The Theory of Machines and Mechanisms

(in 3 volumes)

Pergamon Titles of Related Interest

ANTHONY	Engineering Metrology
BAYOUMI	Current Advances in Mechanical Design and Production
DU QINGHUA	Boundary Elements
HAMOUI *et al.*	Mathematical Analysis and its Applications
HARADA	Fluid Control and Measurement
HAYWOOD	Analysis of Engineering Cycles, Worked Problems
JAPAN SOCIETY OF MECHANICAL ENGINEERING	Flow Visualization
KANT	Finite Elements in Computational Mechanics
NIKU-LARI	Structural Analysis Systems, Vols 1-6

Pergamon Related Journals (sample copy gladly sent on request)

Computers and Fluids

Computers and Industrial Engineering

Engineering Fracture Mechanics

Fatigue and Fracture of Engineering Materials and Structures

International Journal of Applied Engineering Education

International Journal of Engineering Science

International Journal of Machine Tools and Manufacture

International Journal of Mechanical Sciences

International Journal of Non-Linear Mechanics

Journal of Applied Mathematics and Mechanics

Mechanism and Machine Theory

Robotics and Computer-Integrated Manufacturing

The Theory of Machines and Mechanisms

Proceedings of the 7th World Congress
17–22 September 1987, Sevilla, Spain

(in 3 volumes)

Editors

Emilio Bautista, *ETSII Madrid, Spain*

Javier Garcia-Lomas, *ETSII Sevilla, Spain*

Alfredo Navarro, *ETSII Sevilla, Spain*

Associate Editors

Jaime Dominguez, *ETSII Sevilla, Spain*

Justo Nieto, *ETSII Valencia, Spain*

Volume 3

Gears and Transmissions
Control Systems
Machines and Systems Design
Cams
Teaching Methods
Rotor Dynamics
Biomechanisms
History
Rational Mechanics

PERGAMON PRESS

OXFORD · NEW YORK · BEIJING · FRANKFURT
SÃO PAULO · SYDNEY · TOKYO · TORONTO

U.K.	Pergamon Press, Headington Hill Hall, Oxford OX3 0BW, England
U.S.A.	Pergamon Press, Maxwell House, Fairview Park, Elmsford, New York 10523, U.S.A.
PEOPLE'S REPUBLIC OF CHINA	Pergamon Press, Room 4037, Qianmen Hotel, Beijing, People's Republic of China
FEDERAL REPUBLIC OF GERMANY	Pergamon Press, Hammerweg 6, D-6242 Kronberg, Federal Republic of Germany
BRAZIL	Pergamon Editora, Rua Eça de Queiros, 346, CEP 04011, Paraiso, São Paulo, Brazil
AUSTRALIA	Pergamon Press Australia, P.O. Box 544, Potts Point, N.S.W. 2011, Australia
JAPAN	Pergamon Press, 8th Floor, Matsuoka Central Building, 1-7-1 Nishishinjuku, Shinjuku-ku, Tokyo 160, Japan
CANADA	Pergamon Press Canada, Suite No. 271, 253 College Street, Toronto, Ontario, Canada M5T 1R5

First edition 1987

Library of Congress Cataloging-in-Publication Data

World Congress on the Theory of Machines and Mechanisms (7th: 1987: Sevilla, Spain)
The theory of machines and mechanisms.
1. Mechanical engineering—Congresses.
I. Bautista, Emilio. II. García-Lomas, Javier.
III. Navarro, Alfredo. IV. Title.
TJ5.W56 1987 621.8 87–14770

British Library Cataloguing in Publication Data

The Theory of machines and mechanisms: proceedings of the 7th world congress, 17–22 September 1987, Sevilla, Spain.
1. Machinery
I. Bautista, Emilio II. García-Lomas, Javier
III. Navarro, Alfredo
621.8 TJ145

ISBN 0-08-034815-7

Printed in Great Britain by A. Wheaton & Co. Ltd., Exeter

Foreword

The Seville Congress is the seventh in a series of international congresses on the Theory of Machines and Mechanisms. The first was held in Varna (Bulgaria) in 1965, the second in Zakopane (Poland) in 1969, the third in Kupari (Yugoslavia) in 1971, the fourth in Newcastle (UK) in 1975, the fifth in Montreal (Canada) in 1979, the sixth in New Delhi (India) in 1983.

During the Zakopane Congress, the International Federation for the Theory of Machines and Mechanisms was officially constituted. This Federation took as one of its responsibilities the organization of the successive congresses, changing the denomination from "International" to "World" Congresses. This denomination may have seemed rather ambitious at that time; it appears much less so today.

The booklet collecting the abstracts of the Zakopane Congress contains a total of 81 papers. The USSR was present with 19 papers; USA with 14 papers; Japan with 2; the UK with 1. All the other papers came from six European countries (Bulgaria, Czechoslovakia, the German Democratic Republic, the German Federal Republic, Poland, Romania). For the Seville Congress the number of accepted papers is 417, the countries of origin of the authors total 34, and all the continents are represented: Africa, North and South America, Asia, Australia, Europe. The Federation can certainly be pleased with this result and the term "world" can now be used without embarrassment.

The significance of the geographical and numerical development of the congress is, however, determined by the significance of the papers presented. This cannot be measured by simple statistics.

General congresses of this kind do not intend to compete with the specialized symposia on restricted topics, which our Federation also organizes. The purpose of the Congress is to offer a general picture of the research done in the field of machines and mechanisms throughout the world, to promote this research by bringing together junior and senior scientists, and favour the interdisciplinary understanding and cooperation so important today in many projects.

If we read through the proceedings of all the IFToMM congresses we have good reason to feel that our goals have been achieved. From congress to congress, in a diversified and lively series of papers, we can trace the development, through the exposition of its scientific foundation, of all modern mechanical engineering. The collection of the proceedings has itself become an important tool for reference and for the training of researchers.

Results like these are obtained through the contribution of many persons. Today our thanks go particularly to our Spanish colleagues, whose generous efforts have made this seventh congress possible.

Giovanni Bianchi
President of IFToMM

Preface

For centuries Seville has been a melting-pot of cultures coming from all over the world. Open to every thought flow, this city played an important role in the development of the science, the technology and the mentality that allowed the discovery of the New World. Once again people from all over the world have met in Seville during the Seventh World Congress of the International Federation for the Theory of Machines and Mechanism (IFToMM) on September 17-22, 1987 at Seville University, Spain.

The Congress Proceedings contain 417 papers contributed from 38 countries, and the Special Lectures presented by internationally renowned engineers. The technical sessions cover various fields of machines and mechanisms and in particular, Kinematic Analysis and Synthesis, Dynamics of Machines and Mechanisms, Gearing and Transmissions, Rotor-Dynamics, Vibrations and Noise in Machines, Biomechanisms, Tribology, Robots, Manipulators and Man-Machine Systems, Computer-Aided Design and Optimization, Industrial Applications for Special Machines and Mechanisms and Experimental and Teaching Methods.

Historically IFToMM came into being on September 27, 1969, during the Second International Congress for the Theory of Machines and Mechanisms at Zakopane, Poland as a result of recommendations brought forward by an International Coordinating Committee. The Coordinating Committee was created at the suggestion of the Bulgarian delegation, participating in the First International Congress on Machines and Mechanisms held at Varna, Bulgaria in September 1965. The countries represented in the inaugural assembly at Zakopane included Australia, Bulgaria, Czechoslovakia, German Democratic Republic, German Federal Republic, Hungary, India, Italy, Netherlands, Norway, Poland, Romania, United States, Soviet Union and Yugoslavia. The Spanish National Council for the Theory of Machines and Mechanisms was formally accepted as a regular Country member of IFToMM in 1979 at Montreal, Canada during the Fifth Congress. The Sixth Congress was held in 1983 at New Delhi, India.

In the past, IFToMM has played a major role in the unification of scientists working in the field of machines and mechanisms. Through its sponsorship of international meetings every four years, it has influenced new developments, it has enabled contacts between colleagues to be renewed, and in general, provided the incentive to develop new and productive friendship. It is sincerely hoped that the Seventh World Congress will continue to act as the catalyst for the advancement of new ideas and concepts in our field.

The Organizing Committee for the Seville Congress has been generously supported by many bodies. In particular, the Executive Council of IFToMM and the Spanish Administration have contributed financial assistance. Seville University and the Engineers Association of Seville have contributed immensely to the success of this undertaking by providing the facilities and resources required in several aspects of the organizational activities, as well as a number of professional services essential to the organization of this Congress. To all these societies, institutions and individuals too numerous to single out, the Organizing Committee wishes to express its sincere thanks and appreciation.

The Spanish Council for the Theory of Machines and Mechanisms extend a warm welcome to all delegates and their guests and wish you all a pleasant and successful meeting.

On behalf of the Organizing Committee
J. Nieto Nieto

Sponsors

The International Federation for the Theory of Machines and Mechanisms (IFToMM)

Comite Espanol del IFToMM (CEIFToMM)
(Spanish Committee of IFToMM)

Comision Asesora de Investigacion Cientifica y Tecnica (CAICYT)
(Consultant Commission for Scientific and Technical Research)

Consejeria de Educacion y Ciencia de la Junta de Andalucia
(Board of Education and Science of the Government of Andalusia)

Direccion General de Industria, Energia y Minas
de la Consejeria de Economia y Formento de la Junta de Andalucia
(General Directorate of Industry, Energy and Mines
of the Board of Economy and Promotion of the Government of Andalusia)

Instituto de Fomento de Andalusia (IFA)
(Institute for Promotion of Andalusia)

Diputacion Provincial de Sevilla
(County Council of Seville)

Ayuntamiento de Sevilla
(City Council of Seville)

Colegio Oficial de Ingenieros Industriales de Andalucia Occidental
(Official Institution of Engineers of Western Andalusia)

Colegio Oficial de Ingenieros Industriales de Madrid
(Official Institution of Engineers of Madrid)

Universidad de Sevilla
(University of Seville)

Universidad Politecnica de Madrid
(Technical University of Madrid)

Universidad Nacional de Educacion a Distancia
(Spanish Open University)

Escuela Tecnica Superior de Ingenieros Industriales de Sevilla (ETSII)
(Engineering School of Seville University)

Departamento de Ingenieria Mecanica de la Universidad de Sevilla
(Mechanical Engineering Department of Seville University)

Departamento de Ingenieria Mecanica de la Universidad Politecnica de Madrid
(Mechanical Engineering Department of the Technical University of Madrid)

Exposicion Universal Sevilla 1992
(1992 Seville World Exposition)

IBM Espana S.A.

Compania Espanola de Petroleo S.A. (CEPSA)

La Cruz del Campo S.A.

Monte de Piedad y Caja de Ahorros de Sevilla

Monte de Piedad y Caja de Ahorros de Almeria

Ingenieria y Diseno S.A.

Landis & Gyr Espanola S.A.

Fosforico Espanol S.A.

Sociedad Andaluza Para el Desarrollo de la Informatica y la Electronica S.A. (SADIEL S.A.)

Fuerzas Electricas del Noroeste (FENOSA)

Organising Committee

Honorary President

G Bianchi
President IFToMM, Politecnico di Milano, Italy

Honorary Chairmen

L Maunder
University of Newcastle-upon-Tyne, UK
B Roth
Standford University, California, USA

International Programme Committee

E Bautista (Chairman), ETSII Madrid, Spain
S H Crandall, MIT Cambridge, Massachusetts, USA
J Dominguez, ETSII Sevilla, Spain
T Hayashi, Tokyo Intstitute of Technology, Japan
G Kunad, Technische Hochshule Otto Von Guerick, Magdeburg, DDR
L Maunder, University of Newcastle-upon-Tyne, UK

Executive Organizing Committee

J Nieto (Chairman), ETSII Valencia, Spain
J Dominguez (Secretary), ETSII Sevilla, Spain
J Garcia-Lomas (Treasurer), ETSII Sevilla, Spain
E Bautista, ETSII Madrid, Spain
A Navarro, ETSII Sevilla, Spain
M Cabrera, ETSII Sevilla, Spain

Members

J C Guinot, Université Pierre et Marie Curie, France
H Rankers, Delft University of Technology, the Netherlands
K N Gupta, Indian Institute of Technology, Delhi, India
D W Pessen, Technion, Haifa, Israel
A Rovetta, Politecnico di Milano, Italy
M O M Osman, Concordia University, Montreal, Canada
R Chicurel, Instituto de Ingenieria and DEPFI, UNAM, Mexico
Z A Parszewski, University of Melbourne, Australia
L Belivagna, ABCM, Rio de Janeiro, Brasil
M S Konstantinov, Central Laboratory for Manipulators and Robots, Sofia, Bulgaria
K Luck, Technische Universitat Dresden, German Federal Republic
Tao Hengxian, Chinese Mechanical Engineering Society, Beijing, China
Y C Tsai, National Sun Yat-Sen University, Taiwan, China
L Pust, Academy of Sciences, Prague, Czechoslovakia
Z Terplan, Technische Universität für Schwerindustrie, Hungary.
Z Zivkovic, Masinski Facultet, Univerziteta u Beogradu, Yugoslavia
J Oderfeld, Technical University, Warsaw, Poland
N I Manolescu, Institul Politechnic Bucuresti, Rumania
A P Bessonov, USSR Academy of Sciences, Moscow, USSR
A Seireg, University of Wisconsin, USA
T C Warren, The Royal Society, London, UK
S Fujil, Japan Society of Precision Engineering, Tokyo, Japan
T Leinonen, University of Oulu, Finland

Contents of Volume 3

Contents of Volume 1

Contents of Volume 2

Gears and Transmissions

Mechanische Vorrichtungen Fuer Antrieb von Systemen mit Robotern

Iw. P. DIMITROFF, CHR. IW. GROSEFF, D. D. ANDREEV AND An. Iw. DOBREWA

Forschungslabor für Zahnradgetriebetechnik, Technische Hochschule "A. Kantschev" – Russe, 7004 Russe, Str. Komsomolska 8, VR Bulgarien

Kurzfassung. Anliegen dieses Beitrages ist der mechanische Antrieb von Robotern sowie von unterschiedlichen Vorrichtungen in der Produktion und Steuerung welche mit Robotern ausgerüstet sind. Es werden unterschiedliche Getriebearten betrachtet die für eine adaptive Verwirklichung von Rückkopplung zwischen dem jeweiligen Datenangeber und dem mechanischen Wirkungsglied dienen können. Es werden Gesichtspunkte angegeben die auf einen Entscheidungstreffen bei dem Auswahl eines Getriebeartes gerichtet sind.

Schlüsselwörter. Power transmission; gear; harmonic-drive

EINLEITUNG

Die Steuerungs- oder Servomechanismen finden ihre breite Anwendung in den Konstruktionen von Robotern sowie in den unterschiedlichen Systemen mit Robotern wie Transportmittel, Überwachungssysteme und CNC-gesteuerte Werkzeugmaschinen. In diesen Systemen findet eine breite Anwendung das Prinzip der Rückkopplung durch Datenangeber welche die Geschwindigkeit und Lage der Arbeitsorganen überwachen und die Steuerung der Servomotoren regeln. Zur Verwirklichung der erforderlichen Drehgeschwindigkeit oder zum Verändern des Drehmomentes ist jeweils eine Kopplung der Servomotoren der Steuerungseinrichtung mit einem mechanischen Antrieb erforderlich. Wenn wir von den Forderungen ausgehen welchen solche robotisierte Systeme genügen sollen, dann läßt sich die Auswahl der zum Steuerungsmechanismus gekoppelten mechanischen Vorrichtung nach folgenden grundlegenden Merkmalen machen: Lehrgang und Unverformbarkeit, Abmessungen, Massenträgheitsmoment, optimale Übersetzungszahl, Sicherheit und Wirkungsgrad (Bruchmann, 1985).

Die Arbeits- und Qualitätsmerkmale der mechanischen Antriebe hängen mit dem angewendeten kinematischen Prinzip sowie mit der Konstruktion zusammen. Von den langjährigen theoretischen und experimentellen Untersuchungen welche an dem Forschungslabor für Zahnradgetriebetechnik an der TH "A.Kantschev" - Russe durchgeführt wurden, ist die Schlußfolgerung zu ziehen, daß Zahnradgetriebe deren Kinematik auf der Basis der sich im Eingriff befindenden außen- und innenverzahnten Zahnräder mit kleiner Zähnezahldifferenz beruht, am besten den von den Steuerungsmechanismen gestellten Forderungen genügen. Zu diesen Getriebearten zählen: das exzentrische (zykloidale, oder auch trochoidale genn.) Getriebe, das harmonic-drive Getriebe, das radiale Kugelgetriebe und das exzentrische K-H-V-Umlaufrädergetriebe.

EXZENTRISCHE (ZYKLOIDALE) GETRIEBE

Bild 1. zeigt seine Konstruktion wobei 1- Exzentrik, 2 - außenverzahntes Umlaufrad, 3 - Rolle, 4 - Sonnenrad, 5 - Radiallager und 6 - fest an der Abtriebswelle befestigte Klauen sind. Die Zahnflanken der Umlaufräder haben die Form einer Trochoide, und diese des Sonnenrades haben zylindrische Form. Zwangsläufig verlangt das die Anwendung von Sonderwerkzeugen welches nun zu begrenzter Anwendung dieser Art von Getrieben führt. Die Klauen 6 welche eine Wälzbewegung in den Bohrungen der Umlaufräder durchführen, betreiben die Abtriebswelle. Die Bohrungsachsen und die Klauen sind auf Kreise mit gleichen Durchmessern verteilt und exzentrisch aneinander auf den Achsabstand verschoben. Ein Drehwinkel $\psi^1_{abtr.}$ welcher die Genauigkeit der Positionierung bestimmt wird auf grund des Eingriffsspieles, des Spieles Klauen-Bohrung und der Torsionsdeformation der Abtriebswelle berechnet. Bei einer genaueren Berechnung von $\psi^1_{abtr.}$ werden auch die Verschiebungen mitberechnet welche an der Stellen des Eingriffes Klauen-Bohrung verursacht sind.

Bild 2. veranschaulicht die Verteilung des Spiels $j=f(r°)$ für einen Satz mit $i=35$. Die Verteilung des Spiels für das Paar Klauen-Bohrung ist auf dem Bild 3. für einen Satz mit $i=71$ gezeigt (Lobastoff, 1971)

Bei einem exzentrischen Abtrieb bewegt sich der Win kel $\psi^1_{abtr.}$ unter Arbeitsbelastung zwischen 0.3° und 1.0° und vor allem dieser Tatsache ist seine kleineren Anwendung in den Steuerungsvorrichtungen zurückzuführen.

Andererseits ist dieses Getriebe mit kleinen Abmessungen, weist einen großen Wirkungsgrad ($\eta=0.85$ bis 0.92), kann Übersetzungszahlen von 9 bis 80 verwirklichen und mittelmässige Drehmomente übertragen. Dieses Getriebe ist für Vorrichtungen ohne große genauigkeit der Positionierung geeignet.

HARMONIC-DRIVE GETRIEBE

Der Eingriff bei diesem Getriebe kommt infolge der elastischen Deformation eines der Zahnräder zustande und diesem Hauptmerkmal der Konstruktion ist der Name zurückzuführen. Bild 4. veranschaulicht die Konstruktion des Getriebes wobei 1 - Antriebswelle, 2 - elastisches Zahnrad, 3 - innenverzahntes Zentralrad, 4 - elastisches Lager, 5 - Wellengenerator, 6 - Abtriebswelle.

Das elastische Getriebe gewährleistet eine große Genauigkeit bei der Ubertragung von Bewegungen, welche nur wenig von dem Schrittfehler auf dem Zahnfußkreis abhängt. Dies ist auf der großen Zahl der gleichzeitig und in den beiden Zonen im Eingriff kommenden Zähnen zurückzuführen. Andererseits weist dieses Getriebe eine veränderliche Nachgiebigkeit auf, welche von der Belastung abhängt und im gewissen Maßen die Genauigkeit der Positionierung verschlechtert. Der Spielraum der Übersetzungszahl ist i=80 bis 260. Die untere Grenze ist von der Dauerfestigkeit des elastischen Zahnrades begrenzt, die obere von der Herstellungsmöglichkeiten eines Werkzeuges mit sehr kleinem Modul.

Die Veränderung des Wirkungsgrades $\eta=f(i)$ ist auf dem Bild 5. dargestellt (Harmonic-drive, 1981).

Das harmonic-drive-Getriebe ist geeignet für Vorrichtungen die eine große Positionierungsgenauigkeit bei ständiger Arbeitsweise verlangen.

RADIALES KUGELGETRIEBE

Dieses Zahnradgetriebe stellt eine Kombination zwischen dem exzentrischen und dem harmonic-drive-Getriebe dar. Seine Konstruktion ist auf dem Bild 6. veranschaulicht wobwei 1 - Antriebswelle, 2 - Kugelkörper, 3 - Rollen, 4 - Wellengenerator, 5 - elastisches Lager, 6 - Abtriebswelle sind. Die Kugelkörper sind in den radialpositionierten Bohrungen auf der Abtriebswelle eingesetzt.

Der Wirkungsgrad des Getriebes schwänkt in den Gränzen $\eta=0.75$ bis 0.85.

Dieses Getriebe ist für Vorrichtungen geeignet welche eine große Genauigkeit der Positionierung bei Belastungen bis zu 100 Nm verlangen.

EXZENTRISCHES K-H-V-UMLAUFRÄDERGETRIEBE

Die Konstruktion dieses Getriebes ist auf dem Bild 7. veranschaulicht wobwei 1 - Exzentrik, 2 - Umlaufrad, 3 - Sonnenrad, 4 - Rollenlager, 5 - Klauen, 6 - Abtriebswelle sind.

Die Zahnflanken sind mit der Form einer Evolvente. Die Zähnezahldifferenz zwischen dem Sonnenrad und dem Umlaufrad beträgt $Z_{\Sigma}=1$ bis 5. Die Hauptschwierigkeit bei der Entwicklung von Setzen dieser Art ist das Vermeiden der Interferenz zweiten Grades. Dieses kann auf einer der folgenden Methoden erreicht werden:

1. Annehmen einer Zahnflankenkorrektur $X_{\Sigma}>0$ bei einem Winkel des Bezugsprofiles des Werkzeuges $\alpha_t=20°$ und Zahnkopfkoeffizienten $h_a^*=1$. Dieses ist mit einer Verschiebung der aktiven Strecke der Eingriffslinie von dem Eingriffspunkt, welches mit einem Erhöhen der Geschwindigkeit des Berührungspunktes, der Verluste in der Eingriffszone und mit Minderung des Wirkungsgrades verbunden ist (die Kurve 2 auf dem Bild 8).

2. Das Verkleinen des Zahnkopfkoeffizienten - $h_a^* = 0.6$ bis 0.8. In diesem Fall wird die Länge der aktiven Sträcke der Eingriffslinie vermindert welches zwangsläufig zu einem theoretischen Überdeckungsgrad $\varepsilon_r < 1$ führt. Der wirkliche Überdeckungsgrad ist höher als der theoretische un beträgt meistens $\varepsilon_r=4$ bis 5 bei $\varepsilon_r=0.8$ (Groseff, 1984; Dimitroff, 1983).

3. Veränderung des Winkels des Bezugsprofiles α_t Bei $\alpha_t=35°$ ($\alpha_t=30°$) ist die Wahrscheinlichkeit von Interferenz sehr gering. Eine andere posizive Erscheinung ist daß die aktive Sträcke der Eingriffslinie durch den Eingriffspunkt verläuft welches zum Erhöhen des Wirkungsgrades führt (die Kurve 1 auf dem Bild 8.).

Zum Entfernen des Eingriffsspieles werden zwei gleiche Sonnenräder eingesetzt die aneinander auf einen geringen Winkel relativ verschoben sind. Dann kommen die entgegengesetzten Flanken des Umlaufrades jeweils mit unterschiedlichen Sonnenrädern in Berührung welches zum Entfernen des Spieles führen kann.

Zum Erhöhen der Zahl von gleichzeitig arbeitenden Klauen werden elastischen Klauen eingesetzt (Bild 9.) oder elastischen Ringe die zwischen dem Rollenlager und dem Umlaufrad montiert sind (Bild 10.). Durch diesen Bauelemente werden die Fehler von der Fertigung und Montage sowie die Achs- und Winkelverschiebungen ausgeglichen und Lärm sowie Schwingungen vermindert.

Für das exzentrische K-H-V-Umlaufrädergetriebe bewegt sich der Winkel $\psi^1_{abtr.}$ bei Arbeit unter Belastung in den Gränzen 0.01° bis 0.02°.

Die Übersetzungszahl kann in den Gränzen i=12 bis 125 verändert werden. Die unterschiedlichen Übersetzungszahlen werden bei denselben Abmässungen des Getriebes verwirklicht welches von großer Bedeutung ist zum Vermeiden von Resonanzerscheinungen in dem gesamten System.

Der Wirkungsgrad verändert sich in den Gränzen von 0.85 bis 0.96 und hängt von der Übersetzungszahl ab (Bild 11.).

Von den dargestellten Ausführungen ist ersichtlich, daß die exzentrischen K-H-V-Umlaufrädergetriebe für solchen Steuerungsmechanismen geeignet sind, die große Genauigkeit bei der Positionierung verlangen und sowohl bei unveränderlichen als auch bei veränderlichen Belastungen arbeiten.

LITERATUR

Bruchmann, K (1985) Große Übersetzungen bei kleinem Bauvolumen. Techno-Tip, 3, 64-78.

Dimitroff, Iw. u.a. (1983) Das Wellgetriebe - eine Möglichkeit zur Optimierung von Antrieben. Referate Zahnradgetriebe. Dresden, Teil 2, 257-265.

Groseff, Chr. Über das Bestimmen des Flankenspieles bei Innenverzahnung mit kleiner Zähnezahldifferenz. (auf Bulgarisch). Wiss. Zeitschrift der TH "A.Kantschev" - Russe, 1984.

Harmonic-drive Getriebe. (1981). Katalog BRD, 1981.

Judin, W.A., Lobastov, W.K. (1971). Über die Theorie der Entwicklung "reeller" Umlaufrädergetriebe mit büchsenartigem Eingriff außerhalb des Eingriffspunktes. (auf Russisch). Verlag "Nauka", 83-85.

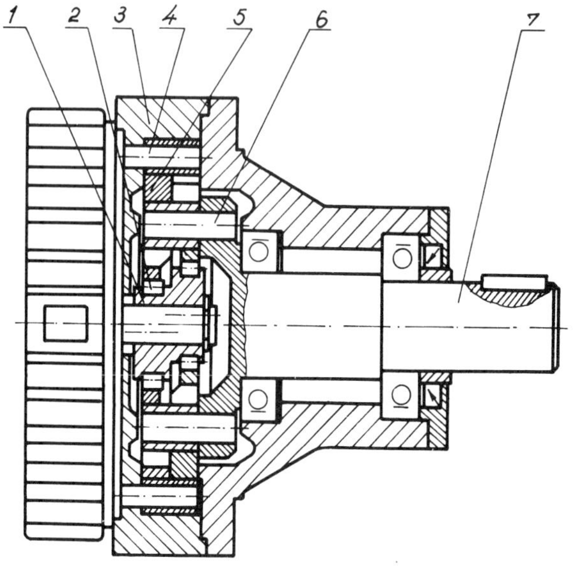

Bild 1

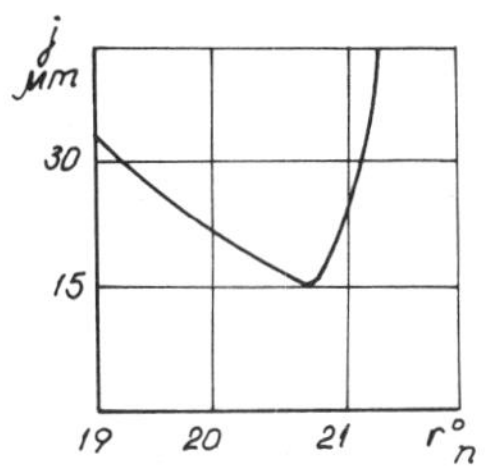

Bild 2

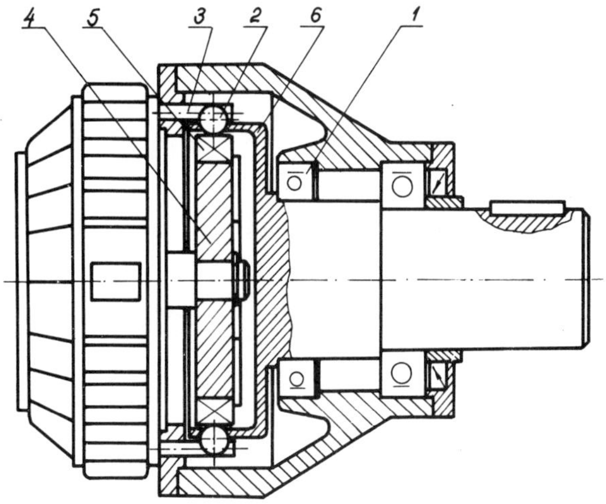

Bild 6

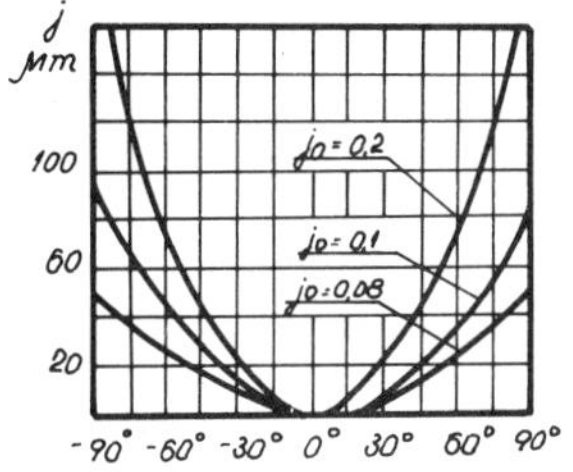

Bild 3

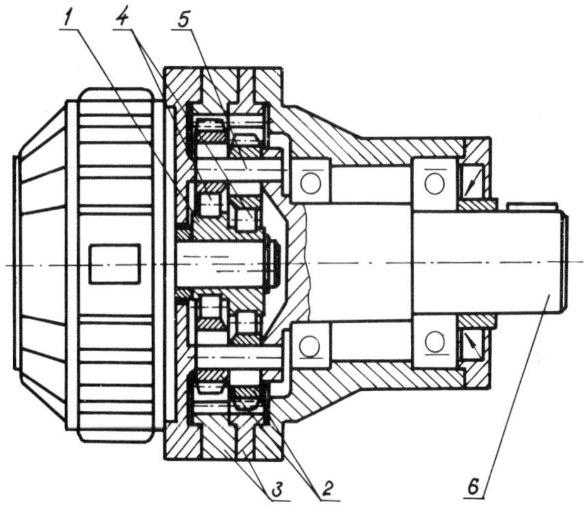

Bild 7

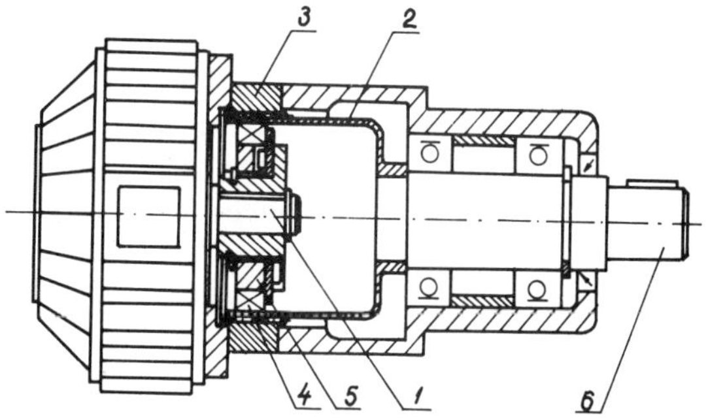

Bild 4

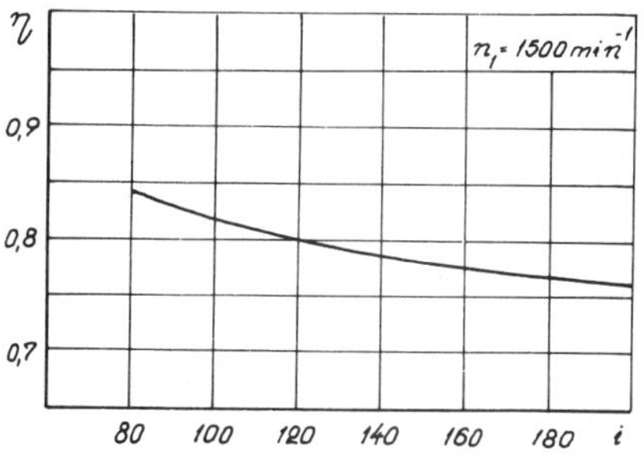

Bild 5

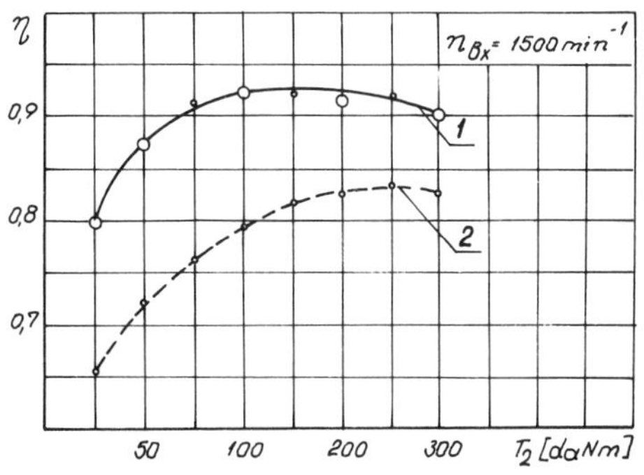

Bild 8

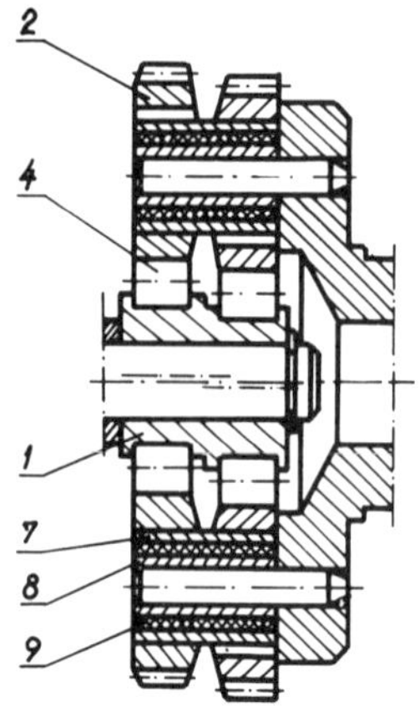

Bild 9

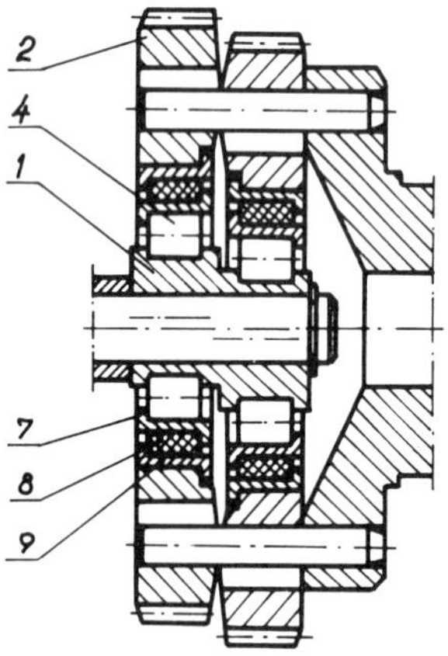

Bild 10

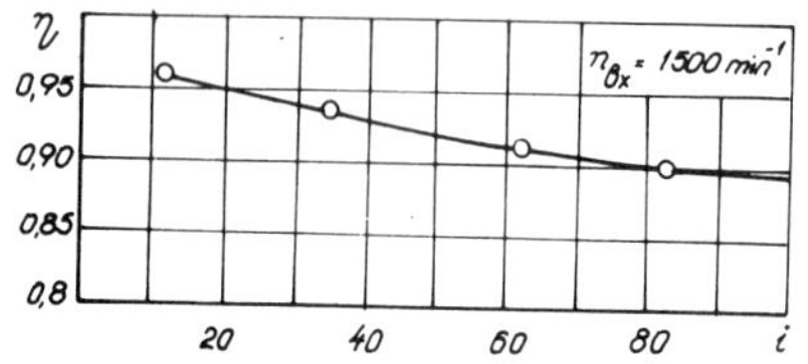

Bild 11

Механизмы управления или сервомеханизмы широко применяются в различных видах роботизированных системах:манипуляторах, промышленных роботах,робокарах,следящих системах и металлорежущих станках с програмным управлением.

Для реализации необходимой частоты вращения или для редукции крутящего момента ходится серводвигатели механизма управления агрегатировать с механическими переда чами. Исходя из требований,предъявляемых к роботизированным системам,выбор механических передач,включенных в механизмах управления,следует производить на основе следующих основных показателей:

1.Мертвый ход и неподатливость-критерии, определяющие точность позиционирования рабочих органов механизма.

2.Компактность-необходимо,чтобы передача имела небольшой объем и легко монтировалась и демонтировалась к приводимым в движение модулям.

3.Массовый инерционный момент-определяет стабильность работы механизма.

4.Оптимальное передаточное число-зависит от требуемой частоты вращения или от передаваемого крутящего момента.Его определение следует производить с учетом режима работы и вида механизма.

5.Надежность и к.п.д.

Теоретические и экспериментальные исследования,проводимые в течении многих лет в ОНИЛ по редукторостроению при ВТУ"А.Кынчев Русе показали,что передачи типа K-H-V удовлетворяют наилучшим образом эти требования. К этому виду передач относятся: эксцентриковая K-H-V передача с трохоидным профилем зубьев,волновая K-H-V с эвольвентным профилем зубьев,эласто передача и радиально-шариковая передача.

В настоящей работе приведен анализ различных видов передач. На основе их качеств определены области приложения передач в связи с их включением в роботизированные системы.

Meshing Theory and Methods of Forming Vaulted-toothed Wheels

H. I. MARINOV, M. B. PAMUKSCHIEV, D. I. ZAFIROV AND T. P. PETKOVA

A. Kiinscheva Higher Technical Institute, Department of the Theory of Mechanisms and Machines, Rousse, Bulgaria

Резюме. Опыт использования конусных и гипоидных передач с коническими зубчатыми колесами со спиральными зубьями доказывает большие преимущества локализованного контакта по высоте и длине зубьев. Спиральная форма зуба используется и при формообразувании цилиндрических зубчатых колес. Цилиндрические зубчатые передачи со спиральными (арочными)зубьями вместе с плавностью при работе, большей поверхностью и объемной прочностью обладают и еще одним ценным преимуществом - нечувствительность при изменении расположения осей. В работе мы делимся нашим опытом, объесняя некоторых особеностей, связанныи с картиной зацепления арочных зубчатых колес. В работе доказывается существование арочных передач двух линий зацепления. Исследована адаптивность этих передач по отношению к изменению межосевого расстояния. Указана область использования арочных передач с новой геометрией.

Keywords. Power transmission; gear; circular zerol gear.

Практически могут быт реализированы несколько методов формообразования профиля зубьев арочных зубчатых колес - метод непрерывного деления, осуществленный спиральной дисковой Фрезой (Фиг. 1), метод перерывного деления, осуществленный торцевой резцовой головкой (Фиг. 2) и метод непрерывного деления, реализированный торцевой резцовой головкой и тангенциальной подачей инструмента.

При первом методе можно реализировать высокая производительность (Marinov, 1982), но используется очень сложная инструментальная поверхность - эвольвентный спиральный геликоид. Для получения сопряженных профилей зубьев в данном случае необходимо использовать две инструментальные поверхности с одинаковыми параметрами и различными по направлению спиралями. Поэтому исходные контуры должны входить один в другой как шаблон и контрашаблон. Это приводить к резкому повышению номенклатуры инструментов.

Для устранения этого недостатка мы разработали метод формообразования зубьев арочных зубчатых колес одним инструментом. Чтобы избежать интерференцию профилей зубьев при зацеплении предлагается дообработать шестерню при помощи допольнительного вертикального смещения инструмента (Marinov, 1985a) (Фиг. 3). Таким способом вогнутые стороны нарезанных зубьев шестерни 1 дообрабатываются в верхней части, а выпуклые - в нижней.

Уравнения направляющих кривых профиля зубьев в неподвижной системе координат Oxyz (Фиг. 3):

$$\begin{aligned} y^{(1)} &= R_{\Phi}\cos(\nu_1+\delta)+\frac{m}{2}\nu_1\cos\nu_1 \\ z^{(1)} &= R_{\Phi}\sin(\nu_1+\delta)+\frac{m}{2}\nu_1\sin\nu_1-\frac{m}{2} \end{aligned} \qquad (1)$$

$$\begin{aligned} y^{(2)} &= R_{\Phi}\cos(\nu_2-\delta)-\frac{m}{2}\nu_2\cos\nu_2 \\ z^{(2)} &= R_{\Phi}\sin(\nu_2-\delta)-\frac{m}{2}\nu_2\sin\nu_2+\frac{m}{2}+\Delta z \end{aligned} \qquad (2)$$

где ν_j $(j = 1,2)$ - угловой параметр текущей точки режущей кромки;

δ - угол подъема спирали.

Отсуствие интерференции гарантируется равенствами:

$$\begin{aligned} z^{(1)} &= z^{(2)} = b_w/2 \\ y^{(1)} &= y^{(2)} \end{aligned} \qquad (3)$$

Из (1), (2) и (3) получаем:

$$\begin{aligned} &R_{\Phi}\sin(\nu_1+\delta)+\frac{m}{2}\nu_1\sin\nu_1-m/2=b_w/2 \\ &R_{\Phi}\sin(\nu_2-\delta)-\frac{m}{2}\nu_2\sin\nu_2+m/2+\Delta z=b_w/2 \\ &R_{\Phi}\cos(\nu_2-\delta)-\frac{m}{2}\nu_2\cos\nu_2 = \\ &= R_{\Phi}\cos(\nu_1+\delta)+\frac{m}{2}\nu_1\cos\nu_1 \end{aligned} \qquad (4)$$

При решении системы уравнений (4) определяется необходимое смешение Δz, обеспечивающее отсуствие интерференции в направляющих кривых зубьев.

Главное преимущество метода формообразовании профиля зубьев перерывным делением - использование технологичного инструмента - торцевой резцовой головки с конусной произ-

водящей (инструментальной) поверхностью.
Пусть в системах координат, неподвижно связанных с заготовками (поворачивающиеся соответственно на углы Ψ_1 и Ψ_2) определенны координаты точек вогнутых сторон зубьев зубчатого колеса 1, выпуклой стороны зубьев зубчатого колеса 2 и орты нормали (Marinov, 1985b).
На Фиг. 4 показана схема зацепления зубчатых колес. Введены следующие системы координат: $\Sigma_j (j = 1, 2)$ - координатные системы, неподвижно связанные соответственно с нарезанными зубчатыми колесами 1 и 2 и Σ_o - неподвижная система координат.
В неподвижной системе координат для координат текущей точки контакта Q имеем:

$$x_1^{(o)} = x_1 \cos\varphi_1 - y_1 \sin\varphi_1 = x_1^{(o)}(u_1, \nu_1, \varphi_1)$$
$$y_1^{(o)} = x_1 \sin\varphi_1 + y_1 \cos\varphi_1 - r_{w1} = \qquad (5)$$
$$= y_1^{(o)}(u_1, \nu_1, \varphi_1)$$
$$z_1^{(o)} = z_1 = z_1^{(o)}(u_1, \nu_1, \varphi_1)$$

$$x_2^{(o)} = x_2 \cos\varphi_2 + y_2 \sin\varphi_2 = x_2^{(o)}(u_2, \nu_2, \varphi_2)$$
$$y_2^{(o)} = -x_2 \sin\varphi_2 + y_2 \cos\varphi_2 + r_{w2} = \qquad (6)$$
$$= y_2^{(o)}(u_2, \nu_2, \varphi_2)$$
$$z_2^{(o)} = z_2 = z_2^{(o)}(u_2, \nu_2, \varphi_2)$$

где x_j, y_j, z_j - координаты текущей точек, определенны в системой Σ_j;
φ_1 и φ_2 - соответствующие углы поворота двух поверхностей (двух зубчатых колес);
r_{w1} и r_{w2} - радиусы начальных окружностей зубчатых колес;
u_j и ν_j - криволинейные координаты поверхностей.

Аналогично для координат орты нормали получаем:

$$e_{x1}^{(o)} = e_{x1} \cos\varphi_1 - e_{y1} \sin\varphi_1 = e_{x1}^{(o)}(u_1, \nu_1, \varphi_1)$$
$$e_{y1}^{(o)} = e_{x1} \sin\varphi_1 + e_{y1} \cos\varphi_1 = e_{y1}^{(o)}(u_1, \nu_1, \varphi_1) \qquad (7)$$
$$e_{z1}^{(o)} = e_{z1} = e_{z1}^{(o)}(u_1, \nu_1, \varphi_1)$$

$$e_{x2}^{(o)} = e_{x2} \cos\varphi_2 + e_{y2} \sin\varphi_2 = e_{x2}^{(o)}(u_2, \nu_2, \varphi_2)$$
$$e_{y2}^{(o)} = -e_{x2} \sin\varphi_2 + e_{y2} \cos\varphi_2 = e_{y2}^{(o)}(u_2, \nu_2, \varphi_2)$$
$$e_{z2}^{(o)} = e_{z2} = e_{z2}^{(o)}(u_2, \nu_2, \varphi_2) \qquad (8)$$

Условие касания двух поверхностей при передачи движения выражается системой уравнений (Litvin, 1968):

$$x_1^{(o)}(u_1, \nu_1, \varphi_1) = x_2^{(o)}(u_2, \nu_2, \varphi_2)$$
$$y_1^{(o)}(u_1, \nu_1, \varphi_1) = y_2^{(o)}(u_2, \nu_2, \varphi_2) \qquad (9)$$
$$z_1^{(o)}(u_1, \nu_1, \varphi_1) = z_2^{(o)}(u_2, \nu_2, \varphi_2)$$
$$e_{x1}^{(o)}(u_1, \nu_1, \varphi_1) = e_{x2}^{(o)}(u_2, \nu_2, \varphi_2)$$
$$e_{y1}^{(o)}(u_1, \nu_1, \varphi_1) = e_{y2}^{(o)}(u_2, \nu_2, \varphi_2) \qquad (9)$$
$$e_{z1}^{(o)}(u_1, \nu_1, \varphi_1) = e_{z2}^{(o)}(u_2, \nu_2, \varphi_2)$$

Из(5), (6),(7), (8) и (9) после соответствующих преобразований получаем:

$$\nu_1 = \nu_2 = \nu$$
$$u_1 = u_2 = u$$
$$\Psi_1 - \varphi_1 = -(\Psi_2 - \varphi_2) = \eta \qquad (10)$$
$$A\cos\eta - B\sin\eta = 0$$
$$A\sin\eta + B\cos\eta = a_w$$

где $A = R_1 - R_2 + \Psi_2 r_2 - \Psi_1 r_1$;
$B = r_1 + r_2 - (R_1 - R_2)\operatorname{cotg}\alpha$;
R_j - радиусы в среднем сечении резцовых головок;
r_j - радиусы делительных окружностей зубчатых колес;
α - угол профиля производящей поверхности.

Если обозначим с $\Delta R = R_1 - R_2$ и $a = r_1 + r_2$ (делительное межосевое расстояние), то из последних двух уравнений в системы (10) получаем:

$$\Delta R^2 \operatorname{cotg}^2\alpha(1 + \operatorname{cotg}^2\alpha\cos^2\nu) - 2a\Delta R\operatorname{cotg}\alpha + a^2 - a_w^2 = 0 \qquad (11)$$

Аналогичная зависимость между основными параметрами инструментальных поверхностей и зубчатой передачей существует и в арочных передачах с продольным циклоидальным профилем зуба (получен при тангенциальной подаче супорта):

$$\frac{\operatorname{cotg}^4\alpha(1-\cos^2\nu)(u_1\sin\alpha + R_1)^2\ \Delta R}{[r_o\cos\nu - (u_1\sin\alpha + R_1)]^2} = a_w - (\Delta R\operatorname{cotg}\alpha + a)^2 \qquad (12)$$

где r_o - радиус производящей окружности;
u_1 - линейный параметр производящей поверхности.

Из уравнении (11) и (12) видно, что если ν_o есть некоторое решение, то существует и решение $-\nu_o$. Это показывает, что арочные зубчатые колеса с продольным круговым и циклоидальным профилем зубьев могут иметь одну линию зацепления (при $\nu_o = 0$) и две линии зацепления (при $\nu_o \neq 0$).
Обычно зона контакта исследуется экспериментально определением формы и размеров пятна контакта. Наглядное представление о существовании двухлиний контакта дает нам характер распределения зазоров между зубьями в процессе зацепления.
На Фиг. 5 показаны находящиеся в зацеплении в данный момент два зуба 1 и 2. Цилиндричной поверхностью с радиусом R_x и осью

совпадающей с геометрической осью зубчатого колеса 1, пересекаем сопряженные поверхности зубьев в точках их касания.
В неподвижной системе координат Σ_o определяем расстояние между точками T_1 и T_2, лежащими на линиях пересечения поверхностей зубьев с цилиндрической поверхностью (Фиг. 5).
Расстояние h по длине зубьев соответственно с продольным круговым и циклоидальным профилем показаны на Фиг. 6 а,б. Из фигуры видно, что действительно существуют две контактные линии и зазор в арочных передачах изменяется плавно, при этом в концах зуба сильно увеличивается, что является благоприятным условием для избежания контакта зубьев по ребру. Зазоры между зубьями при циклоидальном профиле при равных других условиях по длине зуба очень малы.
Интерес в данном случае представляет и определение передаточного отношения зубчатого механизма с двумя линиями контакта.
Для определения параметров движения φ_1 и φ_2 в явном виде используем систему (9), из которы после преобразования уравнений получаем:

$$\Psi_1 = \eta + \varphi_1 \qquad (13)$$

$$\eta = \text{arctg}\frac{\Delta R\text{cotg}^2\alpha\cos\nu}{a - \Delta R\text{cotg}\alpha} \qquad (14)$$

$$u = \frac{R_1(\sin\alpha + \text{cotg}\alpha\cos\alpha\cos\nu)}{\cos\nu} - \frac{r_1\sin\alpha(\eta + \varphi_1)}{\cos\nu} \qquad (15)$$

Зависимость (15) можно представить в виде:

$$u = p_u + q_u\,\varphi_1 \qquad (16)$$

где

$$p_u = \frac{R_1(\sin\alpha + \text{cotg}\alpha\cos\nu) - \eta r_1\sin\alpha}{\cos\nu}$$

$$q_u = -r_1\sin\alpha\ /\cos\nu$$

Из уравнения зацепления угловой параметр Ψ_2 выражаем так:

$$\Psi_2 = p_{\Psi 2} + q_{\Psi 2}\varphi_1 \qquad (17)$$

где

$$p_{\Psi 2} = \frac{R_2(\sin\alpha + \text{cotg}\alpha\cos\alpha\cos\nu)}{r_2\sin\alpha} - \frac{p_u\cos\nu}{r_2\sin\alpha}$$

$$q_{\Psi 2} = q_u\cos\nu\ /r_2\sin\alpha$$

Окончательно после замещения (17) в третьем уравнении системой (10) и диференцирования по времени получаем:

$$d\varphi_2 = \frac{r_1}{r_2}\,d\varphi_1 \qquad (18)$$

или

$$i_{12} = \frac{d\varphi_1}{d\varphi_2} = \frac{r_2}{r_1} = \text{const}$$

Последнее выражение показывает еще одно существенное преимущество арочных зубчатых передач с двумя линиями зацепления - передаточное отношение не зависит от изменения межосевого расстояния.
Сделанные исследования показывают некоторые существенные эксплоатационные преимущества арочных передач - возможность использования передач с повишеной несущей способностью (две контактные линии), адаптивность к изменению межосевого расстояния и взаимного расположения осей колес и др.
Эти преимущества являются важной предпоставкой для широкого внедрения этого зацепления при тяжелонагруженный зубчатых передачах.

LITERATURA

Litvin, F.L. (1968). Teoria zubschatiih zaceplenii. In L.V. Korosteliov (Ed.), Nauka, Moskva.

Marinov, H.I., D.I. Zafirov, O.L. Alipiev, M.B. Pamukschiev (1982). Visokoproizvoditelen metod za obrabotvane na arkoidni zabni kolela sas spiralno-diskova freza. Machinostroene, 11, 500-502.

Marinov, H.I., D.I. Zafirov, O.L. Alipiev, T.P. Petkova (1985). Metod za obrabotvane na arkoidni zabni predavki s nadlajen lokalizovan kontakt. AS 68939/RA. 22.02.1985.

Marinov, H.I., M.B. Pamukschiev (1985). Varhu analiza na arkoidni predavki s dve linii na kontakt. Peti nacionalen kongres po teoretischna i prilojna mehanika, Varna, t. 2, 195-200.

Die Erfahrung bisheriger Anwendung von schräg- und bogenverzahnten Kegelrädern beweist entscheidend die großen Vorteile einer günstig über die Höhe und Breite des Zahnes lokalisierten Kontaktfläche.
Eine bogenartige Verzahnung wird auch bei der Formbildung von Stirnzahnrädern angewendet. In dieser Arbeit wird unsere Erfahrung bei der Untersuchung des gesamten Bildes der Wechselwirkungen bei der Verzahnung und dem Eingriff von bogenverzahnten Rädern.
Bewiesen wird die Möglichkeit zum Bilden von Zahnradpaarungen mit zwei Eingriffslinien. Die durchgeführten Untersuchungen zeigen manche von den bedeutenden experimentellen Vorteilen der Zahnradsätzen dieser Art:
- erhöhte Tragfähigkeit,
- eine gewisse Anpassungsfähigkeit bei kleinen Veränderungen des Achsabstandes sowie bei Achsversetzungen u.a.

Diese Vorteile sind eine wichtige Voraussetzung für höhere Anwendung dieser Art der Verzahnungen bei den hochbelasteten Zahnradsätzen.

Фиг. 1

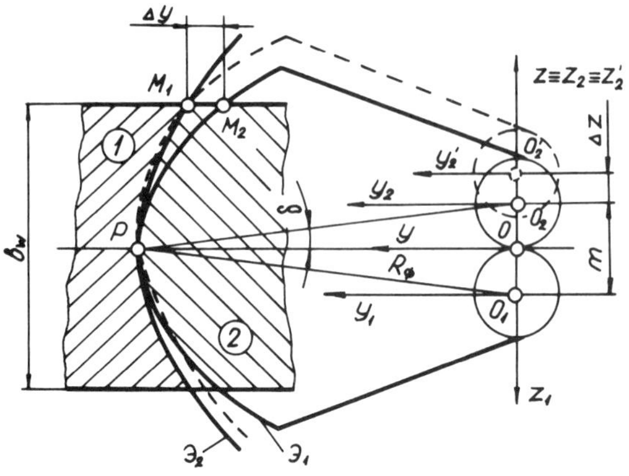

Фиг. 3

Фиг. 2

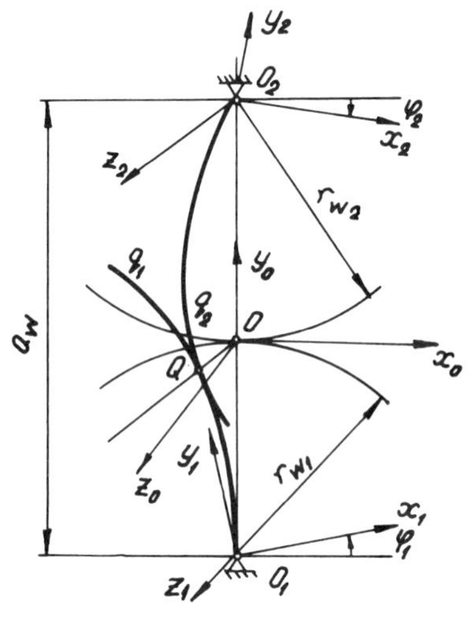

Фиг. 4

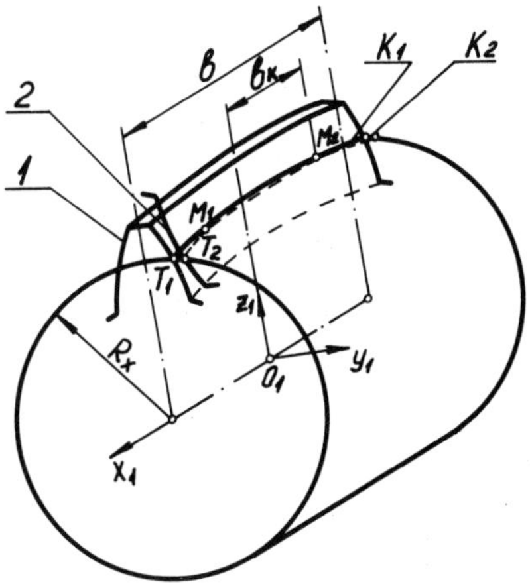
2
1
в
вк
K1
K2
M1
M2
T1
T2
z1
y1
O1
Rx
x1

Фиг. 5

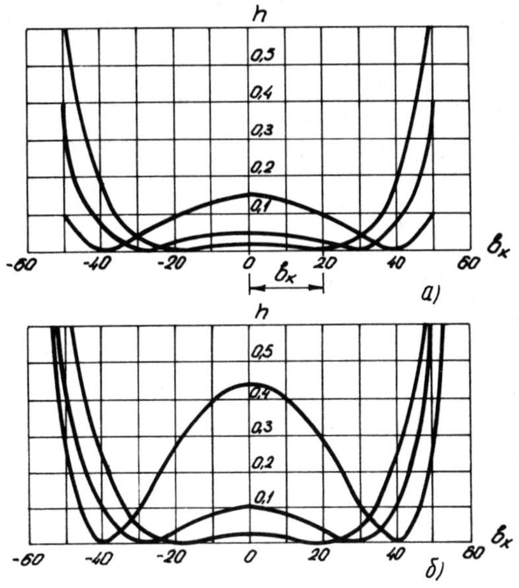
h
0,5
0,4
0,3
0,2
0,1
-60
-40
-20
0
20
40
60
вx
вк
a)
h
0,5
0,4
0,3
0,2
0,1
-60
-40
-20
0
20
40
60
вx
б)

Фиг. 6

Some Aspects of the Geometric and Technological Synthesis of Hypoid and Spiroid Gearings

K. MINKOV, V. ABADJIEV AND D. PETROVA

Institute of Mechanics and Biomechanics, Bulgarian Academy of Sciences, 1090 Sofia, P.O. Box 373, Bulgaria

Abstract. Hypoid and spiroid gearings transforming a rotation between arbitrarily skew axes are considered. The stages of the total design of the geometry and of the technology are briefly discussed. Estimation criteria for choosing an optimal variant are taken into account. The experience of the authors in constructing CAD/CAM systems on modular principle and in their applications in industry is presented.

Keywords. Hypoid and spiroid gearings; modelling; optimisation; standardisation; quality control; computer software; computer-aided design.

INTRODUCTION

The hypoid and spiroid gearings with orthogonally skew axes are well examined. They have already found large application in industry. A lot of articles concerning their geometry and technology as well as calculation instructions of some leading firms are published.

The non-orthogonally skew-axes hypoid and spiroid gearings are of special interest from both theoretical and practical viewpoint. Their mathematical theory is still being developed. These mechanisms are more complicated for design, production and control than the orthogonally ones but they have a lot of merits, such as high loading capacity and higher efficiency, small dimensions and weight, high reduction and universality. These merits destine these gearings as gears of the future. The creation of a good theory and computer programs for automated optimizing design of the construction and technology of non-orthogonally skew-axes hypoid and spiroid gearings will contribute to their application in industry. We have directed our attention to this purpose in recent years I2,3,4,6,7,10,11,12,13,14I.

Two basic stages of the synthesis of hypoid and spiroid gearings can be formulated: a)a calculation of an optimal geometry and construction of the blancs, teeth and tools, and b)a calculation of the optimal technological parameters for generating tooth surfaces.

GEOMETRIC SYNTHESIS

The synthesis of the basic (primary) surfaces of the blancs and of the teeth is of great importance for the design, manufacture and service use of the hypoid and spiroid gearings. The functional qualities of the gearing depend to a great extent on the choice of the form, measures and position of these unreal surfaces. The necessary and sufficient conditions for a choice of the primary pitch surfaces B_1 and B_2 are: a) their geometric axes coincide with the axes 1-1 and 2-2 of rotation of the pinion (worm) 1 and of the gear 2, b)B_1 and B_2 are in contact in a point P, called a pitch point, and have a common tangent plane in this point, and c)B_1 and B_2 should be choosen in such a way that the technical realization of the gearing would be simple enough. The necessary and sufficient conditions for a choice of a basic surface B_Σ of the teeth are:a)B_Σ is rigidly connected with B_1 or B_2 and passes through the point P, b)B_Σ is a regular surface which ensures a technical realization of the two tooth surfaces Σ_1 and Σ_2 and is such that a normal contact exists, and c)the condition $\bar{n}.\bar{v}_{12} = 0$ should be satisfied in the point P, thus realizing a given angular velocity ratio $i_{12} = \omega_1/\omega_2$. Here $\bar{n}$ is the normal vector to B_Σ and $\bar{v}_{12}$ is the sliding velocity vector. Two basic problems of the synthesis of the primary pitch surfaces exist a)a choice of their form, and b)a calculation of their dimensions. Following mentioned above, B_1 and B_2 are to be specified as cones (degenerated as cylinder and discs in some particular cases) (Fig. 1). Ten geometric parameters determine the measures and mutual displacement of B_1 and B_2 I3I: a - the offset, Σ - the shaft angle δ_i - the primary pitch cone angle, r_i, θ_i, a_i - semi-polar coordinates of the point $P \in B_i$ (i=1,2). The dependences between these parameters can be obtained,for example, by the necessary and sufficient conditions for the pitch point P contact of B_1 and B_2 I2,3I, which are the coincidence of the radius-vectors $\bar{r}_1$ and $\bar{r}_2$ and the collinearity of the unit normal vectors $\bar{m}_1$ and $\bar{m}_2$ to the surfaces B_1 and B_2, respectively, in the point P. The set of the two vector equations is equivalent to a set of 5 independent transcendental equations. Consequently, a unique solution exists if 5 parameters among these 10 are specified. Taking into account strengh and constructive requirements, a , Σ , r_2, a_2 and δ_2 are given concret values (i.e. the gear 2 is "given") when designing hypoid gearings. Then the following algorithm is obtained:

$$(1)\quad \mathrm{tg}\,\theta_1 = a/((a_2 + r_2\,\mathrm{tg}\,\delta_2)\sin\Sigma)$$
$$\sin\mu = \sin\theta_1 \sin\Sigma/\cos\delta_2$$
$$m = -\sin\delta_2\cos\Sigma$$
$$n = \cos^2\delta_2\cos^2\mu + \sin^2\delta_2$$
$$k = \cos^2\delta_2\cos^2\mu - \cos^2\Sigma$$
$$(2)\quad \sin\delta_1 = (m + \sqrt{m^2 + m.n})/n$$
$$(3)\quad \sin\theta_2 = \sin\theta_1.\cos\delta_1/\cos\delta_2$$
$$(4)\quad r_1 = (a - r_2\sin\theta_2)/\sin\theta_1$$
$$(5)\quad a_1 = (r_1\cos\theta_1\cos\Sigma + r_2\cos\theta_2)/\sin\Sigma$$

Some of the 5 independent parameters can be fixed (for example a , Σ and r_2). Then by means of a discret variation of the rest of them a lot of variants can be generated and examined, and the optimal one can be choosen I3I. Equations (1)-(5) are invariant with respect to the indices 1 and 2. Thus, if the parameters a , Σ , r_1, a_1 and δ_1 are given (i.e. the spiroid worm is "given" as we proceed in the case

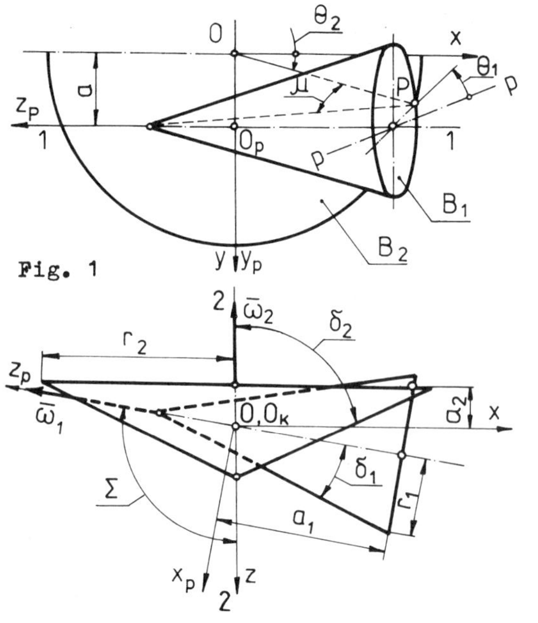

Fig. 1

of spiroid gearings I4I) then the rest 5 parameters can be determined by the same relations just upon changing the indices. Relations (1)-(5) are the most general ones.All particular cases conform to them. The tooth surfaces of the spiroid gear are generated by Oliver's second principle. That is why the geometric synthesis of the worm tooth flancs, which are linear conic helicoids, is of essential interest.Moreover, the geometric parameters of these helicoids have influence on the primary pitch cones synthesis. The equations of a right-hand conic convolute helicoid in the coordinate system of the worm are I4I

$$x_1 = r_0\cos\vartheta \pm (u\sin\xi - p_t\vartheta)\sin\vartheta$$
$$(6)\quad y_1 = r_0\sin\vartheta \mp (u\cos\xi - p_t\vartheta)\cos\vartheta$$
$$z_1 = p_s\vartheta \pm u\cos\xi$$

Here r_0 is the radius of the basic cylinder of the worm, p_s is the axial lead per radian, p_t is the lead (perpendicular to the axis 1-1) per radian in a plane which is tangent to the basic cylinder, ξ is the obtuse angle between the generatrix of the helicoid and the axis of the worm. The two sets of equations corresponding to the upper and lower signs describe the two tooth flancs of the helical tooth. The equations of an archimedian conic helicoid or of an involute one are obtainable from (6) upon assuming $r_0 = 0$ or $p_s + (r_0 \pm p_t)\mathrm{ctg}\,\xi = 0$, respectively.

We proceed now with the estimation criteria for variants.

Critical contact. Such a contact is observed between the tooth surfaces Σ_1 and Σ_2 when the contact lines have a common point (node) or an envelope. Then the common normal vector $\bar{n}_{cr}$ is called a critical normal vector. The condition for existence of a critical contact is I5I

(7) $\bar{a}.(\bar{n}_{cr} \times \bar{\varrho}_g \sin\Sigma - a\,\bar{n}_{cr} \cos\Sigma)=0$

where $\bar{a}=\overline{O_kO_p}$ is the offset vector, $\bar{\varrho}_g=\overline{O_gP}$ (g=k,p) is the radius-vector of P. Generally, the problem for avoidance of damaging critical contact is reduced to a choice of a such orientation of Σ_1 and Σ_2 that the normal vector $\bar{n}_i$ to Σ_i and the critical vector $\bar{n}_{cr}$ are not collinear in the zone of action. For example, the conditions for non-existence of a critical contact in any contact point in the case of archimedian orthogonal spiroid gearing are I15I

(8) $a/p_1 - M/r_a < i_{12} < a/p_1 + M/r_a$

Here M is the minimal mounting displacement starting from the offset line of the gearing, r_a is the maximal external diameter, and $p_1 = p_s \pm p_t \mathrm{ctg}\,\xi$.

<u>Favourable transmission of the normal forces</u>. The pressure angle α_{21}, formed between the normal force vector $\bar{P}_n$ and the circumferential velocity vector $\bar{v}_2$ of the point $P \in 2$ depends on the mutual displacement of B_i and Σ_i. This angle has considerable influence on the transmission forces conditions and on the efficiency of the gearing. It should be smaller than a definite value $I\alpha_{21}I$ for to ensure avoiding wedging.

(9) $\alpha_{21} = \arccos(\bar{n}_1 \cdot \bar{v}_2 / |\bar{n}_1| . |\bar{v}_2|) \leq I\alpha_{21}I$

$I\alpha_{21}I$ depends on the material, the lubrication and the working conditions of the concret mechanism I9,14I.

<u>The sliding velocity</u> between the tooth surfaces directly influences the intensity of the gearing and the tendency to scoring. Its value should be in conformity with the concret operational conditions of the gearing I4,14I.

<u>Minimizing the bearing reactions</u>. The quantative distribution of the axial and radial components of the transmissed normal force on the bearings is exclusively essential for heavily loaded hypoid and spiroid gearings. Requirements for minimum of some forces may be imposed, for example, the radial force on the axis of one of the gears, the two radial forces, the axial force on the axis of one of the gears, the two axial forces, the radial and the axial forces on the first axis or on the other one I2,4I.

The following criteria are related to the method of generation of Σ_1 and Σ_2.

<u>Optimal mean radius of the cutter</u>. A limit curvature $\varkappa_s$ of the mean tooth spiral of the gear is determined in the case of hypoid gearings. This curvature should ensure the same meshing conditions for both tooth surfaces I10I. Having in mind the assumption that the relative velocity vector $\bar{v}_{r2}$ of the point $P \in \Sigma_2$ coincides with the vector $\bar{v}_{12}$, then

(10) $\varkappa_s = -\bar{v}_{12} \cdot ((\bar{\omega}_1 - \bar{\omega}_2) \times \bar{n}_{cr}) / |\bar{v}_{12}|^2 = \frac{1}{R_s}$

The minimal mean radius R_s serves for the choice of standard nominal cutter diameter.

<u>Standard module of the spiroid gearing</u>. The practical realization of the tooth surfaces demands considerable expences due to the producing of the tool, i.e. the hob. The measures of the hob are functions of the axial module of its helical teeth. The decrease in the nomenclature of cutting tools involves the necessity of the design of a spiroid gearing, and a hob, resp., with a standard axial module I9I.

(11) $m_x = \frac{2r_1 \cos\delta_1 \cos\beta_1 \sqrt{r_1^2+r_o^2}}{\sqrt{1-\cos^2\delta_1 \cos^2\beta_1}\,(r_1 \mathrm{tg}\,\delta_1 + \sqrt{r_1^2+r_o^2})}$

Here β_1 is the helix angle of the worm. Requirements treating the coefficient of loss of friction, the efficiency at P, the avoidance of points of interference, etc. are taken into account also.

OPTIMIZING TECHNOLOGICAL SYNTHESIS

This synthesis connects directly the geometrical and technological parameters of the gearing with the technological parameters for generation of the tooth surfaces. In the case of hypoid gearings Σ_1 and Σ_2 should be in contact in a choosen contact point $M \neq P$. Besides, the instant angular velocity ratio $i_{12}=\omega_1/\omega_2$ at M is equal to the given ratio $u=z_1/z_2$. Definite characteristics of the profil mismach and of the direction of the contact line are required, These requirements are described by parameters which control the law of motion $\varphi_2(\varphi_1)$ of the gear and the form, measures and behaviour of the tooth contact pattern I1,6,14I. The control of the pattern in the contact point field is realized by means of mathematical modelling of the permanent contact of Σ_1 and Σ_2 when assuming one of them (Σ_2 in most of the cases) to be kept and the other (Σ_1) to be modified by a change of the generating parameters of its cutting and of the dimensions of the tool surfaces. The parameters of the conical imaginery generator that define the technological parameters of the shaping of Σ_1 are solutions of non-linear system I14I. The optimal solution is found upon cheking the results obtained for series of values of the optimizing contact parameters. This multivariant analysis (TCA) is impossible without special software and computers.

The technological synthesis of the spiroid gearings treates the synthesis of the hob and the relative motion between the gear blanc and the hob I4,9I.

CAD/CAM SYSTEM

This system includes a calculation of: a)the optimal geometric and constructive parameters of the gearing according to the criteria, b)the settings of the machines and tools (in the case of hypoid gearing), c)the geometric parameters of the hob (in the case of spiroid gearing) I4,6,7,14I. A generalized block-scheme is shown in Fig.2. The input data are: x_j^I, x_j^F and Δx_j - the initial and final values and steps of the "internal" geometry parameters; C_n - constructive and kinematic parameters; A_i and A_m - data arrays for cutting machines and tools; q_k - optimizing parameters; Q_G and Q_C -quality criteria concerning the geometry of B_1, B_2 and B_Σ, and the contact of Σ_1 and Σ_2, resp. The output information is written in tables containing the necessary and sufficient data for producing Σ_1 and Σ_2, recommendations, etc.

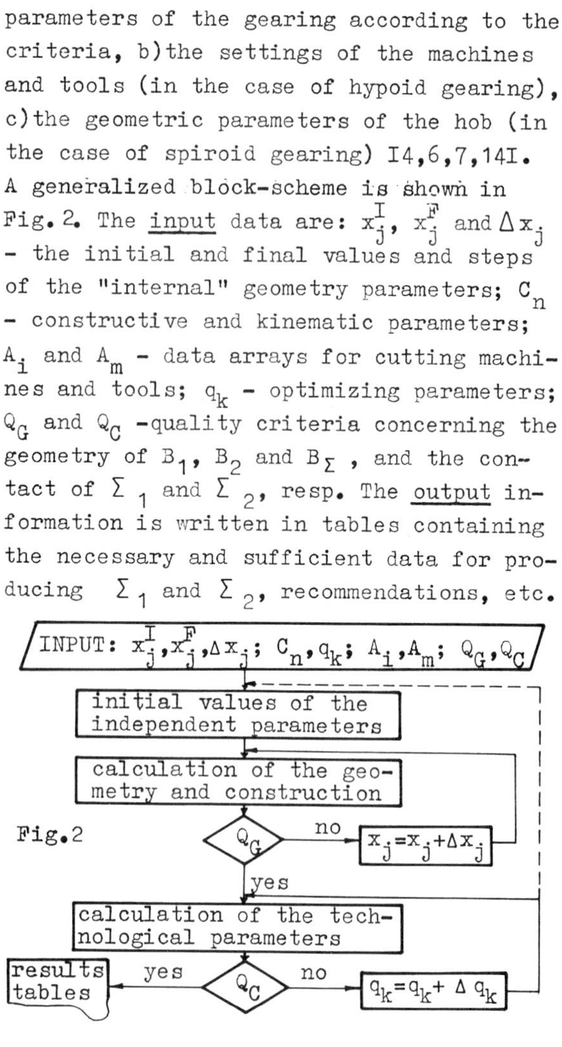

Fig.2

CONCLUSION

The constructed algorithms and program systems have already been successfully applied in industry. The created gearings are pattented for inventions I11,12,13I.

REFERENCES

1. Minkov, K. (1971). Synthesis and analysis of hypoid gears. Thesis Ph.D., Leningrad, 176 pp. (in Russian)
2. Litvin, F.L., K.M.Petrov, and V.A.Ganshin (1974). The effect of geometrical parameters of hypoid and spiroid gears on its quality characteristics. J.Eng. of Ind.-ASME, Vol.96,S. B-1,pp.330-334
3. Minkov, K. (1975). Basic synthesis of surfaces of mechanisms with arbitrary crossed axes. J. Theor.&Appl. Mech.,3, Sofia, 61-74 (in Russian)
4. Abadjiev, V. (1984). On the synthesis and analysis of spiroid gearings. Thesis Ph.D., Sofia, 158 pp. (in Bulgarian
5. Minkov, K. (1973). Avoidance of usual knot contact points of gearings. Proc. of 2nd Congr. Theor.&Appl. Mech., Vol.1 Varna, Bulgaria, pp. 430-439
6. Minkof-Petrof, K. (1983). Automated technological synthesis of spatial gears. Proc. of 6th World Congr, of IFToMM. Vol. 2, New Delhi, pp. 807-810
7. Abadjiev, V., and D.Petrova (1985).Program system for designing of spiroid gears. Proc. of Akt. probl. mas. elem. i konstr, Vol.1, Ohrid,Yugoslavia,65-76
8. Abadjiev, V., and D.Petrova (1985). Optimization of the geometric characteristics of spiroid gearings by computers. Proc. of 3rd conf."Gear Trains'85", Varna, Bulgaria (in Bulg.)
9. Abadjiev, V., and D.Petrova (1985). Automated design of spiroid gears with standard module. Proc. of 5th Congr. of Theor.&Appl. Mech., Vol.2,Sofia,111-116
10. Minkov, K. (1985). Synthesis of the parameters of the tool used for cutting hypoid gears. Proc. of 3rd conf."Gear Trains'85",Varna, Bulgaria 9in Bulg.)
11. Abadjiev, V., and K.Minkov (1983).Tooth mechanism. Pattent Bulgaria, No 36455
12. Abadjiev, V., K.Minkov, and D.Petrova (1984). Helicon tooth mechanism.Pattent Bulgaria, No 66191/11.07.1984
13. Abadjiev, V. (1984). Hyperbolic tooth mechanism.Pattent Bulgaria,No 66192
14. Minkov, K. (1986). Mechano-mathematical modelling of hyperbolic gears. Thesis Dr.Sc., Sofia, 303 pp. (in Bulgarian)
15. Abadjiev, V.,and D.Petrova (1986).Some results of the investigation of the conditions for non-existence of ordinary nodes on the contact lines of spiroid gearings. J.Theor.&Appl. Mech.,4,Bulg.

РЕЗЮМЕ

В работе изложен подход, на основе которого осуществлен оптимальный геометрический и технологический синтез гипоидных и спироидных передач с произвольно скрещивающимися осями. Специальное внимание уделено критериям для оценки вариантов базовых поверхностей заготовок и зубьев. Созданные САД/САМ системы успешно применены в болгарской промышленности. Ряд высокоредукционных передач защитены авторскими свидетельствами.

Performance Analysis of UHMWPE Forged Spur Gears

R. GAUVIN*, Q. X. NGUYEN*, H. YELLE* AND R. D. BOURBEAU**

*Mechanical Engineering Department, Ecole Polytechnique de Montréal, P.O. Box 6079, Station "A", Montreal, Quebec, Canada H3C 3A7
**Solidur, 200 Plum Industrial Court, Pittsburg, PA 15239, USA

Abstract. Forging of ultra high molecular weight polyethylene (UHMWPE) has been done. Three different compounds were used and the performance tests show that even small percentage of additive could affect the results. Another process where the resin powder is heated in a radio frequency oven and then compression molded is also investigated. In all cases, the RF heating technique produced the best of all results.

Keywords. Plastics gears; plastics industry; power transmission; manufacturing process; heat systems.

INTRODUCTION

Ultra high molecular weight polyethylene (UHMWPE) is a semi-crystalline thermoplastic with unique properties. Its outstanding combination of abrasion, low temperature impact and chemical resistance and a low coefficient of friction makes it a good choice in many applications. The polymer is supplied as a fine powder. Because it has an extremely high molecular weight which translate into exceptionnaly long molecular chains with a high degree of entanglement, it is not process by the usual thermoplastics processing technique such as injection molding and single screw extrusion, but rather by compression molding and ram extrusion.

According to commercial literature, the percentage of applications fulfil by the various processes are the following: 60% by compression molding, 35% by ram extrusion, 3-4% by twin screw extrusion and less than 1% by injection molding.

For gear manufacturing, the most common process is to machine a billet cut from a ram extruded rod, as shown in Fig. 1a. For diameters larger than 15 cm, the blank is generally cut from a compression molded thick sheet. Parts can also be compression molded directly from powder in a heated mold but the cycle time is extremely long.

In literature, the solid phase forming of a three-dimensional part is often refer to as forging [1,2]. In this paper, we refer to forging as being a process where a three-dimensional part is formed into shape, in a mold or a die, by the application of pressure on preheated solid blank [3,4]. It is schematically shown in Fig. 1b. An interesting survey on this process as been published in reference [5]. The technique has several advantages over the previous one for reasons such as shorter cycle time and better mechanical properties. Little work has been published to date for the manufacture of gears with that process [3,6]. In a previous work [7], it was shown that when applied to UHMWPE gears manufacturing, a significant load bearing capacity improvement could be expected.

More recently [8], a process using radio frequency heating was proposed and is schematically described in Fig. 1c. Its main advantage over the previous two, being to skip the production of a semi-finished product. As oppose to the classical compression molding process where the powder is heated in a compression mold, here, the polymer is heated in a radio frequency oven which reduced drastically the heating time and then transferred into a compression mold to produce the part.

In this paper, an extensive evaluation of forged gears obtained with various UHMWPE Compound is presented. Preliminary data for gears molded with R-F heated resin are also reported and compared with forged gears.

MATERIAL AND GEAR GEOMETRY

Four different compounds where used in this study. Three for the forging process and one for the R-F heating and compression molding process. Their respective formulation and the corresponding gear manufacturing process as well as their abreviation used in this article are listed in Table 1. The GUR-413 is a Hoescht resin having a density of 0.934 gr/cm^3 and an average molecular weight $M_W \simeq 4.0 \times 10^6$. The so call 9M is an experimental formulation with an average molecular weight in the order of 9 millions.

TABLE 1 Identification of the Various Compounds Used with the Two Processes

COMPOUND IDENTIFICATION	FORMULATION	PROCESS
Zo	GUR-413 Zinc oxyde added	Forging
Zs	GUR-413 Zinc Sterate added	Forging
9M	UHMWPE $M_W \simeq 9. \times 10^6$	Forging
R-F	GUR-413 Carbon Black added	Radio-Frequency Heating

For forging, ram extruded rods were used to machine billets. Machining was decided to remove any skin effect and also to obtain a specific billet size. The dimension of the billet was choosen to have a forming ratio (F.R.) of 50% defined by:

$$R.F. = [(D_g - D_b) \div D_b] \; 100\% \qquad (1)$$

D_g is the gear pitch diameter and D_b the billet diameter. The geometry of the billet and the finished forged gear are shown in Fig. 2. They have a 20° pressure angle, a module (m) of 7.6 and 30 teeth.

For R-F heating, Carbon Black was used as a conductive agent and dry blended with the powder resin. Carbon Black was chosen because of its high electrical conductivity. Compression molding of the R-F heated powder blend was done in the same tooling used for forging to assure a constant geometry for all the gears.

EQUIPMENT AND PROCESS DESCRIPTION

The two processes under study are schematically represented in Fig. 1b and 1c.

Forging

In the forging process, billets are cut from extruded rods and heated in a convective oven at a proper temperature. The billet is then transfer and centered into the mold cavity where pressure is applied at a constant speed of 80 cm/min to forge the gear. Pressure is maintained for a certain period of time, call the dwell time (τ) to allow the material to cool down and the part to retain its shape.

Optimization of the processing parameters was done in previous work [7,9]. Based on that, the following processing parameter were choosen in this study:

Dwell pressure	P : 200 MPa
Mold temperature	Tm: 35° C
Billet temperature	Tb: 150° C
Dwell time	τ : 1 min
Ram speed	v : 80 cm/min

It was experimentally verified that a heating time of two hours in an oven settle at 150° C was long enough to evenly heat the billet at that temperature. Considering that the melting point of the perfect polyethylene crystal is 141.5° C [10], it is reasonable to assume that the polymer is completely molten at 150° C. However, it does not flow easily at that temperature and has a rubber like behavior.

The mold cavity (forging die) is shaped as an internal gear and the punch as an external gear. Since the polymer remains relatively solid, flash is not a problem unless the clearance between the punch and the cavity is over 0.125 mm. Of primary importance for the strenght of forged gears, is the tooth root radius. A small root radius will creates intensive shear during the forming process and promotes the formation of small cracks under the surface an parallel to it [6]. As shown in Fig. 3 a full fillet radius was used to minimised that problem. The dotted line on that figure represents the geometry of a hub cut gear.

Radio Frequency Heating and Molding

This process, here called the R-F process, is described in Fig. 1c. First, the powder compound is heated in a radio frequency oven and then transfer into the mold cavity where pressure is applied to form the part. The Radio Frequency oven used is a flat platen electrodes type having a power of 5 Kw and a fixed frequency of 55 MHz. Because of its high temperature, the polymer powder stick together and can easily be transferred from the oven to the mold cavity.

This process is still under investigation and has not been fully optimized yet. Nevertheless, the following processing conditions gave good results and were used for the gears tested in this work:

Dwell pressure	P : 200 MPa
Mold temperature	Tm: 110° C
Powder surface temperature	Tp ≃ 150° C
Heating time in the R-F oven	Tr: 2.5 min
Dwell time	τ : 4 min

The processing parameters are highly dependant on the oven power and frequency as well as the distance choosen between the electrodes. The mass of the compound is also important. To avoid over heating of the polymer, the residence time in the oven (Tr) must be carefully controlled. The R-F oven does not heat the compound evenly and for the conditions choosen the polymer temperature varies from 150° C at the surface to approximately 165° C to 170° C in the center.

To get a good surface finish, it was found that a much higher mold temperature (110° C) had to be used compared to forging (35° C). Along with that higher mold temperature, a longer dwell time (4 min) was needed to allow for enough cooling. Nevertheless, total processing time from powder form to finished part is in the order of 6 to 7 min, which is a tremendous improvement over the other processes.

RESULTS AND DISCUSSION

Shrinkage

Before testing gears performances, their quality and shrinkage had to be evaluated. In a forged gear, shrinkage is a complex phenomenon. It is a combination of diameter variation, similar to a disk, and tooth deformation which can be visualized as swelling at the tooth base and shrinkage at the top. To measure the quality of a forged gear, a gear rolling tester was used. This apparatus measures the variation of the center distance between the tested gear and a master gear, when they mesh without backlash. To obtain a gear that perfectly fits the mold geometry, a zero shrink material was cast in the cavity.

The results for the mold and a typical gear are schematically shown in Fig. 4. Each curve is the result of the excentricity and the tooth to tooth composite error of the measured gear. The difference between the maximum and the minimum value is called the total composite variation (TCV) of a gear. For shrinkage evaluation, reference diameters for the mold (D_{RM}) and for the forged gears (D_{RG}) were defined. As shown in Fig. 4, these reference diameters are taken at the mid value of the total composite variation. They are in vicinity of the pitch diameters.

The shrinkage of forged gears is then defined as the following:

$$S = [(D_{RM} - D_{RG}) \div D_{RM}] \; 100\% \qquad (2)$$

It is an evaluation of the percentage of variation of the forged gear geometry with respect to the mold geometry.

Results obtained with several forged gears are given in Table 2. The average values and standard deviations are fairly small which is a good indication of the gears quality and the optimization of the processing parameters.

As mentioned previously, the R-F oven process is not fully optimized yet and the shrinkage values are higher but quite satisfactory. Measurements done on 10 gears gave an average value of 1.64%, a maximum of 2.16%, a minimum of 1.3% and a standard deviation of 0.3%.

TABLE 2 Shrinkage of Forged Gears

MATERIAL	UHMWPE Zo	UHMWPE Zs	UHMWPE 9M
NUMBER OF GEAR	10	10	10
MEAN VALUE	0.0132%	- 0.0005%	0.76%
STANDARD DEVIATION	0.154 %	0.0974%	0.199%
MAXIMUM VALUE	0.392 %	0.13 %	1.05 %
MINIMUM VALUE	- 0.065 %	- 0.2 %	0.39 %

Performance test

Performance test of all gears were conducted on a four square testing bench with recirculating powder at a speed of 2000 rpm, against a steel pinion. All tests were run without lubrication at room temperature.

For simplicity, the rooth bending stress was computed using the Lewis equation without any correction factors:

$$\sigma \quad \frac{Wt}{m\ F\ Y} \qquad (3)$$

where:

σ = rooth bending stress (MPa)
Wt = total transmitted tangential load (N)
m = gear module (mm)
F = tooth width (mm)
Y = Lewis tooth form factor for the load applied at pitch point. In this case = 0.606

In all, 55 gears were tested for the four compounds, representing some 1500 hours of testing. Figures 5, 6 and 7 show the results for the forged gears while Fig. 8 shows the results for the R-F oven heated compound. Each dot represents a life test on one gear. As can be seen non-failed gears were stop after 10^7 cycles (84 hours) of continuous running. The solid lines on these figures represents the best guest fit of the data points.

Figure 9 is a summary of the four previous figures. As one can notice the Zo and Zs compound have similar behavior for high load and short life. The long life loading capacity of the Zs compound is somewhat higher. This is probably because the Zinc Sterate is an internal lubricant which reduces the coefficient of friction and help keep the heat generation at a lower level.

The 9 Millions molecular weight compound has a higher short life strenght but a somewhat smaller one at long life that the previous two compounds. This material having a stronger memory, the gear teeth do not retain their geometry as well under stress an temperature combination. This increases the interference in the gear mesh which is probably the reason for a lower capacity at long life.

It is interesting to note that the UHMWPE - Carbon Compound process with the R-F oven has the best performances of all the compounds studied. This is probably partly due to a slight lubricating effect of the carbon black but is certainly a sign of a good bounding between the UHMWPE particles.

CONCLUSION

The results of this study show that forging can produce good and long lasting gears. It also indicates that additives, even in very small percentage can have a significant effect on the performance of dry UHMWPE gears. Finally, it demonstrate the feasibility of the R-F heating technique and its capacity to produce very good parts at a much cheaper cost.

ACKNOWLEDGEMENTS

The authors wish to acknowledge the excellent support of Mr. B. MacDonald from Robco Inc. This work was partly funded by the National Research Council of Canada, grant A-1608 and the Quebec Government FCAR program, grant EQ-3361.

REFERENCES

[1] Werner, A.C. and J.J. Krim (1968). Forging High Molecular Weight Polyethylene. SPE Journal, Vol. 24, No 12, pp. 76-79.
[2] Kraufer, H. and P. Kristukat (1980). Behavior of Partially Cristalline Thermoplastics when Compression Stretched and Attainable Properties, Using POM as Example. Kunststoffe 70.
[3] Kulkarni, K.M. (May 1979). Review of Forging, Stamping and Other Solid-Phase Forming Processes. Polymer Eng. and Sci., Vol. 19, No 7, pp. 474-481.
[4] Kulkarni, K.M. (July 1974). Forging of Rigid Crystalline Plastics. U.S. Patent 3,825,648.
[5] Haudin, J.M. and E. Weynant (Mars 1980). Le forgeage des polymères, synthèse bibliographique. Ecole Nat. Sup. des Mines.
[6] Brezina, M., P. Bertrand and P. Wiser (1980). Dimensional Stability and Fatigue Resistance of Acetal Homopolymer Gears Forged in the Solid State. Int. Power Transmission and Gearing Conf., San Francisco, Paper 80-C2/DET-106.
[7] Gauvin, R., Q.X. Nguyen, J.P. Chalifoux, and H. Yelle (March 1986). Le forgeage, une solution pour augmenter la capacité de charge des engrenages en UHMWPE. 2nd World Congress on Gearing, Vol. 2, pp. 3-11, Paris.
[8] Miller, B. (March 1981). New Twist in Plastics Processing Melting with Radio Waves. Plastics World, pp. 99.
[9] Gauvin, R., Q.X. Nguyen and J.P. Chalifoux (1986). Shrinkage Analysis of Forged UHMWPE Parts. Proc. of 44th ANTEC, pp. 1017-1024, Boston.
[10] Wunderlich, B. and C.M. Cormier (1967). Heat of Fusion of Polyethylene. J. Polym. and Sci., A-2, 5, pp. 987-988.

RESUME

Le polyéthylène à très haut poids moléculaire (UHMWPE) est un thermoplastique semi-cristallin. Son exceptionnelle résistance à l'usure et à l'impact, de même que son excellente résistance aux produits chimiques en font un choix judicieux pour plusieurs applications d'engrenages.

Contrairement aux autres thermoplastiques, il est très difficile à mettre en oeuvre par extrusion à vis ou par injection à cause de son poids moléculaire très élevé ($M_W > 3.0 \times 10^6$). Généralement, les engrenages qu'on en fabriquent, sont taillés dans des barreaux cylindriques produits par extrusion à piston ou encore, ils sont moulés sous presse à partir de la résine en poudre. Ces deux procédés sont très longs. Pour contourner ce problème, cet article propose deux procédés. Le premier est le forgeage de billettes extrudées et préalablement chauffées dans un four à convection. Le deuxième consiste à chauffer un mélange de résine en poudre dans un four à hautes fréquences pour ensuite mouler directement l'engrenage.

Les essais sont faits sur trois composés forgés et un autre chauffé par hautes fréquences et moulé. Les tests de performances sont conduits à 2000 rpm sans lubrification et à température ambiante. Les résultats montrent une performance semblable pour les trois composés forgés et une performance nettement supérieure pour le composé chauffé par hautes fréquences et moulé.

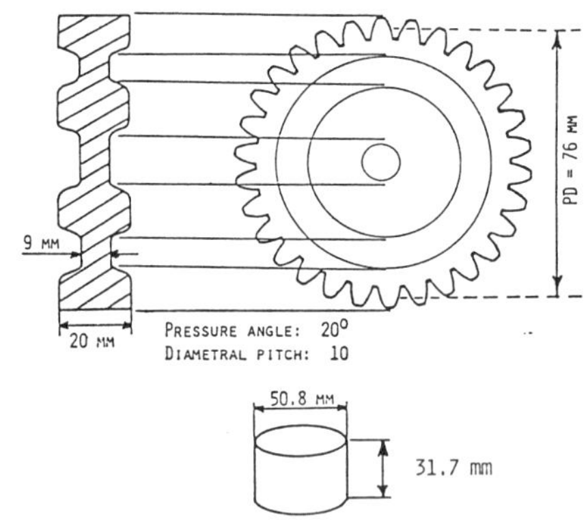

Fig. 2. Billets used and the final shape of the forged gear.

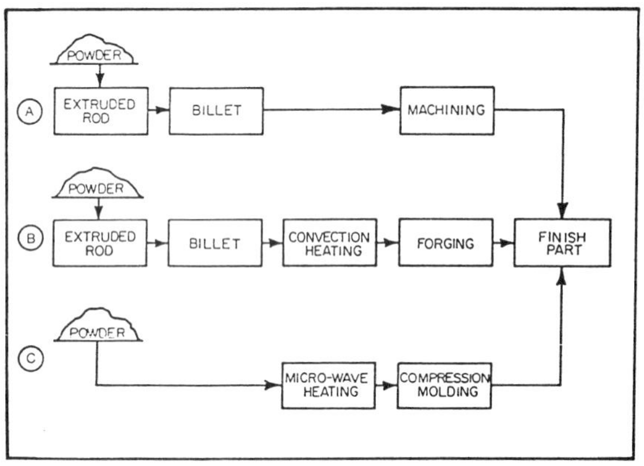

Fig. 1. Processing techniques for UHMWPE

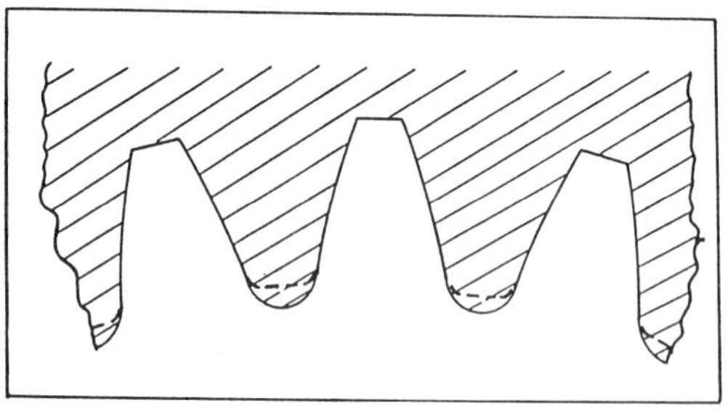

Fig. 3. Geometry of the mold cavity. The dotted line represents the geometry of a hub cut gear.

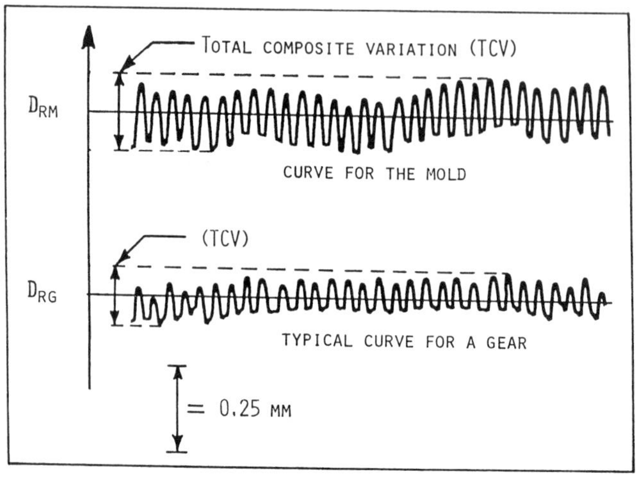

Fig. 4. Center distance variation measured on a gear rolling tester.

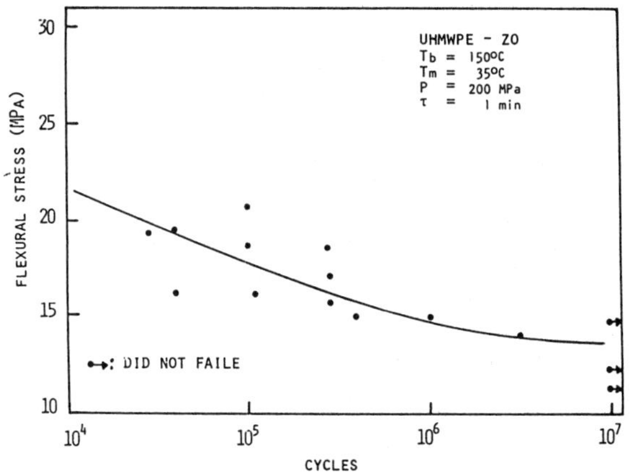

Fig. 5. Root bending fatigue strength of UHMWPE-Zo gears meshing with a steel pinion, without lubrication.

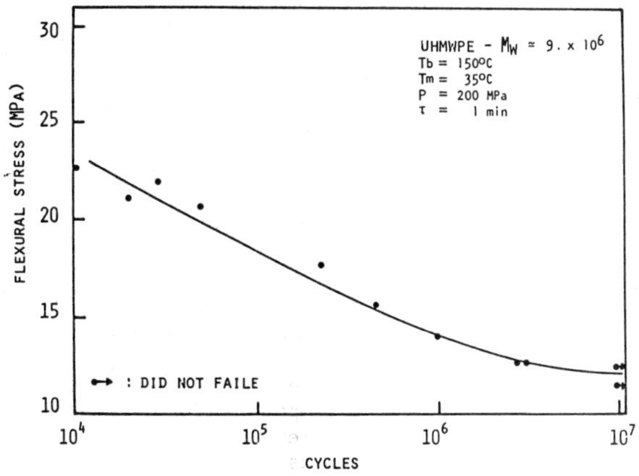

Fig. 7. Root bending fatigue strength of UHMWPE-$M_W \simeq 9. \times 10^6$ gears meshing with a steel pinion, without lubrication.

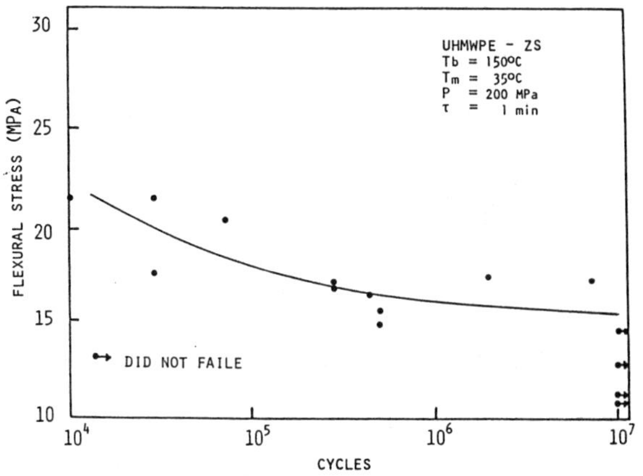

Fig. 6. Root bending fatigue strength of UHMWPE-Zs gears meshing with a steel pinion, without lubrication.

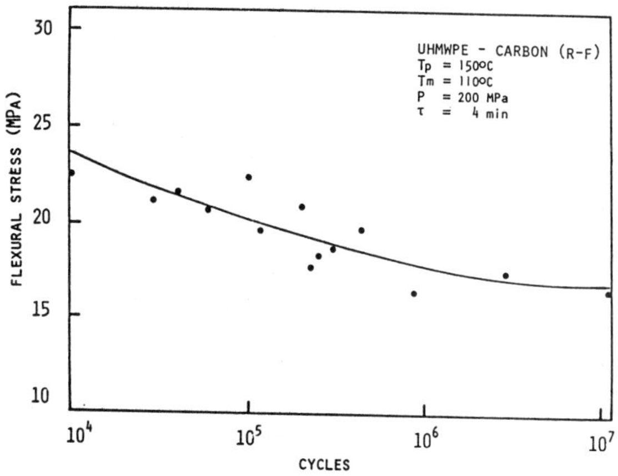

Fig. 8. Root bending fatigue strength of UHMWPE-Carbon gears, process with R-F, meshing with a steel pinion, without lubrication.

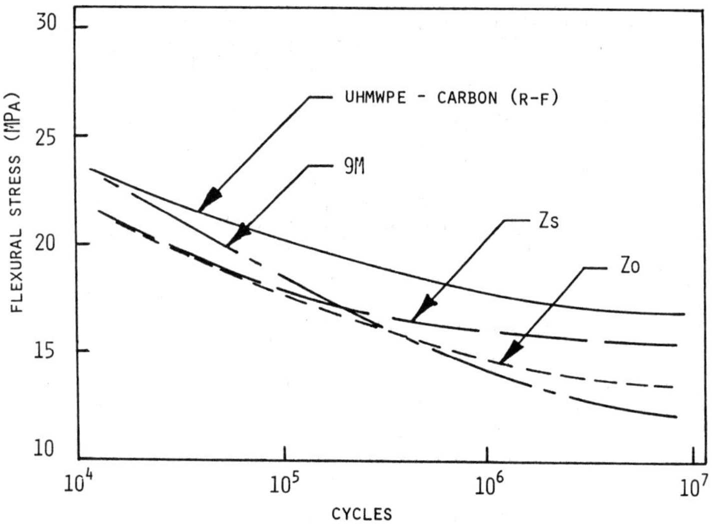

Fig. 9. Comparison of the performances of the four compounds and two processes studied.

Flank Temperature Profile for Thermoplastic Spur Gears

R. LIU*, D. J. BURNS** AND H. YELLE***

*Department of Mechanical Engineering, Xian Institute of Metallurgy and Construction Engineering, Xian, Shaanxi Province, PRC
**Department of Mechanical Engineering, University of Waterloo, Ontario, Canada
***Department of Mechanical Engineering, Ecole Polytechnique de Montréal, Québec, Canada

Abstract. Analytical calculating techniques have suggested the presence of two peaks of temperature on the flank of thermoplastic spur gears. An experimental technique using an infrared pyrometer has permitted to detect and measure these two peaks.

Keywords. Hysteresis; plastics industry; stress control; power transmission; gears.

INTRODUCTION

As with metal gears, thermoplastic (plastic) gears may fail by fatigue cracking at the root of the tooth; by tooth surface fatigue or by excessive wear. However, plastic gears are much more sensitive to temperature and attention must be given to the heat generated by surface friction and hysteresis. Heat buildup in the tooth may accelerate one of the failure mechanisms mentioned above or be so severe that thermal softening occurs. This leads to either local scuffing on the tooth flank or, in extreme cases, to large tooth deformation and shear failure.

A previous paper [1] outlined a method for estimating the temperature rise in plastic spur gears and for assessing whether gear failure is likely to occur by thermal softening rather than root bending fatigue. This previous analysis indicated that the most important heat source is friction. Other heat sources are hysteresis heat produced by cyclic contact between the teeth and by cyclic bending of the teeth.

Figure 1 illustrates the relative magnitude of the friction and hysteresis energy losses on the flanks of a typical gear pair made of a 24 tooth nylon gear driving a steel pinion. Details of the calculations used as the basis for Fig. 1 are given in reference [2]. These heat sources are only acting during 5 to 7 per cent of each revolution of the gear. For the remainder of the revolution, the teeth are not loaded and travel in air or oil mist. The most important heat removal mechanism during this non loaded phase is forced convection. A previous paper [1] explains how finite element and finite difference techniques can be used to calculate the quasi static equilibrium temperature distribution along the flank of plastic spur gears. Figures 2(a) and (b) show this theoretical temperature distribution for 16 pitch, 36 teeth acetal gear pairs at 52.8 N/mm and 5.6 m/s and at 37.1 N/mm and 3.8 m/s respectively. Temperature variations of this magnitude along the flank cause a very significant variation in the yield strength, S_y, along the flank. This also causes a variation in modulus which in turn influences the value of the contact stress, σ_h, at any point.

In a previous paper [1], it was argued that thermal softening and local scuffing is likely to occur if at any point on the flank the hertzian contact stress exceeds the local yield strength. This condition is illustrated on Fig. 2(a). When the speed and the load are reduced for the same gear pair (Fig. 2(b)), σ_h no longer exceeds S_y at any point along the flank. It will be noticed in Fig. 2(a) and 2(b) that the temperature variation along the flank shows two maxima that coincide with two peaks in the distribution of the heat generated by friction (Fig. 1). For the higher load and speed case, the conditions are so severe that the analysis indicates that scuffing damage may also occur near the tip as well as near the root. Indications of two damage zones have been obtained by the authors in tests on various plastic gears. A particularly severe example is shown in Fig. 3. The pinion shown was Ultra High Molecular Weight Polyethylene (UHMWPE) with 25 teeth and a module of 5.08 mm (DP = 5).

Various authors have attempted to measure the temperature distribution along the tooth flank. The late professor Takanashi and professor Shoji used thermal sensitive paint on nylon/steel gear pairs and found "that irrespective as to whether a nylon-6 gear is operated as driver or follower, higher temperature regions appear near the pitch point on its teeth". An examination of the photograph in their paper [3] suggests that the highest temperature may be below the pitch point, but there is nothing to support the two peak prediction discussed above.

A much more sensitive method for monitoring flank temperature is to use an infrared pyrometer. Yousef, Burns and McKinley [4] were the first to use this technique with thermoplastic gears, but they only recorded a mean flank temperature. Gauvin, Girard and Yelle [2,5] made a much more detailed study of flank temperature using an infrared pyrometer. They did not detect the two peak temperature profile illustrated on Fig. 2(a) and 2(b); however, an examination of their experimental technique indicated that they used a pyrometer orientation that would tend to mask the second peak. Therefore, their experiments have been repeated using better pyrometer orientations. This paper explains the modified experimental technique and presents temperature profiles for nylon gears that were driving or driven by a nylon or a metal gear.

TESTED GEARS AND TECHNIQUE

TABLE 1 lists the combinations of materials and gear ratio studied. These materials were chosen because it was desirable to have some data that could be compared directly with the earlier data obtained by Gauvin, Girard, Yelle [2,5]. A four-square gear described in [2] was used.

The apparatus used for temperature measurements was a Barnes Engineering "Spectral Master" infrared research radiometer, model 12-550, serial W2202. The radiometer consists of a sensing head that collects radiation emitted by a remote object and focusses it on an Indium - Antimonide (In-Sb) detector cooled by liquid nitrogen.

Figure 4 shows schematically the pyrometer oriented towards the working flank of a driving tooth shortly after disengagement. With this pyrometer, it was possible to reduce the spot size to about 1.5 mm diameter by setting the front face of the sensing head approximately 600 mm from the pitch point P (distance UP in Fig. 4). As the flank moves in front of the pyrometer, this distance changes by about one per cent for the gears listed in TABLE 1. With this spot size and the temperature range of interest (20°C to 80°C), the pyrometer resolution is better than 0.1°C [6].

Figure 4 shows that as the gear rotates, the pyrometer looks at points B', D, P, C, Q, B and B'. For convenience, the pyrometer is shown moving relative to the gear, in particular positions (1), (2) and (3) where the pyrometer spot is at B, P and Q respectively. The angle between the radiometer axis and the normal to the tooth flank, known as the emittance angle θ, varies from point to point. This variation may lead to erroneous estimates of flank temperature. To establish acceptable values for θ, a radiating source was held at temperatures of 52°C, 56°C, 64°C or 69°C and inclined at various angles θ to the pyrometer. Figure 5 shows the radiance readings of the pyrometer as θ was increased from 0 to 90 degrees. Figure 5 shows that variations in θ are unimportant as long as θ is less than about 65 degrees.

The maximum value for the angle θ is a function of the setting angle γ of the pyrometer, see Fig. 4. Gauvin, Girard and Yelle [5] used an angle γ of 0 degree. When γ is zero and the pyrometer is looking at points on the tooth flank below the pitch circle, the angle θ approaches 90 degrees, as illustrated by the hoken line tooth on Fig. 4. Therefore, Fig. 5 shows that their temperature distribution were incorrect over part of the tooth flank and that they would not detect the double peak mentioned earlier.

As the angle γ is increased from zero, θ_{max} first reduces and then increases. Also, the angle γ cannot be more than a few degrees above zero because the following tooth will mask the lowest meshing point on the tooth flank being monitered. For the gears tests reported herein, γ was set between 2 and 3 degrees thereby keeping θ_{max} below 50 degrees.

Figure 6 shows how θ varied in the present tests as the measuring point moved along the tooth flank from B' and B, see Fig. 4. Figure 6 also shows a typical plot of pyrometer output against measuring position on the tooth flank. Pyrometer output was related to measuring positions on the tooth flank by generating a reference signal. This signal was obtained by gluing a strip of aluminum foil to the top of a tooth to reflect a reference light beam into the pyrometer. There are three peaks on the pyrometer output shown on Fig. 6. The peaks due to gear interaction were not sensitive to small changes in γ within the specified setting of 2 to 3 degrees. The peak due to the reflection of the reference signal from the aluminum foil was sensitive to small changes in γ as shown schematically for γ_1, γ_2 and γ_3 on Fig. 6. This sensibility made it easy to identify the peak due to the reference signal and to calibrate the flank positions.

Figure 7 is a photograph of an oscilloscope display of the pyrometer signal. In this case, aluminum strips were placed on two adjacent teeth so both of the profiles show three peaks. Oscilloscope photographs of this type have been used to generate the temperature distributions shown in Fig. 8 and 9. Temperature data is shown for a nylon/nylon (Fig. 8) and a nylon/cast-iron gear pairs (Fig. 9). Data is shown for the nylon gear acting as the driving or driven gear. Both figures were obtained for a tangential load of 20 N/mm and various tangential speeds V. The effect of varying the tangential load and holding the speed constant at 6.65 m/s was also investigated but the data is not presented herein because of space constraints. In all cases, a double peak (excluding the reference signal) was observed in the temperature distributions and these peaks occurred at or near to the locations expected theoretically.

As shown in Fig. 1, the maximum energy losses in a gear pair occurs at the beginning of engagement. Therefore, the maximum peak in the temperature distribution for the flank of the driving gear occurs below the pitch circle whereas the maximum peak for the driven gear occurs above the pitch circle. This difference between driving and driven gear temperature distributions is very clear on Fig. 8 and 9 and becomes more evident as the pitch line velocity increases. This was also true in tests where the pitch line velocity was held constant and the load was increased. It will be noticed that the highest peaks were obtained in the experiments where the nylon gear was driving the cast-iron gear. This arrangements would not normally be used because the tip corner of the metal gear tends to scrape the flank of the plastic gear at the beginning of engagement. This is shown clearly in Fig. 3 for a UHMWPE pinion driving a steel gear.

In the experiments reported herein there was a tendency for the plastic gears running against the cast-iron gear to have higher flank temperatures than those in a plastic/plastic gear pair. This difference is believed to be directly related to the coarse surface finish on the cast-iron gear; improving the surface finish of the metal gear should reduce or reverse this difference because of the reduction in the friction losses and the much higher thermal conductively of metal surfaces.

Figures 8 and 9 show that at this load and range of speeds, the maximum flank temperature rise was between 10°C to 50°C whereas the root temperature rise was only 5°C to 10°C. There is ample experiment evidence [1] that this load level is unlikely to lead to a root bending fatigue failure in less than 10^7 cycles. Therefore, the design concern is whether local flank temperature rises as higher 50°C will lead to excessive wear or scuffing. At present, there is no simple method for answering this question. As mentioned earlier, the authors have some experimental evidence that excessive wear or scuffing may occur if local hertzian stress exceeds the local yield strength. This requires further study now that a more satisfactory technique has been developed for assessing local flank temperatures. Of course, the actual peak flank temperatures must be somewhat higher than those recorded just after disengagement.

REFERENCES

[1] Yelle, H., Yousef, S., Burns, D.J. (1981). Root bending fatigue and thermal softening of acetal thermoplastic spur gears. Fatigue 1981, Warwick, England, 24-27 mars 1981, pp. 261-270.
[2] Gauvin, R., Girard, P., Yelle, H. (1983). Prediction of the peak temperature on the surface of thermoplastic gear teeth. Paper no P149.03, AGMA Fall Technical Meeting, Montreal, Canada, October 17-19, 1983.
[3] Takanashi, Shoji. (1975). Bulletin of Japan Soc. Mech. Engrs., Vol. 6, No 75, p. 73.
[4] Yousef, S., Burns, D.J., McKinley, W. (1973). Techniques for assessing the running temperature and fatigue strength of thermoplastic gears. Mechanism and Machine Theory, Vol. 8, pp. 175-185.
[5] Gauvin, R., Girard, P., Yelle, H. (1980). Investigation of the running temperature of plastic/steel gear pairs. ASME Int. Power Transmission and Gearing Conf., San Francisco, Calif., August 18-21, 1980, Paper no 80-C2/DET-108.
[6] Instruction Manual of Spectral Master Infrared Research Radiometer, Model 12-550, Barnes Engineering, 30 Commerce Rd, Stamford, Conn., U.S.A.

RESUME

Des études analytiques déjà publiées par les auteurs suggéraients la présence de deux pointes de température sur le profil de la denture d'engrenages cylindriques droits en matières plastiques: une à la tête et l'autre près du pied de la dent. Les programmes de mesure expérimentaux avaient, jusqu'ici, échoué à mettre en évidence l'existence de ces deux pointes.

Cet article décrit en détail la technique expérimentale qui a permis de détecter ces deux pointes avec un pyromètre optique infrarouge et de mesurer leur amplitude. Les mesures effectuées ont permis de découvrir que l'emplacement relatif de la pointe de température la plus importante est inversé selon que l'engrenage est mené ou menant. Les mesures rapportées mettent en évidence l'effet de la vitesse et de la charge pour un module (5.08 mm, 25 dents) et un matériau (nylon) dans des conditions de fonctionnement sans lubrification.

TABLE 1 Test Gears

Gear pair	Driving gear Material	Driving gear Tooth no	Driven gear Material	Driven gear Tooth no	Module (DP)	Face width mm (in)
A	Nylon 6-6	25	Nylon 6-6	25	5.08 (5)	12.7 (0.5)
B	Nylon 6-6		Cast-iron			
C	Cast-iron		Nylon 6-6			

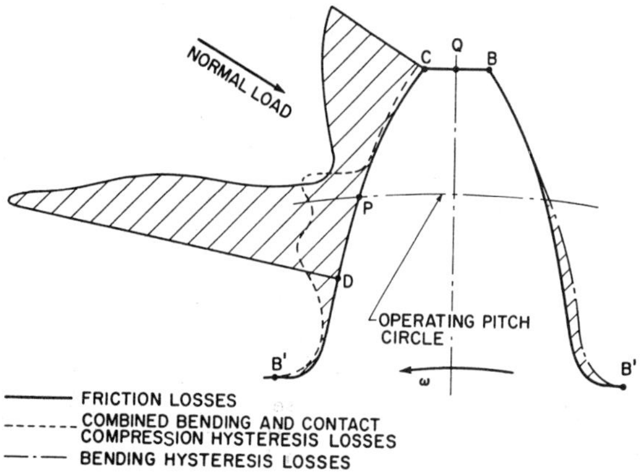

Fig. 1. Total friction and hysteresis energy losses in a plastic/steel gear pair shown on the driving plastic tooth.

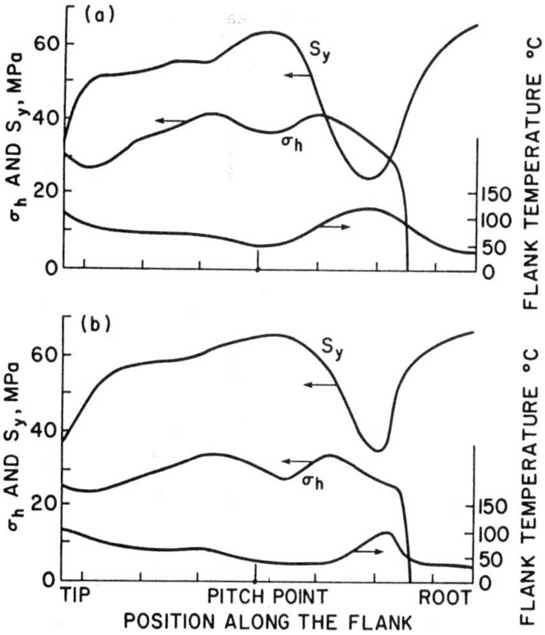

Fig. 2. Variation of contact stress and temperature along the flank for pairs of 1.58 mm module (16 DP) acetal gears.

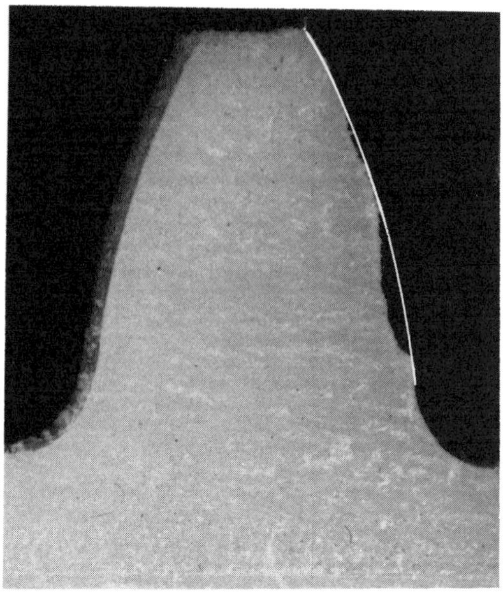

Fig. 3. Flank damage zones at mid face of a U.H.M.W.P.E. pinion.

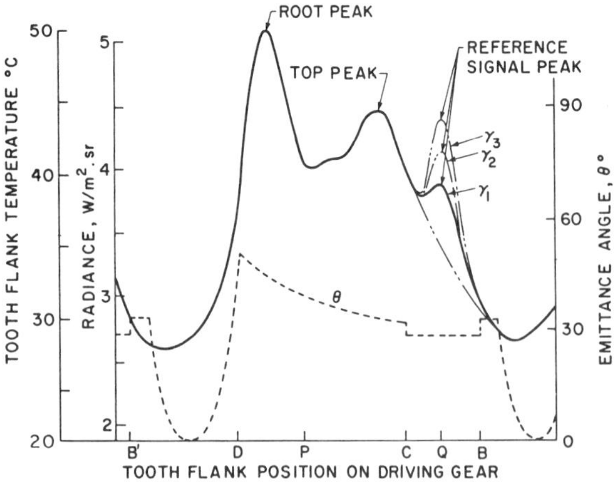

Fig. 6. Variation of emittance angle and radiance (flank temperature) along a tooth flank.

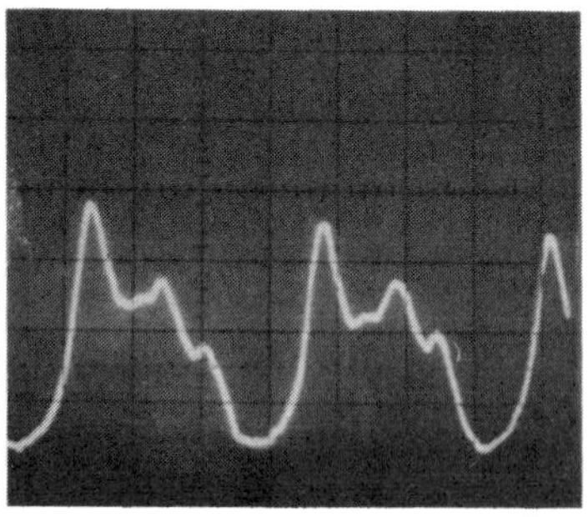

Fig. 7. Typical oscilloscope record of pyrometer output.

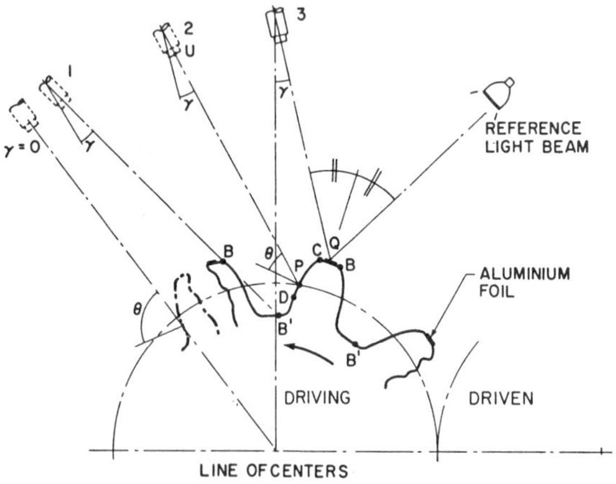

Fig. 4. Relative orientation of the pyrometer, tooth flank of interest and the reference light beam.

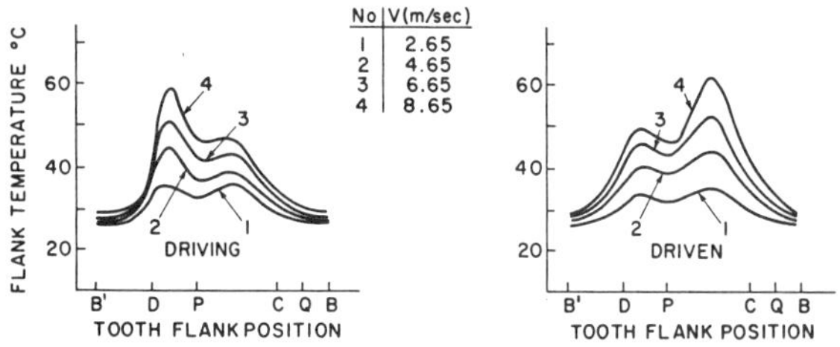

Fig. 8. Effect of pitch line velocity (V) on flank temperature on the driving and driven gears of a nylon/nylon gear pair.

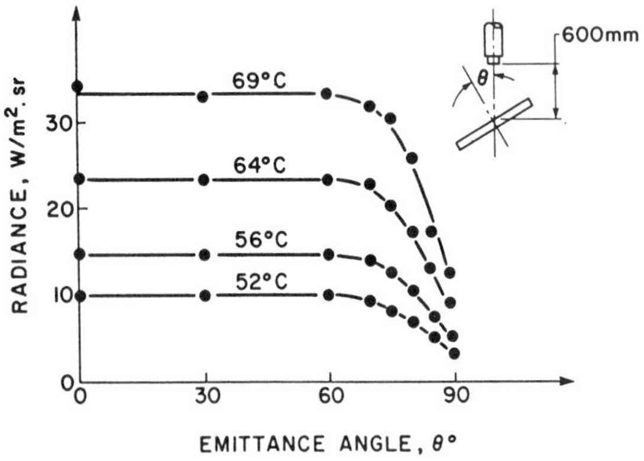

Fig. 5. Effects of emittance angle on pyrometer reading.

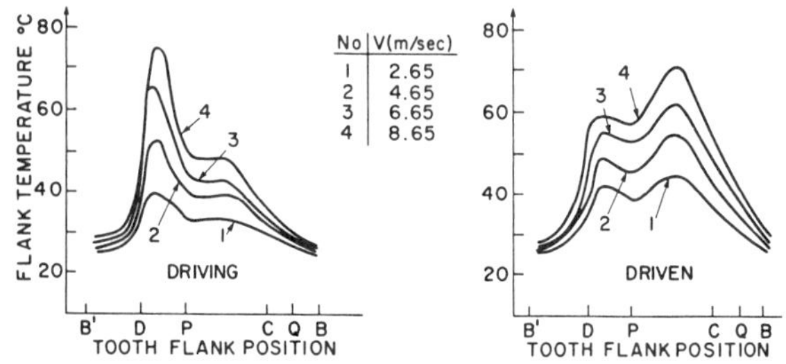

Fig. 9. Effect of pitch line velocity (V) on flank temperature of a nylon gear driving or driven by a cast-iron gear.

Etude Analytique de L'Energie Perdue et de Sa Repartition dans les Engrenages en Plastique

D. KOFFI*, H. YELLE** AND R. GAUVIN**

*Département d'ingénierie, Université du Québec à Trois-Rivières, Québec, Canada

**Département de génie mécanique, Ecole Polytechnique de Montréal, Québec, Canada

Résumé. Une méthode analytique a été développée pour calculer les températures de surface et globale de la dent d'un engrenage. Cet article discute la méthode développée et fait une étude de l'effet de la charge, du module et du coefficient de transfert de chaleur sur la distribution de la température.

Keywords. Heat transfer; plastics industry; power transmission; gears; hysteresis.

INTRODUCTION

Le calcul des engrenages en plastique porte, en général, sur deux aspects: la résistance en fatigue de flexion au pied de la dent et la résistance en fatigue de surface sur le flanc. Reconnaissant que la tenue des matériaux plastiques aux sollicitations mécaniques est fortement affectée et limitée par la température, il devient important de pouvoir évaluer celle-ci aux endroits stratégiques, soit au pied et sur le flanc de la dent.

Quoique l'on ait essayé plusieurs fois de mesurer la température des engrenages en plastique en fonctionnement [1,2,3,4], ce n'est que récemment qu'une technique satisfaisante semble avoir été mise au point [5]. Les auteurs ont développé une méthode analytique pour calculer la température sur la surface et à l'intérieur d'une dent. La méthode de calcul fait appel aux résultats de travaux sur le partage de la charge [6], la génération et la répartition de l'énergie perdue [7] et le coefficient de transfert de chaleur par convection forcée entre une dent et l'air ambiant [8].

Cet article discute la méthode de solution des équations de transfert de chaleur par les différences finies. Après une analyse du mouvement des dents dans une paire d'engrenages, on présente et discute les résultats de calculs effectués en faisant varier la charge et le module. A la fin, on compare des températures mesurées par thermocouple dans le centre de la dent avec des valeurs calculées.

Les calculs et les mesures présentés dans cet article sont pour une paire d'engrenages cylindriques droits en plastique ayant un angle de pression de 20°, 30 dents chacun et tournant à une vitesse de 4 m/s sans lubrification. Pour les calculs, on a utilisé un coefficient de frottement dynamique de 0.1, un coefficient de perte par hystérésis tg δ = 0.03 et un module d'élasticité dynamique de 1.0 GPa.

THEORIE

L'étude de la distribution de température est basée sur la résolution des équations de convection-diffusion en régime permanent. L'étude est faite en deux dimensions car il a déjà été montré [8] que la température varie peu dans la direction axiale; l'effet du rayonnement est aussi négligé. L'équation générale de transmission de chaleur s'écrit:

$$\frac{\delta^2 T}{\delta x^2} + \frac{\delta^2 T}{\delta y^2} = 0 \qquad (1)$$

où T est la température d'équilibre.

Pour solutionner l'équation (1) par la méthode des différences finies, il faut créer un maillage suivant les axes x et y comme montré à la figure 1. Le secteur d'engrenage étudié fait 360/Z degrés (Z étant le nombre de dents) et se prolonge jusqu'à une hauteur de dent sous le cercle de pied. Le maillage fait N_x+2 noeuds en x et N_y+2 noeuds en y.

Lors de la solution de l'équation (1), on suppose que le coefficient de conductibilité thermique k du matériau est constant dans toutes les directions, que la distribution du coefficient de transfert de chaleur par convection forcée h est celle trouvée par Akozan [8] et que les conditions aux frontières sont:

- Sur le flanc non chargé:

$$Ak\frac{\delta T}{\delta n} + gP = hA\,(T - T_a) \qquad (2)$$

où A est la surface de transfert de chaleur, n est la direction normale à la surface de transfert, gP est le taux de dégagement de chaleur totale par cycle dans la région de ces noeuds et T_a est la température ambiante.

Le taux de dégagement de chaleur totale se détermine suivant la relation:

$$qP = qF + qh_f + qh_c \qquad (3)$$

où qF désigne le taux de chaleur de frottement par cycle et qh_f et qh_c, les taux de chaleur d'hystérésis de flexion et de contact respectivement.

La valeur de la chaleur d'hystérésis de flexion ou de contact s'obtient comme suit:

$$qh_{f,c} = \frac{\sigma^2_{f,c}}{tE} \times \frac{\pi}{2} \times \frac{tg\,\delta}{1+tg^2\delta} \times V \qquad (4)$$

où: tg δ = facteur de perte du matériau de la dent

E = module élastique

$\sigma_{f,c}$ = contrainte de flexion ou de contact
V = volume de l'élément étudié
t = durée d'un cycle

Dans son étude, Akozan a calculé des coefficients de transfert par convection en quatre points distincts identifiés à la figure 1 par h_1, h_2, h_3 et h_4. Sur le tiers inférieur de la hauteur de la dent, la valeur du coefficient de transfert h dans l'équation (2) est obtenue par interpolation linéaire entre les valeurs de h_4 et h_1, soit h_{4-1}. Sur les deux tiers supérieurs, l'interpolation est effectuée entre h_1 et h_2 pour donner h_{1-2}.

- Sur le sommet de la dent:
  dans la zone de ces noeuds, h dans l'équation (2) devient h_{2-3}.

- Sur le flanc chargé:
  l'équation (2) s'applique encore avec, sur les deux tiers supérieurs de la dent, h_{3-4} et, sur le tiers inférieur, h_{4-1}.

- Sur les limites droite et gauche du secteur:
  en ces points, sous le cercle de pied, la convection est nulle. Par symétrie, on impose une égalité de température et de flux de chaleur point par point, soit:

$$T\Big]_{i,j} = T\Big]_{i',j'} \tag{5}$$

$$\frac{\delta T}{\delta n}\Big]_{i,j} = \frac{\delta T}{\delta n}\Big]_{i',j'} \tag{6}$$

  où i, j, i' et j' sont les coordonnées suivant x et y respectivement des noeuds qui doivent correspondent.

- Sur la limite inférieure du secteur:
  à une hauteur de dent sous le pied, on considère un gradient de température linéaire, soit:

$$T_{(i,\ N_y+2)} = 2T_{(i,\ N_y+1)} - T_{(i,\ N_y)} \tag{7}$$

  où i varie de 2 à N_x+1.

- Pour les noeuds aux coins supérieurs droit et gauche:
  la loi de Kirelhof s'applique à ce noeud comme suit:

$$\sum_i Kk_i \frac{\delta T}{\delta n} + \sum_i Kc_i\ (T_a - T) + gP = 0 \tag{8}$$

  où Kk_i et Kc_i sont les conductances en conduction et en convection respectivement de l'élément suivant la direction i.

- Pour les noeuds internes:
  l'équation (8) s'applique toujours sauf que $Kc_i = 0$.

Les conditions frontières décrites ci-haut sont les mêmes que celles de Wang et Cheng [9] , modifiées pour limiter l'étude à une hauteur de dent sous le cercle de pied.

La procédure de solution consiste à supposer d'abord une distribution de température initiale. Une nouvelle distribution est ensuite calculée à l'aide des équations caractéristiques et puis comparée à la précédente. L'écart maximum toléré entre deux distributions successives pour arrêter l'itération est de 0.05°C.

MOUVEMENT DES DENTS

La figure 2 montre la position relative des dents d'un engrenage menant (1) et mené (2) au début (S/p_n < 0) et à la fin (S/p_n > 0) du contact sur la ligne d'action. L'axe identifié S/p_n représente la distance S entre le point de contact sur la ligne d'action et le point primitif P, normalisée par rapport au pas normal de base p_n. Ainsi, en se référant à cet axe, une valeur de $S/p_n < 0$ indique que la dent sur l'engrenage mené fait contact en un point situé au-dessus du cercle primitif et que la dent sur l'engrenage menant fait contact en un point situé au-dessous du cercle primitif.

Les points marqués de 1 à 4 sur les dents menées et menantes de la figure 2 réfèrent aux endroits auxquels sont calculés les coefficients de transfert de chaleur par convection forcée h_1, h_2, h_3 et h_4 respectivement. La figure 2 indique que les coefficients sur le flanc chargé de la dent menée sont h_3 et h_4, alors les coefficients sur le flanc chargé de la dent menante sont h_1 et h_2.

Les coefficients h_1, h_2, h_3 et h_4 utilisés dans cet article sont donnés au tableau 1. On remarque, au tableau 1, une certaine variation entre h_1, h_2, h_3 et h_4 pour un même module. Les comparaisons intéressantes à faire sont h_3 avec h_2 (à la tête de la dent menante et menée respectivement) et h_4 avec h_1 (au pied de la dent menée et menante respectivement). Il s'avère ainsi que h_3 est systématiquement inférieur à h_2 alors que $h_4 > h_1$ pour les modules 3.18 et 2.12 et $h_4 < h_1$ pour les modules 5.08 et 2.54. En moyenne, cependant, comme l'indique les colonnes marquées d'un * dans le tableau 1, le transfert de chaleur est systématiquement plus important sur le flanc chargé de la dent menante (h_1 et h_2) que sur le flanc de la dent menée (h_3 et h_4).

RESULTATS ET DISCUSSION

La figure 1 montre, sous la forme d'isothermes, la distribution de température calculée par l'ordinateur. Cette figure montre que la température à l'intérieur de la dent est plutôt modeste (environ 60°C dans la dent et 50°C sous le cercle de pied) même si elle est assez élevée sur le flanc chargé (95°C). Au cours de la solution, le module d'élasticité du matériau utilisé pour calculer les déformations de la dent est supposé constant partout. De la figure 1, on voit que c'est une simplification grossière. Ce premier résultat peut cependant servir comme point de départ pour une analyse plus minutieuse. Le même raisonnement s'applique aussi au facteur de perte tg δ du matériau.

Les figures 3, 4 et 5 présentent la température d'équilibre calculée à la surface de la dent sur le flanc chargé en fonction de la distance normalisée S/p_n du point de contact. Ainsi que l'on pouvait s'y attendre, la figure 3 indique que la température sur la surface de la dent augmente avec la charge totale mais à une puissance un peu inférieure à 1. Ceci s'explique par la méthode utilisée pour calculer la charge sur chaque dent: lorsque la charge totale augmente, les dents fléchissent plus et la fraction de la charge totale reprise par chaque dent diminue [6]. Comme gP, équation (3), est une fonction de la charge sur la dent, la température calculée par l'équation (2) n'est pas directement proportionnelle à la charge.

Pour le module 2.54 mm représenté à la figure 3, le tableau 1 indique une valeur de h_1 plus grande que celle de h_4, ce qui se traduit par une température près du pied de la dent menée (2) plus grande que celle près du pied de la dent menante (1), pour toutes les charges. Pour la même raison, la température à la tête de la dent menée est plus élevée que celle à la tête de la dent menante. La faiblesse relative des coefficients h_3 et h_4 vis-à-vis h_2 et h_1 se traduit par une chute plus lente de la température à partir du pied vers la tête sur la dent menée comparativement à la dent menante.

La figure 4 montre l'effet du module sur la température à la surface de la dent. D'abord, pour chaque module prit individuellement, on remarque que les valeurs relatives des coefficients h_1, h_2, h_3 et h_4 du tableau 1 ont le même effet que signalé plus haut. Ainsi, pour le module de 2.12 mm, le tableau 1 indique que $h_4 > h_1$ et, en conséquence, la température au pied de la dent menée est plus basse que celle au pied de la dent menante.

Un effet particulier du module, cependant, est d'augmenter considérablement le gradient de température entre la tête et le pied de la dent. Ainsi, à la figure 4, on mesure des gradients maximums de 10, 12, 17 et 22°C pour les modules de 5.08, 3.18, 2.54 et 2.12 mm respectivement. La gradation des gradients maximums en fonction du module pourrait s'expliquer par la baisse graduelle de la moyenne des valeurs des coefficients h_2 et h_1 sur la dent menante et h_3 et h_4 sur la dent menée, si ce n'est que cette moyenne a son maximum pour un module de 3.18 mm (voir tableau 1). Ainsi, un raisonnement qui s'appuierait exclusivement sur les valeurs des coefficients h du tableau 1 conclurait-il non seulement à un gradient de température entre la tête et le pied plus grand pour le module de 5.08 mais aussi à une inversion des courbes de températures des modules 5.08 et 3.18. En d'autres mots, suivant l'ordre régressif de la moyenne des valeurs de h_1 et h_2 et de h_3 et h_4 du tableau 1, on devrait rencontrer, en partant du bas de la figure 4, les courbes pour les modules 3.18, 5.08, 2.54 et 2.12 mm. Cette inversion des modules 5.08 et 3.18 s'explique par le fait que, dans le calcul de la charge sur chaque dent, le module apparaît à une puissance 0.22 [6]. Pour les grosses dents (modules grossiers), c'est le module qui est le facteur déterminant et les courbes apparaissent dans l'ordre du module croissant. Mais lorsque le module diminue, il perd graduellement de son influence au profit des coefficients h qui redeviennent le facteur prédominant.

La figure 5 montre les températures mesurées par thermocouples aux points 1, 2, 3 et 4 de la figure 1 sur l'axe de symétrie de la dent, au centre de la face d'un engrenange menant. Sur la figure 5 apparaissent également les températures calculées aux mêmes points et pour les mêmes charges. Loins de la source de chaleur par frottement et hystérésis, au point 4, les valeurs mesurées et calculées sont en très bon accord. L'intersection de la courbe calculée de T_4 à charge nulle se trouve à une température d'environ 15°C, ce qui est près des températures ambiantes mesurées durant les essais, soit de 18°C à 20°C. En ce qui concerne les températures calculées aux points 1, 2 et 3, la concordance avec les mesures est moins bonne. Aussi, les calculs semblent faire converger la température de ces trois points à une valeur de 30°C à charge nulle. Les points 1, 2 et 3 étant placés dans une zone près des sources de chaleur par frottement et hystérésis, l'explication peut provenir d'un déplacement de l'axe neutre dû aux grandes déformations et de là un plus grand dégagement d'énergie par hystérésis que celui calculé. Il est possible aussi, à faible charge, que les thermocouples aient tendance à sous-estimer la température à cause de l'effet de puits de chaleur.

CONCLUSION

Cette étude démontre l'influence du coefficient de transfert de chaleur et de sa variation le long du profil de la dent. L'emplacement du flanc chargé par rapport à l'écoulement de l'air ou d'un brouillard d'huile autour de la dent ne semble pas négligeable. Les grands modules et les lourdes charges ont aussi un effet non négligeable.

D'une façon générale, le modèle analytique prédit les mêmes tendances que les mesures expérimentales avec thermocouples. Les divergences signalées indiquent qu'il est nécessaire de raffiner le modèle. L'emplacement des pointes de température sur le flanc des dents menées et menantes est cependant en accord avec les prédictions théoriques [4] et des mesures expérimentales rapportées dans la référence [5].

REMERCIEMENTS

Les auteurs tiennent à remercier les organismes gouvernementaux canadiens et québécois qui ont soutenu financièrement ce projet au cours des années.

TABLEAU 1 Coefficients de transfert de chaleur par convection forcée, h, $W/m^2/s$ [8] $Z = 30$ dents; $\alpha = 20°$; $V = 4$ m/s

	Module, mm							
	5.08	*	3.18	*	2.54	*	2.12	*
h_1	595	548	645	623	510	525	395	413
h_2	500		600		540		430	
h_3	395	475	500	585	435	455	340	380
h_4	555		670		475		420	

$$* \quad \frac{h_1 + h_2}{2} \quad \text{et} \quad \frac{h_3 + h_4}{2}$$

REFERENCES

[1] Cornelius, E.A., Budich, I.W. (1970). Investigation of gears of acetal resins. Konstruktion, Vol. 22, No 3, pp. 103-116.

[2] Takanashi, Shoji. (1975). Bulletin of Japan Soc. Mech. Engrs., Vol. 6, No 75, p. 73.

[3] Yousef, S.S., Burns, D.J., McKinlay, W. (1973). Techniques for assessing the running temperature and fatigue strength of thermoplastic gears. Mechanisms and Machine Theory, Vol. 8, pp. 175-185.

[4] Gauvin, R., Girard, P., Yelle, H. (1983). Prediction of the peak temperature on the surface of thermoplastic gear teeth. Paper no P149.03, AGMA Fall Technical Meeting, Montreal, Canada, October 17-19, 1983.

[5] Liu, R., Burns, D.J., Yelle, H. (1987). Flank temperature profile for thermoplastic spur gears. Seventh World Congress on the Theory of Machines and Mechanisms, Sevilla, September 17-22, 1987.

[6] Koffi, D., Gauvin, R., Yelle, H. (1985). Heat generation in thermoplastic spur gears. The Fourth International Power Transmission and Gearing Conference, Cambridge, Mass., October 10-12, 1984, Journal of Mech., Trans. and Automation in Design, March 1985, Vol. 107, pp. 31-37.

[7] Yelle, H., Koffi, D., Gauvin, R., Ghamraoui, M. (1986). Effet du matériau du pignon sur les performances d'un engrenage en plastique. 2e Congrès Mondial des Engrenages, Paris, Vol. 2, Mars 1986, pp. 59-68.

[8] Akozan, M. (1982). Etude expérimentale du coefficient de transfert de chaleur par convection pour les engrenages cylindriques droits en thermoplastiques. Mémoire de maîtrise, Ecole Polytechnique de Montréal.

[9] Wang, K.L., Cheng, H.S. (1981). A numerical solution to the dynamic load, film thickness and surface temperatures in spur gears. Part I analysis. ASME Journal of Mechanical Design, Vol. 103, pp. 177-187.

ABSTRACT

Plastic gears are often used in applications requiring stress calculations at the root and on the flank. The strength of the material, in both cases, is greatly affected by temperature. This paper discussed an analytical method by finite differences to calculate mean and flank temperature of plastic gears. The method uses a load sharing technique to calculate the energy losses and variable forced convection heat transfer coefficients [6,7,8]. The effects of the load, the module and the heat transfer coefficients are discussed.

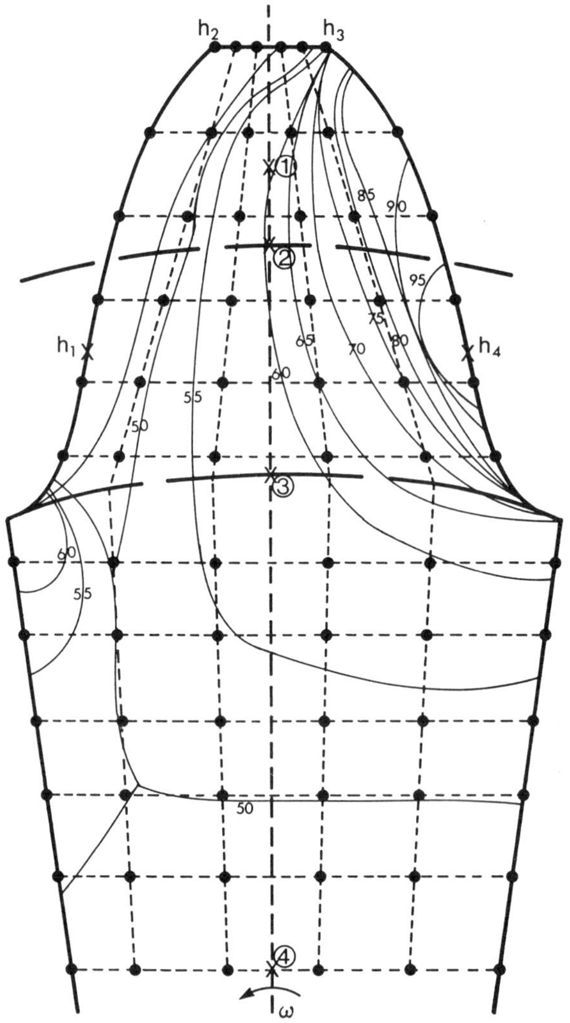

Fig. 1. Géométrie du secteur d'engrenage étudié et isothermes d'une solution. Dent menée; W_o = 15.4 N/mm; m = 2.54 mm; T_a = 20°C.

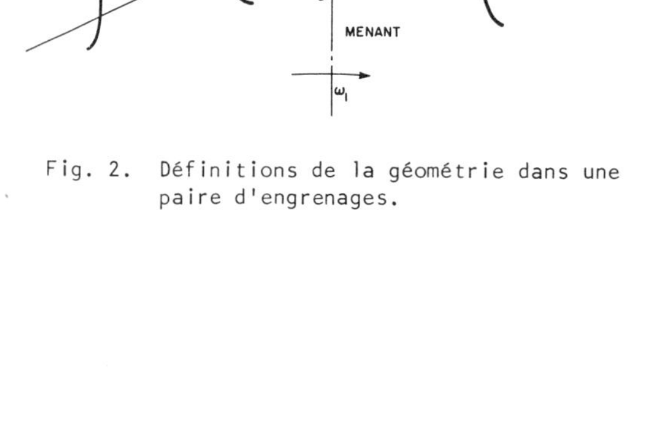

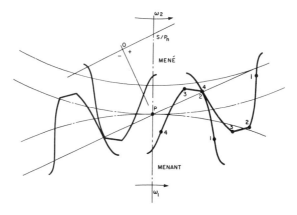

Fig. 2. Définitions de la géométrie dans une paire d'engrenages.

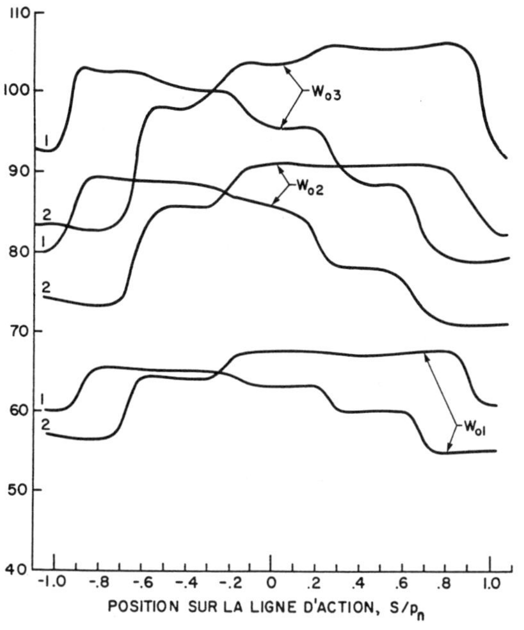

Fig. 3. Effet de la charge totale transmise W_o sur la température calculée sur la surface de la dent. W_{o1} = 9.3 N/mm; W_{o2} = 15.4 N/mm; W_{o3} = 19 N/mm; Z_1 = Z_2 = 30 dents; m = 2.54 mm; T_a = 20°C.

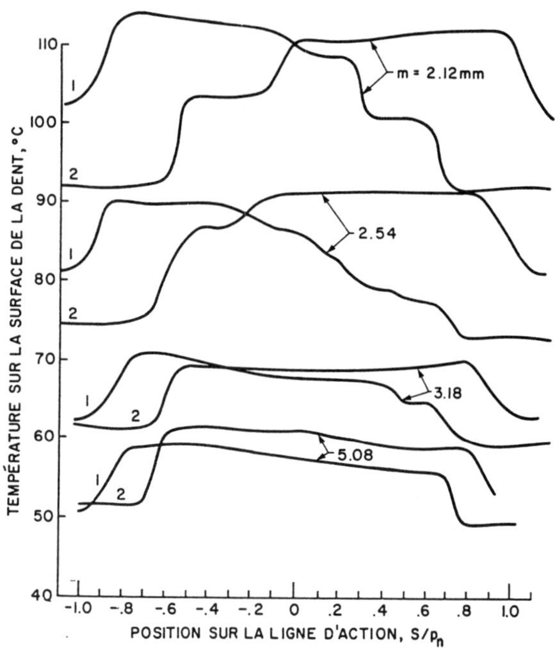

Fig. 4. Effet du module sur la température calculée sur la surface de la dent. W_o = 15.4 n/mm; Z_1 = Z_2 = 30 dents; T_a = 20°C.

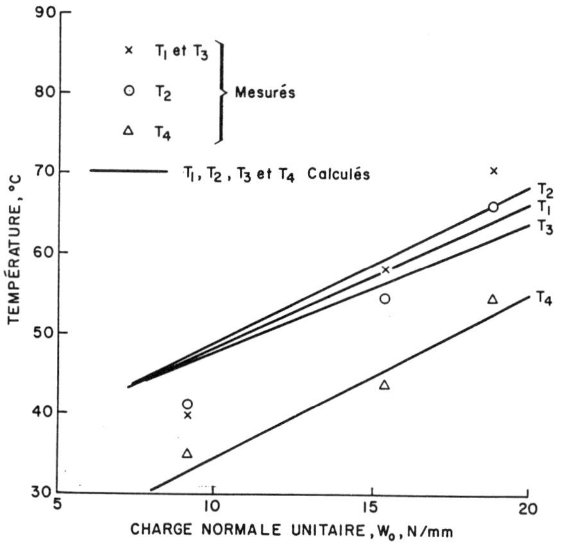

Fig. 5. Comparaison des températures mesurées par thermocouples et calculées au centre de la dent sur l'engrenage mené. Z_1 = Z_2 = 30 dents; m = 2.54 mm; T_a variable entre 18 et 25°C.

Dynamic Response of a Helical Geared Shaft System Subjected to Random Support Excitations

S. V. NERIYA, R. B. BHAT AND T. S. SANKAR

Department of Mechanical Engineering, Concordia University, 1455 de Maisonneuve Blvd., West Montreal, Quebec, Canada H3G 1M8

Abstract. The response of a helical geared shaft system subjected to random support excitations is investigated. Support excitations occur, for instance, on board moving vehicles. These excitations are in general random in nature and the response can be obtained using a statistical analysis. In this paper, a helical geared shaft system is modelled using finite elements and the coupling between torsion, flexure, axial motion and rotation about a diameter is considered. The geared rotor system is excited by a displacement type of support excitation which is the output of a filter, the input to which is a Gaussian, stationary process with a white noise type of PSD. The excitation is assumed to be in the vertical direction only and the excitations through the supports are assumed to be uncorrelated. Results for the response power spectral densities are presented for two kinds of filters.

INTRODUCTION

Rotating shafts are indispensable in mechanical power transmission systems. Dynamic responses of such rotating shafts must be clearly understood for their design. As is very common in power systems, gears are often employed to transmit power between shafts.

In a geared rotor, some of the sources of excitation are mass unbalance, geometric eccentricity and errors of manufacture, all of which originate in the gear pair itself. However, the geared rotor system may be subjected to external forms of excitation such as that from the supports, for instance when on board various types of vehicles. If the support excitations are random in nature with considerable power distributed over a frequency range, the system will respond at those frequencies. Moreover, if one of the system natural frequencies of the geared rotor coincides with the excitation frequencies, the resulting response may be of concern.

In a system of rotating shafts using spur gears, the torsional and flexural motions are coupled due to the presence of mating gears. The coupling of the torsional and flexural motions occurs in two distinct forms, 1) 'force coupling' and 2) 'dynamic coupling'. The predominant form is the 'force coupling'. This is due to the lateral mesh forces generated by torsional oscillations of the shaft and is of the order of geometric eccentricity, ε. Lund [1] considered the force coupled vibration of a geared train of rotors. He carried out the rotor analysis for torsional and lateral vibrations separately and subsequently coupled them through impedance matching at the gear meshes. Also, he obtained the dynamic response and studied the stability of the system. The second form of coupling is the 'dynamic coupling' which occurs due to the angular acceleration of the gears. Dynamic coupling terms are of the order ε^2 [2]. Iida et al [3] studied a spur geared system and calculated the response due to mass unbalance and geometric eccentricity. The geared shaft system was described as a 4 DOF lumped mass model where the driven shaft was considered flexible in bending and the driving shaft was considered rigid. Neriya et al [4] studied a simple spur geared shaft system which was discretised using finite elements. The driving and driven shafts were considered flexible in bending and the flexibility of the mating teeth was considered in the analysis. The frequency response and the orbital behaviour were obtained by normal mode analysis.

Helical gears are often used in industrial applications. There is a 'force coupling' between the torsional, rotational (about the diametral axis) flexural and axial motions of the helical gear pair. Daws [2] studied three dimensional vibration in a helical geared rotor system. He obtained the forced vibration response due to mesh errors. Kacukay [5] used a 8 DOF model to study a pair of helical gears and obtained the response and dynamic tooth loads. Kiyono et al [6] also studied a pair of helical gears and showed that from the viewpoint of dynamic load, the rotational vibration plays a more important role than the longitudinal one. Kiyono et al [7] also carried out an the experimental investigation of their analysis in [6].

Several investigators have also studied rotors subjected to random support excitations. Lund [8] carried out response spectral density analysis of rotor systems due to stationary random excitations of the base, considering excitations only in the vertical direction. Tessarzik et al [9] analysed the turbo-rotor responses due to external random vibrations. The rotor-bearing system was treated as a linear, three mass model and experimental results were found to compare well with calculations of amplitude power spectral density for the case where the vibrations were applied along the rotor axis. used. Subbiah et al [10] obtained the amplitude PSD of a simple rotor subjected to random support excitations using modal analysis methods. The excitations were assumed to be stationary and Gaussian with a white noise type of PSD.

In this paper, the geared rotor system is modelled using finite elements and the coupling between torsion flexure, axial motion and rotation about a diameter is considered. The geared rotor system is subjected a displacement type of support excitations which is the output of a linear filter, the input to which is a Gaussian, stationary random process with a white noise type of PSD. This excitation is assumed to be in the vertical direction only and the excitations through the supports are assumed to be uncorrelated. The PSD of the response process is then obtained using a statistical analysis.

NOMENCLATURE

$\bar{c}_t$	average flexural damping of the gear tooth in the normal direction.
$c_{x1}, c_{y1}, c_{\theta 1}, c_{\phi 1}$	lumped damping at the driving gear in the x,y, θx & θy directions.
$c_{x2}, c_{y2}, c_{\theta 2}, c_{\phi 2}$	lumped damping at the driven gear in the x,y, θx & directions.
C_m, C_ℓ	lumped torsional damping at the motor and dynamo.
$[C]$	generalised global damping matrix.
$\{F\}$	generalised force vector.
$\{F\}_R$	Random support excitation force
$[H(j\omega)]$	Frequency response function matrix
I_1	moment of inertia of the driving gear.
I_2	moment of inertia of the driven gear
J_1	moment of inertia of the motor.
J_2	moment of inertia of the dynamo.
$\bar{k}_t$	average flexural stiffness of the gear tooth in the normal direction.
k_{xx}	stiffness of the rolling contact bearing in the x y & z directions
k_{yy}	
k_{zz}	
$k_{x1}, k_{y1}, k_{\theta 1}, k_{\phi 1}$	lumped stiffnesses at the driving gear in the x,y, θ_x and θ_y directions.
$k_{x2}, k_{y2}, k_{\theta 2}, k_{\phi 2}$	lumped stiffnesses at the driven gear in the x, y, θ_x and θ_y directions.
$[K]$	generalised global stiffness matrix.
m_1	mass of the driving gear.
m_2	mass of the driven gear.
$[M]$	generalised global mass matrix.
N	number of degrees of freedom of the system.
n	no of teeth in the driven gear.
$\{q\}$	generalised rotor displacement vector.
$\{q_r\}$	relative displacement vector
$\{q_s\}$	support displacement vector
r_1	base circle radius of the driving gear.
r_2	base circle radius of the driven gear.
$[S_F(\omega)]$	matrix of force PSD
$[S_{qr}(\omega)$	matrix of relative displacement PSD
$[S_{qs}(\omega)]$	matrix of support displacement PSD
T_m	input torque.
T_ℓ	output torque.
$[\gamma]$	diagonal damping matrix.
β	helix angle.
δ_n	displacement of the gear tooth in the normal direction.
$\varepsilon(t)$	static transmission error.
$[\kappa]$	diagonal stiffness matrix.
λ_j	jth eigenvalue.
$[\mu]$	diagonal mass matrix.
$\{\sigma(t)\}$	normalized force vector.
$\{\phi_j\}$	jth eigen vector.

ANALYSIS

In the present work, the geared rotor system is modelled using finite elements [11-12]. In general, the rotor finite element considers gyroscopic effects in addition to rotating inertia and shear deformation effects. However, the gyroscopic terms are skew symmetric in nature and hence a simple normal mode analysis is not possible. A modal analysis employing biorthorganility relations [13] is capable of considering the skew symmetric gyroscopic terms also, however, for simplicity a normal mode analysis was resorted to in the present study by neglecting the gyroscopic effects. The finite element discretisation of the helical geared shaft system is carried out as follows. The shafts are divided into beam elements and each node has 6 degrees of freedom. The element mass and stiffness matrices are obtained from the consistent formulation and are then assembled to form the global mass and stiffness matrices. The elemental mass matrix is the sum of translational and rotational mass matrices. These are obtained from [10]. The concentrated masses and inertias (including transverse moment of inertia) due to the motor, dynamo, gears etc. are introduced into the appropriate locations in the global mass matrix. The stiffnesses of the rolling contact bearings are included in the analysis. The coupling between torsion, flexure, axial motion and rotation about a diameter of the gear obtained from a study of the dynamics of the helical gear mesh (see Appendix I) are introduced into the appropriate locations in the global stiffness matrix.

The equations of motion for a N degree of freedom geared shaft system subjected to support excitations can be expressed as

$$[M]\{\ddot{q}_r\} + [C]\{\dot{q}_r\} + [K]\{q_r\} = \{F\}_R \quad (1)$$

where $\{q\}$ = generalised rotor displacement vector

$\{q_s\}$ = support displacement vector

$\{q_r\}$ = $\{q - q_s\}$, is the relative displacement vector.

and $$\{F\}_R = -[M]\{\ddot{q}_s\} \quad (2)$$

The homogeneous form of Eq. (1) neglecting damping is solved to obtain the eigen values λ_i and eigen vectors $\{\phi_i\}$ of the system.

Expressing the response $\{q_r\}$ in terms of the modal coordinates $\{p\}$ as

$$\{q_r\} = [\phi]\{p\} \quad (3)$$

where $[\phi]$ is the modal matrix formed using the eigenvector $\{\phi_i\}$

Using Eq, (3) in (1) and premultiplying by $[\phi]^T$ we get,

$$[\mu]\{\ddot{p}\} + [\gamma]\{\dot{p}\} + [\kappa]\{p\} = \{\sigma(t)\} \quad (4)$$

where
$$[\mu] = [\phi]^T [M] [\phi]$$
$$[\kappa] = [\phi]^T [K] [\phi] \quad (5)$$
$$\{\sigma(t)\} = [\phi]^T \{F\}_R$$

Now $$\{p(t)\} = [H(\omega)]\{\sigma(t)\} \quad (6)$$

where $$H_i(\omega) = \frac{1}{(\kappa_i - \mu_i\omega^2) + j(2\zeta_i\omega_i\mu_i\omega)} \quad (7)$$

and ζ_i is the modal damping ratio.

The PSD matrix of the response can be obtained as

$$[S_{qr}(\omega)] = [\phi][H(j\omega)][\phi]^T [S_F(\omega)][\phi] \times [H(-j\omega)]^T [\phi]^T \quad (8)$$

The PSD matrix of the random excitation force $[S_F(\omega)]$ can be expressed in terms of the PSD matrix of the displacement support excitation $[S_{qs}(\omega)]$ as

$$[S_F(\omega)] = \omega^4 [M][S_{qs}(\omega)][M]^T \quad (9)$$

The displacement support excitation is assumed as the output of a linear filter, the input to which is an excitation in the vertical direction only. This excitation is an uncorrelated Gaussian stationary process with a white noise type of PSD.

NUMERICAL RESULTS

The details of the helical geared shaft system used to obtain the numerical results are given in Table 1. The pedestals through which the support excitations are transmitted to the geared rotor system are assumed to be flexible in the x,y, & z directions. A finite element discretisation of the geared shaft system considered for the numerical example is shown in Fig. 3. The details of the beam elements are given in Table 2. The system natural frequencies in the range 0-100 Hz are given in Table 3. The details of the linear filter are given in Appendix II.

The normalized PSD of relative amplitude in the y direction at the driven gear location (DOF #6) is shown in Fig. 4. The base excitation is assumed to be the output of a first order linear filter, the input to which is an uncorrelated white noise type of PSD only in the vertical direction, and this is plotted on the x axis. The PSD plots shows peak response at all the elastic modes within the frequency range. The maximum response is seen at modes within the frequency range. The maximum response is seen at modes 2 & 3 and is about 3.16×10^{-3} m²/Hz.

A similar PSD plot as in Fig. 4 using a second order filter is shown in Fig. 5. The behavior of the response is the same as in Fig. 4, but the magnitude is lower for all the modes. The maximum response is found to be around 3.16×10^{-5} m^2/Hz.

The normalized PSD of relative amplitude in the y direction at the driving gear location (DOF #19) is shown in Fig. 6. The base excitation is assumed to be the output of a first order linear filter, the input to which is an uncorrelated white noise type of PSD in the vertical direction. The PSD plot shows peak response at all the elastic modes within the frequency range. The maximum response is observed in the 3rd mode and corresponds to about 3.16×10^{-3} m^2/Hz.

A similar PSD plot as in Fig. 6 using a second order filter is shown in Fig. 7. The behavior of the response is the same as in Fig. 6, but the magnitude is lower for all the modes. The maximum response is found to be around 1.78×10^{-4} m^2/Hz.

The normalized PSD of relative amplitude in the y directions at the bearing locations (DOF #2 and 24) are shown in Figures 8 & 9. The base excitation is assumed to be the output of a first order linear filter, the input to which is an uncorrelated white noise type of PSD in the vertical direction. The responses are as expected, much smaller than the response at the gear locations. The behavior of the response for the bearing location (DOF #2) on the driven shaft is the same as that for the driven gear location (Fig. 4 & 5), but the maximum response is 1×10^{-10} m^2/Hz. The behavior of the response for the bearing location (DOF #24) on the driving shaft is the same as that for the driving gear location (Fig. 6 & 7) but the maximum response is 1.78×10^{-7} m^2 Hz.

ACKNOWLEDGEMENTS

The work was partially supported by grants 7104 and 1375 from the Natural Sciences and Engineering Research Council of Canada.

REFERENCES

1. Lund, J.W., "Critical Speeds, Stability and Response of a Geared Train of Rotors", Journal of Mechanical Design, July 1978, Vol. 100, pp. 535-539.

2. Daws, J.W., "An Analytical Investigation of Three-Dimensional Vibration Gear- Coupled Rotor Systems", Ph.D. Dissertation, Virginia Polytechnic and State University, 1979.

3. Iida, H., Tamura, A., Kikuch, K. and Agata, H., "Coupled Torsional-Flexural Vibration of a Shaft in a Geared System of Rotors", Bulletin of the JSME, Vol. 23, No. 186, December 1980, pp. 2111-2117.

4. Neriya, S.V., Bhat, R.B. and Sankar, T.S., "Coupled Torsional-Flexural- Vibration of a Geared Shaft System using Finite Element Analysis", Shock and Vibration Bulletin, June 1985.

5. Kucukay, F., "Dynamic Behaviour of High Speed Gears", I Mech E Conference Publications, "Vibrations in Rotating Machinery", C305/84, 1984, pp. 81-90.

6. Kiyono, S., Aida, T. and Fujii, Y., "Vibration of Helical Gears, Part 1 - Theoretical Analysis", Bulletin of the JSME, Vol. 21, No. 155, May 1978.

7. Kiyono, S., Aida, T. and Fujii, Y., "Vibration of Helical Gears, Part 2 - Experimental Investigation", Bulletin of the JSME, Vol. 21, No. 155, May 1978.

8. Lund, J.W., "Response Characteristics of a Rotor with Flexible Damped Supports", Symposium of International Union of Theoretical and Applied Mechanics, Lyngby, Aug. 12-16, 1974, pp. 319-349.

9. Tessarzik, J.M., Chiang, T., and Badgley, R.H., "The Response of Rotating Machinery to External Random Vibration", Journal of Engineering for Industry, Trans. ASME, Vol. 96, No. 2, May 1974, pp. 477-489.

10. Subbiah, R., Bhat, R.B. and Sankar, T.S., "Response of Rotors Subjected to Random Support Excitations", Journal of Vibration, Acoustics Stress and Reliability in Design, Oct. 1985, Vol. 107, pp. 453-459.

11. Nelson, H.D., and McVaugh, J.M., "The Dynamics of Rotor-Bearing Systems using Finite Element Approach", Journal of Mechanical Design, Trans. ASME, Jan. 1980, pp. 158-161.

12. Nelson, H.D., "A Finite Rotating Shaft Element using Timoshenko Beam Theory", Journal of Mechanical Design, Trans. ASME, Vol. 102, Oct. 1980, pp. 793-803.

13. Bhat, R.B., Subbiah, R., Sankar, T.S., "Dynamic Behavior of a Simple Rotor with Dissimilar Hydrodynamic bearings by Modal Analysis", Journal of Vibration, Acoustics, Stress and Reliability in Design, Trans. ASME, Vol. 107, April 1985, pp. 267-269.

TABLE 1

Details of the Rotor System Under Study

E	2.0×10^{11} N/m^2
G	0.8×10^{11} N/m^2
I_1	0.08485 kg m^2
I_2	0.01142 kg m^2
J_m	0.459 kg m^2
J_L	0.02148 kg m^2
$\bar{k}_t$	2.0×10^9 N/m
k_{xx}	8.0×10^9 N/m
k_{yy}, k_{zz}	8.0×10^8 N/m
m_1	11.36 kg
m_2	6.13 kg
n	50
r_1	0.1 m
r_2	0.05 m
e_{rav}	6×10^{-4} m (variable)
β	10°

TABLE 2

Details of the Rotor Elements

Element No.	Length, m	Diameter, m	Mass per Unit Length kg/m
1	0.5	0.02	2.51
2	0.6	0.02	2.51
3	0.5	0.03	5.665
4	0.6	0.03	5.665

TABLE 3

System Natural Frequencies 0-100 Hz

Mode No.	System Natural Frequency, Hz
1	zero
2	22.84
3	42.73
4	62.43
5	70.95

APPENDIX I

A schematic representation of a helical geared shaft system is shown in Fig. 1. Figure 2 shows the pair of helical gears in mesh. Each gear of the pair has motions in the x, y, θx and θy directions. If we consider a lumped mass model of the geared shaft system shown in Fig. 1, we can write the equations of motion as

$$m_1 \ddot{y}_1 + k_{y1} y_1 + c_{y1} \dot{y}_1 + \bar{k}_t \delta_n \cos\beta + \bar{c}_t \dot{\delta}_n \cos\beta = [\bar{k}_t \varepsilon(t) + \bar{c}_t \dot{\varepsilon}(t)] \cos\beta \tag{1}$$

$$m_2 \ddot{y}_2 + k_{y2} y_2 + c_{y2} \dot{y}_2 + \bar{k}_t \delta_n \cos\beta + \bar{c}_t \dot{\delta}_n \cos\beta = [\bar{k}_t \varepsilon(t) + \bar{c}_t \dot{\varepsilon}(t)] \cos\beta \tag{2}$$

$$I_1\ddot{\theta}_1 + k_{\theta1}(\theta_1 - \theta_m) + c_{\theta1} \dot{\theta}_1 + \bar{k}_t \delta_n \cos\beta\, r_1 + \bar{c}_t \dot{\delta}_n \cos\beta\, r_1 = [\bar{k}_t \varepsilon(t) + \bar{c}_t \dot{\varepsilon}(t)]\, r_2 \cos\beta \tag{3}$$

$$I_2\ddot{\theta}_2 + k_{\theta2}(\theta_2 - \theta_\ell) + c_{\theta2} \dot{\theta}_2 + \bar{k}_t \delta_n \cos\beta\, r_2 + \bar{c}_t \dot{\delta}_n \cos\beta\, r_2 = [\bar{k}_t \varepsilon(t) + c_t \dot{\varepsilon}(t)]\, r_2 \cos\beta \tag{4}$$

$$m_1 \ddot{x}_1 + c_{x1} \dot{x}_1 + k_{x1} \dot{x}_1 + \bar{k}_t \delta_n \sin\beta + \bar{c}_t \dot{\delta}_n \sin\beta = [\bar{k}_t \varepsilon(t) + \bar{c}_t \dot{\varepsilon}(t)] \cos\beta \tag{5}$$

$$m_2 \ddot{x}_2 + c_{x2} \dot{x}_2 + k_{x2} \dot{x}_2 + \bar{k}_t \delta_n \sin\beta + \bar{c}_t \dot{\delta}_n \sin\beta = [\bar{k}_t \varepsilon(t) + \bar{c}_t \dot{\varepsilon}(t)] \sin\beta \tag{6}$$

$$(\frac{I_1}{2})\, \ddot{\phi}_1 + k_{\phi1} \phi_1 + c_{\phi1} \dot{\phi}_1 + \bar{k}_t \delta_n \sin\beta\, r_1 + \bar{c}_t \dot{\delta}_n \sin\beta\, r_1 = [\bar{k}_t \varepsilon(t) + \bar{c}_t \dot{\varepsilon}(t)] \sin\beta\, r_1 \tag{7}$$

$$(\frac{I_2}{2})\, \ddot{\phi}_2 + k_{\phi2} \phi_2 + c_{\phi2} \dot{\phi}_2 + \bar{k}_t \delta_n \sin\beta\, r_2 + \bar{c}_t \dot{\delta}_n \sin\beta\, r_2 = [\bar{k}_t \varepsilon(t) + \bar{c}_t \dot{\varepsilon}(t)] \sin\beta\, r_2 \tag{8}$$

$$J_m \ddot{\theta}_m + k_{\theta1}(\theta_m - \theta_1) + c_m \dot{\theta}_m = T_m \tag{9}$$

$$J_L \ddot{\theta}_L + k_{\theta2}(\theta_\ell - \theta_2) + c_\ell \dot{\theta}_\ell = T_\ell \tag{10}$$

$$\delta_n = (y_1 + y_2 + r_1 \theta_1 + r_2 \theta_2) \cos\beta + (x_1 + x_2 + r_1 \phi_1 + r_2 \phi_2) \sin\beta \tag{11}$$

The equations of motion (1) to (10) are simplified using (11) and put in a matrix form from which we obtain the stiffness terms coupling the torsional, rotational (about the diametral axis), flexural and axial motions.

a. Stiffness Matrix

$$\begin{Bmatrix} y_1 \\ y_2 \\ \theta_1 \\ \theta_2 \\ x_1 \\ x_2 \\ \phi_1 \\ \phi_2 \end{Bmatrix} \xrightarrow{\bar{k}_t} \begin{bmatrix} A_1 & A_1 & A_1 r_1 & A_1 r_2 & A_2 & A_2 & A_2 r_1 & A_2 r_2 \\ & A_1 & A_1 r_1 & A_1 r_2 & A_2 & A_2 & A_2 r_1 & A_2 r_2 \\ & & A_1 r_1^2 & A_1 r_1 r_2 & A_2 r_1 & A_2 r_1 & A_2 r_1^2 & A_2 r_1 r_2 \\ & & & A_1 r_2^2 & A_2 r_2 & A_2 r_2 & A_2 r_1 r_2 & A_2 r_2^2 \\ & & & & A_3 & A_3 & A_3 r_1 & A_3 r_2 \\ & \text{Symmetric} & & & & A_3 & A_3 r_1 & A_3 r_2 \\ & & & & & & A_3 r_1^2 & A_3 r_1 r_2 \\ & & & & & & & A_3 r_2^2 \end{bmatrix}$$

where $A_1 = \cos^2\beta$
$A_2 = \cos\beta \sin\beta$
$A_3 = \sin^2\beta$

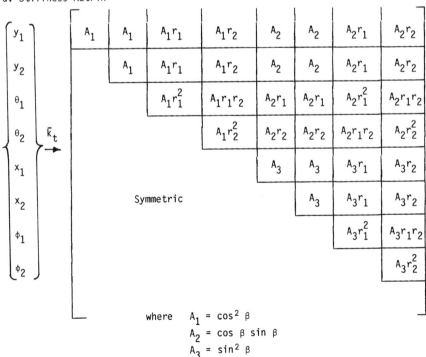

APPENDIX II

Output Y(t), Input X(t)

Order	Governing Equation	$\lvert H(j\omega)\rvert^2$
First	$\alpha \dot{Y} + Y = X$	$\dfrac{1}{1 + \alpha^2\omega^2}$
Second	$\alpha \ddot{Y} + \beta\dot{Y} + Y = X$	$\dfrac{1}{(1 - \alpha\,\omega^2)^2 + (\beta\omega)^2}$

$\alpha = \beta = 1.0$

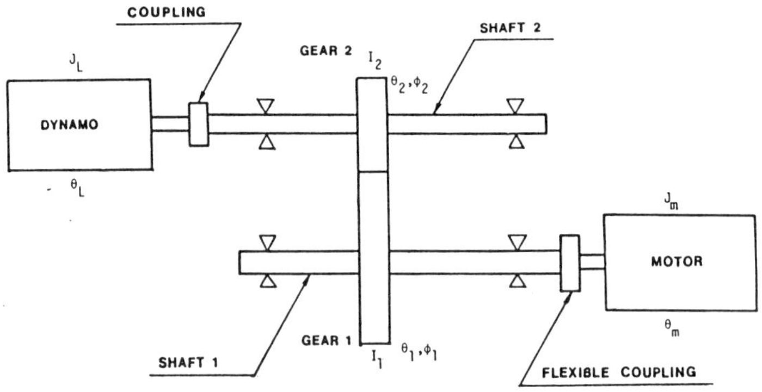

Fig. 1. A Simple Helical Geared Shaft System

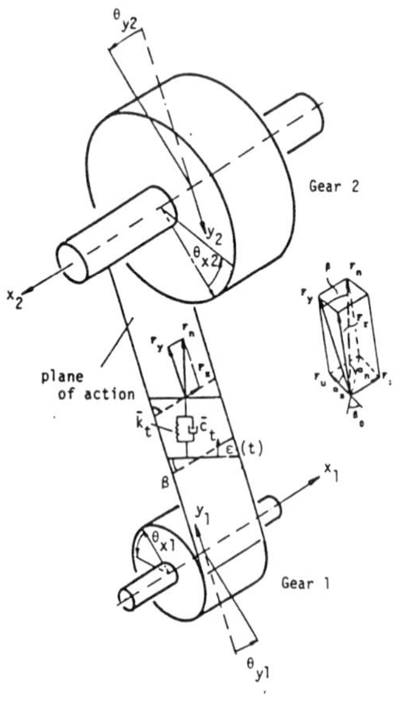

Fig. 2. A Pair of Helical Gears in Mesh

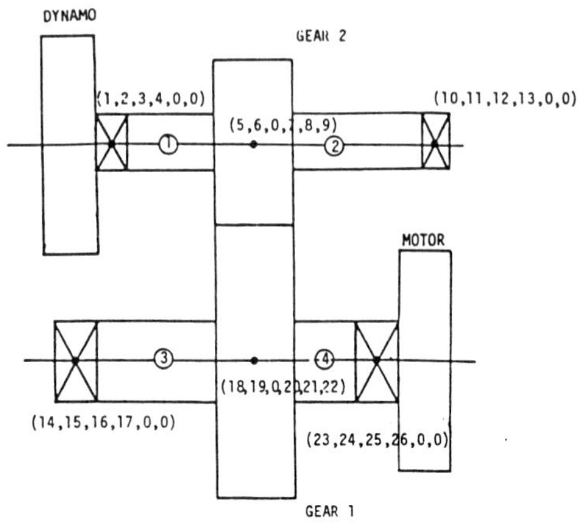

Fig. 3. Finite Element Discretisation of the Helical Geared Shaft System

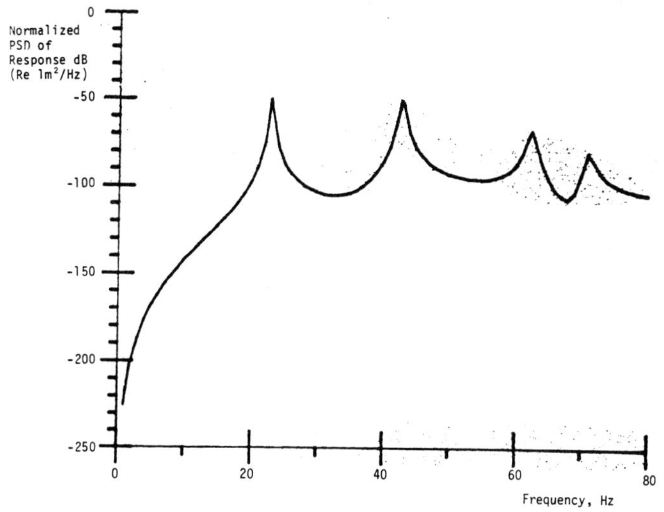

Fig. 4. Normalized PSD of relative amplitude in the y direction at the driven gear location (DOF #6) against the frequency of excitation. First order filter used.

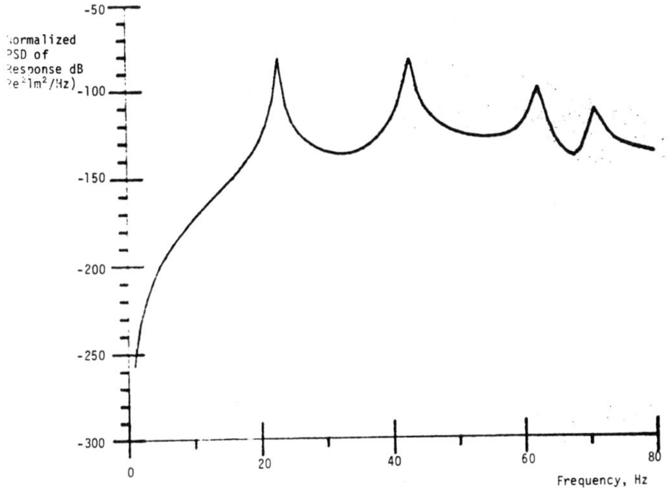

Fig. 5. Normalized PSD of relative amplitude in the y direction at the driven gear location (DOF #6) against the frequency of excitation. Second order filter used.

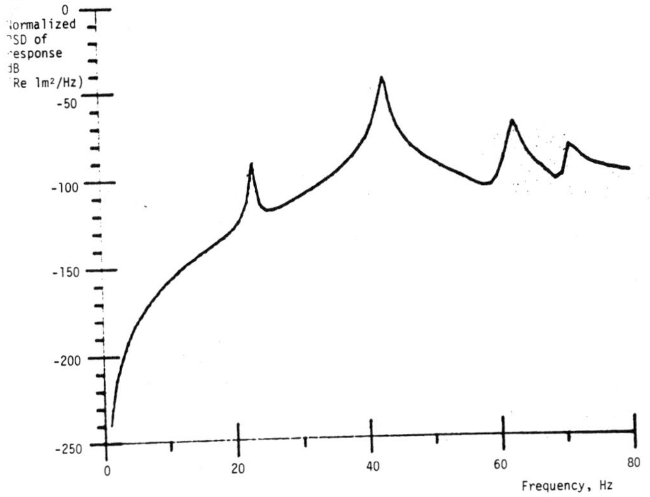

Fig. 6. Normalized PSD of relative amplitude in the y direction at the driving gear location (DOF #19) against the frequency of excitation First order filter used.

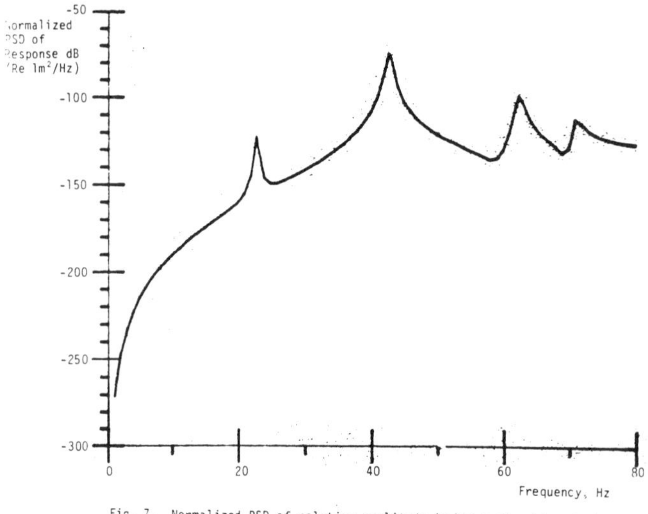

Fig. 7. Normalized PSD of relative amplitude in the y direction at the driving gear location (DOF #19) against the frequency of excitation. Second order filter used.

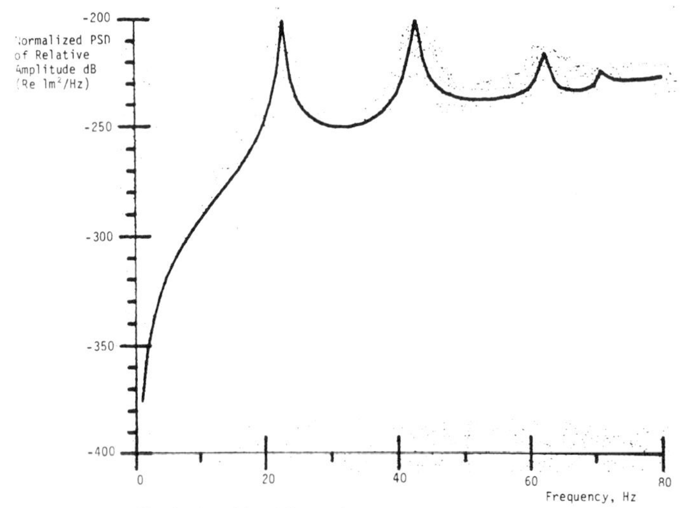

Fig. 8. Normalized PSD of relative amplitude in the y direction at the bearing location (DOF #2) against the frequency of excitation. First order filter used.

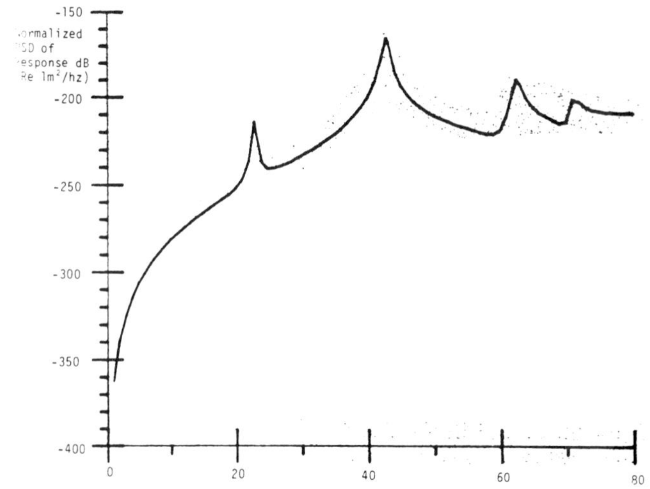

Fig. 9. Normalized PSD of relative amplitude in the y-direction at the bearing location (DOF #24) against the frequency of excitation. First order filter used.

A Study on the Influence of the Distance Between Pitch and Cutting Planes in W-N Spiral Bevel Gears

C. GOSSELIN AND L. CLOUTIER

Department of Mechanical Engineering, Laval University, Ste-Foy Quebec, Canada G1K 7P4

Abstract. Because of various practical advantages (interchangeability, simplicity, insensitivity to center distance changes), the involute profile is universally accepted as the basis for gearing mechanisms. However, its load capacity is limited because of the convex-convex contact in the transverse plane on external gear pairs. E. Wildhaber designed a helical gear system using circular profiles [1926] with convex-concave punctual contact. M. L. Novikov's experiences [1959] indicated a load capacity of 3 to 4 times that of involute gears of the same dimension. It has been shown that these gears could be cut on Gleason's type of spiral bevel gear generators [Ginzburg, 1959 & Litvin, 1985]. However, there are important factors in the use of the cutting tool that will affect its cutting conditions and life. In this paper, the cutting conditions are linked to the geometry of the cutter and a comparison is made with an application on Gleason Spiral bevel gears.

Keywords. Computer graphics; machining; bevel gears; optimization.

SYMBOLS

ϕ	:	design pressure angle;
S	:	angular position on circular profile cutter edge;
S_{min}	:	angular position on cutter edge at fillet of tooth;
S_{max}	:	angular position on cutter edge at tip of tooth;
R_1	:	cutter profile radius;
DR_1	:	distance between pitch and cutting planes;
L_i	:	craddle travel increment;
f	:	generated flat on tooth flank;
f_e	:	generated flat on tooth flank, projected on cutter edge;
S_p , S_g	:	contact point angular position on cutter edge (pinion, gear);
R_{p1}, R_{g1}	:	cutter profile radius (pinion, gear)
DR_{p1}, DR_{g1}	:	distance between pitch and cutting planes (pinion, gear).

INTRODUCTION

For the following explanation, a conventional straight edge cutter will be used; a parallel will be extended to circular cutters afterwards. Figure 1 presents the basic elements for generation. The pinion and the gear rotate about their axes, X_3 for the pinion, Y_3 for the gear. Their pitch cones roll at a given ratio. The pitch plane of the generating wheel rolls on the pitch cones of the pinion and the gear; its axis of rotation is colinear with Z_1, its pitch plane lies in the Z_2 Z_3 plane.

If the pinion and gear teeth are cut by congruent generating wheels, the traces left on the pinion and the gear as they rotate together are conjugate surfaces. At any moment, there exists a common curved line of contact which is the projection of the axis of instantaneous rotation on the generating surface [Baxter, 1977].

In practice, the generating wheel is replaced with a rotating craddle and the tooth is reproduced by a rotating cutter, mounted on the craddle, whose edge and motion duplicate the shape of the generating wheel teeth.

Normally, the pinion and the gear cutters are given different radii to ensure some longitudinal mismatch that will reduce their sensitivity to assembly errors : therefore, they are not conjugated longitudinally. The generated part is called the work.

The cutter teeth are usually evenly distributed on the circumference. Since the cutter speed and teeth are finite, the generated profiles are made of facets caused by the rotation of the work between two adjacent cutter teeth; these facets (or flats) correspond to the amount of craddle travel L_i, between two cutter teeth, projected on the cutter edge (fig. 2). The amount of craddle travel depends on the size of the work, the speed ratio of the craddle to the work, the cutter speed and the cutter blades pitch [Baxter, 1977].

CIRCULAR CUTTERS

Figure 3A presents the basic geometry of the pinion and gear circular cutters. They are arranged such that the generated profiles will contact at point "P" for the chosen pressure angle ϕ, under no fabrication or assembly error (fig. 3B).

In the basic configuration, the pinion tooth is all addendum while the gear tooth is all dedendum, a situation dictated by the cutter geometries : for conjugate motion to occur between the cutter and the generated profile, the common normal during generation must pass through the center of instantaneous rotation. Therefore (fig. 3A), no conjugate point can exist under the pitch plane (the plane tangent to the pinion and gear pitch cones) for a convex tooth in the normal plane, the pinion in this case, nor above the pitch plane for a concave tooth.

CIRCULAR CUTTER DESIGN : GENERAL PRACTICE

Figure 4 depicts the usual geometry adopted in designing the cutters. The pinion cutter profile radius is centered on the pitch plane whereas the gear's is at a distance DR_{g1} below to ensure some profile radius mismatch. Such a radius difference is necessary to prevent the pinion and gear profiles from contacting over their full height, thereby reducing the sensitivity of the pair to assembly and manufacturing errors. The resulting contact is punctual.

DR_{g1} is the distance between the pitch plane and the cutting plane (the plane in which the cutter profile radius center R_{g1} lies) of the gear. The value of DR_{g1} is based on the difference between the pinion and gear cutter radii and the desired working pressure angle :

$$DR_{g1} = [R_{g1} - R_{p1}] \, SIN(\phi) \qquad (1)$$

EFFECT OF THE DISTANCE BETWEEN THE PITCH AND CUTTING PLANES

The fundamental law of conjugate motion states that conjugate contact occurs only when the common normals to the profiles in presence pass through the center of instantaneous rotation (CIR). As generation progresses (fig. 5A), cutter and generated profile obey the law : the contact normals at S_5, S_4 ... S_0 pass through the CIR in turn, sustending distances L_{4-5}, L_{3-4} ... L_{0-1} on the pitch plane as the cutter moves along the tooth flank. As evidenced by the figure, the sustended distances L_{4-5}, L_{3-4} ... L_{0-1} on the pitch plane vary for constant sustended angles S_{4-5}, S_{3-4} ... S_{0-1}.

Distances L_{4-5}, L_{3-4} ... L_{0-1} are defined by equation 2 :

$$L_{i,i+1} = \frac{DR_1}{TAN\,(S_i)} - \frac{DR_1}{TAN\,(S_{i+1})} \qquad (2)$$

As can be seen, from figure 5A and equation 2, the sustended distances L_{4-5}, L_{3-4} ... L_{0-1} are affected by the value of DR_1 : they cannot become constant for any value of DR_1.

It can also be seen that if the value of DR_1 is zero, the cutter will contact the generated profile over its full height (since all the normals pass through the CIR at the same time), and unfavorable condition for surface finish and tool life.

Choosing a constant value, L_i, for L_{4-5}, L_{3-4} ... L_{0-1} (fig. 5B), it is found that the sustended angles S_{4-5}, S_{3-4} ... S_{0-1} vary, beeing rather large near the tip of the tooth and decreasing towards the fillet : this behaviour indicates that the cutter/generated-profile contact width will decrease as the cutter blade progresses along the tooth flank.

Sustended angles S_{4-5}, S_{3-4} ... S_{0-1}, for a given craddle travel L_i, are obtained from :

$$S_{i,i+1} = TAN^{-1} \left| \frac{DR_1 \; TAN(S_i)}{DR_1 - L_i \; TAN(S_i)} \right| - S_i \qquad (3)$$

The length of the sustended arc is given by :

$$f_{i,i+1} = R * (S_{i,i+1}) \qquad (4)$$

It is clear from equations 3 & 4 and figure 5 that no value of DR_1 will yield constant values of sustended angles S_{4-5}, S_{3-4} ... S_{0-1}.

It is also clear that the cutter profile radius R_1 center should be located at a distance below the pitch plane in order to minimize the cutting width near the tip of the tooth, ensuring better surface finish and longer tool life.

DR_{p1} AND DR_{g1} : INFLUENCE ON THE CUTTING PROCESS

F.S. Dikhtyar [1959] investigated methods for profiling hobs to produce parallel axis Novikov gears. The principal conclusions reached are that "when the cutting plane coincides with the pitch plane, the rack profile fully coincides with the fillet and the side clearance angle decreases to zero, a situation which unfavorably affects surface finish and hob life. Consequently, the section of the cutter profile next to the generating line should be constructed on a tangent to the hob profile at an angle of 5-10 degrees ...". "... The pitch plane of generation must be somewhat above the cutting plane, as a result of which a better tool rack tooth profile is obtained, with satisfactory side clearance angles. The accuracy of the component profile is sufficiently high and the mating profiles are run in fairly quickly".

Such a relief angle can be obtained, without modification of the tool profile, by locating its profile radius center at a distance below the pitch plane.

Baxter [1977] established that a certain amount of "flattening" of the generated tooth profile, along the line of constant roll angle, could be tolerated : the larger the flat, the shorter the generating cycle and the higher the output of the generator. He established operating values that satisfied both the output and quality criteria. Baxter's "flat" cross-widths, obtained from generating conditions such as craddle to work speed ratio, cutter feed and speed and cutter teeth pitch (table II), are :

f = 0,00456" for the pinion,

f = 0,00744" for the gear.

In a plane parallel to the cutter edge, these values become :

f_e = 0,00473" for the pinion,

f_e = 0,00772" for the gear.

The craddle travel increments L_i producing such flat widths (fig. 2, eqns 3 & 4) are :

L_i = 0,01236" for the pinion,

L_i = 0,02018" for the gear.

Applying the same approach to circular arc gears, one obtains sustended arcs instead of flats on the generated profiles.

Figure 6 presents the results of the calculated sustended arc values for the pinion depicted in table I, using different values of DR_{p1}. Table II presents the cutting speeds and feeds used. The value of DR_{p1} dictates the relief angle, S_{min}, at the root of the tooth :

$$S_{min} = TAN^{-1} \left| \frac{DR_{p1}}{R_{p1}} \right| \qquad (5)$$

It is clear from figure 6 that the sustended arcs for circular cutters (with cutting plane close to the pitch plane) are far from the flats obtained by Baxter with standard straight edge cutters; however, they can be of the same order of magnitude by properly choosing the location of the profile radius center. These sustended arcs represent the part of the cutter edge in contact with the work at a given instant. Since, with involute gears, the contact between pinion and gear moves across the tooth, it is important to obtain a proper flat arrangement on the teeth such that a smooth movement can be obtained even with coarse cutter feeding (large flats).

The problem is different with circular profile gears : the contact moves along the tooth; therefore, as long as the flats are along the tooth flank, no detrimental situation is likely to occur. However, from a cutting point of view, it is better to remove a thin width of material along the cutter motion.

TABLE I W-N Pair Summary

	Gear		Pinion
Number of teeth	29		17
Pitch angle (deg)	59.62		30.38
Mean cone dist. (in)		4.569	
Face width (in)		1.496	
Spiral angle (deg)		36.000	
Pressure angle (deg)		22.500	
Addendum (in)	0.3510		0.0000
Dedendum (in)	0.0000		0.3510

TABLE II Generation Summary

	Gear		Pinion
Cutter radius (in)		4.500	
Profile radius (in)	0.7781		0.8559
Cutter blades per side	10		20
Blade pitch (in)	2.830		1.410
Cutter feed (ft/min)	150		150
Cutter speed (in/sec)	30		30
Cradle roll (rad/sec)	0.0379		0.0466

Figures 7, 8, 9 and 10 present the results of a computer simulation of the pinion cutting process. The relief angles (S_{min}) are 0.5, 5.0, 10.0 and 15.0 degrees respectively, for the generating conditions given in Table II. Five sequential "snapshots" (from a to e) of each process were taken : they show the pinion cutter leaving the tooth at the heel (left) and the width of material cut during the pass.

It is clear, from figure 7, that a cutter profile radius R_{p1} centered on the pitch plane (0.5 degree relief angle) presents adverse cutting conditions near the tip of the tooth : the beginning of the cut is very wide (although short) on the cutter edge and tooth flank; this affects tool life and surface finish. Moreover, the descent of the cut trace is extremely steep : the cutter edge does not work evenly.

However, referring to figures 8, 9 and 10, improvement to the cutting process can be obtained by locating the cutting plane below the pitch plane. The flat cross-width is not affected significantly; rather, the height of the cut on the cutter edge near the tip of the tooth is reduced and the slope of the cut trace becomes more progressive along the tooth. The cutter edge works in an evenly manner.

Figure 11 presents the cutting process of the spiral bevel pinion described in Table III and referenced by Baxter. As shown, the traces cross-widths in Figures 7 to 11 are comparable; the difference lies in the slope of the trace, thus the width of the cutter edge in contact with the work at a given moment.

TABLE III Gleason Pair Summary

	Gear		Pinion
Number of teeth	29		17
Pitch angle (deg)	59.62		30.38
Mean cone dist. (in)		4.569	
Face width (in)		1.496	
Spiral angle (deg)		36.000	
Pressure angle (deg)		22.500	
Dedendum angle (deg)	4.40		2.67

DR_{p1} AND DR_{g1}: EFFECT ON THE DESIGN PRESSURE ANGLE

As the cutting plane is located below the pitch plane to improve the cutting conditions, the design pressure angle must be increased in order to keep the tooth geometry balanced. It has been shown that, under assembly errors, the contact point will shift either towards the tip or the fillet of the tooth, but with essentially the same sensitivity [Cloutier & Tordion, 1967].

Therefore, the working pressure angle (fig. 12) should be at mid-point between the tip angle S_{max} and the relief angle S_{min}, since the contact can move as much towards the tip as towards the fillet.

However, increasing the pressure angle increases the radial component on the bearings. Practically, the relief angle S_{min} at the fillet of the tooth should not exceed 10 degrees, with a working pressure angle in the order of 30 degrees.

CONCLUSION

It has been shown that when circular arc gears are generated with their profile radius centered on the pitch plane, adverse cutting conditions are present : the width of the cut is very large near the tip of the tooth, and decreases sharply towards the fillet, resulting in an uneven use of the cutter, with expected poor surface finish and tool life.

This situation can be improved by locating the cutting plane somewhat below the pitch plane; the cutting operation is evened and better tool life and surface finish can be obtained.

There is, however, a limit as to how much the cutting plane can be located below the pitch plane without increasing the pressure angle beyond practical limits. It is suggested to keep the relief angle below 10 degrees and to select the pressure angle at mid-point between the tip and fillet angles.

REFERENCES

Baxter, M.L. (October 1977). Lattice Contact in Generated Spiral Bevel Gears, ASME paper 77-DET-63.

Cloutier, L., Tordion, G.V. (1967). Méthode générale d'analyse du contact des engrenages du type Wildhaber-Novikov aux axes quelconques. Bulletins No 51, 52, 53, SEIE, Paris.

Dikhtyar, F.S. (1959). Hobs for Novikov Gears, Vestnick Masch., No 9.

Ginzburg, Y.G. (1959). Roues dentées et vis sans fin : quelques problèmes sur la théorie, le calcul et la production, Chap. 3 : Engrenages coniques avec un nouveau système d'engagement, Nikolchin, Mashgiz, Moscou.

Litvin, F.L., Coy, J.J. (1985). Generation of spiral Bevel Gears with Zero Kinematical Errors and Computer Aided Simulation of their Meshing and Contact, Proceedings, Computers in Engineering 1985, Vol. 1, ASME, pp. 335-339.

Novikov, M.L. (1959). Nouveau système d'engrenages, Moscou.

Wildhaber, E. (1926). Helical Gearing, United States Patent Office, 1, 601, 750, Oct. 5.

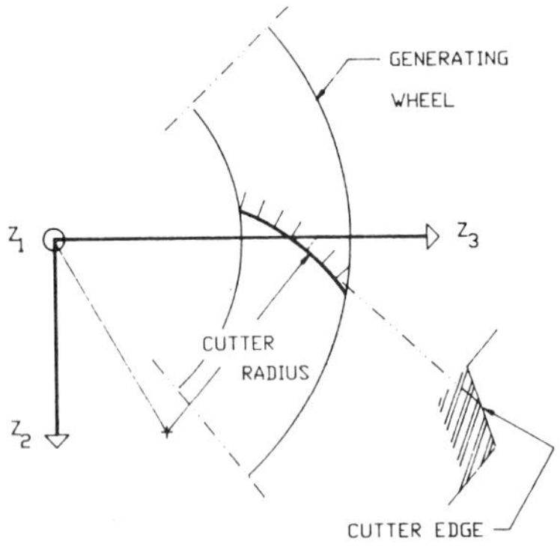

Fig. 1.a Basic elements for generation.

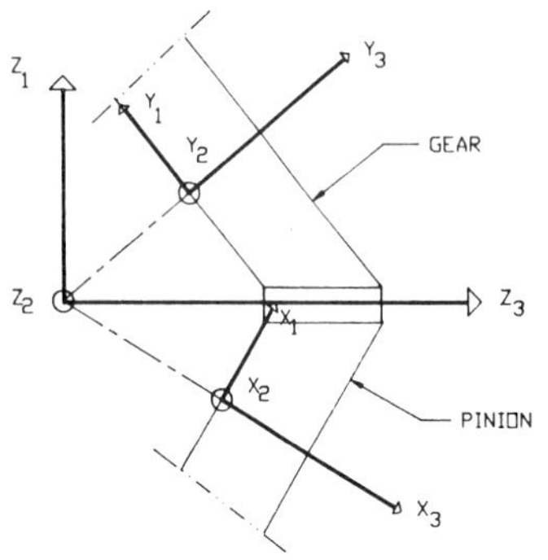

Fig. 1.b Basic elements for generation.

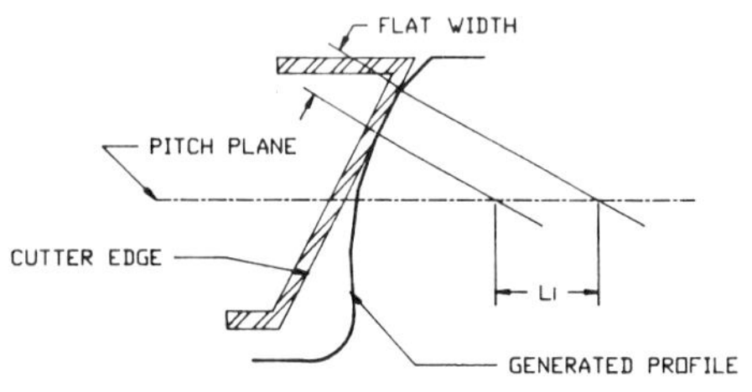

Fig. 2 Profile flat.

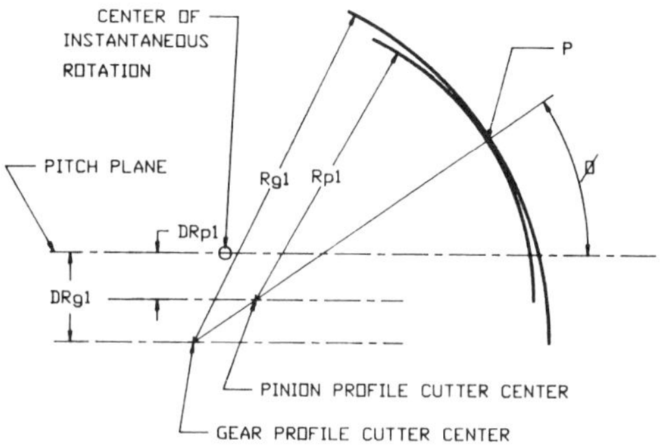

Fig. 3.a Pinion & gear, circular cutter geometries.

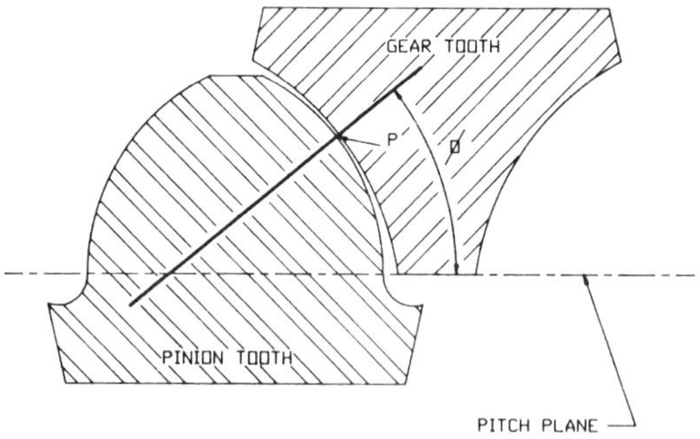

Fig. 3.b Pinion & gear, circular cutter geometries.

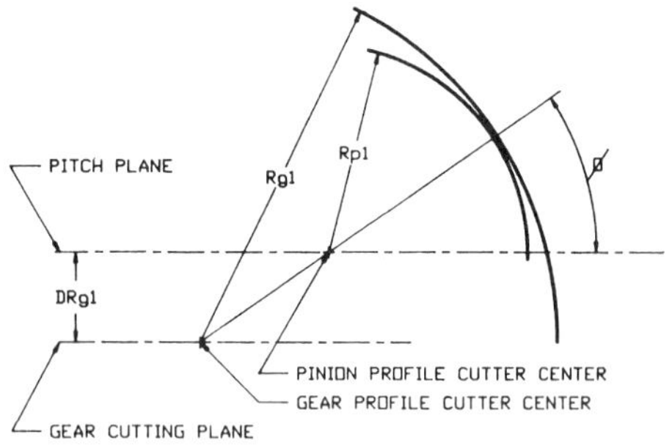

Fig. 4 Circular cutter design : general practice.

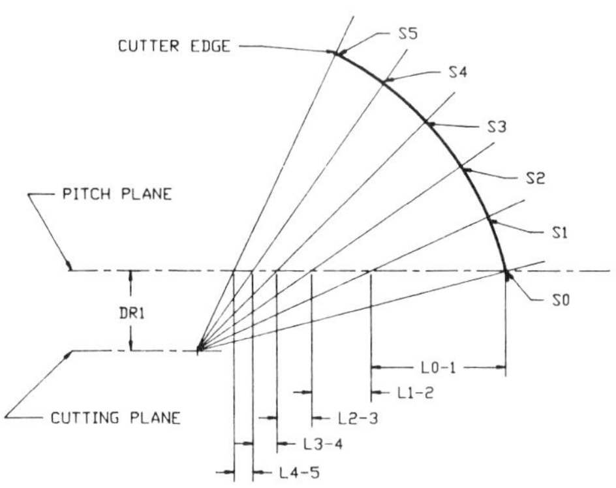

Fig. 5.a Circular cutters sustended arcs.

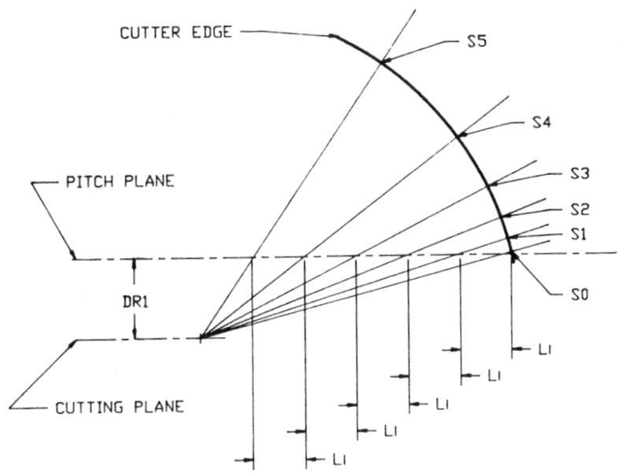

Fig. 5.b Circular cutters sustended arcs.

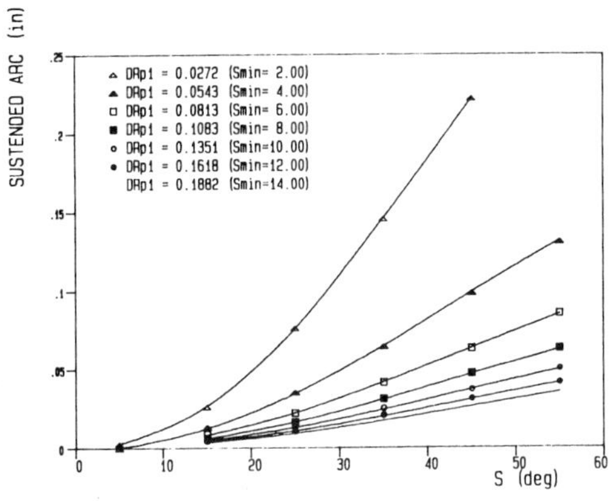

Fig. 6 Influence of DR_{p1} on pinion cutter sustended arc.

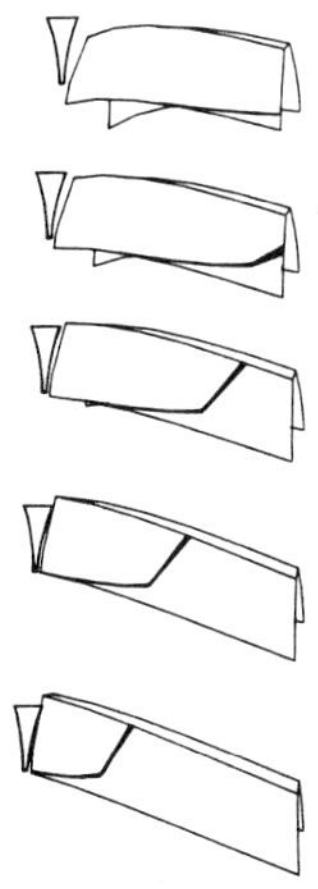

Fig. 7

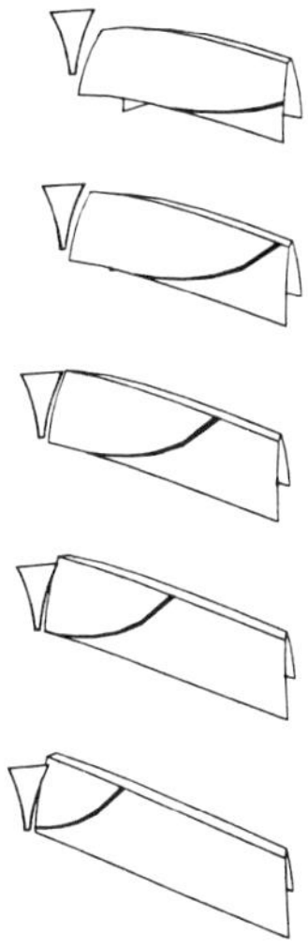

Fig. 8

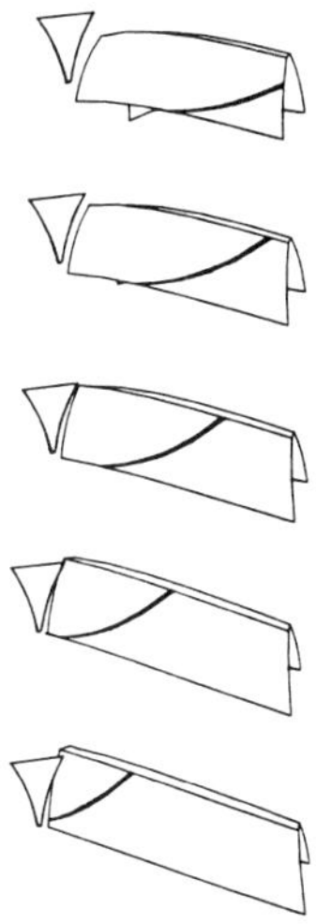

Fig. 9

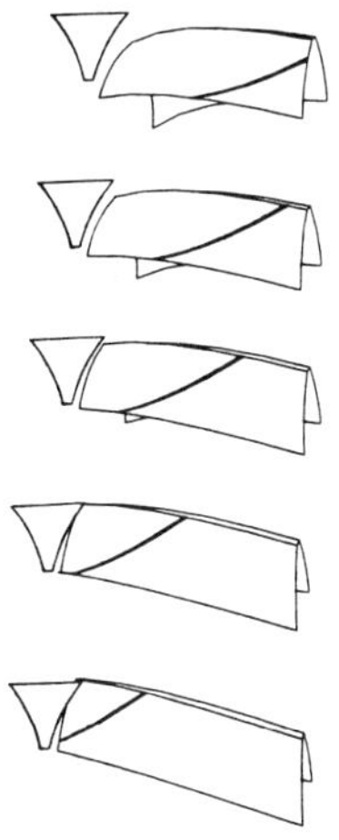

Fig. 10

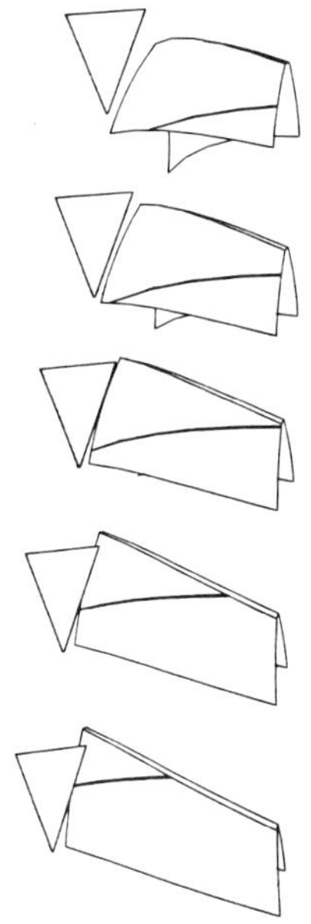

Fig. 11

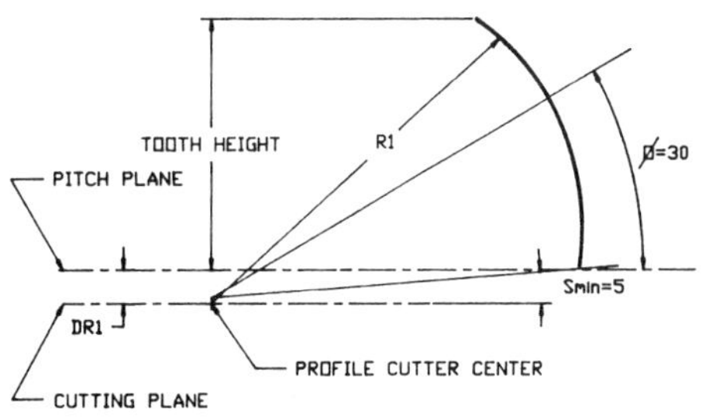

Fig. 12.a Influence of DR_{p1} on tooth balance.

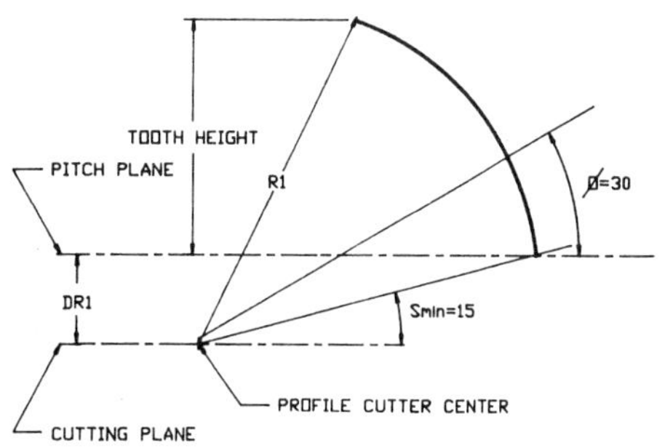

Fig. 12.b Influence of DR_{p1} on tooth balance.

Einfluss der Radkörpergestaltung auf die Tragfähigkeit der Verzahnung

H. LINKE, M. SENF AND A. KÖRTING

Sektion Grundlagen des Maschinenwesens, Technische Universität Dresden, DDR

Zusammenfassung. Durch die genauere Ermittlung der Last- und Spannungsverteilung bei elastisch gestalteten Radkörpern können die Zuverlässigkeit wesentlich erhöht und Tragfähigkeitsreserven erschlossen werden.
Es gelang, ein wenig aufwendiges Verfahren zu entwickeln und aus den bisher vorliegenden Ergebnissen Schlußfolgerungen für die Konstruktion zu ziehen. Das Verfahren geht von der Bestimmung der räumlichen Nennspannung mit Hilfe eines Schalenberechnungsprogrammes aus und ermittelt näherungsweise die örtliche Spannung mit Hilfe des Singularitätenverfahrens an Segmenten bei ebener Betrachtungsweise.

Keywords. Gear unit, Gear rim, Computer programming, Numerical methods, Load distribution, Stress distribution

EINLEITUNG

Größere Zahnräder werden nicht mehr als Vollräder ausgeführt. Meist wird ein relativ dünner Zahnkranz mit der Nabe durch einen oder zwei Stege verbunden. Durch diese Ausführung kann eine wesentliche Materialeinsparung erreicht werden. Bei der Wärmebehandlung ergeben sich andere Abkühlbedingungen und höhere Kernhärten. Wesentlich wird aber auch die Beanspruchung beeinflußt. Durch den elastischen Zahnkranz ergibt sich eine geänderte Last- und Spannungsverteilung. Die Zahnfußspannung wird von der Spannung im Zahnkranz bedeutend mitbestimmt, sodaß Brüche dann ausgehend vom Zahnfuß durch den Zahnkranz verlaufen können.
Es sind rechnerische Untersuchungsergebnisse zur Beanspruchung elastisch gestalteter Radkörper bekannt, die meist auf der Basis der Finiten Elemente gewonnen wurden (Oda, 1981; Fritsch, 1984). Diese Ergebnisse liegen bisher nur für wenige Ausführungen vor, da die FEM-Rechenprogramme sehr laufzeitintensiv sind und die Bereitstellung großer Datenmengen erfordern.

Durch Auswertung von Meßergebnissen (Winter, 1983) liegen für die Anwendung aufbereitete, aber ebenfalls noch keine erschöpfenden Ergebnisse vor. Um weitere Aussagen zu erhalten, wurden Berechnungen mit einem weniger aufwendigen Verfahren durchgeführt. Über die erhaltenen Ergebnisse berichtet dieser Beitrag. [1]

BERECHNUNGSMETHODE

Das grundsätzliche Vorgehen bestand darin, daß zunächst mit einem relativ wenig aufwendigen Verfahren das Belastungs-Verformungsverhalten des aus Kranz-Steg-Nabe bestehenden Rades bestimmt wird. Hierzu fand ein Schalenberechnungsprogramm (ROSCHA) auf der Basis des Übertragungsmatrizenverfahrens Anwendung (Hellmann, 1977). Unter Berücksichtigung dieses Elastizitätsverlaufes wurde dann die Breitenlast-

[1] Die Untersuchungen wurden an der TU Dresden für den VEB Kombinat Getriebe und Kupplungen Magdeburg (ASUG) durchgeführt.

verteilung und Nennspannungsverteilung (σ_F/Y_S) im Zahnfuß bestimmt (Mitschke, 1982). Die Berechnung der örtlichen Zahnfußspannung erfolgte mit Hilfe des Singularitätenverfahrens (Brechling, 1969; Linke, 1978; Linke, 1983). Dazu wurde ein Segment zugrunde gelegt, bei dem die Schnittkräfte durch die genannten Berechnungen bekannt sind.

Die in diesem letzten Schritt erfolgte Zurückführung der Ermittlung der örtlichen Spannung für den räumlichen auf einen ebenen Belastungsfall ergibt hier erfahrungsgemäß nur vernachlässigbare Abweichungen.

Die Lastverteilung und die Verteilung der Spannungen wurden aus einer anteiligen Überlagerung von 3 Lastfällen bzw. Verformungsanteilen (konstant, linear veränderlich, parabolisch veränderlich) berechnet (Bild 1). Dabei wurde der abgeschnitten gedachte Zahn ebenso belastet wie der Radkörper und aus der erforderlichen Gleichheit der Verformungen für eine gegebene äußere Belastung die Anteile der 3 zugrundegelegten Reaktionen bestimmt.

Bild 2 zeigt das benutzte Rechenprogramm, das aus einer Kopplung des erwähnten Schalenberechnungsprogrammes und Programmen zur Ermittlung der Lastverteilung und der örtlichen Zahnfußspannung besteht. Die örtliche Zahnfußspannung wird näherungsweise aus dem infolge der Tangentialkraft sich ergebenden Anteil σ_{FZ} und aus dem infolge der Radialkraft sich ergebenden Anteil σ_{FK} ermittelt.

Eine Überlagerung kann punktweise in der Zahnfußkurve entsprechend Gleichung (1) erfolgen.

$$\sigma_F = \sigma_{FZ} \pm \sigma_{FK} \qquad (1)$$

Unter der als Näherung geltenden Voraussetzung, daß die Lage der Maxima beider Spannungsanteile gleich ist bzw. nicht wesentlich voneinander abweicht, kann die überlagerte Spannung für eine konstante Lage ermittelt werden.

Durch Bezug der örtlichen Spannungsanteile für eine mittlere Lage, z.B. bei $\theta = 60^\circ$, auf die entsprechende Nennspannung ergeben sich die Spannungsverhältnisse (Spannungskonzentrationsfaktoren) Y_{SFt} und Y_{SB} (Gl. 2 und 3). Im einzelnen wird der aus der Tangentialkraft folgende Anteil der örtlichen Zahnfußspannung bei $\theta = 60^\circ$ auf die üblicherweise berechnete Zahnfußbiegenennspannung bei $\theta = 30^\circ$ und der aus der Radialkraft (durch ein reines Biegemoment angenähert) folgende Anteil, ebenfalls bei $\theta = 60^\circ$, auf die Zahnkranzbiegenennspannung bezogen.

$$Y_{SFt} = \frac{\sigma_{FZ}\,(\theta = 60^\circ)}{\sigma_{FZn}\,(\theta = 30^\circ)} \qquad (2)$$

$$Y_{SB} = \frac{\sigma_{FK}\,(\theta = 60^\circ)}{\sigma_{FKn}} \qquad (3)$$

Die überlagerte örtliche Spannung F im Zahnfuß ergibt sich damit zu

$$\sigma_F = \sigma_{FZn} \cdot Y_{SFt} \pm \sigma_{FKn} \cdot Y_{SB} \qquad (4)$$

σ_{FZn} - Zahnfußbiegenennspannung

Y_{SFt} - Spannungskonzentrationsfaktor infolge Tangentialkomponente der Zahnkraft

σ_{FKn} - Zahnkranzbiegenennspannung

Y_{SB} - Spannungskonzentrationsfaktor infolge Radialkomponente der Zahnkraft

BERECHNUNGSERGEBNISSE

Bild 3 zeigt die berechnete Lastverteilung bei drei verschiedenen Ausführungen des Radkörpers mit Kontaktlinienabweichung. Es ist bemerkenswert, daß es bei einer elastischen Ausführung des Radkranzes zu einer Minderung der Beanspruchung kommt. Bei einem Zweistegrad wird die Beanspruchung gegenüber dem Einstegrad bei Kontaktlinienabweichung wieder erhöht, da die Zahnenden steifer sind.

Untersuchungen zur Lage der Spannungsmaxima für die Beanspruchung durch die Tangentialkomponente und durch die Radialkomponente (reines Biegemoment) sind im Bild 4 dargestellt. Generell liegen die Maxima im untersuchten Bereich über $\theta = 30^\circ$, also tiefer im Zahngrund als die durch die Tangente bei $\theta = 30^\circ$ fixierte Lage.

In Abhängigkeit vom Kerbparameter $2\,\varrho_{Fn}/s_{Fn}$ zeigen die Bilder 5 und 6 die Ergebnisse zur Ermittlung der Spannungsverhältnisse (Kerbfaktoren) für die beiden Belastungsanteile bei verschiedenen Zahnkranzdicken. Obwohl nach den Vorlagen für Y_{SFt} und Y_{SB} für dünnere Zahnkränze die Spannungskonzentration sinkt, ergeben sich trotzdem steigende resultierende Spannungen, da die Nennspannungen wesentlich stärker ansteigen.

Zu einer maßgeblichen Verringerung der Zahnfußbiegespannung auf der bei Vollrädern gefährdeten Zugseite des Zahnes kann es bei breiten und sehr elastisch gestalteten Radkörpern kommen.
Der Abbau der Spannung auf der Zugseite und die Erhöhung auf der Druckseite werden durch die Druckkomponente der Zahnkraft und der daraus bedingten Biege(-Druck)-Spannung hervorgerufen.

LITERATUR

Brechling, J.: Zur Berechnung der Spannungsverteilung im festen Körper im Fall des ebenen Spannungs- und Verformungszustandes infolge äußerer Belastung; Maschinenbautechnik 18 (1969) 12

Fritsch, M.; Hagedorn, A.: Analyse eines Mühlen-Zahnkranzes mit Hilfe der Finiten-Element-Methode; Antriebstechnik 23 (1984) Nr. 5

Hellmann, V.: Die statische und dynamische Berechnung beliebig belasteter Rotationsschalen nach verschiedenen Theorien; Diss. A, TU Dresden, Sektion Grundlagen des Maschinenwesens, 1977

Linke, H.: Ergebnisse und Erfahrungen bei der Anwendung des Singularitätenverfahrens zur Ermittlung der Spannungskonzentration an Verzahnungen; Wiss. Zeitschrift der TU Dresden 27 (1978) 3/4

Linke, H.: Spannungskonzentration bei Verzahnungen; Maschinenbautechnik 32 (1983)

Mitschke, W.: Untersuchungen des Einflusses der Radkörpergestaltung auf die Tragfähigkeit der Verzahnung und die Beanspruchung des Radkörpers; Diss. A, TU Dresden, Sektion Grundlagen des Maschinenwesens, 1982

Oda, Satoshi: Stress Analysis of Thin Rim Spur Gears by Finite Element Method. Bulletin of the JSME, Vol. 24, No. 193, July 1981

Winter, H.; Podlesnik, B.: Zahnsteifigkeit von Stirnradverzahnungen, Antriebstechnik 22 (1983) Nr. 3 u. 5; 23 (1984) Nr. 11

Summary. Reliability can be improved considerably and reserves of the load-carrying capacity can be made available by the precise determination of the load and stress distribution in elastically designed gear rims.
A technique with a low degree of expenditure was developed and conclusions for design were drawn from the available results. The technique is based on the determination of the triaxial nominal stress with the aid of a shell calculation program and it determines approximately the local stress with the aid of the singularity method at segments with a two-dimensional approach.

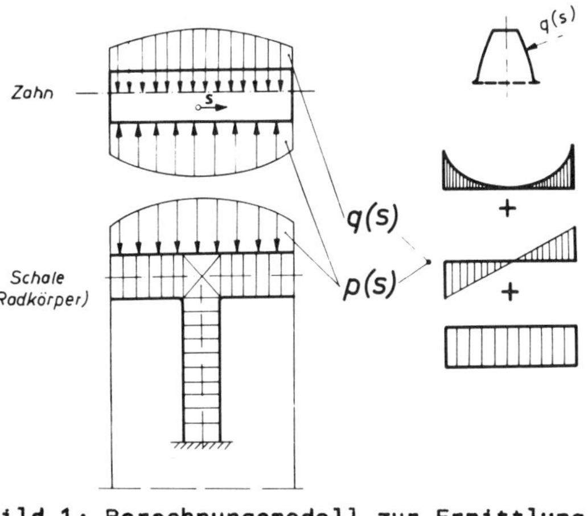

Bild 1: Berechnungsmodell zur Ermittlung der Last- und Spannungsverteilung

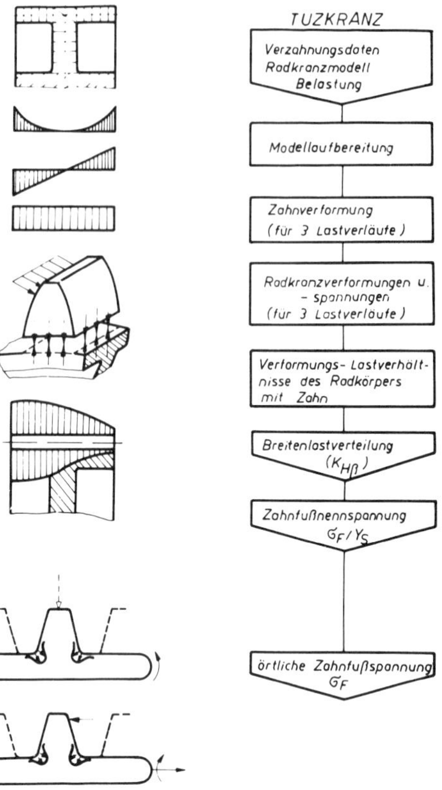

Bild 2: Rechenprogramm TUZKRANZ

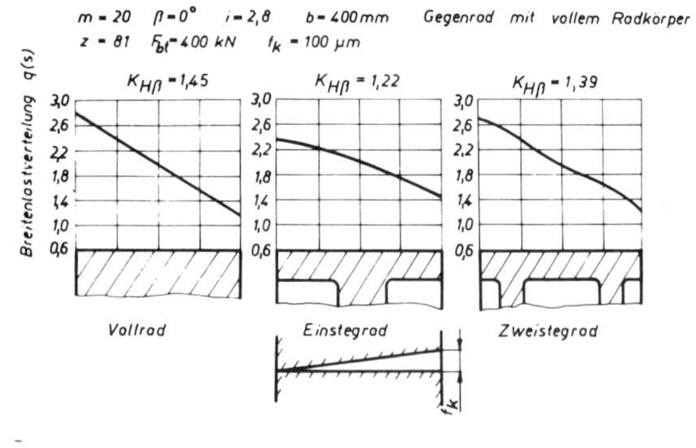

Bild 3: Lastverteilung beim Einsteg-, Zweisteg- und Vollrad

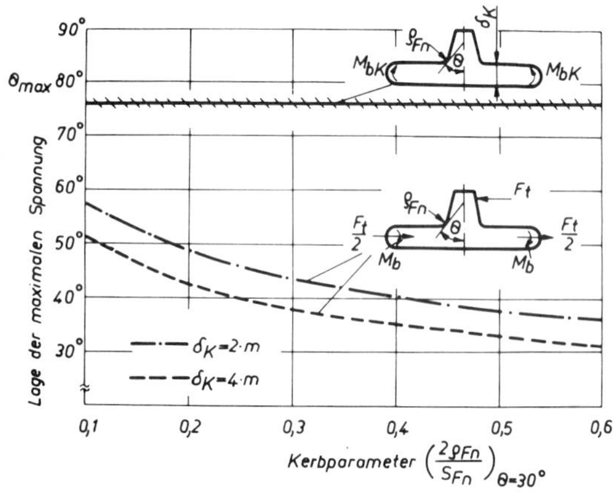

Bild 4: Lage der maximalen Spannungen infolge Tangential- und Radialkraftkomponente

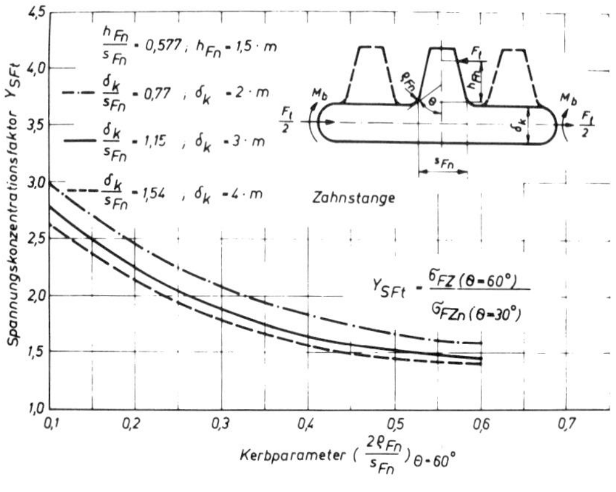

Bild 5: Spannungskonzentrationsfaktor Y_{SFt} infolge Tangentialkraftkomponente

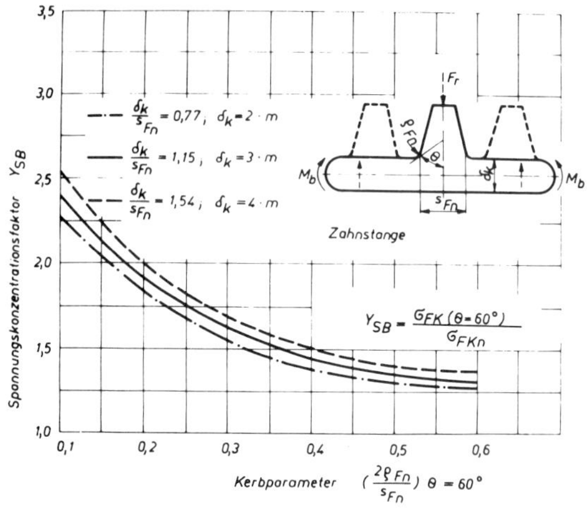

Bild 6: Spannungskonzentrationsfaktor Y_{SB} infolge Radialkraftkomponente

Design Aspects of Self-Locking Lifting Mechanisms with Power Screws

C. SAMÓNOV

Postal address: c/o Dr. James E. Haefely, Hoehenweg 7, CH-4102 Binningen, Switzerland

Abstract. The principal disadvantage of self-locking power screws being used as a lifting mechanism is their inherent very low efficiency. But if the thrust bearing of a power screw has an appreciable friction torque, then this torque can be used as the restraining torque required to prevent rotation of the NON self-locking power screw and consequent lowering of the supported load. Therefore, a mechanism consisting of a NON self-locking power screw and a plain sliding contact axial bearing can be made self-locking. The over-all efficiency of such mechanism will be several times greater as compared with the efficiency obtainable from use of a self-locking power screw. The illustrative example indicates that the over-all efficiency of a self-locking lifting mechanism with NON self-locking power screw can be as high as 48 % which very favourable compares with 14 % efficiency in case a self-locking power screw is being used.

Keywords. Lifting mechanism; power screw; self-locking action; mechanical efficiency; trapezoidal screw thread; buckling of columns; plain axial bearings.

INTRODUCTION

Power screws are widely used machine elements, especially in cases where a self-locking action is required. Unfortunately, self-locking power screws must have helix angle of thread smaller than the friction angle of the thread flanks, and consequently the resulting efficiency of lifting the applied load will be very low, in the order of 10 % to max. 14 %.

High-efficiency multi-start power screws with helix angle being several times greater than the friction angle would of course be NON self-locking. Therefore, in cases when a self-locking mechanisms are required, the NON self-locking screws must be used in conjunction with some locking mechanism, such as one-way clutch, automatic brake, or some similar device, in addition to anywhy required axial bearing, which in this case should be preferably of the rolling contact variety. But it should be borne in mind that the abovementioned safety device and rolling contact axial bearing would cost more in the initial outlay and also in maintenance as compared with the simple NON self-locking power screw with plain axial bearing, and consequently can be economically justified only in special cases when the low power consumption is the most important consideration.

The other solution to obtain an inexpensive and more efficient self-locking mechanism, as compared with a self-locking screw, would be to use a NON self-locking multi-start screw and at the same time to utilize the frictional torque of a plain sliding-contact axial bearing as a restraining torque to prevent rotation of the NON self-locking power screw and consequent lowering of the applied load. Such a mechanism will have the advantage of being simple and inexpensive, requiring very little maintenance, and woul be eminently suitable for an infrequent lifting of wide range of loads.

The purpose of this paper is to describe a rational design procedure for the determination of principal parameters corresponding to the highest feasible efficiency of a self-locking mechanism consisting of a NON self-locking power screw and a plain sliding-contact axial bearing. The mathematical model of such mechanism is made up of several interrelated and interdependent problem which are discussed in following chapters. The multiplicity of parameters, variables and equations makes the design procedure complex and time-consuming. Consequently, tha practical application can be economically feasible only by using a computer and prepared design program.

BUCKLING OF POWER SCREWS

In design of power screws loaded in compression it is important to ensure that lateral buckling will not occur. Neglecting the stiffening effect of screw thread, the power screw can be considered as being a column of round section with diameter equal to the root diameter d_3 of screw thread and the length L. But in order to account for the end restraints of such column, the effective length L_{ef} shall be used in calculations:

$$L_{ef} = \nu L \tag{1}$$

where ν is the effective length factor, which is defined as the distance between the inflection points in the elastic line exhibited by the deflected neutral axis of the column at the instant of buckling (Popov, 1968). The values of the effective length factor (which is termed by some authors as the end fixity coefficient) are given in Table 1.

TABLE 1 Effective Length Factor for Buckling of Columns

Both ends firmly fixed	0.5
One end firmly fixed, the other end on a ball	$\frac{1}{2}\sqrt{2} = \sim 0.7$
Both ends on balls	1
One end firmly fixed, the other end free	2

The buckling of columns occurs, depending on the slenderness ratio $\lambda = L/i$, in one of the two regions: elastic or in-elastic. The critical compressive stress at the instant of buckling in the elastic region is given by Euler equation, which is a second order hyperbola:

$$(\sigma_{cr})_{Eu} = \frac{(\pi)^2 E}{(\lambda_{ef})^2} \tag{2}$$

where $(\lambda)_{ef}$ is the effective slenderness ratio:

$$(\lambda)_{ef} = \frac{L_{ef}}{i} \tag{3}$$

and i is the radius of gyration:

$$i = \sqrt{\frac{I_{ax}}{A}} \tag{4}$$

for the round section:

$$i = \frac{d}{4} \tag{5}$$

The mathematical solution of column buckling in the in-elastic region has been developed by Engesser, and nowadays usually being used in structural engineering. But in mechanical engineering, where anywhy a high factor of safety is required, an approximate and more conservative solution by means of a common parabola, as has been suggested by J.B. Johnson, is usually preferred. Johnson parabola has its apex on the vertical σ - axis and is, as depicted in Fig. 1, tangential to Euler hyperbola.

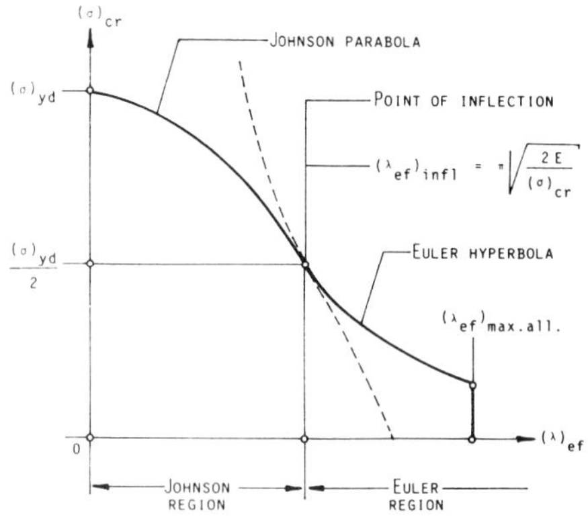

Fig. 1. Buckling curve for columns

The equation of Johnson parabola is:

$$(\sigma_{cr})_{Jo} = (\sigma)_{yd} - (\sigma_{yd})^2 \frac{(\lambda_{ef})^2}{4(\pi)^2 E} \tag{6}$$

It should be noted that the inflection point on the buckling curve, where Johnson parabola is tangent to Euler Hyperbola, is the cross-over point between the regions of in-elastic and elastic buckling. The coordinates of the point of inflection can be easily determined and are:

$$(\sigma_{cr})_{infl} = \frac{(\sigma)_{yd}}{2} \tag{7}$$

$$(\lambda_{ef})_{infl} = \pi \sqrt{\frac{2E}{(\sigma)_{yd}}} \tag{8}$$

Compressive design stress of course shall be less than the critical stress in buckling, therefore:

$$(\sigma)_{des} = \frac{(\sigma)_{cr}}{(n)_b} \tag{9}$$

where $(n)_b > 1$ is the factor of safety in buckling.

In order to avoid a trial and error solution for the determination of an appropriate design region of buckling, it is necessary to know the effective length of column corresponding to the said point of inflection. Using the above given equations and firstly eliminating the column diameter, and then solving for the effective length, the following equation is finally obtained:

$$(L_{ef})_{infl} = \frac{1}{(\sigma)_{yd}} \sqrt{\pi E (n)_b F_a} \tag{10}$$

It is now clear that if $L_{ef} > (L_{ef})_{infl}$ then column is in the elastic Euler region; otherwise column is in the Johnson in-elastic region of buckling.

Should a maximum allowable slenderness ratio be prescribed, then the length which corresponds to the point on Euler hyperbola where $(\lambda)_{ef} = (\lambda_{ef})_{max.all.}$ is:

$$(L_{ef})_{max.all.} = \frac{(\lambda_{ef})^2_{max.all.}}{2\pi} \sqrt{\frac{(n)_b F_a}{\pi E}} \tag{11}$$

For determination of the column diameter d_3 the following equations have been developed.
In the in-elastic (Johnson) region:

$$(d_3)_{Jo} = 2\sqrt{\frac{(n)_b F_a}{\pi (\sigma)_{yd}} + \frac{(\sigma)_{yd} (L_{ef})^2}{(\pi)^2 E}} \tag{12}$$

At the point of inflection

$$(d_3)_{infl} = \sqrt{\frac{8 (n)_b F_a}{(\sigma)_{yd}}} \tag{13}$$

In the elastic (Euler) region:

$$(d_3)_{Eu} = \sqrt[4]{\frac{64 (n)_b F_a (L_{ef})^2}{(\pi)^3 E}} \tag{14}$$

In the region where $(\lambda)_{ef} = (\lambda_{ef})_{max.all.}$:

$$(d_3)_{(\lambda_{ef})max.all.} = \frac{4 L_{ef}}{(\lambda_{ef})_{max.all.}} \tag{15}$$

SLIDING CONTACT AXIAL BEARING

Sliding contact axial bearings with thin-film lubrication are usually calculated in accordance with the uniform wear theory, which is based on the assumptions that:

1. the amount of wear per unit of area per unit of time remains constant over the whole area of the bearing.
2. the coefficient of friction is is independent of speed.

Therefore:

$$p\,v = \text{const.} \tag{16}$$

From the equation of static equilibrium:

$$F_a = \int_{D_i}^{D_o} p\, \partial A \tag{17}$$

the following two basic equations are obtained:

$$F_a = 0.5\,\pi D (D_o - D_i) \tag{18}$$

$$p = \frac{2\,F_a}{\pi\,D\,(D_o - D_i)} \qquad (19)$$

According to Eq. (19) the pressure distribution is hyperbolic and has its maximum and minimum values at the inside and outside bearing diameters respectively

$$(p)_{max.} = \frac{2\,F_a}{\pi\,D_i\,(D_o - D_i)} \qquad (20)$$

$$(p)_{min.} = \frac{2\,F_a}{\pi\,D_o\,(D_o - D_i)} \qquad (21)$$

Pressure distribution and corresponding bearing diameters are shown in Fig. 2.

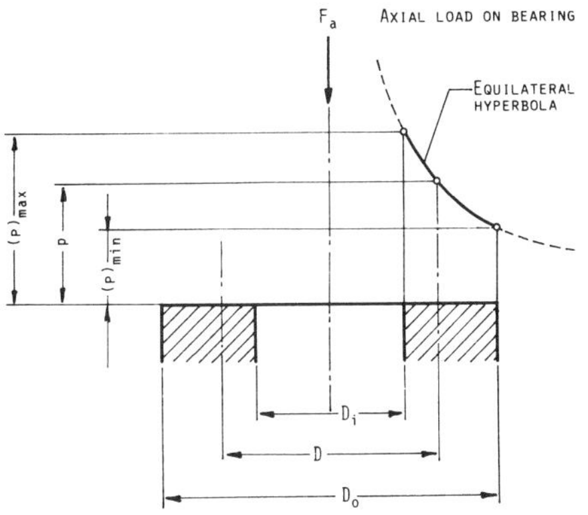

Fig. 2. Pressure distribution in plain sliding contact axial bearing.

As there are two unknowns but only one equation, the problem of determination of inside and outside diameters is ambigious. Moreover, according to Eq. (20) for every assumed value of the inside diameter there always is a corresponding outside diameter:

$$D_o = \frac{2\,F_a}{\pi\,D_i\,(p)_{max.}} + D_i \qquad (22)$$

But determination of the inside diameter for the given outside diameter has two solutions:

$$D_i = \frac{D_o}{2} \pm \sqrt{\frac{(D_o)^2}{4} - \frac{2\,F_a}{\pi\,(p)_{max.}}} \qquad (23)$$

and is feasible only if:

$$\frac{(D_o)^2}{4} \geq \frac{2\,F_a}{\pi\,(p)_{max.}} \qquad (24)$$

In order to obtain the smallest and consequently less expensive bearing, it is necessary to optimize it for the minimum outside diameter. Using the Eq. (22) and optimizing it by the rules of elementary calculus, the following equations are obtained:

$$(D_i)_{min.} = \sqrt{\frac{2\,F_a}{\pi\,(p)_{max.all.}}} \qquad (25)$$

$$(D_o)_{min.} = 2\sqrt{\frac{2\,F_a}{\pi\,(p)_{max.all.}}} \qquad (26)$$

And the frictional torque:

$$T_B = (\mu)_B\,\frac{(D_o + D_i)}{4}\,F_a \qquad (27)$$

THREAD OF POWER SCREWS

Screws with a single-start trapezoidal thread have been standardized by various national standards, and also by the ISO. But there are, to the authors knowledge, neither national nor international standards for the multi-start power screws.

The relationship between pitch diameter and the pitch of the thread is one of the most important parameters and is based on the many years of experience with use and manufacture of power screws. Therefore, it would be reasonable to assume, at least for the purpose of this paper, that the basic dimensions of the multi-start trapezoidal thread would be the same as for the single-start thread.

For the convenience of computer programming, the tabulated numerical values have been replaced by empirical equations which allow a continuous calculation of all geometrical parameters for the given root Dia. of screw thread. These equations have been based on ISO standards 2901...2904 and are valid for the root diameters between 15.5 mm and 274 mm.

Pitch of thread as a function of root diameter:

$$p = 1.6\sqrt{(d_3 + 2\,a_c) + 5} - 3.8 \qquad (28)$$

Pitch diameter as a function pich:

$$d_2 = \frac{1}{2.56}\,(p^2 + 7.6\,p + 1.64) \qquad (29)$$

Pitch diameter as a function of root diameter:

$$d_2 = 0.8\sqrt{(d_3 + 2\,a_c) + 3.26} + (d_3 + 2\,a_c) - 1.58 \qquad (30)$$

Pitch diameter as a function of outside Dia.:

$$d_2 = d - 0.8\sqrt{d + 7.06} + 2.22 \qquad (31)$$

Outside diameter as a function of pitch diameter:

$$d = d_2 + 0.5\,p = d_2 + 0.8\sqrt{d_2 + 5} - 1.9 \qquad (32)$$

Root diameter as a function of pitch diameter:

$$d_3 = (d_2 - 2\,a_c) - 0.8\sqrt{d_2 + 5} + 1.9 = d - 0.5\,p - 2\,a_c \qquad (33)$$

DETERMINATION OF POWER SCREW PARAMETERS

The screw torque required to RAISE the load is considered as being POSITIVE and can be written (Dubbel, 1981; Hancke, 1955) as:

$$T_{screw\ raise} = \frac{(d)_2}{2} \cdot \frac{\cos(\alpha)_n \tan(\beta) + (\mu)_{th}}{\cos(\alpha)_n - (\mu)_{th} \tan(\beta)}\,F_a \qquad (34)$$

The equation of torque which must be applied to screw in order to LOWER the load (Dubbel, 1981; Hancke, 1955) is:

$$T_{screw\ lower} = \frac{(d)_2}{2} \cdot \frac{\cos(\alpha)_n \tan(\beta) - (\mu)_{th}}{\cos(\alpha)_n + (\mu)_{th} \tan(\beta)}\,F_a \qquad (35)$$

In the Eqs. (34) and (35), (β) denotes the helix angle, and $(\alpha)_n$ is the flank angle of trapezoidal thread in the section normal to helix line and can be approximated by the following Eq. (36) :

$$\cos(\alpha)_n = \frac{1}{\sqrt{1 + \cos^2(\beta)\tan^2(\alpha)}} \quad (36)$$

(α) is the flank angle of trapezoidal thread in the axial section, and (β) denotes the helix angle:

$$\tan(\beta) = \frac{(1)_{th}}{\pi (d)_2} \quad (37)$$

$(1)_{th}$ denotes the LEAD of the multi-start thread and is:

$$(1)_{th} = p\, N_{st} \quad (38)$$

Of course, the number of starts N_{st} must be an integer. Therefore, an adjustment of the calculated p and/or $(1)_{th}$ will be required.

It should be noted that the sign of Eq.(35) depends on the sign of numerator. If in the numerator of Eq. (35) :

$$\cos(\alpha)_n \tan(\beta) < (\mu)_{th} \quad (39)$$

then the Eq.(35) will have a NEGATIVE sign, which indicates that the screw thread is SELF-locking, and consequently a NEGATIVE lowering torque (with its direction opposite to the RAISING torque) must be applied to screw in order to lower the load. On the other hand, if:

$$\cos(\alpha)_n \tan(\beta) > (\mu)_{th} \quad (40)$$

then the screw thread is NON self-locking, and therefore a POSITIVE RESTRAINING torque (with the same direction as the raising torque) must be applied to prewent SELF-ROTATION of screw and consequent LOWERING of the load.

The efficiency of screw thread :

$$(\mu_{th})_{\substack{screw\\raise}} = \frac{(1)_{th} F_a}{2\pi T_{\substack{screw\\raise}}} = \frac{(1)_{th}}{\pi (d)_2} \cdot \frac{\cos(\alpha)_n - (\mu)_{th}\tan(\beta)}{\cos(\alpha)_n \tan(\beta) + (\mu)_{th}} \quad (41)$$

is a function of the helix angle (β), and increasing with the increase in (β). Consequently, the NON SELF-locking power screws are much more efficient in RAISING the load as compared with SELF-locking screws.

In cases when a SELF-LOCKING lifting mechanism is required, it is feasible in order to increase mechanical efficiency to use a NON SELF-LOCKING power screw with the necessary RESTRAINING TORQUE provided by the frictional torque of the plain sliding contact axial bearing. But in order to make such lifting mechanism, consisting of NON SELF-LOCKING power screw and a sliding contact axial bearing, secure in SELF-LOCKING, the frictional torque of axial bearing shall always be greater than the POSITIVE RESTRAINING torque required to prevent SELF-LOWERING of the load. In other words, a factor of safety in self-locking $(n)_{s.l.}$ shall be used in calculations :

$$T_B = (n)_{s.l.}\left[T_{\substack{screw\\lower}}\right] \quad (42)$$

It should be borne in mind that the coefficient of friction μ is a function of many variables, as for instance: material, surface finish, lubrication and the relative speed of surfaces in contact. Therefore, in the absence of reliable data for the coefficient of friction, it is advisable to use the MINIMUM expected values of μ for calculation of screw LOWERING torque, and of frictional bearing torque as well. And the MAXIMUM expected values of μ shall be used for determination of the efficiency of mechanism.

As usually only the NOMINAL values of (μ) are known, the expected extreme (minimum and maximum) values can be estimated using an empirical coefficient of scatter (s) defined as shown below:

$$s = \frac{(\mu)_{max.}}{(\mu)_{min.}} \quad (43)$$

$$\frac{(\mu)_{max.}}{(\mu)_{nom.}} = \frac{(\mu)_{nom.}}{(\mu)_{min.}} = \sqrt{s} \quad (44)$$

Therefore:

$$(\mu)_{max.} = (\mu)_{nom.}\sqrt{s} \quad (45)$$

$$(\mu)_{min.} = \frac{(\mu)_{nom.}}{\sqrt{s}} \quad (46)$$

Using the minimum values of the coefficient of friction in Eq.(42) and developing it, the following design equation for determination of the helix angle (β) is obtained:

$$F_a \frac{(\mu_{nom.})_B}{\sqrt{(s)_B}} \cdot \frac{(D_o + D_i)}{4} = F_a \frac{\cos(\alpha)_n \tan(\beta) - \dfrac{(\mu_{nom.})_{th}}{\sqrt{(s)_{th}}}}{\cos(\alpha)_n + \dfrac{(\mu_{nom.})_{th}}{\sqrt{(s)_{th}}}\tan(\beta)} \cdot \frac{(d)_2}{2} \quad (47)$$

Equation (47) solved for $\tan(\beta)$ yields:

$$\tan(\beta) = \frac{A^* \cos(\beta)_n + B^* (d)_2}{C^* (d)_2 \cos(\alpha)_n - D^*} \quad (48)$$

where :

$$A^* = (D_o + D_i)(\mu_{nom.})_B \sqrt{(s)_{th}} \quad (49)$$

$$B^* = 2(n)_{s.l.} (\mu_{nom.})_{th} \sqrt{(s)_B} \quad (50)$$

$$C^* = 2(n)_{s.l.} \sqrt{(\mu_{nom.})_B (\mu_{nom.})_{th}} \quad (51)$$

$$D^* = (D_o + D_i)(\mu_{nom.})_B (\mu_{nom.})_{th} \quad (52)$$

The unknown parameters $(\alpha)_n$ and (β) are found by iteration of the Eqs. (36) and (48). The lead of the multi-start screw thread is now :

$$(1)_{th} = (d)_2 \tan(\beta) \quad (53)$$

and the corresponding number of starts is:

$$N_{st} = \frac{(1)_{th}}{p} \quad (54)$$

The value of the pitch of thread is obtained by firstly determining the root diameter of screw as required for stability in buckling, and then calculating pitch (p) using Eq. (28). The number of starts must be an integer, therefore the calculated theoretical values of lead and pitch must be accordingly adjusted. After such an adjustment the required pitch diameter shall be re-calculated using the Eq.(55) obtained by solving Eq.(48) for $(d)_2$:

$$(d)_2 = E^* + \sqrt{(F^*)^2 - F} \quad (55)$$

where :

$$E^* = \frac{\left[C^* (1)_{th} - \pi A^*\right] \cos(\alpha)_n}{2\pi B^*} \quad (56)$$

$$F^* = \frac{(1)_{th} D^*}{\pi B^*} \quad (57)$$

CHECKING OF CALCULATED PARAMETERS

Total torques applied to a lifting mechanism consisting of a NON SELF-LOCKING power screw and a plain sliding contact axial bearing are as stated below.
Maximum POSITIVE torque required to RAISE the load

$$\left[T_{\substack{raise\\total}}\right]_{max} = \left[T_B\right]_{max} + \left[T_{\substack{screw\\raise}}\right]_{max} \quad (58)$$

Maximum NEGATIVE torque required to LOWER the load:

$$\left[T_{\substack{lower\\total}}\right]_{max} = \left[T_{\substack{screw\\lower}}\right]_{min} - \left[T_B\right]_{max} \quad (59)$$

Factor of safety in SELF-LOCKING :

$$(n)_{s.l.} = \frac{\left[T_B\right]_{min}}{\left[T_{\substack{screw\\lower}}\right]_{max}} \quad (60)$$

The region in which buckling will occur is determined by the point of inflection calculated in accordance with the Eq.(10). Factor of safety in buckling in Johnson (in-elastic) region is obtained by first solving Eq.(12) for $(n)_b$, and then calculating the existing factor of safety for the root diameter $(d)_3$. In the Euler (elastic) region the procedure is the same, but instead of Eq.(12) Eq.(14) shall be used.

Safety in combined bi-axial loading in compression and torsion is calculated in accordance with Mises-Hencky distortion energy theory of failure in yield:

$$(n)_\sigma = \frac{(\sigma)_{yd}}{(\sigma)_{red}} = \frac{(\sigma)_{yd}}{\sqrt{(\sigma)^2 + 3(\tau)^2}} \quad (61)$$

Should the factor of safety in yield failure be less than required, then the min. required root Dia. is the one and only real and positive root of the following Eq.(62), obtained from Eq.(61) :

$$(d_3)^6 - (d_3)^2\left[\frac{4F_a(n)_\sigma}{\pi(\sigma)_{yd}}\right]^2 - 768\left\{\frac{(n)_\sigma\left[T_{\substack{raise\\total}}\right]_{max}}{\pi(\sigma)_{yd}}\right\}^2 = 0 \quad (62)$$

Total efficiency of a mechanism consisting of NON SELF-LOCKING screw and a sliding contact axial bearing:

$$(\eta)_{\substack{total\\raise}} = \frac{(1)_{th} F_a}{2\pi T_{\substack{total\\raise}}}$$

$$= \frac{\tan(\beta)}{\dfrac{\cos(\alpha)_n \tan(\beta) + (\mu)_{th}}{\cos(\alpha)_n - (\mu)_{th}\tan(\beta)} + (\mu)_B \dfrac{(D_o + D_i)}{2(d)_2}} \quad (63)$$

In order to determine the lowest total efficiency, the maximum values of the coefficients of friction for screw thread and axial bearing shall be used in Eq. (63). And the maximum feasible efficiency will corresond to the minimum values of the said coefficients of friction.

ILLUSTRATIVE EXAMPLE

Design of a lifting mechanism consisting of power screw and plain sliding contact axial bearing.

The given design input data is summarized in Table 2. Two design variants have been calculated: one using single start self-locking power screw, and the other one with multi-start NON self-locking power screw.

TABLE 2 Design Input Data

Description	Symbol and Value
Axial load on screw	F_a = 100 kN
Length of Screw	L = 2 500 mm
Effective length factor	ν = 0.8
Maximum allowable slenderness ratio	$(\lambda_{ef})_{max}$ = 180
Young modulus of elasticity	E = 207 GPa
Tensile yield strength of screw material	$(\sigma)_{yd}$ = 324 MPa
Maximum allowable bearing pressure	$(p)_{max}$ = 5 MPa
Screw thread profile	ISO, trapezoidal
Flank angle of thread	α = 15°
Nominal coefficient of friction for axial bearing.	$(\mu_{nom})_B$ = 0.08
Nominal coeff. of friction for flanks of thread	$(\mu_{nom})_{th}$ = 0.12
Coefficient of scatter for axial bearing	$(s)_B$ = 1.5
Coefficient of scatter for flanks of thread	$(s)_{th}$ = 1.8
Factor of safety in buckling	$(n)_b$ = 6.0
Factor of safety in self-locking	$(n)_{s.l.}$ = 1.1
Factor of safety in combined bi-axial loading	$(n)_\sigma$ = 3.0

TABLE 3 Calculated Output Data Which is Common to Single-Start and Multi-Start Screws

Description	Symbol and Value
Effective length of screw	L_{ef} = 2 000 mm
Effective length at the point of inflection	$(L_{ef})_{inf}$ = 1 948 mm
Buckling region	Euler (elastic)
Effective length corresponding to max. allowable slenderness ratio	$(L_{ef})_{max.all.}$ = 4 953 mm
Outside diameter of bearing	D_o = 226 mm
Inside diameter of bearing	D_i = 112 mm
Pitch diameter of thread	$(d)_2$ = 75.5 mm
Root diameter of thread	$(d)_3$ = 70.0 mm
Outside (nominal) diameter of thread	d = 80.0 mm
Pitch of thread	p = 9.0 mm
Pressure at the inside Dia. of axial bearing	$(p)_{max}$ = 4.99 MPa
Pressure at the outside Dia. of axial bearing	$(p)_{min}$ = 2.47 MPa
Effective slenderness ratio	$(\lambda)_{ef}$ = 114.3
Calculated factor of safety in buckling	$(n)_b$ = 6.02

For either of the two design variants the calculations have been carried out twice: first with the maximum and then with the minimum expected values of the coefficient of sliding friction for screw thread and axial sliding contact bearing.

The results of calculations using a prepared computer program are collated to Tables 3 and 4. Table 3 contains all parameters which are common to two design variants. Table 4 on the other hand compares the two design variants.

TABLE 4 Calculated Design Output Data Which Depends on the Number of Starts and the Coefficient of Friction

Parameters	Single-Start $N_{st} = 1$, $p = 9$ mm, $(l)_{th} = 9$ mm		Multi-Start $N_{st} = 6$, $p = 9$ mm, $(l)_{th} = 54$ mm	
Coefficients of friction	$(\mu_B)_{max} = 0.0980$ $(\mu_{th})_{max} = 0.1610$	$(\mu_B)_{min} = 0.0653$ $(\mu_{th})_{min} = 0.0894$	$(\mu_B)_{max} = 0.0980$ $(\mu_{th})_{max} = 0.1610$	$(\mu_B)_{min} = 0.0653$ $(\mu_{th})_{min} = 0.0894$
$(\alpha)_n$	14.98970°		14.64202°	
β	2.17300°		12.82565°	
T_{screw} raise	+ 777.327 N m	+ 494.517 N m	+ 1546.175 N m	+ 1234.396 N m
T_{screw} lower	− 482.879 N m	− 205.579 N m	+ 222.832 N m	+ 449.935 N m
T_B	± 827.928 N m	± 551.952 N m	± 827.928 N m	± 551.952 N m
T_{total} raise	+ 1605.255 N m	+ 1046.469 N m	+ 2374.103 N m	+ 1786.348 N m
T_{total} lower	− 1310.807 N m	− 757.530 N m	− 605.096 N m	− 52.017 N m
$(\eta)_{screw}$ raise	0.184272	0.289655	0.555847	0.696241
$(\eta)_{total}$ raise	0.089232	0.136879	0.362005	0.481114
$(\sigma)_{red}$	48.781 MPa	37.410 MPa	66.356 MPa	52.781 MPa
$(n)_{\sigma}$	6.642	8.661	4.883	6.139
$(n)_{s.l.}$	absolute	absolute	3.175	1.104

CONCLUSION

It is very instructive to compare the efficiencies and power requirements of lifting mechanisms made with multi-start or single-start power screws. For instance, if it is required to lift a load of 100 kN at the speed of 0.1 m/s, then according to Table 4 the required input power for a mechanism with multi-start screw will be 21 kW in case of maximum efficiency, and 28 kW for minimum expected efficiency. Should the single-start power screw be used, then the power demand shall be 73 kW or 112 kW respectively. These figures are speaking for themselves.

Long live the multi-start power screws !

REFERENCES

Dubbel, H. (1981). Mechanik, 1.11 Reibung. In W. Beitz, (Ed.), Dubbel Taschenbuch für Maschinenbau, 14th ed., Springer, Berlin. pp. 129-130.

Hancke, A. (1955). Anzugsmoment, Reibungswert und Verspannkraft bei hochfesten Schrauben. Draht, vol. 6, No. 3, 86-93.

Popov, E. P. (1968). Elastic Buckling of Columns with Different End Restraints. In Introduction to Mechanics of Solids, 1st ed., Prentice-Hall, Englewood Cliffs. p. 530.

ZUSAMMENFASSUNG

Der Hauptnachteil der SELBSThemmenden Bewegungsschrauben ist ihr sehr niedriger Wirkungsgrad. Aber falls das Spurlager als ein Gleitlager ausgeführt wird, dann kann sein Reibungsmoment, um die Selbstdrehung der NICHT-selbsthemmenden mehrgängigen Bewegungsschrauben unter Last zu verhindern, ausgenützt werden. Deswegen es ist möglich, einen Mechanismus, bestehend aus einer NICHT-selbsthemmenden Schrauben und einem Gleitspurlager, als SELBSThemmend auszuführen. Der Gesamtwirkungsgrad eines solchen Mechanismus wird auf alle Fälle vielmal grösser sein als bei der Benützung einer SELBSThemmenden Schrauben. Das berechnete Beispiel zeigt, dass der Gesamtwirkungsgrad von solchen SELBSThemmenden Mechanismen, bestehend aus NICHT-selbsthemmenden Schrauben und und einem Gleitspurlager, sogar 48 % hoch sein könnte, was im Vergleich mit 14 % bei der Verwendung von SELBSThemmenden Schrauben viel günstiger ist. Wird es z.B. notwendig, eine Last von 100 kN mit einer Geschwindigkeit von 0.1 m/s zu heben, dann ist die erforderliche Leistung 21 kW (min. Reibungswerte für Gewinde und Spurlager) oder 28 kW (max. Reibungswerte), für einen SELBSThemmenden Mechanismus mit NICHT-selbsthemmenden Schrauben. Dagegen, bei der Verwendung von SELBSThemmenden Schrauben, erhöht sich die erforderliche Leistung auf 73 kW, bzw. auf 112 kW.

Es leben die mehrgängigen Bewegungsschrauben !

Vergleich genormter Tragfähigkeitsberechnungen für Stirnrader nach AGMA 218.01, DIN 3990, ISO/DIS 6336 und TGL 10545

T. HÖSEL

Technische Universität München, Forschungsstelle für Zahnräder und Getriebebau

KURZFASSUNG: Anhand von Berechnungen ausgeführter Getriebe werden die gültigen Normen nach AGMA, ISO/DIN und TGL/RGW zur Zahnfuß- und Flankentragfähigkeit miteinander sowie mit den Normen verglichen, die sie ersetzen. Durch die verschiedenartige Erfassung der Verzahnungsgeometrie ergeben sich teils beachtliche Unterschiede in der Größe der errechneten Spannungen oder dem Tragfähigkeitsgewinn, den die Verwendung von Schräg- statt Geradverzahnung bzw. von Hoch- statt Normalverzahnung bringt. Durch Einbezug der zulässigen Beanspruchungen für vergütete und einsatzgehärtete Zahnräder wird gezeigt, in welchem Ausmaß für das gleiche Getriebe nach den Normen sich unterschiedliche übertragbare Drehmomente ergeben.

ABSTRACT: In calculations of practical gears the actual AGMA, ISO, DIN and Comecon Standards are compared to each other as well as to their own replaced old versions. As gear geometry is taken into account in different ways remarkable differences in the calculated stress level is observed. For a constant stress level the possible increase in torque transmission is also different for the different calculation methods using helical instead of spur gears or high addendum instead of normal addendum gears. Introducing allowable stresses according to the standards for through hardend and case hardened steels it can be shown that for the same gear the different standards lead to different transmittable torques.
This paper will be presented in English.

1. Einleitung

Zur Berechnung der Grübchen- und Fußtragfähigkeit sind neue Normen erstellt worden. In den USA erschien AGMA 218 im Dezember 1982 [3] als Ersatz für [5, 6].

1987 wird DIS 6336 als Draft International Standard erstmals veröffentlicht. In der BRD wird die in allen wesentlichen Punkten hiermit übereinstimmende neue DIN 3990 (Ausgabe 1987) [2] erscheinen, die eine Weiterentwicklung der Ausgabe von 1970 [7] ist. In den osteuropäischen Ländern wird an einem RGW-Standard [4] gearbeitet, der in den Grundformeln gleichfalls mit [1] bzw. [2] übereinstimmt und die DDR-Norm TGL 10545 [8] ablösen wird. Die alte DIN 3990, Ausgabe 1970 und die TGL 10545 sind weitgehend übereinstimmend und basieren auf dem Rechenverfahren von G. Niemann [9].

Für eine vergleichende Tragfähigkeitsberechnung sind in Zukunft vor allem Unterschiede zwischen den Rechenverfahren nach [3] und [1,2] bedeutsam. Ein Vergleich der "alten" mit den "neuen" Rechenverfahren ist für die Getriebehersteller und Betreiber wichtig, damit die mit den "alten" Verfahren gewonnenen Erfahrungen in die "neuen" übertragen werden können. In den Grundformeln sind die neuen und alten AGMA-Standards [3] und [5,6] nur für die Berechnung der Flankentragfähigkeit bei Schrägverzahnung geringfügig verändert. Zwischen Niemann [9] und den Normen [7,8] bestehen Unterschiede bei Fuß und Flanke. Zwischen alter und neuer DIN 3990 wurde die Zahnfußberechnung grundlegend geändert, d.h. statt der früheren "Nennspannung" wird jetzt eine "wirkliche Zahnfußspannung" verwendet.

2. Getriebestufen und Annahmen für die Vergleichrechnungen

Zugrunde gelegt wird die Getriebestufe (3) in Bild 1. Sie gehört zu einem Hubwerksgetriebe mit vergüteten Zahnrädern. Mit den gleichen Verzahnungsabmessungen könnte sie aber gleichfalls als einsatzgehärtete Stufe in Industriegetrieben eingesetzt werden. Hauptabmessungen: Achsabstand a = 710, Zahnbreite, b = 295 mm, Zähnezahlen z_1=21, z_2=66, Flankenwinkel α_n=20°, Normalmodul m_n= 16, Schrägungswinkel ß=10°, Profilverschiebung x_1 = 0.207 --> (A). Um Geometrieeinflüsse aufzuzeigen, wird hieraus eine "Hochverzahnung" entwickelt --> (B). Bei konstanten Kopfkreisen wie in (A) entsteht eine Geradverzahnung (C) mit ß=0 nur durch m_n = 16.2468 und weiterhin eine geradverzahnte Hochverzahnung (D) mit Kopfkreisen wie in (B), Bild 2. Die Zahnprofile im Stirnschnitt zu (C) und (D) unterscheiden sich nur minimal von denen zu (A) und (B).

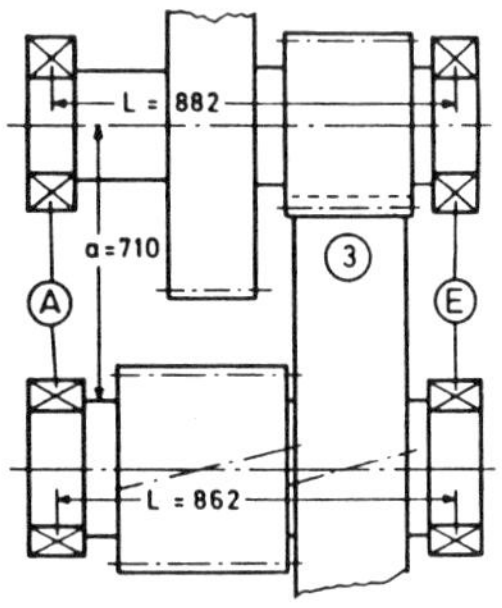

Bild 1: Getriebestufe 3 für Vergleichsrechnungen. Verzahnung A.

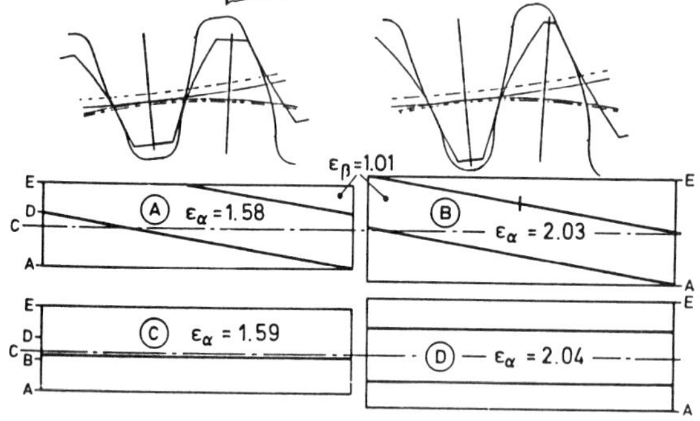

Bild 2: Stirnzahnprofile, Berührlinien im Eingriffsfeld (minimale Gesamtlänge) $\varepsilon_\alpha, \varepsilon_\beta$ = Profil-, Sprungüberdeckung

Die Rechenverfahren [1...9] enthalten Faktoren zur Erfassung äußerer und innerer Zusatzkräfte (Anwendungs- und Dynamikfaktor) sowie der Lastverteilung über Zahnbreite und Zahnhöhe ("Tragbild- oder Lastverteilungsfaktor). Bei AGMA [3] und DIN/ISO [1,2] entsprechen z.B. $C_a \leftrightarrow K_A$, $1/C_v \leftrightarrow K_V$ und $C_m \longleftrightarrow K_{H\beta} \cdot K_{H\alpha}$. Sind diese Einflüsse durch Messungen an der betrachteten Getriebestufe bekannt, sind die betreffenden Faktoren in summa in allen Rechenverfahren gleichgroß einzusetzen. Bei den nachfolgenden Vergleichsrechnungen können diese Faktoren deshalb entfallen, d.h. sämtlich 1.0 gesetzt werden. Bleibt das Antriebsdrehmoment T_1 konstant, dann zeigen Vergleichsrechnungen zwischen [1]...[9] nur die Unterschiede in den Spannungen bzw. Pressungen auf, die allein durch die fest vorgegebenen Vorschriften zur Erfassung der geometrischen Abmessungen bedingt sind.

Der Vergleich der übertragbaren Drehmomente ergibt sich, mit den den Verfahren [1]...[9] jeweils zugehörigen "zulässigen Beanspruchungen". Sie werden bei AGMA direkt als s_{ac} und s_{at} angegeben.

Bei den übrigen sind sie das Verhältnis "Dauerfestigkeitswert/Mindestsicherheit", z.B. $\sigma_{HP}=\sigma_{Hlim}/S_{min}$ bei [1,2]. Für die Vergleiche werden folgende Annahmen gemacht:

- o Hochwertiger Vergütungsstahl, HV = 300
- o hochwertiger Einsatzstahl, HV = 720
- o optimale Wärmebehandlung und Fertigung
- o d.h. zulässige Werte bzw. Dauerfestigkeiten entsprechend der "Obergrenze" gepaart mit der angegebenen Mindestsicherheit.
  V = vergütet (z.B. 34 CrNiMo 6)
  E = einsatzgehärtet (z.B. 17 CrNiMo 6).

Mangels Angabe werden für DIN alt die Mindestsicherheiten von Niemann verwendet. Die u.g. "zulässigen Werte" gelten gleichermaßen für Gerad- und Schrägverzahnung.

zul. Spannung		Flanke	Zahnfuß in N/mm²
D_n = DIN neu:	V	845/1.0	690/1.4
	E	1650/1.0	1040/1.4
A_n = AGMA neu:	V	931	325
	E	1551	482
A_a = AGMA alt:	V	931	325
	E	1551	448
D_a = DIN alt:	V	$710/\sqrt{1.3}$	314/1.8
	E	$1600/\sqrt{1.3}$	516/1.8
T = TGL 10545	V	750/1.5	216/1.5
	E	1785/1.5	342/1.5
N = Niemann	V	$752/\sqrt{1.3}$	310/1.8
	E	$1880/\sqrt{1.3}$	492/1.8

Die in den Grundformeln übereinstimmenden Verfahren DIS 6336, DIN 3990 (1987) und RGW-Standard werden mit ISO/DIN abgekürzt.

3. Zahnfußspannungen - Geometrieeinflüsse

Bild 3 zeigt oben in welchem Ausmaß die Zahnform mit x_1=0.2 allein durch Änderung von x_1 bei x_1+x_2 = konst änderbar ist. Die Zahnfußspannungen s_t und σ_F zu AGMA "alt" und "neu" sowie zu "DIN alt" und TGL 10545 sind jeweils gleich groß. Die Spannungen s_t nach AGMA liegen zahlenmäßig über σ_F zu DIN alt, verlaufen aber tendenziell gleich: Erhöhung der Profilverschiebung x_1 senkt die Fußspannung am Ritzel. Der Einfluß der maximalen Werkzeug-Kopfabrundung ist relativ klein und nur wenig von der Profilverschiebung abhängig.

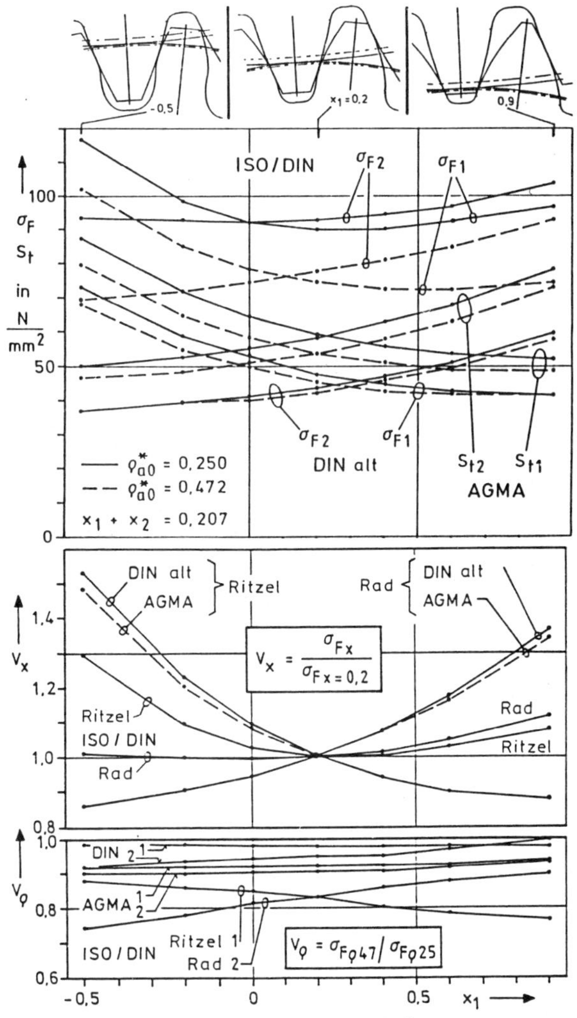

Bild 3: Einflüsse auf Zahnfußspannung Zahnprofile und Spannungen. Werkzeugkopfhöhe h_{a0} = $1.25 \cdot m_n$ mit Kopfabrung ϱ_{a0} = $0.25 \cdot m_n$ und ϱ_{a0} = $0.47 \cdot m_n$

Die Spannungen σ_F nach ISO/DIN sind wesentlich größer als nach "DIN alt" ("Faktor um 2,0") und auch deutlich größer als s_t nach AGMA. Die veränderten geometrischen Abhängigkeiten zeigen die Verhältnisse V_x

V_x = Spannung bei x_1 / Spannung bei x_1=0.2

und analog V_ϱ zum Einfluß von ϱ_{a0}. Im allgemein verwendeten Profilverschiebungsbereich (x_1=-0.5 und +0.9 sind sehr extreme Grenzwerte) ändert sich die Zahnfußspannung σ_F nach ISO/DIN nur unwesentlich mit x. Dagegen senkt die große Werkzeugkopfabrundung σ_F vergleichsweise stark, wobei die Senkung deutlich von x abhängt.

4. Vergleich der Flankenpressungen und Zahnfußspannungen

Die Verfahren [1]...[8] verwenden zur Beurteilung der Flankenbeanspruchung die "Hertzsche Pressung". Bild 4 zeigt die errechneten Werte als Balkendiagramm mit der Abszisse rechts. (Die Stribeckpressung gemäß Niemann [9] wurde in p_{Hertz} umgerechnet). Für die Verzahnung A...D und die Annahmen nach Abs. 2 sind als Punkte gleichfalls die Zahnfußspannungen mit der Abszisse links eingetragen. Das Drehmoment T_1 ist in allen Fällen konstant. Die unterschiedlichen Pressungs- bzw. Spannungswerte für dieselbe Verzahnung sowie die Relationen z.B. zwischen (a) und (B), oder die Tatsache, daß bei "DIN alt" $\sigma_{F1} > \sigma_{F2}$, aber bei "ISO/DIN" $\sigma_{F2} > \sigma_{F1}$ vorliegt, beruhen allein auf der verschiedenen Berücksichtigung der Verzahnungsgeometrien in [1]...[9].

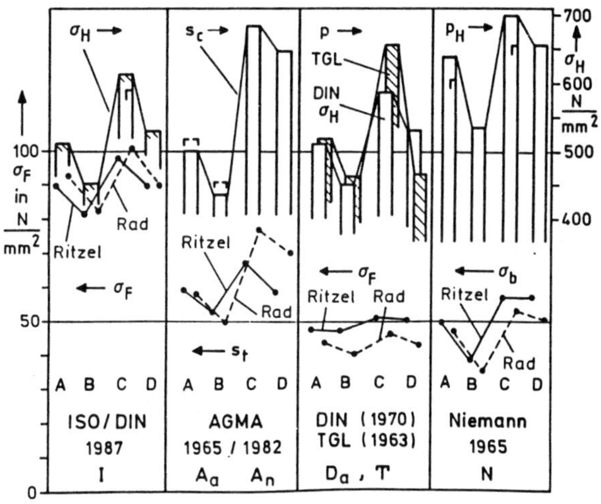

Bild 4: Flankenpressungen und Fußspannungen. Verzahnungen A...D nach Abs.2. T_1 = 26000 Nm, n_1 = 153 min^{-1}

Als V_H sind in Bild 5 oben die aus Bild 4 resultierenden Erhöhungsfaktoren angegeben, um die das Drehmoment T_1 jeweils erhöht werden kann, wenn statt der Geradverzahnung (C) die Schrägverzahnung (A) verwendet wird und in beiden Fällen die gleiche rechnerische Hertzsche Pressung vorliegt. Weiterhin sind die rechnerischen Erhöhungen für den Übergang auf Hochverzahnungen mit $\varepsilon_\alpha > 2.0$ angegeben.

Explizite Hinweise auf die Anwendbarkeit der Verfahren auf Hochverzahnungen mit $\varepsilon_\alpha > 2$ fehlen in [1]...[9], lediglich [2] enthält einen Hinweis zur Fußspannungen. Zur Flanke ist [8] mehrdeutig interpretierbar.

Hervorstechend sind die verschiedenen Steigerungen für Übergang C ⟶ A. Niemann [9] (als Basis für "DINalt") ergibt eine Drehmomentsteigerung auf 120%, DIN alt 133% und bei ISO/DIN sind rechnerisch 144% zulässig, um bei ß=0 und ß=10° die gleiche Pressung zu erhalten. AGMA alt und neu ergeben gar 176 und 188%. Eine derartige Steigerung erscheint dem Autor als sehr unrealistisch und die Steigerung Niemann - DINalt - ISO/DIN, d.h. "Geometrie-Bewertung" 1960 - 1970 - 1987 als nicht unproblematisch. (D) (B) ergibt ähnlich große Steigerungen.

Die sich bezüglich gleicher Zahnfußspannung ergebenden Drehmomentsteigerungen V_F sind demgegenüber klein. In Bild 5 ist jeweils der kleinere Wert für Ritzel oder Rad angegeben.

5. Vergleich der "übertragbaren Drehmomente"

Die sich aus Bild 4 mit den zulässigen Beanspruchungen nach Abs. 2 ergebenden "zulässigen oder übertragbaren Drehmomente T_{1z}" zeigt Bild 6. Es sei nochmals betont, daß diese Momente nur für die genannten Annahmen gelten: d.h. die Faktoren für Zusatzkräfte, Tragbild etc. sind in allen Verfahren gleich groß und die Gesichtspunkte zur Festlegung der "Mindestsicherheiten" sind gleichwertig. Die schraffierten Balken gelten für "vergütete Verzahnungen", wo T_{1z} ausnahmslos von der Flanke begrenzt wird. F und H markieren bei "Einsatzhärtung", ob die Fußspannung (σ_F) oder die Flankenpressung (σ_H) die Begrenzung ist. Die darüber liegende Markierung H und F zeigen die entsprechenden Momente an, die wegen dem "Gegenkriterium" aber nicht ausnutzbar sind.

Bei "Einsatzhärtung" ergibt DIN neu, bei "Vergütet" AGMA das höchste rechnerisch übertragbare Moment für die gleiche Verzahnung. Der Abfall für DIN alt und Niemann ist teils beträchtlich. Bei "Einsatzhärtung" verschieben sich teils auch die Begrenzungen F und H. In Bild 5 unten ist das Verhältnis $V_{T1} = T_{1zul} / T_{1zul}$ DIN 3990 neu aufgetragen, das die relative Zuordnung der einzelnen Verfahren besser erkennen läßt. Bei der Geradverzahnung (C) ergeben bei "vergütet" DIN neu und AGMA praktisch das

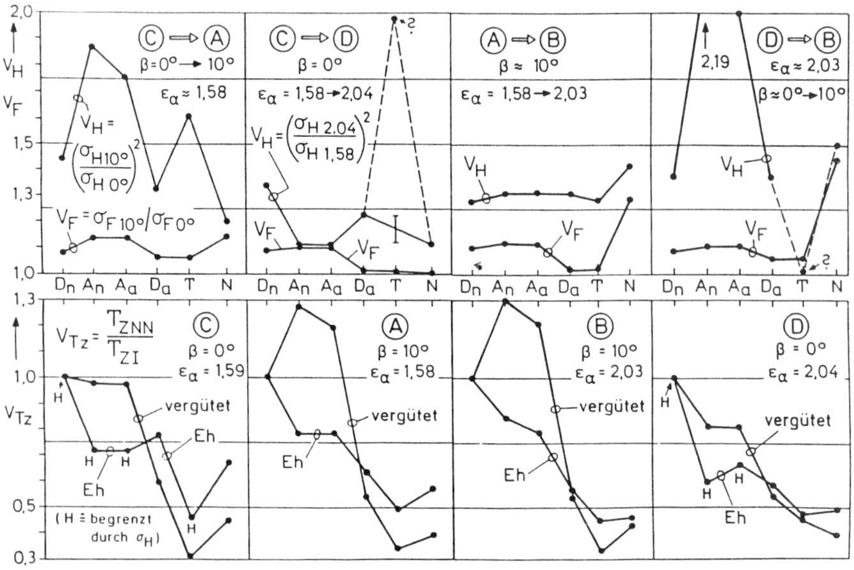

Bild 5 oben: Drehmomenterhöhungsfaktoren V_H und V_F für Übergang von Gerad- auf Schrägverzahnung bzw. von Normal- auf Hochverzahnung bei konstanter Pressung und Fußspannung.
unten: Vergleich der übertragbaren Drehmomente für die "zulässigen Beanspruchungen" nach Abs. 2. Verzahnungen A...D wie Bild 4.

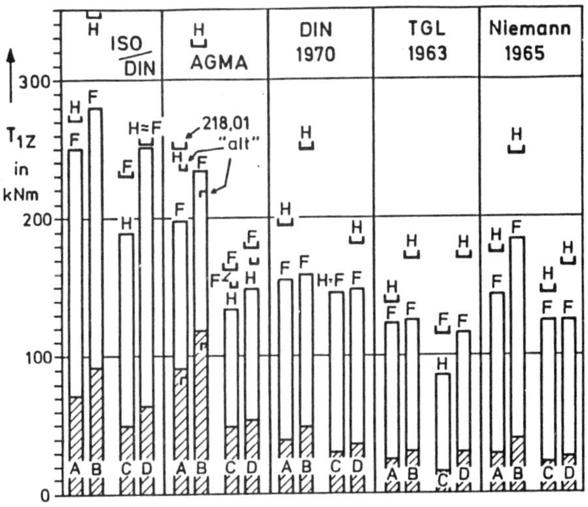

Bild 6: Übertragbare Drehmomente für "zulässige Beanspruchungen" nach Abs. 2

gleiche T_{1zul}, bei "Einsatzhärtung" fällt AGMA auf etwa 72% von DIN neu. Bei "Einsatzhärtung" und Schrägverzahnung (A) liegen die Verhältnisse etwa ähnlich, während bei "Vergütet" AGMA neu und alt 128% bzw. 120% von DIN neu ergeben. Die Hochverzahnungen B und D zeigen ähnliche Relationen. Der teils starke Abfall von DIN alt und Niemann als Basis für DIN 3990 Ausgabe 1987 sollte Anlaß sein, bei der Übertragung der mit den alten Verfahren gewonnenen Erfahrungen auf die neue DIN 3990 sehr sorgfältig zu sein.

Eine Zukunftsaufgabe für die internationale Normung sollte es sein, die Berücksichtigung der Verzahnungsgeometrie besser anzugleichen, vgl. Übergang Gerad- und Schrägverzahnung.

Anmerkung: Für Berechnungen und Zeichnungen der Zahnprofile wurde das FVA-Stirnradprogramm [10] verwendet, das der Autor im Auftrag erstellt hat für die
Forschungsvereinigung Antriebstechnik,
Postfach 710109, 6000 Frankfurt 71./BRD
Der Tragfähigkeitsteil umfaßt z.Z. 15 Rechenverfahren. Der Geometrieteil erfaßt auch "Sonderderverzahnungen". Die Herstellwerkzeuge können in Dateien gespeichert und diese Werkzeugdateien auch nach verwendbaren Werkzeugen abgesucht werden.

Schrifttum

[1] DIS 6336: Calculations of Load Capacity of Spur and Helical Gears (1987). ISO

[2] DIN 3990: Tragfähigkeitsberechnung von Stirnrädern Teil 1, 2, 3, 5, 11, Ausgabe 1985. DIN-Institut 5000 Köln/BRD

[3] AGMA 218.01: Pitting Resistance and Bending Strength of Spur and Helical Involute Gear Teeth. Dec 1982. AGMA Arlington, Virginia, USA

[4] RGW-Standard (Entwurf Okt. 84): Stirnradpaare-Festigkeitsberechnungen. VEB Rationalisierung, 8020 Dresden / DDR

[5] AGMA 210.02 (Jan. 65), AGMA 220.02 (Aug.66): Strength and Surface Durability of Spur Gear Teeth. Siehe [3]

[6] AGMA 211.02 (Feb.69), AGMA 221.02 (July 65): Strength and Surface Durability of Helical and Herringbone Gear Teeth.

[7] DIN 3990: Tragfähigkeitsberechnung von Stirn- und Kegelrädern, Blatt 1-9. Dezember 1970. Siehe [1].

[8] TGL 10546: Richtlinien zur Tragfähigkeitsberechnung von Stirnrädern mit Außenverzahnung. November 1963. VEB Fachbuchverlag 7031 Leipzig/DDR.

[9] Niemann, G.: Maschinenelemente Bd. 2. Springerverlag 1965

[10] Hösel, Th.: Benutzeranleitung zum FVA-Stirnradprogramm. FVA-Forschungsheft 209 (1986). FVA, 6000 Frankfurt 71/BRD.

Determination of the Weakest Section of Spur Gear Teeth Subject to Bending

ST. A. MAVROMMATIS

Department of Mechanical Engineering, University of Patras, Patras, Greece

Реферат. Известно что формирующая корень зуба кривая, под названием fillet (галтель), является результатом трохоидального пути, который проходит центр дуги закругленности зуба производящей рейки, при зубонарезании. В представляемой работе возникновение галтели рассматривается вследствии поворота делительной прямой производящей рейки, вокруг делительной окружности изготовляемого зубчатого колеса. При таком рассмотрении стал возможным вывод формул при помощи, которых можно определить позицию каждой точки галтели в полярных координатах в зависимсти от угла вышеупамянутого поворота, это в свою очередь дало возможность расчета ширины зуба на разных уровнях гальтели. Величина и местоположение опасного сечения определяются при отыскивании максимальных напряжений, применяя формулу защемленной балки с учетом коэффициента концентрации напряжений. С помощью разработанной методики и эвм были расчитаны в зависимости от числа зубьев зубчатого колеса, численные значения всех величин, входящих в расчетную формулу.

Keywords. Gear tooth geometry; fillet geometry; gear teeth endurence;

ВВЕДЕНИЕ

Общепринято, что каждый зуб зубчатого колеса работает в качестве защемленной балки (рис.1) для которой можно написать следующие уравнеия:

a) Для максимальных растягивающих напряжений

$$\sigma_\varepsilon = \frac{F_n}{b\,S_\sigma}\left(\frac{6h\cos\beta}{S_\sigma} - \sin\beta\right)k_t \qquad (1)$$

b) Для максимальных сжимающих

$$\sigma_\vartheta = \frac{F_n}{b\,S_\sigma}\left(\frac{6h\cos\beta}{S_\sigma} + \sin\beta\right)k_t \qquad (2)$$

В уравнениях (1) и (2) обозначено:

b - длина зуба (ширина зубчатого венца)

k_t - коэффициент концетрации напряжений.

Обозначения других величин согласно рис.1.

Трудность применения вышеуказанных уравнений в инженерной практике заключается в определении:

1. Величины опасного сечения S_σ.
2. Величины h т.е. расстояние этого сечения от точки приложения нагружающей силы F_n.
3. Коэффициента k_t, когда он расчитывается согласно [7] по формуле:

$$k_t = H + \left(\frac{S_\sigma}{\rho_f}\right)^L \cdot \left(\frac{S_\sigma}{h}\right)^M \qquad (3)$$

Где H,L и M величины зависящие от угла зацепления α.

В этой формуле как видно кроме величин S_σ и h входит и величина радиуса кривизны галтелн ρ_f на уровне сечения S_σ.

Для преодаления этих трудностей были предложены разные коэффициенты, которые косвенным путем учитывают вышеперечисленные величины. Более распространённые из них a) Lewis Form Factor [2] b) Geometry Factor, который предлогает AGMA (American Gears Manufactures Association) [4] c) Коэффициент формы зуба применяемый в основном в Советской техничеэской литературе [5]. Но если сравним для конкретного случая результаты расчета с применением этих коэффициентов устанавливаем, что они различны, факт, который создает сомнения о точности численных значений упамянутых коэффициентов.

В последние годы с развитием и широким распространением эвм для решения той же задачи, находит применение методика "Finite Elements" [9]. Однако должны отметить, что эта методика дает удовлетворительные результаты при предположении, что имеем заданное, точное очертание профиля зуба и особенно у основания его в области галтели. Известно, что геометрия галтели зависет не только от геометрии режущего инструмента, но и от числа зубьев нарезаемого колеса. Эти зависимости мало исследованы поэтому, эта методика не может дать еще какое-то обобщенное решение. Цель этой работы, сформулировать методику для определения прямым путем значения всех величин, входящих в уравнениях (1),(2) и (3).

ГЕОМЕТРИЯ ГАЛТЕЛИ

На рисунке 2 показывается процесс формирования профиля зуба, когда режущий инструмент соответствует производящей рейке. Известно [3], что точка B, которая является центром округления радиуса r ножки зуба инструмента, в процессе резки описывает трохоидальную траекторию а огибающая кривая последовательных положений дуги GH формирует галтель.

Рисунок 3 показывает формирование, как трохоиды,так и кривой галтели как результат поворота делительной прямой MN, режущей рейки, над делительной окружностью нарезаемого колеса. Для усиления наглядности увеличен размер AB рисунка 2 и заштрихован круговой сектор по дуге GH радиуса r. При предположении,что исходная позиция прямой AB будет та, которая совпадает с радиусом колеса на основание рисунка 3,могут быть сформулированы уравнения:

$$R_f = \sqrt{(C_1D)^2 + R^2 - 2R(C_1D)\cos\beta_2} \tag{4}$$

$$\gamma_f = \arccos\frac{R_f^2 + R^2 - (C_1D)^2}{2RR_f} - \beta_1\frac{180}{\pi} \tag{5}$$

где $$\cos\beta_2 = \frac{(AB)}{\sqrt{(R\beta_1)^2 + (AB)^2}}$$

Нужно отметить, что размер C_1D соответсевует радиусу кривизны галтели в точке, где оканчиваетя радиус R_f. Этот размер в будущем будем обозначать ρ_f. Он согласно рис.3 равняется:

$$\rho_f = \sqrt{(R\beta_1)^2 + (AB)^2} + r \tag{6}$$

Если будут выражены $R = mz/2$, $AB = m\ell_B$, $R_f = m\ell_f$, $\rho_f = m\ell_{\rho f}$ и $r = m\ell_r$, где m - модуль зацепления и z - число зубьев колеса, после нескольких алгебраических преобразований можно написать:

$$\ell_f = 0{,}5\sqrt{z^2 + 4\ell_{\rho f}^2 - c_1 z} \tag{7}$$

$$\gamma_f = \arccos\frac{2z - c_1}{4\ell_f} - \frac{180}{\pi}\beta_1 \tag{8}$$

$$\ell_{\rho f} = (2\ell_r + c_2)/2 \tag{9}$$

где $$c_1 = 8\ell_B\ell_{\rho f}/c_2 \quad \text{и} \quad c_2 = \sqrt{(z\beta_1)^2 + 4\ell_B^2} \tag{10}$$

Как видно из этих уравнений при заданных ℓ_r, ℓ_B (величины, которые характеризуют режущий инструмент), а также z, величины ℓ_f, γ_f и $\ell_{\rho f}$ определяющие геометрию галтели, являются функциами угла β_1. Этот угол в процессе создания галтели меняется от нуля до какого-то максимального значения, которое обозначим β_o. Значение β_o можно определить по рисунку 4. На этом рисунке показывается часть зуба производящей рейки, которая находится ниже делительной прямой и рассматривается движение рейки по направлению стрелки Г. При этом движении точка H, которая сопрягает дугу радиуса r с прямолинейным участком зуба рейки, пересечет линию зацепления n-n в точке c. Рассматривая в дальнейщем точку c, точкой зуба колеса с которым взаимодействует зуб рейки, устанавливается, что она определеяет и границу, где сопрягаются кривая галтели с кривой эвольвенты. Угол β_o, на который поварачивется колесо при вышерассмотренном движении рейки, равносилен с искомым углом β_1 на, который поварачивалась-бы рейка в обращенном движении. Согласно схеме рис.4 можно написать:

$$\beta_o = \arccos\frac{R_o^2 + (R - L_H)^2 - (Hc)^2}{2R_o(R - L_H)}$$

где: $$R_o = \sqrt{r_b^2 + (r_b\,\mathrm{tg}\alpha - L_H/\sin\alpha)^2}$$

$$Hc = L_H/\mathrm{tg}\alpha - r\cos\alpha - \eta/2 \quad \text{и} \quad r_b = R\cos\alpha$$

Если примем во вниманий, что обычно размер L_H равняется модулю тогда угол β_o можно выразить уравнением:

$$\beta_o = \arccos\frac{z(z-4) + c_3}{\sqrt{z(z-4) + (2:\sin\alpha)^2}\cdot\sqrt{(z-2)^2 + 4(\ell_r\sin\alpha)^2}} \tag{11}$$

где:

$$c_3 = 2\left[1 + (1:\sin^2\alpha) + (\ell_r\cos\alpha)^2 - (1/\mathrm{tg}\alpha - \ell_r\cos\alpha - c_4/2)^2\right] \tag{12}$$

$$c_4 = \eta/m = \pi/2 - 2(\ell_B\sin\alpha + \ell_r)/\cos\alpha \tag{13}$$

Уравнения (11) до (13) показывают, что при заданной геометрии производящей рейки и заданном числе зубьев нарезаемого колеса значение $\beta_o = \mathrm{const}$.

Для нахождения последовательных позиций точек галтели,в уравнениях (8) и (10) угол β_1 заменяем выражением $\varkappa\beta_o$ где коэффиуиенту $\varkappa$ даем значения от нуля до единицы.

ГЕОМЕТРИЯ ЗУБА

Из рис.3 видно, что на уровне точки, где оканчивается радиус R_f, толщина зуба если её обозначим S равняется двум хордам дуги ε. Поэтому:

$$S = 2R_f\sin\frac{180\varepsilon}{\pi R_f} \tag{14}$$

Где ε, согласно схеме, имеет значение:

$$\varepsilon = R_f(\beta_z - \gamma_f) - \frac{\eta}{2} \tag{15}$$

Угол β_z соответсвует углу между осью зуба и соседней впадиной, $\beta_z = \pi : z$. Буквой η обозначено расстояние между центрами радиусов r.

Выражая $S = m\delta_f$ и принимая во внимание уравнения (7) (8) и (13) можно написать формулу:

$$\delta_f = 2\ell_f \sin\left[\frac{180}{z} - \arccos\frac{2z-c_1}{4\ell_f} + \frac{180}{\pi}(\beta_1 - \frac{c_4}{2\ell_f})\right] \tag{16}$$

δ_f - безразмерная величина характеризующая толщину зуба на уровнях соответствующих радиусам R_f.

Если вместо R_f примем расстояние данного уровня от делительной окружности и обозначим, это расстояние Y_f, тогда имеем:

$$Y_f = mz/2 - R_f \quad \text{или} \quad \upsilon_f = z/2 - \ell_f \tag{17}$$

где: $\upsilon_f = Y_f : m$

При заданной геометрии режущей рейки и заданном z, изменяя угол β_1 от нуля до β_o, получаем с помощью уравнения (16) значения величины δ_f на уровнях соответствующих υ_f, вычисленному по уравнению (17).

Для того, чтобы начертить часть зуба выше галтели можно применить формулу:

$$S' = 2R_\kappa \sin\frac{180}{\pi}\frac{S_\kappa}{2R_\kappa} \tag{18}$$

Где:

R_κ - радиус колеса, который оканчивается на какойто точке к, эвольвенты зуба.

S_κ - дуга радиуса R_κ между эвольвентами сторон зуба.

S'_κ - хорда соответсвующая дуге S_κ.

Согласно [6] для вычисления S_κ имеем:

$$S_\kappa = 2R_\kappa\left[\frac{1}{z}(\frac{\pi}{2} + 2\xi tg\alpha) + inv\alpha - inv(\arccos\frac{mz\cos\alpha}{2R_\kappa})\right] \tag{19}$$

ξ - коэффициент коррекции. Для стандартных зубчатых колес $\xi = 0$.

Если и здесь заменим $R_\kappa = m\,\ell_\kappa$ и $S_\kappa = m\delta_\kappa$ будем иметь:

$$\delta_\kappa = 2\ell_\kappa \sin\frac{180}{\pi}\left[\frac{\pi}{2z} + inv\alpha - inv(\arccos\frac{z\cos\alpha}{2\ell_\kappa})\right] \tag{20}$$

При заданном z в уравнение (20) независимой переменной является величина ℓ_κ, которая меняется от $\ell_\kappa = \ell_f$, когда ℓ_f соответствует углу $\beta_1 = \beta_o$, до $\ell_\kappa = z/2 + 1$ соответсвующей окружности вершин колеса.

На рисунке 5 показываются профили зубьев, полученные с помщью эвм на основании выведенных формул. Для сравнения даются и фотографии зубьев нарезанных червячной фрезой по ГОСТу 9324-60. Конфигурации профилей, полученных теоретически, совпадают с действительными.

ОПАСНОЕ НА ИЗГИЕ СЕЧЕНИЕ

Формулу (1) можно преобразовать:

$$\sigma_\varepsilon / K = \Lambda_\varepsilon \tag{21}$$

где

$$\Lambda_\varepsilon = \left(\frac{6(\upsilon_\sigma + \upsilon_1)\cos\beta}{\delta_\sigma^2} - \frac{\sin\beta}{\delta_\sigma}\right)k_t \tag{22}$$

в формулах (21) и (22) обозначено:

$K = F_n/bm$, $\upsilon_\sigma + \upsilon_1 = h/m$, и $\delta_\sigma = S_\sigma/m$

Величина υ_σ определяет расстояние опасного сечения характеризуемого велиучиной δ_σ от делительной окружности, в то время как υ_1 характеризует расстояние точки встречи с осью зуба направления усилия F_n, до той-же окружности.

Из уравнения (21) становится очевидным, что напряжение σ_ε принимает максимальное значение, когда Λ_ε станет максимальным. Однако с самого начала нужно отметить, что решающую роль на значение величины Λ_ε играет выбор точки приложения силы F_n на поверхности зуба. Примем для нашего исследования случаи показанный на рис. 1 т.е. когда F_n действует на вершину зуба. Теоретически в этом положении зуба в зацеплении находится и другая, следующая пара зубьев и часть F_n воспринимается следующим зубом. Следует отметить однако, что из-за неточности изготовления зубчатого колеса не исключен случай отсуствия контакта между зубьями второй пары и тогда рассматриваемая схема нагружения становится наиболее опасной. Не случайно, что в подавляющем большинстве случаев, расчеты на изгибную прочность ведутся именно для такого вида нагружения.

Влияние на Λ_ε величин β и υ_1

На основании рис.6 можно написать следующие уравнения:

a) Для угла β:

$$\beta = tg\beta_\kappa - \pi/2z - inv\alpha \tag{23}$$

где

$$tg\beta_\kappa = \sqrt{(z+2)^2 - (z\cos\alpha)^2}/z\cos\alpha$$

b) для υ_1:

$$\upsilon_1 = 0{,}5\left[(z+2)(1 - \gamma tg\beta_\kappa) - z\right] \tag{24}$$

где: $\gamma = S_\kappa/2R_\kappa$

R_κ - радиус здесь окружности вершнн, $R_\kappa = m(z+2)/2$

S_κ - дуга радиуса R_κ расчитываемая по формуле (19).

Как видно из уравнении (23) и (24) величины β и υ_1, входящие в фармулу (23), при заданном угле зацепления α, завнсят только от числа зубьев колеса z.

Влияние коэффициента k_t

Как уже отметили коэффициент концетрации напряжений k_t расчитывается по формуле (3). Для $\alpha = 20^0$, согласно [8] эта формула конкретизируется:

$$k_t = 0{,}18 + \left(\frac{S_\sigma}{\rho_f}\right)^{0{,}16} \cdot \left(\frac{S_\sigma}{h}\right)^{0{,}45}$$

или $$k_t = 0{,}18 + \left(\frac{\delta_\sigma}{\ell_{\rho f}}\right)^{0{,}16} \cdot \left(\frac{\delta_\sigma}{\upsilon_\sigma + \upsilon_1}\right)^{0{,}45} \qquad (25)$$

Нужно здесь подчеркнуть, что величина $\ell_{\rho f}$, которая характеризует радиус кривизны галтели на уровне δ_σ, является переменной величиной, значение которой расчитывается формулой (9). Это когда ряд авторов, например [1], [10], считают ее постоянной. Как видно из формулы (25), значение k_t уведичивается, чем ниже находится опасное сечение (больше δ_σ и меньше $\ell_{\rho f}$). Конкретное знацение он принимает в процессе определения велнчин δ_σ и υ_σ.

Отыскиьание величин δ_σ и υ_σ

Эти величины определяют толщину зуба в опасном сечении и расстояние его от делительной окружности. Нахождение их и дает ответ на поставленную задачу. Они представляют собой одну из пар значений δ_f и υ_f для, которых значение Λ_ε становится максимальным и находятся решением уравнения (22) для каждого конкретного числа зубьев z для которого β и υ_1 постояны. Практически это делается изменением коэффициета $\varkappa$ при угле β_o. При современной вычислительной технике лучше всего это сделать составлением соответствующей программы и решением на эвм. В таблице 1 даются результаты полученные таким путем. В программу, данные относящиеся к геометрии режущего инструмента (червячной фрезы) вошли, $\alpha = 20^0$ $\ell_r = 0{,}4$ $\ell_{AB} = 1{,}25$ (согласно ГОСТу 9324-60).

ТАБЛИЦА 1

z	κ	υ_σ	υ_1	β^0	δ_σ	$\ell_{\rho f}$	k_t	Λ_ε	Λ_Θ
18	0,49	0,94	0,78	30,30	1,79	1,86	1,19	2,98	3,65
19	0,50	0,94	0,79	29,89	1,82	1,88	1,20	2,94	3,59
20	0,51	0,94	0,79	20,51	1,84	1,89	1,20	2,90	3,54
21	0,52	0,94	0,79	29,16	1,85	1,91	1,21	2,87	3,50
23	0,53	0,95	0,79	28,54	1,89	1,93	1,21	2,81	3,43
25	0,54	0,95	0,80	28,00	1,92	1,95	1,22	2,77	3,36
27	0,55	0,96	0,80	27,53	1,95	1,96	1,23	2,73	3,31
30	0,56	0,97	0,80	26,92	1,98	1,98	1,23	2,68	3,25
35	0,57	0,98	0,81	26,10	2,03	1,99	1,24	2,63	3,17
40	0,58	0,98	0,81	25,46	2,07	2,01	1,25	2,59	3,11
50	0,59	1,00	0,82	24,52	2,12	2,02	1,26	2,53	3,02
70	0,60	1,01	0,82	23,36	2,18	2,03	1,27	2,47	2,93
100	0,60	1,03	0,83	22,43	2,24	2,02	1,28	2,43	2,87
150	0,59	1,04	0,83	21,67	2,28	1,99	1,29	2,40	2,82

В этой таблице дается значение и величины Λ_Θ, которая соответствует сжимающим напряжениям согласно формуле (2).

Из данных таблицы характерно:

1. Положение опасного сечения находится на расстояние от делительной окружности равное $0{,}94 \cdot m$ для $z = 18$, увеличивающееся до $1{,}04 \cdot m$ для $z = 150$.
2. Ширина его принимает значения от $1{,}79 \cdot m$ для $z=18$ доходящее до $2{,}28 \cdot m$ для $z=150$. (Более интенсивное увеличение величины δ_σ с ростом z по сравнению с υ_σ обьясняет более высокую прочность зубьев на больших z, что отражается и в значениях Λ_ε и Λ_Θ).
3. Угол β с ростом z уменьшается от $30{,}30^0$ для колеса 18 зубьев до $21{,}67^0$ для колеса 150 зубьев.
4. Величина коэффициента k_t растет хотя и медленно с увеличением z и принимает значения от 1,19 до 1,30.

ЗАКЛЮЧЕНИЕ

В результате изложенной работы можно предложить для расчета зубьев на изгиб следующую последовательность:

1. Пишем уравнение защемленной балки под видом:

$$\sigma = \frac{F_n}{bm}\left(\frac{6(\upsilon_\sigma + \upsilon_1)\cos\beta}{\delta_\sigma^2} \mp \frac{\sin\beta}{\delta_\sigma}\right)k_t \qquad (26)$$

2. Преобразуем его:

$$\text{a)}\ \sigma_\varepsilon = \frac{F_n}{bm}\Lambda_\varepsilon, \qquad \text{b)}\ \sigma_\vartheta = \frac{F_n}{bm}\Lambda_\Theta \qquad (27)$$

Уравнение (27a) применяется для нахождения растягивающих напряжений а уравнение (27b) дл сжимающих. Значения величин Λ_ε и Λ_Θ принимаются в зависимости от числа зубьев z расчитываемого колеса с таблицы 1, где для справки даются и значения всех велицин определяющих их.

Здесь нужно отметить, что в зависимости от материала из которого изготовляется зубчатое колесо, могут быть более опасными в одних случаях растягивающие напряжения, в других сжимающие [10].

3. Нагрузающую силу F_n определяем по формуле:

$$F_n = \frac{F_t}{\cos\alpha} K' \qquad (28)$$

Где:

F_t - окружное усилие соответствующее устанавившемуся режиму работы и зависящее от передаваемой мощности.

K' - коэффициент являющийся произведением ряда коэффициентов, учитывающих такие факторы, как перенагрузку, неравномерное распределение нагрузки по длине зуба, всякие динамические факторы и.т.д

Значение этих коэффициентов в зависимости от условий работы передачи, даются в технической литературе.

Литература

1. Baronet C.H.,G.V.Tordion "Exact Stress Distribution in Standard Gear Teeth and Geometry Factors". Journal of Engineering for Industry. Nov.1973.
2. Black P.H. and O.Eugence Adams, Jr "Machine Design" p.352 Mc Graw-Hill copyright 1968.
3. Buckingham E. "Analytical Mechanics of Gears" p.49 Dover Puplications 1949.
4. Deutshman A. W. Michels and C.Wilson "Machine Design" p.548. Macmillan 1975.
5. Кудрявчев В.Н. "Детали Машин" p.276 Л. Машиностроение. 1980.
6. Mavrommatis A.S."Theory of Mechanisms and Machines" vol.2 p.300 Patras 1984 (In Greece).
7. Mitsiner R.G., H.H.Mabie "The Determination of the Lewis Form Factor and the AGMA Geometry Factor J for External Spur Gear Teeth". Journal of Mechanical Design. Jen.1982.
8. Shigley J.E., Larry D.Mitchell "Mechanical Engineering Design" p.600 Mc Graw-Hill series 1983.
9. Wilcox L., W.Coleman "Application of Finite Elements to the Analysis of Gear Tooth Stresses". Journal of Engineering for Industry. Nov.1973.
10. Winter H., M.Hirt "The Measurements of Actual Strains at Gear Teeth, Influence of Fillet Radius on Stresses and Tooth Strength". Journal of Engineering for Industry. Febr.1974.

DETERMINATION OF THE WEAKEST SECTION OF SPUR GEAR TEETH SUBJECT TO BENDING

Abstract. Practical calculations of the bending stress at the root of spur gear teeth are based on the use of a coefficient variously known as Lewis form factor, geometry factor or coefficient of tooth form. This coefficient expresses in the cantilever beam formula the influence of tooth geometry on the bending strength.

Since the coefficient used by various authors are not equivalent, the purpose of this work is to formulate a method for the direct and precise determination of the tooth thicness at the weakest section and the distance of this section from the point of application of the load for spur gear teeth made by the hobbing process.

The fillet portion of the tooth profile of such gears is a product of the trochoidal path traced by the center of the tip rounding of the rack shaped cutter. In this work, the generation of the fillet was investigated with reference to the rotation of the rack pitch line around the pitch circle of the gear being cut and was succeeded in expressing the position of different points of the fillet in polar coordinates as a function of the angle of this rotation. This made possible the calculation of the tooth thickness at various levels, from the root of the tooth to the point of blending of the fillet and involute portions of the profile. It also made possible the precise drawing of tooth profiles of hobbed gears, for different numbers of teeth. The weakest section was determined for the case when the load acts on the tip of the tooth. The size and position of this section was estimated by searching for the biggest stress in the cantilever beam formula, taking simultaneonsly in to account the stress concentration factor.

Using this methodology, all quantities which enter in the cantilever beam formula were calculated and tabuluted as functions of the number of teeth.

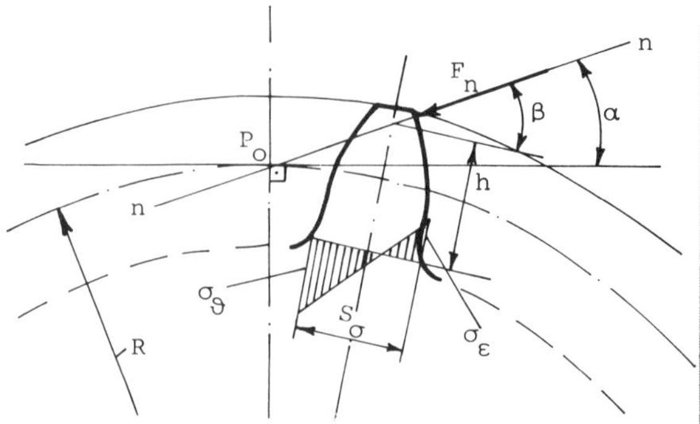

Рис.1. Схема нагружения зуба

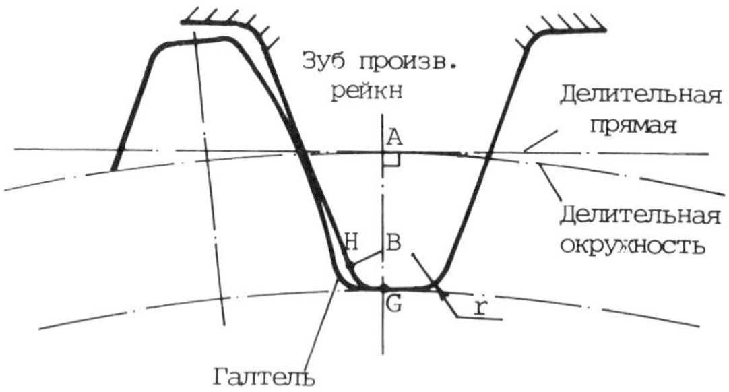

Рис.2. Формирование зуба производящей рейкой при зубонарезании

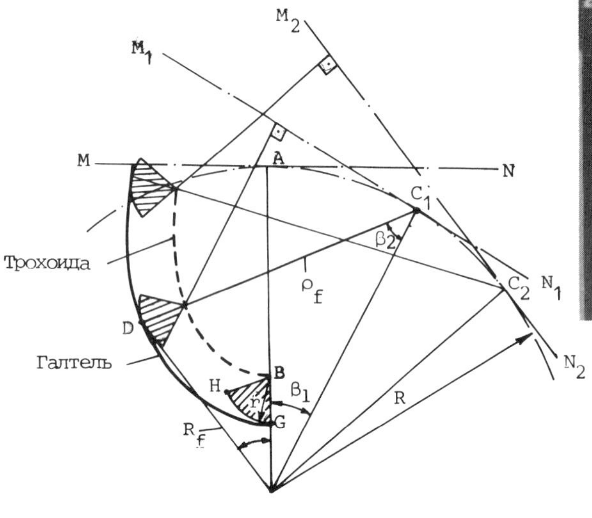

Рис.3. Схема образования трохоиды и галтели

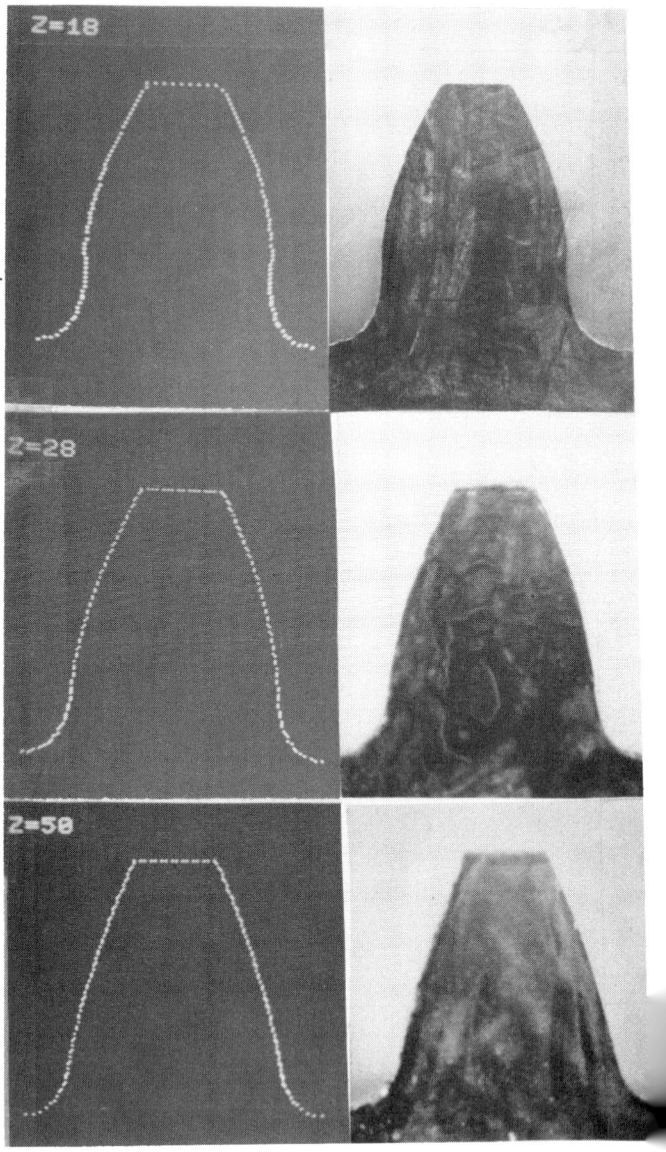

Рис.5. Сопоставление конфигураций зубьев расчетных и действительных

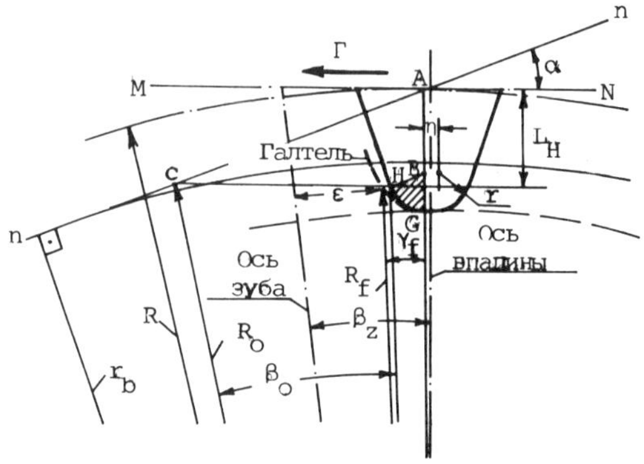

Рис.4. К определению величин β_o и ε.

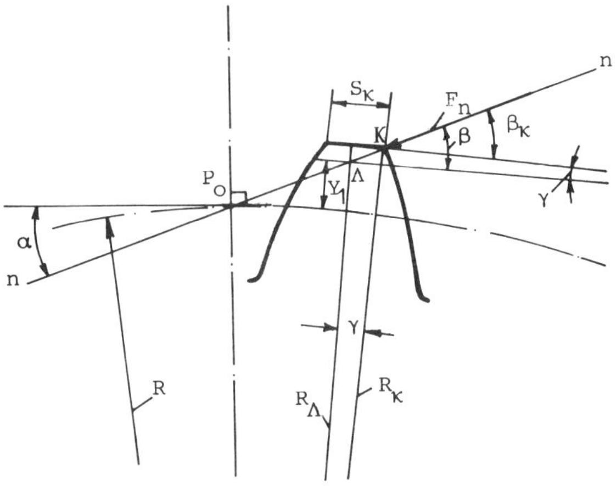

Рис.6. К определению угла β и величины Y_1

The Further Development of the Sizing of Epicyclic Gear Trains

Z. TERPLÁN AND Á. DÖBRÖCZÖNI

Institute of Machine Elements, Technical University of Heavy Industry, 3515 Miskolc, Hungary

Abstract. Of the numerous types of epicyclic gear trains the so-called normal type in which the planet gears engage simultaneously with the externally toothed sun and the internally toothed ring gear is most frequently used. Another variety is represented by the twin planet gear, by means of which the transmission limits may be extended. With this type, one planet gear engages with the sun gear, and the other with the ring gear. Both types are statically uncertain mechanisms, which means that load distribution between the planet gears is not uniform, further, that load is unevenly distributed along the contacting tooth generatrices as well. Effects of geometrical design, bearing rigidity, manufacturing and mounting errors, elasticity of the structural material, input moment, etc. are investigated by the authors. In the gear train investigated, influences of about 150 input parameters were analysed. Research results: Selection of the efficient parameters, and investigation of their effects.

Keywords. Epicyclic gear trains; longitudinal load distribution; load distribution between planet gears; sizing.

INTRODUCTION

All over the world, in the fields of transport, traffic, agriculture, and the general engine design, the epicyclic gear trains for reducing, or increasing, rpm are making headway. Their main advantage consists in that they enable the power to be transmitted to be broken down into several arms. This breaking down of power into a number of arms may also be observed with conventional tooth gear drives. Unfortunately these efforts are effective only if the distribution of power brings the desired result. The phenomenon of unequal load distribution does not relate to the load distribution between planet gears alone. If more than one pair of teeth are meshing in the same engagement, different loads will develop at the contacts of each of the pairs of teeth, with the load distribution being not uniform along the contacting tooth generatrix either.

According to experience, two ways offer themselves for attaining uniform load distribution: increase of manufacturing accuracy; utilization of the natural properties of the structural material, that of steel, elasticity.

For the sizing for fatigue of active tooth surfaces, use of the factors KH_α and KH_β which characterize the transversal load distribution, and longitudinal load distribution, respectively, is prescribed by various standards. These standards do not yield identical results, however. In epicyclic gear trains it is the most heavily loaded contact sun gear-planet gear that determines the load capacity of the whole gear train, i.e., when determining the standard load, the factor also makes its appearance besides the factors KH_α and KH_β [1] . Since $KH_\alpha>1$, $KH_\beta>1$, and $\Omega>1$, it may be asumed that the standard load calculated with their product would increase an unjustifiable amount. Since one and the same elastic construction is concerned, the factors are obviously interrelated, mutually influencing each other; that is, if their values are selected independently of one another, it is to be feared that an inaccurate result might be obtained, with the gear train becoming oversized.

The parameters that influence the contact of the tooth surfaces, the elastic deformation of the elements, and, hence, uneven load distribution, may be

of constructional nature: design and deformation of the gear bodies and rims of the sun gear, planet gear, and ring gear; deformation of the planet carrier and the planet shaft; deformations of the bearings design of the gear couplings driving the sun or ring gear,

of technological nature: tooth directional errors, inaccurate seating of the sun and planet gear on the shaft, shaft positional errors caused by positional errors of the bearing bores, positional errors manifesting themselves in the mutual arrangement of the planet gears, positional errors of the central gears (sun and ring gear), manufacturing errors of the gear couplings or other joining elements of the central gears, etc.

THE THEORETICAL MODEL

For the purposes of the investigation, a mathematical model describing the geometric relations and elastic deformations of a twin epicyclic gear train (Fig. 1.) was developed and applied to a computer. Based on the model, the meshing positions of the teeth, the backlashes between the theoretically contacting tooth generatrices, and the load distributions along the line of contact may be calculated in six fields of contact.

The mathematical model - based on K. I. Zablonsky's theory [2] - exists in the form of a linear algebraic equation system describing the simultaneous displacement of the contact points of the meshing teeth; it contains 150 independent variable parameters. Inclination of the teeth is calculated on the basis of a cantilever fixed plate model [3] , and contact deformation of the teeth is computed on the basis of the elastic half-space deformation. The model functions in the form of a program written in Fortran language.

INFLUENCE OF SOME DESIGN AND TECHNOLOGICAL PARAMETERS

Influence of the point of application of torque. Depending on the distances L3(1)... ...L3(4) at which the disks for torque transmission, are situated from the gear surfaces, as well as on how the sun gears, planet gears, and ring gears are arranged in relation to each other, different load distribution take place. With conventional gear drives it may be said in general that load distribution is more uniform if the points of torque application are situated on the opposite end faces of the meshing gears.

In the case of an epicyclic gear train, different relations have to be taken into account because of the asymmetric deformation of the planet carrier, the inclination of the planet shafts, and the design of the engaging central gears.

TABLE 1

L3(1) (mm)	L3(2) (mm)	$KH_{\beta 1}$
120	120	1,68
0	120	2,09
0	0	2,16
120	0	1,74

L3(3) (mm)	L3(4) (mm)	$KH_{\beta 4}$
120	120	1,53
0	120	1,70
0	0	1,75
120	0	1,58

Effect of the elasticity of bearings. According to TABLE 2,several paarings of bearings were investigated, from time to time also assuming an absolutely rigid bearing. Influence of a few bearings having identical and different spring constant ZC was also examined. It is evident that, there is no reason for installing uniform planet bearings.

TABLE 2

ZC(1,1)(mm/N)	ZC(2,1)(mm/N)	$KH_{\beta 1}$	$KH_{\beta 4}$
$2\cdot 10^{-6}$	$2\cdot 10^{-6}$	1,681	1,704
$1\cdot 10^{-6}$	$2\cdot 10^{-6}$	1,961	1,862
$2\cdot 10^{-6}$	$1\cdot 10^{-6}$	1,354	1,48

Effect of tooth directional errors. It is seen in the figures showing the factor for the load distribution along the contacting tooth generatrices that for both the sun gear (Fig. 2.) and the ring gear (Fig. 3.) a favourable compensating effect can be obtained with tooth directional errors HN of given directions.

Effect of a positional error occurring at the left-side end of a single planet gear in the case of a sun gear rigidly fitted with bearings. It is assumed there exists a 0,2 mm excentricity error at paces of 30°. Owing to the way the sun gear is gripped, the backlashes on the individual planet gears are different, and it is to be expected that one of the gears might be overloaded in relation to the others. If the factors KH_{β}, Ω , and RK = $KH_{\beta}\cdot\Omega$ are plotted in a polar diagram, it becomes evident which of the planet gears will have to bear maximum load in a given position of the error if engaged with the sun gear whilst the two other planet gears become completely relieved, and conversely, load is greatest on the gears 2 and 3 when the gear 1 is running idle (Fig. 4.). In the contact field belonging to the ring gear, conditions are similar to those mentioned above, yet values for RK are smaller.

Effect of a positional error occurring at the left-hand side end of a single planet gear shaft in the case of a sun gear swivelon a gear coupling. Since the position of only the planet gear 1 does change there develop, due to the sun gear's alignment, additional obliquites in the fields 1...3, the field 3 will change owing to the error, with the load distribution curves of the fields 5 and 6 being nearly in agreement with one another (Fig. 5.). It is seen in the polar diagram for the contact field 1 (Fig. 6.) that although the planet gear is everywhere in engagement during the complete turn of the error, there is a marked change of the factor KH with, however, the factors and RK nowhere reducing to zero.

Effect of ring gear positional error. An excentricity of 0,02 mm is assumed at the left side end of the geometrical axis of the ring gear. Variation in the load distribution is investigated whilst this error is being turned around at paces of 15°. Owing to the rigidly - without clutch - clamped sun gear, an appreciably high load may occur on one of the planet gears in relation to the others (Fig. 7.). It is apparent that, if the error falls in the field of contact, its effect is much greater than it would be if the error had a direction perpendicular to the field of contact. If the sun gear swivels on a gear coupling, the factor RK = $KH_\beta \cdot \Omega$ serving as the basis for the standard loading takes on a smaller value (Fig. 8.).

COMPARATIVE ANALYSIS OF THE EFFECTIVITY OF PARAMETERS THAT DETERMINE LOAD DISTRIBUTION

Unequality of load distribution may be reduced by the use of efficient parameters. Consequently, it has to be decided which of the 150 independent parameters that describe the general model influence appreciably the load distribution, i.e., which of them are efficient.

Starting from a great number of independent parameters, statistical methods were used to select the efficient ones. The constructional and technological parameters were randomly varied within the intervals obtained by using the random number generator of a computer. After calculating the load distributions for a great number of epicyclic planet variants, the efficient parameters were selected by the random weights method [4]. For each variant the factors KH_β and were calculated in each field of contact then of the factors RK found in three fields each of both the sun and the ring gear the best one was selected, and the KH_β and Ω values belonging to it were marked with a solid circle, and the non-standard ones with a blank circle in Figs. 9. and 10. related to the sun gear and the ring gear, respectively. Looking for a linear relationship between the factors Ω and KH_β it may be established that for that reference gear in which the sun gear rigid bearing support in Figs. 9. and 10., for that reference gear in which the sun gear swivels on a gear coupling

$$\Omega \approx 1,681 - 0,165\ KH_\beta$$

in the sun gear meshings,

$$\Omega \approx 1,627 - 0,166\ KH_\beta$$

in the ring gear meshings.

That is, it may be stated that Ω reduces with increasing KH_β and vice versa.

The calculations showed that of the 150 parameters describing the epicyclic gear train, only 30 were efficient at one time, the effect of the others on load distribution being negligible if varied withing given intervals.

With almost every variety the breadth dimensions BS(1)...BS(4) of the gears are efficient, their increase entailing an increase in RK in most cases. The co-ordinates L3(1)...L3(4) of the moment application points of the gears are effective for several varieties. In general, every kind of positional error and tooth directional error could be encountered among the efficient parameters.

CONCLUSIONS

On the basis of numerous computer-aided experiments on epicyclic gear trains with twin planet gears, the conclusion may be drawn that a relationship exists between the factor KH_β of the longitudinal load distribution and the factor Ω of the load distribution between the planet gears, i.e., if KH_β increases, Ω decreases, and vice versa. By using the efficient parameters selected by mathematical statistical methods, the longitudinal load distribution and the load distribution between planet gears may be greatly improved.

REFERENCES

1 Terplán, Z. (1974). Dimensionierungsfragen der Zahnrad-Planetengetriebe. Akadémiai Kiadó, Budapest.

2 Zablonsky, K. I. (1977). Zubchatue peredachi. Technika, Kiew.

3 Wellauer, E. J., and Seireg, A. (1960) Bending Strength of Gear Teeth by Cantilever-Plate Theory. Transactions of ASME, Series B, vol.82, No. 3. 213-222.

4 Hartmann, K., and Lezki, E., and Schäfer, W. (1974). Statistiche Versuchplanung und -auswertung in der Stoffwirtschaft. VEB Deutscher Verlag für Grundstoffindustrie, Leipzig.

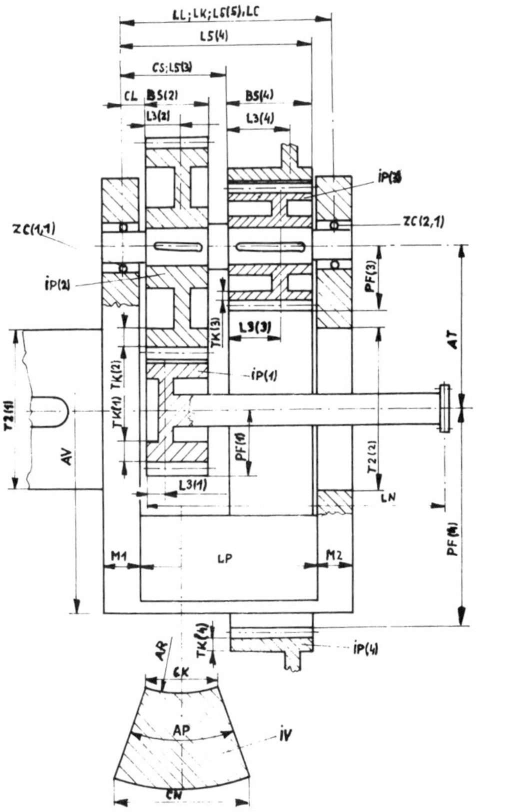

Fig. 1.

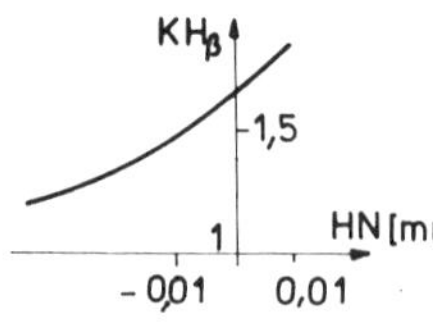

Fig. 2.

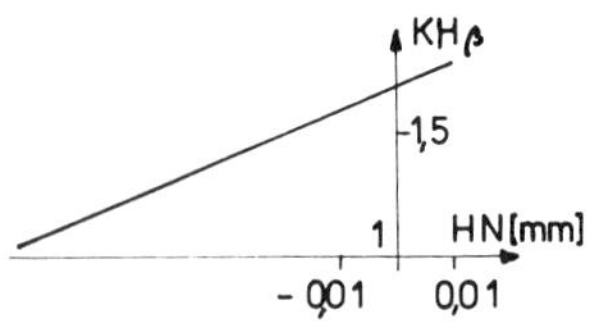

Fig. 3.

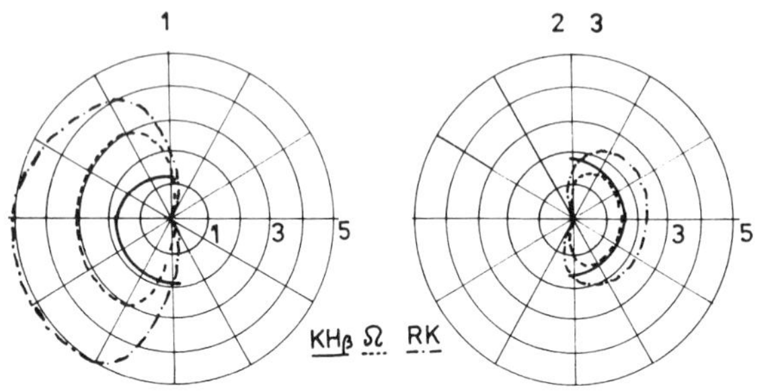

Fig. 4.

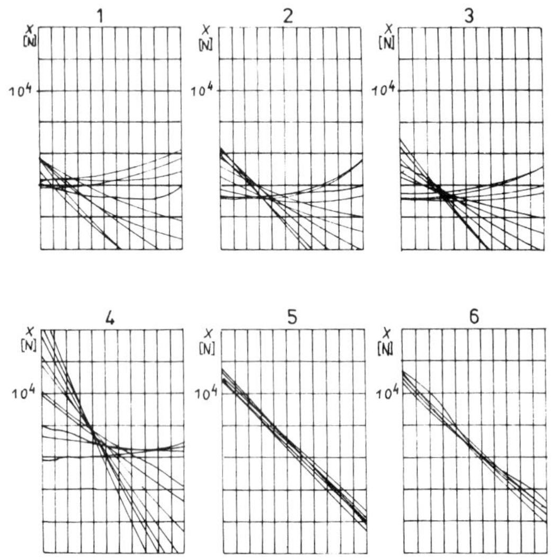

Fig. 5.

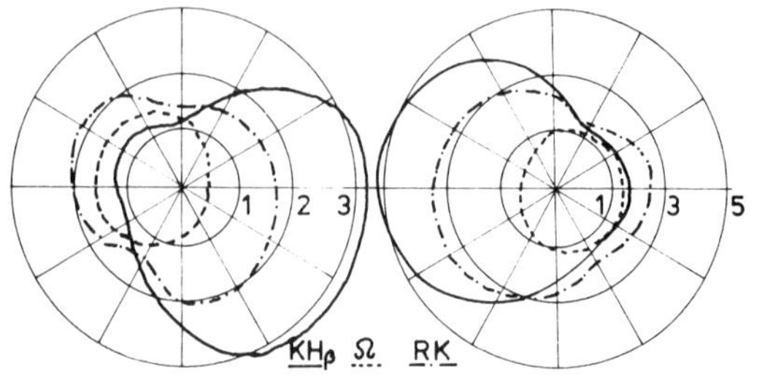

Fig. 6.

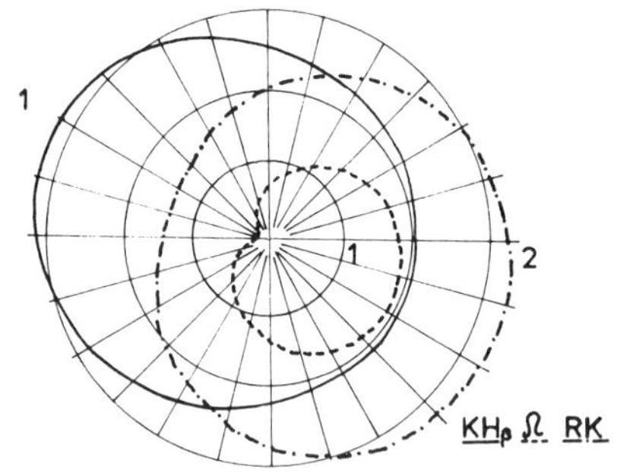

Fig. 7.

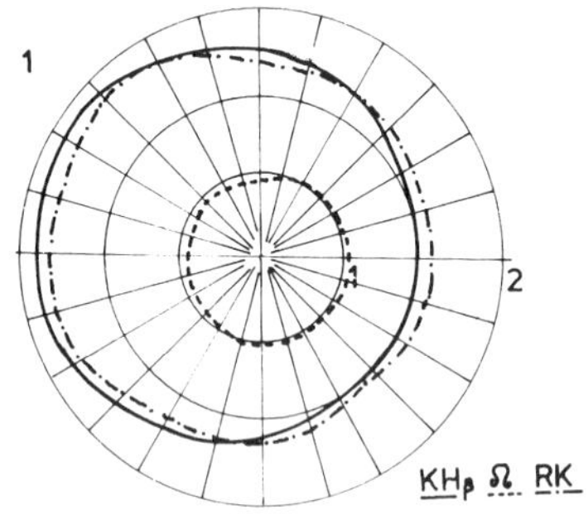

Fig. 8.

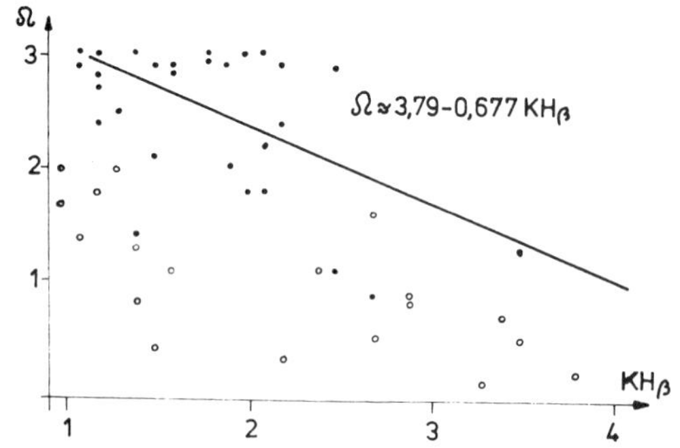

Fig. 9.

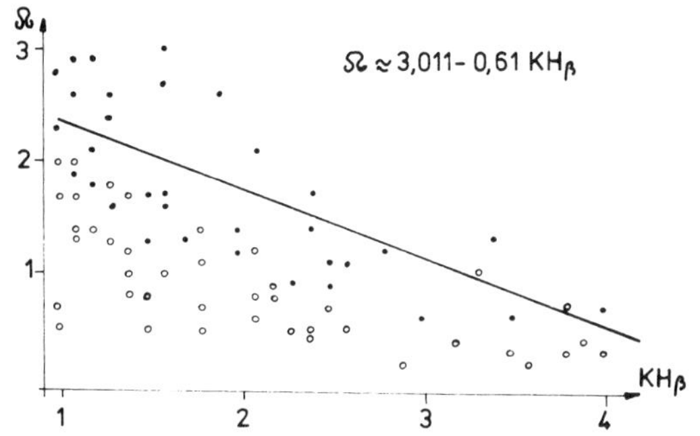

Fig. 10.

Investigation of the Transmission Characteristics of Plastic Involute Fine Mechanism Gears

T. NAGARAJAN AND G. V. N. RAYUDU

Department of Mechanical Engineering, Indian Institute of Technology, Madras-600 036, India

Abstract. Plastic gears are finding increased applications in fine mechanism due to certain specific advantages over metal gears. They are easily manufactured generally by injection moulding and hence are economical. But the inherent dimensional instability in injection moulding process affects the transmission characteristics of gear drives viz., instantaneous angular transmission efficiency (η_{ω}), and instantaneous torque tranmsission efficiency (η_M). In this article investigations have been carried out to find the manufacturing errors, deformation and wear on the transmission characteristics of plastic involute gears (module between 0.2 mm and 0.5 mm) in mesh with metallic mates like brass and suitable recommendations have been made for the use of polyacetal plastic gears (POM) in Fine mechanism.

Keywords. Plastic gears, Instantaneous angular and torque efficiencies, Involute, Clock gears, polyacetal gears.

INTRODUCTION

Gear materials extend over a wide range from ferrous alloys and light metals to nonmetallic plastics. The technology of plastics offers following advantages in Fine mechanism gearing: light weight, silent operation, vibration damping, wear resistance, non-corrosiveness, little or no need for lubrication, suitability for injection moulding and low cost. Chief among the disadvantages are the difficulty of machining them accurately and their temperature instability. However the plastics are suitable for variety of consumer products and instrument packages with proper selection and application. Since the signal transmission is the main function in precision gearing, strength and duraility are considered to be secondary importance. However, the main five types of applications of gears in fine mechanism are listed below under two heads [Roth 1964; Glaser 1974].

I Constant velоctiy ratio gearing (η_{ω} = constant)

1. Measuring trains (e.g.) dial indicator
2. Positioning trains or minimum backlash gearing (e.g) position meter drives, microscope drives
3. Power transmission type (e.g.) sewing machine, hand drills.

II Permissible variation of velocity within one pitch length (η_{ω} = variable)

4. Going or running trains (e.g.) watches and clocks
5. Counter gear units (e.g.) energy and gas meter drives.

Of these, the positional devices are demanding constant instantaneous velocity ratio whereas the going trains require constant instantaneous torque ratio. In positional devices such as potentiometers or resolvers, used in robots, the accuracy of these devices are limited to the accuracy of mating gears. In deciding the torque control of various links of the robots [Koren 1983], the accuracy of torque is dependent on the type of gearing used. In the soft servo system used in robot axes,correction forces from the axis motor are varied in proportion to the position error of the axes concerned. With the possibility of soft servo system, it is still possible to use plastic gears for hybrid type of robots incorporating both position and torque correction simultaneously.

For gearing, the widely used thermoplastic materials are polycetals or polyoxymethlene (POM), Polymids (PA) and Polyurethane. Of these Acetal homopolymers (e.g.) Delrin and acetal co-polymers (e.g.) hostaform, are widely recommended in Fine mechanism, because they retain greater precision compared with other plastics. Mating plastic with plastic results in maximum dimensional variations. The use of making one of them a metal mate can significantly reduce the above problems. Hence the investigation of transmission characteristics injection moulded plastic gears (POM) with metal gears (brass) are necessary with reference to the allowable manufacturing errors, the deformation of teeth and the wear of plastic gears. In this respect, Roth [1973] has studied wear aspects of plastic gears with different material combination for the gear modules greater than 0.8 mm and E. Nill [1977] has studied transmission behaviour of plastic gear pairs. For the practical investigation, the straight spur gears of involute profile in the module range of 0.2 to 0.5 mm (supplied by various clock manufacturers in West Germany) were used. The number of teeth vary from 10 to 80. The quality of plastic gears (POM) are in the range No.10 .. 12 according to DIN 3960-3967 and DIN 58405. The brass mate is of quality 7 .. 9 and brass conforms to DIN 17660 (usually denoted as MS 58). Most of the gear profiles are of involute in nature according to DIN 867. The various tests setups used in the investigations have been developed in Uhren Technik Institute (UTI), Stuttgart, West Germany. Some important definitions relating the transmission behvaiours of two mating gears are given in the reference Roth (1964) and the same symbols are used in this report.

The effect of manufacturing errors

The dimensional variations of injection moulded plastic components are mainly due to parameters like the moulding conditions, the working cycle for the component, the material characteristics like crystal structure, internal stresses, ageing effects, orientation of crystals and the shape of the press tools. To find out the variation of the parameters like module (m), pressure angle (α_o), addendum radius (r_K) correction factor (x), etc., in injection moulded plastic involute gears, theoretical profiles are drawn with the aid of computer and compared with actual profiles (Fig.1). The results of the profile comparison show that addendum circle radius is reduced and the normal pressure angle is increased compared to the design parameters. The effect of torque transmission by change of pressure angle is shown in Fig.2, and the effect of addendum diameter reduction is shown in Fig.3. The increase in pressure angle and reduction in tip radius will try to reach the cross-over line of the η_M vs. ϕ_1 curve and thereby affecting adversely the instantaneous torque efficiency (η_M) endangering kinematic accuracy. The angular transmission errors ($\Delta\phi_2$) are mainly due to the deviations in the involute profile and the pitch errors due to shrinkage. The shrinkage of the plastics affects adversely the angular transmission behaviour, because the conjugate action of the gears is not satisfied since the involute profile has errors in its geometrical form, after injection moulding. This is showns in Fig.4.

The effect of the deformation

Due to the inherent characteristics of the thermo-plastic plastics, the gear tooth will have reversible and irreversible deformation even on the lightest load applied intermittantly. The permanent plastic deformation increases as the frequency of the loading cycle increases. A number combination of gear pairs are used to investigate the central deformation angles ($\Delta\phi$). These are shown in Figs.5 and 6. These tests are conducted with the load being applied at the rate of 120 cycles/hour. In the case of a plastic (POM) gear with brass, as the number of teeth increases, the central deformation angle reduces, since the contact ratio increases and the load is more evenly distributed.

The torque and the angular transmission behaviours of the plastic/ brass gear pair are summarised in Figs.7 and 8. In the case of involute gears, the tip contact is unavoidable due to deformation (since contact ratio $\varepsilon_o \geqslant 1$). This tip contact affects the transmission behaviour, especially the angular transmission. But the elastic deformation increases the contact ratio, thereby the instantaneous torque efficiency (η_M) is improved at the same time the instantaneous angular efficiency (η_ω) is adversely affected.

Effect of wear

Generally, plastic gear materials are some what unpredictable and required life testing, to establish performance for specific applications and conditions. Fig.9 shows wear-depth in mm in plastic gears run against brass gears. The tests are conducted without any lubricant medium in between mating gears, since plastics are basically self-lubricating. The wear depth is measured by microslicing of plastic gears after embedding with suitable moulding compound. (Specifix - commercial name). Fig.10 and 11 show the η_M curve and $\Delta\phi_2$ curve for a specimen with a particular wear depth respectively and compare the same with gears under newstand. The effect of wear on the transmission characteristics will be more, if the worn out flanks completely alter the geometry itself. Since the plastic deformations are set in, the torque transmission efficiency improves. But the angular and torque transmission behaviours are somewhat resembling that of a circular-arc profile since the permanent plastic deformation and wear on the teeth appears to be more as a circular-arc rather than the involute nature.

CONCLUSIONS

Due to deformation, the contact ratio increases, thereby improving the instantaneous torque efficiency (η_M) but at the same time the angular transmission is affected as shown in Figs.7 and 8; Tip contacts are unavoidable - thereby increase of wear.

Initial wear subjected to a limit, improves the transmission behaviour but increased wear leads to poor angular transmission behaviour. The recommended loading for POM/brass gear pairs is given in Figs.12 and 13 and this is compared to the recommended value by Nill and Glaser [1977] for POM/POM gear pairs (these values for an allowable addendum tip crest contact factor of f_k=0.1 m).

The safe limit of F_u/b can be increased upto 2N/mm in case of POM/brass combination instead of 1.5 N/mm in case of POM/POM combination. In the case of wear studies, the allowable contact pressure can be increased for POM/brass pairs upto 50...

60 N/mm^2 to keep a wear depth of t_{vmax} = 10 μm after 10^6 cycles. The corresponding load on the tooth is 2N/mm breadth. The safe limits of correction factors and tolerances are further investigated and recommended by Nagarajan [1976,1977] for plastic gears.
Hence if one gear of the pair is metallic, like brass the performance is better instead of both being made of polyacetal. The recommended limits should be carefully adhered to in usng plastic gears otherwise the transmission characteristics will not be satisfactory.

Acknowledgements

The author acknowledges Prof. G. Glaser, Prof. F. Assmuss and E. Nill of Institut für Uhren Technik und Feinmechanik, University Stuttgart, West Germany for sparing the test setups and various West German industries for donating the plastic and brass gears.

References

Glaser, G., (1974) Lexikon der Uhrentechnik, Verlag W. Kemptar, Ulm.
Koren,Y.,(1983) Computer Control of Manufacturing Systems - Industrial Robots,McGraw Hill Book Co.,pp.221-247.
Nagarajan, T.,(1980) Investigation of the transmission characteristics of plastic involute fine mechanism gears, Diss. I.I.T., Madras.
Nagarajan, T., (1976) Untersuchung verschiedener Evolventenprofile auf deren Eignung für die Herstellung aus Kunststoff, Uhrentechnik (UT), Nr.4, pp.12-20.
Niemann, G. and Roth. (1964) K.H., Zahnformen und Getriebeeigenschaften bei Verzahnungen der Feinwerktechnik Feinwertechnik, 68, (1964), Nr.9, pp.344 .. 357 (Part I), Nr.10, pp.409.. 427 (Part II).
Nill, E.,(1977) Die Übertragungseigenschaften spritz-gegossener Kunststoffzahnräder der Feinwerktechnik, Diss. Tech. Hochschule, Stuttgart, Betreuer Dr. G. Glaser.
Roth, K.H., (1973) Verschliess-untersuchungen an Kunststoffzahnrädern der Feinwerktechnik, VDI Berichte 195 pp.1...8.

Die Übertragungseigenschaften evolventen Kunststoff Zahnräder der Feinwerktechnik

In der Feinwerktechnik werden in Zunehmendem Masse Zahnräder aus Kunststoff verwendet. Die Herstellung der Zahnräder erfolgt im Spritzguss verfahren. Diese Arbeit ist ein Versuch, die Einflusse der Herstellung, des Verformung - und Abriebsverhaltens bei Evolventen verzahnungen aus Polyacetal (POM) mit Messing zahnrädern, auf deren Übertragungseigenschaften darzustellen. Die Modulegrösse der Zahnräder betragt m = (0.2 ... 0.5)mm. Durch die Verformung wird der Überdeckungsgrad erhöht, die daraus resultierenden Winkelübertragungsfehler vergrössern jedoch die Laufgeräusche der Verzahnung. Der mittelere Übertragungswert von η_M wird kleiner. Er kann mit dem Abrieb zunehmen, so dass die Momentenübertragung sogar besser wird als die der theoretishem Verzahnung. Als Grenze der Verwendbarkeit von Evolventenverzahnungen kann die Kopfkantenüberdeckung f_k angesehen werden. Fur die Untersuchungen wurden verschiedene Geräte von UTI , Stuttgart, West Germany benützt.

****

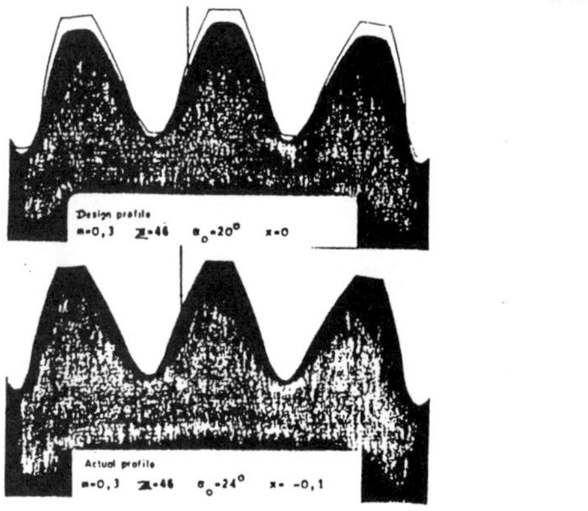

FIG.1.COMPARISON OF PROFILES

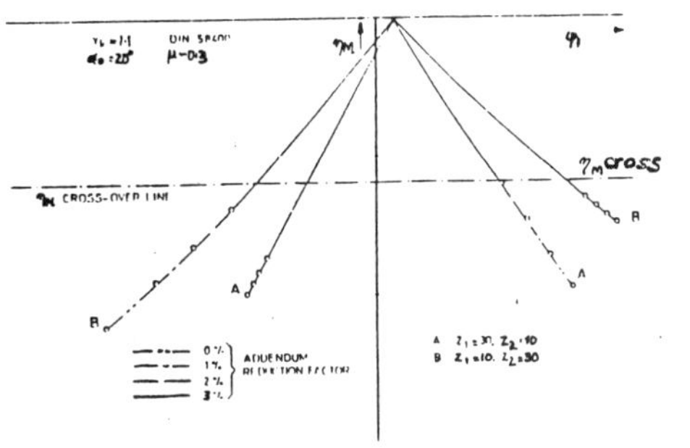

FIG.3. EFFECT ON ADDENDUM DIA.REDUCTION ON TORQUE EFFICIENCY

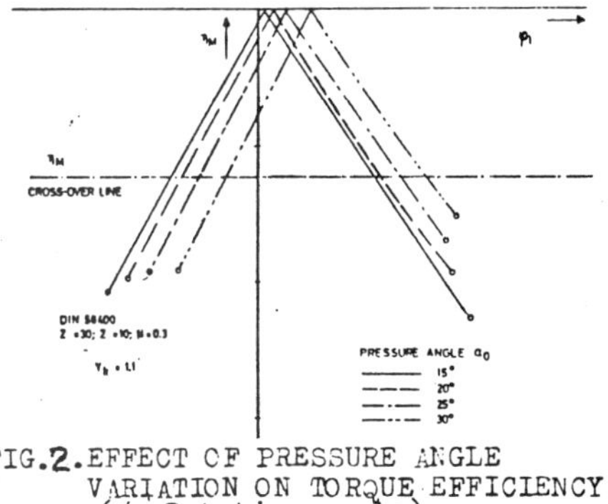

FIG.2.EFFECT OF PRESSURE ANGLE VARIATION ON TORQUE EFFICIENCY ($\emptyset_1$-Rotation angle)

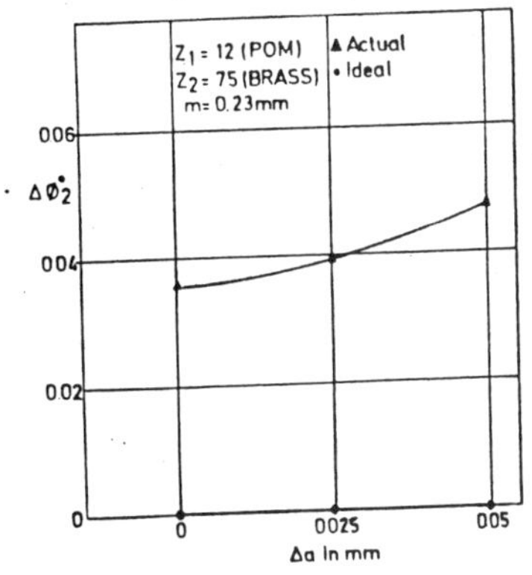

FIG.4.EFFECT OF SHRINKAGE ON $\Delta\emptyset_2$ ($\Delta\emptyset_2$-Angular transmission error)

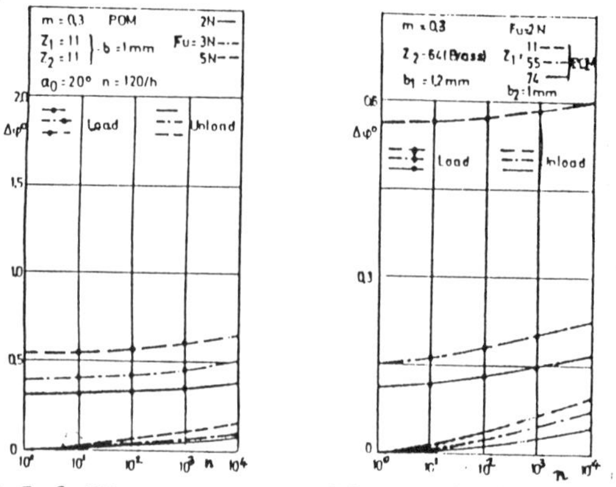

FIG.5-6. DEFORMATION ANGLE Vs NUMBER OF CYCLES($\Delta\emptyset$-Central deformation angle)

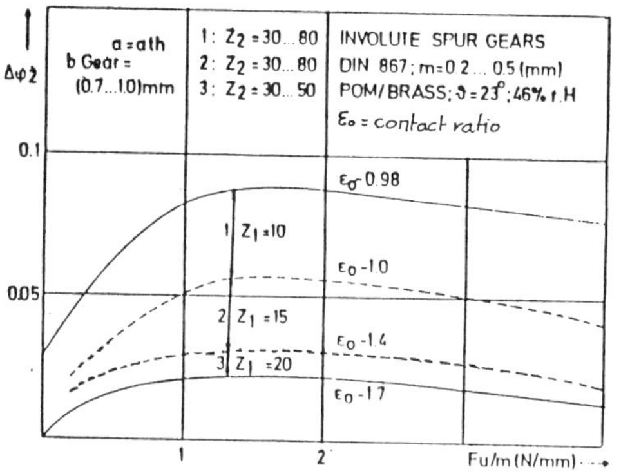

FIG.7. EFFECT OF LOAD ON $\Delta\emptyset_2$

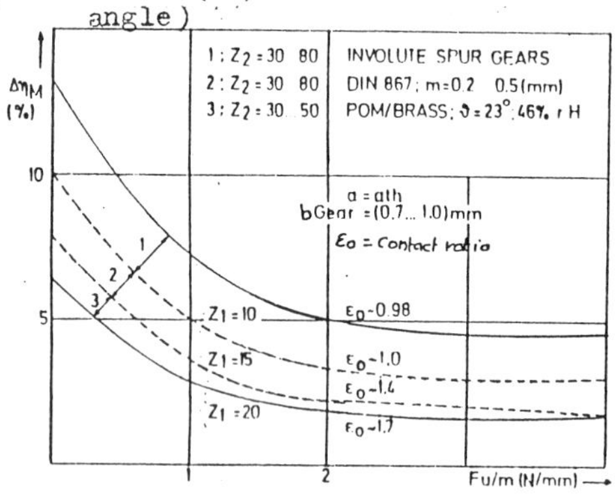

FIG.8. EFFECT OF LOAD ON $\Delta\eta_M$

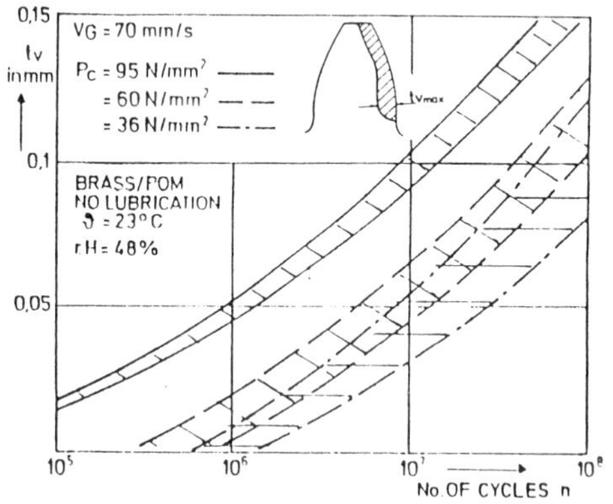

FIG.9. WEAR DEPTH WITH DIFFERENT LOAD

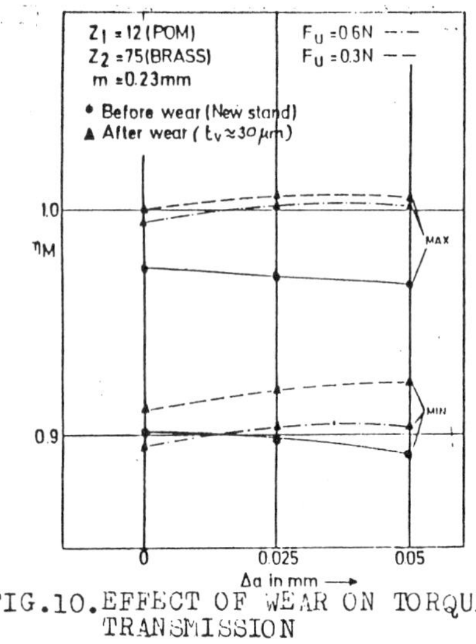

FIG.10. EFFECT OF WEAR ON TORQUE TRANSMISSION

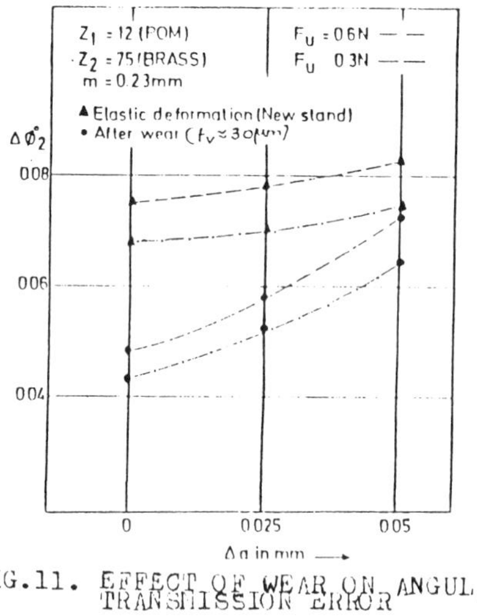

FIG.11. EFFECT OF WEAR ON ANGULAR TRANSMISSION ERROR

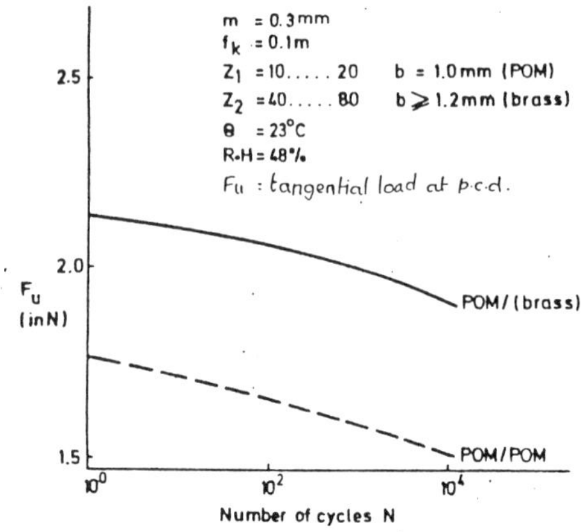

FIG.12. RECOMMENDED LOAD FOR POM/BRASS COMBINATION (f_k = 0.1m)

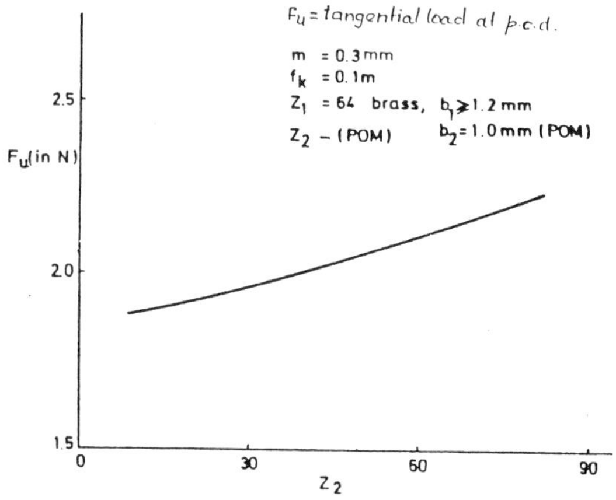

FIG.13. RECOMMENDED LOAD WITH DIFFERENT TEETH

Geared Rotary Actuator: A Way of Providing a Power Hinge with Large Speed Ratio and Minimum Envelope

F. DE BONO

Dipartimento di Meccanica, Politecnico di Torino, Torino, Italy

Abstract. The main features concerning performance and design of geared rotary actuators, often referred to as "power hinges", are discussed; it is pointed out which relations should be used to determine numbers of teeth and number of planets, in order to obtain a high speed ratio with an easy assembly procedure. The main structural features that concern the elements of the actuator are studied, in particular referring to gear design. Efficiency expressions are calculated and some considerations are developed to evidence influence of single meshing efficiencies, numbers of teeth and speed ratio on the overall efficiency and on the non reversibility condition.

Keywords. Actuator; spur gear; speed ratio; correction; efficiency.

A major problem of several actuation systems of new advanced aircrafts is to fit within very limited envelopes while maintaining a large load capability. In those applications where the actuation system consists of a high speed motor (hydraulic, pneumatic or electric) followed by a speed reducer, a device recently developed to minimize volume and weight is a specially configured geared reducer generally referred to as "geared rotary actuator". This device is now present in a few advanced military and civil aircraft and its use is expanding as the knowledge increases on how to design and manifacture it.

KINEMATIC AND GEOMETRICAL CONSIDERATION

A rotary actuator essentially is (Fig. 1-2) an epicyclic gear reducer consisting of a sun gear, driving compound planet gears meshing with two lateral fixed ring gears connected to the structure and a central gear connected to the actuating element.

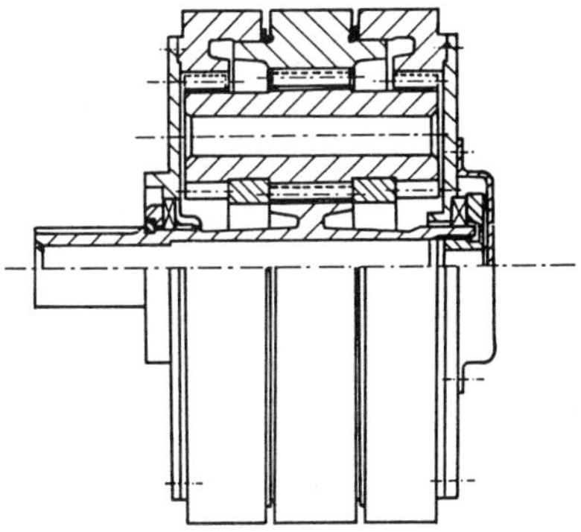

Fig. 1 Real representation of a rotary actuator

TABLE 1 Nomenclature

i=operating centre distance
m_0=nominal modulus
m_{ij}=operating modulus of gears i and j
n_s=number of planet gears
t_c=clearance between the outside circles of adjacent planets
t_0=tooth thickness at the outside radius
x_i=correction factor
x_{ij}=sum of corrections for gears i and j
z_i=number of teeth of gear i
α_0=nominal pressure angle
α_{ij}=operating pressure angle of gears i and j
η_{ij}=efficiency when power flows from 1 to 2
σ_b=tensile bending stress
σ_H=maximum hertz contact stress
$\tau_{ij}=\Omega_j/\Omega_i$=speed ratio

The actuator planetary gear train provides the function of large torque amplification and speed reduction; this is achieved by having similar, but slightly different ratios of planet teeth to ring

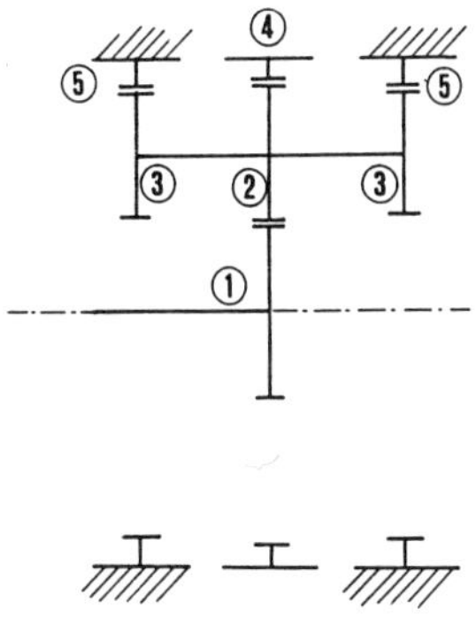

Fig. 2 Schematic representation of a rotary actuator

gear teeth for the centre and outer sections (Fig. 3). In consideration of the high performance that this device must provide, even the choice of the number of teeth must be very accurate, in order to obtain high speed ratio, with a great load supporting capability and an easy assembly procedure. Therefore some conditions must be satisfied:

$z_4=z_5$ (or $z_3=z_2$) (1)
$z_1, z_4=n_s$(whole number) (2)
$z_2=z_3+1$ (or $z_4=z_5+1$) (3)

Equations (1) and (2) enable to have planets equally spaced and to ease the assembling procedure; in fact , if planet gears are machined with three teeth aligned, a correct assembling is simple. At first , with a suitable tool,all planets are put around the sun, so that they are equally spaced, with the three aligned teeth arranged externally on the conjunction of sun and planet centres. Then the whole gear unit is inserted in the central ring, oriented so that the three aligned teeth of the planets can slip on the corresponding hollows in the ring. In this way we have only to verify previously that, with the chosen corrections, there is no interference in the insertion between central ring teeth and the two teeth adjacent to the aligned ones in the lateral planet gears . At last the two lateral rings, arranged as the central one, are inserted. If, instead of Eq.(2), we followed the traditional epicyclic gear drive condition (Henriot, 1968):

$z_1+z_4=n_s$(whole number) (2')

we should have machined,to make assembly possible, every planetary with a different angle between central and lateral gears; this solution is of course no cost effective.

Also Eq. (1) is necessary for assembling reasons, if we want to have a one piece planet. On the other hand, this condition involves that the gears must be non standard. In this way, as it is shown in Figure 3, the components of the rotary actuator are actually operating according to three different modules and associated pitch circles.

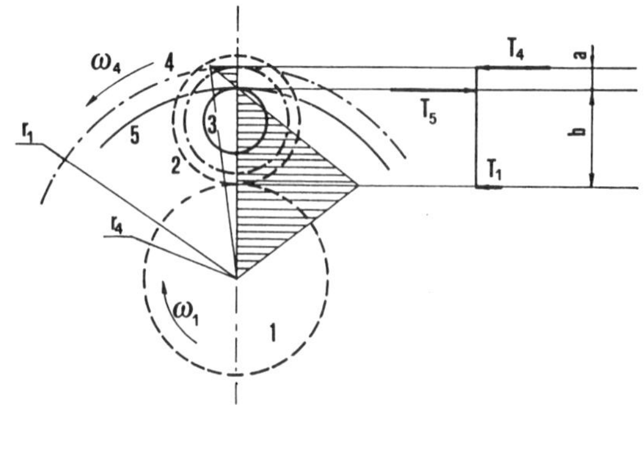

Fig. 3 Kinematics and lever model of a rotary actuator

Therefore, the operation of a rotary actuator may be easily represented by a simple lever model with the "sun gear" arm much longer then the "central ring gear" arm; it is so possible to obtain the reduction ratio:

$$\tau_{14} = -\frac{a\ r_1}{b\ r_4} = \frac{-z_1/z_2}{z_1/z_2 + z_5/z_3}[(z_2/z_4)(z_5/z_3)-1] \qquad (4)$$

Assuming, according to Eq.(1): $z_4 = z_5$, Eq. (4) becomes:

$$\tau_{14}=-(z_2/z_3-1)/[1+(z_5/z_1)(z_2/z_3)] \qquad (5)$$

The ratio z_5/z_1 may be considered quite constant for a chosen value of n_s. In fact, to have a large load capability, clearance between the outside circles of adjacent planets is very small, therefore, with great approximation, we can put :

$$z_5/z_1 \approx \frac{1-\sin(\pi/n_s)}{1+\sin(\pi/n_s)} = \text{const.} \qquad (6)$$

Consequently, τ_{14} depends mainly from z_2/z_3 and in particular the speed ratio increases much closer is z_2/z_3 to the unity. This explains Eq. (3), that corresponds, for a certain z_3 value, to make the speed ratio maximum. On the other hand, if the condition of Eq.(3) is assumed, we will have higher speed ratios for higher values of z_3. Introducing Eq. (3) and (6) into Eq. (5), we find out a qualitative relation between τ_{41}, n_s and z_3 (Fig. 4).

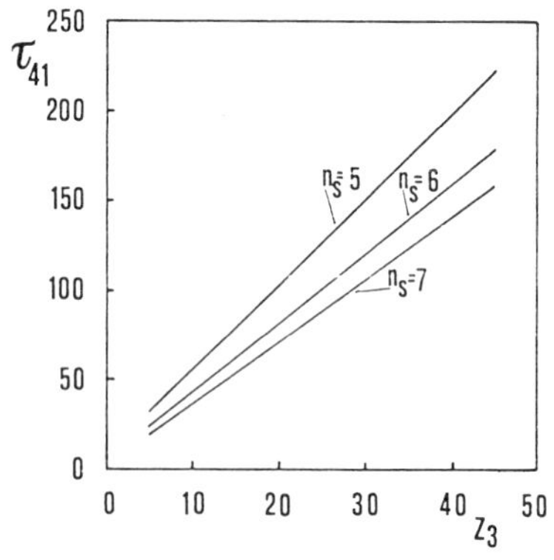

Fig. 4 Speed ratio against z_3 for different n_s

In conclusion, it is difficult to indicate a way to find out analytically numbers of gear teeth; Figure 4 may be helpful, but the choice is strongly limited by Eq. (2). Moreover, as it will be pointed out afterwards, z_i values may influence significantly other important mechanical features as efficiency.

STRUCTURAL FEATURES

The lever model previously introduced in order to explain the large speed reduction, clearly shows that the planet gear meshings with the central and lateral ring gears undergo the highest loads, therefore planet gear corrections must be carefully chosen .

According to the overall dimension of the device,

m_0 and m_{24} are chosen and x_{24}, x_{35} and x_{12} values are calculated, using the following relations, that can be obtained from the equality of the centre distances (Eq. 7), the equality of the basic radii (Eq. 8) and from the meshing condition for non-standard gears (Eq. 9, 9', 9")

$$i=(z_1+z_2)m_{12}=(z_4-z_2)m_{42}=(z_5-z_3)m_{53} \quad (7)$$

$$m_{ij}=m_0 \cos\alpha_0/\cos\alpha_{ij} \quad (8)$$

$$x_{12}=x_1+x_2=(\text{inv}\alpha_{12}-\text{inv}\alpha_0)(z_1+z_2)/(2\text{tg}\alpha_0) \quad (9)$$

$$x_{42}=x_4-x_2=(\text{inv}\alpha_{42}-\text{inv}\alpha_0)(z_4-z_2)/(2\text{tg}\alpha_0) \quad (9')$$

$$x_{53}=x_5-x_3=(\text{inv}\alpha_{53}-\text{inv}\alpha_0)(z_5-z_3)/(2\text{tg}\alpha_0) \quad (9'')$$

As planet theet are the most critically stressed, x_2 and x_3 values are chosen and x_1, x_4 and x_5 values are obtained from Eq. (9,9',9"). Then stress analysis can be performed following traditional calculation methods (BSS, AGMA, etc.). In this way the best x_i values may be obtained. When the complete design of gears is available, it can be useful to verify these calculations with numerical methods (FEM, BEM). It must be also pointed out that not only bending stresses, but also hertz contact stress should be considered carefully; in fact, this device is the last element of an actuating system and so it must support high load at very low speed. In these conditions a correct lubrication is rather difficult , also using special gear lubricants.

In the choice of x_i , not only stress, but also "geometrical" conditions must be taken into account, as:

- -correct values of contact ratio and tooth thickness at the outside radius
- -right clearance between the outside circles of adjacent planets
- -no interference in the assembling procedure between lateral planet gears and central ring.

TABLE 2 Stress and Geometrical Variation in Lateral Planet Gears[1] for Different i and x_3

i (mm)	x_{53}	x_3	σ_b (MPa)	σ_H (MPa)	t_o (mm)	t_c (mm)
33.959	6.45	0.35	1393	3211	0.481	0.661
		0.30	1398	3282	0.541	0.706
		0.25	1403	3359	0.598	0.916
33.659	7.90	0.35	1465	3141	0.481	0.130
		0.30	1460	3213	0.541	0.320
		0.25	1457	3286	0.598	0.530

[1] $z_1=18$, $z_2=14$, $z_3=13$, $z_4=z_5=48$, $n_s=6$, $m_0=2$, $\alpha_0=25$, Torque on ring gear 4 =11275 Nm, $\eta_{14}=0.6$)

Table 2 shows how different centre distances and corrections influence stress and "geometrical parameters". An increasing of the centre distance, if compatible with the overall dimensions, has positive effect on stress and clearance between planets. For a chosen centre distance, positive corrections reduce stress, in particular hertz contact stress may be reduced by working far from base circles. On the other hand high values of x_2 and x_3 greatly reduce tooth thickness and clearance between the outside circles of adjacent planets. For these reasons generally x_i values should be the highest compatible with an acceptable tooth thickness at the outside radius and clearance between planets.

If x_2 and x_3 are chosen in order to optimize the internal meshing, also the external meshing parameters are automatically determined. This may bring some problems , in particular contact ratio may be too low; in this case a positive effect may be obtained by increasing the head radius of the sun, if a small reduction of clearance between the outside radius of the sun gear and the root radius of the planet is allowed.

A particular feature of the rotary actuators, which contributes significantly to minimize their envelope, is the absence of a real planet carrier as in a traditional epicyclic gears reducer. Two rings support radial loads, but the planets are circumferentially completely free, thus providing high flexibility and a better load distribution; on the other hand this solution involves precise and accurate planet design. Under load the planets support mainly two kinds of loads: a bending moment and a torque. If the bending stiffness of the structure is too large or too small, this can produce a nonuniformity of the load distributions in meshing , with stress concentration that may occur in particular points where also torsional loads are high. To check if a final design is correct it is necessary to use FEM analysis in order to verify that the overall deformation under load is acceptable.

Planet design affects also the overall torsional stiffness of the actuator, that is mainly due to the torsional and the bending stiffness of the planets. A correct evaluation of the actuator stiffness is particularly important in order to study the dynamic behaviour of the whole actuating system to which the rotary actuator is connected. In fact, in consideration of its high speed ratio, this device is generally the most flexible element of the system and so it influences significantly the frequency of the vibrations during transients.

As the name "power hinge" puts in evidence, a geared rotary actuator may also support large shear loads; these loads are transferred from the actuating ring to the fixed one and consequently to the structure by means of a sliding coupling. For these reasons the central ring is highly stressed, particularly where torque and shear loads act together.

POWER LOSS AND EFFICIENCY

The main loss of mechanical power in geared rotary actuators is determined by friction in gear meshing; it is so possible to express actuator efficiency as a function of speed ratio and efficiency of each couple of meshing gears:

$$\eta_{14}=\frac{\tau_{14}}{\dfrac{-(z_1/z_2)/\eta_{12}[(z_2/z_4)/\eta_{24}\,(z_5/z_3)/\eta_{53}-1]}{(z_1/z_2)/\eta_{12}+(z_5/z_3)/\eta_{53}}} \quad (10)$$

$$\eta_{41}=\frac{\dfrac{-(z_1/z_2)\eta_{21}[(z_2/z_4)\eta_{42}\ (z_5/z_3)\eta_{35}-1]}{(z_1/z_2)\eta_{21}+(z_5/z_3)\eta_{35}}}{\tau_{14}} \quad (11)$$

Eq. (10) and (11) may be obtained with the equivalent lever method suggested by De Bona and Jacazio (1986); in this way it was also possible to evidence easily the relations between efficiencies, speed ratio and single mashing efficiency. If the speed ratio increases, the efficiency decreases, particularly η_{41}. For high speed ratio the efficiency is not affected significantly by η_{12} , on the other hand η_{24} and η_{53} have the same strong influence, in particular, especially for very high values of τ_{41}, a small decrease of η_{24} or η_{35} produces a large decrease of η_{14} and η_{41} .

These relations are confirmed by experiments; in geared rotary actuators manufactured by Microtecnica, efficiency was evaluated and the measured values always showed a good correspondence with the theoretical values obtained from Eq. (10) and (11).

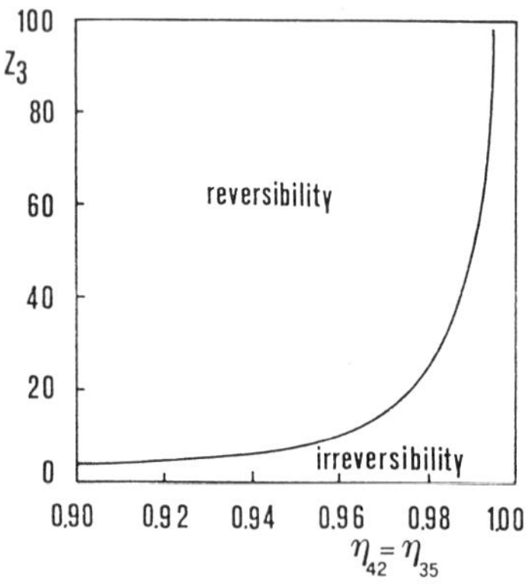

Fig. 5 Irreversibility condition

If we put in Eq. (11) $\eta_{41}=0$, we obtain the irreversibility condition:

$$(z_2/z_4)(z_5/z_3)=1/(\eta_{42}\eta_{35}) \quad (12)$$

Remembering Eq. (1),(2) and supposing $\eta_{35}=\eta_{42}$ it's possible to find (Fig. 5) a relation between z_3 and η_{42} for irreversibility: higher values of z_3 make irreversibility more easy to be obtained.

CONCLUSION

Geared rotary actuators are mechanical devices that provide a power hinge with large speed ratio and minimum envelope. Such remarkable performance is pursued by means of a multiple epicyclic gear reducer which essentially operates as a simple lever with one arm much longer then the other. In order to get a satisfactory result design must be very accurate, especially regarding structural analysis of the most critical elements, as the planets, that are not supported by a real planet carrier, and the central ring that supports both torque and shear loads. As this paper points out, although with some degree of approximation, the relation between design parameters (number of teeth, corrections, etc.) and actuator performance (load capability, speed ratio, efficiency, etc.) are already clear; therefore at present rotary actuator use is expanding, especially in aircraft application, where a high power to weight ratio is at a premium.

ACKNOWLEDGEMENT

This work was supported by by Microtecnica S.p.A, Torino, Italy.

REFERENCES

De Bona, F., G. Jacazio (1986). Un nuovo metodo per il calcolo del rendimento dei ruotismi epicicloidali che evidenzia l'influenza dei rendimenti delle singole coppie di ruote. Atti dell' 8 congr. naz. dell' Associazione Italiana di Meccanica Teorica e Applicata,II, 809-813.

Henriot, G.(1968). Traitè theorique et pratique des engranages, Tome I, Dunod , Paris.

AUSZUG

Die hauptsächlichen Charakteristiken der Zahnräderaktuatoren werden illustriert, deren Anwendung im Luftschiffwesen, dank ihrem günstigen Leistung-Gewicht Verhältnis,schon begonnen ist.

Sie können ein hohes Ubersetzungsverhältnis mit kleinem Raumbedarf liefern, mit der Möglichkeit auch Schubkräfte zu ertragen, wie wirkliche motorisierte Scharniere.

Das wird mittels eines mehrstufigen epyzikloidalen Zahradgetriebe erlangt, das im wesentlichen als ein Hebel mit grösserem Triebkraftarm als Widerstandsarm arbeitet.

Die optimalen Auswahl Zähle der Zähne und der Satelliten wird bestimmt, um ein hohes Übersetzungsverhältnis mit hohem Wirkungsgrad und leichtem Zusammenbau zu gewinnen. Die wichtigsten Festigkeitberechnungen für die mehr beanspruchten Elemente werden auch entwickelt.

An Experimental Analysis of the Jumping Phenomena in Timing Belts

N. HAGIWARA, Y. ONOE AND M. FURUDONO

Hitachi Ltd., Mechanical Engineering Research Laboratory, Kandatsu 502, Tsuchiurashi, Ibaraki 300, Japan

Abstract. This report is an experimental analysis of the jumping phenomena in timing belts widely used in office automation (OA) machines. One of the design policies that should be used to avoid jumping is the ease of engagement between a belt and a pulley. This takes into consideration that the test results show that the higher initial tension, the lower belt flexibility and so on work to increase the jumping limit torque. Higher shaft rigidity also increases the limit torque. The important fact is that the bending natural frequencies f and f_0 (one at setting) have two linear relations with torque Q and Qcr (limit value); $f=aQ+f_0$, $f_0=a(Qcr-Q_0)$. This report suggests that if the gradient 'a' would be analytically predicted, the jumping phenomena would be clarified.

Keywords. Power transmission; vibration measurement; timing belt; belt jumping; natural frequency.

INTRODUCTION

Timing belts have become very popular in many fields. Their original application was to automobiles but recently their use has been extented to OA machines. The reason for their wide use is their nonslip transmission to distant areas with angular precision at fairly high speeds. OA machines usually use XL or MXL type timing belts under various conditions; these include the variety of support flexibility, initial setting tension and wrapping angle between a belt and a pulley and so on. From this background, the load distribution, the strength and transmission characteristics of timing belts have been analysed by Gerbert,1978; Koyama, 1979; Naji,1983; Kagotani,1982; Funk,1982 and others. But the dynamical characteristics of jumping phenomena have rarely been reported. In addition to the timing belt itself, the extent of jumping occurence depends on such factors as its torque capacity and the shape. However perhaps it is most influenced by its conditions of use.
The purpose of this report is to make clear experimentally both the fundamental dynamical characteristics of the timing belt under operating conditions and the causes of jumping.

Type of timing belt This report deals with only XL-U and XL-G types shown in Table 1. The XL-U type is made of polyurethane and has two kinds of tensile members; kevlar fiber (polyamide fiber) (XL-UK), and steel wire (XL-US). The XL-G is made of polychloroprene with a glass cord tensile member. All the belt shapes are basically identical in most area such as in tooth pitch. However they do differ in tooth height and belt thickness. The XL-U type have special structures between each tooth called noses where tensile members are not covered by polyurethane, due to the necessities of belt production.

Experimental Equipment
The experimental apparatus has two pulley shafts as overhanging structures. The reason is that OA machines usually use timing belts near the end of a shaft and outside of a flame to transport and handle papers etc.. One of the axes is a driving shaft with a motor and a torque meter, the other is a driven

Table 1 Timing Belt Type

Type		Shapes and sizes (mm)	Materials
XL	G	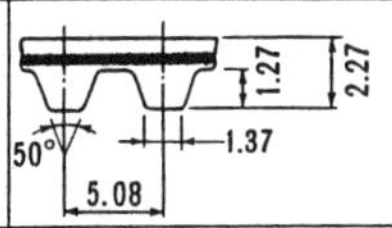	<Rubber> Polychloroprene <Tensile member> Glass fiber (G)
	U	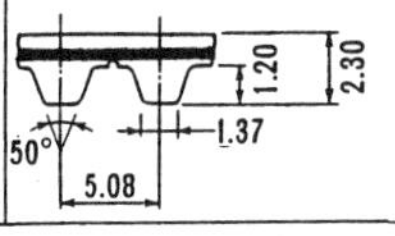	<Rubber> Polyurethane <Tensile member> Steel (US) Kevlar (UK) Tetron

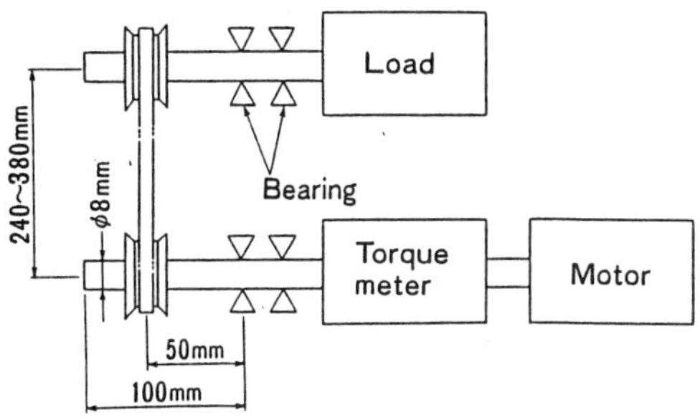

Fig.1 Experimental apparatus

axis with a powder clutch motor producing electric magnetic force as load torque. The test can change the overhanging length, pulley center distance L, initial setting tension T_0 and load torque Q. In this experiment, tension T is always measured by a belt bending natural frequency f by the relation as with a wire,

$$f = 1/2L \cdot T/m \quad (1)$$

where m is belt mass per unit length.

BASIC JUMPING PHENOMENA

The jumping phenomena occur at a certain load torque and always at the beginning of the engagement of a driven pulley as shown in Fig. 2. Even if a belt tooth rides on the tip of a pulley tooth, the belt tooth can not always maintain its similar state from the beginning till the end of engagement. Figure 3 shows the history of increasing load torque and the repetition of jumping occurence. It suggests that the jumping phenomena continue to occur and disappear alternatively and at an almost constant interval.

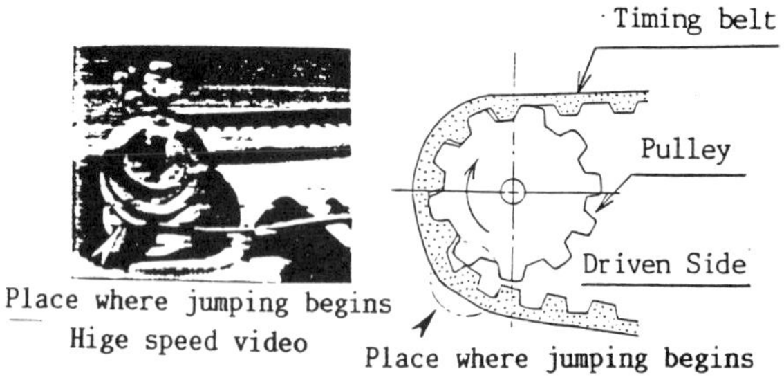

Fig.2 A jumping phenomenon

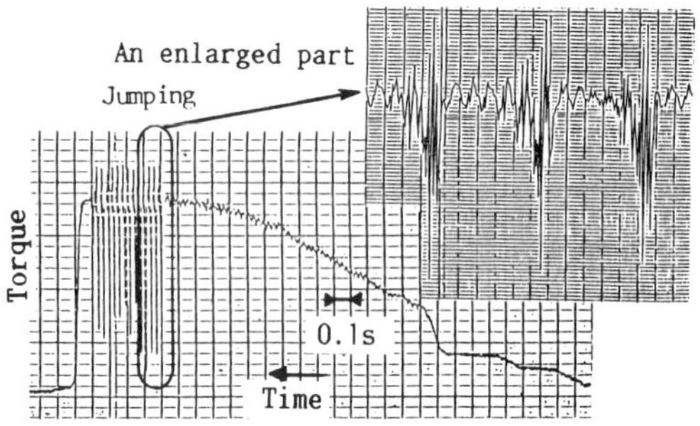

Fig.3 Torque change during jumping

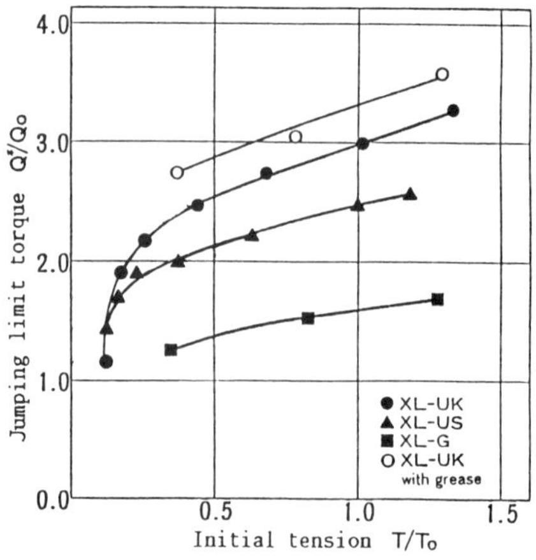

Fig.4 Comparison of jumping limit torque with belt type

The Effect of Belt Types, Tension and Friction

Figure 4 shows the differences of jumping limit torque due to belt type with or without grease lublication as a parameter for initial setting tension. At room temperature, the XL-UK (Kevlar fiber tensile) is best and the lubrication effect improve the jumping limit. As a belt tension increases, the limit torque increases. When belt tension is very low, jumping limit torque declines drastically and belt type or lubrication have no effect on it.

Belt Bending Stiffness

Figure 5 shows the relationship between belt bending stiffness and jumping limit torque under the same belt setting tension condition. The belt bending stiffness K is defined to measure the natural frequencies fc of a canti-lever as indicated in Fig. 5.

$$k = 3EI/L_2^3 \quad (2)$$

$$fc = \lambda^2/(2\pi L_2^2) \cdot EI/m, \quad \lambda = 1.875 \quad (3)$$

L_2: Overhanging length as shown in Fig.5

Nondimensional parameters are used as the axes of Fig. 5 in reference to the kevlar belt data (XL-UK). It is found that the limit torque of jumping is almost proportional to belt bending flexibility. The bending stiffness esentially corresponds to young modulus E, because L_2 and I are almost the same in any belt type.

Considering all the above facts concerning the jumping phenomena i.e., the jumping occurence at the beginning of the engagement of a driven pulley, low friction and belt flexibility, it can be concluded that the design policy for avoiding jumping occurences should be to ease engagement between the belt and pulleys.

Influence of Shaft Stiffness

Figure 6 shows the limit torque of jumping measured by changing the overhanging position where the belt is set. This result indicates that the limit torque increases linearly with shaft bending rigidity Ks.

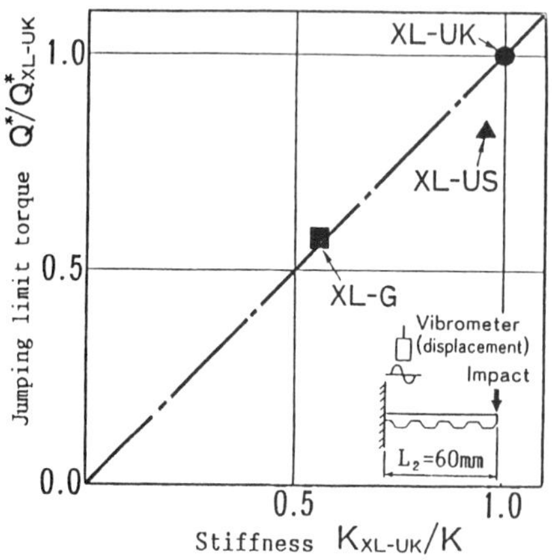

Fig.5 Relationship between belt bending stiffness and limit torque

$$Ks = 3EI/L_0{}^2L_1(1+L_0/L_1) \qquad (4)$$

L_1 ; distance between two bearings.

Further, the relation 'Q = Ks • A' can exist and the 'Acr = Qcr/Ks' corresponds to a displacement and is constant from Fig. 6. When a displacement increases beyond a certain same critical amplitude Acr, a belt jumping occurs. The belt tension effect shows that the linear relation changes nearly in parallel in the direction that the limit torque increases.

Influence of center Distance

Jumping limit torque is influenced by the pulley center distance as shown in Fig. 7. The distance increases, the limit torque decreases. The curve representing the relations moves in parallel as the distance changes. The numbers written in Fig. 7 are the belt bending natural frequencies at the initial belt set. The same setting frequency shows the approximately equal level of the limit torque. These frequencies will be explained in the next section.

The Shaft Radial Force from a Belt

A timing belt works to load on the belt supporting shaft according to belt tension. The measured shaft force changes gradually at the range of the relatively small load torque and increases steeply near the limit of jumping torque as shown in Fig. 8. This fact is unexpected because the shaft radial force is thought to be usually constant. In the usual case, the shaft is sufficiently rigid. That is to say, the shaft flexibility causes the change of shaft force. The level of force increase depends on the initial tension. When the initial tension is smaller, the change is more significant. When the belt tension is higher, the shaft force does not change significantly. The shaft rigidity works to increase the jumping limit torque in contrast to belt flexibility. This means that the shaft rigidity maintains the condition of ease of engagement between a belt and a pulley.

The broken line shown in Fig. 8 indicates the enveloped curve as the jumping limit torque. The curve corresponds to the one described in Fig. 4.

Belt Bending Natural Frequency at Setting

A very important fact was uncovered, that is, that the setting natural frequency has a linear relation with the jumping limit torque as shown in Fig. 9, described as

$$f_0 = a_1(Qcr-Q_0) \qquad (5)$$

From this fact, the relationship between the initial setting tension and the limit torque can be written as

$$T_0 = m(2Lf_0)^2$$
$$= 4L^2a_1{}^2m(Qcr-Q_0)^2 \qquad (6)$$

The center distance L has an influence on the parameter Q_0, but not on 'a_1'. With the case of longer center distance, the natural frequency should be set higher. That is to say, higher tension to keep the same limit torque. The 'a_1' may be influenced by the belt type and shaft stiffness.

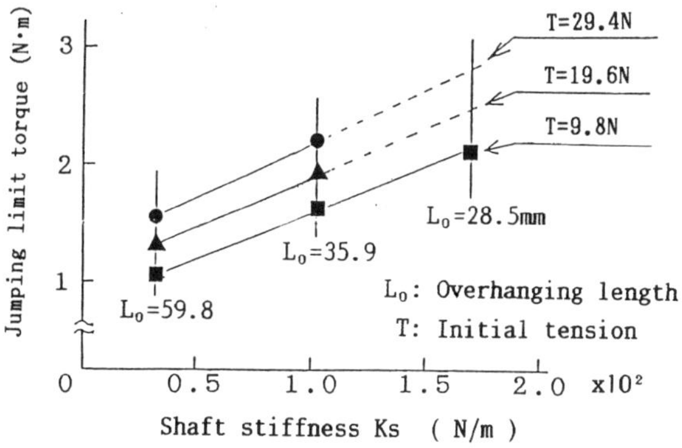

Fig.6 Jumping limit torque influenced by shaft stiffness

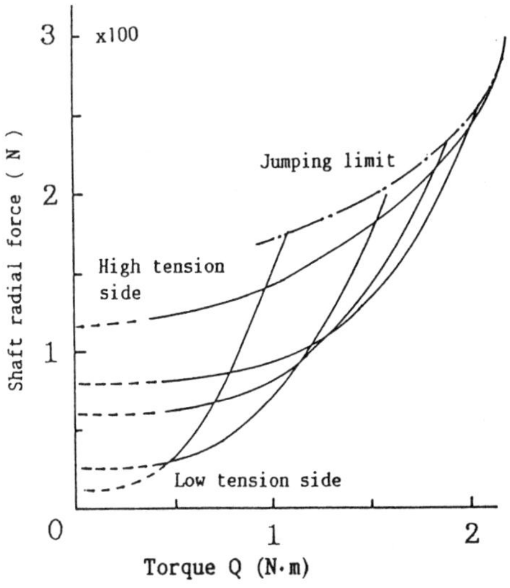

Fig.8 Relationship between shaft radial force and load torque

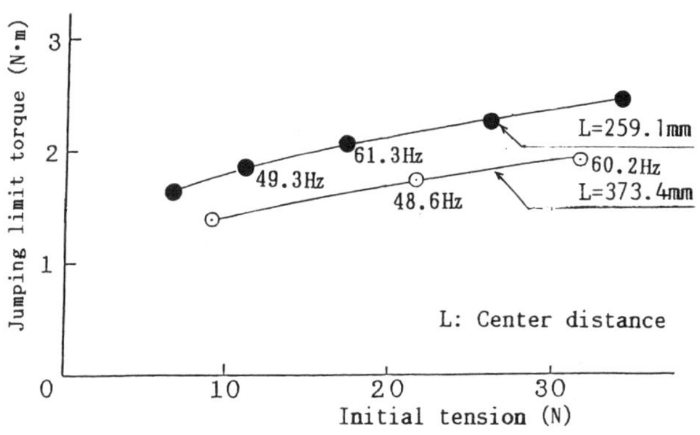

Fig.7 Influence of pulley center distance on jumping torque

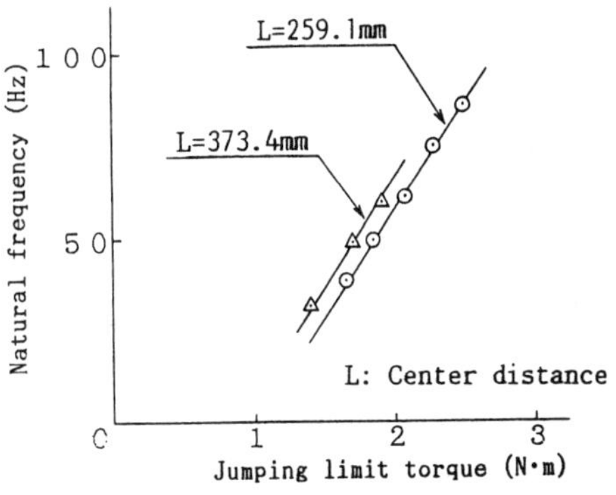

Fig.9 Relationship between belt natural frequency at setting and limit torque

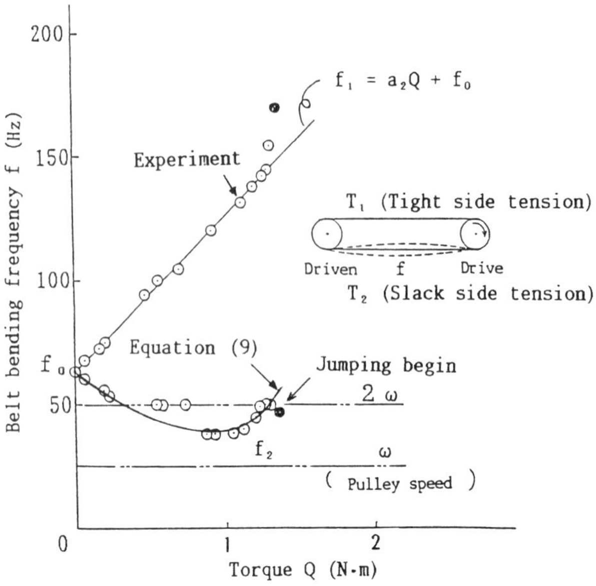

Fig.10 The change of Belt natural frequencies due to load torque (at constant speed)

BELT NATURAL FREQUENCIES

To clarify the relationship between load torque and belt natural frequencies, the belt is hit to measure its frequencies by changing load torque under the condition of constant belt speed. Figure 10 shows the measured belt frequencies for both the tight and slack sides. It is very important that the natural frequency of the tight side f_1 increases almost in proportion to load torque and can be formulated as follows,

$$f_1 = a_2 Q + f_0 \tag{7}$$

The natural frequency of the slack side f_2 can be described,

$$\text{as} \quad Q = (T_1 + T_2) \cdot R \qquad R\text{: pulley radius} \tag{8}$$

$$f_2{}^2 = a_2{}^2 Q^2 + (2a_2 f_0 - 1/4L^2 mR)Q + f_0{}^2 \tag{9}$$

The 'f_2' predicted by Equation (9) is shown as a real line in Fig.10 and coincides quite well with the experiment. The coincidence between the two coefficients a_1, a_2 is another interesting fact.

$$a = a_1 = a_2 \tag{10}$$

This is one of the typical dynamic characteristics of jumping phenomena.

Considering the above facts, the relationship between the shaft radial force and the load torque described in Fig. 8 can be formulated as follows,

$$F = T_1 + T_2$$
$$= 8L^2 a^2 mQ^2 + (16L^2 amf_0 - 1/R)Q + 8L^2 mf_0{}^2 \tag{11}$$

The equation indicates that the shaft radial force is in proportion to the square of the load torque Q and is influenced by the setting tension (natural frequency f_0).

CONCLUSIONS

The jumping phenomena occurring in timing belts widely used in OA machines are investigated by a fundamental experiment. Several new results have been obtained as follows and can indicate design method to avoid jumping.

1. One of the design policy for avoiding jumping should be to ease engagement between a belt and a pulley.
2. The lower flexibility of the timing belt and the higher initial setting tension work to increase the jumping limit torque.
3. As the shaft rigidity increases and the center distance of the pulleys decreases, jumping limit torque increases.
4. It is an important fact that the bending natural frequency f of the belt tight side changes almost linearly in proportion to the load torque Q and can be described as $f = aQ + f_0$.
The initial setting natural frequency f_0 has a linear relation to the limit torque Qcr with the same gradient 'a' as follow,
$f_0 = a(Qcr - Q_0)$
5. As indicated bythe above results, the radial directional force of a pulley shaft increases relatively to the square of load torque Q.
The gradient 'a' is a very important parameter to predict jumping analytically.

Finally, the autors would like to express their appreciation to assistant professor T.Koyama, Osaka Inst. Tech. and also assistant professor M.Kagotani, Osaka Ind. Uni. for their instructions and suggestions. We would also like to thank Mr. I.Misaki, sub-manager and Mr. H.Kato,engineer of Hitachi Ltd. Asahi Work for their support and suggestions.

REFERENCES

Gerbert,G.,et al. (1978). Load distribution in timing belts, Trans. ASME, J.Mech.Des. 100, 2, 208.

Koyama, T., et al. (1979). A study on strength of toothed belt, Bull. Jpn. Soc. Mech. Eng.July 1979, 22, 169, 982 July, 1980, 23, 181, 1235.

Naji, M.R. and Marshek, K. M. (1983). Toothed belt-load distribution, Trans. ASME, J. Mech. Transm. Autom. Des. 105, 339.

Kagotani, M., et al. (1983). A study on transmission characteristics of toothed belt drives, Bull. Jpn. Soc. Mech. Eng., Jan. 1983, 26. 211, 132, July 1983, 26. 217, 1238.

Funk, W. and Koester, L. (1982). Power transmission of timing belts, ASME Paper 82-DET-113.

Zusammenfassung. Der vorliegende Bericht ist über experimentelle Untersuchungen von Sprung-Erscheinungen an Zahnriemen wie sie in Büromaschinen weit verbreitet sind. Um Sprung Erscheinungen zu vermeiden sollte eine leichte Verzahnung gewährleistet sein. Test Ergebnisse weisen darauf hin, dass höhere Vorspannkraft, hohere Riemen-Steifigkeit usw. das Sprung-Erscheinungs-Grenzmoment erhöhen. Höhere Achsen-Steifigkeit wirkt ebenfalls in diesem Sinne. Eine wichtige Tatsache betrifft die linearen Zusammenhänge von Biege-Eigenfrequenzen f (bzw. f_0 im Stillstand) und Drehmoment Q (bzw. Grenzwert Qcr): $f = aQ + f_0$, $f_0 = a(Qcr - Q_0)$. Dieser Bericht legt nahe, dass eine analytische Berechnung des Gradienten 'a' zum Verständnis der Sprung Erscheinungen beitragen würde.

Fatigue Failure and Dynamic Performance of Surface-Hardened Gears

A. YOSHIDA*, K. FUJITA**, K. MIYANISHI*** AND D. KONISHI†

*Department of Applied Mechanics, Okayama University, Okayama 700, Japan

**Department of Mechanical Engineering, Okayama University of Science, Okayama 700, Japan

***NTN Toyo Bearing Co. Ltd., Tokyo 141, Japan

†Department of Mechanical Engineering, Tsuyama Technical College, Tsuyama 708, Japan

Abstract. Induction-hardened, case-hardened and nitrided gears whose hardened depths are close to the optimum hardened depths for surface durability were tested using a power circulating gear fatigue testing machine, and their fatigue strengths, failure modes and dynamic characteristics such as tooth root strain, vibration and noise were elucidated. The fatigue strength of the case-hardened gear was highest followed by that of the nitrided and the induction-hardened gears. In the cases of the case-hardened and the nitrided gears, the tooth breakage due to spalling near pitch point and the tooth breakage due to bending fatigue at tooth fillet occurred, and the changes of the dynamic characteristics during the fatigue process were slight. In the case of the induction-hardened gear, the tooth breakage due to pitting was the dominant failure mode, and the changes of the dynamic characteristics during the fatigue process were remarkable.

Keywords. Gear; induction-hardening; case-hardening; nitriding; fatigue strength; failure mode; tooth root strain; vibration; noise.

INTRODUCTION

Recently, surface-hardened gears such as induction-hardened, case-hardened and nitrided gears are often used as power transmission gears with an increase of operating speed and load carrying capacity of machines. It is customarily said that the load carrying capacity of surface-hardened gears is restricted by the bending strength rather than the surface durability. However, the failure mode of surface-hardened gears has to be investigated in detail from an aspect of the strength design of surface-hardened gears. On the other hand, it is important from not only an aspect of the strength design of gears but also an aspect of the foreknowledge of gear failure to clarify the relation between the tooth profile change and the changes of dynamic characteristics such as tooth root strain, vibration and noise during the fatigue process of gears.

In this report, in order to elucidate the fatigue strength and failure mode of the surface-hardened gears and the change of dynamic performance of the gears during the operational fatigue process, induction-hardened, case-hardened and nitrided gears whose hardened depths are close to the optimum hardened depths for surface durability were fatigue-tested using a power circulating gear testing machine. The fatigue strength of each gear was given and the failure mode of each gear was classified by some detailed observations. The relation between the tooth profile change and the changes of dynamic characteristics such as tooth root strain, vibration and noise during the fatigue process was discussed for each failure mode.

TEST GEAR

Dimension, material, heat treatment, finishing and accuracy of test gears are given in Table 1. Test gears IG, CG and NG are induction-hardened,

TABLE 1 Gear Data

Item	Gear			Pinion
Specimen mark	IG	CG	NG	CP
Module	4			
Nominal pressure angle	20°			
Number of teeth	26			19
Addendum modification coefficient	0			0.2925
Tip circle diameter mm	112.00			86.34
Center distance mm	91.5			
Face width mm	10	8		14
Contact ratio	1.413			
Material	SCM440	SCM420	SACM645	SCM420
Heat treatment	Induction -hardening	Case -hardening	Nitriding	Case -hardening
Tooth surface finishing	Reishauer grinding	Maag grinding		
Accuracy*	Class 1	Class 0	Class 1	Class 0

* JIS B 1702

case-hardened and nitrided spur gears respectively. Test pinion CP mating with those test gears is a case-hardened spur gear. These test gears and pinion were cut with a protuberance hob, and thereafter they were surface-hardened and ground. Only test gear NG was surface-hardened after grinding. Materials of test gears IG, CG and CP are chromium molybdenum steel and material of test gear NG is aluminimum chromium molybdenum steel. Figure 1 shows the hardness distributions of test gears. Judging from Fig.1, the effective hardened depths of gears IG and CG are 1.3 mm and 0.75 mm respectively, and the total hardened depth of gear NG is 0.8 mm. The surface hardnesses of gears IG, CG and NG are Hv640, Hv856 and Hv1030 respectively. Relative radius of curvature at working pitch point of these test gear pairs is 8.52 mm. The hardened depths of gears IG, CG and NG are nearly equal to the optimum

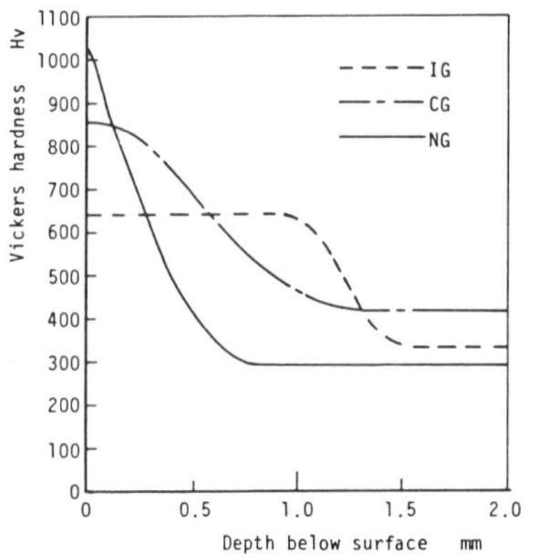

Fig.1 Hardness distributions of test gears.

hardened depths (Refs.1,2 & 3) for surface durability at that relative radius of curvature. The effective hardened depth and the surface hardness of test pinion CP are 1.8 mm and Hv820 respectively. In these experiments, test gears IG, CG and NG were taken as objects of failure, and number of teeth of test gears and pinion is prime each other for avoiding a meshing of specific teeth.

EXPERIMENTAL PROCEDURE

The gear fatigue testing machine employed in these experiments is of power circulating type (Ref.4). The fatigue tests were performed under the combinations of gear IG with pinion CP, gear CG with pinion CP and gear NG with pinion CP. The rotational speed n_2 of gears IG, CG and NG was 1800 rpm. EP gear oil (Viscosity: 214.2 mm^2/s at 311 K, 17.78 mm^2/s at 372 K) was pressure fed into the engaging side of gear pairs at a rate of 1.0 L/min and the oil temperature was adjusted to 313±5 K. A vibration transducer was fixed on this test rig. The operation of this test rig was stopped automatically when the vibration increase due to the tooth failure was detected.

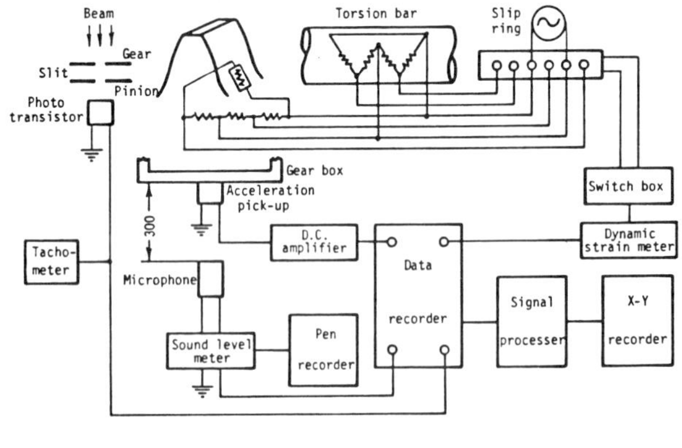

Fig.2 Block diagram of measuring system.

Figure 2 shows a block diagram of measurement system of noise, vibration and tooth root strain of the gears. The measuring methods of noise, vibration and tooth root strain are the same as described in a previous report (Ref.5).

EXPERIMENTAL RESULTS AND DISCUSSION

Fatigue failure and strength

The relationship between the number N_2 of cycles to gear failure and the Hertzian stress p_{max} at pitch point or the actual tooth bending stress σ_t (Ref.6), that is, p_{max}-N_2 curve or σ_t-N_2 curve is presented in Fig.3 for each gear. In Fig.3, the tooth failure modes decided from the observation results by naked eye, light microscope and scanning electron microscope are also given.

The fatigue life and the fatigue limit of gear CG are largest followed by those of gears NG and IG. The Hertzian stresses at 10^8 cycles life of gears CG, NG and IG are 1800, 1700 and 1350 MPa respectively.

In Fig.3, the tooth breakage due to bending fatigue at tooth fillet is presented by B, and in this case the appearance of breaking crack is shown as A in Fig.4. The tooth breakages due to surface fatigue failures which are spalling and pitting near pitch point are presented by S→B and P→B in Fig.3, and in these cases the appearance of breaking crack is shown as B in Fig.4. P in Fig.3 indicates a case that the test rig was automatically stopped by the pitting whose area percentage was more than 5 %, though the tooth breakage did not occur yet.

Figure 5 shows an example of pit and pitting cracks occurred near pitch point of a tooth of gear IG. It can be understood from this figure that the pitting cracks are initiated from the tooth surface. Figure 6 shows an example of spalling crack occurred in a tooth of gear NG. The spalling crack is initiated beneath the tooth surface near pitch point and runs parallel to the tooth surface in the single tooth contact region. The depth of this spalling crack coincided with the depth where the amplitude (Ref.2) of the ratio of reversing orthogonal shear stress to Vickers hardness became maximum. The appearances of the spalling cracks occurred in gears

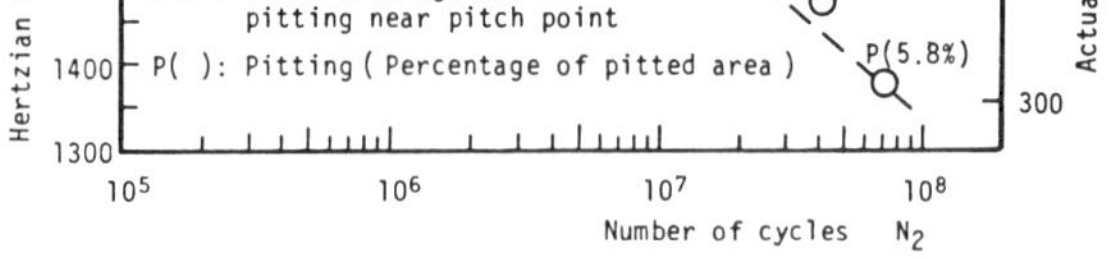

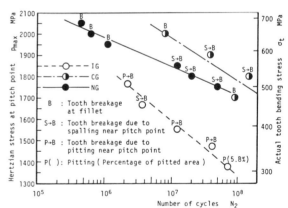

Fig.3 p_{max} - N_2 curves.

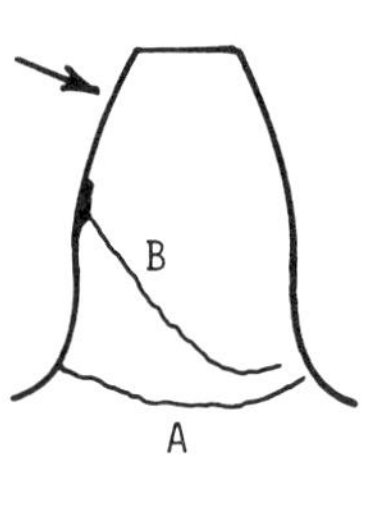

Fig.4 Appearance of tooth breakage.

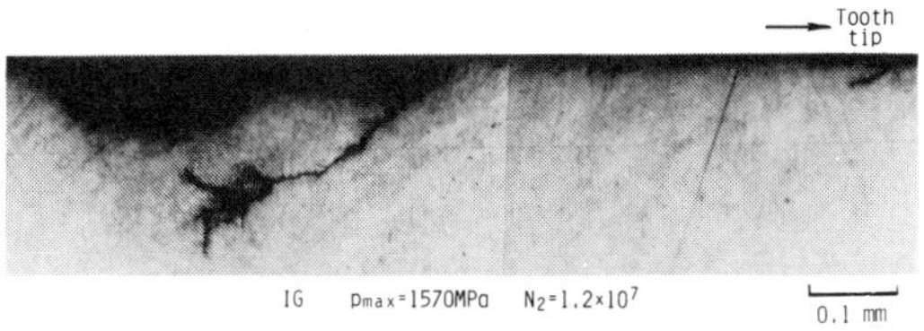

Fig.5 Pit and pitting cracks.

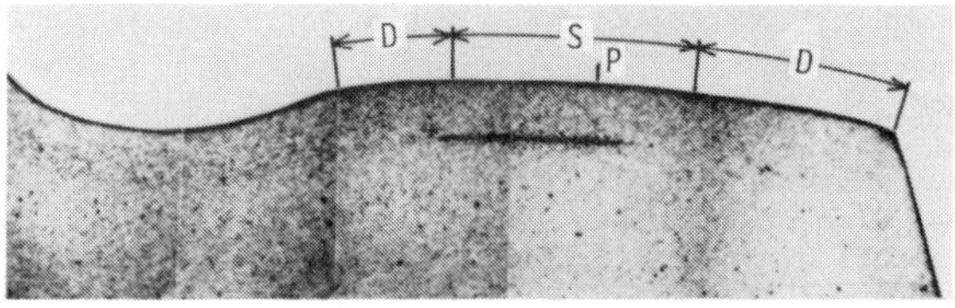

P:Working pitch point D:Double tooth contact S:Single tooth contact
NG p_{max}=1800MPa N_2=2.0x10^7

Fig.6 Spalling crack.

CG and IG had also the same tendency as shown in Fig.6.

Concerning Fig.3, in the cases of gears CG and NG, under the higher stress levels the tooth breakage due to bending fatigue at tooth fillet occurs and under the lower stress levels the tooth breakage due to spalling occurs except one example. In the case of gear IG, the tooth breakage due to surface fatigue failure that is spalling or pitting occurs. Thus, concerning all the surface-hardened gears employed in these experiments, under the lower stress levels near the fatigue limits the tooth breakage due to surface fatigue failure that is spalling or pitting occurred. It can be understood from this that at the strength design of the surface-hardened gears the surface durability becomes the most important problem.

Tooth profile change and dynamic characteristics

The measured results of tooth profile, sound pressure, vibration acceleration and tooth root strain during the fatigue process of the gear are shown in Figs.7, 8 and 9 with the results of sound and vibration power spectra.

Figure 7 is the result for a case of the tooth breakage due to pitting near pitch point under a Hertzian stress of 1570 MPa in a gear pair IG/CP. In this case, at the early stage of the fatigue test, a mild wear occurs near the tooth tip and root, and a running-in effect may be expected. At this stage the tooth root strain profile is hardly changed qualitatively compared with that at the initial stage. Thereafter the tooth profile is gradually deteriorated. At the number N_2 of cycles 1.2x10^7, that is, just before the tooth breakage of the gear, the dedendum flank is heavily worn near the pitch point of the gear and the tooth root strain decreases near the transition position from the double tooth contact to the single tooth contact. The sound pressure and the vibration acceleration profiles change periodically at each tooth meshing period T_Z. The power spectra of the frequency analyses are shown concerning the sound pressure level (SPL) and the vibration acceleration level (VAL). The frequency at which these values reach a peak is a multiple of the tooth meshing frequency f_Z and the tendencies in both SPL and VAL power spectra agree with each other. The higher frequency components of the SPL and VAL tend to disappear under the running-in effect at the early stage of the test and appear again at the number N_2 of cycles 1.2x10^7 when the tooth profile deteriorates heavily.

Figure 8 is the result for a case of the tooth breakage due to spalling near pitch point under a Hertzian stress of 1800 MPa in a gear pair NG/CP. Concerning the tooth profile change during the fatigue process of the gear, at the early stage of

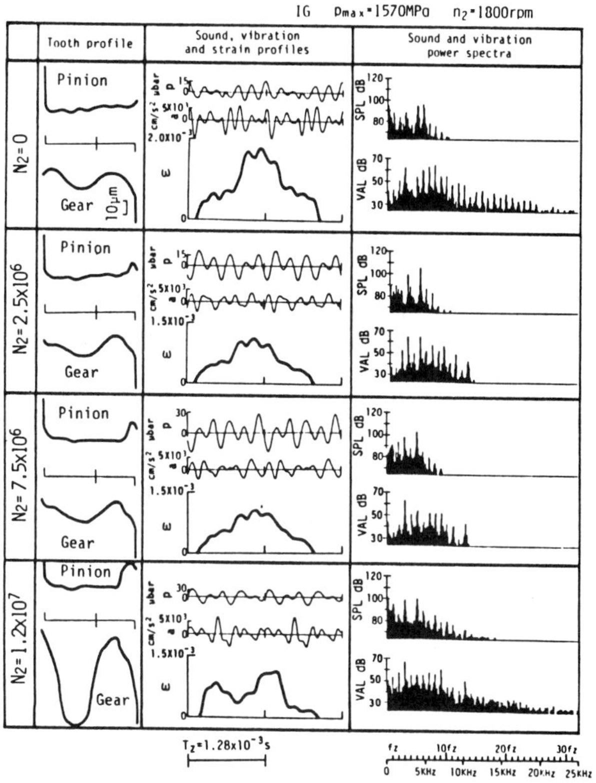

Fig.7 Tooth profile change and dynamic characteristics.

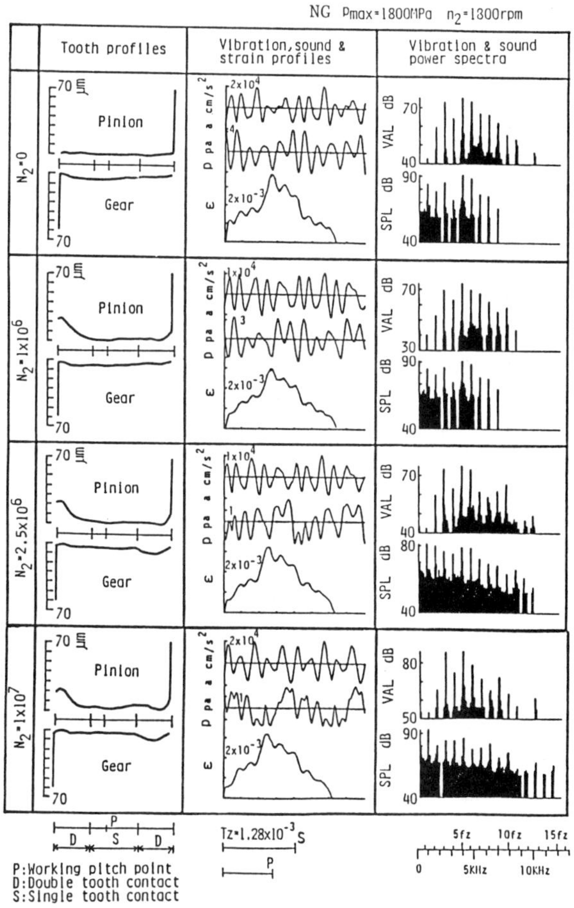

Fig.8 Tooth profile change and dynamic characteristics.

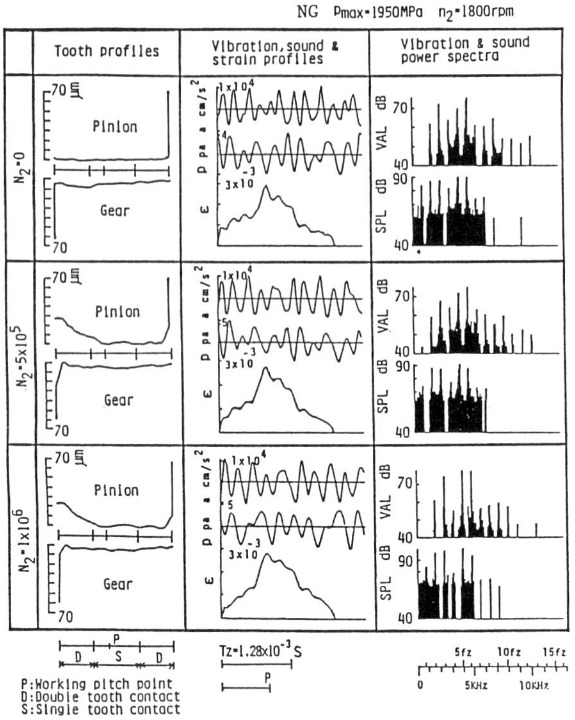

Fig.9 Tooth profile change and dynamic characteristics.

the fatigue test, a wear occurs near the tooth tip and root of the pinion, and thereafter the change of tooth profile of the gear and the pinion is not remarkable until the tooth breakage. Corresponding to this the tooth root strain profile is not changed remarkably during the fatigue test and this tendency is different from the tendency for the case of the tooth breakage due to pitting shown in Fig.7. The power spectra of VAL and SPL are hardly changed at the early stage of the fatigue test, and after the number N_2 of cycles 2.5×10^6 the higher frequency components of the VAL and the SPL tend to appear slightly. The results for the cases of the tooth breakage due to spalling in gear pairs CG/CP were almost the same result as shown in Fig.8.

Figure 9 is the result for a case of the tooth breakage due to bending fatigue at tooth fillet under a Hertzian stress of 1950 MPa in a gear pair NG/CP. The tendency in this case shows almost the same tendency as the case of the tooth breakage due to spalling shown in Fig.8. However, the higher frequency components of the VAL and the SPL do not appear even at the final stage of the fatigue test. The results for the cases of the tooth breakage due to bending fatigue at tooth fillet in gear pairs CG/CP were also the same tendency as shown in Fig.9.

As mentioned above, in the case of the tooth breakage due to pitting, the changes of dynamic characteristics such as tooth root strain, vibration and noise during the fatigue process of the gear are remarkable, but in the cases of the tooth breakage due to bending fatigue at tooth fillet and the tooth breakage due to spalling, the changes of those dynamic characteristics are not remarkable. This suggests that the tooth breakages due to bending fatigue and spalling are unexpected fracture macroscopically. Therefore, it cam be said that the case-hardened and the nitrided gears have to be employed cautiously.

CONCLUSIONS

Induction-hardened, case-hardened and nitrided gears whose hardened depths are close to the optimum hardened depths for surface durability were fatigue-tested, and the fatigue strength, the failure mode and the dynamic performance of each gear were discussed. The main results are as follows:

1. The fatigue strength of the case-hardened gear was highest followed by that of the nitrided and the induction-hardened gears. In the cases of the case-hardened and the nitrided gears, under the higher stress levels the tooth breakage due to bending fatigue at tooth fillet occurred and under the lower stress levels the tooth breakage due to spalling occurred. In the case of the induction-hardened gear, the tooth breakage due to pitting was the diminant failure mode.
2. In the cases of the pitting and the tooth breakage due to pitting the changes of the dynamic characteristics such as tooth root strain, vibration and noise during the fatigue test of the gear were remarkable, but in the cases of the tooth breakages due to bending fatigue and spalling the changes of those dynamic characteristics were not remarkable.

ACKNOWLEDGEMENTS

This investigation was supported financially in part by the Scientific Research Fund of The Japanese Ministry of Education, Science and Culture to which the authors express their gratitude. The authors are indebted to Sumitomo Metal Industries, Ltd. for the supply of test gears.

REFERENCES

1. K.Fujita, A.Yoshida and K.Nakase; Bull.JSME; Surface Durability of Induction-Hardened 0.45 % Carbon Steel and Its Optimum Case Depth; 1979; Vol.22, No.169; p.994-1000.
2. K.Fujita, A.Yoshida, T.Yamamoto and T.Yamada; Bull.JSME; The Surface Durability of the Case-Hardened Nickel Chromium Steel and Its Optimum Case Depth; 1977; Vol.20, No.140; p.232-239.
3. A.Yoshida, K.Fujita, K.Miyanishi, O.Torii and K.Higashi; Trans.JSME; Effects of Hardened Depth and Relative Radius of Curvature on Surface Durability of a Nitrided SACM645 Steel Roller; 1986; Vol.52, No.476; p.1394-1401.
4. K.Fujita, A.Yoshida and K.Akamatsu; Bull.JSME; A Study on Strength and Failure of Induction-Hardened Gears; 1979; Vol.22, No.164; p.242-248.
5. K.Fujita, A.Yoshida and K.Ota; Proc. 6th World Congress on TMM; On the Possibility of Early Detection of Gear Tooth Failure by Noise and Vibration; 1983; Vol.I; p.668-672.
6. T.Aida and Y.Terauchi; Trans.JSME; On the Bending Stress of Spur Gear; 1961; Vol.27, No.178; p.862-868.

Zusammenfassung. Induktionsgehärtete, einsatzgehärtete und nitrierte Zahnräder, deren Härtetiefen fast der optimalen Härtetiefe für Flankentragfähigkeit entsprechen, wurden mit einer FZG-Verspannungsprüfmaschine mit Leistungskreislauf geprüft, und ihre Dauerfestigkeit, Getriebeschädenformen und dynamische Charakteristika wie Zahnfussspannung, Vibration und Geräusch wurden experimentell erklärt. Die Dauerfestigkeit der einsatzgehärteten Zahnräder war am höchsten, gefolgt von der Dauer festigkeit der nitrierten Zahnräder und der induktionsgehärteten Zahnräder. Bei einsatzgehärteten Zahnrädern und nitrierten Zahnrädern erfolgten Zahnbrüche unter hohen Belastungen infolge von Biegungsermüdung an den Zahnfussrundungen, und unter niedrigen Belastungen Zahnbrüche durch Abplatzen. Bei induktionsgehärteten Zahnrädern war Zahnbruch wegen Grübchenbildung die dominierende Fehlerform. Bei Grübchenbildung und bei Zahnbruch wegen Grübchenbildung waren die Veränderungen der dynamischen Charakteristika wie Zahnfussspannung, Vibration und Geräusch während der Ermüdungsversuche am Zahnrad bemerkenswert, aber bei Zahnbrüchen durch Biegungsermüdung und durch Abplatzer waren die Veränderungen dieser dynamischen Charakteristika gering.

A Study of Strength Design Methods of Plastic Gears (Calculation Method of Fatigue Strength Due to Strain Energy)

N. TSUKAMOTO*, T. TAKI** AND N. NISHIDA***

*Chiba Institute of Technology, Narashino City, Japan
**Ishikawajima-Harima Heavy Industries Co. Ltd., Kōtōku, Tokyo, Japan
***Faculty of Engineering, Nagasaki University, Nagasaki, Japan

Abstract. Existing design methods for plastic gears are essentially based on those developed for steel gears. Recently, several problems have been pointed out in relation to these design methods because there exist several experimental evidences to which the existing methods can not be applied. One of these problems is that the increase of revolution speed of plastic gears brings about the elongation of their lives. Another one is that most of fractures of gear teeth in plastic gears occur near the pitch point. Tsukamoto, one of co-workers, has performed many experiments and has proposed an entirely new strength formula, including the product of the contact period of teeth and the loading torque, based on experimental results. The present study is aimed at establishing a practical gear strength design formula by developing Tsukamoto's theory. In this report, a calculation method of the fatigue strength of plastic gears due to strain energy is proposed.

Keywords. Plastics; gear; plastic gear; nylon gear; fatigue strength; strain energy.

INTRODUCTION

Many strength design methods for plastic gears are proposed like those for steel gears. The items which must be examined in the strength design of plastic gears are the bending strength of the tooth and wear resistance of the tooth flank. The weaker one of these two items must be strong enough against the load.

The basic formula for bending strength is Lewis' formula. This is derived from the stress theory for a cantilever. This formula gives the strength defined by the stress at the tooth root. For wear resistance, there is no practical theory or no practical design formula. Hence, the load-carrying capacity to wear has been decided from the relation between the load and the amount of wear which was obtained experimentally.

The above mentioned design philosophy for plastic gears is fundamentally the same one that is applied to steel gears. Recently, several problems have been pointed out in relation to the design philosophy and the design formula for plastic gears. One of these problems is that the increase of revolution speed of plastic gears brings about the elongation of their lives in spite of dry operation. Another one is that most of the fractures of plastic gear teeth take place near the pitch point, not at the tooth root where the fracture is theoretically expected to appear.

Although Takanashi (1981) used the design method of steel gears as that of plastic gears, he pointed out that the temperature of the tooth flank of plastic gears was highest near the pitch point and that this fact had to be considered in the strength design of plastic gears.

Tsukamoto (1981), one of co-workers, has proposed a new strength design method for plastic gears and contributed to the development of strength design methods for plastic gears. In this Tsukamoto's method, the product of the meshing period of a gear tooth and the torque is used as a design parameter. This Tsukamoto's theory for plastic gear strength differs from other existing theories in including the consideration of the meshing period of gear teeth. However, the physical definition of this theory has been slightly vague.

The aim of this study is to develop furthermore this Tsukamoto's theory and to get a practical strength design formula for plastic gears.

EXPERIMENTS

The essential points of the experiments which have been performed by Tsukamoto (1979, 1981, 1982) are as follows.

Testing Apparatus for Plastic Gears

Two torque testing apparatus (power circulating types), a large sized one and a small sized one, are used. In the large sized one, the center distance between the driving and the driven shaft can be changed from 190 mm to 300 mm. In the small sized one, the distance is fixed at 135 mm.

Tested Plastic Gears and Their Mates

In this test, four types of test gears shown in Table 1 were tested. The materials of driving gears and driven gears were MC-nylon and steel (S45C) respectively. All gears were cut by hobbing. The backlash of gears was given by increasing the depth of cut by 0.075 module for type I, type II and type III of MC-nylon gears and by 0.04 module for steel gears. For type IV gears, it was given by increasing the center distance by 0.8 mm in comparison with the standard distance. The precision of these gears ranged from 4th to 5th grade in JIS B1702. Every gear had a disk like shape with the thickness which was equal to the face width.

Measurement of Tooth Temperature

The tooth temperature of MC-nylon gears and steel gears was measured by a chromel-alumel thermocouple. The thermocouple was fixed at the bottom of a hole with a diameter of 1.2 mm, which was bored from the side of gear to the center of face width as shown in Fig.1. The output voltage of the thermocouple was taken out through slip rings.

Experimental Conditions

The experiment was performed under the condition of the driving torque ranging from 7.35 Nm (0.75 kgf-m) to 88.2 Nm (9.0 kgf-m) and the number of revolutions of driving shaft ranging from 500 to 1500 rpm and dry operation.

RELATION BETWEEN STRAIN ENERGY AND FATIGUE STRENGTH

The effective method for evaluating the strength of plastic gears has apparently not been published. In polymeric materials, the hysteresis loss due to repeated stress causes the rise of material temperature which reduces the fatigue strength of materials (Fujiwara, 1973; Nielsen, 1966). The relation between the fatigue strength of polymeric materials σ_A and the temperature of the materials T given by Nielsen (1966) is

$$\log \sigma_A = A + B / T \qquad (1)$$

where A and B are constants.
The frictional heat caused by the slip between meshing tooth flanks and the temperature of the room must be considered as heat sources in addition to the hysteresis loss. We can not adopt Eq.(1) as it stands because we do not know the relation between the temperature of gears and the heat, which is generated within materials or supplied to materials, such as the heat due to hysteresis loss, the frictional heat and the heat from the environment. Now let us discuss a different

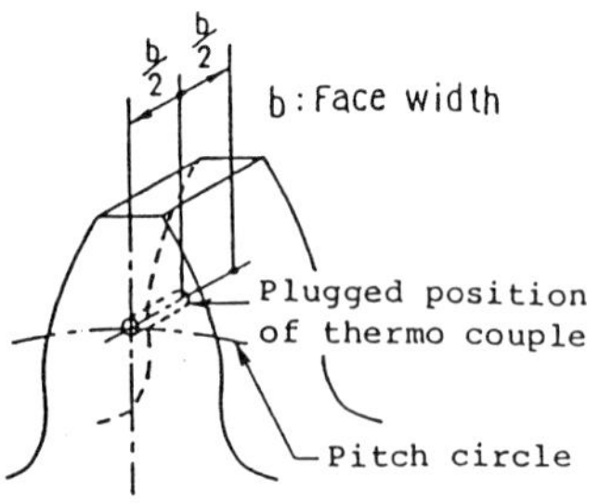

Fig. 1. Measuring position of gear tooth temperature.

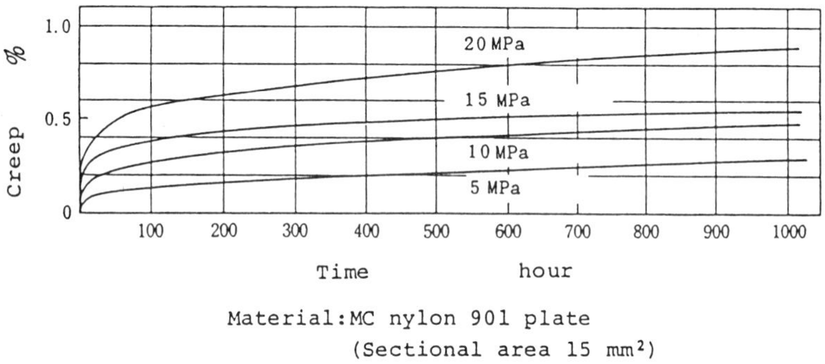

Fig. 2. Creep in compression (at 20°C).

TABLE 1 Specification of Gears

Gear type	Type I		Type II		Type III		Type IV	
	Driving gear	Driven gear	Driving gear	Driven gear	Driving gear	Driven gear	Driving gear	Driven gear
Module	3		4.5		5		5	
Pressure angle	20°		20°		20°		20°	
Number of teeth	30	60	25	35	17	37	34	57
Pitch circle dia.	90	180	112.5	157.5	85	185	170	285
Face width	10	10	10	10	10	10	10	10
Center distance	135		135		135		228.3	
Clearance coef.	0.25							

N.B.) A unit of module, pitch circle diameter, face width and center distance is mm.

method for estimating the fatigue strength of plastic gears, namely a method for estimating it by means of strain energy.

Strain energy U is expressed in terms of stress and strain as follows.

$$U = \int_0^{\varepsilon} \sigma \, d\varepsilon \qquad (2)$$

In viscoelastic materials such as polymeric materials, the relation between stress and strain is influenced by the rate of strain. The stress-strain curve in the short contact period of tooth flanks must be clarified experimentally for various plastic materials in future. In this study, the relation between the stress, the strain and the contact period is estimated, as the first step, from the creep curve of plastic gear materials (MC-nylon 901) (Nippon Polypenco Company,1984) shown in Fig.2. From Fig.2, this relation can be expressed approximately by the following equation.

$$\varepsilon = C_1 \sigma t^{1/4} \qquad (3)$$

From Eqs.(2) and (3), the strain energy U in Eq.(2) is rewritten as follows.

$$U = \frac{1}{2} C_1 \sigma^2 t^{1/4} \qquad (4)$$

Most damages of plastic gear teeth observed in Tsukamoto's experiments are fractures near the pitch points and they are caused by the cracks generated at these points. From these results, we can adopt the idea (Tsukamoto, 1979) that the initiation of the crack is caused by the fatigue of the tooth flank due to stress and the fracture takes place when the crack extends to some degree. So, we can replace σ in Eq.(4) with Hertz's contact stress P_H at the pitch point. Consequently, Eq.(4) is rewritten as follows.

$$U = \frac{1}{2} C_1 P_H^{\,2} t^{1/4} \qquad (5)$$

Hertz's contact stress P_H(Pa) at the pitch point and contact period t(s), during which the contact point of tooth flanks passes through the contact width due to the Hertz's contact stress P_H at the average meshing speed on the tooth flank, can be obtained from the following formulas.

$$P_H = 0.564 \sqrt{q E \frac{2(Z_1 + Z_2)}{Z_1 Z_2 m \sin\alpha}}$$

$$t = \frac{60b}{n Z m}$$

where q is the load per unit face width, E is the equivalent Young's modulus, Z_1 and Z_2 are numbers of gear teeth in driver and driven gears respectively, m is the module, α is the pressure angle, 2b is the contact width due to Hertz's contact stress at the pitch point and n is the number of revolutions (rpm).

$$1/E = (1-\nu_1^2)/E_1 + (1-\nu_2^2)/E_2$$

where E_1, ν_1 and E_2, ν_2 are Young's moduli and Poisson's ratios for driver and driven gear materials respectively.

The constant C_1 in Eq.(5) is indefinite now. Therefore, we divide the both side of Eq.(5) by the constant C_1 and obtain U/C_1. Fig.3 shows the result of Tsukamoto's experiments arranged in U/C_1. Although the data are rather scattered, the relation between U/C_1 and the total number of repeated load (total number of revolutions) N can be expressed roughly as follows.

$$U/C_1 = -38 \log N + 500 \qquad (6)$$

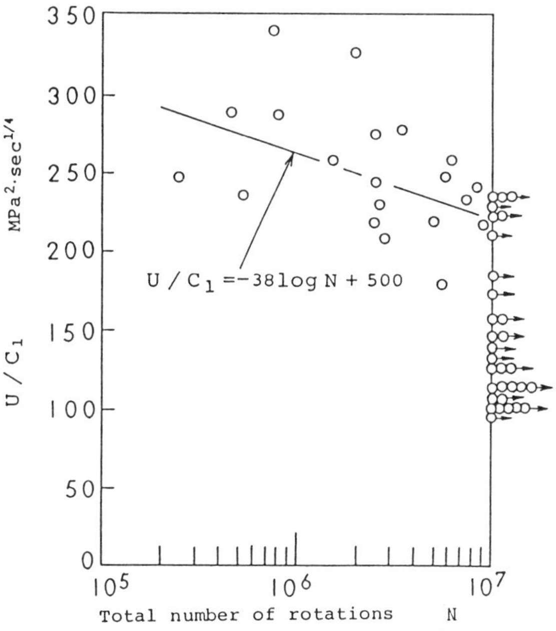

Fig. 3. Relation between the fatigue strength and U/C_1.

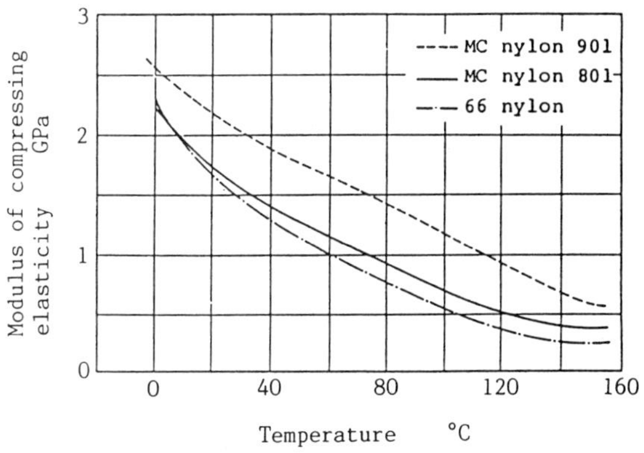

Fig. 4. Relation between temperature and modulus of elasticity in compression.

THE REMAINING PROBLEMS

The strength design equation for plastic gears proposed in this report includes some indefinite factors such as the value of the constant C_1 and the influence of the frictional heat between tooth flanks. In plastic materials, the changes of their Young's modulus (Fig.4) and creep characteristics (Fig.5) due to temperature are very large and also their mechanical properties strongly depend upon their water absorptions. Therefore, the influence of the environment, where plastic gears are used, to the mechanical properties of plastic materials should be also considered in the strength design equation.

CONCLUSIONS

We can considered that fatigue failures take place when the accumulated strain energy in the materials owing to repeated stress has exceeded their own threshold levels. In this report, we applied the same idea to plastic gears and proposed the calculation method for fatigue strength of plastic gear teeth based on the strain energy. In viscoelastic materials such as polymeric materials, the fatigue strength evaluated by the strain energy is greatly influenced by the period in which one gear tooth is subjected to the load because the creep is quite different in the loaded period. The distinctiveness of this report is that the relation between the stress, the strain and the loaded period was roughly approximated by Eq.(3) derived from the creep characteristics (Fig.2). In this report, we could not indicate the value of the constant C_1 and also could not introduce the factors into our design equations concerning the environmental conditions such as room temperature and water absorption. We hope we can answer these questions in future.

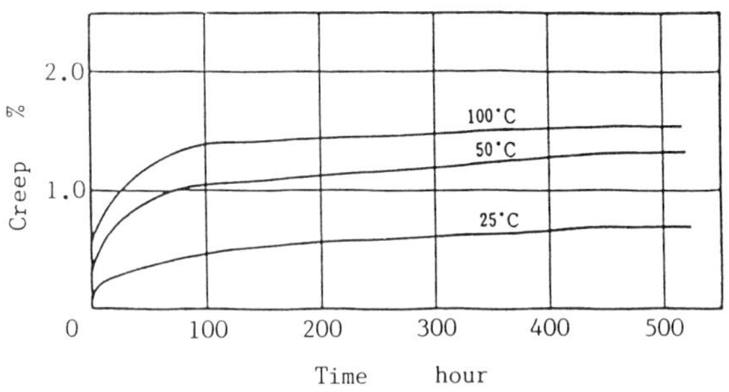

Material : MC nylon 901 plate (Sectional area 15 mm²)

Fig. 5. Influence of temperature to creep in compression (at 10 MPa).

We would like to thank Research Assistant Hiroki Maruyama, Chiba Technical Institute, for his help in arranging the data of this report.

References

Fujiwara, S. and others (1973). Tensile fatigue properties of FRP. Trans. JSME, 39, 805-812.

Nielsen, L. E. (1966). In S. Ogino (Japanese translator), Mechanical Properties of Polymers, Kagakudonin, Tokyo. Chap. 9, pp. 220-222.

Nippon Polypenco company (1984). Technical Report of MC-nylon, Nippon Polypenco Company, Tokyo. No.4, pp.4-21.

Takanashi, S. (1981). Friction and wear of plastic materials used for gears. J. JSLE, 26, 310-315.

Tsukamoto, N. (1979). Investigation about the strength of plastic gears (1st report, the strength of nylon gears which have counter-crowning). Bull. JSME, 22, C, 1685-1692.

Tsukamoto, N. (1981). Investigation about the strength of plastic gears (2nd report, abrasion of the nylon gear for power transmission, meshing with the steel gear). Bull. JSME, 24, C, 872-881.

Tsukamoto, N. (1981). Investigation about the strength of plastic gears (3rd report, strength of the nylon gear meshing with the steel gear which was cut by semitopping hob). Bull. JSME, 24, C, 2194-2204.

Tsukamoto, N., Yano, T. and Sakai, H. (1982). Noise and transmission efficiency under deformation of tooth form of nylon gear. Bull. JSME, 25, C, 1465-1474.

Реферат. Методы конструирования пластмассовой шестерни в основном применяются, основываясь на методах конструирования стальной шестерни на прочности. Однако, в последнее время поднимались несколько вопросов к этим методам потому, что существующие методы не могут приненять к практике. Важнейшие вопросы из них состоют в том, что даже при использовании пластмассовой шестерни без смазки повышение ее скорости вращения способствует увеличению ее срока службы, и что не всегда возникают поломки ее зубка в части его корня. Цукамото, один из авторов, продвинул на шаг впереди теорию прочностного конструирования пластмассовой шестерни и предложил совершенно новый метод с использованием величины произведения контактного периода и вращающего момента в пластмассовой шестерни. Эта теория оригинальна и отличается от существующих теории по прочности и износу пластмассовой шестерни, но физические определения недостаточно точны. Настоящая работа ставит своей целью дальнейшего продвижения вперед теорию Цукамото и оформления практического выражения прочностного конструирования пластмассовой шестерни.

On Analysis and Prediction of Machine Vibration Caused by Gear Meshing

AIZOH KUBO

Department of Precision Mechanics, Kyoto University, Kyoto 606, Japan

A new method to analyze or predict the state of machine vibration owing to gear meshing is introduced: that is a method to reduce a forced vibrational system with parametric exciting terms to a quasi-equivalent forced vibrational system without parametric exciting term. Using this method, the vibrational exciting force due to gear manufacturing and alignment error and to periodical change of tooth mesh rigidity is integrated into one total vibrational exciting force E_V which acts at tooth meshing position. Using E_V, the machine vibration owing to gear meshing can be analyzed by conventional methods of vibration analysis for systems without parametric excitation. The solution of this approximation method for the estimation of gear vibration and that of the original exact method were compared with measured results of the vibrational level and of the frequency spectra of gear vibration to show the scope of application of this simple approximation method.

Key words: Gear, Vibration, Vibration Analysis, Approximate Solution

1. Introduction

There have been a lot of research on noise, vibration and dynamic loading of involute spur gears. Among these investigations, only those concerning the relative rotational vibration of a spur gear pair whose shafts have flexible enough twisting rigidity and solid enough transverse rigidity have shown good agreement between theoretical results and measured results. It is the present status that the analytical results of gear vibration in multi-stage gear drive or in a system including driving and driven machines do not agree well with the measured results. The main causes of this status are: 1) The vibrational system of gear drive in actually used machinery in industry is much more complicated than that of a pair of gears which is the objective model in most of gear vibration research. The application of the results shown by most of gear vibration research until now to the vibrational problem of actual machines is therefore very difficult. 2) An elaborate expression of gear vibration in actual gear drive represents a set of very complicated simultaneous differential equations with parametric exciting terms, and it is not easy to get a solution to it. This report proposes a method to break down such a status of present research on gear vibration.

2. Natureof 2nd Order Differential Equation with Parametric and Forced Exciting Term

Let us consider the following differential equation:

$$\xi'' + 2\zeta\xi' + \kappa(\tau)\cdot\xi = C + \Psi(\tau) \qquad (1)$$

where $\kappa(\tau)$ is a periodic function whose average value is unity, and $\Psi(\tau)$ is a periodic function whose average value is zero, and they are expressed by

$$\left.\begin{aligned} \kappa(\tau) &= 1 + \sum_{m=1}^{\infty}(\alpha_m \cos m\omega\tau + \beta_m \sin m\omega\tau) \\ \Psi(\tau) &= \sum_{m=1}^{\infty}(\eta_m \cos m\omega\tau + \gamma_m \sin m\omega\tau) \end{aligned}\right\} \qquad (2)$$

Now considering the differential equation (3) which is obtained by substituting into the parametric exciting term of the equation (1) its average value of unity,

$$\xi'' + 2\zeta\xi' + \xi = C + \Psi(\tau) \qquad (3)$$

The vibration expressed by eq.(1) or (3) whose forced term $\Psi(\tau)$ corresponds to a vibrational excitation due to pressure angle error of a pair of spur gears, and $\kappa(\tau)$ which changes in rectangular form or sinusoidal form is dealt with. $\Psi(\tau)$ and $\kappa(\tau)$ change in the tooth meshings period T_z. Figure 1 (a) and (b) show the degree of difference between the solutions of eq.(1) and eq.(3), when the condition of the constant C changes from unity to zero. Figure 2 (a) and (b) show the difference of the effective values of vibrational acceleration calculated by eqs.(1) and (3). On the ordinate, the nondimensional speed ω is taken. When the value of C is unity, the solutions of eq.(1) for both the cases of rectangular $\kappa(\tau)$ and sinusoidal $\kappa(\tau)$ differ very much from the solution of eq.(1), in which C is a constant of unity. On the contrary, when the value of C is zero, the solutions of these three cases of equation are quasi-equal ones: especially, when $\kappa(\tau)$ changes sinusoidally, the

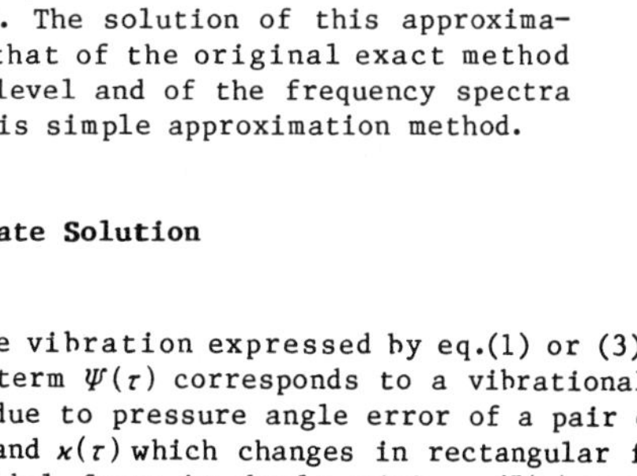

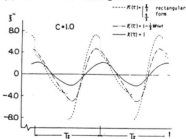

(a) For a case, when the time average value of the forced term is unity

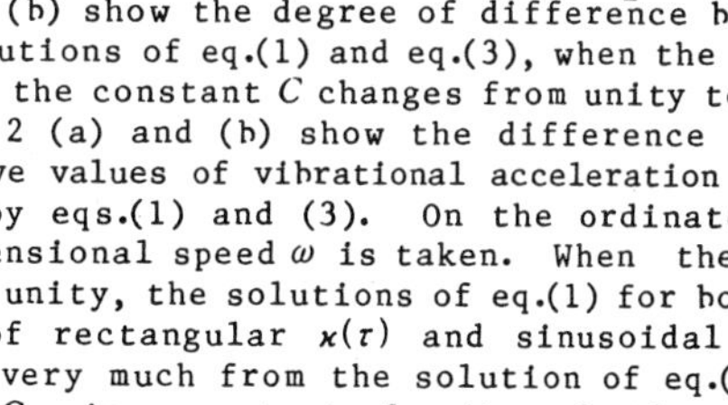

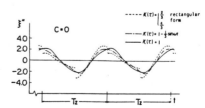

(b) For a case, when the time average value of the forced term is zero

Fig.1 Influence of parametric exciting term on the wave form of steadystatesolution of the 2nd orderdifferential equation, (ω=1.0)

solution of eq.(1) in the region of small ω values comes closer to the solution eq.(3) than in the case, in which $\kappa(\tau)$ changes to rectangular form.

It is admitted that when the time average value of the right term of the second order parametric excited differential equation is zero, the solution of this equation can be well approximated by the solution of differential equation with constant coefficient which is the time average value of parametric excited function $\kappa(\tau)$. This fact is well explained by matrix analysis of eqs.(1) and (3),/1/-/3/. The fact mentioned above is valid also for the case of simultaneous differential equations.

3. Total Vibrational Excitation due to Gear Meshing

Relative circumferential vibration of a pair of gears is expressed by the following equation /3/-/5/ :

$$m\ddot{x}+2\zeta\sqrt{m\sum_i k}\,\dot{x}+\sum_i k_i x=W+\sum_i k_i e_i \quad \cdots\cdots(4)$$

where

$$x=x_1-x_2,\quad x_1=r_{g1}\theta_1,\quad x_2=r_{g2}\theta_2$$

r_{g1}, r_{g2}: base circleradii of driving and driven gears respectively

θ_1, θ_2: rotational angles of driving and driven gears respectively

$$m=m_1m_2/(m_1+m_2),\quad m_1=J_1/r_{g1}^2,\quad m_2=J_2/r_{g2}^2$$

J_1, J_2 : polar moments of inertia of driving and driven gears respectively

ζ : damping coefficient

k_i : composite tooth stiffness of a tooth pair i

W : transmitting normal load

e_i : composite gear error of a tooth pair i

$\sum_i$: summation concerning the number of simultaneous tooth pairs in meshing

Using the constant value $\bar{K}$ which is the time average value of $\sum_i k_i$, and setting

$$x_s=\frac{W}{\bar{K}},\quad \xi=\frac{x}{x_s},\quad \omega_e=\sqrt{\frac{\bar{K}}{m}},\quad \tau=t\omega_e$$
$$\kappa(\tau)=\frac{\sum_i k_i}{\bar{K}},\quad \phi(\tau)=\frac{\sum_i k_i e_i}{W},\quad '=\frac{d}{d\tau} \quad \cdots\cdots(5)$$

the equation (4) is transformed into a nondimensional form of

$$\xi''+2\zeta\sqrt{\kappa(\tau)}\,\xi'+\kappa(\tau)\xi=1+\phi(\tau) \quad \cdots\cdots(6)$$

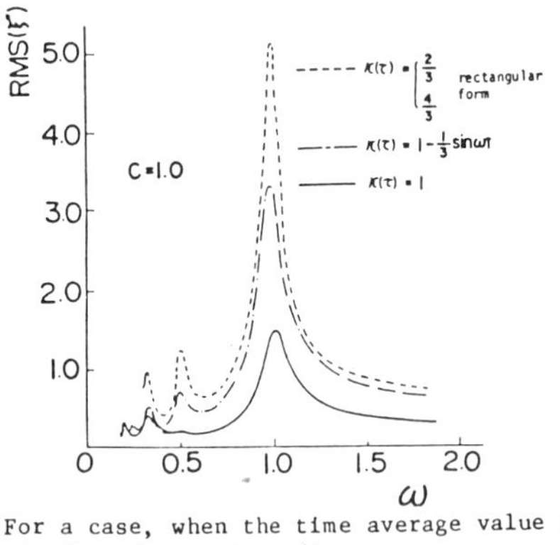

(a) For a case, when the time average value of the forced term is unity

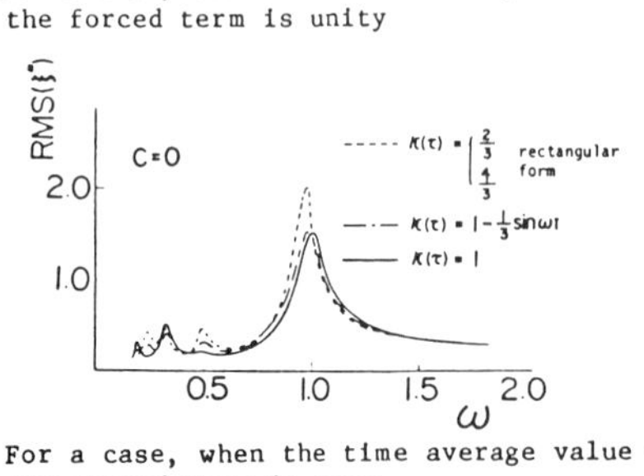

(b) For a case, when the time average value of the forced term is zero

Fig.2 Influence of parametric exciting term on the rms value of the acceleration of steady state solution of the 2nd order differential equation

As the time average value of is unity, setting

$$\kappa^*=\kappa(\tau)-1,\quad \phi^*=\phi(\tau)-\bar{\phi},\quad \xi^*=\xi-1-\bar{\phi} \quad \cdots\cdots\cdots\cdots(7)$$

the equation (6) becomes

$$\xi^{*\prime\prime}+2\zeta\sqrt{\kappa(\tau)}\,\xi^{*\prime}+\kappa(\tau)\xi^*=\phi^*-\kappa^*(1+\bar{\phi}) \quad \cdots\cdots(8)$$

The biggest difference between eq.(8) and eq.(6) is that the time average value of the right term of differential eq.(8) is zero. Using the result of the former section, the solution of eq.(8) is therefore well approximated by the solution of the following equation:

$$\xi^{*\prime\prime}+2\zeta\xi^{*\prime}+\xi^*=\phi^*-\kappa^*(1+\bar{\phi}) \quad \cdots\cdots\cdots\cdots(9)$$

The right term of the equation (9) is therefore recognized as the total excitation of the relative circumferential vibration of a gear pair which is due to manufacturing and alignment errors of gears and periodical change of tooth mesh rigidity.

The same result as mentioned here can be obtained, when the number of gear stages is increased or gears have more moment of inertia on their driving or driven shaft, or when gear vibration in multi-directions, e.g. the vibration of a pair of involute helical gears shown in Fig.3, /6/-/8/, is dealt with /1/.

The total vibrational excitation $[\phi^*-\kappa^*(1+\bar{\phi})]$ can be obtained by statical analysis, without solving differential equations, only from composite gear error and tooth mesh stiffness change of a gear pair at each gear stage: it suffers no influence from gears of other stages or from driving and driven machines. This fact makes all the vibration analysis of gear system much easier.

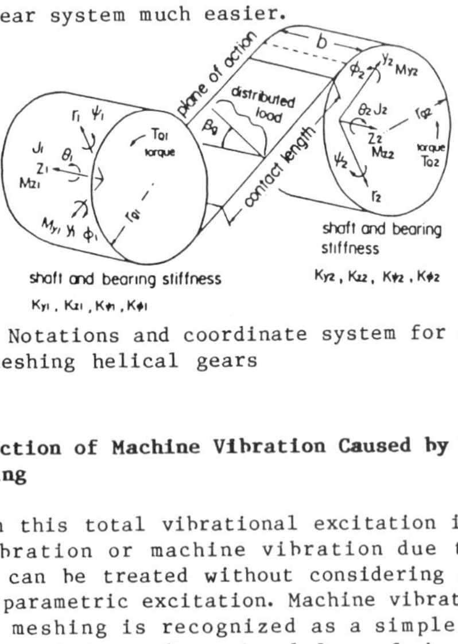

Fig.3 Notations and coordinate system for a pair of meshing helical gears

4. Prediction of Machine Vibration Caused by Gear Meshing

When this total vibrational excitation is used, gear vibration or machine vibration due to gear meshing can be treated without considering the effect of parametric excitation. Machine vibration due to gear meshing is recognized as a simple forced vibration and the nondimensional form of the forcing term is $[\phi^*-\kappa^*(1-\bar{\phi})]$ which acts at the position corresponding to tooth meshing. The method of calculating this nondimensional total vibrational excitation $[\phi^*-\kappa^*(1+\bar{\phi})]$ and a rough estimating method of its tooth meshing frequency component are reported in /6/ and /9/10/ respectively.

Figure 4 shows the changing state of rotational delay x as a function of transmitting load W schematically. When gears have a manufacturing and/or alignment error, tooth stiffness shows a strong nonlinearity as a function of transmitting load, because tooth flank begins to contact at one point on a definite geometrical contact line and contact length or zone grows on the geometrical contact line, when the transmitting load increases from zero to a definite value.

Considering the state of point A in Fig.4, the average value of mesh stiffness $\bar{K}$ of gear teeth is an inverse of the value of gradient of the line which connects the origin and the point A. This value $\bar{K}$ is different from the stiffness value K_A at the transmitting load corresponding to the point A.

The vibrating part F_d of the dynamic loading which acts on gears is expressed by

$$F_d = K_A(x - \bar{x}) \quad \cdots\cdots (10)$$

where $\bar{x}$ is the average value of x and given by

$$\bar{x} = x_s + \frac{\sum_i k_i e_i}{\bar{K}} \quad \cdots\cdots (11)$$

When the transmitting load is large enough, stiffness value K_A takes almost the same value as the stiffness value K_0 of error free gears. Substituting K_A by K_0, the following equation is obtained:

$$F_d = W\left(\frac{K_0}{\bar{K}}\right)\xi^* \quad \cdots\cdots (12)$$

Comparing eq.(12) with eq.(9), it is understood that a machine with power transmission gears has the vibrational source at a position of gear meshing, the amplitude of which is given by

$$E_v = W\left(\frac{K_0}{\bar{K}}\right)[\phi^* - \kappa^*(1+\bar{\phi})] \quad \cdots\cdots (13)$$

The value E_v is recognized as the total vibrational excitation due to gear meshing.

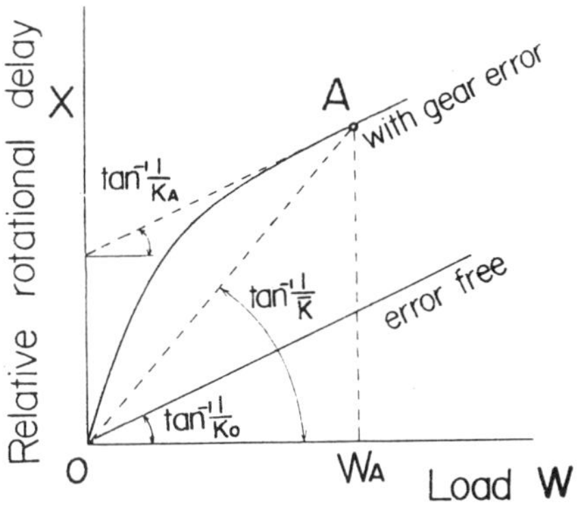

Fig.4 Schematic figure for the run of relative rotational delay of the driven gear to the driving gear as a function of transmitting load

5. Comparison with Measured Vibration of Gears

For a case of relative circumferential vibration of a pair of gears, the solution of eq.(4) is approximated by the solution of the constant coefficient differential equation, which is obtained by replacing $\sum_i k_i$ in eq.(4) with the average stiffness K_0 and $W + \sum_i k_i e_i$ with E_v. Here the solution of this constant coefficient differential equation be called "Simple Solution", while the solution of eq.(4) "Exact Solution". The acceleration of circumferential vibration of spur gears was measured, and compared with two kinds of predicted values: one from the exact solution and the other from the simple solution /11/. The ordinate of Fig.6 takes the effective value of vibrational acceleration and the abscissa the driving speed of gears. The effective value of acceleration obtained from the simple solution and that from the exact solution somewhat differs to each other, but the degree of disagreement is far more smaller than the degree of disagreement between the measured value and the value from exact solution which is generally approved as the model to express the state of actual vibration of spur gears. This fact indicates, that the state of gear vibration is as well estimated by using the simple solution method as by exact solution method.

Figure 7 shows the comparison of frequency construction of vibrational acceleration among that from measurement, that from exact solution and that from simple solution, when the transmitting load was changed in three different stages under constant driving speed of gears. It is recognized, that the state of frequency construction from measurement, that from the exact solution and that from the simple solution correspond qualitatively fairly well each other. And concerning to the degree of correspondence of the results from two estimating methods to the measured result, the method using the simple solution has not shown any inferiority to that using the exact solution.

6. Conclusions

The following results are obtained by this investigation:

The solution of a parametric excited 2nd order differential equation with forced term(s) and a set of such equations can be well approximated by a solution of forced 2nd order differntial equation(s) of constant coefficients, when its variable(s) and constants are so transformed that the time average value(s) of the forced term(s) becomes zero. The new constants which substitute the parametric exciting functions are their time average values. The vibration of gears in service is excited by (a) forced term(s) due to gear manufacturing and alignment errors and by (b) parameric excited term due to periodical change of tooth mesh stiffness with progress of gear meshing. But by applying the procedure mentioned above, it is possible to have only one total vibrational excitation E_v at tooth meshing point, and the influence of the periodical change of tooth mesh stiffness can be left out of consideration. The total vibrational excitation can change its magnitude, when the gear system becomes complicated or when the vibrations in multi directions are dealt with: but the nature or characteristics of the total vibrational excitation do not change. When the vibrational system of a machine is decided by some means, the machine vibration due to gear meshing can be treated by the conventional method of vibrational analysis with use of this total vibrational excitation.

Table 1 Dimensions of test gears

Tooth form	Normal
Module	4 mm
Press.angle	20 deg.
Tooth number	42 / 42
Tooth width	7.74 / 8.10 mm
Add.modf.factor	0 / 0
Center distance	169.32 mm
Contact ratio	1.40

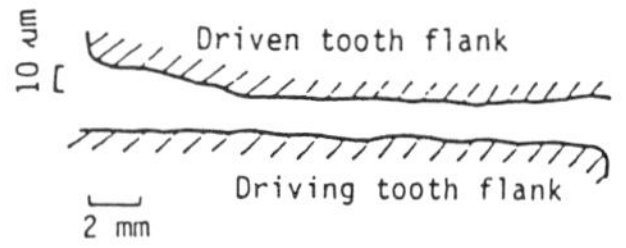

Fig.5 Tooth form of test gears

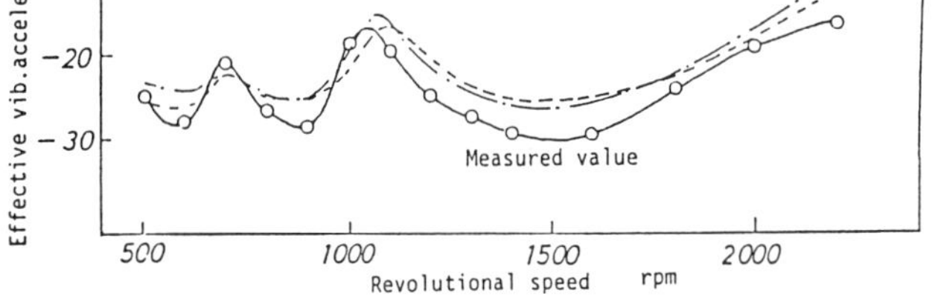

(a) Contact ratio $\varepsilon_\alpha = 1.40$, specific normal load $w_N = 64$ N/mm

Fig.6 Difference between measured level of spur gear vibration and the levels estimated after exact method and simple method

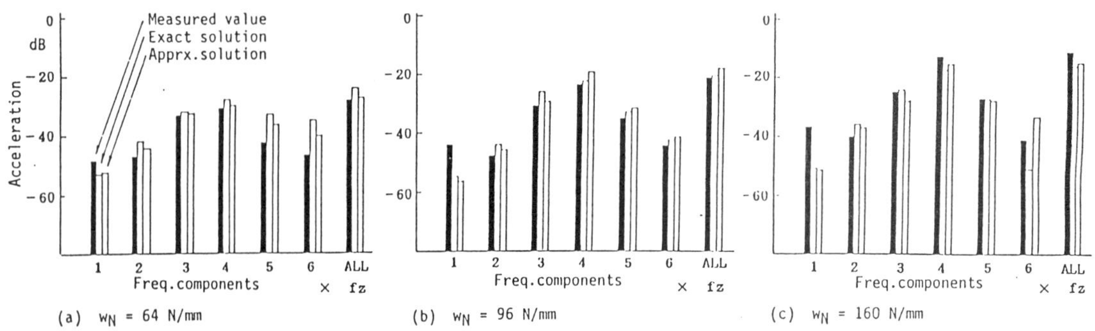

Fig.7 Comparison between vibrational spectra obtained by measurement and those estimated by exact method and by simple method

References

/1/ Kubo,A., etal., Trans.JSME,No.52-476, 1986 April, C, p.1420

/2/ Kiyono,S., et al., JSME Proc.No.830-8, 1983 Aug., p.323

/3/ Kubo,A., et al., JSME Proc.No.850-3, 1985 Aug., p.109

/4/ for example: Utagawa,M., Jourl.JSME, 1958 March, p.296

/5/ Kubo,A., et al., Trans.JSME, No.43 Vol.371, 1977 July, p.2771

/6/ Kubo,A., et al., Trans.JSME, No.46 Vol.401, 1980 Jan., p.86

/7/ Kiyono,S., et al., Trans.JSME, No.43 Vol.373, 1977 Sept., p.3556

/8/ Kiyono,S., Trans.JSME,(C) No.47 Vol.416, 1981 April, p.516

/9/ Kubo,A., Proc.JSME Int'l Symp.Gear., 1981 Tokyo, C-34

/10/ JSME Report of RC-58 Committee, 1983 June, Section 5.3

/11/ Kubo,A., et al., Trans.JSME, No.52-479, 1986 July, C, p.1992

Zusammenfassung

Maschinenvibrationen, die von lastuebertragenden Zahnraedern verursacht werden, werden an den im Eingriff befindlichen Zahnflanken durch dynamische Lastwechsel hervorgerufen. Ursache des dynamischen Lastanteils sind in erster Linie die Schwingungen eines Zahnradpaares relativ zur Umfangsrichtung, die durch eine Differentialgleichung zweiter Ordnung mit Stoerfunktion und zeitlich abhaengiger Federkonstante beschrieben werden koennen. Die Loesung dieser Gleichung kann jedoch nicht mit einem einfachen Ausdruck dargestellt werden. Bisher wurden hauptsaechlich Untersuchungen veroeffentlicht, die fuer Vibrationsbetrachtungen von Zahnradpaaren nur einfache Schwingungsmodelle anbieten: Die Zahnraeder werden als zweidimensional angesehen, d.h. als geradverzahnt ohne Lagen- und/oder Fertigungsfehler in Achsrichtung, ausserdem werden die Zahnradachsen als unsteif gegenueber Torsion, aber steif gegenueber Biegung behandelt. Die tatsaechlichen Schwingungsverhaeltnisse fuer die meisten Getriebe oder Maschinen mit Zahnraedern sind jedoch weitaus komplexer. Daher ist die direkte Uebertragung der Ergebnisse dieser Arbeiten auf die praktische Anwendung aeusserst problematisch. Allgemein ist es nicht einfach, den Einfluss von Zahnradabmessungen sowie von Fertigungs- und Lagenfehlern auf das Schwingungsverhalten von Zahnraedern wie auch Maschinen mit Zahnraedern zu untersuchen.

In der hierliegenden Arbeit wird eine neuartige Methode zur Untersuchung und zur Abschaetzung der Eigenschaften von Maschinenvibrationen, hervorgerufen durch im Eingriff befindliche Zahnraeder, vorgestellt, genauer gesagt ein Weg zur Rueckfuehrung eines Schwingungssystems mit Stoerfunktion und zeitlich abhaengiger Federkonstante auf ein praktisch gleichwertiges System mit Stoerfunktion ohne Zeitabhaengigkeit der Federkonstante. Mit dieser Methode werden die schwingungserzeugenden Kraefte, die aus Fertigungs- und Lagenfehlern sowie periodischem Wechsel der Zahnsteifigkeit im Eingriffspunkt resultieren, in einer schwingungserregenden Kraft E_v zusammengefasst, die an der Kontaktstelle der Zahnflanken wirksam ist. Mit dieser Kraft E_v koennen die Maschinenschwingungen mit herkoemmlichen Untersuchungsmethoden fuer Schwingungsanalysen ohne zeitliche Abhaengigkeit der Federkonstante behandelt werden. Es werden Eigenschaften und Genauigkeit der Aufloesung von Schwingungssystemen mit Zahnraedern anhand der Naeherungsloesung mit E_v besprochen. Ausserdem werden Resultate dieser neuen Naeherungsmethode und der herkoemmlichen genauen Berechnung, also der numerischen Loesung der Differentialgleichung mit zeitlich abhaengiger Federkonstante und Stoerfunktion, mit dem gemessenen Schwingungsniveau und dem Frequenzspektrum von Zahnradschwingungen verglichen und die Moeglichkeiten, die in der Anwendung dieser einfachen Naeherungsberechnung liegen, aufgezeigt. Ueber den Weg der Berechnung der Ersatzkraft E_v ist es moeglich, den Einfluss von Faktoren wie Abmessungen, Fertigungs- und Lagenfehlern auf die Laufeigenschaften von Zahnraedern abzuschaetzen. Ohne Differentialgleichungen loesen zu muessen kann die Ersatzgroesse E_v als statische Kraft bestimmt werden, sofern Abmessungen der Zahnraeder, Fertigungs- und Lagenfehler, sowie Arbeitsbedingungen bekannt sind. Daher duerfte die Groesse E_v ein gefragter Kandidat bei der Einstufung der Zahnraeder in Gueteklassen sein, die in direktem Zusammenhang mit deren dynamischem Verhalten steht.

Third Order Contact Analysis and Optimal Synthesis of Tooth Surfaces

X. C. WANG

Department of Mechanical Engineering, Xi'an Jiaotong University, Xi'an, PRC

Abstract. A new means of analysing the contact situations of bevel and hypoid gearing was developed in this paper, with which the 2- and 3-order contact characterestics at a reference point may be studied in details and expressed in parameter form, hence an objective function is easy to be obtained to characterize the contact quality. Based on this means an optimal synthesis is carried out. The advantages of the optimizing method presented in this paper is that: 1) the position of the reference point on gear tooth flank may be arbitrarily appointed; 2) the 2-order contact characterestics of the gearing may be prescribed and may keep unchanged during the optimizin process.

Keywords. Machine; gear; contact analysis; spiral bevel gear; hypoid gear.

INTRODUCTION

The contact situation of spiral bevel and hypoid gearing are usually analysed with tooth contact analysis (TCA) program developed by Gleason works. Though the results of TCA are fairly evident, the operation of TCA are rather numerous, and the final contact quality depends on the subjective judgement of the operator to a certain extant, hence a real optimal synthesis is hard to be achieved. Professor Litvin has investigated the synthesis of tooth surfaces on the supposition that all the 3-order derivatives of the tooth surface equal zero and the contact point path (CPP) on the surface is a geodetic line in the local sense. But that supposition is unsuited to the actual gears, besides, in his work all the 3-order contact characterestics except for the geodetic curvature of CPP are not concerned. In this paper almost all the 3-order contact features are concerned, which includes the rate of change of the 2-order contact characterestics with respect to the angle of rotation of the pinion or to the arc length of CPP, and the changing rates with respect to the moving distance, s^*, of the center point of the contact pattern along the lengthwise line of the gear face under V and H check. Hence the criterion for the machine setting scheme is obtained, and the base of optimal synthesis is erected.

REFERENCE POINT ON GENERATING AND GEAR TOOTH FLANK

Suppose the reference point on gear flank (Σ^G) is M (Fig.1),

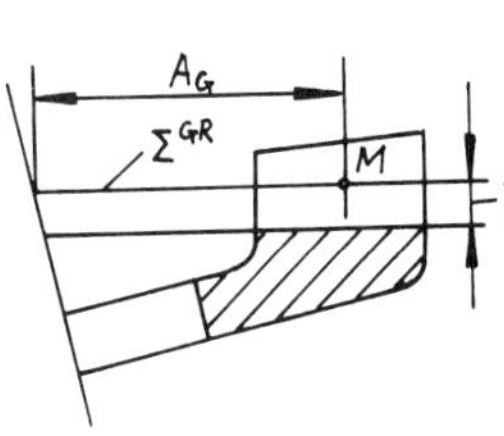

Fig. 1. Reference point on Σ^G.

through M make a cone parallel to the root cone of the gear, which is named reference cone (Σ^{GR}). The pitch cone distance, A_G, and the sector between the root cone and Σ^{GR}, b_G, are determined by the position of M. On the other hand, if a set of A_G and b_G is given, the position of M is settled. The generating flank Σ^C is a cone formed by rotating cutting edges of the cutter, and the reference point on Σ^C is the point which mating with M on Σ^G. It is easy to be obtained from the position of M on Σ^G.

GEOMETRIC PARAMETERS OF THE GEAR TOOTH FLANK

The structure of Σ^C is very simple, the principal directions of which are the generatrix and the latitude of the cone, respectively. Let $\bar{e}_1^c$ and $\bar{e}_2^c$ denote the unit vectors of the principal direction of the reference point on Σ^C, and $\bar{n}^c = e_1^c \times e_2^c$ the outward normal, the principal curvatures and their partial derivatives with respect to the length of the lines of curvature may be written as

$$K_1^c = 0, \quad K_2^c = \pm \cos\Phi_c / r, \tag{1}$$

$$K_{1,1}^c = K_{1,2}^c = K_{2,2}^c = 0, \; K_{2,1}^c = \mp (K_2^c) \tan\Phi_c \tag{2}$$

where r denotes the radius of the cutter cone at the reference point, Φ_c the pressure angle of the cutter. The upper sign of "±" and "∓" suit for the convex side of the gear tooth, and the lower the concave side. The geodetic curvatures of the lines of curvature of Σ^c will be

$$\rho_1^c = K_{1,2}^c/(K_1^c - K_2^c), \; \rho_2^c = K_{2,1}^c/(K_1^c - K_2^c) \tag{3}$$

To each point on Σ^c ($\bar{e}_1^c, \bar{e}_2^c, \bar{n}^c$) composes an orthogonal frame, which moves in accordance with the movement of the point on Σ^c. Let $\tilde{\omega}^{c1}$ and $\tilde{\omega}^{c2}$ denote the projections of the displacement of the contact point on $\bar{e}_1^c$ and $\bar{e}_2^c$, respectively, we get

$$\tilde{d}\bar{e}_1^c = \tilde{\omega}_1^{c2}\bar{e}_2^c + \tilde{\omega}_1^{c3}\bar{n}^c \tag{4}$$

$$\tilde{d}\bar{e}_2^c = -\tilde{\omega}_1^{c2}\bar{e}_1^c + \tilde{\omega}_2^{c3}\bar{n}^c \tag{5}$$

$$\tilde{d}\bar{n}^c = -\tilde{\omega}_1^{c3}\bar{e}_1^c - \tilde{\omega}_2^{c3}\bar{e}_2^c \tag{6}$$

where

$$\tilde{\omega}_1^{c2} = \rho_1^c\tilde{\omega}^{c1} + \rho_1^c\tilde{\omega}^{c2}, \tag{7}$$

$$\tilde{\omega}_1^{c3} = K_1^c\tilde{\omega}^{c1}, \quad \tilde{\omega}_2^{c3} = K_2^c\tilde{\omega}^{c2}. \tag{8}$$

Let $\bar{a}^C$ denote the unit vector of the axis about which Σ^C rotates, and ϕ^C the angle of rotation of Σ^C. Thus, in the fixed coordinates

$$d\bar{e}_1^c = \tilde{d}\bar{e}_1^c + \bar{a}^c\times\bar{e}_1^c\, d\phi^c \tag{9}$$

$$d\bar{e}_2^c = \tilde{d}\bar{e}_2^c + \bar{a}^c\times\bar{e}_2^c\, d\phi^c \tag{10}$$

$$d\bar{n}^c = \tilde{d}\bar{n}^c + \bar{a}^c\times\bar{n}^c\, d\phi^c \tag{11}$$

When the moving frame on Σ^C moves a distance ds along the instantaneous contact line,

$$\frac{\partial^2\bar{n}^c}{\partial s^2} = -\frac{\partial\omega_1^{c3}}{\partial s^2}\bar{e}_1^c - \frac{\partial\omega_2^{c3}}{\partial s^2}\bar{e}_2^c - \frac{\omega_1^{c3}}{\partial s}\frac{\partial\bar{e}_1^c}{\partial s} - \frac{\omega_2^{c3}}{\partial s}\frac{\partial\bar{e}_2^c}{\partial s} \tag{12}$$

Let $\bar{a}^G$ denote the unit vector of the gear axis, ϕ^G the angle of rotation of the gear, and $u_{CG} = d\phi^C/d\phi^G$.

Keep in mind the supposition that $\partial\tilde{\omega}^{c2}/\partial\phi^c = 0$, we get

$$\frac{\partial^2\bar{n}^c}{(\partial\phi^G)^2} = -K_{1,1}^c\left(\frac{\tilde{\omega}^{c1}}{\partial\phi^G}\right)^2\bar{e}_1^c - K_1^c\frac{\partial\tilde{\omega}^{c1}}{(\partial\phi^G)^2}\bar{e}_1^c - K_1^c\frac{\tilde{\omega}^{c1}}{\partial\phi^G}\frac{\partial\bar{e}_1^c}{\partial\phi^G} + \dot{u}_{CG}\bar{a}^c\times\bar{n}^c + u_{CG}\bar{a}^c\times\frac{\partial\bar{n}^c}{\partial\phi^G} \tag{13}$$

Using the moving frame on Σ^G to express $\partial^2\bar{n}^c/\partial s^2$ and $\partial^2\bar{n}^c/(\partial\phi^G)^2$, respectively, the follows may be written

$$\frac{\partial^2\bar{n}^c}{\partial s^2} = -\frac{\partial\omega_1^{G3}}{\partial s^2}\bar{e}_1^G - \frac{\partial\omega_2^{G3}}{\partial s^2}\bar{e}_2^G - \frac{\omega_1^{G3}}{\partial s}\left(\frac{\omega_1^{G2}}{\partial s}\bar{e}_2^G - \frac{\omega_1^{G3}}{\partial s}\bar{n}^c\right) - \frac{\omega_2^{G3}}{\partial s}\left(-\frac{\omega_1^{G2}}{\partial s}\bar{e}_1^G - \frac{\omega_2^{G3}}{\partial s}\bar{n}^c\right) \tag{14}$$

$$\frac{\partial^2\bar{n}^c}{(\partial\phi^G)^2} = -\frac{\partial\tilde{\omega}_1^{G3}}{(\partial\phi^G)^2}\bar{e}_1^G - \frac{\partial\tilde{\omega}_2^{G3}}{(\partial\phi^G)^2}\bar{e}_2^G - \frac{\tilde{\omega}_1^{G3}}{\partial\phi^G}\frac{\partial\bar{e}_1^G}{\partial\phi^G} - \frac{\tilde{\omega}_2^{G3}}{\partial\phi^G}\frac{\partial\bar{e}_2^G}{\partial\phi^G} + \bar{a}^G\times\frac{\partial\bar{n}^c}{\partial\phi^G} \tag{15}$$

where $\bar{e}_1^G$ and $\bar{e}_2^G$ denote the unit vectors of the principal directions of Σ^G. Spreading Eq. (14) and (15) according to the components of $\bar{e}_1^G$ and $\bar{e}_2^G$ gives a system of 4 linear equations. From which the patial derivatives, $K_{i,j}^G$, of the principal curvatures of Σ^G with respect to the arc length of the lines of curvature may be obtained. The geodetic curvatures of the lines of curvature may be obtained as well as ρ_i^c.

GEOMETRIC PARAMETERS OF THE PINION TOOTH FLANK

The 2-order parameters of the pinion face are determined by the mating gear tooth parameters and the prescribed 2-order contact characterestics. The later are listed as follows: 1) the angle, between the unit tangent, t_G, of CPP on Σ^G and the unit tangent, $\bar{\tau}_1$, of the profile line (Fig.2);

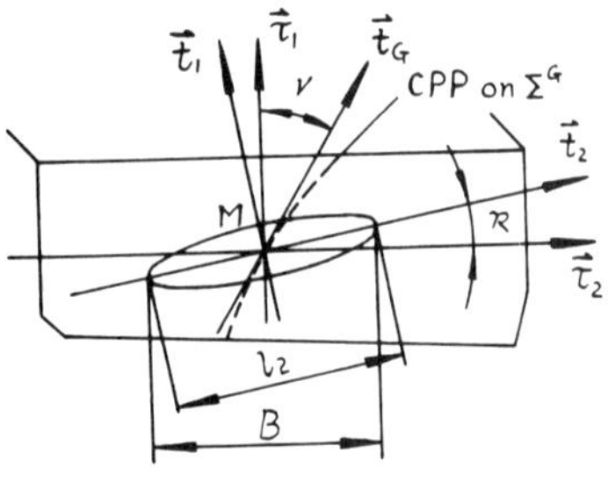

Fig.2. Prescribed contact characterestics

2) the projection length, B, of the length, l_2, of the instantaneous contact field on the unit tangent, $\bar{\tau}_2$, of the lengthwise line; 3) the angular acceleration, $\dot{u}_{GP} = d^2\phi^G/(d\phi^P)^2$, of the gear with respect to the pinion.

Suppose that $d\phi^P/dt \equiv 1$ and $u_{GP} = d\phi^G/d\phi^P$ equals the theoretical value i^{GP} when the contact point coincide with M. the speed of the contact point moving on Σ^G will be

$$ds^G/d\phi^P = \bar{n}\cdot\bar{q}^{PG}/\bar{t}_G\cdot\bar{p}^{PG} \tag{16}$$

where

$$\bar{p}^{PG} = \bar{V}^{PG}\cdot\nabla\bar{n}^G - \bar{\omega}^{PG}\times\bar{n} \tag{17}$$

$$\bar{q}^{PG} = -\dot{u}_{GP}\bar{a}^G\times\bar{R}^G + \bar{\omega}^{PG}\times\bar{V}^G - i\bar{a}^G\times\bar{V}^{PG} \tag{18}$$

In which $\nabla\bar{n}^G$ denotes the curvature tensor of Σ^G, $\bar{\omega}^{PG}$ the relative angular velocity between the pinion and the gear, $\bar{V}^{PG}$ the relative velocity between the tooth surfaces. The velocity of the contact point moving on Σ^P is

$$d\bar{r}^P/d\phi^P = \bar{t}_G\, ds^G/d\phi^P - \bar{V}^{PG} \tag{19}$$

Set

$$ds^P/d\phi^P = |d\bar{r}^P/d\phi^P|, \quad \bar{t}_P = (d\bar{r}^P/d\phi^P)/(ds^P/d\phi^P), \tag{20}$$

the derivative of the unit normal, $\bar{n}^{GP}$, of the relative curvature surface with respect to s^P may be expressed as

$$\frac{\widetilde{d\bar{n}}^{GP}}{ds^P} = \bar{t}_p \cdot \nabla \bar{n}^G - \left(\frac{ds^G}{d\phi^P} \bar{t}_G \cdot \nabla \bar{n}^G - \bar{\omega}^{GP} \times \bar{n} \right) \Big/ \frac{ds^P}{d\phi^P} \tag{21}$$

Hence

$$\varkappa = -\tan^{-1}\left[\left(\frac{\widetilde{d\bar{n}}^{GP}}{ds^P} \cdot \bar{\tau}_2 + K^{GP} \bar{t}_p \cdot \bar{\tau}_2 \right) \Big/ \left(\frac{\widetilde{d\bar{n}}^{GP}}{ds^P} \cdot \bar{\tau}_1 + K^{GP} \bar{t}_p \cdot \bar{\tau}_1 \right) \right] \tag{22}$$

Let δ denote the thickness of the marking compound used to check the gear bearing, and the area where the distance between the mating surfaces is less then δ will become the contact field. Thus

$$K_2^{GP} = 8\delta \cos^2 \varkappa / B^2 \tag{23}$$

From Eq.(22) and (23) $\varkappa$ and K_2^{GP} may be obtained by iteration. The other principal relative curvature is

$$K_1^{GP} = \bar{\tau}_1 \cdot (\widetilde{d\bar{n}}^{GP}/ds^P) / \bar{\tau}_1 \cdot \bar{t}_p \tag{24}$$

Having got K_1^{GP}, K_2^{GP} and $\varkappa$, the 2-order parameters of Σ^P are fixed.

Having defined the machine setting according to the 2-order geometric parameters of Σ^P, the 3-order parameters of Σ^P can be determined as well as the method used for Σ^G.

THIRD-ORDER CONTACT ANALYSIS

Using both the moving frames on Σ^P and Σ^G to express the derivatives of $\bar{n}$ and the position vectors, $\bar{R}^P$ and $\bar{R}^G$, in the fixed coordinates, respectively, we get

$$\frac{d\bar{R}^P}{d\phi^P} = \frac{\tilde{\omega}^{P1}}{d\phi^P} \bar{e}_1^P + \frac{\tilde{\omega}^{P2}}{d\phi^P} \bar{e}_2^P + \bar{a}^P \times \bar{R}^P \tag{25}$$

$$\frac{d\bar{R}^G}{d\phi^P} = \frac{\tilde{\omega}^{G1}}{d\phi^P} \bar{e}_1^G + \frac{\tilde{\omega}^{G2}}{d\phi^P} \bar{e}_2^G + \frac{d\phi^G}{d\phi^P} \bar{a}^G \times \bar{R}^G \tag{26}$$

$$\frac{d\bar{n}^P}{d\phi^P} = -K_1^P \frac{\tilde{\omega}^{P1}}{d\phi^P} \bar{e}_1^P - K_2^P \frac{\tilde{\omega}^{P2}}{d\phi^P} \bar{e}_2^P + \bar{a}^P \times \bar{n} \tag{27}$$

$$\frac{d\bar{n}^G}{d\phi^P} = -K_1^G \frac{\tilde{\omega}^{G1}}{d\phi^P} \bar{e}_1^G - K_2^G \frac{\tilde{\omega}^{G2}}{d\phi^P} \bar{e}_2^G + \frac{d\phi^G}{d\phi^P} \bar{a}^G \times \bar{n} \tag{28}$$

Considering that $d^2\bar{R}^P/(d\phi^P)^2 = d^2\bar{R}^G/(d\phi^P)^2$ and $d^2\bar{n}^P/(d\phi^P)^2 = d^2\bar{n}^G/(d\phi^P)^2$, a system of 4 linear equations may be made up, and from which $d\tilde{\omega}^{P1}/(d\phi^P)^2$, $d\tilde{\omega}^{P2}/(d\phi^P)^2$, $d\tilde{\omega}^{G1}/(d\phi^P)^2$ and $d\tilde{\omega}^{G2}/(d\phi^P)^2$ can be derived. Then, the geodetic curvature of CPP on Σ^G may be expressed as

$$G^G = \left(\frac{\tilde{\omega}^{G1}}{d\phi^P} \frac{d\tilde{\omega}^{G2}}{(d\phi^P)^2} - \frac{\tilde{\omega}^{G2}}{d\phi^P} \frac{d\tilde{\omega}^{G2}}{(d\phi^P)^2} \right) \Big/ \left(\frac{ds^G}{d\phi^P} \right)^3 - \frac{\tilde{\omega}^{G2}}{d\phi^P} \Big/ \frac{ds^G}{d\phi^P} \tag{29}$$

The higher-order angular acceleration of the gear with respect to the pinion will be

$$\frac{d^3\phi^G}{(d\phi^P)^3} = \frac{\left(\bar{\omega}^{GP}, \frac{d^2\bar{R}^P}{(d\phi^P)^2}, \bar{n}\right) + 2\left(\bar{\omega}^{GP}, \frac{d\bar{R}^P}{d\phi^P}, \frac{d\bar{n}}{d\phi^P}\right) + \bar{V}^{GP} \cdot \frac{d^2\bar{n}}{(d\phi^P)^2} + 2\dot{u}_{GP}\left[\left(\bar{a}^G, \frac{d\bar{R}^P}{d\phi^P}, \bar{n}\right) - \left(\bar{a}^G, \bar{R}^G, \frac{d\bar{n}}{d\phi^P}\right)\right]}{(\bar{a}^G, \bar{n}, \bar{R}^P)} \tag{30}$$

And the rate of change of the instantaneous contact field length with respect to the arc length of CPP ON Σ^G will be

$$\frac{dl_2}{ds^G} = -\frac{1}{2} \frac{l_2}{K_2^{GP}} \frac{dK_2^{GP}}{d\phi^P} \Big/ \frac{ds^G}{d\phi^P} \tag{31}$$

During V and H check, the center point of contact moves along the lengthwise line of the gear face, and the derivatives of the position vectors and the unit normal will be

$$\frac{d\bar{R}^P}{ds^*} = \frac{\tilde{\omega}^{P1}}{ds^*} \bar{e}_1^P + \frac{\tilde{\omega}^{P2}}{ds^*} \bar{e}_2^P + \frac{d\phi^P}{ds^*} \bar{a}^P \times \bar{R}^P \tag{32}$$

$$\frac{d\bar{R}^G}{ds^*} = \bar{\tau}_2 + \frac{d\phi^G}{ds^*} \bar{a}^G \times \bar{R}^G \tag{33}$$

$$\frac{d\bar{n}^P}{ds^*} = -K_1^P \frac{\tilde{\omega}^{P1}}{ds^*} \bar{e}_1^P - K_2^P \frac{\tilde{\omega}^{P2}}{ds^*} \bar{e}_2^P + \frac{d\phi^P}{ds^*} \bar{a}^P \times \bar{n} \tag{34}$$

$$\frac{d\bar{n}^G}{ds^*} = \bar{\tau}_2 \cdot \nabla \bar{n}^G + \frac{d\phi^G}{ds^*} \bar{a}^G \times \bar{n} \tag{35}$$

In order to get the values of $\tilde{\omega}^{P1}/ds^*$, $\tilde{\omega}^{P2}/ds^*$, $d\phi^P/ds^*$ and $d\phi^G/ds^*$, a system of 4 linear equations is needed. Let $\bar{\xi}_1$ and $\bar{\xi}_2$ denote the unit vectors of the vertical and horizontal directions of V and H displacement, respectively,

$$\frac{d\bar{R}^P}{ds^*} = \frac{d\bar{R}^G}{ds^*} + \frac{dV}{ds^*} \bar{\xi}_1 + \frac{dH}{ds^*} \sec\gamma\, \bar{\xi}_2 \tag{36}$$

where γ denotes the pitch angle of the pinion. Set $\bar{\xi}_3 = \bar{\xi}_1 \times \bar{\xi}_2$, and from the numerical production of Eq.(36) with $\bar{\xi}_3$ we get the first linear equation; from $d\bar{n}^P/ds^* = d\bar{n}^G/ds^*$ two others may be obtained. The final is derived from the supposition that the transmission ratio of the gearing keep unchanged at the center point of contact, which can be written as

$$(\bar{a}^P, \frac{d\bar{R}^P}{ds^*}, \bar{n}) + (\bar{a}^P, \bar{R}^P, \frac{d\bar{n}}{ds^*}) = i\left[(\bar{a}^G, \frac{d\bar{R}^G}{ds^*}, \bar{n}) + (\bar{a}^G, \bar{R}^G, \frac{d\bar{n}}{ds^*})\right] \tag{37}$$

Having got those values, dV/dS^* and dH/dS^* may be worked out.

After the displacement dV, dH of the gear axis relative to the pinion, the statement $d\bar{R}^P/d\phi^P = d\bar{R}^G/d\phi^P$ and $d\bar{n}^P/d\phi^P = d\bar{n}^G/o$ are still valid. Hence a system of linear equations containing $d(\tilde{\omega}^{P1}/d\phi^P)/dS^*$, $d(\tilde{\omega}^{P2}/d\phi^P)/dS^*$, $d(\tilde{\omega}^{G1}/d\phi^P)/ds^*$, $d(\tilde{\omega}^{G2}/d\phi^P)/dS^*$ is generated from Eq.(25),(26),(27) and (28). Having got the values, the rates of change of the 2-order contact characterestics with respect to S^* may be obtained as follows: the rate of change of the angle between the direction of CPP on Σ^G and the profile line will be

$$\frac{dv}{ds^*} = \left[\frac{\tilde{\omega}^{G1}}{d\phi^P}\frac{d}{ds^*}\left(\frac{\tilde{\omega}^{G2}}{d\phi^P}\right) - \frac{\tilde{\omega}^{G2}}{d\phi^P}\frac{d}{ds^*}\left(\frac{\tilde{\omega}^{G1}}{d\phi^P}\right)\right] \Big/ \left(\frac{dS^G}{d\phi^P}\right)^2 - G_L + \frac{\tilde{\omega}_1^{G2}}{dS^*} \tag{38}$$

where G_L denotes the geodetic curvature of the lengthwise line on ^{G}; the rate of change of the angular acceleration of the gear relative to the pinion is equal to

$$\frac{d\dot{u}_{GP}}{ds^*} = \frac{(\bar{\omega}^{GP}, \frac{d}{dS^*}(\frac{d\bar{R}^P}{d\phi^P}), \bar{n}) + (\bar{\omega}^{GP}, \frac{d\bar{R}^P}{d\phi^P}, \frac{d\bar{n}}{dS^*}) + \bar{V}^{GP} \cdot \frac{d}{dS^*}(\frac{d\bar{n}}{d\phi^P}) - (\bar{a}^P, \frac{d\bar{R}^P}{dS^*}, \frac{d\bar{n}}{d\phi^P}) + i(\bar{a}^G, \frac{d\bar{R}^G}{dS^*}, \frac{d\bar{n}}{d\phi^P}) + \dot{u}_{GP}\left[(\bar{a}^G, \frac{d\bar{R}^G}{dS^*}, \bar{n}) + (\bar{a}^G, \bar{R}^G, \frac{d\bar{n}}{dS^*})\right]}{(\bar{a}^G, \bar{n}, \dot{\bar{R}}^G)} \tag{39}$$

and the rate of change of the length of the contact field major axis can be written as

$$\frac{dl_2}{ds^*} = -\frac{1}{2}\frac{l_2}{K_2^{GP}}\frac{dK^{GP}}{dS^*} \tag{40}$$

THE OPTIMAL SYNTHESIS PROCESS

The practical calculation process is as follows:

First, calculate the geometric parameters of the gear tooth surface at the reference point, and the 2-order parameters of the pinion flank.

Second, tentatively define a set of freely chosen machine setting parameters of the gear cutting machine and calculate the other parameters and the 3-order geometric parameters of the pinion tooth surface.

Third, calculate the 3-order contact characterestics between the gear and pinion tooth surfaces, and set up a weighted objective function according to the relative importance of the influnce of the 3-order contact parameters to the engagement performance.

Finally, a direct search method is used to minimize the objective function by means of changing the freely chosen parameters for machine setup.

CONCLUSION

In order to get a fairly comprehensive understanding of the engagement performance of bevel and hypoid gearing, only one point on the path of contact, the reference point, is needed where the 3-order contact analysis is carried out. And by means of optimal synthesis, it is sure that gearings with high quality of contact characterestics may be manufactured effectively.

REFERENCE

Litvin, F. L., and Y. Gutman (1981). A method of local synthesis of gears grounded on the connections between the principal and geodetic curvatures of surfaces. Trans. ASME., Vol.103, No.1, Jan., 114-125.

DIE ANALYSE DER DRITTEN ORDNUNG UND DIE OPTIMALE SYNTHESE VON FLANKEN

Die Beruehrverhaeltnisse der Kegelraeder mit Bogenverzahnung und Hypoidraeder werden allgemein durch das Programm der Kontakt-analyse von Flanken (TCA) analysiert, das von Gleason-Gesellschaft entwickelt wird. Obwohl die Resultate von TCA anschaulich ist, gibt es eine Masse Operationen und ist die Kontakt-qualitaet von subjektiven Aussagen des Operators abhaengig, so dass es schwierig ist, zu einer wirklichen optimalen Synthese zu gelangen. In diesem Aufsatz wird eine neue analytische Methode der Beruehrverhaeltnisse von diesen Zahnraeder dargestellt, mit der die Kontakt-charakteristiken der zweiten und dritten Ordnung an Bezugspunkten zu untersuchen sind und daraus ergeben sich eine Folge in Parameterform. Dann ist eine objektive Funktion, die Kontakt-qualitaet beschreibt, lecht zu erhalten. In bezuk auf dieser Methode wird eine Optimale Synthese vorgestellt, deren Vorteile es ist.

— Stellung der Bezugspunkten an Flanken unschraenkt zu geben sein,
— die Kontakt-charakteristiken der zweiten Ordnung von Zahnradgetrieben vorauszubestimmen und waehrend des Prozess der Optimierung unveraendert zu halten sein.

The Width of Instantaneous Contact Locus and Formula of the Contact Stresses of Model JB2940-81 W-N Gear

CHEN CHEN-WEN, CHEN RONG-ZENG AND CHEN SHI-CHUN

Department of Mechanical Engineering, Harbin Institute of Technology, Harbin, PRC

Abstract. Drill a small hole and embed a wiring at the initial contact point on middle of the tooth lenth, measure electricallg through on-off action the turning angle of the gear which can be converted into the width of instantaneous contact locus. The formula may be derived through regression analysis of the different widths of contact locus obtained from different helical angles and loads of the gears, then the formula of contact stresses may be drived consequently.

Keywords. Power transmission; W-N Gear; Contact stress; Pressure measuement; Machine tools.

The form of instantaneous contact locus on tooth surface of the W-N gear is still complicated after its being run-in sufficiently. As fig. 1a shown, the width $2l_y$ is different along the tooth elevation (i.e. with the different pressure angle α_y). The dimension of $2l_y$ depends not only on the load, but also on the local equivalent radius of curvature ρ_y, therefore the instantaneous contact locus is wider near the pitch culinder and narrower at the tooth top or tooth root.

At initial contact point before its being run-in, the pressure angle α for[3] model JB2940-81 W-N gear is 24° with local equivalent radius of curvature ρ(fig. 1.a,b). Relation between pressure angle α_y (or α) and equivalent radius of curvature ρ_y (or ρ) is as follows:

$$\bar{\rho}_y = \frac{\rho_y}{m_n} = \frac{z_1}{2\sin\alpha_y \sin^2\beta \cos\beta} \cdot \frac{u}{1+u} \qquad (1)$$

Any length with bar on its top represents the ratio of that length to the module m_n, and becomes non-dimensional hereafter.

Suppose that the contact stress along tooth length (or x axis) is distributed elliptically as fig. 1.c shown. The tooth surface still wears slowly in operation after running-in. Assume that the tooth profile is always in form of an arc in the coure of wear, i.e. the instantaneous contact line on the normal plane of tooth surface as an arc if without considering elastic deformation, and the center of the arc is at its pitch point. The original center of tooth profile moves a distance Δ (fig. 1.e) along the pitch circle due to the result of wear, then the worn-out depth at tooth elevation α_y is to be $\Delta \cdot \cos\alpha_y$. Again suppose the worn-out depth is derectly proportional to the local frictional work, then

$$\frac{\int_t \mu \sigma v\, dt}{\Delta \cdot \cos\alpha_y} = \text{constant} \qquad (2)$$

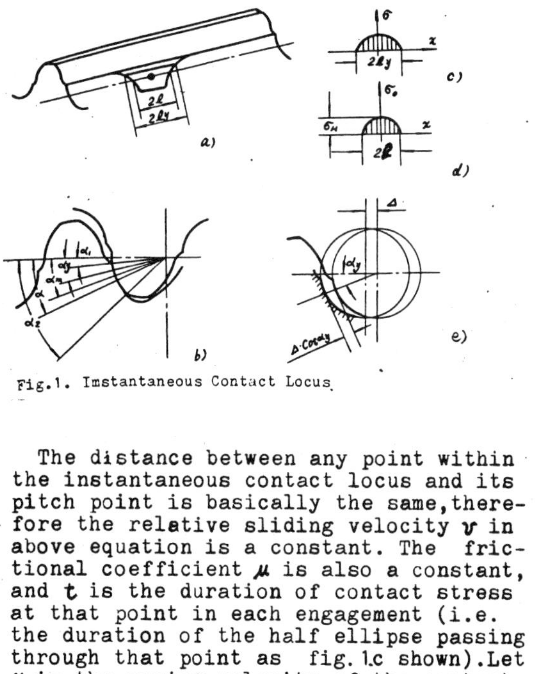

Fig.1. Imstantaneous Contact Locus

The distance between any point within the instantaneous contact locus and its pitch point is basically the same, therefore the relative sliding velocity v in above equation is a constant. The frictional coefficient μ is also a constant, and t is the duration of contact stress at that point in each engagement (i.e. the duration of the half ellipse passing through that point as fig. 1.c shown). Let u is the moving velocity of the contact locus along tooth length. i.e. $u = dx/dt$ = constant, then

$$\int_t u \sigma\, dt = \int_{-l_y}^{+l_y} \sigma\, dx$$

therefore $\quad \frac{1}{\cos\alpha_y} \int_{-l_y}^{+l_y} \sigma\, dx$ = constant

Assume that the max. value of contact stress σ_0 at initial contact point (at $\alpha = 24^\circ$) is σ_H, then (fig. 1.d)

$$\frac{1}{\cos\alpha_y} \int_{-l_y}^{+l_y} \sigma\, dx = \frac{1}{\cos 24^\circ} \int_{-l}^{+l} \sigma_0\, dx = \frac{\pi l \sigma_H}{2\cos 24^\circ} \qquad (3)$$

Also assume that: the tooth profile arc radius ρ_m after running-in is the mean value of original radi of concave and convex tooth profiles, that is

$$\rho_m = (\rho_a + \rho_f)/2 = 1.355\, m_n \quad \text{(when } m_n = 3.5\sim 6\text{)}$$

If the frictional force is neglected, then the tangential force F_{nt} and radial force F_r at the pitch cylinder on the normal plane induced by the stress on whole contact locus are as follows respectively:

$$F_{nt} = \int_{\alpha_2}^{\alpha_1}\int_{-l_y}^{+l_y} \cos\alpha_y\, \sigma \rho_m\, dx\, d\alpha_y = \int_{\alpha_2}^{\alpha_1} \frac{1.355\pi}{2\cos 24^\circ}\, l\, m_n \sigma_H \cos^2\alpha_y\, d\alpha_y$$

$$= 1.165\, l\, m_n \sigma_H \left[\alpha_2 - \alpha_1 + \tfrac{1}{2}\sin 2\alpha_2 - \tfrac{1}{2}\sin 2\alpha_1\right] \qquad (4)$$

$$F_r = \int_{\alpha_2}^{\alpha_1}\int_{-l_y}^{+l_y} \sin\alpha_y\, \sigma \rho_m\, dx\, d\alpha_y = 1.165\, l\, m_n \sigma_H \left[\sin^2\alpha_2 - \sin^2\alpha_1\right] \qquad (5)$$

Therefore the direction of the resultant force (i.e. the mean pressure angle α_m) on normal plane (fig. 1.b) may be obtained as:

$$\alpha_m = \tan^{-1}\frac{F_r}{F_{nt}} = \tan^{-1}\frac{\sin^2\alpha_2 - \sin^2\alpha_1}{\alpha_2 - \alpha_1 + \frac{1}{2}\sin 2\alpha_2 - \frac{1}{2}\sin\alpha_1} \qquad (6)$$

Here α_1 and α_2 as fig. 1.b shown, α_1 is the tip angle on basic ract for processing requirement, for model JB2940-81 W-N gear, when $m_n = 3.5\sim 6$ $\alpha_1 = 9°19'30''$. α_2 depends upon equivalent number of convex teeth Z_v'

$$\alpha_2 = \sin^{-1}\frac{\bar{h}_a Z_v' + \bar{h}_a^2 - \bar{\rho}_m^2}{\bar{\rho}_m Z_v'} \qquad (7)$$

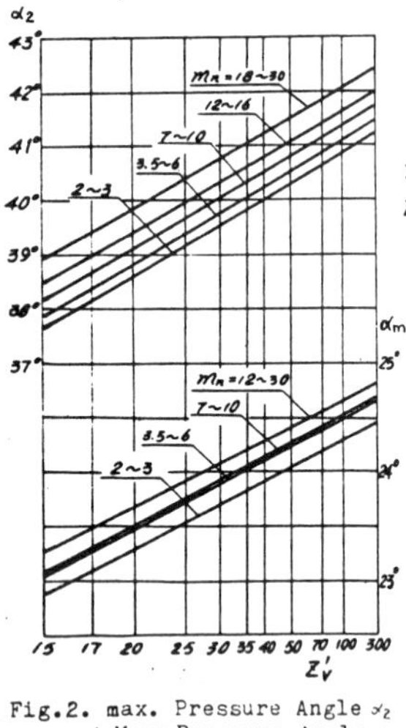

Fig.2. max. Pressure Angle α_2 and Mean Pressure Angle α_m

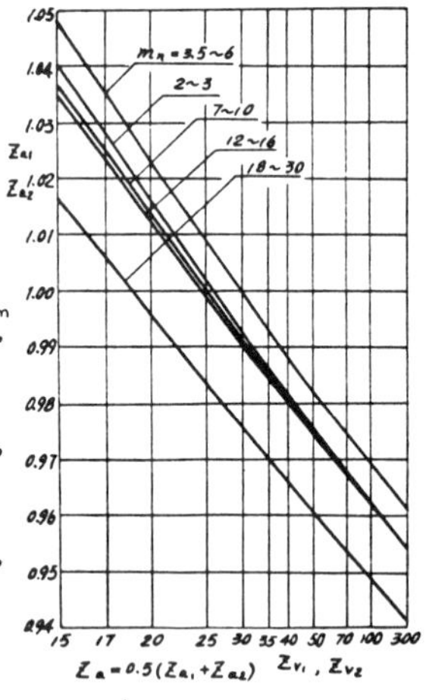

Fig.3. Coefficient of Chordal Length of Contact Line

The calculated α_m and α_2 from equation (6) and (7) are plotted on fig. 2. with different equivalent convex teeth number Z_v'. It can be seen that after running-in the direction of the resultant pressure on tooth surface in not 24° but α_m. The value of α_m which varies not much with Z_v' may be taken still as 24°.

μ_ε is the whole number in value of overlap ratio ε (= b/p_x), when the decimal part of ε is small, there are $2\mu_\varepsilon$ whole contact loci being loaded most of time,

The normal pressure on each contact locus is

$$F_n = \frac{F_t}{2\mu_\varepsilon \cos\beta \cos 24^\circ} \quad \text{or} \quad F_n = \frac{T_1}{0.9135\,\mu_\varepsilon m_n Z_1} \qquad (8)$$

From equation (4) we may get the relation between the formal pressure F_n, with of instantaneous locus $2\bar{l}$ and max. contact stress σ_H at initial contact point.

$$F_n = \frac{F_{nt}}{\cos 24^\circ} = \frac{0.5496}{Z_a}\, 2\bar{l}\, m_n^2 \sigma_H \qquad (9)$$

The constant 0.5496 in equation (9) is derived from equation (4) accounding to $m_n = 3.5\sim 6$, $Z_v' = 30$. For different m_n and Z_v', the value of F_n must be corrected with "coefficient of chordal length of contact line along tooth height Z_a". Z_a is the ratio of Fn and Fn at $Z_v' = 30$, $m_n = 3.5\sim 6$, then

$$Z_a = \frac{\bar{\rho}_m^*}{\bar{\rho}_m}\cdot\frac{\alpha_2^* - \alpha_1^* + \frac{1}{2}\sin 2\alpha_2^* - \frac{1}{2}\sin 2\alpha_1^*}{\alpha_2 - \alpha_1 + \frac{1}{2}\sin 2\alpha_2 - \frac{1}{2}\sin 2\alpha_1} \qquad (10)$$

Plot Z_a calculated from equation(10)on fig.3. For a gear couple, coefficient is the mean value of Z_{a1} and Z_{a2} being determined by Z_{v1} and Z_{v2}.

The width of instantaneous contact locus $2\bar{l}$ is obtained by measurement. The experiment was done on the locked-in torque gear testing stand with center distance of 120mm. The test gears with module of 4 were made of tempered 40Cr steel (HB=250-270). Other parameters are listed in table 1. Before measuring their $2\bar{l}$, test gear couples have been sufficiently run-in.

Drill a small hole of 0.5mm in diameter at the initial contact point (here $\alpha = 24°$) of convex part on the middle portion of tooth length; drill it through until the root of next tooth space. A special drill jig is used. Since the equivalent teeth numbers Z_{v1} of the seven driving gears lay between 28.4-31.1, and the form of normal teeth space are basically same, so the same drill jig is used for accunate location of the small hole.

Embed a 0.35mm painted wiring and fill the hole up with glue for insulation. The lead from cavity bottom of neighbouring tooth is connected to the universal electric meter through collecter rings, the other terminal of universal electric meter is connected to the gear box. Level the lead at hole opening and run-in the gear couple again shortly, then clean the lubricating oil and measure electrically the turning angle between on and off; finally convert it into the "on" width along tooth length and minus the width of lead end along tooth length. We get the width $2\bar{l}$ of instantaneous contact locus.

Fig. 4 was developed from the pitch cylinder of driving gear. At point A, We have drilled a small hole and embed a wiring. When contact locus approach point A, there is an another contact locus on the concave surface of next tooth whose border is at the distance of j from the approach end. In order to ensure two whole contact loci being loaded at that time, must $j > 0$. If g_n is the distance between two pitch points for convex and concave surface of corresponding flank on basic rack tooth profile, f is the distance between convex surface of drilled tooth and concave surface of next tooth on normal plane as shown in fig.4, then (when $m_n = 3.5 \sim 6$)

$$g_n = 0.5\pi m_n + C_a + C_f + \frac{x_a - x_f}{\tan\alpha} = 2.8717\, m_n$$

$$f = \pi m_n - \left[g_n - 2\left(\rho_a + \frac{x_a}{\sin\alpha}\right)\cos\alpha \right] = 2.9665\, m_n$$

$$j = \frac{b}{2\cos\beta} - f\tan\beta - \frac{\pi m_n}{\tan\beta}\gamma - 2l \qquad (11)$$

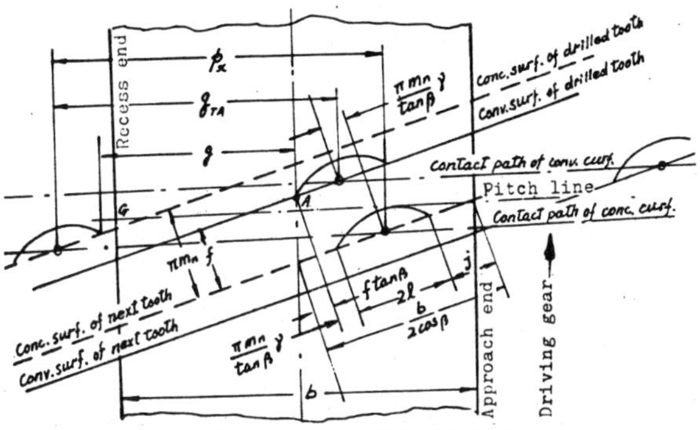

Fig.4. Developed Plane from the Pitch Cylinder

In above equations, γ is "the rate of displacement of contact loci on two neighbouring teeth along tooth length" [2] for model JB 2940-81 , $\gamma = 0.0859$. Symbol of parameters of basic rack: ρ_a . ρ_f , C_a , C_f , x_a and x_f are all according to reference [3] .

For seven test gear couples under different load as listed in table 1, the value of j is always greater than zero, so the second contact locus is also complete; including constant 2 in decominator of equation (8) to determine Fn is reliable. But for test gear couple with β being 33°33′26″, according to equation (11), $2l$ can not exceed 6.1mm, otherwise while first contact locus approach point A, the second contact locus has not complete enter, therefore most load is only 7297 N for this test gear conple.

In addition, in order to ensure the contact locus on concave surface of drilled tooth has been completely recede from contact, must $g > b/2$ as fig.4 shown. If g_{TA} is the axial distance between two center of contact loci of corresponding blank, then (when $m_n = 3.5 \sim 6$)

$$g_{TA} = \frac{g_n}{\sin\beta} - 2\left(\rho_a + \frac{x_a}{\sin\alpha}\right)\cos\alpha\sin\beta = \left(\frac{2.8717}{\sin\beta} - 2.4483\sin\beta\right) m_n$$

$$g = g_{TA} - 2l\cos\beta \qquad (12)$$

For four test gear couples with $\beta \leq 21°2'22''$ under diferent load as listed in table 1, the value g is always greater than $b/2$. For fifth test gear couple with $\beta = 25°50'31''$ while Fn>4500N could not satisfed that condition; for sixth and seventh test gear couples with more $\beta \geq 29°55'35''$ commonly conld not satisfied it. For this reason, measure electrically after thinning-down by filing the convex surface of mating tooth near recess end (point G as shown in fig.4), the length of thinning-down is $(b-2g)/2\cos\beta$ from recess end, i.e. 1.4; 2.9 and 2.6mm for fifth, sixth and seventn test gear couples respectively.

The measured widths of contact loci at different loads Fn and different helical angles β are listed in table 1. For the same test gear, more the loads Fn is more the width $2l$ of the contact locus will be, and their relation showing a straight line on logarithmic coordinates is expressed in the equation as below:

$$2\bar{l} = k_1 F_n^{x_1} \qquad (13)$$

the value of $2\bar{l}$ varies apparently with the equivelent radius of curvature $\bar{\rho}$ (at $\alpha = 24°$). The values $\bar{\rho}$ of seven test gear couple are different and the relation between $2\bar{l}$ and $\bar{\rho}$ under constant load may be expressed as follows:

$$2\bar{l} = k_2 \bar{\rho}^{x_2} \qquad (14)$$

The values x_1 and x_2 are almost same which are of 0.27 approximately.Therefore the equation (13) and (14) may be combined as:

$$2\bar{l} = k_3 (F_n \bar{\rho})^{x} \qquad (15)$$

Equation (15) is applicable for constant m_n only. According to the theorem of similarity, when both the module of elasticity E and Poisson ratio ν keep constant, and $C_p = C_l^2$ (here C_p and C_l are coefficients of load similarity and geometry similarity respectively), or in other words, when Fn and m_n^2 increase with same proportion, then the stress value keeps constant, but $2l$ and m_n vary with same coefficient C_l , so $2\bar{l}$ keeps constant, then

$$2\bar{l} = k\left(\frac{F_n \bar{\rho}}{m_n^2}\right)^{x} \qquad (16)$$

TABLE 1 Parameters of Tested Gears and Width of Contact loci Being Measured

No	β	Z_1	Z_2	b mm	ε	$F_n/2l$ (N/mm)								
1	10°28′31″	27	32	79.5	1.15	5067/9.80	6081/10.8	7297/11.4	8108/11.5	9122/12.4	10136/12	11149/12.8	12163/13.1	
2	14°50′06″	26	32	57	1.16	4210/7.75	5263/8.65	6315/8.70	7297/9.35	8420/9.90	9473/9.80	10525/10.5	11577/10.5	12630/10.6
3	18°11′41″	25	32	46	1.14	4378/7.10	5473/7.50	6568/8.20	7297/8.45	8757/8.55	9852/9.15	10946/9.3		
4	21°02′22″	24	32	41	1.17	3991/6.48	4789/6.78	5587/7.18	6386/7.48	7297/7.88	7982/8.18	8780/8.48		
5	25°50′31″	22	32	33	1.14	3483/5.26	4354/5.78	5225/6.34	6095/6.63	7297/7.02	7837/7.22	8708/7.48	9578/7.55	
6	29°55′35″	20	32	29	1.15	3831/4.87	4798/5.62	5747/5.95	6704/6.32	7297/6.18	7662/6.35	8620/6.81	9578/6.93	10536/7.15
7	33°33′26″	18	32	26	1.14	4257/5.11	5321/5.49	6385/5.79	7297/5.92					

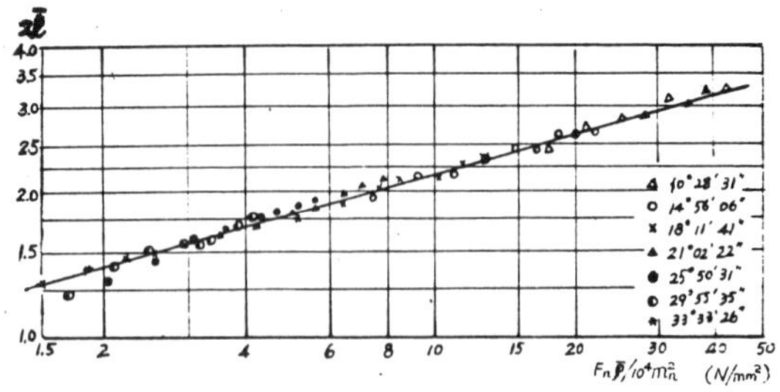

Fig.5. Relationships between ($F_n\bar{\rho}/m_n^2$) and ($2\bar{l}$)

Plot the values of $2\bar{l}$ and ($F_n\bar{\rho}/m_n^2$) on logarithmic coordinates as fig.5 shown, their relation is nearly a straight line. The values of k and x in equation(16) may be rewritten as:

$$2\bar{l} = 0.0975\,(F_n\bar{\rho}/m_n^2)^{0.27} \qquad (17)$$

For W-N gear of JB929-67, the relationships between non-dimensional parameters: ($F_n\bar{\rho}/m_n^2 E'$) and width of instantaneous contact locus $2b$ have been calculated,[1] ($2b$, namely $2\bar{l}$ in this paper), their relation showing a straight line in logarithmic coordinate, i.e. the value of $2\bar{l}$ is depends upon not merely the load F_n/m_n^2 and equivalent radius of curvature $\bar{\rho}$, but also the reduced modulus of elasticity of materials E' as well. Both F_n/m_n^2 and $1/E'$ have same effect on $2\bar{l}$, equation (17) being obtained from experiment only suits forged steel ($E=2.1\times10^5$, $\nu=0.27$, $E'=E/(1-\nu^2)=2.27\times10^5$ N/mm^2). For different meterial whase E' not comform to 2.27×10^5, ($F_n\bar{\rho}/m_n^2$)may be multipled by ($2.27\times10^5/E'$), therefore

$$2\bar{l} = 2.7235\,(F_n\bar{\rho}/m_n^2 E')^{0.27} \qquad (18)$$

From equation (9) and (18), we can get equation (19) and (20),(length in mm and force in N)

$$\frac{\sigma_H\bar{\rho}}{Z_a E'} = 0.67\left(\frac{F_n\bar{\rho}}{m_n^2 E'}\right)^{0.73} \qquad (19)$$

or

$$\sigma_H = 0.67\left(\frac{F_n}{m_n^2}\right)^{0.73}\left(\frac{E'}{\bar{\rho}}\right)^{0.27} \qquad (20)$$

In this paper, the experiment are static and the effect of oil film to the stress distribution is not considered as the calculation of the contact stress of involute gear at present. After the fatigue test on the gear couples for determinating the limiting contact stresses of different materials,[4] the equation (20) may be used as the basic equation in calculation of their contact strength. Although the measurement of the value $2\bar{l}$ and the derivation of contact stress mentioned above are conditional, however the calculation is reliable due to utilization of data obtained from the fatigue test as well as the limiting stress derived under the same condition.[4]

REFERENCES

[1] CHEN Chen-Wen
The Relationships Between Load Parameter ($F_n\bar{\rho}/m_n^2 E'$) of W-N Gear and its Stiffness Parameter, Contact Stress Parameter, Root Stress Parameter Recpectively.
Proceedings of International Symposium on Gearing & Power Transmissions (1981, Tokyo) . Vol. 1. pp.93-98.

[2] CHEN Chen- Wen, CHEN Rong-Zeng, YE Rei- Da
Experimental Study of the Helical Gear Couple With Stepped-Double-Circular-Arc H-2 Type Tooth Profile.
Proceedings of International Symposium on Gearing & Power Transmissions (1981, Tokyo) . Vol. 1. pp.99-104.

[3] YANG Ji-Bin, CHEN Chen-Wen
A preliminary Investigation on the Contact Problem of Helical Gear with Double Circular-arc Tooth Profile
Proceedings of 2nd World Congress on Gearing (1986, Paris) vol.2. pp.19-27.

[4] 陳荣增等
81型双圆弧齿轮的齿面接触疲劳试验研究
哈尔滨工业大学科学研究报告 1986年第7期 PP.1-14

Analysis of Pinion Elastic Deformation and Curve of Axial Modification

T. WANG

Department of Mechanical Engineering, Shanghai Jiao-Tong University, Shanghai, PRC

Abstract. It is necessary for us to be aware of the elastic deformation of a pinion in order to make the axial modification. On the basis of the analysis of pinion with different parameters and teeth by F.E.M., some elastic deformation curves have been obtained and a corresponding theoretic 3-D axial modification curve is derived by the author. The relative values of bending, torsion, compressive deformations of a pinion shaft, the bending deflection of a tooth, and the elastic deformation of teeth contact surface area are given in this paper. Accordingly, some suggestions have been made about the axial modification to the pinion elastic deformation.

Keywords. Elastic deformation; Axial modification; pinion shaft; curves of total bending, twist and compression; Profile modification

INTRCDUCTION

Profile modification is usually made in order to make a gearset run more quietly, reliably and have a longer endurance. The main problem on the profile modification is that we must obtain the elastic deformation value of a loaded pinion and such that we can reasonably determine the value of profile modification. Obviously, it is a very important subject for a heavy load gear.

Many scholars and specialists have made some studies for the above-mentioned problem and have gotten tremendous developments such as the famous American gear specialist Dr. Dudley who has researched the 2-D axial theoretical modification curves of a double-helical pinion shaft (symmetry assembling). Dr. Sigg has raised the 2-D axial theoretical modification curve of a spur pinion. Their work has established a theoretical basis for the profile modification of a pinion.

This article has mainly analyzed the case of a single-helical pinion shaft which is now commonly used. By F.E.M. and an analytical method, some examples of pinion shaft have been calculated and the total elastic deformation curves and a corresponding 3-D theoretical axial modification curve have been made. After the fixed quantity analysis the main factors which have an influence on the elastic deformation of a pinion shaft have been found out.

THE CALCULATING MODEL AND MAIN PARAMETERS

As the example of analytical calculation, two single-helical pinion shafts are selected and their main parmaeters and working loads are given:

Type	A-500	A-400
No. of pinion teeth	22	19
Module (mm)	3	2.75
Normal pressure angle	20°	20°
Helix angle	14°	14°
Modification coefficient	0.412	0.425
Rotation rate (rpm)	1500	1500
Torsion (N.mm)	$53.8\ 10^{3}$	$29.92\ 10^{3}$

In fig. 1. the outline and dimensions of

one pinion shaft (Type-500) have been given. Figure 1. shows the pinion is mounted unsymmetrically relative to the two bearing suppots (fig. 1. "▼" shows the point of bearing support).

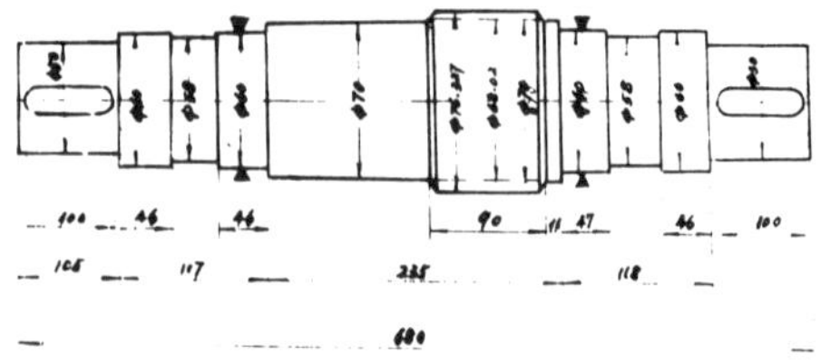

Fig. 1. Dimensions of the Pinion Shaft

For the convenience of analysis, fig. 1. can be simplified further and the calculating model can be formed (see fig. 2.).

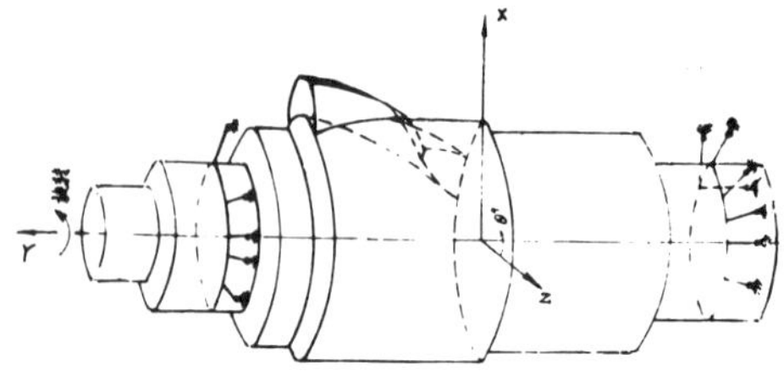

Fig. 2. Calculating Model

The torque is applied to the left end of pinion shaft. The load of tooth surface acts along the normal direction and it is uniformly distributed on the contact line. Four kinds of working load positions have been calculated. The pinion shaft is supported on ball bearings. As a ball bearing can only provide radial reaction force on the loaded half of the ring, so on the other half of the ring the gap is produced. Some specific elements have been used to describe the role of ball bearing suport. The position of these specific elements can be adjusted to a compressional state and then an approximate actual status of the pinion shaft sill be formed. A 3-D Isoparametric element with 8-21 nodes is used to carry out the F.E.M. analysis for increasing the accuracy of the calculation. The pinion shaft has been divided into 356 3-D block elements[1].

THE ELASTIC DEFORMATION OF PINION

According to the results calculated by F.E.M., the elastic deformation of the pinion shaft can be analyzed further.

Bending elastic deformation. Figure 3

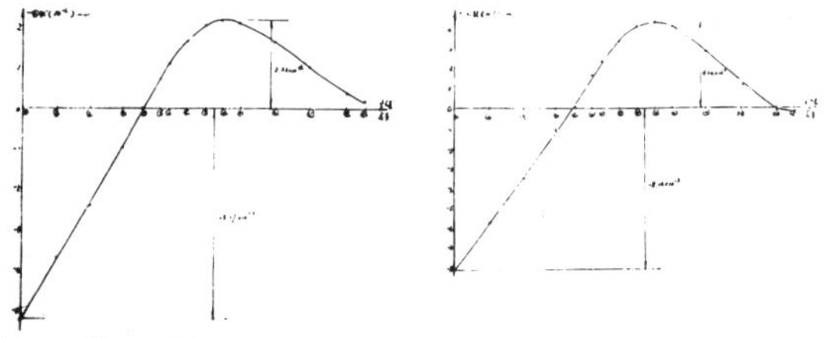

Fig. 3. The Bending Curve

shows the case of bending deflection of a pinion shaft in XOY and XOZ coordinate plans. Figure 3 shows the maximum deflection is 2.19 10^{-3}mm.

Compressive deformation. A compressive curve may be drawn according to the results by F.E.M. (see fig. 4.). Figure 4 shows that the compressive value of a pinion shaft is less than 4.14 10^{-4}mm. This value is much less than the bending deflection. So it may be neglected.

Fig. 4. Curves of Compnession Deformation

Twisting deformation. According to the results of the calculation by F.E.M., a twisting curve of the pinion shaft has been drawn (see fig. 5.). The line B_T'-B_T' is a twisting curve of the pinion shaft in fig. 5. The maximum twisting deformation is 5.32 10^{-3}-7.23 10^{-3}mm. The twisting deformation is 1.72 10^{-3}-2.15 10^{-3} at the maximum bending deflection (section 10) of the pinion shaft.

Curve of total bending, twist and compression. According to the calculated results, a total elastic deformation curve of the pinion shaft (Type A-500) can be drawn (see B'-B' in fig. 6.). Another total elastic deformation curve of the pinion shaft (Type-400) has been shown in fig. 7. In the light of Dr. Sigg's viewpoint the symmtrical line of the total elastic deformation curve is exactly the 3-D theorie-

[1] Slides will show the calculating mesh.

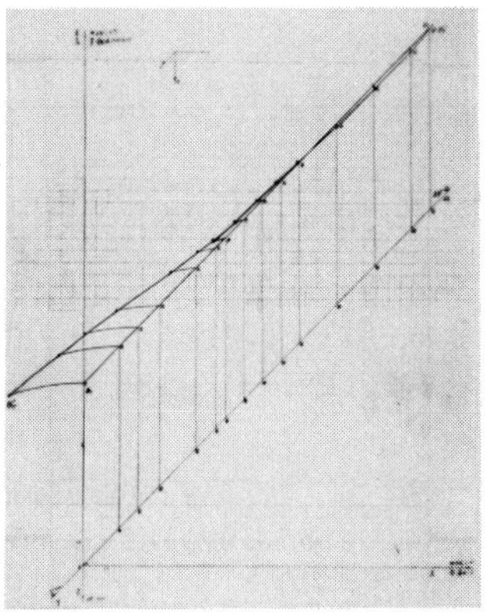

Fig. 5. Curve of Twist of the Pinion Shaft

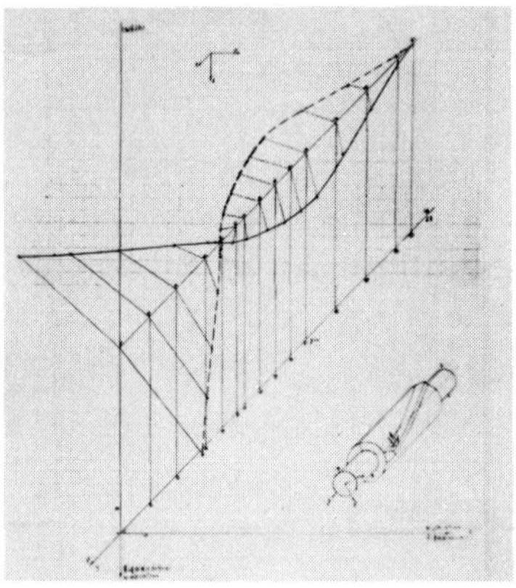

Fig. 6. Curve of Total Twist, Bending and Compression, and 3-D Axial Modification Curve

tical axial modification curve (dashed line in fig. 6. and fig. 7.). Some values of the maximum bending and twisting deformation of the pinion are listed in table 1.

TABLE 1 <u>The Maximum Bending Deflection of the Pinion Shaft and Twisting Deformation</u>

Op. No / Direction / Deformation / Type			A-500	
			Component of maximum deflection mm	Cross section No
N	X (u)		4.22×10^{-3}	10
	Z (w)		1.70×10^{-2}	10
	Y (v)		-4.14×10^{-2}	10
	Torsion	Δ u	-1.64×10^{-3}	10
		Δ w	$-0.6\ \times10^{-3}$	10
		Composition	1.75×10^{-3}	10
		$\Delta\ \beta$	$\pm3.387\times10^{-3}$	7-11

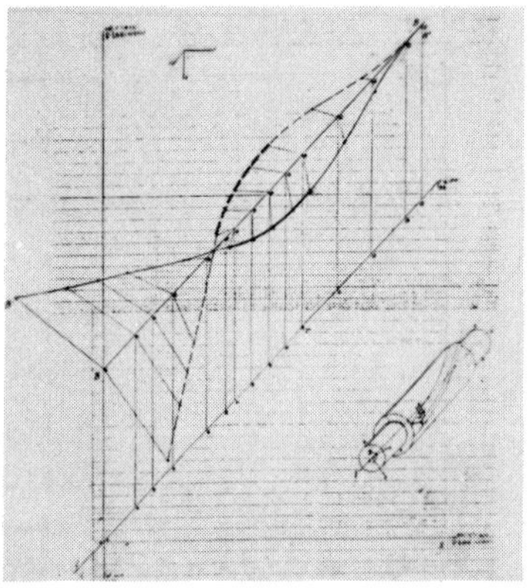

Fig. 7. Curve of Total Twist, Bending and Compression, and 3-D Axial Modification Curve

<u>Bending deformation of a tooth and the local elastic deformation in the contact zone.</u> In order to compare 5 kinds of elastic deformation, the F.E.M. has been used to analyze the bending of a tooth.

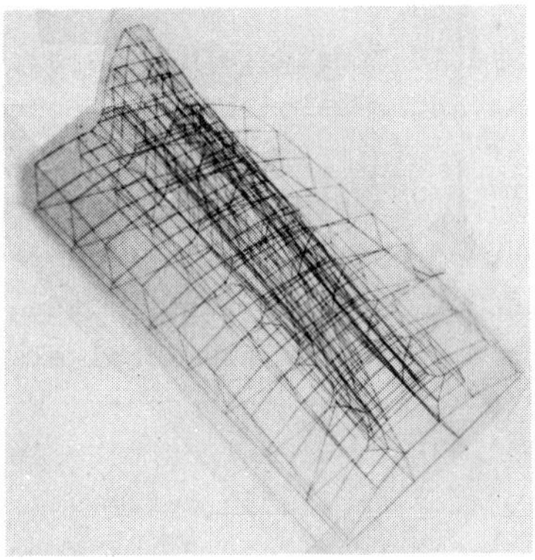

Fig. 8. Calculation Mesh of a Tooth

The results calculated are listed in table 2.

TABLE 2 <u>Five Kinds of Elastic Deformation Values</u>

No	Item	A-500	A-400
1	The maximum deflection of pinion shaft	2.19×10^{-2} mm	2.22×10^{-2} mm
2	The maximum twisting deformation	7.23×10^{-3} mm	$3.91\ 10^{-3}$ mm
3	The conrpressive deformation	4.14×10^{-4} mm	$5.55\ 10^{-4}$ mm
4	The bending deflection of a tooth	1.19×10^{-2} mm	$8.4\ 10^{-3}$ mm
5	The elasto-deformation of surface in contact area	$< 0.775\times10^{-3}$ mm	$< 0.781\times10^{-3}$ mm

The local elastic deformation can be calculated with the formula of the elastic theory

$$S=\Delta S_1+\Delta S_2$$

$$S_1=\rho_1\left[1-\sqrt{1-\left(\frac{X}{\rho_1}\right)^2}\right]$$

$$S_2=\rho_2\left[1-\sqrt{1-\left(\frac{X}{\rho_2}\right)^2}\right]$$

The results calculated are listed in table 2.

Comparsion of 5 elastic deformations. Bending, compressive, and twisting deformation of a loaded pinion shaft, the bending deflection of a tooth, and the local elastic deformation are all listed in table 2.

CONCLUSIONS

According to the analysis of the example, the elastic deformation of the loaded pinion shaft mainly involves: bending deflection of a pinion shaft; torsional deformation of pinion shaft; compressive deformation of a pinion shaft; bending deflection of a tooth, and local elastic deformation in the contact zone. With the F.E.M. and analytical method, the approximate value of the elastic defermation may be solved out. From Tab. 2. we know the bending and torsional deformation of the pinion shaft and the deflection of the tooth have an important role in the elastic deformation of the pinion. So, the bending and torsional deformation of the pinion shaft cannot be ignored while the axial modification is employed; the deflection of a tooth also can't be ignored while the tip or root relief is carried out.

If the helix angle modification is made for the above-mentioned pinion shaft, the value $\Delta\beta$ can be looked up in the table 1. If you want to get a crowned tooth, the crowning value can be evaluated according to the bending deflection of pinion shaft. If total elastic deformation is considered you can take a reference of the fig. 6. or fig. 7., where the total elastic deformation curve and the theoretical profile modification curve have been offered.

REFERENCES

Dudley, D. W. (1954). Special design problems. Practical gear design. McGraw-Hill company, New York. Chap. 8, PP. 317-323

Sigg, H. (1965). Profile and longitudinal corrections on involute gears. AGMA, 109, 16

Wang, T. (1981). Analysis of gear-tooth stress distribution by F.E.M. Journal of Shanghai Jiao-Tong University, 31, PP. 17-38

Wang, T. (1982). Analysis and study of grinding steps of tooth root. Chinese Journal of Mechanical Engineering, Vol. 18, No 3. PP. 93-103

SUMMARY

КОНСПЕКТ

Для модификации профиля зуба необходимо изучать упругие деформации шестеренки. Для этого, занимались анализами и расчетами шестеренок с различными параметрами методом конечных элементов. На основе количественных анализов упругой деформации шестеренок сделали комплексные s^x-координатные кривые упругих деформаций этих шестеренок, и поставили соотвественные теоретические кривые модификаций профилей зубьев. В статье отдельно дали величины сопоставления пятых различных типов деформаций, являющихся деформациями изгиба, кручения и сжатия валов шестеренок, а также, изгибными деформациями зубьев и местными упругими деформациями в области контакта зацепления. Тем самым и указали главные факторы влияющие величины модификации и задали некоторые предложения по модификации профилей зубьев для шестеренок.

Tooth Longitudinal Correction Against Thermal Distortion for High Speed Gears

TAO YANGUANG, LI SHANGWEI, MA XIANBEN AND CHANG KEQIN

Department of Gear Driving, Zhengzhou Research Institute of Mechanical Engineering, Zhengzhou, PRC

Abstract. High speed gear arises nonuniform temperature field across its face width and along its radial direction during operation,thus resulting in nonuniform thermal distortion. A appropriate tooth longitudinal correction must be performed to ensure desired uniform load distribution The paper presents the testing stand for high speed gearing with full scale at full speed, the measuring of temperature field for such gears, the tooth longitudinal correction tests against thermal distortion and practical examples of gearbox utilized the technology of modificating thermal distortion. It also provides a calculation method for determining amount of tooth longitudinal correction against distortion.

Keywords. Test stand; temperature measuring; thermal distortion; tooth longitudinal correction; thermal distortion calculating formulae.

INTRODUCTION

More and more singular helical high speed gears carburized, which have a pitch line velocity not less than 100m/s and transmitted power of 3,000KW and over, are being used in petrochemical industry. Temperature across face width of such gears is in nonuniform during operation. It leads to tooth-contact move towards the leaving end and reduce the operating performance of gears seriously. Therefore, longitudinal flank correction against thermal distortion in high speed gears is a key technology in their manufacturing.

Investigation on high speed gears used in industrial field shows that temperature distribution in such gears mainly depends on their pitch line velocity. At special high speed (such as more than 100m/s) in gearing, tooth contact can also move towards the leaving end obviously under light load, even idling. Based on these priciples, a testing stand (no loading) for high speed gearing has set up. A test series with full scale was performed on a high speed increasing gearbox used in petrochemical industry. It was measured that temperature across tooth face width and at the different radii during test. The amount of longitudinal flank correction against thermal distortion can be established by using formulae based on elasticity theory and theory for heat transfer, depending on the temperature data obtained from the test. The tooth corrected gears have bean tested again. Experiments shows that tooth contact is uniform and that the longitudinal correction is satisfying.

THE TESTING STAND FOR HIGH SPEED GEARS

Accorded with the requirments for temperature measurement and longitudinal flank correction against thermal distortion in high speed gears, a testing stand has been set up. It is composed of AC commutator motor, planet speed-increasing unit, speed-torque sensor and the tested gearbox with full scale. The commutator motor is 150KW in power and can be changed steplessly in speed within 350-1050rpm. The ratio in the planet gearing is about 11. The speed and torque transmitted to the tested gearbox can be measured and shown out by means of the sensor and speed-torque apparatus. Turbine oil flow sensor fixed at lube oil inlet for the tested gearbox can automatically display the flow rate of lube oil with flow indicating-accumulation apparatus. Nylon rope couplings have been used to reduce vibration caused disalignment of multi-axes. This coupling absorbs shock and vibration, dos not require lubrication and can be used conveniently and reliably. The test stand has been equipped with lube oil supply of 250 liter per minute in flow rate and oil pressure of 0.4MPa. The lubricant is turbine oil with kinematical viscosity of 20cSt. The stand can be used to control and measure the following parameters:

Speed, torque and lube oil flow rate entered into the tested gearbox;
Lubricant temperature at inlet and outlet of the tested gearbox and pressure at inlet;
Temperature in the sliding bearings.

THE TESTED HIGH SPEED GEARBOX WITH FULL SCALE

Having considered the complicated factors forming temperature field in high speed gears, it is decided to choose a high speed increasing gearbox supplied with compressor, which reflects practical situation of industrial field as tested unit with full scale. The basic data for the tested unit is as below:

Center distance	250mm
Normal module	3.5mm
Tooth number	93/46
Helix angle	13°20'40"
Working face width	120mm
Pitch diameter	334.533/165.467 mm/mm
Modification coefficients	+0.26/ -0.26
Nameplate rating	4000KW,7132/14420 rpm

Gear material is 14CrNi4, and teeth of both pinion and gear are case carburized to HRc58-62 and ground. The grade of accuracy in gears is 4 according to JB179-83 and the tooth surface finish is 0.8µm. The pinion was mating-ground with gear.

MEASURING TEMPERATURE FIELD FOR HIGH SPEED GEARS

For measuring temperature field, 13 thermocouples have been inserted in pinion teeth and its body across face width at different radii, and 4 thermocouples in gear, as shown in Fig. 1 and Fig. 2 respectively. Thermocouples are with diameter of 0.3mm. They were installed at the measurement point, fed through the gears body and shaft center hole and connected to a coupling driving a Lebow slip-ring assembly, then linked to a dynamic temperature recorder (TR2723, made in Japan) via a special temperature compensating device.

The temperature diagram at pitch circle of pinion is shown in Fig. 3. It can be known from Fig. 3 that temperature distribution across tooth face width is obviously nonuniform when pitch line velocity exceeds 100m/s, that temperature at the leaving end of tooth is highest and that temperature at the leading end is lowest. It can be also seen that temperature difference across tooth face width is 27°c when pitch line velocity reaches 130m/s in the case. Fig. 4 shows radial temperature distribution at three different sections aross face width of pinion. Table 1 gives the temperature data measured in the gear. Compared with temperature data measured in the pinion, it can be discovered that temperature difference across the face width in gear is about as half as that in pinion at various speed.

Table 1.: Temperature data measured in the gear

PLV (m/s)	Temp. at measured point, °c			
	15	16	17	18
95	43.7	49.6	45.1	49.7
105	49.2	57.1	50.3	57.2
115	56.2	67.8	56.9	67.8
130	68.7	83.6	69.8	83.9

Temperature measuring under the following situations was also performed in order to study effect of different contact of tooth flanks on temperature across tooth face width:

To make contact of tooth flanks be over about half face width near the leaving end under static state;

To locate contact of tooth flanks over about half face width near the leading end under static state;

To have teeth corrected in the pinion.

Testing results prove that temperature distribution across tooth face width is not affected by contact position under static state and longitudical flank correction, and that temperature data are identical under all situation aboue. It is still further proved that temperature distribution across face width of high speed gears mainly depends on its pitch line velocity.

LONGITUDINAL CORRECTION TEST AGAINST THERMAL DISTORION

It is still difficult now to theoretically analysize and calculate temperature field of high speed gears, since it is complicated for factors affecting the temperature field in such gears. There fore, the following were assumed before establishing value of longitudinal correction against thermal distortion:

Gear running at high speed is considered as a uniform cylinder placed in a steady temperature field and has a well-distributed thermal source on its outside cylinder surface. Conductivity factor of gear material is considered as constant.

Gear can be divided into numerous thin disk on which axial temperature can be considered as no change.

Based on the assumptions, formulae calculating amount of thermal distortion in gear tooth can be obtained by using knowledge of elasticity mechanics and heat transfer theory.

The temperature distribution in gear which meets the assumptions from heat transfer theory is

$$t=t_c-(t_s-t_c)\frac{r^2}{r_a^2} \qquad (1)$$

where t—temperature at radius r in gear, °c; t_c —temperature at axis of gear, °c; t_s —temperature on outside cylinder surface of gear, c;

r—radius at any point in gear,mm;

r_a—radius of outside circle of gear,mm.

Radial heat expansion of disk which has axial symmetric temperature distribution is:

$$u=(1+v)\frac{a}{r}\int_0^r trdr+\frac{(1-v)a}{r_a^2}\int_0^{r_a} trdr \qquad (2)$$

where u—radial heat expansion at radius r in disk,mm;

v—possion's ratio for the disk material;

a—heat expansion coefficient of disk;

r, r_a—same as in Eq.(1).

Providing ambient temperature is 20 °c, place Eq.(1) into Eq.(2), then radial heat expansion u at any radius r in gear is obtained:

$$u=ar\left\{t_c-20+\frac{t_s-t_c}{4}\left[(1-v)+\frac{r^2}{r_a^2}(1+v)\right]\right\} \quad (3)$$

Noticing that Eq.(3) is obtained under some assumed situation, u must be modified, so radial heat expansion at any radius u becomes:

$$u'=k\cdot u \quad (4)$$

where u'—modified radial heat expansion, mm; k—a modification factor considering assumed conditions, test results and field experiences.
So the normal thermal distortion u_t at transverse tooth from is

$$u_t=u'\sin a_r \quad (5)$$

where a_r—transverse pressure angle at radius r.
Then, the values of longitudinal flank correction against thermal distortion can be determined by using Eq.(3) to (5) and temperature data obtained form tests. Longitudinal pinion correction against thermal distortion of the tested gearbox is shown as in Fig. 5. Tooth lead accuracy in the pinion was checked before and after longitudinal correction, and its curves are shown as in Fig. 6.
The running tests were performed continuously 24 hours for the tested gearbox before and after longitudinal correction respectively to investigate changes on contact position of tooth flanks. An especially resistant tooth bearing dye is painted on tooth surface before each test. Only the dye on areas contacted can be wore off. Before running test without longitudinal correction, contact in teeth under static state is 95% of face width and 65% of tooth depth. It is discovered that contact in teeth under dynamic state is 45% of face width near the leaving end and 50% of tooth depth after running test without longitudinal correction. Before running test with having the pinion longitudinal corrected, contact in corrected teeth under static state is 60% of face width near the leading end and 60% of tooth depth. It is shown that the contact in corrected teeth under dynamic state is 80% arround the middle of face width and 50% of tooth depth. It indicates that contact in longitudinal corrected teeth is uniform, and so is the load distribution

CONCLUSION

It is quite nonuniform for temperature across face width of high speed gears . The longitudinal correction must be reasonably established for such gears which pitch line velocity exceed 100m/s. This is a key technology ensuring gearbox well-operating.

Temperature of such singular helical gear gears changes little over about half the face width near the leading end, rises obviously in another half face width near the leaving end, and reachs its maximum at a point about 1/6 of face width near the leaving end at all speed.

The radial temperature in singular helical gear running at high speed reachs its maximum near the tooth root in the gears body.

Tests show that temperature field of high speed gears mainly depends upon their pitch line velocity and that contact position of tooth flanks and longitudinal correction or not have no effects on temperature field of high speed gears.

The amount of longitudinal correction based on the formulae in the paper is suitable for gears with diameter of not more than 400mm.

REFERENCES

Dudley, D.W. (1984). The kinds and causes of gear failures. In P.Allen-Browne and S.H.Gillams (Ed.), Handbook of practical gear design. Mc Graw-Hill Book Company, New York. pp.7.47-7.51.
Huang Yan. (1982). Themal stress. Engineering elasticity mechanics. Qinghua University Press, Beijing. pp.243-245.
Martinaglia, L. (1973). Thermal behavior of high speed gears and tooth correction for such gears. Mechm and Mach Theory, 8, 293-303.
Wang Buxuan. (1982). Engineering heat and mass transfer. Science Press, Beijing.

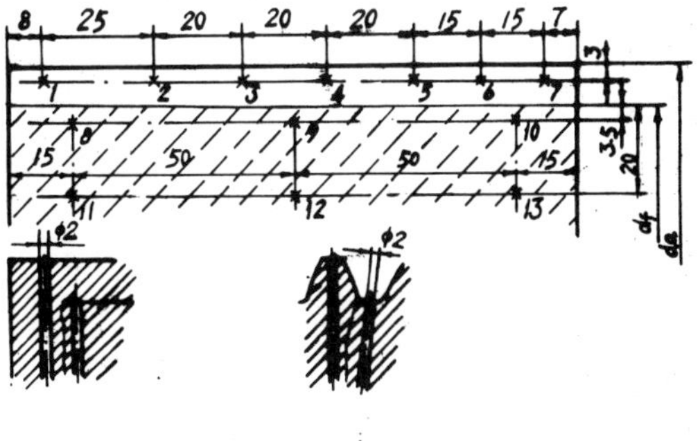

Fig. 1. Location of the thermocouples in the pinion tooth and body

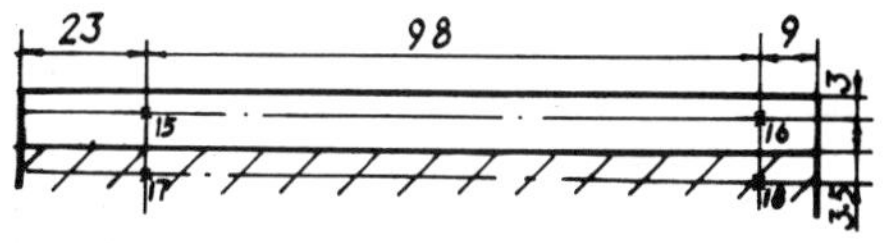

Fig. 2. Location of the thermocouples in the gear tooth and body

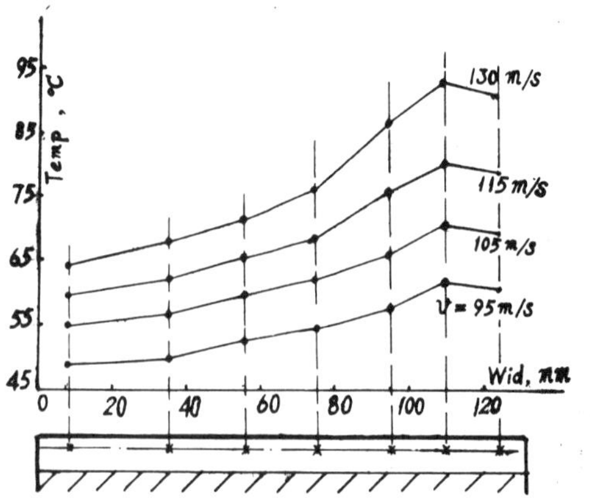

Fig. 3. Curves of temperature across tooth face width measured in the pinion of the tested gearbox at various speed

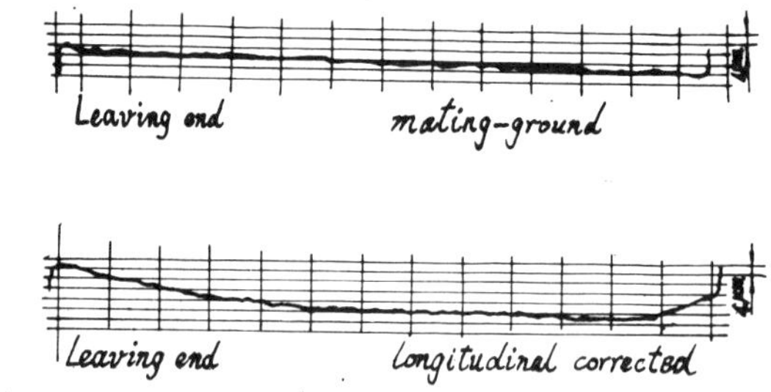

Fig. 6. Curve checked on tooth lead accuracy in the pinion of gearbox

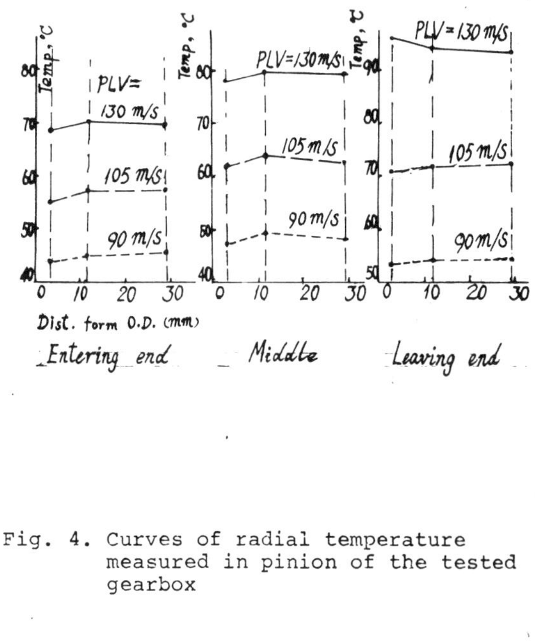

Fig. 4. Curves of radial temperature measured in pinion of the tested gearbox

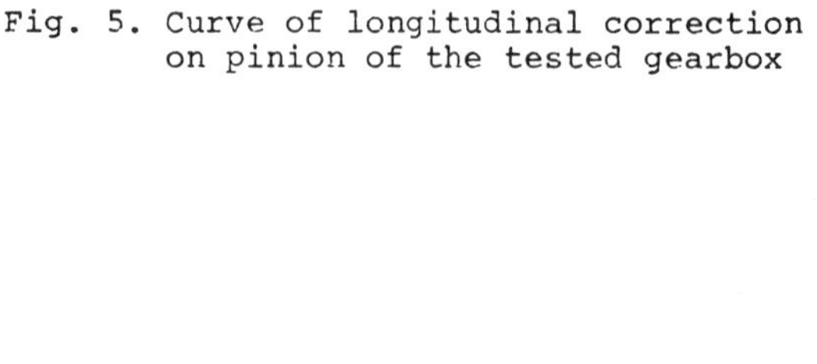

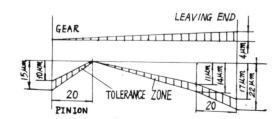

Fig. 5. Curve of longitudinal correction on pinion of the tested gearbox

SOMMAIRE

La table d'essai de l'engrenage de la vitesse élevée, la mesure de température de l'engrenage de la vitesse élevée, l'expérimentation de la modification de la déformation thermale et l'application sur des produits d'engrenages avec la modification de la déformation thermale sont exposées dans ce rapport.
La table d'essai qui a été construite pour la mesure de température de l'engrenage de la vitesse élevée et l'expérimentation de modification, sa vitesse arrive 15000r/min et la vitesse lineaire d'engrenage arrive 130m/s. A la table d'essai, on peut faire des divers mesures et contrôles, il comprend la quantité courante de l'huile de graissage, la température d'huile dans l'entré et la sortie, la pression d'huile dans l'entré, la torque, la vitesse, la température d'arbre, ect.
Dans la vitesse lineaire 95-130m/s, on a mesure la température d'une paire de roue de denture, petite et grande:
Les données de distributions de température de petite roue au long de la direction du longueur de la roue, les donnéesde distributions de température de la petite roue en position de deux différentes longitudes et les données de distribution de temperature de la grande roue de la direction du longueur de la roue. En faisant encore la mesure de température d'engrenage après la modification. Sur les données de distributions de la température mesurées avec les équations de la calculation concernées la thermoconductique et la mécanique d'élasticité, les équations de calculations de la déformation thermale d'engrenage ont été dérivées et la quantité de la modification de déformation thermale d'engrenage a été définie. On a fait d'une essai de la vitesse totale sur la table d'essai avec la roue modifiée, le resultat satisfait a été obtenu.

Engagement Analysis of Involute Harmonic Gearing

SHEN YUNWEN

Department of Mechanical Engineering, Northwestern Polytechnical University, Xi'an, Shaanxi, PRC

Abstract. Starting from the study on theory of engagement of harmonic gearing, this paper derives an equation of the theoretical profile and gives its numerical solution. The error of the technologically effective profile (involute profile) as compared with the theoretical profile is evatuated by means of numerical approximation. Thus it is stricatly proved appropriate to substitute involute profile for theoretical profile. This paper also discusses the effect of involute profile on transmission error of harmonic gearing. Finally, the author presents a new concept of backlash control and solves the problem of selecting geometrical parameters for a given minimum backlash with the aid of an optimization algorithm.

Keywords. Harmonic gearing; wave generator; flexible gear; rigid gear; theoretical profile; technologically effective profile; backlash control.

INTRODUCTION

The harmonic gearing is a new type of mechanical drive which is being developed both in China and abroad. At present, for extending the use of this type of drive in various sectors of Chinese economy, the study of the involute harmonic gearing (i.e. with involute tooth profile) holds immediate significance. Some theoretical analyses on involute harmonic gearing have been done in USSR (Shuvalov, 1974; Volkov, 1976) and other countries. But thus far, no theoretical proof has been given to justify the use of involute profile. A method for selecting the optimal engagement parameters with given backlash has not yet been obtained. The purpose of this paper is to fill in some gaps in the theory of harmonic gearing.

ON STUDY OF THE THEORETICAL PROFILE AND TECHNOLOGICALLY EFFECTIVE PROFILE

As is well known, in its general form the theoretical profile of harmonic gearing is a mathematical curve which cannot be described by elementary mathematical functions, and its manufacture is too difficult to be achieved. For ease in manufacture and measurement of teeth, we have to substitute an easily machined profile (such as involute profile, etc.) for theoretical profile as the technologically effective profile and theoretical profile. The theoretical profile refers to a conjugate profile which is obtained by the theory of engagement of harmonic gearing. There are two basic problems in the theory of engagement of harmonic gearing (Shen, 1979a). They are: 1. for the given form of original curve $\tilde{C}$ and profile $\tilde{R}$ of flexible gear or $\tilde{G}$ of rigid gear, to find the conjugate profile; 2. for the given profiles $\tilde{R}$ and $\tilde{G}$, to find the form of original curve $\tilde{C}$. This paper discusses only problem 1. If the profile of the flexible gear is taken by involute and the original curve $\tilde{C}$ is given, then the profile $\tilde{G}$ of rigid gear can be obtained.

Without sacrificing universality, we'll examine the case of fixed wave generator. As shown in Fig. 1, let the fixed coordinate system $\{XOY\}$, mobile coordinate systems $\{x_1 o_1 y_1\}$ and $\{x_2 o_2 y_2\}$ be connected with the wave generator, flexible gear and rigid gear respectively. If the flexible gear is rotated clockwise with a given angular velocity ω_1, it will cause the rigid gear to move in the same direction. In this instance, a tooth of flexible gear displaces with point c on the original curve $\tilde{C}$ along the surface of wave generator, and rotates an angle μ relative to the radius vector of point c. Consequently, the coordinates of origin o_1 (i.e. point c) of $\{x_1 o_1 y_1\}$ in polar-coordinate system (ρ, φ_1) may be represented as:

$$\begin{cases} \rho = r_m + w \\ \varphi_1 = \varphi + v/r_m \end{cases} \qquad (1)$$

where ρ — the polar radius of original curve in polar-coordinate system; r_m — the radius of center line of flexible ring in a undeformed state; w, v — radial and tangential displacement of points on center line of flexible ring; φ — angle of rotation of underformed end plane of flexible gear from axis Y, positive when measured clockwise, $\varphi = z_2 \varphi_2 / z_1$; φ_2 is the angle of rotation of rigid gear. z_1 and z_2 are the number of teeth of flexible and rigid gear respectively.

In coordinate system $\{x_1 o_1 y_1\}$ the equations of right-side involute profile of flexible gear are written as

$$\begin{cases} x_1 = r_1 [-\sin(u_1 - \theta_1) + u_1 \cos\alpha \cos(u_1 - \theta_1 + \alpha) \\ y_1 = r_1 [\cos(u_1 - \theta_1) + u_1 \cos\alpha \sin(u_1 - \theta_1 + \alpha) - r_m \end{cases} \qquad (2)$$

where r_1 — pitch circle radius of flexible gear; u_1 — parameter; α — the shaped angle of the initial outline of the rack; θ_1 — the half central angle corresponding to the pitch-circle tooth thinkness.

According to the theory of envelopes, the conjugate profile $\tilde{G}$ of rigid gear will be enveloped by the family of curves of the involute profile $\tilde{R}$ of flexible gear in relative motion. If $\varsigma = \Phi - (u_1 - \theta_1)$, $\lambda = \Phi - (u_1 - \theta_1 + \alpha)$, $\Phi = \gamma + \mu$, $\gamma = \varphi_1 - \varphi_2$ and $\mu = |\dot{w}| / r_m$, then the theoretical profile G may be expressed by a system of equations in coordinate system $\{x_2 o_2 y_2\}$, that is

$$\begin{cases} x_2 = r_1 (\sin\varsigma + u_1 \cos\alpha \cos\lambda) - r_m \sin\Phi + \rho \sin\gamma \\ y_2 = r_1 (\cos\varsigma - u_1 \cos\alpha \sin\lambda) - r_m \cos\Phi + \rho \cos\gamma \\ r_1 \dot{\Phi} (\cos\alpha \sin\alpha + u_1 \cos^2\alpha) - r_m \dot{\Phi} [\sin(\varsigma - \Phi) \\ \quad + \cos\alpha \sin(\Phi - \lambda) + u_1 \cos\alpha \cos(\Phi - \lambda)] \\ \quad - \dot{\rho} [\cos\alpha \cos(\lambda - \gamma) - \cos(\varsigma - \gamma) \\ \quad + u_1 \cos\alpha \sin(\lambda - \gamma)] - \rho \dot{\gamma} [\cos\alpha \sin(\lambda - \gamma) \\ \quad - \sin(\varsigma - \gamma) - u_1 \cos\alpha \cos(\lambda - \gamma)] = 0 \end{cases} \qquad (3)$$

where μ is measured relative to radius vector ρ , positive when clockwise, Φ is the angle between the axes of $\{x_1 o_1 y_1\}$ and $\{x_2 o_2 y_2\}$, and $\dot{\Phi}$, $\dot{\gamma}$, $\dot{\rho}$, $\dot{w}$ are the derivatives of Φ, γ, ρ, w with respected to .

For this manner of expressing $\tilde{G}$, only numerical solution can be obtained. As it is evident that the third equation of expression (3) is actually an implicit function in the form of $F(u_1, \varphi) = 0$. Taking a reasonable number of values of $u_1 \in [u_{g1}, u_{a1}]$ (where u_{a1}, u_{g1} are the parameters corresponding to the addendum and the starting of the involute profile of $\tilde{R}$), the value of φ may be found by means of the Muller's method. After the substitution of values of φ in other equations, we obtain the coordinates (x_2, y_2) of theoretical profile $\tilde{G}$ in $\{x_2 o_2 y_2\}$.

For example, for a harmonic gearing having $z_1 = 200$, $z_2 = 202$, $\alpha = 20^\circ$, m=0.5 mm and shift factor of flexible gear $x_1 = 3.0$, if $\tilde{C}$ is given by the deformed curve of center line of circular ring under the action of four forces with $\beta = 30^\circ$ (where β is the angle between the acting force and the major axis of $\tilde{C}$) (Shen, 1979b), $r_m = 50.375$ mm, the maximum radial displacement $w_o = 0.5$ mm and the wall thickness of flexible gear $\delta = 0.9$ mm, then, according to expression (3) the numerical solution of the theoretical profile $\tilde{G}$ can be found and shown in Fig.2.

In order to use the involute profile as the technologically effective profile, we must give the best fit to theoretical profile $\tilde{G}$ with an involute profile $\tilde{G}_w$ by means of a numerical approximation method. Thus, the appropriate engagement parameters are determined. To ensure the non-intersection of $\tilde{G}$ and $\tilde{G}_w$ in the approximating process, $\tilde{G}_w$ always lies at the right-side of $\tilde{G}$. Therefore, the mean error $\varepsilon = \sum_{k=1}^{n} d_k/n$ must be minimum, in which d_k is the distance of the kth corresponding points between $\tilde{G}$ and $\tilde{G}_w$. It is evident that d_k may be represented by $d_k = f(x_2, u_{2k})$, where x_2 and u_{2k} are the shift factor of $\tilde{G}_w$ and the parameter of the point k respectively. It now becomes a problem to find the conditional extreme value satisfying minε and $d_k \geqslant 0$. For this reason, we select the one-dimensional search method and carry out the one-side approximation from positive side. The results calculated by computer show that the best fit will be obtained if $x_2 = 2.66696$. In this case, $\varepsilon = 5.37187 \cdot 10^{-4}$ mm, and the maximum error $d_{kmax} = 7.70516 \cdot 10^{-4}$ mm, it is about 1/30 as much as the tolerance of tooth profile for 7th degree gear.

It can be seen that the involute profile may accurately approximate the theoretical profile if the shift factor of involute profile is properly selected. Thus, strict proof has been provided for the reasonability of substituting involute profile for theoretical profile.

THE EFFECT OF INVOLUTE PROFILE ON TRANSMISSION ACCURACY OF THE HARMONIC GEARING

In meching process there are two contact states of teeth:

In normal contact. When the flexible gear moves along the surface of the fixed wave generator, on the assumption that the section of deformed flexible gear are all still a plane, it may be considered that the instantaneous rotating centres of sections lie along the evolute $\tilde{S}$ of the original curve $\tilde{C}$ (Fig.3). Let the contact point be k, then according to Willis' theorem, the instantaneous pitch point P is the point of intersection of common normal passing through the point k and the extended line of $\overline{OO}_s$. Thus the instantaneous velocity ratio at this engaged position will be

$$i_M = \overline{PO}/\overline{PO}_s = \overline{PO}/(\overline{PO} - a_c) \tag{4}$$

where the value of $a_c = \overline{OO}_s$ may be found by differential geometry. And the value of $\overline{PO}$ may be found by the geometrical relationship of internal gearing with involute profile. Thus

$$\begin{cases} \overline{PO} = m z_2 \cos\alpha / 2\cos\alpha_k \\ \alpha_k = \alpha_k' + \Delta\alpha_k' \end{cases} \tag{5}$$

and

$$\begin{cases} \alpha_k' = \arccos(m\cos\alpha / a_c') \\ \Delta\alpha_k' = \arccos \dfrac{a_c^2 + a_c'^2 - (r_c - r_m)^2}{2 a_c a_c'} \\ a_c = \overline{OO}_{j1} = \sqrt{r_m^2 + \rho^2 - 2 r_m \rho \cos\mu} \end{cases} \tag{6}$$

in which if $a_c' > a_c$, the value of $\Delta\alpha_k'$ is negative; otherwise, the value of $\Delta\alpha_k'$ is positive; r_c is the radius of curvature of the point c on $\tilde{C}$.

By inserting Eq.(5) into Eq.(4), the instantaneous velocity ratio i_M at this engaged position can be determined, and then the transmission error may be evaluted by $\Delta i = i_M - z_2/z_1$.

In tip edge contact. In general, the instantaneous velocity ratio i_{Mb} in tip edge contact can only be estimated approximately, and it will be

$$i_{Mb} = \frac{r_{a\varphi}}{r_{a\varphi}(1 + \dot{v}/r_m) + (r_{a1} - r_m)\dot{\mu} + \dot{w}\,\mathrm{tg}\,\alpha_{P1}} \tag{7}$$

where α_{P1} is the profile angle of flexible gear; $\dot{w}, \dot{v}, \dot{\mu}$ are the derivatives of w,v,μ with respect to φ ; $r_{a\varphi}$ is the polar radius of the addendum of flexible gear; r_{a1} is the radius of addendum circle of flexible gear.

When the contact point k lies on the active part of profile, the contact of a pair of teeth belongs to the normal contact. Otherwise, the contact belongs to tip edge contact. This contact state is due to elastic deformation of flexible gear. The state of normal contact may be determined by the inequality

$$r_{a2} \leqslant \sqrt{X_k^2 + Y_k^2} \leqslant r_{g2} \tag{8}$$

where r_{a2} — the diameter of the addendum circle of rigid gear; r_{g2} — the radius of involute terminal point on profile of rigid gear; X_k , Y_k — the coordinates of the contact point of rigid gear in coordinate system $\{XOY\}$.

Calculations show that the normal meshing region, with $x_1 = 3.075$ and $x_2 = 2.746$ for the two-wave drive having the above parameters, is about 4.75°, and the errors of instantaneous velocity ratio are less than 0.5%. It is obvious that the effect of involute profile on transmission error of harmonic gearing is quite small. On the basis of backlash requirement, we can directly select the shift factor by the aid of electronic computer, providing theoretical basis for study of backlash control.

THE BACKLASH CONTROL OF INVOLUTE HARMONIC GEARING

To ensure normal operation of harmonic gearing, the condition of noninterference in engagement must be statisfied, that is, the circumferential backlash in arbitrary engaged position should be $j_t > 0$. Consequently, avoiding interference and controlling backlash are two sides of the same problem.

But the effect of backlash on operating performance of harmonic gearing is considerable. The backlash demands are different for harmonic gearing enployed in various kinds of service. Therefore the backlash must be appropriately controlled in design.

The backlash control is essentially that the selected parameters of engagement, wihle satisfying a series of constraint conditions, make the difference $|j - j_c|$ minimum. j_c is a given backlash based on operating requirements, and in general, $j_c = 0$. Therefore, this problem can be solved by constrained optimization. For a power transmission, it is necessary to check the backlash j_{cb} at the end point of the arc of action.

Let the objective function be

$$f(x) = \begin{cases} |j_t(x) - j_c| & \text{if } j_c \neq 0 \\ j_t(x) & \text{if } j_c = 0 \end{cases} \qquad (9)$$

where $j_t(x)$—the circumferential backlash for an abitrary position of engagement; $x = \{x_1, x_2, x_3, x_4\}$, and the components x_1, x_2, x_3, x_4 of x are respectively parameters x_1, x_2, φ, h_n (where h_n—working depth of teeth).

In this problem the inequality constraints $g_i(x) \geqslant 0$ ($i = 1,2,\ldots,k$) are as follows:

1. To prevent the overlapping interference of profile, in the first quadrat we have

$$g_1(x): \begin{cases} X_{M2}(x_2, x_3, x_4) - X_{M1}(x_1, x_3, x_4) \geqslant 0 \\ Y_{M1}(x_1, x_3, x_4) - Y_{M2}(x_2, x_3, x_4) \geqslant 0 \end{cases}$$

where X_{M1}, Y_{M1} and X_{M2}, Y_{M2} are the coordinates of corresponding points on $\tilde{R}$ and $\tilde{G}$ at arbitrary engaged position (Shen, 1979a).

2. According to the condition for avoiding transition curve interference, then

$$g_2(x) = [r_{g2}(x_2) - r_{g1}(x_1)] - (x_4 + w_o) > 0$$

3. The working depth of teeth should be less than the allowable limit working depth, that is

$$g_3(x) = m(0.5z_1 + x_1 + 1) - r_{g1}(x_1) - x_4 > 0$$

4. The working depth of teeth should be greater than module m, that is

$$g_4(x) = x_4 - 0.5[r_{g2}(x_2) - r_{g1}(x_1) + m - w_o] > 0$$

5. To ensure a certain radial clearance between addendum of flexible gear and dedendum of rigid gear, we give

$$g_5(x) = r_{f2}(x_2) - [r_{a1}(x_1, x_4) + w_o + 0.2m] \geqslant 0$$

6. The addendum width should be greater than 0.25m, that is

$$g_6(x): \begin{cases} s_{a1}(x_1, x_4) - 0.25m \geqslant 0 \\ s_{a2}(x_2, x_4) - 0.25m \geqslant 0 \end{cases}$$

7. To ensure full tooth disengagement in minor axis, the constraint function may be represented by

$$g_7(x) = [r_{a2}(x_2, x_4) + 1.09w_o] - r_{a1}(x_1, x_4) > 0$$

8. All $j_t(x)$ values must be within the arc of action, that is

$$g_8(x) = u_{M2}(x_1, x_3, x_4) - u_{a2}(x_2, x_4) \geqslant 0$$

In above inequality constraints, $r_{g1}(x_1)$ is the radius of the starting point of involute profile of flexible gear.

In this problem, we adopt the following ways:

1. To make the conditional extreme value change into common extreme value, let

$$j_t(x) = \begin{cases} j_t(x) & \text{if } x \in L \\ N & \text{if } x \bar{\in} L \end{cases}$$

where $N \gg j_t(x)$, L is the feasible region.

2. The initial point x^0 is chosen by means of uniformly distributed false random numbers, so that the global extreme value may be found. Of course, $x^0 \in L^0$ (L^0 is the set of all interior points in the feasible region L and not empty) and should be in convex region of f(x).

3. We use the Hooke-Jeeves' method so as to find the extreme value of f(x).

By using this method, we have designed a power harmonic gearing in servo system. Its parameters are: $z_1 = 200$, $z_2 = 202$, $m = 0.5$mm, $\alpha = 20^\circ$, $\beta = 30^\circ$, $w_o = 0.5$mm, $\delta = 0.9$mm. Manufacture of teeth of flexible gear is by hobbing, and the teeth of the rigid gear are machined by pinion cutter (its number of teeth $z_o = 60$, shift factor $x_o = 0$). This gearing demands that $j_c = 0$ and $j_{cb} = 0.030 \pm 0.005$mm. The results, computed by digital computer, are: $x_1 = 2.080087$, $x_2 = 2.52038$, $h_n = 0.827873$mm, $r_{a1} = 51.7884$mm, $r_{f1} = 50.7254$mm, $r_{a2} = 51.6087$mm, $r_{f2} = 52.4620$mm, $j_{t\,min} = 6.92597 \cdot 10^{-5}$mm, $j_{tb} = 0.032577$mm.

CONCLUSION

1. This paper not only provides a numerical method of analysis on the theory of engagement of the harmonic gearing, but also strictly proves appropriate to use $\alpha = 20^\circ$ involute profile in harmonic gearing. In this case the involute profile merely needs to be reasonably shifted. It is for promoting the extensive use of the involute harmonic gearing that its theoretical basis has been found in this paper. Moreover, the basic principle described in this paper is also applicable to harmonic gearings using other original curves.

2. In this paper the backlash control offers a practical method for selecting optimal engagement parameters for a given minimum backlash in design. Not only in controlling minimum backlash, but also in control of backlash at the border of acting arc may this method be used. The theoretical analysis and design calculations show that the engagement parameters selected by controlling backlash are satisfactory. For selection of parameters of spline output and design of dual drive, this method is similarly effective.

REFERENCES

Shen, Y.W. (1979a). A few problems in the study of engagement of involute harmonic gears. Gear Journal, 1, China, 17-27.

Shen, Y.W. (1979b). A numerical method on study of engagement of harmonic gearing and its geometrical calculation. Research Report (NPU), No.871, China.

Shuvalov, S.A. (1974). Calculation of harmonic drives with due regard to compliance of the components. Russian Engineering Journal, No.6, 46-51.

Volkov, D.P., and A.F. Krainev (1976). Harmonic Gear Drives, Tekhnika, Kiev.

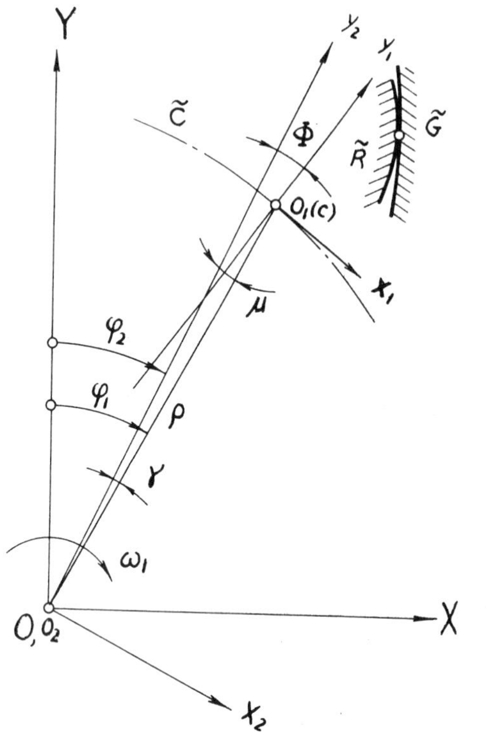

Fig. 1. Relationships of coordinates systems for finding theoretical profile.

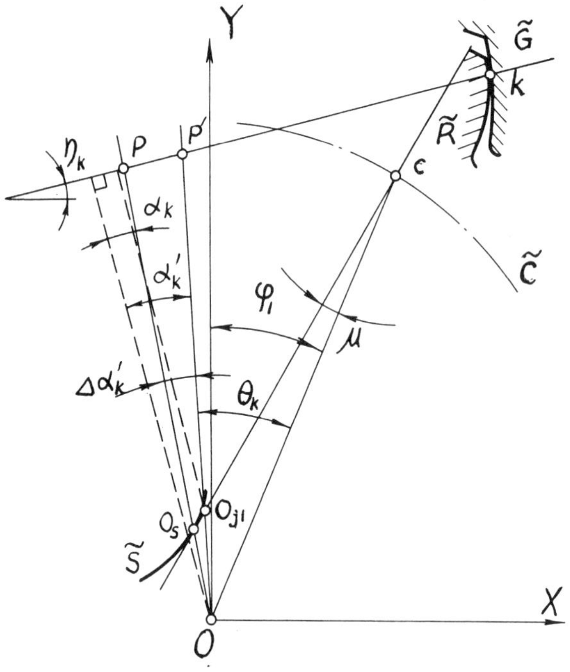

Fig. 3. The results of the numerical solution of theoretical profile $\tilde{G}$.

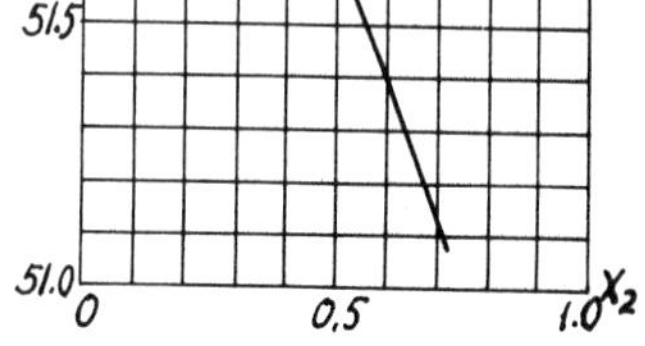

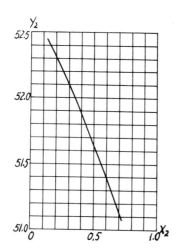

Fig. 2. A sketch for calculating instantaneous velocity ratio.

The Connections Between Gradient Angles at Mating Points of Conjugate Gear Tooth-Profiles and its Applications in Planar and Crossed Helical Gearing

CAO CUN-CHANG

Nanchang Gear Research Institute of Nanchang Gear Factory, Nanchang, PRC

Abstract. In finding a mating gear tooth-profile in plane gearing, perhaps the profile-normal-method is the simplest one to be used. There are three angles at mating point of given tooth profile 1, namely gradient angle γ_1, pressure angle ψ_1 and rotation angle φ_1. Regarding the connections between γ_1 and γ_2 and ψ_1 and ψ_2 are yet unpublished. This paper reveals their connections, also studies the pressure angle Ψ_2 when pitch radius of gear 2 takes R_2 instead of r_2 and the rotation angle Φ_3 of conjugate gear 3. In practice, the given tooth profile is often simple in form. For example, the profile of internal gear of pin gearing is only a circle, but its mating pinion tooth-profile is rather complex, and the profile of generating tool for pinion becomes the most complex one. Shannikov (1959) and M.B.S.P. (1979) have given sets of lengthy and complex formulas for this cutter. This paper shows that the above-mentioned rack cutter profile may be solved without differentiation and can be simplified to one-tenth of theirs and have been verified by actual cutting. When the mating gear becomes a rack, its gradient angle is $\gamma = \gamma_1 + \varphi_1$. If using this rack as a medium in solving crossed helical gear problem, we can obtain equations of the three-dimentional coordinates of the contact points in fixed space.

Keywords. Plane gear problem; profile-normal method; Gradient angle; Pressure angle; Angle connections; New pin-gearing; Rack cutter profile; Crossed helical gearing; Contact point coordinates.

THREE ANGLES AT POINT OF GIVEN PROFILE

In solving a gear mesh problem in plane gearing, perhaps the profile-normal method (Litvin, 1968) is the simplest one to be used. In Fig. 1, O_1 is the center of pitch circle of radius r_1, pitch point P lies on the center line y_1y_1' of two gears. S_1S_1' is the given tooth profile. $M_1(x_1,y_1)$ is any point at S_1S_1'. If M_1 becomes a contact point, its normal n_1n_1' must pass through pitch point P, that is the point P_1 must rotate through an arc $\widehat{PP_1}$ to arrive at point P. The angle P_1O_1P is called the rotation angle φ_1. The line T_1T_1' is tangent to the pitch circle at P_1 and angle $T_1P_1n_1$ is called the pressure angle ψ_1. The line t_1t_1' is tangent to profile at point M_1, making an angle t_1Nx_1 with x_1-axis

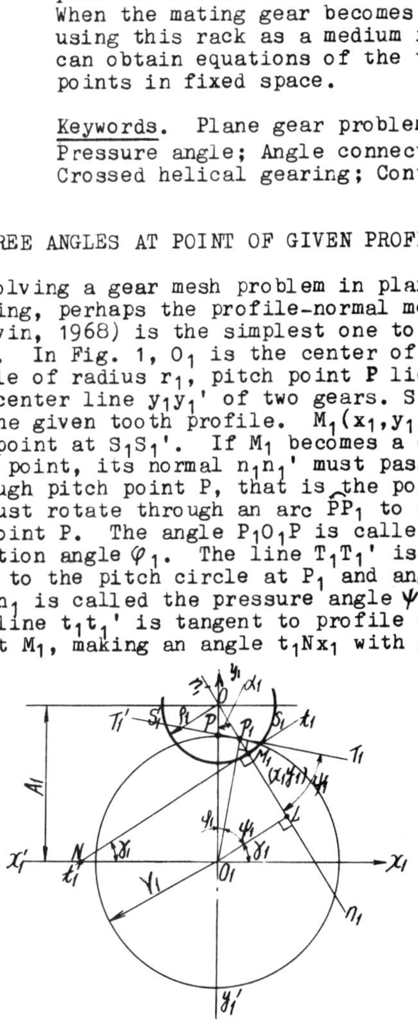

Fig. 1. The three-angles at given point M_1 and coordinates of profile S_1S_1'.

and is called the gradient angle γ_1. Angles φ_1, ψ_1 and γ_1 are three angles at point M_1 of profile S_1S_1'. According to the method of profile-normal

$$\tan \gamma_1 = \frac{dy_1}{dx_1} \tag{1}$$

$$\cos \psi_1 = \frac{x_1 \cos \gamma_1 + y_1 \sin \gamma_1}{r_1} \tag{2}$$

$$\varphi_1 = 90^\circ - (\gamma_1 + \psi_1) \tag{3}$$

CONNECTIONS BETWEEN EACH PAIR OF THREE ANGLES

On the mating gears, the connection between a pair of rotation angles is

$$\varphi_2 = (z_1/z_2)\,\varphi_1 \tag{4}$$

and is well-known, but what are the connections between γ_2 and γ_1, and ψ_2 and ψ_1 are yet unpublished. This paper reveals their connections:

$$\gamma_2 = \gamma_1 + \varphi_1 \pm \varphi_2 = \gamma_1 + (1 \pm \frac{z_1}{z_2})\,\varphi_1 \tag{5}$$

("+" for external, "-" for internal contact) and

$$\psi_2 = \psi_1 \tag{6}$$

and gives proofs as follows:

Proof of Connection Between a Pair of Gradient Angles

Transforming relative positions of mating points at tooth profiles may be represented by formulas of transformation of coordinates. In external gearing:

$$\left.\begin{aligned} x_2 &= x_1\cos(\varphi_1+\varphi_2) - y_1\sin(\varphi_1+\varphi_2) + A\sin\varphi_2 \\ y_2 &= x_1\sin(\varphi_1+\varphi_2) + y_1\cos(\varphi_1+\varphi_2) - A\cos\varphi_2 \end{aligned}\right\} \quad (7)$$

Differentiating (7) with respect to x_1, we have

$$\frac{dy_2}{dx_2} = \tan(\gamma_1+\varphi_1+\varphi_2) = \tan\gamma_2$$

$$\gamma_2 = \gamma_1+\varphi_1+\varphi_2 = \gamma_1 + (1+\frac{z_1}{z_2})\varphi_1 \quad (5)$$

This is the proof of formula (5) in the case of external contact. If internal contact is used, we change the sign of φ_2 with minus.

If rack-pinion contact is used, then $z_2 = \infty$, and $\varphi_2 = 0$, we have

$$\gamma_2 = \gamma_1 + \varphi_1 \quad (5)$$

Proof of Connection Between a Pair of Pressure Angles

In the formula (3), we have

$$\varphi_1 = 90^\circ - (\gamma_1 + \psi_1) \quad (3)$$

and

$$\varphi_2 = -[90^\circ - (\gamma_2 + \psi_2)] \quad (3')$$

We put minus sign before the right side of formula (3') because we take φ_2 in the direction opposite to φ_1 in the external contact. From formula (5), we have

$$\gamma_2 = -\psi_1 + \psi_2 + \gamma_2$$

Hence, we have proven formula (6)

$$\psi_1 = \psi_2$$

In like manner, we can prove

$$\psi_1 = \psi_2 \quad (6)$$

in the internal contact. In the rack and pinion contact, $\varphi_2 = 0$, from formula (3) we have $\psi_2 = 90^\circ - \gamma_2$. Furthermore,

$$\psi_1 = 90^\circ - \gamma_2 = \psi_2 \quad (6)$$

These are proofs of formula (6) in different cases.

Radius of Generating Circle R_2 Differs from Pitch Circle r_2

On the design of generating tool 3, in order to avoid undercut or interference of generated gear 2, it is often to take generating circle radius R_2 different from pitch circle radius r_2. Using Φ_2, Ψ_2 and Γ_2 for φ_2, ψ_2 and γ_2 in this case. As tooth profile (x_2, y_2) is not changed with a change from pitch radius r_2 to generating radius R_2, so we have

$$\Gamma_2 = \gamma_2 = \gamma_1 + \varphi_1 \pm \varphi_2 = \gamma_1 + (1 \pm \frac{z_1}{z_2})\varphi_1 \quad (8)$$

But $$\cos\Psi_2 = \frac{x_2\cos\gamma_2 + y_2\sin\gamma_2}{R_2} = \frac{r_2}{R_2}\cos\psi_1 \quad (9)$$

and $\gamma_2 = 90^\circ - (\psi_2 + \varphi_2)$

$$\Gamma_2 = 90^\circ - (\Psi_2 + \Phi_2)$$

For $\gamma_2 = \Gamma_2$

So $\psi_2 + \varphi_2 = \Psi_2 + \Phi_2$

Thus, we obtain

$$\Phi_2 = \varphi_2 + (\psi_2 - \Psi_2) = \varphi_2 + \psi_1 - \Psi_2 \quad (10)$$

Example of Application in Plane Gearing

In a new type pin-gearing, the tooth profile of internal pin gear tooth is a circular arc S_1S_1' with radius ρ_1 and center O (Fig. 1). O_1 is the center of pitch circle with pitch radius r_1, $OO_1 = A_1$, center O and O_1 all lie on the y_1-axis. $M_1(x_1, y_1)$ is a point of tooth profile, α_1 is the angle made by line OM_1 with y_1-axis. Take α_1 as a parameter, we have the equation of pin-tooth profile as follows:

$$\left.\begin{aligned} x_1 &= \rho_1\sin\alpha_1 = \rho_1\sin\gamma_1 \\ y_1 &= A_1 - \rho_1\cos\alpha_1 = A_1 - \rho_1\cos\gamma_1 \end{aligned}\right\} \quad (11)$$

The three angles at point $M_1(x_1, y_1)$ are

$$\gamma_1 = \alpha_1 \quad (12)$$

$$\cos\psi_1 = \frac{A_1}{r_1}\sin\gamma_1 = \frac{A_1}{r_1}\sin\alpha_1 \quad (13)$$

$$\varphi_1 = 90^\circ - [\alpha_1 + \cos^{-1}(\frac{A_1}{r_1}\sin\alpha_1)] \quad (14)$$

Find the tooth-profile of mating pinion. This is an example of internal contact, formulas of coordinates transformation are

$$\left.\begin{aligned} x_2 &= x_1\cos(\varphi_2-\varphi_1) + y_1\sin(\varphi_2-\varphi_1) - A\sin\varphi_2 \\ y_2 &= -x_1\sin(\varphi_2-\varphi_1) + y_1\cos(\varphi_2-\varphi_1) - A\cos\varphi_2 \end{aligned}\right\} \quad (15)$$

where $\varphi_2 = \frac{z_1}{z_2}\varphi_1$

A=center distance

$z_1, z_2 (= z_1 - 1)$=number of teeth of internal pin gear and pinion in planetary gear reducer, (or gear pump.)

Substituting x_1 and y_1 into equation (15) we have:

$$\left.\begin{aligned} x_2 &= \rho_1\sin\gamma_2 + A\sin(\varphi_2-\varphi_1) - A\sin\varphi_2 \\ y_2 &= -\rho_1\cos\gamma_2 + A\cos(\varphi_2-\varphi_1) - A\cos\varphi_2 \end{aligned}\right\} \quad (16)$$

$$\gamma_2 = \gamma_1 + \varphi_1 - \varphi_2 \quad (4)$$

From equation (16), we see that pinion tooth profile is a parallel of curtate epicycloid.

Find the tooth-profile of rack cutter for pinion with radius of generating circle R_2.

For rack-pinion contact, the coordinates transformation formulas are

$$\left.\begin{aligned} x &= x_2\cos\Phi_2 - y_2\sin\Phi_2 + R_2\Phi_2 \\ y &= x_2\sin\Phi_2 + y_2\cos\Phi_2 - R_2 \end{aligned}\right\} \quad (17)$$

Substituting x_2 and y_2 of equation (16) into equation (17), we have

$$\left.\begin{aligned} x &= \rho_1\sin\Gamma + A_1\sin(\varphi_2-\Phi_2-\varphi_1) - A\sin(\varphi_2-\Phi_2) + R_2\Phi_2 \\ y &= -\rho_1\cos\Gamma + A_1\cos(\varphi_2-\Phi_2-\varphi_1) - A\cos(\varphi_2-\Phi_2) - R_2 \end{aligned}\right\} \quad (18)$$

where $\Gamma = \gamma_2 + \Phi_2$ (5')

$$\gamma_2 = \gamma_1 + \varphi_1 - \varphi_2 \quad (5)$$

$$\Phi_2 = \varphi_2 + \psi_1 - \Psi_2 \quad (10)$$

$$\Psi_2 = \cos^{-1}\left(\frac{r_2}{R_2}\cos\psi_1\right) \quad (9)$$

$$\psi_1 = \cos^{-1}\left(\frac{r_2}{r_1}\sin\gamma_1\right) \quad (13)$$

From formul s (18) it can be seen that the profile of rack cutter is a parallel of curtate cycloid. The above equations (18) may be further simplified if we make $R_2 = \frac{r_2}{r_1}A_1$, then $\Phi_2 = \varphi_2 - \varphi_1$, $\Psi_2 = 90^\circ - \gamma_1$, $\Gamma = \gamma_1$, and the equations of rack profile become:

$$\left.\begin{aligned} x &= \rho_1\sin\gamma_1 - A\sin\varphi_1 + \frac{A_1}{z_1}\varphi_1 \\ y &= -\rho_1\cos\gamma_1 - A\cos\varphi_1 - \frac{A_1}{z_1}z_2 \end{aligned}\right\} \quad (19)$$

These are the simplest equations of rack tooth profile and can be generated by mechanism of linkage devized by Cao (1975).

On literatures Shannikov (1959) & M.B.S.P. (1979), sets of equations of rack cutter profile are deduced by way of differentiation and they are so lengthy and complex, that they are quite tiresome for practical use. We have simplified to about one-teeth of them by the help of connections between gradient angles and pressure angles with no differentiation.

Use of Medium Rack Method and Connection Between Gradient Angles in Solving Crossed Helical Gearing Problems

The rack can mesh with any tooth profile of spur or helical pinion. The rack meshes with pinion in line contact. In crossed helical gearing, two pinions contact the medium rack in two lines of contact intersecting at a common point. This point is called the point of contact. Let us determine this point in fixed space.

The rack and pinion have contact action in the direction perpendicular to axis of pinion, i.c. they contact on the transverse plane (Buckingham, 1936), so we can calculate the profile of rack-tooth on the transverse plane of helical pinion. If the given work is a spur gear, then the medium rack is also a spur rack, and the method of solution is just the same as in plane gearing. If the given work is a helical gear, then the medium rack is a helical rack, after finding out the transverse profile of rack-tooth, we must also determinate the three-dimentional coordinates $(-x_p, y_p, -z_p)$ of contact point in fixed space as follows:

$$-x_p = y_p \tan\gamma = y \tan\gamma \quad (20)$$

$$y_p = y \quad (21)$$

$$-z_p = -x_p \tan\beta_1 = y \tan\gamma \tan\beta_1 \quad (22)$$

where y = the ordinate of profile of the medium rack-tooth, as given in (19)
β_1 = the helix angle of the given work.
γ = the gradient angle of profile of the medium rack-tooth ($=\tan^{-1} dx/dy$)

Here γ is calculated by means of the connection between the gradient angles of $\gamma = \gamma_1 + \varphi_1$ instead of differentiation. As γ_1 and φ_1 are all known, so they can be calculated directly. Proofs of formulas (20) to (22) of coordinates of contact point in fixed space are given in Cao's paper (1984).

CONCLUSIONS

1. Connection between gradient angles at mating points of conjugate gears is established at first time in this paper.
2. This connection is so simple in form that the solution of plane gearing may be greatly simplified by use of them.
3. Formulas for determing modified values of Ψ_2 and φ_3 are also given when the pitch radius r_2 of gear 2 is changed to generating radius R_2 to avoid undercut or interference with gear 3, the latter being a generating cutter.
4. An example of rack cutter profile design for trochoidal pinion is included to show how this method being superior to those with differentiation.
5. Another example extends the application of connection of gradient angles into crossed helical gearing and gives simple formulas of coordinates of contact points in fixed space.
6. The use of gradient angle will greatly simplify equation of any plane parallel curve.

REFERENCES

Buckingham, E. (1936). Analytical Mechanics of Gears 1st. ed. McGraw-Hill. pp.143-144.

Cao Cun-chang, (1975). Principle of angle transformation at mating points of conjugate tooth profiles and its application in solving planar complex problems. (in Chinese). 2nd. Meeting of the Mechanical Transmission Institute of CMES.

Cao Cun-chang, (1984). Direct method to calculate the three dimensional coordinates of contact points on tooth profiles of crossed helical gears and its applications. (in Chinese). 4th Meeting of the Mechanical Transmission Institute of CMES.

Litvin, F.L. (1968). The Theory of Gearing 2nd ed. (in Russian). Nauka. pp.47-49.

M.B.S.P. (Machinery Bureau of Sichuan Province) (1979). Design Handbook of Complex Cutters. (in Chinese). pp.787-793.

Shannikov, V.M.(1959). In N.I.Kolchin(Ed) Collected Works of Gear and Worm Transmission. (in Russian). Mashgiz. pp.91-94.

DER ZUSAMMENHANG ZWISCHEN TANGENTENWINKELN AN KONTAKTPUNKTEN DER ZAHNPROFILE DER KONJUGIERTEN RÄDER UND SEINE ANWENDUNGEN AUF DIE STIRN- UND SCHRAUBENZAHNRADGETRIEBE

Kurzfassung. Für Suchung der Verzahnungsprofile bei ebenen Zahnradgetrieben ist die Profilnormal Methode wahrscheinlich die einfachste gewendete Methode. Es gibt drei Winkel am Eingriffspunkt der gegebenen Zahnprofile. nämlich der Tangentenwinkel γ_1, der Eingriffswinkel ψ_1 und der Drehwinkel φ_1. Bezüglich der Zusammenhänge zwischen drei Winkeln am Eingriffspunkt der Zahnprofile 1 und 2 ist die Formel $\varphi_2=(z_1/z_2)\varphi_1$ bekannt, aber die Zusammenhänge zwischen γ_1, γ_2 und ψ_1, ψ_2 sind noch nicht veröffentlicht. In dieser Arbeit sind die Zusammenhänge bzw. die Formeln des Tangentenwinkels $\gamma_2=\gamma_1+\varphi_1\pm\varphi_2$ und $\psi_2=\psi_1$ ("+"für außere Kontakte, "-"für innere Kontakte). Der Eingriffswinkel ψ_2 und der Drehwinkel ϕ_3 des konjugierte Rades 3, das das Verzahnungsprofil 3 zum Kontakt mit Profil 2 bringt, dabei statt des Teilkreis Radius r_2 des Rades 2 mit Wälzkreis Radius R_2, sind auch untersucht. Solche Änderung des Radius ist oft sinnvoll bei der Konstruktion des Schneidzeugs, um die Unterschneidung oder die Interferenz mit Rad 2 zu vermeiden. Unter normalen Verhältnissen sind die Zahnprofile einfach, um die Zähne leicht zu erzeugen. Zum Beispiel, bei Zapfengetrieben ist das Profil des Innenrades nur ein Kreis. Das Profil des konjunktive Rades ist aber ziemlich kompliziert, das eine Kurve parallele zum gestreckte epitrochoid ist und das Profil des Schneidzeugs noch komplizierter. In Referenze (Shannikov,1959:B.M.S.P.,1979) ist eien Reihe der komplizierten Formel für das Schneidzeug gegeben. Diese Arbeit zeigt, das obenerwähnte Stangenschneidzeugprofil Kann gelöst werden, ohne Differential zu verwenden. Im Vergleich zu (Shannikov,1959; B.M.S.P.,1979),die Formel für das Zahnstangenprofil kann zu Zehntel verkürzt werden und hat schon durch Schnitt-Test geprüft werden. Wenn das Verzahnungsrad geht zu eine Zahnstange,ist der Tangentenwinkel $\gamma=\gamma_1+\varphi_1$. Bei Lösungen der Probleme der Schraubenradgetriebe mit dieser Zahnstange als Medium, bekommen wir einige Gleichungen zur Beschreibung der 3-D Koordinatenwerte der Kontaktpunkte im bestimmten Raum. Die Kontaktpunkte können durch die Koordinatenwerte Y der Zahnprofile für die Mediumzahnstange, des Tangentenwinkels γ und des Spiralwinkels β_1 direkt dargestellt werden. Die Formeln zur Berechung der Tangentenwinkel und Engriffswinkel sind ausführlich hergeleitet.

The Analysis of Meshing in the Worm Gears With the Worm Made with the Rotary Head of Circular - Concave Profile

K. J. WOŹNIAK

Institute of Machine Design, Technical University of Łódź, Poland

Abstract. An equation of the tool contact line with the worm, and an equation of meshing of the worm and the worm - wheel have been determined. To illustrate the usefulness of these equations, an example of the analysis of gear meshing with the gear axis specing equal to 100 mm and gear ratio of 12,6 has been given. The contact line shape of the worm and the worm - wheel in the process of meshing has been given. And on this basis, it has been found out that the head diameter does not affect the conditions of the worm and the worm - wheel meshing. It has also been found that the meshing conditions of the gear with the worm made with the rotary head are the same as the meshing conditions of the gear with the worm of circular - convex profile. The above conclusions have been supported by experimental investigations.

Keywords. Worm gears; rotary head; contact line; analysis; axis meshing.

THE APATIAL CONTACT OF TOOL AND WORM

Modern trends in the cylindrical worm gear development tend towards the application of curvilinear profile worms which comprise, among other things, a circular-convex profile. Gears of such a profile have a number of advantages. Among them are good conditions of creating fluid friction in meshing and an increased resistance to pitting. These problems were the subject of the paper presented at the 5th World Congress of Mechanism and Machine Theory in Montreal.

The helical surface of such worms is made with the help of a disk - type grinding wheel, which is not efficient machining. The achievement of good effects by a number of western firms, e. g. Gleason, Klingelnberg, in the machining of herdened teeth of bevel gears allows one to suppose that the method will be applicable in the production of worms as after - machining. Therefore, an analysis of the worm gear with the worm made with the rotary head of circular - concave profile has been presented in the paper.

The shape of the cutting edges of the head tools is circular - concave about the radius ρ . The coordinates of the profile circle centre amount a and b, Fig. 1. The tool is set at the angle γ_p in realtion to the worm axis, of the helix line lift on the worm reference cylinder. The condition of the tool contact with the worm expressed by zeroing of the scalar product of the normal unit vector to the tool surface and the realitve velocity vector of the tool and the worm after transformations has the form:

$$tg\,\vartheta = \frac{(A_n - a)\cos\psi + A_n + p\,ctg\,\gamma_p}{b\cos\psi + (A_n ctg\,\gamma_p + p)\sin\psi}$$

where: ϑ – the tool profile parameter
A_n – the distance between the tool axis and the worm axis
ψ – the angele of rotation of tool
$p = H/2\pi$ – the helix line parameter
H – the worm lead

The contact line of the worm with the worm – wheel is the geometrical place of the contact points in which the relative velocity vectors and the vectors of the normal to the worm surface are reciprocally perpendicular. The equation of meshing has the following form after the transformations.

$$-i_{21}(Q\eta - R\alpha)\cos\Theta - i_{21}(P\eta + R\delta)\sin\Theta + P\alpha + Q\delta + ARi_{21} = 0$$

where:

$$P = \sin\vartheta\sin\psi\cos\gamma_p + \cos\vartheta\sin\gamma_p$$

$$Q = -\sin\vartheta\cos\psi$$

$$R = \sin\vartheta\sin\psi\sin\gamma_p - \cos\vartheta\cos\gamma_p$$

$$\alpha = (\rho\sin\vartheta - A_n + a)\cos\psi - A_n$$

$$\eta = -(\rho\sin\vartheta - A_n + a)\sin\psi\sin\gamma_p + (\rho\cos\vartheta - b)\cos\gamma_p - p\beta$$

$$\delta = (\rho\sin\vartheta - A_n + a)\sin\psi\,\gamma_p + (\rho\cos\vartheta - b)\sin\gamma_p$$

$$\Theta = \beta - \varphi_1$$

β – the angle of rotation of the worm in the machining process.
φ_1 – the angle of rotation of the worm in the meshing process.
A – the worm gear axis spacing.
i_{21} – gear ration expressed by the relation of the angular velocity of the worm – wheel to the angular velocity of the worm.

The above dependences will be used to carry out an analysis of meshing of the worm gear of the axis spacing A = 100 mm and the gear ratio $i = 1/i_{21} = 12{,}6$. The calculations of the axial profile of the worm made with the head of the diamater of 70 and 200 mm with the circular – concave profile of the radius $\rho = 25$ mm were not essentionally different form the circular – convex profile of the radius $\rho = 25$ mm.

Figure 2 shows the contact line shape in the process of meshing of the worm – wheel and the worm made with rotary heads of the diameters given above. No effect of the head diameter upon the position of the contact line in the gear is observed.

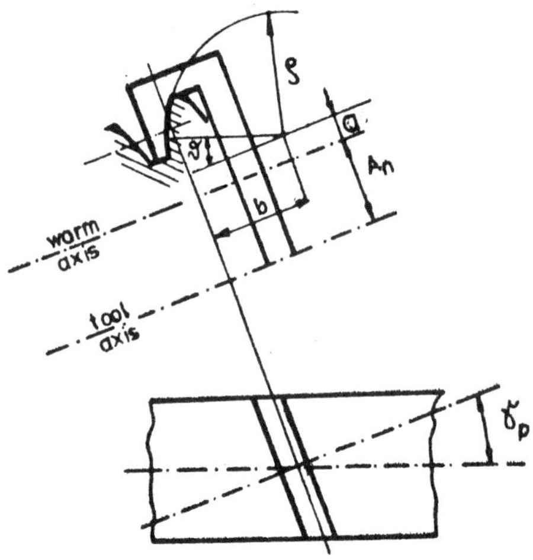

Fig. 1. Setting of the rotary head.

Moreover, the contact line shape of the gear with the worm made with the rotary head of circular – concave profile is the same as the contact line shape of the gear with the worm having a circular – convex profile in the axial section. As experimental investigations show, the efficiency of the gear with the worm made with the head amounts to 90% and is of the same order as the afficiency of the gear with the worm of circular – convex profile in the axial section.

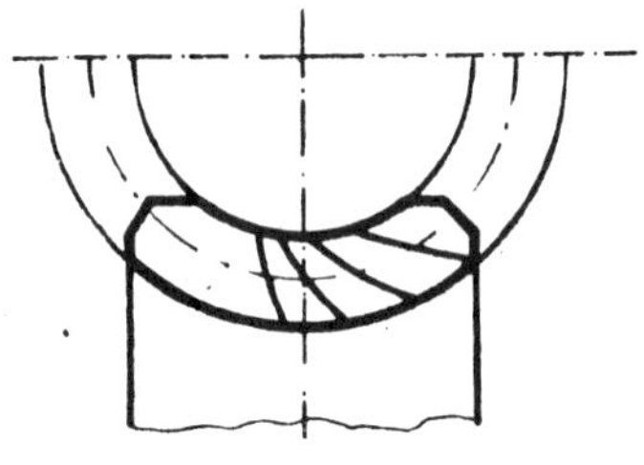

Fig. 2. Contact line shape of the worm – wheel and worm

THE PLANE CONTACT OF TOOL AND WORM

The tool contact line with the worm was a spatial line, which made the helix surface of the worm not to have, in any section a planar line which would make checking of the worm profile easier. Now, such setting of the tool in realtion to the worm mechined will be shown, in which their contact will be presented in the form of the planar line corresponding to the circular profile of the head tool edges. This requires setting of the head axis, in realtion to the worm axis, at the angle γ_n different from the angle γ_p of the worm helix line lift on the reference cylinder. It is known from the theory of meshing that, in the process of worm machining with the rotary tool, there are two, the so called, meshing axes which are intersected by the normal to the tool surface at the contact point of the tool with the work - piece. For the rotary head, one of these axes is its own axis, and the position of the other depends on the angle between the axes of the head and the worm. The position coordines of the other axis will be determined below. The relative movement of the tool and the worm can be presented in two ways:

1/ as the helical movement characterized by relative velocity vectors ω and $p\omega$
2/ as the sum of the component movements around the meshing axis I-I and II-II with the velocities ω^{I} and ω^{II}, Fig. 3.

Thus, we have

$$\omega = \omega^{I} + \omega^{II}$$

$$m_o(\omega^{I}) = m_o(\omega^{II}) = p\,\omega$$

where:
$$m_o(\omega^{I}) = r^{I} \times \omega^{I}$$
$$m_o(\omega^{II}) = r^{II} \times \omega^{II}$$

The vector r^{I} has the coordinates $-A_n,0,0$ and the vector r^{II} has the coordinates $X^{II},0,0$. From the solution of the equation we shall find the position of the other meshing axis, determined by the coordinate X^{II} and the angle β_n that it forms with the worm axis.

$$x^{II} = p/\mathrm{tg}\,\gamma_n \quad ; \quad \mathrm{tg}\,\beta_n = p/A_n$$

where: γ_n - the angle between the tool the worm axes.

If the angle $\gamma_n = \gamma_p$, then the coordinate $X^{II} = r_r$, i, e, the other meshing axis, lies at the altitude of the reference radius of the worm, Fig. 3. If the other meshing axis is set to intersect the tool profile circle centre, that is the coordinate $X^{II} = a$, then all the normals to the tool surface at the points of contact with the worm will intersect both the meshing axesand then the tool profile /circle/ will be the contact line of the tool and the worm. The position of the other meshing axis and the position of the tool will be determined in the following way:

$$X^{II} = a; \; b = 0; \; \mathrm{tg}\gamma_n = p/a.$$

The helical surface of the worm will have a circular - convex profile /easy to be tested/ in the section established at the angle γ_n, and the tool will have a circular - concave profile inits axial plane. For the helix surface made in this way, the equations that describe it are obridged.

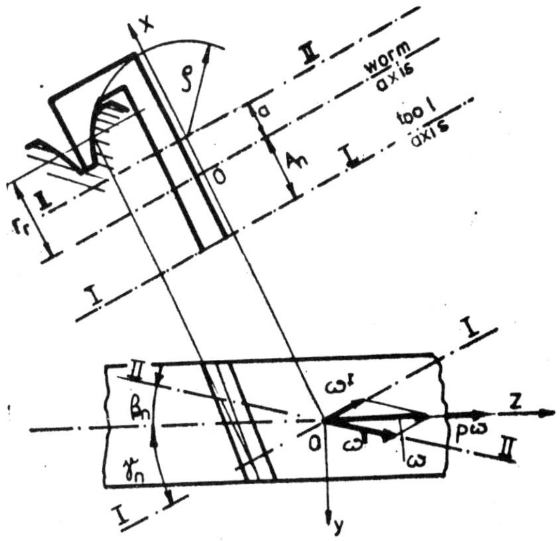

Fig. 3. Meshing axes in the process of the worm making with the rotary head.

The condition of the tool and the workpiece for $b = 0$ and $\mathrm{tg}\,\gamma_n = p/a$ is brought to $\psi = 0$, that is the contact line of the head and the worm does not depend on the angle of rotation of the tool. The condition of the worm and worm - wheel teeth contact in the meshing process is reduced to the form:

$$\mathrm{tg}\,\vartheta = \sin\gamma_p\,\mathrm{tg}\,\Theta + \frac{\cos\gamma_p}{p\beta}\left(\frac{p}{i_{21}\cos\Theta} - \frac{A}{\cos\Theta} - a\right)$$

The shape of the worm tooth in the axial

section, made with the rotary head set at the angle γ_n=arc tg p/a is slightly different from the worm tooth made with the rotary head at the angle γ_p of the helix lift on the reference cylinder. For the gear of parameters given earlier /the axis spacing A = 100 and the gear ratio i = 12,6/ the profile angle changes from 20° to 26°. This is the reason the contact line shape of the worm and the worm - wheel slightly changes. Figure 4 shows the contact line shape of the gear with the worm of the profile angle of 25°. As the results of the experimantal investigations show, the efficiency of such a gear amounts to 90% and is of the same order as the efficiency of the gear made with the rotary head set at the angle γ_p of the helix lift on the reference cylinder, and of the same order as the efficiency of the gear with the worm of circular - convex profile in the axial section.

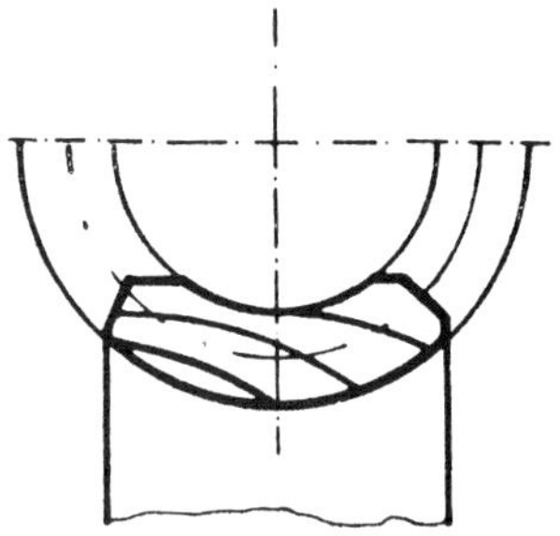

Fig. 4. Contact lines of the worm gear with the worm of circular - convex profile with the profile angle of 25°

CONCLUSION

To sum up, it can be stated that shaping of the helical surface of the circular - convex worm with the rotary head creates good conditions of the worm and worm - wheel contact in the meshing process. The efficiency of worm gears with the worm made with the rotary head whose axis is set, in relation to the worm axis, at the angle of helix lift on the reference cylinder, or the head axis is set in such a way that the contact of the tool and the workpiece is circular, is of the same order as the efficiency of the worm gear with the worm having a circular - convex profile in the axial section.

REFERENCES

Woźniak, K. J. /1979/. Worm gears of increased setzure and pitting resistance. Proceedings of the Fifth World Congress on Theory of Machines and Mechanisms. Montreal.

The Use of Gears With Small Number of Teeth in Driving Industrial Robots

V. MERTICARU, P. TIVLEA AND C. OPRIŞAN

Mechanical Engineering Faculty, Polytechnic Institute, Jassy, Romania 6600

Abstract. The paper presents the use of gears with small number of teeth in driving systems of industrial robots. Some references one made concerning the electromechanical drivers of industrial robots with direct current motors and step by step electric motors. The paper presents the advantages of the working made up of direct current motors or step by step electric motors, and reducing gear with gears with small number of teeth. Some accomplished pinions with small number of teeth and servo - motor-reducing gear are presented in this paper and in our photographs. We make some references concerning the precision for these working units and concerning the necessity to use systems for the compensation of the gear backlash. The compensation methods for gear backlash are presented with constructive solutions. The computation relations describing the design and working of these systems are logically made up and synthetically presented.

Keywords. a.c.motors; torque amplifiers; d.c.motors; error compensation; electric motors; electromechanical drivers; manipulation; power transmission; robots; servomechanisms; servomotors; torque control.

INTRODUCTION

Both the direct current and the step by step motors are more and more used in the driving of the industrial robots. These motors working very well at high speeds have electrical and mechanical advantages namely a precise setup in a large rotation speeds rate, a reduced rotor inertia, small overall dimensions.

But these motors transmit reduced torque. Good results are achieved when they are associated with torque multipliers (speed reducing gears), some driving units for industrial robots being thus obtained. These units must have high ratios within small overall dimensions, high output, durability and kinematic precision, reduced cost.

If for these units speed-reducing gears with small number of teeth are used, the above mentioned advantages are fulfilled.

CYLINDRICAL GEARS WITH VERY SMALL NUMBER OF TEETH

The usual cylindrical gears have small ratios ($i \leq 6,3$). The decrease of the number of teeth at the driving gear up to $z=1$, allows the accomplishment of high ratios within small dimensions. These gears have some advantages, namely, correct gearing at deviations from the distance between the axes, simple processing, high output, small cost. These gears are used for small and medium power transmissions.

Specific Problems for Design, Generation and Gearing

The decrease of the number of teeth has the following negative effects: bottom clearance, sharpening of the tooth head, inadequate contact degree.

To avoid the bottom clearance, the minimum coefficient of the profile displacement is computated:

$$x_{nmin}=h^{*}_{an} - z \sin^2\alpha_t/(2 \cos\beta), \qquad (1)$$

To avoid the sharpening of the tooth, the tooth head width is calculated:

$$s_{an}=d_a((\pi+4x_n tg\alpha_n)/2z-inv\alpha_{ta}+inv\alpha_t)\cos\beta_a$$

$$s_{an} \geq 0,5m_n \text{ for hardened teeth,}$$

$$s_{an} \geq 0,3m_n \text{ for improved teeth} \qquad (2)$$

The gearing continuity is ensured when the contact degree is overunitary. The required minimum value of the total contact degree is ensured by the overlap ratio.

Accomplishments

Gear cutting is done by the rolling method with standard cutting tools. Gears with small number of teeth are presented in Fig.8, Fig.9 and Fig.10. Some speed-reducing gears and servomotor-reducing gears are presented in Fig.11, Fig.12, Fig.13 and Fig. 14.

DRIVING UNITS FOR INDUSTRIAL ROBOTS

Some driving units for industrial robots have been obtained with the mentioned electric motors and two-stage speed re - ducing gears with $z_1=2$ and $z_3=3$. In Figure 1 a driving unit made up of a speed reducing gear with a transmission ratio i=5o4 and a direct current motor is presented. Figure 2 presents a driving unit made up of a speed reducing gear with a transmission ratio i=621, and a direct current motor with disk rotor and axial air gap. A two-output driving unit made up of two speed reducing gears with a transmission ratio i=648 and a step by step motor with a step angle $\theta_p=7^o$ is shown in Fig.3. For the last unit it has been imposed that the rotation angle in the speed reducing gear θ_r, due to the backlash between the teeth flanks, at the changing of the rotation direction, should not exceed the step angle ($\theta_r \leq \theta_p$).

METHODS AND SYSTEMS FOR THE COMPENSATION OF THE BACKLASH BETWEEN FLANKS

The compensation of the backlash between flanks has been achieved by using some special constructive systems. The backlash between the teeth flanks can be compensated by the variation of the distance between axes, by using systems with eccentric bushing as in Fig.4 and in Fig.12, or the systems with case in translation as in Fig.5.

The presented chart in Fig.4, allows the calculus of the rotation angle α_{r1} of the eccentric bushing,

$$\alpha_{r1}=\mathrm{arc}\ \cos(1/(2(a_w-k)k)((a_w(u-1)+k).(a_w(u+1)-k)-k^2)$$

where $u=(\cos\alpha_{tw})/\cos(\alpha_{tw}-(2(j_n-j)/d_b)$, (3).

The chart from Fig.5 allows the calculus of the axial displacement Δa_w of the case in translation,

$$\Delta a_w = a_w(1-u) \qquad (4).$$

The stratification gear in the frontal plane as in Fig.6, Fig.7 and Fig.11 also allows the compensation of the backlash between the flanks. The relative rotation angle α_{r2} of two neighbour layers is calculated :

$$\alpha_{r2} = 2(j_n - j)/d_b \qquad (5).$$

The relative axial displacement of two neighbour layers is calculated :

$$\Delta d = (j_n - j)/\sin\beta \qquad (6).$$

CONCLUSIONS

The gears with small number of teeth give better results in the manufacture of the torque multipliers (speed reducing gears). A gear transmission with small number of teeth and a direct current motor or a step by step motor forms an efficient unity for driving industrial robots. The mechanical systems for the compensation of the backlash between the flanks represent solutions with high fiability in producing the driving units for the industrial robots.

The mechanisms laboratory of the Polytechnic Institute of Jassy has accomplished and tested such driving units with very good results .

NOTATIONS

α_n, α_t-normal and transverse pressure angle of rack
α_{ta}-transverse pressure angle corresponding to diameter d_a
α_{tw}-transverse pressure angle corresponding to diameter d_w
a_w-center distance
β-reference pitch helix angle of gear
Cw-rolling circle
Cb-base circle
d_a-tip diameter
d_b-base diameter
d_w-rolling diameter
h^*_{an}-reference coefficient of the addendum
j_n-normal backlash
j- accepted backlash
k-eccentricity
m_n-module in the normal plane
s_{an}-tooth top sharpening

REFERENCES

Merticaru, V., and V. Atanasiu (1979). Optimum design of mechanisms with spur and helical gears and involute profile. Proc. of the 5-th World Congr. on the TMM, vol II, ASME, New York, pp.114o-1143.
Merticaru, V., and P. Tivlea (1985). Torque amplifiers for industrial robots. Mechanisms with gears with small number of teeth. Bul. Inst. Polith., Iaşi, Constr. de maşini, pp.7-13.
Waldron, K.J., and A. Kumar (1979). Develop - ment of theory of errors for manipulators. Proc. of the 5-th World Congr. on the TMM, vol.I, ASME, New York, pp.821-826.

Résumé. Le texte présente l'emploi des engrenages à pignons avec un nombre réduit de dents dans les systemes d'actionnement pour les robots industriels.

On fait des références sur l'actionnement électromécanique des robots industriels avec moteurs à courant continu et moteurs pas à pas. On présente les avantages qu'ont les untés d'actionnement pou robots formées de moteurs à courant continu ou de moteurs pas à pas et de réducteurs à pignons avec un nombre réduit de dents. On présente de manière succinte les problèmes importants à concernant l'étude, l'usinage et l'engrenage de ces pignons .

Quelque pignons avec un petit nombre de dents et servomotoréducteurs exécutés sont présentés aussi dans les photos. On fait des références sur la précision de ces unités d'actionnement et sur la nécessité de l'emploi de systèmes de compensation du jeu entre les dents d'un engrenage. Les méthodes de compensation de jeu entre les dents d'un engrenage sont suivies de la présentation de solution de construction. Les relations de calcul qui présentent la construction et le fonctionnement de ces systèmes sont logiquement déduits et présentés synthétiquement .

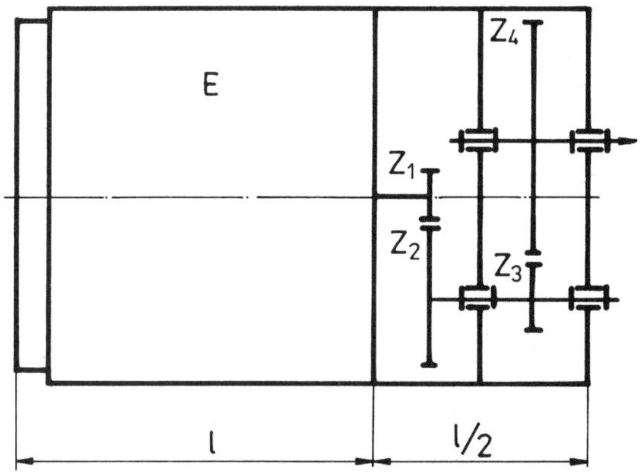

Fig. 1 .Servomotor-reducing gear, i=504

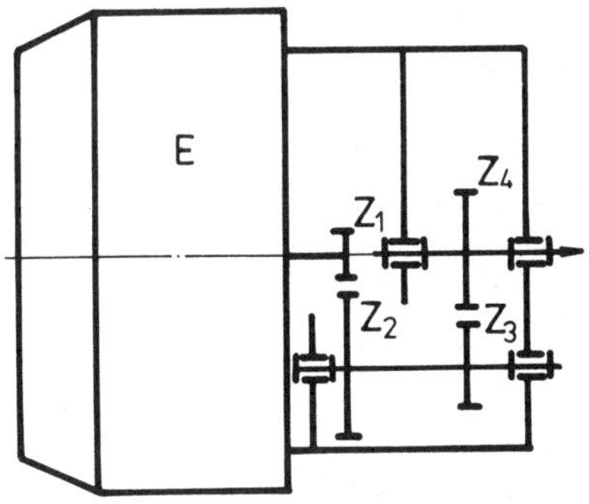

Fig. 2 .Servomotor-reducing gear, i=621

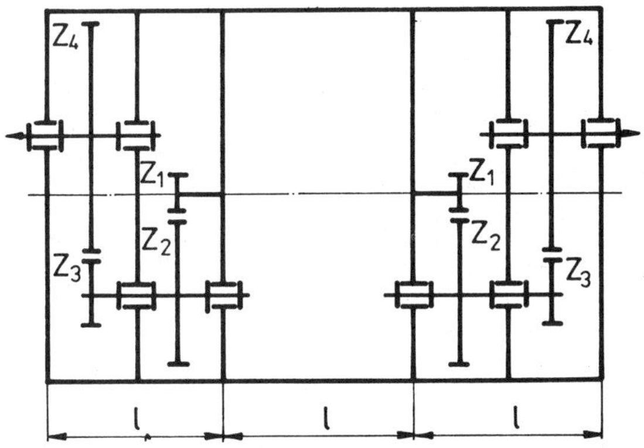

Fig. 3 .Servomotor-reducing gear with two-output, i=648

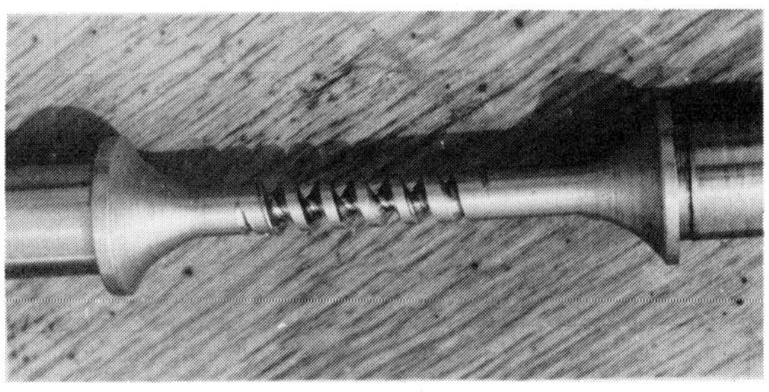

Fig. 8 . Pinion , z = 1

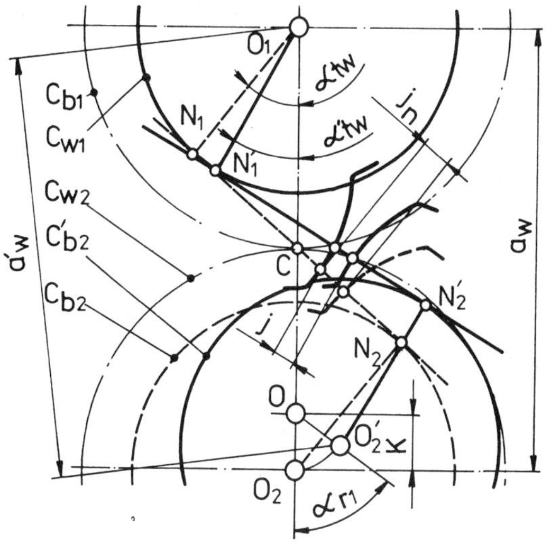

Fig. 4 .The compensation of backlash with eccentric bushing

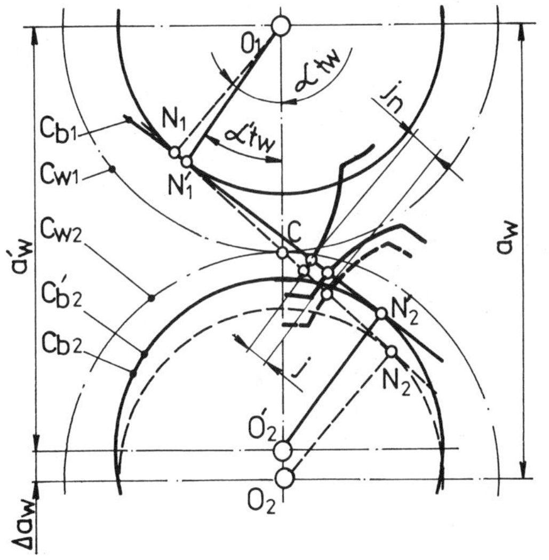

Fig. 5 .The compensation of backlash with case in translation

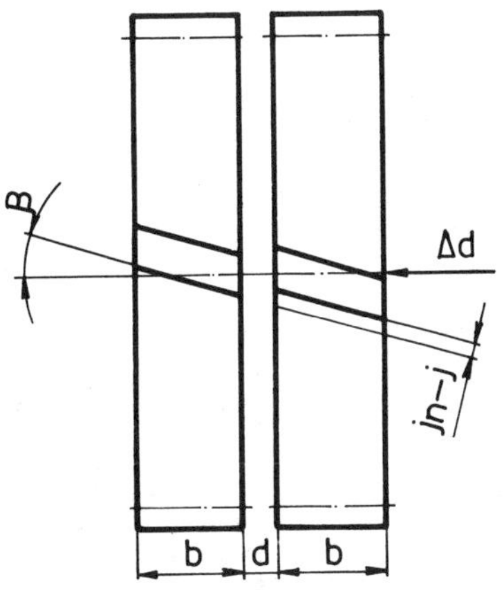

Fig. 7 .The compensation of backlash with relative axial displacement

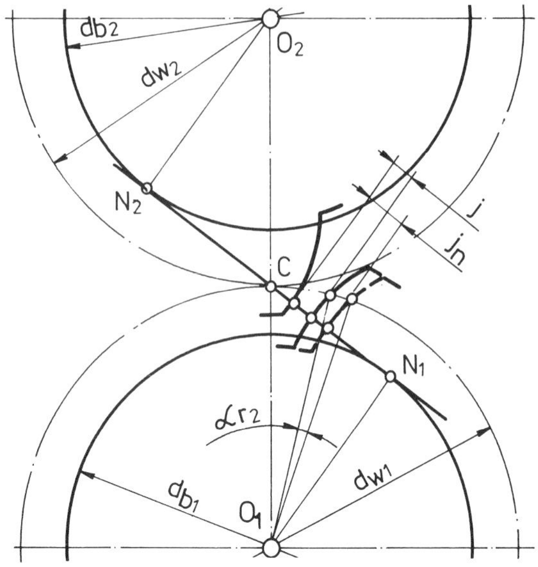

Fig. 6 .The compensation of backlash through relative rotation

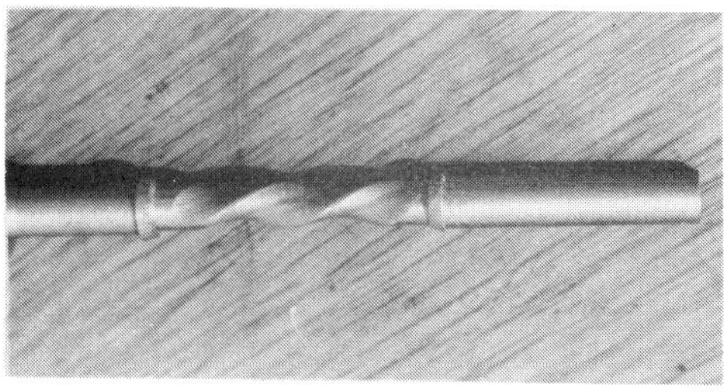

Fig. 9 . Pinion , z = 2

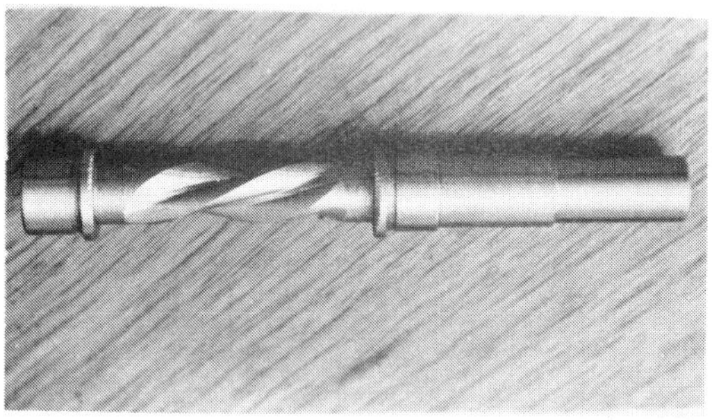

Fig.lo. Pinion , z = 3

Fig.11. Compensation system with rotation

Fig.12.Compensation system with eccentric bushing

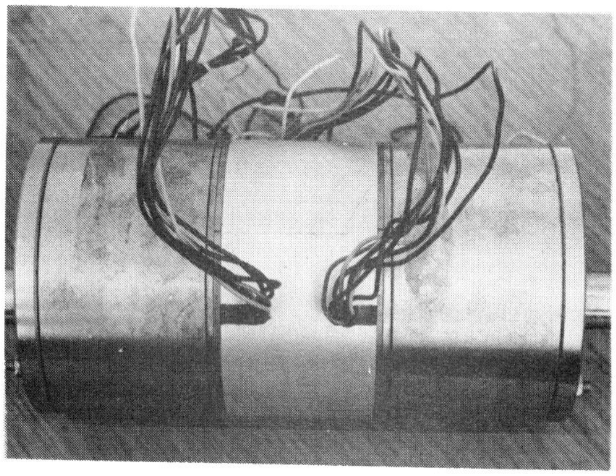

Fig.13 . Servomotor -reducing gear

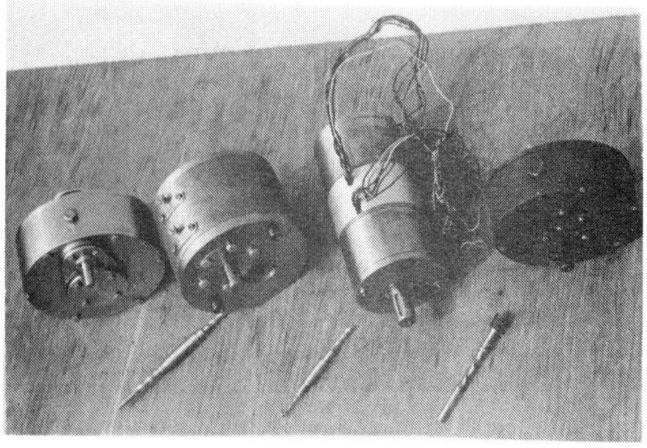

Fig.14.Reducing gears and servomotor - reducing gears

Possibilities of Extending the Gear Wheels Application Range

I. SZÉKELY AND A. DALI

Department of Mechanisms and Machine Parts, the Polytechnical Institute of Cluj-Napoca, Romania

Abstract. The paper points out the distribution laws governing the profile displacement coefficient, permitting to obtain toothed wheels and gearings of special characteristics. As a result the parctical application of gear mechanisms is being extended. Some of the possible cases, worth mentioning are the following: the involute bevel gears that are being used in the construction spur-bevel gears with gears with parallel axes in which the play between the flanks may be adjusted. The value of the load-carrying capacity of harmonic drive in which the harmonic gear wheel teeth are cut in undeformed state. Variable pitch spur gears are basis for variable transmission rate spur gearing assemblies. Discrete speed variators consisting of spur gear wheels having the same external diameters but different numbers of teeth.

Keywords. Variable speed gear; Beveloid gear; Special gears; Harmonic drive; Textile industry; Machine tools; Vehicles.

INTRODUCTION

It is well known that by adequately selecting the geometrical form of gear wheel tooth of involute profile the service life and load carrying capacity of gear assemblies may be inreased without changing the material. The desired shape of tooth flank may be obtained while the gear wheel is being machined, indicating the profile displacement coefficient (i. e. by correcting). But it is less known that by adequately selecting the profile displacement coefficient one can obtain special gears which when used in the construction of some machines may improve their performances. In this respect it is worth mentioning that the machining of corrected gear wheels does not require additional expenses, and does not alter or complicate the technological process. It does not require special tools or machine tools either. In the present paper the authors set into evidence the laws of profile displacement coefficient variation permitting us to obtain some gear wheels and gear assemblies with special characteristics, a fact resulting in the extension of gear mechanisms utilization range. Some examples of practical application of gear mechanisms are given that could be of great interest to machine designers.

EXAMPLES OF UTILIZATION OF PROFILE DISPLACEMENT VARIATION LAWS.

The profile displacement coefficient x varies liniarly over the tooth width (Fig.1) having maximum value on the right hand side of front plane $+x_1$, and minimum value on the left of front plane $-x_1$, in midplane it is equal to zero.

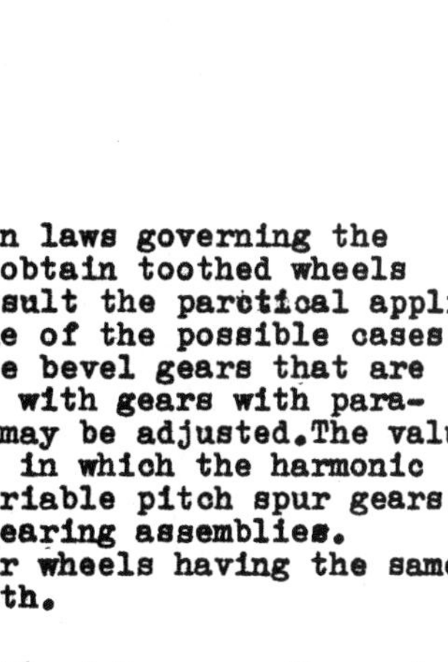

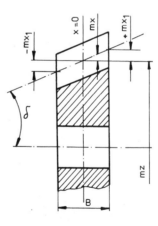

Fig. 1. Linear distribution of profile displacement coefficient over the tooth width.

Székely (1960) thoroughly studied the dependence of gear wheel elements, comb tool machined, on relative motion character between cutting tool and blank, whence the distribution law of profile displacement coefficient over the tooth width is clearly derived. Observing these imposed conditions in working out the gear wheel, an involute bevel gear is obtained (Fig.2) resembling an ordinary bevel gear, at first sight, but essentially differing from it, having the same modulus and the same tooth height in each section perpendicular to the axis. Likewise the flanks of these teeth have the form of some helical-involute surfaces one "e" inclined towards the right and the other "f" inclined towards the left.

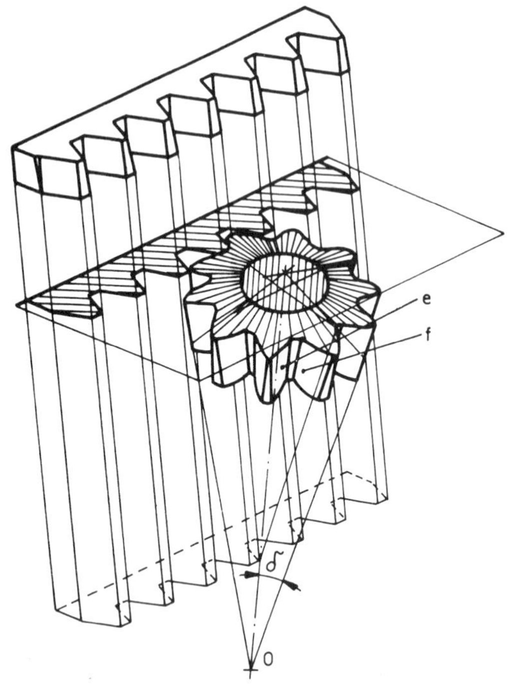

Fig. 2. The relative position between rack tool and the blank during work. The teeth flanks are helical-involute surfaces with vaious inclinations.

In another paper (Székely 1961,a) has deduced the formulas of tooth side flanks determination starting from a cutting method by standard comb cutting tools.

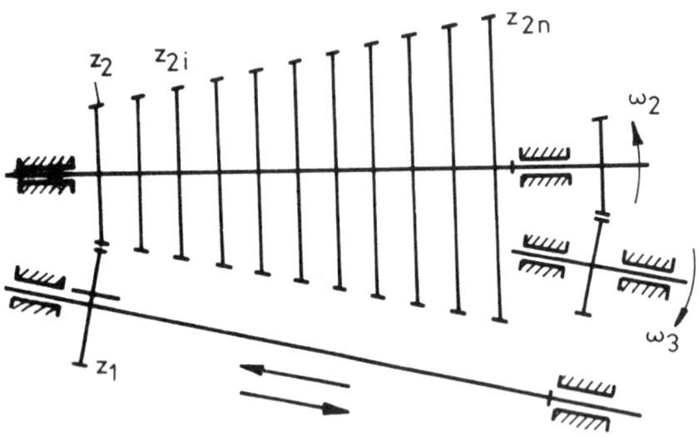

Fig. 3. Discrete speed variators with beveloid gears.

The involute bevel gear wheels are employed in some cylindrical-bevel gear assemblies used in the construction of discrete speed variators with gear wheels (Fig. 3) in the form of a conical toothed wheel assembly. Such a gear assembly provides gear shifts while running under load. In the same paper (Székely 1961,a) showed the additional conditions to be observed in assembling the toothed conical block so that shifting from a gear step to another be possible while in motion under load. We notice that the difference in the number of teeth between two neighbouring wheels of the conical gear assembly may be 2 or 4. By properly selecting the angle δ and the profile displacement coefficient x_1 we provide equality between the external, respectively, the internal diameters of neighbouring toothed wheels in their front meshing areas. Consequently an alignment of the teeth along a generator of the conical gear block is obtained making possible for the sliding pinion to pass from one step to another, along two generators, by means of a logical coupling mechanism. The operating principle of the logical coupling mechanism is described in (Székely, 1961, b). The involute bevel gear wheels may be used in the construction of spur gears with parallel axes in which the play between flanks may be adjusted (Fig. 4). When the technological process calls for small angles between bevel gear axes (5°-25°) some difficulties are encountered in ordinary gear wheel processing, some special machine-tools and cutting tools being required. In such cases the utilization of involute bevel gear wheels is justified as they are processed on the same machine tools as the spur gears. This will result in remarkable economy.

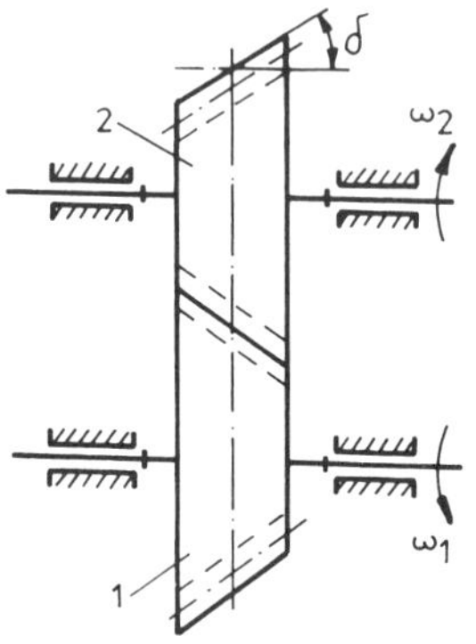

Fig. 4. Parallel axes gear assembly consisting of beveloid gear wheels in which the side clearence may be adjusted.

The bevel-involute gear wheels having small vertex angles are used for aircraft propeller pitch variator gear box construction, as well as in the construction of motor boat main drives. In cases when the straight toothed spur gears have great widths and are cantilever mounted onto the shafts they may be replaced with involute bevel gear wheels with angles at vertices of 1° - 3°. We have, as a result, a spur gear set having an advantageous distribution of load over the tooth width. The gear assemblies of cement mills, in which the pinion was replaced with a bevel-involute gear wheel of angle 1°, had twice the service-life of ordinary gear assemblies. Similarly load-carrying capacity of harmonic toothed wheels in undeformed state (Fig. 5) is improved by applying a linear distribution of profile supplementary displacement coefficient Δx over the width of tooth. In this case, a bevel-involute gear wheel is obtained having a very small vertex angle (0.5° - 1°). Dali (1982) showed that the harmonic gear wheel with involute teeth required for its processing anadequate profile correction to be applied.

It is however necessary, to use supplementary correction coefficient, linearly distributed, in order to increase the load carrying capacity of harmonical drive.

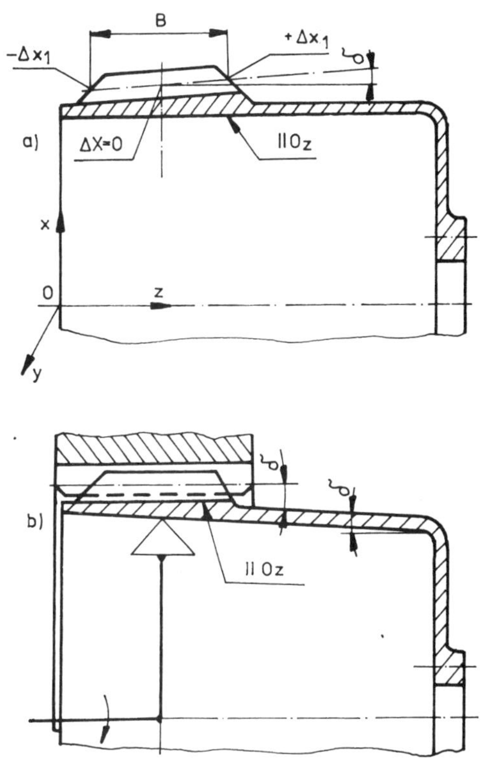

Fig. 5. Harmonic bevel-involute gear wheel

During harmonic drive operation the harmonic wheel (whose teeth were cut in involute-conical shape) is deformed by the generator. The teeth of this wheel take advantageous positions in the meshing zone. As a result the possibility that the teeth mesh on the edge of rigid spur gear (Fig.5,b) was eliminated.
The profile displacement coefficient has a constant value over the tooth width but varies from tooth to tooth, thus creating a gear wheel in which the teeth widths measured over pitch line have different values (Fig.6). As a result, the teeth of a same gear will have different forms, some having the bases thickned and others thinned. The gear assembly consisting of an ordinary spur gear plus a gear wheel of circular form with variable pitch is called circular spur gear. In his doctoral thesis (Müller, 1976) investigates all the types of circular gear assemblies providing a variable transmission ratio and he also gave some appropriate geometrical and kinematic calculation methods. The circular spur gear assemblies are used to achieve some variable transmission ratios (in accordance with some prescribed rules) within a complete rotation of the driven wheel while the driving wheel has constant angular speed. The average transmission ratio is mentioned to be a constant magnitude.

The circular spur gear set is employed for generating vibrations used in various technological processes such as, automatic machine feed, positioning, transportation and sorting of workpieces by means of vibration, testing some measuring apparatus by vibration etc. The variable pitch spur gearing was also applied in leveling chipping speeds in shaping machines. In the case when the profile displacement coefficient varies both with tooth width and from tooth to tooth of the same wheel a quasivariable involute-bevel gear wheel is obtained.

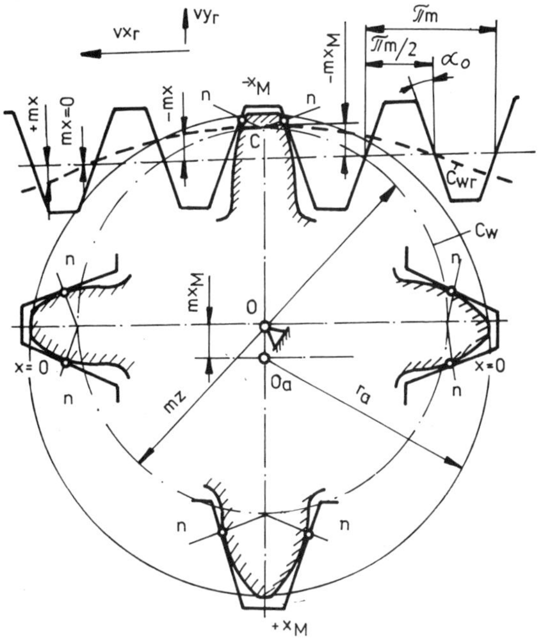

Fig. 6. Cycloid spur gear with quasi variable pitch

Distributing the profile displacement coefficients in accordance with well-determined rules over more gear wheels mounted on the same shaft (forming a spur gear assembly) a set of toothed wheels (about nine in number) are obtained, having the same external diameter but different number of teeth (Fig.7).
Such a spur gear assembly is used in the construction of discrete speed variators, having a reduced adjusting range, but very fine velocity steps. Such a speed variator (Fig.6,a) was used in the drive of warp sizing machines in textile industry having adjusting ranges of 1,2 and 9 velocity steps (Fig.7,a).
In their paper (Csulak and Székely, 1962) investigated more constructive types of speed variators consisting of several cylindrical toothed assemblies and more sliding wheels. One of the types can achieve 5329 velocity steps over an adjusting range of 2.o7.

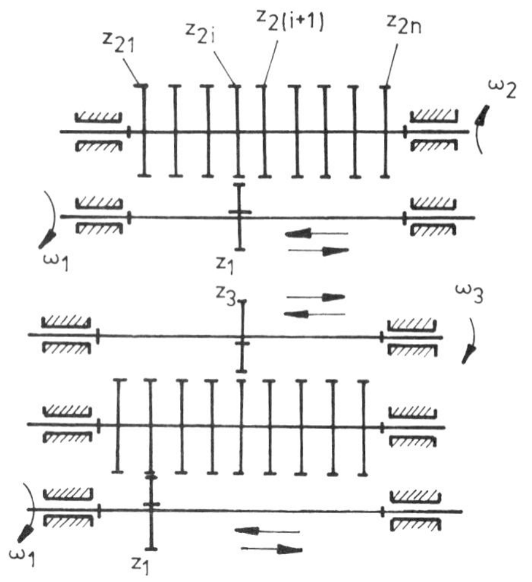

Fig. 7. Discrete speed variator having a reduced adjusting range.

CONCLUSIONS

The paper emphasizes the manner in which the profile correction (displacement) coefficient influences the shape of tooth, the shape of the gear wheel and the type of gear assembly.
The examples taken from technological processing offer a wide opportunity to designers who on the basis of the remarkable achievements of the late 2oth century, could create new machines and mechanisms with high performances.

REFERENCES

Csulak, A., Székely, I. (1962) Mecanisme cu roţi dinţate pentru obţinerea unui mare număr de rapoarte de transmitere diferite. In: Studii şi cercetări de mecanică aplicată, Nr.4, 989-999.

Dali, A., (1982) Contribuţii la geometria, cinematica şi calculul de rezistenţă a angrenajelor cu elemente dinţate deformabile. Teză de doctorat. Institutul politehnic Cluj-Napoca.

Müller, A., (1976) Roţi dinţate cilindrice şi cilindro-conice cu raport de transmitere variabil. Teză de doctorat. Institutul politehnic Cluj-Napoca.

Székely, I., (196o) Dependenţa elementelor roţii dinţate prelucrate cu cuţit pieptene de caracterul mişcării relative dintre sculă şi semifabricat. In: Buletinul ştiinţific al Institutului politehnic Cluj-Napoca, 33o-335.

Székely, I., (1961) Studiu asupra blocului conic dinţat compus din roţi dinţate conico-evolventice. In: Studii şi cercetări de mecanică aplicată. Editura Academiei R.S.R. Bucureşti, 1o57-1o74.

Székely, I. (1961) Variator de viteză cu roţi dinţate cu raport de transmitere reglabil sub sarcină. In: Metalurgia şi construcţia de maşini, Nr.2, 147-15o.

POSSIBILITES D'EXTENSIONS DU DOMAINE DES APPLICATIONS DES RUES DENTEES

Resume

Dans le présent article les lois de repartition de coefficient de modification des profils des dents permottent d'obtenir des roues dentées et engrenage a caracteristiques speciales. Par la suite, les applications du mécanismes comportant des roues dentées preuvent être etendues à d'autres domaines Voilà quelques uns des ces cas possibles:

- si à la places des engrenage cylindrique on utilise des engrenages conique-évolventique le jeux entre les profils peut ȇntre éliminé,
- on agrendit la capacite portante des engrenages harmoniques si la denture est réalisée de manière conique-évolvente,
- on peut realiser des roues dentees à pes quasi variable où le rapport de transmission est modifie,
- on peut obtenir des variateurs discretes de tours utilisant des roues ayant le même diamètre extérieur mais un nombre de dents differents.

Computerized Modelling of Quasi – Involute Gears

V. CHIOREANU* AND M. CHIOREANU**

*Manufacturing Center, SMA. Sighetu Marmației, Jud. Maramures, Romania
**Faculty of Mechanics, Institutul Politechnic Cluj – Napoca, Romania

Abstract . By generalizing the geometry of crossed axes gearings and their component gears , the author introduces the term quasi - involute, which includes the whole family of space gearings , having a point - shaped contact , and gears manufactured by involute tools . After defining , then , a series of geometric and kinematic elements which are specific for this new computing conception , we find the mathematical expression of the general quasi - involute gear tooth flank , and we also elabotate the corresponding numeric model , applied in a high efficiency sub - program .

Keywords . Numerical Methods ; Mathematical programing ; Computer aided gear design ; Quasi - Involute gears modeling .

INTRODUCTION

In paper [1, 2] the author demonstrates that the general quasi - involute gears , with crossed axes may be of three types : cylindric (C) ; conic - involute (CE) ; hyperbolic (HP) ; as in the following five possible combinations : C – C ; C-CE ; CE-CE ; C-HP ; CE-HP ; The first three types are space involute gears , having generally a point shaped contact . The last two types (C-HP ; CE-HP) are monoparametric , with instant contact , lineary , which in modern constructions are liable to corrections . Taking into account this observation and the fact that the gears of this gearing are kinematically generated by the help of involute tools (of the rack type ; comb knife ; or the whell type of the Fellow or Skiwing method .) the author suggests the general term of qoasi-involute attributed both to the general gearing with crossed axes , and to its gears .

By convention the constituent gears of the quasi - involute gearing will be identified by indices k (k=1,2 , ...) and defined through : number of teeth Z_k ; conicity angle δ_k (fig.1.) ; inclination

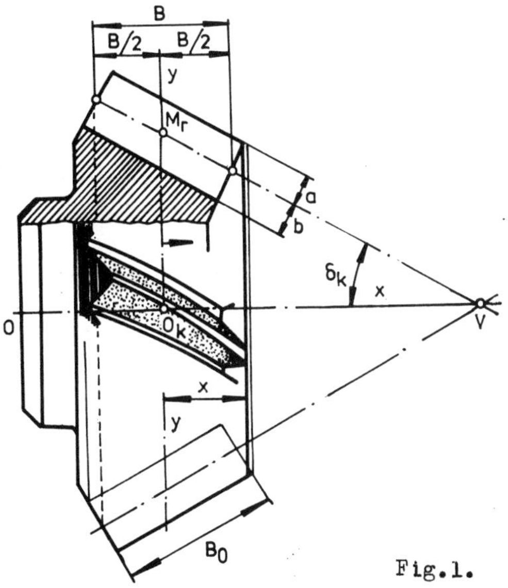

Fig.1.

of toothing β_k and the nominal profile ξ_k movement (the module being considered unitary) . Gear of rank k has the dividing cylinder of radius given by the known formula :

$$R_d = Z_k / 2 \cos \beta_k \tag{1}$$

In the reference plane we define the reference circle of radius :

$$R_k = R_d + \xi_k + d_k \tag{2}$$

where d_k is the correction by which we

assure the gearing without clearance between the teeth . For limiting the semifabricated part and the tooth flank generating process we take into account the variation of external radius R_x and the internal one r_x , of the gear function of x . From where :

$$R_x = R_k + \frac{a - x\sin\delta_k}{\cos\delta_k} \; ; \; r_x = R_k - \frac{b + x\sin\delta_k}{\cos\delta_k} \quad (3)$$

As the breadth of the gear is limitted at quota B , the variable in relation (3) will have a meaning between limits :

$$\left.\begin{aligned} x &= \pm\left(\frac{B}{2}\cos\delta_k + \sin\delta_k\right) \text{ pentru } R_x \\ x &= \pm\left(\frac{B}{2}\cos\delta_k - \sin\delta_k\right) \text{ pentru } r_x \end{aligned}\right\} \quad (4)$$

This means that the maximal and minimal diameter of the gear will have the expressions :

$$\left.\begin{aligned} D_M &= 2\left(R_k + \frac{B}{2}\sin\delta_k + \cos\delta_k\right) \\ D_m &= 2\left(R_k + \frac{B}{2}\sin\delta_k - \cos\delta_k\right) \end{aligned}\right\} \quad (5)$$

THE ELEMENT OF REFERENCE

We admit that regardless the method of generating the gears of the quasi - involute gearing , the element of reference is plane S , (fig.2.) of the right flank of the generating rack having a standard normal profile , defined in the

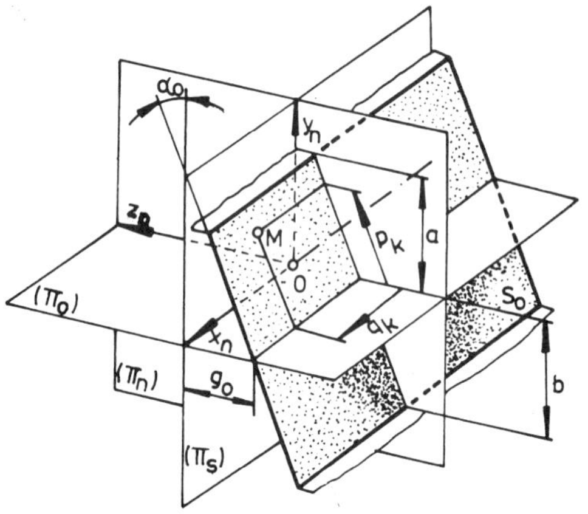

Fig.2.

system of reference Σ_n by the vector radius and the normal by relations :

$$X_n = \begin{bmatrix} q \\ p\cos\alpha_0 \\ k_1(k_2 p\sin\alpha_0 - g_0) \end{bmatrix} = \begin{bmatrix} x_n \\ y_n \\ z_n \end{bmatrix} = X_n(p,q) \quad (6)$$

in which : - p and q - are the independent parameters of the surfaces ; - k_1 and k_2 - coeficients ; - g_0 - is the semi thickness of rack tooth ; - α_0 - angle of normal profile : $\alpha_0 = 20^0$.

The normal vector N_n attached to relations (6) has constant projections

$n_{xn} = 0$; $n_{yn} = tg\,\alpha_0$; $n_{zn} = -k_1 k_2$

As the conicity angle δ_k and the inclination of tooth β_k are particular elements of the gear , we considered them to be introduced in the process of calculus in the same time with the rack . For this purpose , the system Σ_n will rotate twice : first around the axis $O_n y_n$ with angle β_k (fig.3 a) after the transformation :

$$T_{\beta n} = \begin{bmatrix} \cos\beta_k & 0 & -\sin\beta_k \\ 0 & 1 & 0 \\ \sin\beta_k & 0 & \cos\beta_k \end{bmatrix} \quad (7)$$

then around the axis $O_c z_c$ (fig.3 b) with angle after the transformation :

$$T_{c\beta} = \begin{bmatrix} \cos\delta_k & \sin\delta_k & 0 \\ -\sin\delta_k & \cos\delta_k & 0 \\ 0 & 0 & 1 \end{bmatrix} \quad (8)$$

The transformation from Σ_n in Σ_c will be obtained from the $T_{c\beta} T_{\beta n}$ produce , after matrix :

$$T_{cn} = \begin{bmatrix} \cos\beta_k\cos\delta_k & \sin\delta_k & -\sin\beta_k\cos\delta_k \\ -\cos\beta_k\sin\delta_k & \cos\delta_k & \sin\beta_k\sin\delta_k \\ \sin\beta_k & 0 & \cos\beta_k \end{bmatrix} \quad (9)$$

Applying matrix (9) for vector (6) we obtain for the normal the expression of flank S_0 related to the system of reference Σ_c , under the form of :

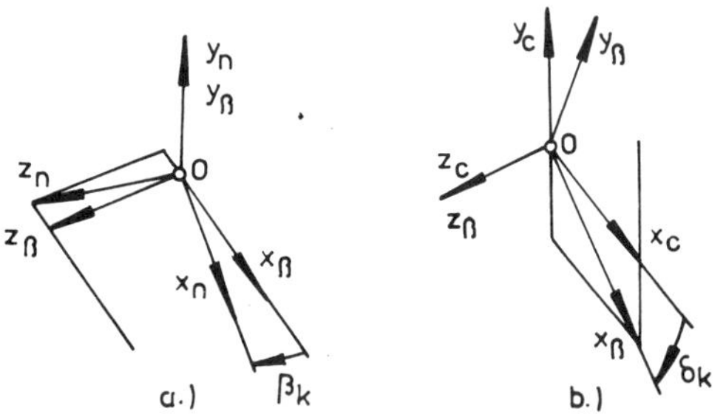

Fig.3.

$$X_c = \begin{bmatrix} [q\cos\beta_k - k_1(k_2 p\sin\alpha_0 - g_0)\sin\beta_k]\cos\delta_k + p\cos\alpha_0\sin\delta_k \\ [q\cos\beta_k - k_1(k_2 p\sin\alpha_0 - g_0)\sin\beta]\sin\delta_k + p\cos\alpha_0\cos\delta_k \\ q\sin\beta_k - k_1(k_2 p\sin\alpha_0 - g_0)\cos\beta_k \\ 1 \end{bmatrix} = \begin{bmatrix} x_c \\ y_c \\ z_c \\ 1 \end{bmatrix} = X_c(p,q) \quad (10)$$

$$N_c = \begin{bmatrix} tg\,\alpha_0\sin\delta_k + k_1 k_2\sin\beta_k\cos\delta_k \\ tg\,\alpha_0\cos\delta_k - k_1 k_2\sin\beta_k\sin\delta_k \\ k_1 k_2\cos\beta_k \\ 0 \end{bmatrix} = \begin{bmatrix} n_{xc} \\ n_{yc} \\ n_{zc} \\ 0 \end{bmatrix} = \text{const} \quad (11)$$

Regrouping the expressions of X_c vertor's projections by help of the coeficients :

$$a_x = -k_1 k_2\sin\alpha_0\sin\beta_k\cos\delta_k + \cos\alpha_0\sin\delta_k \; ;$$
$$a_y = k_1 k_2\sin\alpha_0\sin\beta_k\sin\delta_k + \cos\alpha_0\cos\delta_k \; ;$$
$$a_z = k_1 k_2\sin\alpha_0\cos\beta_k \; ;$$

$$\left.\begin{aligned} b_x &= \cos\beta_k\cos\delta_k \; ; & c_x &= k_1 g_0\cos\delta_k\sin\beta_k \\ b_y &= -\cos\beta_k\sin\delta_k \; ; & c_y &= -k_1 g_0\sin\delta_k\sin\beta_k \\ b_z &= \sin\beta_k \; ; & c_z &= -k_1 g_0\cos\beta_k \end{aligned}\right\} \quad (12)$$

the expression of vector radius (10) will become :

$$X_c = \begin{bmatrix} a_{xp} & +b_{xq} & +c_x \\ a_{yp} & +b_{yq} & +c_y \\ a_{zp} & +b_{zq} & +c_z \\ & 1 & \end{bmatrix} = \begin{bmatrix} x_c \\ y_c \\ z_c \\ 1 \end{bmatrix} = X_c(p,q) \qquad (13)$$

The profile angle of the rack has the expression :

$$\operatorname{tg}\alpha_k = \frac{\operatorname{tg}\alpha_0 \cos\delta_k}{\cos\beta_k} - k_1 k_2 \operatorname{tg}\beta_k \sin\delta_k \qquad (14)$$

If we impose that angle δ_k must keep the value α_0, between δ_k and β_k we obtain the relation of dependence :

$$\beta_k = 2\operatorname{arc\,tg}\left[\sin\delta_k \frac{1-\cos\alpha_0}{(1+\cos\delta_k)\sin\alpha_0}\right] \qquad (15)$$

GENERATION OF QUASI - INVOLUTE TOOTHING

While generating the flank S_k (fig.4) of quasi - involute gear we take as known: - the generating rack S_0, defined by relations (11) , (12) and (13) ; - division radius given by relation (1) ; - profile movement ξ_k ; - quotas of semifabricated part , acording to fig.1. and relation(5). The process of generating is kinematic and the flank S_k is determined as the geometric locus of the points in which we satisfy the gearing equation :

$$Q_u(p,q,u_k) = V_r N_c = 0 \qquad (16)$$

Where : V_r is the relative speed between surfaces S_0 and S_k ; U_k - the parameter of the relative speed ; - the normal given by the relation (11) .

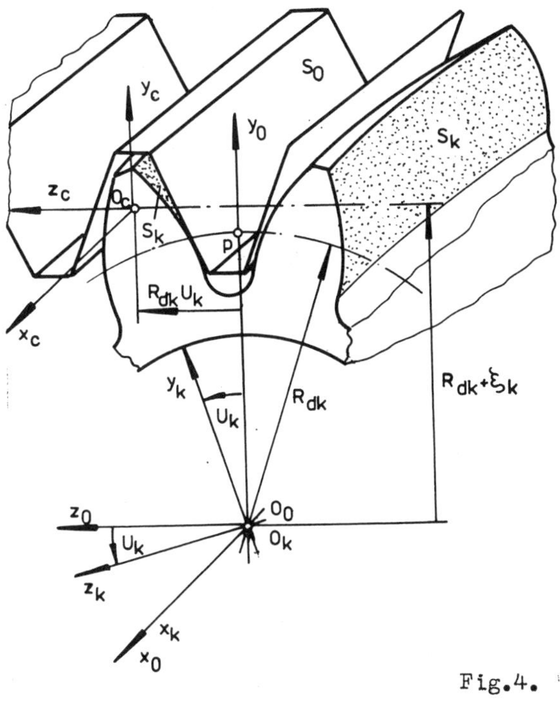

Fig.4.

For developing this equation , between the reference sysrems Σ_c and Σ_k, of the rack , respectively of the gear , accor - ding to the scheme of **figure** 4 , we determine the transfer matrix :

$$T_{oc} = \begin{bmatrix} 1 & 0 & 0 & 0 \\ 0 & 1 & 0 & R_d+\xi_k \\ 0 & 0 & 1 & R_d U_k \\ 0 & 0 & 0 & 1 \end{bmatrix} \qquad (17)$$

$$T_{ok} = \begin{bmatrix} 1 & 0 & 0 & 0 \\ 0 & \cos U_k & -\sin U_k & 0 \\ 0 & \sin U_k & \cos U_k & 0 \\ 0 & 0 & 0 & 1 \end{bmatrix} \qquad (18)$$

and the inverse matrix of these , after which :

$$T_{kc} = T_{ko} T_{oc} =$$

$$= \begin{bmatrix} 1 & 0 & 0 & 0 \\ 0 & \cos U_k & \sin U_k & (R_d+\xi_k)\cos U_k + U_k R_d \sin U_k \\ 0 & -\sin U_k & \cos U_k & -(R_d+\xi_k)\sin U_k + U_k R_d \cos U_k \\ 0 & 0 & 0 & 1 \end{bmatrix} = T_{kc}(U_k) \qquad (19)$$

in which : the system Σ_0 is considered to be fix . The operator of the relative speed , [1] in this application becomes :

$$\Omega_u = T_{ok}\,\partial T_{kc}/\partial U_k = \Omega_u(U_k) \qquad (20)$$

Which together with the vector (13) will define the relative speed as :

$$V_r = \Omega X_c = \begin{bmatrix} 0 \\ z_c + U_k R_d \\ -y_c - \xi_k \end{bmatrix} = \begin{bmatrix} V_x \\ V_y \\ V_z \end{bmatrix} = V_r(U_k) \qquad (21)$$

The condition of gearing is expressed by the scalar produce between vectors (21) and (11) obtaining the equation :

$$Q_u(p\ q) = (z_c + U_k R_d) n_{yc} - (y_c + \xi_k) n_{zc} = 0 \qquad (22)$$

with the solution :

$$U_k = \left[(y_c+\xi_k)\frac{n_{zc}}{n_{yc}} - z_c\right] / R_d - U_k(p,q) \qquad (23)$$

Introducing solution (23) in matrix T_{kc} we obtaing the defined flank univocally by relations :

$$X_k = T_{kc} X_c = \begin{bmatrix} x_c \\ (y_c+\xi_k)(\cos U_k + \sin U_k \operatorname{ctg}\alpha_k) + R_d \cos U_k \\ -(y_c+\xi_k)(\sin U_k - \cos U_k \operatorname{ctg}\alpha_k) - R_d \sin U_k \end{bmatrix} =$$

$$= \begin{bmatrix} x_k \\ y_k \\ z_k \end{bmatrix} = X_k(p,q) \qquad (24)$$

$$N_k = \begin{bmatrix} n_{xc} \\ n_{yc}\cos U_k + n_{zc}\sin U_k \\ n_{zc}\cos U_k - n_{yc}\sin U_k \end{bmatrix} = \begin{bmatrix} n_{xk} \\ n_{yk} \\ n_{zk} \end{bmatrix} = N_k(p,q) \qquad (25)$$

in which α_k is given by the expression(14) Relations (24) and (25) represent the mathematic model of flank S_k . For determining flank S_k by the help of the computer, we prepared the subprogramme SPR-1 , rendered in the logic scheme from figure 5 , which functions on the basis of parameters p , q , i , j , k , x .

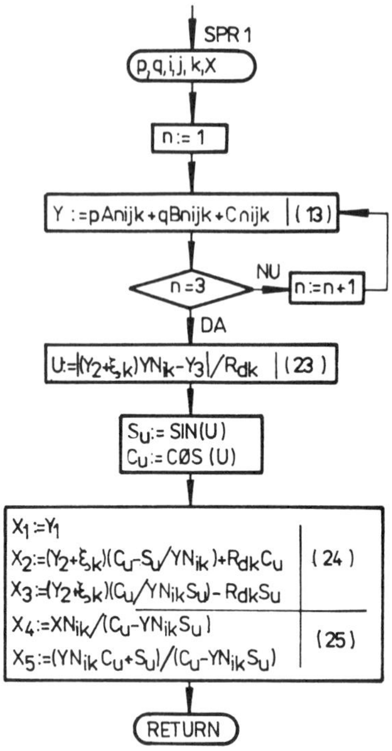

Fig.5.

After determining vector (13) , the SPR-1 (fig.5) procedure goes on to the calculation of position parameter $\bar{U}$, using for this purpose formula (23) . Then after shortenings in the assignments block (1) we apply formulae (24) and (25) completing matrix X_L ; (L = 1,5) , which is at its turn the output argument of the programme , defining numerically the current point , corresponding to the concrete numerical values of arguments p and q .

CONCLUSION

Taking into account figure 5 we observe that the calculus algorithm of the general quasi - involute tooth flank is finally reduced to a remarcably simple numerical model , univocally defined in any situation of usage field of these gears . From the practice of numerical research we know that the necessary computing time used for realizing programme SPR-1 is 10^{-3} S , which permits intensive usage of this programme in very cheap conditions .
The numeric model of quasi - involute toothing , shown in the logical scheme of figure 5 , was successfully used for realizing practical formulae of calculus .

REFERENCES

Chioreanu , V. (1983) . Doctorate paper The Study of Gearings made up by Conic-Involute Gears . Institutul Politehnic Cluj - Napoca Romania .

Szekely , I. , J. , Bocian , and V.Chioreanu . General Quasi - Involute Gear Toothing . International Symposium on Gearing and Power Transmissions .Tokyo (1981) Vol.1.paper a.5.p.23 - 28 .

Chioreanu , V. (1983) . Determination of quota over the balls in conic - involute gears by numerical modeling . Procedings of the Sixth World Congress on theory of Machines and Mechanisms New Delhi , p. 932 - 935 .

Chioreanu , V. , M. Chioreanu . (1984) Numerical Study of Limits in General Quasi - Involute Gear Toothing . International Symposium on Design and Synthetis . Tokyo .

Chioreanu , V. , M. Chioreanu . (1985) . Computer Analysis of Quasi - Involute Gearings . Fourth IFToMM International Symposium on Linkages and Computer Aided Design Methods . Bucharest , Romania , Paper 7 , p. 51 - 62.

ABSTRACT

By generalizing the geometry of crossed axes gearings and their component gears , the author introduces the term quasi - involute , which includes the whole family of space gearings , having a point - shaped contact , and gears manufactured by involute tools .
After defining them , a series of geometric and kinematic elements which are specific for this new computing conception we find the mathematical expresion of the general quasi - involute gear tooth flank, and we also elaborate the corresponding numeric model , applied in a high efficiency sub - program .
The numeric model uses for realizing practical formulae of calculus regarding ;
- determination of quota over the balls ; - calculus of limits in under - cutting and sharpening of tooth ;

Tragfähigkeitsberechnungen von Zahnrädern. Bestimmung der Faktoren Y_F und $K_{F\beta}$

E. MIRIŢĂ AND GH. RĂDULESCU

Lehrstul für Maschinenelemente, Polytechnischer Institut, Bucharest, Romania

Ubersicht. Arbeit geht von einer Analyse des aktuellen Stadiums im Tragfähigkeitsberechnungen von Zahnrädern an. Es macht sich die Bemerkung bezüglich der großen Zahl von Berechnungsfaktoren und der heteroklitischen Karakter des physikalischen Modellen, die zu deren Bestimmung benutzt sind. Dies hat zur Folge eine Erschwerung des Konstruktionsberechnungen und stellt die Präzision in die Frage. Eine Verbesserung dieser Situation, durch eine einheitliche Betrachtung - mit der Methode finiter Elemente - des Formfaktors Y_F und des Breitenfaktors für die Spannungsverteilung $K_{F\beta}$, ist es möglich. Prezisierung der Berechnungsstrukturen, der Strukturierung und der Arbeitsweise werden erfasst. Die ermittelten Faktoren werden im Vergleich mit anderen Werten gezeigt.

Schlüsselwörte. Zahnrädern; Tragfähigkeit; Zahnfußspannungen; Methode finiter Elemente; Zahnformfaktor; Breitenfaktor.

EINLEITUNG

Die Bestimmung des efektiven Spannungen im Zahnfuß ist von großer Bedeutung für die Tragfähigkeitsberechnungen der Zahnradgetrieben. Das Problem ist komplex und schwierig, weil das Verhältnis der Dimensionen den Zahn als einen massiver Körper bestimmen, und die Fußausrundung, die nach sich eine Kerbwirkung einziet, große Geometrieveränderungen hat. Alles reflektiert sich ins Struktur des Rechnungsbeziehungen für die Zahnfußspannungen - Beziehungen die einen sehr großen Zahl von Einflußfaktoren enthälten. Oft werden diese Einflüße bei sehr verschiedenen Modellen und Rechnungshypothesen ermittelt. Dies ist der Fall auch für die genormten Rechnungsmethodologien. So, werden Spannungen im Zahnfußausrundung mit einer Beziehung dieser Typ bestimmt (ISO/DP 6336/III):

$$\sigma_F = \frac{F_t}{bm} Y_F\, Y_S\, K_{F\beta}\, K_{F\alpha}\, K_A\, K_V \qquad (1)$$

wo:

- F_t - Umfangkraft am Teilkreis;
- b - Zahnbreite; m - Modul;
- Y_F - Formfaktor;
- Y_S - Spannungskorekturfaktor;
- $K_{F\beta}$ - Breitenfaktor für die Zahnfußbeanspruchung;
- $K_{F\alpha}$ - Einflußfaktor für die Eingriffsphase;
- K_A, K_V - Dynamische Faktoren.

Der Formfaktor Y_F wurde im Tragfähigkeitsberechnungen in 1892 von W.Lewis eingeführt. Lange Zeit wurde dies Faktor ausschließlich auf Grund der Biegspannungen auf Modell eines eingespannten Balken bestimmt. Es wurden deshalb Korekturfaktoren für Kerb - (Y_S) und Breitenlastverteilungwirkung ($K_{F\beta}$) ins Beziehung (1) eingeführt.

Erlangung einer analytischen Lösungen des räumlichen Spannungszustandes im Zahnfuß ist es nicht möglich. Die Kerbwirkung wurde aber durch eine Reihe von approximativen Lösungen im Rahmen des Elastizitätstheory bestimmt (Linke, 1983; Ustinenko, 1972). Zur Berücksichtigung auch der Breitenlastverteilung, Zahn wird als eine Platte im Plattentheory betrachtet.

Diese Situation ist nicht befridigend, weil so bestimmten Y_F, Y_S und $K_{F\beta}$ - Faktoren zusammen ins Beziehung (1) eingehen während Anforderungen an Genauigkeit eine einheitliche Betrachtung der räumlichen Spannungszustände verlangen.

In letzter Zeit, die Methode finiter Elemente erlaubt so eine Betrachtung für einen großen Zahl von Zahn- und Zahnradgeometrien. In vorliegenden Arbeit die Methode finiter Elemente wird für Bestimmung des Y_F und $K_{F\beta}$ - Faktoren verwendet.

ZAHNFUSSPANNUNGENBERECHNUNG ENTSPRECHEND EINER GLEICHMASSIGEN LASTVERTEILUNG

Die gleichmässige Lastverteilung über die Kontaktlänge ist der Idealfall der Zahnbelastung. Die Hypothese einer ebenen Verformungszustand entspricht gut dieser Situation. Die so Berechneten Spannungen können zum Bestimmung den Y_S oder $Y_F Y_S$ - Werten dienen. Gleichzeitig sind sie Vergleichspannungen für Abschätzung der Wirkung einer ungleichmässigen Lastverteilungen.

Bestimmung des physisches Modelle und der Strukturierung

Bild 1 zeigt den untersuchten physischen Modell. Abmessungen h_c und s_c sind so groß genommen, damit sie die ganze Ver - spannungszone erfassen können. In diesen Fall kann man das Einspannen als Randbedingung nehmen.

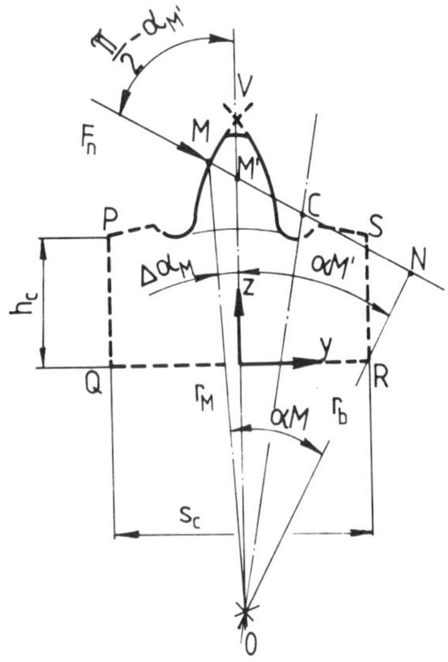

Bild 1. Physisches Modell der Zähne.

Strukturierung der Zähemodelle wurde manuell durchgeführt. Ein Beispiel ergibt sich im Bild 2.

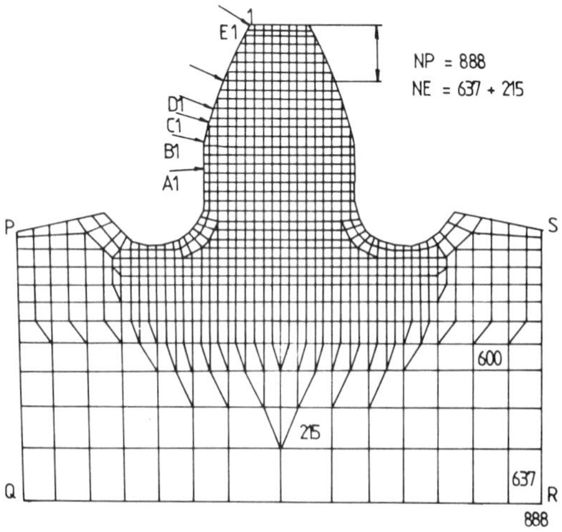

Bild 2. Beispiel von Strukturierung.

Als Rechnenprogram wird SAP IV (Bathe, 1973) benützt.

Auswertung der Rechnerergebnisse

Ein schwieriges Problem beim Abschätzung der Rechnerergebnisse ist die räumliche Spannungszustände zu equivalierung. Schliesslich machte sich die Option um die Verwendung der maximalen Normalspannungentheory. Einen Beweisgrund dazu war auch die Tat, daß für einen großen Zahl von Dauerbruchsektionen, Mikrophotographien im Rastermikroskop einen überwiegend fragilen Relief zeigten. Also:

$$\sigma_{Fmax,gl} = \sigma_1 \qquad (2)$$

$\sigma_{Fmax,gl}$ - die maximalen örtlichen Spannungen im Fußausrundung bei einer gleichmäßigen Lastverteilung über die Kontaktlänge.

Im Bild 3. sind die Spannungsverteilungen am Oberfläche des Zahn z = 17 dargestellt.

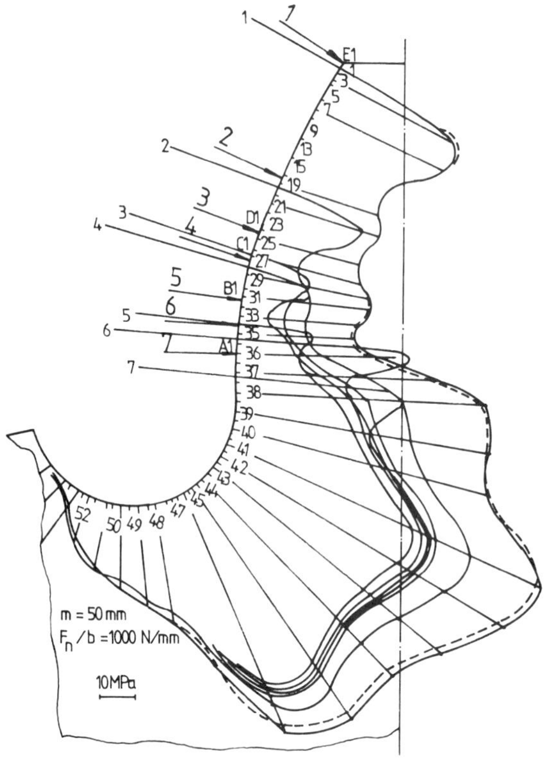

Bild 3. Spannungsverteilungen (σ_1) am Oberfläche des Zahn z=17.

Bestimmung der Formfaktoren - Y_F .

Die Formfaktoren werden aus den maximalen örtlichen Spannungen entsprechend einer gleichmäßigen Lastverteilung über die Kontaktlänge, gemäß Beziehung (3), bestimmt.

$$Y_F = \sigma_{Fmax,gl}\ b\ m\ /\ F_t \qquad (3)$$

Die so ermittelten Faktoren werden auch die Kerbwirkung der Fußausrundung berücksichtigen und damit Spannungskorekturfaktor Y_S wird beiseide gelasst. Man betrachtet daß so eine Arbeitsweise die Berechnungen vereinfachen und die Präzision steigern. Es ist das Zeit gekommen, wenn wir den alten Modell des gespannten Balken verlassen können. Tafel 1 zeigt die berechneten Faktoren im Vergleich mit anderen Werten.

TAFEL 1 Y_F Faktoren

	z			
	17	50	90	∞
Eigene Werte	4,98	4,08	4,03	3,80
Baronet, 1973	-	3,69	4,21	4,15
Chabert, 1974	4,12	3,38	-	-
ISO/DP 6336/III	4,58	3,94	3,92	4,06
Tobe, 1979	4,44	3,80	3,76	3,72
Ustinenko, 1972	4,27	3,74	3,74	3,77

ISO Werte bedeuten tatsächlich die Erzeugnisse $Y_F Y_S$.

ZAHNFUSSPANNUNGENBERECHNUNG ENTSPRECHEND EINER UNGLEICHMASSIGEN LASTVERTEILUNG

Bestimmung des physisches Modelle, der Strukturierung und der Arbeitsweise

Damit die Wirkung ungleichmäßigen Lastverteilungen erfasst wird, werden räumliche Modelle der Zähne verwendet. Das Erfordernis an Erlangung , am räumlichen Modelle, einer Lösungen erstrebenswert mit dieselben Präzision wie auf ebenen Modelle, führt zu einen zu großen Zahl von Knoten und Elementen. In dieser Situation, damit man die Rechnerzeiten zu vermindern, macht sich die Option für Lösung des Problems in zwei Etappen. In der ersten Etappe, mit

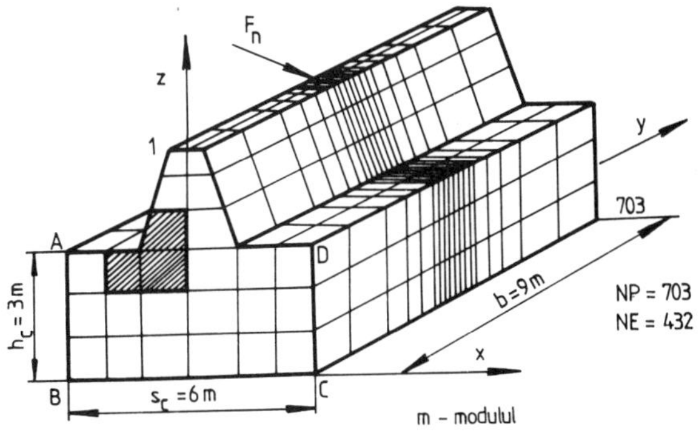

Bild 4. Räumliche Struktur für Berechnung der Verschiebungen (z =∞).

eine grobe Strukturierung (Bild 4.), im Hinblik des Zahnfußspannungen, aber genugend präzis hinsichtlich der Knotenverschiebungen, werden die Verschiebungen des Knoten ermittelt.

Im zweiten Schritt, auf Unterstrukturen des Fußausrundung fein strukturiert (Bild 5.), werden die Spannungen berechnet. Für "Entladung" dieser Unterstrukturen werden die Verschiebungen des erstens Etappe verwendet. Das Programm SAP IV ermöglicht so eine Betrachtung durch Verwendung eines speziellen Randelement.

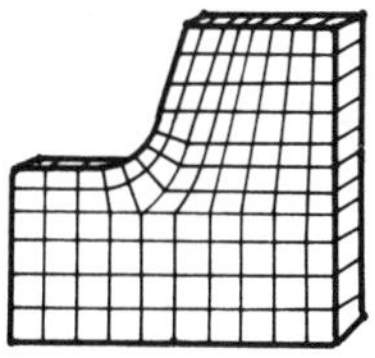

Bild 5. Unterstruktur für Spannungenberechnung (z =∞).

Betrachtungen über die Lastverteilungen

Für die durchgeführten Rechnungen machte sich die Voraussetzung einer linearen Gesetz der Lastverteilung über die Kontaktlänge. Diese Annahme beruht sich auf einen großen Zahl von theoretischen und experimentellen Untersuchungen und entspricht gut dem gewöhnlichen gehärteten Getrieben mit $b/d < 1$.

Das Verhältnis b/m, ein wichtiger Parameter für die Lastverteilung, nimmt in dieser Versuche vier Werte: 5; 9; 12 und 17.

Die liniar verteilten Kräfte entsprechen bei vier Breitenlastverteilungsfaktoren: $K_{H\beta}$ = 1,25; 1,5; 1,75 und 2. Gleichzeitig würden die Zähne z = 17; 50 und ∞ (b/m = 9), auch mit konzentrierten Kräfte am Mitte und am Ende des Zahnes entladet. Für alle analysirte Situationen Kraft wirkt am Kopfe des Zahnes und am Abstand von 0,6 Modul von diesen.

Bestimmung der $K_{F\beta}$ - Faktoren

Die errechneten Spannungszustände dienen zum Bestimmung der $K_{F\beta}$ - Faktoren, als Verhältnis zwischen dem maximalen örtlichen Spannungen bei einen ungleichmäßigen Lastverteilung - $\sigma_{Fmax,ugl}$ - und den maximalen örtlichen Spannungen bei gleichmäßigen Lastverteilung.

$$K_{F\beta} = \sigma_{Fmax,ugl} / \sigma_{Fmax,gl} \qquad (4)$$

Tafel 2 zeigt einige der so berechneten K - Faktoren im Vergleich mit anderem Werten. Die Rechnungbeziehungen die am Grunde dieser Werte stehen sind:

$$K_{F\beta} - 1 = (K_{H\beta} - 1)[1 - 0{,}54e^{\alpha}] \qquad (5)$$
$$\alpha = -0{,}1(b/m - 4)$$

$$K_{F\beta} = K_{H\beta}^{f/(f+1)}; \; f = 0{,}27 \; b/m \qquad (6)$$

$$K_{F\beta} = 2a - (2a - 1)e^{-0{,}64 \, F}; \qquad (7)$$
$$F = (K_{H\beta} - 1)/(a + 1)^{1,2};$$
$$a = 0{,}27 \; b/m$$

$$K_{F\beta} = K_{H\beta}^{N} \; ; \; N = \frac{(b/h)^2}{1 + (b/h) + (b/h)^2} \qquad (8)$$
$$h = 2{,}25m$$

TAFEL 2 $K_{F\beta}$ - Faktoren

	$K_{H\beta}$	b/m 5	9	12
Eigene Werte (Gl.4) z = 17	2	1,34	1,61	1,72
	1,75	1,25	1,45	1,55
	1,5	1,16	1,30	1,38
	1,25	1,07	1,15	1,21
Biger, 1979 (Gl.5)	2	1,21	1,28	1,34
	1,75	1,16	1,22	1,26
	1,5	1,15	1,15	1,18
	1,25	1,06	1,08	1,10
Reschetov, 1969 (Gl.6)	2	1,49	1,63	1,70
	1,75	1,38	1,49	1,53
	1,5	1,26	1,33	1,36
	1,25	1,14	1,17	1,19
Schleifer, 1967 (Gl.7)	2	1,35	1,52	1,59
	1,75	1,26	1,38	1,42
	1,5	1,16	1,24	1,26
	1,25	1,07	1,11	1,12
ISO/DP 6336/III (Gl.8)	2	1,21	1,70	1,76
	1,75	1,65	1,53	1,58
	1,5	1,12	1,36	1,39
	1,25	1,06	1,18	1,20

ZUSAMMENFASSUNG

Die Formfaktoren - Y_F und die Breitenverteilungsfaktoren - $K_{F\beta}$ wurden mit der Methode finiter Elemente ermittelt. Die Zähnemodelle und die Strukturierung wurden so genommen, daß, ohne zu großen Aufwand im Rechnerzeiten, genugend präzise Lösungen finden.

Der Formfaktor wird aus maximalen Spannungen im Fußausrundung gemäss ebener Verformungzustände (Zahn infiniter Breite unter gleichmäßigen Lastverteilung) ermittelt. Die so ermittelten Faktoren berücksichtigen auch die Kerbwirkung der Fußausrundung.

Die Breitenfaktoren für die Zahnfußberechnungen - $K_{F\beta}$ werden als Verhältnis der maximalen Spannungen bei ungleichmäßigen Lastverteilungen und der maximalen Spannungen bei gleichmäßigen Lastverteilung ermittelt.

Die berechneten Faktoren sind im Vergleich mit anderen Werte vorgestellt. Natürlich gibt es Unterschiede, die als erste Erklärung die große Varietät der Methoden haben. Die durgeführeten Untersuchungen haben auch die wesentliche Einflüße der Form (durch Zähnezählen) und der Eingriffsphase gezeigt.

LITERATUR

Baronet,C.N. und Tordion,G.V. (1973). Exact stress distribution in standard gear teeth and geometry factors. Trans. ASME, B95, 1159-1163.

Bathe, K.J., E. Wilson und F. Peterson (1973). SAP IV. A structural analysis program for static and dynamic response of linear systems.

Birger, I.A. und andere (1979). Raszet na prozinosti detalei maschin. Maschinostroienie. Moskwa.

Chabert, G. und andere (1974). An evaluation of stresses and deflection of spur gear teeth under strain. Trans. ASME, B96, 85-93.

Linke, H. (1983). Spannungskonzentrazion bei Verzahnungen. Maschinenbautechnik, 4, 174-179

Reschetov, D. und M. Schleifer (1969). Ob ozenke wliania konzentrazii nagruski na raspredelenii napreajenii w galteli suba subceatogo kolesa. Maschinovedenie, 3, 66-74.

Schleifer, M.A. (1967). Ozenka wliania konzentrazii nagruski na napreajenia w galteli suba subceatogo kolesa. Vestnik Maschinostroenia, 8, 38-41.

Tobe,T., M. Kato und K. Inoue (1979). True stress and stiffness of spur gear tooth. Proceeding of the Fifth World Congress on Theory of Machines and Mechanisms. 1105-1108.

Ustinenko, V.L. (1974). Napreajennoe sostoianie subiew zilindriceskih preamosubih koles. Maschinostroienie, Moskwa.

ISO/DP 6336/III. Basic Principle for the Calculation of Load Capacity of Spur and Helical Gears; Calculation of Tooth Strength.

In this paper the finite element method is applied to compute the true stress at the root of involute spur gear tooth. The computed stresses are used to determinate the form factor Y_F and the longitudinal distribution factor for the stress calculation $K_{F\beta}$ (according to ISO/DP 6336/III).

The form factor Y_F are obtined from the maximum tensile stress on the root of tooth under uniformly distributed load (plane strain). They including the stress concentration effect of the root form.

The factor $K_{F\beta}$ are determinate as ratio between the maximum stress for a nonuniformly distributed load and the maximum stress for uniformly distributed one.

The parameters taked into account are the tooth form (throught z), the ratio b/m, nonuniformity of load distribution ($K_{H\beta}$) and the engagement phases.

Optimising the Ratio Series of Common-Type and Planetary Gear-Boxes

S. BOBANCU

University of Braşov, R. S. Romania

Optimisation of the real transmission ratio series (ratio series series of real numbers) of a gear-box with (gear-box multi-stage transmission) a view to accomplishing the maximum possible corespondence with the initially considered theoretical series is achieved by the series connection of an additional constant transmission-ratio gear mechanism and a gear-box. An exhaustive procedure is proposed in the paper for the computation of the transmission ratio of the additional gear mechanism.

1. The paper deals with the following structural problem (for exemple [1,2] the structure of a multi-stage transmission): given (fig.1) a sequence (of transmission ratios) $b_1 < b_2 < \dots < b_n$, determine the sequence fo real numbers $t_1, t_2, \dots, t_k$, so that the elements of the set $\{t_i b_j\}$ with $i=1\dots k$, $j=1\dots n$, may be arranged in a sequence to best approsimate a previosly given ideal sequence $a_1 < a_2 < \dots$ $\dots < a_{kn}$.

2. If any one of the numbers

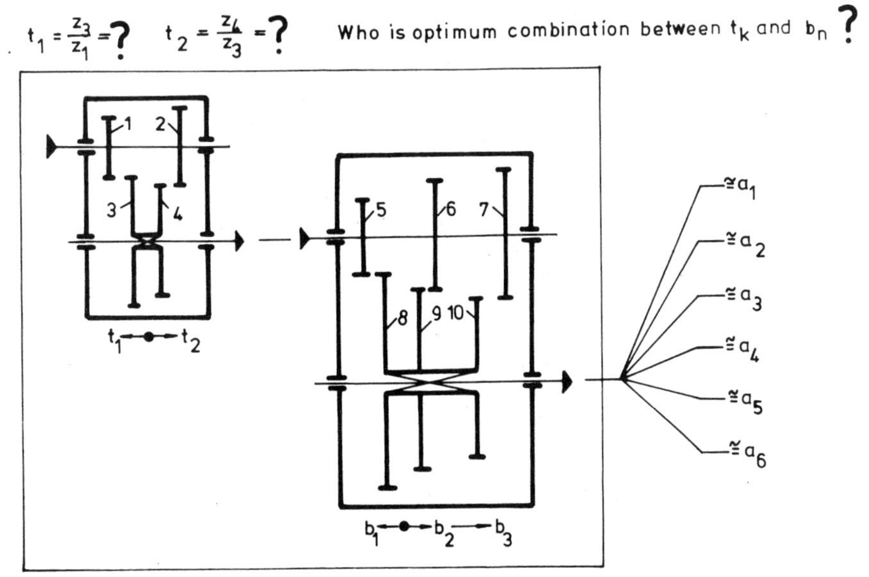

$b_1 = \frac{z_8}{z_5}$; $b_2 = \frac{z_9}{z_6}$; $b_3 = \frac{z_{10}}{z_7}$; theoretical series a_{kn} : GIVEN

Figure 1

$$d_p(x,y)=\begin{cases}(\sum_i |x_i-y_i|^p)^{\frac{1}{p}}, & 1\le p<\infty,\\ \max_i |x_i-y_i|, & p=\infty,\end{cases}$$

is propoved as the distance between two sequences $x=(x_i)_i$ and $y=(y_i)_i$, then the problem, as stated, consist in the determination of the numbers $t_1^o,\dots,t_k^o$ and a permutation W^o to attain the value $m(p)=\min m_W(p)$ with $m_W(p) = \min\{|a_{W(1)}-t_1b_1|^p+\dots+|a_{W(n)}-t_1b_n|^p+\dots+|a_{W((k-1)n+1}-t_kb_1|^p+\dots+|a_{W(kn)}-t_kb_n|^p\}^{1/p}$ with respect to $t_1,\dots,t_k\in[0,\infty)$ with W covering the set $\sum_{kn}$ of all permutations $\{1,2,\dots,kn\}$

3. The paper demonstrates that:

(I) the minimum of m(p) is reached;

(II) the numbers $t_1^o,\dots,t_k^o$ are solely determined for $p\in(1,\infty]$ and for p=1 they describe a closed interval each [3];

(III) m(p) is a increasing function in the variabile $p\in[1,\infty]$;

(IV) $\min\{m_W(p):W\in\sum_{kn}\}=\min\{m_W(p):W\in\sum^*_{kn}\}$ where $\sum^*_{kn}$ is the set of those permutations W for which $W(1)<\dots<W(n)$; $W(n+1)<\dots<W(2n);\dots;W((k-1)n+1)<\dots<W(kn)$;

(V) the number of the elements of the set $\sum^*_{kn}$ is equal to $(kn)!/(k!(n!)^k)$;

(VI) given a number $m>0$, a necessary condition (and sufficient one in case of $p=\infty$) for $m_W(p)\le m$ is the satisfaction of the following inequalities

$$\max_{1\le i\le n}\frac{a_{W(\ell n+I)}-m}{b_i}\le\min_{1\le i\le n}\frac{a_{W(\ell n+i)}+m}{b_i}$$

with $\ell = 0, 1, \dots, k$.

4. In terms of computer time, (IV) assures a substantial saving, for instad effectuating the B_W subrutine of $m_{W(p)}$ for all (kn)! possible permutations of $\sum_{kn}$, it is sufficient to take into account the $(kn)!/(k!(n!)^k)$ permutations of $\sum^*_{kn}$ (for example, for n=4, k=3 the number of permutations can be reduced from 479oo16oo to 5775).

5. The result (VI) also increases the rapidity of obtaining the final solution because those permutations $W\in\sum^*_{kn}$ wich the calculation of $m_W(p)$ is still necessary, can be tested by T (see infra).

6. A subrutine A^* is used for the description of the permutations of $\sum^*_{kn}$, the subrutine B_W (see tab.1) being also presented for $p\in\{1, 2, \infty\}$. The T - test verifies the inequalities (VI), in which m is m_W, calculated for the first permutation $W^o\in\sum^*_{kn}$.

7. The procedures proposed are illustrated by the following exemples:

Exemple 1) :

Given: n = 3 ;

k = 2;

$b_1= 3$;

$b_2= 5$;

$b_3= 8,1$;

$a_1 = 1,2$;

$a_2 = 1,5$;

$a_3 = 1,7$;

$a_4 = 2,5$;

$a_5 = 2,7$;

$a_6 = 4$;

we obtain for $p = \infty$

$t_1 = 0,3636$

$t_2 = 0,4968$

which involue the following approximations

$t_1b_1 = 1,0909 \cong 1,2 = a_1$;

$t_2b_1 = 1,4906 \cong 1,5 = a_2$;

$t_1b_2 = 1,8181 \cong 1,7 = a_3$;

$t_2b_2 = 2,4844 \cong 2,5 = a_4$;

$t_1 b_3 \cong 2{,}9454 = 2{,}7 = a_5$.

Table 1

Subrutine B_W

CRITERION-FUNCTION:

- absolute case:
- percentage case:

$$E_p = \begin{cases} \left(\sum |a_n - tb_n|^p \right)^{\frac{1}{p}}, \\ 100\left(\sum \left| \frac{a_n - tb_n}{a_n} \right|^p \right)^{\frac{1}{p}}, \end{cases} \quad t \in R, \quad p \in [1, \infty] .$$

GIVEN: $k=1,\ b_1,\ b_2,\ \ldots,\ b_n,\ a_1,\ a_2,\ \ldots,\ a_n$.

$$\min E_p \Leftarrow t_{optimum} = ?$$

p=1	p=2	p=∞
- $p=1$	- $p=2$	- $p=\infty$
- calculate series $\frac{a_i}{b_i}$	- $t=\frac{\sum(a_q b_q)}{\sum(b_q^2)}$	- calculate series $c_q=\frac{a_i+a_j}{b_i+b_j}$ with $i \neq j$ and $i,j \in \{1,2,\ldots,n\}$
- arrange creasing series $\frac{a_q}{b_q}$	- cal. $E_2(t)$	- cal.series $h_q=\|a_i-c_q b_i\|$
- calculate sum $c_s = -\sum b_q$		- max $h_q \Rightarrow q \Rightarrow t_\infty = c_q$
- cal.series $c_q=c_s+2b_q$		- min $E_\infty(t)=\max h_q$
- if $c_{q-1} < 0$ and $c_q > 0$ $t=\frac{a_q}{b_q}$		
- if $c_{q-1} = 0$ $t = \left[\frac{a_q}{b_q}, \frac{a_{q+1}}{b_{q+1}}\right]$		
- calculate min $E_1(t)$		

$t_2 b_3 \cong 4{,}0248 = 4 = a_6$

Exemple 2):

Given: $n = 4$;

$k = 1$ (fig. 2);

$b_1 = 2{,}9166$;

$b_2 = 2{,}4000$;

$b_3 = 1{,}9565$;

$b_4 = 1{,}7000$;

$a_1 = 3{,}4450$;

$a_2 = 2{,}8300$;

$a_3 = 2{,}2150$;

$a_4 = 1{,}6000$;

we obtain for $p = \infty$

$t = 1{,}1561$

wich involue the following approxima - tions

$tb_1 = 3{,}3719 \cong 3{,}4450 = a_1$;

$tb_2 = 2{,}7747 \cong 2{,}8300 = a_2$;

$tb_3 = 2{,}2619 \cong 2{,}2150 = a_3$;

$tb_4 = 1{,}1618 \cong 1{,}6000 = 4_4$.

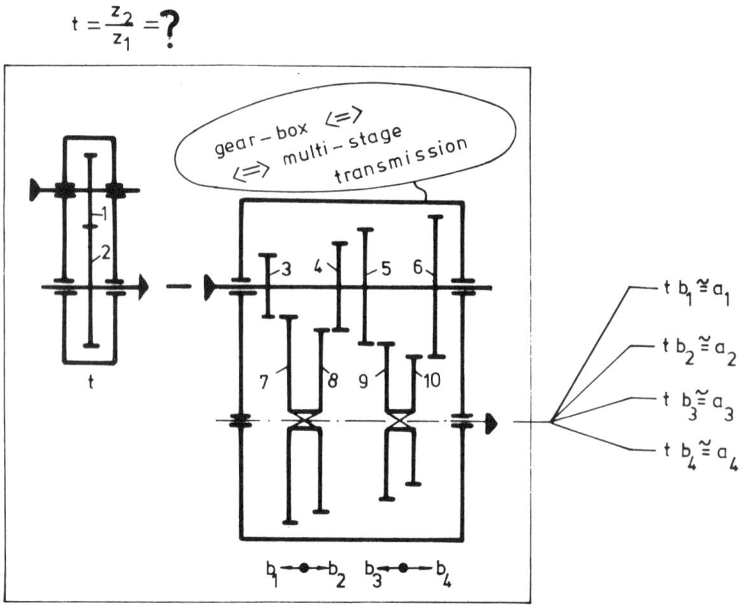

$b_1 = \frac{z_7}{z_3}$; $b_2 = \frac{z_8}{z_4}$; $b_3 = \frac{z_9}{z_5}$; $b_4 = \frac{z_{10}}{z_6}$; theoretical, series a_n: GIVEN

Figure 2

8. As for as multi-stage transmissions (particular case k=1 - figure 2), are concerned, the authors recommend operate with p = ∞, since in this way the dispersion field of the deviations $|a-tb|$ narrows down to its minium.

REFERENCES

1. Bobancu,S., Contribuţii la analiza şi sinteza structurală şi numerică a mecanismelor cu roţi dinţate cu axe mobile, Rezumatul tezei de doctorat,Universitatea din Braşov,1977.

2. Bobancu,S. Finisarea seriei de rapoarte a unor scheme cu grupuri planetare în raport cu o serie dată, Proc.SYROM'77, vol. III,p.69...78,Bucharest, R.S.Romania, 1977.

3. Bobancu,S., Benko,I., Criteriu şi algoritm pentru aproximarea în raport cu un şir impus, a şirului de rapoarte al oricărei cutii de viteze,Proc.SYROM'77 vol.III,p.57...68,Bucharest,R.S.Romania, 1977.

4. Bobancu,S., Les criteres absolus et de pourcentage,et leur solution en tant que problemes de programmation lineare, a l'optimisation de la suite des raports de toute transmis - sion a plusieurs echelons,Annuaire des ecoles superieurs, Mechanique technique technique, vol. XII,livre 1, p.19...25, Bulgaria,1977.

L'OPTIMISATION DES SERIES DES RAPORTS DES TRANSMISIONS MULTI-RAMPES PLANETAIRES ON A ROUES À DENTS À AXES FIXES

- Résumé -

Le modiffication de la series de raports reeles d'une transmision multirampe,pour une meilleure approximation d'une serie théoretique s'obtient par la conexion type serie d'un reducteur ou amplificateur a rapport de transmision fixé.Dans l'ouvrage on presente une procedure exhaustive pour le calcul de ce rapport, quand on connais les series reale et theoretique.

Error Calculation of Tredgold's Approximation

F. MONTOYA* AND F. ROMANO**

*Applied Mechanic Department, University of Valladolid, Spain
**E.T.S.I.I., Polytechnique University of Valencia, Spain

Abstract. The limit number of teeth calculation of bevel gears made by a crown-rack is usually made by Tredgold´s aproximation, projecting the teeth spherical curves intersection over the back cone. By unfolding this cone, this spheric al problem is reduced to plane one. A general procedure is exposed in this paper , to calculate the exact limit number of teeth to avoid undercut, and it is applied to the particular case above exposed comparing results.

Key words. Gears; Undercut

INTRODUCTION

The theoric bases of this work are developed in [1], and here will only be used their results.

The generating surface of a crown rack is a plane which contains the pitch cone apex of the wheel (the parameters of the generating surface are refered to with the sub index 3, and the ones from the wheel with the 1).

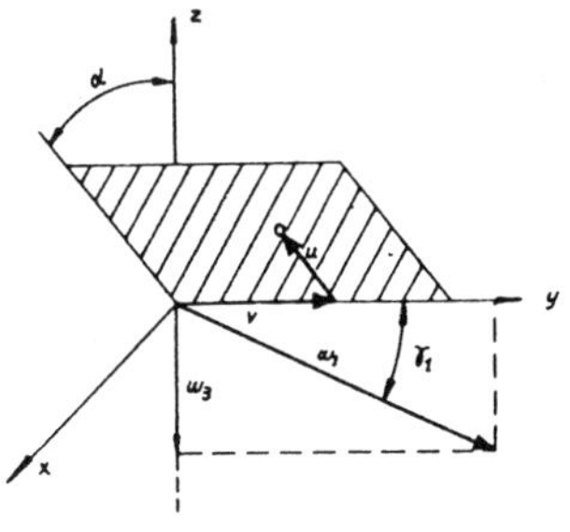

Being γ_1 the pitch-cone angle, the speeds relation between the generating surface / and the wheel is:

$$\underline{w}_3 = \underline{w}_1 \cdot \cos {}_1 \quad (1)$$

GENERATING SURF. GEOMETRY

The generating surface plane will be defined by the u, v, parametric coordinates , being its equation in the initial position of the Fig., (the variables refered to the initial position have asterisk):

$$\underline{r}^* = u \text{ sen}\alpha \, \underline{i} + v \, \underline{j} + u \cos\alpha \, \underline{k} \quad (2)$$

The normal vector, for the initial position is:

$$\underline{n}^* = -\cos\alpha \, \underline{i} + \text{sen}\alpha \, \underline{k} \quad (3)$$

being null the normal gradient tensor, $\nabla \underline{\underline{n}}$.

TURN ANGLE OF CONTACT

The turn angle, from the initial position, so the u, v, generating surface point is making contact is calculated in [1], and its value, for this particular case is:

$$\tan \phi_3 = -u/(v \text{ sen}\alpha) \quad (4)$$

For instance, the plain points u=0, are contact points for the initial position, where the turn angle is 0. In another instant the contact lines are the straight lines:

$$u/v = -\text{sen}\alpha \cdot \tan\phi_3 \quad (5)$$

being Φ_3 a constant.

CUSP EDGE

Being $\underline{v}_i$ the velocity of the contact point of the wheel (i=1), or of the generating / surface (i=3), and:

$$\underline{q} = \underline{w}_3 \times \underline{v}_1 - \underline{w}_1 \times \underline{v}_3 \qquad (6)$$

$$\underline{v}_3 = (\underline{w}_3 - \underline{w}_1) \times \underline{n} - \underline{\underline{\nabla_3 n}} \cdot (\underline{v}_3 - \underline{v}_1) \qquad (7)$$

the condition for a cusp is:

$$\underline{v}_3 \cdot (\underline{v}_3 - \underline{v}_1) + \underline{n} \cdot \underline{q} = 0 \qquad (8)$$

Taking account that:

$$[\Phi_3] = \begin{bmatrix} C & S & 0 \\ -S & C & 0 \\ 0 & 0 & 1 \end{bmatrix} \qquad (9)$$

and $\underline{r}_3 = \underline{\Phi}_3 \cdot \underline{r}^*_3$, $\underline{n} = \underline{\Phi}_3 \cdot \underline{n}^*$, $\underline{v}_i = \underline{w}_i x \underline{r}_i$,

substituting these values into (8), the condition for a cusp in the conjugate point of u,v, is:

$$\cos^2\alpha \cdot \mathrm{sen}^3\phi - \mathrm{sen}\phi + \tan\gamma_1 \cdot \tan\alpha = 0 \qquad (10)$$

being Φ the turn angle (4). As this relation doesn´t depend upon u, v, this means that the cusp edge is a contact line.

TREDGOLD´s APPROXIMATION ERROR

So as to calculate the teeth limit number the limit condition will be that the tooth foot be the cusp edge, so:

$$-u\cos\alpha = m = 2R_1/z_1 = 2v\,\mathrm{sen}\gamma_1/z_1 \qquad (11)$$

from (4):

$$\tan\Phi = 2\,\mathrm{sen}\,\gamma_1 / (z_1\,\mathrm{sen}\,\alpha\,\cos\alpha) \qquad (12)$$

Equations (10) and (12) establish the relation between the teeth limit number versus the pitch cone angle. If the turn angle is smaller than 5°, the previous equations can be simplified, resulting:

$$z_1 = 2.\cos\gamma_1/\mathrm{sen}^2\alpha \qquad (13)$$

which are the teeth limit number according Tredgold´s aproximation.

In the next table the exact and the approximate teeth limit number are calculated, versus the wheel pitch-cone angle:

γ_1	z Tred.	z real
10°	16.84	16.75
15°	16.51	16.32
20°	16.07	15.71
25°	15.50	14.91
30°	14.81	13.95
35°	14.01	12.76
40°	13.10	11.30
45°	12.09	9.39

CONCLUSION

The teeth limit number calculation of bevel gears made by a crown-rack, according Tredgold´s approximation is valid until / pitch-cone angles of 30° and this approximation is conservative.

RESUME

Pour calculer le nombre exact limit de / dents dans engrenages concourants a denture droite, on a utilisè une mèthode analytique. Les resultats ont etè comparès / avec ceux obtenis par moyen l´aproximation de Tredgold, en etant celui valide pour / semiangles de le cône primitive puisque / 30°. Le resultats de cette aproximation / ont conservatives.

REFERENCES

[1] Montoya F. (1985). Fundamentos de la Geometría de los Engranajes. ETSII , Valladolid. Spain.

[2] Bucckingham E. (1963). Analytical Mechanics of Gears. Dover Publications. N.Y.

Equalizing Blok's Flash Temperature in the Extremities of the Contact Line, as an Optimization Criterion for Spur Gears

C. RIBA I ROMEVA

Departament of Mechanical Engineering, Universitat Politècnica de Catalunya, Spain

Abstract. In the course of research on the optimal geometry of spur gears, starting from the intrinsic limitations of the gearing --that is, not from those imposed by normalized tools-- we analized the geometrical consequences of tooth fatige breakeage, fatigue pitting of the surface, and scuffing of the teeth. In relation to scuffing prevention, the author considered Blok's flash temperature theory, and took the equalizing of the flash temperature in the extremities of the contact line as an optimization criterion. The resulting geometric condition (the relative situation of the extremities in the contact line) is the same as that which results from equalizing the Almen PV factor, and also from equalizing the specific slinding in the extremities of the contact line. The coincidence of geometrical condition resulting from these last two criteria (equalizing the Almen PV factor, and equalizing specific slinding) was well known. The coincidence with the equalizing of Blok's flash temperature, analized here, gives new possibilities to this important design criterion for spur gears.

Keywords. Optimization; spur gear; scuffing; scoring; flash temperature.

INTRODUCTION

The determination of spur gear geometry, as a function of load capacity and running conditions required, means taking into account several design criteria to prevent gear failures --frequently contradictory--, about which we have to find a compromise solution.

Today, the following criteria are widely accepted in the designing of spur gear:

1) The prevention of fatigue pitting on gear teeth surfaces --often the most limiting criterion-- is that which leads to the determination of the main dimensions of the gear: center distance, facewidth, and also the pitch angle.

2) The prevention of fatigue breakage of gear teeth is the criterion that permits us to determine the size, and as far as possible, the form of the teeth.

3) The prevention of scuffing failure is the criterion that facilitates the situation of the contact zone inside the line of action T_1T_2, and hence, the size of tip diameters of pinion and wheel.

Refering to scuffing failure, we analized *Almen's PV factor* (Ref. 1) and *Blok's flash temperature* criteria (Ref. 2).

In the course of research on the optimal geometry of spur gears, starting from the intrinsic limitations of the gearing --that is, not from those imposed by normalized tools-- the author takes into account the scuffing prevention criteria just mentioned

Starting from a simplified model, we can consider that the scuffing is of the greatest severity in the extremities of the path of contact (points A and E in figure 1). Hence, the geometric condition that minimizes the danger of this damage corresponds to the equalizing of either Almen's PV factor, or Blok's flash temperature.

(Later we shall analize the influence of the load sharing that occurs in the zone ends of the contact line between two successive mesh contact points (double-pair contact), and the consequences of the introduction of the sharing factor X_Γ as defined in ISO/DP 6336 IV, Ref. 5).

It is well known (Ref. 4) --and the author has corroborated (Ref.6)-- that the equalizing of Almen's PV factor in the ends of the contact line (points A and E) gives the same geometric condition as that imposed by the equalizing of specific sliding.

When we studied the new geometric condition that results from the equalizing of Blok's flash temperature at the same points (A and E), we found that it is again the same condition as the two last mentioned criteria: equalizing specific slidings, and equalizing Almen's PV factors. That conclusion, although it was established for intrinsic geometry, preserves its validity for gears made with normalized tools.

The object of this paper is, thus, to present the above-mentioned result, which in our opinion reinforces this criterion for spur gear design, and also to analize the influence of the sharing factor X_Γ (as defined in ISO/DP 6336 IV) on minimizing the danger of scuffing damage, taking Blok's flash temperature hypothesis.

GEOMETRIC DEFINITIONS

Starting from the well known parameters of gear theory:

Z_1	number of teeth in pinion
Z_2	number of teeth in wheel
a	centre distance
α	pressure angle
$u = Z_2/Z_1$	gear ratio
ω_1	angular velocity of pinion (rad/s)
$\omega_2 = \omega_1/u$	angular velocity of wheel (rad/s)

we define, taking the line of action T_1T_2, as a length unit, two adimensional factors (we call them *curvature factors*), according to the following equations (see figure 1):

$$T_1T_2 = a.\sin(\alpha) \quad \text{line of action length}$$

$$\rho_{TE} = \rho_{1E}/T_1T_2 \quad \text{E-curvature factor}$$
$$\rho_{TA} = \rho_{2A}/T_1T_2 \quad \text{A-curvature factor} \qquad (1)$$

factors related to the gamma-parameters defined in the ISO/DP 6336 IV by means of the following expressions:

$$\rho_{TE} = (1+\Gamma_E)/(1+u) \qquad \Gamma_E = (1+u).\rho_{TE} - 1$$
$$\rho_{TA} = (u-\Gamma_A)/(1+u) \qquad \Gamma_A = u - (1+u).\rho_{TA} \qquad (2)$$

The components of velocity, normal to the path of contact, at the end points A and E are:

$$v_{1E} = \rho_{TE}.a.\sin(\alpha).\omega_1 \quad \text{pinion at E}$$
$$v_{1A} = (1-\rho_{TA}).a.\sin(\alpha).\omega_1 \quad \text{pinion at A}$$
$$v_{2E} = (1-\rho_{TE}).a.\sin(\alpha).\omega_2 \quad \text{wheel at E}$$
$$v_{2A} = \rho_{TA}.a.\sin(\alpha).\omega_2 \quad \text{wheel at A} \qquad (3)$$

and the sliding velocity moduli at these two same points are:

$$v_{gE} = |\, v_{1E} - v_{2E} \,|$$
$$v_{gA} = |\, v_{1A} - v_{2A} \,| \qquad (4)$$

From the former definitions, the specific slidings for each point and each profile, are defined as follows (Refs. 4 and 6):

for the pinion, at point E:

$$g_{s1E} = \frac{v_{gE}}{v_{1E}} = \frac{(1+u).\rho_{TE}-1}{u.\rho_{TE}} \qquad (5)$$

for the pinion, at point A:

$$g_{s1A} = \frac{v_{gA}}{v_{1A}} = \frac{(1+u).\rho_{TA}-u}{u.(1-\rho_{TA})} \qquad (6)$$

for the wheel, at point E:

$$g_{s2E} = \frac{v_{gE}}{v_{2E}} = \frac{(1+u).\rho_{TE}-1}{1-\rho_{TE}} \qquad (7)$$

for the wheel, at point A:

$$g_{s2A} = \frac{v_{gA}}{v_{2A}} = \frac{(1+u).\rho_{TA}-u}{\rho_{TA}} \qquad (8)$$

The maximum values of specific sliding occurs in the respective teeth feet, that is, at point A for the pinion, and point E for the wheel. Hence, establishing the equality:

$$g_{s1A} = g_{s2E} \qquad (9)$$

we obtain the geometric condition resulting from equalizing the specific sliding at the end points of the path of contact:

$$\frac{\rho_{TE}-1}{\rho_{TE}} = u^2 \frac{\rho_{TA}-1}{\rho_{TA}} \qquad (10)$$

SCUFFING. BLOK'S HYPOTHESIS.

Scuffing is damage in tooth surface that occurs rapidly and suddenly when the teeth are submitted to a hard working conditions, specially to high contact pressure and great sliding velocity, and in general case, presents its maximum severity in the end profile zones, that is, in the tip and foot of the teeth.

The scuffing phenomenon --which mechanism is not yet fully understood-- nevertheless seems to be a consequence of the breaking down the oil-film between the contact surfaces, caused by friction heat, the formation microwelds, and immediate tearing off due to the relative velocity and the inertia.

In 1937, Blok (Ref. 2) established the hypothesis from which the oil-film breakage occurs when the contact temperature surpasses an admissible value. Blok distinguishes two sumands in the *contact temperature* T_c:

a) The *bulk temperature* T_b, that is the temperature of tooth surfaces before meshing, the value of which changes slowly in time.

b) The *flash temperature* T_{fla}, that is the local and instantaneous rise of temperature that occurs in the contact zone due to the friction heat.

In this paper, where we propose to study the evolution of the contact temperature T_c along the path of contact --and hence, we do not consider the absolute values--, our interest will only be in the flash temperature variation along the mesh.

The Blok hypothesis (Ref. 3) gives the following equation for the flash temperature at any point of contact C:

$$T_{fla} = 0.62\ \frac{f}{b_u}\ \sqrt[4]{\frac{F_{nu}^3 . E_r}{\rho_c}}\ \left|\sqrt{v_{1c}} - \sqrt{v_{2c}}\right| \qquad (11)$$

where the symbols have the following significance:

- F_{nu} normal load per unit width (N/mm)
- f coefficient of friction
- b_u thermal contact coefficient $(N/(mm.s^{1/2}.{}^{\circ}C)$
- E_r reduced modulus of elasticity (N/mm^2)
- ρ_c reduced radius of curvature (point C) (mm)
- v_{1c} pinion velocity normal to T_1T_2 (point C) (mm/s)
- v_{2c} wheel velocity normal to T_1T_2 (point C) (mm/s)

The adopted criterion consisting in the equalizing of the flash temperature at both ends of the contact line, A and E, gives the following equality:

$$T_{flaA} = T_{flaE} \qquad (12)$$

Introducing in this equality the expression of each one of its parameters, all the terms turn out to be constant for our purpose, except the reduced radii of curvature and the velocities of pinion and wheel. The equality becomes:

$$\frac{\left|\sqrt{v_{1E}} - \sqrt{v_{2E}}\right|}{\sqrt[4]{\rho_{rE}}} = \frac{\left|\sqrt{v_{1A}} - \sqrt{v_{2A}}\right|}{\sqrt[4]{\rho_{rA}}} \qquad (13)$$

where the reduced radii of curvature are defined as follows (see parameters in figure 1):

$$\frac{1}{\rho_{rE}} = \frac{1}{\rho_{1E}} + \frac{1}{\rho_{2E}} = \frac{\rho_{1E} + \rho_{2E}}{\rho_{1E} . \rho_{2E}}$$

$$\frac{1}{\rho_{rA}} = \frac{1}{\rho_{1A}} + \frac{1}{\rho_{2A}} = \frac{\rho_{1A} + \rho_{2A}}{\rho_{1A} . \rho_{2A}} \qquad (14)$$

Introducing in the former equality (13) the expressions of reduced radii of curvature (14), and the teeth contact velocities (3), after simplifying, we arrive at:

$$\frac{\sqrt{u.\rho_{TE}} - \sqrt{1 - \rho_{TE}}}{\sqrt[4]{\rho_{TE}.(1 - \rho_{TE})}} = \frac{\sqrt{\rho_{TA}} - \sqrt{u.(1 - \rho_{TA})}}{\sqrt[4]{\rho_{TA}.(1 - \rho_{TA})}} \qquad (15)$$

Squaring twice successively and simplifying, the former expression becomes:

$$u^2 . \frac{\rho_{TE}}{1-\rho_{TE}} + \frac{1-\rho_{TE}}{\rho_{TE}} = \frac{\rho_{TA}}{1-\rho_{TA}} + u^2 . \frac{1-\rho_{TA}}{\rho_{TA}} \qquad (16)$$

Finally, establishing that $A=(1-\rho_{TE})/\rho_{TE}$ and $B = (1-\rho_{TA})/\rho_{TA}$, and after resolving the resulting equation, we obtain two solutions: $A = u^2.B$ and $A = 1/B$. From the first, we obtain the same relation as that for the equalizing of the specific slidings and also the equalizing of Almen PV factors (compare equations 6 and 17):

$$\frac{1 - \rho_{TE}}{\rho_{TE}} = u^2 . \frac{1 - \rho_{TA}}{\rho_{TA}} \qquad (17)$$

The second solution is trivial: $\rho_{TE}+\rho_{TA}=1$ and means that all the path of contact is reduced into the pitch point C.

EFFECTS OF LOAD SHARING.

As we have said above, in the end zones of the path of contact (segments AB and DE, in figure 1), there is a sharing of the load between two consecutive pairs of teeth (double-pair contact).

The ISO/DP 6336 IV, establishes four types of load-sharing diagram along the contact line, according to the class of tooth profile and the use (see respectively figures 2, 3, 4 and 5):

a) Speed-reducing gear with profile modification.
b) Speed-increasing gear with profile modification.
c) One-ratio gear with profile modification.
d) Gear without profile modification.

The sharing factor X_Γ affects the normal load F_{nu} that acts over the teeth, and multiplies flash temperature by a value of $X_\Gamma^{3/4}$.

In consequence, given the diagrams established in ISO/DP 6336 IV, it is not possible, in the general case, to solve the ploblem of optimizing the flash temperature simply by applying the geometric condition (17) in the ends of the contact line.

However, we can see that the geometric condition (17) applied at any two points whatever of the path of contact, gives the equalizing of flash temperature at them, if and only if the corresponding sharing factors X_Γ have the same value.

Proceeding on this reasoning, we have examined the following solutions for the flash temperature optimization.

a) For a speed reducing gear with profile modification, we establish the geometrical condition (17) at the highest single-pair contact points (B and D in figure 1). For a low gear ratio (ϵ below than 1.2), the maximum flash temperature values (equalized, in this case) occur in the same highest single-pair contact points. For the greater gear ratio, the maximum flash temperature (here not yet equalized) appears in the zones AB and DE (see example in figure 2), but these temperatures hardly ever exceed the optimal flash temperature that we can obtain from an exact calculation by more than 15 % . In the zone AB, corresponding to the pinion head, the greater growth of the flash temperature wHen we go away from the pitch point C, is compensated by the greater attennuation of the sharing factor X_Γ in the speed-reducing gear, from which the flash temperature in the two zones AB and DE tend to level out.

b) For a speed-increasing gear with profile modification, the lower attennuation of the sharing factor X_Γ corrsponds to the Wheel head zone DE, and the flash temperatures tend to separate. In this case, the establishing of the geometrical condition to the highest single-pair contact points B and D gives a result far away from the exact optimal solution (see example in figure 3). The same occurs if we take into acount the end contact line points A and E.

c) For a one-ratio gear with profile modification, the symmetry gives the optimal solution for all the cases (see example in figure 4).

d) Finally, for the gears without profile modification, as is logical, the sharing factor effect in the extremities of the path of contact (points A and E, where $X_\Gamma = 1/3$) on the flash temperature are more important than the former cases; moreover, the symmetry of the sharing diagram according to ISO/DP 6336 IV (see figure 2), had led us to establish the geometrical condition (17) at the end contact line points A and E. For a gear ratio ϵ greater than 1.2, the maximum flash temperatures reached hardly ever exceed the optimal flash temperature that we can obtained from an exact calculation by more than 15 % . If the gear ratio is lower than 1.2, the flash temperatures in the highest single-pair contact points B and D prevails, and the geometrical conditions (17) should be applied in them (see exemple in figure 5)

CONCLUSIONS

Specific sliding, Almen's PV factor, and Blok flash temperature, in spite of being conceptualy and dimensionaly different, when we impose the equalizing at two points in the path of contact of a spur gear, give the same geometric condition (9) or (17) between their respective positions on the contact line, and hence, they are equivalent from a design point of view

The geometric condition (17) is only valid if the sharing factors X_Γ at the two considered points have the same value.

From the point of view of flash temperature, the geometric condition (17) gives a good approximation for gear optimization if: a) we apply it at the single-pair contact points, in speed-reducing gear with profile modification; b) we apply it in the extremities of the path of contact, in a gear without profile modification. However, it gives a poor result when we apply it to the speed-increasing gear with profil modification.

REFERENCES

1. Almen, J.D. (1935) Factors influencing the durability of spiral-bevel gears for automobiles, Automotive Industries, nov. 23, pp. 676-701.

2. Blok, H. (1937) Les temperatures de surfaces dans des conditions de graissage sous extreme pression, Proc. Second World Petrol Congress, Paris, 3, pp. 471-486.

3. Blok, H. (1970) The postulate about the constancy of scoring temperature, in "The interdisciplinary approach of concentrated contacts", NASA SP-237, pp. 153-248

4. Henriot, G. (1979) Traité théorique et pratique des engrenages, Vol. I, Dunod, Paris

5. ISO/DP 6336 IV (draft of international standard), Calculation of Load Capacity of Spur and Helical Gears; Calculation of Scuffing Load Capacity.

6. Riba i Romeva, C. (1976) Estudi de la influència de la configuració geomètrica dels dentats sobre la capacitat de càrrega dels engranatges rectes. (Study about the influence of geometric configuration of the teeth on load capacity of spur gears). Escola Tècnica Superior d'Enginyers Industrials de Barcelona (Universitat Politècnica de Catalunya), doctoral thesis.

Resume. Dans le cours d'une recherche sur la géométrie optimale pour les engrenages droits en partant des limitations intrinsèques des dentures --c'est à dire, non de celles imposées par les outils normalisés-- on analysa les conséquences géometriques derivées de la résistence à la rupture de fatigue des dents, les piqûres sur les dentures, et le grippage. En relation avec cette dernière détérioration, l'auteur considéra la théorie des temperatures éclair de Blok et établit l'égalization de cette temperature en les extrêmes de la ligne de conduite comme un critérium d'optimization. La condition géometrique résultante (la position relative de les extrêmes de la ligne de conduite) c'est la même que celle résultant de l'égalization des facteus PV d'Almen en ces points, aussi bien que pour les glissements spécifiques. La coïncidence de les conditions géometriques resultantes de l'égalization des facteurs PV d'Almen et égalization des glissements spécifiques était déjà bien conue. Cependant, la nouvelle coïncidence avec l'égalization de la temperature éclair de Blok, analysée dans ce travail, donne des nouvelles possibilités à cet important critérium de dessin pour les engrenages droits.

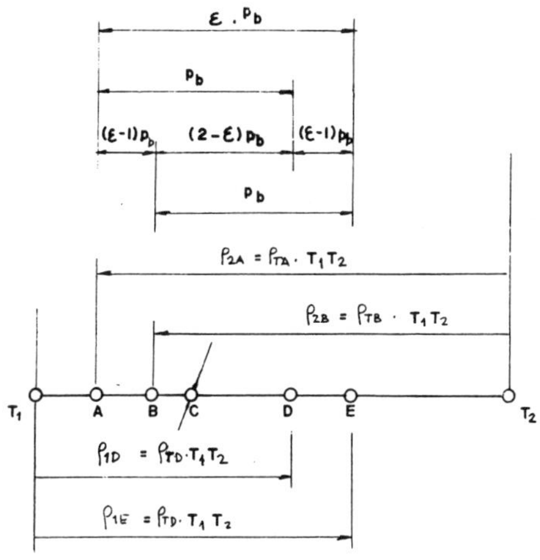

FIGURE 1. Geometric definition of curvature factors.

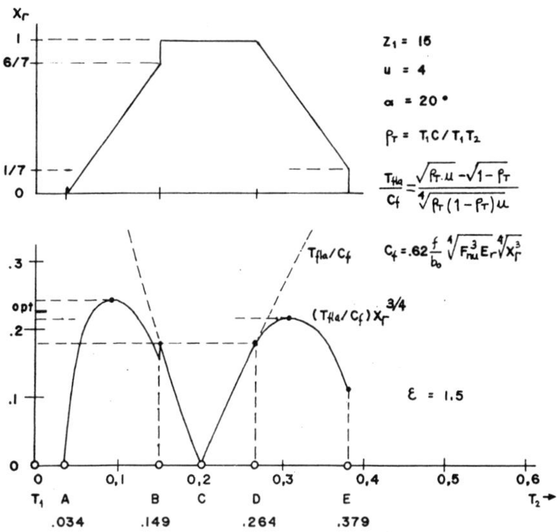

FIGURE 2. Speed-reducing gear with profile modification

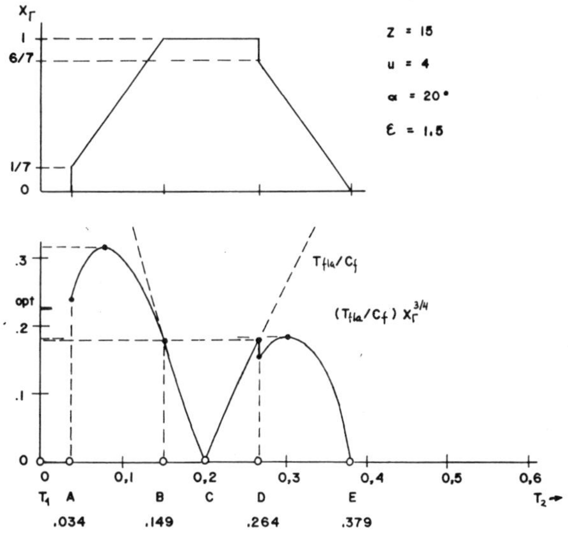

FIGURE 3. Speed-increasing gear with profile modification

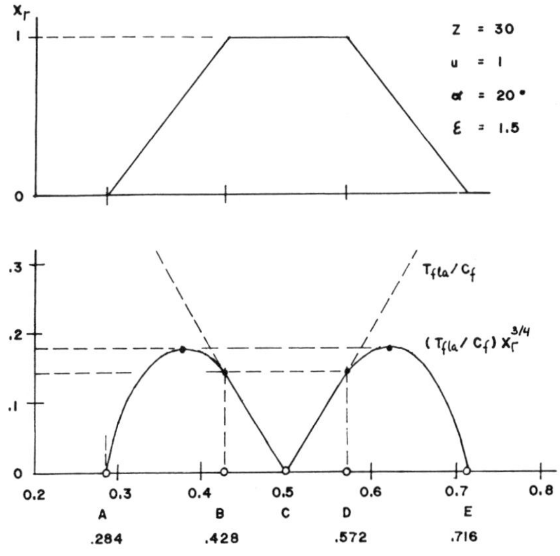

FIGURE 4. One-ratio gear with profile modification

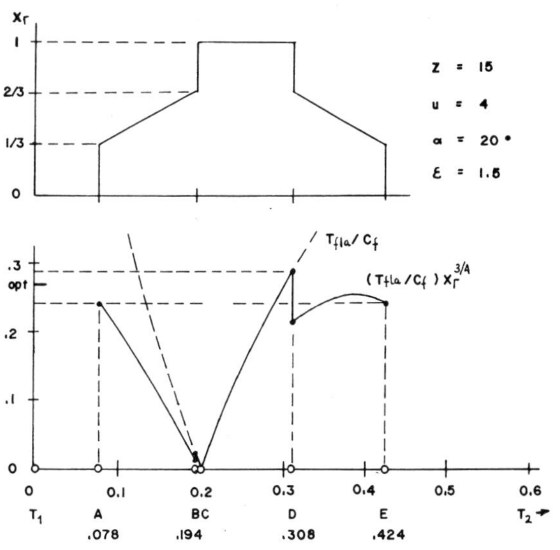

FIGURE 5. Gear without profile modification

A Matrix-Based Method to Analyse Complex Mechanical Transmissions

A. HEDMAN

Division of Machine Elements, Chalmers University of Technology, S-412 96 Göteborg, Sweden

Abstract. A method for analysing general mechanical transmission systems is presented. This method is adapted for computer calculations. The transmission system is considered as being built up of different transmission components, such as gear and belt transmissions, bearings, clutches, and epicyclic gear trains. The shafts of these components are coupled in nodes to form a transmission network. This is very similar to electric circuits and meshes of finite element analysis. Relationships between the angular velocities and between the torques of the shafts of each component can be formulated easily. These relationships are arranged in matrix form. Prescribed values of torques and angular velocities (e.g. power input or stationary shafts) must be included before the system of equations can be solved. The angular speed and the torque of every shaft of the transmission system are thus obtained. Then the power flows can be calculated. The method can handle speed and torque losses of several types. The method is thus convenient for calculating the efficiency of complex transmission systems.

Keywords. Power transmission; matrix algebra; computer applications; network analysis; transmission system; gear train; epicyclic gear train; CVT.

1. INTRODUCTION

During the last few years interest for using continuoulsly variable transmissions (CVTs) has increased. CVTs offer the advantage of an infinite number of transmission ratios. In, for instance, motor vehicles a CVT makes it possible to, at each instant, follow the curve of minimum specific fuel consumption of the engine. Since there is a continuous change of transmission ratio, the acceleration will be very smooth with no "jerky" gear changes.

Unfortunately, the efficiency η of a CVT is not very good compared to a "normal" gear transmission. Another disadvantage is that CVTs cannot normally be made as small as a comparable gear transmission.

A method frequently used to reduce the effect of those disadvantages of CVTs is called power branching or split power design (cf. Beckmann, 1968; Dorey and McCandlish, 1986; Findeisen, 1981; Gackstetter, 1966; Hedman, 1985; Jarchow, 1967; Pichard and Besson, 1977; Polder, 1969; White, 1976; Yu and Beachley, 1984) The principle of this is shown schematically in Fig. 1. Two more advanced designs are shown in Fig. 2.

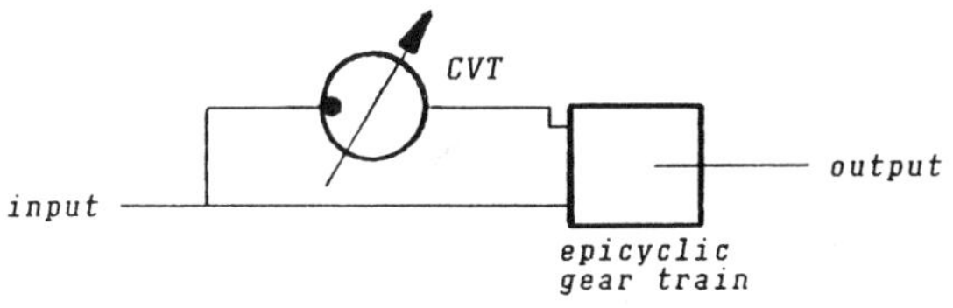

Fig. 1. Power branching (schematic)

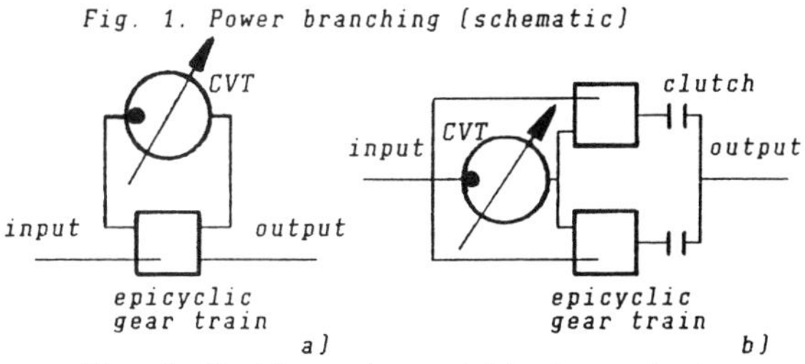

Fig. 2. Continuously variable transmission system with advanced power branching. The output of design b) can be coupled to either of the epicyclic gear trains. The total transmission ratio range can hence be extended.

Much research work on split power transmissions has been presented. In most of these studies the analysis has been carried out in roughly the same way. The power flows, efficiencies, etc. have been obtained by means of manual algebraical eliminations of the relationships between the torques and rotational speeds of the shafts of the transmission system. These eliminations have been valid only for specific transmission configuration(-s) that has (have) been studied.

Some more general methods of handling the equations for torques and angular speeds have been presented (e.g. Pichard and Besson, 1977; Polder, 1969; Wojnarowski, 1976). The equations are in these methods handled in a systematic way which can easily be implemented on a computer. Unfortunately they all demand that each epicyclic gear train in the network has one planet carrier, exactly two sun gears, and exactly three shafts. If the losses are neglected, an epicyclic gear train with more than three shafts can be regarded as a number of coupled epicyclic gear trains with three shafts. It can, however, be shown that this cannot be done when the losses are to be considered (Hedman, 1985; Mägi, 1974).

A general method of analysis of transmission networks that can handle the losses in epicyclic gear trains with more than three shafts will be presented in this paper. The paper is based on an earlier report by the author (Hedman, 1985) with some additions. The method is capable of analysing any kinematically possible configuration of a transmission system. It is adapted for computer calculations. When the transmission system has been described, equations for speeds and torques can be generated and solved automatically by the computer.

2. NOTATION

A matrix is denoted by two bars: $\bar{\bar{M}}$. A vector (i.e. a column matrix) is denoted by one bar: $\bar{\omega}$.

d	Belt pulley diameter
i	Transmission ratio
$\bar{\bar{M}}$	Matrix
N_e	Number of shafts of an epicyclic gear train
N_n	Number of nodes in the transmission system
N_s	Number of shafts of the transmission system
N_T	Number of torque relationships
N_β	Number of boundary conditions
N_ω	Number of angular velocity relationships
n	Number of a node
p	Planet wheel
R	Fixed-carrier ratio of epicyclic gear train
s	Slip factor
$T, \bar{T}$	Torque, torque vector
T_l	Drag torque in a uni-directional clutch, torque loss in a bearing or in a seal
z	Number of teeth
$\beta, \bar{\beta}$	Boundary value, boundary value vector
η	Efficiency
κ	Efficiency factor
$\omega, \bar{\omega}$	Angular velocity, angular velocity vector

SUBSCRIPTS

$1,2,\ldots$ i,j,k	Refer to the shaft or wheel of the same number
a	Arbitrarily chosen reference shaft
e	Epicyclic gear train, epicyclic rolling contact transmission
T, ω	Refer to torque and angular velocity

SUPERSCRIPT

T	Transposed

3. COMPONENTS

3.1 ASSUMPTIONS, NOTATIONS, AND DEFINITIONS

The different components that can be put together to form a transmission system will now be presented. First some assumptions and notations will be introduced:

1) In many transmission systems the different shafts are parallel. Therefore, the network that describes the transmission system can be outlined so that the axes of most of the shafts will be horizontal, see Fig. 3.

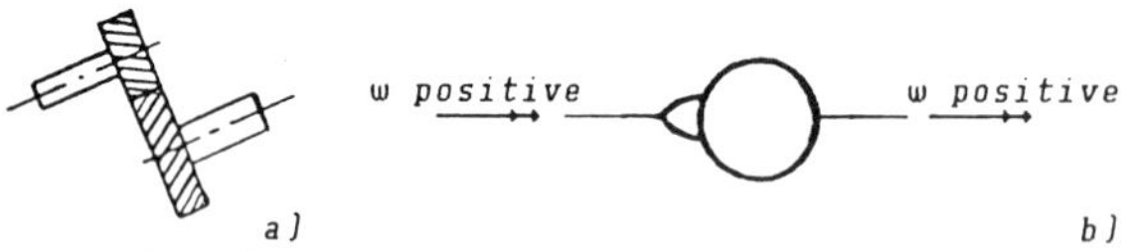

a) b)

Fig. 3. A very simple transmission system, a gear transmission. a): sketch, b): network form with the axes of the shafts horizontal and the chosen positive sense of rotation.

2) For the shafts that are placed according to notation 1), one sense of rotation is chosen as positive, see Fig. 3 b).
3) A torque acting on a shaft is considered positive if it tends to drive the shaft in the positive sense of rotation.
4) In some transmission systems not all of the shafts are parallel as discussed in notation 1). This will be the case when there are bevel gearings in the transmission system. The sense of rotation of such a shaft depends on the location of the bevel gear wheels, see Fig. 4. How the positive sense of rotation of "non-horizontal" shafts can be found is shown by Hedman (1985).

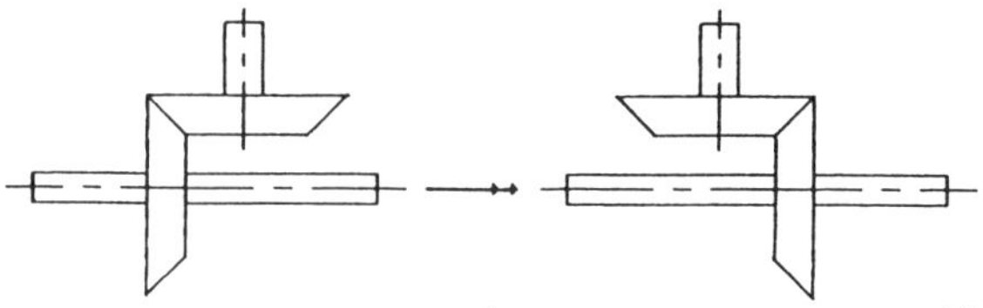

a) b)

Fig. 4. Two locations of bevel gear wheels.

5) Both torque <u>and</u> velocity losses are considered.
6) Only stationary cases are considered.
7) Transmission networks with hydrokinetic transmissions, i.e. hydrodynamic torque converters and hydrodynamic couplings, are not considered here. There is, however, a similar method of analysis for such transmission systems (Andersson, 1982; Hedman, 1985).

3.2 COUPLING AND CLUTCH

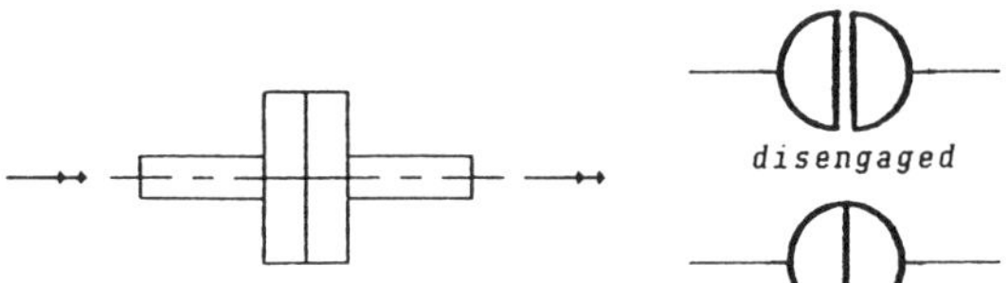

Fig. 5. Coupling and clutch. a): sketch with chosen positive senses of rotation and torque, b) and c): simplified network representations.

For a coupling or an engaged clutch of Fig. 5 there are the following relationships:

$$\omega_1 - \omega_2 = 0 \tag{1}$$

$$T_1 + T_2 = 0 \tag{2}$$

A disengaged clutch is equivalent to two free shaft ends:

$$T_1 = T_2 = 0 \tag{3}$$

3.3 UNI-DIRECTIONAL CLUTCH

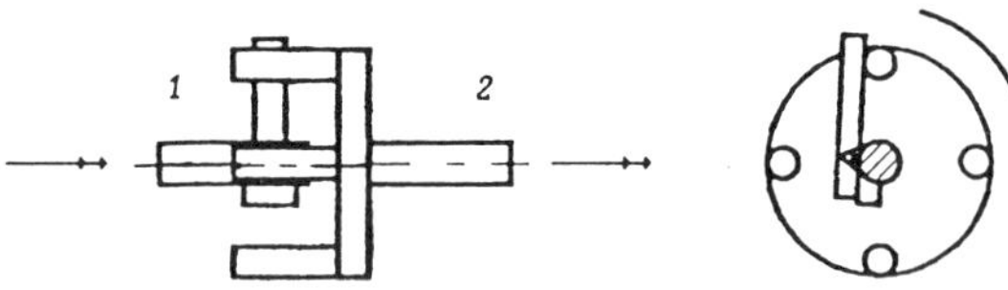

Fig. 6. Uni-directional clutch: sketch with chosen positive senses of rotation.

A uni-directional clutch either works as a coupling ("locked") or as a disengaged clutch ("overrunning" or "unlocked" uni-directional clutch). Normally, the user only has a vague indication of whether the uni-directional clutch is locked or unlocked. Thus, it cannot be treated as a "normal" clutch. For the locked uni-directional clutch in Fig. 6 the relationships are the same as for a coupling, i.e. equations (1) and (2) together with the condition:

$$T_1 \geq 0 \tag{4}$$

An unlocked uni-directional clutch has the following relationships:

$$\left.\begin{array}{l} \omega_1 \leq \omega_2 \\ T_1 = -T_l \\ T_2 = T_l \end{array}\right\} \tag{5}$$

Here, T_l is a possible drag torque. A "reversed" form of the clutch of Fig. 6 is shown in Fig. 7.

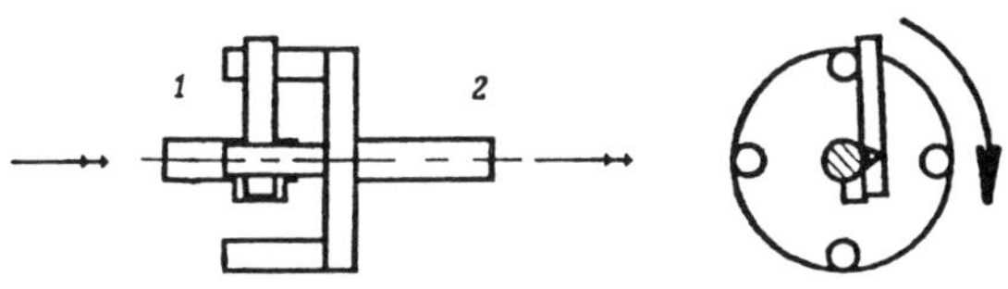

Fig. 7. Uni-directional clutch: sketch with chosen positive senses of rotation.

The relationships corresponding to equations (1), (2), (4), and (5) are easily obtained by studying the differences between Fig. 6 and Fig. 7. The results for a locked clutch are:

$$\left.\begin{array}{l}\omega_1 - \omega_2 = 0\\ T_1 + T_2 = 0\\ T_1 \leqslant 0\end{array}\right\} \quad (6)$$

and for an unlocked one:

$$\left.\begin{array}{l}\omega_1 \geqslant \omega_2\\ T_1 = T_l\\ T_2 = -T_l\end{array}\right\} \quad (7)$$

An algorithm to determine whether the uni-directional clutch is locked or unlocked is as follows:

1. Assume that the uni-directional clutch is locked, i.e. equations (1) and (2) are valid (or (6) if it is of the same type as Fig. 7).
2. Calculate the angular velocities and the torques using this assumption. If the resulting transmission system is kinematically impossible, then the assumption was incorrect.
3. If the inequality in (4) (or (6)) is satisfied, then the above assumption was correct, otherwise the uni-directional clutch is unlocked; calculate the angular velocities and the torques again using equation (5) (or (7)).

3.4 TRANSMISSIONS WITH TWO SHAFTS

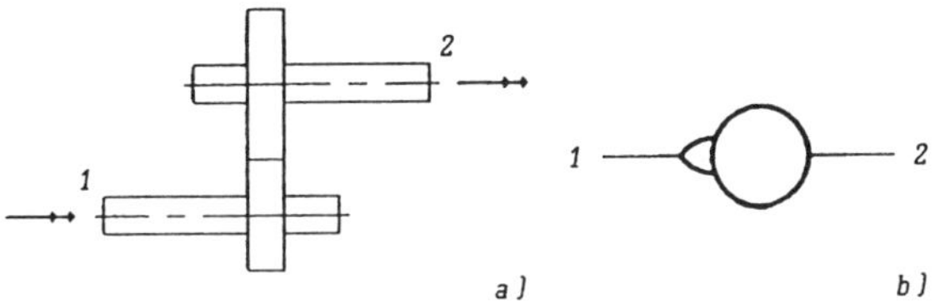

Fig. 8. A gear. a): sketch with chosen positive senses of rotation (and torque), b): simplified network representation.

There are a lot of types of transmissions with two shafts, e.g. gear, belt, hydrostatic, and rolling contact transmissions. CVT designs are usually of the three latter types. Introduce the transmission ratio i. If the losses are neglected, the angular speed and torque relationships are given by:

$$\omega_1 - i\omega_2 = 0 \quad (8)$$

$$iT_1 + T_2 = 0 \quad (9)$$

For a gear transmission the transmission ratio is (z denotes the number of teeth):

$$i = -z_2/z_1 \quad (10)$$

The equation is also valid for internal gear transmissions <u>if</u> the number of teeth of its internally cogged gear wheel is negative, cf. Mägi (1974). The transmission ratio of timing belt and chain transmissions is:

$$i = z_2/z_1 \quad (11)$$

Introduce the (pitch) diameter d of the (V-)belt pulley of a traction belt transmission. Then a nominal, non-slip, transmission ratio i can be defined for such transmissions:

$$i = d_2/d_1 \quad (12)$$

In practice, torque losses always occur. The equation (9) will then be changed. If power is transmitted from shaft 1 to shaft 2, then:

$$\left.\begin{array}{l}T_1\omega_1 > 0\\ i\eta_{T12}T_1 + T_2 = 0\end{array}\right\} \quad (13)$$

Here η_{Tij} is the <u>t</u>orque efficiency of the transmission when power is transmitted from shaft i to shaft j. If the power is transmitted from shaft 2 to shaft 1, then:

$$\left.\begin{array}{l}T_1\omega_1 < 0\\ iT_1 + \eta_{T21}T_2 = 0\end{array}\right\} \quad (14)$$

Gear, timing belt, and chain transmissions have no speed losses. Most other types of two-shaft transmissions have speed losses. This is also called slip. Denote the slip by s. Then the angular speed relationship for traction belt transmissions is:

$$\left.\begin{array}{l}T_1\omega_1 > 0\\ (1-s_{12})\omega_1 - i\omega_2 = 0\end{array}\right\} \quad (15)$$

$$\left.\begin{array}{l}T_1\omega_1 < 0\\ \omega_1 - (1-s_{21})i\omega_2 = 0\end{array}\right\} \quad (16)$$

The quantities (1-s) can also be called speed efficiencies, η_ω. The values of the loss quantities η_T and s (or η_ω) must be obtained. Experimental values can be used as well as values of experience. Another possibility is to use values found by mathematical modelling of the specific transmission type. For gear transmissions Dorey and McCandlish (1986) have presented different models for the torque efficiency. Gerbert (1972) has shown how both the slip and the torque efficiency of a traction belt transmission can be obtained. Mägi (1974) has shown how the slip and the torque efficiency of a general traction drive can be calculated.

Belt and chain transmissions can have more than two shafts. A generalization to such cases is given by Hedman (1985).

An algorithm to determine the direction of the power flow in a two-shaft transmission is:

1. Calculate the torques and angular speeds neglecting the losses, i.e. use equations (8) and (9).
2. Determine preliminary values of the slips and the torque efficiencies using those values of the angular velocities and the torques.
3. If $T_1\omega_1 > 0$ then equations (13) and (15), otherwise (14) and (16), are to be used.
4. Calculate the angular velocities and the torques again using the correct pair of those equations.
5. Go back to point 2 if better accuracy is desired, or if the sign of $T_1\omega_1$ has changed.

3.5 BEARING AND SEAL

Fig. 9. Network representations. a): bearing, b): seal.

In bearings and seals a loss torque of magnitude T_l is always counterdirected to the rotation:

$$\omega_1 - \omega_2 = 0 \quad (17)$$

$$\omega_1 > 0: \quad T_1 + T_2 = T_l \quad (18)$$

$$\omega_1 < 0: \quad T_1 + T_2 = -T_l \quad (19)$$

The value of T_l of a seal is normally regarded as constant. It can also be taken as dependent on the square of $\omega_1=\omega_2$ if the seal itself rotates. The magnitude of T_l for journal bearings can be considered as proportional to the speed of the shaft. For ball and roller bearings it is normally both speed and torque dependent. Expressions for the loss torque of both journal bearings and ball and roller bearings have been given by Dorey and McCandlish (1986). They have also stated that oil churning losses can be regarded as proportional to the speed of the shaft. These losses can hence be treated as bearing or seal losses.

In some designs a bearing or a seal is mounted between two rotating concentric shafts. This is similar to an unlocked uni-directional clutch. Equations (5) or (7) can then be used. It is not unusual that a bearing or a seal is mounted at the end of a shaft. Appropriate equations are then:

$$\omega > 0: \quad T = T_l \tag{20}$$

$$\omega < 0: \quad T = -T_l \tag{21}$$

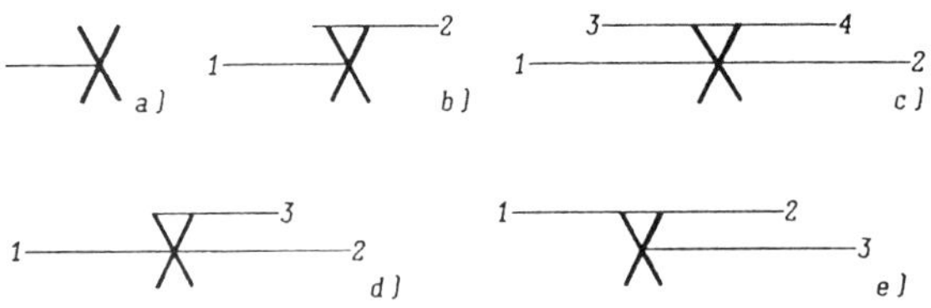

Fig. 10. Different bearing (or seal) arrangements (schematic), a): a shaft end, b): two concentric shaft ends, c): two concentric shafts, d) and e): a shaft and a concentric shaft end.

Two other possible bearing or seal arrangements with concentric shafts are shown in Fig 10 c)-e). In Fig. 10 c), there are two concentric shafts. In Fig. 10 d) and e), one of those shafts has been substituted by a shaft end. Appropriate equations for an arrangement of Fig. 10 c) are:

$$\omega_1 - \omega_2 = 0 \tag{17}$$

$$\omega_3 - \omega_4 = 0 \tag{22}$$

$$\omega_1 > \omega_3: \quad \left.\begin{aligned} T_1 + T_2 &= T_l \\ T_3 + T_4 &= -T_l \end{aligned}\right\} \tag{23}$$

$$\omega_1 < \omega_3: \quad \left.\begin{aligned} T_1 + T_2 &= -T_l \\ T_3 + T_4 &= T_l \end{aligned}\right\} \tag{24}$$

Equation (17) is also valid for arrangements like those in Fig. 10 d) and e). The torque equations can be obtained from equations (23) and (24) by omitting the T_4-terms. It can easily be shown, that the other arrangements can also be derived from the general type of Fig. 10 c).

To determine the losses, an algorithm like the one for transmissions with two shafts can be used.

3.6 EPICYCLIC GEAR TRAIN

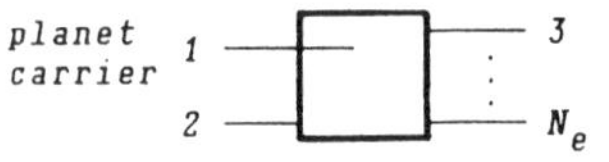

Fig. 11. Network representation of an epicyclic gear train.

An epicyclic gear train of Fig. 11 consists of one planet carrier (denoted 1) and N_e-1 sun gear shafts 2,3,...N_e ($N_e \geq 3$). The planet wheel(s) may be configurated in many ways. In Fig. 12 a) there is only one, "single", planet wheel. The epicyclic gear train in Fig. 12 b) has two directly connected planet wheels. This can also be called a "multiple" or "double" planet wheel. Fig. 12 c) shows an epicyclic gear train with three planet wheels. There is one double planet wheel A and B and one single planet wheel C. The wheels B and C are in mesh.

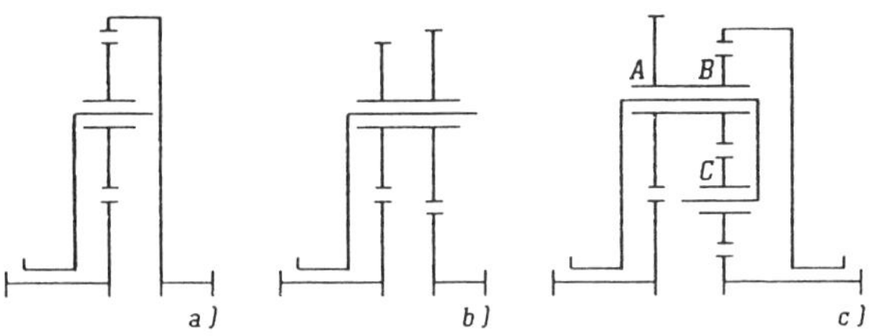

Fig. 12. Examples of planet wheel configurations. a): a single, b): a double planet wheel, c): a double and a single planet wheel in mesh.

The speed relationships are given by:

$$\frac{\omega_i - \omega_1}{\omega_a - \omega_1} = R_{ia} \qquad i = 2,3,\ldots N_e \tag{25}$$

"a" is an arbitrary sun gear shaft. The "fixed-carrier ratio" R_{ia} is thus the transmission ratio between shaft i and shaft a <u>when</u> $\omega_1=0$. It can thus be expressed by ratios of the numbers of teeth of the wheels, cf. equation (10). By definition R_{aa} equals unity. Equation (25) can be rewritten:

$$\omega_1(R_{ia}-1)-\omega_a R_{ia}+\omega_i=0 \qquad i=2,3\ldots a-1,a+1\ldots N_e \tag{26}$$

The torque equilibrium of the whole epicyclic gear train is:

$$\sum_{i=1}^{N_e} T_i = 0 \tag{27}$$

If the losses are neglected, then the total input power equals the total output power:

$$\sum_{i=1}^{N_e} T_i \omega_i = 0 \tag{28}$$

which combined with equations (25) and (27) yields:

$$\sum_{i=2}^{N_e} R_{ia} T_i = 0 \tag{29}$$

Assume that there is only one planet wheel, single or multiple. Neglect the bearing losses of this planet wheel. Then equation (29) can be altered to take account for the gear mesh losses:

$$\sum_{j=2}^{N_e} \kappa_{Tj} R_{ja} T_j = 0 \tag{30}$$

where

$$\left.\begin{aligned} \kappa_{Tj} &= \eta_{Tjp} \\ T_j(\omega_j-\omega_1) &> 0 \end{aligned}\right\} \qquad j = 2,3,\ldots N_e \tag{31}$$

$$\left.\begin{aligned} \kappa_{Tj} &= 1/\eta_{Tpj} \\ T_j(\omega_j-\omega_1) &< 0 \end{aligned}\right\} \qquad j = 2,3,\ldots N_e \tag{32}$$

Here η_{Tjp} is the efficiency of power (or torque) transmission from shaft j to the planet wheel, etc. An algorithm similar to the one of section 3.5 can be used to determine the power flows and losses.

If there is more than one (multiple) planet wheel in the epicyclic gear train, a more accurate method of analysis is required. This is also the case when the bearing losses of the planet wheel(s) are to be considered. Mägi (1974), has presented how a very accurate analysis of the losses in a general epicyclic gear train can be done.

In transmissions where high angular speeds are involved, epicyclic rolling contact transmissions are sometimes used. An example of this is gas turbines.

Some rolling contact CVT designs are also of epicyclic type; a classification of those is made by Brännare (1986). Mägi's method for epicyclic gear trains has been extended to cover epicyclic rolling contact transmissions (Hedman, 1985).

4. MATRIX-BASED NETWORK ANALYSIS THEORY

With the definitions and assumptions in chapter 3, a transmission system can be composed and defined. The shaft ends of the components of the transmission system are connected in nodes, see Fig. 13.

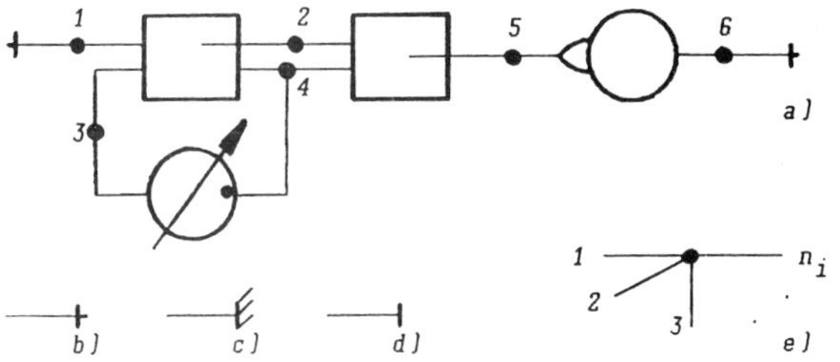

Fig. 13. a): A transmission system network, the components are connected to nodes, b): a general boundary shaft, c): a stationary shaft, d): a free shaft, e): a node "i" with n_i shafts.

All the n_i shafts of node i $(i=1,2,\dots N_n)$ have the same angular speed ω_i. Torque equilibrium yields:

$$\sum_{j=1}^{n_i} T_j = 0 \qquad i = (1,2,\dots N_n) \tag{33}$$

Power input and output are symbolized by "boundary shafts", see Fig. 13 b). These boundary shafts can also be stationary shafts and free shaft ends, Fig. 13 c) and d). Both stationary and free shafts are examples of boundary conditions enforced on boundary shafts. A boundary condition is a prescribed value of either an angular velocity of a node or a torque of a shaft. The boundary conditions are thus of the following forms:

$$T_i = -\beta_{Ti} \qquad (\beta_{Ti} \longrightarrow \quad \longrightarrow T_i) \tag{34}$$

$$T_i = \beta_{Ti} \qquad (\beta_{Ti} \longrightarrow T_i \longrightarrow \quad) \tag{35}$$

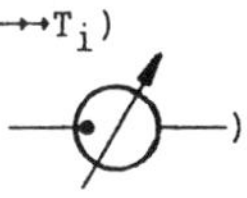

$$\omega_j = \beta_{\omega j} \tag{36}$$

The transmission system must not be "locked" kinematically. This means that all the shafts, except possible stationary ones, must be able to rotate. Additionally, there must not be any identical transmissions coupled in parallel. As an example, a V-belt transmission with two belts must be considered as if it was only one belt. This can be explained in the following way. The whole belt transmission will transmit a certain amount of power. The distribution of this power between the two belts cannot be determined by means of equilibrium equations. Since this method is based on equilibrium equations, the power distribution cannot be calculated. Polder (1969) has implicitly shown how both kinematically locked parts of a transmission network and parallel transmissions can be detected.

The equations for the angular velocities can be written in matrix form:

$$\bar{\bar{M}}'_\omega \bar{\omega} = 0 \tag{37}$$

$$(\bar{\omega})^T = \{\omega_1, \omega_2, \dots \omega_{N_n}\} \tag{38}$$

$\bar{\bar{M}}'_\omega$ is a matrix of the size N_n columns and N_ω rows. N_ω is the number of angular velocity equations obtained from the different components (of chapter 3) or from (possible) pairs of nodes which are directly interconnected by a shaft. For every such pair there is an equation of the type $\omega_i - \omega_j = 0$ since the angular velocities are equal. The connecting shaft can and will in the following be implemented as a coupling. The difference $N_{\beta\omega} = N_n - N_\omega$ is the number of angular speed boundary conditions that has to be enforced on the nodes. Sometimes there is more than one speed boundary value to be enforced. The first boundary value can be enforced on any node. This can however not be done with the other boundary values. Take as an example a gear transmission. Suppose that the rotational speed of the node, to which one of its shafts is coupled, has been prescribed. Then, the speed of the node, to which the other shaft is coupled, must not be prescribed. This is because that speed is already determined by the first speed via the speed relationship of the gear transmission. If $\bar{\bar{M}}'_\omega$ is "filled" with the $N_{\beta\omega}$ boundary value equations, (37) changes to:

$$\bar{\bar{M}}_\omega \bar{\omega} = \bar{\beta}_\omega \tag{39}$$

where the elements of the vector $\bar{\beta}_\omega$ are zero except for those corresponding to the boundary conditions. Solving equation (39) yields the angular velocities of the transmission system.

Now, let

$$N_s = \sum_{i=1}^{N_n} n_i \tag{40}$$

be the total number of shafts of the transmission system. Neglect the losses. Then, the equations for the torques of those shafts can be written:

$$\bar{\bar{M}}'_T \bar{T} = 0 \tag{41}$$

$$(\bar{T})^T = \{T_{1,1}, T_{1,2}, \dots T_{1,n_1}, T_{2,1}, \dots T_{N_n, n_{N_n}}\} \tag{42}$$

The first subscript is the node number and the second is the number of the shaft in that node. $\bar{\bar{M}}'_T$ is a matrix of the size N_s columns and N_T rows. N_T is the number of torque equations obtained from the components and the nodes (33). Let $N_{\beta T} = N_s - N_T$ be the number of torque boundary conditions to be enforced on shafts. When this number is larger than one, only the first torque boundary condition can be enforced on any shaft, as was the case for speed boundary values. "Fill" $\bar{\bar{M}}'_T$ with the torque boundary conditions to give:

$$\bar{\bar{M}}_T \bar{T} = \bar{\beta}_T \tag{43}$$

The torques of the shafts of the transmission system are then obtained when equation (43) is solved.

The matrices $\bar{\bar{M}}_\omega$ and $\bar{\bar{M}}_T$ in equations (39) and (43) are in general not symmetric. The eigenvalues and eigenvectors are then in general not real. If any part of the transmission system would have been kinematically locked, then the determinant of the matrix $\bar{\bar{M}}_\omega$ would be zero. The same applies for both matrices if there would be any identical parallel transmissions in the transmission system.

The losses can be considered in accordance with the algorithm of section 3.4. Preliminary values of the torques and angular speeds are obtained from the solution of equations (39) and (43). Those values can be used to achieve the loss parameters. Then $\bar{\bar{M}}_\omega$, $\bar{\beta}_\omega$, $\bar{\bar{M}}_T$ and $\bar{\beta}_T$ can be updated and the systems of equations can be solved again. This procedure can be repeated until the accuracy is satisfactory.

When the "loss-considered" angular velocities and torques of the transmission system are known, the efficiency, etc., can be calculated. The analysis of the specific transmission is then completed.

5. CONCLUSIONS

This paper presents a computer oriented method of analysis for mechanical transmission systems. The transmission system is composed by "simple" transmissions, eg. belt and gear transmissions and epicyclic gear trains, and other components such as bearings and couplings. For each such transmission or component, equations between the torques and between the angular velocities of its shafts are obtained. The transmission network is formed by coupling the shafts of those transmissions and components in nodes. This will give additional coupling equations. It is assumed that the transmission system is not kinematically locked, i.e. the shafts must be able to rotate. The system of equations are arranged in matrix form. Prescribed torques and angular velocities (this also includes stationary and free shafts), are then comprehended. The system of equations can then be solved.

Both torque and speed losses can be handled. The losses are determined in an iterative way. Approximate values and directions of the power flows are obtained by analysing the transmission system when the losses are neglected. It is then possible to estimate the losses. The system of equations can then be adjusted with respect to the losses and be solved again. This procedure can be repeated to achieve high accuracy. The resulting torques and angular velocities can be used to calculate efficiencies, power losses etc. The matrix arrangement of the equations makes the analysis rapid. The method is well-suited for analysis of both hypothetic types of transmission system configurations and existing designs. The very tedious analysis work of complex mechanical transmission systems is facilitated.

6. REFERENCES

Andersson, S. (1982). On hydrodynamic torque converters. Diss., Lund Technical University.

Beckmann, K. (1968). Wirkungsgrad und Stellbereich von Planetengetrieben mit Überlagerungszweig in Abhängigkeit vom Verhältnis der Überlagerungsleistung zur Gesamtleistung. *Konstruktion*, 20, No. 9, 358-364.

Brännare, G. (1986). On rolling variators. Diss., Division of Machine Elements, Chalmers University of Tecnology, Göteborg.

Dorey, R.E., and D. McCandlish (1986) The modelling of losses in mechanical gear trains for the computer simulation of heavy vehicle transmission systems. *Proceedings, International Conference on Integrated Engine Transmission Systems*, Bath, *IMechE Conference Publications* 1986-7, 69-82.

Findeisen, D. (1981). Gleichförmig übersetzende Getriebe stufenloser Übersetzungsänderung: Gegenüberstellung von mechanischer und fluidtechnischer Energieübertragung, Teil 2. *Konstruktion*, 33, No.1, 15-24.

Gackstetter, G. (1966). Leistungsverzweigung bei der stufenlosen Drehzahlregelung mit vierwelligen Planetengetrieben. *VDI-Z.*, 108, 210-214.

Gerbert, B. G. (1972). Force and slip behaviour in V-belt drives. *Acta Polytechnica Scandinavica, Mech. Eng. Ser.*, No. 67, Helsinki.

Hedman, A. (1985). A method to analyse mechanical transmission systems. Report No. 1985-11-08, Division of Machine Elements, Chalmers University of Technology, Göteborg.

Jarchow, F. (1967). Überlagerungsgetriebe für stufenlose Drehzahl- und Drehmomentenwandlung in Kraftfahrzeugen. *VDI-Berichte*, No. 105, 69-78.

Mägi, M. (1974). On efficiencies of mechanical coplanar shaft power transmissions. Diss., Division of Machine Elements, Chalmers University of Technology, Göteborg.

Pichard, J., and B. Besson (1977). Proposition d'un modele general representatif des transmissions de puissance. Exemples d'applications. *Ingeniéurs de l'Automobile*, No. 4, 183-191.

Polder, J. W. (1969). A network theory for variable epicyclic gear trains. Diss., University of Technology, Eindhoven.

White, G. (1976). Compounded two-path variable ratio transmissions with coaxial gear trains. *Mechanism and Machine Theory*, 11, 227-240.

Wojnarowski, J. (1976). The graph method of determining the loads in complex gear trains. *Mechanism and Machine Theory*, 11, 103-121.

Yu,D., and N. Beachley (1984). Alternative split-path transmission designs for motor vehicle application. *Proceedings, First International Symposium on Design and Synthesis*, Tokyo.

Eine Matrizen-basierte Methode zur Analyse gekoppelter Getriebe.

Kurzfassung. Eine Analysenmethode für generelle Getriebesysteme wird vorgestellt. Diese Methode ist für Computerberechnung angepasst. Das Getriebesysteme ist aus verschiedenen "Teilgetrieben", wie zum Beispiel Zahnradgetribe, Riemengetribe, Lager, Kupplungen und Umlaufgetribe, aufgebaut. Gleichungen zwischen den Drehgeschwindigkeiten und zwischen den Drehmomenten sind einfach aufzustellen. Als ein einfaches Beispiel kann das Verhältnis der Drehgeschwindigkeiten der Wellen eines Zahnradgetriebes im Verhältnis der Zahnzahlen der Zahnräder ausgedrückt werden. Die Wellen der Teilgetriebe werden in Knotenpunkte gekuppelt. Ein Netz ergibt sich daraus in Ähnlichkeit mit elektronischen Kreisen und Netzen der Methode der Finiten Elemente. Die Gleichungen zwischen den Drehmomenten und Drehgeschwindigkeiten der verschiedenen Wellen bilden eine Gleichungssysteme, die in Matrizenform aufgestellt wird. Vorgeschriebene (bekannte) Werte gewisser Drehmomente und Drehgeschwindigkeiten müssen eingeschlossen werden, bevor die Gleichungssysteme gelöst werden kann. Die Drehgeschwindigkeiten und Drehmomente der sämtlichen Wellen werden dadurch erhalten. Die Methode kann Verluste verschiedener Typen behandeln. Erst wird die Gleichungssysteme unter Vernachlässigung der Verluste gelöst. Die so erhaltene Drehgeschwindigkeiten und Drehmomente werden dann zur Schätzung der Verlustkennwerte (Wirkungsgrade der Teilgetriebe usw.) verwendet. Die Gleichungen können dann in bezug auf die Verluste korrigiert werden, und die Gleichungssysteme kann wieder gelöst werden. Dieses iterative Verfahren kann wiederholt werden, bis erwünschte Genauigkeit erreicht worden ist. Die Methode ist also zur Berechnung verschiedener Eigenschaften, wie zum Beispiel Gesamtübersetzung und Gesamtwirkungsgrad, bei gekoppelten Getrieben geeignet. Die Analyse solcher Getriebe ist wesentlich vereinfacht worden.

A New Graph Theory Representation for the Topological Analysis of Planetary Gear Trains

D. G. OLSON, A. G. ERDMAN AND D. R. RILEY

Productivity Center, Mechanical Engineering Department, University of Minnesota, Minneapolis, Minnesota, USA

Abstract—Graph theory has been demonstrated to be useful during the conceptual, or type synthesis phase of mechanism design. For the particular class of mechanisms known as planetary gear trains, the "conventional" graph representation has some inadequacies for certain steps within the type synthesis process. This paper addresses these inadequacies, and proposes a new graph representation which enables these steps to be performed in a straight-forward manner. In the process, some new questions are raised regarding the criteria for which a particular planetary gear train may be either eliminated or considered as a candidate for further design.

Introduction

Type synthesis can be viewed as consisting of two separate steps: topological synthesis and topological analysis [1]†. Topological synthesis is the determination of distinct kinematic chains (topologies), perhaps by exhaustive enumeration of a given class of mechanisms (with a given distribution of links and joints). Topological analysis refers to the analysis of a given topology to determine all of the distinct ways of assigning the fixed link and the input and output links, and evaluation of the topology based on the designer's functional requirements.

There have been a number of studies concerned with topological synthesis of geared kinematic chains, with particular attention paid to those that could be used as planetary gear trains [2–8]. However, there has been a considerable lack of attention given to the subsequent topological analysis steps; namely, assigning the ground link and identifying potential input and output links. There has also been some confusion relating to the isomorphism question, that is, what it means for two planetary gear trains to be distinct.

This paper will briefly review the established graph-theory-based approaches for topological synthesis, and introduce a new type of graph representation for planetary gear trains in which multiple joints are explicitly represented. This enables the subsequent topological analysis steps to be performed in a systematic, unambiguous fashion.

Planetary Gear Trains

Gear trains are typically used to transmit a constant angular velocity ratio between two shafts. There is an enormous variety of possible gear train configurations for such applications as automobile transmissions, gas turbine engines, etc.

A gear train is referred to as an "ordinary gear train" if the shafts of the gears are fixed in space [9]. It is often possible to achieve the same overall gear ratio with a gear train of smaller size and weight, if a "planetary" arrangement is used [10].

A *planetary gear train* (PGT) consists of one or more central "sun" gears with "planet" gears revolving around them in such a way that points on the planet gears trace out epicyclic curves. For this reason, PGT's are sometimes called "epicyclic gear trains" or "epicyclic drives." Each meshing pair of gears has associated with it a link called the *gear carrier* or *arm*, which ensures that the distance between the centers of the two gears remains constant.

Figure 1a shows a "three-dimensional" functional representation of a planetary gear train. A more conventional functional sketch is shown in Figure 1b. The planet gears are typically doubled or tripled in a symmetric fashion around the sun gears to prevent imbalance and for load sharing. Topologically and kinematically, the additional planet gears are irrelevant, so they are not shown. Also, the sizes of the gears and whether they mesh internally or externally is of no interest during the type synthesis phase. Further, there is no need at this stage for separate identification of the three different gears that are all part of link 1. For this example, link 5 is the arm for all three gear pairs.

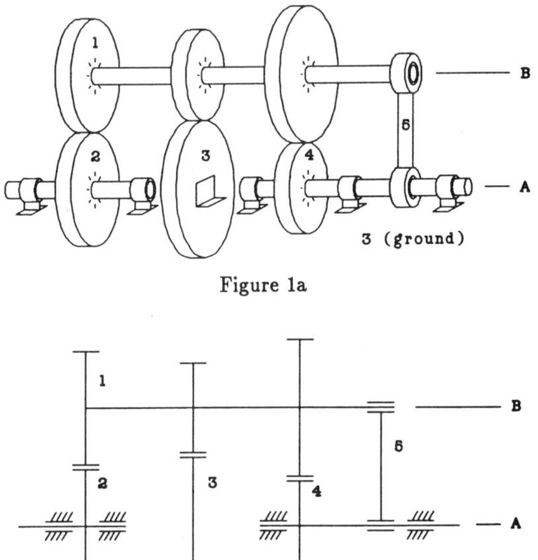

Figure 1a

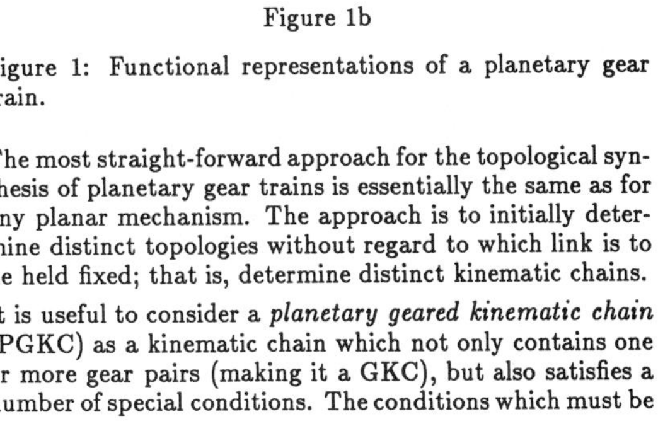

Figure 1b

Figure 1: Functional representations of a planetary gear train.

The most straight-forward approach for the topological synthesis of planetary gear trains is essentially the same as for any planar mechanism. The approach is to initially determine distinct topologies without regard to which link is to be held fixed; that is, determine distinct kinematic chains.

It is useful to consider a *planetary geared kinematic chain* (PGKC) as a kinematic chain which not only contains one or more gear pairs (making it a GKC), but also satisfies a number of special conditions. The conditions which must be

†Numbers in brackets denote references at the end of the paper.

met by a GKC in order for it to qualify as a PGKC are most easily expressed in terms of the graph theory representation.

Graph Theory Representation

Since the early 1960's, the graph representation has been used as an abstract model of kinematic chains and mechanisms, to aid in the creative stages of mechanism design [4–8,11–16]. In the graph representation, the links of the mechanism are represented as *vertices* and the joints (pairs) are represented as *edges*. Two vertices are connected by an edge if the corresponding links are connected to each other by a joint.

The usefulness of the graph representation during the conceptual phase of mechanism design lies in its relative simplicity and conciseness. There are one-to-one correspondences between the vertices of the graph and the links, between the edges of the graph and the joints, and between the circuits of the graph and the "kinematic circuits" [11] of the kinematic chain. The graph, in the form of its vertex-vertex adjacency matrix, is also the basis of algebraic techniques for the topological analysis of kinematic chains and mechanisms [12–15].

In the late 1960's, the graph representation was first used to represent kinematic chains containing gear pairs [4]. Figure 2 shows the graph representation of the planetary gear train of Figure 1 with a number of enhancements to represent additional topological information. Gear pairs are represented by dashed edges and revolute pairs by solid edges. The ground link is indicated by a circle drawn around the vertex representing the ground link. Each revolute edge is labeled according to the location of its axis in space, and each geared edge is labeled with the number corresponding to the vertex representing the gear carrier.

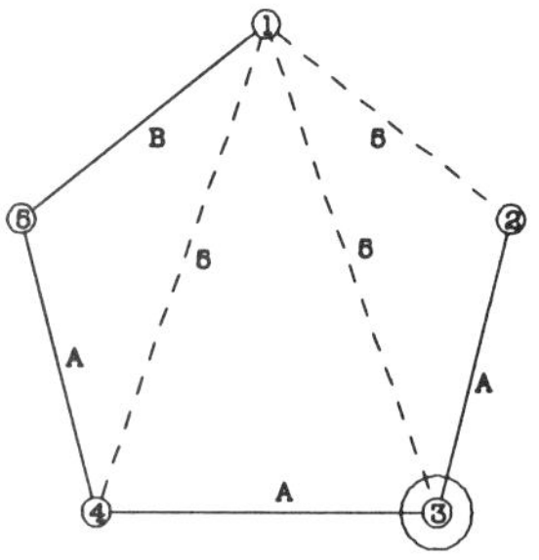

Figure 2: Graph representation of PGT of Figure 1.

The rules which must be met for a graph to represent a legitimate planetary gear train have been well documented [4–8]. In part, they are as follows:

Rule 1: The subgraph obtained by deleting all of the geared edges is a tree. This means that the number of vertices is one more than the number of revolute edges, and that there are no circuits consisting exclusively of revolute edges.

Rule 2: Any geared edge added to this tree forms a unique circuit (called a *fundamental* circuit, or *f*-circuit).

Rule 3: Associated with each *f*-circuit (i.e., geared edge) there is a *transfer vertex* such that all of the revolute edges on one side of the transfer vertex are at the same level, and all edges on the other side are at a (single) different level. The transfer vertex represents the gear carrier.

Implicit in the statement of these rules (as with most other uses of graph theory in kinematics) is that all joints are considered to be simple joints (i.e., connecting only two links). However, a consequence of rule 3 is that a number of the revolute joints must have a common axis (must be at the same level). If the gear train of Figure 1 is drawn from the "side," so that all the links are drawn in the same plane, it is clear that all revolute joints at the same level are in fact the same (multiple) joint (see Figure 3).

Figure 4 shows a graph representation of the planetary gear train of Figure 3 which differs from the one in Figure 2. The fact that the same gear train can be represented by different graphs has resulted in difficulty in establishing the isomorphism between gear trains, and has led researchers to define a number of different graph representations in an effort to establish a "canonical" one.

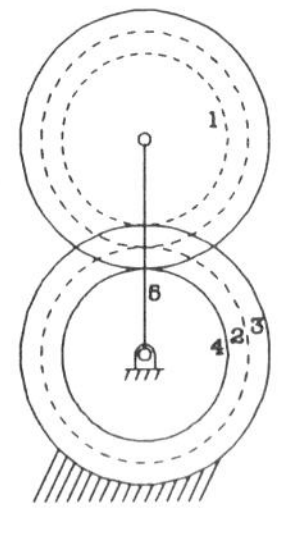

Figure 3: "Side" view of PGT of Figure 1.

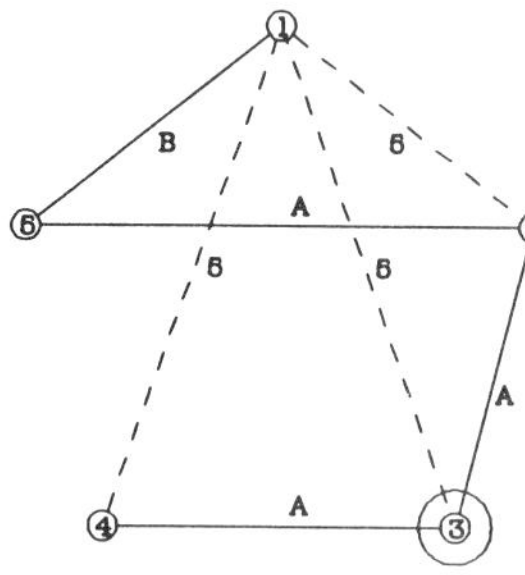

Figure 4: Alternative graph representation of PGT of Figure 1.

Structural vs. Rotational vs. Displacement Isomorphism

The planetary gear trains represented by the graphs of Figures 2 and 4 are said to be "stucturally non-isomorphic" [7], suggesting that the "structural" graph of Figure 4 is actually representing the distinct PGT of Figure 5 (note the different location of link 5). However, for the purposes of type synthesis, there is no need to distinguish between the PGT's of Figures 1 and 5. The fact that there is no essential difference between the two gear trains is reflected by the fact that the *rotation graph* and the *displacement graph* are the same for each, that is, the two are *rotationally isomorphic* and *displacement isomorphic*†.

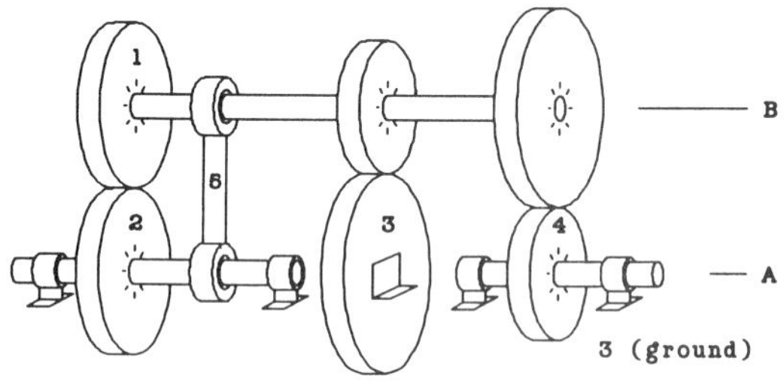

Figure 5: PGT which is structurally non-isomorphic, but displacement isomorphic to PGT of Figure 1.

The rotation graph was originally defined as the graph obtained from the structural graph of the PGKC, by deleting all of the revolute edges (and transfer vertices if incident to only revolute edges) [5]. Hence, the geared edges are the only edges represented; they are labeled with their associated transfer vertex. The displacement graph was defined to be the same as the rotation graph, but with each vertex also labeled with the level of the revolute edge that is incident to it in the original graph. If a vertex is incident to revolute edges at more than one level, the level associ-

†Although the definitions of these terms can be easily extended to apply to PGT's (for which a fixed link is identified), their use in [5,7,8] and the remainder of this paper is confined to PGKC's.

ated with the f-circuit is used. The terms rotation graph and displacement graph were introduced based on their respective relationships to the derivation of the rotational and linear displacement equations [5].

Ravisankar and Mruthyunjaya recently defined the rotation and displacement graphs in such a way that the vertices representing gears are connected directly to their associated transfer vertex with revolute edges [7]. This results in a displacement graph which is itself *a* structural graph. (Freudenstein called this graph the "canonical graph" [5].) It is also pointed out in [7], that for some PGKC's with more than 5 links, the rotation graph can contain circuits consisting exclusively of revolute edges. This will sometimes occur when a link consists of several gears not all associated with the same transfer vertex.

For a distinct rotation graph, the levels can be assigned in a number of different ways resulting in a number of distinct displacement graphs, each of which could have the actual revolute edges drawn in a variety of ways (resulting in a number of distinct structural graphs). The usual approach for topological synthesis is to determine, for a given number of vertices and degrees-of-freedom, all of the possible non-isomorphic rotation graphs, and then, for each one, determine the number of distinct ways of assigning the levels (i.e., determine all possible non-isomorphic displacement graphs) [5,7,8].

There is no need to go a step further and determine all of the different ways of drawing the revolute edges (i.e., determine distinct structural graphs), because there is no meaningful distinction to be made between structural graphs derived from the same displacement graph.

Using the definition of rotation graphs from [7], it is possible to determine algebraically whether two rotation graphs are non-isomorphic by simply comparing the coefficients of the characteristic polynomial of the vertex-vertex adjacency matrix, using suitably chosen values to represent the different types of edges (for example, 1's for geared edges, 2's for revolute edges)†. Because each geared edge is connected by revolute edges directly to its associated trasfer vertex, there is no need to represent the transfer vertex numbers in the adjacency matrix, as there would be using the original definition of rotation graph.

It is somewhat more involved to detect isomorphism between displacement graphs derived from a rotation graph by assigning levels to the revolute edges. The method introduced in [7] was to compare the characteristic coefficients of a vertex-vertex adjacency matrix defined by 1's for gear pairs, 2's for revolute edges at level A, 3's for revolute edges at level B, etc. However, since the labels for the levels are completely arbitrary, the numbers used for the different levels would have to be permuted in all possible ways to be certain that two PGKC's are distinct.

Coincident-Joint Graph

It is possible to simplify the preceeding discussion, and also facilitate a straight-forward topological analysis approach by defining a new graph representation called the *coincident-joint graph*‡. Referring to Figure 4, vertex 2 is adjacent to vertex 3 and vertex 2 is adjacent to vertex 5, therefore vertex 3 is adjacent to vertex 5. In fact, links 2, 3, 4, and 5 are all adjacent to each other; therefore, in the coincident-joint graph, those four vertices are all adjacent to each other, forming what is called a *complete subgraph*, as shown in Figure 6.

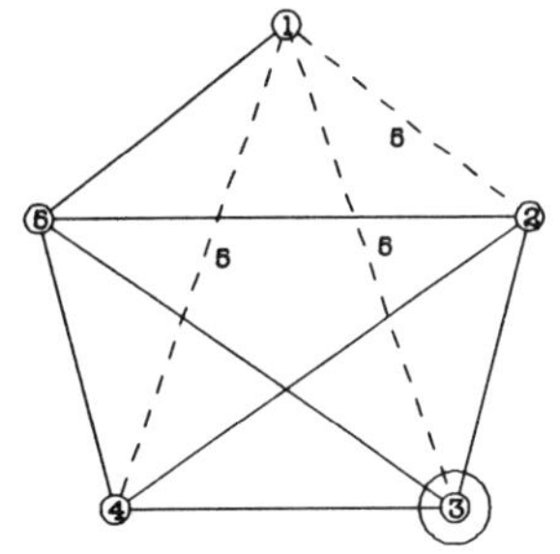

Figure 6: *Coincident-joint* graph representation of PGT of Figure 1.

The coincident-joint graph characterizes each displacement non-isomorphic PGKC. For detection of displacement isomorphism among coincident-joint graphs, there is no need to explicitly represent in the adjacency matrix, either the levels associated with each revolute edge, nor the transfer vertices associated with each geared edge.

As an example, Figure 7a shows one of the twenty-six distinct rotation graphs with six vertices and four gear pairs, as enumerated in [7,8]. Two possible ways of assigning the levels of the revolute edges are shown in the displacement graphs of Figures 7b and 7c. To show that the two are indeed non-isomorphic, the two graphs are transformed into their coincident-joint graphs (Figures 7d and 7e), their corresponding adjacency matrices (Figures 7f and 7g), and finally the characteristic polynomials of the adjacency matrices (Figures 7h and 7i). Because the coefficients of the characteristic polynomials are not the same, it can be concluded that the two PGKC's are displacement non-isomorphic.

Topological Analysis

For a given PGKC, the topological analysis steps leading to a distinct PGT can be performed in a straight-forward manner based entirely on the coincident-joint graph. The first step is to determine, for a given PGKC, all of the distinct ways of assigning the fixed link, that is, determining all of the unique *kinematic inversions* of the PGKC.

For the PGKC represented by the coincident-joint graph of Figure 6, it is quite obvious using inspection that assigning either link 2, 3, or 4 as the fixed link results in the same PGT—vertices 2, 3, and 4 are represented identically in terms of their adjacencies. It is not possible to reach that conclusion based on the structural graphs of Figures 2 or 4. The coincident-joint graph is an unbiased representation, which makes it possible to use it as the basis for implementing topological analysis techniques on a digital computer.

The algebraic method for determining all of the unique inversions of a pin-jointed kinematic chain introduced in [14] can be applied directly to PGKC's represented by their coincident-joint graphs. The basis of the method is the fact that the structural identity of each link is reflected by the diagonal terms of the powers of the adjacency matrix.

During the process of calculating the characteristic polynomial using Bocher's formula [13], it is necessary to first determine the powers $[M_{vv}]^k, (k = 1, 2, \ldots n)$. The physical significance of the elements of these matrices (denoted here by a^k_{ij}) is known from graph theory: a^k_{ij} is the number of paths from vertex i to vertex j of length k§ [14].

For two different choices of the ground link (say i and j) to be topologically distinct (resulting in different mechanisms), it is *necessary* for the two sequences of numbers $a^k_{ii}, (k = 1, 2, \ldots n)$ and $a^k_{jj}, (k = 1, 2, \ldots n)$ to be different. This condition has not yet been proven to be *sufficient*, but

†The characteristic polynomial was once thought to be unique for the graph of every distinct pin-jointed kinematic chain. Recent studies however, have revealed a few cases with ten or more links, where distinct chains yield the same characteristic polynomial [15,16]. It seems very likely that the characteristic polynomial *is* unique for the distinct rotation graphs of GKC's, though no thorough studies are known to have been done. Nonetheless, it is always possible to state that if the characteristic polynomials of two graphs are different, then the graphs are definitely non-isomorphic.

‡A similar graph and a "generalized matrix notation" for representing multiple-jointed kinematic chains (with all revolute joints) was introduced in [13].

§If the adjacency matrix has entries other than 0 and 1, the elements represent the total "value" of the paths [13].

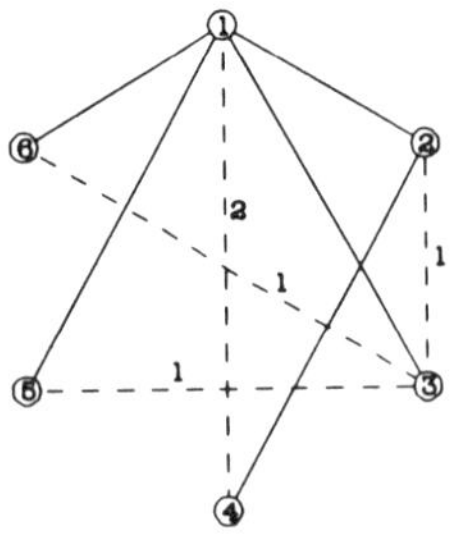

Figure 7a

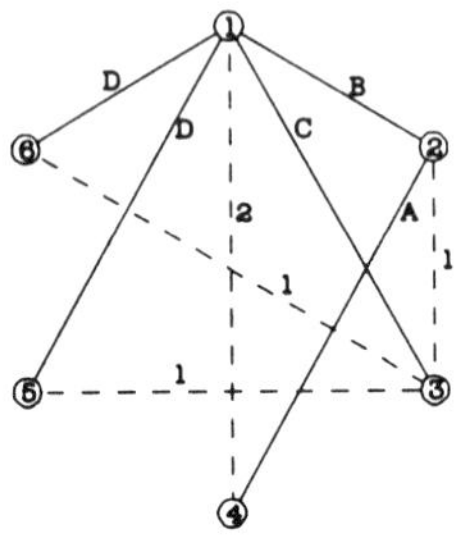

Figure 7b

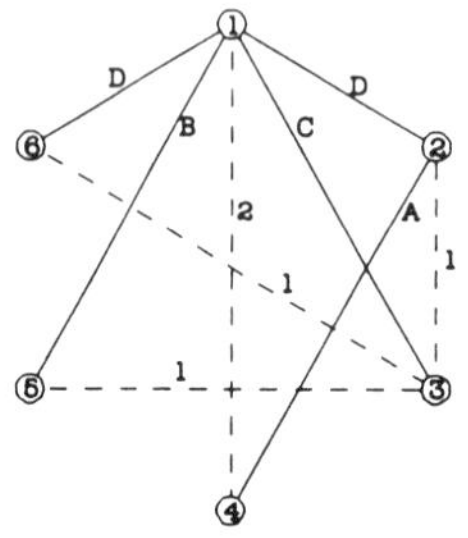

Figure 7c

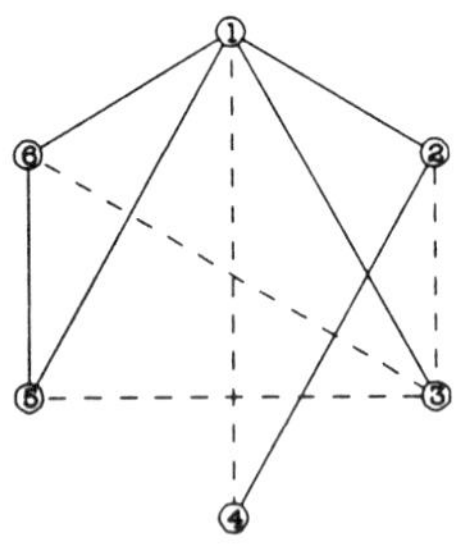

Figure 7d

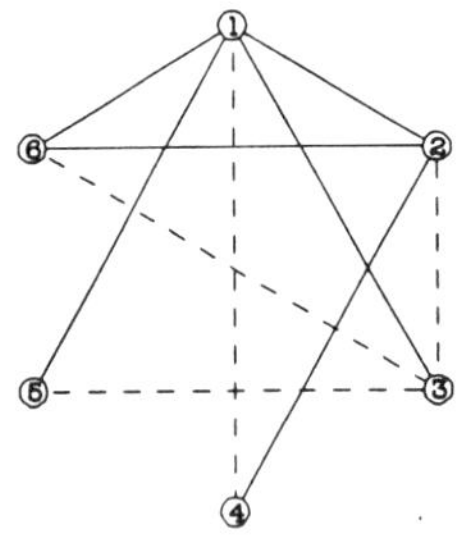

Figure 7e

$$[M_{vv}] = \begin{bmatrix} 0 & 2 & 2 & 1 & 2 & 2 \\ 2 & 0 & 1 & 2 & 0 & 0 \\ 2 & 1 & 0 & 0 & 1 & 1 \\ 1 & 2 & 0 & 0 & 0 & 0 \\ 2 & 0 & 1 & 0 & 0 & 2 \\ 2 & 0 & 1 & 0 & 2 & 0 \end{bmatrix}$$

Figure 7f

$$[M_{vv}] = \begin{bmatrix} 0 & 2 & 2 & 1 & 2 & 2 \\ 2 & 0 & 1 & 2 & 0 & 2 \\ 2 & 1 & 0 & 0 & 1 & 1 \\ 1 & 2 & 0 & 0 & 0 & 0 \\ 2 & 0 & 1 & 0 & 0 & 0 \\ 2 & 2 & 1 & 0 & 0 & 0 \end{bmatrix}$$

Figure 7g

$$P(x) = x^6 - 28x^4 - 52x^3 + 75x^2 + 212x + 92$$

Figure 7h

$$P(x) = x^6 - 28x^4 - 52x^3 + 43x^2 + 84x - 4$$

Figure 7i

Figure 7: Example illustrating the use of the coincident-joint graph for detection of displacement isomorphism.

no known counter-examples have been found [14].

For the coincident-joint graph of Figure 7d, the six sequences of numbers are shown assembled in the matrix below, with each column corresponding to a link. It can be seen that distinct PGT's will result by fixing links 1, 2, 3, 4, or 5—fixing link 6 is topologicaly identical to fixing link 5. By comparison, all of the links are distinct for the PGKC represented by the coincident-joint graph of Figure 7e.

$$\begin{bmatrix} 0 & 0 & 0 & 0 & 0 & 0 \\ 17 & 9 & 7 & 5 & 9 & 9 \\ 48 & 16 & 28 & 8 & 28 & 28 \\ 429 & 167 & 189 & 69 & 207 & 207 \\ 2032 & 708 & 1032 & 304 & 1072 & 1072 \\ 13141 & 4741 & 6131 & 1881 & 6485 & 6485 \end{bmatrix}$$

The criteria for rejecting any particular choice of ground link for a given PGKC should take into account the designer's functional requirements. If a gear carrier is selected as the ground link, at least that portion of the gear train will be "ordinary." To avoid this situation, a link which is incident to revolute edges all at the same level should be chosen as the ground link.

As a practical matter, input and output links must be adjacent to the ground link via revolute edges. This will typically rule out a number of possible ground links for which there are fewer than two incident revolute edges, for example, links 3 and 4 for the PGKC represented in Figure 7d. The algebraic procedure used to determine the topologically distinct ground links can also be used to identify distinct choices for the input and output links.

Finally, it is possible to make the following observation: If, for the PGT of Figure 8, link 5 is the input and 2 the output, gear pair 1-4 has no effect on the overall gear ratio. This can be seen in the coincident-joint graph of Figure 6. Edge 1-4 is not incident to any of the vertices central to this inversion (i.e., vertices 2, 3, and 5). Based on this observation, it is likely that virtually all of the eighty distinct displacement non-isomorphic PGKCs with six links and four gear pairs enumerated in [7,8] contain extraneous gears for any given inversion (assuming only one output).

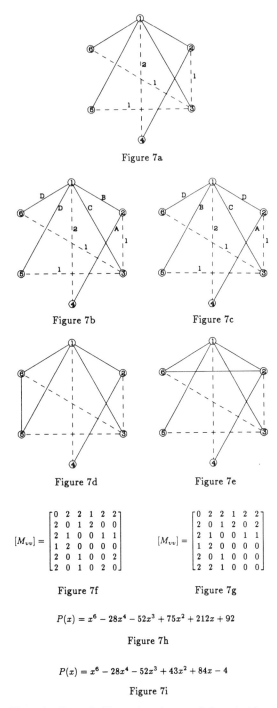

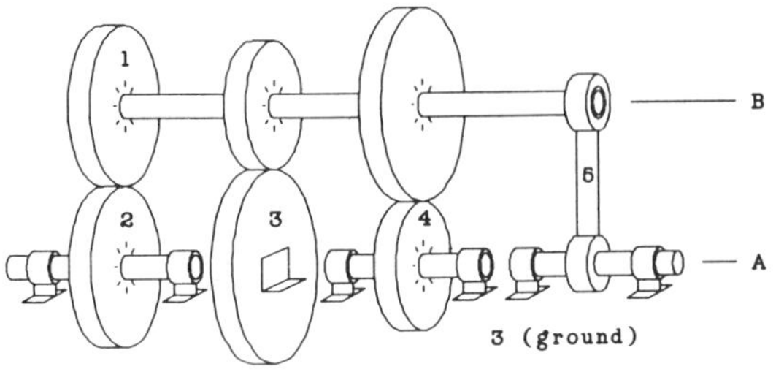

Figure 8: PGT with extraneous gears (gear pair 1-4 has no effect on overall ratio).

Conclusion

The coincident-joint graph representation for PGT's introduced in this paper makes it possible to utilize algebraic topological analysis methods formerly applied only to pin-jointed kinematic chains. The discussion here was limited to planetary gear trains, but the representation can also be used for geared linkages as well as ordinary pin-jointed linkages.

Acknowledgement

The financial support of the National Science Foundation (Grant No. MEA 8120893 and MEA 8409004) is greatly appreciated. The authors would also like to express their appreciation to the Productivity Center at the University of Minnesota for the use of their facilities.

References

1. Olson, D. G., A. G. Erdman, and D. R. Riley: "A Systematic Procedure for Type Synthesis of Mechanisms with Literature Review," *Mechanism and Machine Theory*, Volume 20, pp. 285–295, 1985.

2. Johnson, R. C. and K. Towfigh: "Creative Design of Epicyclic Gear Trains using Number Synthesis," *ASME Transactions, Journal of Engineering for Industry*, Volume 89, pp. 309–314, 1967.

3. Manolescu, N. I. and P. Antonescu: "Structural Synthesis of Planetary Mechanisms Used in Automatic Transmissions," ASME Paper No. 68–MECH–44, 1968.

4. Buchsbaum, F. and F. Freudenstein: "Synthesis of Kinematic Structure of Geared Kinematic Chains and Other Mechanisms," *Journal of Mechansisms*, Volume 5, pp. 357–392, 1970.

5. Freudenstein, F.: "An Application of Boolean Algebra to the Motion of Epicyclic Drives," *ASME Transactions, Journal of Engineering for Industry*, Volume 93, pp. 176–182, 1971.

6. Freudenstein, F. and Yang, A. T.: "Kinematics and Statics of a Coupled Epicyclic Spur-Gear Train," *Mechanism and Machine Theory*, Volume 7, pp. 263–275, 1972.

7. Ravisankar, R. and T. S. Mruthyunjaya: "Computerized Synthesis of the Structure of Geared Kinematic Chains," *Mechanism and Machine Theory*, Volume 20, pp. 367–387, 1985.

8. Tsai, L. W.: "An Application of the Linkage Characteristic Polynomial to the Topological Synthesis of Epicyclic Gear Trains," ASME Paper No. 86–DET–66, 1986.

9. Erdman, A. G. and G. N. Sandor: *Mechanism Design: Analysis and Synthesis, Volume 1*, Prentice Hall, Englewood Cliffs, NJ, 1984.

10. Lynwander, P. *Gear Drive Systems: Design and Application*, Marcel Dekker, NY, 1983.

11. Olson, D. G., T. R. Thompson, A. G. Erdman, and D. R. Riley: "An Algorithm for Automatic Sketching of Planar Kinematic Chains" *ASME Transactions, Journal of Mechanisms, Transmissions, and Automation in Design*, Volume 107, pp. 106–111, 1985.

12. Uicker, J. J. and A. Raicu: "A Method for the Identification of and Recognition of Equivalence of Kinematic Chains," *Mechanism and Machine Theory*, Volume 10, pp. 375–383, 1975.

13. Mruthyunjaya, T. S. and M. R. Raghavan: "Structural Analysis of Kinematic Chains and Mechanisms Based on Matrix Representation," *ASME Transactions, Journal of Mechanical Design*, Volume 101, pp 488–496, 1979.

14. Mruthyunjaya, T. S. and M. R. Raghavan: "Computer-Aided Analysis of the Structure of Kinematic Chains," *Mechanism and Machine Theory*, Volume 19, pp. 357–368, 1984.

15. Mruthyunjaya, T. S. and H. R. Balasubramanian: "In Quest of a Reliable and Efficient Computational Test for Detection of Isomorphism in Kinematic Chains," *Proceedings of the National Conference of Machines and Mechanisms* (India), pp. 53–60, 1985.

16. Sohn, W. J. and F. Freudenstein: "An Application of Dual Graphs to the Automatic Generation of the Kinematic Structures of Mechanisms," *ASME Transactions, Journal of Mechanisms, Transmissions, and Automation in Design*, Volume 108, pp. 392–398, 1986.

EINE NEUE GRAPHENTHEORETISCHE DARSTELLUNG ZUR TOPOLOGIEANALYSE VON UMLAUFGETRIEBEN

Kurzfassung—Es ist bekannt, daß Graphentheorie während der konzeptionellen oder typensynthesen Phase von Getriebekonstruktionen brauchbar ist. Für Umlaufgetriebe, als eine spezielle Klasse von Getrieben, hat die herkömmliche graphische Darstellung einige Unzulänglichkeiten bei bestimmten Schritten der Typensynthese. Diese Arbeit bespricht diese Unzulänglichkeiten und schlägt eine neue graphische Darstellung vor, mit der diese Schritte direkt durchgeführt werden können. Dabei werden neue Fragen aufgeworfen bezüglich der Kriterien, ob ein bestimmtes Umlaufgetriebe als Kandidat für weitere Enswicklung in Betracht gezogen werden kann oder dafür nicht in Frage kommt.

Static and Dynamic Jamming in Self-locking Gearings

R. V. VYRABOV

Moscow Automobil Mechanical Institute, Bolshaya Semionovskaya 38, Moscow 105023, USSR

Аннотация. Передача Twinworm , имеющая высокий КПД прямого хода с одновременной возможностью обеспечения самоторможения при обратном ходе, а при определенном соотношении параметров и так называемого самоторможения второго рода, при котором обратный ход сопровождается заклиниванием, т.е. оказывается невозможным при любых приложенных к звеньям передачи внешних силах, известна уже 2,5 десятилетия. Однако до настоящего времени не было проведено силового анализа процессов статического и динамического заклинивания таких и аналогичных им цилиндрических зубчатых передач. Данная работа направлена на восполнение этого пробела, что должно способствовать лучшему пониманию физики явлений и уточнению их критериев.

Ключевые слова. Самоторможение; оттормаживание; заклинивание; динамическое заклинивание.

ВВЕДЕНИЕ

Особенностью нового класса передач, начало которому было положено созданием Поппером и Пессеном (1960) передачи Twinworm (передача спаренными червяками) является высокий КПД прямого хода с одновременной возможностью самоторможения при обратном ходе. Искомые зависимости между силами и выражения для КПД прямого и обратного ходов в такой передаче авторы получают из рассмотрения ее клиновового аналога (рис. 1), в котором тела 1 и 2 имитируют соответственно червяки с меньшим - φ_1 и большим - φ_2 углами подъема винтовой линии на начальных цилиндрах (рис. 2). С учетом только трения в зацеплении червяков имеем: для прямого хода - при ведущем червяке 1

$$\eta_{12} = \frac{1 + \mathrm{tg}\,\rho / \mathrm{tg}\,\varphi_2}{1 + \mathrm{tg}\,\rho / \mathrm{tg}\,\varphi_1}, \quad (1)$$

для обратного хода - при ведущем червяке 2

$$\eta_{21} = \frac{1 - \mathrm{tg}\,\rho / \mathrm{tg}\,\varphi_1}{1 - \mathrm{tg}\,\rho / \mathrm{tg}\,\varphi_2}, \quad (2)$$

где применительно к передаче под ρ подразумевается приведенный угол трения в контакте витков червяков.

Из формулы (1) установлено, что при близких между собой углах φ_1 и φ_2 ($\varphi_1 < \varphi_2$) КПД прямого хода может быть весьма высоким, а из формулы (2) получено условие самоторможения при обратном ходе:

$$S_1 = \frac{\mathrm{tg}\,\rho}{\mathrm{tg}\,\varphi_1} \geqslant 1 \quad (3)$$

где S_1 - коэффициент запаса самоторможения.

Для обратного хода в клиновом аналоге передачи с $\rho > \varphi_1$ под действием движущей силы F_2, когда осуществление движения требует приложения также движущей-оттормаживающей силы F_1 и к телу 1, из соотношения сил

$$F_1 = F_2 \frac{\sin(\rho - \varphi_1)}{\sin(\varphi_2 - \rho)} \quad (4)$$

авторы устанавливают, что при

$$S_2 = \frac{\mathrm{tg}\,\rho}{\mathrm{tg}\,\varphi_2} \geqslant 1 \quad (5)$$

в передаче возникает заклинивание. Это явление они назвали самоторможением второго рода и соответственно S_2 - коэффициентом запаса самоторможения второго рода.

Поппер и Пессен (1960) считают, что при самоторможении КПД не имеет физического смысла. Поэтому они предлагают в этом случае определять так называемое "отношение между силами", которое, однако, представляют как отношение работ, совпадающее по форме с выражением для КПД, и отмечают, что это отношение "может принять любое отрицательное значение и даже может быть больше единицы". По-видимому, это послужило первопричиной последующего некорректного обращения с КПД как критерием самоторможения.

Панюхин (1979) в работе, посвященной самоторможению в эвольвентной косозубой цилиндрической передаче с очень большим углом наклона зубьев и с заполюсным зацеплением, получает аналогичные по структуре с (2) и (4) выражения для КПД обратного хода и для момента оттормаживания, на основании которых формулирует следующие признаки самоторможения:

обычное самоторможение $\eta_{21} < 0$, (6)

самоторможение второго рода $\eta_{21} \geqslant 1$ (7)

СТАТИЧЕСКОЕ ЗАКЛИНИВАНИЕ

Неравенства (6) и (7) в совокупности создают ложное представление, будто обычное самоторможение и самоторможение второго рода имеют в своей основе различную природу. Между тем при корректном обращении с КПД общепринятое условие самоторможения (6) является единственным условием, в пределе охватывающим и второй род самоторможения. Поскольку КПД есть отношение мощностей, то даже отрицательный КПД, являющийся признаком невозможности совершения полезного действия, нельзя получать как результат формальной математической операции по подстановке в исходную формулу некоторого набора значений соответствующих параметров, обеспечивающих самоторможение. Необходимо, чтобы комбинация этих параметров проистекала из условия возможности осуществления движения звеньев, когда только и можно говорить о соотношении мощностей на выходе и входе. Так, если условно рассматривать оттормаживающую силу как "отрицательную силу полезного сопротивления" на выходе механизма при обратном ходе, то отрицательное числовое значение КПД представляет собой взятое со знаком минус отношение мощности "отрицательной силы полезного сопротивления" к мощности движущей силы, стремящейся вызвать обратный ход, совместное действие которых обеспечивает равновесное движение механизма в процессе оттормаживания.

Рассмотрим с этих позиций правомерность подстановки в формулы (2) и (4) значений ρ, превышающих φ_2. Для этого проанализируем изменение сил в клиновом аналоге передачи Twinworm (см. рис. 1-2) с $\rho > \varphi_1$, нагруженном силой F_2, под действием возрастающей оттормаживающей силы F_1. Такой анализ в вышеперечисленных работах отсутствует, а он должен способствовать лучшему пониманию физики явления заклинивания.

Если к телу 2 самотормозящегося при обратном ходе механизма приложить силу F_2 (рис. 3), то без учета трения в опорах тел равновесие установится при неполной силе трения, которая обусловит расположение полной реакции между телами 1 и 2 перпендикулярно к направляющей тела 1. Только при этом условии тело 1, свободное от внешней силы F_1, будет находиться в равновесии. На рис. 3 показаны треугольник действующих на тело 2 сил F_2, $\mathring{F}_{21}$ и $\mathring{F}_{23}$ и силы $\bar{F}^{\circ}_{12} = -\bar{F}^{\circ}_{13}$, действующие на тело 1. Приложение к этому телу постепенно возрастающей силы F_1, стремящейся вызвать оттормаживание механизма, первоначально сопровождается увеличением неполной силы трения и поворотом и возрастанием полной реакции $F_{21}(F_{12})$, происходящим как по причине увеличения силы трения, так и, главным образом, вследствие уменьшения угла между F_{21} и F_{23} при неизменной силе F_2. Одновременно возрастают реакции направляющих. Это изменение сил сопровождается изменением деформаций в зоне контакта и соответствующими им относительными микроперемещениями тел как тангенциальными (предварительным смещением), так и нормальными.

При $\rho < \varphi_2$ состояние покоя (отсутствие макроперемещений тел) будет сохраняться, пока угол между полной реакцией $F_{21}(F_{12})$ и перпендикуляром к плоскости контакта тел не достигнет угла трения ρ, т.е. пока сила трения не достигнет полной силы трения. Лишь после этого начнется оттормаживание механизма. Треугольники сил, действующих на тела 1 и 2 при оттормаживании, также показаны на рис. 3. Из этих треугольников из условия равенства $F_{21} = F_{12}$ получаем известную формулу (4), причем как из построений на рис. 3, так и из формулы (4) ясно, что чем меньше разность углов φ_2 и ρ, тем при больших реакциях $F_{21}(F_{12})$, F_{23}, F_{13} и силе F_1, а, согласно формуле (2), и при большей абсолютной величине отрицательного КПД будет происходить оттормаживание. В пределе же, когда угол ρ стремится к φ_2, оттормаживание, требующее беспредельно большой силы F_1, которая приводит к неограниченному возрастанию и всех реакций при конечной силе F_2, становится неосуществимым, т.е. происходит заклинивание.

Если $\rho > \varphi_2$, то картина сил при заклинивании остается такой же, как при $\rho \to \varphi_2$: оно происходит при угле между полной реакцией $F_{21}(F_{12})$ и перпендикуляром к плоскости контакта тел 1 и 2, приближающемся к φ_2, и, значит, при неполной силе трения. Следовательно, угол трения больший, чем φ_2, физически не может быть реализован, как бы ни был велик запас по заклиниванию согласно формуле (5). Поэтому подстановка в формулу (2), как и (4), угла трения $\rho > \varphi_2$, при котором движение обратного хода в механизме при любом соотношении сил F_1 и F_2 невозможно, является некорректной. Ведь нельзя же определять отношение мощностей сил, действующих на реальный механизм с трением в условиях его статического равновесия. Итак приходим к заключению, что пределом использования отрицательного КПД как критерия самоторможения является случай $\rho \to \varphi_2$, когда $\eta_{21} \to -\infty$.

ЗАПАС САМОТОРМОЖЕНИЯ

Таким образом, отрицательный КПД, изменяющийся от нуля до $\eta_{21} \to -\infty$, охватывает всю область самоторможения, включая и предельный случай - заклинивание при оттормаживании. При этом важно подчеркнуть, что абсолютная величина отрицательного КПД, принимаемого за критерий самоторможения, дает в общем случае более полное представление о запасе самоторможения, чем используемый для этой цели коэффициент запаса S_1, определяемый по формуле (3). Применение такого коэффициента запаса заимствовано из теории наклонной плоскости, винтовой пары и обычной червячной передачи, где эта характеристика однозначно определяет запас самоторможения. Между тем запас самоторможения, например, в передаче Twinworm зависит не только от того, насколько коэффициент S_1 больше единицы, но еще и от величины угла φ_2, возрастая с уменьшением φ_2.

ДИНАМИЧЕСКОЕ ЗАКЛИНИВАНИЕ

Представляет интерес рассмотрение, как подчеркивают Поппер и Пессен (1960), особого свойства - передачи Twinworm с самоторможением второго рода — немедленно блокироваться в момент отключения привода вращения входного червяка 1, если на валу выходного червяка 2 имеется небольшой маховик, благодаря инерции которого червяк 2 стремится вращаться быстрее, чем если бы он приводился червяком 1. Очевидно, что речь идет о заклинивании передачи при выбеге в режиме оттормаживания, названном Вейцем (1969) динамическим заклиниванием. Но поскольку при определенном соотношении параметров оно возникает и в обычной - ортогональной червячной передаче, для которой не может быть выполнено условие так называемого самоторможения второго рода, то можно предположить, что и в передаче Twinworm возникновение заклинивания при выбеге в режиме оттормаживания вовсе не требует обязательного выполнения условия (5), являющегося необходимым для заклинивания передачи при оттормаживании в статике.

Рассмотрим механизм динамического заклинивания при выбеге передачи Twinworm на основе силового анализа передачи, выполняя его для клинового аналога передачи, для которого картина сил более наглядна и который вместе с тем, при условии учета для каждого тела приведенных (к точке соответствующего червяка, совпадающей с полюсом зацепления передачи P (см. рис. 2)) массы и силы сопротивления, динамически эквивалентен передаче. Очевидно, что к вышеназванной точке каждого червяка приводятся, кроме собственных, еще массы и силы сопротивления кинематически связанных с ними звеньев: для первого червяка - звеньев привода, для второго - звеньев исполнительного механизма или выходного органа, если таковые имеются.

На рис. 4 показаны реакции в контакте тел $F_{12}(F_{21})$ в самотормозящемся клиновом механизме, т.е. при $\rho > \varphi_1$, с приведенными массами тел m_1 и m_2 и приведенными силами сопротивления F_{c1} и F_{c2}. При заданном направлении скоростей тел V_1 и V_2 картина сил соответствует выбегу в режиме оттормаживания, при котором в связи с тем, что червяк 2 обладает большей инерцией, чем червяк 1, то по сравнению с прямым ходом, когда червяк 1 был ведущим (см. штриховое изображение тел при том же направлении их скоростей), червяки приходят во взаимодействие противоположными сторонами своих витков.

Условием выбега в режиме оттормаживания является выполнение неравенства

$$\frac{a_{1Fc1}}{a_{2Fc2}} > \frac{\sin \varphi_2}{\sin \varphi_1}, \qquad (8)$$

где составляющие ускорений тел, обусловленные только внешними для механизма силами сопротивления, соответственно равны

$$a_{1F_{C1}} = \frac{F_{C1}}{m_1}\ ; \quad a_{2F_{C2}} = \frac{F_{C2}}{m_2}\ . \qquad (9)$$

Составляющие ускорений тел, вызванные действием полных реакций в контакте, согласно схеме сил, соответственно равны

$$a_{1F_{12}} = \frac{F_{12}\sin(\rho-\varphi_1)}{m_1}; \quad a_{2F_{21}} = \frac{F_{21}\sin(\varphi_2-\rho)}{m_2}, \qquad (10)$$

причем полные ускорения тел определяются как

$$a_1 = a_{1F_{C1}} + a_{1F_{12}}\ ; \quad a_2 = a_{2F_{C2}} + a_{2F_{21}}\ , \qquad (11)$$

а их отношение, как следует из показанного на рис. 4 плана ускорений (равное отношению скоростей), равно

$$\frac{a_1}{a_2} = \frac{\sin\varphi_2}{\sin\varphi_1}\ . \qquad (12)$$

Отметим, что в самотормозящемся механизме все вышеописанные слагаемые ускорений характеризуют замедление тел и, следовательно, отрицательны. Однако, поскольку они имеют одинаковый знак, а нас интересует соотношение между их величинами, то знак ускорений во внимание не принимается.

Подстановка в равенства (11) выражений (9) и (10) с учетом отношения (12) приводит к выражениям

$$a_1 = \frac{F_{C1} - F_{C2}\dfrac{\sin(\rho-\varphi_1)}{\sin(\varphi_2-\rho)}}{m_1 - m_2\dfrac{\sin\varphi_1}{\sin\varphi_2}\cdot\dfrac{\sin(\rho-\varphi_1)}{\sin(\varphi_2-\rho)}}\ , \qquad (13)$$

$$F_{21} = F_{12} = \frac{F_{C1}m_2\sin\varphi_1 - F_{C2}m_1\sin\varphi_2}{m_1\sin\varphi_2\sin(\varphi_2-\rho) - m_2\sin\varphi_1\sin(\rho-\varphi_1)} \qquad (14)$$

Условие получения конечных значений a_1 и $F_{21}(F_{12})$ с учетом формул (10) можно записать в виде

$$\frac{\sin(\rho-\varphi_1)}{m_1} \Big/ \frac{\sin(\varphi_2-\rho)}{m_2} = \frac{a_{1F_{12}}}{a_{2F_{21}}} < \frac{\sin\varphi_2}{\sin\varphi_1} \qquad (15)$$

Если неравенство (15) выполняется, то, поскольку его знак противоположен знаку неравенства (8), вытекающего из условия задачи, то равенство (12), без выполнения которого невозможно согласованное движение тел, удовлетворяется при конечных значениях ускорений $a_{1F_{12}}$ и $a_{2F_{21}}$, а следовательно, и конечной реакции $F_{21}(F_{12})$. При этом, чем ближе между собой левые и правая части неравенства (15), тем большими должны быть сила $F_{21}(F_{12})$ и составляющие ускорений $a_{1F_{12}}$ и $a_{2F_{21}}$, чтобы удовлетворилось равенство (12). В пределе же, когда неравенство (15) стремиться превратиться в равенство, то при заданном условием задачи неравенстве (8) выполнение равенства (12) требует неограниченного роста реакции в контакте тел и определяемых ею составляющих их ускорений. В противном случае торможение тела 1 должно было бы происходить интенсивнее, чем это позволяет ему тело 2, что, очевидно, исключено. Таким образом, если неравенство (15) стремится превратиться в равенство, то при жестких звеньях $F_{21}(F_{12}) \to \infty$, $a_1 \to \infty$ и $a_2 \to \infty$. Это означает, что выбег такого механизма в режиме оттормаживания приводит к мгновенному заклиниванию. При этом угол трения ρ может быть весьма далеким от определяемого неравенством (5). Так, для механизма с $\varphi_1 = 5^\circ$, $\varphi_2 = 10^\circ$ при $m_2 = 5m_1$ и соотношении сил F_{C1} и F_{C2}, удовлетворяющем условию (8), заменяя в формуле (15) знак неравенства на знак равенства, подстановкой указанных данных находим, что критический угол трения $\rho_{кр}$, достижение или превышение которого обусловливает возникновение динамического заклинивания, равен $6^\circ 26'$.

Соотношению приведенных масс $m_2 = 5m_1$ при $\varphi_1 = 5^\circ$, $\varphi_2 = 10^\circ$ и $U_{12} = 1$, если использовать формулы перехода $J_1 = m_1 \tau_{w1}^2$, $J_2 = m_2 \tau_{w2}^2$ и учесть, что при $U_{12} = 1$ имеем $\tau_{w1}/\tau_{w2} = \sin\varphi_2/\sin\varphi_1$ соответствует соотношение приведенных моментов инерции червяков $J_2 = 1{,}25 J_1$. Если принять $m_2 = 10m_1$, чему соответствует $J_2 = 2{,}5 J_1$, то критический угол трения, приводящий к динамическому заклиниванию, оказывается равным $5^\circ 50'$. Из этих примеров следует, что если в самотормозящейся передаче Twinworm с $U_{12} = 1$ условие (8) выполняется за счет превышения J_2 над J_1, то угол $\rho_{кр}$ оказывается значительно меньшим угла φ_2 и с ростом отношения J_2/J_1 все более приближается к φ_1. Лишь в случае, когда J_1 значительно больше J_2, а неравенство (8) удовлетворяется за счет требуемого для его выполнения превышения M_{C1} над M_{C2}, угол $\rho_{кр}$ может оказаться близким к φ_2.

ЗАКЛЮЧЕНИЕ

Обычное самоторможение и самоторможение второго рода имеют в своей основе одинаковую природу. Второй род самоторможения является предельным случаем проявления обычного самоторможения. Они характеризуются единым критерием – отрицательным КПД, который в пределе, когда его абсолютная величина беспредельно возрастает, охватывает и второй род самоторможения.

Обеспечение после отключения привода вращения ведущего звена, при определенном соотношении масс звеньев, мгновенной остановки рассматриваемого класса передач не требует выполнения условия самоторможения второго рода.

ЛИТЕРАТУРА

Вейц В.Л. (1969). Динамика машинных агрегатов Л., Машиностроение, 1969, с. 245-249.

Панюхин В.И. (1979). Самотормозящиеся зубчатые передачи. Вестник машиностроения, № 2, 22-24.

Popper B., Pessen D.W. (1960). The Twinworm Drive-A Self-Locking Worm Gear Transmission of High Efficiency. Transactions of the ASME, Ser. B., vol. 82, N 3, 191-199.

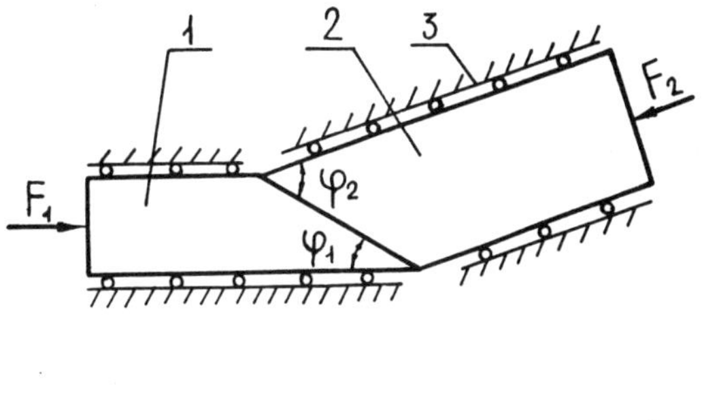

Рис. 1. Клиновой аналог передачи Twinworm

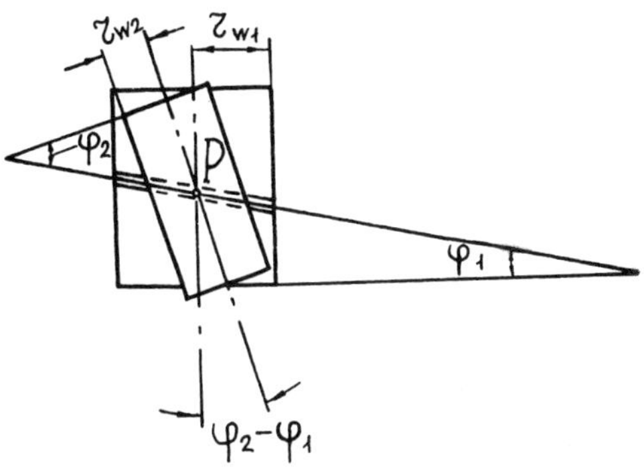

Рис. 2. Схема передачи Twinworm

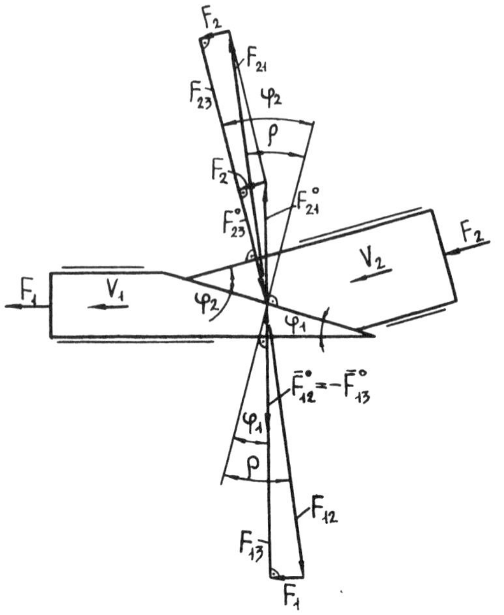

Рис. 3. Картина сил при оттормаживании клинового аналога самотормозящейся передачи Twinworm

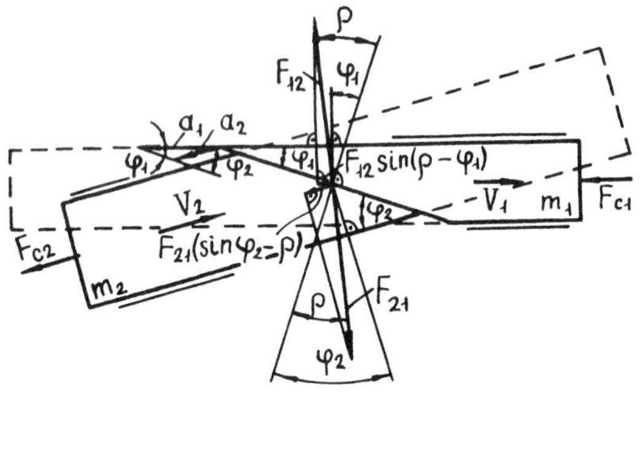

Рис. 4. Картина сил при выбеге в режиме оттормаживания клинового аналога самотормозящейся передачи Twinworm

SUMMARY

During the last decades a new class of self-locking gearings emerged, the first of which was the Twinworm gearing. They have high efficiency ratio in the direct direction of movement, and at certain relationships of elements provide so-called self-locking of the second type which means that jamming takes place during delocking. In connection with the new properties of these gearings in some papers various criteria of self-locking were proposed which create false illusion as if the usual self-locking and self-locking of the second type are of different nature. It is shown in the paper on the base of analysis of forces in the process of jamming of self-locking gearing during delocking in static condition, that the self-locking of the second type is the extreme case of the usual self-locking and that they are characterised by the same criteria-negative efficiency numerical value of which at jamming tends to negative infinity. In addition on the base of analysis of forces in the process of dynamic jamming during run-out of the self-locking gearing in the delocking state the physical reason of this phenomena is explained and it is shown that in the analysed class of gearings the dynamic jamming takes place at friction angles that are much less than those which are needed for the jamming of the gearing during delocking in the statics i.e. for the self-locking of the second type.

Einfluß der Werkseugsgeometrie, der Zaehnezahl z und des Profilverschiebungsfaktors x auf den Fussrundungsradius ρ_F

D. STAMBOLIEV

Fakultät für Maschinenwesen der Universität in Skopje, Yugoslavia

ZUSAMMENFASSUNG: Im vorliegende Arbeit wird eine Analyse über der Einfluss der Zahnform des Hobelkamm,bzw. des Wälzfräser,sowie der Zähnezahl z und des Profilverschiebungsfaktors x auf den Fussrundungsradius ρ_F, durchgeführt.
Bei dem festgestellt ist dass, bei eine Herstellung nach des Fellows-Verfahren, hat die Zahnform,vor allem der Zahnkanteradius des Werkzeug kein Einfluss an der Fussrudungradius des Zahnrades.

Dagegen diesem,bei eine Herstellung mit einem Hobelkamm bzw. Wälzfräser,die Hauptgrössen des Zahnprofils h_{ao} und ρ_{ao} zusammen mit die Zähnezahl z und des Profilverschiebungsfaktors x, ein besonder Einfluss auf dem Fussründungsradius ausüben, und auf folgende Weise: mit eine Vergrösserung der Zähnezahl und mit eine positive Profilverschiebung nimmt der Fussrundungsradius ρ_F schnell ab, und zum Zahnkantenradius des Profils des Werkzeugs ρ_{ao}, strebt, was ungünstig auf der Fusstragrähigkeit einfliesst. Deshalb, bei Zahnräder mit relativ grosse Zähnezahl z soll nicht positive Profilverschiebung verwenden, und auf diese Fall darf Werkzeuge mit je grössere Zahnkantenradius ρ_{ao} anwenden, um die Spannungskonzentration je kleinere wird.

EINLEITUNG

Wie bekannt ist (nach DIN 3972) die Kopfabrundung am Zahnstangenwerkzeugen ab 0,2.m_n bis $\rho_{ao\,grenz}$ variiert (Bild 1), wo

$$\rho^*_{ao\,grenz} = \frac{h^*_{ao}-1}{1-\sin\alpha_n} \qquad (1)$$

ist.

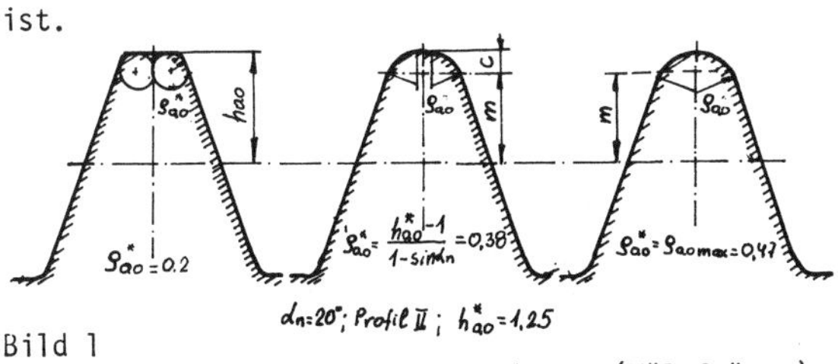

Bild 1
Kopfabrundung am Zahnstangenwerkzeug (Wälzfräser)

Auf diese Weise,der Teil der Flanke,der die aktive Evolvente ausbildet,gekürzt wird nicht (Bildmitte). Durch weitere Vergösserung des Kopfradius bis zur vollen Abrundung ($\rho^*_{ao\,max}$ = 0,47, für h^*_{ao} = 1,25), rechts im Bild,wird bereits ein Teil der aktiven Werkzeugflanke weggeschnitten.Dadurch es möglich ist,dass die aktive Flanke im Zahnfuss des Werkstückes nicht voll ausgebildet wird, was zur Eingriffstörungen im Getriebe führt.Dises hat Ungleichförmigkeiten der Bewegung und frühzeitiges Zerstören der Zahnräder zur Folge.

Im Gegensatz zum Wälzfräsen,bei die Ausbildung der Zahnfussausrundung beim Wälzstossen mit Schneidrad, die Grösse der Fussausrundung immer grösser ist. Deshalb werden die Kopfecken nur im Sonderfällen abgerundet, i.a. werden sie nur leicht gebrochen. Deswegen wird die nächste Analyse nur mit die Herstellung mit Wälzfräsen umfassen.

Ausserdem mit diese Analyse werden auch die Zahnräder mit verschiedene Zähnezahl,ausgeführte mit einige karakteristische Werte des Profilverschiebumgsfaktors x, sowie mit gewisse Werte der Verzahnungsgeometrie: h_{ao} und ρ_{ao}, umfassen.

KINEMATISCHE UNTERLAGEN

An Bild 2 sind die Eingriffsverhältnisse zwischen abgerundetem Wälzfräser und Werkrad mit Profilverschiebung beim Fräsen der Zahnfussausrundung dargestellt. Der Mittelpunkt B beschreibt beim Abwälzen eine verschlungene, allgemeine Evolvente, die im Zahnfussgebiet mit der Zahnfussausrundung identisch ist.

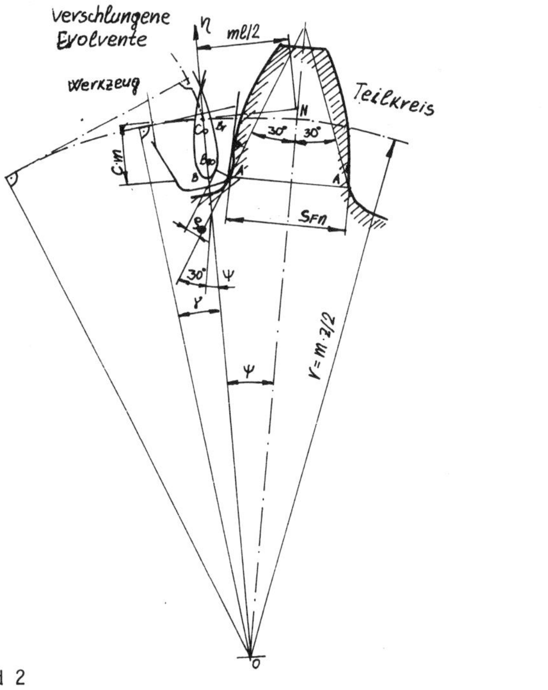

Bild 2
Relative Laufbahn des Mittelpunkt B der Werkzeugkopfabrundung für ein Fell $x < h^*_{ao} - \rho^*_{ao}$.

Um eine Feststellung der Ausrundung von Fussausrundungsradius im Berechnungsquerschnitt ρ_F, es nötig ist einigen Hilfsgrössen vorausbestimmen. Im diesem Sinne, als erste Hilfsgrösse einführt sich der Abstand

$$G=h^*_{ao}-\rho^*_{ao}-x \qquad (2)$$

Für der Abstand G drei Fällen möglich sind, von welcher am häufig in Anwendung ist:

$$G > 0; \quad x < h^*_{ao} - \rho^*_{ao}$$

Im diesem Fall der Mittelpunkt der Werkzeugkopf-abrundung B eine werschleugene,allgemeine Evolvente beschreibt.

Für $\quad G = 0; \quad x = h^*_{ao} - \rho^*_{ao}$

der Punkt B ausführt eine geschweifte,allgemeine Evolvente. Das eintritt für $x > 1$, was relativ selten ist.

Als andere,sehr wichtige Grösse ist der so genannt Abrollwinkel ϕ, dessen Ausdruck durch die transzendente Gleichung gegeben ist:

$$\frac{z_n}{2}.\phi = G.\cot g(30^o + \phi) + H \tag{3}$$

Für der Grenzwert des Radius ρ_{ao} grenz, der Parameter H durch die folgende Gleichung gegeben ist:

$$H = \frac{h^*_{ao}.(1+\sin\alpha_n)-1}{\cos\alpha_n} + \frac{\pi}{4} \tag{4}$$

Für

$$G = 0; \qquad \phi = \frac{2}{z_n}.H$$

Im Fortsetzung wird der Ausdruck für des Radius $\rho^*_F = f(\rho_{ao}, h^*_{ao}, z, x)$ ausgeführt werden in der Berührpunkt 30° - Tangente an dem Fussrundungsgebiet des Zahnprofils des Werkrades.

Wegen eine Vereinfachung wird zuerst der Fall gebetracht wird wenn der Wälzfräser nicht abgerundet ist ($\rho_{ao}=0$) und wenn x=0 (Bild 3).

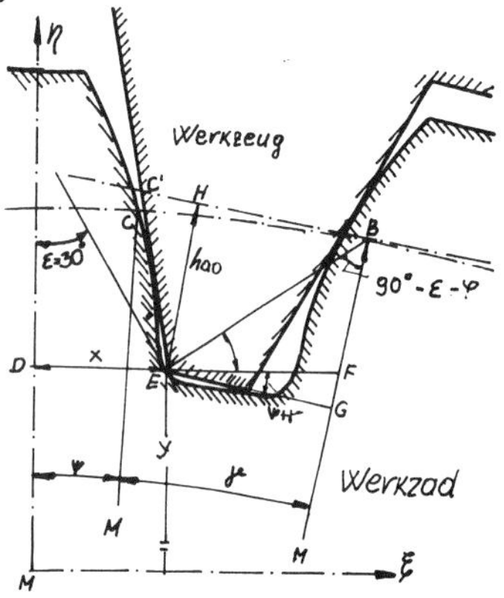

Bild 3. Eingriffsverhältnisse zwischen nicht abgerundetem Wälzfräser und Werkrad bei der Ausbildung der Zahnfussausrundung.

Die Koordinaten ξ und η dieser Evolvente beziehen sich auf das rechtwinklige Koordinatensystem, dessen Nullpunkt der Werkradmittelpunkt und dessen η - Achse eine Zahnmittellinie ist.
Da beim Verzahnen der Wälzpunkt B auf dem Wälzkreis d_w wandert und damit sich der Hilfswinkel γ ändert, sind ξ und η von γ abhängig:

$$\xi = f(\gamma); \quad \eta = f(\gamma)$$

Die Summe des Winkels γ und des Zahndicken-Halbwinkel ψ ist gleich dem Abrollwinkel ϕ, so dass kann man schreiben:

$$\gamma + \psi = \phi \tag{5}$$

Aus das Bild 3 folgt:

$$\psi = \overline{DF} - \overline{EF} \quad \text{und} \quad \eta = \overline{MF}.\cos(\psi + \gamma)$$

Die Grössen $\overline{DF}$, $\overline{EF}$ und $\overline{MF}$ werden aus die trigonometrische Verhältnisse sich bestimmen, auf welche Probleme werden wir uns nicht anlassen. So, auf Grund der Gleichung 5 und mit eine Reihe matematische Operationen bekommen sich für die Koordinaten dieser Evolvente folgende Ausdrücke:

$$\xi = (\frac{z}{2} - h^*_{ao})\sin\phi - (\frac{z}{2}\gamma - h^*_{ao}.\text{tg}\alpha_o).\cos\phi \tag{6}$$

$$\eta = (\frac{z}{2} - h^*_{ao})\cos\phi + (\frac{z}{2}\gamma - h^*_{ao}.\text{tg}\alpha_o).\sin\phi \tag{7}$$

Der Krümmungsradius der Zahnfusskurve ergibt sich dann zu:

$$\rho^*_o = \frac{(\xi'^2 + \eta'^2)^{3/2}}{\xi'.\eta'' - \xi''.\eta'} \qquad (\text{für } \rho^*_{ao} = 0) \tag{8}$$

mit $\xi' = \frac{d\xi}{d\gamma}$ und $\eta' = \frac{d\eta}{d\gamma}$ sowie $\xi'' = \frac{d^2\xi}{d\gamma^2}$, $\eta'' = \frac{d^2\eta}{d\gamma^2}$

Beim Wälzfräsen mit einem am Kopf abgerundeten Werkzeug mit dem Abrundungsradius ρ^*_{ao} und mit eine Profilverschiebung, ändert sich die Form der Fusskurve und damit ihre Koordinaten. Es wird dann:

$$\xi = (\frac{z}{2} - h^*_{ao} + x + \rho^*_{ao}).\sin(\psi + \gamma) - $$
$$- \left| \frac{z}{2}.\gamma - (h^*_{ao} - x - \rho^*_{ao})\text{tg}\alpha_o - \frac{\rho^*_{ao}}{\cos\alpha_o} \right|.\cos(\psi+\gamma) \tag{9}$$

$$\eta = (\frac{z}{2} - h^*_{ao} + x + \rho^*_{ao}).\cos(\psi + \gamma) + $$
$$+ \left| \frac{z}{2}\gamma - (h^*_{ao} - x - \rho^*_{ao})\text{tg}\alpha_o - \frac{\rho^*_{ao}}{\cos\alpha_o} \right|.\sin(\psi+\gamma) \tag{10}$$

Durch die Kopfabrundung wird die Fussausrundung als Aequidistante im Abstand ρ_{ao} zur Bahn des Abrundungsmittelspunktes ausgebildet. Die Grösse der Fussausrundung deswegen beträgt:

$$\rho_F = \rho_o + \rho_{ao} \tag{11}$$

Bestimmend die Ausdrücke für ξ' und η' sowie für ξ'' und η'' aus die Gleichungen 9 und 10, der Ausdruck für des Mittelpunktradius der Werzeugkopfabrundung ergibt sich dann zu:

$$\rho^*_o = \frac{\left|(\frac{z}{2}.\gamma - \frac{G.\sin\alpha_o + \rho^*_{ao}}{\cos\alpha_o})^2 + G^2\right|^{3/2}}{(\frac{z}{2}.\gamma - \frac{G.\sin\alpha_o + \rho^*_{ao}}{\cos\alpha_o})^2 + (\frac{z}{2} + G).G} \tag{12}$$

in dem der Parameter G ist durch die Gleichung 2 gegeben.

Speziell Fall:

für $\quad G = 0; \quad x = h^*_{ao} - \rho^*_{ao} \approx 1$

dann

$$\rho^*_o = \frac{z}{2}.\gamma - \frac{\rho^*_{ao}}{\cos\alpha_o} \tag{13}$$

In die Gleichungen 12 und 13 bestimmt sich der Winkel γ durch das Verhältnis 5, mit Bestimmung des Winkels ϕ im voraus, doch wird der Zahndicken-Halbwinkel durch folgende Gleichung gegeben:

$$\psi = \frac{s}{d} = \frac{\pi + 4.x.\text{tg}\alpha}{2.z} \tag{14}$$

Auf Grund der angegebenen Gleichungen, und für ein breit Bereich der Zähnezahlen z, Profilverschiebungsfaktors x sowie für die übliche Werte der Grössen h^*_{ao} und ρ^*_{ao}, im Fortsetzung wird eine Berechnung von entsprechende Werte des Mittelpunktradius der Werkzeugkopfabrundung ρ^*_o (Gl.12) sowie des Fussrundungsradius im Berechnungsquerschnitt ρ^*_F (Gl.11) des Werkrades durchgeführt.

BERECHNUNGSERGEBNISSE

Nach die Angaben im vorgehend Punkt werden die ergegebene Ergebnisse tabelarisch geordnet, aber wegen Raumsparen (des Vortrages), werden die Tabellen

weggelassen, doch im nächste Bilder (Nr 4 bis 8) werden die Ergebnisse ($\rho^*_F=f(z,x,h^*_{ao},\rho^*_{ao})$ grafisch dargestellt und im Fortsetzung auch ausgelegt. Die Interessente können konkrete Tabellen bei Verfasser bekommen.
Sonst, werden Zahnräder mit folgenden Angaben gebetrachtet:

Tabelle

h^*_{ao}	ρ^*_{ao} →	0,2	0,25	0,304	0,38
1,2			24	23	
1,25		33	34		36
1,3			44		

Die Zahlen in Inneren der Tabelle kennzeichnen der Grösse bzw. die Masse des Werkzeugzahnes.
Bezüglich der Zähnezahlen werden die Zahnräder mit z=14; 18; 25; 40; 58 und 68 geanalysiert ohne und mit Profilverschiebung, so dass, die grössere Verschiebung wird bei die Zahnräder mit kleinere Zähnezahl geverwendet.

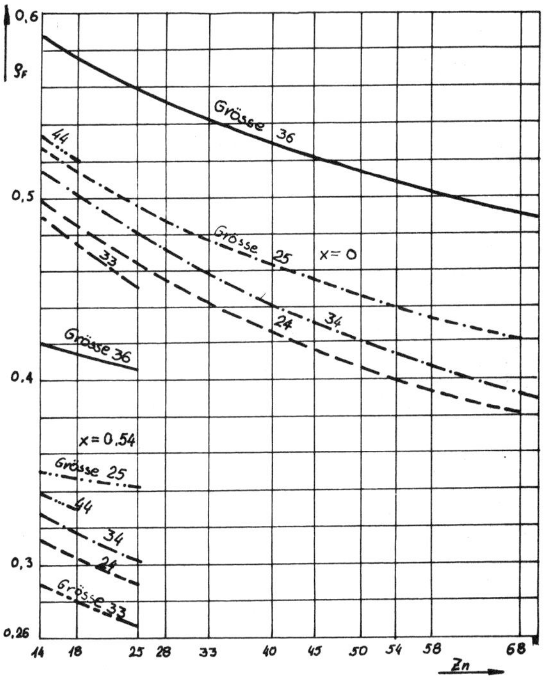

Bild 4

So, an den Bild 4 wird der Verlauf von Faussrundungsradius im Berechnungsquerschnitt ρ^*_F in Abhängigkeit von Zähnezabl z=14 bis 68, ohne und mit Profilverschiebung x=0,54 (für z bis 25) und mit verschiedene Grössen h^*_{ao} und ρ^*_{ao}, dargestellt. Als erst kann man bemerken die Situation, dass erstreckt sich die Garbe der Linien von die Zahnräder mit eine Profilverschiebung erheblich niedriger im Vergleich mit die Linien für die Zahnräder ohne Profilverschiebung, was ungünstiger ist. Beide Garben der Linien zeigen eine schnelle Senkung im Bereich der kleinere Zähnezahl. Bezüglich der Kopfhöhe des Werkzeugs h^*_{ao}, bei andere gleiche Bedingungen (ρ^*_{ao},z,x), erreicht sich grössere Fussrundungsradius ρ^*_F mit eine Werkzeugs mit grössere Zahnkopfhöhe h^*_{ao} (bei 1,3 bezüglich 1,2). Selbstverständlich, dass am grösste Radius ρ^*_F kann man erreichen mit einem Kopfkanten Rundungsradius des Werkzeugs $\rho^*_{ao}=\rho^*_{ao\,grenz}$ (die Linie Nr 36, erstreckt sich erheblich oberhalb der Linie Nr 33, bei die der Radius ρ^*_{ao}=0,2).

An die Bilder 5 bis 7, wird, für die betrachtenden Zahnräder, die Änderung des Fussrundungsradius ρ^*_F im Abhängigkeit von Profilverschiebungsfaktors x der Räder mit z=14, 18 und 25, dargestellt. Im alle Fälle, mit Vergrösserung des Profilverschiebungsfaktors x, kann man bemerken eine jähe Senkung des Wert des Radius ρ^*_F, was bezüglich der Fusstragfähigkeit, heisst eine grössere Ungünstigkeit (wegen grössere Spannungskonzentration). Die Linienordnung dieselbe ist, doch, im Vergleich mit die Zahnräder mit grössere Zähnezahl, ihre Dichtigkeit am grösste ist bei die Zahnräder mit kleinere Zähnezahl (klein Einfluss der Höhe h^*_{ao},(vergleich die Bilder 7; 5 und 6). Auch im diese Darsteillung erstreckt sich die Linie (Nr 36) mit am grösste Radius ρ^*_{ao}=0,38 bemerkbar oberhalb anderen Linien.

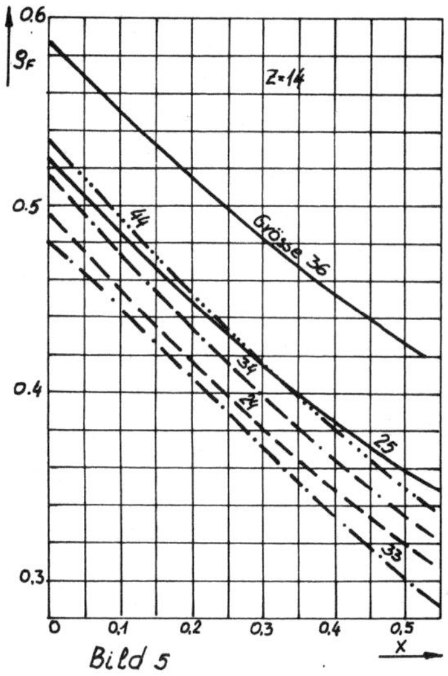

Bild 5

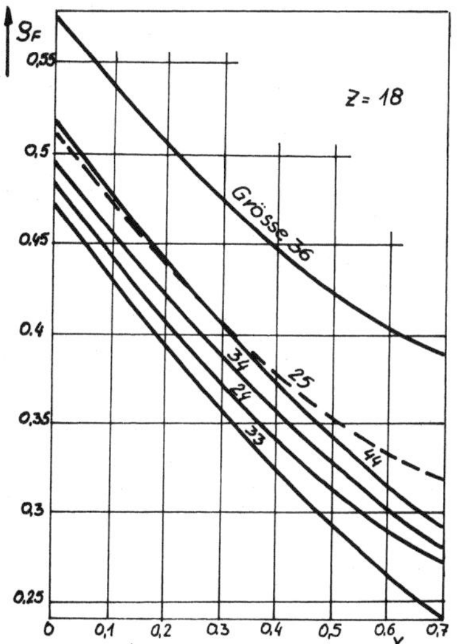

Bild 6

Das Bild 8 zeigt den Verlauf des Radius ρ^*_F im Abhängigkeit des Profilverschiebungsfaktors x für eines Zahnrad mit z=40, ausgeführt mit ρ^*_{ao}=0,25 und h^*_{ao}=1,2 und 1,25. Aus des Diagramm geht hervor dass der Radius ρ^*_F etwas grösser ist bei die Zahnräder ausgeführte mit h^*_{ao} = 1,25 (Linie Nr.34), was im Übereinstimmung mit die vorangehende Bilder (Nr 4-7) ist. Hier noch bemerkbar ist, dass mit eine Erhöhung des Profilverschiebungsfaktors x bis 0,8 und mehr, nähert sich der Radius ρ_F^* zu dem Kopfkanten Abrundungsradius der Werkzeug $\rho_{ao}^*=0,25$, was ungünstig ist.

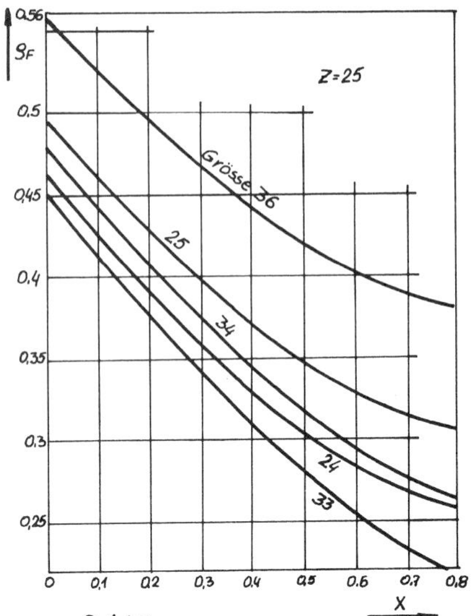

Bild 7

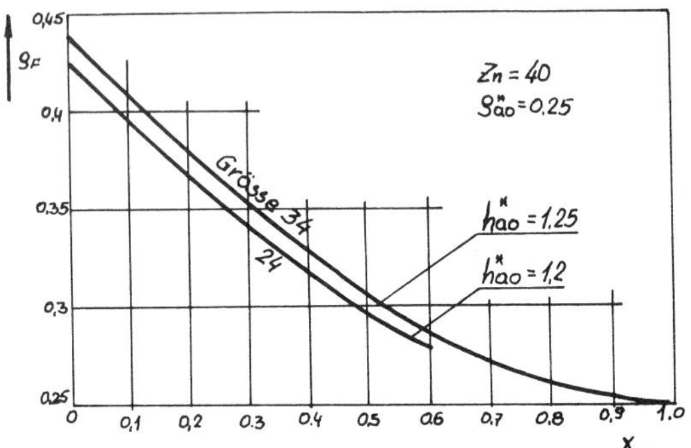

Bild 8

FOLGERUNGEN

- Für dieselbe Werte der Grössen h_{ao}^*, ρ_{ao}^* und x, mit Vergrösserung der Zähnezahl z, nimmt der Fussrundungsradius ρ_F^* ab (Bild 4), so dass, im Bereich der kleinere Zähnezahl, die Senkung jähe ist.
- Für dieselbe Werte der Grössen h_{ao}^*, ρ_{ao}^* und z, mit Erhöhung des Profilverschiebungsfaktors x, nimmt der Fussrandungsradius ρ_F^* ab (Bild 4, vergleichen die Werte für x=0 und für x=0,54).
- Mit Erhöhung des Profilverschiebungsfaktors x, lässt sich die Garbe der Linien herab in dem Bereich der kleinere Werte des Radius ρ_F^*. Das heisst dass, der Verringerung des Radius ρ_F^* mehr einfliesst die vergrösserte Profilverschiebung als ein Zuwachs der Zähnezahl, besonders im Bereich der grössere Zähnezahl.
- Für dieselbe Werte der Grössen h_{ao}^*, z und x, mit Vergrösserung des Radius ρ_{ao}^* nimmt der Wert des Radius ρ_F^* degressiv zu, oder anderes gesagt, mit eine Vergrösserung des Kopfkanten Abrundungsradius ρ_{ao}^* verringert sich der Mittelpunkradius des Werkzeugs ρ_p^* (Punkt B, Bild 2, Gl.12), besonders bei die Zahnräder mit grosse Profilverschiebung.

RÉSUMÉ

Dans le travial ci-présenté on a fait une analyse de l'influence de la forme, des dimensions du profil du dent d'outil crémaillère, ainsi que du nombre des dents z et le coefficient de déport x sur le rayon de courbure du flanc de raccord ρ_F^*.

On a constaté que, en appliquant la methode Fellow, lors de la fabrication, la forme de dent, surtout le rayon à la tête de dent de l'outil, n'a pas une influence sur le rayon de courbure du flanc de raccord.

Au contraire, pendant une fabrication avec une crémaillère ou avec une fraise-mère, les dimensions du dent (h_{ao}^* et ρ_{ao}^*) et le nombre des dents z, ainsi que le coefficient de déport x influencent d'une maniére essentielle sur le rayon de courbure du flanc ρ_F^* de sorte que par une augmentation du nombre des dents z et par un coefficient de déport positif le rayon de courbure du flanc ρ_F^* diminue rapidement et tend vers le rayon à la tête de dent de l'outil ρ_{ao}^*. La conséquence en est une influence négative sur la contrainte de flexion au pied de la section critique.

Il en suit la constatation que chez les roues d'engrenage avec un nombre de dents plus grand il ne faut pas appliquer un déport de denture positif mais employer un outil à rayon à la tête de dent ρ_{ao}^* plus grand pour que la concentration de la contrainte ait une influence minimum.

LITERATURVERZEICHNESS:

1. Dudley, D.W. - Winter H.: Zahnräder, Springer, Verlag, 1961
2. Stamboliev, D.: Einfluss der Verzahnungsgeometrie auf der Fusstragfähigkeit der Stirnräder, Dissertation, Univerzität Skopje, YU, 1978.
3. Niemann, G. und Winter, H.: Maschinenelemente Band II und III, Springer Verlag, 1983.
4. DIN Normen: 3960, 3972, 3990, Teil 1 - 5, 1985.

Computer Aided Optimisation of Planetary Gears

E. VETADŽOKOSKA AND H. IVANOVSKI

Mechanical Engineering Department, University of Skopje, Skopje, Yugoslavia

РЕЗЮМЕ. Предлагается алгоритам для проектирования оптимальных планетарных передаточных механизмов с применение ВМ. Алгоритам дает возможность проанализыровать большое число возможных решении выполняющих поставлених требовании ограничении, а врезультате получается механизам с наименьшими габаритами и оптимальными геометрическими параметрами (числа зубьев, моделеи, ширина шестерен). Предложенныи алгоритам, кроме того, обеспечивает контроль выносливости и договечности зубьев и подшибников. Применением алгоритма можно минимизыровать и другие зависьмости (осевое растояние, окружная скорость сателлитов ширина колеса), а также для анализа существующих механизмов.

ВВЕДЕНИЕ

При проектировании современых машин нужно стремиться к получению компактных механизмов с увеличенной долговечностью и уменьшению вредного влияния но окружающую среду. При проектировании планетарных механизмов нужно учитывать - его назначение, експлоатационные условия, качество обработки, материала, габаритов, себестоимость и сл. Кроме того, планетарные механизмы должны удовлетворить ряд производственных и эксплоатационных требовании, а так-же отдельные условя зацепления, ограничивающие число зубеев шестерен находящиеся в одновременном зацеплении.

Так-как возможно получение большего числа вариантных решении, выполняющих поставление требования, определение оптимального передачного механизма обыкновенном способом скоро невозможно.

АЛГОРИТАМ ДЛЯ ПРОЕКТИРОВАНИЯ ОПТИМАЛЬНОГО ПЕРЕДАТОЧНОГО МЕХАНИЗМА

При проектировании оптимального механизма, показанного на рис.1, потребуем выполнение следующих услови:

1. реализации передаточного отношения i
2. кпд. больше допускаемого $\eta > \eta_{min}$
3. максимальное число оборотов сателитов должно бить в допускаемом диапазоне
4. число зубьев $z_{min} > z_L^i > z_{max}$
5. число сателитов должно быть n_w
6. допускаемое отступление передаточного отношения Δi
7. Передаточное отношение u должно удоволетворить

$$U_{min} < U < U_{max}$$

8. условие монтажа

$$\frac{z_b + z_a}{n_w} = c$$

9. условие соседства

$$z_g + h_g \leq (z_a + z_g)\sin(\frac{\pi}{n_w})$$

10. услови соосности

$$z_a + 2z_g = z_b$$

11. Коэффициент запаса прочности в отношении контактного давления боков зубьев больше допускаемого

$$A_1 \frac{z_1^2 . m^3 . b^2}{Ft} \frac{Z_L^2 \cdot Z_R^2 \cdot Z_V^2 \cdot Z_N^2 \cdot Z_W^2}{K_A \cdot K_V \cdot K_{H\alpha} \cdot K_{H\beta} \cdot Z_H^2 \cdot Z_\varepsilon^2} \sigma^2_{Hlim} \geq S_{Hmin}$$

12. Коэффициент запаса прочности от напряжения корена зубца больше допускаемого

$$A_2 \frac{m^2 . z^2 . b^2}{Ft} \frac{Y_X \cdot Y_N \cdot Y_\delta \cdot Y_R \cdot Y_{st}}{K_A \cdot K_V \cdot K_{F\alpha} \cdot K_{F\beta} \cdot K_{Fa} \cdot Y_{Sa} \cdot Y_\varepsilon \cdot Y_\beta} \sigma_{Flim} \geq S_{Fmin}$$

13. Минимального обьема механизма

$$V = A_3 b\left[m^2(z_a+2)^2 K_a + m^2(z_b+2)^2 K_b + \right.$$
$$\left. + n_w m^2 (z_g+2) K_g\right] = V_{min}.$$

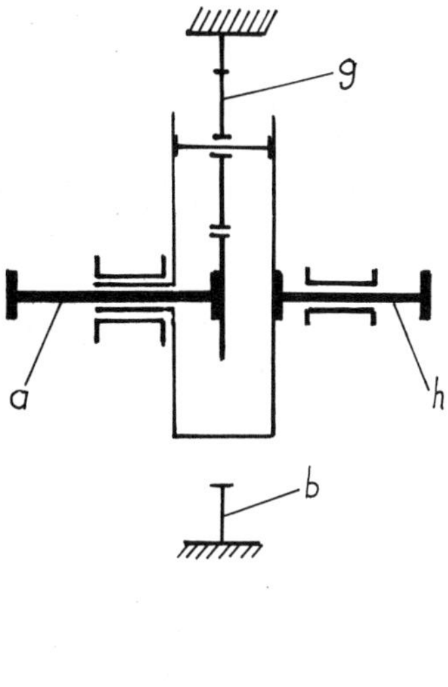

Рис.1. Схема планетарнои передачи

ОПИСАНИЕ АЛГОРИТМА

Блок-схема алгоритма приведена на рис.2. Входные данные: передаточное отношение i, которое нужно реализовать; крутящии момент T ведущего звена; срок службы передачи T_L вид напрузки; степень точности KVL; шероховатость поверхностеи зубьев Ra; материал VM; способ упрочняющеи обработки; относительная ширина зубьев ; число сателитов n_w; допуск на передаточное отношение Δi; $x_1 = x_2 = 0$; $\beta = 0$; U_{min}; U_{max}; минимальныи коэффициент запаса S_{Hmin}, S_{Fmin}.

Блок 1 - чтение входны данных

Блок 2 - вычисление оборотов и крутяших моментов всех звеньев

Блок 3 - число циклов перемен нипряжении N_{cj} и коэффициентов долговечности Z_N

Блок 4 - Ориентировочное значение допускаемых контактных напряжении $[\sigma_H]$

Блок 5 - коэффициенты неравномерности распределенилянепрузки в зацеплениях $K_{\varepsilon H}$ и коэффициент неравномерности распределения нагрузки между сателлитами Ω

Блок 6 - вычисление ориентировачного значения диаметра делительнои окружности солнечного колеса от условия что определяющим является контактное давление

Блок 7 - коэффициент долговечности Y_N и допускаемые значения напряжении изгиба на переходнои поверхности зуба

Блок 8 - вычисление z_{1max} из условия равнопрочности по напряжениям изгиба и контактным напряжениям, вычисление числа зубьев всех колес и контроль условия монтажа и соседства.

Блок 9 - определение уточненных значении m, z, d_1 кооректировка размера b, вызвинного округлением до стандартного m и целого значения, z_1

Блок 10 - определение коэффициентов K_A, K_V, $K_{H\alpha}$, $K_{H\beta}$, Z_E, Z_ε, Z_H, Z_L, Z_R, Z_V, Z_W.

Блок 11 - определение допускаемых контактных напряжении $[\sigma_{Hj}]$ расчетных контактных напряжении σ_H; при нев полнение условия $\sigma_H \leq [\sigma_H]$ увеличиватся расчетная ширина b.

Блок 12 - определение коэффициентов Y_{Fa}, Y_ε, Y_{sr}, Y_{Sa}, Y_σ, Y_x, Y_R: расчетное σ_F и допускаемое напряжения изгиба $[\sigma_F]_j$ при невыполние условия $\sigma_F \leq [\sigma_F]_j$ уменьшается число зубьев

Блок 13 - выбырается тип подшипника; проверяется не превышает ли число замен подшипников за полныи срок службы редуктора допускаемое значение

Блок 14 - несущая способность подсишипника сателита лимитирует габаритные размеры планетарнои передачи. В этом случае следует определить диаметар сателлита, при котором может быть подобран подшипник требуемои долговечности.

Блок 15 - вычисляется наружныи диаметар подшипника

Блок 16 - определение σ_H; σ_{Hlim}, σ_F; σ_{Flim} колеса b и подбор материала

Блок 17 - определение размеры для контроля

18 - определение КПД

19 - В случае когда совокупность параметров удовлетворяст поставленные условия производится селекция вариантных решении для определения наименьшего габарита передачи. Для минимизации функции V, определения шага, направления и линии деиствия перемены вектора геометрических характеристик, используются методы нелинеиного программирования. Для определения обьема передачи, обьем каждого колеса вычисляется уз предпоставки что колесо цилиндер с диаметром равным диаметру d_a выступов колеса и высотои равнои ширине колеса; K - коэффициент заполнения, равныи отношению обьема зубчастого колеса K обьему цилиндра с диаметром d_a и высотои b.

Блок 20 - печатаются результаты наименшего обьема, числа зубьев, величин модулеи и ширину каждого зубчатого колеса оптималbнои передачи.

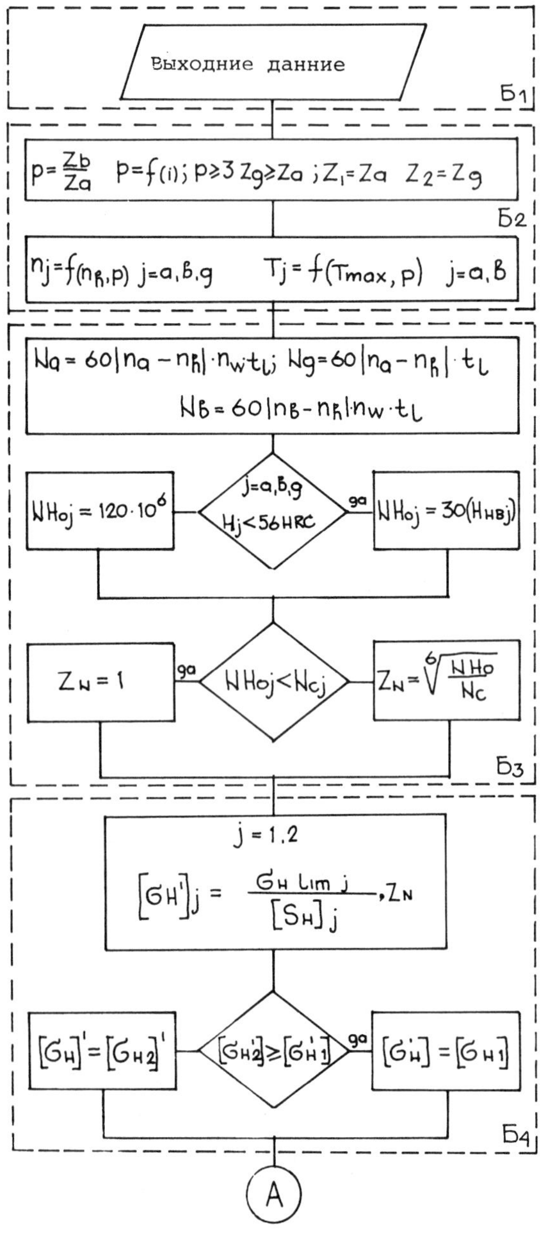

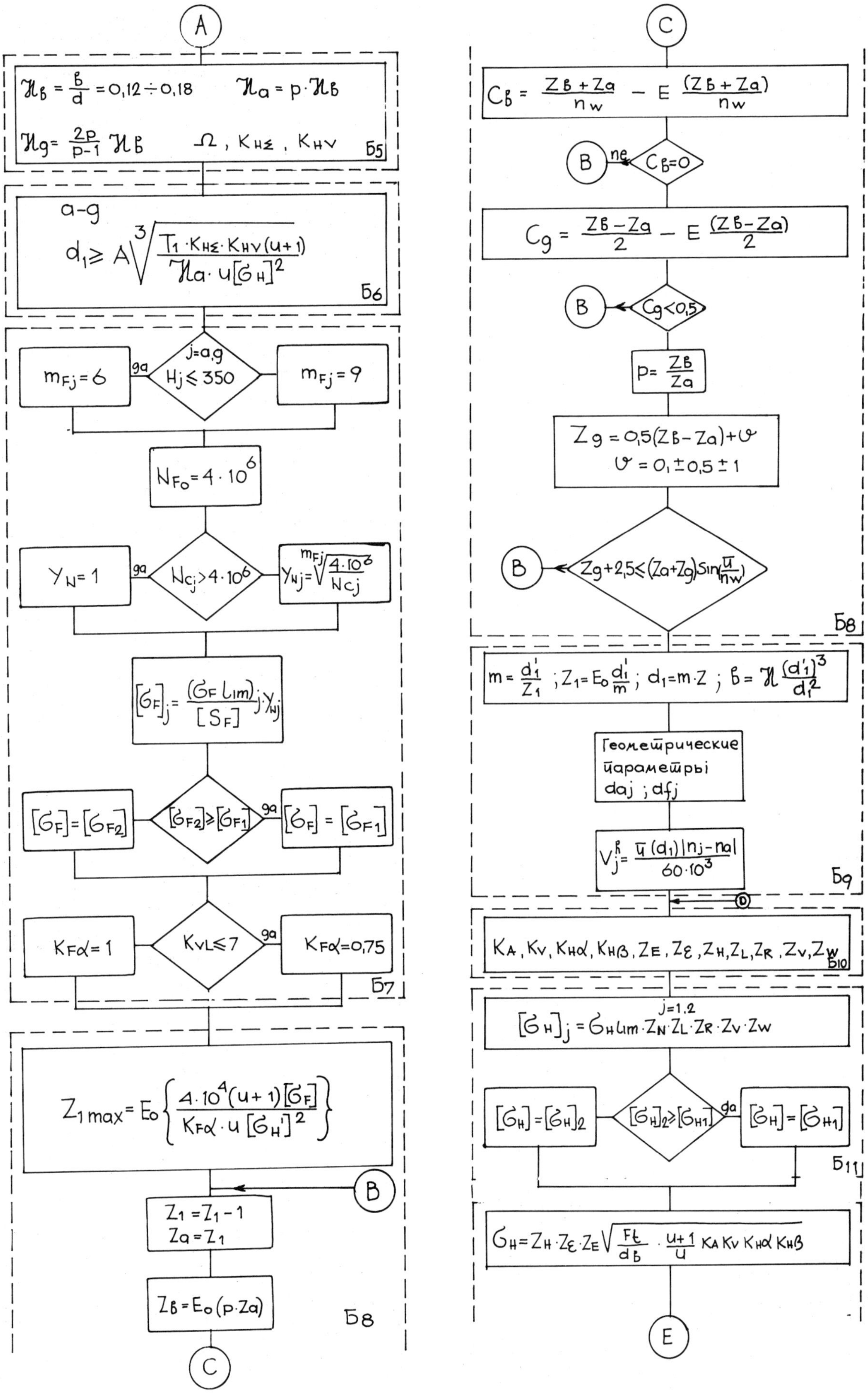
A
Б5
Б6
a-g
j=a,g
Hj ≤ 350
mFj=6
mFj=9
ga
NFo=4·10^6
Ncj>4·10^6
YN=1
Б7
Kvl≤7
KFα=1
KFα=0,75
Б8
B
Z1=Z1-1
Za=Z1
ZB=Eo(p·Za)
C
ne
CB=0
Cg<0,5
Б9
Геометрические параметры
daj ; dfj
D
Б10
KA, Kv, KHα, KHβ, ZE, Zε, ZH, ZL, ZR, Zv, Zw
j=1,2
Б11
E

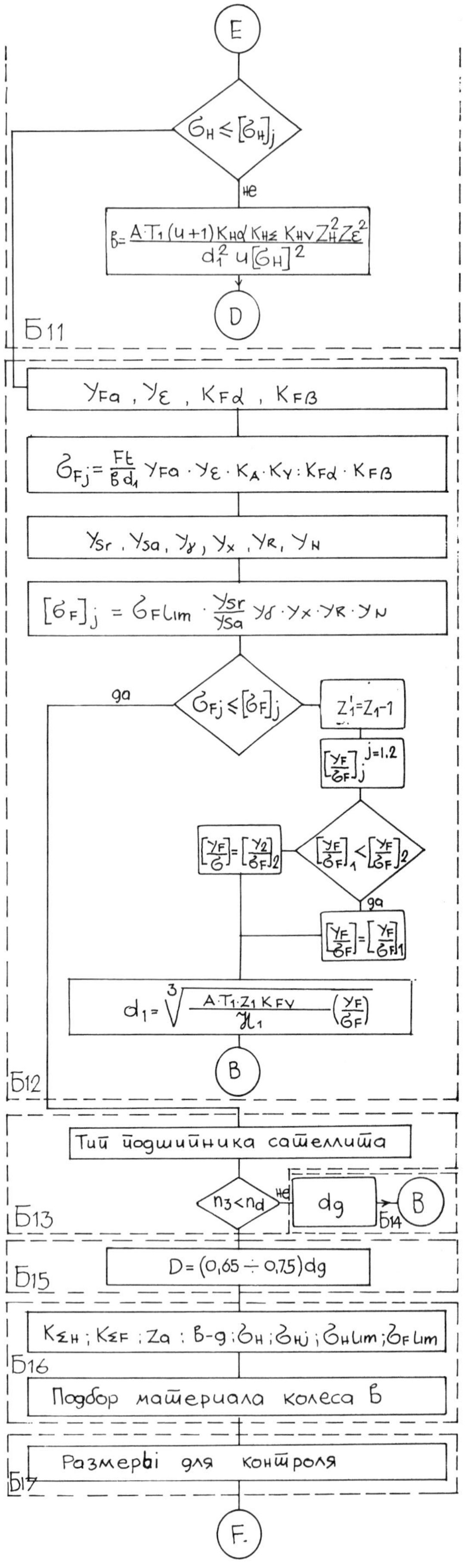

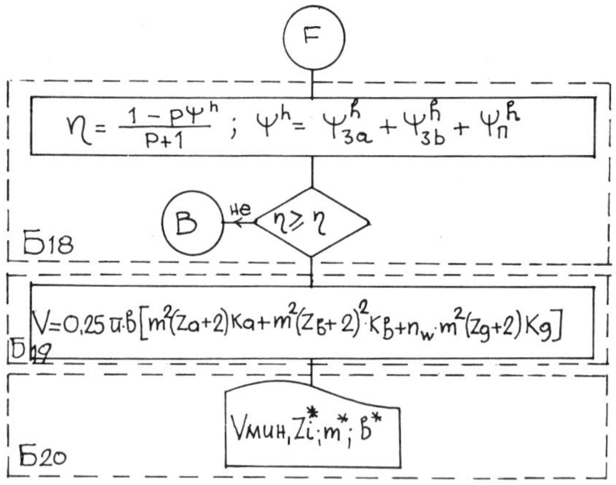

Рис.2 Блок схема алгоритма

ЗАКЛЮЧЕНИЕ

Основное преимущество предлагаемого алгоритма состоит в точности, скорости выбора и правильнои селекции вариантных решении. Алгоритам дает возможность автоматизировать проектирование планетарных передач, а можно его использовать и для анализа влияния конструкционных параметров на несущуя способность (ширина, шереховатость, передаточное отношение, осевое растояние). Этот алгоритам может быть основои для развития алгоритма для вышекритериумскои оптимизации, в которои одновременно будут минимизироватся выше обьективных функции.

SUMMARY

A planetary gearbox represents the main assembly within almost all kinds of machines in the sense of function and economic value. Henceforte a special attention should be taken to its design. Becouse of very complex calculations a great number of constrains and a large number of solution variants the design of an optimal planetary gearbox using hand graph methods is impossible.
The paper presents an algorithm for computer aided optimization design of planetary gearbox with one degree of freedom. The algorithm can be used for an analisys of all solution variants as well as for fast and accurate selection of optimal solution i.e a gearbox with minimal volume and optimal basic constructional parameters as number of teeth, modules and widths of the gear wheel rims. It can be used also for development of an algorithm for multicriterion optimization of planetary gears which minimizes simultaneously more objective functions (the width of the gearbox, the distance between axes) and determines the best compromise solution.

ЛИТЕРАТУРА

1. В.Кудрявцева и Ю.Курдяшева. Планетарние передачи. Машиностроение, 1977 г.
2. Е.Ветацокоска, Х.Ивановски, Об оптимизации планетарных передач с внутреным зацеплением IV IFTooM, Bukurešt, 1985
3. M.A.Depster, Introduction to Optimization Methods, London, 1981.
4. E.Vetadžokoska: Optimization of planetary gear trains with one degree of freedom, International AMSE Conference, 1986, Naples, Ijaly

Analysis of the Effect of Defects in the Manufacture and Assembly of Planetary Gears

V. B. DJOKIĆ AND A. V. ANDRIJAŠIN

Абстракт. Рассмотрены основные погрешности изготовления и монтажа планетарной передачи, влиякщие на распределение нагрузки между сателлитами. Определены функции и параметры их распределения. Предложена методика, позволяюща на стадии проектирования планетарной передачи оценить влияние допусков изготовления на неточность положения профилей контактирующих зубьев в зацеплениях сателлитов с центральными колесами.

Важна слова. Нагрузка; допусков; зазор-натяг; радиально биение; закон распределения.

ВВОД

В планетарных передачах погрешности изготовления и монтажа деталей приводят к неравномерному респределению нагрузки между сателлитами. Поэтему для выравнивания нагрузки в конструкциях передач применяется ряд конструктивных мероприятий (Кудравцев, 1977; Решетов, 1979; Руденко, 1980). Эффективность применения в передаче уравнительных средств определяется степенью выравнивания нагрузки между сателлитами и зависит от конструкции уравнительного средства, а также от величины неточности положения профилей контактирующих зубьев в зацеплениях сателлитов с центральными колесами, которую необходимо компенсировать уравнителем.

ВЛИЯНИЕ ОСНОВНЫХ ПОГРЕШНОСТЕЙ ИЗГОТОВЛЕНИЯ И МОНТАЖА

Влияние основных погрешностей изготовления и монтажа деталей планетарной передачи на неточность положения профилей контактирующих зубьев в зацеплениях сателлитов с центральными колесами рассмотрим на примере планетарной передачи, выполненной по схеме рис. 1. Ведущим звеном передачи является центральное колесо с внешними зубьями a, установленное на подшипниках в крышке 1, ведомым - водило h, закрепленное на шлицах вала 3. В водиле на осях установлены сателлиты g. Централ ное колесо с внутренними зуб ми b закреплено неподвижно в корпусе 2. С цел в влени размеров, отклонени котор х от номинал н х учит ва тс при нахождении неточности положени профиле зуб ев, детали передачи показани в сме енном положении.

За основную ось передачи, относительно которой определяются смещтения основных звеньев, принимаем ось посадочной поверхности BV подшипников вала водила.

Для оценки неточности положени рабочих профилей контактирующих зубьев расмотримо зазоры-натяги в зацеплениях сателлитов с центральными колесами в планетарной передаче с произвольным числом сателлитов. Если погрешности изготовления деталей отсуствуют, то после приложения незначительного крутящего момента к центальному колесу с внешними зубьями a при остановленных водиле h и центральном колесе с внутренними зубьями b, выберутся зазоры и зубья всех сателлитов вступят в контакт с соответствующими зубьями центральных колес. При такой теоретической схеме находятся зазоры-натяги в зубчатых зацеплениях сателлитов, для определения которых погрешности изготовления проецируются на линии зацепления. Положительный знак проекции соответствует условному натягу, т.е. условному внедрению рабочих профилей зубьев, а отрицательный - зазору между рабочими профилями зубьев при отсуствии моментов на звеньях передачи. Зазор-натяг, измеряемый по нормали к рабочим профилям зубьев в точке их контакта зависит от конструкции передачи, допусков на узготовление деталей и сумарного про влени погрешностей при сборке. Поскольку погрешности изготовления и монтажа носят случайный характер, то случайными будут и зазоры-натяги в зубчатых зацеплениях сателлитов с центральными колесами.

Смещение e_{na} посадочной поверхности A подшипников центрального колеса a относительно радиальными биениями цилиндрических поверхностей G относительно BV, E относительно D, и A относительно E. Смещтение e_{na} вызывает зазор-натяг в зацеплениях j-того сателлита (рус. 2):

$$\delta_{eaj} = e_{na} sin(\psi_a + \alpha_{wa} - \gamma_h - \gamma_j) \qquad (1)$$

где ψ_a и γ_h - углы, определяющие смещение e_{na} и положение водила в неподвижной системе кординат XOY, связанной с корпусом передачи,

$\gamma_j = 2\pi(j-1)/n_w$ - угловая координата сателлита с номером j,

n_w - число сателлитов планетарной передачи.

Смещтение e_h оси водила оснотильно основной оси передачи обусловлено радиальным биением шлицев вала водила и вызывает зазор-натяг в зацеплениях:

$$\delta_{ehj} = -2e_h \cos\{(\alpha_{wa} + \alpha_{wb})/2\} \times$$

$$\times \sin\{\psi_h + (\alpha_{wa} - \alpha_{wb})/2 - \gamma_j\} \qquad (2)$$

где ψ_h - угол, определяющий смещение e_h

в системе координат X_1OY_1, вращающейся вместе с водилом передачи.

Смещение e_{gj} отверстия под ось j-того сателлита в водиле относительно номинального положения вызывает зазор-натяг в зацеплениях:

$$\delta_{egj} = -2e_{gj}\cos\{(\alpha_{wa}+\alpha_{wb})/2\} \times$$
$$\times \sin\{\psi_{gj}+(\alpha_{wa}-\alpha_{wb})/2\} \qquad (3)$$

где ψ_{gj} - угол, определяющий смещение e_{gj} отверстия под ось j-того сателлита.

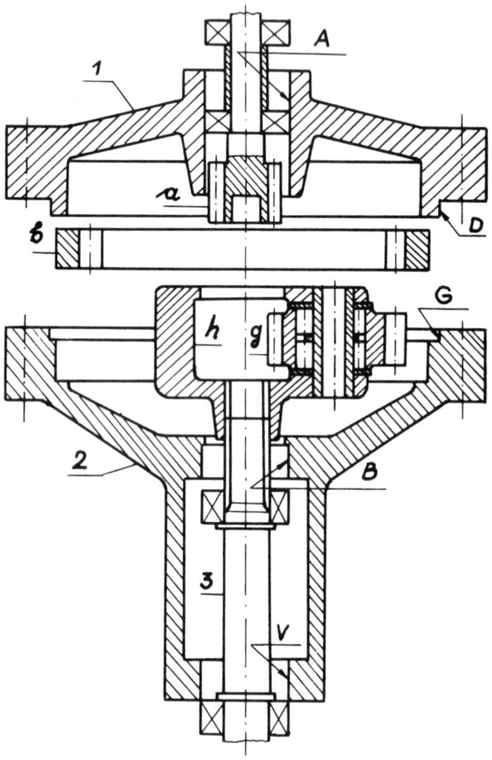

Рис. 1. Монтажная схема планетарной передачи

Разность $\Delta\delta_j$ диаметрального зазора в подшипниках j-того сателлита и среднего диаметрального зазора по всем подшипникам сателлитов планетарной передачи вызывает зазор-натяг в зацеплениях:

$$\delta_{\delta j} = -\Delta\delta_j(\cos\alpha_{wa}+\cos\alpha_{wb})/2 \qquad (4)$$

Разность Δs_j толщины зубьев j-того сателлита и среднего значения толщтин зубьев сателлитов планетарной передачи вызывает зазор-натяг в зацеплениях:

$$\delta_{sj} = \Delta s_j(\cos\alpha_{wa} + \cos\alpha_{wb})/2 \qquad (5)$$

Радиальное биение зубчатого венца определьется относительно его рабочей оси вращения (Марков, 1977). Эксцентриситет зубчатого колеса равен:

$$e = F_r/2$$

где F_r - радиальное биение зубчатого венца.

Радиально биение зубчатого венца центрального колеса a относительно посадочной поверхности A подшипников вызывает зазор-натяг в зацеплениях:

$$\delta_{aj} = e_{ra}\sin(\Theta_a+\alpha_{wa}+\gamma_a-\gamma_h-\gamma_j) \qquad (6)$$

где e_{ra} - эксцентриситет центрального колеса a,

Θ_a - фазовый угол, определяющий положение эксцентриситета в системе координат XOY в исходном положении (при $\gamma_h = 0$), при неподвижном централ ном колесе b:

$$\gamma_a - \gamma_h = \gamma_h z_b/z_a$$

Радиальное биение зубчатого венца центрального колеса b относительно посадочной поверхности BV подшипников вала водила вызывает зазор-натяг в зацеплениях:

$$\delta_{bj} = e_{rb}\sin(\Theta_b-\alpha_{wb}-\gamma_h-\gamma_j) \qquad (7)$$

где e_{rb} - эксцентриситет центрального колеса b,

Θ_b - фазовий угол, определяющий положение эксцентриситета в системе координат XOY

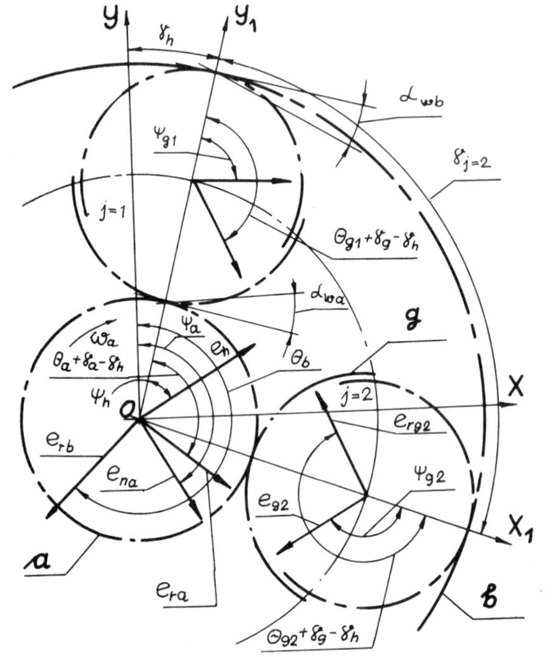

Рис. 2. Схема для определения зазоров-натягов в зацеплениях састеллитов с центральными колесами

Радиальное биение зубчатого венца j-того сателлита относительно оси его вращения вызывает зазор-натяг в зацеплениях:

$$\delta_{gj} = -2e_{rgj}\cos\{(\alpha_{wa}+\alpha_{wb})/2\} \times$$
$$\times \sin(\Theta_{gj}+\gamma_g-\gamma_h+(\alpha_{wa}-\alpha_{wb})/2\} \qquad (8)$$

где e_{rgj} - эксцентриситет j-тогс сателлита

Θ_{gj} - фазовий угол, определяющий положение эксцентриситета j-того састеллита в исходном положении. При неподвижном колесе b:

$$\gamma_g - \gamma_h = -\gamma_h z_b/z_g.$$

Зазор-нат г Δ_j в зацеплени х j-того сателлита с централ н ми колесами вл етс суммо рассмотренн х в ше сжагаем х:

$$\Delta_j = \delta_{eaj} + \delta_{ehj} + \delta_{egj} + \delta_{\delta j} + \delta_{sj} + \delta_{aj} + \delta_{bj} + \delta_{gj} \qquad (9)$$

Величина зазора-натяга Δ_j характеризует неточность положения рабочих профилей зубьев в зацеплениях сателлитов с центральными колесами относительно теоретического, которое они занимали бы при отсуствии погрешностей изготовления и монтажа деталей и моментов на звеньях передачи.

Можно выделить постоянную, низко и среднечастотные составляющие зазора-натяга. Постоянная составляющая, обусловленная погрешностями e_h, e_g, Δ_δ и Δ_s не зависит от угловых положений зубчатых колес или водила. Низкочастотная составляюща , определяемая погрешностями e_{na} и e_{rb} изменяется с частотой вращтения водила. Среднечастотная составляюща , обусловленная погрешностями e_{ra} и e_{rg} связане с частотой вращения зубчатых колес a и g относительно водила.

Известно (Кудравцев, 1977; Цутович, 1984) что на величину передаваемой нагрузки j-тым сателлитом оказывает влияние только часть полного зазора-натяга, которая равна разности между зазором-натягом в зацеплениях j-того сателлита с центральными колесами и средней величиной зазоров-натягов в зацеплениях всех сателлитов, т.е. зазор-натяг:

$$\Delta'_j = \Delta_j - (1/n_w) \sum_1^{n_w} \Delta_j \qquad (10)$$

Величина зазора-натяга Δ'_j характеризует неточность положения рабочих профилей контактирующих зубьув в зацеплениях сателлитов с центральными в планетарной передаче, работающей под нагрузкой, относительно теоретического, которое они занимали бы при отсуствии погрешностей изготовления и монтажа деталей.

Определим зазор-натяг Δ'_j через погрешности изготовления и монтажа деталей планетарной передачи. Подставля в уравнение (10) выражение зазора-натяга (9) через его составляющие (1-8), и учитывая, что

$$\sum_1^{n_w} \sin\gamma_j = \sum_1^{n_w} \cos\gamma_j = 0$$

после преобразовани , получим:

$$\Delta''_j = e_{na}\sin(\psi_a + \alpha_{wa} - \gamma_h - \gamma_j) - 2e_h\cos\{(\alpha_{wa} + \alpha_{wb})/2\} \times \sin\{\psi_h + (\alpha_{wa} - \alpha_{wb})/2 - \gamma_j\} + e_{ra}\sin(\Theta_a + \alpha_{wa} + \gamma_a - \gamma_h - \gamma_j) + e_{rb}\sin(\Theta_b - \alpha_{wb} - \gamma_h - \gamma_j) + B_j - (1/n_w)\sum_1^{n_w} B_j \qquad (11)$$

где

$$B_j = -2e_{gj}\cos(\alpha_{wa} + \alpha_{wb})/2 \times \sin(\psi_{gj} + (\alpha_{wa} - \alpha_{wb})/2) + (\Delta s_j - \Delta\delta_j)(\cos\alpha_{wa} + \alpha_{wb})/2 - 2e_{rgj}\cos(\alpha_{wa} + \alpha_{wb})/2) \times \sin(\Theta_{gj} + \gamma_g - \gamma_h + (\alpha_{wa} - \alpha_{wb})/2)$$

Как следует из уравнения (11), если одинаковы смещтения отверстий под оси сателлитов и углы, определяющие смещения, толщины зубьев сателлитов и зазор в подшипниках, эксцентриситеты сателлитов и их фазовые углы, то данные погрешности изготовления не вызывают неравномерного распределения нагрузки между сателлитами и исключаются при определении зазора-натяга Δ'_j.

Эта особенность используется на практике при специальных методах сборки передач (синфазная установка сателлитов, подбор подшипников по зазорам, установка сателитов с одинаковой толщиной зубьев и т.д.).

Определим закон и параметры распределения зазора-натяга Δ'_j, для чего рассмотрим плотности и параметры распределений независимых случайных погрешностей изготовления и монтажа деталей. Для погрешностей изготовлени Δs и $\Delta\delta$ принимаетс нормалдный закон распределения, а для погрешностей e_{na}, e_h, e_{ra}, e_g, e_{rb}, e_{rg} - закон распределения Рэлея, при этом случайные углы ψ_a, ψ_h, ψ_g, Θ_a, Θ_b и Θ_g - равномерно распределенными в интервале $(0,2\pi)$ (Куцокон 1971).

Плотность распределения неотрацительной случайной величины e, распределенной по закону Рэлея и соответствующей смещению или эксцентриситету, равна:

$$f(e) = (e/K_e^2)\exp(-e^2/2K_e^2), \quad e \geqslant 0 \qquad (12)$$

где

K_e - параметр распределения.

Проекции случайного вектора, длина которого распределена по закону Рэлея, на оси прямоугольной системы координат при угле ψ, определяющем положение вектора в данной системе координат, равномерно распределенном в интервале $(0,2\pi)$, являются независимыми случайными величинами, распределенными по нормалнму закону, при этом математическое ожидание их равно нилю, а K_e и K_e^2 - представляют среднее квадратическое и дисперсию нормального распределения (Куцоконь, 1971).

Используя известные теоремы теории вероятностей, определим, что зазор-натяг Δ' распределен по нормальному закону, математическое ожидание его равно нулю, а дисперсия равна:

$$D(\Delta') = K_{ena}^2 + 2(1+\cos(\alpha_{wa} + \alpha_{wb})) \times K_{eh}^2 + K_{era}^2 + K_{erb}^2 + (n_w - 1)/n_w(2(1+\cos(\alpha_{wa} + \alpha_{wb}))(K_{eg}^2 + K_{erg}^2) + 1/4(\cos\alpha_{wa} + \cos\alpha_{wb})^2 (D(s) + D(\delta))), \qquad (13)$$

где

K_{ena}, K_{eh} K_{eg} - параметры распределений погрешностей изготовления, обусловленных смещением осей посадочной поверхности подшипников центрального колеса a, водила и отверстий под оси сателлитов;

K_{era}, K_{erb}, K_{erg} -параметры распределений эксцениситетов зубчатых колес, связанных с радиальными биениями зубчатых венцов;

$D(s)$ $D(\delta)$ - дисперсии погрешностей изготовления, обусловленных разнотолщинностью зубьев и разнозазорностью подшипников сателлитов.

При определении составляющих дисперсии зазора-натяга принимается для погрешностей, распределенных по нормальному закону, поле допуска равным $6K_e$, а для погрешностей, распределенных по закону Рэлея $3{,}44K_e$. Исходя из этого, определим:

$$K_{ena} = \Delta e_{na}/3{,}44 \;;\quad K_{eh} = F_{rh}/6{,}88 \;;$$

$$K_{era,b,g} = F_{ra,b,g}/6{,}88 \;;$$

$$K_{eg} = \Delta e_g/3{,}44 \;;\quad D(s) = T_c^2 / 36 \;;$$

$$D(\delta) = \delta_\Delta^2 / 36 \qquad (14)$$

где

Δe_{na} - наибольше смещение посадочной поверхности подшипников центрального колеса a относительно основной оси передачи,

F_{rh} - допуск на радиальное биение шлицев вала водила,

$F_{ra,b,g}$ - допуск на радиальное биение зубчатых венцов центральных колес и сателлитов,

Δe_g - допуск на смещение отверстий под оси сателлитов в щеках водила от номинального положения,

δ_Δ - допуск на зазор в подшипниках сателлитов и

T_c - допуск на толщину зубьев сателлитов.

Для данной передачи смещение e_{na} является суммой независимых случайных векторов, углы между которыми случайные величины, равномерно распределенные в интервале $(0,2\pi)$, а длины векторов представляют эксцентриситеты соответствующих поверхностей и распределены по закону Рэлея. Суммарное смещение также будет распределено по закону Рэлея, наибольше значение которого определим из выражения:

$$\Delta e_{na} = 1/2 \;\; (F_{rG}^2 + F_{rE}^2 + F_{rA}^2)^{1/2} \qquad (15)$$

где

F_{rG}, F_{rE}, F_{rA} - допуски на радиальное биение поверхностей G, E и A.

Необходимые данные для расчета дисперсии зазора-натяга определяются по рабочим чертежам планетарной передачи.

ЗАКЛЮЧЕНИЕ

Дисперсия зазора-натяга $D(\Delta')$ позволяет оценить возможную неточность положения рабочих профилей контактирующих зубьев в зацеплениях сателлитов с центальными колесами относительно теоретического, которое они занимали бы при отсутствии погрешностей изготовления и монтажа деталей, а также выявить относительное влияние допусков изготовления деталей на эту неточность. Очевидно, чем больше величина дисперсии, тем больше ожидаемая величина неравномерности распределения нагрузки между сателлитами. Необходимо отметить, что величина дисперсии зависит от числа сателлитов планетарной передачи: с увеличением числа сателлитов она возрастает при неизменных остальных параметрах.

Таким образом, предлагаемая методика позволяет на стадии проектирования планетарной передачи оценить возможную неточность положения профилей контактирующих зубьев в зацеплениях сателлитов с центральными колесами, в вит относител ное вли ние допусков изготовлени детале на ту неточност , чтоб при возможности изменит их дл снижени неравномерности распределени нагруски между сателлитами.

ЛИТЕРАТУРА

Кудрявцев, В.Н. (1977). Справочник. Машиностроение.

Куцоконь, В.А., С.Г. Малошевский (1971). Применение теории вероятностей при проектировании механизмов приборов. Машиностроение.

Марков, А.Л. (1977). Измерение зубчатых колес. Машиностроение.

Решетов, Л.Н. (1979). Самоустанавливающиеся механизмы. Альбум конструкций. Машиностроение.

Цитович, И.С. (1984). Нагруженность зубчатых зацеплений планетарных передач. Минск. ИНДМАШ АН БССР.

Power Losses in High Speed Spur and Helical Gears

V. SIMON

Faculty of Technology, University of Novi Sad, Novi Sad, Yugoslavia

Abstract. In high speed spur and helical gears the main power losses are due to EHD lubrication and the pumping effect of gear mesh.

To determine the EHD power loss a full thermal elastohydrodynamic analysis of lubrication for all the instantaneously engaged teeth is carried out. The viscosity variation with respect to pressure and temperature and the density variation with respect to pressure is included.

The teeth entering in mesh are pumping out the mixture of air and oil from the tooth gaps of the mated gear. A method for the calculation of the corresponding power loss is developed.

The effects of gear geometry and operating conditions on power losses is discussed. Based on the obtained results, by regression analysis equations are derived for the calculation of power losses.

Keywords. Power transmission; EHD lubrication; air pumping; partial differential equations; numerical methods; computer application.

INTRODUCTION

The power losses in high speed gears are due to EHD lubrication, pumping effect of gear mesh, windage, and bearing losses.

A number of papers is published suggesting equations for the calculation of power losses in oil film due to EHD lubrication (Gu, 1973; Wellauer, Holloway, 1976; Martin, 1981; Anderson, Loewenthal, 1982; Matsumoto and co-workers, 1985). All these equations are derived using the results of EHD analysis of lubrication for rollers with line contact, usually omitting the thermal effect on oil viscosity. The special geometry and kinematics of helical gears make such a calculation doubtful. The full thermal EHD analysis of involute gears and worm gears is treated in (Sato, Takanashi, 1981; Simon, 1981, 1985).

The teeth entering in mesh are pumping out the mixture of air and oil from the tooth gaps of the mated gear. At high speed gears, as it will be shown, the corresponding power loss is of the same range as that due to EHD lubrication. For this kind of power losses no results can be found in literature.

For the calculation of bearing losses reliable methods exist.

The results presented in (Anderson, Loewenthal, 1981) show that the windage losses are negligable in comparison with other power losses.

In this paper methods are developed for the calculation of power losses due to thermal elastohydrodynamic lubrication and due to the pumping effect of gear mesh. By applying the corresponding computer programs the influence of gear parameters and operating conditions on power losses is investigated and the obtained results are discussed. Based on the obtained results, by regression analysis equations are derived for the calculation of power losses.

EHD LUBRICATION OF SPUR AND HELICAL GEARS

The pertinent equations governing the pressure and temperature distributions, and the oil film shape are the Reynolds, elasticity, energy, and Laplace's equations (Simon, 1981, 1985).

The general Reynolds equation for EHL of point contacts

$$\frac{\partial}{\partial x}\left(F_2\frac{\partial p}{\partial x}\right)+\frac{\partial}{\partial y}\left(F_2\frac{\partial p}{\partial y}\right)=-\frac{\partial}{\partial x}\left[\frac{F_3}{F_o}(U_1-U_2)\right]-\frac{\partial}{\partial y}\left[\frac{F_3}{F_o}(V_1-V_2)\right]+\rho(W_1-W_2) \qquad (1)$$

The full energy equation is given by

$$\rho c_p\left(u\frac{\partial T}{\partial x}+v\frac{\partial T}{\partial y}+w\frac{\partial T}{\partial z}\right)-k_o\left(\frac{\partial^2 T}{\partial x^2}+\frac{\partial^2 T}{\partial y^2}+\frac{\partial^2 T}{\partial z^2}\right)=\alpha_T T\left(u\frac{\partial p}{\partial x}+v\frac{\partial p}{\partial y}\right)+\eta\left[\left(\frac{\partial u}{\partial z}\right)^2+\left(\frac{\partial v}{\partial z}\right)^2\right] \qquad (2)$$

The equation governing the heat transfer in the gear teeth is Laplace's equation

$$\frac{\partial^2 T_l}{\partial x_l^2} + \frac{\partial^2 T_l}{\partial y_l^2} + \frac{\partial^2 T_l}{\partial z_l^2} = 0 \qquad (3)$$

The elasticity equation which determines the composite normal elastic displacement of the teeth surfaces is given by

$$d(x,y) = \frac{2(1-\mu^2)}{\pi E} \int_{x_{min}}^{x_{max}} \int_{y_{min}}^{y_{max}} \frac{p(X,Y)}{\sqrt{(x-X)^2+(y-Y)^2}} dX\, dY \qquad (4)$$

The film thickness is defined by expression

$$h(x,y) = h_{min} + d(x,y) + s(x,y) \qquad (5)$$

where

h_{min} = minimum film thickness due to hydrodynamic effects

$s(x,y)$ = separation due to the geometry of the teeth surfaces.

The viscosity variation with respect to pressure and temperature, and the density variation with respect to pressure are represented by

$$\eta = \eta_o e^{\alpha p - \beta (T - T_o)} \qquad (6)$$

$$\rho = \rho_o \left(1 + \frac{\alpha_1 p}{1 + \beta_1 p}\right) \qquad (7)$$

The friction factor is defined by the ratio of the frictional force to the load

$$f_{EHD} = \frac{\sqrt{F_{Tx}^2 + F_{Ty}^2}}{W} \qquad (8)$$

The appropriate boundary conditions of Reynolds, energy, and Laplace's equations are described in (Simon, 1981, 1985).

The Reynolds, elasticity, energy, and Laplace's equations represent a highly nonlinear integrodifferential system. The method used in the solution of these equations is a finite difference method. A three-dimensional mesh is applied. The intervals used to divide the coordinates along the oil film are irregular. They decrease gradually as they approach the pressure peak. The application of such a non uniform mesh reduces considerably the computational time. Automatic mesh generation and decrease of grid size due to actual pressure distribution is introduced.

The systems of equations, obtained by finite difference approximations, are solved by succesive-over-relaxation method.

By applying the corresponding computer program the influence of radius of relative tooth surface curvature in normal section (ρ_n), helix angle (β_o), addendum modification factors (x_p), face width (b_f), crownings (c_r), length of crownings (ℓ_{cr}), oil viscosity (η_o), minimum oil film thickness (h_{min}), pitch velocity (v_p), and supplied

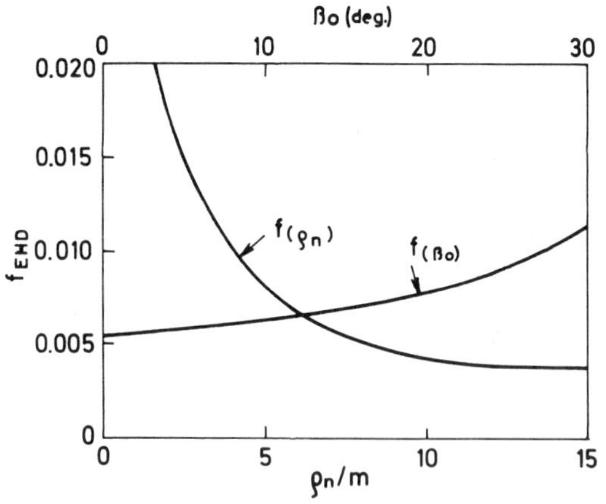

Fig. 1. EHD power losses for relative tooth surface curvature and helix angle variations.

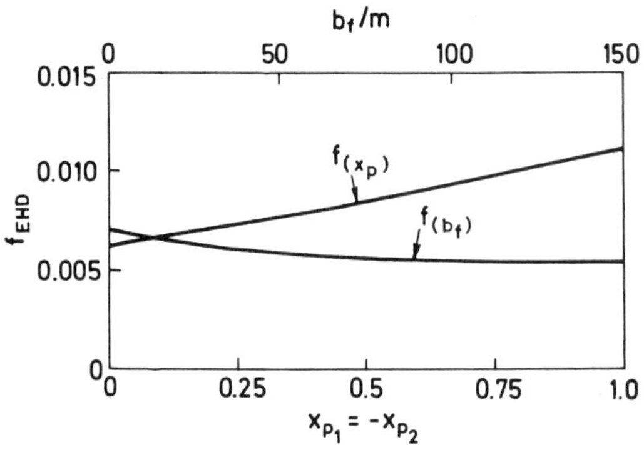

Fig. 2. EHD power losses for addendum modification factor and face width variations.

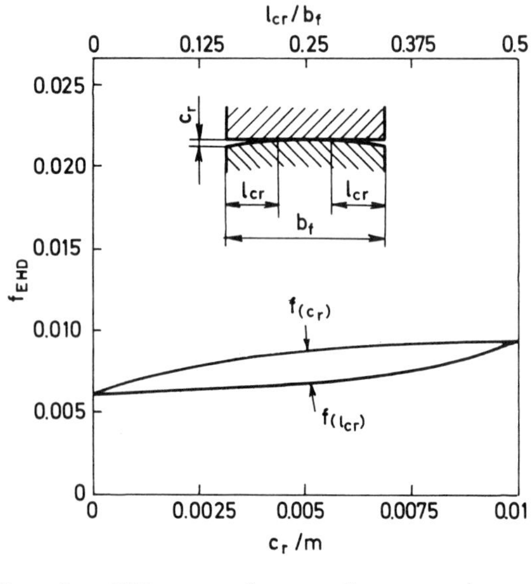

Fig. 3. EHD power losses for crowning and crowning length variations.

oil temperature (T_o) on EHD power losses is investigated. The obtained results are shown in Figs. 1-4. It can be seen that the radius of relative tooth surface cur-

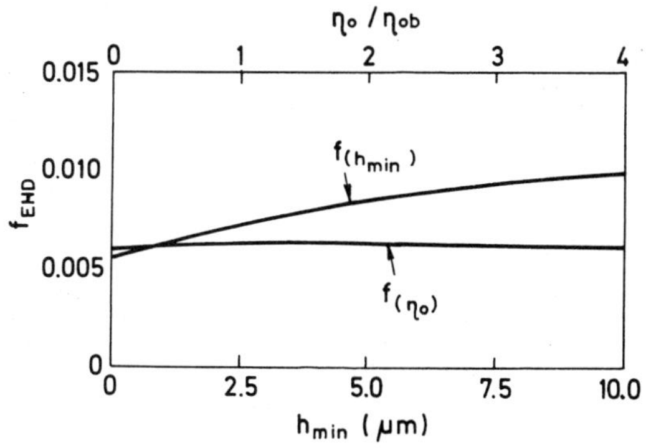

Fig. 4. EHD power losses for minimum oil film thickness and viscosity variations.

vature has the strongest effect on the factor of power losses (f_{EHD}), i.e. the rise of the radius causes a sharp drop in power losses. A considerable effect on the power losses have the helix angle, addendum modification factors, minimum oil film thickness, and crownings. The investigations show that the influence of oil viscosity, pitch velocity, and supplied oil temperature on EHD power losses is negligible.

POWER LOSSES DUE TO PUMPING EFFECT OF GEAR MESH

The teeth entering in mesh are pumping out the mixture of air and oil from the tooth gaps of the mated gear. The governing equations are the continuity (9) and momentum (10) equations. The compressibility of the mixture is included (11).

$$-\frac{\partial}{\partial t}\int_V \rho dV = \int_A \rho v dA \qquad (9)$$

$$\frac{\partial u}{\partial t} + u\frac{\partial u}{\partial x} + v\frac{\partial u}{\partial y} = X - \frac{1}{\rho}\frac{\partial p}{\partial x}$$

$$\frac{\partial v}{\partial t} + u\frac{\partial v}{\partial x} + v\frac{\partial v}{\partial y} = Y - \frac{1}{\rho}\frac{\partial p}{\partial y} \qquad (10)$$

$$\frac{p_1}{p_2} = \left(\frac{\rho_1}{\rho_2}\right)^{\kappa} \qquad !11)$$

A special finite difference method is used to solve this system of equations. The "control volumes" are obtained by dividing any of tooth gaps, instantaneously engaged with the teeth of the mated gear, on an equal number of parts in the lengthwise direction of teeth.

The power loss due to the pumping effect is calculated by the following equation, based on (Saad, 1966)

$$P_{pump.} = \frac{p}{\kappa-1}\left|\frac{d\Delta V}{dt}\right|\left[1+(\kappa-2)\,2^{(1+\frac{1}{\kappa})}\left(\frac{p_o}{p}\right)^{\frac{1}{\kappa}}\right] \qquad (12)$$

By applying the corresponding computer program the effects of gear tooth numbers (z_1, z_2), dedendum (h_f), helix angle (β_o), fillet radius (r_{fil}), face width (b_f), backlash with mate (b_ℓ), pitch velocity

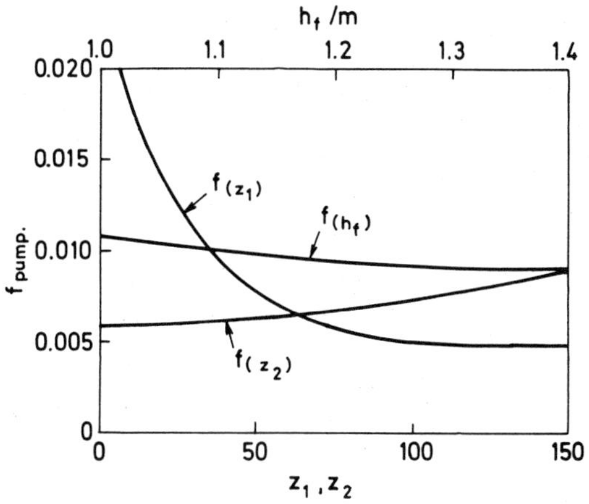

Fig. 5. Effects of gear tooth numbers and dedendum on pumping power losses.

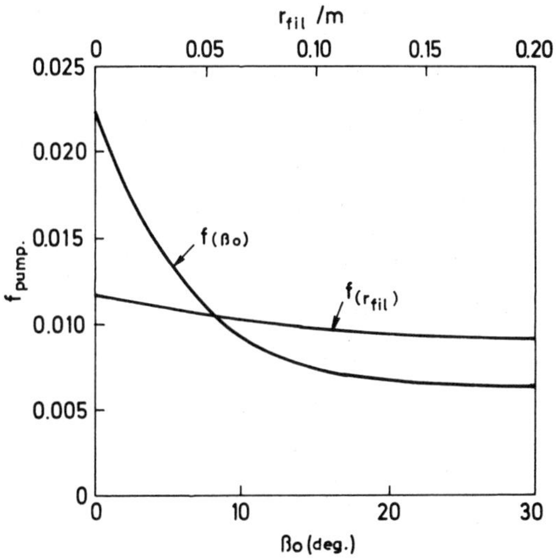

Fig. 6. Effects of helix angle and fillet radius on pumping power losses.

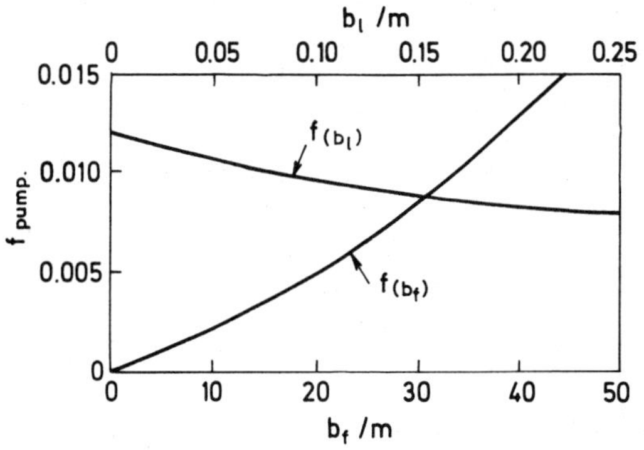

Fig. 7. Effects of face width and backlash with mate on pumping power losses.

(v_p), and percentage of oil in air-oil mixture on power loss factor $f_{pump.} = \frac{P_{pump.}}{P}$ is investigated. The obtained results are shown

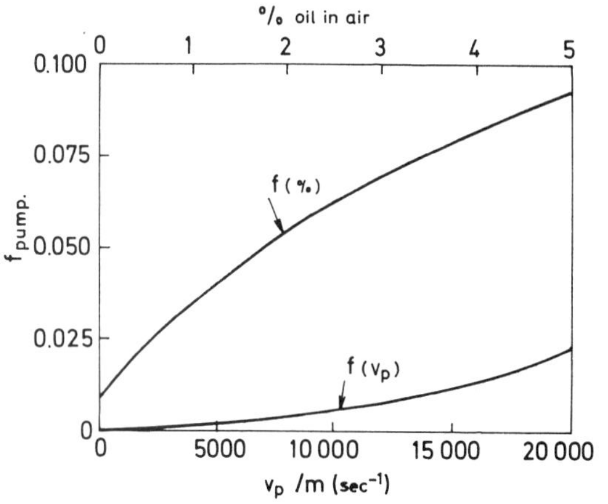

Fig. 8. Effects of gear speed and percentage of oil in the mixture on pumping power losses.

in Figs. 5-8. The conclusion can be made that most of the parameters have a strong effect on the power losses due to pumping effect. Only the influence of dedendum and fillet radius is moderate. The investigations also show that the power losses are almost constant through the meshing cycle.

TEST RESULTS

For the justification of the developed methods and the obtained results a high speed gear set is tested.

By applying a testing power of 147 kW, a 6.04 kW power loss in gear mesh is measured. By calculation, a 2.05 kW power loss in the oil film and a power loss of 3.71 kW due to the pumping effect is obtained, in total 5.76 kW. The difference between the measured and calculated power losses is 4.6 per cent.

CONCLUSIONS

Methods are developed for the determination of power losses due to EHD lubrication and due to pumping effect of gear mesh.

On the basis of the obtained results it can be concluded that the power losses can be considerably reduced by increasing the tooth numbers and decreasing the addendum modification factor, minimum oil film thickness, and crownings.

ACKNOWLEDGMENTS

This project was supported by the National Science Foundations of USA and Yugoslavia through funds made available to the U.S.--Yugoslav Joint Board on Scientific and Technological Cooperation.

The author wish to thank the University of Pennsylvania where partially the research was done. Particular thanks are due to Dr. B. Paul and W. Pizzichil for support and valuable suggestions made.

REFERENCES

Gu, A. (1973). Elastohydrodynamic Lubrication of Involute Gears. J. Eng. Ind., 95, 1164-1170.

Wellauer, E. J., and G. A. Holloway (1976). Application of EHD Oil Film Theory to Industrial Gear Drives. J. Eng. Ind., 98, 626-634.

Martin, K. F. (1981). The Efficiency of Involute Spur Gears. J. Mech. Des., 103, 160-169.

Anderson, N. E., and S. H. Loewenthal (1982). Design of Spur Gears for Improved Efficiency. J. Mech. Des., 104, 767-774.

Matsumoto, S., S. Asanabe, K. Takano, and M. Yamamoto (1985). Evaluation Method of Power Loss in High-Speed Gears. Proc. JSLE Int. Trib. Conf., Tokyo, 1165-1170.

Sato, M., and S. Takanashi (1981). On the Thermo-Elastohydrodynamic Lubrication of the Involute Gear. Int. Symp. Gearing, Tokyo, paper b-10.

Simon, V. (1981). Elastohydrodynamic Lubrication of Hypoid Gears. J. Mech. Des., 103, 195-203.

Simon, V. (1981). Load Capacity and Efficiency of Spur Gears in Regard to Thermo--EHD Lubrication. Int. Symp. Gearing, Tokyo, paper b-9.

Simon, V. (1985). Thermoelastohydrodynamic Analysis of Lubrication of Worm Gears. Proc. JSLE Int. Trib. Conf., Tokyo, 1147-1152.

Saad, M. (1966). Thermodynamics for Engineers, Prentice-Hall, pp. 52-54.

ENERGIEVERLUSTE BEI GERAD-UND SCHRÄGSTIRNRÄDER GROSSER DREHZAHLEN

Bei Stirnräder mit Evolventenverzahnun mit Gerad-und Schrägzähnen die mit grossen Drehzahlen arbeiten, entstehen die Hauptenergieverluste durch die Reibung in Ölfilm der sich zwischen den Zähnen im Eingriff formiert, und durch das Auspressen der Mischung von Luft und Öl aus den Zahnlücken der gepaarten Räder.

Die Energieverluste im Ölfilm sind mit der termoelastohydrodinamischen Analyse der Schmierung bei allen Zähnen die im Augenblick im Eingriff sind bestimmt. Dabei ist die Abhängigkeit der Viskosität des Öls vom Druck und der Temperatur, und der Öldichtheit vom Druck in Bezug genommen.

Der Radzahn presst beim Eintritt in den Eingriff die Mischung von Luft und Öl aus der Zahnlücke des Radpaares. Es wurde die Methode zum Ausrechnen entsprechender Energieverluste ausgearbeitet.

Auf Grund so ausgearbeiteter Teorie sind entsprechende Computerprogramme aufgestellt. Mit diesen Programmen ist der Einfluss der Zahnradparameter und der Betriebverhältnisse auf die Energieverluste geprüft.

Control Systems

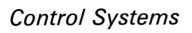

Response Potentials of the Powered Servodrive Technologies

P. DRANSFIELD, S. ORR AND E. S. P. TAN

Department of Mechanical Engineering, Monash University, Australia

Abstract: The paper is concerned with the absolute and the relative response capabilities of pneumatic, electrical and hydraulic servodrives using state-of-the-art technology and working under similar loading conditions.

Keywords: Actuators, D.C. motors, dynamic response, hydraulic systems, nonlinear control systems, pneumatic countrol, position control, robots, servomechanisms.

INTRODUCTION

This paper is concerned with controlled variable-state machines

- o which have to overcome impeding loads as they control the position and/or velocity of a payload.
- o which are intended to receive and respond satisfactorily to deliberate control inputs, and which have to maintain acceptable performance in the face of unpredicted load disturbances.
- o for which the speeds and shapes of their responses to inputs is just as important as is the final state achieved following those inputs.

The impeding loads referred to can have inertive (also inductive), compliant (capacitive), resistive, gravitational or other external forces such as wind loading etc. components.

Machines in this category include aircraft control systems, some machine tools, some lifting and transfer machines, and industrial robots. They usually have low-energy control systems interfaced with high-energy actuation systems. The control system provides the logic, sensing, feedback, and decision making. The actuation system(s) provide the muscle to put the control decision into effect.

The paper is concerned primarily with the actuation systems. These can be electromechanical (AC or DC motors, with gear reduction and/or mechanical linkages to provide required motions and speeds), hydraulic (cylinder, or motor with or without gear reduction and/or mechanical linkage), or pneumatic (cylinder, or motor with gear reduction and mechanical linkage). Control, for all three actuator technologies, is increasingly electronic.

The accuracy of the new desired final state of payload position or velocity following a control input is vitally important. The present paper is concerned, however, with the manner and the speed of the change of state. It is considered that "designing" this manner and speed of change-of-state is a major task for the designer of high-quality machines.

To change the state of an impeding load requires energy in a suitable form. To change state within a specific time frame, and in a specific manner within that time frame requires control of the flow rate of energy. That is, the power flowing from source to payload has to be controlled and predictable.

Power flow modelling techniques, and power bond graphs in particular are based on the fundamental fact that it is the instant power exchanges taking place between machine and system components which dictate the dynamic responses taking place. The bond graph dynamic modelling procedure, followed by digital solution of the model, provides the designer of powered systems or machines with a particularly relevant means of designing response. It is particularly relevant because, like most modern powered systems and machines, it is modular, allowing local changes for trial purposes, and repeated re-use in other machines or systems of developed component models. Its use of R (resistive), C (compliance or capacitive), and I (inertive or inductive) concepts to classify all dynamic effects allows direct compatibility of approach to modelling in all three of the powered actuation technologies (electromechanical, hydraulic, pneumatic) or mixtures of them. The good system and machine actuator designer should have all three technologies in his repertoire so that the best for a particular application can be sought via dynamic modelling and simulation.

The paper describes some of the results from a study of comparable pneumatic, electrical, and hydraulic servodrives operating on the same load system. The results given are from digital simulations. Initial experimental measurements tend to confirm the simulations.

PNEUMATIC MOTOR SERVO

Fig 1(a) illustrates the electronically-controlled air-motor powered servodrive. The objective is to control the motion and the final position of the mass load. The strategy is to control mass speed by opening one to four of the on-off solenoid valves, depending upon the size of the position error. That is, if the input command requires the load to go from an initial position to a new position, the initial error is large, fast response is required, and all four valves are

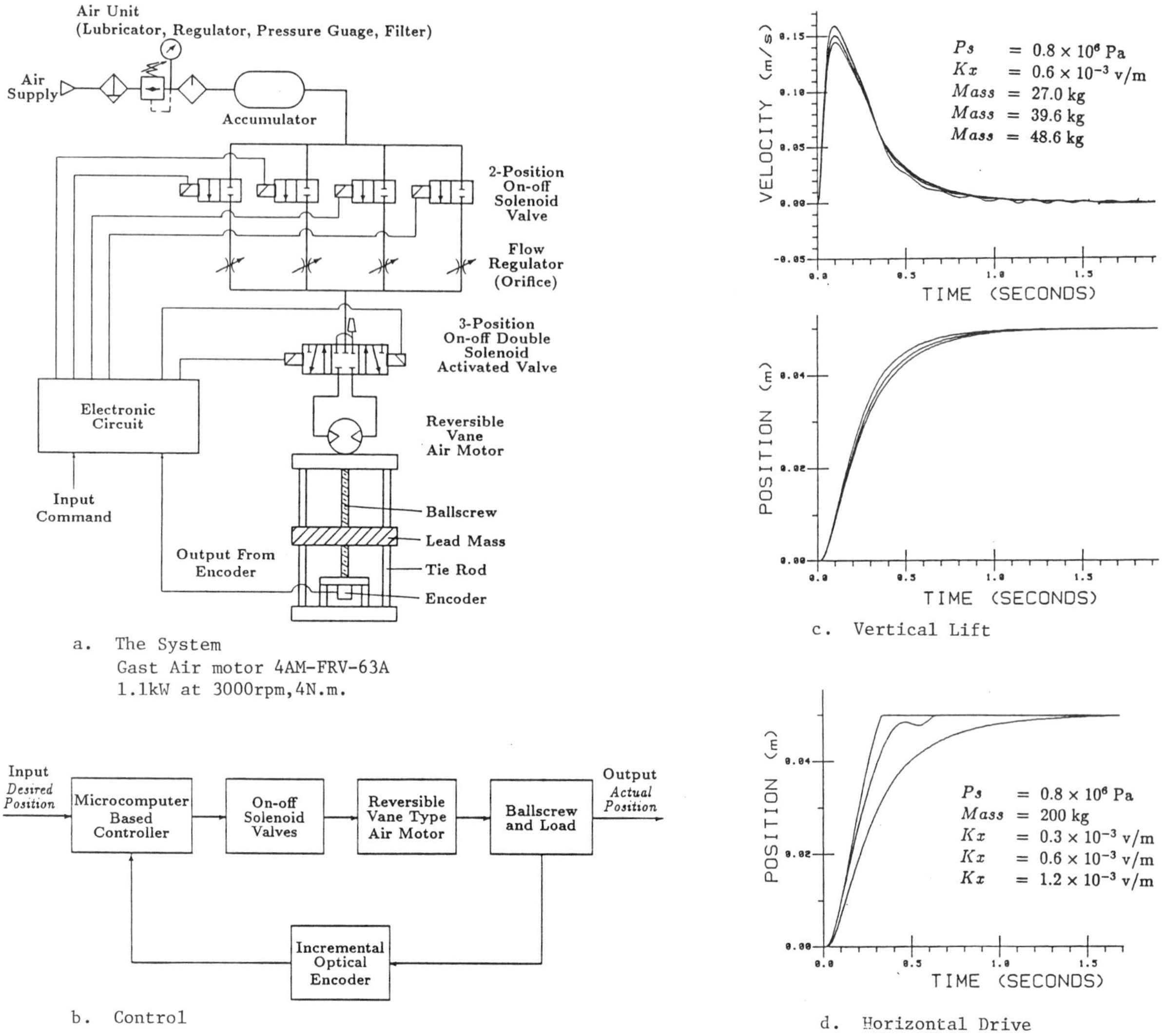

a. The System
Gast Air motor 4AM-FRV-63A
1.1kW at 3000rpm,4N.m.

b. Control

c. Vertical Lift

d. Horizontal Drive

FIG. 1 PNEUMATIC SERVODRIVE

opened to provide maximum flow-rate of air to the air motor. As the load approaches the new desired state, the flow valves are closed discreetly until final approach is made via one open valve, the orifice for which is finely adjusted. The double solenoid operated three-position valve merely dictates the direction of rotation of the air motor, or holds it stationary. The encoder allows position and velocity feedback.

Fig 1(b) illustrates the system control.

The system bond graph, equation set model, and coefficeints are described in Tan (1).

Fig. 1(c) shows velocity and displacement responses of the system to step input command for three values of load mass, for upward vertical motion of 50mm. Supply air pressure Ps is 800 kPa, and the position feedback coefficient Kx is 0.6E-3 v/m. The responses are smooth and well-shaped. The response time of about 1 second is satisfactory for many industrial applications. The effect on response of varying the load mass is minor, which is highly desirable.

The simulation was used to study the effects on system response of various values of supply pressure, control valve orifice area, position feedback coefficient, velocity feedback, etc. The effect of whether the load motion was vertical up or down or horizontal was also studied. Fig. 1(d) shows the responses when the load is 200kg and the motion horizontal for three values of position feedback coefficient. The sharpness of the top edge of the fastest response needs further investigation.

ELECTRICAL MOTOR SERVO

Fig. 2(a) illustrates the load system being driven by a brushless D.C. servodrive with pulse-width-modulation control. The component characteristics are described briefly with the diagram.

Fig. 2(b) illustrates the control strategy.

Fig. 2(c) shows load velocity and position responses obtained via digital simulation of a bondgraph derived model of the system. The load mass was 200kg. The two step input commands were for change of load position vertically upward 25mm, and 150mm respectively. The four curves on each diagram are for each of four values of position feedback coefficient. The position

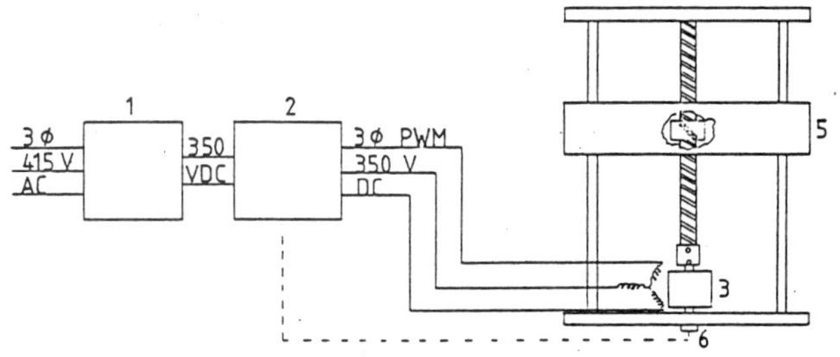

1. Isolation Transformer
2. Analogue Controller
3. Brushless DC Motor (Moog 1.5 kw at 2000 rpm, Peak torque 63.2 Nm)
4. Ball Screw
5. Mass load
6. Position Resolver (incorporated in motor)

a. System

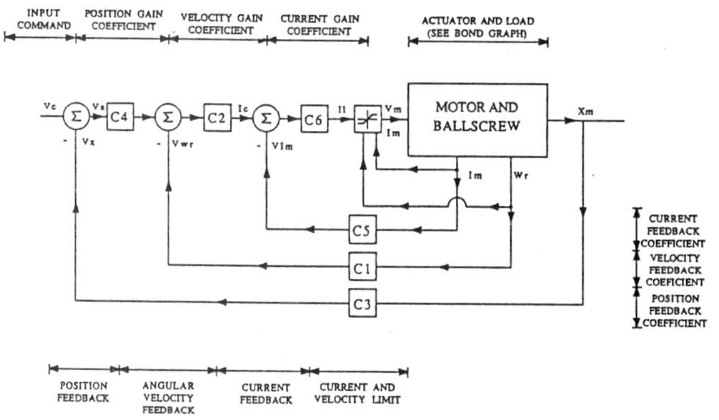

b. Control

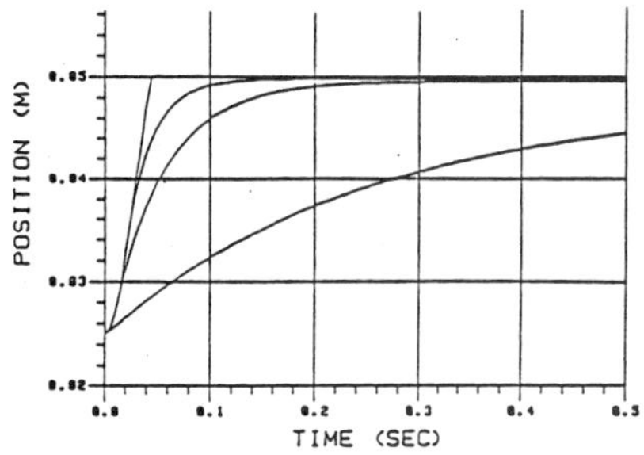

c. Response, 25mm

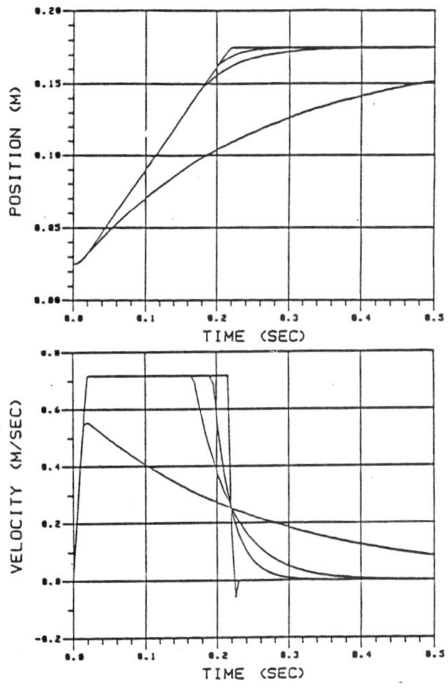

d. Responses, 150mm

FIG. 2 ELECTRIC SERVODRIVE

responses are fast, stiff, and well shaped. In particular, the zero-overshoot rise time of 0.05s for the fastest case (highest value of position feedback coefficient for well-shaped response) for the large input is impressive. The corresponding maximum value of velocity, 0.7m/s, was reached in 0.02s without overshoot. Orr (2) gives full details of the system, the model and coefficients, and the results.

HYDRAULIC SERVOS

Fig. 3(a) illustrates the hydraulic drive system with a modern laminar-fit double-acting cylinder providing the thrust. The cylinder can be replaced by a hydraulic motor with the ballscrew arrangement of the pneumatic and electric motor drives. When the cylinder is being used, position state is sensed by an LVDT. The hydraulic motor drive incorporates the rotary position encoder used with the pneumatic and electric motor drives.

Figs 3(b) and (c) show some of the simulation results for the load being driven upwards.

Fig. 3(b) shows responses of the cylinder-driven system for large (150mm) and small (25mm) step input commands, each for the same four values of the feedback gain coefficient. System coefficients had been chosen to give the fastest well-shaped response.

Fig. 3(c) shows the response of the hydraulic motor-ball screw driven system under the same condition as for Fig. 3(b). The responses are somewhat slower, but are of the same general shapes.

COMMENTS

The simulation results show that satisfactorally shaped controlled load-position responses can be achieved with all three drive technologies. In the hydraulic case the controlled motion can be achieved using either a hydraulic motor with the ball-screw system, or a direct-drive hydraulic cylinder. The pneumatic cylinder direct-drive has not been examined.

The "best" of the results shown are

- o the pneumatic motor servo with the largest position feedback coefficient, which responds to step position change of 50mm in 0.3s smoothly and without overshoot, when driving a load of 200kg horizontally (Fig. 1(d)).
- o the brushless D.C. electric motor servo with the largest position feedback coefficient, which responds to step position change of 25mm in 0.05s and 150mm in 0.21s, driving a 200kg load vertically upward (Fig. 2(c)).

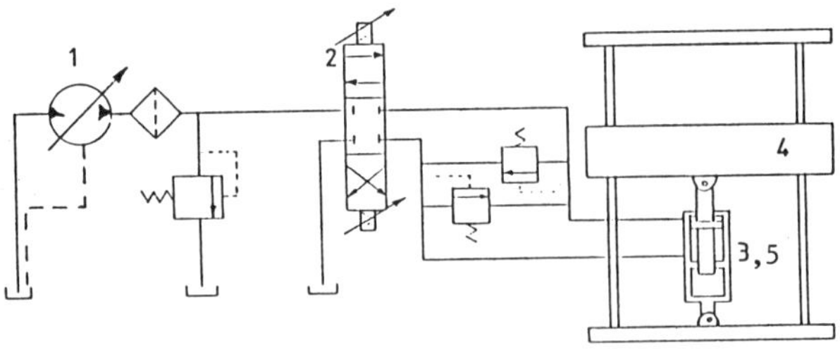

a. The System

1. Pump
2. Super-high response servovalve, 35MPa, 19ℓ/min
3. Laminar fit cylinder Effective area 3.6cm^2, 35MPa Moog Model 853
4. Mass Load
5. Volvo motor F11-5 35MPa, 5.4kW at 2000rpm, 25.6N.m

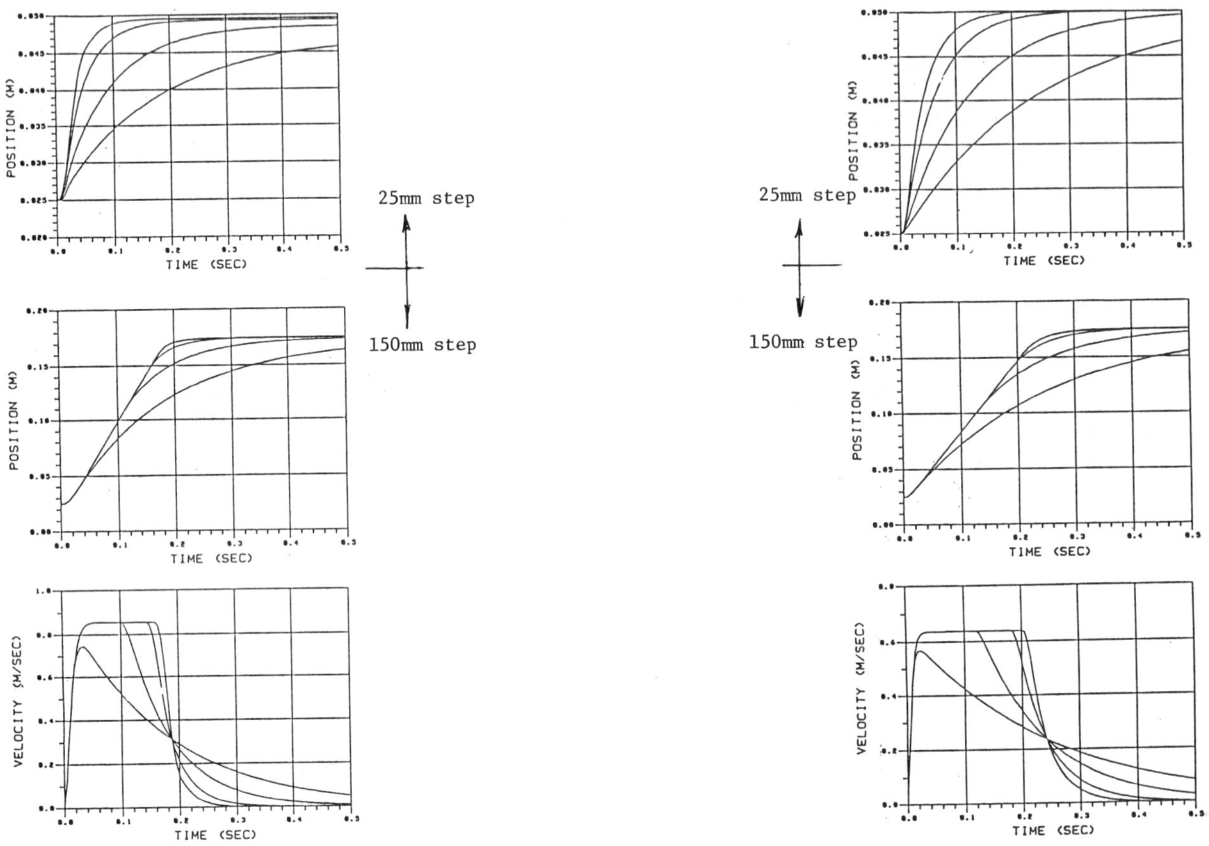

b. Cylinder Driven

c. Motor Driven

FIG. 3 HYDRAULIC SERVODRIVES

- the hydraulic motor servo with largest feedback coefficient which drove a load of 200kg upwards 25mm in 0.2s and 150mm in 0.3s (Fig. 3(c)).
- the hydraulic cylinder driven servo with the largest feedback coefficient, which drove a load of 200kg upwards 25mm in 0.14s and 150mm in 0.23s (Fig. 3(e)).

In all four of these cases the load was 200kg and the input command a step. All responses were smooth, without oscillation, and without overshoot. The motion was vertically upward for the electrical and the two hydraulic drives but was horizontal for the pneumatic drive. Fig. 1(c) showed the present pneumatic servo's performance for vertical lift.

CONCLUSION

The results given are from a comparative study of modern pneumatically, electrically, and hydraulically powered position-control servo technologies. The study is well advanced but not yet complete. Initial experiments tend to validate the digital simulation results presented.

ACKNOWLEDGEMENT

The authors thank the Harold Armstrong Memorial Fund for its financial assistance to the work.

REFERENCES

1. Tan, E.S.P., On Air Motor Servodrives, M.Eng.Sci. Thesis, Monash University, 1987.

2. Orr, S., On Electric and Hydraulic Servodrives, M.Eng.Sci. Thesis, Monash University, 1987.

Mechatronics

M. S. KONSTANTINOV*†, S. P. PATARINSKI*, Z. M. SOTIROV†
AND L. G. MARKOV*

*Department of Instrumentation and Robotics, Institute of Mechanics and Biomechanics, Bulgarian Academy of Sciences, 1113 Sofia, Bulgaria
†Department of Robotics, Higher Institute of Mechanical and Electrical Engineering, 1156 Sofia, Bulgaria

Abstract. The concept of Mechatronics (MECHAnics & elecTRONICS), proclaimed by the Japanese researchers during the days of the 9th ISIR (Tokyo, 1979), is considered in three different aspects: (i) as a methodology for developing simple, reliable and inexpensive technological automation; (ii) as a basis for trajectory planning in the continuous path control of manipulation robots; and (iii) as a generalized approach (treating both mechanics and control of motion simultaneously) in mechanisms design. Examples are given, illustrating the basic ideas and the possible applications.

Keywords. Mechatronics; robots; trajectory planning; mechanisms design; joint motions decoupling.

INTRODUCTION

An even cursory review of the modern Theory of Machines and Mechanisms (and especially, as applied to Robotics) shows the tremendeous progress made during the past two decades. This progress is very closely related and due to the development and mutual infiltration with Control Theory and Microelectronics. Significant illustrations are:

(i) the snake-like robots of Hirose (1985) with a great number of actively driven and passive joints;

(ii) the micro-manipulators of Hayashi (1985), utilizing memory-shaped alloys;

(iii) the walking machines of McGhee and Iswandhi (1979), Raibert (1981), Hirose (1984), Bessonov (1985), Hayashi (1985) and others;

(iv) the dexterous hands (three fingers with three degrees of mobility each), developed at Stanford University (Roth, 1985);

(v) the joint motions decoupling in parallelogram-type manipulation robots (Markov and others, 1985), (Patarinski and Markov, 1985).

Results of this type were the reason for the Japanese robotsman to proclaim the concept of Mechatronics (a symbiosis of MECHAnics & elecTRONICS) during the days of the 9th ISIR in Tokyo. Since 1979 there is an evergrowing interest to Mechatronics, but it has been considered from a more or less applicational point of view so far (Pelecudi, 1983; Hirose, 1985; Minoru, 1985; Bögelsack, 1985) allowing a purposeful combination of the developments in Mechanics and Control with the modern products in Microelectronics and the new technologies. Noway Mechatronics finds an universal acceptance while its implementation is being effected by the absence of a general theoretical basis.

In the present paper three particular examples are considered, illustrating the three basic principles of Mechatronics:

(*) the principle for functions distribution among mechanics and control of mechanisms and machines;

(**) the principle for decomposition of the 3D motion in the space and in the time;

(***) the principle for common treatment of mechanics and control in the product analysis, synthesis ans design.

Further on the following working definition of Mechatronics will be used (not pretending for either generality or exhaustiveness):

Mechatronics is a field of modern science and engineering dealing with mechanics and control of technological automation (incl. various types of manipulation and mobile robots, programmable handling devices and peripherals) from a common theoretical stand. The main approach of Mechatronics is the general treatment of mechanics and control of programmable automation simultaneously in the course of its analysis, synthesis and design. The industrial application of Mechatronics is based on the achievements of Microelectronics, new technologies and Production Engineering.

In the following paragraphs the three principles of Mechatronics, stated above, are discussed using simple, illustrative examples.

FUNCTIONS DISTRIBUTION: CLASSICAL MECHANISMS

Consider the following problem: Given some

plannar (open or closed) curve c. A mechanism must be synthesised, so that the trajectory of a selected point, located on one of its links, coinsided with c (eventually - within a prescribed deviation). A great number of problems of this type are well known and completely solved in the classical Theory of Machines and Mechanisms and it is not necessary to discuss any details here.

In Fig. 1 a particular curve is shown together with the kinematic scheme of one of the possible mechanisms (Eschenbach and Tesar, 1968), realizing it with respect to the trajectory of a selected point P from link 2. The trajectory of P has two singular points: S_1 and S_2, where the projection of its velocity v_P on the x axis is zero. It is clear, that the trajectory of P is always one and the same, independently from the law of motion $q_1 = q_1(t)$ of the driving link (link 1) - here and further on t stands for the current time.

If now an additional requirement is imposed on the velocity v_P and acceleration a_P of motion (as shown for example in Fig. 2), the driving link should move properly to provide the desired velocity and acceleration profiles of P. The real motion will be performed under model uncertainties and parameter inaccuracies, that implies the use of respective (position and velocity) feedbacks - in other words, control system and position/velocity sensors are called for. In this way, the very simple four-bar mechanism of Fig. 1 can perform complicated manipulation tasks, including synchronization with a moving conveyor.

What should be specially emphasised here is, that the trajectory of motion (as a plannar or spatial, in the general case, curve) is fully determined and completely depends on the kinematic structure and geometric parameters (the mechanics) of the mechanism, while the characteristics of motion (velocity, acceleration, etc.) are defined by the control system. This is an illustration of the first principle of Mechatronics - the principle for functions distribution among mechanics and control.

Concluding this paragraph we shall point out, that automatic devices of the type described are comparatively simple, highly reliable and very cheap, while in the same time they can be applied in rather complicated tasks, often requiring at first sight universal robots - this is the reason to call them manipulating devices with built-in kinematic intelligence (Konstantinov and others, 1985).

DECOMPOSITION OF THE 3D MOTION: TRAJECTORY PLANNING IN ROBOTICS

Denote by x the vector, comprizing three linear p and three angular ϕ components, uniquely determining respectively the position and the orientation of the robot's end-effector in the 3D space. Consider an example of assembly automation (Fig. 3), when a shaft with a key (denoted by 1) must be properly inserted into a second shaft with a groove key (denoted by 2). The problem (trajectory planning) is to find a smooth-enough trajectory $f^*(x)$ in the 3D space along which shaft 1 to be moved for insertion into shaft 2.

The trajectory planning can be naturally sub-divided into two stages:

- task trajectory definition - $f^* = f^*(x)$;

- parametrization with respect to the time.

The task trajectory definition can be accomplished independently for the position and the orientation of the end-effector - i.e. $f^*(x)$ can be represented as two mutually independent functions: $f_1^*(p)$ and $f_2^*(\phi)$, as shown below.

Assuming the positional trajectory $f_1^*(p)$ as a third order spline, its coefficients must be determined so that:

a) it passed through the initial and the final positions of the shaft one (represented in Fig. 3 by the unit vectors $\bar{e}_1^o$, $\bar{e}_3^o$ and $\bar{e}_1^f$, $\bar{e}_3^f$ respectively).

b) $\bar{e}_3^o$ and $\bar{e}_3^f$ coinsided with the tangent vectors to $f^*(p)$.

The orientational trajectory $f_2^*(\phi)$ can be then defined providing $\bar{e}_3$ being currently tangent to $f_1^*(p)$ and rotating shaft 1 around the axis $\bar{e}_3$ so that $\bar{e}_1$ coinsided with $\bar{e}_1^f$ when O coinsides with O_f. Thus both $f_1^*(p)$ and $f_2^*(\phi)$ are uniquely determined (Fig. 4).

At the second stage the positional and the orientational trajectories can be parametrized (Patarinski, 1986), as $p = p^*(t)$ and $\phi = \phi^*(t)$ under prescribed linear $v = v^*(t)$ and angular $w = w^*(t)$ velocities of the end-effector. Because of the smoothness of both functions $f_1^*(\circ)$ and $f_2^*(\circ)$, they are explicitly resolvable with respect to anyone of their arguments. Expressing from there the linear velocity $\dot{p}$, having in mind the relation between the derivative $\dot{\phi}$ and the angular velocity of the end-effector (Wittenburg, 1977), and equalizing to $v^*(t)$ and $w^*(t)$ respectively, differential equations are obtained from where the required parametrization $x^*(t)$ can be accomplished (for details see (Patarinski, 1986)).

It is important to note here, that the trajectory planning, as performed above, is based on splitting up the task into geometry (desired trajectory in the 3D space) and parametrization of this trajectory with respect to the current time. This is an illustration of the second principle of Mechatronics.

COMMON TREATMENT OF MECHANICS AND CONTROL: JOINT MOTIONS DECOUPLING

A common solution in Industrial Robotics is to use parallelogram-type designs to locate the actuators near the stand, thus reducing the loads of the links, improving robot dynamics and decreasing the energy consumption. Unfortunately, due to the transmissions used, the joint motions become mechanically coupled (a few actuators must be operated in a coordinated manner to provide a prescribed displacement of only one joint), that may create certain difficulties related to the additional computational burden imposed on the control.

If these two inherently conflicting requirements are considered along at the early stage of robot analysis, synthesis and design, a proper compromise might be worked out, as for example in (Patarinski and Markov, 1985) where joint motions decoupling is achieved by suitable mechanical arrangement (Markov and others, 1984). Solution of this kind was implemented in the two armed robot ARO, developed at the Department of Instrumentation and Robotics, Institute of Mechanics & Biomechanics for assembly in the relative space. The robot has two identical hands with five degrees of mobility each and its schematics is shown in Fig. 5.

This is an example of the third principle of Mechatronics for common treatment of mechanics and control in product analysis, synthesis and design. The basic idea is to clearly state all existing problems, to estimate the costs for their solution by purely mechanical, hardware or software means, and to select the most effective solution. In this way, the third principle of Mechatronics is opposed to the standard engineering practice today to solve the design problems consequtively and independently with respect to mechanical design, control hardware and software.

CONCLUSIONS

In the paper the basic principles of Mechatronics, defined as an interdisciplinary field of modern science and engineering dealing with mechanics and control of technological automation from a common theoretical stand, are considered. The following three principles are formulated on an heuristic basis:

° functions (trajectory generation and velocity (acceleration, etc.) control) distribution among mechanics and control;

°° decomposition of 3D motion in the space and in the time (creating tasks for trajectory generation and velocity/acceleration/etc. control); and

°°° common treatment of mechanics and control in the product analysis, synthesis and design (solving the problems in the most effective manner by mechanical and/or control means).

Thus the three principles are in a perfect harmony with each other.

Examples are given, illustrating the application and effectivity of Mechatronics' ideas.

REFERENCES

Bessonov, A. (1985). Some problems in mechanics of walking machines. Talk at 3rd Int. Summer School on Pract. Rob. - PRACTRO'85. Pamporovo, Oct. 1-6.

Bögelsack, G. (1985). Development of TMM in Instrumentation - Mechatronics. Proc. IFToMM Symp. "Progress in TMM - Robotics" Varna, Oct. 7-9 (in German).

Eschenbach, P. W., and D. Tesar. (1968). Optimization of four-bar linkages satisfying four generalized plannar positions. Trans. ASME, Paper No. 68-Mech-30.

Hayashi, T. (1985). Miniature mechanisms for realizing a miniature robot. Talk at 32nd Annual Design Eng. Conf., Chicago, March 11-14.

Hirose, S. (1984). A study of design and control of a quadruped walking vehicle. The Int. J. Robotics Research, Vol. 3, No. 2.

Hirose, S. (1985a). Snake-like robot. Talk at 3rd Int. Summer School on Pract. Rob. - PRACTRO'85. Pamporovo, Oct. 1-6.

Hirose, S. (1985b). Mechatronics. Proc. IFToMM Symp. "Progress in TMM - Robotics" Varna, Oct. 7-9.

Konstantinov, M. S., V. Mihailov, Ch. Bechev and T. Tonchev. (1985). Manipulator for welding in the relative space. Proc. 5th Nat. Congress Theor. Appl. Mech., Vol. 4, Varna, Sept. 23-29 (in Bulgarian).

Markov, L., M. S. Konstantinov, V. Zamanov, and V. Mihailov. (1984). Industrial robot with rotational joints. Bulg. Patent No. 37888 (in Bulgarian).

McGhee, R. B., and G. I. Iswandhi. (1979). Adaptive locomotion of a multilegged robot over rough terrain. IEEE Trans. Syst. Man Cybern.. Vol. SMC-9, No. 4.

Minoru, U. (1985). Mechatronics. J. Inst. Elec. Eng. Japan, Vol. 105, No. 11 (in Japanese).

Patarinski, S. (1986). Some problems in kinematics of manipulation robots. Proc. 5th Symp. Inf. Cont. Problems in Manuf. Tech. Robotics and FMS, Suzdal, April 22-25.

Patarinski, S., and L. Markov. (1985). Joint motions decoupling in anthropomorphic robots. Proc. 5th Nat. Congress Theor. Appl. Mech., Vol. 4, Varna, Sept. 23-29.

Pelecudi, C., and others. (1983). Merotechnika. Proc. Simp. Nat. Roboti Ind., Vol. 2, Bucharest (in Roumanian).

Raibert, M. H. (1981). Dynamic stability and resonance in a one legged hopping machine. Proc. CISM-IFToMM Symp. Theory and Practice Robots and Manipulators. Amsterdam, Elsevier.

Roth, B. (1985). Recent results in regard to the theory and the use of dexterous mechanical hands. Talk at 3rd Int. Summer School on Pract. Rob. - PRACTRO'85. Pamporovo, Oct. 1-6.

Wittenburg, J. (1977). Dynamics of Systems of Rigid Bodies. Stuttgart, B. G. Teubner.

МЕХАТРОНИКА

Резюме. Общей характеристикой современных манипуляционных роботов является то, что их механические системы и системы управления проектированы последовательно и независимо друг от друга. Нетрудно объяснить эту ситуацию, но она не является естественной, так как робототехника интердисциплинарное направление науки и технологии на грани пересечения нескольких классических наук. В докладе обсуждается концепция МЕХАТРОНИКИ /МЕХАника и елекТРОНИКА/, предложенная японскими специалистами по робототехнике во время 9-того Международного симпозиума по промышленным роботам /Токио, 1979г./, в трех разных аспектах: а/ как методология разработки простой, надежной и дешевой технологической автоматизации; б/ как базис для планирования траекторий в контурном управлении универсальными промышленными роботами; и в/ как общий подход рассматривания механики и управления едновременно в процессе проектирования. Даны примеры, илюстрирующие основные идеи и возможные применения Мехатроники.

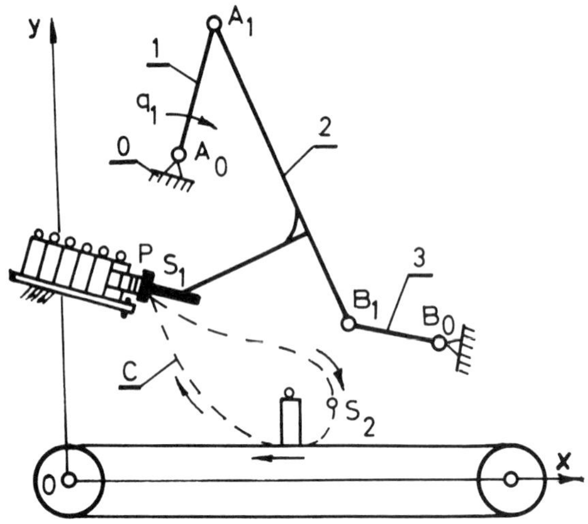

Fig. 1. Kinematic scheme of a simple four-bar mechanism

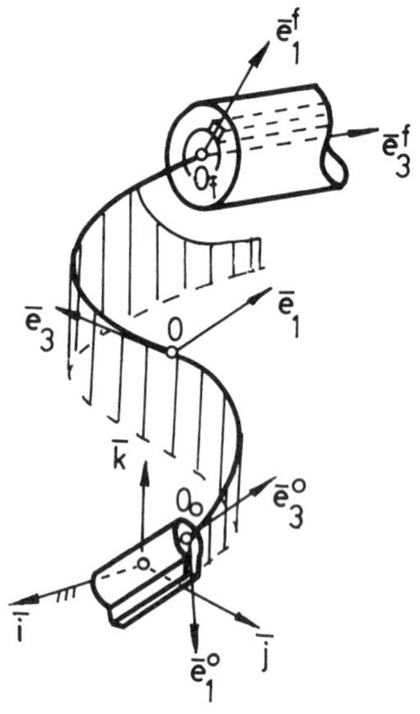

Fig. 3. Trajectory planning

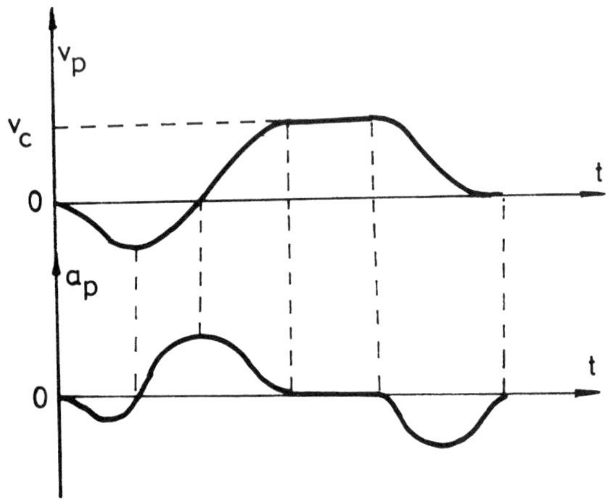

Fig. 2. Desired velocity and and acceleration of P

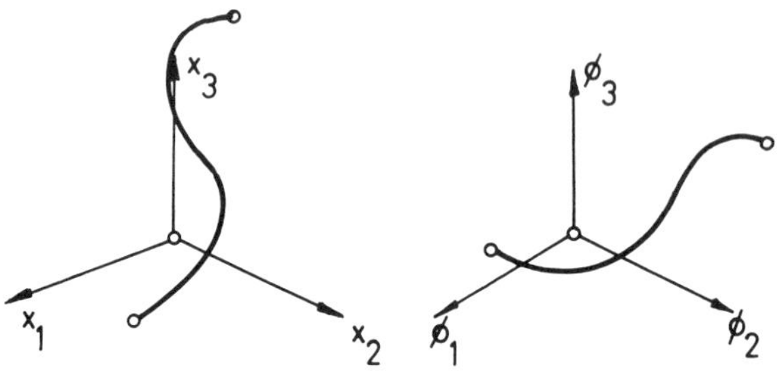

a) positional trajectory $f_1^*(p)$ b) orientational trajectory $f_2^*(\phi)$

Fig. 4. Desired trajectory definition with respect to position and orientation of the end-effector

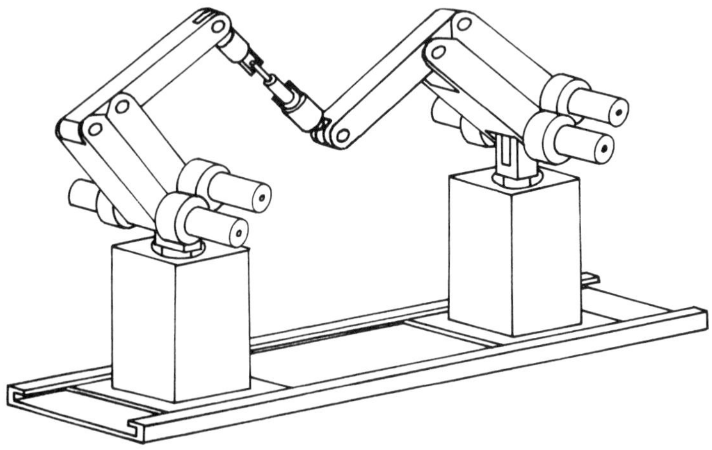

Fig. 5. The two-armed assembly robot developed at the Department of Instrumentation and Robotics

Low Cost Electronic Fuel Control Concept for Small Gas Turbine Engines

T. KREPEC, M. KREPEC AND A. I. GEORGANTAS

Department of Mechanical Engineering, Concordia University, Montreal, Canada H3G 1M8

ABSTRACT

Low cost fuel control has been developed from a diesel fuel injection pump with the help of an electronic controller. The conversion occured by connecting all of the pump section outputs into a single fuel exit, by introducing a pressure equalizer and by bringing the pump metering valve under the management of a digital actuator. The fuel control unit was tested on a test bench under computer control to evaluate the fuel delivery characteristics versus engine speed and the dynamic response during a transient process. The preliminary tests confirmed the feasibility of that project. Due to the very low cost of diesel injection pump produced in very large quantities, the cost of the fuel control unit for a small gas turbine engine could be cut substantially when using this concept.

KEYWORDS

Gas turbines, electronic, fuel control, injection pump.

INTRODUCTION

The highly non-linear characteristic of fuel delivery, as required by the gas turbine engines due to the inherent characteristic of the air compressor, created always difficulties in development of fuel control units, particularly for small engines where the cost of more sophisticated FCU becomes more detrimental. The first and the most logical solution was a variable displacement fuel pump controlled by a system of bellows adjusting the pump output proportionally to the air mass flow rate delivered by the compressor (Watson, 1948). Simple schematic of such system is shown in Fig. 1. In case of a reciprocating pump the number of pump cylinders and the duration of the pumping stroke had to be high enough so that a semicontinuous fuel flow could be obtained to the injectors, as required by the gas turbine engine.

Another solution was the use of a constant displacement fuel pump which produced the fuel flow rate being proportional to the speed of the engine. Then, the fuel flow was modulated using a metering valve actuated by a bellows provided with a compressor pressure signal. To reduce the pressure force acting on the metering valve, a relatively small differential pressure was created across the metering valve using a sophisticated bypass valve (Pratt & Whitney Ltd, 1977). Such system is schematically presented in Fig. 2. In both systems described above however, it was difficult to obtain the optimum fuel delivery characteristic, particularly in the engine acceleration mode where it is important to minimize the occurance of a compressor stall. Special complex devices were invented to fulfill these requirements.

CONCEPT BASED ON INJECTION PUMP

With the recent development of electronic controls as well as sensors and actuators, new opportunities emerged for low cost fuel control systems for small gas turbine engines. One of them is particularly attractive when adapting a commercially available low cost variable displacement hydraulic pump which is produced in large quantities. Such pumps are built for fuel injection in the diesel engines by the million every year. In this paper a proposal of such electronically controlled fuel control unit for small gas turbine engines will be presented based on a converted diesel injection pump. The concept is based on the following assumptions:

1) The diesel injection pump is inexpensive, for it is produced in very large quantities.
2) It is already developed for engines with high vibration level and reliability.
3) It is designed and manufactured for high fuel pressures and precise delivery schedules with similar speed governing options.

There are two major difficulties in adapting a diesel injection pump to a small gas turbine engine. First, the pulsating character of fuel flow must be overcome by a special flow equalizing system. Second, the requirement for a non-linear fuel delivery characteristic of the acceleration and the deceleration schedules must be taken into consideration. In the electronic type control, however, the characteristic could be easily shaped using a digital actuator under microprocessor control operating the metering valve of the injection pump.

The requirement for the characteristic change is shown in Fig. 3, where the original diesel injection pump delivery line is shown as the dose of fuel injected during one revolution of the pump versus the engine speed. For the gas turbine engine, however, the fuel delivery line has to follow the pressure curve of the compressor which corresponds approximately to the air mass flow rate, as shown in Fig. 3.

Three fuel delivery schedules can be distinguished as typical for a gas turbine engine:

1) Acceleration schedule for maximum fuel delivery.
2) Governing schedule to maintain a preset speed range when the load of the engine changes (similar to that in diesel engine).
3) Deceleration schedule to maintain the minimum fuel flow and avoid flame-out of the engine.

The governing, as well as the steady state fuel schedule can be produced by the speed governor, either electronic or mechanical, which is already available in the diesel injection pump. However, the acceleration and deceleration schedules would be produced exclusively by an electronic controller.

MODIFICATIONS INTRODUCED

To obtain the highest economic gain from that project the least expensive diesel injection pumps were considered for conversion. The modifications should be made as follows:

-exits from all sections should be combined into one single output,
-discharge rate should be increased,
-flow pulsations should be reduced,
-pump metering system should be modified.

Similar attempt was made by Robert Bosch who successfully converted his "in line" diesel injection pump to an hydraulic pump (Robert Bosch, GMBH, 1964). In this paper, the modifications of two rotary diesel injection pumps are described to convert them into fuel control units for small gas turbine engines. The pumps converted were
1) CAV-DPA 6-section pump produced in England.
2) Stanadyne-DB2 8-section pump produced in USA.
Both pumps have similar rotary distributors and fuel metering systems, as shown in Fig. 4.

The first modification was a circular groove machined on the rotor to connect the exit ports into one single outlet, as shown in Fig. 4,a.

The possibility of fuel discharge rate increase was next investigated. A mathematical model of the pump metering system was created and applied to the CAV pump shown in Fig. 4.

The following flow equations were considered:

-Fuel delivered by the transfer pump:

$$Q_t = C\ N\ \eta_v \qquad (1)$$

Fuel flow through regulating valve, metering valve and rotor feeding orifices; general equation:

$$Q = F\mu\sqrt{\frac{2}{\rho}\Delta P} \qquad (2)$$

Fuel flow continuity for the transfer pump, metering valve and regulating valve:

$$Q_t = Q_m + Q_r \qquad (3)$$

Because of only small pressure variations in the metering system, fuel was considered incompressible. Therefore: $Q_m = Q_f = Q$, and the metering pressure could be evaluated from (2) & (3) as follows:

$$P_m = \frac{(F_m\mu_m)^2\ P_t + (F_r\mu_r)^2\ P_r}{(F_f\mu_f)^2 + (F_m\mu_m)^2} \qquad (4)$$

where:

Q_t = fuel delivery rate of transfer pump

$Q_{r,m,f}$ = fuel flow rate in regulating valve, metering valve or feeding orifice

$F_{r,m,f}$ = flow area in regulating valve, metering valve or feeding orifice

$\mu_{r,m,f}$ = flow coefficient for regulating valve, metering valve or feeding orifice

$P_{t,r,m,f}$ = pressure in transfer pump, regulating valve, metering valve or rotor

N = rotational speed of transfer pump

C = discharge volume of transfer pump

η_v = volumetric efficiency of transfer pump

ρ = density of fuel

Having the value of P_m, the Q value was obtained from (2). Then, it was possible, for given speed and metering valve position to find the dose of fuel entering the pump chamber during the feeding period. This was done by integrating the instant fuel flow over the time of opening of the feeding orifice in the rotor. The fuel output so calculated is shown in Fig. 5.

The above calculation method was also used to predict the impact of two design changes introduced to increase the discharge volume of the CAV pump. First, a chamfer was made at the entry to the feeding ports in the rotor. (Fig. 4,b). Second, the shape of the ports was changed to rectangular (Fig. 4,c). The discharge characteristic obtained after these modifications is shown in Fig. 6.

As the next step, the delivery valve retraction stroke was reduced and that increased the pump discharge rate by 20-25%. Figure 6 also shows the desired fuel flow characteristic, as required for a small gas turbine engine. It was provided by controlling the metering valve position with electronic control system, as described below. The pump conversion was completed by the introduction of a flow equalizer which provides the pump with a continuous fuel discharge, as shown in Fig. 7.

ELECTRONIC CONTROL

The proposed electronic control block diagram is shown in Fig. 8. It consists of an "On-board" microcomputer interfacing a linear digital actuator operating the metering valve of the pump and also a speed sensor installed on the shaft of the engine. Because the digital actuator provides a "rigid" relationship between the position of the metering valve and the digital signal received from the computer, the control system was preliminarily made as an open loop system.

The above described electronic control was incorporated into the prototype of the modified injection pump. A linear actuator Airpax (Philips) type K92121 was used with Northstar Horizon microcomputer which was interfacing the actuator and the speed sensor through a home made interface unit. During the tests also the fuel flow and the actuator (metering valve) position were monitored by a home made data acquisition system. An example of the fuel delivery characteristic obtained with the CAV pump on a computer controlled test bench is shown in Fig. 9.

The fuel delivery schedule recorded during the tests was obtained as expected. Any required changes of that schedule could be easily introduced through reprogramming of the memory. Also, any other changes due to the variations in the fuel temperature or the compressor air pressure could be made based on the signal from the proper transducer.

To ensure about the proper dynamic response of the electronic fuel control unit fitted with a digital actuator, a fast transient process was investigated, as shown in Fig. 10. The unit was run on a test bench, first at selected steady state conditions to stabilize the flow at the governing mode. Then, the speed of the test bench has been reduced rapidly by the computer during 1 sec (as recorded on the graph). The change in the metering valve position and in the fuel flow rate has been also recorded. As can be seen on the graph (Fig. 10), the changes in the metering valve position and in the flow follow almost instantly that of the speed at the beginning of the transient process; the delay of 0.2 sec at the end of the process can also be observed in fuel flow stabilization.

The discussed electronic fuel control system has also the great advantage that it allows to neglect, to some extent, the manufacturing deviations in the mechanical part of the injection system. This is possible because the characteristic of each pump can be corrected individually during the calibration process and the specific program for particular unit can be written in the PROM (Programmable Read Only Memory) of the microcomputer (Kimberley, J.A., 1982). Recalibration, if required, is also possible in case of engine reconditioning and is being followed by reprogramming of the PROM.

The fuel control unit based on the converted fuel injection pump CAV type DPA did undergo a short 30 hours endurance run on a test bench using the jet fuel with lower viscosity than the diesel fuel. After disassembly, no visible traces of wear have been found on the pump components. It has to be underlined that the converted pump was operating with fuel pressure about 8 times lower than the original fuel injection pump used on a diesel engine.

REFERENCES

Kimberley, J.A. and J.R. Voss (1982). A Mid-Range Fuel Injection Pump. SAE Paper No. 820448, 143-144.

Pratt and Whitney Canada Ltd. (1977). PT6A-50 Descriptive Notes, 95-96.

Robert Bosh GNBH (1964). Eigengetriebene Forderpumpe. VDT-AKP 420/2, Blatt 1.1.5.1964.

Watson, E.A. (1948). Fuel Systems for the Aero-Gas Turbine. Proceedings of the Institute of Mechanical Engineers, 158, 187-208.

RESUME

La caractéristique non linéaire du débit d'air du compresseur d'une turbine à gaz a toujours causé des difficultées dans l'adaptation des systèmes à combustible pour ces moteurs. C'est seulement récemment que la situation s'améliore grâce au contrôle électronique qui permet le changement du débit du combustible d'une façon très libre. Cela donne une possibilité d'utilisation des pompes à injection des moteurs diesel qui sont produits en grandes quantitées et à coût peu élevé, pour les incorporer dans les systèmes à combustible pour les petites turbines à gaz. Deux pompes à injection ont été converties grâce aux modifications suivantes:

1) Toutes sorties de la pompe sont reliées par un passage intérieur afin d'obtenir une sortie unique du combustible.
2) Les pulsations du débit du combustible sont diminuées par l'introduction d'un égalisateur de pression.
3) La soupape de dosage est placée sous le contrôle d'un actuateur digital qui ajuste le débit de la pompe.

Toutes ces modifications ont démontré qu'une telle unité de contrôle à combustible est capable de produire une caractéristique désirable pour une petite turbine à gaz.

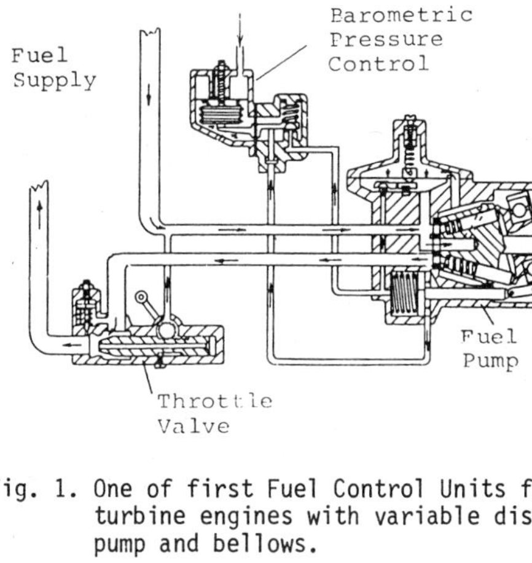

Fig. 1. One of first Fuel Control Units for gas turbine engines with variable displacement pump and bellows.

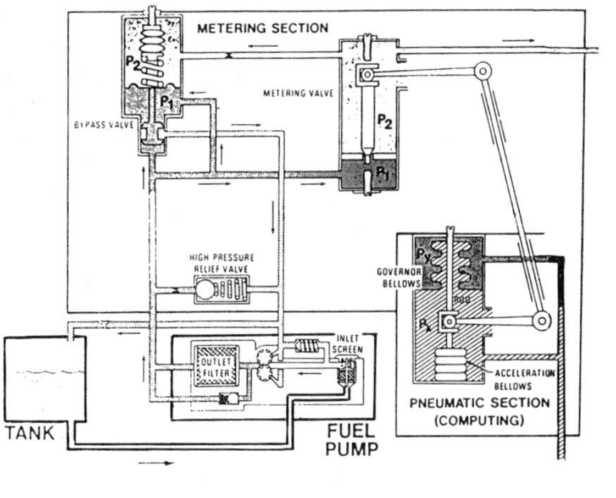

Fig. 2. Bendix Fuel Control Unit with constant displacement pump and flow modulation by metering valve (Courtessy of Pratt & Whitney Aircraft of Canada Ltd.).

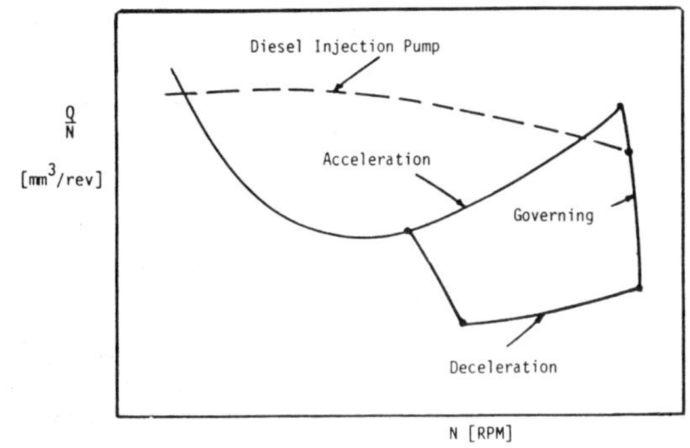

Fig. 3. Pump delivery characteristics before and after conversion.

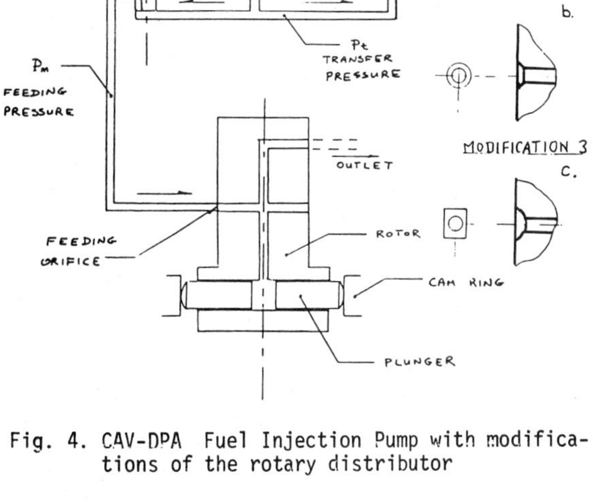

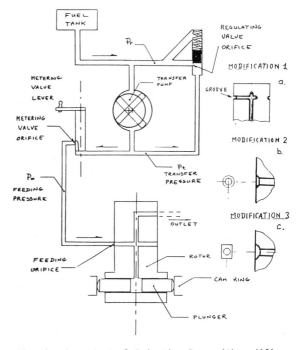

Fig. 4. CAV-DPA Fuel Injection Pump with modifications of the rotary distributor

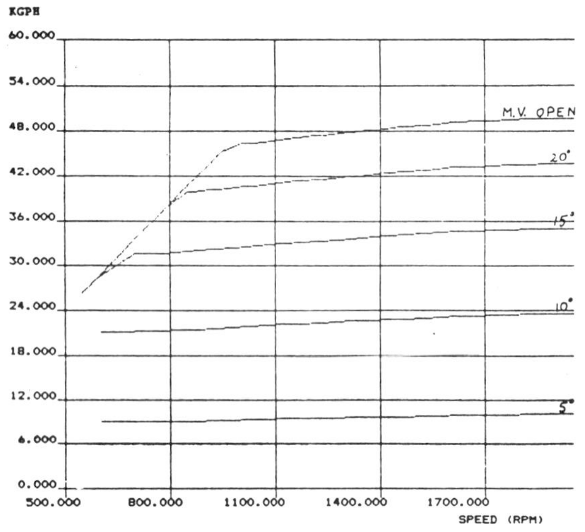

Fig. 5. Fuel flow calculated versus speed of the pump and versus metering valve position.

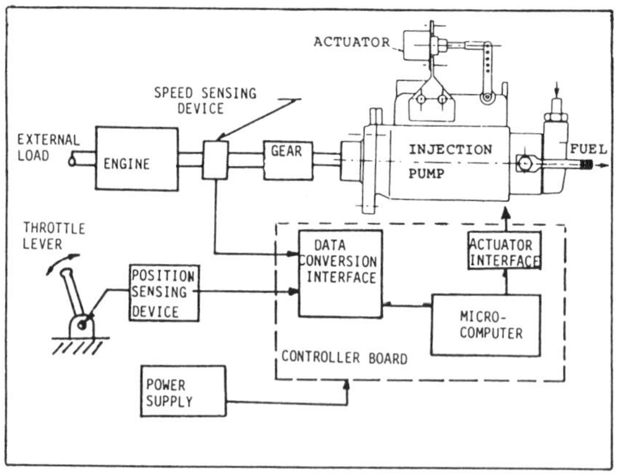

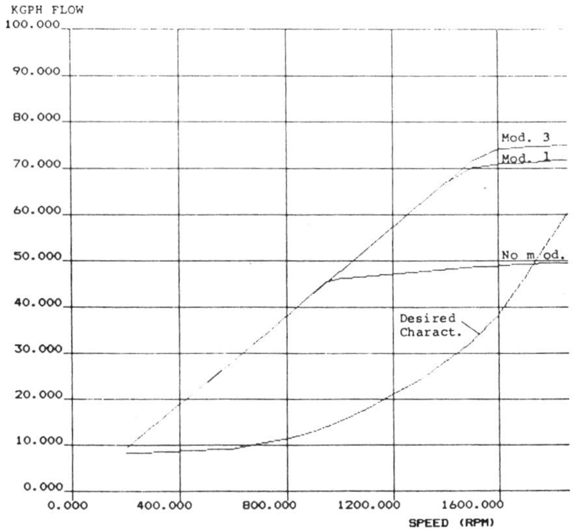

Fig. 6. Pump characteristics after modifications compared to desired output of the FCU.

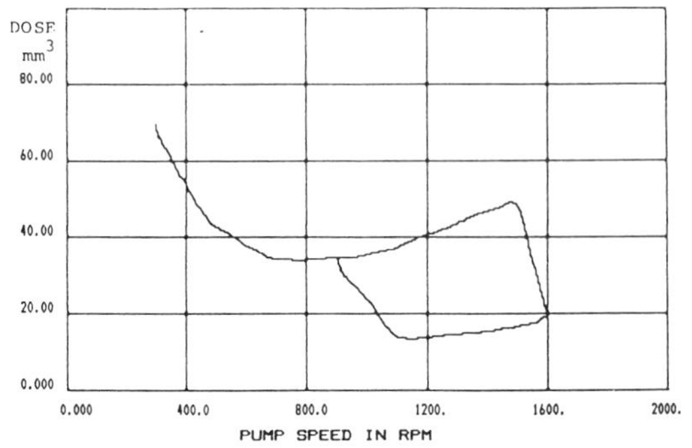

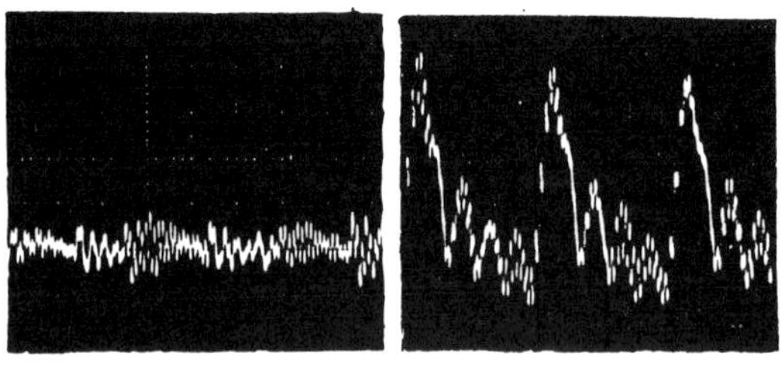

Fig. 7. Fuel pressure pulsations obtained from the converted Fuel Injection Pump with pressure equalizer (left) and without equalizer (right). The pressure scale is the same for both cases.

Fig. 8. Schematic of electronic controller used in the converted Fuel Control Unit.

Fig. 9. Fuel delivery schedule for the Fuel Control Unit with electronic control as recorded versus speed on a test bench under computer control.

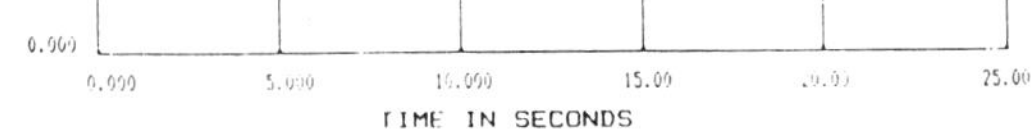

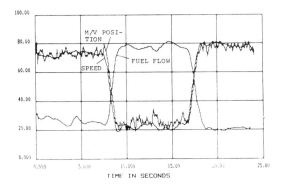

Fig. 10. Transient process for converted Fuel Control Unit recorded versus time on a test bench under computer control.

Electronical Loading-rate Limiter

J. VINCENT, S. JURAJ AND B. TIBOR

TU-Košice, Švermova 9, 041 87 Košice, Czechoslovakia

Abstract. In the process of lifting burden, the system of elasticity in the lifting machine vibrates so that the dynamical forces arised in all parts of the crane. The result of the analysis confirm the deffects of mechanical loading-rate limiters, whose exceed the value for examination of stability by ČSN. $\Delta F_c = 0{,}33\ Q_m\ g$, for non-permissible lifting process with the free rope system force increments reach the value $\Delta F_c = 0{,}52\ Q_m\ g$, so the necessity of the new loading-rate limiter conception was confirmed. The dynamical character of the lifting process requires two differents parts of the determining and elaborating unit of the electronical limiter - the statical and the dynamical ones. The sugested conception of the electronical loading-rate limiter protects crane from deffects and registers actual condition of the burden. It is also very reliable.

Keywords. Building tower crane; Dynamic stability; Modelling; Partial differential equations; Analog computer applications; Loading-rate limiter.

INTRODUCTION

In the process of lifting burden the system of elasticity in the lifting machine vibrates so that the dynamical forces arised in all parts of the crane. The total force in the lifting rope and in the system of the arm rope is greater than the statical burden. In the static condition the calculation of the bearing structure of the lifting machine the effect of dynamics in lifting of the burden is expressed by the coefficient of dynamics ψ.

$$\psi = \frac{F_{dyn} + F_{stat}}{F_{stat}} = 1 + \frac{F_{dyn}}{F_{stat}} \qquad (1)$$

The dynamical coefficient ψ in this case is a function $\psi = f(\dot{x}_1$ - velocity, c_1 - stiffness of the mecanical system in the rope and crane, manner of the operation, type of control or manipulation of the crane . Introducing the dynamical coefficient however doe s not provide the possibility of getting real view of the force ratios in the lifting rope and in the system of the arm rope.

According to what type of structure is taken into consideration it is necessary to know the real force ratios in the rope system of the lifting machine of the restrictor bearing capacity. This helps us too in order to determine the criteria for permitable function of the machines. Because of this it was necessary to examine the influence of dynamical forces to the electronical loading-rate limiter's activity, to perform analysis of dynamical forces by means of theoretical calculations, thus to determine the real course of force quantities.

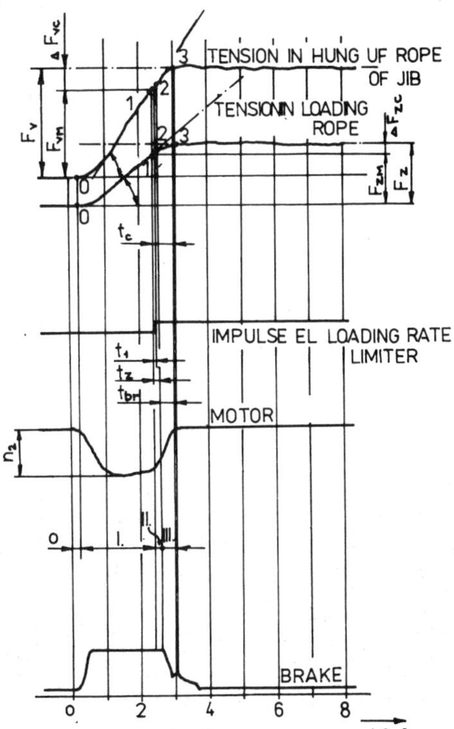

Fig. 1

The process from imparting the impulses to switching off dangerous motions up to the complete stopping mechanism of the lifting drum may be divided into segmented time intervals. At the moment of imparting the impulse to the switching off, point 1 in figure 1, the controlling electrical circuit will be interrupted and so the driving electrical motor will be switched off too. After the time interval t_1 the impulse is given to the electromagnetic restrictor. The braking time t_z is possible to get at point 2. The total stopping of the time of the lifting mechanism drive will be achieved after the time interval $t_c = t_z + t_{br}$, point 3. The magnitude of time interval t_{br} depedns on the velocity of rotation and on the moment of inertia of individual parts of lifting mechanism, which are in motion. Within the time interval $t_c = t_z + t_{br}$, forces in the lifting rope as well as in the hanging rope of the arm will increase in the value $\Delta F_{zc} + \Delta F_{yc}$ according to figure 3. The resultant value of pulling force in the burden rope will be:

$$F_v = F_{zn} + \triangle F_{zc} \quad (2)$$

in the hanging rope of the arm:

$$F_v = F_{vm} + \triangle F_{vc} \quad (3)$$

The described part in the whole process of lifting the burden is marked as the action of overtaking the driving machine of the crane.

SUPLEMENTARY MECHANICAL SYSTEM OF THE BUILDING TOWER CRANE

In the selection of the supplementary mechanical system of the lifting machine it is necessary to start with certain simplifications, which are not real in practice. The real system of the lifting machine is for mathematical purposes substitued by a mechanical model. The aim of drawing this mechanical models is to get by means of simple calculations results that ought to approximate as close as we wish to the real results.

To investigate the influence of dynamical forces and the mass of inertia of the lifting mechanism on the function of restrictor bearing activity there was used the third mass system according to fig. 2.

The process of lifting will be carried out in 3 phases:

1. phase: To roll up the existing free lifting rope.
2. phase: The rope is stretched. The pulling force in the rope will be equal to the magnitutde of the force of gravity of the burden, $F_z = Q_B\, g$.
3. phase: The burden is lifted up from the ground.

Individual movements are expressed with the help of coordinates x_1 , x_2 , $\varphi_3 = x_3\, r$ which correspond to the motion of shifting or rotation of individual parts of the structure in the following cases:

A/ In the region of lifting the burden.
B/ In the region of its own structure.
C/ In the region of lifting mechanism.

According fig. 2 the following equations hold:

$$\sum F = 0, - m_1\ddot{x}_1 - m_1 g + F_z = 0$$
$$\sum F = 0, - m_2\ddot{x}_2 - F_2 + F_z = 0 \quad (4)$$
$$\sum M_R = 0, - I_3\ \ddot{\varphi}_3 + F_3\, r - F_z = 0$$

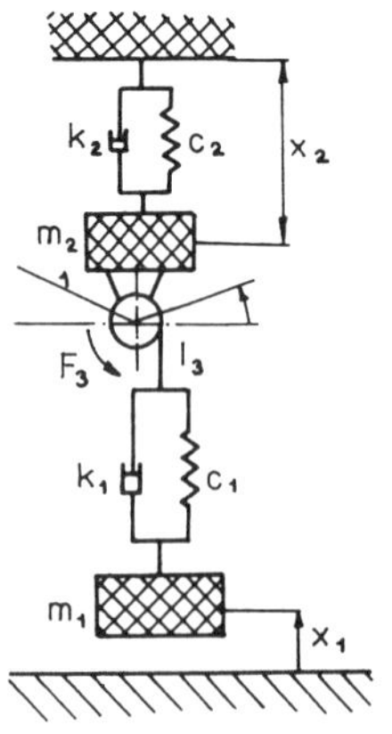

Fig. 2

The above given motion equations are simply solved by means of analogous computer. It is sometimes very complicated and pretentious to determine the particular model parameters but is has been published many times and also it was introduced in the research report at our department.

ANALOGOUS CALCULATING CRANE SCHEMA

In fig. 3 there is illustrated the calculating scheme for programming the mathematical model of the crane together with the braking system of the lifting mechanism. It is a combination of three fundamental calculating circuits describing:

- drive of the lifting mechanism
- the steel structure of the crane
- the burden

and the independently realised brake circuit. In this calculating scheme there are given two types of potenciometers - functional and control ones.

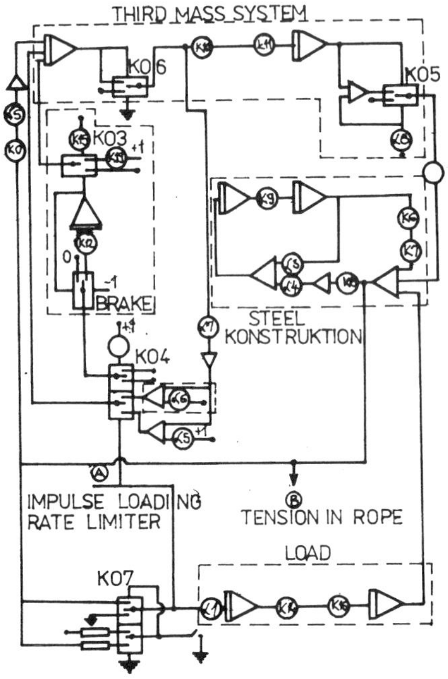

Fig. 3

Functional potenciometers are marked with the letter α ; they consider the constants of the differentied equations of the motion and characterise the crane parameters as a functional dependence.

$\alpha_3 = f(c_1 , c_2)$ $\alpha_4 = f(m_2 , m_1)$

$\alpha_5 = f(I_3 , m_1)$ $\alpha_6 = f(F_o)$

$\alpha_7 = f(F_o ; \dot{\varphi}_3 , c_1)$ α_B - adjusting the length of the free rope

Values for calculation of particular mass and stiffness parameters for various crane types are calculated on the basis of product technical documentation and experimental measuring results. The control potenciometers are marked with a capital letter K and a numbers. They prevent from saturation of the computer network.

It is especially necessary here to mention importance of the potenciometer K15 by means of which it is possible to adjust the value of braking time. Entry signal to the analogous model of the loading-rate limiter is introduced as a final signal from the result of mathematical model of the crane.

Integration I_1 creates dynamical restriction and comparator K01 secure the decision in the dynamical part of the system. Comparator K02 serves as a statical part of the loading-rate limiter. In the case of letting the switching off signal to the comparator K04 and comparator K04 will let the brake to function. Signal for the creation of statical restriction is created by the trans-ducer.

ANALYSIS OF THE ELECTRONICAL LOADING-RATE LIMITER'S ACTIVITY

According to the scheme in fig. 3 a functional model of the loading-rate limiter was prepared whose function was examined by means of the entry signal taken from the supplimentar and mechanical model of the tower crane. In order to judge without lias its reliability and safety in operational conditions and to apply its shortcomings, if any, to the prototype production, there was parallel connection of both mechanical loading-rate limiter and electronical loading-rate limiter during the measurements on MB crane.

One of the most important parts of the experimental measurements was to determine the switching off points, curve of restriction, to find the time of reaction, its shortening during the application of the electronical loading-rate limiter.

Entry signal of the pulling force was registered as the secondary signal with the system of the hanging rope of the arm from a place according in fig. 6, drawn through the amplifier of the operator TDA 6 to the determining part of limiter. Measurements were carried out for different lifting service conditions and especially the statical restrictions were positioned in every lifting process.

The value of the pulling force in the hanging rope of the arm was set up according to the motion of the smooth character of lifting with the II degree of velocity; curve b in fig. 4. Lifting with higher degree of velocity, curve c , and lifting with the free rope, curve d , are considered as non-permissible operation. Im lifting with the 1 st degree of velocity the values of statical restrictions PM and PG were set up in a manner and also for electronical limiter might react in $t_p = 0{,}1$ s later. The angle of the integrated curve in dynamical restriction α_{pE} was chosen for the purpose of safety:

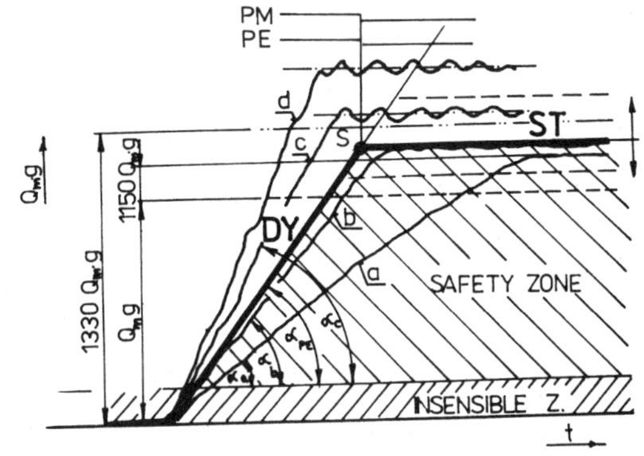

Fig. 4

$$\alpha_a < \alpha_{PE} < \alpha_b \qquad (5)$$

Considering this fact the value of the switching off force F_{PE} has decreased, while $F_{PE} = F_{PM} - \Delta F_{PE}$ that made the elctronical limiter PE to react in t_{PEb} earlier than the mechanical one, which had an effect in the following equation:

$$F_{PE} + \Delta F_c < F_{PM} + \Delta F_c \qquad (6)$$

While lifting the burden decreasing of switching off forces F_{PE} against the F_{PM} and all the more shortening the reaction time t_{PEc} will be manifested in more expressive manners, while the value $F_{PE} + \Delta F_c$ will not exceed the value of the statical restriction.

According to the above analysis and experimental measurements came to the conclusion that the electronical loading-rate limiter could register exactly the process of lifting with regard to the lifting velocity thus preventing from the higher velocity value in the rope system without excedd the permissible value F_{st} .

CONCLUSION

a/ The result of this analysis and experimental measurements confirm the deffects of mechanical loading-rate limiters, whose pulling increments during switching off period exceed the value in the stable condition, prescribed for examination of stability by ČSN 27 0142 $\Delta F_c = 0{,}33\ Q_m\ g$; for non--permissible lifting process with the free rope system force increments reach the value $\Delta F_c = 0{,}52 \cdot Q_m \cdot g$; so the necessity of the new loading-rate limiter conception was confirmed.

b/ The dynamical character of the lifting process requires two differents parts of the determining and elaborating unit of the electronical limiter - the statical and the dynamical ones. The dynamical restriction is determining partly by the creation of the increasing curve in the process of integration; partly by determining the statical restriction curves for the concrete type of tower crane.

c/ The suggested conception of the electronical loading-rate limiter protects the crane from deffects and registers actual condition of the burden. It is also very reliable.

The electronical loading- rate limiter made in Chechoslovakia, on the basis of the fundamental research results of our department is protected by the ownership of the author's patent N° 201.792. Its overall workmanship is documented in the next figures.

REFERENCES

Jasaň, V. (1984). Influence constructional parameters on working quality of cranes. Conference, High Tatras.

Jasaň, V. (1985). Dynamics of crane bridge modeling. Conference, High Tatras.

Sinay, J. and Bugár, T. (1982). Electronical loading--rate limiter. Conference, High Tatras.

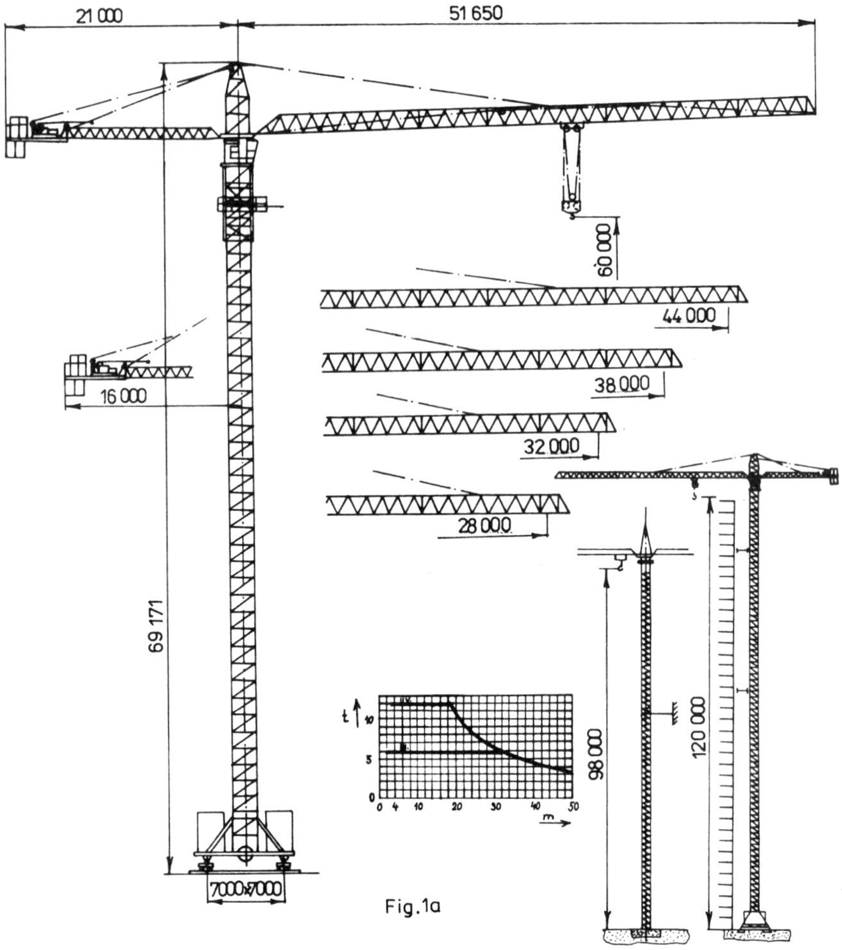
21 000
51 650
60 000
44 000
38 000
32 000
28 000
16 000
69 171
98 000
120 000
7000x7000

Fig. 1a

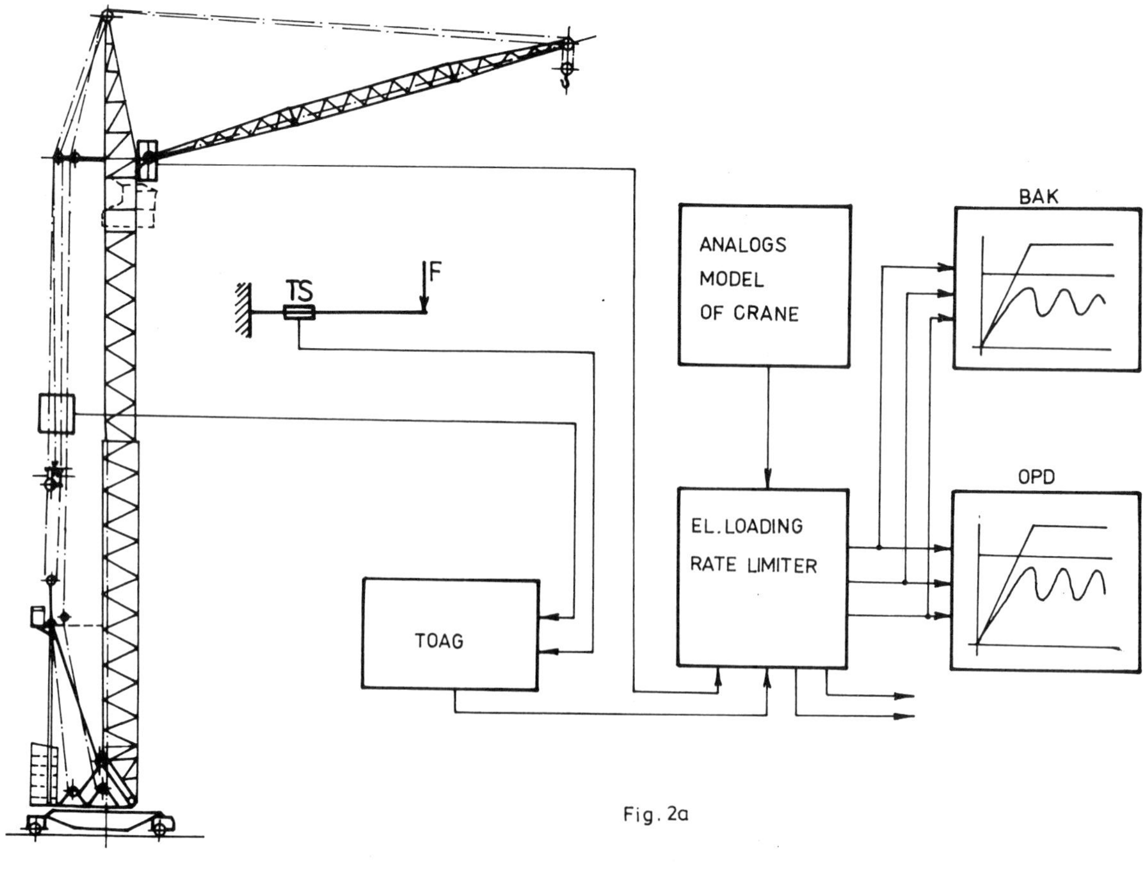
TS
F
ANALOGS
MODEL
OF CRANE
BAK
EL. LOADING
RATE LIMITER
OPD
TOAG

Fig. 2a

Energetically Optimal Control of Industrial Robots

M. Valášek

Department of Automatic Control, Faculty of Mechanical Engineering, Czech Technical University of Prague, Prague, Czechoslovakia

Abstract. A new very efficient algorithm of the synthesis of energetically optimal trajectory of industrial robots (so called method of suboptimal approximations) is described. Especially the advantageous property of the algorithm is the complete applicability of each approximation, i.e. every approximation satisfies all kinematic constraints, restrictions of servo's dynamics, collision avoidance to obstacles etc. The convergence of this algorithm has been proved and has been very rapid in practice. The most advantageous performance index of optimality is from many points of view the minimum energy consumption. This algorithm enables us to realize the program optimal control of industrial robots. It calculates after several iterations very good approximation of optimal control which is completely applicable and the robot is driven along this optimal trajectory by a classical feedback control (CNC) or by the feedback control with dynamic compensation.

Keywords. Optimal control; robots; energy control; programming control; maximum principle; optimal trajectory; feedback control with dynamic compensation; positional control; path tracking control.

INTRODUCTION

Industrial robots driven electrically with a feedback control loop (and even then hydraulically) create highly nonlinear dynamic systems the control of which in real time is a difficult problem of the control theory. From the theoretical point of view it is a classical problem of optimal control. A system is to be controlled from the initial state to the terminal one so as to minimize some performance index of optimality. But the synthesis of the optimal control as a feedback control has not been till now successful because of high nonlinearity with couplings among the joints and of strict constraint conditions at the terminal states (Luh, 1983a, 1983b). Otherwise only the computation of the time dependance of optimal control is possible only by numeric iterations convergence of which has been questionable (Valášek, 1983). So the conventional control of today's robot (Luh, 1983a, 1983b; Paul, 1981; Brady, 1983) is universally realized by the program CNC control, i.e. controller generates the desired corner points of the path and a conventional feedback control (linear servo) for each joint stabilizes the robot state according to the varying desired one. The trajectory of motion being the input to the control system is either numerically fed into the system or furnished through "teaching by doing".

The mass use of robots for more difficult application (i.e. the surroundings of the robot is dynamic, the accuracy and complexity of robot's positioning exceeds the human possibilities in "teaching by doing" and/or it is necessary a quick change of robot's job according to the production change by many robots in one moment) need the synthesis of suitable trajectories of industrial robots by a computer without direct interaction with the robot. This necessity of the synthesis of robot's trajectories by a computer enables us to use an alternative approach towards optimal control, i.e. the computation of the optimal control and the optimal trajectory for the desired robot's movement and the realization of this optimal trajectory by the conventional CNC control. Then, it is a program optimal control of an industrial robot. We regard a suitable trajectory as the trajectory which is convenient from the technological point of view and which is admissible for the dynamics of joint servos and which minimizes the performance index of optimality. The energy consumption is a very advantageous performance index.

A very efficient method of suboptimal approximations for solving this problem has been developed (Valášek, 1984). This method computes from admissible control and admissible trajectory of robot (i.e. satisfying constraint conditions, restrictions of servo's dynamics, obstacles in space etc.) next admissible control and trajectory on which the performance index of optimality is lower Further the synthesis of various optimal trajectory of industrial robots is formulated, the advantages of energy consumption performance index and the control structure of industrial robots with dynamic compensation are included and the method of suboptimal approximations is descriebed.

SYNTHESIS OF OPTIMAL TRAJECTORY

According to different constraints we can distinguish several problems of trajectory synthesis:

A. Positional control. The initial and terminal positions of an industrial robot are given and the trajectory between them is not specified. In this way the problem has an infinite number of solutions. The ambiguity of solutions can be used for the minimization of some performance index. We speak about positional optimal control of robots.

B. Path tracking control or program control or continuous path control. The trajectory along which the industrial robot is to move is specified. The trajectory is usually given only by pure geometric shape of the path and so the problem has even here an infinite number of solutions determined by a possible choice of the velocity profile along the specified path. The ambiguity can be used again for the minimization of some performance index. We speak about path tracking optimal control of industrial robots.

Both problems can be solved with redundant degrees of freedom (Valášek, 1984). Only the solution of positional optimal control without redundant degrees of freedom will be described here.

Among many performance indexes of optimality the most advantageous one from several points of view is the energy consumption. Other possible performance indexes of optimality are the minimization of motion time, of average performance effort, of values of reaction forces etc. The minimization of motion time is a natural performance index, but it has a lot of disadvantages. At the numerical solution there can appear degenerated cases with infinite number of solutions which are eliminated only by the more or less strong nonlinear couplings between robot's joints. During the movement according to this index all actuators have their maximum output. Their energy consumption is also maximum. And at the same time the robot must wait at the final position for the time between the minimum one necessary for the movement and the time resulting from the technological requirements on the robot's movement. During this movement the actuator torque is being switched from the maximum driven value in one direction to the opposite one and this causes the impacts in the robot gears and the efforts of the robot actuators and robot construction. By the others performance indexes these disadvantages are decreased in a different degree. But only the performance index of energy consumption fluently removes all these disadvantages. The degenerated cases cannot arise because among robot's joints there is always a stable coupling through the performance index. By this movement there is obviously the minimum energy consumption and the time of the movement which is chosen can completely conform to the technological requirements. From the temporal point of view we don't lose anything because the chosen time of robot's movement can be shortened to the minimum time. Then the control for the minimum energy consumption is equal to the control for the minimum motion time. The impacts at robot's actuators and the extremal efforts of robot's construction don't occur because the time behaviour of robot's actuator torques are continuous.

ROBOT CONTROL STRUCTURE WITH DYNAMIC COMPENSATION

The dynamic equations of an industrial robot with electric servos can be compactly written (Brady, 1983)

$$\mathbb{n}_n = \mathbb{I}(\mathbb{s})\ddot{\mathbb{s}} + \dot{\mathbb{s}}^T\mathbb{C}(\mathbb{s})\dot{\mathbb{s}} + \mathbb{V}(\mathbb{s})\dot{\mathbb{s}} + \mathbb{F}(\mathbb{s})\dot{\mathbb{s}} + \mathbb{g}(\mathbb{s}) \qquad (1)$$
$$\mathbb{n}^- \leq \mathbb{n}_n \leq \mathbb{n}^+$$

where $\mathbb{n}_n$ is the n-dimensional vector of constant joint torques, $\mathbb{I}(\mathbb{s})$ is the nxn generalized inertia tensor, $\mathbb{C}(\mathbb{s})$ is the nxnxn generalized tensor in the formulation of the Coriolis and centrifugal forces, $\mathbb{V}(\mathbb{s})$ is the nxn generalised tensor in the formulation of viscous friction forces, $\mathbb{F}(\mathbb{s})$ is the nxn generalized tensor in the formulation of velocity actuator torque dependency such as the back EMF of electric motors, $\mathbb{g}(\mathbb{s})$ is the n-dimensional vector of gravity forces, $\mathbb{s} = [s_1, \ldots, s_n]^T$ is the n-dimensional vector of robot's joint coordinates, $\mathbb{n}^+ = [n_1^+, \ldots, n_n^+]^T$ and $\mathbb{n}^- = [n_1^-, \ldots, n_n^-]^T$ are maximum and minimum constant torque limits. This description can be easily transferred to the state description with variables

$$x_{1i} = s_i \qquad i = 1, \ldots, n$$
$$x_{2i} = \dot{s}_i \qquad (2)$$
$$u_i = \frac{n_{ni} - \frac{n_i^+ + n_i^-}{2}}{\frac{n_i^+ - n_i^-}{2}}$$

and then

$$\dot{x}_{1i} = x_{2i} \qquad k = 1,2; \quad i,j = 1, \ldots, n$$
$$\dot{x}_{2i} = f_i(u_m, x_{kj}) = \sum_{m=1}^{n} H_{im}(x_{kj})u_m + L_i(x_{kj}) \qquad (3)$$

with the control variable restriction

$$|u_i| \leq 1 \qquad i = 1, \ldots, n \qquad (4)$$

Now we use the method of the change of dynamic properties of a system by the transformation of input variables (Valášek, 1982) for the derivation of the new robot control structure with dynamic compensation. To the system (3) we choose another dynamic system with desirable dynamic properties

$$\dot{y}_{1i} = y_{2i} \qquad (5)$$
$$\dot{y}_{2i} = \sum_{m=1}^{n} K_{im} w_m$$

with inputs w_m and outputs y_{ki}. We require the agreement of outputs of both systems $y_{ki} = x_{ki}$. From this according to (Valášek, 1982,1984) we derive the transformation equations between input variables u_i and w_m

$$u_i = Q_i(w_m, x_{kj}) = \qquad (6)$$
$$= \left(\left[H_{im}(x_{kj})\right]^{-1}\left[\sum_m K_{im} w_m - L_i(x_{kj})\right]\right)_i =$$
$$= \sum_{m=1}^{n} M_{im}(x_{kj}) w_m - N_i(x_{kj})$$

If these equations (6) are realized in real time the original dynamic system would have new dynamic properties given by equations

$$\dot{x}_{1i} = x_{2i} \qquad i = 1, \ldots, n \qquad (7)$$
$$\dot{x}_{2i} = \sum_{m=1}^{n} K_{im} w_m$$

with new inputs w_m. If we use the equations (7) for the calculation of the control variable behaviour on any prescribed trajectory and apply it by (6) to the robot control, the industrial robot would move along the prescribed trajectory only in the case that equations (3) are exact and complete dynamic model of the controlled robot. It cannot be practically satisfied. Therefore we must use the feedback principle for the compensation of uncomplete identification. The suggested control structure is in Fig.1. (for one degree of freedom). The control system (CS)

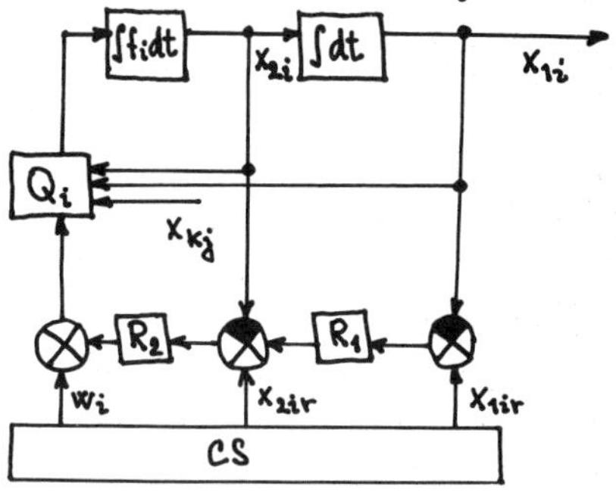

Fig. 1. Robot control structure with dynamic compensation

generates the control variable w_i and also the corresponding required position x_{1ir} and velocity x_{2ir}. These required values are compared with the real values of the robot and the differencies are transferred by the controllers R_1, R_2 to the correction of the control variable $\tilde{w}_i$ which is transformed in the compensator Q_i according to (6) to the real control variable u_i. The properties of this control structure are better compared with traditional ones even by the uncompleteness of (6). This control structure is able to realize robot's motion along the dynamically acceptable trajectory quite exactly and thus to realize the program optimal control of industrial robots.

METHOD OF SUBOPTIMAL APPROXIMATIONS

We have given the boundary conditions of motion

$$x_{1i}(0) = x_{1i0} \quad x_{1i}(T) = x_{1if} \qquad t\in[0,T] \quad (8)$$
$$x_{2i}(0) = x_{2i0} \quad x_{2i}(T) = x_{2if}$$

and we have to synthetise the optimal trajectory between these positions on which the energy consumption would be minimal

$$E = \int_0^T \Big(\sum_{i=1}^{n} e_i u_i^2\Big) dt \qquad (9)$$

where weighting coefficients e_i express contributions of particular control variables to the total energy consumption (9). For the synthesis of optimal trajectory we use the robot system (7) after the transformation. It is possible to show by the generalized optimum principle of V.F.Krotov (Krotov, 1973) that the solutions for both system's description (3),(7) are the same (Valášek, 1984). For the system (7) we assemble the hamiltonian

$$\mathcal{H} = -\sum_{i=1}^{n} e_i Q_i^2(w_m, x_{kj}) + \sum_{i=1}^{n}\Big(\psi_{1i} x_{2i} + \psi_{2i}\big(\sum_m K_{im} w_m\big)\Big) \qquad (10)$$

and the corresponding conjugate system of differential equations

$$\dot{\psi}_{1j} = \sum_{i=1}^{n} e_i 2\Big(\sum_m M_{im} w_m - N_i\Big)\Big(\sum_m \frac{\partial M_{im}}{\partial x_{1j}} w_m - \frac{\partial N_i}{\partial x_{1j}}\Big) \qquad (11)$$
$$\dot{\psi}_{2j} = -\psi_{1j} + \sum_{i=1}^{n} 2 e_i \Big(\sum_m M_{im} w_m - N_i\Big)\Big(\sum_m \frac{\partial M_{im}}{\partial x_{2j}} w_m - \frac{\partial N_i}{\partial x_{2j}}\Big)$$

and the condition of hamiltonian maximum

$$\sum_{i=1}^{n} e_i \sum_m M_{im} M_{il} w_m = \frac{1}{2}\sum_{i=1}^{n} \psi_{2i} K_{il} + \sum_{i=1}^{n} e_i N_i M_{il} \qquad (12)$$

This special form is the basis for the derivation of the method of suboptimal approximations. Let us have an admissible control as the J-th approximation $w_m^J(t), x_{1i}^J(t), x_{2i}^J(t)$ which satisfies the boundary conditions (8) and even further constraints. We substitute it into (11) and (7) and because of the linearity of (7) we get from the solution of (11) the nonlinear time function with the superposition of the linear solution of (7)-(12) determined by the choose of initial values $\psi_{ki}(0)$. After the comparison with the boundary conditions (8) we obtain the linear algebraic equations for the unknown initial values

$$x_{2jf} = x_{2j0} + \sum_i \Big(\psi_{1i}^J(0) \sum_m K_{jm} F_1^J(t) + \psi_{2i}^J(0) \sum_m K_{jm} F_2^J(t)\Big) + \sum_m K_{jm} F_3^J(t)$$
$$x_{1jf} = x_{1j0} + \sum_i \Big(\psi_{1i}^J(0) \sum_m K_{jm} \int_0^T F_1^J(t)dt + \psi_{2i}^J(0) \sum_m K_{jm} \int_0^T F_2^J(t)dt\Big) + \sum_m K_{jm} \int_0^T F_3^J(t)dt \qquad (13)$$

Hence we get the J*-th approximation $w_m^{J*}(t)$, $x_{1i}^{J*}(t), x_{2i}^{J*}(t)$ which is admissible for (8). Now we define

$$w_m^{J+1} = (1-\alpha) w_m^J + \alpha w_m^{J*}$$
$$x_{2i}^{J+1} = (1-\alpha) x_{2i}^J + \alpha x_{2i}^{J*} \qquad (14)$$
$$x_{1i}^{J+1} = (1-\alpha) x_{1i}^J + \alpha x_{1i}^{J*}$$

If α is constant, equations (14) define a new approximation satisfying (7) and (8). The unknown α is determined from one-dimensional minimization of energy consumption

$$E^{J+1} = \int_0^T \Big(\sum_{i=1}^{n}\Big(\sum_m M_{im}(x_{kj}^{J+1}) w_m^{J+1} - N_i(x_{kj}^{J+1})\Big)^2\Big) dt \qquad (15)$$

because (15) is just a function of α and for $\alpha\in[0,1]$ the (J+1)-th approximation is varying from J-th to J*-th approximation. The change of energy consumption in the neighbourhood of J-th approximation is (Valášek, 1984)

$$\delta E = -2\int_0^T \alpha \sum_{i=1}^{n} e_i \Big(\sum_m M_{im}(x_{kj}^J)(w_m^{J*} - w_m^J)\Big)^2 dt \qquad (16)$$

Equation (16) proves that if the linear theory is valid these approximations lead to the decrease of energy consumption. The validity of linear theory guarantees that the change of energy consumption $E^{J+1} - E^J$ is given by the first differential of optimum functional. Its validity is guaranteed by a small variation from J-th approximation to (J+1)-th approximation which can be ensured by the choice of sufficiently small α in (14). On the basis of (16) we can prove the Theorem (Valášek, 1984): The sequence of approximations (14) is either finite and then its last point satisfies the necessary conditions of Pontryagin's maximum principle, or is infinite and then its limit point satisfies the necessary conditions of Pontryagin's maximum principle.

The above described approximations can be further modified for consideration of further constraints (4), state constraints and etc. Let the J-th approximation satisfy the constraints. If the whole J-th approximation is on the boundary of constraints and J*-th approximation leads into the region where some constraint is not fulfilled, then the value of performance index cannot be decreased. On the contrary let the (J+1)-th approximation satisfy the constraints for some α_1 but not for α_2, and let both α decrease the performance index. We construct a new (J+1)-th approximation. Equation (16) is also valid for α as a time function and for every positive $\alpha(t)$ it leads to the decrease of the performance index. Instead of (14) we set

$$\begin{aligned} w_m^{\overline{J+1}}(t) &= (1-\alpha(t))w_m^J(t) + \alpha(t)w_m^{J*}(t) \\ x_{2i}^{\overline{J+1}}(t) &= (1-\alpha(t))x_{2i}^J(t) + \alpha(t)x_{2i}^{J*}(t) + \\ &\quad - \int_0^t \frac{d\alpha(t)}{dt}(x_{2i}^{J*}(t) - x_{2i}^J(t))dt \\ x_{1i}^{\overline{J+1}}(t) &= (1-\alpha(t))x_{1i}^J(t) + \alpha(t)x_{1i}^{J*}(t) + \\ &\quad -2\int_0^t \frac{d\alpha(t)}{dt}(x_{1i}^{J*}(t) - x_{1i}^J(t))dt + \\ &\quad + \int_0^t\int_0^t \frac{d^2\alpha(\tau)}{d\tau^2}(x_{1i}^{J*}(\tau) - x_{1i}^J(\tau))d\tau dt \end{aligned} \tag{17}$$

We substitute from the solution of (11)-(12) and from (17) to (7) and for given $\alpha(t)$ we obtain a system of linear algebraic equations for $\psi_{ki}^J(0)$ which together with (12) determine (J+1)-th approximation by (17). By the choice of $\alpha(t)$ between $\alpha(t)=\alpha_1$ and $\alpha(t)=\alpha_2$ (e.g. in the form of a spline function) we easily find the (J+1)-th approximation which satisfies the constraints and decreases the performance index below $\alpha(t)=\alpha_1$. The other possibility is to set $w_m^{\overline{J+1}}=w_m^{Jb}$, where w_m^{Jb} is the control of the movement along the boundary of the region of constraints, on the time interval where the (J+1)-th approximation fails to satisfy the constraints. If we set

$$\alpha(t) = \frac{w_m^{Jb} - w_m^J}{w_m^{J*} - w_m^J} \tag{18}$$

then for $\alpha(t) > 0$ this construction decreases the performance index (Valášek, 1984).

The described method of suboptimal approximation has been applied to several robot's models and control tasks and the practical convergence on the computer has been very rapid. The decrease of energy consumption has been approximately quadratic and the number of approximations necessary for the change of the performance index only in per centes has been about five.

CONCLUSION

The developed method of suboptimal approximations is a very efficient method of the synthesis of robot's energetically optimal trajectory. Together with the new robot control structure with dynamic compensation it enables to realize program energetically optimal control of industrial robots.

REFERENCES

Brady, J.M., and etc. (1983). Robot Motion: Planning and Control. MIT press, Cambridge.

Krotov, V.F., and V.I. Gurman (1973). Methods and Problems of Optimal Control. Nauka, Moskva. In Russian.

Luh, J.Y.S., (1983a). An anatomy of industrial robots and their controls. IEEE Trans. on Automatic Control, 28, 2, 133-153.

Luh, J.Y.S., (1983b). Conventional controller design for industrial robots - a tutorial. IEEE Trans. Systems Man and Cybernetics, 13, 3, 298-316.

Paul, R.P., (1981). Robot Manipulators: Mathematics, Programming and Control, MIT Press, Cambridge.

Valášek, M., (1982). The change of dynamic properties of a system by the transformation of input variables. Acta Polytechnica, 10, II, 4, 159-164. In Czech.

Valášek, M., (1983). Energetically suboptimal and program control of industrial robots in real time. Automatizace, 26, 12, 296-300. In Czech.

Valášek, M., (1984). Synthesis of Optimal Trajectory of an Industrial Robot. Ph.D. Thesis, Faculty of Mechanical Engineering, Czech Technical University of Prague, Prague. In Czech.

ENERGETISCH OPTIMALE STEUERUNG VON INDUSTRIELLEN ROBOTERN

Ein neuer, sehr effektiver Algorithmus (Methode der Suboptimalapproximation) für die Optimaltrajektoriensynthese von industriellem Roboter wird geschrieben. Aus einem zulässigen Steuerungsdurchlauf und aus einer zulässigen Robotertrajektorie (d.h. erfüllende kinematische Grenzenbedingungen, Begrenzung der Stellantriebdynamik, Ausweichen den Hindernissen im Raum usw.) rechnet der Algorithmus den weiteren zulässigen Steuerungsdurchlauf und die weitere zulässigen Trajektorie aus, an welcher das Optimalitätskriterium niedriger ist. Die Konvergenz des Algorithmus wurde bewiesen und hat sich als sehr schnell gezeigt. Für den grössten Vorteil des Algorithmus ist die vollständige Verwendbarkeit jeder Iteration angesehen. Die Energieverbrauch ist das vorteilhafte Optimalitätskriterium. Dieser neue Algorithmus zussamen mit Steuerungsstruktur des Roboters mit dynamischer Kompensation ermöglicht die programmierte Optimalsteuerung vom industriellen Roboter zu realisieren. Ebenfalls wird eine neue Steuerungsstruktur des Roboters mit dynamischer Kompensation beschrieben.

Stability Charts for a Man-machine System

V. D. TRAN* AND G. STÉPÁN**

*Department of Precision Mechanics and Applied Optics, Budapest University of Technology, H-1525 Budapest, Hungary
(Hanoi Institute of Technology, Hanoi, Vietnam)
**Department of Engineering Mechanics, Budapest University of Technology, H-1525, Budapest, Hungary

Abstract. This work presents the general stability investigation of a man-machine system. The system is described by means of a fourth-order differential difference equation. The stability of its trivial solution is analysed with the help of a new necessary and sufficient condition. This study results in stability charts in parameter spaces chosen in a proper way. The parameters of the human controller and those of controlled technical object (which is a slow dynamic system) are separated. The knowledge of the stability regions helps the designer to determine the paremeters of the controlled system and to predict whether the control will be a simple, difficult or impossible exercise for a human being. The stability charts are useful both in analysis and synthesis problems of such systems.

Keywords. Man-machine systems; delays; dynamic stability; stability criteria; human factors.

INTRODUCTION

When a human operator is a link of a complex system he establishes a connection with technical objects according to the technology in question. The prediction of the behavior and the stability of these systems is an interesting and important problem because the quality and reliability of the system operation depend not only on the controlled machine but also on the operator's knowledge, practical experience, discrimination and reflex. The present work investigates the stability of such a system. Stability charts are constructed with the help of a new method. These charts give a clear geometrical picture about the influences of all relevant technical and human parameters from stability point of view. In the engineering practice, the charts can be used well for the synthesis of man-machine systems.

THE SYSTEM MODEL

In our practical case we have been interested in the manual control of slow dynamic systems like vehicle and ship steering, crane handling or space-ship manipulation. As the control of the system in question was not too complex the case of a supervisory control was neglected and the block-diagram of Fig.1 was assumed. Of course, the transfer function Y(s) of the controlled system and the human transfer function H(s) have to be described in a proper way using the relevant parameters.

In the case of crane handling, when a human operator manipulates an object with a hydraulic servomechanism, the transfer function Y(s) can be simplified as follows

$$Y(s) = \frac{1}{J\, s^2(T^2s^2 + 2\zeta Ts + 1)}, \tag{1}$$

where T, ζ describe the main properties of the servomechanism and J is the inertia of the controlled object. In the given case where stability is investigated J, T and ζ can be assumed as relevant parameters and the compressibility of the hydraulic fluid and the hysteresis backlash of the servosystem are disregarded meanwhile the controlled object is supposed to be spring-damper free.

It is far more difficult to approximate the human transfer function H(s) and to choose its relevant parameters because the human operator adapts himself to the controlled system. The different approaches of the human controller's model are summarized in the excellent paper of McRuer (1980). From the point of view of system stability, it is allowed to take into consideration only the linear part of the model and to neglect the so-called remnant. The complexity of the transfer function H(s) depends on the accuracy with which the reproduction of the controller's characteristics is attempted. However, a fairly large number of phenomena can be accounted for by

$$H(s) = H_1(s)\cdot H_2(s)$$

$$H_1(s) = K\frac{T_L s+1}{T_I s+1}, \quad H_2(s) = \frac{e^{-\tau s}}{T_N s+1} \tag{2}$$

(see McRuer et al. (1957)). H_2 represents some basic inherent human limitations such as the reaction time delay τ and the lag time T_N of the neuromuscular system while H_1 represents human equalization characteristics (see Kleinman et al. (1970)). As the stability charts will show the relevant human parameters are the delay τ, the lead time constant T_L and the operator gain K. The parameters T_N and T_I have no significant role in system stability thus they are assumed to be zero in this case.

On the basis of the block-diagram of Fig.1 and the formulae (1) and (2) it is easy to construct the differential equation for the variable x. After a time transformation with respect to the time delay τ it has the form

$$x^{(IV)}(t)+a_1\dddot{x}(t)+a_0\ddot{x}(t)+b_1\dot{x}(t-1)+b_0x(t-1)=0 \tag{3}$$

where

$$a_1=2\frac{\zeta\tau}{T},\quad a_0=\left(\frac{\tau}{T}\right)^2,\quad b_1=\frac{K}{J}\,\frac{\tau^3 T_L}{T^2},$$

$$b_0=\frac{K}{J}\,\frac{\tau^4}{T^2}. \tag{4}$$

From mathematical point of view our aim is to investigate the asymptotic stability of the trivial solution x = 0 of the retarded differential difference equation (RDDE) (3) in Lyapunov sense.

STABILITY INVESTIGATION

There are several stability criteria for RDDEs in the special literature (see Bellman and Cooke (1963) or Hale (1977)). However, because of the great number of the parameters, a new method has been applied in this case.

Let us consider the characteristic function of (3)

$$D(\lambda)=\lambda^4+a_1\lambda^3+a_0\lambda^2+b_1\lambda e^{-\lambda}+b_0e^{-\lambda}$$

and the functions

$$M(y)=\mathrm{Re}D(jy)=y^4-a_0y^2+b_1y\sin y+b_0\cos y, \tag{5}$$

$$S(y)=\mathrm{Im}D(jy)=-a_1y^3+b_1y\cos y-b_0\sin y \tag{6}$$

where $j=\sqrt{-1}$, $y\in R^+$.

Theorem 1. The x = 0 solution of (3) is asymptotically stable iff

$$S(y_k)\neq 0,\ k=1,\dots,m \text{ and} \tag{7}$$

$$\sum_{k=1}^{m}(-1)^k \operatorname{sign}S(y_k)=2 \tag{8}$$

where $y_1\geq y_2\geq \dots y_m\geq 0$ are the real nonnegative zeros of M(y).

This theorem is a special form of a general necessary and sufficient condition proved in the paper of Stépán (1979).

Theorem 2. If the x = 0 solution of (3) is asymptotically stable then

$$b_0>0 \quad\text{and}\quad b_1>b_0. \tag{9}$$

As this necessary condition of stability is practically always fulfilled (see(4)) in our control system we do not deal with its proof here. The construction of the stability charts can be based on the case $a_1=0$ (i.e. the daping ratio ζ is negligible) and b_0 is small positive value.

Theorem 3. Let us consider the RDDE (3) with $a_1=0$ and $b_0\to +0$. The x = 0 solution is asymptotically stable iff (see the shaded areas of Fig. 2):

$b_1>0$ and $b_1<-(\pi/2)^3+(\pi/2)a_0$ and

$b_1<(3\pi/2)^3-(3\pi/2)a_0$ or

$b_1<-(5\pi/2)^3+(5\pi/2)a_0$ and $b_1<(7\pi/2)^3-(7\pi/2)a_0$ ... etc.

Proof. If $a_1=0$, $b_0=+0$ in (6) the positive zeros of S(y) are $\pi/2+i\pi$, i=0,1,2,.... On the basis of (9) it is easy to determine the sign variation of S(y). Thus, conditions (7) and (8) of theorem 1 are fulfilled iff

$M(0)>0$ and $M(\pi/2)<0$ and
$M(3\pi/2)>0$ or
$M(3\pi/2)<0$ and $M(5\pi/2)<0$ and $M(7\pi/2)>0$
etc. Formula (5) shows that these conditions are equivalent to the inequalities in theorem 3. □

In the case of $a_1>0$, $b_0>0$ explicit conditions cannot be obtained any more and simple computer programs have to be used to determine the boundaries of the stability regions in the parameter space on the basis of the implicit equations

$$M(y)=0,\quad S(y)=0 \tag{10}$$

and conditions (7) and (8) of theorem 1.

STABILITY CHARTS

By means of formulae (4), the stability conditions for the mathematical parameters can be transformed into conditions for the human and technical ones. Fig.3 shows the shaded stability regions on the plane of the human parameters τ and T_L for different values of damping ζ. With the help of a simple extremum calculation in (10) the optimal values T_{Lopt} and τ_{opt} can be determined when τ and T_L have the greatest value respectively .Figure 3 shows that these values depend only slightly on damping. Thus, if we use $\zeta=0$ and the approximation

$$KT^2/J\ll 1$$

which is true in technical cases then we get the simple formulae

$$T_{Lopt}=\sqrt{(3J)/(2K)},\quad \tau_{opt}=\pi T/\sqrt{2}. \tag{11}$$

Some points of this stability chart were controlled by means of the Hall-Nyquist diagram of the open loop transfer function

$$Y_0(s)=\frac{K}{J}\,\frac{(1+T_Ls)e^{-\tau s}}{s^2(T^2s^2+2\zeta Ts+1)}$$

(see Fig.5a). It shows the good stability properties of system at the optimal human parameters (11). Fig.4 shows the stability chart on the plane of the technical parameters J and T. The stability regions have been determined by a computer program based on theorem 1.

CONCLUSIONS

If we are interested in the human influence we get conclusions from the stability chart (τ, T_L) of Fig. 3. The operator can find the optimal lead time T_{Lopt} only after a long practice. Thus, he is a beginner in system controlling, he has to concentrate very intensely on doing his work with good reflexes with a small value of τ close to τ_{opt}. On the other hand if $\tau \to 0$ and T_L is great the case of human fluster is detected which causes instability again. After sufficient practice when $T_L \approx T_{Lopt}$, the operator reflexes may be slow. However if the delay is too great then the system can be stabilized only after a long transience (see Fig. 5b). It is also clear that the operator's behavior does have to contain the property of a lead compensator because the system is unstable if $T_L \to 0$.

As regards the technical parameters, an optimal interval of the time constant T can be determined by means of the vertical asymptotes of the boundary curves of stability regions in Fig. 4. The horizontal asymptote gives a simple formula for the necessary value of the inertia. It is quite natural that a great value of inertia J is good from stability view-point.

Though these conclusions have been got from a strongly simplified model omiting the complexity of the human behavior, we think that the influence of the relevant parameters on system stability can be analysed well in this way.

REFERENCES

McRuer, D. (1980). Human Dynamics in Man-Machine Systems. Automatica, Vol. 16, pp 237-253. Pergamon Press Ltd. Printed in Great Britain.

McRuer, D. T. and Krendel, E. S. (1957) Dynamic Response of Human Operators. WADC-TR-56-524.

Kleinman, D.L., Baron, S. and Levison, W.H. (1970). An Optimal Control Model of Human Response (I-II). Automatica, Vol. 6, pp 357-369 and 371-383. Pergamon Press. Printed in Great Britain.

Bellman, R., Cooke, K.L. (1963). Differential-Difference Equations. Academic Press, New York.

Hale, J.K. (1977). Theory of Functional Differential Equations (Applied Mathematical Sciences 3). Springer-Verlag. New York-Heidelberg-Berlin.

Stépán, G. (1979). On the stability of linear differential equation with delay. Coll. Math. Soc. J. Bolyai. 30. QTDE. North Holland, pp 971-984.

RÉSUMÉ

Cet article présente l'étude de la stabilité générale d'un système homme-machine. Le système est décrit au moyen d'une équation différentielle-différence du quatrième ordre. La stabilité de sa solution triviale a été analysée à l'aide d'une nouvelle condition nécessaire et suffisante. Cette étude conduit à des cartes de stabilité dans les espaces des paramètres choisis d'une manière propre. Les paramètres de l'homme-contrôleur et ceux de l'objet technique commandé (qui est un système dynamique lent) se sont séparés. La connaissance de la région stable aide l'inventeur dans la détermination des paramètres du système commandé et dans la prédiction si la commande soit simple, difficile ou non traitable pour un homme. Les cartes de stabilité sont utiles à la fois poure les problèmes d'analyse et de synthèse des tels systèmes.

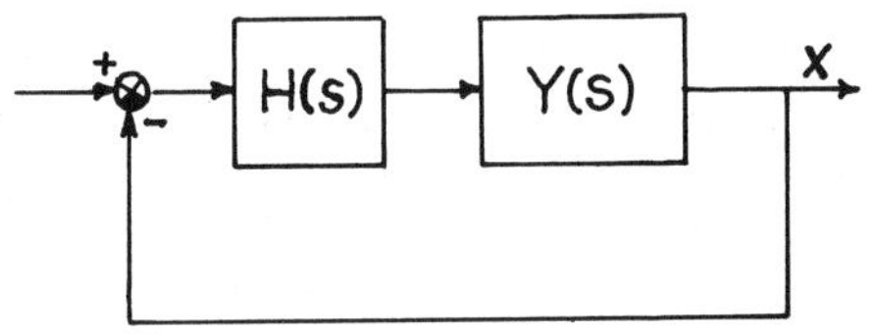

Fig. 1. Block-diagram

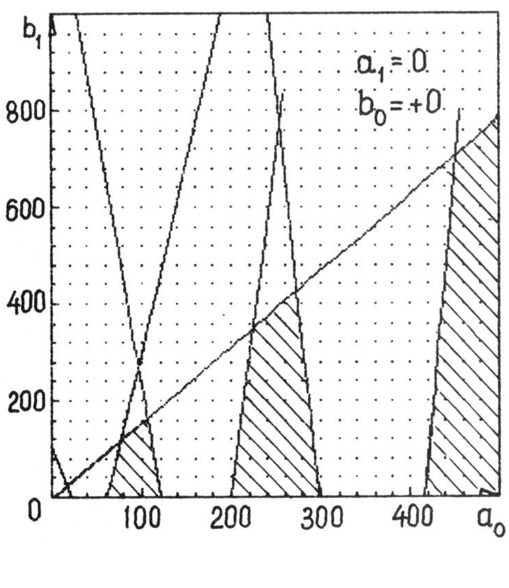

Fig. 2. Stability chart for mathematical parameters

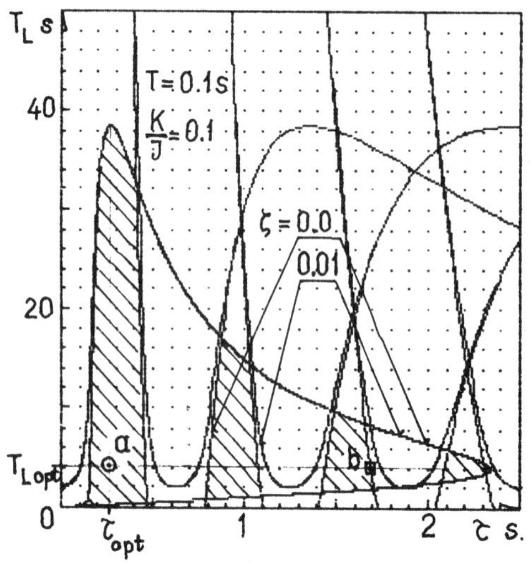

Fig. 3. Stability chart for human parameters

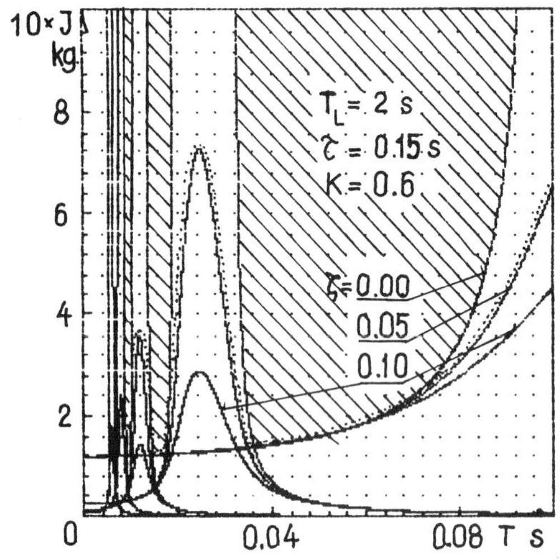

Fig. 4. Stability chart for technical parameters

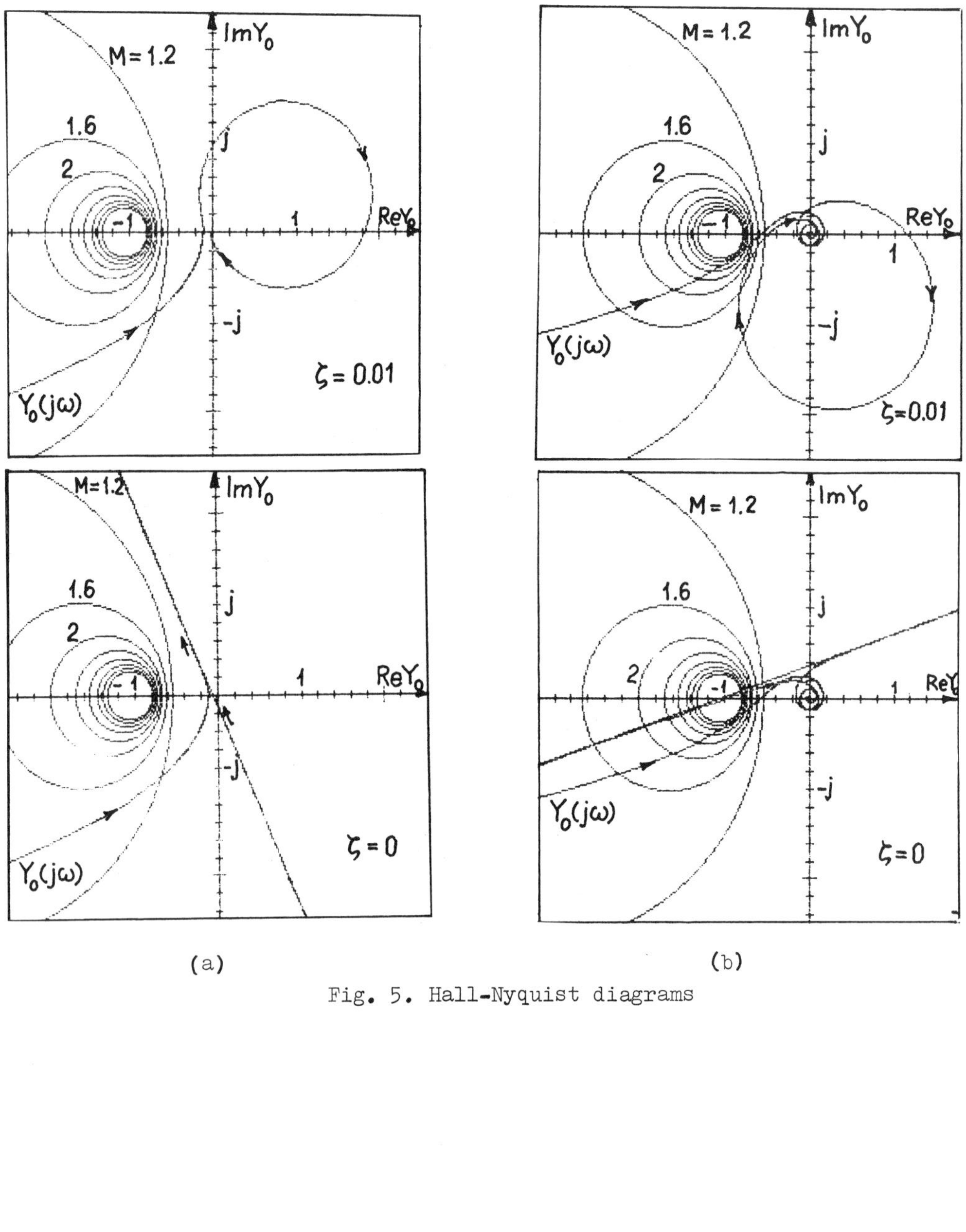

(a) (b)

Fig. 5. Hall-Nyquist diagrams

Consideration of Human Error While Determining Optimum Inspection Interval of a Standby System

C. CHOWDHURY

Department of Precision Mechanics and Applied Optics, Egry J. u. 1; E. ép. III. em. 1; Budapest-1521, Hungary

Abstract. In order to determine the optimum inspection interval of a standby system a model has been worked out, where three types of human errors are considered. For the purpose of optimization a complex state variable has been developed which is a functional derived from the reliability of the system, cost of operating it and the market conditions.

Keywords. Inspection interval; standby system; human error.

INTRODUCTION

System reliability can be improved by providing a standby unit, which plays an important role when failure of an operating unit is costly and/or dangerous. These systems are periodically tested in order to ensure their availability in need. The possibility of human error during testing and maintenance of such systems is obviously present. Numerous studies have addressed the problem of determining optimum test intervals. Chay and Mazumdar (1975) developed a model for determining the test interval which maximizes the availability of certain standby systems of k-out-of-n:G configuration. Sherwin (1979) developed a model which is used to determine optimum inspection schedules of condition maintained items according to certain cost criteria. But these models have not considered the human error effects. The inclusion of human error in the development of appropriate inspection plans and procedures has only recently been undertaken. Apostolakis and Bansal (1977) considered the effect of a single type of human error on the availability of certain k-out-of-n:G configured redundant standby system. In McWilliam and Martz (1980), the effects of two types of human error on the steady stete availability of a system are considered.

In this paper, in order to determine the optimum inspection interval of a standby system a new model is worked out, where three types of human error are incorporated. The possibility that a bad standby system is judged as good (type A human error), possibility that a good standby system is inadvertently left in a bad state after inspection or the standby system fails during inspection (type B error) and the possibility that a good standby system is judged as bad (type C human error), will be considered.

DESCRIPTION OF THE MODEL

It is assumed that the basic and the standby systems are not necessarily of same type. The system is represented with the help of twelve states. The corresponding probabilities are X_i (i=1,2,3,...,12). Twelve different states are defined as follows:

S_1: Basic system is operating, standby system is in an operable state. System is functioning.

S_2: Basic system is not working, but the standby system is successfully switched in and thus the system is functioning. Detection of the cause of failure of the basic system has started.

S_3: Good standby system is under inspection and the basic system is operating. Thus the system is functioning.

S_4: Bad or judged bad standby system is under detection of failure and the basic system is operating. The system is functioning.

S_5: Bad standby system is under inspection and the basic system is operating. System is functioning.

S_6: Standby system is not available, because it is in a failed state. But it is wrongly assumed that the system is in an operable state. The basic system is working. Thus the system is functioning.

S_7: After detection of the fault of the basic system it is undergoing repair Standby system is operating. The system is functioning.

S_8: After detection of the fault of standby system it is undergoing repair. Basic system is operating. The system is functioning.

S_9: Basic system is not working as per specification (e.g. in case of a manufacturing unit system is manufacturing faulty products), but this has not been noticed as yet and thus the standby system has not been switched in. The system is not functioning as per specification.

S_{10}: Basic system is not working as per specification, but this has not been noticed as yet. Standby system is not in an operable state. The system is not functioning as per specification.

S_{11}: Standby system is not working as per specification, but this has not been noticed as yet. Basic system is not in an operable state. The system is not functioning as per specification.

S_{12}: Neither the basic system nor the standby system are operable. The system is not functioning.

In Fig. 1, the transition diagram of the model is shown - Reference for Fig. 1:

$$A_2 = -\left(\frac{\omega_1(S)}{S} + \frac{\omega_2(S)}{S} + \frac{1}{S}\frac{\tau_1}{\tau+\tau_1} + \frac{\omega_H(S)}{S}\cdot\frac{\tau}{\tau+\tau_1}\right);$$

$$B_2 = -\left(\frac{\alpha_1}{S} + \frac{\omega_4(S)}{S} + \frac{\omega_5(S)}{S}\right);$$

$$C_2 = -\left(\frac{\omega_1(S)}{S} + \frac{\omega_2(S)}{S} + \frac{1}{S\tau_1} + \frac{P_B}{\tau_1 S}\right);$$

$$D_2 = -\left(\frac{\alpha_2}{S} + \frac{\omega_1(S)}{S} + \frac{\omega_2(S)}{S}\right);$$

$$E_2 = -\left(\frac{\omega_1(S)}{S} + \frac{\omega_2(S)}{S} + \frac{1}{S\tau_1}\right);$$

$$F_2 = -\left(\frac{\omega_1(S)}{S} + \frac{\omega_2(S)}{S} + \frac{1}{S}\frac{\tau_1}{\tau+\tau_1}\right);$$

$$G_2 = -\left(\frac{\omega_4(S)}{S} + \frac{\omega_5(S)}{S} + \frac{\mu_1}{S}\right);$$

$$H_2 = -\left(\frac{\mu_2}{S} + \frac{\omega_2(S)}{S} + \frac{\omega_1(S)}{S}\right);$$

$$I_2 = -\frac{\delta}{S}$$

$$J_2 = -\frac{\delta}{S}$$

$$K_2 = -\frac{\delta}{S}$$

Using Laplace transformations the following set of differential equations are derived that represent the system:

$$\dot{X}_1(t) = \frac{1-P_C}{\tau_1}\cdot X_3(t)+\mu_1\cdot\varepsilon_3\cdot X_7(t)-(\omega_1(t)+ \omega_2(t)+\frac{\tau_1}{\tau+\tau_1}+\omega_H(t)\cdot\frac{\tau}{\tau+\tau_1})\cdot X_1(t)+\mu_2\cdot X_8(t) \quad (1)$$

$$\dot{X}_2(t) = \omega_2(t)\cdot\varepsilon_1\cdot X_1(t)+\delta\cdot\varepsilon_2\cdot X_9(t)-(\alpha_1+\omega_4(t) + \omega_5(t))\cdot X_2(t) \quad (2)$$

$$\dot{X}_3(t) = \frac{\tau_1}{\tau+\tau_1}\cdot X_1(t)-(\omega_1(t)+\omega_2(t)+\frac{1}{\tau_1}+\frac{P_B}{\tau_1})\cdot X_3(t) \quad (3)$$

$$\dot{X}_4(t) = \frac{P_C}{\tau_1}\cdot X_3(t)+\frac{1-P_A}{\tau_1}\cdot X_5(t)-(\alpha_2+\omega_1(t) + \omega_2(t))\cdot X_4(t) \quad (4)$$

$$\dot{X}_5(t) = \frac{P_B}{\tau_1}\cdot X_3(t)+\frac{\tau_1}{\tau+\tau_1}\cdot X_6(t)-(\omega_1(t)+\omega_2(t) + \frac{1}{\tau_1})\cdot X_5(t) \quad (5)$$

$$\dot{X}_6(t) = \omega_H(t)\cdot\frac{\tau}{\tau+\tau_1}\cdot X_1(t)+\frac{P_A}{\tau_1}\cdot X_5(t) - (\omega_1(t)+\omega_2(t)+\frac{\tau_1}{\tau+\tau_1})\cdot X_6(t) \quad (6)$$

$$\dot{X}_7(t) = \alpha_1\cdot X_2(t)-(\omega_4(t)+\omega_5(t)+\mu_1)\cdot X_7(t) \quad (7)$$

$$\dot{X}_8(t) = \alpha_2\cdot X_4(t)-(\mu_2+\omega_2(t)+\omega_1(t))\cdot X_8(t) \quad (8)$$

$$\dot{X}_9(t) = \omega_1(t)\cdot X_1(t)-\delta\cdot X_9(t) \quad (9)$$

$$\dot{X}_{10}(t) = \omega_1(t)\cdot X_3(t)+X_4(t)+X_5(t)+X_6(t) + X_8(t))-\delta\cdot X_{10}(t) \quad (10)$$

$$\dot{X}_{11}(t) = \omega_4(t)\cdot X_2(t)+\omega_4(t)\cdot X_7(t)-\delta\cdot X_{11}(t) \quad (11)$$

where

P_A = probability of human error that a bad standby system is undetected upon inspection;

P_B = probability of human error that a good standby system is inadvertently left in a bad state or the standby system fails during inspection;

P_C = probability of human error that a good standby system is judged as bad;

τ_1 = inspection time (i.e. time required for accomplishing the inspection activities);

τ = time between inspections;

δ = reciprocal of the average time that is required to notice that the system is not working as per specification;

$\omega_1(t)(\omega_4(t))$ = failure rate of the basic (standby) system when it is still working but not as per specification;

$\omega_2(t)(\omega_5(t))$ = failure rate of the basic (standby) system (complete break down of the basic (standby) system);

$\omega_3(t) = \omega_1(t) + \omega_2(t)$;

$\omega_6(t) = \omega_4(t) + \omega_5(t)$;

$\omega_H(t)$ = failure rate of the standby system in cold state (usually much smaller than $\omega_5(t)$)

α_1 = reciprocal of the average time to detect the cause of failure;

α_2 = reciprocal of the average time to diagnose failure during inspection time;

μ_1 = average repair time required for repairing the basic system;

μ_2 = average repair time required for repairing the standby system.

ε_1 = Reliab. of successful switching of standby;

ε_2 = reliability of the switch responsible for switching off the basic system and connecting the standby system;

ε_3 = reliability of the switch responsible for switching off the standby system;

While developing the differential equations of the model the following initial conditions are used:

$X_1(0) = 1$

and $X_j(0) = 0$ where $j = 2,3,\ldots 12$

Reliability of the system can be represented with the help of following equation:

$$R(t) = \sum_{i=1}^{8} X_i(t), \quad (12)$$

The probability that the system functions not as per specification:

$$S(t) = X_9(t) + X_{10}(t) + X_{11}(t), \quad (13)$$

The probability that the system is completely down:

$$X_{12}(t) = 1- \sum_{i=1}^{11} X_i(t), \quad (14)$$

OPTIMISATION

For the purpose of determination of the optimum inspection interval of a standby system a complex state variable is developed which is a functional of economic factors, market conditions and reliability of the system. It is expressed as follows:

$$K(t_m) = \int_0^{t_m} \dot{K}(t)dt , \quad (15)$$

where

$$\dot{K}(t) = \frac{dK(t)}{dt} = k_p(t)\cdot X_9(t)+X_{10}(t)+X_{11}(t)) + $$
$$+k_{d1}(t)\cdot X_2+k_{d2}(t)\cdot X_4(t)+k_{r1}(t)\cdot X_7(t) +$$
$$+k_{r2}(t)\cdot X_8(t)+k_i(t)\cdot(X_3(t)+X_5(t)) -$$
$$-k_F(t)\cdot R(t) \quad (16)$$

where

$R(t)$ = reliability of the system;

$k_p(t) = dK_p(t)/dt$; $K_p(t)$ = proportional cost;

$a'(t) = dA'(t)/dt$; $A'(t)$ = revenue;

$k_F(t) = \frac{dF(t)}{dt}$; $F(t) = A'(t)-K_p(t)$ = residual revenue;

$k_{d1}(t)(k_{d2}(t))$ = cost of detection of failure of the basic (standby) system;

$k_{r1}(t)(k_{r2}(t))$ = cost of repair of the basic (standby) system;

$k_i(t)$ = cost of inspection;

The upper limit of the integration of Eq. (15) can be estimated with the help of a formula developed in a simple four states model in Szász (1977). The approximate value of t_m can be expressed as follows:

$$t_m \approx \frac{\lambda^2(1+\varepsilon)+\lambda(1+\varepsilon)\cdot(\alpha_1+\mu)+\alpha_1\cdot\mu}{\lambda^3+\lambda^2(\alpha_1+\mu)+\lambda\cdot\alpha_1\cdot\mu(1-\varepsilon)} \quad (17)$$

where λ = reciprocal of the mean time between failures of the basic system;

$\varepsilon = (\varepsilon_1+\varepsilon_2+\varepsilon_3)/3$

$\mu = (\mu_1+\mu_2)/2$

The optimum inspection interval has to minimize $K(t_m)$ of Eq. (15). A computer programme has been developed to compute $K(t_m)$ for any value of τ and to find the value $\tau = \tau^*$ which minimizes $K(t_m)$ of Eq. (15).

CONCLUSION

In order to determine the optimum inspection interval of a standby system, a model has been worked out, where three types of human errors are considered. The possibility that a bad standby system is judged as good (type A human error), the possibility that a good standby system is inadvertently left in a bad state after the inspection or the standby system fails during inspection (type B error) and the possibility that a good standby system is judged as bad (type C human error), are considered. In the attempt to determine the optimum inspection interval of a standby system the literature either ignore the residual revenue or discuss it with marginal importance. The residual revenue is to be deducted from the cost suffered due to inspection and repair activities. The resultant value should be minimized in order to ascertain the optimum inspection interval of a standby system.

With the help of the model discussed in this paper human error effects can be explicitly accounted for in ascertaining optimum inspection interval of a standby system. For the purpose of determining the optimum inspection interval of a standby system a complex state variable has been developed which is a functional derived from the reliability of the system, cost of operating it and the market condition.

REFERENCES

Apostolakis, G.E., and P.P. Bansal (1977): Effect of human error on the availability of periodically inspected redundant systems, IEEE Trans. Reliab., Vol. R-26, No. 3, pp. 220-225.

Chay, S.C., and M. Mazumdar (1975): Determination of test intervals in certain repairable standby protective systems, IEEE Trans. Reliab., Vol. R-24, August, pp. 201-205.

McWilliams, T.C., and H.F. Martz (1980): Human error considerations in determining the optimum test interval for periodically inspected standby systems, IEEE Trans. Reliab., Vol. R-29, No.4, October, pp. 305-310.

Sherwin, D.J. (1979): Inspection intervals for condition-maintained items which fail in an obvious manner, IEEE Transactions Reliab., Vol. R-28, No. 1, pp. 85-89.

Szász, G. (1977): Rendszerek megbizhatóségi függvényének meghatározása elektronikus számitógéppel, Finommechanika-Mikrotechnika, 16. évf., pp. 140-142.

ZUSAMMENFASSUNG

Die Zuverlässigkeit eines Systems kann dadurch erhöht werden, indem man eine Reserveeinheit einfügt. Diese Einheit spielt eine grosse Rolle wenn der Ausfall einer Betriebseinheit teuer und/oder gefährlich ist. Ein typisches Beispiel dafür ist die Anwendung eines Notgenerators in Krankenhäusern. Die Reserveeinheiten werden aktiviert wenn Notzustände eintreten und ihre Funktion ist, die Folgen dieser Kritischen Zustände zu verhindern. Diese Systeme werden periodisch getestet, um ihre Gebrauchsfähigkeit im Notfall zu sichern. Während der Inspektion können auch die einzelnen Komponenten getestet werden. Die Möglichkeit von menschlichen Fehlern kann natürlich beim Testen und Betrieb dieser Systeme nicht ausgeschlossen werden.

Zur Festlegung der optimalen Inspektionsintervalle wurde ein Modell entwickelt, wobei drei Typen von menschlichen Fehlern berücksichtigt wurden. Zum Ziel der Festlegung der optimalen Inspektionsintervalle eines Notsystems wurde eine komplexe zustandsbedingte Variable entwickelt, welche eine Funktionale ist, die von der Zuverlässigkeit des Systems, den Betriebskosten und Marktbedingungen abgeleitet wurde.

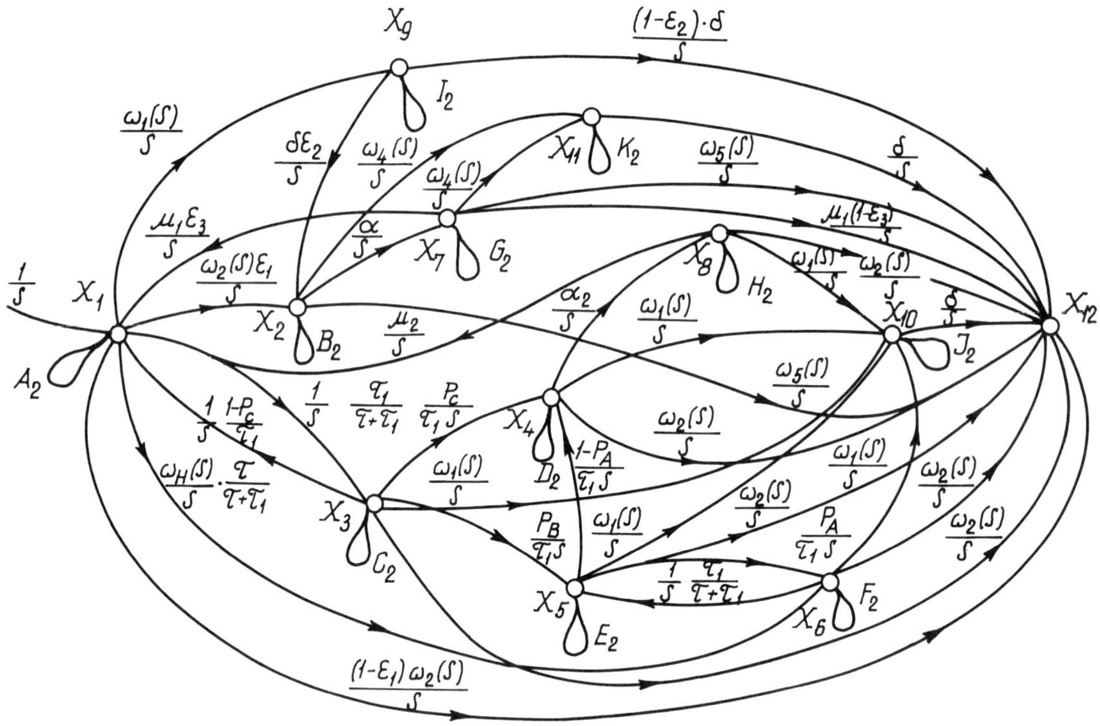

Fig. 1. Transition diagram of the model

Control Mechanism for a High Precision Straight Motion System

H. AOYAMA, I. WATANABE AND A. SHIMOKOHBE

Research Laboratory of Precision Machinery and Electronics, Tokyo Institute of Technology, 4259 Nagatsuta, Midori-ku, Yokohama 227, Japan

ABSTRACT

This paper introduces a control mechanism for correcting straight motion errors of an air slide table. In this mechanism, position and vibration of the slide table are measured and feedbacked to piezoelectric translators which control air pads supporting the table. As an example, position of a steel cubic block floated by air film is controlled by using the mechanism. In the experiment, positioning accuracy of 0.01μm is obtained.

Keywords: straight motion, air slide, precision positioning, vibration reduction, air pad, piezoelectric translator.

INTRODUCTION

In a diamond turning machine or a coordinate measuring machine, accuracy of machining or measuring depends on how slide tables of XYZ axes deviate from an ideal straight motion. Even if an air slide is used, motion errors arise because of profile errors of bases as well as change in air gap by load disturbance. Moreover small vibration occurs because of air flow fluctuation. In order to realize a high precision straight motion, it is necessary not only to correct such motion errors which correspond to each degree of freedom, i.e. vertical and horizontal position errors, pitch, roll and yaw angle errors but also to eliminate such small vibration.

The purpose of this study is to develop a control mechanism which is capable of correcting straight motion errors of an air slide and removing its small vibration. The mechanism consists of air pads, a piezoelectric translator and a capacitance micrometer. As an example, a steel cubic block is supported by five air pads and one of them is driven by the piezoelectric translator. Displacement and small vibration of the block is measured by the capacitance micrometer and feedback control system is designed so as to achieve both precison positioning and vibration reduction.

CONTROL MECHANISM

SYSTEM PRINCIPLE

Figure 1-(a) illustrates an air slide system schematically. Generally the slide table is supported by the pressurized air film and can slide without mechanical contact. Approximately the air film may be considered as a spring-damper system. Thus the thickness of the air film tends to change by external force and the table vibrates by air fluctuation. For correction of the thickness change of the air film and reduction of vibration without contacting the table, control mechanism shown Fig.1-(b) is suggested. The slide table is supported by the air film of the air pad which can be moved by a piezoelectric translator. Change of the thickness of the air film and vibration of the block are measured by a sensor and the translator is controlled so as to correct the thickness change and to eliminate the vibration.

MECHANICAL CONFIGURATION

Figure 2-(a) and (b) illustrates an equivalent system to the control mechanism shown in Fig.1-(b). A steel cubic block(135 mm each side and 7 Kg mass) is supported by the air film of five circular air pads which surround the block. Regulated air of 0.8MPa is supplied to the air pads. An air pad under the block is put on a piezoelectric translator. The translator provides expansion of 20μm when a voltage of 1000V is supplied. The bandwidth(-3dB) of the translator is about 220Hz. The supporting mechanism composed by the air pads and the combination of the air pad and the piezoelectric translator provide an actuation system free from mechanical friction.

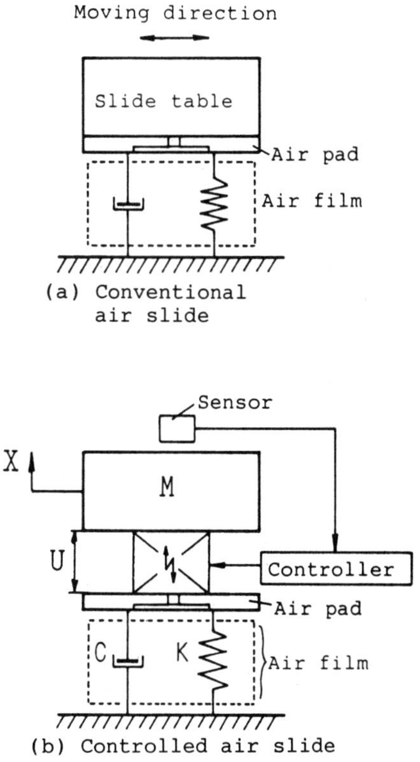

Fig.1 Schematics of air slide

Displacement of the block is measured by a capacitance micrometer(A.D.E. Microsense 3026A) set above the block. These are installed on the surface plate which is supported by pneumatic vibration isolators.

STEADY-STATE CHARACTERISTICS

Small amplitude vibration of the block is shown in Fig.3. It is considered that this block is beaten by pressurized air flowing through the gap between the block and air pad. It should be noted that this vibration limits motion accuracy of the block. Thus for such system, a method of reducing such vibration must be considered.

DYNAMIC CHARACTERISTICS

In schematic diagram of the system shown in Fig.2-(b), the relation between displacement of the block(x), that of piezoelectric translator(u) and disturbance(d) can be written as follows:

$$M\ddot{x}+C\dot{x}+Kx=C\dot{u}+Ku+d \qquad (1)$$

where M is mass of the block, C is damping coefficient, K is stiffness of air film.

The transfer function from U to X is described as follows:

$$\frac{X(s)}{U(s)}=\frac{Cs+K}{Ms^2+Cs+K} \qquad (2)$$

In addition, performance of the piezoelectric translator and the amplifier for driving it is assumed:

$$\frac{U(s)}{E(s)}=\frac{K_p}{1+Ts} \qquad (3)$$

where E is input voltage to the amplifier.

Thus the open loop transfer function W(s) of the system is given and represented by F(s) and G(s) in Fig.5.

$$W(s)=\frac{X(s)}{E(s)}=\frac{K_p}{1+Ts}\cdot\frac{Cs+K}{Ms^2+Cs+K}\cdot K_s$$

$$=F(s)\cdot G(s) \qquad -(4)$$

$$F(s)=\frac{K_p(Cs+K)}{1+Ts}, \quad G(s)=\frac{K_s}{Ms^2+Cs+K}$$

where performance of the sensor is assumed as a proportional element(Ks).

Measured and calculated dynamic responses are shown in Fig.4. The responses display a prominent peak at 112Hz. This is the resonance of the mechanical system consisting of the block mass and the stiffness of air film.

CONTROL SYSTEM DESIGN

Block diagram of the control system is shown in Fig.5. To improve the controller performance, a cascade element F1 and a feedback element F2 are considered. F1 and F2 involve proportional(P), integral(I) and derivative(D) elememts.
Namely F1 and F2 are represented,

$$F_1(s)=\alpha_1 s+\beta_1+\gamma_1/s \qquad (5)$$

$$F_2(s)=\alpha_2 s+\beta_2 \qquad (6)$$

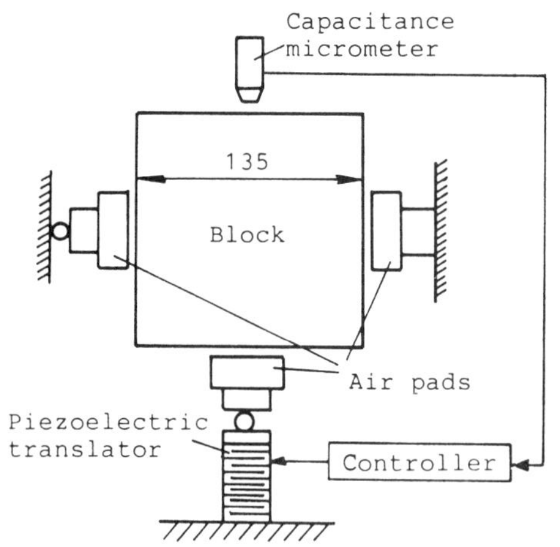

(a) Example system

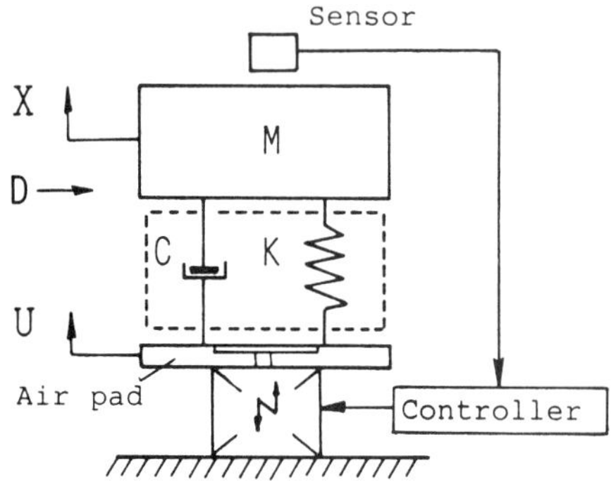

(b) Schematic diagram

Fig.2 Control mechanism

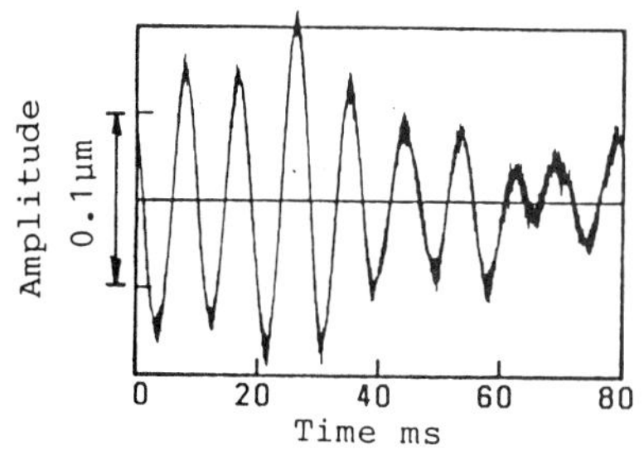

Fig.3 Vibration

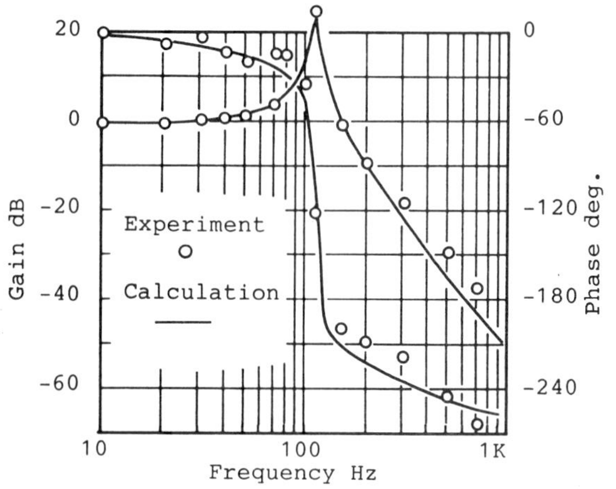

M=7 Kg, K=3.46×10^6 N/m, C=2.96×10^2 Ns/m, Kp=2.2 μm/v, Ks=1.9 v/μm.

Fig.4 Dynamic characteristics of system

From the block diagram, the transfer functions from reference input Xr(s) to controlled variable X(s) and from disturbance input D(s) to X(s) are given,

$$\frac{X(s)}{Xr(s)}=\frac{F_1(s)\cdot F(s)\cdot G(s)}{1+F(s)\cdot G(s)\cdot (F_1(s)+F_2(s))} \quad (7)$$

$$\frac{X(s)}{D(s)}=\frac{G(s)}{1+F(s)\cdot G(s)\cdot (F_1(s)+F_2(s))} \quad (8)$$

Eqs.(7) and (8) mean that the sum of F1+F2 can be designed so that response to disturbance become minimum and then F1 can be determined so that response to reference become desirable.

EXPERIMENT

DISTURBANCE

In the experiment, three types of controller are considered.

(1)PID type controller.

$$(F_1=\alpha_1 s+\beta_1+\gamma_1/s,\ F_2=0)$$

(2)I-PD type controller.

$$(F_1=\gamma_1/s,\ F_2=\alpha_2 s+\beta_2)$$

(3)PID-D type controller.

$$(F_1=\alpha_1 s+\beta_1+\gamma_1/s,\ F_2=\alpha_2 s)$$

(In actual system, incomplete D element $\frac{\alpha_1 s}{1+T_d s}$ is used, where $T_d=8\times10^{-5}$s)

Parameters of each controller are determined so as to make the effect of disturbance as small as possible and the parameters is shown in Table 1. Figure.6 shows vibration of the block in the steady-state under control. The vibration shown in Fig.2 has been reduced to the value close to the noise level of the sensor(Fig.6-a). This indicates that the actuation system has positioning resolution of about 0.01μm. Figure 7 shows results of FFT. Vibrations are reduced very well after control. Remained prominent peaks at 50, 100 and 150Hz come from noises of the sensor.

Controller	Parameters	
PID	$\alpha_1=9.3\times10^{-4}$ $\beta_1=0.38$ $\gamma_1=187.2$	$\alpha_2=0$ $\beta_2=0$
I-PD	$\alpha_1=0$ $\beta_1=0$ $\gamma_1=187.2$	$\alpha_2=9.3\times10^{-4}$ $\beta_2=0.38$
PID-D	$\alpha_1=5.9\times10^{-4}$ $\beta_1=0.38$ $\gamma_1=187.2$	$\alpha_2=3.4\times10^{-4}$ $\beta_2=0$

Table. 1

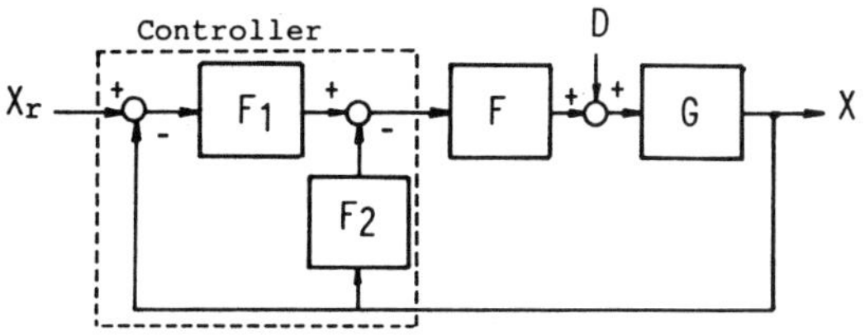

Fig.5 Block diagram of system

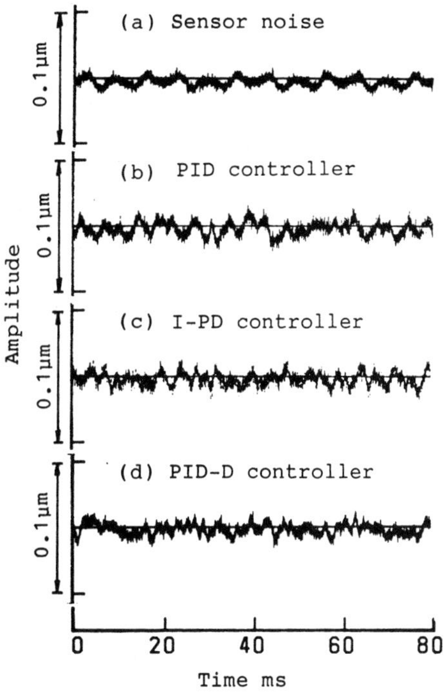

Fig.6 Vibration under control

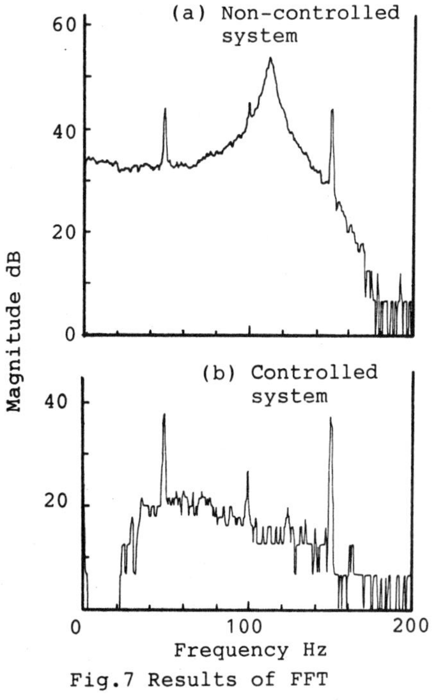

Fig.7 Results of FFT

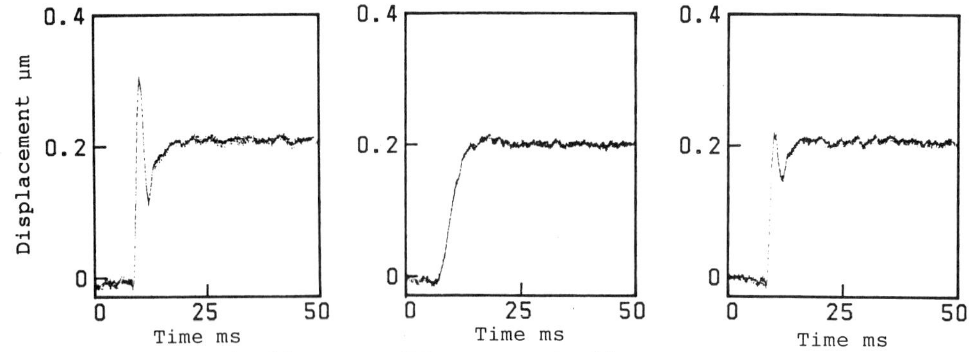

(a) PID controller (b) I-PD controller (c) PID-D controller

Fig.8 Indicial responses

REFERENCE

Indicial responses for reference input to the system are shown in Fig.8-a, b and c. Different responses are obtained with PID, I-PD and PID-D controllers, though the performances for disturbance are similar. The response of the system with PID controller has too large overshoot and that of the system with I-PD is slow.
The optimum performance is achieved by using PID-D controller. Figure.9 shows frequency characteristics of the system using PID, I-PD and PID-D controllers. Bandwidth of about 500 Hz is obtained when using PID-D controller.

CONCLUSIONS

An example system of control mechanism which is developed for correction of straight motion errors of air slide is introduced. This paper is summarized as follows:

1) The control mechanism consists of the combination of air pads and piezoelectric translator and can function as a non-contact actuation system.
2) In this control mechanism, PID-D controller is more useful for improving both performances for reference input and disturbance.
3) The experiments show that vibration by disturbance of the system is reduced to about 0.01μm and dynamic response of about 500 Hz in bandwidth is obtained by using PID-D control system.

This mechanism is to be applied to an ultra high precision straight motion system, which has motion accuracy of 0.01μm and 0.01 arc sec over 1000mm travel.

ZUSAMMENFASSUNG

Eine Kontrollvorrichtung ist untersucht für die Korrektion der Fehler des gerade Bewegung in einer Luftführungslager. In der vorrichtung, Position und Vibration der lager sind gemessen und rückkoppelt zum piezoelektrishen Übersetzer um die Luftlagen zu treiben. Zum Beispiel, Position der kubiche Block, aus die Stahl gemacht, ist kontrollt und die erreichte Genauigkeit ist 0.01μm.

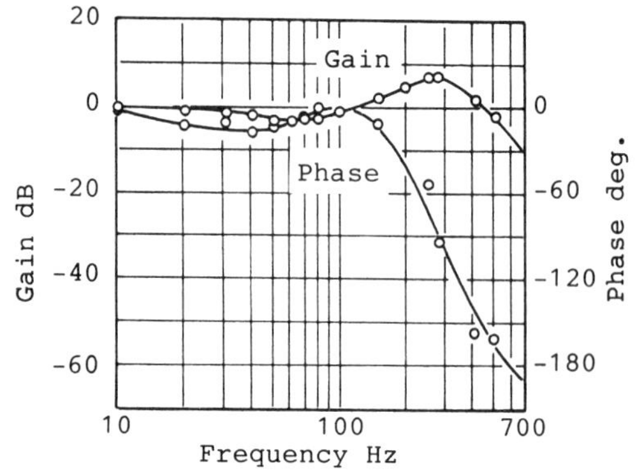

(a) PID controller

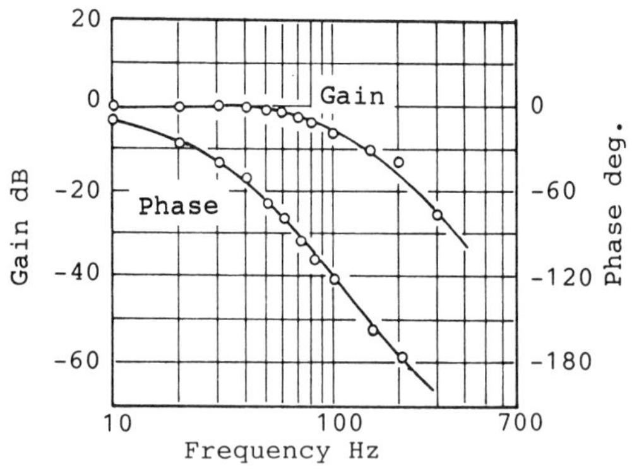

(b) I-PD controller

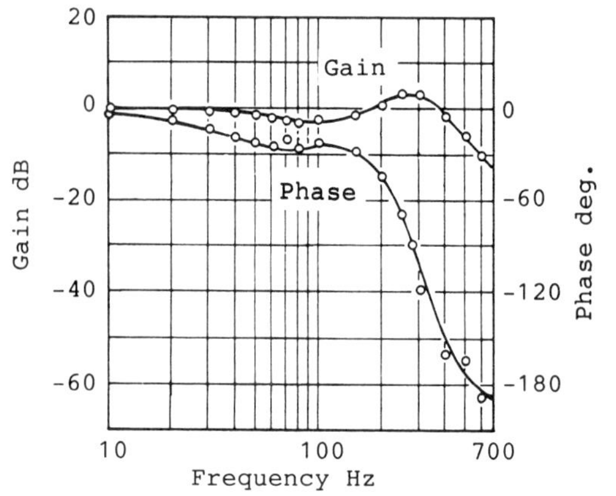

(c) PID-D controller

Fig.9 Frequency characteristics

Commensurate Positioning of a Stepmotor Actuated Stewart Platform

G. P. RATHBUN AND G. R. DUNLOP

Mechanical Engineering Department, University of Canterbury, Christchurch, New Zealand

Abstract. This paper describes the use of a Stewart Platform for six axis machining. The Stewart Platform has considerable potential as a rigid cheaply constructed NC milling machine. The development involved path synthesis in six dimensions, multiaxis stepper motor control, software engineering, and operational testing of the complete machine. Verification of both hardware and software design was obtained by machining several test shapes from urethane foam. However, the platform's movement envelope was found to be comparatively small. Optimization of the geometry to increase the rigidity and the range of movement is necessary if the Platform is to be used commercially.

Keywords. Attitude control; direct digital control; machine tools; microprocessors; multivariable control systems; robots; stepping motors.

INTRODUCTION

The Stewart Platform shows much promise as a rigid, cheaply constructed milling machine with six degrees of motion freedom. However only five are required for full capability machining. A Stewart Platform (c.f. Fig. 1), was originally constructed for research into the manipulation of objects during automated assembly (McCallion and Truong, 1979) and has been adapted for milling operations.

Each of the platform's six legs comprises a lead screw and nut. The nut is rotated by a stepper motor via a Hooke's joint located on the base. Attached to each motor is an open loop "Actuator Control Unit" (ACU). Each ACU has two inputs: a pulsed signal commanding the motor to move one step in the direction given by the logic level of the direction signal. A Z80 based microcomputer controls the machine via a specially constructed "Machine Control Unit" (MCU) which produces the signals required by the six ACUs.

The functions required to operate the platform as a milling machine were:
a) A path synthesis procedure such that the motion of on object attached to the platform is specified using simple geometric commands.
b) Electronic hardware in the MCU that enables the computer to simultaneously control six stepper motors.
c) A multiple pulse stream control algorithm for correct control of the MCU by the computer.
d) Software to create a data manipulation, storage and communication structure for the computer to use in transforming the input geometric data into platform movements.

Each of these requirements is discussed in the following sections.

PATH SYNTHESIS

The movement of the platform in six space (denoted by SS) is analyzed in this section. Six space is the geometric space in which a body can move relative to a fixed coordinate system. A point C within this space can be described by:

$$C = (U,\ V,\ W,\ \phi,\ \theta,\ \psi) \tag{1}$$

Where U, V, W, describe the spatial position of the point with respect to the fixed coordinate system, and ϕ, θ, ψ describe the Euler rotations of the point with respect to the fixed coordinate system. The orientation of the point is specified by the orientation of a movable coordinate system (the platform) to which the point is attached by a fixed vector $\underline{p}$. (Vectors in the fixed and movable coordinate systems are denoted by upper and lower-case characters respectively.) Thus the movement of a fixed point $\underline{p}$ on the platform is now considered where $\underline{p}$ represents a point on a workpiece or the cutting point of a tool attached to the platform. A typical movement of the platform is described by: "Move the point $\underline{p}$ on the platform from a starting position of C_s to a finishing position of C_f". This is shown in Fig. 2 as $\underline{p}$ moves on a straight line from C_s at the upper left to C_f at the lower right. Note that the orientation of $\underline{p}$ in the finishing position is different from that in its starting position. Therefore, not only the U, V, W, but also the ϕ, θ, ψ terms in C_s and C_f must differ during the movement.

The position of the platform is fully specified by the vector $\underline{T}$ and the matrix [R], giving the position and orientation of the movable coordinate system i.e. the platform. Thus for a point $\underline{p}$ on the platform positioned at C in SS, the values of $\underline{T}$ and [R] must be derived so that the platform may be correctly located.

The analysis starts by assuming the fixed and movable coordinate systems are originally coincident. The new orientation of $\underline{p}$ is obtained by operating on the movable coordinate system by [R]. Thus [R]$\underline{p}$ gives the orientation of $\underline{p}$ in the fixed coordinate system. [R] is the Rodriguez matrix, a 3x3 matrix derived from the three Euler angles of C. Then $\underline{p}$ is translated to position it at $\underline{P}$ which is given by the three spatial coordinates of C. To obtain $\underline{T}$, it is apparent from Fig. 2 that

$$\underline{P} = \underline{T} + [R]\underline{p} \tag{2}$$

or:

$$\underline{T} = \underline{P} - [R]\underline{p} \tag{3}$$

Point to point control of the platform between any two points in SS can thus be achieved by specifying the starting and finishing values of $\underline{T}$ and [R].

Path Control

The actual motion of the platform from C_s to C_f is controlled by varying the six leg lengths in a six dimensional space called Leg Space (LS). Point to point motion in SS is accomplished by a one to one mapping of C_s and C_f points into LS and the control system provides a movement in LS from the starting to the finishing position.

A linear movement in LS will result in an curved path in SS due to a nonlinear mapping function. The deviation from the correct path in SS can be reduced by introducing equally spaced way points or "Path Nodes" along the path in SS, through which the movement in LS is forced to pass. Movement between Path Nodes is along Path Vectors. The separation of the Path Nodes in LS is a function of the SS to LS transformation, and the required path accuracy. The true separation function for the Stewart Platform is extremely complex and was not derived. Instead, a simple geometric approximation based on the movement error of one leg is used.

The Path Nodes in SS between C_s and C_f are generated using "basic path control". This specifies that the path of a controlled point $\underline{p}$ be via a straight line if translations occur, that any orientation changes occur about an axis of rotation maintaining a constant direction, and that velocities be constant, i.e. linear interpolation in SS.

Assuming N Path Vectors are required to span the path from C_s to C_f, then N-1 intermediate Path Nodes C_k must be generated, each with an associated data pair $(\underline{T}_k,[R]_k)$ as derived below.

Positional interpolation: For $\underline{P}$ to be on a straight line between $\underline{P}_s$ to $\underline{P}_f$ we require that

$$\underline{P}_k = \underline{P}_s + k/N(\underline{P}_f - \underline{P}_s) \qquad (4)$$

Rotational Interpolation: With the platform at the origin the initial and final orientations of $\underline{P}$ are

$$\underline{P}_s = [R]_s\underline{p} \qquad (5)$$

and

$$\underline{P}_f = [R]_f\underline{p} \qquad (6)$$

The k'th orientation of $\underline{p}$ is given by:

$$\underline{P}_k = [R]_k\underline{p} \qquad (7)$$

To obtain $[R]_k$, first derive an operator $[R]_{sf}$ that operates on $[R]_s$ to produce $[R]_f$. Once $[R]_{sf}$ is obtained, it is manipulated to create an operator $[R]_{sf}{}^{k/N}$ that allows a variable degree of rotation to be specified about a fixed axis of rotation. To obtain $[R]_{sf}$ first rearrange (5):

$$\underline{p} = [R]_s{}^T[R]_s\underline{p} \qquad (8)$$

since $[R]^{-1} = [R]^T$. Combining (6) with (8) yields:

$$\underline{P}_f = [R]_f[R]_s{}^T[R]_s\underline{p} \qquad (9)$$

or

$$\underline{P}_f = [R]_{sf}[R]_s\underline{p} \qquad (10)$$

where

$$[R]_{sf} = [R]_f[R]_s{}^T \qquad (11)$$

$[R]_{sf}$ maps $\underline{P}_s$ onto $\underline{P}_f$ through Euler rotations ϕ,θ and ψ as shown in method a) in Fig. 3. These parameters do not prescribe the desired mapping path. However, the [R] matrix can be reformulated into the equivalent Euler matrix (c.f. Thompson, 1969) whose parameters specify the mapping to be a rotation of α radians about a line given by the direction cosines (l,m,n) as in method b) of Fig. 3. This specification conforms with the requirements of basic path control as the rotation can be applied in N steps of α/N. Therefore

$$[R]_{sf}{}^{k/N} = \text{FUNCTION}(l,m,n,k\alpha/N) \qquad (12)$$

and

$$[R]_k = [R]_{sf}{}^{k/N}[R]_s \qquad (13)$$

Once $\underline{P}_k$ and $[R]_k$ are defined, the position of the platform origin $\underline{T}_k$ can be obtained from (3):

$$\underline{T}_k = \underline{P}_s + k/N(\underline{P}_f - \underline{P}_s) - [R]_k\underline{p} \qquad (14)$$

Each successive data pair $(\underline{T}_k,[R]_k)$, is mapped, via a transformation algorithm onto a corresponding point $\underline{L}_k$ in LS and thence into a corresponding six element Actuator State vector $\underline{AS}_k$ which has units of motor steps i.e.

$$(\underline{p},\ C_k) \dashrightarrow (\underline{T}_k,\ [R]_k) \dashrightarrow \underline{L}_k \dashrightarrow \underline{AS}_k \qquad (15)$$

THE MACHINE CONTROL UNIT

The MCU contains all the special electronic hardware necessary to interface the computer to the machine. The MCU operates on data contained in a "Machine Control Data Block". The MCDB is sent from the computer and is converted into the six concurrent step signals and the six direction signals required by the ACUs. Each MCDB produces an elemental movement of the platform called a System Vector. Several System Vectors span one Path Vector between two adjacent Path Nodes. The method of operation of the MCU is shown in Fig. 4.

The MCU produces six pulse streams that can be accurately and independently controlled as regards their pulse frequency (Dunlop, 1986). This is achieved by passing a common frequency through six digital frequency dividers called Relative Speed Frequency Divider Groups (RSFDGs). Each $RSFDG_i$ sends motor i step pulses and is loaded with a divisor called the Relative Speed Divisor (RSD_i) from the MCDB. The common input frequency to all RSFDGs is generated by a high speed oscillator (frequency F) and a frequency divider called the Overall Speed Frequency Divider. This divider is loaded with a Speed Divisor (SDIV) from the MCDB. Thus each RSD_i provides individual control of each motor pulse frequency V_i, and SDIV provides an overall control as given by:

$$M_i/t = V_i = F/(RSD_i.SDIV) \qquad (16)$$

where M_i and t specify the required step movement and time allowed for motor i.

Movement along a System Vector is monitored to detect the end of the vector so that a new System Vector can be started. The pulses output to each motor are counted by Counter Comparator Groups (CCGs) which are initialized to zero at the start of every System Vector. When the CCG count equals a preprogrammed TARGET value corresponding to the end of the System Vector, a pulse is sent to the computer to initiate the start of a new System Vector. The direction logic levels are set by the MCU and are constant for each System Vector.

The MCU configuration is variable via software commands from the computer. The particular configuration used is now described: Position monitoring utilizes an incremental approach. At the end of each System Vector the six CCGs contain the number of steps moved by each motor over that vector. The computer reads this data and updates the actuator positions. All the pulse manipulation groups are then rapidly reprogrammed for the next System Vector so that movement is continuous. Rapid is less than the time for a motor to rotate 5% of one step. Rapid reprogramming occurs when the relevant data is prestored in registers on the MCU and is activated by a few commands from the computer.

The pulse streams due to any MCDB are created so that at the end of the System Vector, the last pulses are coincident. This ensures continuity of movement across System Vector Nodes. The coincidence requirement is found to limit the speed of movement, but this requirement is relaxed so as to increase platform speed. The coincidence requirements and the tradeoffs are discussed in the next section.

MULTIPLE PULSE STREAM GENERATION

The path synthesis algorithms produce successive Actuator State vectors $\underline{AS}$ and a corresponding time t to achieve each state. The difference between each successive $\underline{AS}$ pair is a movement vector $\underline{M}$. The control system generates 6 pulse streams and associated direction signals so as to achieve the movement $\underline{M}$ in the time t. As t is constant for all motors during $\underline{M}$ then equation (16) implies that the RSD values are infinitely variable i.e. all final pulses arrive exactly t seconds after the pulse trains were started. However, the RSDs and the SDIV are finite integer values so round off errors lead to unequal values of t for each motor, particularly when small RSD values are used. Two approaches are used to overcome this problem:

The Common Factor theory (CF): Each RSD_i is equated to a product of the required step values as follows:

$$RSD_i = M_1.M_2.M_3.M_4.M_5.M_6./M_i \qquad (17)$$

Incorporation of common factors allows RSDs to be calculated without truncation. Thus all the last pulses arrive simultaneously. However, the above equation shows that for even small step values Mi, the RSD_i will be very large and require a large number of bits to represent them. A practical upper limit of F=5MHz divided by large RSD numbers yields low step rates. The step rate is increased by subdividing the movement vector $\underline{M}$ so that the RSDi are smaller. By way of illustration consider all motors to moving N steps at speed V steps/sec. The equation relating subdivision size and the speed is as follows:

$$N = (F/V{*}SDIV)^{1/5} \qquad (18)$$

The number of MCDBs are minimized by maximizing N. Hence the data storage required is reduced.

Controlled Pulse Arrival Error theory (CPAE): It is apparent the CF theory imposes a severe restraint on the size of the System Vector. However, a small variation in the arrival time of the last pulse generated by an MCDB is acceptable if the variation is less than some maximum (usually less than 5% of one pulse time so that stepper motor synchronization is maintained). The maximum round off error for determining Mi from (16) is less than unity. This yields the following relation which can be compared to the equation (18) for the CF theory.

$$N = k_i.F/(V_i.SDIV) \qquad (19)$$

where N is the maximum number of steps along a subdivision of M_i and k_i denotes the pulse arrival error for motor i (taken as 5% of one step). Graphical evaluation of (17) and (18) is shown in Fig. 5. The graphs indicate that the CF theory is preferred for only a few actuaors, but the CPAE system is superior for the six actuator system used. Thus the CPAE theory is used to calculate the required RSD values given $\underline{M}$ and t. However, given the common input frequency F of 5 MHz, the CPAE theory still cannot generate enough steps per System Vector to span a typical Path Vector. Thus it is necessary to linearly divide each Path Vector into a number of smaller equal vectors called System Vectors which can be generated by one MCDB.

THE SOFTWARE

The software environment is shown in Fig. 6. The steps taken from entering the movement commands to executing the final milling operation can be traced: First the source file, containing geometric and path control commands is created and stored using an editor. The COMPILE program then converts the source file into a run file comprised of MCDBs. This conversion uses the equations developed in the preceding sections. Once the RUN file is created, it is used by the CCS program to provide controlled motion of the platform.

DISCUSSION

The efficacy of the system was tested by performing milling operations with the platform and assessing the accuracy of the resulting objects. The main accuracy tests involved milling "flat" surfaces at various inclinations and examining the resulting surfaces to detect any irregularities. An accuracy of +/- 0.2 mm was expected due to several problems: Hooke's joint errors, free play in the 8 joints of each leg, and measurement inaccuracies. Flat surface milling gave a flatness better than 0.2mm thus indicating correct control system function at fixed orientations.

Control over all six degrees of motion freedom of the platform was tested by cutting a conical shape using an inclined and expanding helical path. Movement envelope restrictions limited the path to one turn of the helix. The path was produced as expected, thus indicating the path synthesis algorithms were functioning correctly.

CONCLUSION

The correct operation of all aspects of the path synthesis algorithms and control systems has been tested and verified by experimental milling operations as indicated. The movement envelope of the machine was found to be too restrictive for useful five axis machining. Thus, if the Stewart Platform is to be further developed for milling, its geometry must be optimized to create an extended movement envelope.

REFERENCES

Dunlop, G.R. (1986). Multiaxis Step Motor Control. To be published in Trans IMC.

Fichter, E. F. (1968). A Stewart platform based manipulator: general theory and practical construction. Int. J. Robotics Res. 5 #2 157-182.

McCallion, H. & Truong, P. D. (1979). The analysis of a six degree of freedom work station for mechanized assembly. Fifth World congress IFToMM, Montreal, 1, 611-616.

Thompson, E. H. (1969). An Introduction to the Algebra of Matrices with some Applications. Adam Hilger, London, p150.

POSICION CONMENSURADA DE UNA PLATAFORMA 'STEWART' ACTUADA POR UN MOTOR 'STEP'

La plataforma Stewart es bien conocida por su aplicación en el campo de simuladores de vuelo en aviación, donde provee aceleración angular en tres dimensiones y aceleración lineal en tres dimensiones. Su uso en máquinas herramientas o manipuladores robóticos es menos conocido y es el tema de este artículo. La plataforma Stewart es una superficie plana que se posiciona en el espacio por medio de coordenadas "roll", "pitch" and "yaw". Seis barras de longitud variable conectan la plataforma a una base sólida. Las longitudes de las seis barras de conexión se calculan en función de las coordenadas espaciales por medio de seis ecuaciones no-lineales.

El sistema descrito en este articulo utiliza un microcomputador de uso general para calcular las transformaciones de coordenadas y para actuar seis motores de 'step' que varian las longitudes de las seis barras de conexión. Cuatro circuitos auxiliares VLSI son añadidos al microcomputador para controlar los accionamientos de los seis motores. El resultado es un sistema de seis ejes controlado numéricamente, relativamente simple y económico, basado en un microcomputador comercial de propósito general.

Control de punto a punto se obtiene facilmente en seis ejes por medio de un solo microcomputador. Control del recorrido en seis ejes no se obtiene facilmente por medio de un solo microcomputador. Sin embargo, la adición de cuatro circuitos VLSI auxiliares permite controlar la aceleración conmensurada, la posición y deceleración por un solo microcomputador. La implementación del sistema está descrita en este artículo, junto con un ejemplo específico de aplicación a una máquina fresadora.

El espacio de seis dimensiones constituido por las seis longitudes de barras de conexión se llama "espacio de seis barras" y el espacio formado por las tres posiciones y las tres coordenadas angulares se llama 'espacio de seis posiciones'. La posición conmensurada dentro del 'espacio de seis barras' está relacionada con el recorrido en el 'espacio de seis posiciones' por medio de un sistema de ecuaciones no-lineales. Por tanto, control preciso de recorrido en el 'espacio de seis posiciones' es extremadamente complejo. La complejidad puede ser reducida especificando puntos de ruta en el recorrido de seis dimensiones y 'mapping' cada punto en el espacio de seis barras. Interpolación lineal entre esos puntos de ruta en el espacio de seis barras da posición conmensurada en el espacio de barras. Sin embargo, el recorrido normalmente se desvía con respecto al recorrido deseado en el espacio de posición. Pero el recorrido pasa por los puntos de ruta en los seis espacios. La desviación en el espacio físico se puede reducir reduciendo los intervalos entre puntos de ruta. Varios ejemplos se incluyen en el artículo, junto con algunas formas geométricas obtenidas en la máquina fresadora. Esas formas geométricas ayudaron a mejorar la precisión del sistema y algoritmos.

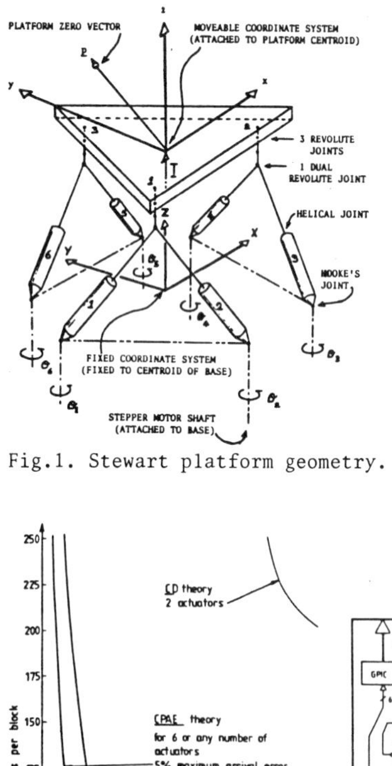

Fig.1. Stewart platform geometry.

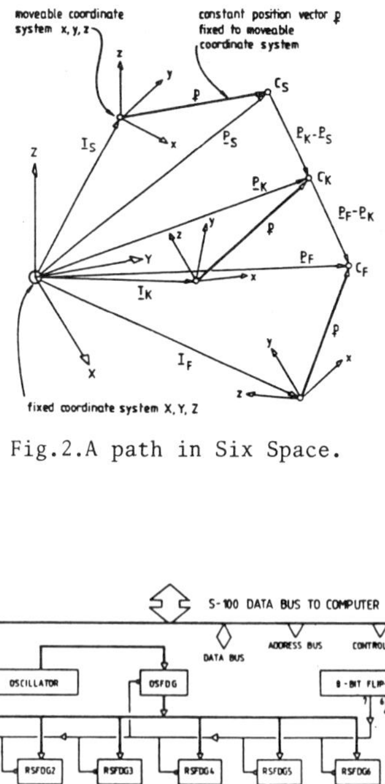

Fig.2.A path in Six Space.

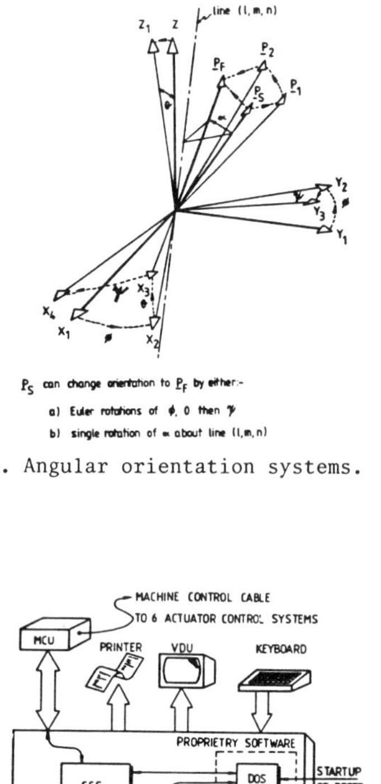

Fig.3. Angular orientation systems.

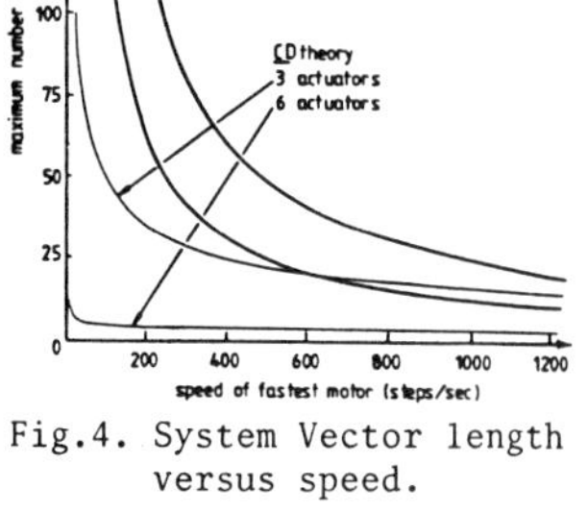

Fig.4. System Vector length versus speed.

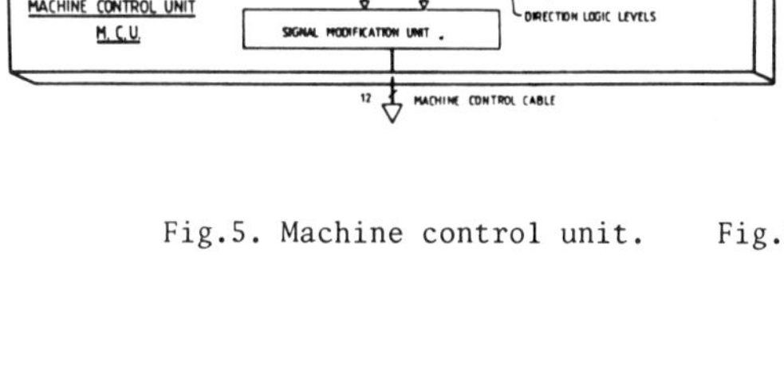

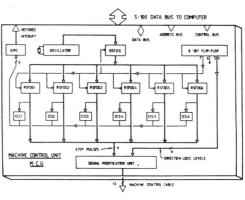

Fig.5. Machine control unit.

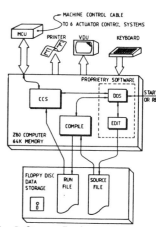

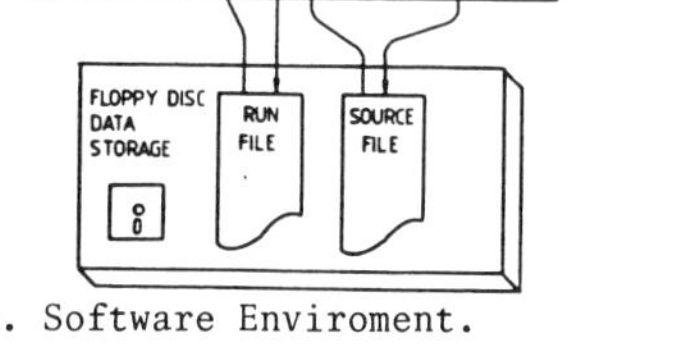

Fig.6. Software Enviroment.

Receptance Synthesis by Means of Block Diagrams

H. HOLKA

Department of Mechanical Engineering, Federal University of Technology, Minna, Nigeria

Abstract. The paper presents the synthesis of the receptance by using the block diagrams. This method could by applied to the active method of vibration control. In the first part of this paper a formulas of connections of two sub-system are derived when their receptance are given. In the second part the same relations are presented but by means of the method of block diagrams. Results of calculations have been demonstrated on the diagrams for specific example.

Keywords. Vibration control; block diagrams; receptance; control system synthesis; discrete systems; feedback; martix algebra.

INTRODUCTION

One of the very useful methods for representing control system are block diagrams. It can be used also when we want to exchange the structure of the system. In both cases the place of the new element or some elements and their interdependence are very much clear in the whole system.
Bishop (1972) investigated complex systems consisting of certain number of continuous and discrete systems connected with one another, assuming that their receptance is known. The receptance method is very comfortable specially when some of the sub-system are given from experiments, because the analytical methods are very complicated and inaccurate.
The author noticed that this method would be excellent for active systems testing if analytical connection could be replaced by graphical methods.

RECEPTANCE SYNTHESIS BY MEANS OF ANALYTICAL METHOD

Two sub-system "B" and "C" as well their receptance " β " and " γ " are given. The forces F_1 and F_2 are acting on the sub-system "B" and also Q_1, Q_2 on sub-system "C", Fig. 1.

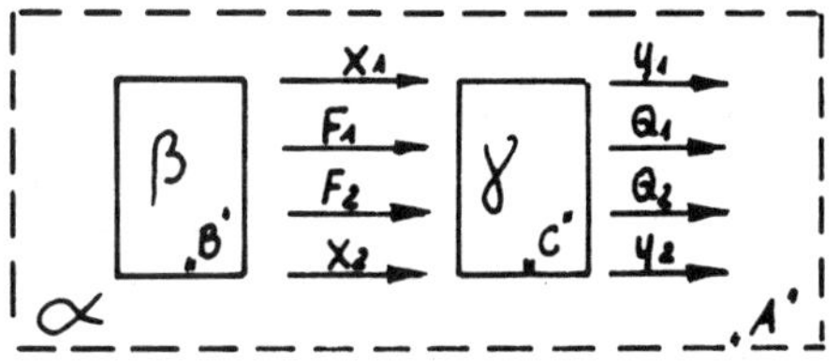

Fig. 1. Connection of two sub-system along two coordinates.

Let us mark $x_{1,2}$ and $y_{1,2}$ as the displacements of the points where the forces $F_{1,2}$ and $Q_{1,2}$ are acting respectively. We can determine the receptance " α " of system "A" after connecting sub-systems by coordinates x_1, x_2. We can write for the sub-system "B".

$$\begin{Bmatrix} X_{1B} \\ X_{2B} \end{Bmatrix} = \begin{bmatrix} \beta_{11} & \beta_{12} \\ \beta_{21} & \beta_{22} \end{bmatrix} \begin{Bmatrix} F_{1B} \\ F_{2B} \end{Bmatrix} \tag{1}$$

where X - matrix of displacements,
β - receptance matrix,
F - matrix of forces.

For the sub-system "C" we have:

$$\begin{Bmatrix} y \\ X_c \end{Bmatrix}_{1,4} = \begin{bmatrix} \gamma_1 & \gamma_2 \\ \gamma_3 & \gamma_4 \end{bmatrix}_{4,4} \begin{Bmatrix} Q \\ F_c \end{Bmatrix} \tag{2}$$

The conditions of connection of the two above mentioned sub-system are

$$\{X_B\} = \{X_c\} = \{X\} \tag{3a}$$

$$\{F_B\} + \{F_c\} = \{F\} \tag{3b}$$

We have to make transformation Eqs. (1),(2) and conditions (3) in order to obtain the following forms

$$\begin{Bmatrix} y \\ X_c \end{Bmatrix} = \begin{bmatrix} \alpha_{11} & \alpha_{12} \\ \alpha_{21} & \alpha_{22} \end{bmatrix} \begin{Bmatrix} Q \\ F \end{Bmatrix} \tag{4}$$

From Eq. (2) we have

$$y = \gamma_1 \cdot Q + \gamma_2 \cdot F_c \tag{5a}$$

$$X_c = \gamma_3 \cdot Q + \gamma_4 \cdot F_c \qquad (5b)$$

Substituting Eq. (3a) into (5b) and next multiplication both side of this Eq. by γ_4^{-1} we get

$$F_c = \gamma_4^{-1} \cdot X - \gamma_4^{-1} \cdot \gamma_3 \cdot Q. \qquad (6)$$

Then we substitute Eq. (6) into (5a) and we get

$$y = (\gamma_1 - \gamma_2 \cdot \gamma_4^{-1} \cdot \gamma_3)\, Q + \gamma_2 \cdot \gamma_4^{-1} \cdot X \qquad (7)$$

After it we multiply both side of Eq. (1) by β^{-1}

$$\beta^{-1} \cdot X_B = F_B. \qquad (8)$$

The sum of Eq. (6) and Eq. (8) gives

$$X = (\gamma_4^{-1} + \beta^{-1})^{-1} \cdot \gamma_4 \cdot \gamma_3 \cdot Q + (\gamma_4^{-1} + \beta^{-1})^{-1} \cdot F \qquad (9)$$

Substituting Eq. (9) into Eq. (7) we get

$$y = \{\gamma_1 - \gamma_2 \gamma_4^{-1} [\gamma_3 - (\gamma_4^{-1} + \beta^{-1}) \gamma_4^{-1} \gamma_3]\} Q + \gamma_2 \gamma_4^{-1} (\gamma_4^{-1} + \beta^{-1}) F \qquad (10)$$

and therefore if we take Eqs. (9), (10), and (4) we can write

$$\alpha_{11} = \gamma_1 - \gamma_2 \cdot \gamma_4^{-1} [\gamma_3 - (\gamma_4^{-1} + \beta^{-1}) \gamma_4^{-1} \cdot \gamma_3], \qquad (11a)$$

$$\alpha_{12} = \gamma_2 \cdot \gamma_4^{-1} (\gamma_4^{-1} + \beta^{-1})^{-1}, \qquad (11b)$$

$$\alpha_{21} = (\gamma_4^{-1} + \beta^{-1})^{-1} \cdot \gamma_4^{-1} \cdot \gamma_3, \qquad (11c)$$

$$\alpha_{22} = (\gamma_4^{-1} + \beta^{-1})^{-1}, \qquad (11d)$$

Where matrix γ have the form:

$$\begin{bmatrix} \gamma_1 & \gamma_2 \\ \gamma_3 & \gamma_4 \end{bmatrix} = \begin{bmatrix} \begin{bmatrix} \gamma_{y_1-y_1} & \gamma_{y_1-y_2} \\ \gamma_{y_2-y_1} & \gamma_{y_2-y_2} \end{bmatrix} & \begin{bmatrix} \gamma_{y_1-x_1} & \gamma_{y_1-x_2} \\ \gamma_{y_2-x_1} & \gamma_{y_2-x_2} \end{bmatrix} \\ \begin{bmatrix} \gamma_{x_1-y_1} & \gamma_{x_1-y_2} \\ \gamma_{x_2-y_1} & \gamma_{x_2-y_2} \end{bmatrix} & \begin{bmatrix} \gamma_{x_1-x_1} & \gamma_{x_1-x_2} \\ \gamma_{x_2-x_1} & \gamma_{x_2-x_2} \end{bmatrix} \end{bmatrix} \qquad (12)$$

Let us calculate for example receptance α_{22} in the points of connection of sub-systems. Matrix α_{22} have the size two by two and can be written as

$$\alpha_{22} = \begin{bmatrix} \alpha_{x_1-x_1} & \alpha_{x_1-x_2} \\ \alpha_{x_2-x_1} & \alpha_{x_2-x_2} \end{bmatrix} \qquad (13)$$

Solving Eq. (11d) we obtain

$$\alpha_{x_1-x_1} = \frac{\beta_{11}(\gamma_{11}\gamma_{22} - \gamma_{12}^2) + \gamma_{11}(\beta_{11}\beta_{22} - \beta_{12}^2)}{(\beta_{11} + \gamma_{11})(\beta_{22} + \gamma_{22}) - (\beta_{12} + \gamma_{12})^2} \qquad (14)$$

$$\alpha_{x_2-x_2} = \frac{\beta_{12}(\gamma_{11}\gamma_{22} - \beta_{12}\gamma_{12}) + \gamma_{12}(\beta_{11}\beta_{22} - \beta_{12}\gamma_{12})}{(\beta_{11} + \gamma_{11})(\beta_{22} + \gamma_{22}) - (\beta_{12} + \gamma_{12})^2} \qquad (15)$$

$$\alpha_{x_2-x_2} = \frac{\beta_{22}(\gamma_{22}\gamma_{11} - \gamma_{12}^2) + \gamma_{22}(\beta_{22}\beta_{11} - \beta_{12}^2)}{(\beta_{11} + \gamma_{11})(\beta_{22} + \gamma_{22}) - (\beta_{12} + \gamma_{12})^2} \qquad (16)$$

In the forms (14), (15) and (16) we assume that $\alpha_{x_1 x_2} = \alpha_{x_2 x_1}$ as well $\beta_{12} = \beta_{21}$ and $\gamma_{12} = \gamma_{21}$.

RECEPTANCE SYNTHESIS BY MEANS OF BLOCK DIAGRAMS

We are looking for receptance connection from Fig. 1. The block diagrams of receptance sub-system "B" and "C" before connection are presented on Fig. 2.

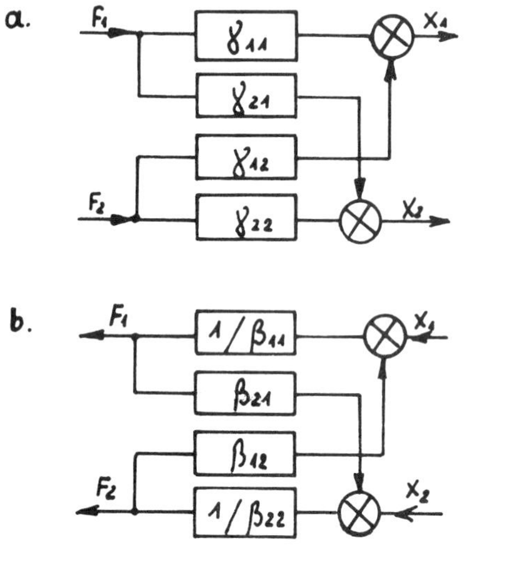

Fig. 2. Block diagrams of receptance of the sub-system "B" and "C" before connection.

As we can see, the block diagram of sub-system "B" is modified in order to preserving condition of connection.
We can get the general scheme of connection from which all interesting us receptance is presented on Fig. 3.

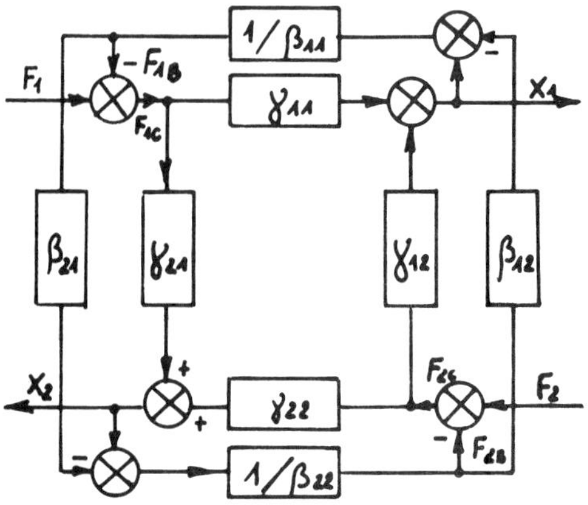

Fig. 3. Construction of block diagram.

For example we would like to find the receptance α_{11}. In the first step we have to transform the block diagram shown in Fig. 3 to the block diagram which is convenient for another transformation, Fig. 4.

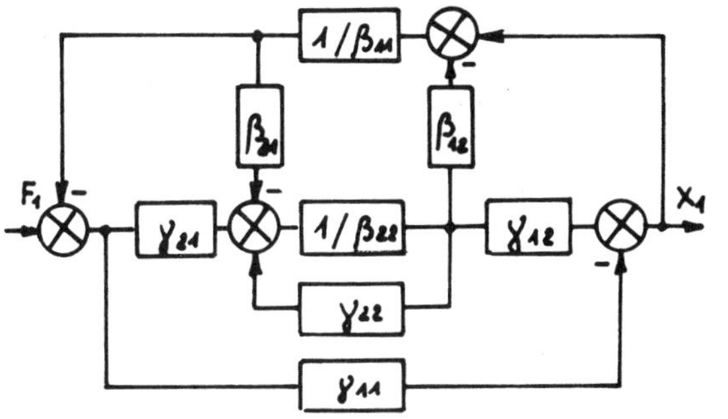

Fig. 4. Reduction of block diagram of receptance α_{11} shown in Fig. 3.

After the transformation the minor feedback loop enclosed the blocks $1/\beta_{22}$ and γ_{22} is reduced to a single block and the block diagrams β_{21} and β_{12} are interchanged. As a result we get a block diagrams shown in Fig. 5.

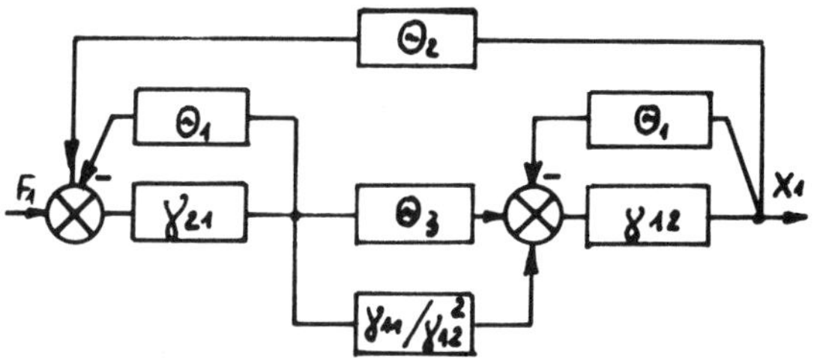

$$\Theta_1 = \frac{\beta_{12}}{\beta_{11}(\beta_{22} - \gamma_{21}) - \beta_{12}^2} = \frac{\beta_{12}}{D}$$

$$\Theta_2 = \frac{\beta_{22} - \gamma_{22}}{D}; \quad \Theta_3 = \frac{\beta_{11}}{D}$$

Fig. 5. Reduction of block diagram of receptance α_{11} shown in Fig. 4.

This scheme is simple to reduce and it gives the overall transfer function of the system corresponding to Eq. (14).

EXAMPLE

In order to illustrate the connection of receptance by means the block diagram we are presented example. Let us consider the system presented in Fig. 6 (Holka, 1984). In this case the second mass has been joined with the first in order to eliminate the vibration of the main system. The first mass has the following parameters: $m_1 = 25$ kg, $k_1 = 50000$ N/m, a = 1 m, b = 3 m, $I_1 = 50$ kgm^2, l = 0,5 m, P = 500 N.

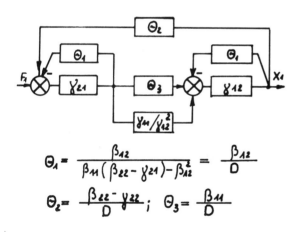

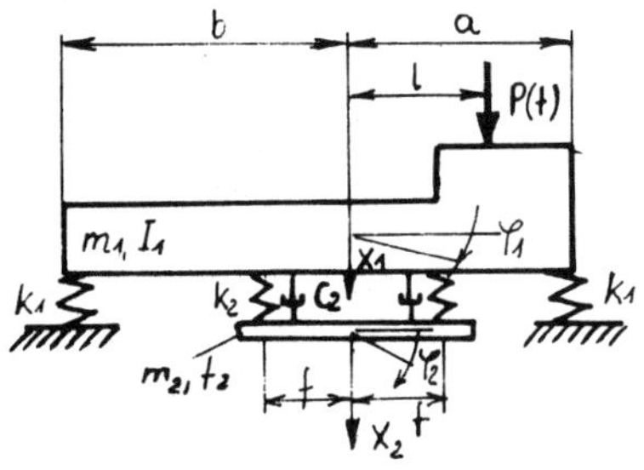

Fig. 6. Vibration system with 4-degrees of freedom

It is the case of two points connection, therefore is obligatory general scheme from Fig. 3. From this Fig. we can see that x_1 and φ_1 are zero when

$$\gamma_{11} = \gamma_{22} = \gamma_{12} \tag{17}$$

The second sub-system has the following characteristic matrix $U(i\omega)$

$$U_c(i\omega) = \begin{bmatrix} 2k_2 & 0 & -2k_2 & 0 \\ 0 & +2k_2f^2 & 0 & -2k_2f^2 \\ -2k_2 & 0 & +2k_2 - m_2\omega^2 & 0 \\ 0 & -2k_2f^2 & 0 & +2k_2f^2 - I_2\omega^2 \end{bmatrix} \tag{18}$$

While the characteristic matrix of the main system has the form:

$$U_B(i\omega) = \begin{bmatrix} 2k_1 - m_1\omega^2 & -k_1(b-a) \\ -k_1(b-a) & +k_1(a^2+b^2) - I_1\omega^2 \end{bmatrix} \tag{19}$$

From Eqs. (18), (19) we can get the particular receptances.

$$\beta_{rn} = \frac{\Delta_{rn}}{\Delta_B} \qquad \gamma_{rn} = \frac{\Delta_{rn}}{\Delta_C} \tag{20}$$

Where Δ_{rn} is the minor of martix $U(i\omega)$
Δ - is the determinator of martix $U(i\omega)$

From Eqs. (18), (20) we can see that conditions - Eq. 17 can be satisfied when

$$2k_2 - m_2\omega^2 = 2k_2f^2 - I_2\omega^2 = 0 \tag{21}$$

Using the notation $\omega_{x_2}^2 = 2k_2/m_2$ and $\omega_{\varphi_2}^2 = 2k_2f^2/I_2$ where ω_{x_2} and ω_{φ_2} are free vibrations of mass m_2 and assuming

$$I_2 = m_2\varrho_2^2 \qquad \varrho_2^2 = f^2 \tag{22}$$

it can be written

$$2k_2\left(1-\frac{\omega^2}{\omega_{x_2}^2}\right)=2k_2f^2\left(1-\frac{\omega^2}{\omega_{x_2}^2}\right)=0 \qquad (23)$$

and now we can see, that

$$X_1 = \Upsilon_1 = 0 \quad \text{when} \quad \omega = \omega_{x_2} \qquad (24)$$

It is evident that the absorber is used when the main system is in resonance. Therefore the last conditions have the forms

$$\omega = \omega_1 = \omega_{x_2} \quad \text{or} \quad \omega = \omega_2 = \omega_{x_2} \qquad (25)$$

Where $\omega_{1,2}$ are free vibrations of the main system. The frequency $\omega_{1,2}$ can be solved with the formula

$$\omega_{1,2}^2=\frac{1}{2}\left[\frac{2k_1}{m_1}+\frac{k_1(a^2+b^2)}{I_1}\mp\sqrt{\left[\frac{2k_1}{m_1}+\frac{k_1(a^2+b^2)}{I_1}\right]-\frac{4k_1(a+b)^2}{m_1 I_1}}\right] \qquad (26)$$

Substituting the date we find $\omega_1 = 53,6$ and $\omega_2 = 105,4$. The ratio m_2/m_1 is assumed 0,2 and from this the mass $m_2 = 0,5$. From conditions of tuning we have $k_2 = 7190$ N/m. We assume f = 0.25 and from this we get $I_2 = 0,3125$.
The general solution have the forms

$$X_r(i\omega)=\sum_{i=1}^{n}\alpha_{rn}(i\omega)\cdot F_n(i\omega) \qquad (27)$$

where individual receptances we calculate from Eqs. (14 - 16) and Eqs. (18), (19). The calculations were made by digital computer. Fig. 7. presents the vibrations of the coordinate X_1 and Υ_1 of the main mass after attaching an absorber. The curves confirm calculations. For the resonance frequency, the magnitudes of X_1 and Υ_1 are equal zero if $c_2 = 0$.

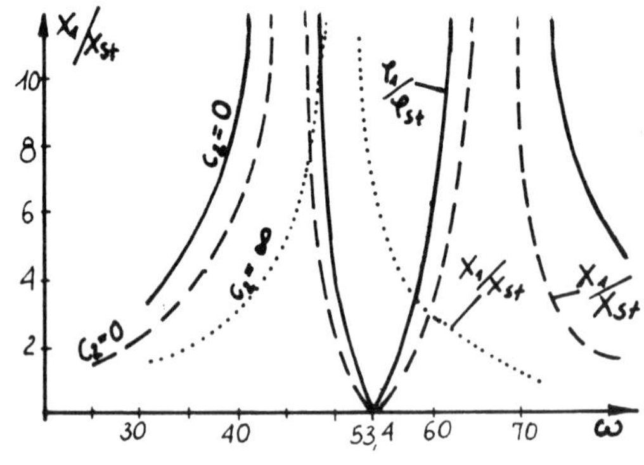

Fig. 7. Amplitudes of the main mass of Fig. 6.

The active method of vibration control could also be used now, in this case we apply again the block diagrams shown above

CONCLUSION

1. Receptance Synthesis by means of block diagrams describe the structure in very transparent way.
2. The block diagrams method permits the change of the structure of the dynamic system by connection of new active or passive elements.
3. All the blocks of the system can be received by theoretical or experimental methods.

REFERENCES

Bishop, R.E.D. and G.M.L. Gladwall (1972). In W.N.T. (Ed.). Macierzowa analiza drgan, Warsaw.

Holka, H. (1984). The Dynamic Vibration Absorber For The Main System with Two Degrees of Freedom. International Symposium on Design and Synthesis, Tokyo, 849 - 852.

An Improved Model of an Electro-hydraulic Servo-Valve

S. BAYSEC* AND J. REES JONES**

*Department of Mechanical Engineering, Middle East Technical University, Gaziantep, Turkey
**Department Mechanical, Marine and Production Engineering, Liverpool Polytechnic, UK

Dept. Mechanical, Marine & Production Engineering, Liverpool Polytechnic, England.

Summary. This paper presents a mathematical model of an electro-hydraulic, two stage servo-valve. The model is of the fourth order in linear elements which is considerably lower than current trends and offers simplicity. The emphasis has changed to non-linear factors within the valve and this has had the result of very accurate modelling, a fact which is validated by experimental work. The paper describes the formulation to include various factors such as saturation, hysteresis, flapper and spool valve flow and presents the digital computer results in comparison with experimentally obtained results.

Keywords. Hydraulic systems; frequency and step response; dynamic modelling; servomechanisms.

INTRODUCTION

Hydraulic servo-drives are being extensively used to power industrial robots and machine tools because of their high power-to-weight ratio and ability of fast start, stop and speed reversals with high response control elements. In the design and development of hydraulic systems, including the servo-drives, a computer simulation enables the designer to see the actual behaviour of the machine without actually building it. A simulation program which produces results in good fit to that of a real system provides the designer with the most valuable tool, the computer aid in design.

Attempts by manufacturers to model the dynamics of a hydraulic servo-valve have generally been confined to a description in terms of linear differential equations. These may be of order as high as eight, but they suggest that empirical models of first, second and third order with experimentally evaluated constant coefficients, are accurate enough for most engineering purposes. These linear models, which are briefly described within the paper, are simple and easy to solve, but their results deviate from that of a real valve considerably at high frequency and high amplitude control signals due to the non-linearities existing. These non-linearities occur at the magnetic torque motor as electromagnetic saturation and hysteresis, at the flapper valve assembly as the physical limitations on the motion of mechanical parts and at the variation of stiffness of elastic, mechanical elements within the valve.

DESCRIPTION OF THE SERVO-VALVE

The servo-valve studied, shown in Fig. 1 is a well known and documented commercially available type, where the control of fluid flow between the hydraulic power source and the actuator is done by a conventional 4-way spool valve. The spool position and the actuator load determines the rate of fluid flow through the spool stage. The spool is positioned by a fast response flapper valve. The two ends of the spool are exposed to the variable pressures P_1 and P_2 of the boost chambers, which are adjusted by a double nozzle flapper valve. Nozzles are fixed to the valve body and the motion of a flapper extending between the two opposing nozzles varies the flow area of the nozzle exits and hence the boost chamber pressures. The flapper is actuated independently by an electrical torque motor, which applies an external, orbitrarily controllable torque on the armature-flapper assembly. Any variation on the magnetic torque causes the flapper to displace, varying the initially equal nozzle exit areas. This in turn varies the pressures P_1 and P_2 at the upstream of the nozzles. Difference in these pressures force the spool to move in the direction of the nett force. While moving, the spool deflects the feedback wire connected to the flapper, gradually creating a counter torque against the magnetic torque. Motion continues until the two torques become equal and the flapper comes to rest at midway between the nozzles, making P_1 and P_2 equal.

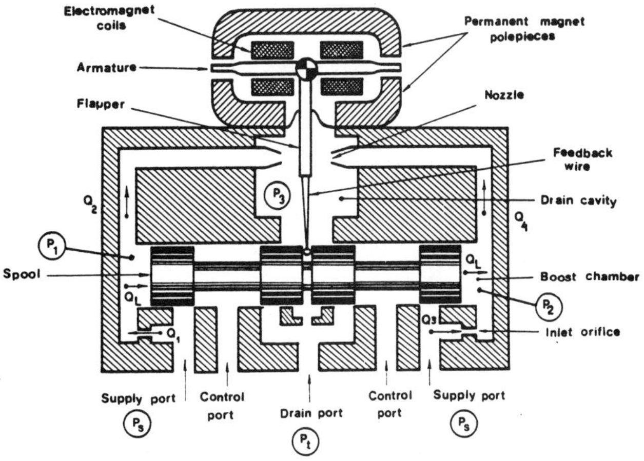

Fig. 1. Schematic structure of a two stage electro -hydraulic servo-valve. Pressures are shown in circles.

MODELLING

The valve contains two inertia bearing elements, the spool and the flapper, which move independently from each other. The deviation of their equation of motion requires the formulation of all the forces and torques acting on the spool and the flapper assembly.

Forces on the spool

i) Hydraulic Control Forces:

Assuming an incompressible fluid in turbulent flow with a constant supply pressure and using the notation of Fig. 1,

$$\begin{aligned} Q_1 &= C_{do} A_o \sqrt{\tfrac{2}{\rho}(P_s - P_1)} \\ Q_3 &= C_{do} A_o \sqrt{\tfrac{2}{\rho}(P_s - P_2)} \\ Q_2 &= C_{df}\pi D_N (x_{fo}-x_f) \sqrt{\tfrac{2}{\rho} P_1} \\ Q_4 &= C_{df}\pi D_N (x_{fo}+x_f) \sqrt{\tfrac{2}{\rho} P_2} \end{aligned} \qquad (1)$$

assuming that the drain cavity pressure P_3 is zero. The continuity of flow, assuming no leakage requires:

$$Q_L = Q_1 - Q_2 = Q_4 - Q_3 = A_s\dot{x} \qquad (2)$$

where:

- C_{do} : Discharge coefficient for fixed inlet orifices (dimensionless)
- C_{df} : Discharge coefficient for the variable orifices at the exit of the nozzles (dimensionless)
- P_s : Supply pressure (N/mm^2).
- P_1, P_2: Boost chamber pressures (N/mm^2).
- A_o : Area of the fixed inlet orifice (mm^2).
- ρ : Mass density of the hydraulic fluid $\simeq 8.3 \times 10^{-10}$ $N.sec^2/mm^4$.
- x_f : Instantaneous displacement of the flapper from its null position at midway between the nozzles (mm).
- x_{fo} : gap between a nozzle and the flapper when the flapper is at its null position (mm).
- x : Spool displacement from its null position (mm).
- Q : Fluid flowrate (mm^3/sec).
- D_N : Nozzle diameter (mm).
- A_s : Spool end area (mm^2).

Substituting equations (1) into (2) yields:

$$P_1^2[(Z_1^2+Z_2^2K^2)^2] + P_1[2\{\dot{x}^2(Z_1^2-Z_2^2K^2) - Z_1^2P_s(Z_1^2+Z_2^2K^2)\}] + [(\dot{x}^2-Z_1^2P_s)^2] = 0 \qquad (3a),$$

or

$$P_2^2[(Z_1^2+Z_2^2L^2)^2] + P_2[2\{\dot{x}^2(Z_1^2-Z_2^2L^2) - Z_1^2P_s(Z_1^2+Z_2^2L^2)\}] + [(\dot{x}^2-Z_1^2P_s)^2] = 0 \qquad (3b)$$

where

$$Z_1 = C_{do}\frac{A_o}{A_s}\sqrt{\frac{2}{\rho}} \qquad [mm^4/(N.sec^2)]^{\frac{1}{2}}$$

$$Z_2 = C_{df}\pi\frac{D_N}{A_s}\sqrt{\frac{2}{\rho}} \qquad [mm^2/(N.sec^2)]^{\frac{1}{2}}$$

- K : Instantaneous gap between nozzle 1 and flapper (mm).
- L : Instantaneous gap between nozzle 2 and flapper (mm).

From Eq. 3, the boost pressures are:

$$P_1 = \frac{-B_1 + \text{sign}(\dot{x})\sqrt{\text{DISC 1}}}{2A_1} \qquad P_2 = \frac{-B_2 - \text{sign}(\dot{x})\sqrt{\text{DISC 2}}}{2A_2} \qquad (4)$$

where

A_1, B_1, DISC 1: first two coefficients and the discriminant of P_1 Eq. 3a.
A_2, B_2, DISC 2: first two coefficients and the discriminant of P_2 Eq. 3b.

Equations 3 and 4 define the boost pressures associated with an arbitrary flapper discplacement from null. In many valve designs, the drain cavity is kept at a high pressure to provide stability in flapper motion. In the case of constant P_3, obtainable by a relief valve between the cavity and the drain line, Eqn. 3a and b become

$$\begin{aligned} &P_1^2[(Z_1^2+Z_2^2K^2)^2] + P_1[2\{\dot{x}^2(Z_1^2-Z_2^2K^2) - (Z_1^2 + Z_2^2K^2)(Z_1^2P_s+ Z_2^2K^2P_3)\}] + \\ &[(\dot{x}^2-Z_1^2P_s)^2 + Z_2^2K^2P_3(Z_2^2K^2P_3+2\dot{x}^2+2Z_1^2P_s)] = 0 \\ &P_2^2\,[Z_1^2 + Z_2^2L^2)^2] + P_2(2\{\dot{x}^2(Z_1^2 - Z_2^2L^2) - (Z_1^2 + Z_2^2L^2)(Z_1^2P_s + Z_2^2L^2P_3)\}] + \\ &[(\dot{x}^2-Z_1^2P_s)^2 + Z_2^2L^2P_3(Z_2^2L^2P_3+2\dot{x}^2+2Z_1^2P_s)] = 0 \end{aligned} \qquad (5)$$

and the solution of these quadratics using Eq. 4 gives the boost pressures. The widely used techinique of maintaining a high exit cavity pressure is to place a fixed orifice between the cavity and the drain line. This increases the difficulty of getting a closed form solution for boost pressures, since P_3 is not constant. The following conditions must exist in addition to equations 1 and 2:

$$Q_2 + Q_4 = Q_5 \qquad Q_5 = C_{dro} A_{dro}\sqrt{\frac{2}{\rho}P_3} \qquad (7)$$

where C_{dro} and A_{dro} are the discharge coefficient and area of the drain orifice respectively. Solution can be obtained by iteration on P_3.

ii) Bernoulli Forces:

Bernoulli forces along the spool axis is given as:

$$F_b = \rho\, Q\, v \cos\Theta \qquad (8)$$

where θ is the angle of the fluid jet leaving the high pressure spool compartment from the spool axis [1]. An empirical definition of the jet angle by Von Mises is given in [6], showing that it approaches to 69 degrees. v is the velocity of flow.

iii) Feedback spring and damping forces:

On the spool, there is a force exerted by the feedback spring and viscous damping, given by

$$F_s = -k_F x \qquad (9)$$
$$F_d = -c\dot{x} \qquad (10)$$

where k_F and c are spring stiffness and damping coefficient respectively. At high control currents, the flapper may deflect till it touches the tip of a nozzle. With this supporting condition, the stiffness of the armature, flapper and feedback wire assembly changes, hence k_F is not constant, but a function of flapper displacement. The spring is therefore non-linear.
The spool moves under the action of these forces according to:

$$\ddot{x} = [(P_2-P_1)A_s + F_b + F_s + F_d]/M_s \qquad (11)$$

where M_s is the mass of the spool.

Forces on the Flapper

i) Magnetic Torque:

An armature magnetised by an electromagnetic coil is integral with the flapper. Upon the application of a control current to the coil, the armature becomes magnetized and produces a magnetic torque with the interaction between the permanent magnet pole pieces shown in Fig. 1. A detailed analysis of the magnetic torque formed is given in [1]. In all valve designs the armature motion is limited to a small range in the middle of the pole pieces and the torque produced is in direct proportion to the control current as:

$$\tau_m = K_t\, i \qquad (12)$$

where K_t : Torque motor gain in N.m/Amp.
i : Control current in Amperes.

In [2], [4], [5], [9], eq. 12 is used to characterise the magnetic torque, where a saturation-free operation is assumed. But ferromagnetic materials show a saturation in their magnetic induction in strong magnetic fields. Also, these materials tend to resist magnetic induction, which is called the *electromagnetic hysteresis*. In an alternating magnetic field, induced magnetization follows a loop around the magnetization curve as seen in Fig. 2. Hysteresis loop is a complicated function, and so, empirical definitions of the loop have been widely used.

In this paper a definition and model of a hysteresis loop is composed of straight lines near to the origin and by second order curves at the saturation region. Further it is composed of two regions called *climbing-up* and *climbing-down* zones as shown in Fig. 2. The magnetization curve passes along the middle of the zones formed. Intercepts N_1 and N_2 and the slope K_t define the two linear border lines of the loop. Two ellipses, reflected diagonally about the origin are defined with their centres at C_1 and C_2 (Fig. 2) to represent the transition into saturation. The coordinates of C_1 and C_2 and the saturation torque τ_{max} define the ellipses which are tangential to the mid-line separating the zones. The point where current reverses is a measure of the residual magnetism and determines the path to follow within the loop after the current reversal. For an increase in current i, the torque moves towards a straight line within the *climbing-up* zone through an elliptic path tangent to it. An increase of current after reversing at point A (Fig.2) causes the torque function to follow the border line. Similarly, if the reversal is at origin, torque increases along the magnetization curve, and a reversal at B moves the torque towards the mid-line. *Climbing-down* occurs in a similar manner in the *climbing-down* zone.

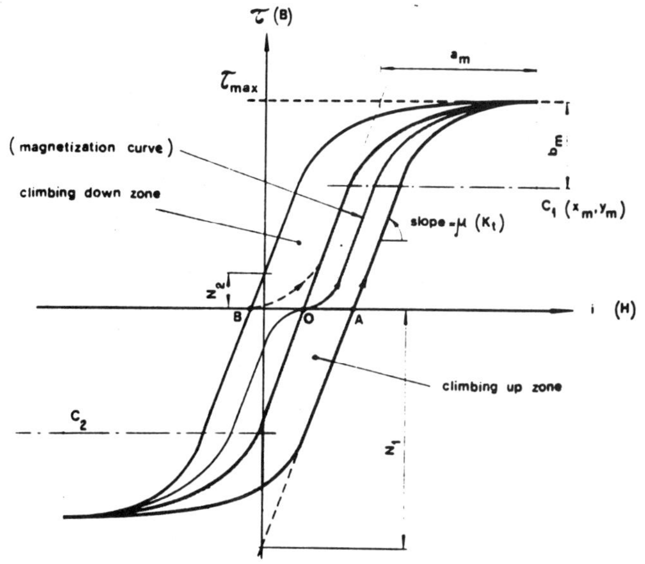

Fig. 2. Magnetic hysteresis loop

Since the location of reversal point is important, the instantaneous current value has to be monitored continuously and at each reversal, the new path that the torque-current function takes must be determined. To estimate the point of reversal, a parabola may be fitted onto the last three data points once a fall in current is detected and the extremum of the parabola is the reversal point. However, in this instance a discrete first order change in direction was found to be effective.

While testing, the armature of the valve under test may be demagnetized initially.

ii) Torque due to the impinging jets of fluid:

The torque exerted on the flapper by two opposing jets of fluid is:

$$\tau_h = [(P_1-P_2)A_N + 4\pi C_{df}{}^2\{K^2(P_1-P_3)^2 - L^2(P_2-P_3)^2\}]\, R_F \qquad (13)$$

where A_N : nozzle crossectional area = $\pi D_N^2/4$ in mm^2
R_F : distance between the axis of the nozzles and the flapper pivot.

This torque increases with the flapper displacement from null, hence acts as a negative spring. In [6] a similar formulation is used.

iii) Restoring spring and damping torques:

The restoring torque on the flapper from the feedback spring is:

$$\tau_s = -k_s \theta \tag{14}$$

where k_s is the equivalent torsional spring constant. Since the flapper is completely in hydraulic fluid, a damping effect must be considered. This is represented by

$$\tau_d = -C_d \dot{\theta} \tag{15}$$

where C_d is the equivalent damping coefficient.

iv) Torques due to contact restrictions:

The flapper is restricted to move in a very small range by the nozzles. If the flapper is displaced x_{fo} from null, it touches the tip of one of the nozzles [6], [10] and comes to rest. This nonlinearity is modelled as a flapper bearing against stiff springs at the end of its motion range, as:

$$\tau_\ell = -k_\ell [x_f - \text{sign}(x_f) \cdot (x_{fo})] R_f \text{ for } x_f > x_{fo} \text{ and } x_f < -x_{fo}$$

$$\text{and } \tau_\ell = 0 \qquad \text{for } -x_{fo} < x_f < x_{fo} \tag{16}$$

where k_ℓ is a high spring stiffness.

The flapper moves under the action of these torques according to:

$$\ddot{\theta} = (\tau_m + \tau_h + \tau_s + \tau_d + \tau_\ell)/I_f \tag{17}$$

where I_f is the mass moment of inertia of the flapper.

OTHER EMPIRICAL LINEAR SERVO-VALVE MODELS

In most servo applications, the load response is slower than the valve and valves are operating at the D.C. end of their bandwidth. Responses of valves operating under this condition can be suitably modelled by linear differential equations as summarised.

i) For valves of the type and size under consideration, operating up to 10 degrees phase lag, which roughly corresponds to 8-10 Hertz, the spool displacement can be assumed to be a linear function of the input control current as

$$x = C_1 i \tag{18}$$

ii) Among the two hydraulic amplifiers in the valve, the flapper stage has a natural frequency of 5-10 times that of the spool stage and so the spool lag starts to influence the valve response at higher excitation frequencies. For frequencies up to 45 degrees phase lag, the valve response can be represented by a first order model as

$$\dot{x} = \frac{Q_L}{A_s} \tag{19}$$

where

$$Q_L = C_2 i$$

and both flapper and spool are regarded massless.

iii) At progressively higher frequencies a higher order model is required. A second order model, resulting from a combination of both spool and flapper stages, which are assumed to be of first order appears as

$$\dot{x}_f = C_3 i$$
$$Q_L = C_4 x_f \tag{20}$$
$$\dot{x} = Q_L/A_s$$

which describe the valve response up to 90 degrees phase lag.

iv) A third order model, assuming an inertia for the flapper and zero mass for the spool can describe the valve response up to 150-160 degrees phase lag using

$$\ddot{x}_f = (C_5 i/I_f) R_f$$
$$Q_L = C_6 x_f \tag{21}$$
$$\dot{x} = \frac{Q_L}{A_s}$$

where $C_1 \ldots C_6$ are experimentally determined coeffients.

A COMPARISON OF RESULTS OF SIMULATION AND REAL VALVE RESPONSE:

To test the servo-valve models against a real valve, frequency and step response tests were carried out on a Moog E076-102 valve. The spool displacement was monitored by eddy-current proximity pick-ups fitted to the boost chambers. The step response of the actual spool to equal current increments, up to its rated current, is shown in Fig. 3.a. The envelope of the response suggests a saturation in spool velocity, which becomes constant at high control currents. Load flow Q_L is a linear function of flapper displacement x_f. A saturation in Q_L therefore implies a saturation in flapper displacement, that is flapper coming to rest at the limits of its motion. In [6] and [10] this nonlinearity is given a special consideration. A step response of the digital model including this non-linearity (eq. 16) is very similar to that of the real valve as seen in Fig. 3.b. Fig. 3.c shows the response of the third order linear model proposed by the manufacturers of the valve.

Figure 4.a shows the experimentally obtained frequency-response of the same valve to different amplitudes of control current. Fig. 4.c shows the response of the 3rd order model proposed by the manufacturers. The different responses associated with different amplitudes show that the real valve is a non-linear system.

The third order model does not exhibit these non-linear characteristics. The frequency-response of a 4th order model which did not include hysteresis and drain orifice is shown in Fig. 4.b. The predominant non-linearity here is the restriction of the flapper motion by the nozzles. Fig. 4.d shows the frequency-response of the 4th order model including the nonlinear spring stiffness. Some correspondence with the experimental result is apparent in the ordered separation of the frequency response curves of Fig. 4.b. However a significant improvement in the correlation with the experimental result is seen to be achieved only with those features included in the model for Fig 4.d.

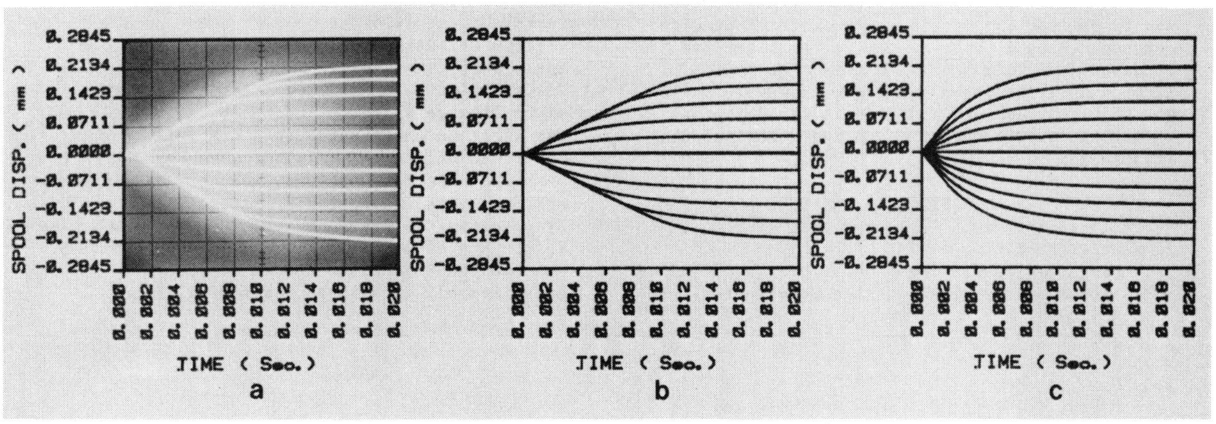

Fig. 3. a) Step response of a MOOG E076-102 Servo-valve. Response recorded is spool displacement.
b) Step response of the digital simulation with saturation at flapper stage.
c) Step response of the third order linear model proposed by MOOG Inc.

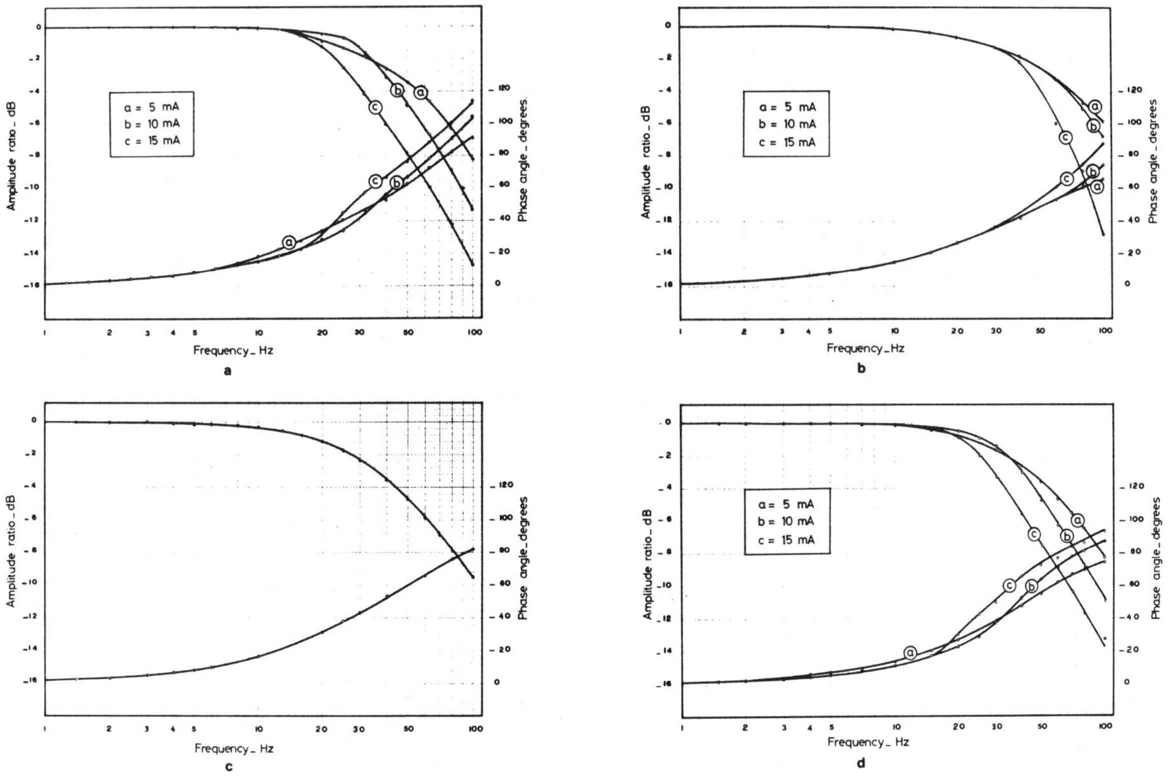

Fig. 4. a) Frequency of a MOOG E076-102 servo-valve. Response recorded is the spool displacement. b) Frequency response of a theoretical 4th order model. c) Frequency response of a 3rd order linear model proposed by Moog Inc. d) Frequency response of a 4th order model with a non-linear spring stiffness and a flapper motion restrictor.

CONCLUSION

The paper has described the development of a refined but efficient model of a servo-valve for use in digital computation. It exhibits precise correspondence with an acutal valve at the low frequency end in common with the other models with which it has been compared. However, it is at the high frequency end that it stands out as by far the most accurate model and this includes frequencies up to 100 Hz and beyond.

REFERENCES

1) Merritt, H.E.
Hydraulic control systems.
Library of Congress Catalog Card Number: 66-28759.
John Wiley & Sons, Inc. - -967.

2) Guillon, M.
Hydraulic Servo-Systems - Analysis and Design.
Suggested UDC No: 621-522.
Butterworth & Co. Ltd., 1969.

3) McCloy, D., Martin, H.R.
The Control of Fluid Power - Analysis and Design.
ISBN 0-585312-135-4.
Ellis Horwood Limited - 1980.

4) Nikiforuk, P.N., Ukrainetz, P.R., Tsai, S.C.
Detailed Analysis of a Two-Stage, Four-way Electro-hydraulic Flow Control Valve.
Journal of Mechanical Engineering Science, Vol. 11, nO. 2. pp. 168-174. - 1969.

5) Martin, D.J., Burrows, C.R.
Dynamic Characteristics of an Electro-Hydraulic Servovalve.
Journal of Dynamic Systems, Measurement and Control Transactions of the ASME.
December 1976. pp. 495-406.

6) Lebrun M. and Scavarda, S.
Simulation of the Nonlinear Behaviour of an Electro-Hydraulic Exciter.
Simulation, October, 1979. pp. 127-142.

7) Schachte, John J., Bollinger John G.
Analogue Computer Simulation Techniques for Analysis of a Machine Tool Hydraulic Drive.
National Conference on Fluid Power, 1972, pp. 248-261.

8) Conrad, R., Jannerup, O., Trostmann, E. Toft Fensvig, A.
On Elements of a Diagnostic System for Identifying Servo-valve Parameters.
Proceedings of National Conference of Fluid Power 33rd Annual Meeting. Vol 31, pp.252-268 - 1977.

9) Wang, P.K.C.
Mathematical Models for Time Domain Design of Electro-Hydraulic Servomechanisms.
AIEE Winter General Meeting - 1961.

10) Zaborszky, J., Harrington, H.J.
A describing function for Multiple Nonlinearities Present in Electro-Hydraulic Control Valves.
AIEE Winter General Meeting - 1957.
pp. 183-190.

Zusammenfassung

Dieser Bericht stellt ein mathematisches Modell eines elektrohydraulischen, zqeistufigen Servoventils dar. Das Modell gohört zu der vierten Gliederordnung, was erheblich weniger ist als zur Zeit üblich - und was Unkompliziertheit anbietet. Betont werden nicht-lineare Faktoren innerhalb des Ventils und das führt zu einem sehr genauen Modell, was experimentell bestätigt wird. Der Bericht beshreibt die Formulierung, welche verschiedene Faktoren einschliesst: z.B. Sättigung, Hysteressis, "flapper and spool valve flow". Knappe und spuleventil fluss. Die von Computer ermittelten Ergebnisse werden mit den experimentall erhaltenen Daten verglichen.

Operational Experience with the Adaptive Suspension Vehicle

K. J. WALDRON*, C. A. KLEIN*, D. PUGH**, V. J. VOHNOUT**, E. RIBBLE**, M. PATTERSON** AND R. B. McGHEE***

*Department of Mechanical Engineering, Ohio State University, Columbus, Ohio 43210, USA
**Adaptive Machine Technologies, Inc., Columbus, Ohio 43210, USA
***Department of Computer Science, Naval Postgraduate School, Monterey, California 93943, USA

Abstract The Adaptive Suspension Vehicle is a legged locomotion system which uses computer coordination and advanced sensing to operate in rough terrain. This presentation will review the results of a program of testing and software installation which has occupied the calendar year 1986. The paper will also contain a review of the major features of the mechanical and electronic hardware architectures, and the control and coordination software.

Keywords Legged locomotion; system development; multi-limbed coordination.

INTRODUCTION

The Adaptive Suspension Vehicle, shown in Figure 1, is a walking machine which was mechanically completed in June 1985. Since that time the machine has undergone mechanical and electronic hardware development, and installation and testing of computer software. At the time of writing three of six designed operational modes have been installed and tested. The machine has operated in the laboratory on level surfaces but, as of the date of writing, has not been operated outdoors on unstructured terrain. It is expected that installation of the remaining software modes will be completed in Autumn 1986, and that outdoor operation on unimproved terrain will be demonstrated before the end of 1986.

In this paper, results from the testing of the vehicle at the date of writing will be summarized. It is expected that, in the conference presentation, it will be possible to report on operation of the machine in the complete planned set of operational modes.

HARDWARE

The machine has been described in detail in several references (Waldron and McGhee 1986 a,b, McGhee and Waldron 1985) In brief, it is a walking vehicle which has six legs each of which has a normal operating length of slightly more than two meters. The body of the machine is approximately five meters long and the machine is about two and one half meters wide. Its weight is about two thousand kilograms, and it is designed to carry a 250 kg payload, together with an operator, and to be completely self-contained. It is powered by a motorcycle engine of about 65 kilowatt peak power via a storage flywheel with a quarter of a kilowatt hour of energy storage. Power distribution to the actuators is via a two stage hydraulic system with a hydrostatic output stage.

The system is coordinated and controlled by means of a computer system consisting of 12 INTEL 86/30 single board computers, together with two custom built, micro-coded, special purpose processors.

The special purpose processors are based on high speed computation chips supplied by AMD Corporation. This system is coupled by two multi-busses and a parallel data link to form a multi-processing computer (McGhee et al. 1985). An especially configured operating system called GEM was developed for this type of system (Schwan et al. 1985) The machine carries a scanning range finder based on an infra red laser which was built by ERIM (Zuk et al. 1985). Other sensing systems include position, velocity, and differential force sensing on all 18 of the hydraulic actuators, and sensing of the swash plate positions on the hydrostatic pumps powering the actuators. The operator commands system velocity and direction by means of a six degree of freedom joystick system, with fore-aft rate, lateral rate and heading rate on the main joystick, roll and pitch are controlled by one of two mini-joysticks mounted on the head of the main joystick, and walking height by the other mini-joystick. The operator also has a keyboard allowing him to select functions from those available in the computer system.

One of the special purpose processors is used to provide enhanced computational speed for the transformation of the data from the scanning range finder into a vehicle based cartesian coordinate system, and the insertion of that data into an elevation map in the machine's memory. The scanning range finder scans a 128 x 128 pixel field at two frames a second producing a data stream of 32,000 eight bit words per second. Thus this processing load is difficult to handle without specially designed hardware. The second special purpose processor is used to enhance coordination computations allocating commanded forces to each of the feet which are on the ground at any given time. This problem has been discussed in references (Waldron 1986, McGhee and Orin 1983), and also represents a substantial computational burden.

SOFTWARE

It is planned to install six operational modes in the machine. Each of these modes is a software module which resides as an upper level above a much larger software system which handles basic control coordination and system monitoring functions. At the time of writing the total amount of code in the system was somewhat in excess of one mega-byte of compiled code. The functions of the operational modes are as follows:

Utility

This mode is used for powering up the system, shutting down the system, system checkout, dumping of data for diagnostic purposes, and other utility functions.

Precision Footing

This mode will be used in very difficult conditions to permit the operator to directly designate foot movements, or body position and attitude movements. It is the most manual of the designed operational modes. However, since it involves coordination of 18 actuated degrees of freedom to produce a maximum of 6 output degrees of freedom, namely the degrees of freedom of body movement, it may be seen that a substantial level of computer coordination is still involved. In the precision footing mode, the operator may either control body position and attitude with 6 degrees of freedom, using the joystick and mini-joystick, or he may select any of the six feet and control foot position in foot coordinates using the 3 degrees of freedom on the mini-joysticks.

Close Maneuvering

This mode is intended for turn in place, backing, lateral or slow forward movements. The operator may command any or all of the six degrees of freedom of body motion provided by the joystick. Phasing of the legs, and foothold selection is completely automatic. A major difference between this operational mode and the higher speed operational modes is that the scanning range finder is not utilized. On level ground, in principle, it is possible to operate at fairly high speed in this mode. In irregular terrain, it will be necessary to operate more slowly since the system must rely on sensing of the contact of the foot with the ground via the differential pressure transducers on the actuators.

At the time of writing the utility, precision footing, and close maneuvering modes of operation have been installed on the machine and thoroughly tested. The software for the remaining operational modes was still under development.

Follow-the-Leader

This mode is a adaptation of the precision footing mode to facilitate operation in severe terrain. The operator will select footholds to be used by the front feet. The middle and rear feet will be placed either in the footprints of the front feet or directly alongside those feet. The advantage of the follow-the-leader gait is that it relieves the operator of the necessity of monitoring the locations of the middle and rear feet. This is significant problem since those feet are not readily visible from the operator's cabin. The follow-the-leader gait also minimizes the number of quality footholds which the operator must identify in difficult conditions. Animals have been observed to make frequent use of follow-the-leader gaits and it is surmised that the reasons for which they do this are related to those stated here. This gait will be used, in particular, to negotiate large obstacles.

Terrain Following

The terrain-following mode is the most completely automated operational mode of the system. In this mode maximal use will be made of all sensor information to allow the machine to move at about 5 kilometers per hour in quite severe terrain. The scanning range-finder data will be used in the selection of footholds and the determination of the return heights of the feet in this mode. A free gait algorithm will be used, in which stability potential is maximized every time the machine makes a movement. That is, no predetermined gait pattern will be used, but rather, whenever the machine selects the next leg to be placed, it will do so in order to maximize future stability while being constrained only to those footholds identified from the rangefinder data as being safe.

Cruise/Dash

This is the highest speed operating mode of the machine. There is no computational difference between the cruise and dash modes of operation. However, there will probably be changes in the mechanical system parameters for operation in the dash mode. Specifically, it is likely that the flywheel speed will increased to its maximum, and it is possible that the primary hydraulic system pressure will be increased. In this mode a patterned gait or set of gaits, specifically wave gaits, will be employed to reduce the coordination computational load. It is likely also that the use of the scanner data will be diminished to selection of return heights, without control of the positioning of the feet. The cruise mode is intended for efficient locomotion over substantial distances in relatively easy terrain. The designed cruise speed is 8 kilometers per hour. Dash speed is expected to be about 12 kilometers per hour.

The software for the machine is programmed in Pascal and compiled on INTEL development systems prior to downloading into the machine. At present, downloading of the software requires about 10 minutes. Installation of a bubble memory, expected shortly, will eliminate this delay.

COORDINATION AND CONTROL

The legs of the machine are treated as force generators by the coordination software. A mode switching scheme is used to control each leg in a force control mode when the foot is on the ground and in a position-velocity control mode when the foot is lifted off the ground. Thus, the primary coordination algorithm is a force distribution algorithm. Computation cycle time is about 50 ms.

Control of the vehicle's position and attitude is achieved by means of an inertial guidance package mounted in the body of the machine. This unit contains a vertical gyroscope, rate gyroscopes on all three axes, and accelerometers on all three axes. The inertial package senses vehicle displacement at high bandwidth and, in combination with the operator's commands via the joystick, generates vehicle body position error signals which are converted in the control software into a desired body force, and then to commanded forces at the legs. Drift due to integration, and instrument error, is corrected at low bandwidth by means of leg position data.

This control scheme makes it unnecessary to use a detailed model of leg dynamics. A simple look-up table of reflected inertia at each actuator as a function of position is used in the leg coordination software. A model of leg compliance is used when correcting inertial system drift via leg position information.

MECHANICAL SYSTEM

The machine is completely self-contained. Power is provided by a 55 kw motorcycle engine. The engine is coupled to an energy storage flywheel of 0.25 kw hr capacity. The flywheel is designed to operate at up to 12,000 rpm, although normal operating speed is about 9,000 rpm. Power from the flywheel is distributed via a two-stage hydraulic system. The primary stage is a pressure-regulated, valve controlled system. The secondary is hydrostatic. The swash plates of the variable displacement pumps which control flow to the actuators are positioned by small rotary actuators in the primary system.

The vehicle legs are planar pantograph mechanisms hinged to the top of the body frame about axes parallel to the longitudinal axis of the body. The leg configuration can be seen in Figure 1.

TEST PROGRAM

At the time of writing the installation of mode software, and the test program have not been completed. The utility, precision footing, and close maneuvering software has been installed and tested. The machine has been operated with an operator in the cab, and also remotely using a communication cable. The special purpose processors have been added to the computer to overcome computational problems leading to excessively long cycle times. A recurring problem has been interaction between the computational cycle times and natural frequencies of the vehicle. However, it has proved to be possible to control this problem by shifting the sampling rates, as appropriate. The coordination software computes several steps ahead to ensure that the vehicle has a safe stopping capability before attempting any maneuver. This has lead to computation cycle time problems. It has been necessary to perform considerable tuning on the software for this, and other reasons. The inertial control system has performed well except for recurring hardware problems, particularly gyroscope dropout. Again, considerable tuning was necessary to determine the best sampling rates, both for the main system and for the low frequency correction of drift.

At the time of presentation it is expected that it will be possible to review the completed test program and the operation of all software modules.

SUMMARY

This is, necessarily, a brief review of the system architecture, and operational experience of the Adaptive Suspension Vehicle. More extensive descriptions of various aspects of the system are to be found in the references cited. Likewise, detailed descriptions of some of the algorithms used are to be found in references.

REFERENCES

K.J. Waldron and R.B. McGhee, "The Adaptive Suspension Vehicle," IEEE Control Systems Magazine, December 1986.

K.J. Waldron and R.B. McGhee, "The Mechanics of Mobile Robots," Robotics, Vol. 2, No. 2, June 1986, pp. 113-122.

R.B. McGhee and K.J. Waldron, "The Adaptive Suspension Vehicle Project: A Case Study in the Development of an Advanced Concept for Land Locomotion," Unmanned Systems, Vol. 4, No. 1, Summer 1985, pp. 34-43.

D. Zuk, F. Pont, R. Franklin and M. Dell'Eva, "A System for Autonomous Land Navigation," Proceedings of Active Systems Workshop, Naval Postgraduate School, Monterey, California, November 1985.

R.B. McGhee, D.E. Orin, D.R. Pugh and M.R. Patterson, "A Hierarchically Structured System for Computer Control of a Hexapod Walking Machine," Theory and Practice of Robots and Manipulators, Ed. A. Morecki, G. Bianchi and K. Kedzior, Kogan Page 1985, pp. 375-382.

K. Schwan, T. Bihari, B.W. Weide and G. Taulbee, "GEM: Operating System Primitives for Robots and Real-Time Control Systems," Proceedings of 1985 IEEE International Conference on Robotics and Automation, St. Louis, March 25-28, 1985, pp. 807-813.

K.J. Waldron, "Force and Motion Management in Legged Locomotion," IEEE Journal of Robotics and Automation, December 1986.

R.B. McGhee and D.E. Orin, "A Mathematical Programming Approach to Control of Joint Positions and Torques in Legged Legged Locomotion Systems," Theory and Practice of Robots and Manipulators, Ed. A. Morecki and K. Kedzior, Elsevier, 1977, pp. 225-232.

RESUMÉ

C'est, de neccesité, un compte rendu bref de la configuration du système, et de nos expériences quand l' Adaptive Suspension Vehicle fut operé. On peut trouver les descriptions plus detaillés des traits divers du système dans les articles qu'y sont cités. Les algorithmes utiliseés sont aussies décrit dans les articles que sont cités.

Figure 1: The Adaptive Suspension Vehicle

Tool Path Error Control, for End Milling of Microwave Guides

D. K. ANAND, J. A. KIRK AND M. ANJANAPPA

Mechanical Engineering Department, University of Maryland, College Park, MD 20742, USA

ABSTRACT

The research reported here is centered around high accuracy end milling operations with particular interest on the use of a magnetically suspended spindle for controlling the tool path error. Specifically, this study is concerned with the quantification and control of tool path error for low horsepower applications in end milling operations using a magnetically controlled spindle. The dynamic tool path error is the difference between the required and actual path of the workpiece relative to the tool. This error is a combination of tool deflection and machine dynamics and degrades the precision of the machined parts. Thin ribbed microwave guides are used for all experimental work. In such machining, part accuracy is important and the surface finish must be burr-free to avoid microwave signal attenuation. The cutting regime is chosen such that tool path errors dominate but without the onset of chatter. The overall research effort supported by the National Science Foundation is a cooperative effort between the University of Maryland, Cincinnati Milacron, Magnetic Bearings, Inc., The Westinghouse Corporation, and The National Bureau of Standards.

Keywords: Magnetic bearings, machining, error compensation, manufacturing processes

INTRODUCTION

End milling to produce thin ribbed components such as microwave guides is one of the major operations in the electronic and aircraft industries. Shown in Figure 1 is a typical aluminum microwave guide which is used in mobile radar applications. The part shape is composed of repeating sections of extremely thin ribs with "X" shaped openings which go completely through the part base. The economical production of thin ribs has proven troublesome because of the difficulty of controlling part tolerances and surface finish while maintaining high metal removal rates (MRR). A particularly troublesome area has been caused by burrs on the "X" shaped openings. These burrs require the finished piece to undergo a secondary deburring operation resulting in increased production time and costs. Studies have shown that the cost for cleaning and deburring can be quite high and is often unaccounted for in process planning for part production (Gillespie, 1978). To improve production efficiency of thin rib components, and to eliminate the secondary deburring operation, it is desirable to increase spindle speeds and table feeds (i.e., to move toward high speed machining) while maintaining part tolerances and surface finish within acceptable limits. In discussions with Westinghouse, Cincinnati Milacron, Magnetic Bearings Incorporated, and the National Bureau of Standards we have concluded that a magnetic bearing spindle can be retrofitted to existing machine tools and, with modification in feed rate, provide a solution to the accuracy, deburring and MRR problems in thin rib machining. Experience by Westinghouse has shown that the deburring operation can be eliminated if the part is machined at higher surface speeds (i.e., higher spindle speeds) provided that part accuracy is maintained. To achieve this goal control of the tool path error via a magnetic bearing spindle is required.

During high speed machining the forces at the interface of the cutting tool and workpiece can cause the tool to chatter. When chatter occurs the effect can not only degrade surface finish and part tolerance but can also damage the tool. Generally, tool chatter is avoided by controlling both the feed rate and spindle speeds of the tool and does not appear to be a limiting factor in improving the metal removal rates in thin rib machining. This paper addresses the specific problem of identifying and controlling tool path error as it effects dimensional accuracy and surface finish in thin rib machining. Specifically, interest is centered around high speed end milling operations with particular interest on the use of a magnetically suspended spindle to control the tool path error.

TOOL PATH ERRORS

Tool path error in three-dimensional cutting can be represented as shown in Fig. 2. The tool path error in computer numerical control machine tools is defined as the distance-difference between the required and actual tool path.

Tool path error (in the absence of chatter) can be classified into the following four categories, based on the source (Tlusty, 1980; MTTF Report, 1980; Anand, Kirk and Anjanappa, 1986) of the error for each category;

- deterministic position errors
- deformation due to heat sources
- deformation due to weight forces
- deformation due to cutting forces.

These four error sources can cause three types of tool path errors, viz: static deterministic, dynamic deterministic and stochastic. The first three sources are considered deterministic and may be measured using a laser metrology system and then put in the form of error maps. As an example of applying this technique Hocken and Nanzetta of the National Bureau of Standards reports on its successful implementation on a precision coordinate measuring system (Hocken and Nanzetta, 1983) and later on a machining center (Nanzetta, 1984; Simpson, Hocken and Albus, 1982). Once the error measurements for a given machine tool are obtained they must be used to correct the machine tool movement. Based upon discussions with Cincinnati Milacron and other machine tool manufacturers no machine tool manufacturer at present incorporates provisions for position error correction to be included in their controller.

DEFORMATION DUE TO CUTTING FORCES

The deformations due to cutting forces are both static deterministic and stochastic in nature (Anand, Kirk and McKindra, 1977; Tlusty, 1980; Tlusty and Macneill, 1975). These errors are best understood by considering typical thin rib machining with a straight teeth cutter as shown in Fig. 3. After the cut has been completed the thickness of the rib should be a theoretically perfect t_f. However, because of steady state and stochastic cutting forces the final rib shape is ramped in the vertical direction (characterized by the error Δt_{fa}) and has dimensional variations (characterized by the error Δt_{fr}) in the longitudinal direction. The ramp variations, Δt_{fa}, are deterministic and are due to the cantilever deflection of the thin rib (i.e., compliance between the tool and workpiece). The remaining workpiece deflection error is along its longitudinal length and is identified as Δt_{fr}. This error, caused by stochastic fluctuations in the cutting force, cannot be characterized by conventional means and must be treated by stochastic methods. The cutting force errors can therefore be considered as:

a. Deterministic tool path errors due to compliance between the tool and workpiece.

b. Stochastic tool path errors due to variations in the depth of cut.

The errors due to cutting forces show up as position differences between the required and actual tool/workpiece position and result in workpiece shapes which are not perfect. The departure from perfect shape is considered acceptable if workpiece tolerances and surface finish are within user defined limits.

The tool path error due to the static deterministic deformation caused by varying compliance between tool and workpiece is in general a combination of the compliance of tool and workpiece structures. However, in end-milling operations, of thin ribbed parts (e.g., microwave components), the variation in compliance is principally that due to the workpiece. The variation in compliance between tool-workpiece in this situation is therefore due to the variation in workpiece geometry, which is deterministic. No method of correcting these errors is currently available but, through the use of a magnetic bearing spindle, it will be possible to tilt the spindle to compensate for these errors. This method of error minimization is currently under study and will be reported at a later date.

The deformation due to variations in depth of cut is caused by stochastic cutting force variation. These variations are a combination of cutting an imperfect blank shape, and the influence of machine dynamics on the cutting process. Depth of cut errors can be understood by considering the case of end milling as shown in Fig. 4.

The imperfect blank shape consists of a nominal depth of cut which has superposed on it random variations in thickness (x_0). The nominal depth of cut gives rise to steady state cutting forces which act on the tool/workpiece structure to cause deformations. These deformations can be considered static deterministic and may be predicted by applying chip cutting mechanics (Anand, Kirk and McKindra, 1977). The random variations in rough machined blank thickness give rise to stochastic variations in cutting forces which, in turn, give rise to stochastic variations in part shape or, alternatively, cause stochastic tool path error.

The tool path errors due to variations in depth of cut are termed "copying errors" and occur as follows. The random blank error (x_0) is 'copied' on to the machined surface (as x_1) to a reduced scale. The 'rate of copying' of form error can be written as $i = x_1/x_0$, where i varies from 0 to 1 and is a function of μ which is a non-dimensional measure of system stiffness.

In most machining operations $\mu \ll 1$ so that $i \cong \mu$, and the copying error does not propagate between passes. End milling does not follow the normal behavior and a typical value of μ for end milling is about 2.67 for cutting steel (Tlusty, 1980), which gives a value of $i = 0.73$. Therefore for every consecutive pass, the error only reduces by a factor of about 1.4. The cutter teeth are helical in practice and therefore make the relationships more complicated (Tlusty, 1980; Tlusty and Macneill, 1975).

No method of correcting the stochastic errors is currently available but, through the use of a magnetic bearing spindle, it will be possible to translate the cutting tool to minimize these errors. The techniques for this approach to error minimization are currently under study and will be reported at a later time.

Shown in Fig. 5 is a listing of the errors which are applicable to end milling, in general, and the machining of thin rib structures in particular. Deterministic position errors (both static and dynamic) are defined as those repeatable errors which will reoccur when an identical set of input parameters exist on a given machine tool structure. Stochastic errors, on the other hand, are defined as those errors which occur when a random input is presented to the machine tool. The main sources of stochastic error are due to surface roughness of blank and in the cutting process itself. All the errors are further discussed in Anand, Kirk and Anjanappa (1986).

MAGNETICALLY CONTROLLED SPINDLES

The magnetic spindles for use on machine tools are fairly experimental at this time. The only spindles currently available for use on machine tools are developed and built by Societe Mecanique Magnetique (S2M) of France. At present there are three models of magnetic spindles available for milling purposes. These 3 models cover the speed range between 30,000, and 60,000 rpm with a rated horsepower between 20 and 34 (S2M Literature; SKF Report, 1981).

Magnetic spindles consist of a spindle shaft supported by contactless, active radial and thrust magnetic bearings, as is shown in Fig. 6. In operation, the spindle shaft is magnetically suspended with no mechanical contact with the spindle housing. Position sensors placed around the shaft continuously monitor the displacement of shaft in three orthogonal directions. It is particularly important to note that the spindle shaft can be translated up to ±.005 inches and tilted up to 0.5° with no effect on the performance of the spindle system. This unique feature of magnetically controlled spindles can have significant impact in correcting tool path errors.

The unique design of magnetic spindles provides significant advantages over conventional spindles with regard to tool path error correction (Anand,

Kirk and Anjanappa, 1986).

Several investigators have used magnetic spindles (Nimphuis, 1984; Raj Aggarwal, 1984; Schultz, 1984) by retrofitting them on existing machine tools. Their primary focus was to use the magnetic bearing spindle to improve metal removal rate. In the approach suggested in this paper, the many other advantages of using magnetically controlled spindles to improve tool path errors can take precedence over the advantage of high metal removal rate. This approach exploits the full capabilities of the magnetic spindles and will be useful for retrofitting existing machine tools for tool path error minimization.

ERROR MINIMIZATION METHODOLOGY

In general, tool path error consists (Anand, Kirk and Anjanappa, 1986) of machine tool errors and cutting force errors. These errors can be static and dynamic deterministic and/or stochastic. The machine tool static and dynamic deterministic errors can be quantified using a laser metrology system and put in the form of an error map for use in software correction. Cutting force errors are both static deterministic and stochastic and can be minimized by utilizing a magnetically controlled spindle and a control strategy which takes advantage of the spindles ability to tilt and translate, while continuing to rotate at high speeds.

A program to implement an error correction methodology in a vertical machining center has been undertaken. The strategy is to utilize an experimentally determined error matrix of a machine, along with models of cutting force errors, and to implement a corrective control scheme to reduce overall part errors in thin rib machining.

A control scheme as shown in Fig. 7, is the proposed block diagram for control of a magnetic bearing spindle. This scheme, although still being refined, will take the overall machine error matrix and cutting force model data and adjust the spindle location (both translation and tilt) in order to minimize the instantaneous overall tool path errors. This research involves the following tasks:

- develop an expert system to correct deterministic errors using error maps
- develop an expert system for stochastic error correction
- develop and implement control algorithms for controlling magnetically suspended spindles to minimize tool path errors
- validate models and algorithms using a CNC center fitted with a magnetically suspended spindle.

CONCLUSIONS

Tool path errors have been characterized as static deterministic, dynamic deterministic and stochastic. The source of each of these errors is either in the machine tool itself or in the nature of the cutting process. Based on the ability of a magnetic bearing spindle to both translate and tilt an initial control scheme for the magnetic bearing has been presented. Furthermore, it is expected that the long term benefits of this on-going research will lead to the enhancement of accuracy by quantifying and controlling tool path error, development of a control strategy for tool path error minimization in end-milling that is machine independent, and potential for increased MRR.

REFERENCES

Anand, D.K., Kirk, J.A., McKindra, C.D., "Matrix Representation and Prediction of Three Dimensional Cutting Forces", Transactions of ASME, Vol. 99, Series B, Nov. 1977, pp. 828-834.

Anand, D.K., Kirk, J.A., Anjanappa, M., "Magnetic Bearing Spindles for Enhancing Tool Path Accuracy", Advanced Manufacturing Processes, Vol. 1, No. 1, April 1986.

Gillespie, L.K. "Advances in Deburring", Society of Manufacturing Engineers, Dearborn, Michigan 1978.

Hocken, R.J., Nanzetta, P., "Research in Automated Manufacturing at NBS", Manufacturing Engineering, October 1983, p. 68-69.

MTTF Report, October 1980, "Technology of Machine Tools", Volume 1-5.

Nanzetta, P., "Update: NBS Research Facility Addresses Problems in Set-ups for Small Batch Manufacturing", Industrial Engineering, June 1984, pp. 68-73.

Nimphius, J.J., "A New Machine Tool Specially Designed for Ultra High Speed Machining of Aluminum Alloys", High Speed Machining, WAM of the ASME, New Orleans, Louisiana, December 9-14, 1984, pp. 321-328.

Raj Aggarwal, T., "Research in Practical Aspects of High Speed Milling of Aluminum", Technical Report, Cincinnati Milacron 1984.

Schultz, H., "High-Speed Milling of Aluminum Alloys", High Speed Machining, WAM of the ASME, New Orleans, Louisiana, December 9-14, 1984, pp. 241-244.

Simpson, J.A., Hocken, R.J., Albus, J.S., "The Automated Manufacturing Research Facility of the National Bureau of Standards", Journal of Manufacturing System, Vol. 1, NO. 1, 1982, p. 17-32.

SKF 1981 Report, "Active Magnetic Bearing Spindle Systems for Machine Tools", SKF Technology Services, June 1981.

S2M Literature, "Application of Active Magnetic Bearing to Machine Tool Industry".

Tlusty, J., Macneil, P., "Dynamics of Cutting Forces in End Milling", Annals of CIRP, Vol. 24, 1975.

Tlusty, J., "Criteria for Static and Dynamic Stiffness of Structures", Section 8.5, Volume 3, MTTF Report, October 1980.

<u>Résumé</u>: Cette étude concerne la quantification et le contrôle de l'imprécision sur la trajectoire de l'outil dans le cas de fraisage de surface utilisant un mandrin à contrôle magnétique. Des guides à parois minces sont utilisés pour les essais expérimentaux. Dans le cas d'un tel usinage, la précision de la pièce est importante et le fini de la surface doit être très soigné afin d'éviter l'atténuation du signal micro-onde. Le régime de coupe est choisi de telle sorte que l'imprécision sur la trajectoire de l'outil existe sans qu'il y ait vibration de l'outil.

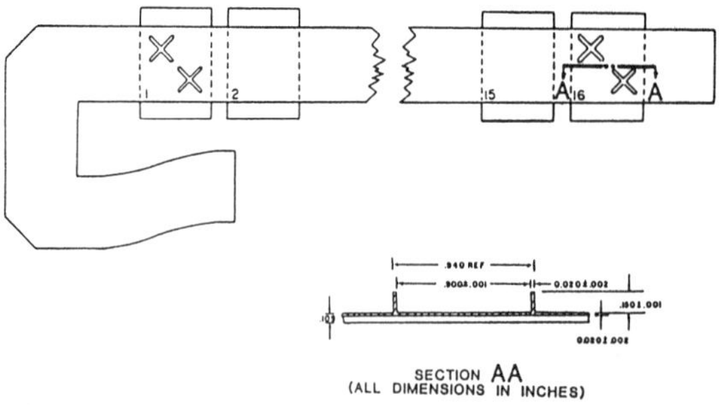

Fig. 1. Microwave guide (Courtesy of Westinghouse Corp.)

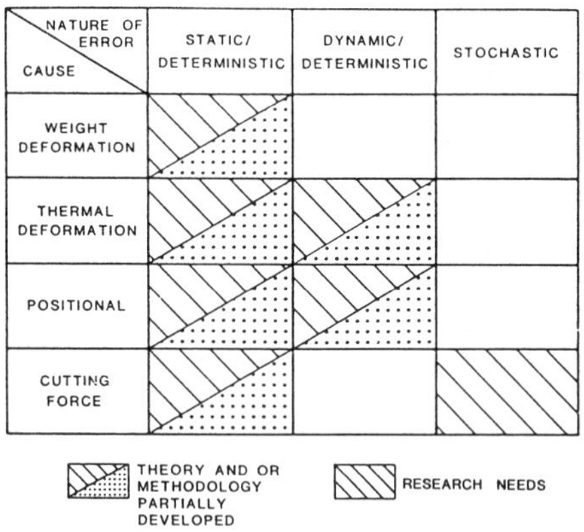

Fig. 5. End milling errors

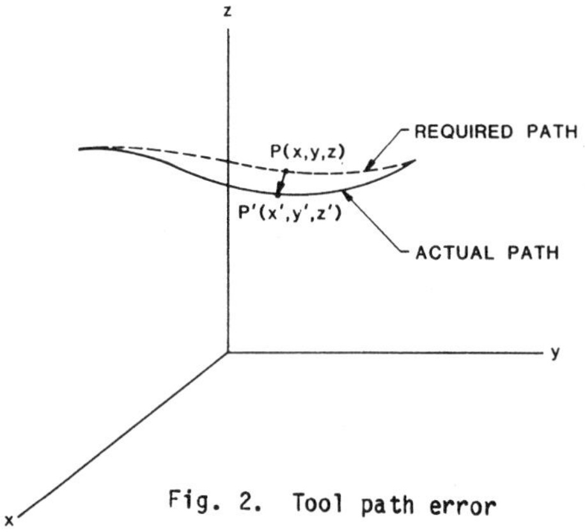

Fig. 2. Tool path error

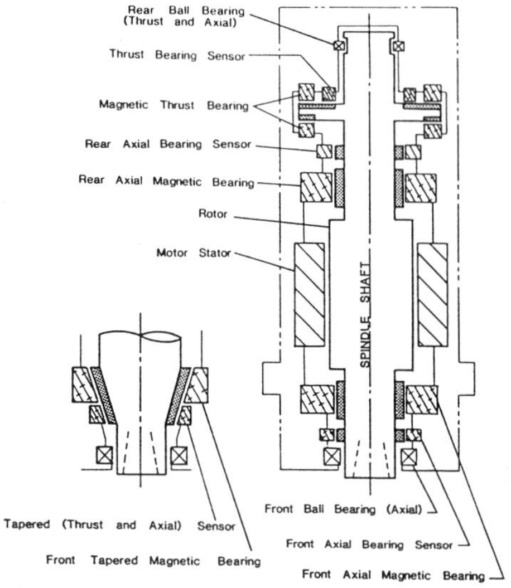

Fig. 6. Magnetic spindle configuration

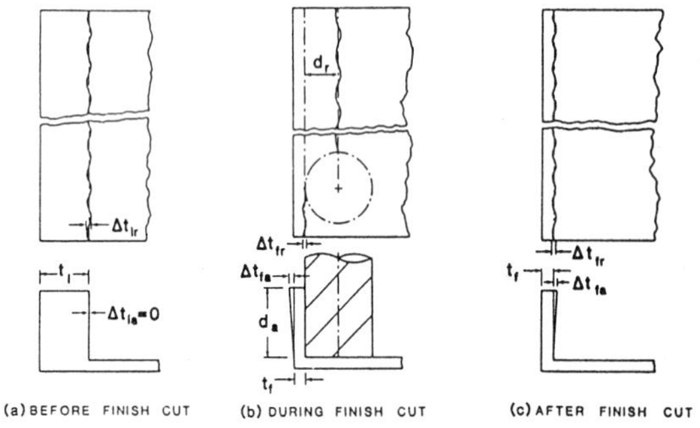

Fig. 3. Thin rib machining

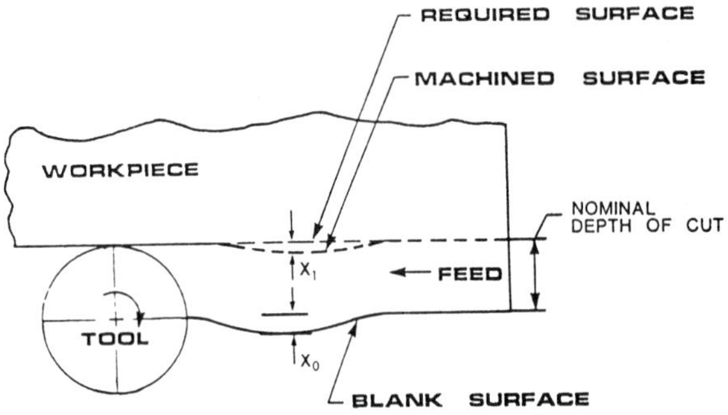

Fig. 4. Variation of depth of cut

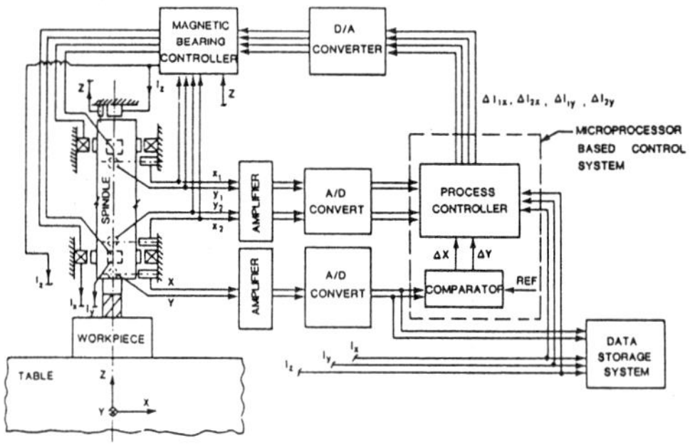

Fig. 7. Schematic diagram of tool path error minimization

Evaluation of Positioning Drive Parameters

G. V. KREJNIN

Institute for the Study of Machines, USSR Academy of Science, Moscow, 101830, USSR

Аннотация. Совершенствование приводов (электрических, гидравлических, пневматических) в результате применения новых материалов, технологий и средств управления сопровождается сближением их свойств, что затрудняет поиск оптимального варианта. Возникает необходимость в общей теории приводов, позволяющей решать задачи анализа и синтеза приводов с единой позиции. Разработанная в рамках такого подхода процедура выбора параметров позиционного привода рассматривается ниже. Она явилась результатом развития исследований автора, распространенных на случай больших перемещений, когда приходится учитывать нелинейные эффекты насыщения в системе и ряд ограничений. Метод выбора параметров охватывает приводы с параллельной и последовательной коррекцией. Приводятся примеры расчета приводов разных типов по единым исходным данным.

Keywords. Positioning drive, control system, parameters evaluation, nonlinear system, transion process, generalized modell

ВВЕДЕНИЕ

В последнее время наблюдается интенсивное совершенствование приводов на основе применения новых материалов, технологий и средств управления. Новые магнитные материалы и конструкции роторов, бесколлекторное распределение, транзисторное управление и др. привели к улучшению массо-габаритных показателей и управляемости электроприводов. Использование современных средств и алгоритмов управления позволило улучшить управляемость гидро- и пневмоприводов, компенсировать вредное влияние сжимаемости рабочего тела. Переход на новые уплотнения обеспечивает снижение утучек даже высоких давлениях, а замена основных деталей более легкими, выполненными из композитных материалов, приводит к снижению массы подвижных частей, уменьшению потребной мощности привода и расхода рабочего тела, а следовательно, к повышению эффективности гидро- и пневмоприводов

Процесс совершенствования приводов сопровождается постепенным сближением их свойств и возможностей, что делает поиск наиболее рационального технического решения в каждом конкретном случае все более трудным. Возникает необходимость в создании и развитии общей теории приводов, позволяющей подходить к анализу и синтезу приводов различных типов с единых позиций, максимально использовать заложенные в каждом приводе возможности.

Рассматриваемая ниже обобщенная процедура выбора параметров позиционного привода (ПП) - результат развития исследований [1] [2] , распространения их на случай больших перемещений, когда приходится выходить за рамки линейной модели и учитывать нелинейные эффекты, в первую очередь эффекты насыщения в системе.

На Фиг. I показаны структурные схемы ПП с параллельной (а) и последовательной (б) коррекцией в контурах управления. В первом случае процесс позиционирования состоит из двух этапов (Фиг. 2а): разгона с выходом на установившуюся скорость и торможения с отработкой заданной позиции. В случае (б) имеет место трапецидальный закон изменения скорости на большей части пути и лишь в конце хода (от точки "о") система активно отрабатывает заданную позицию (Фиг. 2б)

В задачу синтеза ПП входит выбор параметров двигателя, передаточного механизма (редуктора) и параметров контура управления из условия перемещения рабочего органа массой m за заданное время t_n на заданное расстояние $\bar{x}$ при требуемом качестве переходного процесса.

УРАВНЕНИЯ ДИНАМИКИ ПОЗИЦИОННОГО ПРИВОДА

Динамика ПП с параллельной коррекцией на втором этапе позиционирования в линейном приближении описывается безразмерным уравнением

$$\dddot{\xi} + A_1'\ddot{\xi} + A_2'\dot{\xi} + \xi = 1 \qquad (1)$$

$$A_1' = \nu_v + \nu_x \varkappa_d + \nu_x \varkappa_{\ddot{x}}$$
$$A_2' = \nu_x(\varkappa_{\dot{x}} + \varkappa_d \nu_v + 1); \quad A_3' = \nu_x \varkappa_x = 1 \qquad (2)$$

получаемым из исходного введением безразмерного времени $\tau = t/t^*$, где $t^* = (m/A_3)^{1/3}$ и безразмерного перемещения $\xi = x/\bar{x}$; A_3 - коэффициент при x исходного уравнения. Соотношения между безразмерными и размерными параметрами

$$\nu_x = c_x(t^{*2}/m); \; \varkappa_x = K_x t^*; \; \varkappa_{\dot{x}} = K_{\dot{x}};$$
$$\varkappa_{\ddot{x}} = K_{\ddot{x}}/t^*; \; \varkappa_d = K_d(m/t^*); \qquad (3)$$
$$\nu_v = c_v(t^*/m),$$

где ν_x, c_x -жесткость двигателя; $\varkappa_x$, K_x ; $\varkappa_{\dot{x}}$, $K_{\dot{x}}$; $\varkappa_{\ddot{x}}$, $K_{\ddot{x}}$ - коэффициенты усиления контуров по перемещению, скорости и ускорению; $\varkappa_d$, K_d - коэффициенты внутреннего демпфирования привода; ν_v , c_v -коэффициенты жидкостного трения.

Динамика этого же ПП на первом этапе позиционирования описывается уравнением

$$\dddot{\xi} + \bar{A}_1'\ddot{\xi} + \bar{A}_2'\dot{\xi} = B' \qquad (4)$$

где $\bar{A}_1'$ и $\bar{A}_2'$ получаются из (2) при $\varkappa_{\dot{x}} = \varkappa_{\ddot{x}} = 0$. Выражение

$$B' = \nu_x \dot{\xi}^* = 1 - \xi^* - \nu_x \varkappa_x \xi^* \qquad (5)$$

связывает предельную скорость $\dot{\xi}^*$ и координату ξ^*, в которой происходит смена этапов

Динамика ПП с последовательной коррекцией характеризуется постоянством ускорения на участках разгона и большей части участка торможения, а также постоянством скорости на промежуточном участке. В конце процесса когда все усилители выходят из состояния насыщения, начинается активная отработка позиции (Фиг. 2б); на этом этапе динамика ПП описывается уравнением (I), решаемым при ненулевых начальных условиях.

МЕТОДИКА ВЫБОРА ПАРАМЕТРОВ

Параметры ПП с параллельной коррекцией выбираются с помощью безразмерных эталонных переходных характеристик (ЭПХ), получаемых стыковкой решения уравнения (I) с предшествующим ему решением уравнения (4) при учете условия (5). Коэффициенты A_1' и A_2' базового более полного уравнения (I) принимаются равными эталонным значениям $A_{1э}'$ и $A_{2э}'$, которые выбираются по справочным данным 3 из условия воспроизведения переходного процесса желаемого вида. Семейство ЭПХ получается варьированием коэффициентов уравнения (4) (в пределах $\bar{A}_1' \leq A_1'$, $\bar{A}_2' \leq A_2'$) и B′. Имея в виду, что согласно (2) $A_2' \approx \nu_x$, уменьшение $\bar{A}_2'$ означает снижение жесткости двигателя. Этот фактор заметно влияет на характер ЭПХ, что хорошо видно на Фиг. 3а.

Для ПП с последовательной коррекцией при трапецидальном законе изменения скорости мощность двигателя $N_д$ и передаточное отношение i редуктора выбираются известными методами. При выборе параметров системы управления учитывают условия всех этапов движения, в т.ч. последнего (Фиг. 2б), характеризуемого начальным ускорением $\ddot{x}_o$ и скоростью $\dot{x}_o$; точка "о" удалена от конечной точки на относительно малое расстояние $\bar{x}_o$, что позволяет пользоваться линейным уравнением (I). Коэффициенты его выбираются из условия плавного выхода в конечную точку. Требуется также согласовать без размерные начальные условия с действительными и обеспечить заданную длительность t_o последнего этапа.

Выбор параметров управления удобно начать с задания масштаба времени t^*, пользуясь которым можно перейти от действительных начальных условий $\ddot{x}_o$ и $\dot{x}_o$ к безразмерным $\ddot{\xi}_o$ и $\dot{\xi}_o$. Далее по графику Фиг. 3б определяется τ_o с последующим переходом к действительному времени t_o. Пользуясь зависимостями (2) и (3), выбираем коэффициенты K_x, $K_{\dot{x}}$ и $K_{\ddot{x}}$ так, чтобы $A_I' = A_{I\text{э}}'$ и $A_2' = A_{2\text{э}}'$. Поскольку собственно двигатель был выбран раньше, то ν_v, ν_x и $\varkappa_d$ на этом этапе расчета могут считаться известными.

ПРИМЕРЫ РАСЧЕТА

Рассмотрим примеры расчета пневматического и электрического приводов, выполняющих одинаковую задачу перемещения массы m = 40 кг на расстояние $\bar{x}$ = 0,2 м за время t_n = 0,6 с; пневмодвигатель (пневмоцилиндр связан с массой непосредственно, электродвигатель - через редуктор.

Пневмопривод

Выбираем две эталонные кривые (Фиг. 3а), из которых I представляет движение с установившейся скоростью на большей части пути а II - состоит только из участков разгона и торможения. По τ_n и t_n определяем t^* и выполняем переход от $\bar{A}_2' \approx \nu_x$ к размерной жесткости c_x с использованием соотношения (3); далее по c_x находим геометрические параметры пневмоцилиндра.

Эффективная площадь $f_\text{э}$ проходного сечения канала находится по предельной скорости $\dot{x}^*$, связанной t^* и $\bar{x}$ с безразмерной скоростью ξ^*, определяемой из соотношения (5) как $\dot{\xi}^* = B'/\bar{A}_2'$, где B' выбранная величина. Из условий $A_I' = A_{I\text{э}}'$, $A_2' = A_{2\text{э}}'$ и соотношений (2) получаем

$$\varkappa_{\ddot{x}} = (A_1' - \bar{A}_1')/\nu_x; \; \varkappa_{\dot{x}} = (A_2' - \bar{A}_2')/\nu_x; \; \varkappa_x = 1/\nu_x$$

Далее переходим к размерным величинам согласно (3).

ТАБЛИЦА I. Результаты расчета пневмопривода

Параметры		Варианты I	Варианты II
D	, м	0,I50	0,040
$f_\text{э}$	, $м^2$	$22\cdot 10^{-6}$	$6\cdot 10^{-6}$
K_x	, $с^{-I}$	4I	32
$K_{\dot{x}}$	-	I,2	3,4
$K_{\ddot{x}}$	, с	0,0I	0,I

В варианте I, характеризуемом большей жесткостью двигателя, диаметр цилиндра больше (D = 0,I50 м), но одновременно упрощается управление двигателем - коэффициенты $\varkappa_{\ddot{x}}$ и $\varkappa_{\dot{x}}$ относительно малы. Наоборот, в варианте II D = 0,040 м, а коэффициенты $\varkappa_{\ddot{x}}$ и $\varkappa_{\dot{x}}$ много больше, чем в варианте I.

Электропривод

По исходным данным известными методами определен двигатель постоянного тока $N_д$ = 90 Вт с номинальным моментом 0,I43 Нм и частотой вращения 628 $с^{-I}$, который при i = I000 и к.п.д. редуктора 0,8 обеспечивает $t_n \approx$ 0,6 с. Начальные условия последнего этапа $\ddot{x}_o$ = - 3020 $с^{-2}$, $\dot{x}_o$ = 50 $с^{-I}$ и $\bar{x}_o$ = 0,4 рад.

При варьировании масштабом времени t^* в пределах 0,0075 ... 0,0I25 с безразмерные начальные условия $\ddot{\xi}_o$ = -0,4 ... -I,I; $\dot{\xi}_o$ = 0,9 ... I,5, которым согласно Фиг. 3б соответствуют τ_o = 2,2 ... I,0 и t_o = 0,0I6 ... 0,0I3 с. Учитывая, что при выбранном двигателе c_x, K_d и c_v известны, определяем ν_x, $\varkappa_d$ и ν_v согласно (3), находим $\varkappa_x$, $\varkappa_{\dot{x}}$ и $\varkappa_{\ddot{x}}$ из (2) и переходим к размерным величинам по соотношениям (3). В результате имеем K_x = 250 ... 80; $K_{\dot{x}}$ = 5,3 ... I,3 и $K_{\ddot{x}}$ = 0,026 ... 0,004. Уменьшение t^* ограничено возрастанием K_x ... $K_{\ddot{x}}$, а при увеличении t^* отрицательные обратные связи переходят в положительные.

ЛИТЕРАТУРА

1. KREJNIN G.V. Comparative analysis of performance characteristics of various types of drives. "Proc. of the 6-th World Congr. on TMM, Dec.15-20, 1983, New Delhi, India " V1., s. 677-680

2. КРЕЙНИН Г.В. Динамика приводов: модели и оценки. "Second Yugoslav-Soviet Symposium on applied robotics, June 14-15 , 1984, Aranjelovac, Yugoslavia, Proceed." Beograd, 1984, p. 63-69

3. ЯВОРСКИЙ В.Н. и др. Проектирование нелинейных следящих систем. Энергия, Л., 1976, с.206.

CONCLUSION

Extensive application of new materials, technology and electronics in electric, hydraulic and pneumatic drives have as a consequence a narrow gap between various types of drives bringing their perfor - mance characteristics closely together. A problem arose in searching the most sui table technical solution can be solved here using generalized approach covering all types of drives.

In [2] , [3] generalized nondimensional dynamic equations and parameters evaluation procedure in simplest form were presented. Now this approach has been developed to cover the case where large dis - placements take place and nonlinear effects is to be taken into account. Two varieties of control systems are considered : with (a) parallel and (b) series circuits (Fig. 1).

In case (a) eq. (4)at the first and eq.(1) at the second stage (Fig. 2a) are valid together with transitional conditions (5) Evaluation procedure is based upon nondimensional transitional curves (Fig. 3a) generated each as a series solutions of eq. (4) and (1).

In case (b) the movement of the system is represented by the curve shown in the Fig. 2b, where the point "o" notes the beginning of active positioning process. Actuators parameters in this case can be evaluted here using the well-known me - thods.The calculation control system parameters is based upon transition curves represented by eq. (1) generated by non-zero initial conditions. Examples of evaluation procedure applications are shown.

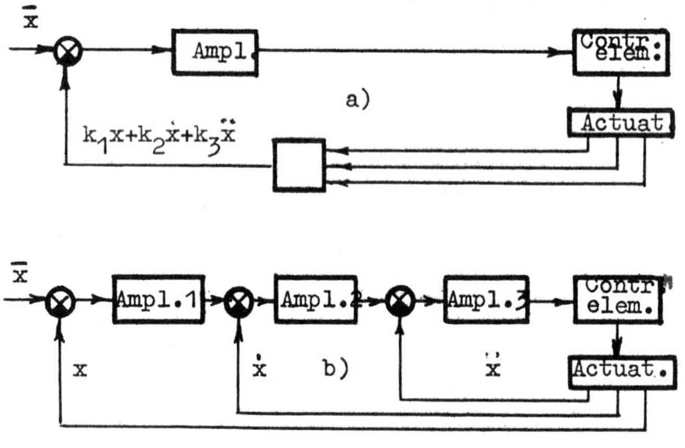

Fig. 1

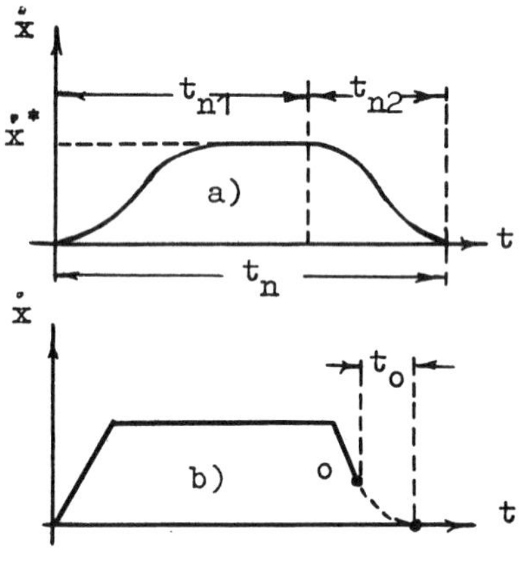

Fig. 2

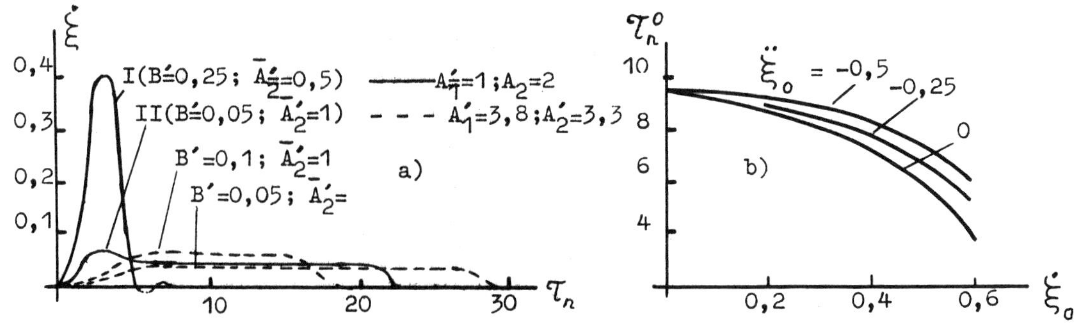

Fig. 3

The Stochastic Analysis of Machines Main Parameters' Fluctuations

N. SALENIEKS*, V. SERGEJEV** AND G. UPITIS

**Department of Mechanical Engineering, Riga Polytechnical Institute, Riga, USSR*

***Institute of Machine Sciences, USSR Academy of Sciences, Moscow, USSR*

Резюме. Техническое состояние машин характеризуется степенью соответствия определяющих параметров их номинальным значениям. Полученные при испытаниях реализации следует подвергать тщательному стохастическому анализу. Изучению подлежат флуктуации, их формирование, свойства детерминированных и случайных составляющих. Стохастический анализ осуществляется путем поочередного применения традиционных и ряда оригинальных методов обработки данных. Статистическая обработка проводится для каждой реализации в отдельности с последующим совместным анализом полученных выборочных статистик.

Ключевые слова. Механическая система, определяющий параметр, случайная функция, реализация, стохастический анализ, демодуляция, спектр.

ВВЕДЕНИЕ

В процессе реального функционирования машин закономерности взаимодействия их элементов претерпевают искажения, что проявляется в поведении определяющих параметров. Изменения определяющих параметров рассматриваются как реализации некоторых составных случайных функций, имеющих начальное значение, тренд и флуктуации - быстропротекающие изменения параметра, вызванные взаимодействием элементов механической системы.

Крупные искажения форм рабочих поверхностей, цикличность взаимодействия элементов, их колебания формируют особенности поведения исследуемого определяющего параметра, выражающиеся в виде некоторых детерминированных изменений. Случайные отклонения форм, блуждание мест контактирования, смещения элементов в поле зазора вызывают случайные флуктуации. При статистической обработке необходимо четко разграничить явно выраженные регулярные изменения параметра и изменения, носящие случайный характер.

СТРУКТУРА ФЛУКТУАЦИЙ

Флуктуации единичного экземпляра представляются в виде составной выборочной функции

$$ {}^{(i)}\varkappa(t) = {}^{i}z(t) + {}^{(i)}c(t), \qquad (1)$$

где ${}^{i}z(t)$ и ${}^{(i)}c(t)$ - детерминированные и случайные флуктуации.

В технических приложениях при объяснении понятия "случайная функция" традиционно основываются на ее представлении в виде совокупности выборочных функций, каждая из которых в отдельности рассматривается как обычная неслучайная функция. Это справедливо для квазидетерминированных функций, у которых каждая реализация описывается неслучайной кривой с однозначно определенными интерполированными и экстраполированными значениями. К этому классу принадлежит составляющая ${}^{i}z(t)$, так как любое ее точечное значение может быть определено однозначно по остальным известным значениям. Но составляющая ${}^{(i)}c(t)$ является выборочной функцией, обладающей случайными свойствами, так как для нее любое точечное значение не может быть однозначно вычислено по остальным ее значениям; могут быть оценены законы распределения вероятностей этих значений. Следовательно, для описания свойств каждой выборочной функции ${}^{(i)}\varkappa(t)$ необходимо выявить вид детерминированной функции ${}^{i}z(t)$ и выборочные статистические оценки вероятностных свойств выборочной функции ${}^{(i)}c(t)$.

АНАЛИЗ РЕАЛИЗАЦИИ ФЛУКТУАЦИЙ

При экспериментальном исследовании машин регистрируют отдельные отрезки определяющего параметра в ряде сеансов наблюдения. Статистическая обработка реализации содержит ряд характерных задач:
- выделение флуктуаций,
- разделение их на составляющие,
- исследование детерминированных флуктуаций,
- статистический анализ случайных флуктуаций.

Разработка процедур статистического анализа носит индивидуальный характер для каждого определяющего параметра, а иногда - даже для каждой реализации. Используют известные статистические процедуры, их модификации, а также прибегают к разработке новых оригинальных методов.

ДЕТЕРМИНИРОВАННЫЕ ФЛУКТУАЦИИ

Вид выделяемых детерминированных флуктуаций определяют при статистическом анализе зарегистрированных данных, руководствуясь представлением о характере формирования изучаемого определяющего параметра, а также некоторыми характерными особенностями поведения зарегистрированной реализации.

Довольно простым является представление детерминированных флуктуаций единичного экземпляра в виде периодической функции с постоянной или переменной структурой - обычно в виде суммы значимых ($j \in \nu_z$) периодических компонент разложения реализации ${}^{i}\varkappa(t)$ в ряд Фурье

$$ {}^{i}z(t) = \sum_{j \in \nu_z} \left[{}^{i}A_j(t) \cos \omega_j t + {}^{i}B_j(t) \sin \omega_j t \right] = \sum_{j \in \nu_z} {}^{i}R_j(t) \cos \left[\omega_j t - {}^{i}\varphi_j(t) \right], \qquad (2)$$

где коэффициенты ${}^{i}A_j(t)$ и ${}^{i}B_j(t)$ или ${}^{i}R_j(t)$ и ${}^{i}\varphi_j(t)$ в общем случае рассматриваются как медленно меняющиеся по наработке детерминированные функции - возрастающие, убывающие или немонотонные. В соответствии с (2) может быть построен дискретный спектр, выделяющиеся значения которого или некоторый их набор могут быть объяснены как особенности функционирования отдельных элементов механической системы.

Однако при физической трактовке гармонического представления (2) следует соблюдать определенную осторожность. При статистической обработке существует возможность получения псевдоспектра, вызванного наличием модуляций у гармонических компонент, присутствием импульсных значений и др. Так, например, в передаточных механизмах часто образуются модулированные по амплитуде и фазе флуктуации. В последующих двух разделах рассмотрено формирование и анализ реализации таких флуктуаций в течение сеанса наблюдения.

ФОРМИРОВАНИЕ МОДУЛИРОВАННЫХ ФЛУКТУАЦИЙ

Амплитудная и угловая (частотная или фазовая) модуляция флуктуаций параметров вращательного движения машин обуславливается переменностью их выборочных кинематических и динамических характеристик: функций мгновенного передаточного отношения передаточных механизмов - геометрических $g(\psi_1)=d\psi_2/d\psi_1$ и временных $q(t)=\omega_2(t)/\omega_1(t)$, связывающих элементарные приращения координат и текущие значения скоростей ведомых и ведущих звеньев (рис.1,а); амплитудно-частотных $|H(\nu,t)|$ и фазочастотных $\Phi(\nu,t)$ характеристик механизмов (рис.1,б), представляемых как линейные нестационарные динамические системы с частотными характеристиками $H(\nu,t)=|H(\nu,t)|\exp[j\Phi(\nu,t)]$, определяющими параметрами которых являются параметр входа $x(t)$ и выхода $y(t)$.

Переменные во времени другие параметры механизмов и параметры движения их звеньев представлены общей моделью $p(t)=\bar{p}+\overset{*}{p}(t)$, где $\bar{p}$ - среднее значение параметра, $\overset{*}{p}(t)$ - флуктуационная компонента, аппроксимируемая тригонометрическим многочленом $\overset{*}{p}(t)=\sum_m p_m\cos(\nu_m t+\varphi_{om})$; p_m, ν_m, φ_{om} - амплитуда, круговая частота и начальная фаза m-ой гармонической составляющей.

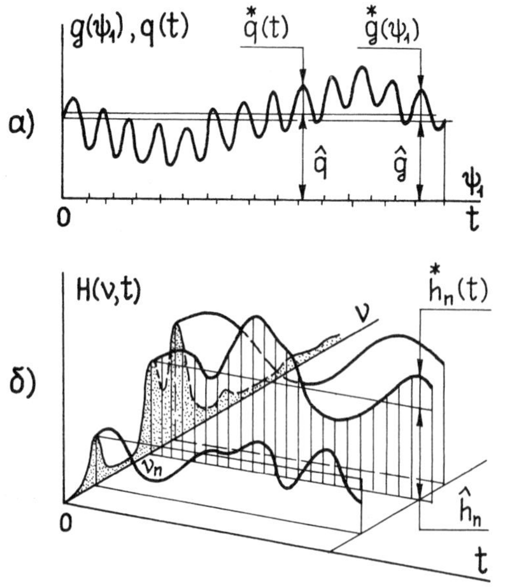

Рис.1. Характеристики передаточного механизма

Амплитудная модуляция флуктуаций обусловлена преобразованиями вида $y(t)=p(t)x(t)$, где в качестве закона модуляции $p(t)$ параметра $x(t)$ могут выступать мгновенное передаточное отношение $q(t)=\bar{q}+\overset{*}{q}(t)$ или амплитудно-частотная характеристика $|H(\nu_n,t)|=\bar{h}_n+\overset{*}{h}_n(t)$. Спектральная структура кинематически или динамически модулированного параметра усложняется возникновением множества модуляционных составляющих:

$$y(t)=\sum_n\Big[\bar{p}+\sum_m p_m\cos(\nu_m t+\varphi_{om})\Big]x_n\cos(\nu_n t+\varphi_{on})=$$

$$=\bar{p}\sum_n x_n\cos(\nu_n t+\varphi_{on})+$$

$$+\frac{1}{2}\sum_n\sum_m p_m x_n\cos\big[(\nu_n-\nu_m)t+\varphi_{on}-\varphi_{om}\big]+$$

$$+\frac{1}{2}\sum_n\sum_m p_m x_n\cos\big[(\nu_n+\nu_m)t+\varphi_{on}+\varphi_{om}\big]. \qquad (3)$$

Значимость отдельных составляющих спектрального разложения определяется коэффициентом глубины модуляции $w_m=p_m/\bar{p}$. Из-за незначительной глубины модуляции $w_{qm}=q_m/\bar{q}\ll 1$ при кинематических преобразованиях параметров движения амплитудная модуляция флуктуаций объясняется, в основном, действием динамических факторов.

Однако кинематическое преобразование может привести к заметным искажениям частотной структуры флуктуаций. Погрешности размеров, форм и взаимной ориентации элементов механизма формируют отклонения мгновенных передаточных отношений, зависящие от координаты ψ_1 ведущего звена и представляемые геометрической характеристикой $g(\psi_1)$. Спектральные составляющие характеристики $g(\psi_1)$ обусловлены эксцентриситетами, овальностями, высокочастотными циклическими погрешностями элементов передачи и распознаваемы по числу циклов в пределах периода функции $g(\psi_1)$. Непостоянство скорости ведущего звена $\omega_1(t)=\bar{\omega}_1+\overset{*}{\omega}_1(t)$, где $\overset{*}{\omega}_1(t)=\sum_{n=1}^{N}\omega_n\cos(\nu_n t+\varphi_{on})$ приводит к искажениям структуры геометрической характеристики $g(\psi_1)$ при ее развертывании на ось времени функционирования:

$$q(t)=g[\psi_1(t)]=\bar{g}+\sum_m g_m\cos\Big[\nu_m\int_0^t\omega_1(\tau)d\tau\Big]. \qquad (4)$$

При последовательном соединении R передаточных механизмов во флуктуациях параметров движения выходного звена проявляется частотно-модулированная структура характеристик мгновенного передаточного отношения всех передач кинематической цепи (рис.2):

$$\dot{\psi}_{R+1}=\frac{d\psi_{R+1}}{dt}=\Big[\prod_{r=1}^{R}q_r(t)-\prod_{r=1}^{R}\bar{g}_r\Big]\,\omega_1(t)\approx$$

$$\approx\sum_{r=1}^{R}\sum_m\Big[A_{rm}\cos\nu_{rm}t+\sum_{\alpha,\beta...\nu}a_{rm}(\alpha,\beta...\nu)\cos(\nu_{rm}+\sigma\Omega)t\Big], \qquad (5)$$

где

$$A_{rm}=\bar{\omega}_{R+1}\frac{g_{rm}}{\bar{g}_r}\prod_{n=1}^{N}B_0\Big(G_r\frac{m}{n}w_n\Big);\quad G_r=\frac{\bar{\omega}_r}{\bar{\omega}_{R+1}};$$

$$a_{rm}(\alpha,\beta...\nu)=\bar{\omega}_{R+1}\frac{g_{rm}}{g_r}B_\alpha(G_r m w_1)B_\beta\Big(G_r\frac{m}{2}w_2\Big)...B_\nu\Big(G_r\frac{m}{N}w_N\Big);$$

$B_n(\varphi)$ - функция Бесселя первого рода; $\alpha,\beta...\nu$ - произвольные целые постоянные, число которых соответствует числу учитываемых гармонических составляющих флуктуаций $\overset{*}{\omega}_1(t)$, $\sigma=\alpha+2\beta+...N\nu$; Ω - частота основной гармоники флуктуаций $\overset{*}{\omega}_1(t)$.

Кинематические возмущения $\ddot{\psi}(t)$, мгновенная частота которых флуктуирует вблизи некоторой из собственных частот системы, возбуждают амплитудно-модулированные флуктуации динамического характера. Другим источником амплитудной модуляции динамических флуктуаций представляется переменность параметров колебательных систем. По этой причине в ряде случаев имеет место также фазовая модуляция флуктуаций.

С учетом модуляционных явлений обширный класс флуктуаций параметров движения машин можно представить в виде модели

$$x(t)=\sum_m A_m(t)\cos[\nu_m t+\varphi_m(t)]+u(t), \qquad (6)$$

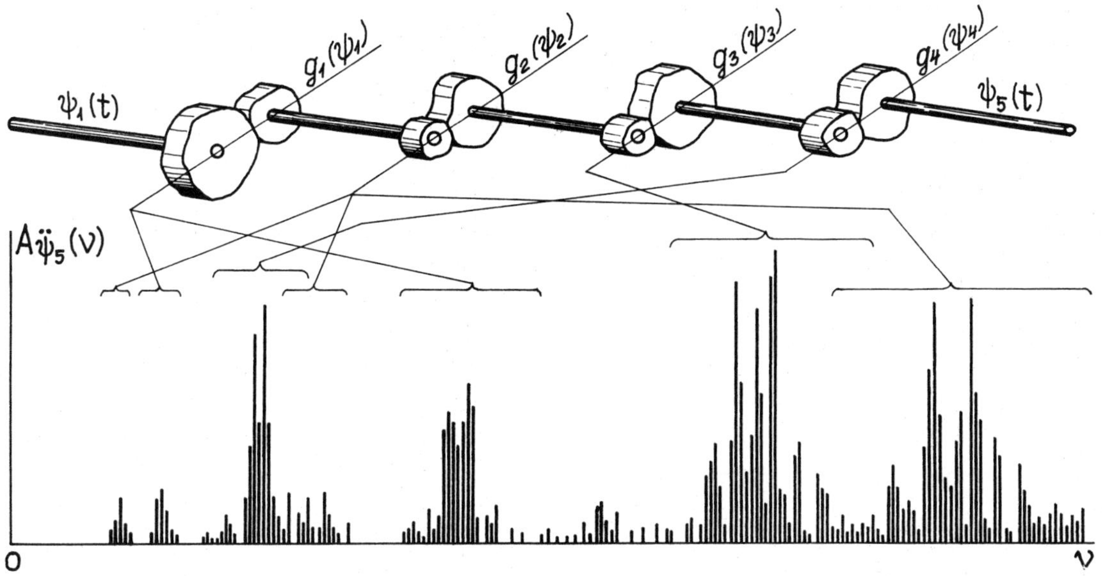

Рис.2. Структура частотно-модулированных флуктуаций углового ускорения

где $A_m(t)$, $\varphi_m(t)$ - законы амплитудной и угловой модуляции флуктуаций с несущей частотой ϑ_m ; $u(t)$ - тренд среднего значения.

АНАЛИЗ МОДУЛИРОВАННЫХ ФЛУКТУАЦИЙ

Практический интерес при анализе флуктуаций со структурой (6) представляет нахождение характеристик $A_m(t)$, $\varphi_m(t)$, ϑ_m; $m = 1, 2, \ldots$, несущих информацию о техническом состоянии объекта исследования. Непосредственное применение известных методов комплексной демодуляции и преобразования Гильберта для решения данной задачи затруднено вследствие составной структуры флуктуаций и отсутствия априорных сведений относительно частотных диапазонов отдельных составляющих.

Модуляционные характеристики флуктуаций могут быть установлены с помощью анализа координат экстремальных точек

$$(x_k; t_k)\ ;\ k = 0, 1 \ldots K-1, \tag{7}$$

принадлежащих колебанию с несущей частотой $\vartheta_M = K\pi/T$, где T - длина исследуемой реализации $x(t)$. Координаты (7) устанавливаются обычными методами поиска экстремумов функции путем аппроксимации экспериментальной реализации аналитическим выражением в виде ряда Фурье.

Другие колебания с несущими частотами $\vartheta_m < \vartheta_M$ образуют тренд среднего значения наиболее высокочастотной компоненты, структуру флуктуации можно представить в виде упрощенной модели

$$x(t) = U_M(t) + A_M(t)\cos[\vartheta_M t + \varphi_M(t)], \tag{8}$$

где

$$U_M(t) = u(t) + \sum_{m=1}^{M-1} A_m \cos[\vartheta_m t + \varphi_m(t)].$$

Закон угловой модуляции $\varphi_M(t)$ устанавливается в следующем порядке:

1) определяются смещения экстремумов модулированного колебания относительно эквидистантных экстремумов стационарного колебания

$$\tau_k = t_k - k\frac{T}{K}\ ;\ k = 0, 1 \ldots K-1; \tag{9}$$

2) для последовательности (9) рассчитываются комплексные коэффициенты $\mathcal{T}(m)$ дискретного преобразования Фурье (ДПФ), позволяющие аппроксимировать рядом Фурье функцию $t(\Phi)$, являющуюся обратной к изыскиваемому закону изменений фазового угла

$$\Phi_M(t) = \vartheta_M t + \varphi_M(t)\ ;$$

3) отклонения фазового угла находятся обращением временных смещений по алгоритму

$$\varphi_M(t) = \vartheta_M \sum_{m=0}^{K-1} \mathcal{T}(m)\exp\left\{jm\left[t - \frac{\sum \mathcal{T}(n)\exp(jnt)}{1 + \sum \mathcal{T}(n)jn\exp(jnt)}\right]\right\}; \tag{10}$$

$$t \in [0; 2\pi].$$

Определение закона амплитудной модуляции $A_M(t)$ включает следующие этапы:

1) строятся разреженные последовательности ординат x_{2n} и x_{2n+1}, $n = 0, 1 \ldots K/2-1$, представляющие либо только максимумы, либо только минимумы флуктуаций;

2) аппроксимируются верхняя и нижняя псевдоогибающие $E_+(t)$, $E_-(t)$ флуктуаций (рис.3) с помощью общего алгоритма

$$E(t) = \sum_{m=0}^{K/2-1} X(m)\exp\{jm[t - \varphi_M(t)/\vartheta_M]\}, \tag{11}$$

где $X(m)$ - коэффициенты ДПФ соответствующей последовательности ординат экстремумов (максимумов или минимумов);

3) искомая характеристика находится усреднением

$$A_M(t) = \frac{1}{2}[E_+(t) - E_-(t)]. \tag{12}$$

Аддитивные низкочастотные компоненты флуктуаций выделяются в виде переменного среднего значения флуктуаций

$$U_M(t) = \frac{1}{2}[E_+(t) + E_-(t)]. \tag{13}$$

Как слудует из модели (8), тренд среднего значения (13) может содержать другие, более низкочастотные составляющие колебательного характера. Поэтому организуются повторные циклы анализа, в которых в качестве анализируемых флуктуаций выступают выделенные рассмотренным способом компоненты $U(t)$. Разложение прекращается, если составляющая (13) не имеет колебательного характера.

СЛУЧАЙНЫЕ ФЛУКТУАЦИИ

Случайные флуктуации в рамках представления (1) исследуются на основе остатка, полученного после выделения из состава реализации ${}^i\varkappa(t)$ детерминированной составляющей ${}^iz(t)$, т.е. путем статистического анализа дискретных значений реализации

$${}^ic(t) = {}^i\varkappa(t) - {}^iz(t). \tag{14}$$

Характерным является представление случайных флуктуаций единичного экземпляра в виде некоторой выборочной случайной функции

$${}^{(i)}c(t) = {}^i\sigma_c(t)\,{}^{(i)}n(t), \tag{15}$$

где ${}^{(i)}n(t)$ - нормированная стационарная случайная функция с ${}^i\overline{n(t)} = 0$, ${}^i\sigma_n^2(t) = 1$ и относительно быстро убывающей корреляционной функцией ${}^ik_n(t;\Delta)$, $\lim_{\Delta\to\infty} {}^ik_n(t;\Delta) = 0$;

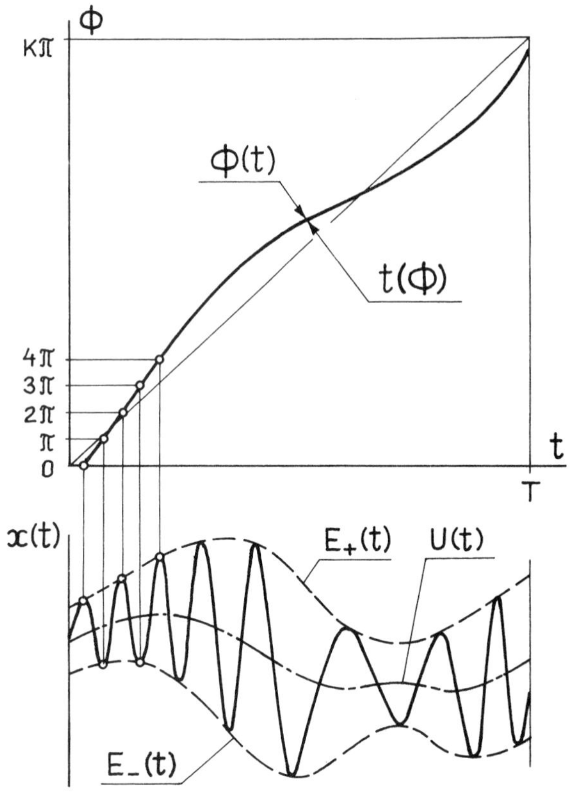

Рис.3. Анализ флуктуаций методом локальных экстремумов

$^{i}\sigma_c(t)$ - среднее квадратическое отклонение случайных флуктуаций - медленно меняющаяся по наработке функция.

В более сложных случаях строят другие вероятностные модели для $^{(i)}c(t)$, позволяющие учесть возможность наличия у реализации нестационарности и меняющегося во времени функционирования структурного состава. При этом применяются модифицированные методы статистического анализа, при создании которых могут быть использованы некоторые выявленные при исследовании детерминированных флуктуаций явления, как, например, имеющие место в передаточных механизмах преобразовательные свойства, закономерности модуляции и др.

SUMMARY

The theoretical relationship between the elements of machine parts has been distorted in real operation conditions. It is assumpted that fluctuations of every main parameter consist of the deterministic and random components, nevertheless the individual character of every item remains unchanged. Realization of the main parameters must be stochastically and thoroughly analysed on the basis of the physical structural model.

The main parameter random sample functions are compound for every item. During the data processing the deterministic and random components must be separated and their properties must be analysed.

The deterministic component has been considered to be a periodic function of a certain type. It is obtained that certain properties like amplitude or phase modulation, impulse-like components etc. are of great importance.

The statistical procedures have been derived for a single realization on the basis of a structural model. The tools in use are both standard and original methods such as impulse-like components detection, decomposition, filtering, demodulation etc.

The random component is non-stationary in common sense. Often it is possible to represent it as one joined to a stationary random function, i.e. as some transformant of a stationary random function.

The main original feature of our method is based on the use of the sample function. The statistical analysis is completed on a single realization contrary to the common use of the full set processing.

When a subset of objects must be analysed then the single item sample functions as Fourier discrete spectrum estimates, distribution function parameter estimates etc. are averaged. The averaging is not completed mechanically but the homogeneity of the subset is tested by the use of methods of the clusters or discriminating analysis.

REFERENCES

Kordonsky, H. B., Salenieks, N.K., and Zviedris, A.V. (1979). Statistical analysis of accuracy and parametrical reliability of machines and mechanisms. Proceedings of the Fifth World Congress on Theory of Machines and Mechanisms. ASME, 1200-1203.

Салениекс Н.К., Сергеев В.И., Упитис Г.В. (1979). Стохастические модели исследования формирования флуктуаций параметров механизмов. - Машиноведение, № 4, с.43-47.

Салениекс Н.К., Сергеев В.И., Упитис Г.В. (1983). Модели модулированных флуктуаций параметров движения передаточных механизмов. - Машиноведение, № 6, с.24-30.

Салениекс Н.К., Сергеев В.И., Упитис Г.В. (1984). Декомпозиция сложных флуктуаций методом локальных экстремумов. - В кн.: Методы исследования динамических систем на ЭВМ.- М.: Наука, с.3-12.

Machines and Systems Design

Extreme Synthesis of Mechanisms

K. Ts. ENCHEV

Higher Technical Institute, "A. Kynčev", Ruse, Bulgaria

Тезисы. В докладе излагаются основные положения специальной методики синтеза многозвенных плоских рычажных механизмов по бесконечно близким положениям, характеризующейся той простотой, которой отличается синтез простых четырехзвенных механизмов. Число бесконечно близких положений, с которыми оперирует методика, в зависимости от сложности синтезированного механизма может достигнуть до пятнадцати. Приведены два примера, в которых синтезированы восьмизвенные механизмы.

ВВЕДЕНИЕ

Синтез плоских рычажных механизмов по бесконечно близким положениям - это классическая область синтеза механизмов, в которой посредством определенной механизмом зависимости (траектории, функции положений и т.п.) аппроксимируется (воспроизводится приближенно) заданная геометрическая зависимость, оперируя с одним узлом аппроксимации высокой кратности. Подход к решаемым задачам в этой области иллюстрирован на фиг. 1, где заданная и реализованная (полученная в результате синтеза) геометрические зависимости представлены в графическом виде. Обозначен и интервал ε_x, в котором вторая кривая должна быть приближена к первой. Параметры механизма определяются так, чтобы в упомянутом интервале эти две кривые имели общую точку U, выбранную подходящим образом, и в этой точке последовательные производные $y', y'', \dots, y^{(m-1)}$ ($m = 2, 3, \dots$) соответствующих двух функций были соответственно равны. При этом ожидается, что насколько кратность m узла аппроксимации U выше, настолько максимальное по модулю отклонение Δ_{max} реализованной зависимости от заданной в интервале ε_x будет меньше, т.е. настолько более высокая степень аппроксимации будет обеспечена.

Фиг. 1

Задачи в рассматриваемой области синтеза механизмов решаются удобно и просто с помощью классического геометрического аппарата Бурместера (Burmester, 1888) и точнее - методами той его разновидности, котороя известна как кинематическая геометрия бесконечно близких положений (Артоболевский, Левитский, Черкудинов, 1959) и которая разработана последователями Бурместера: Мюллером, Альтом, Котельниковым, Геронимусом, Черкудиновым и др. Но, к сожалению, при применении этого аппарата кратность узла аппроксимации ограничена неравенством $m \leq 5$ и это является причиной для низкой и неудовлетворительной степени аппроксимации при более значительных интервалах ε_x, которые требует современная практика. А ограничение это произходит из самой геометрической природы задачи, сформулированной и решенной Бурместером, и поэтому кажется непреодолимым.

К счастью, это только нам кажется. В течение многих лет автор настоящего доклада искал возможности для преодоления ограничения $m \leq 5$ при сохранении произходящих из кинематической геометрии удобств и простоты методов синтеза. В результате этого ему удалось создать (Entchev, 1975; Енчев, 1985) соответствующую искомую методику синтеза плоских рычажных механизмов по бесконечно близким положениям и он назвал ее экстремальным синтезом.

ОБЩАЯ ХАРАКТЕРИСТИКА ЭКСТРЕМАЛЬНОГО СИНТЕЗА

Самое характерное в методике экстремального синтеза является то, что

а) синтез проводится в последовательные и самостоятельно обособленные этапы, после каждого из которых кратность узла аппроксимации (степень достигнутой аппроксимации) увеличивается;

б) на каждом этапе для синтеза применяется аппарат одной и той же структурой и степенью сложности, такими же как структура и степень сложности классического геометрического аппарата Бурместера.

Более конкретно выполняемый таким образом аппроксимационный синтез характеризуется следующим:

1. Максимальное возможное число упомянутых этапов три.
2. На первом этапе синтезируется четырехзвенный механизм (шарнирный четырехзвенник, один из подвижных шарниров которого может быть и бесконечно удаленной точкой). На втором и третьем этапе определяются параметры дополнительной кинематической цепи, добавленной специальным образом к синтезированному на предыдущем этапе механизму и превращающей его соответственно в восьмизвенный и четырнадцатизвенный.
3. В результате проведенного синтеза на данном этапе решаемая задача преобразуется на следующем этапе в задачу для реализации функции положений механизма, обладающей точкой экстремума в положении, соответствующем аппроксимационному узлу. Этот момент ключевой в методике - ему она обя-

зана своим существованием и поэтому с ним связано ее имя.

4. На каждом этапе в сущности синтезируются последовательно лишь одни шарнирные четырехзвенники, представляющие самостоятельно обособленные части механизма. Число этих шарнирных четырехзвенников на данном этапе равно номеру этапа.
5. Последний шарнирный четырехзвенник на каждом этапе синтезируется известным аппаратом классической кинематической геометрии бесконечно близких положений, а все предшествующие его шарнирные четырехзвенники - аппаратом нового типа кинематической геометрии, разработанной автором для нужд экстремального синтеза и названной также экстремальной.
6. На первом этапе кратность узла аппроксимации можно достичь до 5, на втором этапе - до 10 и на третьем этапе - до 15. Число этапов при решении конкретной задачи (а значит и сложность синтезированного механизма) принимается в зависимости от степени аппроксимации, которую необходимо обеспечить.
7. Первый этап соответствует известным до сих пор возможностям аппроксимационного синтеза по бесконечно близким положениям, а второй и третий этапы по своей сущности являются его естественным дальнейшим развитием.

ЭКСТРЕМАЛЬНАЯ КИНЕМАТИЧЕСКАЯ ГЕОМЕТРИЯ

Аппарат кинематической геометрии бесконечно близких положений в рамках экстремального синтеза применяется для решения задачи, которая формулируется следующим образом: найти параметры шарнирного четырехзвенника OABC (фиг. 2) так, чтобы существовало положение его, для которого последовательные производные $\beta', \beta'', \ldots, \beta^{(\mu_1-1)}$ ($\mu_1 = 2,3,4,5$) функции положений механизма $\beta = \beta(\alpha)$ имели заданные значения. Аппарат экстремальной кинематиче-

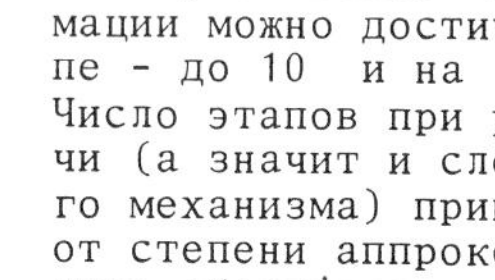

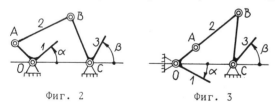

Фиг. 2

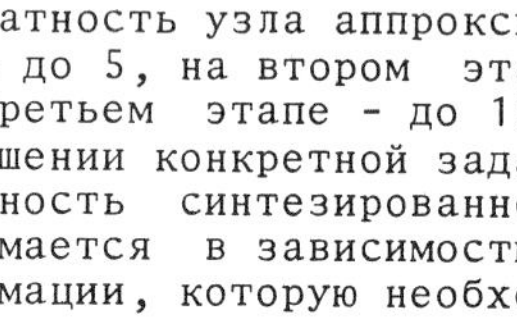

Фиг. 3

ской геометрии создан для решения задачи с такой формулировкой: найти параметры шарнирного четырехзвенника так, чтобы существовало положение его (фиг. 3), для которого выполнены первые μ_2 ($\mu_2 = 2,3,4,5$) из следующих пяти условий -

$$\beta' = 0, \tag{1}$$

$$\beta'' \neq 0, \tag{2}$$

$$-\frac{1}{3}\frac{\beta'''}{\beta''} = k_1, \tag{3}$$

$$-\frac{1}{3.10}\frac{\beta^{V}}{\beta''} - \frac{1}{3}\frac{\beta^{IV}}{\beta''}k_1 = k_2, \tag{4}$$

$$-\frac{1}{3.10.21}\frac{\beta^{VII}}{\beta''} - \frac{1}{3.10}\frac{\beta^{VI}}{\beta''}k_1 + \frac{1}{3}\frac{\beta^{IV}}{\beta''}(3k_1^3 - k_2) = k_3, \tag{5}$$

где значения величин k_1, k_2 и k_3 заданы. Другими словами, при новом типе кинематической геометрии вместо заданных значений производных функции положений требуется реализовать заданные значения $k_1, k_2, \ldots, k_{\mu_2-2}$ специальных аналитических выражений, содержащих эти производные.

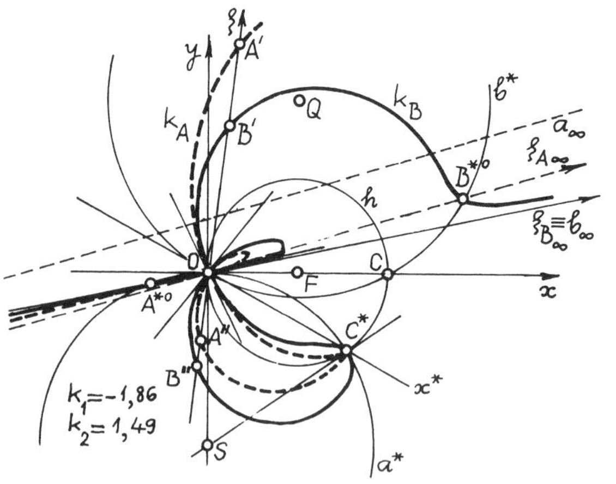

Фиг. 4

Экстремальная кинематическая геометрия аналогична по структуре и степени сложности кинематической геометрии бесконечно близких положений и ее аппарат деформирован определенным образом относительно аппарата последней. На фиг. 4 изображены кривые k_B и k_A (геометрические места точек B и A при $\mu_2 = 4$), которые характеризуют экстремальную кинематическую геометрию и являются аналогами известных кривых Бурместера (кривой круговых точек и кривой центров).

ПРИМЕРЫ

Иллюстрируем последовательность вычислительных операций и возможности экстремального синтеза при помощи двух численных примеров.

Синтез прямолинейно-направляющего механизма. Параметры восьмизвенного механизма из фиг.5 определены так, чтобы траектория q его точки Q воспроизводила приближенно заданную прямую q_0 в окресности заданного узла аппроксимации Q_0. На первом этапе решения задачи аппаратом кинематической геометрии бесконечно близких положений синтезирован шарнирный четырехзвенник OKLM (с неподвижными шарнирами O и M) и обеспечена кратность 4 узла аппроксимации. На втором этапе аппаратами экстремальной кинематической геометрии и кинематической геометрии бесконечно близких положений синтезированы один за другим шарнирные четырехзвенники OABC и CDEF и кратность узла аппроксимации увеличина на 8.

Переход от первого к второму этапу характеризуется следующим. Шарнир M уже синтезированного шарнирного четырехзвенника OKLM освобождается от неподвижности и добавляется новое звено MF с новым неподвижным шарниром F, место которого выбирается произвольно (но не и на прямой LM). Получается пятизвенный механизм OKLMF с двумя степенями свободы, функция положений $\psi = \psi(\varphi)$ которого однако вполне определена условием, что точка Q движется по прямой q_0. Это позволяет, с помощью подходящо выведенных формул, вычислить последовательные производные $\psi', \psi'', \ldots, \psi^{(m-1)}$ этой функции для положения механизма, в котором $Q \equiv Q_0$. При этом, благодаря предварительному синтезу цепи OKLM, получается

$$\psi' = \psi'' = \psi''' = 0, \quad \psi^{IV} \neq 0. \tag{6}$$

Это означает, что в упомянутом положении функция $\psi = \psi(\varphi)$ имеет точку экстремума.

Далее отнимается одна из степеней свободы механизма соединением его конечных звеньев 1 и 5 кинематической цепью ABCDE. Таким образом, появляется шестизвенный механизм OABCDEF, функция положений которого должна аппроксимировать функцию $\psi=\Psi(\varphi)$, т.е. параметры которого необходимо определить так, чтобы реализовать вычисленные производные $\Psi', \Psi'', \ldots, \Psi^{(m-1)}$. В принципе при $m \leq 9$ это возможно, так как функция положений шестизвенного механизма определяется 9 параметрами. Эта чрезвычайно трудная в общем случае задача[1], благодаря точке экстремума функции положений $\psi=\psi(\varphi)$, распадается на две простые задачи для последовательного синтеза шарнирных четырехзвенников OABC и CDEF, из которых состоит шестизвенный механизм. Величины $k_1, k_2, \ldots, k_{\mu_2-2}$

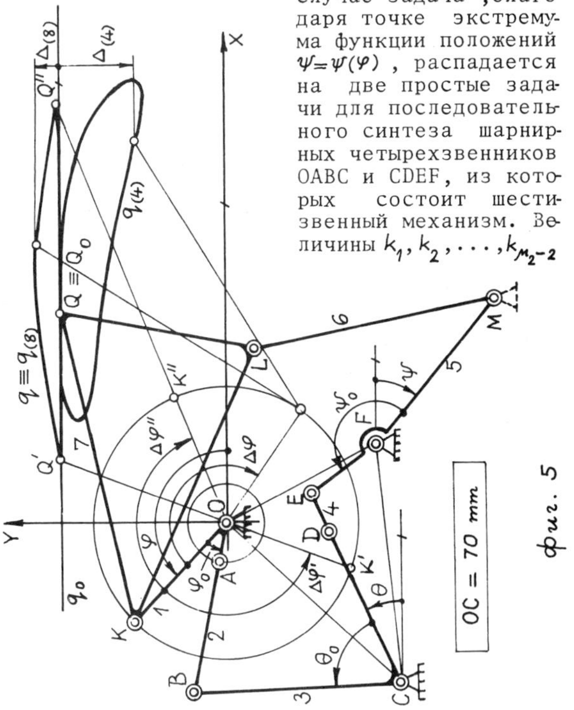

фиг. 5

[см. формулы (1),(2),...,(5)], реализуемые при синтезе шарнирного четырехзвенника OABC, вычисляются предварительно по формулам, в которых участвуют производные $\Psi^{IV}, \Psi^{V}, \ldots, \Psi^{(m-1)}$. Производные $\psi'_\theta, \psi''_\theta, \ldots, \psi_\theta^{(\mu_1-1)}$ функции положений $\psi=\psi(\theta)$ шарнирного четырехзвенника CDEF, подлежащие реализации при его синтезе, вычисляются также предварительно по формулам, которые кроме производных $\Psi^{IV}, \Psi^{V}, \ldots, \Psi^{(m-1)}$ содержат и производные $\theta', \theta'', \theta''', \ldots$ функции положений $\theta=\theta(\varphi)$ уже синтезированного шарнирного четырехзвенника OABC.

В рассматриваемом примере принято $m = 8$ и $\mu_2=\mu_1=4$. Траектория точки Q как точка синтезированного на первом этапе шарнирного четырехзвенника OKLM на фиг. 5 обозначена $q_{(4)}$, а как точка полученного в конечном счете восьмизвенного механизма - $q_{(8)}$. Обозначены и отклонения $\Delta_{(4)}$ и $\Delta_{(8)}$ обеих траекторий от заданной прямой q_0.

На фиг. 6 построены диаграммы для этих отклонений. Непосредственно видно, что благодаря второму этапу (т.е. экстремальному синтезу) степень аппроксимации увеличивается значительно. Например, при допустимом отклонении $[\Delta]$ = 0,5 mm интервал аппроксимации от $\varepsilon_{\varphi(4)} = 77^\circ$ увеличивается на $\varepsilon_{\varphi(8)} = 227^\circ$, т.е. около трех раз.

Параметры механизма в примере вычислены (в милиметрах и градусах) с высокой точностью (до восьмой цифры после десятичной запятой). Построенная на фиг. 6 кривая $\Delta^*_{(8)}$ получена после округления всех параметров до первой

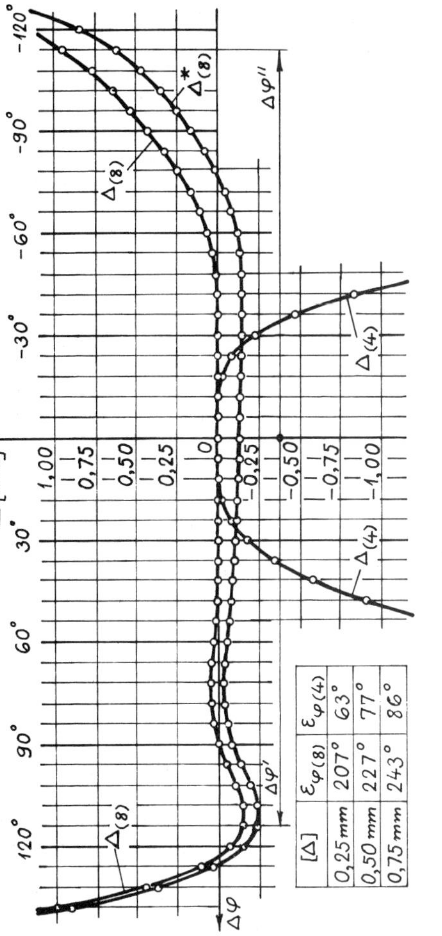

фиг. 6

цифры после десятичной запятой (при округление до второй цифры кривая $\Delta_{(8)}$ не меняет видимо свой вид). Видно, что неточности практической реализации механизма не столь фатальны и что даже в конкретном случае привели к повышению степени аппроксимации (конечно, это случайно).

Синтез уравновешивающего механизма. Второй пример, который иллюстрирован с помощью фиг. 7, 8 и 9, относится к уравновешиванию весов G_1, G_2 и G_3 звеньев шарнирного четырехзвенника $O\hat{A}\hat{B}\hat{C}$ стрелы $O\hat{A}$ портального крана. Приведенной к стреле (к звену 1) момент этих

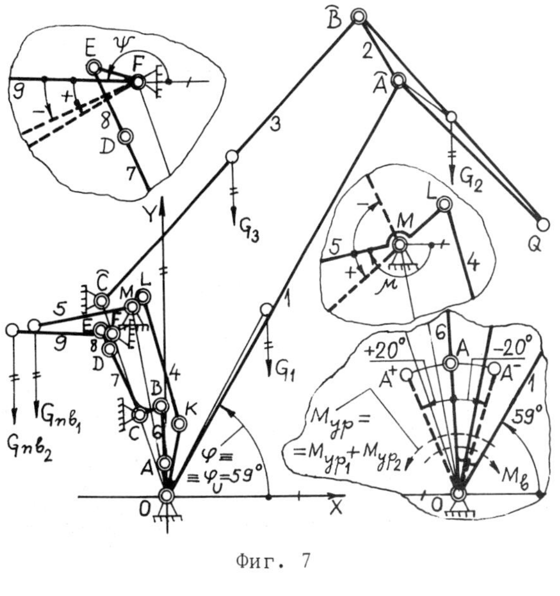

Фиг. 7

[1]Если условия (6) не выполнены, при $m = 9$ нужно решать систему из 26 тригонометрических уравнений, некоторые из которых записываются на целой странице.

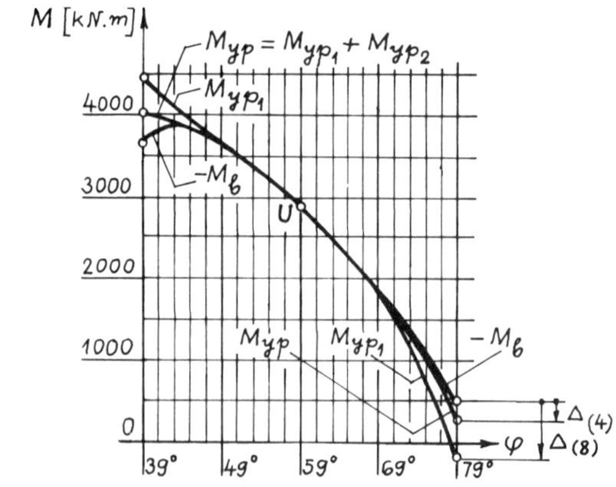

Фиг. 8

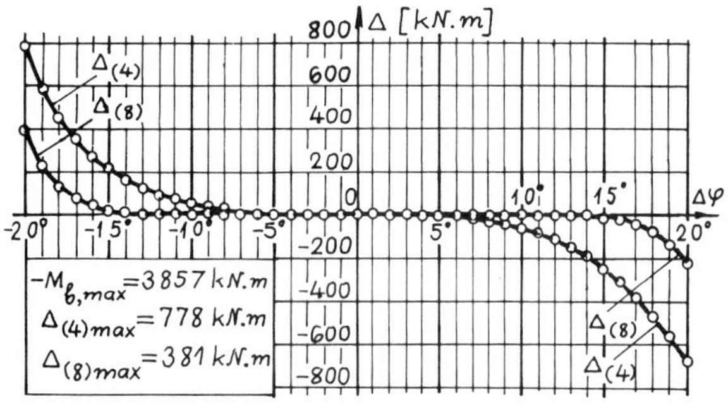

Фиг.9

весов, который подлежит уравновешиванию, обозначен M_b, а приведенные к тому же самому звену моменты уравновешивающих противовесов G_{nb_1} и G_{nb_2} - M_{yp_1} и M_{yp_2}. На первом этапе решения задачи синтезируется основной уравновешивающий механизм - шарнирный четырехзвенник OKLM, а на втором этапе - дополнительный шестизвенный уравновешивающий механизм OABCDEF.

В настоящее время в практике шарнирный четырехзвенник стрелы портального крана уравновешивают с помощью одного шарнирного четырехзвенника OKLM. Однако из примера видно что, при добавлении шестизвенного уравновешивающего механизма OABCDEF степень уравновешенности увеличивается более, чем в два раза, что ведет к экономии энергии в процессе эксплуатации, к повышению надежности в аварийной ситуации и т.п. При этом вычисления показывают, что противовес G_{nb_2}, который необходимо установить дополнительно, в 15 раз меньше, чем основной противовес G_{nb_1}.

Схема рассматриваемого уравновешивающего механизма защищена авторским свидетельством (Енчев, 1979).

ЛИТЕРАТУРА

Артоболевский, И.И., Н.И.Левитский и С.А. Черкудинов (1959). Синтез плоских механизмов. Физматгиз, Москва, стр.1017-1068.

Burmester, L.(1888). Lehrbuch der Kinematik. Leipzig, §§ 249-254.

Entchev, K.Ts. (1975). Fundamental situations of extreme kinematic geometry. Fourth Congress on the TMM, Newcastle upon Tyne(England), vol. 2, 315-320.

Енчев, К.Ц. (1979). Механизъм за уравновесяване на стреловия шарнирен четиризвенник на порталния кран. Авторско свидетелство, № 26067, Н.Р.България.

Енчев, К.Ц. (1985). Екстремен синтез на механизмите. Дисертация. ВТУ "А.Кънчев", Русе (България), 451 стр.

EXTREME GETRIEBESYNTHESE

Zusammenfassung

Das Thema des Vortrags ist aus dem Gebiete der Synthese der ebenen Gelenkgetrieben nach unendlich benachbarten Lagen, d.h. aus dem Gebiete der Synthese, bei der durch einen vom Getriebe bestimmten Zusammenhang (Bahnkurve, Übertragungsfunktion u.a.) eine gegebene geometrische Abhängigkeit approximiert wird (näherungsweise reproduziert wird), wobei mit einem Approximationsknotenpunkt mit hoher Vielfachheit operiert wird. In dem Vortrag werden bis jetzt unbekannte Möglichkeiten zur Durchführung derselben Synthese in aneinanderfolgenden und selbständigen Etappen beschrieben, nach jeder von denen sich der erreichte Approximationsgrad (d.h. die Intervallänge der Annäherung der gewonnenen geometrischen Abhängigkeit an die gegebene) erhöht. Dabei wird in jeder Etappe ein Apparat zur Synthese mit einer und derselben Struktur und mit einem und demselben Schwierigkeitsgrad angewendet, die gleich mit der Struktur und mit dem Schwierigkeitsgrad des klassischen geometrischen Apparats von Burmester sind.

Die höchst mögliche Anzahl der erwähnten Etappen ist drei. In der ersten Etappe wird ein viergliedriges Getriebe synthesiert. In der zweiten und dritten Etappe werden die Kenngrößen der zusätzlichen kinematischen Kette bestimmt, die auf spezielle Weise an das in der vorherigen Etappe synthesierte Getriebe angehängt wird und ihn dementsprechend in acht- und vierzehngliedriges Getriebe umwandelt.

Als Ergebnis der durchgefürten Synthese in einer bestimmten Etappe wird die aufzulösende Aufgabe in der nächsten Etappe in eine Aufgabe zur Realisierung der Übertragungsfunktion umgewandelt, die einen Extrempunkt in der Lage hat, die dem Approximationsknotenpunkt entspricht. Aus diesem Grunde wurde die dargestellte Methodik extreme Getriebesynthese genannt.

In jeder Etappe werden eigentlich nacheinander nur Gelenkvierecke synthesiert, die selbständig abgesonderte Teile des synthesierten Getriebes darstellen. Die Anzahl dieser Gelenkvierecke ist in einer bestimmten Etappe gleich der Nummer der Etappe.

In der ersten Etappe kann die Fachheit des Approximationsknotenpunktes bis 5, in der zweiten - bis 10 und in der dritten - bis 15 erreichen.

In dem Vortrag wird das Wesen der Methodik durch zwei Beispile geklärt. Das Beispil, das mit Hilfe der Abb. 7, 8 und 9 illustriert wird, bezieht sich auf das Balancieren der Gewichte G_1, G_2 und G_3 der Glieder des Pfeilgelenkvierecks $O\widehat{ABC}$ des Hafenwippkranes. Das in dem Getriebeschema bei der zweiten Etappe der Lösung der Aufgabe zusätzlich angehängte sechsgliedrige gleichgewichtschaffende Getriebe OABCDEF verbessert den Gleichgewichtsgrad mehr als 2 mal. Dabei ist das entsprechende zusätzliche Gegengewicht G_{nb_2} 15 mal weniger als die am Anfang angelegte Gegengewicht G_{nb_1}, d.h. als das Hauptgegengewiht.

Theoretical Approach to Problems of Stability of Piping Systems

A. TUČEK*, J. PALYZA* AND J. BARTÁK

*SIGMA Research Institute, K Nouzovu 2090, 143 16 Praha 4, Czechoslovakia
**Department of Mathematics of the Pedagogical Faculty of the Charles University, M. D. Rettigové 4, 116 39 Praha 1, Czechoslovakia

Abstract. The paper investigates the influence of chosen nonlinear elements and the influence of the nonlinearity of the flowing medium on the dynamics of the pipe-line systems. The nonlinearity of the medium is caused by a nonlinear character of the constitutive equation. Other phenomena taken into consideration are the effects of the elongation of the pipe-line system, the effects of the velocity of the flowing medium, the influence of the internal pressure and the heating of the pipe-line. Further, the paper shows the influence of the damping and the already mentioned effects on the stability of the system and on the first eigenfrequency.

Keywords. Pipe-line systems; stability; eigenvalues; nonlinear equations; damping.

INTRODUCTION

The pipe-line systems represent a complicated space structure composed of a set of elements according to Fig. 1. These may be of a linear or nonlinear character. Nonlinearity may be due to both construction properties of the element and the mutual interaction of elements with the flowing medium. The dynamic research is aimed at the investigation of chosen typical phenomena and nonlinearities and their influence on the tuning of the system and the possibility of instable states arising. The application of the results is assumed for Expert systems (ES) and CAD.

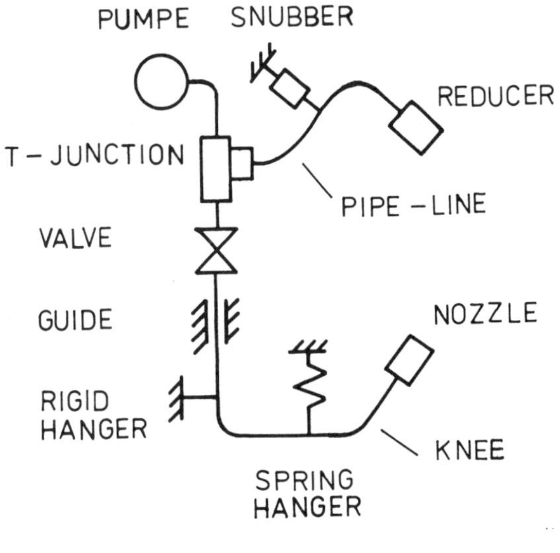

Fig. 1 The scheme of the pipe-line system.

THE DYNAMICS OF NONLINEAR FLOWING MEDIUM

Assume that the flow of the medium posseses a steady state solution characterized by the constant velocity V and the constant pressure P. The small perturbances v, p of this steady state solution will be described by the constitutive and motion equations

$$p_t = E_M v_x^3 \tag{1}$$

$$\rho_M v_t = p_x \tag{2}$$

together with the boundary conditions (see Fig. 2)

$$v(0,t) = A_0 \sin\omega t, \quad p(l,t) = 0.$$

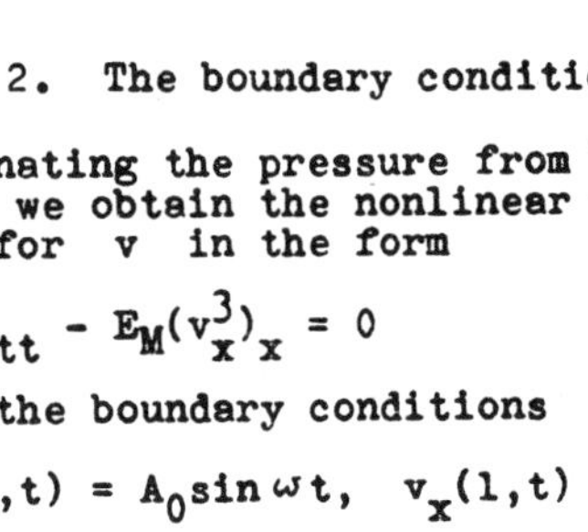

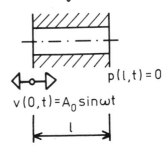

Fig. 2. The boundary conditions.

Eliminating the pressure from both equations we obtain the nonlinear wave equation for v in the form

$$\rho\, v_{tt} - E_M (v_x^3)_x = 0 \tag{3}$$

with the boundary conditions

$$v(0,t) = A_0 \sin\omega t, \quad v_x(l,t) = 0.$$

Applying the Ritz-Galerkin method we obtain the approximate solution in the vicinity of the first eigenvalue in the form

$$v(x,t) = (A_0 + A_1 \sin \Omega x) \sin \omega t, \quad \Omega = \pi/2l,$$

where for the amplitude A_1 we have the equation

$$\frac{4}{9} \frac{c_M^2 \Omega^4}{\omega^2} A_1^3 - A_1 + \frac{4}{3} A_0 = 0 \qquad (4)$$

where $c_M^2 = E_M/\rho_M$.

The dependence of A_1 on the frequency is sketched at Fig. 3. As far as these three solutions are concerned, only the stable solutions are realized in practice. Usually the solution which is the poorest from the point of view of stability corresponds to the middle branch at Fig.3.

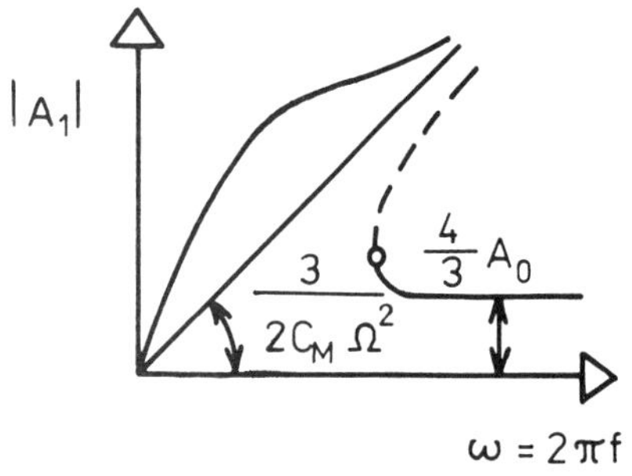

Fig. 3. The amplitude dependence on the frequency.

THE INFLUENCE OF NONLINEAR ELEMENTS

In this section we shall investigate the influence of some parameters on the behavior of the system. Concretely, we shall deal with the influence of the elongation of the pipe-line, the velocity, the pressure and the temperature of the flowing medium.

The Influence of the Elongation of the Pipe-line

The force caused by the elongation of the pipe-line is given under the notation of Fig. 4 by the formula

$$F_E = E_L A_L / 2l \int_0^l u_x^2(x,t)\, dx \, .$$

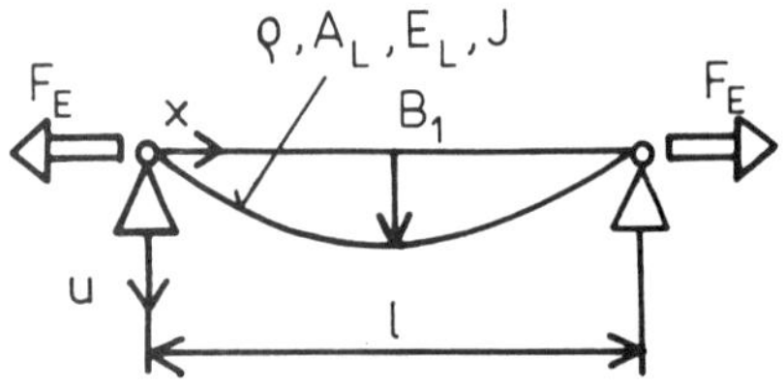

Fig. 4. Elongation of the pipe-line.

The motion equation takes the form

$$u_{tt} + a^2 u_{xxxx} - b \int_0^l u_x^2\, dx \;\; u_{xx} = 0,$$

where $c^2 = E_L/\rho$, $a^2 = c^2 i^2$, $i^2 = J/A_L$ (inertia radius), $b = c^2/2l$, ρ is the density of the pipe-line -medium system. We shall apply the Ritz-Galerkin method and the first approximation will be taken in the form

$$u(x,t) = B_1 \sin \Omega x \sin \omega t, \qquad \Omega = \pi/l.$$

A standard calculus yields the dependence of the first eigenfrequency on the amplitude B_1 of the vibrations in the form

$$(f_1/f_{1L})^2 = 1 + l^2/(6i^2)\ (B_1/l)^2$$

where f_{1L} is the eigenfrequency of the linear problem

$$(f_{1L} = a^2 \Omega^2/\sqrt{2\pi}\ ).$$

The graph of this dependence is sketched at Fig. 5.

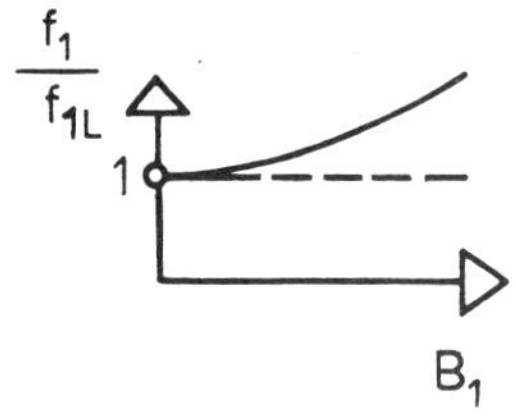

Fig. 5. The dependence of the first eigenfrequency on the amplitude.

The Influence of the Velocity of the Flowing Medium

The force caused by the flowing medium on the pipe-line is considered under the notation of Fig. 6 in the form

$$F_C = \rho_M A_M v_M^2 u_{xx} \, .$$

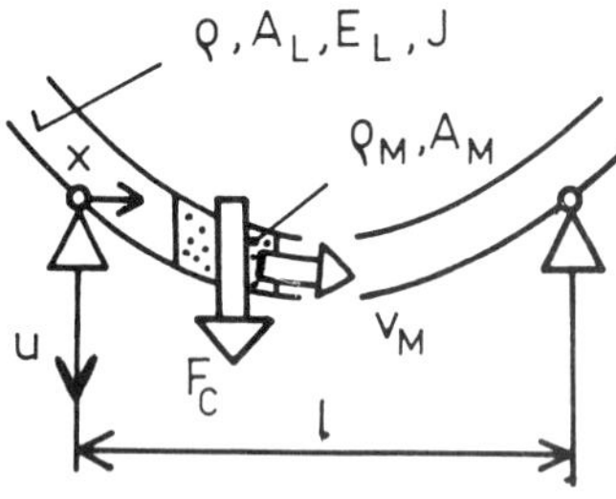

Fig. 6. The influence of the flowing medium.

The motion equation now takes the form

$$u_{tt} + a^2 u_{xxxx} + v^2 u_{xx} = 0 \qquad (5)$$

where $v^2 = k_A/k_\rho \, v_M^2$, $k_\rho = \rho/\rho_M$, $k_A = A_M/A_L$ (A_M is the cross-section of the medium). The solution corresponding to the first eigenfrequency has the form

$$u(x,t) = B_1 \sin \Omega x \sin \omega t \, .$$

Substituting this into the equation (5)

we obtain the dependence of the first eigenfrequency

$$(f_1/f_{1L})^2 = 1 - k_v\,(v_M/c)^2$$

on the velocity v_M (here

$$k_v = k_A/(k_\varrho i^2 \Omega^2)\ ).$$

This dependence is introduced at Fig. 7 where the value of the critical velocity is marked.

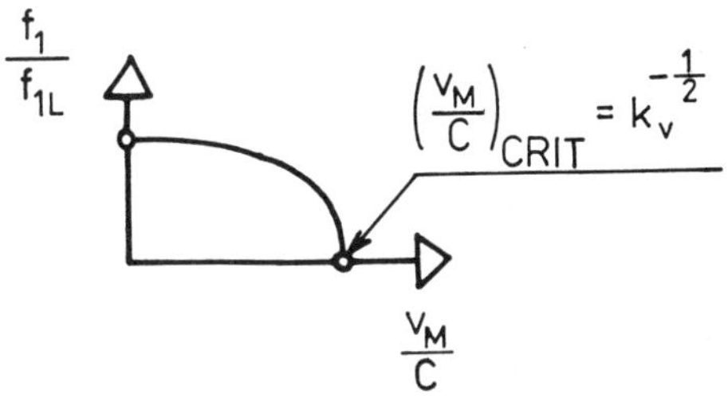

Fig. 7. The dependence of the first eigenfrequency on the medium velocity.

The Influence of the Internal Pressure

The force caused by the internal pressure according to Fig. 8 is given by the formula

$$F_p = p_M\, A_M\, u_{xx}\ ,$$

where p_M stands for the internal pressure of the medium.

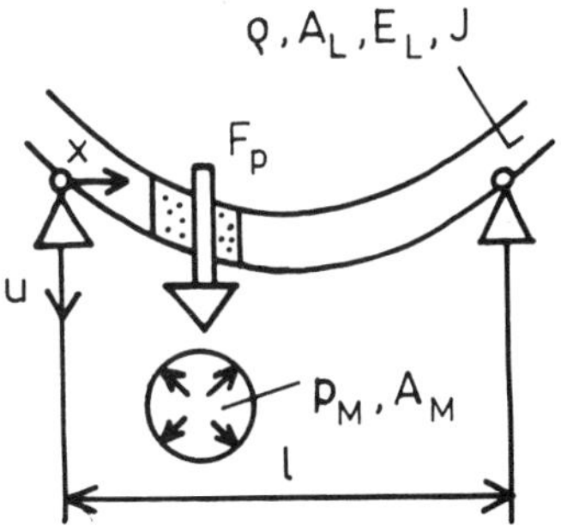

Fig. 8. The influence of the internal pressure.

The corresponding motion equation is of the form

$$u_{tt} + a^2 u_{xxxx} + p u_{xx} = 0 \qquad (6)$$

where $p = k_A/\varrho\ p_M$.

As well as in the last section we obtain the formula

$$(f_1/f_{1L})^2 = 1 - k_p\,(p_M/E)\ ,$$

where $k_p = k_A/(i^2\Omega^2)$. The dependence is introduced at Fig. 9 , where also the critical pressure is introduced.

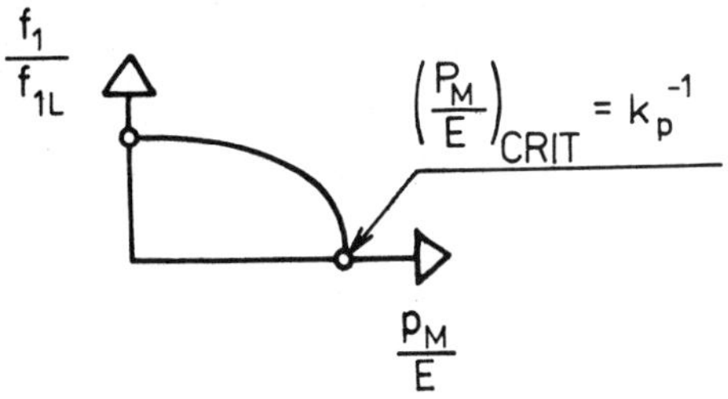

Fig. 9. The dependence of the first eigenfrequency on the pressure.

The Influence of the Temperature of the Pipe-line

The heat dependence of the force can be expressed by the formula

$$F_\tau = \alpha_L\, E_L\, A_L\, \tau_L\, u_{xx}\ ,$$

where α_L is the linear thermal expansion coefficient (see Fig. 10).

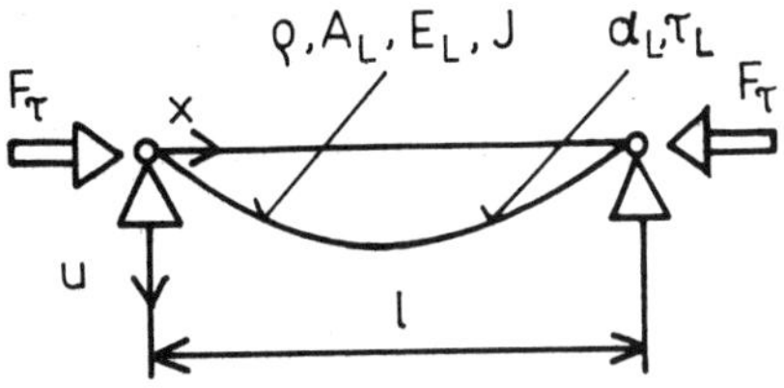

Fig. 10. The influence of the temperature.

The motion equation now reads

$$u_{tt} + a^2 u_{xxxx} + \tau u_{xx} = 0 \qquad (7)$$

where $\tau = \alpha_L E_L/\varrho\ \tau_L$.

This implies the following dependence of the first eigenfrequency on the heat τ_L.

$$(f_1/f_{1L})^2 = 1 - k_\tau\,(\alpha_L \tau_L)\ .$$

This dependence together with the critical heat is introduced at Fig. 11.

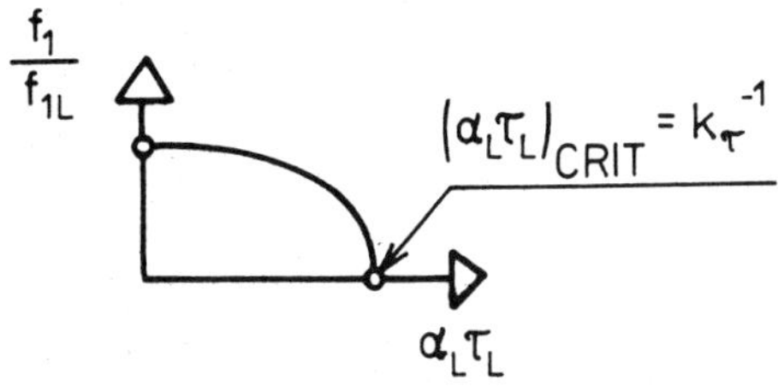

Fig. 11. The dependence of the first eigenfrequency on the heat.

THE STABILITY

For the purpose of the investigation of the stability we add into our consideration the influence of the damping effects, which are always present in a real system and which act in our model through the term u_t. The equations supplemented by the damping term will be written uniformly in the form

$$u_{tt} + a^2 u_{xxxx} + \alpha u_t + \beta u_{xx} = 0 \quad (8)$$

We shall investigate for which values of the parameter β the solution of the equation (8) is under the given damping stable. In the linear case when β is considered a constant, the Fourier method yields the asymptotic stability conditions

$$\alpha > 0, \quad \beta < a^2 \Omega^2 .$$

The region of stability and instability in the variables α, β is sketched at Fig. 12.

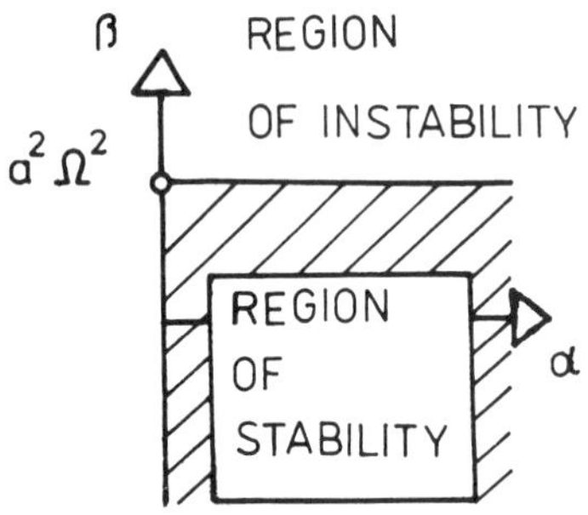

Fig. 12. The stability and instability regions.

REFERENCES

Barták, J. (1976). The Lyapunov Stability of the Timoshenko Type Equation. Časopis pro pěstování matematiky, 101, 130-139.

Barták, J. (1978). Stability and Correctness of Abstract Differential Equations in Hilbert Spaces. Czechoslovak Mathematical Journal, 28, 548-593.

Bondar, N.G. (1971). Nonlinear Autonomous Problems in the Mechanics of Elastic Systems. Izd. Budivelnik, Kiev, 140 pp. (in Russian).

Hayashi, Ch. (1964). Nonlinear Oscillations in Physical Systems. McGraw-Hill Book Company, 392 pp.

Skalický, A. (1978). Nichtlineare Schwingungen des Fördermediums von Kreiselpumpen in Leitungssystem. Proc. of the VIII th International Conference on Nonlinear Oscillations, Prague 1978, 653-658.

Tuček, A. (1978). Application of Spline Function for Solution of Duffing Equation (Finite Element Method in Time). Proc. of the VIII th International Conference on Nonlinear Oscillations, Prague 1978, 707-713.

Резюме. В работе исследуется влияние некоторых избранных нелинейных элементов и влияние нелинейного характера текучей среды на динамику трубопроводов. Нелинейность среды вызвана нелинейностями в уравнении состояния. Далее исследовано влияние удлинения трубопровода, влияние скорости текучей среды, влияние внутреннего давления и нагрева и влияние дампфирования на устойчивость системы и на первую собственную частоту колебаний.

Physikalische Grundoperationen und Funktionsstrukturen Technischer Systeme

R. KOLLER

Fakultät für Maschinenwesen, Rheinisch-Westfälische Technische Hochschule Aachen, DDR

Zusammenfassung: Die große Vielfalt physikalisch-technischer Vorgänge in Maschinensystemen (Maschinen, Geräten, Apparaten) lassen sich auf 7 unterschiedliche physikalische Grundoperationen reduzieren. Aus formalen und praktischen Gründen kann man zu jeder dieser 7 Grundoperationen auch eine entsprechende inverse Grundoperation definieren. Diese 7 Grundoperationen und inverse Operationen lauten: 1. Wandeln - Rückwandeln, 2. Vergrößern - Verkleinern, 3. Richtungändern - Richtungändern (Rückändern), 4. Leiten - Isolieren, 5. Speichern - Entspeichern, 6. Verbinden (Vermischen) - Trennen (Entmischen), 7. Fügen - Teilen. Generell betrachtet, lassen sich die Vorgänge in komplexen Maschinensystemen durch Funktionsstrukturen ("Schaltpläne"), bestehend aus physikalischen-, mathematischen- und logischen Grundoperationen, abstrakt darstellen. Bestehend aus mathematischen- und logischen Grundoperationen deshalb, weil in Maschinensystemen üblicherweise neben physikalischen Vorgängen, auch mathematische und logische Operationen realisiert werden.

Schlagworte: Grundoperationen; physikalische Grundoperationen; Funktionsstrukturen; Grundoperationsstrukturen; technische Systeme.

EINFÜHRUNG UND HYPOTHESEN

Ein wesentliches Ergebnis der Konstruktionsmethodeforschung war die Erkenntnis, daß man die komplexen Vorgänge in Maschinen, Geräten und Apparaten abstrahieren und auf eine relativ kleine Zahl von 7 elementaren Tätigkeiten bzw. Grundoperationen zurückführen kann.
In technischen Systemen kommen Energie-, Stoff- und Signalflüsse vor. Ob ein technisches System als Maschine, Apparat oder Gerät zu bezeichnen ist, hängt davon ab, ob dieses primär dem Zweck dient Energie, Stoff oder Signale umzusetzen oder zu übertragen (leiten).

Die Physik unterscheidet zwischen verschiedenen Energiearten, wie z. B. der mechanischen-, der potentiellen-, elektrischen-, thermischen Energie usw. Ein Energiefluß wird beschrieben durch Angabe der Art der Energie. Im alten Maßsystem wurde die Art der Energie durch die unterschiedlichen Einheiten mechanischer-, elektrischer- oder thermischer Energie gekennzeichnet.
Ein Energiefluß wird ferner durch den skalaren Wert der Energiemenge oder durch Angabe des skalaren Wertes der Energiekomponenten beschrieben. Energiekomponenten sind physikalische Größen, wie beispielsweise die Kraft, der Druck, der Weg, die Spannung, der Strom u. a. Schließlich wird ein Energiefluß bzw. dessen Komponenten durch Angabe der Richtung beschrieben, wenn dies vektorielle Größen sind.
"Energieart", "skalarer Wert und Richtung" der Energiekomponenten sind drei Parameter eines Energieflusses, welche man mit physikalisch-technischen Mitteln verändern kann, andere sind nicht bekannt.

Entsprechend dieser Möglichkeiten hat man das Umsetzen oder Umwandeln einer Energieart in eine andere, als physikalische Grundoperation definiert und diese als WANDELN bezeichnet.

Des weiteren hat man das Ändern des skalaren Wertes einer Energiekomponente ebenfalls als physikalische Grundoperation definiert und dieser den Namen VERGRÖSSERN oder VERKLEINERN gegeben, je nachdem ob die betreffende Größe größer oder kleiner gemacht werden soll.

So wurde auch das Ändern der Richtung einer vektoriellen Größe als physikalische Grundoperation bezeichnet und entsprechend RICHTUNGÄNDERN genannt.

Was kann mit Energieflüssen noch geschehen? Um Energieflüsse überhaupt zu realisieren muß man dafür sorgen, daß Energie von einem Ort A nach einem anderen Ort B fließen kann. Das bedeutet, daß man üblicherweise für eine Leitfähigkeit für eine bestimmte Energieart zwischen A und B sorgen muß. In manchen Fällen ist diese Leitfähigkeit naturgegeben, in anderen nicht. In letzteren Fällen muß der Konstrukteur physikalisch-technische Maßnahmen treffen, daß eine Leitfähigkeit zustande kommt. Ist die Leitfähigkeit für eine Energie naturgegeben, muß etwas getan werden, daß sich diese nicht nach allen Richtungen hin ausbreiten kann, sondern nur auf eine eingegrenzte Bahn von A nach B fließen kann.
Der Energiefluß muß senkrecht zur gewollten Bewegungsrichtung verhindert bzw. isoliert werden.
LEITEN und ISOLIEREN wurden dementsprechend als weitere physikalische Grundoperationen definiert.

Energien zweier Flüsse der selben Energieart können zusammengeführt (addiert bzw. gefügt) werden. Entsprechend wurden die Tätigkeit FÜGEN und die entsprechend inverse Tätigkeit TEILEN als weitere notwendige Grundoperationen definiert.

Auch können zwei Energieflüsse unterschiedlicher Art zusammengebracht werden. Oder es können Energieflüsse nach Art ihrer Energien getrennt werden. Dementsprechend wurde VERBINDEN und TRENNEN als weitere physikalische Grundoperation festgelegt.
Schließlich lassen sich Energieflüsse bzw. Energien speichern oder aus Speichern wieder entnehmen. SPEICHERN wird deshalb ebenfalls als eine physikalische Grundoperation angesehen. Da man zumindest theoretisch jede dieser Operationen wieder rückgängig machen kann, wurde zu jeder der genannten Grundoperationen eine inverse Grundoperation definiert, wie

Wandeln - (Rück)wandeln
Vergrößern - Verkleinern
Richtungändern - Richtungändern
Leiten - Isolieren
Verbinden - Trennen
Fügen - Teilen
Speichern - Entspeichern.

Des weiteren wurden aus praktischen Gründen noch 5 weitere "quasi Grundoperationen" definiert, wie Führen - Nichtführen, Sammeln - Verzweigen, Koppeln - Unterbrechen, Richten - Oszillieren und Emittieren (Quelle) - Absorbieren (Senke). Diese zuletzt genannten Operationen sind keine wirklichen Grundoperationen, vielmehr lassen sich diese durch die Erstgenannten ersetzen; so ist z. B. Richten und Oszillieren durch ein gesteuertes Koppeln und Unterbrechen ersetzbar bzw. realisierbar; Koppeln und Unterbrechen ist letztlich durch alternatives Leiten oder Isolieren (= Schalterfunktion) realisierbar. Bild 1 zeigt die genannte Grundoperation und quasi Grundoperation mit zugeordneten Operationssymbolen.

Für das spätere Verständnis ist es noch notwendig zu wissen, daß diese Grundoperationssymbole außer ihren Ein- und Ausgängen für die Energie, auch noch wenigstens einen Eingang für ein Steuersignal haben können - diese Operationen sind generell als 3-Pole zu betrachten. Realisiert werden diese Grundoperationen, wie im folgenden noch kurz gezeigt wird, durch physikalische Effekte [1]. Die Funktionen physikalischer Effekte (deren Gesetz), beinhalten in den meisten Fällen mehr als nur einen unabhängigen Parameter. Sie enthalten noch weitere Parameter, die man in vielen Fällen zu Steuerzwecken der eigentlich gewollten physikalischen Operationen nutzen kann. Deshalb kann man sich die o. g. Grundoperationen im allgemeinen als 3-Pole vorstellen bzw. mit einem Signal- oder Steuerungsanschluß ausgestattet, annehmen.

Da Signalflüsse, abgesehen von ihren Informationsinhalten, ebenfalls Energieflüsse sind, sind die o. g. Grundoperationen für Signalflüsse ebenso zutreffend, wie für Energieflüsse.

Wie hier aus Umfangsgründen nicht näher ausgeführt werden soll, lassen sich auch für die möglichen Operationen mit Stoffflüssen 7 analoge Grundoperationen angeben [1].

Auch in der Mathematik kennt man eine bestimmte Zahl sogenannter mathematischer Grundoperationen, mit Hilfe deren alle komplexen mathematischen Funktionen aufgebaut werden. Solche mathematischen Grundoperationen sind Addieren - Subtrahieren, Multiplizieren - Dividieren u. a. Auch in der Mathematik gibt es wirkliche Grundoperationen und quasi Grundoperationen. So können z. B. Multiplizieren und Dividieren durch ein mehrfaches Addieren bzw. Subtrahieren ersetzt werden und sind deshalb keine wirklichen Grundoperationen, sondern sinnvolle "Zusammenfassungen" bzw. "Kurzzeichen" für mehrere Grundoperationen.

Auch aus der Entwicklung logischer (Boolescher) Funktionen sind sogenannte Grundoperationen wie die UND-, ODER-Verknüpfung sowie INVERSION bekannt.

Zu den oben genannten physikalischen Grundoperationen gibt es - wie auch bei den mathematischen Grundoperationen - eine Algebra. Diese ist bisher jedoch nur sehr unvollständig erforscht und nicht sehr bekannt. Die wenigen Ansätze können in der Literatur nachgelesen werden [1].

Ein zweites, wesentliches Ergebnis der Konstruktionsmethodeforschung war die Erkenntnis, daß man diese physikalischen Grundoperationen nur durch physikalische-, chemische- oder biologische Effekte oder Effektketten realisieren kann.

Gelingt es, eine technische Aufgabe auf die Realisierung einer solchen physikalischen Grundoperation, -funktion oder Funktionsstruktur zurückzuführen, so lassen sich hierfür alle existierenden Prinziplösungen angeben, indem man alle zu deren Realisierung geeigneten physikalischen-, chemischen- oder biologischen Effekte oder Effektketten angibt. Effektketten braucht man dann, wenn es für eine bestimmte Operation bzw. Funktion keinen Effekt gibt, der diese Operation in einem Arbeitsschritt zu realisieren vermag, oder wenn man zu Ein-Effekt-Lösungen noch alternativ Mehr-Effekt-Lösungen angeben möchte oder angeben muß, weil diese technisch möglicherweise günstigere Eigenschaften besitzen, als die Ein-Effekt-Lösungen. Aufgrund dieses Sachverhaltes besteht ein umfangreiches Arbeitsgebiet der Konstruktionsmethodeforschung darin, die bis heute bekannten physikalischen Effekte (Phänomene) entsprechend den unterschiedlichen physikalischen Grundoperationen zu ordnen und so aufzubereiten, daß diese Konstrukteuren bei der Suche nach neuen Prinziplösungen bequem zugänglich sind [1, 2].

ANWENDUNGSBEISPIELE

Grundsätzlich sind diese Grundoperationen zur Funktionssynthese jeder Art von Maschine, Gerät oder Apparat geeignet. Aus Umfangsgründen kann im folgenden nur ein Beispiel näher behandelt werden.

Bei der Entwicklung verschiedener Maschinen und Geräte stellt sich häufig die Aufgabe, eine intermittierende (aussetzende) Antriebsbewegung zu haben. Will man diese Aufgabe "Erzeugen einer intermittierenden Antriebsbewegung" umfassend lösen und/oder will man die hierfür geeigneten Lösungen systematisch ordnen, so ist es notwendig, nach allen Funktionsstrukturen zu fragen, die geeignet sind, eine aussetzende Bewegungsfunktion zu erzeugen. Zur Lösung dieser Aufgabenstellung ist es notwendig, alle Grundoperationen oder/und deren Strukturen zu betrachten und zu prüfen, ob diese zur Lösung der gestellten Aufgabe geeignet sind.

So kann man sich beispielsweise Systeme vorstellen, die Energie beliebiger Art in Bewegungsenergie wandeln. Mit Hilfe eines Steuersignales kann der Wandelprozeß periodisch oder stochastisch unterbrochen werden, um auf diese Weise einen aussetzenden Bewegungsablauf zu erzielen (s. Bild 2 a).
Eine weitere Möglichkeit, einen aussetzenden Bewegungsablauf zu erzeugen, bietet die steuerbare Operation "Verkleinern". Gemeint ist ein System, das abhängig von der Antriebsbewegung das Übersetzungsverhältnis von Zeit zu Zeit zu Null macht, so entsteht ebenfalls ein intermittierendes Bewegungssystem (s. Bild 2 b).

Eine weitere Möglichkeit intermittierende Antriebssysteme zu entwickeln besteht noch darin, den Energiefluß durch Entkoppeln zeitweise zu unterbrechen (s. Bild 2 c).

Die Funktionsstruktur eines Systems, die eine intermittierende Bewegung dadurch symbolisiert, daß sie die Antriebskraft und Bewegungsrichtung so steuert, daß diese abhängig von dem Abtriebsweg rechtwinkelig (90°) zueinander gerichtet sind, um auf diese Weise einen Stillstand zu erzeugen, zeigt Bild 2 d.

Des weiteren läßt sich eine intermittierende Abtriebsbewegung auch noch dadurch erzeugen, daß man eine gleichmäßige Bewegung in eine oszillierende Bewegung umsetzt und von letzterer mittels gesteuertem Unterbrechen nur die eine Bewegung einer bestimmten Richtung (Halbwelle) auf die Abtriebsseite durchläßt, während man die Bewegung in die entgegengesetzte Richtung unterbricht (s. Bild 2 e).

Bild 2 g zeigt eine der vorangegangenen Funktionsstruktur 2 e sehr ähnliche Struktur, bestehend aus den Operationen Oszillieren und Richten (Gleichrichten). Ein "Gleichrichter" ist ein sich von der Bewegungsrichtung oder Stromrichtung etc. abhängig, sich selbst steuernder Schalter (Unterbrecher).

Auch mit einer solchen Struktur, wie sie Bild 2 g zeigt, sind intermittierende Bewegungsfunktionen zu erzeugen. Durch Überlagerung (Addieren) einer oszillierenden Bewegung und einer gleichmäßigen Bewegung ist die Erzeugung einer intermittierenden Bewegung ebenfalls möglich. Bild 2 g zeigt die entsprechende Funktionsstruktur, bestehend aus einer in Reihe angeordneten Operation Oszillieren und einer Operation Fügen (Addieren).

Schließlich kann man auch noch durch eine Reihenanordnung eines Arbeitsspeichers (Sp1) und eines kinetischen Energiespeichers (Sp2) ein System angeben, welches bei geeigneter Dimensionierung ebenfalls eine intermittierende Bewegung zu erzeugen vermag (Bild 2 g). Hierbei wird die kinetische Energie der bewegten Masse (Sp2) periodisch in einem Arbeitsspeicher (Federspeicher Sp1) gespeichert und von dort wieder auf den kinetischen Energiespeicher übertragen.
Bild 3 zeigt Beispiele zu den einzelnen Funktionen bzw. Funktionsstrukturen. In Bild 3 sind auch noch die üblichen Bezeichnungen der Getriebearten angegeben, die zu den verschiedenen Funktionsstrukturen gehören.

LITERATURVERZEICHNIS

/1/ Koller, R. Konstruktionslehre für den Maschinenbau, Springer-Verlag, Berlin, Heidelberg, New York, Tokyo, 1985

/2/ Roth, K. Konstruieren mit Konstruktionskatalogen, Springer-Verlag, Berlin, Heidelberg, New York, 1982

/3/ VDI-Richtlinie 2721 Schrittgetriebe

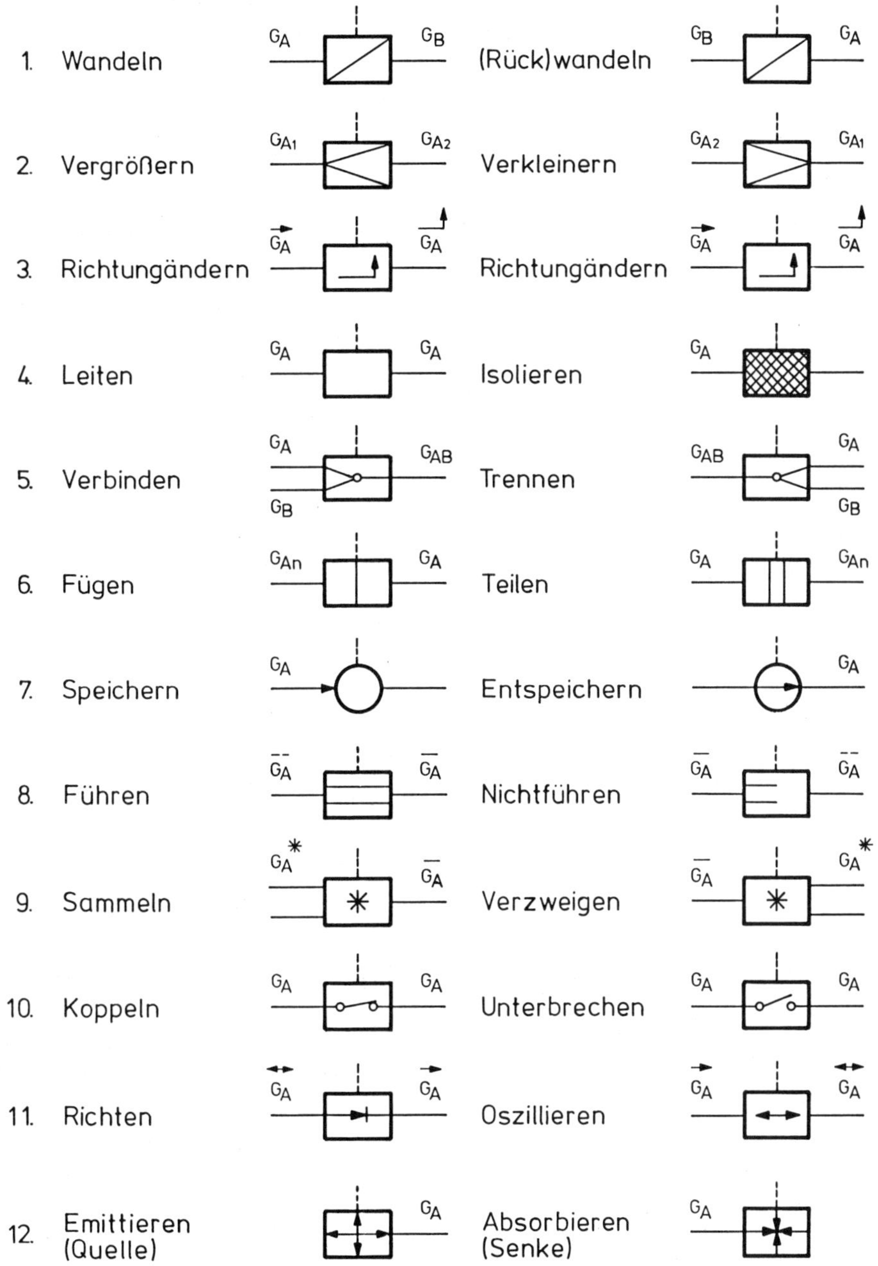

IKT | Physikalische Grundoperationen | Bild 1

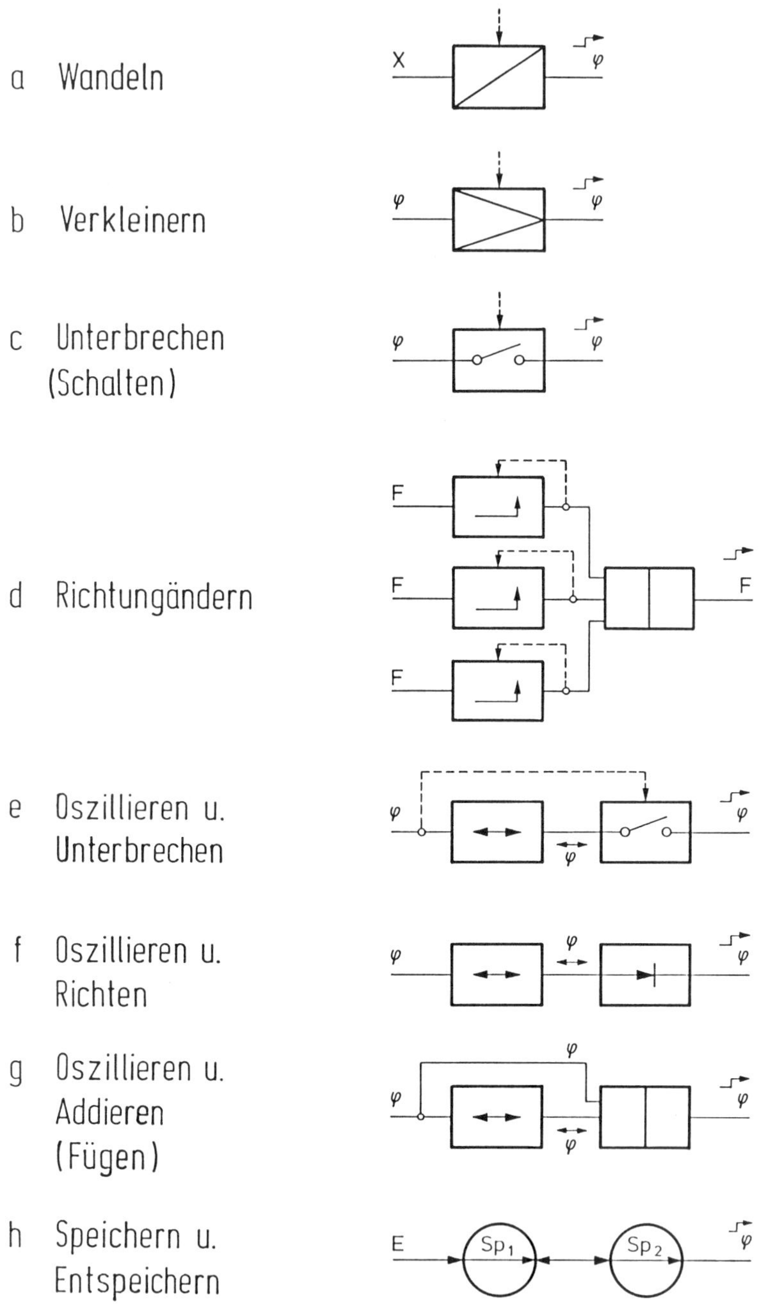

IKT | Grundfunktionsstrukturen von Schrittbewegungssystemen | Bild 2

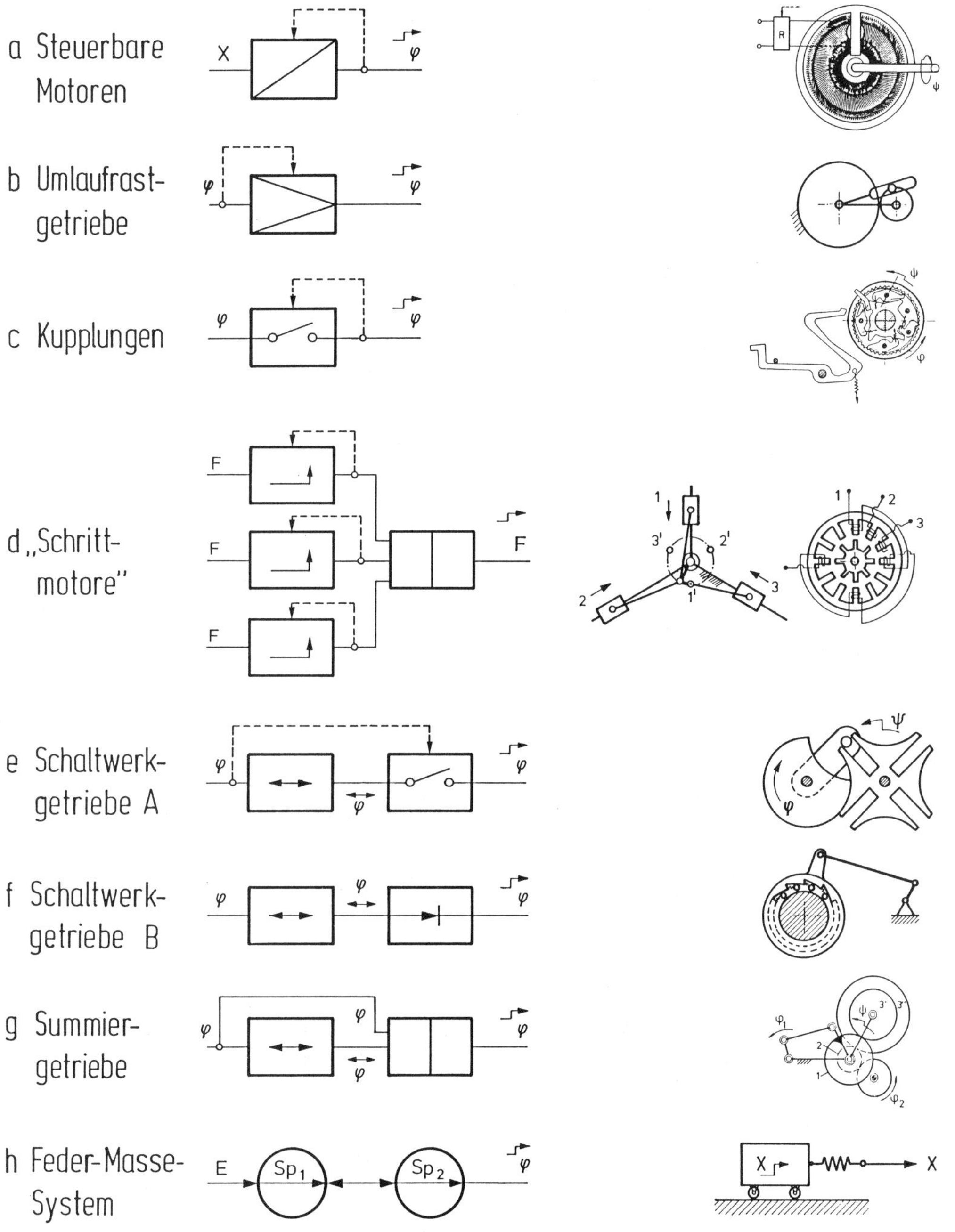

IKT | Systematik der Schrittbewegungssysteme | Bild 3

Hypothesis for Quantitative Functional and Economic Type Synthesis for Mechanisms

H. T. THORAT AND J. P. MODAK

Department of Mechanical Engineering, Visvesvaraya Regional College of Engineering, Nagpur 440 011, India

Abstract. Type synthesis is the selection of best mechanism out of possible alternatives. At present it is done by designer's intuition, previous experience and qualitative comparison. There is a real need of evolving a scientific approach for type synthesis preferably in a quantitative form. The present paper proposes methodologies for quantitative comparison namely Functional Type Synthesis and Economic Type Synthesis for selecting a mechanism having best performance and maximum economy respectively. A new concept of Functional Index is evolved which is a measure of performance of a mechanism; mechanism having best functional index is chosen. Cost Index, total of cost of mechanism per unit task for all aspects, is worked out for each mechanism. Economic type synthesis selects a mechanism which has minimum cost index. The two methodologies are expected to eliminate dependence on intuition and trial and error for type synthesis and give an analytical and more scientific approach for type synthesis.

Keywords. Kinematics; dynamics; mechanism; type synthesis; machine design; modelling.

INTRODUCTION

Type synthesis involves in first place, survey of all possible mechanisms which can perform satisfactorily the desired motion programme, their kinematic and dynamic analysis, comparison on the basis of various aspects of performance and subsequently selecting the best possible mechanism, indicated by comparison. Hain (1967) has divided systematics of mechanism in type synthesis and number synthesis. Most of the literature published (Dobrjanskyj and Freudenstein, 1967; Hain, 1967; Tao, 1964) is on dimensional synthesis and number synthesis, which is essentially in the context of isomorphism of mechanism.

Major problems which have not been dealt with till this time are how to compare amongst and how to select a mechanism. Conventional practice of using intuition, previous experience, and trial and error technique is not a scientific approach. This shortcoming is due to the varied nature of aspects of comparison leading to a qualitative approach. Literature survey reveals that no scientific procedures are reported on type synthesis. This paper presents a pioneering attempt of quantitative comparison for performing type synthesis. Two methods of type synthesis namely Functional Type Synthesis and Economic Type Synthesis are developed.

METHODOLOGIES OF TYPE SYNTHESIS

Reported and Practiced Methods

Tao (1964) has stated that the factors which influence selection are speed, load, reliability and economy. A designer has to use his creative ability, previous experience and keen observation. Hain (1967) observed that only rigorous analysis may not be sufficient and actual building of all mechanisms and testing them may be necessary. Normal practice is to select the mechanism mainly by intuition or by previous experience. Even if the comparison is made, only limited aspects are taken into account. Mechanism selection is not done in some situations where a mechanism, which has been used in similar situations in the past, is adopted without much analysis.

In either case it is necessary that the designer should have a good intuition, vast previous experience and he should be prepared for trial and error approach. Beginner or an inexperienced designer is likely to fail in this task. Also one does not know whether selected mechanism is really the best mechanism.

Above mentioned difficulties arise due to the wide range of aspects considered for comparison. Comparison remains at the qualitative level because units in which various aspects are expressed are different. Selection is possible if only one mechanism is best in all aspects, which is quite unlikely. Selecting the mechanism is not possible when different mechanisms are better in different aspects. Solution to this problem can be obtained if the comparison is quantitative and suitability of mechanism from the angle of various aspects is expressed in the same unit. In view of this line of thinking, two methods of quantification of comparison namely Functional Type Synthesis and Economic Type Synthesis are proposed.

Functional Type Synthesis

Functional type synthesis is suggested

when the best performing mechanism is preferred giving secondary importance to economy. Method of quantification adopted in this case is,comparing the actual performance of the mechanism with ideally desired performance,to get a dimensionless quantity for each aspect. These dimensionless quantities are further modified to get certain number indicating a measure of suitability. Such numbers obtained for various aspects are added to get overall suitability of the mechanism. Suitability is expressed quantitatively and hence selection is very simple.

Economic Type Synthesis

Economic Type Synthesis is recommended when a most economic rather than best performing mechanism is to be selected. In this method each aspect is converted into cost for that aspect. Such costs, calculated for various aspects for each mechanism,are added to get cost as index of suitability of the mechanism. Suitability is expressed in terms of quantity in this method,making the comparison very simple.

METHODOLOGY OF FUNCTIONAL TYPE SYNTHESIS

Various aspects of comparison are divided in three groups namely Input Aspects, Output Aspects and Practical Aspects. Comparison of actual performance with ideal performance can be quantified in two ways. One way is of giving full positive points for ideal performance and zero point for worst performance. The other way can be of assigning zero point for ideal performance and negative (i.e. penalty) points for poor performance even up to infinity. In former case poor performance for some aspects can be covered by good performance for other aspects, whereas in latter case it is not possible and unless other alternative mechanisms are equally poor,may be in some other aspects, the mechanism is eliminated. Latter way of quantification is preferred for this reason.

The comparison of ideal performance and actual performance for an aspect is called here as Aspect Ratio which is a dimensionless quantity. If performance is ideal the aspect ratio is 1 and for non ideal performance it is more than 1 up to infinity. Aspect ratios of individual aspects are discussed below.

Input Aspects

Type of input. Aspect ratio is 1 if electric motor is a drive and it is 2 for any other type of drive. Mechanisms deriving power from main drive of machine also have aspect ratio of 1 and motion modifying and transmitting mechanism is added to the original mechanism.

Input power. This is a very important aspect. Aspect Ratio = Power supplied at Input/Power derived at output.

Input torque. Maximum instantaneous torque at the crank/Average torque over a cycle = Aspect ratio. This aspect accounts for excess dynamic loads arising in the mechanism.

Flywheel requirement. Aspect ratio of this aspect, a measure of smoothness of operation of mechanism is given by Aspect ratio = Fluctuation of energy/Minimum value of fluctuation of energy from all possible mechanisms being compared.

Output Aspects

Static response. Certain motion programme of the output link of mechanism is desired within error limit L for ideal performance or at least within permissible error limit H. For actual error $E < L$, Aspect Ratio = 1, for $E > H$, Aspect Ratio = infinity; and for $L \leq E \leq H$, Aspect Ratio = $(H-L)/(H-E)$.

Dynamic Response. It is desired some times that velocity or acceleration or jerk values of output link should be predetermined; only in such cases aspect ratio is calculated in the same way as static response. Otherwise aspect ratio is 1.

Kinematic Characteristics. Minimum values of velocity, acceleration and jerk are always desirable. Ideally minimum values of these can be obtained only by constant velocity, constant acceleration and constant jerk characteristic of output link respectively. Aspect ratio = 4 - (Ideal minimum velocity/Maximum velocity) - (Ideal minimum acceleration/Maximum acceleration) - (Ideal minimum jerk/Maximum jerk).

Shaking Forces. Shaking forces affect the frame and other mechanisms. Aspect Ratio = Maximum shaking forces/Minimum of maximum shaking forces of all mechanisms being compared.

Vibration Effect. If frequency of highest significant harmonic fh is equal to natural frequency of frame fn then aspect ratio is infinity. For $fn/fh > 2$, Aspect Ratio = 1; For $1 \leqslant fn/fh \leqslant 2$,Aspect Ratio = fh/fn-fh.

Practical Aspects

Space. For main mechanism, Aspect Ratio = Space for mechanism/Minimum of space for all alternative mechanisms. For auxiliary mechanism, Aspect Ratio = Space occupied/ Space provided (subject to Aspect Ratio $\geqslant 1$). If mechanism occupies objectionable space then aspect ratio is infinity.

Weight. If mechanism weight affects the power consumed by other mechanism then aspect ratio is the ratio of increased power consumption of affected mechanism and unaffected power consumption of affected mechanism.

Life. Aspect Ratio = Life of machine/Life of mechanism $\geqslant 1$. For main mechanism, Maximum of life of all mechanisms being compared/Life of the mechanism is the aspect ratio.

Manufacturing Difficulty. Depending upon whether mechanism can be fabricated by ordinary machines or by good machines or by special purpose machines or by most sophisticated machines, aspect ratio is 1 or 2 or 3 or 4 respectively.

Maintenance. Aspect Ratio = (Machine life/ Machine life - Downtime for maintenance of the mechanism).

Cost. Aspect Ratio = Cost of mechanism/ Minimum of cost for all mechanisms being compared.

Weightages. Weightages decide relative importance of each aspect. Weightage of 100 units is given for all primary aspects i.e. input and output aspects except for input power. In the case of input power, variation in aspect ratio is relatively less with the change in power required. In addition to this, as the aspect is very important, weightage of 500 units is allotted for this aspect. Practical aspects which have secondary importance are given weightage of 50 units.

Functional Index

Aspect ratios are converted into functional index for the purpose of comparison. For ideal performance functional index is zero. Functional index is given by

$$\text{Functional Index} = (AR_n - 1) \times W_n \tag{1}$$

where AR_n is the aspect ratio and W_n the weightage of nth aspect. 1 is subtracted from AR_n so as to get zero value for ideal performance and to provide better differentiation of Index near ideal value.

Criterion of Selection

Mechanism with least functional index is to be selected. If the constraints of operation are changed, altogether different mechanism may get selected.

TABLE 1. Aspect Ratios and Weightages

Sr. No.	Aspect	Aspect Ratio Min	Aspect Ratio Max	Weightage
1.	Type of Input	1	2	100
2.	Input Power	1	-	500
3.	Input Torque	1	-	100
4.	Flywheel requirement	1	-	100
5.	Static Response	1	-	100
6.	Dynamic Response	1	-	100
7.	Kinematic characteristics	1	4	100
8.	Shaking forces	1	-	100
9.	Vibration effect	1	-	100
10.	Space	1	-	50
11.	Weight	1	-	50
12.	Life	1	-	50
13.	Manufacturing difficulty	1	4	50
14.	Maintenance	1	-	50
15.	Cost	1	-	50

METHODOLOGY OF ECONOMIC TYPE SYNTHESIS

Only those mechanisms which can satisfactorily perform the function are considered for comparison. Quantification in this method is by calculating cost of mechanism for various aspects. Aspect cost is defined as cost of mechanism due to an aspect. Aspect costs for various aspects are discussed individually. Unit task is defined as some fixed amount of work output in terms of production by machine.

Cost of Mechanism

Type of input, input torque, flywheel requirement, shaking forces, vibration effects, weight, maintenance, manufacturing difficulties and total cost are the aspects of functional type synthesis reflected into this aspect either directly or due to design of the mechanism. Aspect cost = Interest on machine cost per day/ unit tasks performed per day.

Cost of Operation

This aspect accounts for input power and weight aspect of the functional type synthesis. The aspect cost in this case is the operational cost which includes power consumption, labour and consumables per unit task.

Cost Due to Life

This accounts for life aspect of functional type synthesis. Aspect Cost = Cost of mechanism/Unit tasks performed in life of mechanism.

Cost Due to Space

Cost Aspect = Rental of space occupied by mechanism for one day/Unit tasks performed per day.

Cost Due to Quality

Performance of the mechanism is accounted only in this aspect. This aspect is governed by static response, dynamic response and kinematic characteristic. Aspect Cost = Value of standard unit task - Value of produced unit task. Aspect cost in this case can be negative.

Cost Due to Maintenance

Maintenance required is due to some aspects of functional type synthesis namely input torque, kinematic characteristic, vibration effect and inherent need of maintenance. Aspect cost in this case is cost of maintenance including losses due to downtime per unit task performed.

Criterion of Selection

Cost Index is sum of all aspect costs of the mechanism. Mechanism with least cost index is the choice by economic type synthesis. Mechanism selected by economic type synthesis may be different from mechanism selected by functional type synthesis.

HYPOTHETICAL CASE STUDY

Table 2 and Table 3 illustrate the application of proposed theory. Let three mechanisms namely A, B and C fulfil the requirements of the motion programme. The aspect ratios and aspect costs can be worked out for all the aspects enumerated in theory by applying well established techniques of designing the mechanism. The numerical values of aspect ratios and aspect costs are assumed hypothetically. Functional index is calculated using equation 1 and cost index is the sum of aspect costs.

DISCUSSION

Mechanism C is selected by functional type synthesis. Performance of mechanism A is best in 3 aspects and worst in 3 aspects, mechanism B is best in 5 aspects and worst in 4 aspects, whereas mechanism C is best in 2 and worst in 4 aspects. Designer is likely to choose mechanism B as per existing approaches of type synthesis. The quantitative approach presented here has led to selection of mechanism C which otherwise would have been dropped. Economic type synthesis has selected mechanism B. This is a case where indices are close to each other, in many cases they are expected to have large variation.

TABLE 2. Aspect Ratios and Functional Index for the Hypothetical Case

Sr. No.	Aspect	Aspect Ratio		
		A	B	C
1	Type of input	1.00	1.00	1.00
2	Input Power	1.20	1.30	1.15
3	Input Torque	4.00	2.50	3.00
4	Flywheel Requirement	1.00	1.80	2.00
5	Static Response	2.00	1.00	1.60
6	Dynamic Response	1.00	1.00	1.00
7	Kinematic Characteristic	3.30	3.80	2.80
8	Shaking forces	1.00	2.00	1.10
9	Vibration Effect	1.00	1.00	1.00
10	Space	1.00	1.00	1.00
11	Weight	1.00	1.00	1.00
12	Life	1.30	1.00	2.00
13	Manufacturing Difficulty	3.00	2.00	3.00
14	Maintenance	1.10	1.30	1.20
15	Cost	1.20	1.00	1.50
	Functional Index	960	800	780

TABLE 3. Aspect Cost and Cost Index for the Hypothetical Case

Sr. No.	Aspect	Cost (Rs.)		
		A	B	C
1	Cost of Mechanism	60	50	70
2	Cost of Operation	10	11	9
3	Cost due to life	30	15	40
4	Cost due to space	5	5	5
5	Cost due to quality	15	0	5
6	Cost due to maintenance	15	35	20
	Cost Index	135	116	149

CONCLUSION

Two methods of type synthesis namely Functional Type Synthesis and Economic Type Synthesis are evolved. Former method selects a mechanism having best performance, whereas latter selects most economic mechanism. New methodologies use quantitative comparison to reduce possibility of error in type synthesis which otherwise need intuition, experience and trial and error approach. Functional type synthesis and economic type synthesis may select different mechanisms for same set of constraints. Selected mechanisms could be different if set of constraints is changed.

It has been attempted to cover all the aspects, however, few more aspects may need inclusion which can be treated as further development of this subject. Relationship between aspect ratio, actual performance and ideal performance can be modified after studying each aspect carefully. The weightages assigned need to be decided precisely for accurate comparison.

Refining of these two approaches of type synthesis involves in-depth study of each aspect and weightages, supported by exhaustive experimentation. Then these methodologies will prove to be most powerful tool in the hands of machine designers for selecting the appropriate mechanism.

REFERENCES

Dobrjanskyj, L. and F. Freudenstein, (1967), Some applications of graph theory to the structural analysis of mechanisms. Transactions of ASME, 89, Series B, 153-158.

Hain, K. (1967), Applied Kinematics, Mc Graw-Hill, New York, p. 32.

Tao, D.C. (1964), Applied Linkage Synthesis, Addison-Wesley, Massachusetts, p.2.

ABSTRACT IN FRENCH

La synthèse de type du mécanisme est pour choisir le plus meilleur mécanisme des alternatifs possibles. Au présente, c'est éffectué par le dessinateur selon sa pratique, son intuition et la comparaison qualitative. Vraiment dire, c'est nécessaire d'élaborer une approache scientifique, préférablement dans un mode quantitatif, pour la synthèse de type du mécanisme. En cet article, nous proposons les méthodologies de la comparaison quantitative, c'est-à-dire, la synthèse de type - fonctionnelle, et, la synthèse de type - économique, pour choisir un méchanisme donnant le plus meilleur fonctionnement, et, l'économie à maximum, respectivement. À la synthèse de type - fonctionnelle, les mécanismes sont comparé pour, presque touts les aspects fonctionnels. Un nouveau concept d'indice fonctionnel est avancé, ce qui est une mesure du fonctionnement d'une machine. Le mécanisme possedant le plus meilleur indice fonctionnel est choisi. L'indice de côut, c'est-à-dire, le prix total de mécanisme considérant touts les aspects pour unité du travail, est calculé pour chaque mécanisme On croit que l'utilisation des deux méthodologies avancées peut éliminer la dépendance de l'intuition et, d'essai et de l'erreur, pour la synthèse de type du mécanisme donnant une approache analytique, et, scientifique.

Development of Automatic Let-off Mechanism for Shuttle Loom

P. B. JHALA AND S. R. JOSHI

Loom Design Division, Ahmedabad Textile Industry's Research Association, Ahmedabad-380 015, India

Abstract. Optimum tension in the warp threads is essential during weaving of fabric on loom. This is achieved by controlling the amount of warp thread released from the beam by let-off mechanism. The paper presents work on the development of an automatic let-off mechanism particularly for conventional shuttle loom through theoretical analysis and dynamic studies. Optimisation of different design parameters of the mechanism such as co-efficient of friction was carried out using sensitivity analysis technique. The cyclic variation in warp thread tension under dynamic conditions was studied with the help of special measurement techniques. The mechanism provides the automatic warp tension control at low cost and its mechanical simplicity makes it easy to set and maintain.

Keywords. Closed loop systems; Self-adjusting systems; Sensitivity analysis; Dynamic-response; Torque control; Optimal control; Textile industry.

INTRODUCTION

On looms, it is essential to maintain the necessary tension upon the warp threads during weaving to facilitate a clear shed formation for insertion of weft thread and to resist the action of the reed as it moves forward to force the weft thread into the fabric.

Generally, on conventional shuttle looms a weight-lever type of let-off is used, which requires frequent manual adjustments as weaving progresses to maintain constant warp thread tension. In actual mill working conditions, it causes large variations in the warp tension. Incorrect warp tension leads to higher warp thread breakages and the quality of fabric is adversely affected. There are many types of semi-positive as well as positive let-off motions (Foster, 1961) used on automatic looms which control the warp tension, but they are very expensive and difficult to set and maintain by the fitters of the ordinary loomshed.

For designing a simple and robust automatic let-off mechanism for conventional shuttle loom, different parameters were optimised carrying out the sensitivity analysis. The cyclic variations in warp tension under dynamic conditions were studied using special measurement techniques.

THE LET-OFF MECHANISM

A schematic diagram of automatic let-off mechanism developed for conventional shuttle loom at ATIRA[1] is shown in Fig.1. It consists of a brake band wrapped around the brake drum and a bellcrank lever with the backrest at the end of a short arm and a spring at the end of long arm. The bellcrank lever is maintained in equilibrium around the fulcrum by the moments of different forces acting on it.

Theoretical Background

Morrison (1962) has derived an expression for the warp thread tension considering the equilibrium of

1 ATIRA is the authors' institution

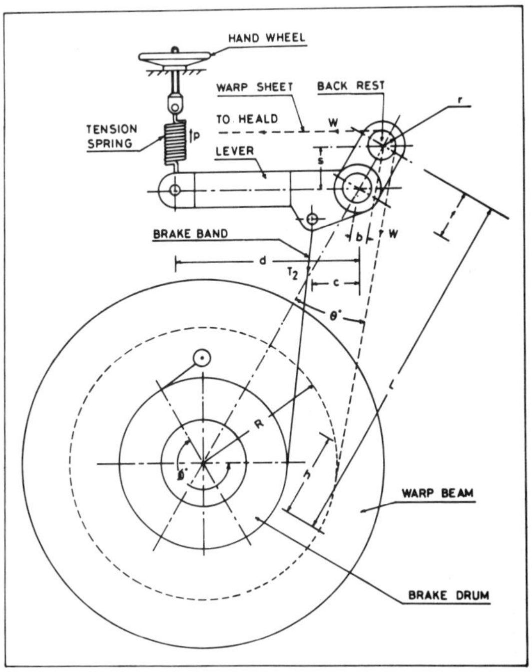

Fig. 1. Automatic Let-off Motion

the bellcrank lever and equilibrium of resisting torque at the brake drum and the torque exerted by the warp threads on the beam. The equation for the warp thread tension W is as follows:

$$W = \frac{\frac{h(e^{\mu\phi}-1)p.d}{c}}{\left(L - \frac{f.h(e^{\mu\phi}-1)}{c}\right)\sin\theta + \left(r + \frac{h(e^{\mu\phi}-1)s}{c}\right)} \quad (1)$$

where μ is the co-efficient of friction between drum and brake liner and p is the preset load on the spring. The other dimensions are as shown in Fig.1.

In the Eq.(1) only θ is variable while all other parameters are constant. To eliminate the effect of θ, on the warp thread tension W, the following condition is to be satisfied:

$$L = \frac{f.h(e^{\mu\phi}-1)}{c} \tag{2}$$

By a proper geometrical configuration of the mechanism and choosing certain values of different parameters, the condition given in Eq.(2) is satisfied in the let-off mechanism, so that warp thread tension remains constant from the start to the finish of the beam.

OPTIMISATION OF DESIGN PARAMETERS

As can be seen from the theoretical analysis carried out in the previous section in designing the automatic let-off mechanism, there are not only dimensional parameters which are to be optimised with the constraints of the loom structure and technological requirements but also other parameters such as co-efficient of friction μ which varies under practical conditions. A sensitivity analysis was carried out to study and to minimise the effect of variation of co-efficient of friction on the warp thread tension.

Sensitivity Analysis

The value of μ co-efficient of friction between brake drum and brake band depends on many other uncontrollable variable factors such as sliding velocity of two surfaces, temperature and roughness of surfaces in contact. Gordeev (1960) has established the direct effect of μ on the warp thread tension W, when μ decreases or increases.

As in the case of negative friction let-off mechanism discussed in the paper, the warp thread length l is released when the driving torque on the beam due to warp thread tension is higher than the braking torque on beam as shown in Fig.2. The beam accelerates during time period t_1, when driving torque is higher than the braking torque and it retards during time period t_2, when the braking torque is higher than the driving torque.

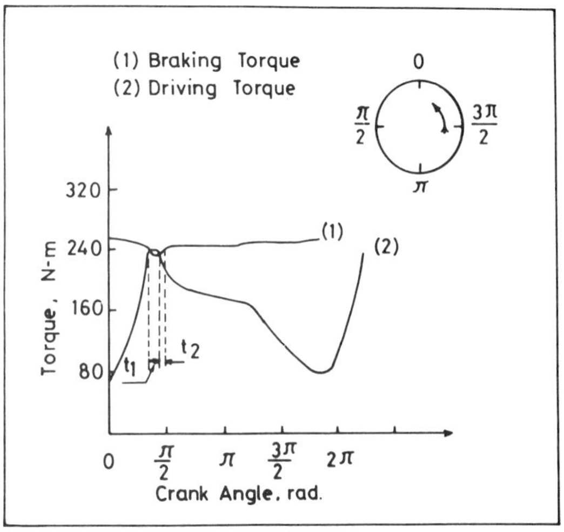

Fig. 2. Patterns of Braking and Driving Torque

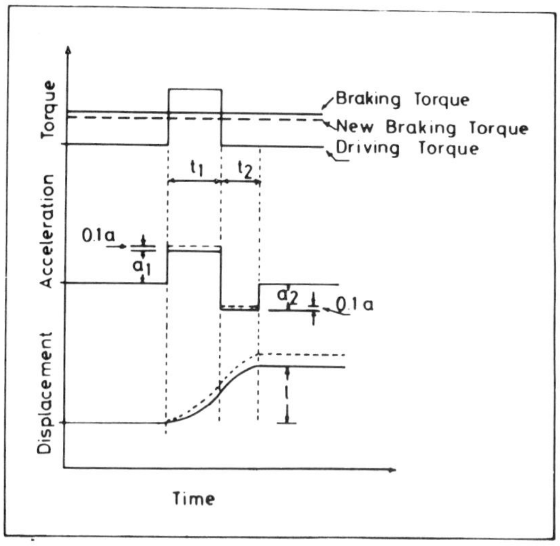

Fig. 3. Diagram of Displacement, Acceleration and Torque of Beam

The acceleration a_1, deceleration a_2 and dispacement (warp length released) l are represented in Fig.3. The warp length released l is given by Eq.(3) as follows:

$$l = 1/2\ at_1^2\ (1/m) \tag{3}$$

where, $a = (a_1+a_2)$ and $m = a_2/a_1 = t_1/t_2$

The acceleration a_1 and deceleration a_2 are assumed constant throughout the time t_1 and t_2 respectively as well as t_1 is constant for a given loom.

When μ changes, Δl is the change in l, warp length released, and m_1 is the new value of m. Then the ratio $\Delta l/l=i$, where i is the let-off variaton factor[2], is given by Eq.(4) as follows:

$$i = \frac{1/m_1 - 1/m}{1/m} \tag{4}$$

In the case of decrease in μ, the changes in the braking torque and accelerations are shown in Fig.3. and the equation for m_1 can be given as follows:

$$m_1 = \frac{a_2-0.1a}{a_1+0.1a}$$

$$= \frac{m-0.1(m+1)}{1+0.1(m+1)} \tag{5}$$

Now, by putting the value of m_1 from Eq.(5) in the Eq.(4), we get the following equation for i:

$$i = \frac{0.1(m+1)^2}{m-0.1(m+1)} \tag{6}$$

Similarly, when μ increases, the equation for let-off variation factor, i can be given as follows:

$$i = \frac{-(0.1)(m+1)^2}{m+0.1(m+1)} \tag{7}$$

2 Absolute value of i is to be considered

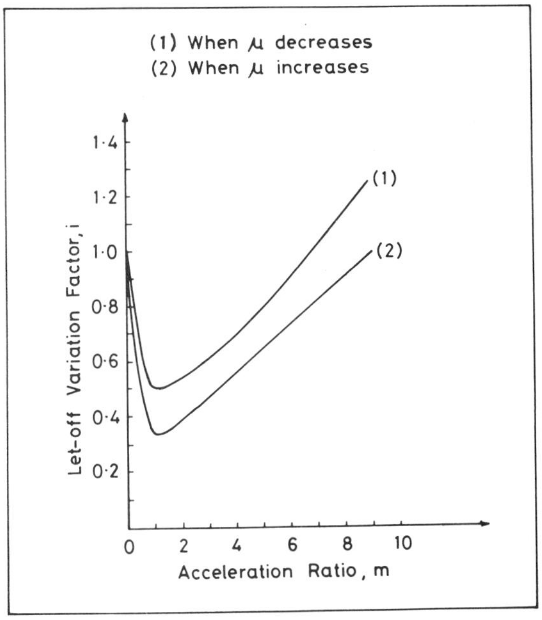

Fig. 4. Relationship between Let-off Variation Factor and Acceleration Ratio

The relationship between let-off variation factor, i and the acceleration ratio, m is given in Fig.4. It shows that in both the cases, the let-off variation factor, i is minimum or warp length released is the least sensitive when m is in the range of 0.7 to 2. In the automatic let-off at full warp beam, m is 1.5 and at empty beam, m=0.82 and these values are within the limit. So, the let-off mechanism is least sensitive to the variation in the co-efficient of friction μ.

EVALUATION OF THE MECHANISM

Initially, the automatic let-off mechanism was evaluated on a conventional shuttle loom in ATIRA pilot mill and was found to regulate the warp tension in a fairly uniform manner from start to finish of the beam.

The effect of cyclic warp tension on the brake band tension (Joshi and Jhala, 1985) was also studied in the pilot mill by pasting strain gauges on the brake band and warp tension transducer. The corresponding signals were graphically recorded on a storage oscilloscope and the oscilloscope traces were photographed which are shown in Fig.5. It shows that the cyclic variation in warp tension is compensated to certain extent by the corresponding changes in the brake band tension.

After successful pilot mill trial, the automatic let-off motion was tried under mill working conditions for long duration. During the trial various data such as warp tension, warp breaks and fabric damage with normal as well as automatic let-off were collected and are presented in Table 1.

It is observed from the data that there is considerable reduction in warp breaks and fabric damage percentage.

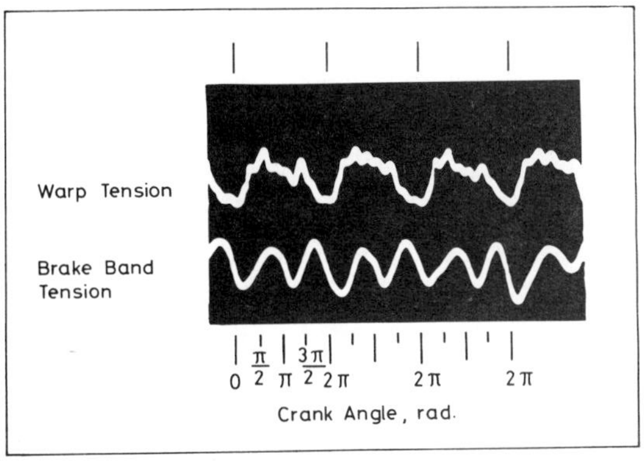

Fig. 5. Patterns of Warp Tension and Brake Band Tension

TABLE 1 Practical Performance of Normal and Auto Let-off

Sort	Coating 2/36s x 13s (Combed) 88 x 48	
Loom Speed	160 rpm	
Loom Width	68" (170 cm)	
Type of Let-off	Normal	Auto
Variation in warp tension, g/thread	23-28	26-28
Warp breaks per loom hour	4.4	3.5
Fabric damage, %	8.1	6.7

CONCLUSIONS

The automatic let-off motion regulates the tension on warp in a fairly uniform manner from start to finish of the beam. It helps in getting the required weft-density and uniform weft spacing there by reducing damage in fabric. It also reduces weaver's work load. It provides the automatic warp tension control at low cost and is easy to set and maintain by the fitters of the ordinary loomshed. It can also be used for automatic shuttle loom.

ACKNOWLEDGEMENT

The authors are grateful to the Director, ATIRA for his kind permission to present this paper at this Congress. They also thank Mr. P.M. Jain and Mr. D.L. Trivedi for the assistance in developmental work, Mr. R.B. Jadhav for trials in the mills. Thanks are also due to Mr. C.S. Vora for building up the instrumentation facilities and Mr. Y.R. Vora for undertaking the arduous jobs of typing on the special mats.

REFERENCES

Foster, R (1961). Positive Let-off Motions. Wool Industries Research Association, Leeds.

Morrison, D (1962). Engineering aspects of power loom design. Proc. of the Symposium on Modern Mechanisms in Textile Machinery, The Institution of Mechanical Engineers, U.K., pp. 16 to 18.

Gordeev, V.A. (1960). On the precision of the action of friction brakes for warp beams. Technology of the Textile Industry, USSR, pp. 96 to 103.

Joshi, S.R., P.B. Jhala (1985). Automatic Let-off Motion for Plain Looms, Ahmedabad Textile Industry's Research Association, Ahmedabad.

ZUSAMMENFASSUNG

Gleichmässige Spannung in den Kettfäden ist notwending bei dem Webprozess auf dem Webstuhl. Dies geschieht durch Kontrollieren der vom Kettbaum freigegebenen Menge des Kettfadens, mit Hilfe einer-Abwickeleinrichtung. Eine Gewicht-Hebeltyp Abwickeleinrichtung wird auf gewöhnlichen Pendelwebstühlen benutzt, die häufige manuelle Einstellungen erfordert wie der Webprozess fortschreitet um konstante Kettfadenspannung zu erhalten. Das bedingt grosse Abweichungen in der Kettfadenspannung in tatsächlichen Betriebsbedingungen. Falsche Kettfadenspannung führt zu höheren Kettfadenbruche und die Qualitat der Gewebe ist verschlechtert. Viele verschiedene Typen von Abwickeleinrichtungen werden auf automatischen Pendelwebstühlen benutzt um die Kettfadenspannung zu kontrollieren aber sie sind sehr teuer und schwierig zum Einstellen durch Schlösser der gewohnlichen Weberei.

Die vorliegende Studie stellt Entwicklungsarbeiten auf einer automatischen Abwickeleinrichtung insbesondere für übliche Webstuhle durch theoretischen-Analysen und dynamischen Studien dar. Eine Optimierung der verschiedenen Auslegungsparametern wurde-mit Hilfe von verschiedenen analytischen Techniken wie Empfindlichkeitsanalyse fur den Reibungskoeffizient erreicht. Die zyklische Variationen in der Kettfadenspannung wurden unter dynamischen Verhältnissen mit besonderen Messtechniken untersucht. Die Abwickeleinrichtung kontrolliert die Kettfadenspannung ziemlich gleichmässig von vollen bis zu leeren Kettbaum und hilft gleichmässige Teilungen und erforderlichen Kettfadendichte zu erreichen. Sie liefert die automatischen Kettfadenspannungskontrolle zu geringen Kosten und ihre-Einfachheit erleichtert die Einstellungen und die Instandhaltung.

Design and Optimization of a Bi-directional Non-reversible Dynamic Brake

L. BORELLO* AND G. JACAZIO**

*Department of Aerospace Engineering, Politecnico, Torino, Italy
**Department of Mechanics, Politecnico, Torino, Italy

Abstract. *The no-back devices serve two different purposes:*
- *they hold the controlled element in a fixed position, independently of the load variations when no change of command is given to the servomechanism of which they are a part;*
- *they act as dynamic brakes and dissipate the excess mechanical energy when the load acting on the controlled element has the same direction of the commanded position change.*

To allow a good operation of the system made up by the no-back and the controlled element, a careful selection of the design parameters must be made, based on the knowledge of how they affect the system performance. To this effect, a detailed analysis of the system was carried out and its performance determined, that allowed to obtain the most suitable combination of the design parameters to get the optimum system performance.

Keywords. *Brakes; actuation systems; servomechanisms.*

INTRODUCTION

The actuation systems used in position servomechanisms often contain a brake which, depending on the application, may serve different purposes.
In many applications the power supply (hydraulic or electrical) is switched off once the controlled item has reached its selected position.
This case is typical of all those applications in which the changes of position demand occur after more or less long intervals of rest. Switching off the power supply provides an advantage in this case for it allows an appreciable energy saving. However, if the system is subjected to a load, some means of maintaining the controlled item stationary must be provided in order to avoid a runaway under load.
A similar case takes place when the system contains provisions to detect a failure and to remove the power supply whenever a failure is recognized. In this case too, a way of maintaining the controlled item stationary under load must be guaranteed.
In some applications the load acting on the actuator may vary from large opposing to large aiding and it may be necessary to provide a means to dissipate excess mechanical energy to prevent an overspeed of the actuator, particularly when the actuator is a hydraulic or electric motor.
Different solutions can be selected for all the cases just outlined. This paper will consider one particular all-mechanical solution consisting of a special mechanical device, that is generally referred to as "no-back", having the following characteristics:
- to allow the flow of mechanical power with very little loss from the servomechanism actuator to the controlled element;
- to act as a brake between the controlled element and the fixed structure whenever the load acting on it tries to overrun the servoactuator.

These favourable characteristics of the no-back devices are, however, balanced by the difficulty of matching their design to the inertia and stiffness characteristics of the controlled element. It is a proven fact that a non correct matching may give rise to an operational instability, with strong limit cycle oscillations, particularly under large aiding loads.
To allow a good operation of the system made up by the no-back device and the controlled element, a careful selection of the design parameter must be made, based on the knowledge of how they affect the system performance. This paper presents a detailed analysis of such a type of systems and its performance is determined to show which is the most suitable combination capable of leading to the optimum system performance. The system usually consists of a rotating device; however, for simplicity, it is represented and modelled as a translating one.

FUNCTIONAL DESCRIPTION

A no-back device is basically a multi-disc brake that is automatically engaged by balls moving on a wedge (Fig.1) whenever any of the following conditions b), c) or d) take place.
The possible operating conditions of a no-back device are the following.
a)- Normal operation with opposing load: the direction of motion is opposite to the external force acting on the driven element (Fig.2).
b)- Standstill: the input is stationary (zero speed) while an external force is acting in either direction on the driven element (Figg.3 or 4).
c)- Normal operation with aiding load: the direction of motion is the same of the direction of the external force acting on the driven element (Fig. 3).

d)- Overload: the force acting on the driven element is opposite and greater than the maximum driving force that the input can provide (Fig.4)

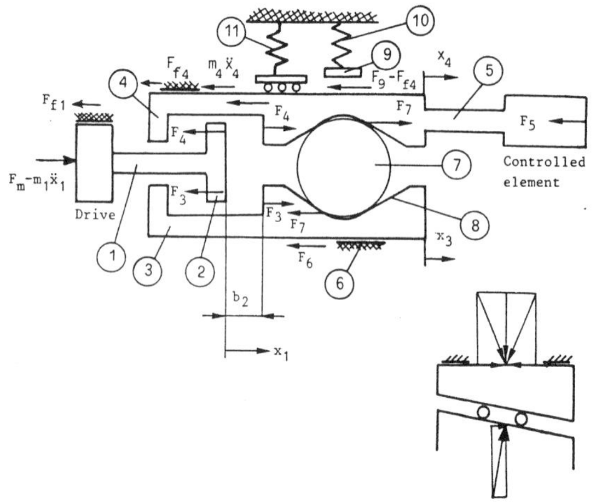

Fig.1 - Definition of the no-back geometry
1= input shaft, 2= teeth, 3= direction wedge disk, 4= driven disk, 5= output shaft, 6= direction brake, 7= balls, 8= ramp groove, 9= main multi-disk brake, 10= loading spring, 11= friction spring.

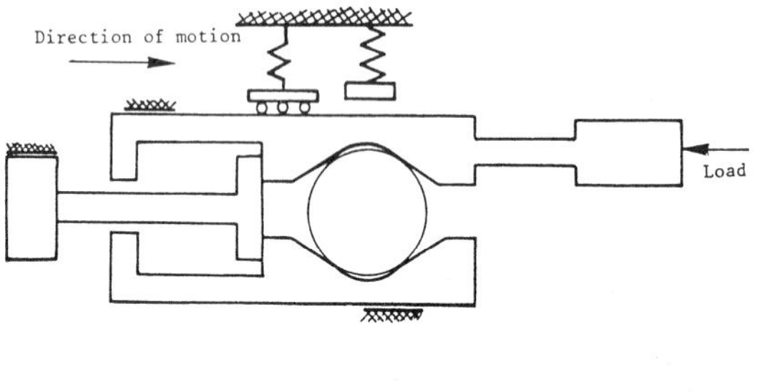

Fig.2 - Operation with opposing load

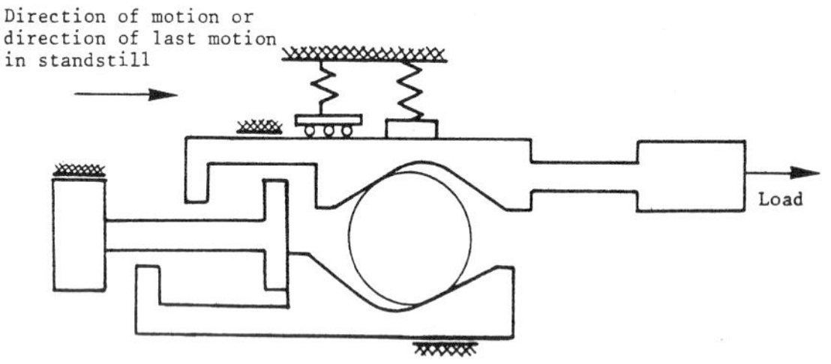

Fig.3 - Operation with aiding load

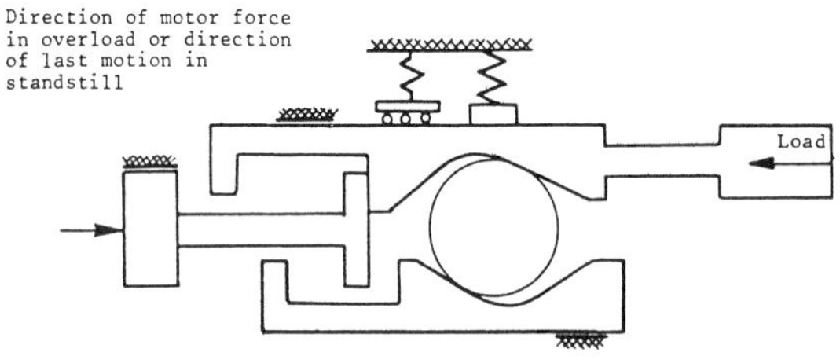

Fig.4 - Operation with overload

MATHEMATICAL MODEL

As a first step for the generation of a mathematical model, it is necessary to establish which are the parameters that affect the values of the forces F_7, F_6, F_9 (Fig.1) and the relationships among them. The force F_7 that the balls transmit to the elements 4 and 3, neglecting the minor effects due to the rolling friction, depends on the differential position x_3-x_4 between the elements 3 and 4 according to the law shown in Fig.5. The value of F_7 is, in fact, determined only by the force of the preloaded spring 11 in the range of x_3-x_4 for which the brake 9 is not engaged ($|x_3-x_4| < b_9$), and grows with x_3-x_4 outside this range because of the further load of the preloaded spring 10. It can therefore be written:

$$F_7 = \frac{2}{\pi} F_{p11} \, tg\,\alpha \, arctg \frac{x_3-x_4}{h_a} + k_{11} tg^2\alpha (x_3-x_4) + \frac{2}{\pi} F_{p10} \, tg\,\alpha \, arctg \frac{T_a}{h_b} + k_{10} \, tg^2\alpha \, T_a \qquad (1)$$

where:

$$T_a = \frac{1}{2} (x_3-x_4-b_9+|x_3-x_4-b_9| + x_3-x_4+b_9-|x_3-x_4+b_9|) = \frac{1}{2}\Big[2(x_3-x_4) + |x_3-x_4-b_9| - |x_3-x_4+b_9| \Big] \qquad (2)$$

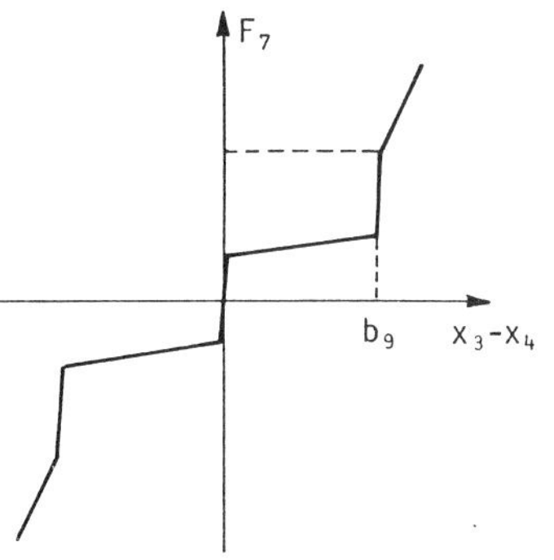

Fig. 5 - Ball force vs. differential position.

Similarly, the passive force F_9 which is on the total acting on the element 4 is the sum of a constant term plus a term that is highly dependent on x_3-x_4. This second term is due to the brake 9 when this is engaged by the preloaded spring 10, according to the law shown in Fig.6 :

$$|F_9| = |F_{f4}| + \frac{2}{\pi} f_9 \, F_{p10} \, arctg \frac{T_b}{h_c} + k_{10} f_9 \, tg\,\alpha \, T_b \qquad (3)$$

where

$$T_b = \frac{1}{2} \Big(x_3-x_4-b_9+|x_3-x_4-b_9| - x_3+x_4-b_9+|x_3-x_4+b_9|\Big) = \frac{1}{2} \Big(|x_3-x_4|-b_9+\big||x_3-x_4|-b_9\big|\Big) \qquad (4)$$

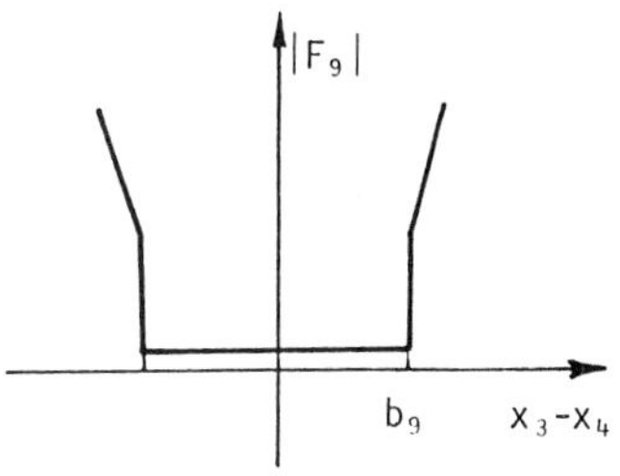

Fig. 6 - Main-brake force vs. differential position.

Also, the passive force F_6, due to the element 6 and felt by the element 3 depends on x_3-x_4 according to the law shown in Fig.7.

$$|F_6| = |f_6 F_{P11}| + k_{11} f_6 \,tg\,\alpha\, |x_3-x_4| + \frac{2}{\pi} f_6 F_{P10} \,arctg \frac{T_b}{h_d} + k_{10} f_6 \,tg\,\alpha\, T_b \quad (5)$$

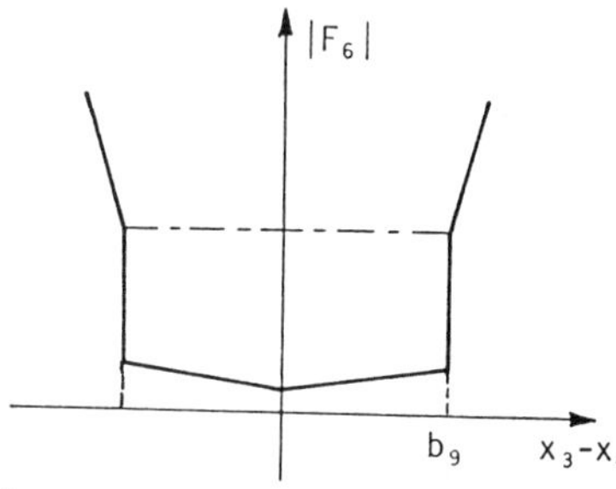

Fig. 7 - Direction-brake force vs. differential position.

It is important to point out that, in order for the no-back to operate, it must always be: $F_6 > F_7$. This means that 3 must never be displaced by 4 trough the balls. This conditions means that:

$$f_6 > tg\,\alpha \quad (6)$$

If not, the sensing brake 6 slips and the main brake 9 can never be engaged. The knowledge of F_6, F_7, F_9 makes it possible to write up the dynamic equilibrium equations for the element 3, for the driving (1 + 2) and for the driven (4 + 5) subsystems. These equations are:

$$F_m - m_1\ddot{x}_1 - F_3 - F_4 - |F_{f1}| \cdot SIGN(\dot{x}_1) = 0 \quad (7)$$

$$F_3 - |F_6| \cdot SIGN(\dot{x}_3) - F_7 = 0 \quad (8)$$

$$F_4 - F_5 - |F_9| \cdot SIGN(\dot{x}_4) - m_4\ddot{x}_4 + F_7 = 0 \quad (9)$$

It must also be noted that:

$$x_1 - b_2 \leq x_3 \leq x_1 + b_2 \quad (10)$$

$$x_1 - b_2 \leq x_4 \leq x_1 + b_2 \quad (11)$$

and that x_3 changes with the time only as a result of element 1 beeing in contact with element 3. If elements 3 and /or 4 are in contact with element 2, it must be: $\dot{x}_1 = \dot{x}_3 = \dot{x}_4$ and $\ddot{x}_1 = \ddot{x}_2 = \ddot{x}_4$; further, the equations (7), (8) and (9) can be changed in the following:

$$F_m - (m_1+m_4)\ddot{x}_1 - (|F_6| + |F_{f1}| + |F_9|) \cdot SIGN(\dot{x}_1) - F_5 = 0 \quad (12)$$

SIMULATION RESULTS

The simulations performed with the mathematical model outlined above yielded interesting results which are now briefly summarized.

a) - If condition 6 is not satisfied there is a total failure for the no-back operation.

b) - Large values of the friction coefficient f_6 of the sensing brake 6 can cause large drag forces when the no-back device is moving both with aiding and opposing loads.

c) - Large values of the static friction coefficient f_9 of the main brake, when related to the wedge angle α, can cause problems while starting against opposing loads.

d) - A large decrease of the friction coefficient f_9 from static to dynamic conditions can give rise to an operating instability with large and fast changes of the speed.

LIST OF SYMBOLS

F_m = motor force
F_3 = force from 2 to 3
F_4 = force from 2 to 4
F_5 = force on 5 produced by the controlled element
F_6 = friction force produced by 6 acting on 3
F_7 = force from 7 to 3 and 4
F_9 = friction force acting on 4 produced mainly by 9 (and by bearings)
m_1 = mass of the system "1 + 2 + upstream portion of the drive + power unit" referred to 1
m_4 = mass of "4 + 5 + downstream portion of the drive + controlled element" referred to 4
x_1, $\dot{x}_1$, $\ddot{x}_1$ = position, speed, acceleration of 1
x_3, $\dot{x}_3$ = position, speed of 3
x_4, $\dot{x}_4$, $\ddot{x}_4$ = position, speed, acceleration of 4
h_a, h_b, h_c, h_d = sufficiently little dimensional (length) constants
b_2 = half - backlash between 2 and 3 or 3 and 4
b_9 = half - amplitude of the differential positions between 3 and 4 for which the brake 9 is disengaged
α = ball ramp angle
F_{P10} = preload - force of the spring 10
F_{P11} = preload - force of the spring 11
k_{10} = stiffness of 10
k_{11} = stiffness of 11
F_{f1} = dynamic friction force on 1
F_{f01} = static friction force on 1
F_{f4} = friction force from 5 to 4

PROJET ET OPTIMIZATION D' UN FREIN DYNAMIQUE DE IRREVERSIBILITE'

Sommaire

Les freins de irreversibilité ont deux different fonctions:
- ils retien l'element controlle' dans una position fixe, independentement de variations du charge, lorsque il n'y a aucum change de la position commandée;
- ils agit comme freins dynamiques et dissipe l'excès de energie mecanique lorsque la force qui charge l'element controllé a la même direction de la variation de la position commandée.

Pour assurer un bon fonctionnement du système constitué du frein et de l'elément controllé on doit faire une accurée optimization de différent paramétre de projet, considerent comme ils agit sur les prestations du système, ce qui est illustré dans cette memoire.

An Active Restriction System for a Hydrostatic Lead Screw

H. MIZUMOTO*, S. OKAZAKI*, T. MATSUBARA* AND M. USUKI**

*Department of Mechanical Engineering, Tottori University, Tottori, Japan
**Nachi-Fujikoshi Corporation, Toyama, Japan

Abstract. To increase the stiffness of the hydrostatic lead screw, we propose a new type of active restrictor. Instead of a diaphragm in a conventional active restrictor, a set of hydrostatic bearings is used for supporting a restriction ring which controls the oil flow rate. The nut of the hydrostatic lead screw with the proposed active restrictor is constructed of two inner sleeve blocks and an outer sleeve. Between these inner and outer sleeves, the restriction ring and the ring-supporting hydrostatic bearings are incorporated. Experimental analysis shows that with this hydrostatically controlled restrictor(HCR), both the thrust and radial stiffnesses of the hydrostatic lead screw can be increased simply by adjusting the oil supply pressure to the ring-supporting hydrostatic bearings. Such an easy adjustment of the restrictor is very difficult for a conventional active restrictor.

Keywords. Machine tools; hydraulic systems; closed loop systems; feedback; hydrostatic lead screw; active restrictor.

INTRODUCTION

The hydrostatic bearing system is essential for a high precision feed drive device. However, the stiffness of the hydrostatic lead screw used in the system is not always enough, because of the limitation of the bearing area (the flank area) of the lead screw. To increase the stiffness of the lead screw, an active restrictor is highly effective. Many active restrictors have been designed for hydrostatic bearings (Moris, 1972; Tully, 1977). According to their mechanism for controlling the restriction gap, active restrictors can be divided into two types: a pocket pressure feedback type, and a working clearance feedback type.

One of the well known active restrictors is a diaphragm controlled restrictor (DCR), where a gap controlling pad is supported by a diaphragm. The pocket pressure fed back to the restrictor moves the gap controlling pad to keep the working clearance on the bearing surface constant. This means that the stiffness of the bearing with DCR can be virtually infinite. To obtain an infinite stiffness bearing, the parameters of the active restrictor, such as the diaphragm stiffness, should be adjusted to an optimum value. However, such an adjustment is difficult because of errors in the manufacturing and the assembling of the restrictor, and there are few reports concerning the hydrostatic lead screw with an active restrictor (Khaimovich, and Migai, 1970).

The mechanism of the active restrictor proposed in the present paper is basically the same as the DCR. Instead of the diaphragm in the DCR, a set of hydrostatic bearings is used for supporting the gap controlling pad. Therefore, this active restrictor can be regarded as a Hydrostatically Controlled Restrictor (HCR). The HCR has already been incorporated into a hydrostatic journal bearing for increasing the bearing stiffness (Mizumoto, and others, 1985). The stiffness of the pad-supporting bearing can be changed by the supply pressure, and the stiffness of the hydrostatic lead screw with an HCR can not only be infinite but also positive or negative. Moreover, as the HCR controls the flow rates to all the pockets of the bearing by one gap controlling pad, it is suitable for being incorporated into the hydrostatic lead screw.

EFFECT OF ACTIVE RESTRICTOR

Figure 1 shows an opposed-pad hydrostatic bearing with a DCR. In the DCR, there is a moving pad which controls the oil flow to the bearing pads. Oil under constant pressure Pp is restricted as it passes through the gaps hd_1 and hd_3 between the moving pad and the opposed surfaces as shown in Fig. 1. Subscripts 1 and 3 represent the loaded side and the unloaded side, respectively. The load carrying capacity W of the opposed-pad bearing is proportional to the pressure difference between the lower and the upper pockets, i.e. $W=Se(P_{11}-P_{13})$, where Se is the virtual bearing area of the bearing pad. This pressure difference $dP=P_{11}-P_{13}$ is fed back to the active restrictor, where the moving pad is supported elastically by a diaphragm. By the feedback pressure difference, the moving pad is displaced to change the restriction gaps by $dhd=dP\cdot Sd/k$; where, Sd is the restriction area of the moving pad, and k is the diaphragm stiffness.

The relationship between W and the eccentricity ε of the working clearance on the bearing surface is given by the following equation:

$$Lf = \frac{\left(1+\frac{Sd\cdot Pp}{k\cdot hd*}Lf\right)^3}{(1-\varepsilon)^3+\left(1+\frac{Sd\cdot Pp}{k\cdot hd*}Lf\right)^3} - \frac{\left(1-\frac{Sd\cdot Pp}{k\cdot hd*}Lf\right)^3}{(1+\varepsilon)^3+\left(1-\frac{Sd\cdot Pp}{k\cdot hd*}Lf\right)^3} \quad (1)$$

where, $Lf=W/(Pp \cdot Se)$ is the load factor and hd^* the restriction gap at non-load, and the ratio of the pocket pressure to the supply pressure is supposed to be $m=P_1/Pp=0.5$.

Now, we define a restrictor constant Vc which represents the effectiveness of the active restrictor.

$$Vc = \frac{Sd \cdot Pp}{k \cdot hd^*} \qquad (2)$$

For a conventional passive restrictor whose restriction gap is constant, k is infinite and Vc=0.

Equation (1) can be approximated by the following equation where the higher order terms of Lf and ε are ignored (the error occurring from this approximation is within 0.5% when Lf and ε are less than 0.1):

$$\varepsilon = \left(\frac{2}{3} - Vc\right) Lf \qquad (3)$$

Equation (3) shows that when Vc<0.67, ε increases proportionally to Lf and the bearing stiffness is positive. At Vc=0.67, ε is zero for any load. This means that the stiffness of the bearing is infinite. When Vc>0.67, the eccentricity is negative, e.g. when the load is applied downward, the bearing shaft displaces upward. In the present paper, such a relationship between ε and Lf is called negative stiffness.

To obtain an infinite stiffness bearing, Vc should be 0.67. However, owing to the errors in the manufacturing and the assembling of the restrictor, it is very difficult for the DCR to adjust Vc to such an optimum value. Further difficulties will occur when incorporating the DCR into a hydrostatic journal bearing or a hydrostatic lead screw, because the number of restrictors to be incorporated should be equal to the number of the independent pockets. To avoid these difficulties in obtaining the infinite stiffness bearing, a hydrostatically controlled restrictor (HCR) has been designed.

A schematic view of an HCR incorporated into a hydrostatic journal bearing is shown in Fig. 2 (Mizumoto and others, 1985). The bearing is constructed of an outer sleeve, an inner sleeve, a restriction ring which corresponds to the moving pad in Fig. 1, a flange, and a shaft to be supported. The pressure of Pp of oil supplied from the right side of the inner sleeve is restricted as it passes through the gap hd between the inner sleeve and the restriction ring. Then, the restricted oil flows into the bearing surface via several supply holes of diameter 4mm to form the lubricating film. Instead of the diaphragm in Fig. 1, a hydrostatic bearing incorporated into the outer sleeve supports the restriction ring. A pressure of Pp^* of oil supplied from the left side of the outer sleeve operates the ring-supporting hydrostatic bearing whose stiffness is proportional to Pp^*.

For the free displacement of the restriction ring, there must be small gaps hd' at both sides of the ring. Through these side gaps, some leakage of the pressurized oil occurs. Here, we can define the leakage factor Fs as the ratio of flow resistance of the restrictor to that of the leak path through the side gaps of the restriction ring; an increase of Fs indicates an increased leakage. The influence of the leakage modifies Eq. (3) as follows:

$$\varepsilon = \left\{ \frac{2}{3}\left(1 + \frac{Fs}{2}\right)^2 - Vc \right\} Lf \qquad (4)$$

For various values of Fs, the effect of Vc on ε is calculated from Eq. (4) and shown in Fig. 3. When leakage does not occur (Fs=0), the stiffness of the journal bearing is infinite at Vc=0.67. As leakage increases (Fs increases), the optimum value of Vc which gives the infinite bearing stiffness increases. Equation (2) indicates that Vc is proportional to the supply pressure Pp, and inversely proportional to k— to the supply pressure on the ring-supporting bearing Pp^*. Consequently, Vc is proportional to the ratio of both the supply pressures Pp/Pp^*. This means that Vc can be adjusted to an optimum value even after incorporating the HCR into the bearing. The influence of leakage can also be compensated by increasing the supply pressure ratio.

The advantages of HCR are summarized as follows:

1. The restrictor constant can be altered by changing the supply pressure ratio.
2. With only one restriction ring, the oil flow to all the oil supply holes can be controlled.

Owing to these advantages, it is easy to incorporate an HCR into a hydrostatic journal bearing, or a lead screw, and to make the bearing stiffness infinite. In the case of the conical bearing and the lead screw, Eq. (4) is applied to the axial and to the radial direction independently.

STRUCTURE OF HYDROSTATIC LEAD SCREW WITH HCR

Figure 4 shows an assembled view of a hydrostatic lead screw with an HCR. The parts of the lead screw are shown in Fig. 5. The specifications of the male screw (9) are as follows: outer diameter 51mm, pitch 16mm, and flank angle 15°. The hydrostatic nut is constructed of two inner sleeve blocks, an outer sleeve (3), a restriction ring (4), and two flanges(6). Each inner sleeve block is constructed of an inner sleeve 1 (1) and an inner sleeve 2 (2). The inner sleeve 1 has two pitches of saw shape threads; the angle of 15° of flank is used as the bearing surface. As shown in Fig. 4, two inner sleeve blocks are arranged so that each bearing surface opposes another. The working clearance is made by an appropriate phase difference between both the helical bearing surfaces. Between the inner sleeve blocks, a spacer ring (7) is inserted so as to form an appropriate restriction gap between the inner sleeve blocks and the restriction ring.

The oil flow in the hydrostatic nut is shown in Fig. 6. A pressure of Pp of oil supplied from the flange is restricted as it passes through the "Restriction region" indicated. Then, the restricted oil flows to the flank and supports the male screw hydrostatically. An O-ring attached to the restriction ring prevents leakage of the pressurized oil. The restriction ring is supported by a hydrostatic radial bearing in the outer sleeve, and by a hydrostatic thrust bearing attached to the innner sleeve 2. A pressure of Pp^* of oil which is independent of the oil for the lead screw operates these ring-supporting bearings. The restrictors of these supporting bearings are slot-restrictors machined on the flange and the inner sleeve as indicated "Restrictor" in Fig. 6.

CHARACTERISTICS OF LEAD SCREW WITH HCR

Radial Stiffness of Lead Screw

The hydrostatic nut was fixed horizontally to a base. Under several supply pressure ratios, the radial displacement of the male screw for a unit load was measured. From the measurement, the radial compliance was calculated as the ratio of the radial displacement to the load. Then, the calculated values were converted into radial compliances at Pp=98KPa, as shown in Fig. 7. In Fig. 7, Cs indicates the compliance of the same lead screw with a conventional passive restrictor, and Co is the intersection of the regression line of the measured values and the axis of the compliance. If leakage does not occur, Co should be equal to Cs. From the discrepancy between Co and Cs, the leakage factor Fs is estimated to be Fs=0.65.

As the supply pressure ratio increases, the radial compliance decreases. At the highest pressure ratio measured, the compliance was about a half of Cs. Therefore, the radial stiffness of the lead screw with the HCR can be twice as large as that of the lead screw without the HCR. However, the infinite stiffness could not be observed. The supply pressure ratio which gives the infinite radial stiffness (zero compliance) was estimated to be Pp/Pp*=15.4, from the intersection of the regression line and the axis of the pressure ratio. This estimated optimum pressure ratio is considerably greater than the designed value, 4.9. The discrepancy between the estimated and designed optimum pressure ratio occurs because of the use of the O-ring which prevents the radial displacement of the restriction ring.

Thrust Stiffness of Lead Screw

The lead screw and the hydrostatic nut were incorporated into a feed drive device, and the male screw was fixed. Then, under several supply pressure ratios, the thrust load was applied to the table of the device and the displacement of the table was measured. From the measurement, the thrust compliances were calculated and are shown in Fig. 8, where, Co and Cs are the same as in Fig. 7. The infinite thrust stiffness (zero compliance) was observed when the supply pressure ratio Pp/Pp*=6.2. As shown in Fig. 7, the radial compliance at Pp/Pp*=6.2 is equal to Cs. Thus, without decreasing the radial stiffness, we can obtaine a lead screw with an infinite thrust stiffness; a desirable characteristic as a lead screw for a high precision feed drive device.

The observed optimum supply pressure ratio, 6.2 is slightly higher than the designed value, 4.7. However, the discrepancy is not so large as in the case of the radial stiffness, because of the reduced influence of the O-ring on the stiffness of the ring-supporting thrust bearing. The negative stiffness was also observed at pressure ratios higher than the optimum value. The leakage factor estimated from the discrepancy between Cs and Co was Fs=0.59, which agrees with the estimation from the radial stiffness.

CONCLUSIONS

The effect of a hydrostatically controlled restrictor (HCR) on the stiffness of a hydrostatic lead screw was analysed, and the following conclusions are obtained:

1. The thrust stiffness of the lead screw can be infinite by adjusting the supply pressure ratio. The negative stiffness can also be realized by relatively high supply pressure ratios.
2. The radial stiffness of the lead screw can be increased, however the infinite radial stiffness can not be observed.
3. Leakage through the sides of the restriction ring affects the stiffness of the lead screw. The influence of leakage can be compensated by increasing the supply pressure ratio.

From the analyses, it is shown that the hydrostatic lead screw with the HCR can be used as the lead screw of the high precision feed drive device.

REFERENCES

Khaimovich, Ya. M., and Yu. A. Migai (1970). Diaphragm-type differential throttle for a closed hydrostatic screw-and-nut transmission. Machines and Toolings, 41, 12-16.

Mizumoto, H., T. Matsubara and M. Kubo (1983). Effective improvements in the design of hydrostatic lead screws. Proceedings of 24th International MTDR Conference, 369-374.

Mizumoto, H., M. Kubo, Y. Makimoto, S. Yoshimochi, S. Okamura and T. Matsubara (1985). A hydrostatically controlled restrictor for an infinite stiffness hydrostatic journal bearing. Journal of the Japan Society of Precision Engineering, 51, 1553-1558.

Moris, S. A. (1972). Passively and actively controlled externally pressurized oil-film bearings. Trans. ASME, Ser.F, 94, 56-63.

Tully, N. (1977). Static and dynamic performance of an infinite stiffness hydrostatic thrust bearing. Trans. ASME, Ser. F, 99, 106-112.

Zusammenfassung. Ein neugeformtes aktives Drosselventil soll vorgeschlagen werden, um die Steifigkeit an der hydrostatischen Leitspindel zu erhöhen. In dem aktiven Drosselventil kontrolliert ein von hydrostatischen Lagern gestützter beweglicher Ring die Ölmenge in die Flanke von der Mutterschraube. Deshalb wird das aktive Drosselventil als "HCR" (Hydrostatically Controlled Restrictor) bezeichnet. Die hydrostatische Mutterneinheit mit dem HCR besteht aus einer äußeren Muffe und zwei inneren. Die inneren Muffe haben je in sich eine Mutterschraube. Der bewegliche Ring wird zwischen die äußeren und die inneren Muffen hineingelegt. Eine Drosselung des in die Flanke eingegossen Öl findet sich in Moment, da es recht zwischen den inneren Muffen und dem Ring fährt. Wird der Ring von den hydrostatischen Lagern in der äußeren Muffe gestützt, so kann er sich in dem Maße von der auf die Leitspindel beförderten Last bewegen. Infolgedessen verändert sich der Abstand zwischen den inneren Muffen und dem Ring zur vermehrten Ölmenge in die belastete Flanke, damit sich dadurch die Leitspindel zurückstoßen und die Steifighkeit der Leitspindel bis zur Unbegrenztheit wird.

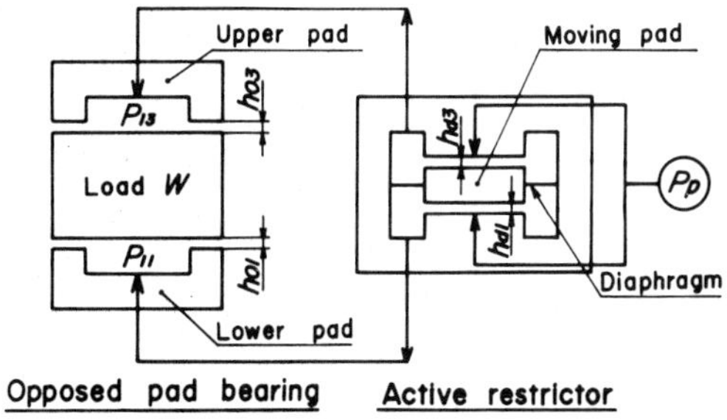

Fig. 1. Opposed-pad Bearing with Active Restrictor.

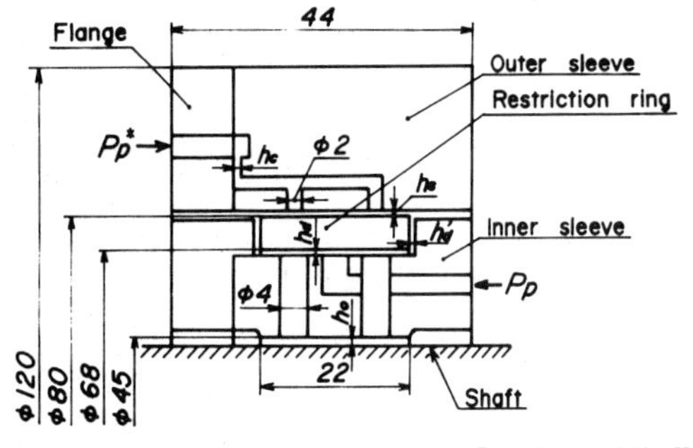

Fig. 2. Magnified view of Journal Bearing with HCR.

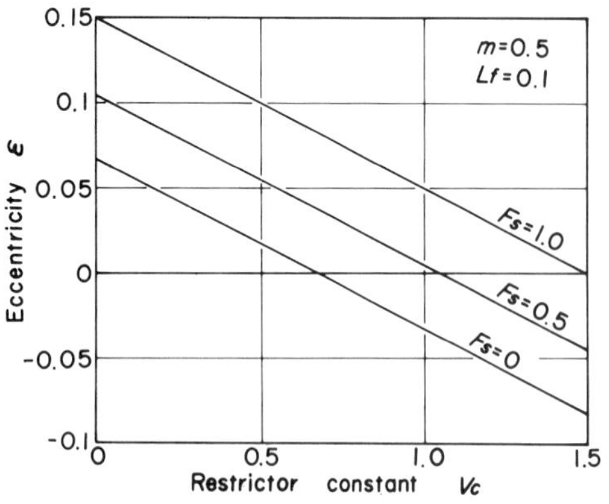

Fig. 3. Influence of Leakage on Eccentricity.

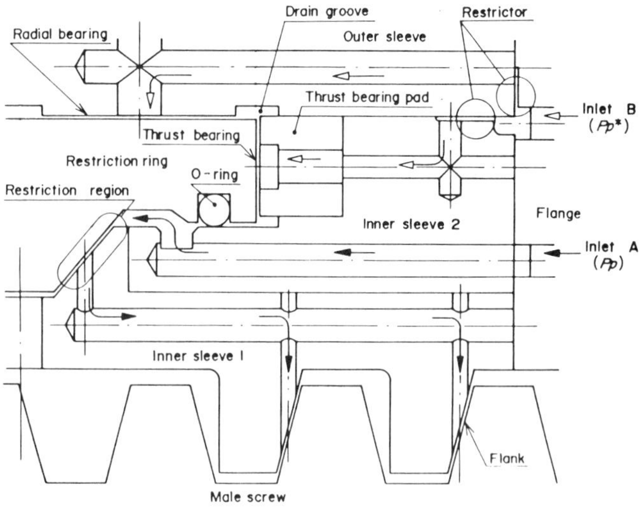

Fig. 6. Magnified view of Lead Screw with HCR.

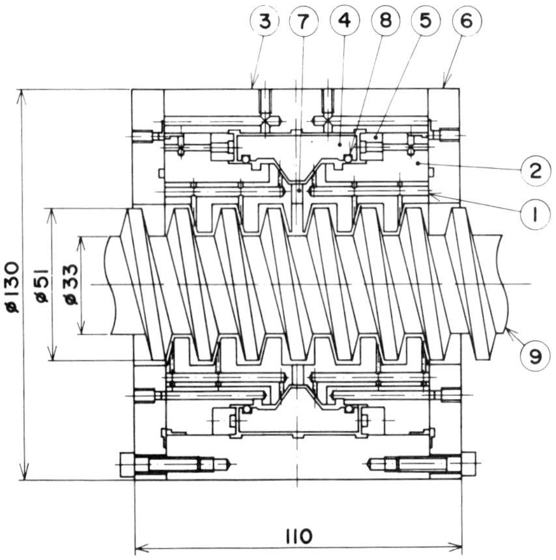

Fig. 4. Hydrostatic Lead Screw with HCR:
1; inner sleeve 1, 2; inner sleeve 2,
3; outer sleeve, 4; restriction ring,
5; thrust bearing pad, 6; flange,
7; spacer, 8; O-ring, 9; male screw.

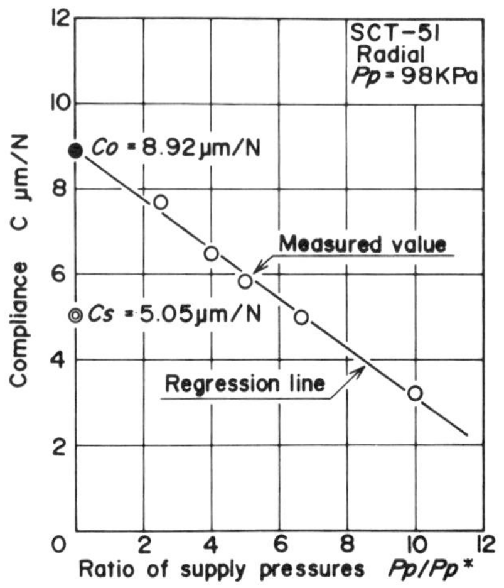

Fig. 7. Radial Compliance of Lead Screw.

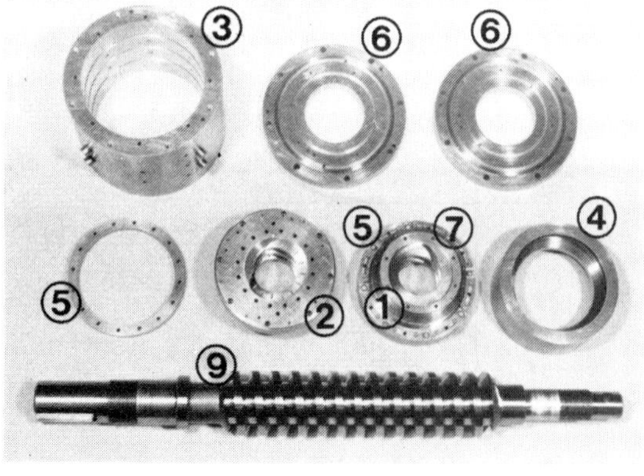

Fig. 5. Parts of Hydrostatic Lead Scrw with HCR (part numbers are the same as in Fig. 4).

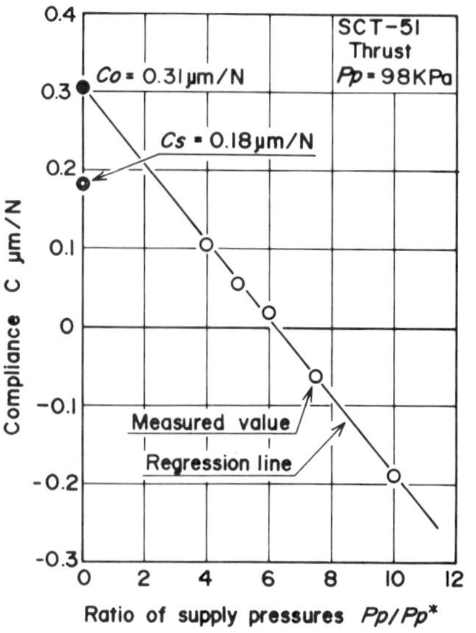

Fig. 8. Thrust Compliance of Lead Screw.

Studies on Aerostatic Lead Screws

T. SATOMI

Faculty of Science and Engineering, Tokyo Denki University, Ishizaka, Hatoyama-machi, Hiki-gun, Saitama 350-03, Japan

Abstract. The purpose of this study is to present theory practically applicable to the design of aerostatic lead screws. Heretofore, the theories on the static rigidity applicable throughout the land clearance between the guiding surfaces were developed, considering that the flow rate conforms to the inherent restrictor theory in the smaller land clearance and to the orifice restrictor theory in the larger land clearance. In the beginning of this study, using a thrust bearing type multi-orifice model with ring shaped plane guiding surfaces corresponding to one-lead screw flanks, the calculating process for the static rigidity is developed. Then a lead screw with thread angle 60 ° mating with a nut consisting of two one-lead internal threads is manufactured on trial and its static rigidity is examined. As the results, the experimental values showed fairly good agreement with the calculated values.

Keywords. Fluidic devices, Pneumatic control equipment, Aerostatic guiding system, Aerostatic lead screws, Air-bearing.

Nomenclature

The following nomenclature is used in the paper

a:a half size of annulus land width
b:a half length between orifices
cf':coefficient of flow rate at inherent restrictor
c2:constant
dw/dh:static rigidity
g:acceleration of gravity
h:land clearance
h1:axial clearance of side A
h2:axial clearance of side B
hm:axial clearance at maximum static rigidity
ha:axial direction clearance at equilibrium point
i,j:variables
K:adiabatic exponent
K1,K2:constants
k:number of orifices
m:number of images
N:amount of orifices
n:polytropic number
p:land pressure
p0:pressure just downstream of orifice pressure at air-pocket
pa:atmospheric pressure
pc:pressure just downstream of inherent restrictor
ps:supply pressure
Qc:mass flow rate from inherent restrictor
Qout:mass flow rate from land clearance into air pocket
Rout:land radius
W1,W2:a load of axial lands
x:radial coordinate on annulus land
y:circumferential coordinate on annulus land
α:half-angle of thread
α':half-angle of thread at right-angle section of thread
β:lead angle
γ_o:specific gravity of air at p0
μ:coefficient of viscosity
ξ:axial displacement
ρa:density of air at atmospheric pressure
ρc:density of air at pc

Introduction

Improvement of manufacturing accuracy on electronic parts, I.C., opto-electronics and others, is expected to be according to new developing electronics, recently. As the aerostatic guiding system has been experimented in machine tools, measuring machines and other guiding systems, the performance was improved. Aerostatic guiding systems are characterized by the followings; ① being free from friction, ② that the fluid is almost infinitely available, ③ no trouble due to air leakage, ④ being little influence from heat, ⑤ that the guiding accuracy does not fail permanently, etc..

The purpose of this study is to present theory practically applicable to the design of an aerostatic guiding system for aerostatic lead screws.

In the first step of this study, the static rigidity is calculated on the annulus guiding surface corresponding to one-lead screw flank. In the second step, the pressure distribution on thread face, lead characteristic, maximum static rigidity, etc. are obtained theoretically, and confirmed experimently on the vertical type lead screws in which the pitch is 32 mm and angle of thread is 60 °. As the result; the experimental values showed fairly good agreement with the theoretical values.

1. Theory

1.1 Assumptions

The following assumptions are made in this investigation.
① the air is regarded as Newtonian fluid, ② laminar flow, ③ compressible, ④ inertia and gravity are negligible, ⑤ pressure and viscosity are uniform along the direction perpendicular to the guiding surface.

1.2 Determination of pressure on the downstream of inherent restrictor

The throttle of an aerostatic guiding system without an air-pocket downstream of orifice becomes a cylindrical inherent restrictor, in which the radius is r0 mm, and the clearance is h μm.
Many small drilled feeder holes(orifices) are drilled on the spiral center of aerostatic screws helicoidal surface.
As practical clearance on the screws surface is expected to be less than 50 μm according to the previous study, influence of inherent restrictor is larger than orifice restrictor in this study. In this case, mass flow rate from inherent restrictor into land clearance is showed by equation (1), and mass flow rate from land clearance into atmosphere is showed by equation (2).

$$Q_c = C'_f \rho_c 2\pi r_0 h \sqrt{\frac{2gKP_0}{(K-1)\gamma_0}\left\{1-\left(\frac{P_c}{P_0}\right)^{\frac{K-1}{K}}\right\}} \tag{1}$$

$$Q_{out} = \left(\frac{C_2 \pi h^3}{6\mu}\right)\left(\frac{n}{n+1}\right)\left(\frac{P_c^{\frac{n+1}{n}}-1}{\ell n(R_{out}/r_0)}\right) \tag{2}$$

Under the condition of continuity of flow, Qc=Qout holds. From this relation, pressure just downstream of inherent restrictor can be calculated. Land radius Rout at screw surface is imaginary, but it is set in annulus land width size according to the result of study which is shown in reference (1).

1.3 Pressure distribution on screw surface

When the screw helicoidal surface can be assumed to be replaced by long nerrow strip piece of land and also when the origin is chosen so as to be between the orifices, which are symmetrically arranged about the origin, pressure distribution of said land is represented by equation (3).

$$\begin{aligned} p = K_1 \Bigg\{ & \frac{1}{2}\sum_{i=1}^{k} \ell n\left(\left(x^2+(y+(2i-1)b)^2\right)\left(x^2+(y-(2i-1)b)^2\right)\right) \\ & + \sum_{j=1}^{m}\Bigg[(-1)^j \frac{1}{2}\sum_{i=1}^{k} \ell n\Big(\big((x-2aj)^2+(y+(2i-1)b)^2\big)\big((x- \\ & 2aj)^2+(y-(2i-1)b)^2\big)\times\big((x+2aj)^2+(y+(2i-1)b)^2\big) \\ & \big((x+2aj)^2+(y-(2i-1)b)^2\big)\Big)\Bigg]\Bigg\} + K_2 \end{aligned} \tag{3}$$

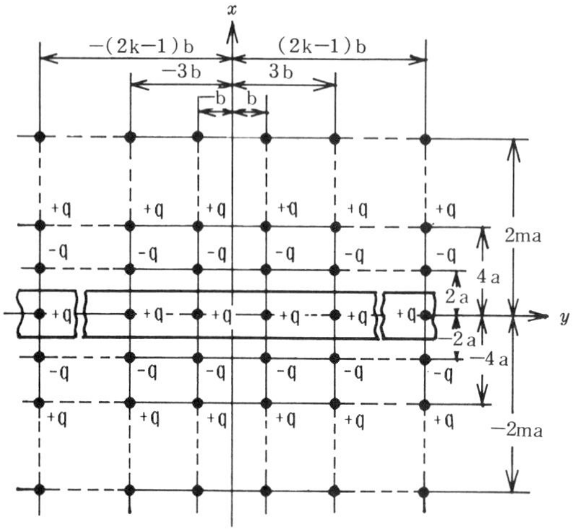

Fig.1 Developed view of land surface in case of multiple orifices

As pressure Pc just downstream of inherent restrictor is calculated by the aforementioned method(showing paragraph 1.2), pressure at the boundary when the origin of coordinate is given such as in Fig.1, becomes p=atmospheric pressure in case coordinate(x,y) is (a,o) and p= pc in case (x,y)=(r0,b). Constant K1 and K2 are determined using these results.

$$K_1 = \frac{1-P_c}{p(a,o)-p(r_0,b)}$$

$$K_2 = 1 - \frac{(1-P_c)\,P(a,o)}{p(a,o)-p(r_0,b)}$$

Pressure p(a,0) AND p(r0,b) are calculated by substituting (a,0) and (r0,b) into equation (3). Generally, the instantaneous pressure distribution on long narrow land can be calculated by equation (3) and the above equations for K1 and K2. In this case, the larger number of images m and the larger number of orifices k yield the better approximation.

1.4 Load component force acting on the screw thread flanks

It is possible to resolve a component force acting perpendicular to the screw thread flanks into three directions, i.e., axial direction, axial perpendicular direction and axial tangential direction, depending on the influence of the angle of the thread and lead angle. In this case, DA is the component force of perpendicular direction on the screw flanks. OA is the component force of axial direction, OE is the component force in the direction perpendicular to the axis and OH is the component force in the direction tangential to the axis. When putting N orifices on the spiral center of one-lead screw flank providing at regular stated intervals, perpendicular load on flanks carried by one orifice is

$$OD = 4 \int_0^a \int_0^b p\,dx\,dy \tag{4}$$

In this case, axial load can be calculated by maltiplying the equation 4 by $\cos\alpha' \cos\beta$.

$$OA = 4 \cos\alpha' \cos\beta \int_0^a \int_0^b p\,dx\,dy \tag{5}$$

And perpendicular load to the axis can be calculated by maltiplying the equation 4 by $\sin\alpha'$.

$$OE = 4 \sin\alpha' \int_0^a \int_0^b p\,dx\,dy \tag{6}$$

Tangential load to the axis can be calculated by maltiplying the equation 4 by $\cos\alpha' \sin\beta$.

$$OH = 4 \cos\alpha' \sin\beta \int_0^a \int_0^b p\,dx\,dy \tag{7}$$

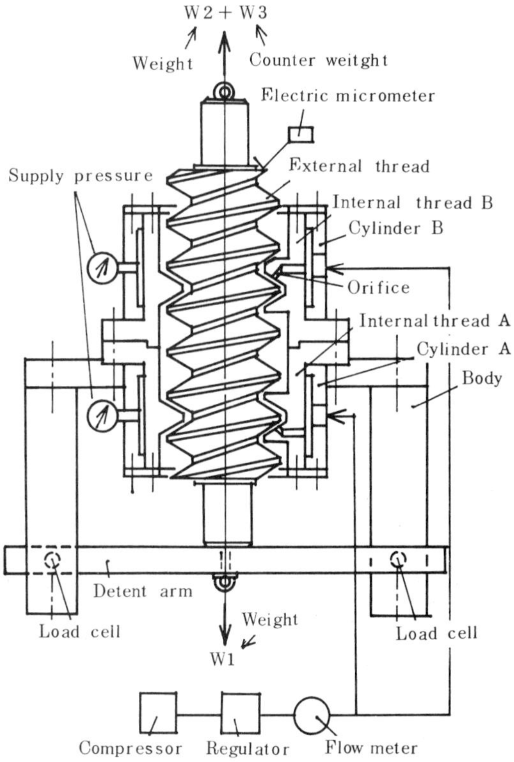

Fig.2 Experimental apparatus for measuring static rigidity

1.5 Calculation of axial static rigidity

As forces in the direction perpendicular to the axis calculated by equation (6) balance each other and yield self-centering action in this experimental equipment, an axial direction of static rigidity can only be treated in this study. Fig.2 shows an outline of the experiment. Compressible air that is derived only on one side of the screw thread flanks is called the single side pressure type and compressible air that is derived on both side of the screw thread flanks called the differential pressure type.

In case of drilling N orifices on these flanks, axial static rigidity can be represented by equation (8).

$$W_1 = 4N \cos\alpha' \cos\beta \int_0^a \int_0^b p\,dx\,dy \tag{8}$$

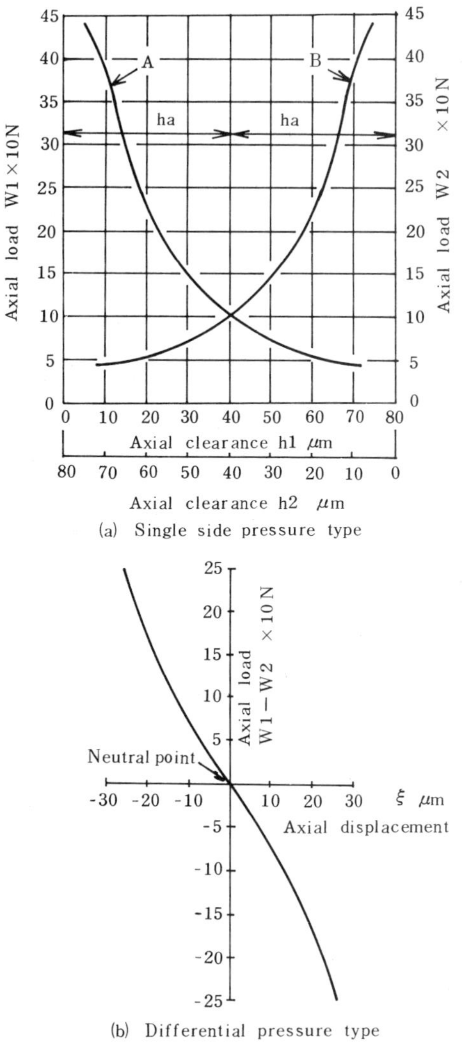

Fig.3 Theoretical consideration on relations between axial displacement and load

Fig.3(a) shows relations between axial clearance and load in this equipment calculated by equation (8). Curve A shows relations h1 and W1, Curve B shows relations h1 and W1. In this case, an intersection point with curve A and curve B is a neutral point. Fig.3(b) shows relations between axial displacement and load(W1-W2) putting origin of coordinate at a neutral point in the differential pressure type. ξ equal h1-h2 in Fig.4(a).

Axial clearance hm is derived maximum static rigidity in case h1 is parameter, and dW1/dh1 shows maximum value in the relations between axial load W1, and axial clearance h1 calculated by equation (8).
Static rigidity in the differential pressure type shows | d(W1-W2)/dξ | a neutral point, and its maximum value is twice as large as maximum dW1/dh1 value at ha=hm.

2. Results of calculation and experiment

Full lines in Fig.4 show the relations between axial displacement ξ and axial load(W1-W2) at differental pressure type. Curve ① shows the theoretical characteristic of maximum static rigidity at clearance hm=17.3 μm, in this case thestatic rigidity is 28.6 N/ μm at the neutral point. Curve ② shows the theoretical characteristic at clearance ha=40 μm, in this case the static rigidity is 7.1 N/ μm at the neutral point. Marks "○", and "△" in Fig. 4 shows the experimental value with the differental pressure type. the "○" expresses an increasing load and the "△" expresses a decreasing load at ha=40 μm. Experimental static rigidity in neutral is 7.1 μm at axial clearance ha=40 μm.
The experimental value shows faily good agreement with the calculated value with differential pressure type. Experimental value corresponding to curve ① could not be measured as the clearance is too narrow.

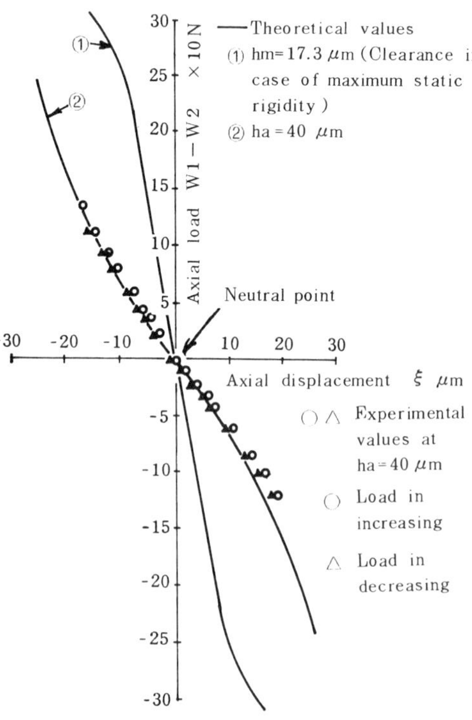

Fig.4 Relations between axial clearance and load in single side pressure type

3. Conclusion

A calculating method on static rigidity of an aerostatic lead screws is represented as follows; when making an opposed arrangement with two one-lead internal thread flanks to an external thread flanks, showing relations with an axial clearance and axial load by applying the calculating method of pressure distribution on a ring shaped plane guiding surface in thread flanks. The aforementioned static rigidity is derived from gradient of these characteristics. and it was ascertained by experiment. As a result, calculated value showed faily good agreement with experimental value.

Reference

1) Tadaatsu Satomi: "Studies on the Aerostatic Guiding System -Theory and Experiments on the Static Rigidity and Stability-", Bull. of JSPE, 17, No.2(19-83)107.

* *

Forschung über aero-statische Vorschubspindel

Tadaatsu Satomi

Fakultät für Naturwissenschaften und Technik, Tokio Denki Universität

Übersicht: Das Ziel dieser Forschungen ist die Gestaltungstheorie für aerostatische Schrauben in ihrer praktischen Anwendung nahe zu bringen.
Erster Abschnitt dieser Abhandlung ist die Forschung über den Drosselbereich. Das Ergebnis ist, die Theorie für die statische Steifigkeit mit der Theorie für die innewohnende Drossel im kleinen Spalt zwischen den Führungsflächen und mit der Theorie für die Mündungs Drossel im grossen Spalt zwischen den Führungsflächen zu erschliessen.
Zweiter Abschnitt dieser Abhandlung ist die Forschüng über die statische Steifigkeit der Ringführungsfläche. Das Ergebnis ist der Berechnungsgang für die statische Steifigkeit der Ringführungsfläche entsprechend mit einer Steigung der Schraubenfläche zu erschliessen. Dritter Abschnitt dieser Abhandlung ist den Flankenwinkel 60 ° und die Schraube 32 mm der Vorschubsspindel versuchs weise zu fabriziern, und zu untersuchen. Diese zwei Führungsmuttern sind mit einer Schraube zu komponieren. Das Ergebnis ist, den berechneten Wert 7.1 N/ μm der statischen Steifigkeit an 40 μm im Spalt zwischen den Führungsflächen und den Maximumrechnungswert 28.6 N/μm der statischen Steifigkeit an 17 μm im Spalt zwischen den Führungsflächen zu gewinnen. Der Experimentelwert ist in erträglich gute Übereinstimmung mit dem berechneten Wert zu weisen.

Considerations on Design of the Actuator Based on the Shape Memory Effect

S. HIROSE, K. IKUTA, M. TSUKAMOTO AND K. SATO

Department of Mechanical Engineering Science, Tokyo Institute of Technology, Tokyo, Japan

Abstract. Shape Memory Alloy (SMA) actuators, for their unique functionality, have been much expected for contribution to the advancement of robot performance. The authors conducted a series of investigations to materialize truly practical SMA actuators in the last several years. Firstly, the basic characteristics of SMA actuators including the shortcomings were evaluated to clarify key problems which might make bottlenecks for practical applications. Secondly, the key problems were substantionally settled by introducing a new actuator construction method named " ξ-array" and a " σ-mechanism" which sharply improved the output characteristics. Thirdly, experimental investigations were conducted to the material properties of SMA and the optimum annealing condition for the actuator material was also found. As regards the application, an active endoscope with 13 mm in diameter 210 mm in length and consists of 5 segments was test produced. The 5 segments are individually controlled by computer. It could be smoothly inserted to the colon by its flexible posture making function of the super miniature actuator system of SMA.

Keywords. Shape memory alloy; servo actuators; robots; annealing temperature; resistance feedback control;active endoscope; fiberscope ;power weight ratio.

INTRODUCTION

Servo actuators using Shape Memory Alloy (SMA), for the unique functionality unavailable by conventional actuators, are expected to be used for designing new advanced robots. Japan had started the studies earlier than others in the world on applying SMA to robot actuators. Accomplishments of foremost studies were reported (Honma, Miwa, Iguchi,1983; Hirose, Ikuta, Umetani, 1983; Hosoda, Fujie, Kojima,1983). Those reports drew attention of the related circles to the potential of SMA as the material of robot actuators, and usefulness was well admitted by many researchers. Nevertheless, SMA has not yet been practically tried for robot actuators.
The authors believe the reason to lie on the lack of comprehensive evaluation of the basic characteristics and construction method of SMA servo actuators (Hirose, Ikuta, Umetani, 1984a).
On this account, investigation was made to clarify not only the advantages but also problems including basic shortcomings and difficulty of handling. Subsequently two new designing principles for settling those issues were introduced and verified for effectiveness by experiment. Also conducted were experimental study on the material properties and optimum annealing conditions as SMA actuators. Finally, an active endoscope making most of the features of SMA suitable for designing super miniature system was test-produced. This paper is going to report the key subjects of the authors' investigation so far conducted.

BASIC CHARACTERISTICS OF SMA ACTUATOR

The basic characteristics of SMA actuators will be discussed in comparison with general actuators commonly used.

The concept of a power weight ratio is a basic evaluation standard to know the performance of an actuator. A new evaluation standard was devised by the authors for robot actuators. The proposed evaluation method is to express the power weight ratio P/M (W/kg) taking account of the power P (W) against unit mass M (kg) in comparison with actuator mass M (kg) as Fig. 1.
Fig. 1 shows the overall trend of actuators so far developed. The trend shows that the larger is the mass, the larger is the power weight ratio, and individual actuators have the mass fields in which they function better, and specific power ratio. The power weight ratio of the SMA actuator in question is stated for one case which was described in details by the authors (Hirose, Ikuta, Umetani, 1986). As a results, the characteristics of the SMA actuator was P/M=200 (W/kg) at about M=0.05 (kg). This falls in the mass field in which electric motors are applicable, in Fig. 1. It is located at position which can increase the P/M by one digit or more. Accordingly, it has power potential exceeding an electric motor at least in terms of power weight ratio. Besides, SMA actuators can be theoretically downsized to super miniature system. An actuator with less than 10^{-2} (kg) of mass which was impossible so far is expected to be realized with considerably good power weight ratio.

The energy efficiency of SMA actuators which are regarded as a kind of thermal engine in terms of its physical principle is rather restricted, for instance, it has the upper limit of Carnot's efficiency. It seems impossible to exceed 10 (%) based on a rough calculation (Wollants and others, 1979). Application is limited because of such inefficiency. To maximize the merit, application should be found in the fields where the energy efficiency problem is negligible. Besides, the response of a SMA actuator cannot be designed too quick because it depends on thermal transfer. Whereas, considerable improvement in this area seems to be possible by introducing such new construction methods as ξ-array to be presented in the later passage.

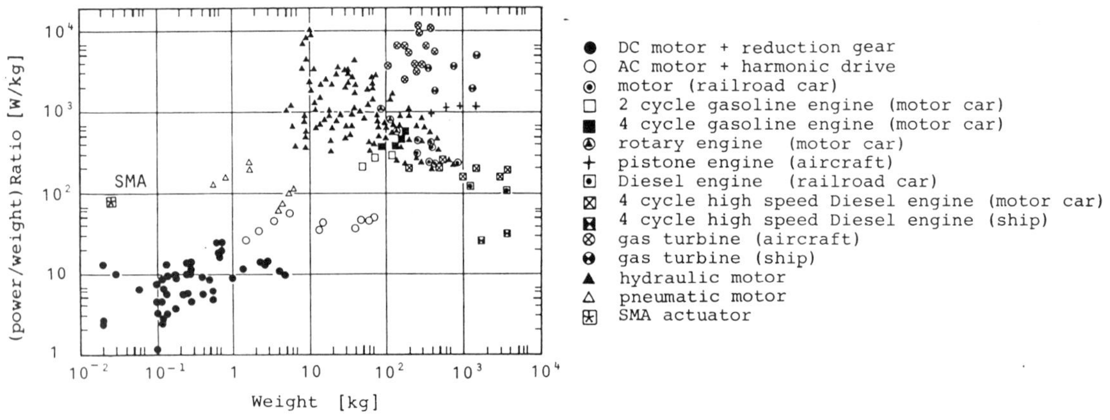

Fig. 1. Power weight ratio vs. weight diagram of actuators

PROBLEMS OF SMA ACTUATORS AND INTRODUCTION OF NEW DESIGNING PRINCIPLE

The authors would like to propose two designing principles to overcome the barriers thwarting SMA actuators from application to practical use.

Introduction of New Construction Method(ξ-array)

The newly proposed method is "to link SMA wires in mechanically parallel, but, in electrically series". That is, mechanical construction is in series as shown by Fig. 3, as previously tried, but, the electrical connection is in series as shown in Fig. 2. The construction is named " ξ-array" as it looks like the Greek letter "ksi". The ξ-array can increase the electric resistance of whole SMA wire so that it can be heated by a power source of high voltage and low current.

This system produces the effects as follows:

1) Diameter of a lead wire can be minimized so that the lead bunch does not restrict the multi-flexible mechanism constructed by a super miniature actuator.
2) Construction of power source and power control circuit is easy.
3) Thin SMA wire increases heat-transfer rate to enhance response.
4) Sensor function of SMA is more efficiently exploited.

The item 4) means the variation of electrical resistance resultant from phase transformation of SMA can be monitored more easily. The electrical resistance of SMA varies by about 10 to 20 (%) as a result of phase transformation, while, the transforming intensity also changes depending on the stress by the effect of stress induced martensite (SIM). Therefore, this characteristic can be used to provide SMA with the function of a sensor for displacement and force. This fact was reported in the past (Honma and others,1983), but, too fine variation of resistance made stable detection difficult so that a SMA actuator construction was unable to make use of the resistance variation. On the contrary, the ξ-array currently proposed can take the absolute values of the resistance itself in large magnitude so that the variation of resistance is expressed in board range, and as the result, can be measured correctly by a simple measuring system. Consequently, it can be applied to feedback signals of a SMA overheat-preventive system, to sharply improve the control characteristic.

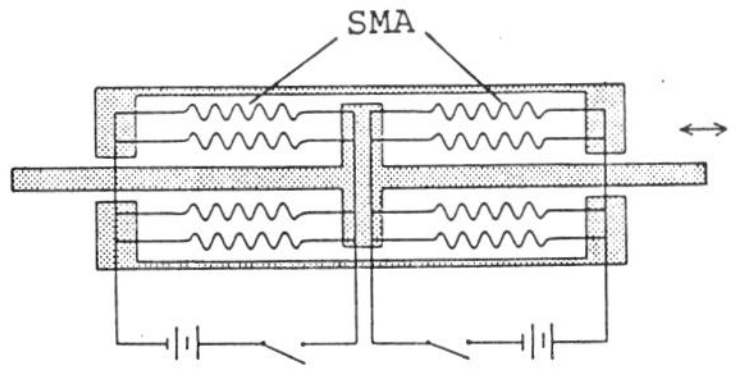

Fig. 2. Conventional parallel array of SMA actuator

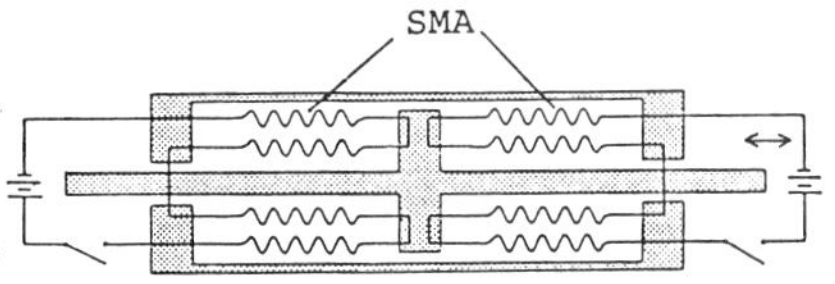

Fig. 3. Proposed ξ-array of SMA actuator

Introduction of σ-Mechanism

Rough modelization is possible to see that SMA is an elastic material which can control the Young Modulus with temperature. Accordingly, when SMA is deformed in some degree, the relation between the deformation and force is governed by Hook's law to first approximation. When it is largely stretched, the preapplied strain is getting large so that large restoring force is generated. On the contrary, when stretch is small, the restoring force is also small. In the conventional actuator mechanism based on a circular pulley as shown in Fig. 4(a), the moment generatable to both left and right directions is equal at the middle of the movable range, but, is significantly different at the both ends of the range. Therefore, the maximum rotating force uniformly generatable on both right/left is distributed in a mountain-shape as shown in Fig. 4(b).

Such uniformity of maximum torque is the characteristic inherent in any systems, e.g., one side of bias spring system, or SMA being in the forms of coil spring or wire. In other words, it is an inherent problem of SMA actuators making use of elastic energy, and produces such difficulties as:

(1) reduction of use efficiency of SMA
(2) reduction of usable moving range
(3) variation of control characteristic due to bending angle.

The maximum torque-uniforming mechanism conceived by the authors is to settle this problem by using a pulley of special configuration at converting the SMA force to torque so that torque generated over whole movable range is made uniform and at the same time movable range is also expanded as shown in Fig.5(b). The pulley has an effect to increase the rotating radius corresponding to the reducing ratio of preapplied strain. This pulley having unique contour is named "σ-pulley" by the authors because of the shape similarity, and the designing method for the configuration is also devised with verification.(Hirose and others, 1984b)

Specifically, the σ-mechanism consists of σ-pulley, non-stretchable belt (or wire) wound thereon and SMA coil spring (or wire) connected to the belt as shown in Fig. 5(a). The output chracteristics of an experimental unit actually introducing the σ-mechanism (maximum torque uniforming mechanism) is shown by solid line in Fig. 6. Although some effects of hysteresis are appearing, it show the above mentioned effect be actually produced when compared with the characteristic curve(broken line) based on a conventional circular pulley.(Hirose and others, 1985)

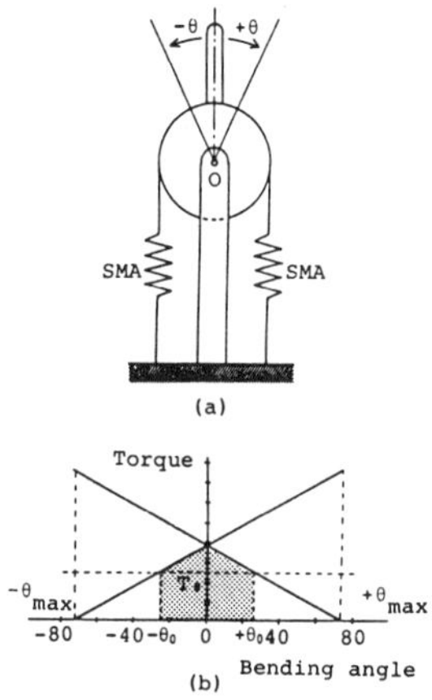

Fig. 4.

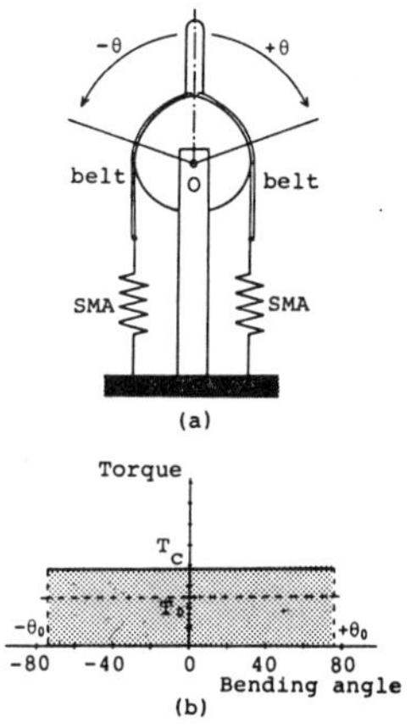

Fig. 5.

Fig. 4. Conventional joint with circular pulley:
(a) structure
(b) produced torques by the antagonistic SMA springs

Fig. 5. Proposed σ-mechanism with σ-pulley
(a) structure
(b) produced uniform maximum torque

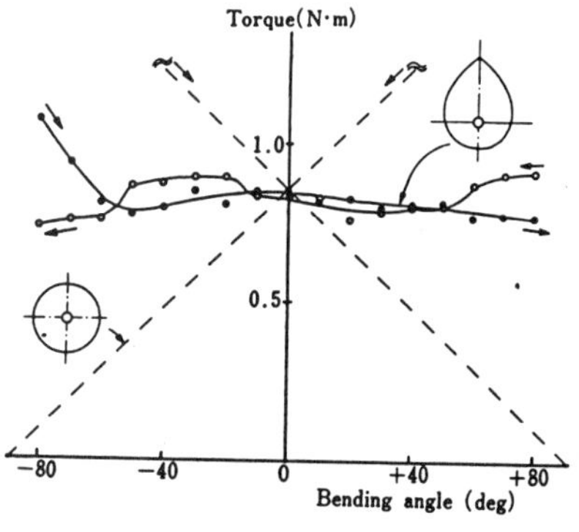

Fig. 6. Experimental results of uniformed maximum torque

Introduction of Antagonistic Type Resistance Feedback Control

As the resistance-monitoring is easy by introducing the ξ-array, introduction of resistance feedback control becomes possible depending on it as feedback signals. This eliminates the needs for such sensors as a potentiometer and encoder, etc. to realize superminiaturization of actuators.

Practically effective is an "antagonistic type resistance feedback control system" making use of servo system in pair-antagonistic manner. Past approaches had never tried such accordant control using resistance of antagonistic type SMA actuators as feedback signals. Therefore, the authors tried an approach which changed λ_1 from 0 to 1 while steadily satisfying the relation of $\lambda_1 + \lambda_2 = 1$, as shown in Fig.7, for λ_1 and λ_2, the ratio of resistance variation of antagonistic SMA.

This approach produced good position-control characteristic over broad range of strain with relatively little hysteresis as shown in Fig. 8. Fig. 9 shows the result of response when cooling air was suddenly shut off as in Fig. 9(a) under certain displacement command. In spite of such drastic change of the cooling condition, the resistance feedback system worked well to simultaneously adjust the heating current as shown in Fig. 9(c) so that the actuator displacement changed little as shown in Fig. 9(b).

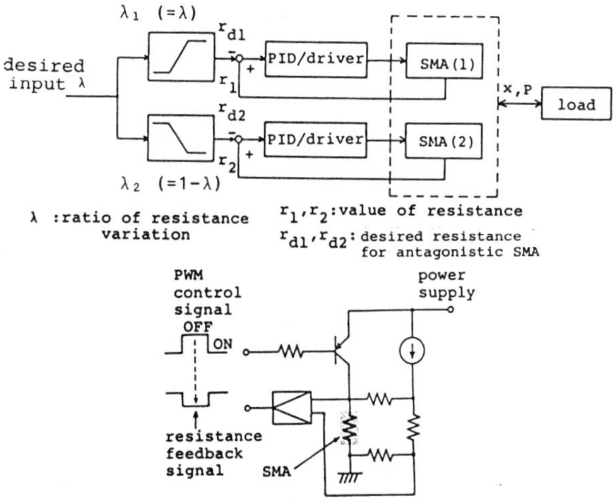

Fig. 7. (a)Antagonistic type resistance feedback control system
(b)PWM controller with SMA resistance detector

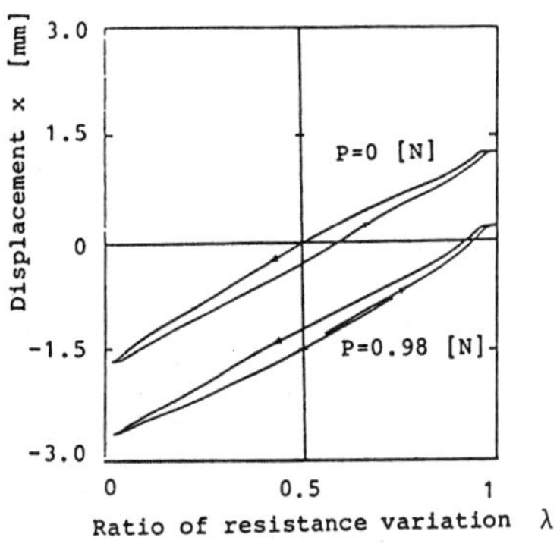

Fig. 8. The experimental results of SMA servo system based on resistance feedback control in two load(P) conditions

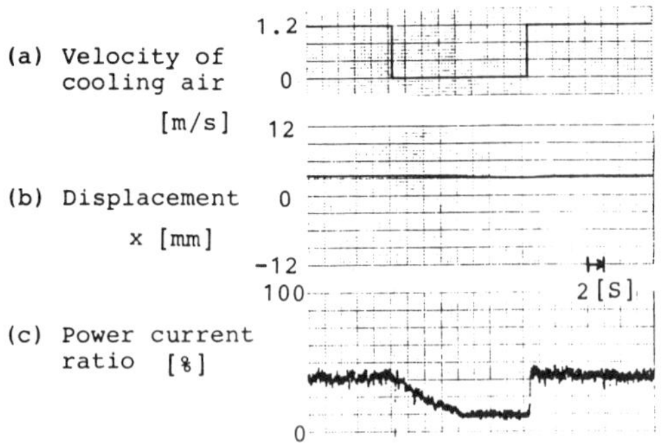

Fig. 9. The response of SMA actuator when the cooling condition changed

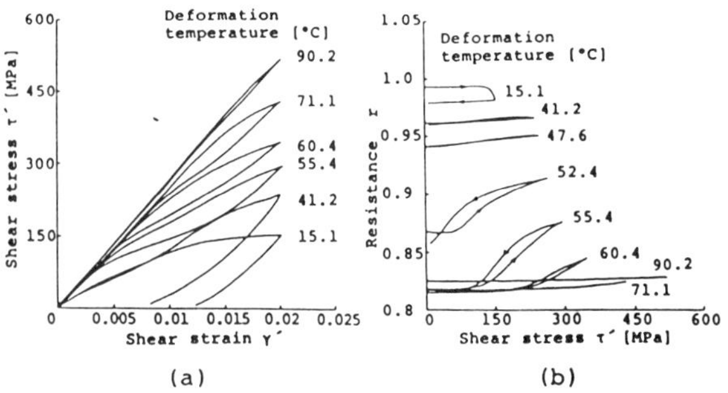

Fig. 10. The result of experiments of SMA coil spring
(a) relations between stress and strain
(b) relatinos between stress and electric resistance

MATERIAL PROPERTIES OF SMA AS SERVO ACTUATORS

The basic properties of SMA in terms of actuator material were not sufficiently presented so far although such information was essential for designing/controlling SMA actuators. The authors constructed property-measuring instruments for SMA coil spring and actuator. Those instruments enabled to systematically collect basic data of SMA and to analyze individual properties and the relations among those properties, with attention paid on phase transformation characteristic. The gist is given below.

Basic Properties of SMA Coil Spring

At first, the basic properties were measured for SMA coil spring which would be most generally used for constructing SMA actuators. Specimen used for measuring the properties were the cold-drawn TiNi SMA (0.2mm of diameter) which was formed to coil spring and subsequently annealed at T°C (T=350, 400, 450 & 480) for t (h) (t=2.5, 2.5, 1 & 1), respectively. The properties were measured mostly by tensile test under constant temperature.

One of such results is respectively shown in a stress(τ)-strain(γ) curve in Fig. 10(a) and a resistance(r) variation curve measured at the same time in Fig. 10(b). Both typical stress of and strain of coil springs were calculated based on the Wahl-modified linear spring theory. The resistance value r was a non-dimensional value $r=R/R_0$, with the value R_0 as the standard when the specimen were cooled from Af point to 0°C.
Summarizing the measurements of Fig. 10(a),(b) and visualizing the mechanical properties and resistance characteristic of SMA coil spring, the authors introduced a new presentation method which showed individual properties by equal strain and equal resistance curves in the stress-strain plane. They are shown in Fig. 11.

In addition, to clarify the effect of phase-transformation on the above mentioned phenomena, stress-temperature phase diagrams were made after processing as follows: (1) force P, deformation δ, strain γ and resistivity ρ by means of the three equations:

$$\gamma = \delta / (\pi n d C^2) \qquad (1)$$

$$\tau = \frac{2C}{\pi d^2} \left(3 + \gamma \frac{d}{d\gamma}\right) P \qquad (2)$$

$$\rho = \frac{d}{2nC} / \left[\left(2 + \gamma \frac{d}{d\gamma}\right) (1/R) \right] \qquad (3)$$

d: strand diameter n: effective winding count
C: spring index

(2) As the mechanical properties and resistance characteristic sharply changed at the phase-transforming point, the stresses at sharp flex of the stress-strain curve and resistivity-stress curve were plotted against individual deforming temperatures, and subsequently, the said information was analyzed based on the qualitative knowledge concerning the known phase-transformation to finally produce the phase diagram. The result is shown in Fig. 12.

Besides, to know the characteristic of SMA servo actuator under the individual annealing conditions, an experiment was conduced for control characteristic by arranging those SMA coil springs (strain range is 0.012 to 0.02) on the antagonistic type resistance control system was applied with several kinds of load, and static characteristic was obtained between input and output (input: λ & output: displacement x). The result is shown by λ-x curve in Fig. 13.

Observations on Characteristic

The actuator characteristic and basic properties of SMA coil springs under different annealing temperatures obtained above are summarized as follows:

As regards SMA coil springs:
(S1) Hysteresis of the stress-strain curve and resistance-stress curve is larger as the annealing temperature is higher.
(S2) Generated stress at the typical strain =0.02 is approx 520 (MPa) at high temperature 20 °C, however, the generated stress is smaller when annealing temperature is lower, as shown in Fig. 11.
(S3) The higher is the annealing temperature, the narrower is the controllable range of the resistance servo system.

As regards resistance servo system:
(A1) The higher is the annealing temperature, the larger is the hysteresis of the static characteristic (λ-x curve) between input and output.
(A2) The higher is the annealing temperature, the larger is the variation resulting from repetition of hysteresis loop of λ-x curve
(A3) Specimen annealed at 400° C generated the largest power.
(A4) Ununiformity of heating/cooling of SMA coil spring, which had been previously reported, was most conspicuous in the specimen annealed at 480°C.

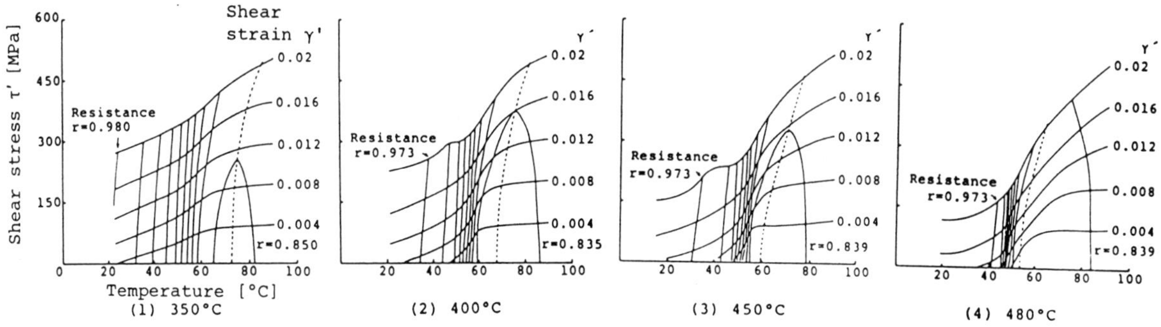

Fig. 11. Experimental results of SMA coil spring in four different annealing conditions expressed by newly introduced general diagram

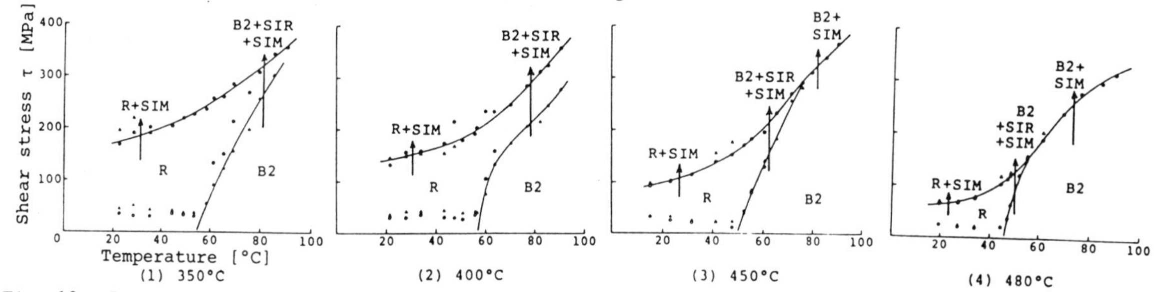

Fig. 12. Derived Temperature-Stress (T-τ) diagrams of the SMA as to four different annealing conditions

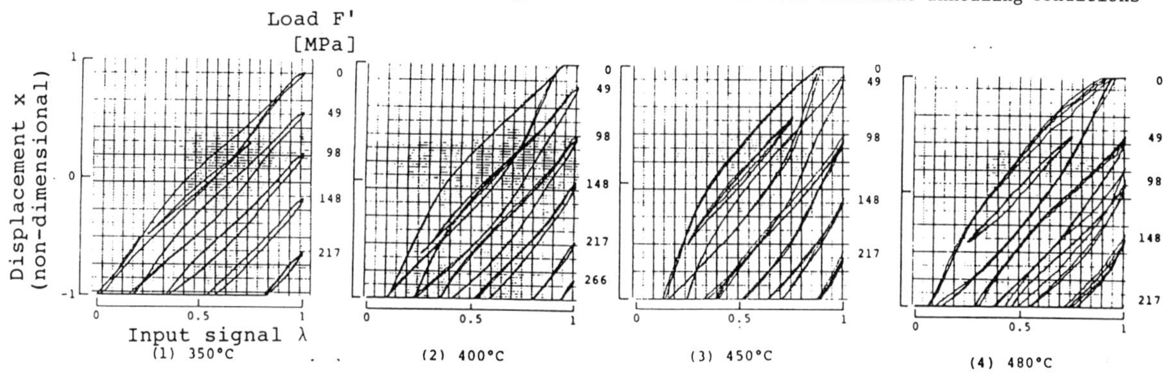

Fig. 13. Input λ (ratio of resistance variation) and output x (displacement) relationship of the SMA servo system based on the electric resistance feedback controll

These characteristics can be explained from the viewpoint of phase transformation. As shown in the stress-temperature phase diagram in Fig. 12, the lower is the annealing temperature rises to 450°C and 480°C, stress induced R-phase (SIR) is not created at temperature higher than 70° C and 50° C, respectively, so that the existence zone of the R-phase is narrowed (Fig. 12 (3)(4)). The phase diagram indicates that the higher is the annealing temperature, the smaller is the SIM generating stress at low temperature (20° C). In this case, increasing hysteresis in the characteristic curves (S1, A1) is estimated to increase the SIM generation. The shape-memory-effect of TiNi SMA has two modes, i.e., B2 phases, and large hysteresis resulting from transformation between martensitic phase (M-phase) and parent phase (B2-phase). The case of low annealing temperature is roughly regarded to be the case of the former, while, the ratio of the latter may increase when the annealing temperature is higher.

A resistance servo system, as basically it makes use of the phase-transformation of the former type, tends to show the characteristic of (S3), while, the M phase, being very soft, is regarded to show the tendency of (S2). The resistance servo system constructed in the current study generated the largest power at 400°C of the annealing temperature. It is because the large power generated at low temperature when annealed 350° C (Fig. 11(1)) cancels the power generated at the construction of the antagonistic type actuator, while, at 450 °C and 480°C of the annealing temperatures, narrow controllable range of the resistance servo (Fig. 11(3)(4)) reduces the generated power.
The specimen annealed at 480 °C produced a lot of M (martensitic) phase so that transformation took place among three phases of R, M and B2. In this case, apart from transformation between two phases, the inter-phase transformation values could not be determined only based on the resistance restriction. This will cause the characteristic of (A2). Easy occurrence of SIM will be the basis of (A4) characteristic.

From the viewpoints of power generation (A3) and controllability (A1,2 & 4), most suitable for resistance servo system will be the specimen annealed at 400 °C, which has broad existence zone of R-phase and phase transforming characteristic producing considerable SIM at low temperature.

Thus clarified above are the relations among phase transformation characteristic, annealing temperature, mechanical properties, resistance property and resistance servo characteristic. The information will be useful for optimum design of SMA servo actuators in the future.

EXPERIMENTAL PRODUCTION OF ACTUAL SCALE MODEL OF ACTIVE ENDOSCOPE

Endoscopes are indispensable in today's medical and industrial fields. One of the bottlenecks is the difficulty of insertion when the path is narrow or of complicated configuration. Especially, improvement is much expected for a fiber-sigmoidscope to abate the discomfort of the patients at insertion. The endoscopes themselves are mostly stiff in construction. Even when a scope has an active-moving section, the section is limited to tip area and the degree of freedom is low (Reynolds and others, 1967). Significant improvement will be possible if so called "active endoscope" is realized. SMA actuator allowed the authors to design and experimentally produce such endoscope in which whole stem could be actively operated.

The SMA actuator is especially suitable for the active endoscope, because;

(1) Miniature actuators system just like muscle can be realized

(2 The constant temperature condition inside intestine is good for the control of the actuator

(3) The energy efficiency and response are not the crucial conditions for the actuator's function

To design the SMA actuator systematic experimental investigations were conducted to the material properties of SMA and the optimum annealing condition for the actuator material. The detail discussions on this problem are shown in our paper to be published (Hirose, Ikuta and Tsukamoto, 1987).

Specifications

The basic design of the active endoscope model was made taking account of its application to a fiber-sigmoidscope. For this purpose, it should have certain mechanical compliance to smoothly pass the sigmoid colon which has the smallest radius of curvature. Also, it should be able to guide a fiberscope. Specifications were set for external diameter and maximum flexing angle to satisfy those needs. The diameter of 13(mm) of the current experimental model is satisfactory in comparison with 10 to 20(mm) of endoscopes on market now. The specifications are summarized on Table 1. According to the plane arrangement of the colon, an endoscope is expected to bend mostly on the plane dividing a human body into front/back sides. In this connection, the model was designed to have five joints comprising four joints with flexibility in the same direction on a plane and one joint at the tip which can bend orthogonally to the direction.

Figure 14 shows newly constructed actual scale model of active endoscope and its maneuver joy-stick.

TABLE 1 Specification of the Constructed 5 Segments Prototype SMA-ACM II

1) Dimension	ϕ 13 × 215 mm
2) Total weight	32 g
(for one segment	6.4 g)
3) Maximum bending angle (for one segment)	60 deg
4) Maximum produced torque (for one segment)	6.9 N m
5) Electric resistance (for one ξ -array)	12 ~ 13 Ω
6) Maximum bending velocity (for one segment)	30 deg/sec
7) Power source (for one ξ -array)	12V-1A
maximum PWM duty ratio	0.33
8) Flow rate of cooling water	0.13 ml/sec

Mechanism of Individual Joints

The driving system, same as the primary model (Hirose and others, 1984b) consists of a stainless steel coil spring which makes main skeleton at the center of a joint and a series of SMA coil spring arranged around it for driving function. In this model, each joint has one degree of freedom, so that a pair of SMA actuators capable of antagonistic motion are arranged in symmetry with respect to the axis, as shown in Fig. 15. The ξ-array construction is sufficiently served by power-supplying lead as thin as less than 0.1(mm) of external diameter including "Teflon" coating. The lead wires are arranged along the main skeleton so that they don't interfere the motion of the joints.

To enhance the response speed and safety, a cooling water tube is arranged as shown in Fig.15. The water flows back and forth in a fine tube (made of silicone rubber) held in coil spring located in the center of the joint to indirectly cool SMA. Based on the knowledge obtained in our investigations for material property of SMA, the driving characteristic is ameliorated by using SMA coil spring of the optimum specifications of Table 2 in terms of annealing temperature and strain range. The sheath (external jacket) is required to pliably bend without buckling. An artificial vein of PTFE serves best for this object.

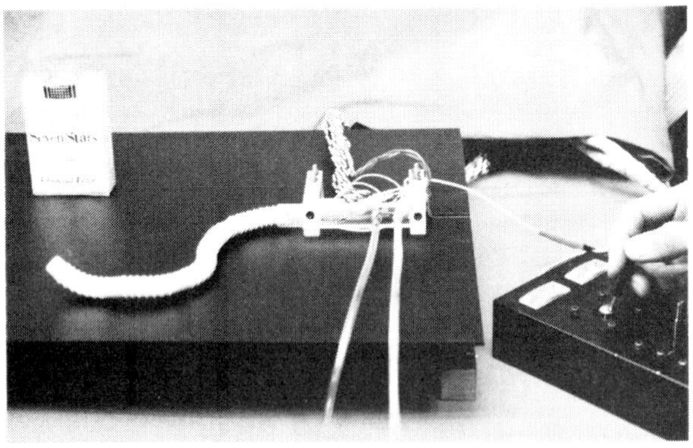

Fig. 14. Produced actual scale model of active endoscope using SMA servo actuators (SMA-ACM II)

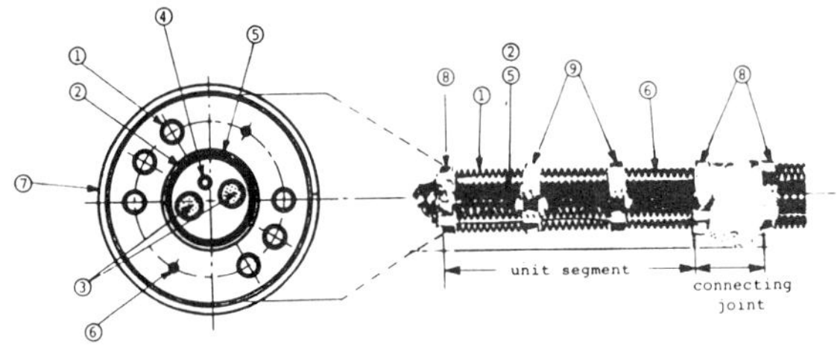

① : SMA coil spring (ξ-array)
② : spinal coil spring
③ : cooling water tubes (counterflow)
④ : fiber scope
⑤ : sheath of spinal coil spring
⑥ : lead wires
⑦ : outer casing (made of artificial vein)
⑧ : side flange
⑨ : intermidiate flange

Fig. 15. Inner constitution of the unit segment of construced active endoscope SMA-ACM II (frontal and side view)

TABLE 2 Specification of SMA actuator used in SMA-ACM II

1) Composition	Ti-50.2at%Ni
2) Annealing condition	350° C 2.5 h
3) Diameter of wire	d = 0.2 mm
4) Diameter of coil spring	D = 1.0 mm
5) Maximum strain	max = 0.06
6) Minimum strain	min = 0.043

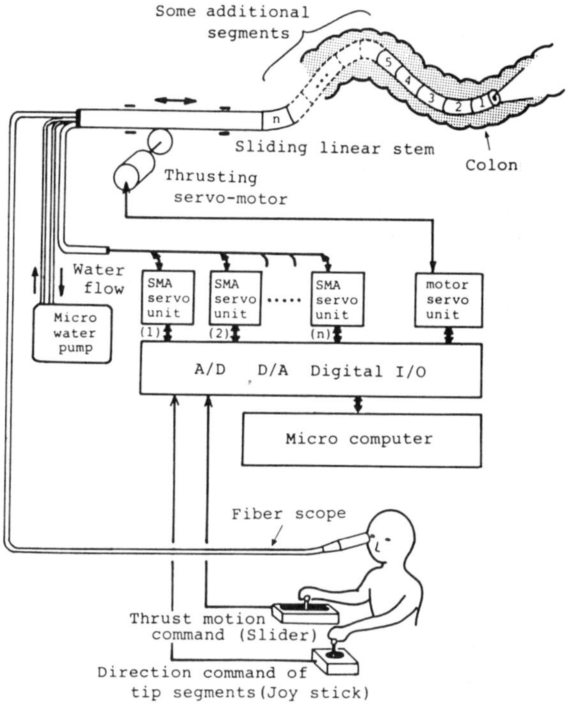

Fig. 16. Total controlling system of the active endoscope using SMA servo actuators with resistance feedback control

Control System

Basic architecture of the control system is shown in Fig. 16. SMA actuators ofindividual joints are microcomputer-controlled via respective resistance measurement/driving circuits. The microcomputer also receives commands from a joystick to control the bending angle of the first and second tip joints and from direct-motion servo system which advances/retracts the whole endoscope.

For insertion into the colon, a medical doctor, watching the images fed by the fiberscope, operates the joystick to manually control the tip two joints and simultaneously pushes in the scope at desired speed by hand. The two kinds of command,i.e., the command for tip joint bending angle given by manual operation and the direct-motion command given by hand at the base of the scope are translated, via the computer, to the shift command. The shift command consecutively transmits the manually-given bending angles at the tip joints, synchronous with the direct-motion speed, to the hind joints. This shift control enables to produce the motion of whole stem pliably following the winding path, instead of the motion by reactive force of the entero-paries.

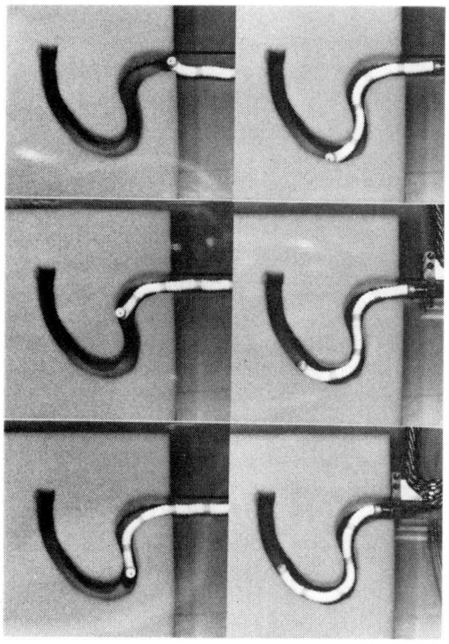

Fig. 17. Sequential motion of the constructed active endoscope as entering into sponge rubber colon model under shift control

The measuring/driving circuits are set for each pair of SMA actuators at each joint. Introduction of ξ-array mechanism is confirmed to enable stable and effective feedback of resistance values to prevent overheat and to keep the control characteristic free of external interference.

Driving Experiment

To verify the basic performance of this device, the driving experiment was conducted in a colon model environment. The active endoscope proved to make pliable motion with about 30 (deg/sec) up to designed maximum angle (60 deg) at the responding speed high enough for the purpose. The shift control characteristic of all joints synchronous with backward/forward motions, was also confirmed to be effective, as exhibited in Fig. 17. The temperature of the sheath during the experiment was approx 30° C in continuous operation under room temperature of approximately 15° C.

In summary of those results, the active endoscope using the SMA actuators is regarded to sufficiently serve the practical object.

Acknowledgment is hereby made to Prof.Y.Umetani of Tokyo Institute of Technology, Dr.Y.Suzuki and Mr. H.Tamura of Central Research Lab, Furukawa Electric Co., Ltd., Tokyo for their valuable advice.

REFERENCES

Hirose.S., K.Ikuta and Y.Umetani. (1983). Study of Servo Actuator Based on Shape Memory Alloy (No.1,No.2). Proc. of 22nd Annual Conf. of Society of Instrument and Control Engineers. pp.543-546 (in Japanese)

Hirose.S., K.Ikuta and Y.Umetani. (1984a). A New Design Nethod of Servo Actuators Based on The Shape Memory Effect. Proc.of 5th RO.MAN.SY. symp. Udine, Italy, pp.339-349

Hirose.S., K.Ikuta, M.Tsukamoto, K.Sato and Y. Umetani.(1984b). Study of Servo Actuator based on Shape Memory Alloy (No.7 Consruction of New Drive Joint) Proc. of 2nd Annual Conf. of the Robotics Society of Japan. pp.127-130 (in Japanese)

Hirose.S., K.Ikuta, M.Tsukamoto, K.Sato and Y. Umetani.(1985a). Study of Servo Actuator based on Shape Memory Alloy (No.8. Test Production of Drive Joint with the property of Uniformed Maximum Torque. Proc. of 24th Annual Conf. of Society of Instrument and Control Engineers (SICE). pp.555-556 (in Japanese)

Hirose.S., K.Ikuta, M.Tsukamoto and Y.Umetani (1985b).Study of Servo Actuator Based on Shape Memory Alloy (No.9 Mesurement of Ralaitionship between Stress-Strian-Electrical resistance of SMA).Proc. of 3rd Annual Conf. of the Robotics Society of Japan. pp.331-334 (in Japanese)

Hirose.S., K.Ikuta and Y.Umetani. (1986). Deveropement of Shape Memory Alloy Actuator (Performance evaluation and introduction of a new design concept). Journal of the Robotics Society of Japan (JRSJ). Vol.4-2, pp.15-26 (in Japanese)

Hirose.S., K.Ikuta and M.Tsukamoto. (1987). Development of Shape Memory Alloy Actuator (Characteristics Measurement of the Alloy and Development of an Active Endoscope). J. of the Robotics Society of Japan (JRSJ). (to be published in Japanese)

Honma,D., Y.Miwa and N.Iguchi. (1983). Application of SME to Digital Control Actuator. Trans. JSME-C, Vol.49(448),pp.2163-2169 (in Japanese)

Hosoda,Y., M.Fujie and Y.Kojima. (1983). Three Fingered Robot Hand by Using SMA. Proc. of 1st Annual Conf. of Robotic Society of Japan. Vol. 1. pp.213-214. (in Japanese)

Reynolds W.E. S.Bazell. A.Brushenko and D.A.Pontarelli. (1967). Fiber Optic Multiple Fibersigmoidoscope. Proc. of 12th SPIE Technical Symposium. Los Angeles, USA. pp.49-53

Wollants.P.,M.D.Bonde, L.Deleay and J.Roos.(1979). Thermodynamic Analysis of the Work Performance of a Martensisic Transformation under Stressed Conditions(Part2). Z.Metallkde. Bd.70. pp.298-304

Zusammenfassung

Es wird erwartet,dass der "Shape Memory Alloy"(SMA) Antrieb, mit seiner einzigartigen Anwendung sehr viel zum Fortschritt der Leistungsfähigkeit der Roboter beiträgt.

Während den letzten Jahren führten die Autoren eine Reihe von Untersuchungen durch um wirklich praktisch anwendbare SMA Antriebe zu konstruieren. Erstens wurden die grundsatzlichen Eigenschaften der SMA Antriebe einschliesslich der Unzulänglichkeiten bestimmt, um die Hauptprobleme abzuklären, die Schwierigkeiten bei der praktischen Anwendung geben könnten.

Zweitens wurden die Hauptprobleme im Wesentlichen gelöst, indem eine neue Methode zur Konstruktion des Antriebs, "ξ-array" und "σ-mechanism" genannt, eingeführt wurde, die die Endeigenschaften merklich verbesserten.

Drittens wurden experimentelle Untersuchungen über die Materialeigenschaften durchgeführt und die optimalen Bedingungen zum Härten des Materials des Antriebes bestimmt.

Für die Anwendung wurde ein aktives Endoskop mit 13 mm Durchmesser und 210 mm Länge, bestehend aus 5 Segmenten, hergestellt und getestet. Die 5 Segmente wurden einzeln mit Hilfe des Computers kontrolliert. Das Endoskop konnte sanft in den Dickdarm dank der beweglichen Halterung, ermöglicht durch das Superminiaturantriebssystem des SMA, eingeführt werden.

Merit Parameters of Transmission in Mechanism Design

HONG-LIANG YIN AND FU-BANG WU

Department of Mechanical Engineering, Shanghai University of Technology, 149 Yan-chang Road, Shanghai 200070, PRC

Abstract. In mechanism design, pressure angle or transmission angle is conventionally regarded as one of the indexes of transmission characters, and it is often presumed that when the pressure angle is equal to 90° or the transmission angle is equal to 0, the mechanism will get jammed. But this traditional concept fails to explain certain problems in mechanisms, especially in the cam-linkage combined mechanisms, as it does not determine whether a mechanism can transmit or not. In this paper, we first taking the four-bar linkage as an example, discuss this problem with a view to mechanical advantage, and then study the synthesis of cam-linkage combined mechanism. Finally, we arrive at a conclusion that in mechanism design we must consider to use the merit factor or merit angle as the accurate rule to determine the jam character of mechanism transmission and not the pressure angle or transmission angle. Only on some simple mechanisms under special action condition of external force, the transmission angle of certain link is just equal to the merit angle of transmission.

Keywords. Mechanism design; mechanical advantage; jam; pressure angle; transmission angle; merit angle; merit factor.

PRESSURE ANGLE, TRANSMISSION ANGLE AND MERIT PARAMETERS OF TRANSMISSION

In mechanism design, pressure angle or transmission angle is conventionally regarded as one of the indexes of transmission characters. According to the usual definition, if the effects of inertia force, gravity force and friction force are not taken into consideration, the acute angle between the direction line of the force that link i pushes to link j and the velocity of action point of this force are known as pressure angle α_{ij} , and the complementary angle of pressure angle as transmission angle γ_{ij} . When $\gamma_{ij}=0$, the mechanism will get jammed. In the discussion of transmission angle, we take into consideration not only the output link, but also all the transmission angles on the kinematic pairs of every driven link. The traditional concept however fails to explain centain problems of mechanisms. For example, the four-bar mechanisms shown in Fig.I, driving moment M_2 acts on crank 2 and load F acts on point G of coupler 3. Then the pressure angle $\alpha_{34}=90°$, which is between the velocity v_c (⊥CD) and the force R_{34}(along CD) pushes to rocker 4 by coupler, thus the transmission angle $\gamma_{34}=0$, but in fact the mechanism will not get jammed. Further in the combined mechanism , some links are often driven by certain bar and cam simultaneously(as link 4 in Fig.3). If one of the pressure angles in the lower pair C or cam pair D of link 4 is equal to 90° , can we determine that the combined mechanism will get jammed?

In this paper, we discuss this problem with a view to mechanical advantage, which is one of the major indexes of transmission characters of mechanism. The mechanical advantage A is indicated by the ratio of output load moment M_o and input driving moment M_i. If we disregard the effects of inertia force, gravity force and friction force, then A is the ratio of angular velocities of input and output links, i.e.

$$A=\left| M_o/M_i \right| = \left| \omega_i/\omega_o \right| \qquad (I)$$

when M_i is given, if $\omega_i/\omega_o=0$, thus $M_o=0$ and $A=0$, this shows the mechanism will get jammed. From this, we discuss the dimensional conditions which have influence in determining the jam characters in mechanism design, i.e. the problem of merit parameters of transmission. In this paper, we first study the basic four-bar mechanism and then approach the combined mechanism.

THE TRANSMISSION ANGLE AND THE MERIT ANGLE OF TRANSMISSION IN FOUR-BAR MECHANISM

In the four-bar mechanism as shown in Fig.I, if load F is given, we sequentially take links 2 and 3 as a freebody, and write the equilibrium equations of forces and moments, then obtain:

$$M_2=\left(Fr_2\left(r_3\sin(\theta_4-\theta_3)\sin(\theta_2-\theta_3-\lambda_F)+r_G\sin(\theta_4-\theta_2)\sin(\lambda_F-\lambda_G)\right)\right)/\left(r_3\sin(\theta_4-\theta_3)\right) \qquad (2)$$

From Eq.(2) it gives that when $\theta_4-\theta_3=0$ or π, the driving moment M_2 will approach to ∞ which is required to overcome a very small load F, i.e. mechanical advantage A=0, thus denotes the mechanism will get jammed. Therefore, the transmission characters of four-bar mechanism be judged by the merit angle of transmission δ, and δ is the acute angle $(\theta_4-\theta_3)$ between links 3 and 4. When $\delta=0$, we can determine the mechanism will get jammed. From Fig.I we can see δ is not equal to the transmission angle γ_{23}.

If the output link of mechanism is link 4 and the load moment M_4 is given(Fig.2), by using similar method we may find the driving moment M_2:

$$M_2=r_2M_4\sin(\theta_2-\theta_3)/r_4\sin(\theta_4-\theta_3) \qquad (3)$$

Under such condition, the merit angle of transmission δ of the mechanism is also the acute angle$(\theta_4-\theta_3)$ between links 3 and 4, and if $\delta=0$, the mechanism will get jammed, but in this case, δ is just equal to γ_{34}.

From this, we know the merit angle of transmission δ is only related to the dimensions of mechanism and is not under the influence of force disposition. But the transmission angle γ will change according to the action condition of external forces.

It is more simple and easier to study the merit angle of transmission by kinematic analysis. In Fig.I, when link 2 is the driving link, we use complex vector method and may obtain the angular velocities $\dot{\theta}_3$ and $\dot{\theta}_4$ of driven links 3 and 4:

$$\dot{\theta}_3=r_2\dot{\theta}_2\sin(\theta_2-\theta_4)/r_3\sin(\theta_4-\theta_3) \tag{4}$$

$$\dot{\theta}_4=r_2\dot{\theta}_2\sin(\theta_2-\theta_3)/r_4\sin(\theta_4-\theta_3) \tag{5}$$

If the merit angle of transmission $\delta=0$(i.e. $\theta_4-\theta_3=0$), from Eq.I, it gives: $A=\dot{\theta}_2/\dot{\theta}_3=0$ or $A=\dot{\theta}_2/\dot{\theta}_4=0$. Thus we can determine that the mechanism will get jammed.

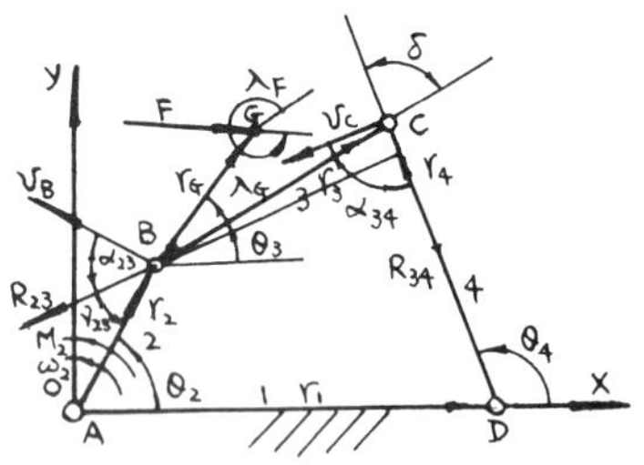

Fig. I.

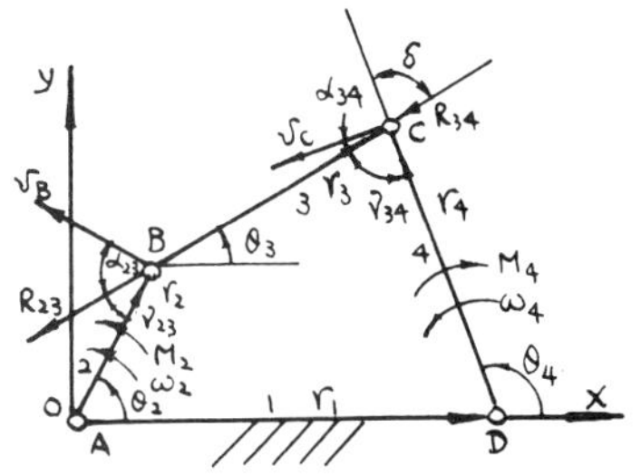

Fig. 2.

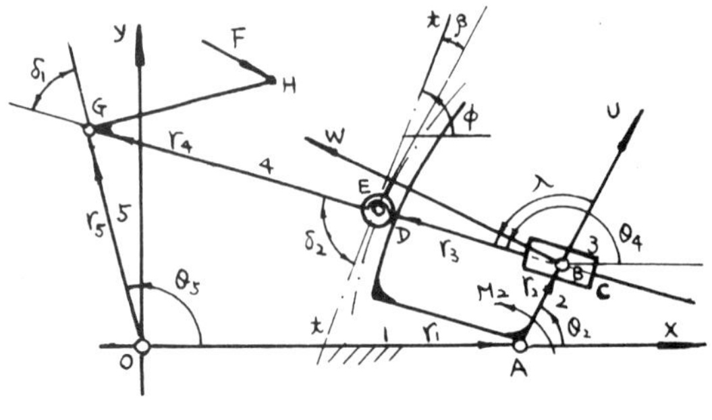

Fig. 3.

If link 4 is the driving link, then obtain;

$$\dot{\theta}_2=r_4\dot{\theta}_4\sin(\theta_4-\theta_3)/r_2\sin(\theta_2-\theta_3) \tag{6}$$

$$\dot{\theta}_3=r_4\dot{\theta}_4\sin(\theta_2-\theta_4)/r\sin(\theta_2-\theta_3) \tag{7}$$

In this case, the merit angle of transmission δ' is the acute angle$(\theta_2-\theta_3)$ between driven links 2 and 3, and if $\delta'=0$, then $A=\dot{\theta}_4/\dot{\theta}_2=0$ or $A=\dot{\theta}_4/\dot{\theta}_3=0$, the mechanism will get jammed.

From this we know, if we take a different driving link in the same mechanism, the merit angle of transmission will be different.

THE MERIT ANGLE AND MERIT FACTOR OF TRANSMISSION IN CAM-LINKAGE COMBINED MECHANISM

Figure 3 shows a cam-linkage combined mechanism, the driving moment M_2 acts on driving link 2, and the load F acts on point H of output link 4. Link 4 is driven by crank 2 through silder 3,and the roller at point E of link 4 is meanwhile pushed by the cam profile D with crank 2 attached. So link 4 is somewhat like a link with variable length. In this link, the length of BE(i.e. r_3) is changable and the length of EG(i.e.r_4) is constant.

Now, we use the method of complex vector to write the configuration equations and take their time derivatives, and then find the angular velocities of driven links 4 and 5:

$$\dot{\theta}_4=\Delta_4/\Delta\ , \quad \dot{\theta}_5=\Delta_4/\Delta \tag{8}$$

and

$$\Delta=r_5(r_3+r_4)\sin(\theta_4-\phi)\sin(\theta_5-\theta_4) -r_5 r_3\cos(\theta_4-\phi)\cos(\theta_5-\theta_4) \tag{9}$$

$$\Delta_4=\begin{vmatrix} \sin(\theta_4-\phi) & r_3\dot{\theta}_2\cos(\theta_4-\phi) & 0 \\ \sin\theta_4 & -r_2\dot{\theta}_2\cos\theta_2 & -r_5\cos\theta_5 \\ \cos\theta_4 & r_2\dot{\theta}_2\sin\theta_2 & r_5\sin\theta_5 \end{vmatrix}$$

$$\Delta_5=\begin{vmatrix} \sin(\theta_4-\phi) & r_3\cos(\theta_4-\phi) & r_3\dot{\theta}_2\cos(\theta_4-\phi) \\ \sin\theta_4 & (r_3+r_4)\cos\theta_4 & -r_2\dot{\theta}_2\cos\theta_2 \\ \cos\theta_4 & -(r_3+r_4)\sin\theta_4 & r_2\dot{\theta}_2\sin\theta_2 \end{vmatrix}$$

ϕ is the inclined angle of tangent line tt at point E of cam theorectical profile in the fixed coordinates xoy.

If Eq.(9) is satisfied with one of the following conditions:

I. $\Delta=0$

2. $\theta_4-\phi=0$ and $\theta_5-\theta_4=\pi/2$

3. $\theta_4-\phi=\pi/2$ and $\theta_5-\theta_4=0$

The mechanical advantage $A=\dot{\theta}_2/\dot{\theta}_4=0$ or $A=\dot{\theta}_2/\dot{\theta}_5=0$, this shows the mechanism will get jammed. Thus we name Δ as the merit factor of this combined mechanism, and its two merit angles are given in the following:

I. δ_1, this is the acute angle $(\theta_5-\theta_4)$ between link 4 and 5;
2. δ_2, this is the acute angle $(\theta_4-\phi)$ between tangent line tt of cam and the line EG of link 4.

If $\delta_1=0$ and $\delta_2=\pi/2$, or $\delta_1=\pi/2$ and $\delta_2=0$, or the relation between δ_1 and δ_2 agrees with $\Delta=0$ of Eq.(9), the mechanism will get jammed. From this we know, the influence on the transmission characters of this cam-linkage combined mechanism is the compound action of two merit angles. In mechanism design, we should not simply consider that δ_1 and δ_2 are larger then their allowable values, for example, when $\delta_1=0$, i.e. r_4 is colinear with r_5, if $\delta_1\neq\pi/2$, the mechanism is still transmissible.

CONCLUSION IN BRIEF

I. When the merit factor of transmission approaches to 0, the velocity ratio of input link and output link turns near 0. If a very slight friction in the mechanism will then cause it get jammed, this indicates the transmission character of mechanism is very poor, and the mechanical advantage is very small. Similarly from the precision analysis of mechanism, we can also find that the output error is greatly enlarged. This signifies the merit factor also represents the sensibility of mechanical error. Consequently, merit angle and merit factor may be jointly known as merit parameters of transmission, and it is used as a major index of transmission characters of mechanism, especially as the accurate rule to determine whether a mechanism can transmit or not.

2. In the analysis and synthesis of mechanism, it is advisable to use the parameters of transmission presented in this paper as the rule of design and check, and not the pressure angle or transmission angle. Only in some simple linkage mechanisms and cam mechanisms and under special action conditions of external force, the transmission angle of certain driven link in these mechanisms is just equal to its merit angle of transmission.

REFERENCES

Shigley, J.E., Uicker, J.J.Jr. (I980). Theory of Machines and Mechanisms. McGraw-Hill Book Company, Inc.

Yin, Hong-liang, (I982). Collections of Translations of Mechanisms. China Mechanical Industry Publishing House.

Yin, Hong-liang, (I985). Synthesis of kinematic adjustable linkages and cam-linkage combined mechanisms in packing machines. Proceedings of The 4th IFToMM International Symposium on Linkage and Computer Aided Design Methods., III-2, 45I-458.

Yin, Hong-liang, and Wu, Fu-bang. (I985). Design of cam-linkage combined mehanisms. Journal of Shanghai University of Technology, 4.

Dynamic Analysis of a Telescopic Crane on an Automotive Chassis

J. KŁOSIŃSKI AND L. MAJEWSKI

Mechanical-Constructional Institute, Branch of Łódź Technical University, Bielsko-Biała, Poland

Abstract. Dynamic analysis of a telescopic crane has been carried out in this work. Two exploitation states have been considered: the state of work and the state of transport. Mathematical models have been designed and methods of their solution discussed. Possibilities of using these models for design-research works, taking into account especially the accuracy of crane work as well as driver-operator's comfort have been presented. Some examples of calculation have been presented.

Keywords. Computer-aided design; cranes; models; Monte-Carlo method; numerical analysis; vehicles; vibration control.

INTRODUCTION

The dynamic analysis of the telescopic crane on an automotive chassis requires taking into account the state of crane while designing its physical and mathematical model. Two exploitation states can occur: the state of work and the state of transport. The first one must assure the user safe and accurate work of crane at the determined, more and more high values of load, radius and link speed. These requirements cause some disadvantageous occurrences, that make the work of crane more difficult and worsen the working conditions for the operator. While in the state of transport it is required that possibly maximum comfort for the driver is assured and disadvantageous movements of the vehicle, that make driver's work more difficult, are avoided.

CRANE ANALYSIS IN THE STATE OF WORK

A computational discrete-continuous model /Fig. 1/ has been worked out in order to test the crane in the state of work. In this model a stiff immobile chassis and a rotary body consisting of the following assemblies:
- stiff rotary platform 1,
- assembly of girders 2,3,4 and 5 connecting the platform to the jib,
- telescopic jib consisting of members 6,7 and 8 with continuous mass and elasticity distribution /coupled elastically with one another/,
- pulley block with hanged material considered as a spheric pendulum

have been separated.

Kinetic-static and dynamic movement equations resulting from Newton's principle have been formulated for these assemblies. No clearance in the nodes, no resistance to motion, negligible weight, inextensibility and ideal elasticity of ropes and minimum displacements of the elements have been assumed.

The equations are as follows:
- for the rotary platform:

$$I\ddot{\gamma} + k_s \gamma = M_{zr} \qquad (1)$$

where: I - platform's moment of inertia in relation to the axis of rotation; k_s - substitute coefficient of stiffness to the torsion for the crane driving assembly; M_{zr} - moment of the forces acting on the platform from the girders, in terms of rotation axis;

- for telescopic jib:

$$\begin{aligned}
&\Sigma P_{jx6} + \sum_{i=6}^{8} m_i \ddot{x}_6 = 0\\
&\Sigma P_{jy6} + \sum_{i=6}^{8} \varrho A_i \int_0^{l_i} \ddot{y}_i \, dx = 0\\
&\Sigma P_{jz6} + \sum_{i=6}^{8} \varrho A_i \int_0^{l_i} \ddot{z}_i \, dx = 0\\
&\Sigma M_{jy6} + \sum_{i=6}^{8} \varrho A_i \int_0^{l_i} \ddot{z}_i x_6 \, dx = 0\\
&\Sigma M_{jz6} + \sum_{i=6}^{8} \varrho A_i \int_0^{l_i} \ddot{y}_i x_6 \, dx = 0\\
&j = 1, 2, \ldots, p_w, \qquad (2)
\end{aligned}$$

where: A_i - area of an i-segment /i=6,7,8/; ϱ - density of material; x_i, y_i, z_i - coordinates of displacements of the i-segment; $P_{jx6}, P_{jy6}, \ldots, M_{jz6}$ - forces and moments acting upon particular jib segment from neighbouring crane assemblies;

- for the pendulum:

$$\begin{aligned}
&P_x^B - S \sin\vartheta \cos\delta = 0\\
&P_y^B - S \sin\vartheta \sin\delta = 0\\
&P_z^B - S \cos\vartheta - m_M g = 0 \qquad (3)
\end{aligned}$$

where: P_x^B, P_y^B, P_z^B - components of all inertia forces, loading the pendulum mass; S - force in pendulum rope; m_M - pendulum mass concentrated in point M;

- for the assembly, which is statically indeterminable, both equation of equilibrium and Menabre's equations have been formulated.

Desides, differential equations of segment vibrations in two planes have been formulated for telescopic jib. It has been assumed that the telescopic segments are prismatic beams with uniformly distributed masses and elasticity, and their joints are elastic. The equations have been formulated with the following additional assumptions: in non-deformed state the axes of jib segments are straight lines; torsional and longitudinal vibrations of the jib are ignored. The equations are as follows:

$$\frac{\partial^2 z_i}{\partial t^2} = -C_{z_i}^4 \frac{\partial^4 z_i}{\partial x_i^4}$$

$$\frac{\partial^2 y_i}{\partial t^2} = -C_{y_i}^4 \frac{\partial^4 y_i}{\partial x_i^4}$$

$$i = 6, 7, 8. \tag{4}$$

and have to fulfil the conditions of continuity of displacements, angles, forces and moments between particular segments of the jib.

Thus 64 conditions have been formulated (Kłosiński, Majewski, 1983). In order to determine the state of crane during the body rotation the solution of equations (4) has been accepted in form of the Prager-Krylow's formulas and applied in the formulas describing conditions of displacements and forces continuity for particular jib segments. Using equations (1),(2)and(3), after some mappings, a non-linear system of 35 equations has been obtained, which can be solved by means of numerical methods for any kinematic and dynamic inputs.

In order to formulate the problem of natural frequencies, the equations have been linearized and accepting periodic solutions in relation to time, the frequencies of natural vibrations of the crane model have been determined as a function of its /selected/ parameters.

CRANE ANALYSIS IN THE STATE OF TRANSPORT

A discrete model of a crane moving on an uneven road has been worked out for the dynamic analysis of this state. In this model /Fig. 2/ the following elements have been separated:

- body 1 based upon massless suspension and wheels with determined stiffness and damping,
- crane jib 2 fastened to the body with elastic-damping joints,
- driver's seat 3 fastened to the body with massless elastic-damping elements.

For the mentioned discrete elements 8 dynamic equations of motion have been formulated. They are as follows:

- for the body:

$$m_1 \ddot{z}_1 + \sum_{i=1}^{5} c_i \cdot \dot{u}_i + \sum_{i=1}^{5} k_i u_i + c_7(\dot{z}_1 - \dot{z}_2) + k_7(z_1 - z_2) = F$$

$$I_{y1} \ddot{\varphi}_1 + \sum_{i=1}^{5} c_i w_{ix} \dot{u}_i + \sum_{i=1}^{5} k_i w_{ix} u_i + c_8(\dot{\varphi}_1 - \dot{\varphi}_2) + k_8(\varphi_1 - \varphi_2) = Fl + h_1(B_1 + B_2 + B_3)$$

$$I_{x1} \ddot{\psi}_1 + \sum_{i=1}^{5} c_i w_{iy} \dot{u}_i + \sum_{i=1}^{5} k_i w_{iy} u_i + c_9(\dot{\psi}_1 - \dot{\psi}_2) + k_9(\psi_1 - \psi_2) = 0 \tag{5}$$

- for the jib:

$$m_2 \ddot{z}_2 + c_7(\dot{z}_2 - \dot{z}_1) + k_7(z_2 - z_1) = 0$$

$$I_{y2} \ddot{\varphi}_2 + c_8(\dot{\varphi}_2 - \dot{\varphi}_1) + k_8(\varphi_2 - \varphi_1) = B_2 h_2$$

$$I_{x2} \ddot{\psi}_2 + c_9(\dot{\psi}_2 - \dot{\psi}_1) + k_9(\psi_2 - \psi_1) = 0 \tag{6}$$

- for driver's seat:

$$m_3 \ddot{z}_3 - c_5 \dot{u}_5 - k_5 u_5 = 0$$

$$m_3 \ddot{x}_3 + k_6(x_3 - L) = B_3 \tag{7}$$

where: m_1, I_{x1}, I_{y1} - mass and moments of inertia of the body; m_2, I_{x2}, I_{y2} - mass and moments of inertia for the jib; m_3 - mass of the seat; u_i, $\dot{u}_i$ - relative displacement and speed of displacement for the points of application of the i-spring and the i-damper /i=1,2,...,5/; w_{ix}, w_{iy} - respective arms of application of forces acting on vehicle body from the suspension and the seat; B_1, B_2, B_3 - d'Alembert's forces; F - dynamic input from the engine. The remaining dimensions, parameters and variables have been shown in Fig. 2.

Equations (5),(6) and (7) have been formulated with the following assumptions: projections of wheel prints on horizontal plane of the road are straight lines, inputs from road unevenness are the functions of time, the engine is located on vehicle's longitudinal axis.

System of equations (5),(6) and (7) has been solved by means of a special numerical programme using Runge-Kutty's method. Solutions have been obtained for different inputs q_i from road unevenness /periodical and random input generated using a computer generator of quasi-random digits and interpolation method using spline functions (Fortuna, 1982)/ and for different parameters of the model.

An example of displacement of chassis 1 for quasi-random inputs q_1 and q_3 has been shown in Fig. 3.

According to obtained results Bode's amplitude characteristics have been worked out and the influence of selected parameters of the model on the swing amplitude has been determined. The analysis has been carried out for both linear and non-linear models of damping and elasticity of vehicle suspension (Kłosiński, Majewski, 1984).

USING OF CRANE MODELS FOR DESIGN PROBLEMS

Both the models have been used for CAD problems. In the first model the influence of the elasticity platform drive k_s and of the joints of jib elements k_{1y}, k_{1z}, k_{2y} etc. on the swing amplitude of the hanged material has been analysed. This influence is related with the correct work of the crane.

The second model has been used in order to select the parameters determining the shape of non-linear characteristics of vehicle suspension. Maximum value of driver's seat displacement while going on a harmonic surface with constant frequency has been accepted as a criterion of model behaviour.

Minimalization of the criterion has been carried out using statistical experiments based on Monte-Carlo method (Zieliński, 1970). For this purpose the criterion values have been calculated for same parameters selected at random from the admissible set describing suspension characteristics. The calculations have been carried out by means of an iterative method using Runge-Kutty's procedure of integration. Then the set of admissible parameter values has been limited to those characteristics of suspension, for which the criterion is of the smallest value, and thereafter the procedure has been repeated. The calculations were repeated until a satisfactory result was obtained.

The calculations were based on technical data of a typical telescopic middle sized crane on an automotive chassis. Dimensions, masses and moments of inertia for the elements have been assumed to be constant and independent in time. The calculations have been carried out on the ODRA-1305 computer.

CONCLUSION

The calculations carried out for a certain type of the automotive crane suggested some changes in the design, especially in the joints of jib segments and vehicle suspension. The last one was very important because of significant swing of the vehicle when going on uneven road.

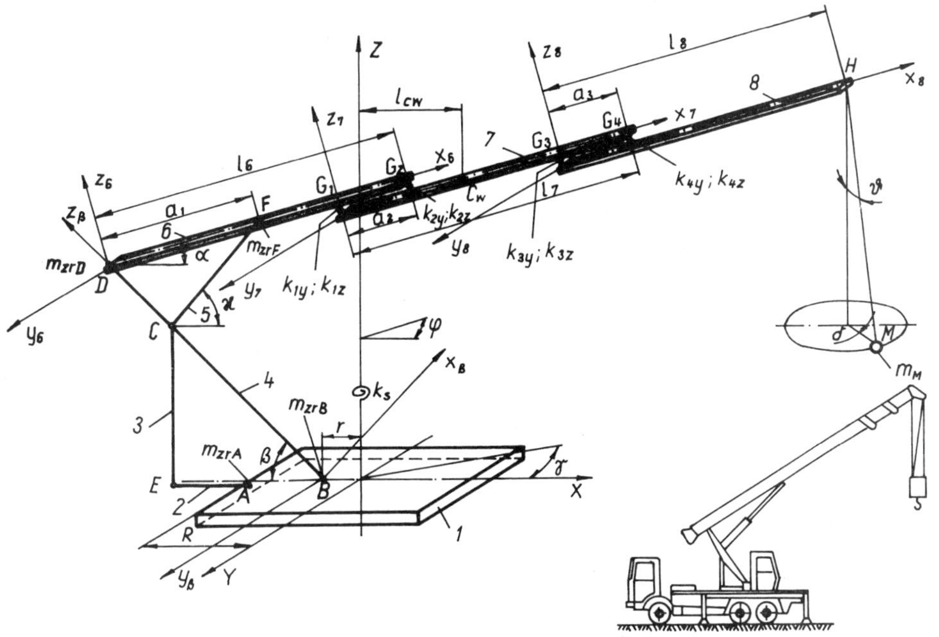

Fig. 1. Model of crane in the state of work

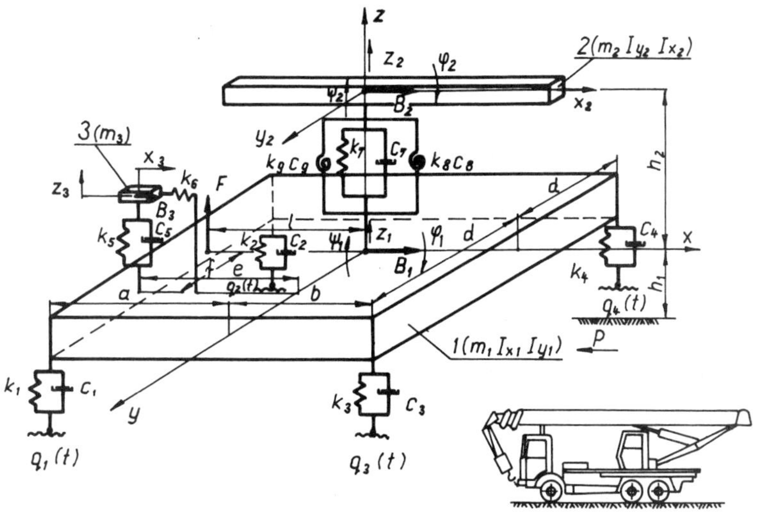

Fig. 2. Model of crane in the state of transport

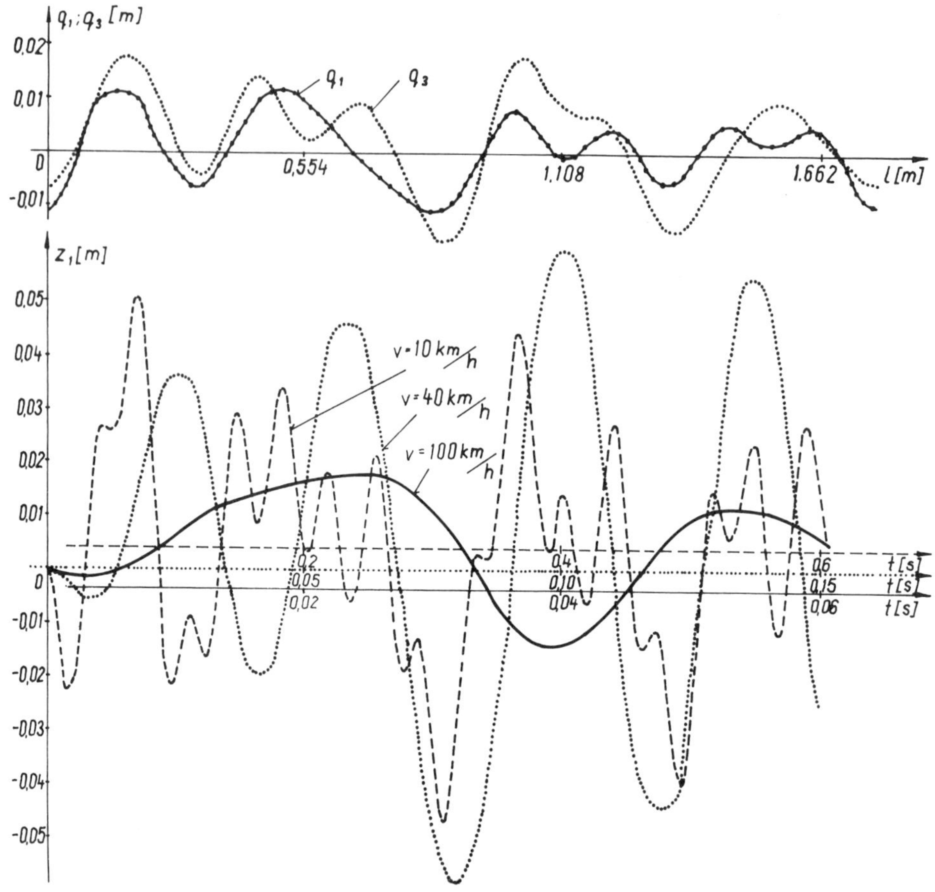

Fig. 3. Time run of selected variables for quasi-random inputs

The main advantages of these numerical simulations are:
- relatively short time of calculation,
- possibility of conversational work with the computer,
- programme flexibility /possibility of changes in mathematical model/.

The authors desire to develope the dynamic analysis of crane, taking into account non-linearities in the joints of the elements and in the suspension, and omitting the restricions related with straight-line going of the vehicle.

REFERENCES

Fortuna, Z., B. Macukow, and J. Wąsowski (1982). Metody numeryczne. WNT, Warszawa. pp. 62-69 and 307-313.

Kłosiński, J., and L. Majewski (1983). An analytic-numerical method for determining the natural frequencies of a discrete-continuous model of a telescopic crane. The Archive of Mechanical Engineering, 30, 59-74.

Kłosiński, J., and L. Majewski (1984). The dynamic analysis of the model of truck crane. In Proceedings of X-th National Conference on Theory of Machines and Mechanisms. ZGPW, Warszawa. pp. 244-249.

Korn, G.A., and T.M. Korn (1983). Matematyka dla prac. nauk. i inżynierów. PWN, Warszawa. Vol. 2, pp. 210-259.

Mitschke, M. (1977). Dynamika samochodu. WKiŁ, Warszawa. pp. 233-348.

Peeken, H., and K. Menniger (1974). Dynamische Kräfte beim Drehen eines Mobilkranes. Fördern und Heben, 24, 1143-1147.

Zieliński, R. (1970). Metody Monte-Carlo. WNT, Warszawa. pp. 223-236.

ZUSAMMENFASSUNG

In der Arbeit wurde die dynamische Analyse eines Teleskopdrehkranes auf dem Fahrgestell durchgeführt. Es wurden zwei Zustände unterschieden und zwar: der Arbeitzustand und der Verkehrszustand. Mathematische Modelle wurden erarbeitet und deren Lösungsmethoden besprochen. Möglichkeiten der Modellsausnutzung für Forschungs- und Entwurfarbeiten wurden dargestellt. Probleme der Genauigkeit der Arbeitbewegungen des Kranes und des Arbeitkomfortes des Fahrers-Operateurs sind inbegriffen. Berechnungsbeispiele wurden beigefügt.

Insight into Computer-Aided Evaluation on Rail Vehicles Dynamics

T. G. IONESCU* AND N. I. MANOLESCU

*Institute for Research and Technological Design in Transports, Bucharest 78341, Calea Grivitei 393, Romania
**Bucharest Polytechnic Institute, Bucharest 77206, Spl. Independenţei 313, Romania

Abstract. Aspects related to updating 4-axle electric locomotives with a view to improving their dynamic behaviour in operation are presented. Following the theoretical modelling of the locomotive, of the rail and of the phenomena of their interaction, extensive experiments and measurements have been developed under specific methodologies. Computer-aided evaluation of the locomotive dynamics on experimental bases uses single input-single output and multiple input-single output signals processing and analysis.

Keywords. Electric locomotives; modelling; dynamic response; computer evaluation; time-domain analysis; frequency response.

INTRODUCTION

The dynamics of railway vehicles, the vehicle-rail interaction and the problem of their improvement are highly complex. The theoretical and experimental research work and the solutions have to fulfil a wide range of sometimes contradictory requirements related to tangent track running, lateral stability, curve negotiation, wheel-rail interaction, response to rail irregularities, derailment, etc.

THEORETICAL MODELLING OF THE VEHICLE-RAIL SYSTEM

The vehicle-rail system may be modelled as a discrete one, formed of a number of rigid masses interconnected by massless links elastic springs and dampers (Fig. 1).

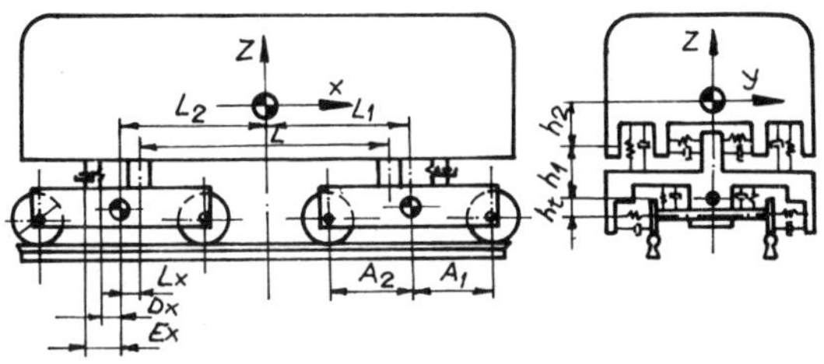

Fig. 1. Functional constructive diagram of the locomotive

The system is subject to certain geometrical, structural and kinematic constraints, mostly non-linear in character. Substituting a linear model for the system (Fig. 2) leads to simplifying the analysis and obtaining useful results (Fallon, 1978).

The irregularities of a homogeneous track segment (Fig. 3) may be defined by three basic processes: a periodic deterministic process represented by the rectified sine wave with wave-length corresponding to the distance between joints, a stationary random process and a modulated periodic deterministic process (Corbin, 1975).

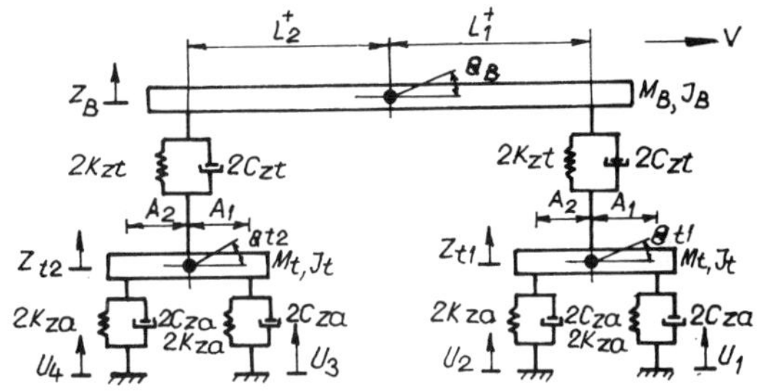

Fig. 2. Mechanical model of the locomotive on a vertical plane

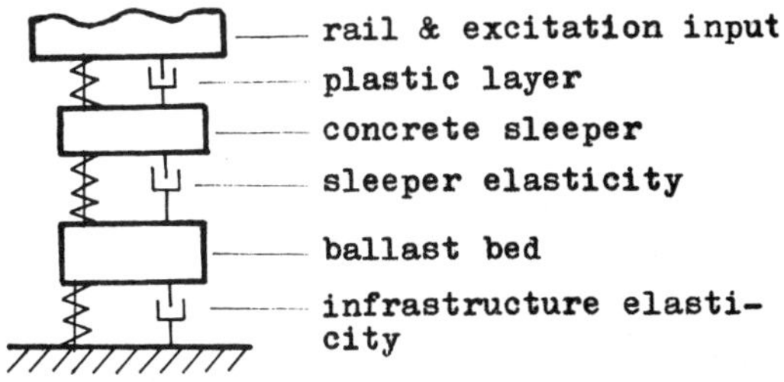

Fig. 3. Vertical mechanical model of the rail

The vertical plane movements of the locomotive may be described function of the mechanical model with 6 degrees of freedom: the vertical displacement (bounce) and pitch displacement of the two trucks and carbody (Ionescu, 1980). Considering the locomotive as a system subjected to the track irregularities input, with a corresponding displacement and acceleration output within the vehicle, the equations of motion may be represented as:

$$[M]\ \ddot{\bar{Z}}(t) + [C]\dot{\bar{Z}}(t) + [K]\bar{Z}(t) = [R]\bar{u}(t) \quad (1)$$

Different techniques, including state variables, Wilson - Θ or Newmark integration methods, Laplace transform, may be alternatively used, computer assisted, to solve the mathematical model in the time or frequency domains (Ionescu, 1985).

EXPERIMENTAL DATA ACQUISITION

The dynamic behaviour of the locomotive towards the rail vibrations and shock inputs has been evaluated in terms of the suspending and damping systems displacement and acceleration vibration responses. Figure 4 sums up the following measurements performed on the vehicle (Ionescu, 1981):

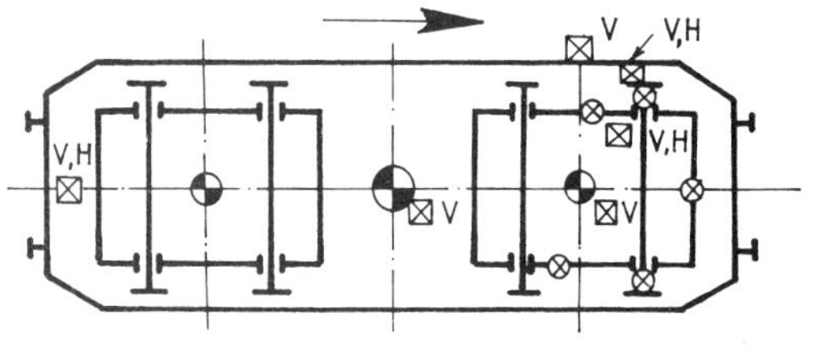

Fig. 4. Lay-out of transducers
⊗ vrd transducers
⊠ vha transducers

1. vertical relative displacement (vrd) between axle box and truck frame, right and left, at the leading wheelset;
2. vrd between truck frame and carbody, at the leading truck, aligned with the secondary suspension, right and left;
3. vrd between truck frame and carbody, at the leading truck, in the middle of the frontal beam;
4. vertical and horizontal accelerations (vha) on one of the axle boxes of the leading wheelset;
5. vha in truck frame above the leading wheelset, on the same side as 4.;
6. vertical acceleration (va) in truck frame under its geometric centre;
7. va in carbody, aligned with the centre of the leading truck, on the same side as 4. and 5. ;
8. va in carbody, under its geometric centre;
9. vha in the driving cab, at the floor level.

All the chosen points have been measured and magnetically recorded simultaneously at speed intervals of 10 km/h up to 120 km/h.

Strain gauge measurements have also been accomplished aiming at determining the distribution and level of strains in the rolling equipment and correlating them with the displacement and acceleration responses.

The experiments have been carried out in simulated conditions on the test rig and then on railway sections in normal conditions of exploitation. The test rig (Fig. 5) has been provided with tangent track and curve sectors for speeds up to 140 km/h as well as with 4 sectors characterized by specially built rail defects. To reveal the locomotive dynamics at shock excitements, a hillock with a maximum vertical deviation on both rails has been arranged.

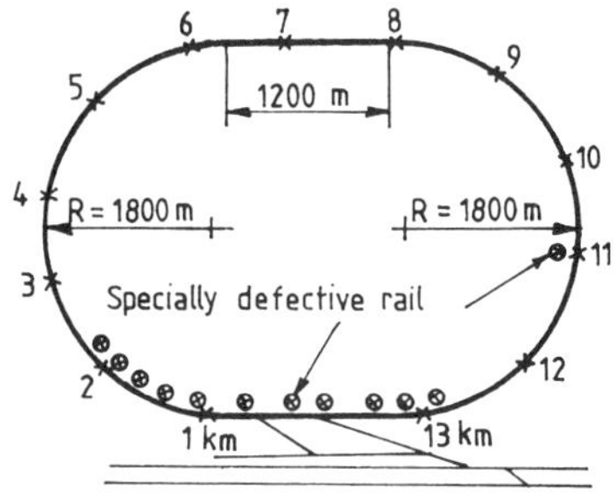

Fig. 5. Diagram of the test rig

The test rig has been measured and its characteristics recorded with a track geometry car.

PROCESSING AND ANALYSIS OF EXPERIMENTAL DATA

Relative Displacement Data

The data analogously recorded on magnetic tape are filtered, transfered onto oscillograph tape and converted by digitalization. The subsequent analysis has been ensured by a desktop computer (HP 9835 A) and interactive programs for digitalization and calculus.

The analized dynamic characteristics of the locomotive function of the running speed refers to primary and secondary suspension displacement transmissibilities, stabilization and delay time, relative speeds, natural periods, logarithmic decrements and damping ratios, quality factors for bounce and bounce and pitch movements, etc.

The analysis of the data defining the test rig track follows almost the same routine, using the recording tape of the track geometry car as the primary source.

Acceleration Data

The recorded signal is low-pass filtered and analogously-numerically converted. The discrete numerical signal obtained is then scaled, memorized and processed by a minicomputer (Coral 4030-PDP compatible). The applied program package, which is modularly built, command-oriented and completely interactive, consists of a set of programs performing time series processing and analysis for single and multiple signals. The main functions of the program package and the recommended processing sequences are presented in Fig. 6 and Fig. 7.

Statistical data analysis module computes the amplitude histogram, specific statistic values and performs a normality test.

Stationarity analysis consists in fitting locally stationary auto regressive (AR) models to non-stationary time series, including system inputs and outputs, by minimum Akaike criterion procedure.

Power spectral densities (PSD) of the system variables are computed from covariance functions by Blackman-Tukey method and by the use of a parametric AR model.

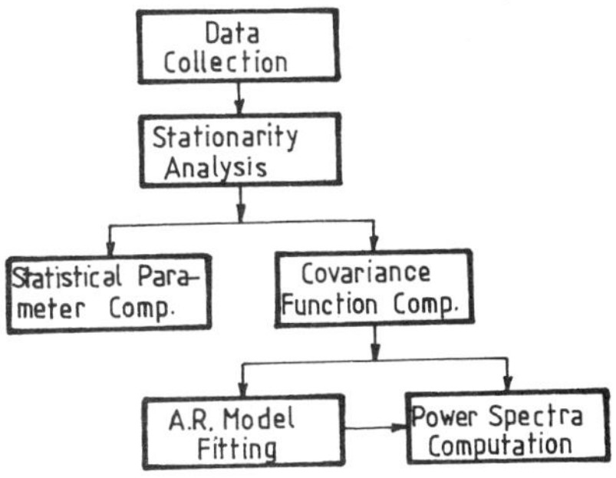

Fig. 6. Single signals processing

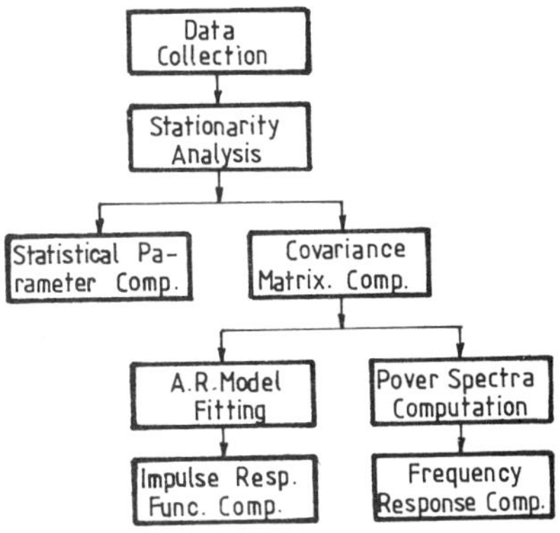

Fig. 7. Multiple signals processing

Frequency response functions (gain and phase) and related functions (simple, partial and multiple coherence, relative error statistics) have been computed from the PSD functions of the system variables.

The computation of a parametric representation of the vehicle-rail system variables can be used to obtain the discrete transfer function as well as the impulse response functions between these variables.

The multiple signals processing is particularly useful when evaluating the characteristics of the rail vehicle as a multiple input-single output system.

RESULTS

Figures 8 and 9 present dynamic displacement characteristics.

Some results regarding the acceleration response signals measured over the hillock at 40 km/h locomotive running speed follow. The statistical characteristics of the axle box vertical acceleration (AVA) signal are (Fig. 10):

max.= 28.506630 m/s^2; min.= -27.708349 m/s^2 ampl.= 56.214981 m/s^2; cent.val.= 0.399140 m/s^2; mean = 1.9093562 m/s^2; for 0.95 confidence, 1.4134003 < mean < 2.4053121; for 0.99 confidence, 1.2568102 < mean < 2.56190; r.m.s. = 6.7162209 m/s^2; std. deviation = 6.4390979 m/s^2; for 0.95 confidence, 6.0852836 < std. deviation < 6.7929573; for 0.99 confidence, 5.9703314 < std. deviation 6.9078644; variance = 41.461982 m^2/s^4; median = 1.7815550 m/s^2; mode = 3.5191307 m/s^2; variance coef.= 3.3723921; skewness = -0.39178935; peakedness = 4.8854404; excess = 1.8854404; sigma skewness = 0.095708586; sigma excess = 0.19053730; Abnormal distribution.

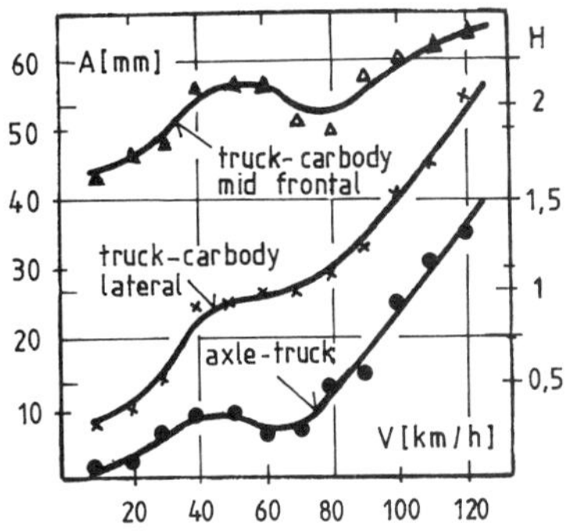

Fig. 8. Relative displacement amplitudes and transmissibilities

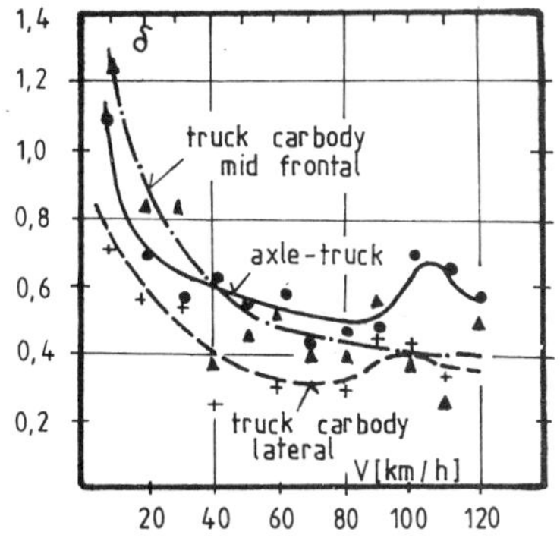

Fig. 9. Logarithmic decrements of the response oscillations

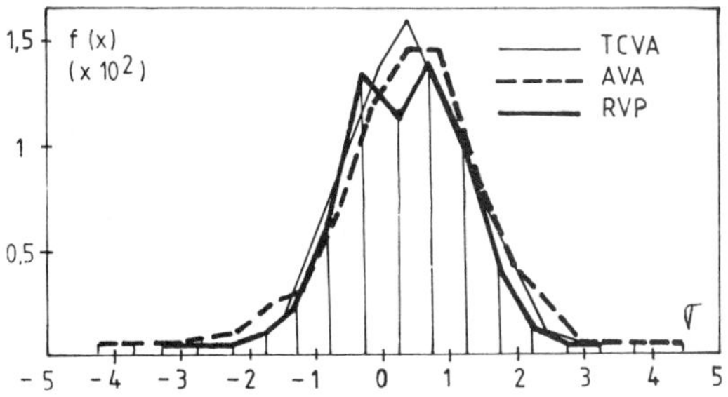

Fig. 10. Amplitude probability histograms (TCVA - truck centre vertical acceleration, RVP - rail vertical profile)

A comparative view of the PSD functions of the AVA signal obtained by Hanning windowing of the sample autocovariance function (Fig. 11) and through the AR parametric model is shown in Fig. 12.

The resulted AR parametric model of the AVA signal is:

$$y(n) = 0.281y(n-1) - 0.214y(n-2) - 0.238y(n-3) + 0.079y(n-4) + 0.149y(n-5) + 0.034y(n-6) + 0.089y(n-7) + 0.072y(n-8) + e(n) \quad (2)$$

with variance = 30.636 for e(n).

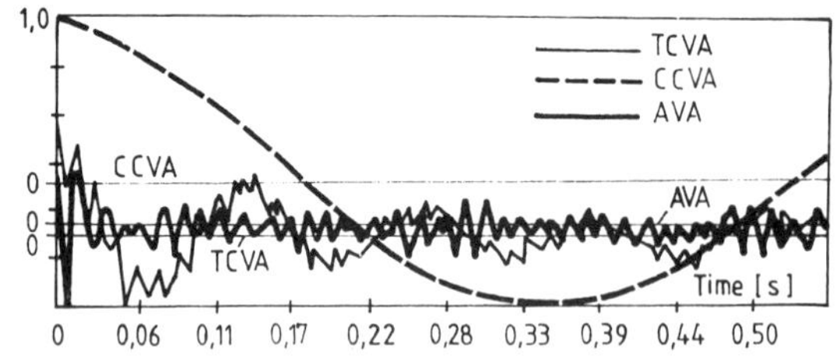

Fig. 11. Autocovariance functions (CCVA - carbody centre vertical acceleration)

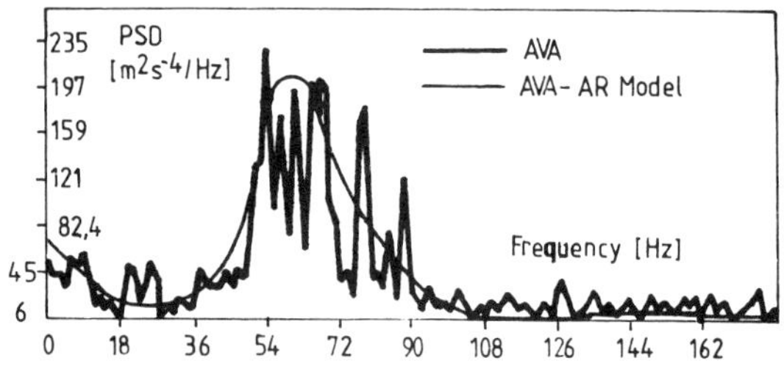

Fig. 12. Power spectra of AVA signal

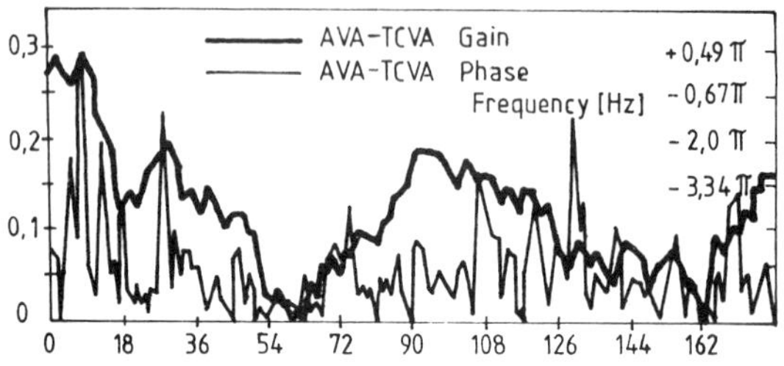

Fig. 13. Frequency response for AVA-TCVA signals

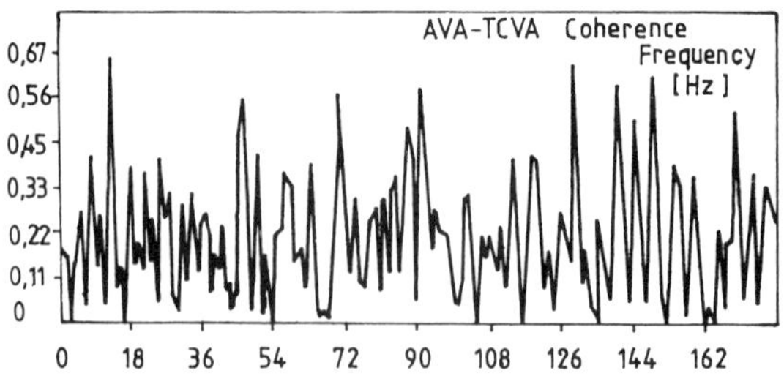

Fig. 14. Coherence function for AVA-TCVA signals

Impulse response function computation has been performed using the multivariate AR parametric model of the considered bivariate time series.

CONCLUSIONS

The theoretical modelling of the locomotive, rail and their interaction has proved adequate, confirming our objectives. The data acquisition under specific

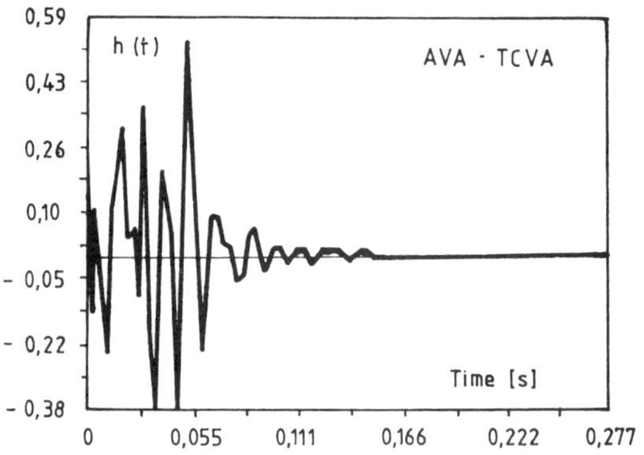

Fig. 15. Impulse response function for AVA-TCVA signals

methodologies in simulated and actual exploitation conditions, has ensured the experimental data base. The results obtained by computer-aided processing of the data base illustrate the potential and flexibility of the automated identification and evaluation techniques. The implementation of this research has led to a better understanding of the functional dynamic characteristics and reliability of the 4-axle electric locomotives, laying the foundation for designing and building new, improved elements.

REFERENCES

Corbin, J. C., V. M. Kaufman (1975). Classifying track by power spectral density. Mechanics of Transportation Suspension Systems, ADM-vol. 2.

Fallon, W. J., N. K. Cooperrider, and E. H. Law (1978). An investigation of techniques for validation of railcar dynamic analyses. Report OR & D/19, Washington.

Ionescu, T. G., and N. I. Manolescu (1980). System analysis and structural criteria concerning Diesel locomotives. ASME Paper 80-DET-105, New York.

Ionescu, T. G., and others (1981). Optimization of the functional dynamic characteristics of the Bo-Bo electric locomotive. Internal Report IRTDT, Buch.

Ionescu, T. G., and T. D. Popescu (1985). Evaluation of Bo-Bo electric locomotive dynamics using time series analysis. Fourth Syrom Symposium, Bucharest. Vol. 4, 65-94.

SOMMAIRE

L'ouvrage a pour objet certains aspects relatifs à la modernisation des locomotives électriques à 4 essieux, avec référence à l'amélioration du comportement dynamique fonctionnel en service de celles ci. Après la simulation et l'analyse théorique de la locomotive, de la voie et de leur interaction, on a procédé à l'exécution d'expérimentations et mesures complexes conformément aux méthodologies établies. L'évaluation assistée par la calculatrice de la dynamique de la locomotive, dans les domaines temps et fréquence, utilise le traitement et l'analyse des signaux correspondants aux grandeurs dynamiques obtenues expérimentalement dans la maniere "une entrée-une sortie", "plusieurs entrées-une sortie".

The Dimensional Synthesis of a Spatial Planetary Mechanism for the Feed of the Wire-Electrode

N. I. MANOLESCU*, ST. ANGHEL** AND I. ANGHEL**

*Polytechnical Institute of Bucharest, Romania
**Subengineers' Institute of Reshitza, Romania

Abstract. The spatial planetary mechanism for the feed of the wire-electrode is used in the construction of the semiautomatic and automatic welding equipments. On the hand of the conclusions from an earlier paper, a method of dimensional synthesis of a spatial planetary mechanism for the feed of the wire-electrode is presented.

Keywords. Welding equipments; steel manufacturing; servomechanisms; dimensional synthesis; spatial planetary mechanism.

INTRODUCTION

The spatial planetary mechanism for the feed of the wire-electrode presents a series of advantages in comparison to other classical mechanisms used in the construction of the semiautomatic and automatic welding equipments. It is recomended in the case of modern techniques of welding in protecting gas medium and under a layer of flow. On the hand of the conclusions from an other paper, the method of dimensional synthesis of the spatial planetary mechanism for the feed of the wire-electrode is shortly presented.

THE MAIN GEOMETRICAL PARAMETERS OF THE MECHANISM

In fig. 1 an axial section through the spatial planetary mechanism for the feed of the wire-electrode is shown. The case in form of the frustrum of a cone (1) has the function of a cam in the translation motion for the sliding pushers (4) to which are fastened the bolts of the rollers (2), materialized by two radial ball-bearings with a slightly concave external face that comes into touch with the wire-electrode (3). The nozzles (6) and (9) guide the wire-electrode between the rollers (2),

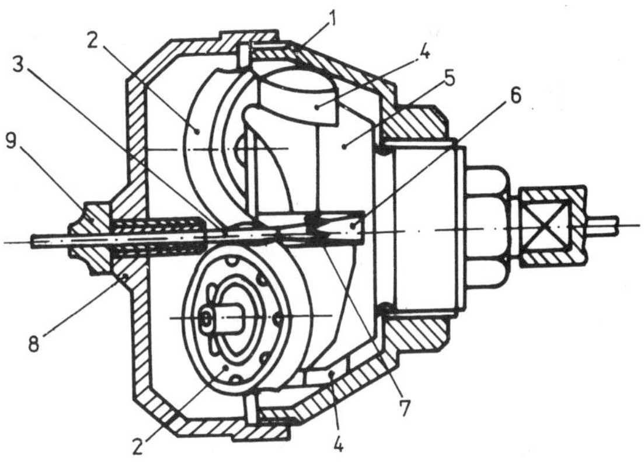

Fig. 1. An axial section through the spatial planetary mechanism

they being changed if the diameter d_e of the wire-electrode is modified. Together with the modification of the wire-electrode, the distance between the rollers (2) also changes by shifting axially the body (5) in which the pushers (4) glide with respect to the cam (1). This adjustment is performed by a helical coupler (screw and nut), acted upon with two spanner-like parts on the special plane portions provided on the body (5) and the cam (1).

The pushers (4) are permanently maintained in contact with the inner side of the cam in shape of a cone-frustrum (1) due to a compression spring (7), and while the mechanism rotates, the centrifugal forces also act and shift the pushers (4) away from each other. The lid (8) protects the mechanism against dirt and has a nozzle (9) in its centre placed to guide the wire-electrode.

By α is marked an angle between a plane normal to the geometrical axis of the mechanism (which also is the axis of the wire-electrode) and the plane of the free rotation of the rollers (2). By β the semiangle at the point of the frustrum-shaped cam (1) is marked, and by d_2-the external diameter of the rollers (2) at the point of contact with the wire-electrode (3). In the following, the notations used in the earlier paper will be maintained, that is n_1 is the speed of revolution, in rot/min, of the planetary head; n_2-the speed of revolution, in rot/min, of the rollers (2); i - the transmission ratio and h - axial shift of the wire-electrode during a whole rotation of the planetary head.

THE DIMENSIONAL SYNTESIS OF THE SPATIAL PLANETARY MECHANISM

To determin the main geometrical dimensions of the spatial planetary mechanism, one starts from the d_e diameter ranges of the wire-electrode, the welding outfit is supposed to work with. The steps between these diameters dictate the interior diameters of the nozzles (6) and (9). If the largest diameter of the wire-electrode is marked with d_{emax}, h_c being the minimal necessary length of the conic portion of the cam (1) then:

$$h_c = d_{emax}/2tg\beta \qquad (1)$$

where the angle β is taken smaller than the critical value β_a, that would lead, due to sliding friction, to a self-blocking of the transmission of motion from cam to the pushers. For the pair of materials made of steel is considered $\beta_a=45^o$, but $\beta=30^o$ recommended as a value convenient both from functional and constructive point of view (reduced size).

The screw-nut couple, interposed between the cam (1) and the body (5) is made up by using a standard finepace p thread and it must allow a relative axial advance H that should surpass by 1...2 mm the magnitude of h_c. The number of relative rotations n_r, at the adjustment of the distance between the rollers (2), will, obviously, be:

$$n_r = H/p \qquad (2)$$

From the standards, ball-bearings will be selected that will serve as rollers; they will be machined so as to obtain a concavity of the external surface of a depth of 0,2...0,3 mm. The external diameter d_2 of the rollers is to be measured in the region of the maximal depth obtained by machining. It is recommended to select ball-bearings of small size to achieve a small size of the whole mechanism.

Using the diagram shown in fig. 2, according to the speed of revolution n_1, in rot/min, that the direct current electric motor of the welding equipment can assure

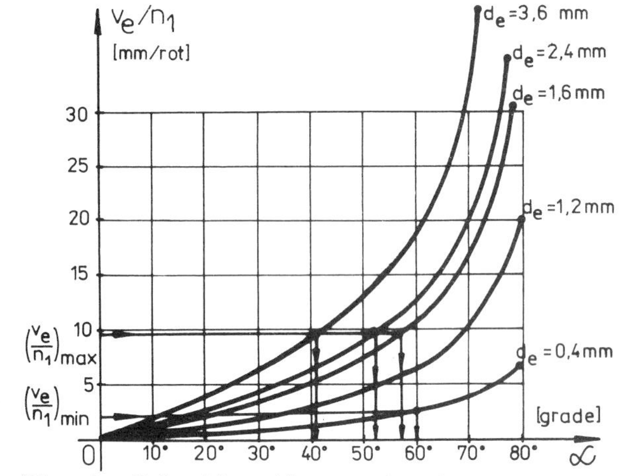

Fig. 2. Selection diagram for the inclination of the rollers

to the planetary head, and choosing the speed interval for the axial feed v_e of the wire-electrode, in mm/min, from the conditions imposed by the technological process of welding (in order to get a high quality welded joint), and taking also d_e into account, the interval is determined within which the angle α may be elected for the inclination of the rollers with respect to a plane, perpendicular to the direction of the feed of the wire-electrode. It is recommended to choose $\alpha=30^o...60^o$, as much as possible towards the middle of the interval determined from the nomogram represented in fig. 2, after the values $(v_e/n_1)_{min}$ and $(v_e/n_1)_{max}$ have been marked on the ordinate, values that determin by horizontal lines the points on the curves d_e, from which verticals may be taken down so as to delimitate a domain within which α can be choosen.

By the help of the value found for the angle α on the corresponding curve d_2/d_e in fig. 3, a point is determined, and the horizontal drawn from that point establishes the value of the transmission ratio i, which allows the calculation of the speed of revolutions n_2 for each of the rollers in rot/min with the equation:

$$n_2 = n_1/i \qquad (3)$$

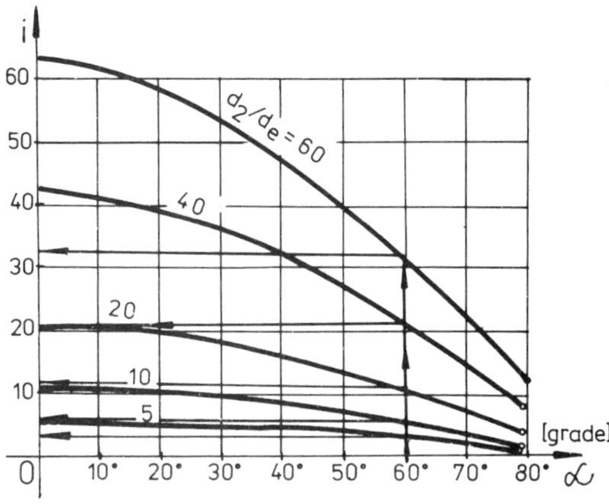

Fig. 3. Diagram for the establishing of the transmission

By means of this value one can determin the durability of the selected ball-bearings and the intervals of their replacement. The other dimensions of the spatial planetary mechanism are to be established constructively and are to be checked for the effort they have to face, by the help of classical dimensioning formulae. It is advisable that the force of the spring, that keeps apart the pushers on which the rollers are fixed, should be as small as possible (1...3 N) in order to reduce the contact strain between the pushers and the cam.

CONCLUSIONS

The method set forth in the paper for dimensional synthesis of the spatial planetary mechanism for the a feed of the wire-electrode provides the designers and those who use such mechanisms with accesible and quick ways to solve the problems implied by the designing of mechanisms of such performance.

After the desing has been accomplished, the values of the kinematical parameters may be calculated with the relations from the another paper, being assured that the mechanism covers the range of values necessary in modern welding processes to which it is applied.

REFERENCES

Anghel, St., I. Anghel and I. Tioc (1985). Aspects Concerning the Kinematics of a Spatial Feed Mechanism of the Wire-Electrode. In The Fourth International Symposium. Theory and Practice of Mechanisms. SYROM'85. Vol. II-1. Bucharest. pp. 19-26.

Manolescu, N.I. (1985). Structural Synthesis of Planar-Came Mechanisms with e=4 elements, c_5=3 Lower Pairs and c_4=2 Higher Pairs. In The Fourth International Symposium. Theory and Practice of Mechanisms. SYROM'85. Vol. I-1. Bucharest. pp. 241-248.

SUMMARY

The spatial planetary mechanism for the feed of the wire-electrode presents a series of advantages in comparison to other classical mechanisms used in the construction of the semiautomatic and automatic welding equipments and is recomended in the case of modern techniques of welding in protecting gas medium under a layer of flow.

The method dimensional synthesis of the spatial planetary mechanism for the feed of the wire-electrode set forth in the paper provides the designers and those who use such mechanisms with accesible and quick ways to solve the problems implied by the designing of mechanisms of such performance.

The Programing in Numerical Control Machine Tools with Five Axes for the Processing of Curved Surfaces

GH. VERTAN*, L. HERZOVI*, ST. ANGHEL** AND M. VERTAN***

**Scientific Research and Technological Engineering Center of Reshitza, Romania*
***Subengineers' Institute of Reshitza, Romania*
****Institute of Calculation Technics Timisoara, Romania*

Abstract. Certain curved surfaces, sometimes known by the coordinates and/or curvatures given for a series of points can be most efficiently processed on NC machine tools. The paper deals with the solving of these problems for the processing of the blades for axial hidraulic turbines on NC machines tools with 5 axes.

Keywords. Turbines; hydraulic turbines; machine tools; NC machine tools; machining; rotor blade of the axial hydraulic turbine.

INTRODUCTION

The curved spatial surfaces, gives by means of coordinates of a multitude of points belonging to the said surfaces, can be processed on numerical controlled (NC) machine tools either by "feeling" a geometrical model (by the help of mechanical a tactile device and, for a higher accuracy, by the use of a LASER probe), or by determining the analytical expression of the surfaces and of the corresponding trajectory of the tool. The use of the analytical expressions is more advantageous compared to the tactile.

This work is a development and generalization of some previous ones and contains the principal programming and machining modality of some curved spatial surfaces on five axles NC like the faces of the Kaplan hydraulic turbine blades. The program that has been worked aut avoids a previous building of the geometrical model of the detail to be machined and uses only the dimensions given in the shop drawing of the detail, aspects that confer to the proposed programme a high efficiency. The validity of these assertions has been fully confirmed by the processing of some hydraulic turbine blades on a 5 axes NC at The Reshitza Machine-building Enterprise using the original programs.

THE SELECTION OF THE INITIAL DATA FOR THE GEOMETRY OF A BLADE

Costumarily, the geometry of an axial hydraulyc turbine rotor blade is given in several initial cylindrical sections, coaxial with the Oz axis of the turbine, fig. 1.a, or in some initial plane sections parallel to the yOz plane, fig. 1,b. These initial cylindrycal or plane sections will be denominated as <u>sections I</u>. For each section I the coordinates of some point and also the curvature elements of the rounded leading edge, fig. 2 are given. If the trailing edge is also

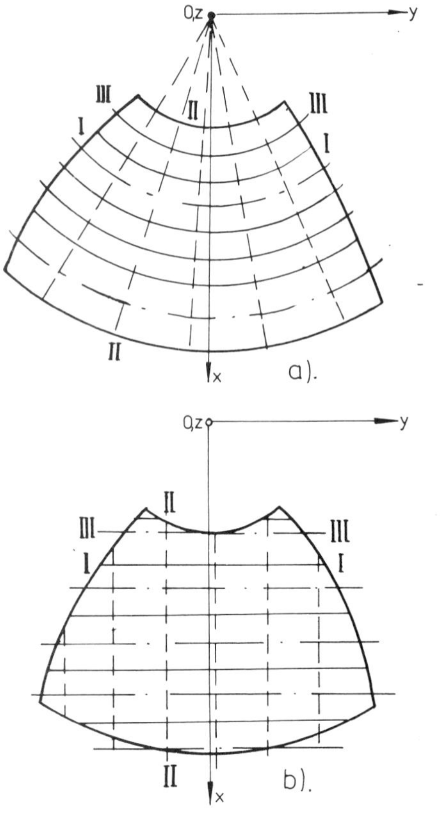

Fig. 1. The selection of the sections through the rotor blade of an axial hydraulic turbine

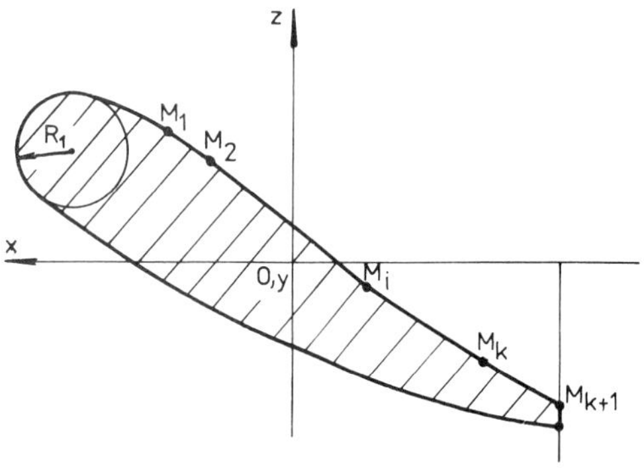

Fig. 2. Sections I or III through the rotor blade

rounded, then it is also necessary to specify the elements that define its curvature.

The <u>sections II</u> are determined, as respectively perpendicular sections on the previously defined sections I. That is, section II will be concurrent planes with respect to the Oz axis, for the case of fig. 1.a, and planes parallel to xOz for the case of fig. 1.b. Sections II will be determined in a convenient number by the help of a suitable calculus program. The sections II intersect a convenient number of surfaces parallel to the initial sections I and so the final <u>sections III</u> are obtained, fig. 1, according to which the curved surface of the blade will be processed.

Each section of type I, II or III at the intersection with the space surface of the blade determines two curves C corresponding to the under side and upper side of the blade. Each curve C is considered a reunion of k arcs of a circle. Each arc has at the centre an angle less then 180°. Some arcs may be degenerated into straight line segments (considered arcs with an infinite radius). These arcs are interconnected, having a common tangent at the connection points. The series of the abscissae of the connection points of the arc of a curve is strictly monotonous.

The analytical relations, that express in a unitary form each of the two C curves of every section through the blade, are obtained by symbolic calculus, using combinations of original functions like:

$$\eta(t) = \frac{1}{2\pi i}\int_{a-i\infty}^{a+i\infty}\frac{e^{tz}}{z}dz, \qquad (1)$$

that, on the ground of the delay theoreme allow to obtain the relation:

$$T(t,\tau_1,\tau_2) = \eta(t-\tau_1)-\eta(t-\tau_2) = \frac{1}{2\pi i}\cdot$$

$$\cdot\left[\int_{a_1-it}^{a_1+it}\frac{e^{(t-\tau_1)z}}{z}dz-\int_{a_2-i\infty}^{a_2+i\infty}\frac{e^{(t-\tau_2)z}}{z}dz\right] =$$

$$= \begin{cases} 0 \text{ for } t<\tau_1, \\ \frac{1}{2} \text{ for } t=\tau_1, \\ 1 \text{ for } \tau_1<t<\tau_2, \\ \frac{1}{2} \text{ for } t=\tau_2, \\ 0 \text{ for } t>\tau_2. \end{cases} \qquad (2)$$

From among those k arcs, the current i arc of a curve C is considered to belong to the circle of equation

$$a_i(x^2+y^2)+b_ix+c_iy+d_i = 0 \qquad (3)$$

where $a_i = 1$ for an arc with a finite radius and, $a_i = 0$ for an arc with an infinite radius (straight line segment).

The ends of the arc i are considered the points $M_i(x_{mi}, y_{mi})$ şi $M_{i+1}(x_{mi+1}, y_{mi+1})$, and by means of the expression (2) and (3) the equation of the arc is obtained in the form

$$T(t, x_{mi}, x_{mi+1}) \cdot \left[a_i(x^2+y^2)+b_i x+c_i y+d_i\right] = 0, \quad (4)$$

which only in the points M_i and M_{i+1} is not valid.

The curve C ist compound of k circle arcs between k+1 points $M_1, M_2, \ldots, M_{k+1}$, than

$$\bigcup_{i=1}^{i=k}\left\{T(t, x_{mi}, x_{mi+1})\left[a_i(x^2+y^2)+b_i x+c_i y+d_i\right]\right\} = 0, \quad (5)$$

which is not true in points M_1 and M_{k+1}.

The solutions found for extreme points are not in the frame of this work.

THE DETERMINING OF THE COORDINATES OF THE END OF THE TOOL AT THE PROCESSING ON 5 AXLE NC MACHINE TOOLS

The diameter D of the front tools does not influence the pich between the machining bands and the accuracy in the case of convex surfaces, but in the case of concave ones there is a close interdependence between the diameter of the tool, the pitch between two succesive passes and the achived accuracy of the machined surface. The working of the blades for axial hydraulic turbines belongs to the second case, the object having also concave surfaces.

To program 5 Axle NC machine-tools in order to work the surfaces of axial hydraulic turbine rotor blades, the surface to be worked was completely divided up in bands, each band being limited by two sections III and considered as a ruled surface. The bredth L of every band is comprised between D and 2D. Each band is processed by at least two passes of the tool. At the first pass the diameter of the tool is maintained tangent to the first section III at the limit of the band (in case of need, tangent to a curve shifted with a constant quantity away from the section III). At the second pass the diameter of the tool is tangent to the second limiting section III of the band (or to a curve shifted away from this). It is considered necessary to have also a third pass, on condition that the point S on the geometrical axis of the tool, at intersection with its frontal plane, should be located on the processed surface in the middle of the band (or on a curve constantly shifted from this middle).

For the points on surfaces with concavities processing bands with a smaller bredth than the diameter of the tool can be taken; then only one single pass of the tool, but in this case the technological correction angle between the axis of the tool and normal in the point of chipping of the surfaces becomes greater, in order to avoid undesired processing that may appear by the action of the periphery of the tool outside the band to be worked.

To determin the tool trajectory of a 5 axle NC machine-tool, it is necessary to calculate the X, Y, Z coordinates of the point S in every point of processing and also two angles (B and C) that give the direction of the axis in space of the space.

Except some small surfaces near the rounded (leading, and, may be, the trailing) edges of the blade, it is admitted that on each ruled band the current generatrix is a straight line segment, comprised in a plane normal to the sections III. The coordinates of the point S in the centre of the frontal tool and considered on the generatrix of the surface rezult like this: at the first pass S is a distance of D/2 from the first section III limiting the band; at the second pass S is at a distance of D/2 from the second section III limiting the band; and as the third pass the point S is in the middle of the generatrix.

THE DETERMINING OF THE DIRECTION OF THE TOOL AXIS ON PROCESSING WITH 5 AXLE NC MACHINE-TOOLS

The plane P_i, i=1,2, is considered, determined by the generatrix of the ruled surface that passes through the point 5 by the tangent to the limiting section III at the end i of the generatrix. For plane P_3 whatever plane comprised in the acute dihedron angle between the plane P_1 and P_2 can be adapted (P_3 can be choosen e.g. as the bisecting plane of the dihedron angle). The generatrix trough S belongs simultaneously to all the planes P_1, P_2 and P_3.

It is admitted that the normal in S to the plane P_3 is the normal in S to the surface to be processed. The direction of the axis of the tool is choosen as shifted with a little angle of $2^o...3^o$ from the normal in S to the worked surface. This technological correction angle allowed to avoid the undesired phenomen of contact with the "back" of the tool.

CONCLUSIONS

The present work permitted to draw up some original calculation program by the help of which the curved surfaces of the blades for axial hydraulic turbines have been processed.

The processing according to cylindrical sections allow higher accuracy, but the processing according to plane sections also provides the necessary precision.

The processing method, set forth in the work, increases technical and econominal efficiency and allows theoretical and applicative developments.

REFERENCES

Vertan, H., Gh., Vertan and M. Vertan(1981). Unterprogramme zur Bearbeitung von gekrümmten Flächen auf numerisch gesteurten Werkzeugmacshinen. In The Third International Symposium. SYROM'81. Bucharest. pp. 419-430.

Vertan, Gh., and H. Vertan (1981). Mathematische Darstellung der Oberfläche von Axialturbinenschaufeln zwecks Bearbeitung auf numerisch gesteuerten Werkzeugmaschinen. In Rev. Roumaine des Sciences Techniques-Mechanique Appliqueé, Tom 26, No.3, Bucharest. pp 481-488.

SUMMARY

The paper presents an original method for the calculation program by the help of which the curved surfaces of the rotorblades for axial hydraulic turbines are processed with on NC machines-tools.

In order to achieve such processing the following is necessary: a) analitical determination of surfaces, known by the coordinates and/or curvature given for a series of points (matrices for car bodies, plane bodies, ship propellers, blades of hydraulic turbines etc.); b) generating cutter paths for NC machine-tools.

The method set forth in the work allows theoretical and applicative developments of great prospect.

On the Dynamic Behaviour of the Matrix Printheads and Its Designing Implications

T. DEMIAN AND C. NITU

Fine Mechanics Department, Polytechnic Institute of Bucharest, Romania

Abstract. The wire displacement periods for printing a dot is annalysed for the three known matrix printheads. The strokes versus time corresponding to these periods are graphically or analytically given, depending on the actuator type. By solving the motion differential equations in dimensionless form, the influence of the different mechanical and electrical parameters upon the dynamic behaviour is examined.

Keywords. Actuators; computer-aided design; dynamic response; printers, solenoids.

INTRODUCTION

Dot matrix printing has become one of the most popular printing techniques, because it allows the user to control the quality of print and it also has an important improvement in the cost performance ratio when is used with impact printers.

There are three generally available types of dot matrix printheads. The first is a tubular solenoid type in which the printwire is attached to a small internal plunjer (fig.1). In the second one, the printwire is accelerated by the movement of an electromagnet rotating armature (fig.2). The third is the so-called magnetic stored energy printhead (fig.3). Magnetic energy is stored by spring held cocked by magnetic field at pole tip gap. Energized coil cancels magnetic field of gap, releasing spring and allowing wire to impact paper at high speed.

PRINTHEAD DYNAMIC BEHAVIOUR SIMULATION

The printwire displacement x vs.time, between its rest position and the place where it compresses the ribbon, paper and platen is shown in fig.4. Different stages of the movement are: t_1-delay time for magnetic force increasing/decreasing; t_2- actuated printwire time; t_3- free printwire travelling time when armature is stopped (it can be missing); t_4- contact time with the paper; t_5 - printwire return time; t_6- settling time.

Dot printing time, $t_p = \sum_{i=1}^{6} t_i$, is particularly determined by t_1 and t_2, when the printwire is energized.

The printhead can operate in crush-mode, when the actuator is energized for the whole printwire displacement, or in ballistic-mode, with a free travelling of

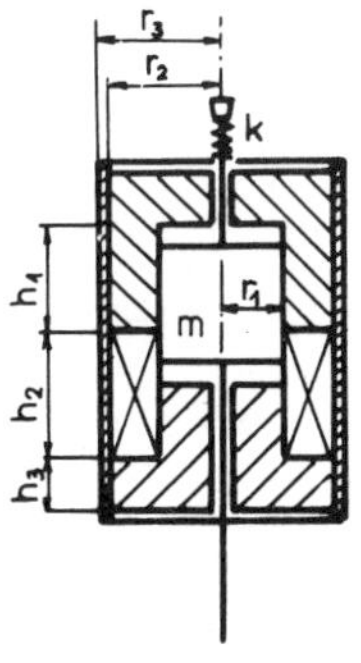

Fig.1 Tubular solenoid with small internal plunjer

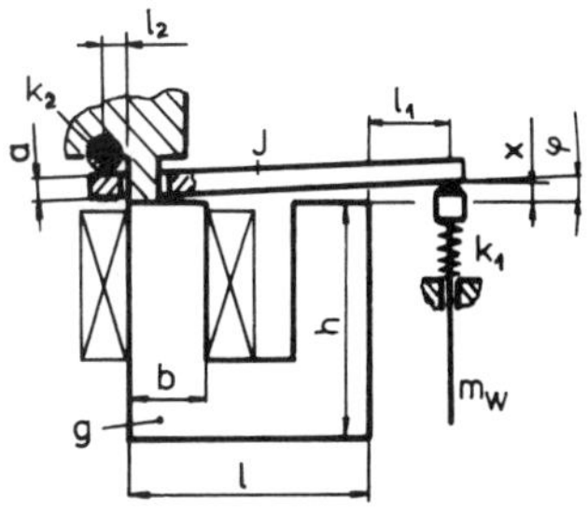

Fig.2 Solenoid with rotating armature

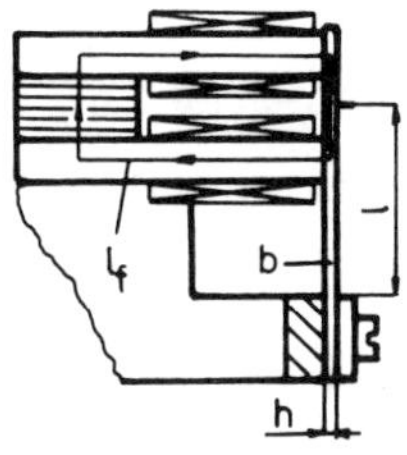

Fig.3 Magnetic stored energy printhead

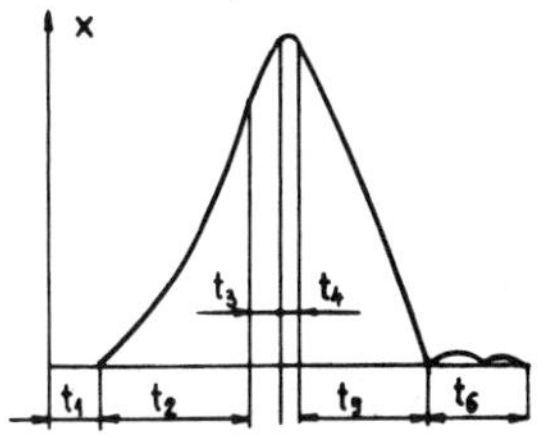

Fig.4 Printwire displacement versus time

the printwire, when the armature is stopped. However the displacement time(t_2+t_3) and the impact velocity v_3 are depending on the magnetic and electric parameters of the actuator and armature and printwire mechanical ones. Someone can obtain them from the armature and printwire motion differential equation, to which are attached the magnetic and electric circuits equations. The core magnetic flux increase can be determined by use of Froelich approximation of the D-C magnetization curve: $B=C_1H(C_2+H)^{-1}$, where: B is flux density; H - field intensity; C_1 - saturation flux density; C_2 - field intensity in the point of maximum permeability (Isobe, 1971).

If a voltage step is applied to the solenoid the dimensionless differential equations of motion and flux, when the leakage is neglected, are:

$$\frac{d^2X}{dT^2} = a_1P^2-a_2X-a_3 \qquad (1)$$

$$\frac{dP}{dT} = b_1-b_2\left[\beta(1-\alpha P)^{-2}+\gamma(1-P)^{-2}+C_1(1-X)/(\mu_0C_2)\right]P \qquad (2)$$

where: X - dimensionless displacement; P - dimensionless flux; T - dimensionless time; μ_0 - air permeability.

The coefficients of these equations depend on electromagnet dimensions and shape and electric parameters of the drive circuit: U - drive voltage; R - coil rezistance; N - number of coil turns; ζ- duration of constant voltage drive pulse. For the solenoid in fig.1, they are:

$$a_1 = \pi(r_3^2-r_2^2)C_1^2\zeta^2/(m\,\delta\, r_1^2);\quad a_2 = k\zeta^2;$$

$$a_3 = F_{res}\zeta^2/(m\delta);\ \alpha = (r_3^2-r_2^2)/\left[(r_1+r_2)h_1\right];$$

$$b_1 = U\zeta(r_3^2-r_2^2)^{-1}(1+RR_1^{-1}N^{-2})^{-1}/(\pi NC_1)$$

$$b_2 = C_2\delta\zeta RN^{-2}r_1^{-2}(1+RR^{-1}N^{-2})/(\pi NC_1)$$

$$\beta=(h_1+h_2)r_1^2/(2\delta h_1h_2)\ln\left[(r_2+r_3)/(2r_1)\right]$$

$$\gamma=(h_1+2h_2+h_3)r_1^2/\left[2\delta(r_3^2-r_2^2)\right].$$

In these equations, δ is maximum air gap; k - spring constant; m - armature and wire mass; F_{res} - friction force between wire, plunjer and their guides; dimensions $h_1, h_2, h_3, r_1, r_2, r_3$ are indicated in fig.1. The influence of the eddy currents is pointed out by R_1(Kallenbach, 1969). If L is the coil inductivity and ρ, the core specific resistance, then:

$$R_1=1.4L\delta\rho\left[2(h_1+h_2+h_3)+r_2+r_3-r_1\right]N^{-2}\mu_0^{-1}r_2^{-2}$$

The dimensionless variables for the solenoid in fig.1 are : $X = x/\delta$, $T = t/\zeta$; $P = \phi/\left[C_1\pi(r_3^2-r_2^2)\right]$.

For the solenoid with rotating armature, the coefficients take the values:

$$a_1 = \alpha C_1^2 agl\zeta^2(l+l_1)^{-2}/(2\mu_0m_r\varphi_0);$$

$$a_2 = k_r\zeta^2/m_r;\ \alpha = a/b;$$

$$a_3 = F_{res}\zeta^2/\left[m_r(l_1+l)\varphi_0\right];$$

$$b_1 = U\zeta(1+RR_2^{-1}N^{-2})^{-1}/(C_1N\,ag);$$

$$b_2 = \alpha RC_2l\varphi_0\zeta(1+RR_2^{-1}N^{-2})^{-1}/(C_1N^2a\,g);$$

$$\beta = (2h+l-2b)/(l\varphi_0);\ \gamma=(a+l-b)/(l\varphi_0).$$

In these equations: $k_r=k_1+k_2l_2^2(l_1+l)^{-2}$ is the equivalent spring constant at the wire; $m_r=m_w+J(l_1+l)^{-2}$ is the equivalent mass at the wire; m_w-wire mass; J - armature moment of inertia; F_{res}-equivalent friction force at printwire. Dimensions $a, b, g, h, l, l_1, l_2, \varphi_0$ are indicated in fig.2. The influence of eddy currents is pointed out by R_2 (Kallenbach,1969):

$$R_2=0.7Ll\varphi_0\rho\, a^2g^2/\left[N^2\mu_0(h+l-b)(a^2+g^2)\right]$$

In this case, the dimensionles variables are $X = x/\left[(l_1+l)\varphi_0\right]$; $T = t/\zeta$; $P=\phi/(ag)$.

At the magnetic stored energy printhead, an exponential magnetic flux variation may be admitted: $\phi = \phi_0(1-e^{-t/\theta})$, where $\phi_0=UA\,\mu_f/(Rl_f)$; $\theta = \mu_fAN^2/(Rl_f)$; l_f- iron core length; μ_f - iron core permeability; R,N - coil rezistance and number of turns; A - core cross section area. The printwire motion equation is:

$$m_r\ddot{x}+kx+(B_0A-\phi)^2/(\mu_0A) = k\,f \qquad (3)$$

where: $k = Ebh^3/(4l^3)$ is spring constant; b,h,l - leaf spring dimensions; E- Young's modulus; $B_0=B_MA_M/A$ is core flux density; B_M, A_M - flux density and cross section area of the permanent magnet; f - spring deflection; m_r - equivalent mass at the printwire.

ANALYSIS OF THE PRINTHEAD DYNAMIC BEHAVIOUR

The printwire motion before impact is described by equations (1) and (2) for the printheads with electromagnetic actuator. These equations were solved by use of Runge-Kutta algorithm for different values of actuator parameters. The influence of the armature dimension modification (fig.5) the influence of voltage step value (fig.6) and duration of drive pulse (fig.7), the influence of spring constant (fig.8) upon the armature displacement and velocity are examined.

As it is expected, when the voltage step and duration of drive pulse increase, the armature displacement and velocity also increase because more energy is supplied. The change of the armature dimensions has an optimum value for α = 0.6, because a large cross section area increases the equivalent mass and a small one leads to a faster magnetic saturation. In the common range, the change of the spring constant value is not significant.

The numerical results point out only the t_1 and t_2 stages from fig.4. In the "balistic" mode, the printwire has an additional displacement before it impacts the paper. The kinetic energy of the wire, at the end of the t_2 stage of motion, decreases during the t_3 stage in order to have

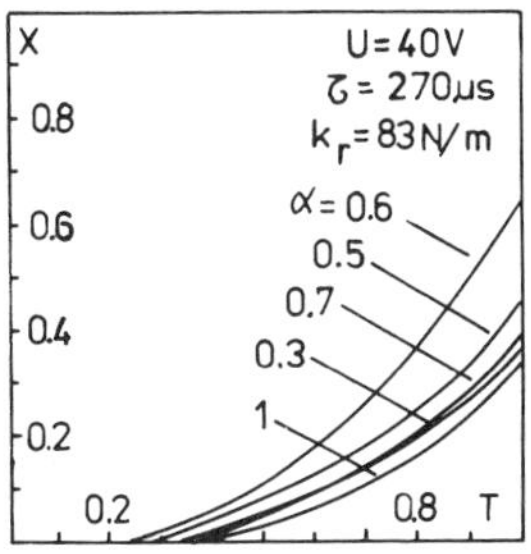

Fig.5 The influence of the armature dimensions upon the printwire displacement

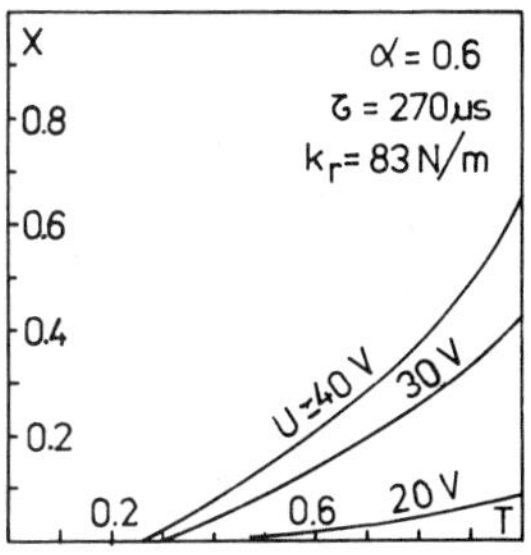

Fig.6 The influence of the drive voltage upon the printwire displacement

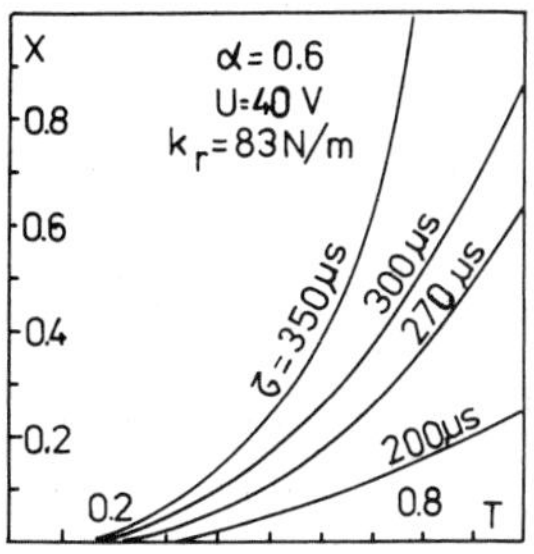

Fig.7 The influence of the drive pulse duration upon the printwire displacement

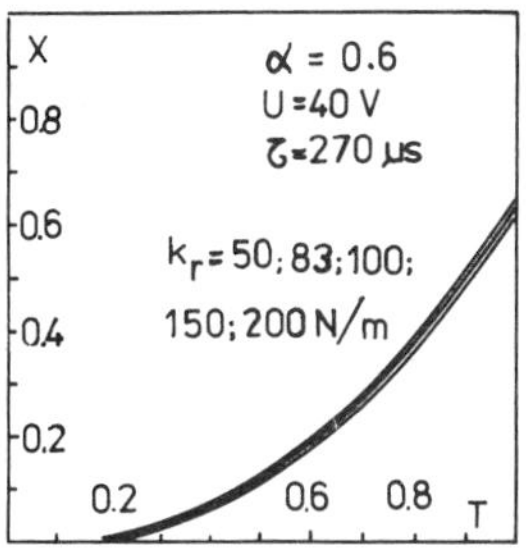

Fig.8 The influence of the spring constant upon the printwire displacement

required value for printing 1 to 6 part forms. The duration t_2 can be analytically calculated with:

$$t_3 = 2\sqrt{m_w/k_1}\cdot\tan^{-1}\frac{v_2\sqrt{\frac{m_w}{k_1}} - \sqrt{\frac{m_w v_2^2}{k_1} - z^2 - \frac{2F_{res}\cdot z}{k_1}}}{z + 2F_{res}/k_1} \qquad (4)$$

where: $z = x-(1+l_1)\varphi_o X_2$ and X_2 - dimensionless displacement at the end of the t_2 stage.

In the case of the magnetic stored energy printheads, by solving the equation (3), the delay time and the motion time can be determined. The delay time is:

$$t_1 = \theta\ln(\phi_o/\sqrt{\mu_o k\ f\ A}) \qquad (5)$$

The motion time $t=t_1+t_2$ is given, for a certain displacement x, by the equation:

$$x = f - \frac{\phi_o^2}{2\mu_o A\theta}\sqrt{\frac{m_r}{k}}\left(\frac{4m}{\theta^2}+k\right)^{-1}\sin\sqrt{\frac{k}{m_r}}\,t + \left[\frac{\phi_o^2}{\mu_o A}\left(\frac{4m}{\theta^2}+k\right)^{-1} - f\right]\cos\sqrt{\frac{k}{m_r}}\,t -$$

$$- \frac{\phi_o^2}{\mu_o A}\left(\frac{4m}{\theta^2}+k\right)^{-1}e^{-2t/\theta} \qquad (6)$$

The equations (5) and (6) show that it is favourable to have a coil with great resistance and large dimensions of the magnetic circuit. The other parameters have also important influence, but they depend on spring resonant frequency, **fatigue** limit of the stress and strain energy.

The period t_4 is the contact time between the printwire and paper and it has a small contribution to the printing time. Experimental researches have pointed out that the contact time is about 20 µs for a maximum impact force of 10 N. The printwire velocity after impact has a diminished value, $v_4 = \varepsilon v_3$, **where** ε is the restitution coefficient. In order to maintain a small contact time, it is necessary to have a small printwire mass a great wire velocity at the impact.

In the case of electromagnetic actuated printheads, the displacement vs.time equation, for the wire return period is:

$$x = x_3 - \varepsilon v_3\sqrt{\frac{m_w}{k_1}}\sin\sqrt{\frac{k_1}{m_w}}\,t +$$

$$+ (x_3 - F_{res}/k_1)(1-\cos\sqrt{\frac{k_1}{m_w}}\,t) \qquad (7)$$

At the magnetic stored energy printhead, upon the spring act the magnetic force and the restored kinetic energy after impact. The return time can be calculated with:

$$t_5 = \int_0^{x_3}\left(\frac{8\phi_o^2 l_f^2}{\mu_o\mu_f A m_r}\ \frac{x}{x-x_3} - \frac{kx^2}{m_r} + \varepsilon^2 v_3^2\right)^{-\frac{1}{2}}dx \qquad (8)$$

The settling time is comparable to the motion ones. It can be diminished by an adequate choosing of spring constants and back stop piece material.

The diagram in fig.4 is experimental confirmed if the printwire strikes a piezoelectric transducer. A dual trace memory oscilloscope shows the drive signal of the solenoid and the transducer output, for a certain distance between the printwire end and the transducer.

CONCLUSION

The overall printhead efficiency is low because only a small part of the input energy is transformed into printwire kinetic energy. This work showed the parameter values and the conditions for optimum efficiency. However the greatest part of the energy is dissipated by heat. The maximum allowed temperature can dramatically limit the calculated solenoid refire rate. It depends on the drive circuit type, the average dot number of a character, the line feed time etc. All of these effects have to be considered when a printhead is designed.

REFERENCES

Kallenbach,E.(1969). Der Gleichstrommagnet. Akademische Verlagsgesellschaft Geest und Portig K.G., Leipzig.Chap.5,pp.162-2o8.

Isobe,M., Uejima,H., Kusu,T.(1971).On the analytical design of high speed printer hammer actuator. NEC Res.and Dev.,22, 35-45.

Endo,K., Fujiwara,S., Woda,M., Agata,K. (1985). Line dot impact printer.NEC Res. and.Dev., 79, 11o-116.

RÉSUMÉ

L'ouvrage étudie les étapes du mouvement de l'aiguille pour imprimer un point, aux trois modeles connus pour les têtes d'impression en mosaïque. Correspondant à ces étapes, on présente, graphique ou analytique, la dépendence course-temps, pour chaque type d'élémente d'actionnement.

Pour l'actionnement à électro-aimant on résoudre les équations différentielles en forme non-dimensionelle et on tire des conclusions au sujet des dimensions optimalles et l'influence des divers paramètres sur le comportement dynamique du système.

Pour l'actionnement avec des ressorts lamellaires, ainsi que pour la course de retour des aiguilles on établit des relations analytiques entre la course et le temps, qui relèvant l'influence des différents paramètres constructifs et dynamiques.

Comme suite de l'étude réalisé on tire de conclusions au sujet de la manière d'élaborer des têtes d'impression en mosaïque.

The Dynamics of the Drive and Positioning Mechanisms with Electrodynamic Linear Engines

V. MĂTIEŞ

The Polytechnic Institute of Gluj-Napoca, Romania

Abstract. The work presents the components of a positioning system of the recording heads on magnetic disks achieved with a liniar electrodynamic engine. Based upon the operational block diagram the system functioning equations corresponding to acceleration, steady run, braking and stop phases are deduced. The equations here deduced point out the parametres which influence the system behaviour in the four phases and the means for increasing its functional performances.

INTRODUCTION

Electrodynamic liniar engines are modern elements, compatible with the numerically controlled technique their development going hand in hand with that of calculation.
These engines are functionally and structurally studied by Olbrich (1973).
Mechanisms achieved with electrodynamic engines have gained a special place especially in applications concerned with the positioning of mechanic receivers in a motion of translation.
The components of such a mechanism used in magnetic disk units are pointed out in Fig.1. Thus, 1 is the liniar engine stator; 2-the movable complex on which recording heads 3 and the mobile rod 4 of the position transducer are fixed; 5-the magnetic disks; 6-the speed transducer; 7-servocontrol.

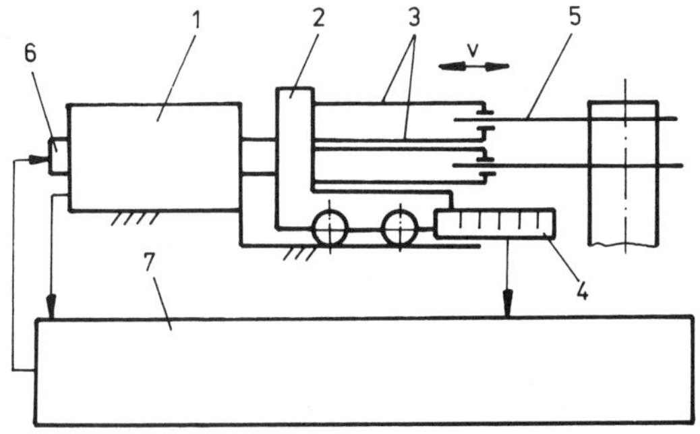

Fig.1. Positioning mechanism with a liniar electrodynamic engine.

The whole complex constitutes a positioning system with a limitation of the current through engine induced and of the mobile complex speed, whose block diagram is shown in Fig.2.
Using the block-diagram, one can write the system functioning equations under the form:

$$U = K(U_i - U_T) \quad , \qquad (1)$$

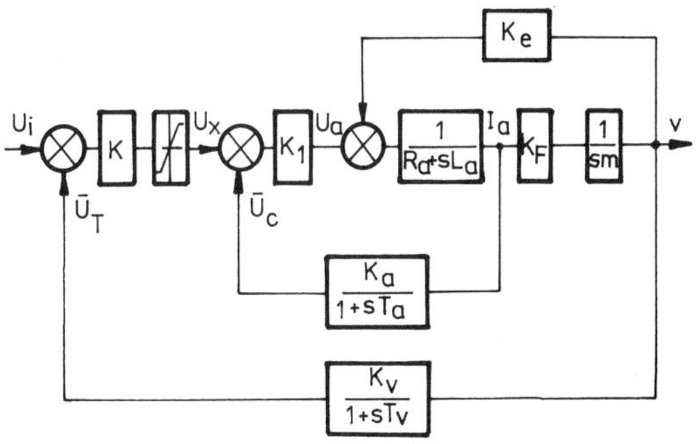

Fig.2. The block-diagram of the positioning system.

$$U_x = f(U) = \begin{cases} U'_H & \text{for} \quad U > U'_H \\ U''_H & \text{for} \quad U < U''_H \\ U & \text{for} \quad U''_H < U < U'_H \end{cases} \qquad (2)$$

$$I_a = \frac{U_x}{K_a} \quad , \qquad (3)$$

$$K_F . I_a = m \frac{dv}{dt} \quad , \qquad (4)$$

$$K_v . v = U_T + T_v \frac{dU_T}{dt} \quad , \qquad (5)$$

where (1) represents the function achived by the error amplifier;
(2) the function achieved by the current limiter.
The notations bear the following significances:

- U_i - reference voltage;
- U_T - feedback signal of the speed transducer;
- K - amplifying factor of the error amplifier;

U_x - output voltage of the current limiter;
U - amplified error;
U'_H - superior threshold of the limiter;
U''_H - inferior threshold of the limiter;
I_a - mobile armature current;
K_a - current feedback constant;
m - mobile complex mass;
v - mobile complex displacement speed;
K_e - electromotor voltage constant;
K_v - speed feedback constant;
T_v - speed loop time constant;
U_c - current transducer feedback signal;
T_a - current loop time constant;
R_a - mobile armature resistance;
L_a - mobile armature inductivity;
U_a - voltage at the terminals of the mobile armature;
K_1 - final amplifier amplification factor.

The operation of these systems is stabile no matter how great the amplification factor K is, such as Mătieş shows (1986). For a positioning cycle, one can distinguish the following stages (Fig.3): acceleration-t_a, stabilized run-t_m, braking-t_f and stop-t_o.

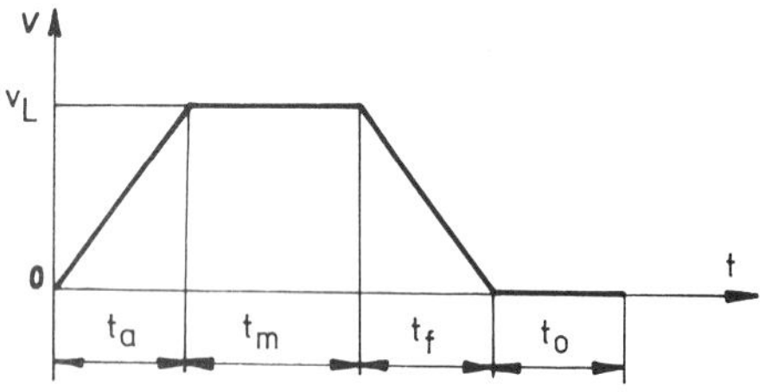

Fig.3. The operational stages of the positioning system.

Based upon the operational equations we can study the behaviour of the system in the four stages.

ACCELERATION STAGE

From (1) one can write:

$$U = K(U_i - U_T) > U'_H \qquad (6)$$

As the difference $U_i - U_T$ is large, U_T being almost negligible, it yields:

$$U \approx KU_i \gg U'_H \,. \qquad (7)$$

The error amplifier enters in saturation and a superior limitation of the voltage U takes place and:

$$U_x = U'_H \,. \qquad (8)$$

Integrating (4), we obtain:

$$v = \frac{K_F I_{a1}}{m} t + C' \,.$$

Constant C' is determined from the initial conditions: t=0, v=0, resulting in C'=0. Thus:

$$v = \frac{K_F I_{a1}}{m} t \,. \qquad (9)$$

Replacing (9) in (5), it yields:

$$\frac{dU_T}{dt} + \frac{1}{T_v} U_T = \frac{K_v K_F I_{a1}}{mT_v} t \,. \qquad (1o)$$

The homogeneous equation solution is written under the form:

$$U_{To} = A_1 e^{-t/T_v} \,. \qquad (11)$$

The special solution is:

$$U_{T_p} = B_1 t + C_1 \,. \qquad (12)$$

From (11) and (12) we obtain:

$$B_1 + \frac{B_1}{T_v} t + \frac{C_1}{T_v} = \frac{K_v K_F I_{a1}}{mT_v} t \,. \qquad (13)$$

After determining constants A_1, B_1, C_1 the general solution of the equation (1o) is:

$$U_{T_1} = \frac{K_v K_F I_{a1} T_v}{m} e^{-t/T_v} + \frac{K_v K_F I_{a1}}{m} t - \frac{K_v K_F I_{a1} T_v}{m} \,. \qquad (14)$$

Consequently, in the acceleration stage the operational equations for the system are:

$$\begin{cases} U = K(U_i - U_{T_1}) \,, \\ U_x = U'_H \,, \\ I_{a1} = \dfrac{U'_H}{K_a} \,, \\ v = \dfrac{K_F I_{a1} t}{m} \\ U_{T_1} = \dfrac{K_v K_F I_{a1} T_v}{m} e^{-t/T_v} + \dfrac{K_v K_F I_{a1}}{m} t - \dfrac{K_v K_F I_{a1}}{m} T_v \,. \end{cases} \qquad (15)$$

RUN STAGE

After the t_a time interval corresponding to the acceleration stage, the error amplifier goes out of saturation,

$$U_x = U = K(U_i - U_{T_2}) \qquad (16)$$

and

$$I_{a2} = \frac{K(U_i - U_{T2})}{K_a} \,. \qquad (17)$$

Replacing (17) in (4) it results:

$$\frac{K_F K(U_i - U_T)}{K_a} = m \frac{dv}{dt}$$

or

$$\frac{dv}{dt} = \frac{K_F K U_i}{mK_a} - \frac{K_F K}{mK_a} U_T \,. \qquad (18)$$

Derivating (5) versus the time, we obtain:

$$K_v \frac{dv}{dt} = \frac{dU_T}{dt} + T_v \frac{d^2 U_T}{dt^2} \,. \qquad (19)$$

From (18) and (19), we can write:

$$\frac{d^2U_T}{dt^2} + \frac{1}{T_v}\frac{dU_T}{dt} + \frac{K_vK_FK}{mK_aK_v}U_T = \frac{K_vK_FK}{mK_aT_v}U_i . \quad (20)$$

Solving the characteristic equation

$$r^2 + \frac{1}{T_v}r + \frac{K_vK_FK}{mK_aT_v} = 0 , \quad (21)$$

we obtain the roots:

$$r_{1,2} = \frac{-\frac{1}{T_v} \pm \sqrt{\frac{1}{T_v^2} - 4\frac{K_vK_FK}{mK_aT_v}}}{2} . \quad (22)$$

The homogeneous equation solution is

$$U_{T_{2o}} = e^{-t/2T_v}(A_2\sin\omega t + B_2\cos\omega t). \quad (23)$$

where

$$\omega = \sqrt{\frac{KK_FK_v}{K_amT_v} - \frac{1}{4T_v^2}} .$$

Constants A_2 and B_2 are determined from the continuity and tangent conditions written under the form:

$$\begin{cases} U_{T_1}(t_a) = U_{T_2}(o) , \\ \left.\frac{dU_{T1}}{dt}\right|_{t=t_a} = \left.\frac{dU_{T2}}{dt}\right|_{t=o} . \end{cases} \quad (24)$$

From (24) one can obtan:

$$A_2 = \frac{2K_vK_FI_{a1}T_v + K_vK_FI_{a1}t_a - mU_i}{2m\omega T_v} ,$$

$$B_2 = \frac{K_vK_FI_{a1}t_a - mU_i}{m} .$$

The general solution of the equation (20) is:

$$U_{T_2} = e^{-t/2T_v}\left[\left(\frac{2K_vK_FI_{a1}T_v + K_vK_FI_{a1}t_a}{2m\ T_v} - \frac{U_i}{2\ T_v}\right)\sin\omega t + \frac{K_FK_vI_{a1}t_a - mU_i}{m}\cos\omega t\right] + U_i . \quad (25)$$

Taking into account (25), from (4) we obtain the expression of the speed:

$$v = v_L + \frac{1}{K_v}e^{-t/2T_v}\left[\left(\frac{2K\ K_FI_{a1}T_v}{4m\omega T_v} + \frac{K_vK_FI_{a1}t_a - mU_i}{4m\omega T_v} - \frac{T_v\omega . K_FK_vI_{a1}t_a}{m} - T_v\omega . U_i\right)\sin\omega t + \left(\frac{K_FK_vI_{a1}t_a}{m} + \frac{K_vK_FI_{a1}T_v}{m} - U_i\right)\cos\omega t\right], \quad (26)$$

where v_L is the imposed speed limit. Equations (16),(17),(25) and (26) describe the functioning of the system in the steady run stage.

BRAKING STAGE

The braking of the engine begins in the moment in which reference voltage level disappears.
If $U_i = 0$ from (1), it yields:

$$U = -KU_{T_3} , \quad (27)$$

$U \ll U''_H$ the limiter comes into function, at its output we obtain:

$$U_x = U''_H , \quad (28)$$

$$I_{a3} = \frac{U''_H}{K_a} . \quad (29)$$

From (4) by integration one gets:

$$v = \frac{K_FI_{a3}}{m}t + C'_3 . \quad (30)$$

As for $t = 0, v = v_L$ then $C'_3 = v_L$ and relationship (30) becomes:

$$v = v_L - \frac{K_FI_{a3}}{m}t . \quad (31)$$

Replacing (31) in (4) we get:

$$\frac{dU_T}{dt} + \frac{1}{T_v}U_T = \frac{K_vK_FI_{a3}}{m}t + \frac{v_L}{T_v}K_v . \quad (32)$$

The solutions of the equation (32) are written so:

$$U_{T3o} = A_3\,e^{-t/T_v} , \quad (33)$$

$$U_{T3p} = B_3t + C_3 . \quad (34)$$

Taking into consideration (34), from (32) it results:

$$B_3 + \frac{B_3}{T_v}t + \frac{C_3}{T_v} = \frac{K_vK_FI_{a3}}{T_vm}t + \frac{v_LK_v}{T_v} . \quad (35)$$

For coefficient identification, constants are determined

$$B_3 = \frac{K_vK_FI_{a3}}{m} \text{ and}$$

$$C_3 = v_LK_v - \frac{K_vK_FI_{a3}T_v}{m} .$$

As $t = 0$, $U_{T3} = U_{T3p} = U_i$,

$$A_3 = U_i + \frac{K_vK_FI_{a3}T_v}{m} - v_LK_v$$

is obtained.

The general solution of the equation (32) is:

$$U_{T3}=(U_i+\frac{K_vK_FI_{a3}T_v}{m}-v_LK_v).e^{-t/T_v}+$$

$$+\frac{K_vK_FI_{a3}}{m}t+v_LK_v-\frac{K_vK_FI_{a3}T_v}{m}.\quad(36)$$

In the braking stage the operation of the system is described by equations (27),(28),(29),(31) and (36).

STOP STAGE

For this stage we write:

$$U_x=U=-KU_{T4},\quad(37)$$

$$I_{a4}=-\frac{KU_{T4}}{K_a}.\quad(38)$$

Having in mind (38),from (4) and (5),we get:

$$\frac{d^2U_T}{dt^2}+\frac{1}{T_v}\frac{dU_T}{dt}+\frac{K_vK_FK}{T_vK_am}U_T=0.\quad(39)$$

The general solution of the equation (39) is:

$$U_{T4}=e^{-t/2T_v}(A_4\sin\omega t+B_4\cos\omega t).\quad(4o)$$

Out of the conditions of continuity and tangency written under the form:

$$\begin{cases}U_{T3}(t_f)=U_{T4}(o),\\ \left.\frac{dU_{T3}}{dt}\right|_{t=t_f}=\left.\frac{dU_{T4}}{dt}\right|_{t=0}\end{cases}\quad(41)$$

it yields:

$$B_4=-\frac{U''_H}{K}\text{ and }A_4=-\frac{2K_FK_vT_vI_{a3}}{m\omega},$$

which lead to

$$U_{T4}=e^{-t/2T_v}(\frac{-2K_FK_vT_vI_{a3}}{m\omega}\sin\omega t-$$

$$-\frac{U''_H}{K}\cos\omega t).\quad(42)$$

if replaced in (4o).
Replacing (42) in (5) we get the speed value:

$$v=\frac{1}{K_v}e^{-t/2T_v}\left[(\frac{U''_H\omega T_v}{K}-\frac{K_FK_vT_vI_{a3}}{m\omega})\sin\omega t-\right.$$

$$\left.-(\frac{U''_H}{2K}+\frac{K_FK_vT_v^2I_{a3}}{m})\cos\omega t\right].\quad(43)$$

Equations (37),(38),(42) and (43) describe the operation of the system in the stop phase.
The form of the equations is a guide to the oscillating periodically damped character of the signals U_{T4},U,I_{a4} and v.

CONCLUSIONS

The equations above deduced allow the determination of the time needed to achieve a positioning cycle and point out the parametres influencing the length of acceleration,braking and stop. This is the basis on which an optimizing of positioning systems with electrodynamic liniar engines is possible.

REFERENCES

Kuo,B.C.,ş.a. (1981) Sisteme de comandă şi reglare incrementală a poziţiei.Ed.Tehnică,Bucureşti.

Mătieş,V., (1986) Dinamica modulelor de translaţie acţionate cu motoare liniare de curent continuu, Lucrările Simpozionului PRASIC-ROBOT'86,Braşov.

Olbrich,O.E.,(1973) Aufbau und Kennwerte elektrodynamischer Linearmotoren als Positionierer fur Plattenspeicher,Feinwerktechnik 77, Heft 4,S 151-157.

On the Stability and Synthesis of Tumbling Vibrating Machines

M. MUNTEANU

Department of Mechanical Engineering, Polytechnic Institute, Bucharest, Romania

Abstract. For some types of tumbling vibrating machines with six degrees of freedom it is necessary to test for stability the mass center rest point. The nonlinear system of differential ecuations for the general case is written and the testing method for stability based on first approximation is outlined. It is found that the machines which have a stiffness symmetry around the central vertical axis of inertia are stable for any design- functional values. The regulating parameter of the vibration generator for prescribed amplitudes and frequency in steady-state is determined. The notion of technological stability is defined and the corresponding criteria for an ordinary type of tumbling vibrating machine are established.

Keywords. Tumbling vibrating machine; vibration generator; dynamic; steady-state vibration characteristics; stability criteria; technological stability

INTRODUCTION

One considers a single mass tumbling vibrating machine driven by vibration generator with unbalanced masses whose rigid working member has six degrees of freedom. The tumbling stationary motion can be harmonic, almost harmonic or periodic. As an example, Fig. 1 shows a tumbling abrasive finishing machine with annular work-bowl, mounted on a set of "n" identical coil springs uniformly distributed on a circle. The working member of some types of tumbling finishing machines, tumbling screens etc. perform small harmonic steady-state vibrations. For such machines, the stationary motion, the paths, velocities and accelerations distribution, the forces transmitted to the foundation and the relative motion of a plate workpiece with sliding friction on the machine working surface have been analysed by Munteanu (1984b 1985) and by Munteanu and Iatan (1985).
In the sequel one analyses the stability based on first approximation and the choice a regulating parameter for some given amplitudes and working frequency in connection with technological stability.

MOTION EQUATIONS. STABILITY

Let us take a fixed coordinate system O'x'y'z' and a moved coordinate system Oxyz connected to the rigid working member whose axes are principal axes of inertia ($J_x=J_1$; $J_y=J_2$; $J_z=J_3$; $J_{xy}=J_{yz}=J_{zx}=0$). The point O being center of mass. The column vector of generalized coordinates is

$$\underline{q}=[\underline{q}'^T\ \underline{q}''^T]^T$$

$$\underline{q}'=[q_1\ q_2\ q_3]^T=[x_0\ y_0\ z_0]^T \qquad (1)$$

$$\underline{q}''=[q_4\ q_5\ q_6]^T=[\varphi\ \psi\ \theta]^T$$

where x_0, y_0, z_0 are coordinates of center of mass and φ, ψ, θ are the Euler's angles specific to vibrations (Krîlov's angles), represented in Fig. 2. The nonlinear differential equations of motion in the case when the resistances in the system can be neglected as it is shown by Munteanu

(1984a) are

$$\begin{bmatrix}[m] & o \\ o & [J]\end{bmatrix}\begin{bmatrix}\ddot{\underline{q}}' \\ \ddot{\underline{q}}''\end{bmatrix} + \begin{bmatrix}o & o \\ o & [B]\end{bmatrix}\begin{bmatrix}\dot{\underline{q}}'^2 \\ \dot{\underline{q}}''^2\end{bmatrix} + \begin{bmatrix}o & o \\ o & [D]\end{bmatrix}\begin{bmatrix}\dot{\underline{q}}'\dot{\underline{q}}' \\ \dot{\underline{q}}''\dot{\underline{q}}''\end{bmatrix} + \sum_i \begin{bmatrix}[k_i] & o \\ o & [\Lambda_i][k_i]\end{bmatrix}\begin{bmatrix}\underline{q}' \\ \underline{q}''\end{bmatrix} +$$

$$+\sum_i \begin{bmatrix}[k_i][\alpha_o]\,\underline{r}^{(i)} \\ [\Lambda_i][k_i][\alpha_o]\,\underline{r}^{(i)}\end{bmatrix} = \begin{bmatrix}[\alpha]\,\underline{F} \\ [\Lambda][\alpha]\,\underline{F}\end{bmatrix} \qquad (2)$$

where $[\alpha] = [\alpha_{nj}]_{n,j=1,2,3}$ $(\alpha_{nj} = \alpha_{nj}(\varphi,\psi,\theta))$ orthogonal matrix of direction cosines and $r_1^{(i)} = x_i$, $r_2^{(i)} = y_i$, $r_3^{(i)} = z_i$ the ith spring point coordinates on moving frame.

$$\dot{\underline{q}}''\dot{\underline{q}}'' = [\dot{q}_4\dot{q}_5 \;\; \dot{q}_5\dot{q}_6 \;\; \dot{q}_6\dot{q}_4]^T$$

$$[k_i] = \begin{bmatrix}k_{ix} & o & o \\ o & k_{iy} & o \\ o & o & k_{iz}\end{bmatrix} \qquad \underline{F} = [F_x \; F_y \; F_z]^T$$

$$[\alpha_o] = [\alpha] - [I_3] \;;\; [J] = [\mu_{ir}]_{i,r=1,2,3} \qquad (3)$$

$$\mu_{ir} = \sum_{k=1}^{3}\sum_{j=1}^{3} J_j \frac{\partial \alpha_{kj}}{\partial \eta_i} \cdot \frac{\partial \alpha_{kj}}{\partial \eta_r}$$

$$\eta_1 = q_3 = \varphi \;;\; \eta_2 = q_5 = \psi \;;\; \eta_3 = q_6 = \theta$$

$$[B] = \left[\frac{\partial \mu_{kj}}{\partial \eta_j}\right]_{k,j=1,2,3} \qquad [D] = [d_{jk}]_{j,k=1,2,3}$$

$$d_{jk} = \frac{1}{2}\left(\frac{\partial \mu_{jp}}{\partial \eta_s} + \frac{\partial \mu_{js}}{\partial \eta_p}\right); \quad \begin{matrix}k=1 \Rightarrow p=1,\; s=2 \\ k=2 \Rightarrow p=2,\; s=3 \\ k=3 \Rightarrow p=3,\; s=1\end{matrix}$$

$$[\Lambda_i] = [\lambda_{jr}^{(i)}]_{j,r=1,2,3} \;;\; \lambda_{jr}^{(i)} = \sum_{p=1}^{3} \frac{\partial \alpha_{rp}}{\partial \eta_j} r_p^{(i)}$$

For the linearized matrix of the direction cosines

$$[\alpha] = \begin{bmatrix}1 & -\theta & \psi \\ \theta & 1 & -\psi \\ -\psi & \varphi & 1\end{bmatrix} \qquad (4)$$

one gets the corresponding linearized differential equations associated to the equations (2)

$$[M]\ddot{\underline{q}} + [K]\underline{q} = \underline{Q} \qquad (5)$$

As it is known, the rest point $\underline{q} = \underline{o}$ of the system (2) is stable in first approximation, if the homogenous linear system associated to (5) has periodic solutions (stable limit cycles). Let us consider the machine from Fig. 1. For all identical coil springs we have

$$k_{ix} = k_{iz} = k_t;\; k_{iy} = k \quad (i=1,2,..n) \qquad (6)$$

Taking into account (3),(4) and (6) the fifth equation of system (5) is

$$J_y\ddot{\psi} + nk_t b^2\psi = o \qquad (7)$$

with the stationary solution $\psi = o$ (for zero initial conditions) and accordingly, the other equations of system (5) are

$$\begin{bmatrix}m & o \\ o & J\end{bmatrix}\begin{bmatrix}\ddot{x}_o \\ \ddot{\theta}\end{bmatrix} + \begin{bmatrix}nk_t & nk_t h \\ nk_t h & nk_t h^2 + nkb^2/2\end{bmatrix}\begin{bmatrix}x_o \\ \theta\end{bmatrix} =$$

$$= \begin{bmatrix}F_{o1}\sin\omega t + F_{o2}\sin(\omega t + \beta) \\ -F_{o1}l_1\sin\omega t + F_{o2}l_2\sin(\omega t + \beta)\end{bmatrix} \qquad (8)$$

$$\begin{bmatrix}m & o \\ o & J\end{bmatrix}\begin{bmatrix}\ddot{z}_o \\ \ddot{\varphi}\end{bmatrix} + \begin{bmatrix}nk_t & -nk_t h \\ -nk_t h & nk_t h^2 + nkb^2/2\end{bmatrix}\begin{bmatrix}z_o \\ \varphi\end{bmatrix} =$$

$$= \begin{bmatrix}F_{o1}\cos\omega t + F_{o2}\cos(\omega t + \beta) \\ F_{o1}l_1\cos\omega t - F_{o2}l_2\cos(\omega t + \beta)\end{bmatrix} \qquad (9)$$

$$m\ddot{y}_o + nk_t y_o = -\varphi[F_{o1}\cos\omega t + F_{o2}\cos(\omega t + \beta)] + \theta[F_{o1}\sin\omega t + F_{o2}\sin(\omega t + \beta)] \qquad (1o)$$

where $F_{o1} = m_{o1}\rho_1$, $F_{o2} = m_{o2}\rho_2$, - statical moments of unbalanced masses.

Since the discriminant of the equation of natural frequencies of systems (9) and (8) can be set up in the positive form

$$(\nu_1^2 - \nu_2^2\varepsilon_2)^2 + \nu_1^4\varepsilon_1^2 + 2\nu_1^4\varepsilon_1 > 0 \qquad (11)$$

where

$$\nu_1^2 = nk_t/m;\; \nu_2^2 = nk/m;\; \varepsilon_1 = h^2/i_g^2; \qquad (12)$$
$$\varepsilon_2 = b^2/2i_g^2;\; \varepsilon_3 = hL/i_g^2;\; i_g^2 = J/m;\; L = l_1 - l_2$$

the machine is stable in first approximation for any design-functional values. It can be seen that in this case the vertical axis of stiffness (Munteanu 1984a) coincide with the principal central axis of inertia and the mass and the stiffness centers lie on this axis. All the machines with these properties lead us to the same conclusions since the stiffness matrices have the same structure. If the two axes do not coincide, one gets a condition relation instead of the positive form (11)

STEADY-STATE VIBRATION CHARACTERIS TICS TECHNOLOGICALLY STABLE

Let us consider again the machine from Fig. 1 for the particular case when $F_{o1} = F_{o2} = m_o\rho$ The stationary solutions of ecuations (8)-(1o) are

$$x_o = R\sin\omega t; \quad \varphi = \phi\cos\omega t;$$
$$z_o = R\cos\omega t; \quad \theta = -\phi\sin\omega t; \qquad (13)$$
$$y_o = -2F_o\phi/nk;$$

Amplitudes expressions are

$$R = \varrho\mu\omega^2 |2(\nu_1^2\varepsilon_2 - \nu_2^2\varepsilon_2 - \omega^2) + \nu_1^2\varepsilon_3| / \Delta \quad (14)$$

$$\phi = \varrho\mu\omega^2 |2\nu_1^2\varepsilon_1 + (\nu_1^2 - \omega^2)\varepsilon_3| / h\Delta \quad (15)$$

$$\Delta = (\nu_1^2\varepsilon_1 + \nu_2^2\varepsilon_2 - \omega^2)(\nu_1^2 - \omega^2) - \nu_1^4\varepsilon_1 \quad (16)$$

$$\mu = m_0/m$$

The statical moment $m_0\varrho$ can be considered a regulating parameter of machine. Appropriate to previous relations we can take $\lambda = \varrho\mu$ as a regulating parameter. In Fig. 3 are ploted the family of curves $R = R(\lambda, \omega)$ for the case

$$p_2 < \omega_R < p_1 \quad (17)$$

where

$$\omega_R^2 = \nu_1^2(\varepsilon_1 + \varepsilon_2) + \nu_2^2\varepsilon_3 \quad (18)$$

is the cancel frequency for R, and

$$p_{1,2} = [\nu_1^2(1+\varepsilon_1) + \nu_2^2\varepsilon_2 \pm \sqrt{[\nu_1^2(1+\varepsilon_1) + \nu_2^2\varepsilon_2]^2 - 4\nu_1^2\nu_2^2\varepsilon_2}\,] / 2 \quad (19)$$

are the natural angular frequencies. If the conditions (17) are not fulfilled, the graphs have a minimum in range $\omega \in (p_1, p_2)$. The family of curves $\phi = \phi(\lambda, \omega)$ has the same shape if the conditions (17) with

$$\omega_\phi^2 = \nu_1^2(2\varepsilon_1/\varepsilon_3 + 1) \quad (2o)$$

are fulfilled. Usually the linear amplitudes, angular amplitudes and working frequency are imposed by technological considerations. Thus, for a given R_0, the exciting frequency can be in the preresonance range ($\omega_1 < p_1$), in the interresonance range ($\omega_2, \omega_3 \in (p_1, p_2)$) or in the postresonace range ($\omega_4 > p_2$) (Fig. 3). Assuming that R_0 and ϕ_0 are given, one obtains the necessary regulating parameter

$$\lambda = R_0\Delta/\omega_1^2[2(\nu_1^2\varepsilon_1 + \nu_2^2\varepsilon_2 - \omega_1^2) + \nu_1^2\varepsilon_3] =$$
$$= \phi_0 h\Delta/\omega_1^2[2\nu_1^2\omega_1 + (\nu_1^2 - \omega_1^2)\varepsilon_3] \quad (21)$$

This means that two amplitudes curves with steady-state points (R_0, ω_1) and (φ_0, ω_1) respectively, have been chosen. From (21) results the condition relationship

$$\gamma_2^2 = p\gamma_1^2 - p \quad (22)$$

with the notations

$$p = 2R_0(2h+L)/\phi_0 b^2; \quad q = 2h(h+L)/b^2$$
$$\gamma_1 = \nu_1/\omega_1 \quad ; \quad \gamma_2 = \nu_2/\omega_2 \quad (23)$$

The two steady-state points (R_0, ω_1) and (φ_0, ω_1) for the determined regulating parameter must fulfilled the technological stability criteria. A vibrating machine is said to be technologically stable if at a certain variation of working frequency the amplitude variation must not exceed an allowable variation. Hence, for a given $\Delta\omega$ of the working frequency ω_1, the condition of technological stability are

$$|\Delta R| \leq |\Delta R a| \quad ; \quad |\Delta\phi| \leq |\Delta\phi_a| \quad (24)$$

Using the derivatives

$$\frac{dR}{d} = R'(\omega); \quad \frac{d\phi}{d\omega} = \phi'(\omega) \quad (25)$$

the conditions (24) can be written as

$$R'(\omega)\Delta\omega\big|_{\omega=\omega_1} \leq \Delta R_a \, ; \, \phi'(\omega)\Delta\omega\big|_{\omega=\omega_1} \leq \Delta\phi_a \quad (26)$$

Setting the derivatives and relationships (21),(22) into (24), the technological stability criteria for amplitudes R_0 and ϕ_0 become

$$M/(NP_1 - S) - 1/U + 1 \leq \delta_R \quad (27)$$

$$M/(NP_2 - S) - 1/V + 1 \leq \delta_\phi \quad (28)$$

where

$$M = \gamma_1^2(1 + \varepsilon_1 + p\varepsilon_2) - q\varepsilon_1 - 2$$
$$N = \gamma_1^2 - 1 \; ; \quad S = \gamma_1^4\varepsilon_1$$
$$P_1 = \gamma_1^2\varepsilon_1(1+p) - q\varepsilon_2 - 1$$
$$P_2 = \gamma_1^2(\varepsilon_1 + \varepsilon_2 p) - q\varepsilon_2 - 1$$
$$U = \gamma_1^2[\varepsilon_2(1+2p) + 2\varepsilon_1] - 2q\varepsilon_2 - 2$$
$$V = 2\gamma_1^2(1+\varepsilon) - 1$$
$$\delta_R = \frac{\Delta Ra}{2R_0}\Big/\frac{\Delta\omega}{\omega_1} \; ; \; \delta_\phi = \frac{\Delta\phi a}{2\phi_0}\Big/\frac{\Delta\omega}{\omega_1} \; ; \; \varepsilon = h/L$$

If $\delta_R = 1$ and $\delta_\phi = 1$, i.e. the percentage of amplitudes variations is twice the percentage of frequency variation, the inequalities (27) and (28) wil take a more simplified form. When $F_{01} \neq F_{02}$; $\beta \neq o$, one can consider three regulating parameters, therefore we can choose the amplitude for the steady-state points from three families of curves with greater possibilities to obtain reasonable criteria of stability.

CONCLUSIONS

The nonlinear differential equations of motion in matrix form (2) is valid for any type of tumbling vibrating machines with a six degrees of freedom working member. It is found that the test for stability is necessary only in the case of dissymmetry of stiffness and/ or dissymmetry of mass.

For the machines which perform small vibrations driven by vibration generators with unbalanced masses, the steady-state vibration characteristics, namely frequency and amplitudes, imposed by technological considerations, can be obtained by the aid of a regulating parameter $m_0\varrho$ given by (21). The requirement of technological stability leads to fulfilling the condition relation (22) and the simultaneous inequalities (27) and (28) for the design - functional values (12).

In discussing the changes in the state of the machine we assume the change of frequency sufficiently slow to allow us to use the amplitude curves, i.e. the relations (14) and (15).

In the paper were treated the case without taking into account the limited capacity of the driving motor. When the moment developed by the motor is introduced into the differential equations, the steady - state vibrations conditions must satisfy the equation of power balance based on the static characteristic of the motor. But the obtained point of dynamic equilibrium must satisfy also some specific stability conditions as it is shown for example in the Munteanu's book (1986).

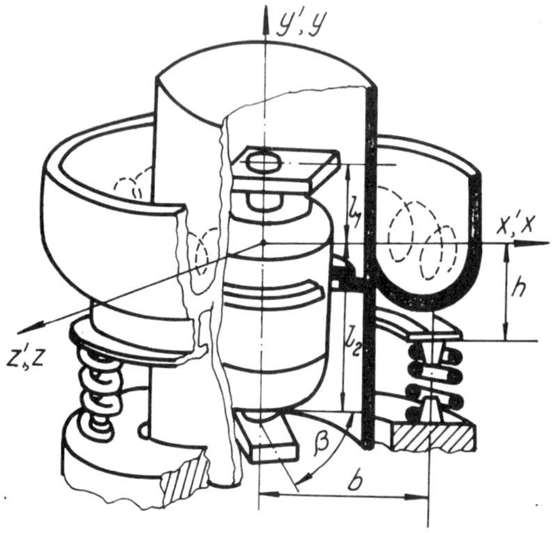

Fig. 1. Tumbling abrasive finishing machine.

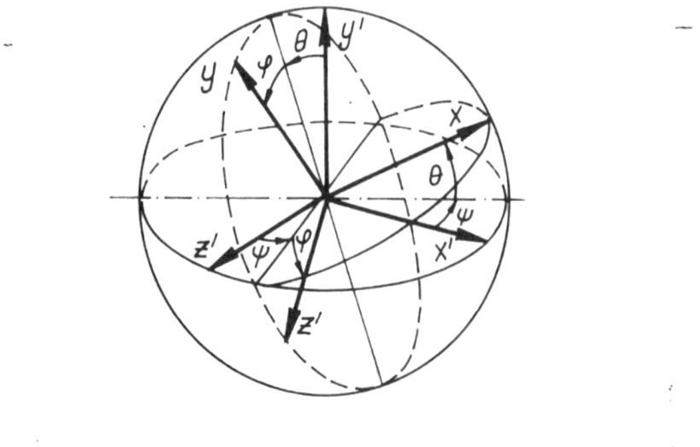

Fig. 2. Euler's angles specific to vibrations

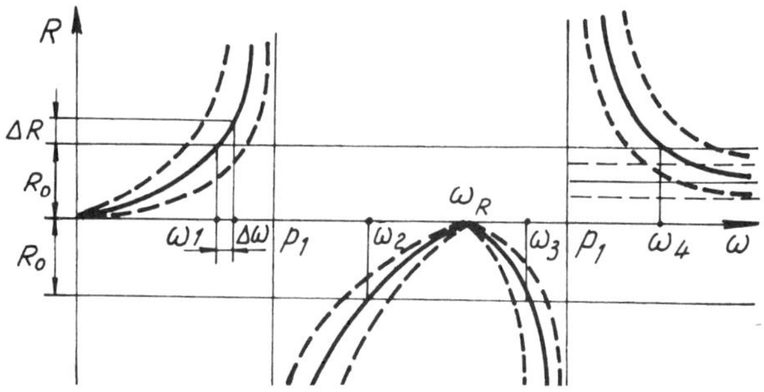

Fig. 3. The amplitude curves $R = R(\lambda, \omega)$.

REFERENCES

Munteanu, M. (1984a). Dynamics of single mass vibrating machines mounted on elastic supports. General case. Bul. Inst. Pol. Bucureşti s. Mecanică XLVII, 95-110

Munteanu, M. (1984b). Dynamics of tumbling vibrating machine driven by vibration generators with unbalanced masses. JSPE -IFTOMM Intern. Symp. Tokyo, Japan

Munteanu, M. and R. Iatan (1985). Dynamics of tumbling abrasive finishing machine with annular work bowl. IFTOMM-SYROM 85 Intern. Symp. Bucharest, Romania, Vol. II-2, 309-316.

Munteanu, M. (1985). Motion with sliding friction of a plate workpiece in a tumbling vibrating machine. IFTOMM-SYROM 85 Intern. Symp. Bucharest, Romania, Vol. II-2, 317-323.

Munteanu, M. (1986). Introducere în dinamica maşinilor vibratoare. Ed. Acad. R. S. Romania, Bucureşti, Cap. VI, 245-268.

SUR LA STABILITE ET LA SYNTHESE DES MACHINES VIBRANTS CHANCELANTES.

Résumé. Pour quelque types des machines vibrants avec le bac de travail rigide qui possède six degrés de liberté, il est souvent necesaire d'examiner la stabilité de la position d'equilibre du centre de gravité. Dans cet travail on a établit le système d'ecuations differentielles non-linéaires pour le cas général et on a indique la méthode concernant l'analyse de la stabilitéen première approximation. On a On a trouve que les machines avec la rigidité symétrique des ressorts par raport à l'axe principale et verticale d'inertie sont bien stables por tout parametres constructifs. Pour les machines vibrants entrainées par generateurs de vibrations a-deux masselotes centrifuges ona determiné un parametre de réglage a condition que les amplitudes et la frequence sont etablies par des considerations technologiques. On a defini la notion de stabilitétechnologique et on a établi les critères correspondants pour une machine vibrant usuelle.

Influence of Dimensional Deviations upon the Position Functions in the Guiding Mechanism of Automotive Front Wheels

I. VIŞA AND P. ALEXANDRU

Department of Machines and Mechanisms, University of Braşov, Romania

Abstract. The space mechanism with link rods guiding the automotive front wheel with contours of the type RSSR - RCS_b^{*} or $RSCS_b$ for the suspension and RSST - SSR or RSSR - SSR for the direction is defined by 25-30 geometrical parameters entailing knowledge of the deviations from the mechanism member draft. Considering that each dimension figure is machined with a deviation from the rated value, in the paper the deviations of major influence are selected from the multitude of deviations and, by deducing the expressions of the position functions to directly include the dimensions figures having deviations, the influences of the deviations on the functions realized by the mechanism are obtained.

Keywords: Automobiles; guidance mechanism; optimization; numerical methods; suspension-steering; vehicles.

DEFINING THE MECHANISM AND POSING THE PROBLEM

Guiding the automotive front wheels is achieved by space mechanisms with link rods comprising the suspension mechanism of the quadrilateral type plus a RSSR-RCS_b shock-absorber or a $RSCS_b$ Mc.Pherson strut (fig.1) and the steering gear of the rack-type (fig.1) or of the central lever type. The mechanism thus obtained is multicontour with a variable construction, functionally adaptable to a great diversity of automotive running conditions. The mechanism is symmetrical to the longitudinal - vertical plane of the motorvehicle, thus allowing for the kinematic study to consider only one half of it with two degrees of mobility, the two independent variables being the stroke L_a of the shock-absorber and the stroke S_c of the gear rack (or the angular stroke of the central lever).

Constant attempts have been made along time for the improvement of these mechanisms. Among the more recently published theoretical works we remark those elaborated by Felzien and Cronin (1981,1983,1985) on the kinematics and optimization of guiding mechanisms with Mc Pherson strut (through the minimization of an objective-function of 7 parameters); Vişa and Alexandru(1981, 1983,1985) on the kinematic analysis of the suspension-steering unit (establishing general analysis relations, synthesis relations for the minimization of the toe angle induced by suspension in automotive steering, the influence of parameters and deviations, optimum values of the main geometrical parameters); Herzog and Kruczynski (1982) on the employment of electronic computing tehniques in kinematic analysis (through a unitary parameter defining diagram), Knapczyk and Kuranowski (1985) on the influence of elasticities

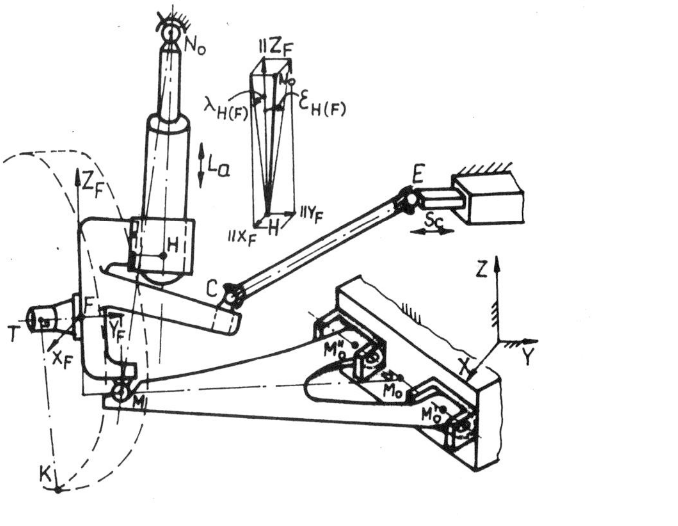

Fig.1. Kinematic-constructional diagram of the guiding mechanism with Mc Pherson strut and gear rack steering.

on the functions of the Mc Pherson mechanism (by considering the link elasticity), Hiller and Woernle (1985) on the analysis of a general guiding mechanism(with the guiding through five points).

We define the mechanism by the geometrical parameters of the component members considered in the reference systems associated with the elements. For the guiding mechanism of the motorcar steering these parameters are (fig.1):

- the positions of the joints M_o',M_o'',N_o,E, given by the coordinates X'_{Mo} , Y'_{Mo} , Z'_{Mo} ; X''_{Mo} , Y''_{Mo} , Z''_{Mo} ; X_{No} , Y_{No} , Z_{No} ; X_E,Z_E, in the central system of the motor-vehicle OXYZ;
- the characteristic lengths of the link

rods 1,4,5 i.e. l_1=MoM; l_4=CE; l_5 = $=2Y_{E(Sc=o)}$;

- the positions of the points M,C,H and T of the stub axle holder 2, given by the coordinates $X_{M(F)}$, $Y_{M(F)}$, $Z_{M(F)}$; $X_{C(F)}$, $Y_{C(F)}$, $Z_{C(F)}$; $X_{H(F)}$, $Y_{H(F)}$, $Z_{H(F)}$; $Y_{T(F)}$, in the system stub axle holder $FX_FY_FZ_F$ (FY_F- the stub axle holder axis, the plane Y_FFZ_F contains the point M);

- the position of the shock-absorber axis given by the angles $\varepsilon H(F)$, and $\lambda H(F)$ in the system $FX_FY_FZ_F$;

- the radius of the front wheel r = Tk.

The right hand half of the guiding mechanism is defined by the same geometrical parameters; the current values may, however, be different owing to dimensional deviations.

We consider that by kinematic synthesis (see the papers mentioned) the optimal nominal values of the geometrical - constructional parameters are obtained. However, due to the dimensional deviations both the static reference (loading) position of the front wheel and the kinematics of the mechanism are altered as to the theoretical case. It is, then, necessary to analyze the influences of the dimensional deviations of the geometrical parameters to determine their influences and to take measures towards a certain machining accuracy and the provision of adjustment devices.

Based on the relations of the kinematic analysis of the wheel guiding mechanism, which are deduced according to the constructional parameters indicated directly on the drafts, the paper analyzes the influence of the dimensional deviations of the constructional parameters in the Mc Pherson strut - rack steering mechanism (fig.1). In the paper the elasticity of links and elements are completely ignored.

POSITION FUNCTIONS IN THE GUIDING MECHANISM

We consider as independent parameters the stroke of the shock-absorber (along the length $L_a = N_oH$) and the stroke the steering rack (S_c) relative to the reference position of the mechanism (rectilinear travel) and as geometrical parameters the values indicated at the preceding point.

The position functions of this complex mechanism that must be known are (fig.2):

a) for rectilinear travel of the motor car:

- variation of the king pin angles $\Delta\alpha(L_a)$ and $\Delta\beta(L_a)$,
- variation of the stub axle angles $\Delta\gamma(L_a)$ and $\Delta\delta(L_a)$,
- variation of the track and base $E(L_a)$, $\Delta A(L_a)$,
- variation of the wheel parallelism $P(L_a)$;

b) for travel of the motorvehicle in curves:

- the turning law $\Theta_e(\Theta_i)$, where $\Theta_e(L_a,S_c)$ and $\Theta_i(L_a,S_c)$-(fig.2),
- the pressure angles in the direction $\beta_{pr}(L_a,S_c)$.

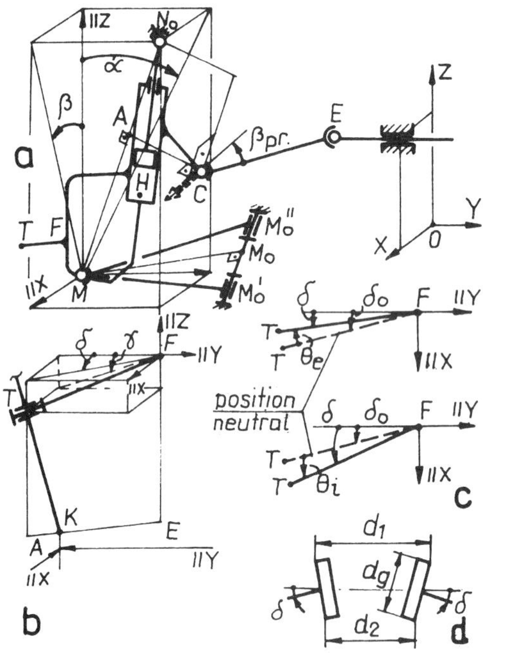

Fig.2. Diagrams and kinematic angles of the Mc Pherson mechanism.

The position functions are established in the OXYZ system attached to the car body (considered fix). In order to obtain these functions the coordinates of a random point P(X,Y,Z) must be determined depending on other three points $P_i(X_i,Y_i,Z_i)$ - i=1,2,3 situated at the known distances $d_i=PP_i$. The computing relations of the coordinates of the point P are obtained by solving the system

$$(X-X_i)^2+(Y-Y_i)^2+(Z-Z_i)^2=d_i^2 \quad (i=1,2,3) \qquad (1)$$

with respect to X,Y,Z, the relation being obtained:

$$X=A_1Z+B_1;\ Y=A_2Z+B_2;\ Z=(-B_3\pm\sqrt{B_3^2-A_3C_3})/A_3, \qquad (2)$$

where

$$A_1=-\frac{1}{C}\begin{vmatrix} Z_2-Z_1 & Y_2-Y_1 \\ Z_3-Z_1 & Y_3-Y_1 \end{vmatrix};\ A_2=-\frac{1}{C}\begin{vmatrix} X_2-X_1 & Z_2-Z_1 \\ X_3-X_1 & Z_3-Z_1 \end{vmatrix};$$

$$B_1=-\frac{1}{C}\begin{vmatrix} E_2-E_1 & Y_2-Y_1 \\ E_3-E_1 & Y_3-Y_1 \end{vmatrix};\ B_2=-\frac{1}{C}\begin{vmatrix} X_2-X_1 & E_2-E_1 \\ X_3-X_1 & E_3-E_1 \end{vmatrix};$$

$$C=\begin{vmatrix} X_2-X_1 & Y_2-Y_1 \\ X_3-X_1 & Y_3-Y_1 \end{vmatrix};\ E_i=\frac{1}{2}(d_i^2-X_i^2-Y_i^2-Z_i^2)\ i=1,2,3;$$

$$A_3=A_1^2+A_2^2+1;\ B_3=A_1B_1+A_2B_2-X_1A_1-Y_1A_2-Z_1;$$

$$C_3=B_1^2+B_2^2-2(X_1B_1+Y_1B_2)-2E_1 .$$

By employing relations (1-3), the current coordinates of the points M(as depending on Mó,Mő,No), C(as depending on M,E,No), H(as depending on M,C,No), F and T(as depending on M,C,H) are determined.

Based on the current coordinates of the points specified above the following functions are determined.

The pivot angles and their variation (fig.2,a):

$$\alpha = \operatorname{arctg}\frac{Y_{No}-Y_M}{Z_{No}-Z_M}\ ;\ \beta = \operatorname{arctg}\frac{X_{No}-X_M}{Z_{No}-Z_M};\quad (4)$$

$$\Delta\alpha = \alpha - \alpha_o\ ;\ \Delta\beta = \beta - \beta_o,$$

where α_o and β_o are reference values computed for the automotive load state(static position).

The stub axle angles and their variation (fig.2,b):

$$\gamma = \operatorname{arctg}\frac{Z_F-Z_T}{Y_F-Y_T}\ ;\ \delta = \operatorname{arctg}\frac{X_T-X_F}{Y_F-Y_T};\quad (5)$$

$$\Delta\gamma = \gamma - \gamma_o\ ;\ \Delta\delta = \delta - \delta_o,$$

where γ_o and δ_o are the values computed for the load position.

The wheel parallelism (fig.2,d) given by

$$P = d_1 - d_2 = 2d_g \sin\delta. \quad (6)$$

The track and axle base variations (fig.3):

$$Y_K = Y_T + r\sin\gamma\cos\delta;\ \gamma' = \operatorname{arctg}(\operatorname{tg}\gamma\cos\delta);$$

$$\Delta E = Y_k - (Y_k)_o; \quad (7)$$

$$X_k = X_T - r\sin\gamma'\sin\delta\ ;\ \Delta A = X_k - (X_k)_o.$$

The pressure angle in the link C (fig.2,a)

$$\beta_{pr} = \arccos\left(\operatorname{sgn}X_{C(F)}\cdot\frac{V_x(x_C-x_E)+V_y(y_C-y_E)+V_z(z_C-z_E)}{l_4\sqrt{V_x^2+V_y^2+V_z^2}}\right),$$

where

$$V_X = \begin{vmatrix} Y_C-Y_A & Z_C-Z_A \\ Y_{No}-Y_M & Z_{No}-Z_M \end{vmatrix};\ V_Y = -\begin{vmatrix} X_C-X_A & Z_C-Z_A \\ X_{No}-X_M & Z_{No}-Z_M \end{vmatrix};$$

$$V_Z = \begin{vmatrix} X_C-X_A & Y_C-Y_A \\ X_{No}-X_M & Y_{No}-Y_M \end{vmatrix}. \quad (9)$$

The toe angle in turn and the turning law (fig.2,c) - turn to the right:

$$\Theta_e = \delta_o - \delta,\ \ \Theta_i = \delta - \delta_o \quad (10)$$

with δ given by (5) but with the coordinate Y_a - independent variable

$Y_E = -(l_5/2 \pm \operatorname{sgn} X_{C(F)}\cdot S_C)$, the "+" sign for Θ_e and the "-" sign for Θ_i.

SELECTION AND INFLUENCE OF DIMENSIONAL DEVIATIONS

The geometrical parameters defining the mechanism fall into two categories: geometrical parameters of the attachment points of the guiding mechanism to the body and the characteristic geometrical parameters of the mechanism elements (stub axle holder, suspension lever, direction tie rod, rack, shock-absorber).

As the dimensional deviations of the attachment points to the body tend to be much greater than the dimensional deviations of the characteristic parameters of the elements, in what follows only the influence of the former will be referred to.

It should be specified that in the range of position dimensional deviations of the points Mó, Mő and No we may include also the deformations due to the elasticity of the corresponding links (the body attachment links being as a rule much more elastic than the other links of the mechanism). To determine the influences of the dimensional deviations upon the position functions qualitatively and quantitatively, a general computer program was elaborated in BASIC, by means of which both the individual influences upon the dimensional deviations and combinations of them were analyzed.

In studying the influences of dimensional deviations kinematically optimized mechanisms were resorted to, in which the nominal values of the geometrical parameters of the points Mó, Mő, No and E were successively altered or combined. Thus, the influences of the deviations were deduced, a part of the conclusions being andvanced below.

The caster angle (β) of the king pin is influenced by the dimensional deviations Z'_{Mo}, Z''_{Mo}, X_{Mo} şi X_{No} (fig.3 a,b).

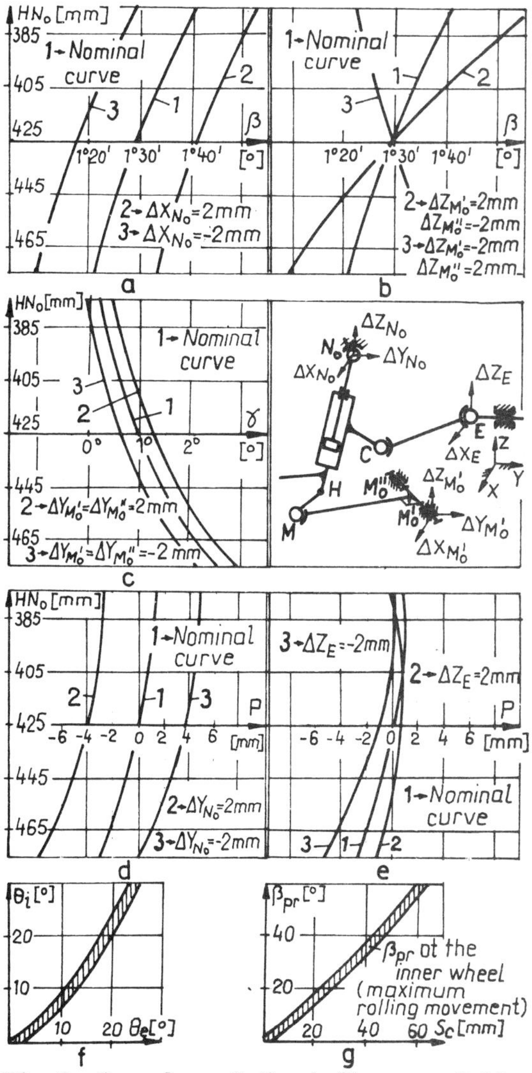

Fig.3. Examples of the influence of dimensional deviations on the position functions.

For the dimensional deviations ΔX_{Mo} and

X_{No} the deviation curves are parallel to the original curve; thus, the static value of the caster angle being influenced. The dimensional deviations $\Delta Z'_{Mo}$ and $\Delta Z''_{Mo}$ modify the mechanism kinematics. The variations $\Delta Z'_{Mo}$ and $\Delta Z''_{Mo}$ of contrary signs amplify the change in carter angle. The most unfavourable case is that in which the differences $\Delta Z'_{Mo} - \Delta Z''_{Mo}$ and $\Delta X_{Mo} - \Delta X_{No}$ have maximum (absolute) values of contrary sign.

The wheel camber angle (γ) is influenced by the dimensional deviations ΔY_{Mo} and ΔY_{No}, the deviation curves being displaced parallel to the original curve (fig.3,c). The dimensional deviations ΔY_{Mo} and ΔY_{No} of different signs increase the change in camber angle.

The wheel parallelism (P) is extremely sensitive to dimensional deviations. Thus, the dimensional deviations $\Delta X'_{Mo}$, $\Delta Y'_{Mo}$, $\Delta X''_{Mo}$, $\Delta Y''_{Mo}$, ΔX_{No}, ΔY_{No}, ΔX_E influence (more or less) the wheel parallelism, while the curves with deviations remain approximately parallel to the original curve.(fig.3,d). However, the dimensional deviations ΔZ_{Mo} and especially ΔZ_E modify the mechanism kinematics (fig.3,e). By combining the dimensional deviations the variation in parallelism may be entirely changed as compared to the original curve.

The track and the axle base are affected by the deviations insignificantly.

The turning laws $\theta_i(\theta_e)$ for the suspension positions during loading, choke, detent and maximum rolling movement, are influenced by the dimensional deviations of the coordinates of the points MQ , Mo , No and E in the same way as the wheel parallelism is influenced. A still greater influence is exercised by the variations $\Delta Y'_{Mo}$, $\Delta Y''_{Mo}$ and ΔY_{No} of the same sign. Under the influence of dimensional deviations the range of the turning laws is widened.(fig.3,f).

Generally, to diminish the influences of dimensional deviations upon parallelism variations (and implicitly upon the turning laws) adjustment devices for the dimension figures Y_E (or l_4) and Z_E are introduced into the mechanism.

The pressure angle (β_{pr}) is influenced in the same way as the wheel toe angles (fig.3,g).

The conclusions here deduced are valid for all guiding mechanisms with Mc Pherson strut and steering rack; the degree of the influences also depends (to a certain extent) upon the nominal values of the geometrical-constructional parameters of the mechanism.

BIBLIOGRAPHY

Alexandru,P. and Vişa,I (1981). Valori optime ale parametrilor geometrici ai mecanismelor de direcţie. Prep. International Symposium IFToMM, Bucureşti, Romania, IV, 5-14.

Alexandru,P. (1983), Allgemeine Gleichungen der Übertragungsgestze der Fahrzeuglenkgetriebe. Proc. of the Sixth Congrres on TMM, New Dehli, India, 369-372.

Cronin,D.L. (1981). Mc Pherson Strut Kinematics. Mechanism and Machine Theory, 16, 631-644.

Felzien,G. and Cronin,D.L. (1985). Steering Error Optimization of the Mc Pherson Strut Automotive Front Suspension. Mechanism and Machine Theory, 20, 17-26.

Herzog,G. and Kruczynschi,F. (1982). EDV - Einsatz ber der Entwicklung von PKW. Kraftfahrzeugtechnik, 4, 108-110.

Hiller,M. and Woernle,Ch.(1985). Bewegungsanalyse einer Fünfpunkt-Radaufhängung. ATZ , 87, 59-64.

Knapczyk,I. and Kuranowski,A. (1985). Analysis of large deflections in Mc Pherson Type Automotive Suspension Spatial Mechanisms. Prep. International Symposium IFToMM, Bucureşti, România, 95-104.

Vişa,I. (1983). Kinematik synthesis of Steering linkages employed in Mc Pherson Suspension type Automobiles. Proc. of the Sixth Congress on TMM, New Dehli, India, 373-376.

Vişa,I. and Alexandru,P. (1985). Influenţele cinematice ale abaterilor dimensionale din mecanismele spaţiale RSSTSSR de ghidare a roţilor directoare ale autoturismelor. Prep. Internaţional Symposium IFToMM, Bucureşti, Romania, 269-280.

Vişa,I. and Alexandru,P. (1985). Einfluss der Massabweichungen im Führungsgetriebe mit Mc Pherson Federbein auf die statische Lage und die Kinematik des gelenkten Kraftfahrzeugrades, Prep. Symposium CONAT, Braşov, România.

L'INFLUENCE DES DÉVIATIONS DIMENSIONNELLES SUR LES FONCTIONS DE POSITION DU MÉCANISME DE GUIDAGE DES ROUES DES VOITURES DE TOURISME

- Résumé -

Le guidage des roues directrices des véhicules de tourisme est réalisé par des mécanismes spatiaux à barres articulées, de structure variable, formés d'un mécanisme de suspension pour chaque roue (de type RSSR au $RSCS_b$) et un mécanisme de direction (de type RSSTSSR on RSSRSSR), qui possède de 4 à 6 contours, pour l'étude-à cause de sa symétrie - étant suffisants 2-3 contours et deux degrés de mobilité: la course de l'amortisseur et la course de la cremaillère. Le nombré des paramètres géometriques qui définissent un pareil mécanisme est d'approximativement 30. Dans l'ouvrage, on déduit les expressions analitiques des fonctions de position de paramètres constructifs du mécanisme, on sélectionne les déviations dimensionnelles, en établissant l'influence des déviations principales sur la position statique de la roue et sur les fonctions de position. Ayant à base les conclusions générales déduites dans le travail, le constructeur est en mesure, dans un cas concret de disposer la réalisation de certaines précisions de construction ou d'introduire des systèmes de réglage qui donnent la possibilité de la compensation de l'effet global des déviations.

*) R - rotation, S - sphere, C - cylinder, S_b - bolt-sphere.

Hydrocentrifugal System for the Regulation of Aeroturbines

A GONZALEZ MARTIN* AND S. BRESO BOLINSHES**

**E.U.I.T.I. de Zamora*
***E.T.S.I.I. de Valladolid*

Abstract. The passive regultaion of the aeroturbines with variable pitch, requires the arrangement of a centrifugal mechanism of inertia, which operates equilibrated, has only one degree of freedom that is, produces the same turning on each blade, only responds by velocity, is damped and performs in a uniform way during a cycle. This can be fulfilled by mechanical systems, though these are very sensible to dis-equilibration and malfunction due to wear and out-of-adjustment, as well as complicated and expensive.

Due to the previous points, we want to analyse the performance of a hydrocentrifugal system that satisfies the described requirements, increases the simplicity and reliability of the same and possibly reduces the costs.

Keywords. Regulation theory; hydraulics systems; turbines; velocity control; fluidics devices.

INTRODUCTION

DISTRIBUTION LAW OF PRESSURE IN A FLUID TURNING WITH HORIZONTAL AXIS

The cylinder in Fig. 1 with horizontal axis normal to the paper, turns full of fluid with angular velocity ω= cte. All the fluid points turn with equal ω angular velocity, there being thus no sliding between the different cylindrical layers.

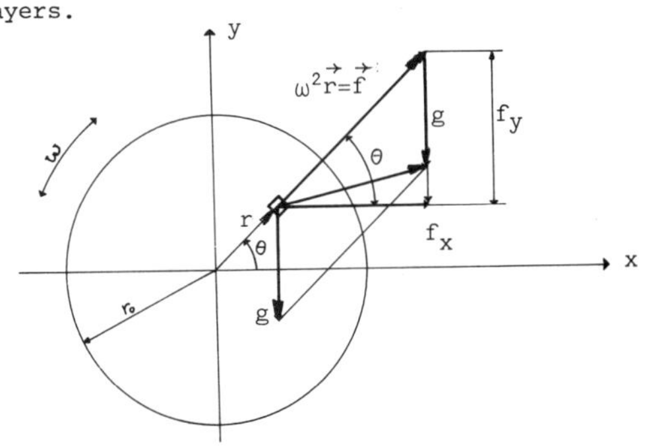

Fig. 1. Diagram of accelerations.

The general equation for the forces balance of a fluid element, exposed to the acceleration of the gravity and to the inertial field $\vec{f}$, regardless of the gravitation one, can be defined:

$$\rho\vec{f} - \vec{\nabla}p^* = 0; \quad \rho = \text{cte.}; \quad p^* = p(x,y,z) + \rho g y$$

$\vec{f} = f(x,y,z)\vec{i} + f(x,y,z)\vec{j} + f(x,y,z)\vec{k}$; in this case the situation being identical in z , the equilibration can be studied in one single plane, xy;

$p^* = p(x,y)$; $\vec{f} = \vec{f}(x,y)$:

$$\text{of } \rho\vec{f} - \vec{\nabla}p^* \begin{cases} \rho f_x - \dfrac{\partial}{\partial x}(p + \rho g y) = 0; \ \rho f_x - \dfrac{\partial p}{\partial x} = 0 \\ \rho f_y - \dfrac{\partial}{\partial y}(p + \rho g y) = 0; \ \rho f_y - \dfrac{\partial p}{\partial y} - \rho g = 0 \end{cases}$$

$$f_x = \omega^2 r\cos\theta \qquad \frac{\partial p}{x} = \rho\omega^2 r\cos\theta$$

$$f_y = \omega^2 r\,\text{sen}\theta \quad \rho\omega^2 r\,\text{sen}\theta - \frac{\partial p}{\partial y} - \rho g = 0; \ \frac{\partial p}{\partial y} = \rho\omega^2 r\,\text{sen}\theta - \rho g$$

The total differential of the pressure is consequently:

$$dp = \frac{\partial p}{\partial x}dx + \frac{\partial p}{\partial y}dy = \rho\omega^2 r\cos\theta \ dx + (\rho\omega^2 r\,\text{sen}\theta - \rho g)dy$$

$x = r\cos\theta$; $dx = \cos\theta\, dr - r\,\text{sen}\theta d\theta$

$y = r\,\text{sen}\theta$; $dy = \text{sen}\,\theta\, dr + r\cos\theta d\theta$

substituting in the total diff.:

$$dp = \rho\omega^2 r\cos\theta\,(\cos\theta dr - r\,\text{sen}\theta d\theta) + (\rho\omega^2 r\,\text{sen}\theta - \rho g)$$
$$(\text{sen}\theta\, dr + r\cos\theta d\theta) = \rho\omega^2 r\cos^2\theta dr - \rho\omega^2 r^2\text{sen}\theta\cos\theta\, d\theta +$$
$$(\rho\omega^2 r\,\text{sen}^2\theta - \rho g\,\text{sen}\theta)dr + (\rho\omega^2 r^2\text{sen}\theta\cos\theta - \rho g r\cos\theta).d\theta =$$
$$= (\rho\omega^2 r\cos^2\theta + \rho\omega^2 r\,\text{sen}^2\theta - \rho g\,\text{sen}\theta)dr + (\rho\omega^2 r^2\text{sen}\theta\cos\theta -$$
$$-\rho\omega^2 r^2\text{sen}\theta\cos\theta - \rho g r\cos\theta)d\theta = (\rho\omega^2 r - \rho g\,\text{sen}\theta)dr - \rho g r\cos\theta d\theta$$

$$dp = (\rho\omega^2 r - \rho g\,\text{sen}\theta)dr - \rho g r\cos\theta d\theta \qquad (1)$$

By doing r = cte. and integrating:

$p = -\rho g r\,\text{sen}\theta + c(r)$; $c(r)$ arbitrary function of r or as well with θ = cte.

$p = \rho\omega^2\dfrac{r^2}{2} - \rho g\,\text{sen}\theta.r + c(\theta)$ " " " θ

Differentiating these two functions, we will get respectively:

$dp = (-\rho g\,\text{sen}\theta + c'(r))\,dr - \rho g r\cos\theta d\theta$

$dp = (\rho\omega^2 r - \rho g\,\text{sen}\theta)dr + (-\rho g r\cos\theta + c'(\theta))\,d\theta$

Identifying anyone of these with (1)

$-\rho g\,\text{sen}\theta + c'(r) = \rho\omega^2 r - \rho g\,\text{sen}\theta \quad c'(r) = \rho\omega^2 r$

Integrating: $c(r) = \dfrac{\rho\omega^2 r}{2}$ + cte

or rather identifying the second form:

$-\rho g r\,\text{sen}\theta + c'(\theta) = -\rho g r\cos\theta c'(\theta) = 0$
$c(\theta)$=cte.

of anyone of the forms:

$$p = \frac{\rho\omega^2 r^2}{2} - \rho g r\,\text{sen}\theta + \text{cte}$$

The determination of the constant depends on the initial conditions. If as in the case of a cylinder, a halted cylinder, it is filled up with a fluid and sealed, having at the inlet a normal atmospheric pressure, manometric pressure at zero, we will find that for

$$\omega = 0;\ \frac{\pi}{2} = \theta\ ;\ r = r_0 \}\quad P = 0 \}\ cte = gr_0$$

$$p = \frac{\rho\omega^2\ r^2}{2} - \rho gr\ sen\theta + \rho gr_0 \qquad (2)$$

Isobar surfaces

The isobar surfaces, which are non-circular cylinders, were reduced to plane lines due to the identity of axial conditions.
On these p=cte.

$$cte = \frac{\rho\omega^2 r^2}{2} - \rho gr\ \ sen\ \theta \quad + cte.,\ \text{being the}$$

$$\text{density}\quad \rho = cte\ ;\ \frac{\omega^2\ r^2}{2} - \ gr\ sen\theta + C = 0$$

One particular isobar for $r = r_0$; $\theta = \frac{\Pi}{2}$

$$\frac{\omega^2\ r_0^2}{2} - gr_0 = -C;\ \frac{\omega^2 r^2}{2} - grsen\theta - \frac{\omega^2 r_0^2}{2} + gr_0 = 0$$

Even though the pressure of that particular line has to be determined through the initial conditions, i.e. through (2).

According to the previous theory, the mechanism we want to design would now be inmersed in the created scalar field of pressures, this mechanism is composed by two inertial masses which are two pistons, with double effect, which will be explained later on. To start and in order to advance the theoretic development, we can suppose that these pistons housed in their corresponding thinwalled cylinders (infinitesmal thickness), hydraulically connected as indicated in figure 2 and perfectly rigid and with absence of friction wetween piston and cylinder.

Connected by the tubes a and b and the whole unit being subject to the internal and external action and being both the tubes and the cylinders thinwalled, these are not subject to mechanical tension due to the pressures.

Excluding the external fluid, the internal distribution of pressures continues the same, thus subjecting the cylinders to the corresponding effort.

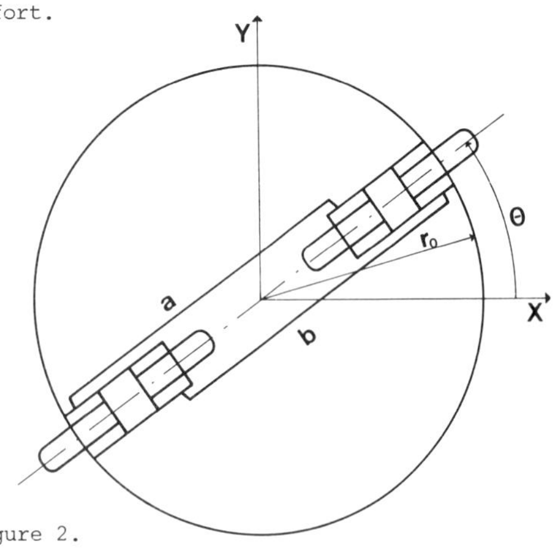

Figure 2.

As to the conception of the mechanism, regarding the positioning of the airscrew blades of the turbines, we will find that the whole unit connected to the hub through the cylinders will rotate, the movement of the pistons activating the blades.

ANALYSIS OF THE FORCES

Once the mechanism being conceived, it is necessary to define the forces on the pistons and cylinders as well as the reactions in the hub, as ω =cte. and relative velocity piston-cylinder, very slight or really null.

Forces in the piston:

a) Owing to the fluid
b) Centrifugal
c) Gravitational

a) The calculation is enormously simplified assuming the approximate hyphotesis that the piston and the rods are situated for each θ, on the corresponding $\underline{\underline{r}}$, being the circular position.

Under these circumstances, the resultant of the pressure actions on the lateral cylinder surfaces is null, being subject to normal effort, piston axis-way, the two plane faces of the circular crown situated on r_1 and r_2.

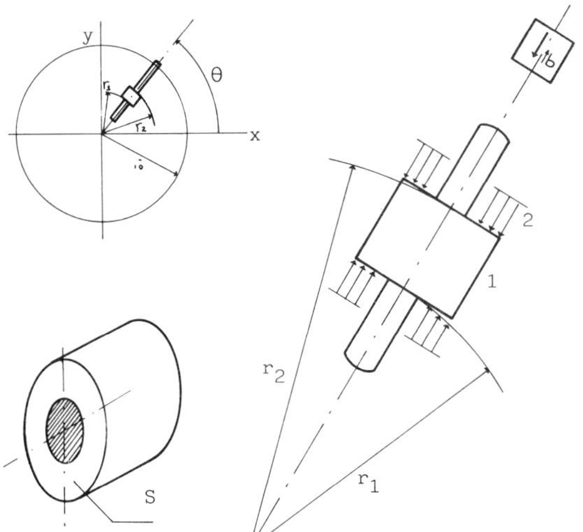

Figure 3.

Pressure force in 1, (figure 3), with initial equation conditions (2); (The initial conditions will, to a start, not influence the final result, if there is no cavitation).

$$\vec{F}_1 = P_1(-\vec{S}) = -\vec{S}\ (\rho\frac{\omega^2\ r_1^2}{2} - \rho gr_1\ sen\theta + \rho gr_0)$$

Force in 2

$$\vec{F} = P_2\vec{S} = \vec{S}\ (\frac{\rho\omega^2 r_2}{2} - \rho gr_2 sen\theta + \rho gr_0)$$

$$\vec{F}_p = \vec{F}_1 + \vec{F}_2 = s\{\rho\frac{\omega^2}{2}\ (r_2^2 - r_1^2) - \rho gsen\ (r_2 - r_1)\}\vec{\sigma}$$

where $\vec{\sigma}$ is a unitary superficial vector in centripetal direction.

$$\vec{F}_p = S\{\rho\frac{\omega^2}{2}\ (r_1 + r_2) - \rho gsen\theta\}(r_2 - r_1)\ \vec{\sigma}$$

$$\vec{F}_p = S\{\rho\omega^2\frac{r_1 + r_2}{2} - \rho g\ sen\theta\}\ (r_2 - r_1)\ \vec{\sigma}$$

* * *

b) Centrifugal: we take any cylindrical form:

$$dF = dm\omega^2 r = \rho\omega^2 r\ dV$$

$$dF_x = \rho\omega^2\ r\ cos\theta dV = \rho\omega^2 x\ dV$$

$$dF_y = \rho\omega^2\ y\ dV$$

Working point of the centrifugal resultant:

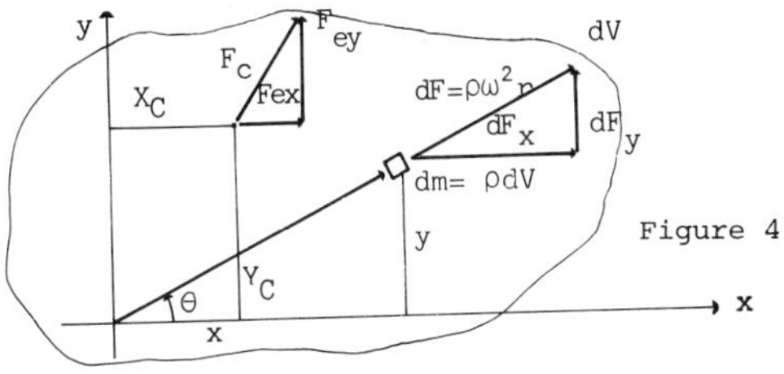

Figure 4.

That is (X_c, Y_e), and F_e result.

Taking moments regarding the centre point:

$F_{ex}\, Y_c = \int_V \rho\omega^2 xy\, dV;\; F_{ey}\, X_c = \int_V \rho\omega^2 xy\, dV$

But $F_{ex} = \int_V \rho\omega^2 x dV\;;\; F_{ey} = \int_V \rho\omega^2 y\, dV$

$\int_V x\, dV$ is the static moment regarding $\underline{\underline{y}}$

$\int_V y\, dV$ " " " " $\underline{\underline{x}}$

and knowing that $\int_V xd = M_T \cdot X_G, \int V\, ydv = M_T\, Y_G$

H_T total mass (area), (X_G, Y_G), gravitation centre $\int_V xy\, dV = I_{xy}$ polar moment of inertia. Hence

$$X_c = \frac{I_{xy}}{M_T\, Y_G}\;;\; Y_c = \frac{I_{xy}}{M_T\, X_G}$$

* * * *

We subject our case to the same simplified hypothesis as in that of the pressure forces, supposing "thin" rod in three sections.

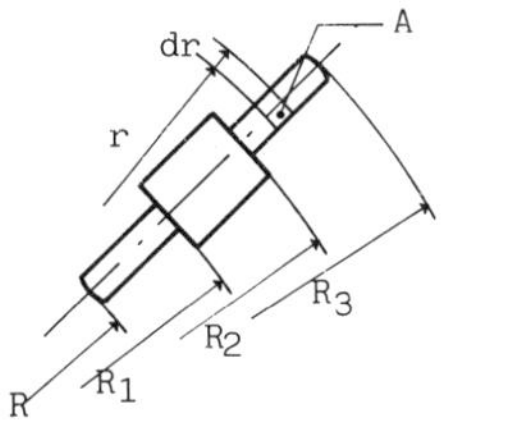

Figure 5.

For one rod

$$dF_c = \rho A dr\, \omega^2 r$$

$$F_c = \int_{R2}^{R_3} \rho A\omega^2 r dr = \rho A\omega^2 \frac{R_3^2 - R_2^2}{2} = \rho A\omega^2 (R_3 - R_2)\, \frac{R_3 + R_2}{2} =$$

$$= M_T\, r_m\, \omega^2$$

$F_c = M_T\, r_m \omega^2;\; r_m$, medium rad.

The piston behaves as a total mass applicated in r_m, if it has transversal symetry, being

$$r_m = \frac{R_3 + R}{2}$$

* * * *

c) Gravitational: the forces appointed up to now have radial direction. In this direction, the gravitational action:

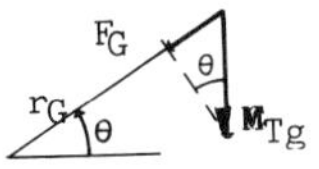

$$F_G = M_T\; g\text{sen}\; \theta$$

TOTAL FORCE EXERCISED ON THE PISTON

We refer to the piston of axis in and to the sense of the unitary vector previously defined:

$$\vec{F}_T = \vec{F}_p + \vec{F}_c + \vec{F}_G = S(\rho\omega^2 \frac{r_1 + r_2}{2} - \rho\; g\text{sen}\,\theta)\; (r_2 - r_1)\,\vec{\sigma} -$$

$$-M_T\, \omega^2 r_m\, \vec{\sigma} + M_T\; g\; \text{sen}\theta\vec{\sigma}$$

$$\vec{F}_T = \{\rho S\; (\omega^2 \frac{r_1 + r_2}{2} - g\text{sen}\theta)\; (r_2 - r_1) - M_T(\omega^2 r_m - g\; \text{sen}\; \theta)\}\vec{\sigma}$$

But: $\rho S(r_2 - r_1) = m_a$ is the fluid mass displaced by the cylindrical ring.

$\frac{r_1 + r_2}{2} = r_{ma}$ medium radius of the ring

Under these conditions:

$$\vec{F}_T = m_a(\omega^2 r_{ma} - g\text{sen}\theta) - M_T(\omega^2 r_m - g\text{sen}\theta)\}\vec{\sigma}$$

In case of the piston of mass M_T having transversal symetry: $r_{ma} = r_m$

$$\vec{F}_T = \{(m_a - M_T)\omega^2 r_m - (m_a - M_T)\; g\; \text{sen}\theta\;\}\;\vec{\sigma}$$

The same force expressed in centrifugal sense, i.e. $\{-\vec{\sigma}\};\; -\vec{F}_T = \vec{F}_{TF}$

$$\vec{F}_{TF} = \{(M_T - m_a)\omega^2 r_m - (M_T - m_a)g\; \text{sen}\theta\}\,(-\vec{\sigma})$$

The mechanism formed that way behaves in the piston in the same way as a conventional mechanism in "dry", with the difference that the mass now is represented by $(M_T - m_a)$

* * * *

The force in the opposite centrifugal cylinder, now according to $\vec{\sigma}$, will be:

$$\vec{F}_{TF\ \text{opposite}} = \{(H_T - m_a)\omega^2 r_m + (M_T - m_a)g\; \text{sen}\theta\}\;\vec{\sigma}$$

Stiffly locked from the hub, are transmitted to this one:

$$\vec{F}_{TF} + \vec{F}_{TF\ \text{opposite}} = 2(M_T - m_a)g\; \text{sen}\theta.\; \vec{\sigma}$$

* * * *

<u>Forces in the cylinder</u>:

a) Owing to the fluid
b) Centrifugal
c) Gravitational

a) Pressure forces: only those reamin that exercise on the circular rings which are the bottoms:

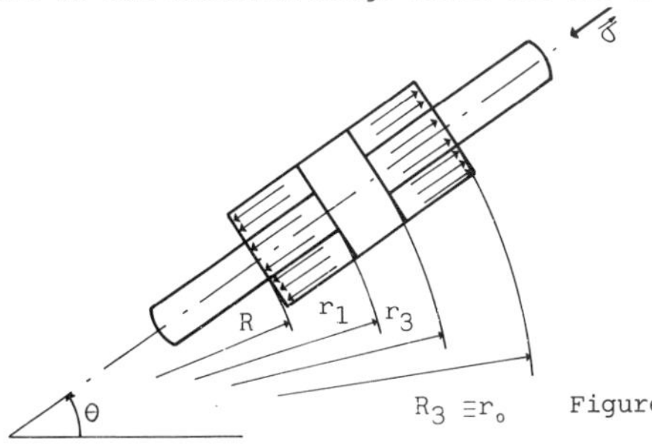

Figure 6.

$$\vec{F}_3 = S(\frac{\rho\omega^2\, R_3^2}{2} - \rho g R_3 \text{sen}\theta + \rho g r_0)\;(-\vec{\sigma})$$

$$\vec{F} = S(\frac{\rho\omega^2\, R^2}{2} - \rho g R \text{sen}\theta + \rho g r_0)\;\vec{\sigma}$$

Adding the two forces in $\{-\vec{\sigma}\}$:

$$\vec{F}_{pc} = S\{\rho\omega^2 \frac{R_3^2 - R^2}{2} - \rho g(R_3 - R)\; \text{sen}\;\theta\}\;(-\vec{\sigma})$$

$$\vec{F}_{pc} = \rho S(R_3 - R)\;(\omega^2 r_m - g\; \text{sen}\theta)\;(-\vec{\sigma});\rho S(R_3 - R) m_{at},$$

Fluid mass "dislodged" by the complete cylindrical crown.

$$\vec{F}_{pc} = m_{at}\;(\omega^2 r_m - g\; \text{sen}\theta)\;(-\vec{\sigma})$$

$$\vec{F}_{pc\ \text{opposite}} = m_{at}\;(\omega^2 r_m + g\; \text{sen}\theta)\;\vec{\sigma}$$

The force transmitted to the hub by the locked cylinders and in radial sense is:

$$\vec{F}_{pc} = 2\; m_{at}\; g\; \text{sen}\; \theta.\; \vec{\sigma}$$

Total force transmitted to the hub in radial sense, by cylinder and piston:

$$\vec{F}_r = 2(M_T - m_a + m_{at})\ g\ \text{sen}\ \theta.\ \vec{\sigma}$$

$m_{at} - m_a$ = mf fluid mass; $\vec{F}_r = 2(M_T+mf)\ g\ \text{sen}\theta\ .\vec{\sigma}$

b) The centrifugal one of the cylinder is compensated by the opposite one.

c) As well the gravitational force (the cylinder´s own weight is considered an integrated part of the hub).

* * * *

Nevertheless, we have not contemplated the tangential forces transmitted to the hub by the mechanism through the cylinders:

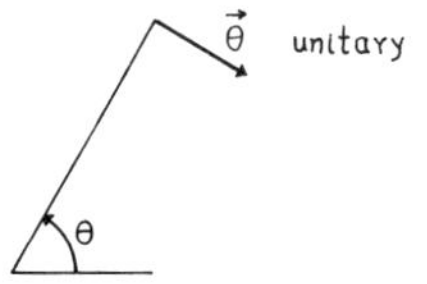

1.- Gravitational of the piston

$$\vec{F}_{1\theta} = 2\ M_{Tg}\ \cos\ \theta.\vec{\sigma}$$

2.- The tangential action of the fluid that would appear integrates the pressure forces on the cylinder surface, is not considered because of the hypothesis regarding the infinite thinness. The force in is only:

$$\vec{F}_{2\theta} = 2mfg\ \cos\ \theta\ .\ \vec{\theta}$$

$$\vec{F} = 2(M_T + mf)\ g\ \cos\ \theta\ .\ \vec{\theta}$$

The remaining force as in previous hypothesis, is now transmitted through the corresponding faces of the cylinders in the pressurized part as seen in the figures 7 and 8.

THE FORCES TRANSMITTED BY THE PISTONS ARE EQUAL

In order to prove this theory, we pass on to the following models

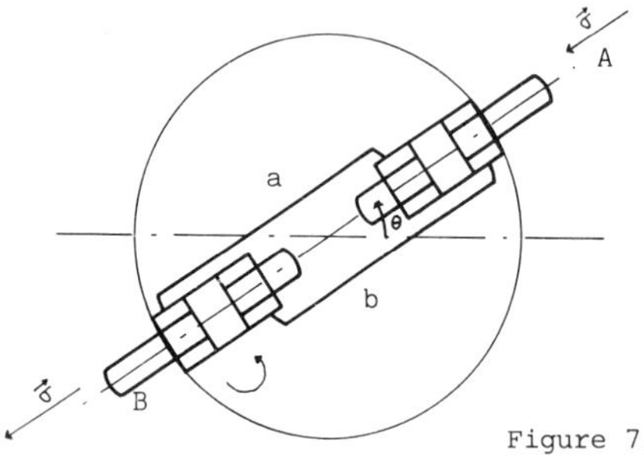

Figure 7.

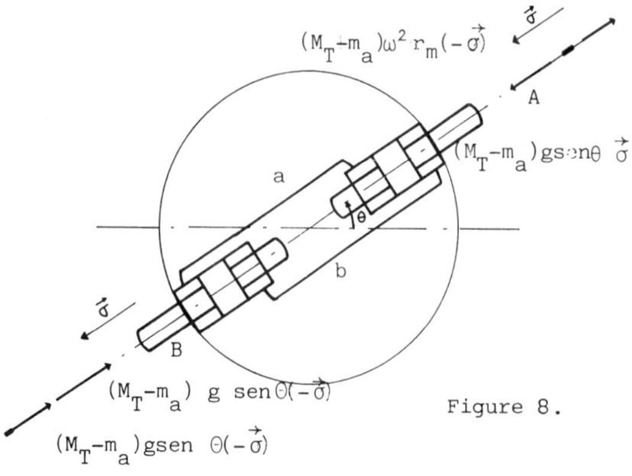

Figure 8.

By turning the cylinder B 180°, Fig. 7, we obtain the equivalent set up as in Fig. 8, where the variable forces due to their own weights (M_T - m_a) are cancelled, remaining a centrifugal one, equal in each piston and this is due to the pressurization of the liquid connected by a; referring to the b one, the hyperstatism requires the fluid stiffness to be very superior to the springs, see fig. 9, in order to confirm previous statements. Anyhow, it is preferrable to study the performace bearing in mind the pressure in a and the volumetric elasticity module of the fluid.

In this case we would not get a critical result if the spring (or springs of the blades) were exactly alike, as anyhow we assured the equality of displacement as in the model of figure 9:

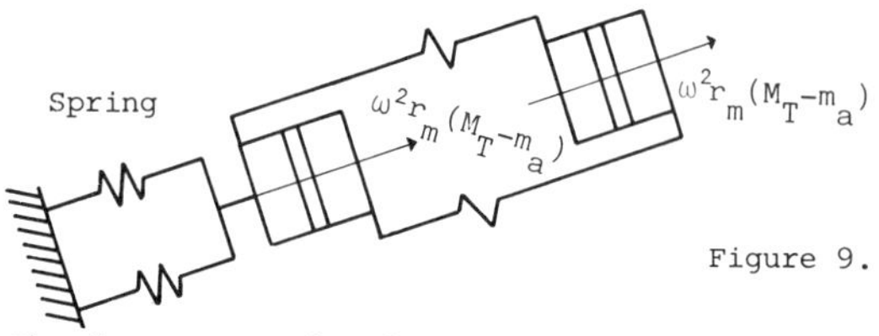

Figure 9.

The force transmitted to the hub is:

$$\vec{F}_r + \vec{F}_\theta = 2(M_T + mf)\ g\ (\text{sen}\theta\ \vec{\sigma} + \cos\theta\ \vec{\theta}\ )$$

but sen$\theta\ \vec{\sigma}$ + cos$\theta\ \vec{\theta}$ is a vertical unitary vector and in a downwards sense, constantly transmitting the weight of the complete system, which absorbed by the bearing, under these circumstances, is turning equilibrated.

RÉSUMÉ

Nous avons voulu développer un système de régulation dont la seule entrée se fait au moyen de la vitesse. C'est un système hydrocentrifuge sur lequel nous avons voulu tester d'une manière analytique l'uniformité de fonctionnement tout au long d'un cycle quand la vitesse de rotation est constante à l'interieur de la turbine, son équilibre et son uniformité de déplacement pour chaque battoir. Pour cela on a obtenu l'échelle de pressions et d'inerties de même que ses actions sur la totalité des parties du mecanisme, l'ayant considéré valide et d'exécution facile. De l'étude ci-dessus réalisée, se dégage de même l'absence de cavitation, et de sa disposition la possibilité d'être amertie a volonté a travers l'étranglement de connexions hydrauliques.

Crack Growth Modelling under Variable-Amplitude Loading

F. ROMANO* AND S. BRESÓ**

*Departamento de Ingeniería Mecánica y de Materiales, Univ. Politécnica, Valencia, Spain
**Instituto Metal-mecánico, Comunidad Valenciana, Spain

Abstract. A mathematical model have been made to predict the crack growth under simple variable-amplitude cyclic loading. The model is based upon the effective Stress Intensity Factor amplitude and crack clousure concepts proposed by Elber (1971). But differently to others models using these concepts, that are based on empirical relationships, this analyses each element behaviour in the crack tip neigbourhood. The Stress Intensity Factor in Unloading is obtained by calculating the residual deformations left at the crack lips by the permanent deformations produced in the plastic zone ahead the crack tip. By analysing the cyclic behaviour of the material ahead the crack tip we calculate the cyclic deterioration variation for constant-amplitude loading and correct the crack growth velocity. This crack growth velocity is calculated by one of the growth law using the Stress Intensity Factor amplitude (Paris, 1962;Forman 1967). To take into account the secuential effects, the "cycle by cycle" calculation method is used. The model describe which is occuring realy in the crack and include the most relevant aspect experimentaly observed on the crack growth in these conditions.

Key words. Crack propagation; Fatigue; Stress; Plastic strain; ciclic loads.

INTRODUCTION

The most crack growth prediction models under variable-amplitude loading (Wheeler, 1971; Willemborg ,1971) are based upon the study of the plastic zone developped ahead of the crack and upon the influence that applied varying loads have on it. This influence is generaly modelliced in an empirical manner, experimentaly finding the model parameters. These fitted parameters have validity only for loading secuences similar to the used for its finding (Romano,1985).

The model we propose avoid this limitation by modelling each element behaviour in the two zones that this behaviour has influence on crack growth velocity.

One of the zones analysed is the plastic zone ahead of the crack tip in wich we characterize the material cyclic behaviour and the plastic deformations created there. The other is the contact zone inmediatly behind the crack tip, in which, because of the residual deformations, the crack clousure ocurs and contact stresses are generated that determine the effective Stress Intensity Factor (SIF) amplitude that produce the crack growth.

PLASTIC ZONE

Size and shape

Although a generaliced agreement on plastic deformation plastic zone size and shape do not exist, the topic has been widely studied from theoretical and experimental view points. The discrepancy on plastic zone shape is not important between theoretical studies and experimental results, and we can say that both monotonic and ciclic zones have lengthened shape under plane strain conditions and more circular under plane stress.

The monotonic zone size in plane strain will be taken between 1.5 and 2 times lower than in plane stress conditions when we mesure it in a crack perpendicular direction. The relation between crack longitudinal and perpendicular direction sizes is aproximately 0.25 for plane strain and 0.8 for plane stress conditions.

The cyclic zone size is between four (Rice, 1967) and ten (Davidson, 1976) times lower than monotonic plastic zone size. The plastic zone size and shape for plane strain and plane stress conditions can be seen in Fig. 1.

Strain distribution

A theoretical strain distribution law in the plastic zone could not been stablished for loading mode I, but Rice (1969) have resolved the problem for mode III loading in plane strain condition and its results used for mode I. According to this solution, the strain distribution is a power law of the distance to the crack tip (r), which exponent is $-1/(1+n)$, where n is the material strain hardening exponent.

Davidson (1976) found the strain distribution experimentaly on the especimen surface (plane stress), beeing the power law exponent of r equal to $-1/2$ in the cyclic zone, in which the strains have more importance.

Cyclic behaviour

The cyclic stress-strain behaviour of a material element placed in the plastic zone can be obtined from the material cyclic stress-strain curve with a good aproximation, according to some authors (Laird, 1977). The cyclic stress-strain curve and a kinematic hardening law, because of the 'Bauschinger effect', will allow as to know the cyclic behaviour of each element while the crack tip approaches it, that is to say, while r decreases in the strain distribution.

The kinematic hardening law choiced is the Mroz (1967) roule. The calculation in each applied load variation will be:

$\epsilon = f(r, SIF)$ Strain distribution curve

$\sigma^{*} = 2 \cdot f(\epsilon^{*}/2)$ Cyclic stress-strain curve and Mroz roule.

beeing $\sigma^{*} = \sigma - \sigma_b$ and $\epsilon^{*} = \epsilon - \epsilon_b$.

The cyclic evolution of an element thus considered can be seen in Fig. 2 for constant-amplitude load cicles and in Fig. 3 when an overload is applied.

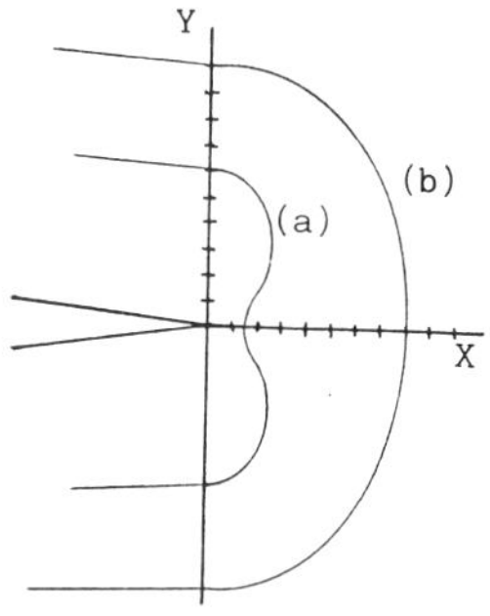

Fig. 1. Plastic zone size and shape for plane strain (a) and plane stress (b) conditions.

Cyclic damage of the element considered

The cycles number of a material element live subjet to plastic and elastic strains can be calculated by the Coffin-Manson (1974) ecuation. Defining the ciclic damage as the inverse of the cicles number that decrease the element live, elimining the damage part produced by elastic strains, because these elastic strains are very little compared with the plastic, and knowing the plastic strain history of an element we can calculate the acumulate damage in every position relative to the crack tip, in particular at the moment that the crack tip reaches its position.

Finding the damage of each element at the moment of its fracture for constant-amplitude loads and knowing the plastic strains history of each element when the load amplitude varies, we will know the growth resistance variation in the material elements affected.

The Fig. 4 show the plastic strain history of an element for constant-amplitude loading and when an overload has been applied. Can be appreciated that the larger plastic strains and therefore the larger damage will be produced when the element considered is placed in the cyclic zone.

The Fig. 5 show how the acumulate damage vary (The growth resistance will vary in the opposite manner) for the material elements affected by the overload while the crack tip reaches them.

Permanent deformations in the plastic zone

The permanent deformations produced at the crack tip are left in the crack lips as residuals deformations. Therefore, it will be necessary to calculate these deformations along the plastic zone and in particular thats produced nearest the crack tip.

The permanent deformations are calculated by integrating the plastic strains along the plastic zone in a perpendicular direction to the longitudinal crack axe. To simplify the calculations we will suppose eliptic shape for the plastic zone and that the strain distribution is maintained in every radial direction from the crack tip.

The permanent deformation obtained is shown in Fig. 6, where can also see the decrease in plastic deformation produced in the plastic zone by the load descent.

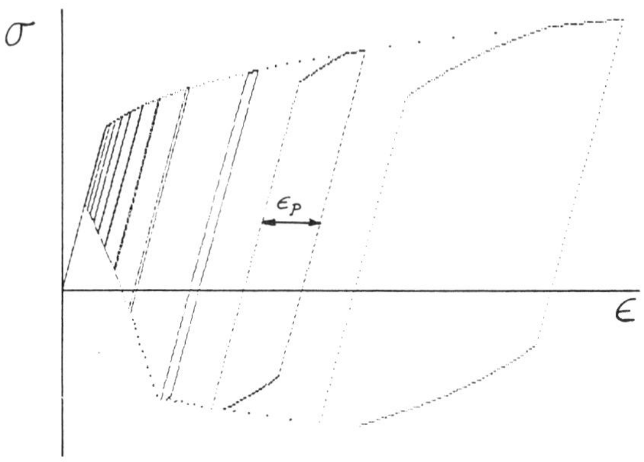

Fig. 2. Cyclic behaviour under constant-amplitude loading.

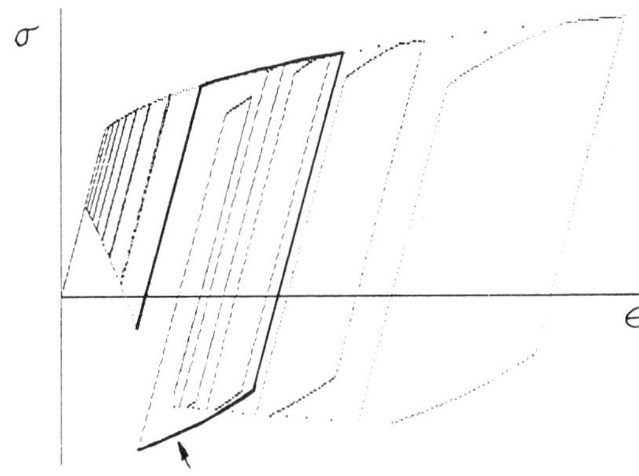

Fig. 3. Cyclic behaviour under overload (↑).

CONTACT ZONE

A new residual deformation is generated by the crack length increment in each load cicle, give by the permanent deformation existing in this plastic zone interval. To decrease the information cuantity to keep, we will compute a new residual deformation distribution when the variaton is significative. When the applied loads vary, it will be reflected in the residual deformations with a magnitude corresponding to the permanent deformation existing in the plastic zone when the crack length increment is produced.

Crack lips theoretical desplacement and interference

The theoretical desplacement of the crack lips at the maximun load is calculated by adding the elastic Westergaard (1939) solution, including the plastic zone in the crack length, and the restriction by the plastic zone (Rice, 1967). In the same manner the desplacement in the load descent will be calculated, taking an account the cyclic zone. The unloading desplacement is obtined by difference.

Residual deformations, before calculated, are present at the crack lips. If these deformations are larger than crack lips desplacement in unloading, a interference zone will exist that will produce the contac stresses neccesaries to the strains take in the interference. These strains remain in the elastic range and, therefore, the contact stresses will be proportional to the interference.

Stress Intensity Factor in unloading

The SIF in unloading will be given by adding the SIF for the minimun load and the SIF for contact stresses at the crack lips. This last SIF will be calculated by applying the SIF expresion for stresses applied on the crack lips given by Paris (1965) and integrating it along the contact zone. Have been checked experimentaly that the overload effect finish when the crack have traversed the plastic zone created by it, therefore, we will take the stresses produced in this zone as the most significants.

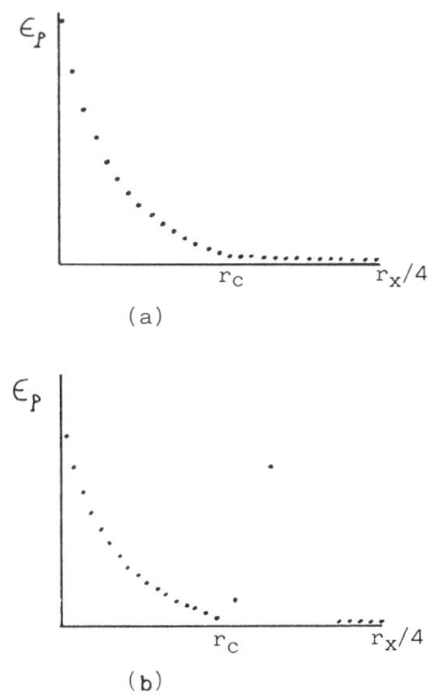

Fig. 4. Plastic strain history of an element for constant-amplitude loading (a) and for an overload (b).

PLANE STRAIN-STRESS TRANSITION

The plane strain or stress condition depends on the relation between the especimen thick and the maximun SIF applied. Microscopicaly, the growth is associate to parallel planes to the crack for plane strain condition and inclined 45° for plane stress. For intermediate conditions the growth is mixed mode. Settling Von Euw (1972) experimental results, percentage of the one or other mode versus the maximun SIF and especimen thick, we can infer the relative influence of both conditions.

For a given especimen thick, while the crack growth and the maximun SIF increase for constant-amplitude load, the condition experience a transition from plane strain to plane stress. The plastic zones change its shape and increase its size according to Fig. 1. The permanent and residual deformations do not vary so much as can be expect relative to the SIF applied because the strain distribution exponent decreases.

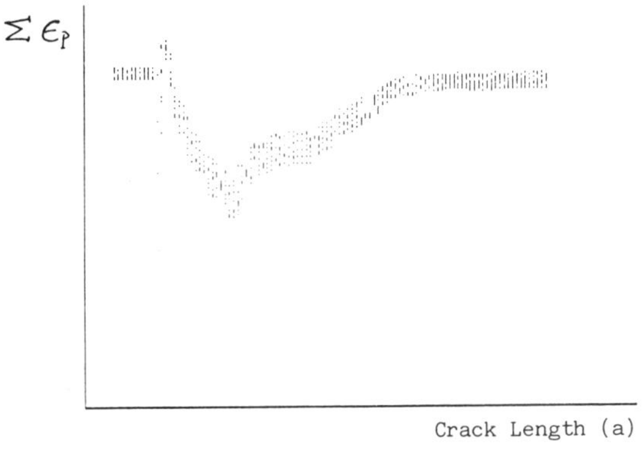

Fig. 5. Acumulate deterioration variation for the elements placed in the overload plastic zone.

MODEL

The crack growth is calculate as a exponential law of the SIF amplitude multiplyed by a correction factor given by the ciclic deterioration. The "cycle-by-cycle" method is used to take an account the secuential effects of each load cycle. For the initial crack length, a residual deformation is supposed, obtined with cyclic loads equal to the first applied cycle, by which the initial residual SIF and stress-strain estate of the elements ahead the crack tip are obtained. For each load cicle the crack growth, residual SIF, cyclic behaviour (Plastic zones size and deterioration factor), permanent deformations and up to dated residual deformation distributions are calculated. Proceeding in a consecutive manner to the programmed cicles, final or critic crack length.

When the crack growth under constant-amplitude loading is simulated, the effective SIF amplitude values obtained are similar to the given by Elber (1971) as a function of the ciclic asymetry (R). The cyclic plastic zone size relative to the monotonic is in the 0.1 - 0.2 range and depends on the cyclic asymetry, being lower for plane stress than for plane strain with constant R.

By varying the load cycle amplitude, the relative size between both zones vary by the variable effective SIF amplitude. The growth velocity is affected by the variations in effecctive SIF amplitude and by the deterioration factor when variable-amplitude loads are applied (Fig. 7).

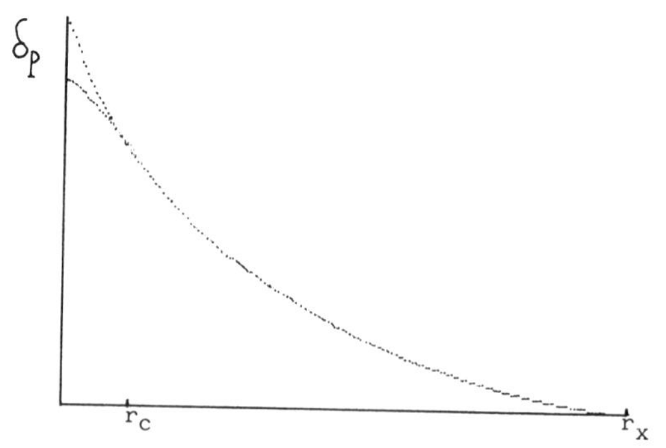

Fig. 6. Permanent deformation in the plastic zone.

CONCLUSION

A model for crack growth prediction is made. This model has an account the interaction effects produced by simple spectra loading. The model analize the material behaviour in the crack tip neigbourhood secuentialy and for each cycle, reproducing cualitative and cuantitatively the phenomena observed experimentaly in especimens subjets to this type of loading. The model describe what is occuring realy in the crack because do not uses empirical relationships.

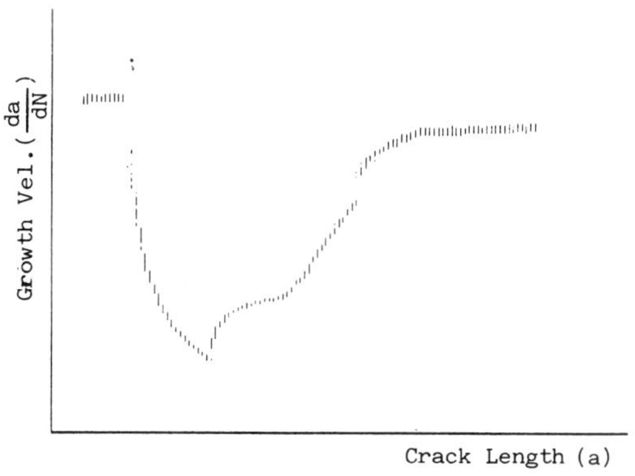

Fig. 7. Crack growth velocity decrease by overload application.

REFERENCES

Davidson D.L. and Lankford J.,Plastic strain distribution at the tips of fatigue cracks,J.Eng.Mat. and Thechnologie,1976,24-29.
Elber W.,The significance of fatigue crack clousure ,ASTM STP 486,1971,230-242.
Forman R.G.,Kearney V.E. and Engle R.M.,Numerical analysis of crack propagation in cyclic loaded structures,ASME Trans.J.B.Eng.,1967,p.459.
Lairz C., ASTM STP 637, 1977, 3-35
Mroz Z.,J.Mech.Phys. of Solids nº 15,1967,163-175.
Paris P.C.,The growth of fatigue cracks due to variations in load,Ph.D.Thesis,Lehigh Univ.,1962.
Paris P.C. and sih G.C.,Stress Analysis of cracks, ASTM STP 381,1965,30-85.
Rice J.R.,The mechanism of crack tip deformation and extension by fatigue,ASTM STP 415,1967.
Romano F.,Breso S. and Alvarez de Ron V.,Aplicabilidad general de los parametros en los modelos de retardo,Anales de Ing.Mec. vol-2,1984.
Von Euw E.F.,Hertberg R.W. and Roberts R.,Delay effects in fatigue crack propagation,ASTM STP 513,1972,230-259.
Westergaard H.,J.of Applied Mech.vol-6,1939,p.a49
Wheeler O.E.,Spectrum loading and crack growth, ASME publ.,1971.
Willemborg J.,Engle R.M. and Wood H.A.,A crack growth retardation model using an effective stress concept,AFFDL-TM-71-1-FBR,1971.

RESUME

On a fait un model matematique pour prédire le croissance des crevasses produits pour charges cycliques avec amplitude variable. Ce model utilise les conceptes de Facteur de Intensité de Tension (SIF) effectif et crevasse fermeture. Mais a différence des otres modeles qui utilisent ces conceptes, que sont foundamentés sur relations empiriques, celui ci analyse la cunduite de chaque élèment du materiel au tours de l'extrême de la crevasse. Le zone analysé est divisé dans deux zones; la zone devant l'extrême et la zone du contac derrière.

Dans la premiérè zone, en modelant la conduite cyclique du matèriel, nous calculons le acroissement du deformation plastique et le détériore cyclique et ses variations quand varient les charges appliqués. Nous calculons aussi la deformatión permanente crée dans la zone plastique que restera sur les bords du la crevasse. Cettes deformations résiduals sont utilisées pour rencontrer le SIF en la descharge. Comme les deformations residuals varient avec les variations de les charges, aussi variera l'amplitude du le SIF.

Pour avoir en cuente les effectes secuencieles on utilise le métode du "cycle-by-cycle" pour le calcul de la vitesse du croissance. Cette vitesse on calcule comme une fonction exponencialle du l'amplitude du le SIF affecté pour le facteur du correction pour le détériore cyclique. Le model décrie ça qui se passe royalement en la crevasse et inclue les aspectes plus relevants que ont eté observé en la croissance des crevasses in cettes conditions.

Instabilities in Calculations of Fatigue Lifetime of Machine Elements by the Local Strain Approach

A. N. ROBLES AND J. G. MARTINEZ

Departamento de Ingenieria Mecánica, E.T.S.I. Industriales de Sevilla, Avda. Reina Mercedes s/n, 41011 Sevilla, Spain

Abstract. Some errors in calculations of plastic strains and stresses at notches, appearing in certain algorithms of wide spread use in making fatigue lifetime predictions of machine elements are analysed. Numerical results referred to a simple load record are evaluated and discussed.

Keywords. Fatigue. Strain. Stress. Cyclic Behaviour.

INTRODUCCTION

The basic assumption in calculating fatigue lifetime of machine elements using the Local Strain Approach is the fact that the crack initiation period in fatigue depends primarily on the edge strains at notches. Therefore the knowledge of the local cyclic stress-strain history at notches is of paramount importance.

In order to perform lifetime predictions, computer programs are developed which simulate this local behaviour handling with the nominal loads on the compoponent, its geometry (usually represented by the stress concentration factor K_t), and the law of the material. The model should be capable of reproducing the hysteresis loops and memory effect displayed by metals when undergoing irregular loads.

Certain procedures used to perform this simulation are analysed in the following pages.

Some rules have been established to predict the stress-strain behaviour based on experimental observations in plain axial test specimens (1,3). They apply to uniaxial stress states and, with some restraints, to three dimensional states (4,5).

The trasients due to cyclic hardehardening or softening is not considered normally, employing the stable stress-strain curve of the material in the calculations. It is usually chosen for such a curve an expression as

$$\varepsilon = \frac{\sigma}{E} + \left(\frac{\sigma}{k'}\right)^{1/n'} \qquad (1)$$

and for the hysteresis loop the same one but twiced

$$\frac{\Delta\varepsilon}{2} = \frac{\Delta\sigma}{2E} + \left(\frac{\Delta\sigma}{2k'}\right)^{1/n'} \qquad (2)$$

This is a close approximation for most of engineering materials (6,7). However, there are some metals in which this approximation is not valid. For example, in gray cast iron, due to its microstructure, the tension and compression behaviour are quite different one from the other, affecting the hysteresis loop shape. There are also certain annealed austenitic stainless steels in which the loops' shape changes markedly with the strain level (8,10).

LOCAL STRESS-STRAIN BEHAVIOUR UNDER IRREGULAR LOADS

In order to obtain the local stress -strain behaviour history at a notch, the cyclic behaviour curve or hysteresis loop is used in the following way (beginning either at a peak or valley in the load sequence)

(1) The curve is always begun in the elastic range.

(2) The curve is followed until a certain control test is satisfied, being this condition one of the proposed rules for calculating plastic strains at notches: Neuber, Hardrath-Ohman, Molki-Glinka, etc. (11,14).

(3) If any of the previous peaks or valleys were reached before the control condition is satisfied, an hyteresis loop is closed, and then the material "remembers" the stress-strain path it was following when interrupted by the strain reversal.

This memory effect is showed in Fig.1. During the excursion 2-3, when the strain reaches 1 again, then the stress-strain path beyond this point is the same as if the excursion 1-2-1 had not occurred (15).

The matter we are dealing with in these pages is whether this change in the strain-stress path must involve another one in the control condition or not.

In particular, consider the Neuber rule, which is one of the most frequently used on account of its simplicity. This relationship, originally derived for sharp notches in shear, essentially states that the theoretical elastic stress concentration factor is equal to the geometric mean of the actual stress and strain concentration factors, i.e.

$$K_t^2 = K_\sigma K_\varepsilon \qquad (3)$$

,which in the case of elastic nominal stresses becomes

$$\Delta\varepsilon\,\Delta\sigma = K_t^2 \frac{\Delta S^2}{E} \qquad (4)$$

This control condition is represented by an equilateral hyperbola in axes whose origin is the same as the one for the behaviour curve. The desired peak falls at the point where both curves intersect, as can be seen in Figs. 2 and 3, referred to the first and second peaks of the load record depicted in Fig. 1.

Two different alternatives arise at point 1 when moving from 2 to 3 (See Fig. 4):

Option (A): The control condition

$$\Delta\sigma\Delta\varepsilon = K_t^2 \Delta S_3^2 / E \qquad (5)$$

can be traced in axes centered at peak number 2.

Option (B): It may be considered, instead, the condition

$$\Delta\sigma\Delta\varepsilon = K_t^2 (\Delta S_3 - \Delta S_2 + \Delta S_1)^2 / E \qquad (6)$$

with origin in O. This means that not only the strain-stress path is changed due to the memory effect, but, rather than that, the criterion's axes are changed as well.

In general, both options lead to different results, since the same curve intersects two different hyperbolae. In option (A) the hyperbola is centered in point 2, while in case (B) it is centered in O.

One of the two options must be chosen in order to make lifetime predictions. Furthermore, to select between them, a stability criterion must be satisfied: given a load record consisting of an arbitrary wave which repeats itself continuously, the ε - σ output should be an hysteresis loop which must be followed periodically, without any peak displacement taking place. That means that given a stable input a stable output must be obtained. It should be noted that transient behaviour has been ruled out since the steady state cyclic curve is used.

There have been developed, and are much used, some methods (16,17) which simulate the cyclic stress-strain curve by means of rectilinear elements with an availability coefficient. They are very efficient in producing the path's change in the ε - σ curve, being able of reproducing these changes without need of keeping memory about all the preceding reversals, just handling the element's availability coefficient at each moment, which is a clear improvement concerning memory storage and time requirement when implemented in the computer.

These methods, however, lose their major advantage (i.e., no need to keep in memory any peak or valley and no need to check for closure of any hysteresis loop), if it is necessary to change the origin of the axes of the control condition to peaks or valleys belonging to non-closed loops, and which should, therefore have been kept in memory. This is to say that the use of such methods which employ the availability coefficient technique only makes sense when Option (A) above is implemented.

However, as it is shown in the next section, Option (A) does not meet with the stability requirement previously discussed: calculations based on (A) result in a shift of the hysteresis loops, giving rise to an effect similar to a "graphical ratcheting".

CALCULATIONS

Both Options (A) and (B) have been implemented in computer programs in order to bring out the differences between them.

Let us consider a MAN TEN steel whose properties are shown in Table 1, where the first set of constants refer to Equation 2 and the second one refer to the ε-N curve (Strain Amplitude - Number of Cycles to Failure), defined by

$$\varepsilon_a = \frac{\sigma_f'}{E} (2N)^b + \varepsilon_f' (2N)^c \qquad (7)$$

and let us take an hypothetical notched element with an elastic stress concentration factor $K_t = 3$.

A simple load record, where clear differences between both options arise, has been chosen to enlighten the problem: a constant amplitude load where the valley - peak excursion is interrupted by a small descent (See Fig. 5).

In Option (B), the stress-strain response is a stable output (Fig. 6), where the respective peaks come together cycle after cycle. Therefore the memory effect is simulated in a correct way with this option.

On the contrary, in Option (A), where no changes are introduced in the control condition, the output turns out to be unstable, i.e., a displacement in the respective peaks is found (Fig. 7 and 8).

The magnitude of this displacement and, hence, the simulation error incurred ,depends heavily on the relative position of the interruption cycle and its amplitude.

The nearer the breaking peak comes to the maximun curvature region (around the cyclic yield point), the bigger the errors become.

These displacements may be positive or negative.

Numerical results obtained in different examples are shown in Tables 2 to 4. (Varying "a" and "b", see Fig. 5. In all the cases $S_{max} = -S_{min} = 300.$ Mpa).

It can be seen that the errors grow as the amplitude becomes smaller due to the particular shape of the control condition (equilateral hyperbola). In the limit case, when the behaviour law became horizontal and the amplitude of the interruption ("a") came close to zero, the error would become infinite.

This fact could account for the improvement in the results obtained in the traditional methods when small cycles are eliminated.

LIFE PREDICTIONS

Notch strain-stress output instabilities arising as Option (A) is used, logically take place as well in lifetime predictions, usually made by using the ε-N curve of the material.

It is found that life predictions about the load record considered here, when based on Option (A), depends on the number of cycles comprising the input blocks, which is physically absurd and causes these predictions to be completely arbitrary.

On the contrary, when Option (B) is applied, this arbitrariness disappears.

In Fig. 9, it is shown the result obtained with the records corresponding to a= 30. Mpa and b= 15. and 250. Mpa (The stress - strain results are given in Fig. 7 and 8). The damage parameter proposed by Smith-Topper-Watson (S.T.W.) (20) and the Palmgren-Miner (21,22) linear rule of damage accumulation have been employed in performing the calculations.

The amount of damage due to the small interruption depends on its location. In agreement with the S.T.W. parameter, the higher the maximun stress of the interruption cycle (the point at which it starts in the record considered here), the bigger the damage it causes. Thus, in the case b= 250. and a=30. Mpa , the following values are obtained:

$$(S.T.W.)_i = \sigma_{max} \frac{\Delta\varepsilon}{2} = 0.089$$

$$D_i = 1/N_i = 1.307 \quad 10^{-9}$$

This value of damage may be neglected if we compare it with that of the overall cycle:

$$(S.T.W.)_o = 3.99$$

$$D_o = 4.263 \quad 10^{-4}$$

In accordance to this, it would be expected that life, considering the overall cycle, without interruption (constant amplitude), were nearly the same as that in the interrupted cycle.

That coincides well with the predictions made by Option (B), as can be seen in Fig. 9. Applying this methodology the stress - strain response becomes stable and the influence of the little interruption cycles may be neglected.

On the contrary, Option (A)'s life predictions are unstable, depending on the number of cycles in the input blocks, and the influence of the small cycles becomes very important: with "a" being 10% of the maximun applied stress, a minimum (with only one cycle in the input block) **error** in terms of life of **50%** is obtained.

No further comment seems necessary.

REFERENCES

(1) Martin J.F, Topper T.H. & Sinclair G. M..Materials Research and Standards, ASTM, Vol.11, No.2, Feb. 1971, 23.

(2) Technical Report No. SR 71-107, Scientific Research Staff, Ford Motor Co. Dearborn, Michigan, August 1971.

(3) Dowling N.E.. Journal of Materials, ASTM, Vol.7, No.1, March 1972, 71.

(4) Dowling N.E.. Transactions of ASME, Vol.105, July 1983. 206.

(5) Mroz Z.. Journal of the Mechanics and Physics of Solids. Vol.15, May 1967, 163.

(6) Morrow J..ASTM STP 378, 1965, 45.

(7) Manson S.S. Expt. Mech. Vol.5, 1965, 193.

(8) Jaske C.E., Mindlin H. & Perrin J.S. ASTM STP 519, 1973, 13.

(9) Abdel-Raouf H., Plumtree A. & Topper T.H..ASTM STP 519, 1973, 28.

(10) Mitchell M.R.. Paper No. 750198, SAE Congress. Detroit. February 1975.

(11) Neuber H.. J. of Appl. Mech. Vol.28, 1961, 544.

(12) Hardrath H.F. & Ohman L. NACA Report No. 1117, Washington, 1953.

(13) Molski K. & Glinka G.. Mat. Science Engng. Vol.50, 1981, 93.

(14) Polak J.. Mat. Science Engng. Vol.61 ,1983 , 195.

(15) Dowling N.E., Brose W.R. & Wilson W. K. Fatigue Under Complex Loading. SAE Advances in Engineering, Vol.6, 55.

(16) Wetzel R.M. Ph.D. Thesis. University of Waterloo, Ontario, Canada, 1971.
(17) Richards F.D., LaPointe N.R. & Wetzel R.M. Paper No. 740278, SAE Congress, Detroit, February 1974.
(18) Nelson D.V. Ph.D. Thesis, University of Stanford, U.S.A., 1978.
(19) Nelson D.V. & Fuchs H.O. Fatigue Under Complex Loading, SAE Advances in Engineering, Vol.6, 163.
(20) Smith K.N., Topper T.H. & Watson R. Journal of Materials, JMLSA, Vol.5, No.4, 1970, 767.
(21) Miner N.A. J. of Appl. Mech., Vol.12, 1945, A-159.
(22) Palmgren A. ZDVDI, Vol. 68, No.14, 1924, 339.

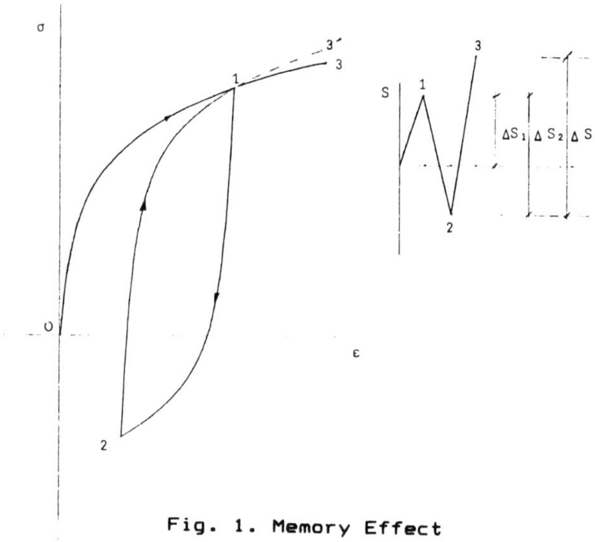

Fig. 1. Memory Effect

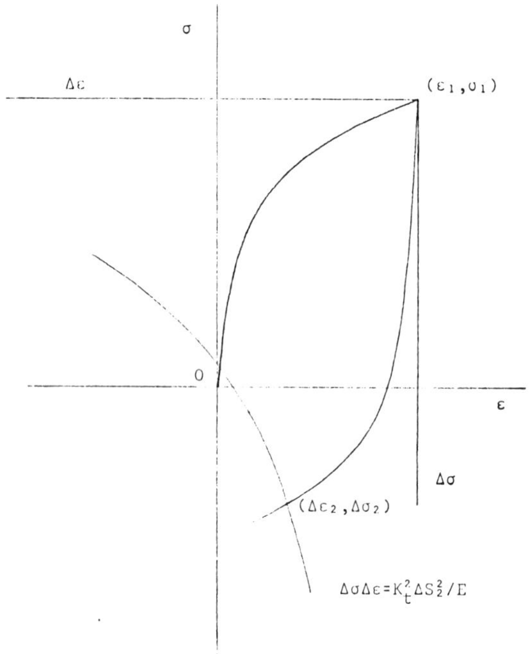

Fig. 3

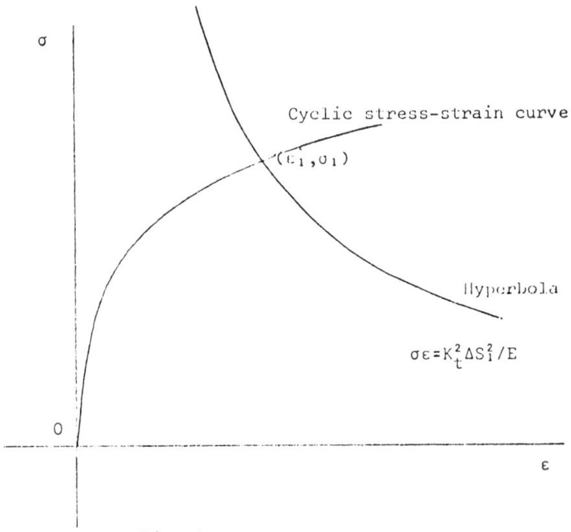

Fig. 2

Fig. 2. & Fig. 3. Application of Neuber's Rule

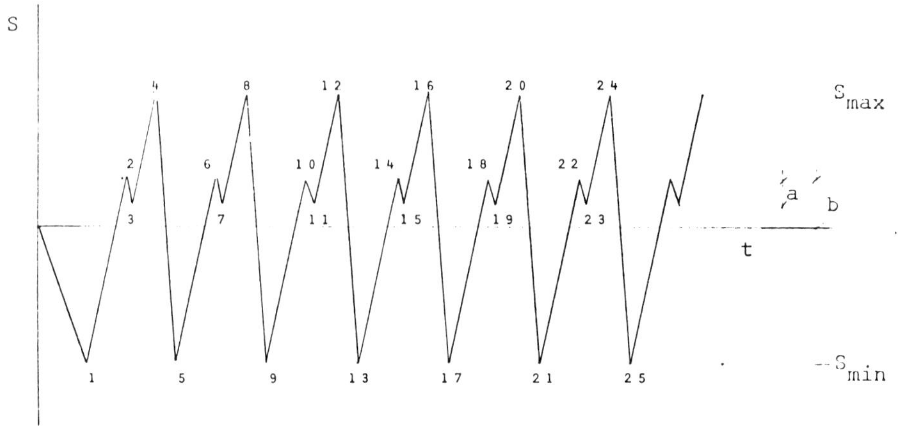

Fig. 5. Load Record Employed in the Simulations

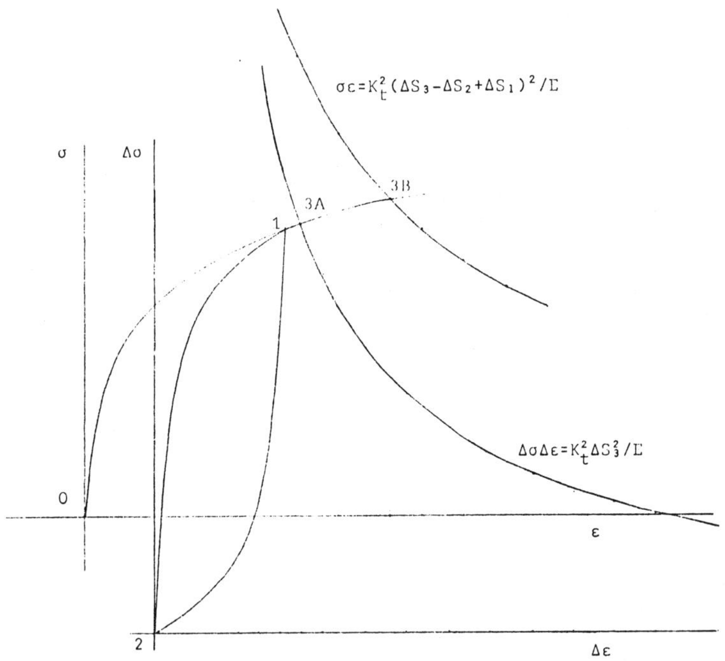

Fig. 4. Comparison between Option (A) and (B)

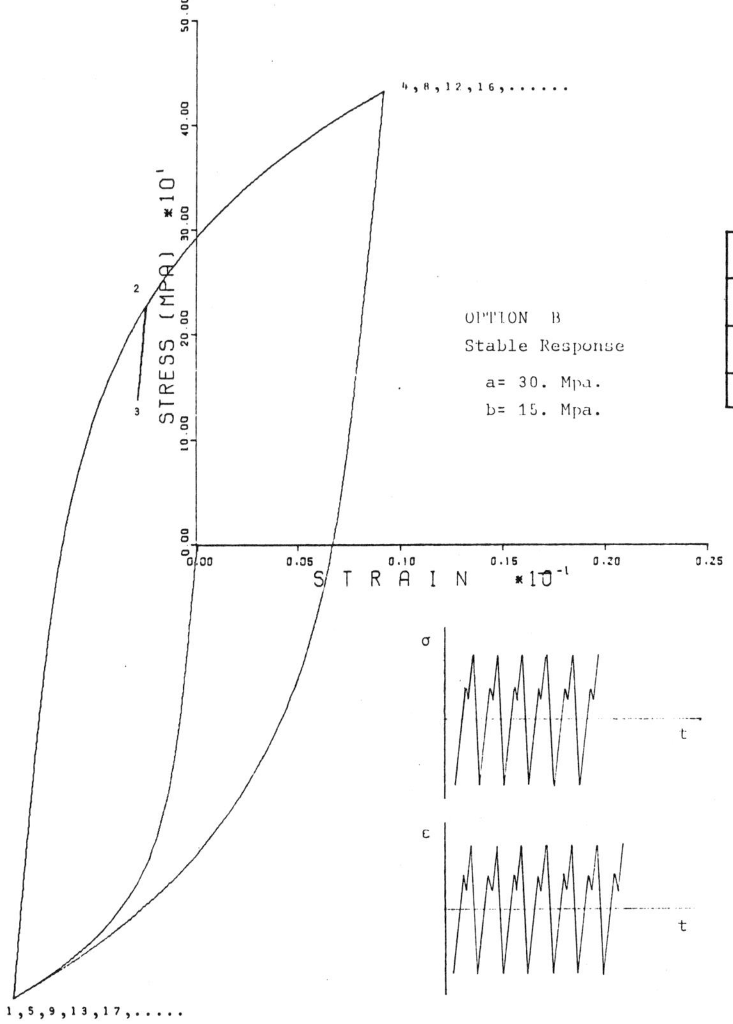

E = 203000 MPa	b = -0.095
σ'_y = 330 MPa	σ'_f = 930 MPa
k' = 1110 MPa	e = -0.47
n' = 0.19	ε'_f = 0.26

Table 1. Fatigue and Cyclic Stress Strain Properties of MAN TEN Steel.

Fig. 6. OPTION (B). Stable Response

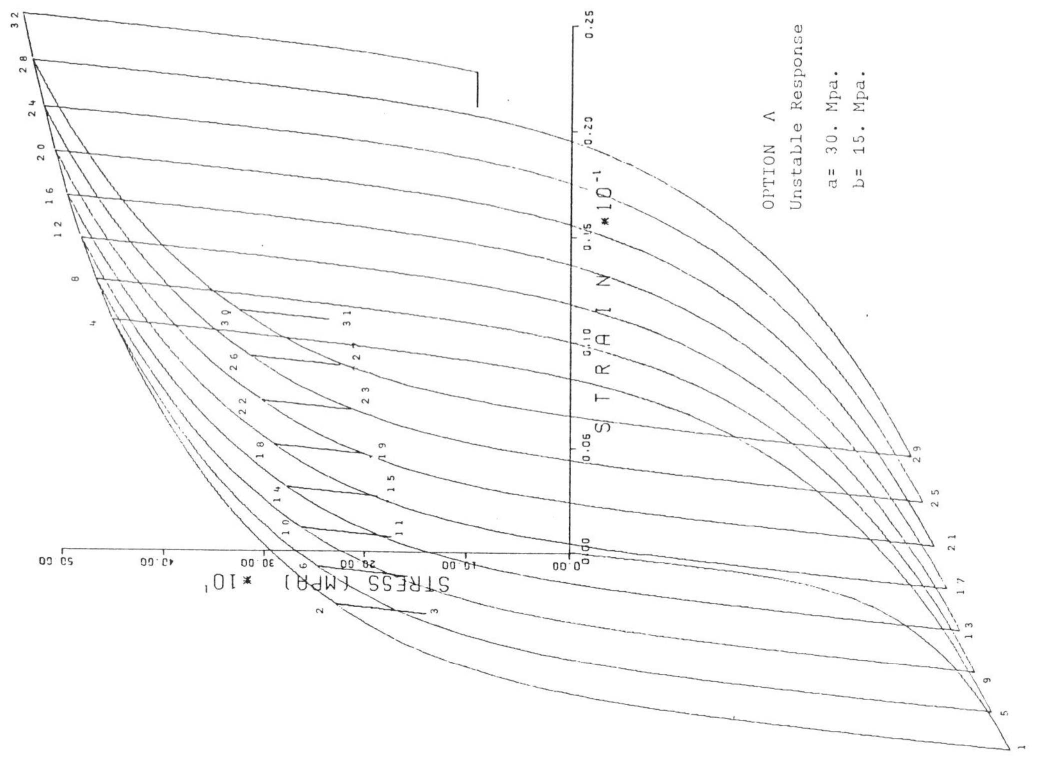

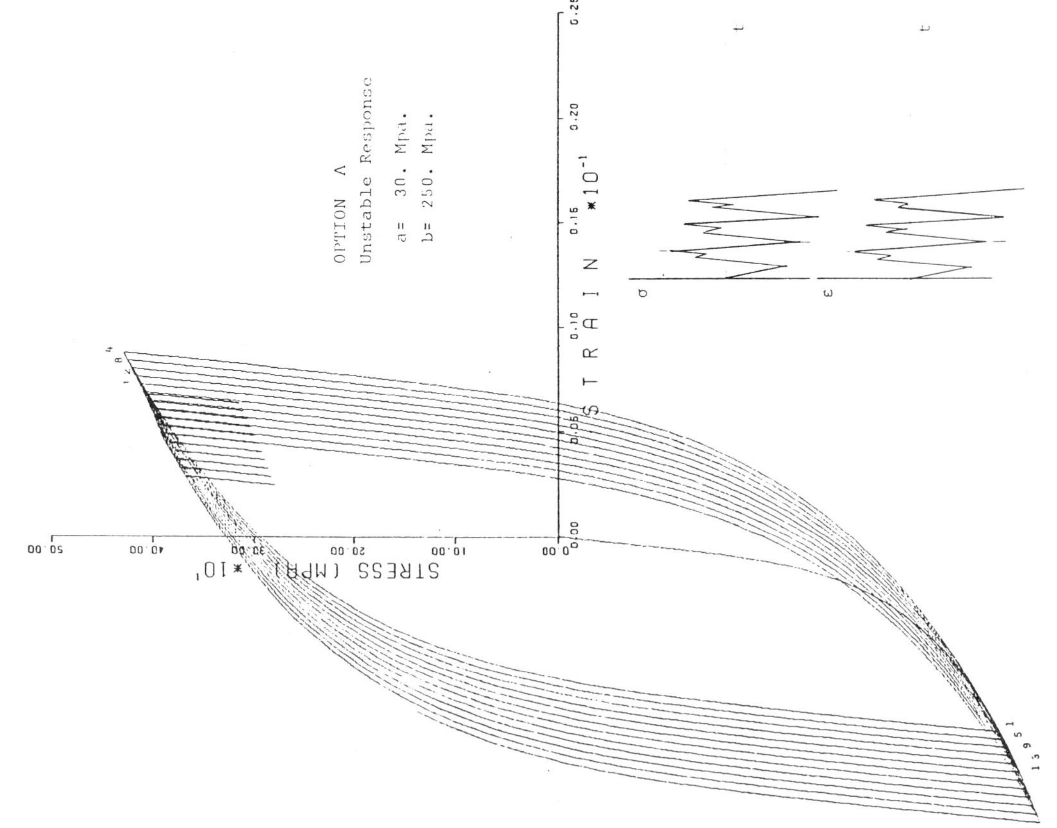

Fig. 7 and 8. OPTION (A). Unstable Response

a= 30. Mpa			a= 10. Mpa			a= 5. Mpa		
b	Δe	ΔS(Mpa)	b	Δe	ΔS(Mpa)	b	Δe	ΔS(Mpa)
-250.	2.93470 E-04	3.182	-250.	6.27831 E-04	6.697	-250.	7.18510 E-04	7.631
200.	1.13708 E-03	11.841	-200.	1.60163 E-03	16.333	-200.	1.72935 E-03	17.536
-150.	1.88732 E-03	19.007	-150.	2.51208 E-03	24.640	-150.	2.78667 E-03	26.165
-100.	2.24387 E-03	22.257	-100.	3.01584 E-03	28.985	-100.	3.23541 E-03	30.828
-50.	2.21697 E-03	22.015	-50.	3.10106 E-03	29.704	-50.	3.35749 E-03	31.840
0.	1.92930 E-03	19.394	0.	2.88724 E-03	27.892	0.	3.17152 E-03	30.295
50.	1.48430 E-03	15.215	50.	2.47770 E-03	24.337	50.	2.78109 E-03	26.981
100.	9.57621 E-04	10.056	100.	1.94492 E-03	19.538	100.	2.25852 E-03	22.388
150.	4.11808 E-04	4.439	150.	1.34083 E-03	13.834	150.	1.65413 E-03	16.823
200.	-8.07452 E-05	-0.890	200.	7.08736 E-04	7.530	200.	1.00545 E-03	10.535
250.	-3.76393 E-04	-4.182	250.	1.11492 E-04	1.220	250.	3.50262 E-04	3.787

Δe : Shift in Strain Between Peaks Numbers 5 and 1
ΔS : Shift in Stress Between Peaks Numbers 5 and 1

Tables 2 to 4. Numerical Results Obtained in Different Examples.

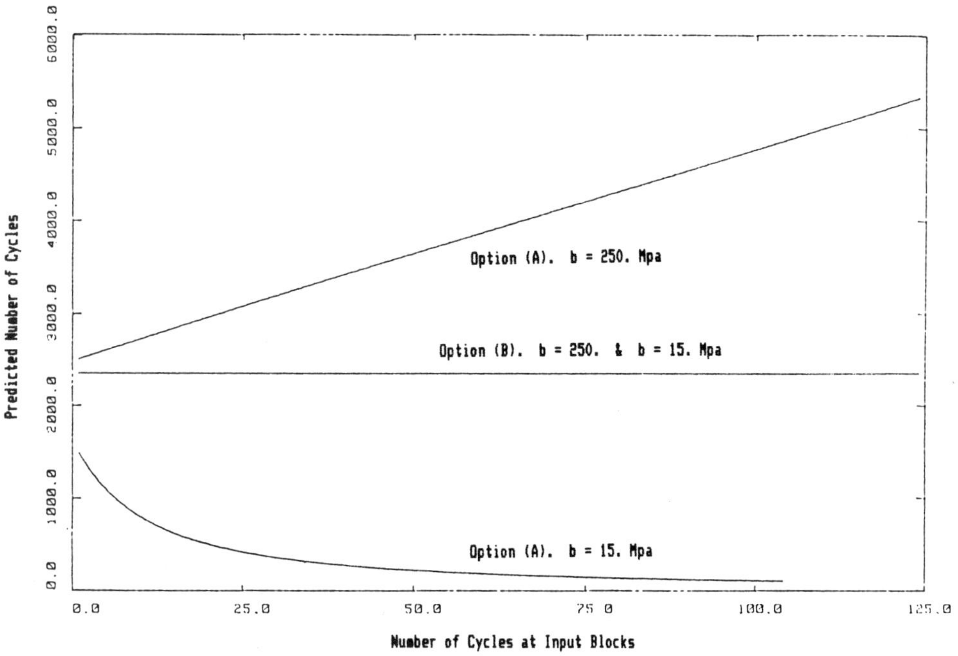

Fig. 9. Lifetime Predictions

INSTABILITÉS DANS LES CALCULS DE LA DURÉE DE VIE A FATIGUE DES ORGANES DE MACHINES AU MÉTHODE DE DÉFORMATION LOCALE

Résumé. On a analysé certaines erreurs dans les calculs des déformations plastiques et dans les contraintes aux entailles apparaisant dans certains algorithmes d'usage très commun dans les estimations du durée a fatigue des éléments de machi nes. Quelques resultats numériques concernant une simple so- llicitation sont évalués et discutés.

Design of Artobolevsky's Lever-gear Piston Mechanism by Curve Matching

T. E. SHOUP

Florida Atlantic University, Boca Raton, FL 33433, USA

Abstract. This paper presents a design method for choosing mechanism geometry of a lever-gear piston mechanism with a double stroke of different lengths. This device is suitable for the design of two stage compressors, two stage pumps, and four cycle engines. Using the kinematic information provided, the designer can select a design with a preselected stroke ratio.

Keywords. Machine theory; gears; linkages; mechanisms; compressor design; synthesis; two stage compressors; double stroke pumps; piston; slider crank.

INTRODUCTION

Because of its unique, double stroke performance, the seven link geared plane mechanism shown in Figure 1 and first presented by Artobolevsky is extremely useful for applications in the design of two stage, positive displacement compressors. This device consists of two gears that drive a vertical piston slide through a binary connecting rod. Because of the 2:1 ratio of sizes for the two drive gears, the piston of this mechanism undergoes a cyclic performance consisting of two strokes of different length as shown in Figure 2. This performance makes the mechanism an ideal candidate for use in the design of two stage compressors, two stage pumps and four cycle engines. Recent research by Shoup into the use of adjustable linkages to augment the design process for variable stroke pumps has utilized the technique of curve matching to provide quick, easy design methods for handling complex problems in a straightforward fashion. Although the original work of Artobolevsky is well documented, it was beyond the intended scope of the work of this original author to look at the use of adjustability of the link geometry as a means of providing design information. Thus it is the purpose of this paper to look at the adjustability of the lever-gear piston mechanism with a view toward finding new, useful design information to augment the original work of Artobolevsky.

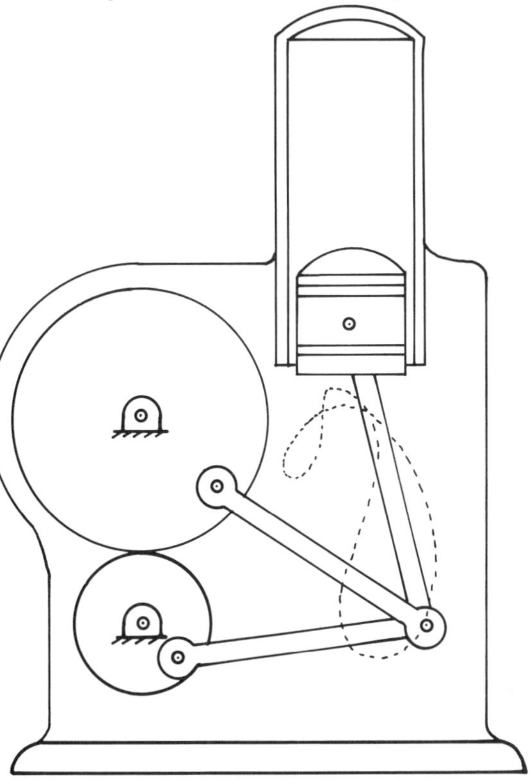

Figure 1. The lever-gear piston mechanism of Artobolevsky.

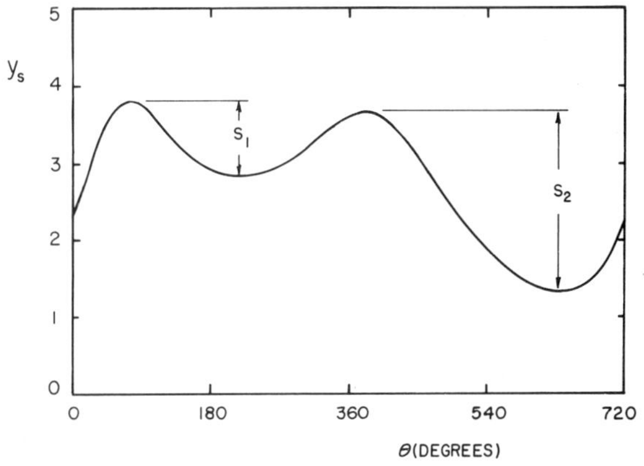

Figure 2. Double stroke performance of the lever-gear mechanism of Artobolevsky.

THE LEVER GEAR PISTON MECHANISM

A kinematic representation of the lever gear piston mechanism is shown in Figure 3. The vector loop equation for loop 1-6 in complex exponential form is as follows:

$$(r_1 + r_2)j + r_4 e^{j\phi} + r_5 e^{j} = r_3 e^{j\theta} + r_6 e^{j} \quad (1)$$

Because of the 2:1 gear ratio the following relationships are valid:

$$\phi = -(r_1/r_2)\theta + AG \quad \text{and}$$

$$r_2 = 2r_1$$

in the first of these relationships the term "AG" is a constant. if the angle "β" is eliminated from equation (1), the following transcendental relationship for the angle "α" can be found:

$$\tan(\alpha/2) = \frac{A \pm [A*A - (C+B)(C-B)]^{0.5}}{(C+B)} \qquad (2)$$

where in this equation:

$$A = 2(r_1+r_2) + 2r_4r_5\sin(\emptyset) - 2r_3r_5\sin(\theta)$$

$$B = 2r_4r_5\cos(\emptyset) - 2r_3r_5\cos(\theta)$$

$$C = r_6^2 + 2(r_1+r_2)\sin(\theta) + 2r_3r_4\cos(\theta-\emptyset)$$

$$- (r_1+r_2)(r_1+r_2) - r_3^2 - r_4^2 - r_5^2$$

$$- 2(r_1+r_2)r_4\sin(\emptyset)$$

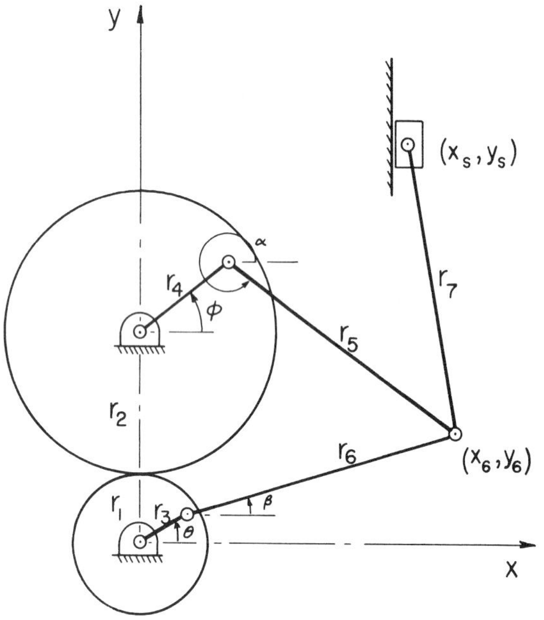

Figure 3. A kinematic representation of the lever-gear piston mechanism.

If, for the analysis of this mechanism, the values of the following parameters are given:

r_1, r_2, r_3, r_4, r_5, r_6, r_7, AG, θ, and x_s

then one can solve for the angle "α". This will allow the user to find the vertical position of the slider piston using the relationship:

$$(x_s - x_6)^2 + (y_s - y_6)^2 = r_7^2$$

For the fundamental mechanism formulated by Artobolevsky the following approximations to the parameters are:

r_1=0.65, r_2=1.3, r_3=0.5, r_4=1.0, r_5 = r_6=2.3,

r_7=2.8, AG = -67.5 degrees, and x_s=1.9

The slider displacement as a function of the input angle θ for this set of geometric parameters is shown in Figure 2. Any selection of geometric parameters other than these will give different performance for the mechanism. In this investigation it was desired to find a suitable adjustment to make to give a variety of choices for the stroke ratio s_2/s_1.

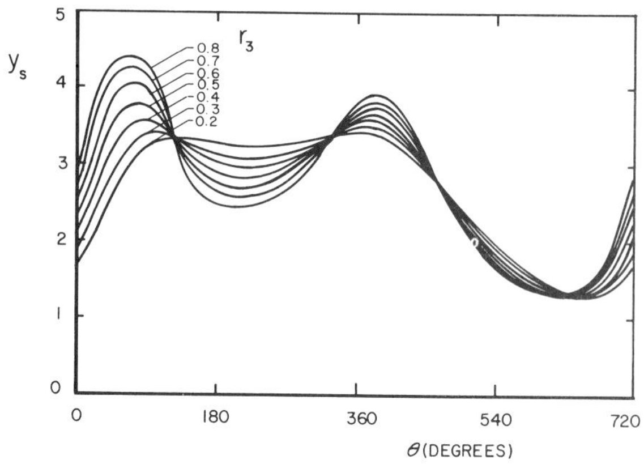

Figure 4. The influence of changes in r_3 on piston displacement. For this case:
r_1=0.65, r_2=1.3, r_4=1.0, r_5 = r_6=2.3,

r_7=2.8, AG = -67.5 degrees, and x_s=1.9

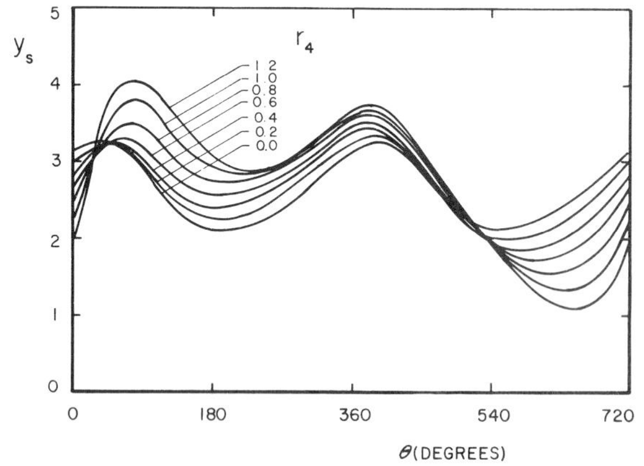

Figure 5. The influence of changes in r_4 on piston displacement. For this case:
r_1=0.65, r_2=1.3, r_3=0.5, r_5 = r_6=2.3,

r_7=2.8, AG = -67.5 degrees, and x_s=1.9

PARAMETER ANALYSIS

The various geometric parameters of the lever gear mechanism were studied using the numerical model and an IBM-AT computer in order to see what geometric dimensions were best suited as adjustment parameters for the design process. The hope in this portion of the study was to find one or two geometric dimensions that could be easily changed to give a wide range of values for the stroke ratio. The results of this sensitivity analysis are now presented.

The drive gears for the mechanism must keep their relative sizes of 2:1 in order to preserve the two stroke cyclic nature of the output. While it would be possible to change the values of r_1 and r_2 in order to adjust the mechanism, this would require replacing both drive gears and could prove rather impractical because of the expense involved if a number of different designs were to be tried.

The output link r_7 could be changed, however it was found that this adjustment leads primarily to a change in the relative position of the slider output and does not have as significant influence on the stroke ratio of the output motion as other parameters.

While the angle "AG" can be adjusted in order to modify the output performance of the mechanism, it was found that modest changes in this geometric dimension soon change the output to the point that the double stroke behavior is lost. Thus this parameter does not appear to be a good candidate for design adjustment for variable stroke performance.

The parameters $r_5 = r_6$ can be adjusted to modify the output of the piston; however, the study undertaken in this research investigation shows that the major influence of varying these parameters is a shift of the average value of the displacement curve without significantly changing the stroke ratio. Thus these two related parameters do not appear to hold much promise as adjustment geometries.

The parameters r_3 and r_4 were found to give the best control over the stroke ratio of any geometry in the mechanism. The influence of changes in r_3 is seen in Figure 4 and the influence of changes in r_4 is seen in Figure 5. Using the mathematical model and the computer program prepared for this research, Figures 6 and 7 were developed. These curves form the basis of the design method presented in the next section.

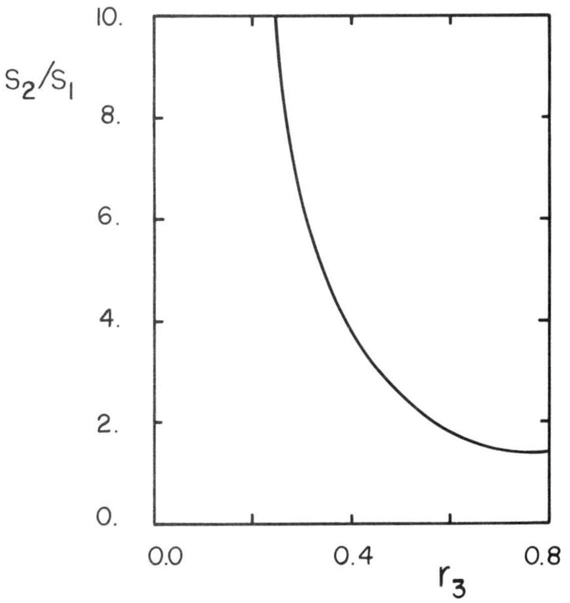

Figure 6. The influence of changes in r_3 on the stroke ratio s_2/s_1. Note that for this case:

r_1=0.65, r_2=1.3, r_4=1.0, r_5 = r_6=2.3,

r_7=2.8, AG = -67.5 degrees, and x_s=1.9

DESIGN METHOD

Figures 6 and 7 can be used to find a design with a particular stroke ratio desired. For example if a stroke ratio of 2.0 is desired, Figure 6 would predict that r_3=5.6 and r_4 = 1.0 would be a good choice. Alternatively, Figure 7 would predict that r_3=0.5 and r_4=0.43 would be a good design choice.

It should be noted from Figures 6 and 7 that values of the stroke ratio greater than 10 are possible, but such designs become impractical because of the way that this ratio is achieved. Also it should be noted that stroke ratios of less than 1.0 are possible if the designer realizes that a 180 degree phase change in the input angle θ would interchange the location of the peaks in the displacement curves. When this happens the stroke ratio is inverted. Thus a stroke ratio of 10. becomes a ratio of 0.1 if this phase relationship is utilized.

Whenever a design choice is selected using Figures 6 or 7 it is wise to utilize the mathematical model to inspect the overall shape of the piston displacement curve. In this way unforseen shapes and difficult transmission angles can be avoided.

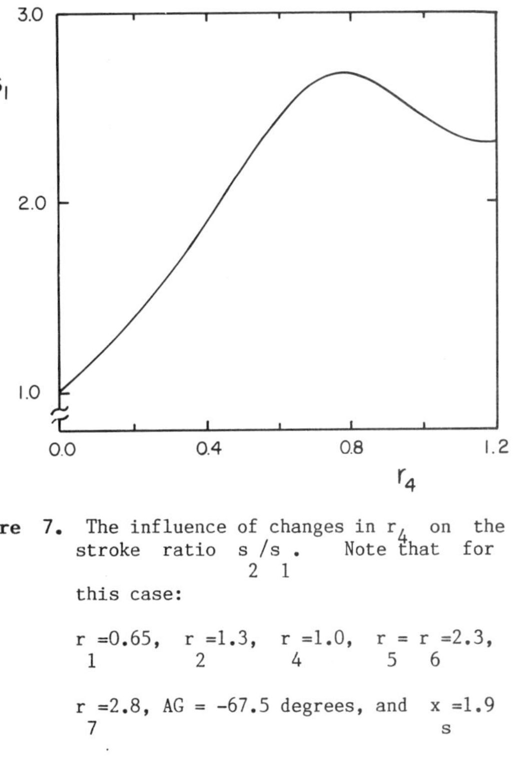

Figure 7. The influence of changes in r_4 on the stroke ratio s_2/s_1. Note that for this case:

r_1=0.65, r_2=1.3, r_4=1.0, r_5 = r_6=2.3,

r_7=2.8, AG = -67.5 degrees, and x_s=1.9

CONCLUSIONS

From the foregoing analysis and design approach, it would appear that Artobolevsky's mechanism continues to hold much promise as a device for special design requirements. The method presented along with the design curves should enhance the utility of this ingenious device created by a highly respected and revered practitioner in the field. Although it is beyond the scope of this investigation, it seems reasonable to expect that a designer might wish to adjust the geometry further to provide not only a desired stroke ration but also a design that has other special geometric features such as equal height peaks, equal height bottom stroke positions, specific sizes for the length of input angle between peaks or minimum positions or specific velocity or acceleration behavior. This design for multiple constraints could be accomplished utilizing an optimization method or by the use of a traditional approach using synthesis equations based on precision positions.

Although it is beyond the intended scope of this study, other mechanisms with multiple stroke behavior could be studied by this same technique. For example, a geared mechanism with a 3:1 ratio of the drive gears might be studied with a view toward finding a family of mechanisms that have triple stroke behavior.

REFERENCES

Artobolevsky, Ivan I. Mechanisms in Modern Engineering Design, A Handbook for Engineers, Designers and Inventors, Volume III - Gear Mechanisms, MIR Publishers, Moscow.

Shoup, T. E. "The Design of an Adjustable Three Dimensional Slider Crank Mechanism," Mechanism and Machine Theory, Vol. 19, 1984, pp. 107-111.

ZUSAMMENFASSUNG

Diese Arbeit beschreibt eine Entwurfsmethode für die Wahl der Geometrie der Mechanik eines Hebel-übersetzten Kolbenmechanismus mit einem doppelten Hub von verschiedenen Längen. Die Methode is geeignet für den Entwurf von Zweistufen-Kompressoren, Zweistufen-Pumpen und Vier-Zyklen-Motoren. Mit hilfe der zur Verfügung gestellten Information über die kinematik kann der Designer einen Entwurf mit vorgewähltem Hubverhältnis auswählen.

Applications on Nonlinear Mechanical Systems for Advanced Machine Elements

E. I. RIVIN

Department of Mechanical Engineering, Wayne State University, Detroit, MI 48202, USA

Summary

The paper describes high performance machine elements based on nonlinear mechanical systems and utilizing their adaptability to changing environment and adjustability by means of internal preloading. Specifically addressed are constant natural frequency and variable stiffness isolators; use of systems with pronounced amplitude dependancy of damping and effective stiffness and double nonlinearity (hardening load-deflection characteristic, and dynamic behavior typical for softening nonlinearity systems); couplings utilizing nonlinear "ideal shaped" elastomeric elements; gears in which sliding on profiles is accommodated by shear deformation in elastomeric coating on teeth; configuration-sensitive key connection. Some test and implementation results are given.

Keywords: non-linear; machine elements; vibration isolator; suspension; coupling; gear; key.

Introduction

Some of the critical requirements for modern machines and mechanisms are: higher operational speeds; higher specific loads; reduced friction losses; improved reliability in harsh environments (e.g., off-road vehicles and round-the-clock unattended production machinery); reduced consumption of critical materials with unreliable supply sources; etc. There are many contradictions in these requirements, such as:

o Reduction of energy consumption in vehicles can be achieved by weight reduction. However, light and strong structures lack structural rigidity. This, together with higher specific loads, leads to excessive deflections and misalignments, resulting in higher friction losses, as well as higher-intensity noise and vibration.

o Reduction of friction losses at high speeds and loads is achieved in high accuracy rolling, as well as in hydrostatic/gasostatic bearing systems. However, this is not the case for limited motions (gears, couplings, U-joints, etc.). Pressurized hydro- or gasostatic systems are not used in such cases due to space/cost constraints. Conventional sliding/rolling bearings cause noise, vibration, and high energy losses which must be dissipated in a limited space.

o Use of advanced materials, such as ceramics, is associated with incompatibility problems due to differences in thermal expansion rates; vastly different behavior under tensile, compression, bending, and impact loading; different sensitivity to stress concentration; etc.

o Vibration/noise problems are exacerbated by such factors as clearances/backlashes in power-transmitting connections; low internal damping in some light-weight materials, such as aluminum; and frequent regime changes (especially in vehicles and production machines), thus preventing use of tuned anti-vibration means.

o Absence of light power transmission elements (gears, couplings, etc.) for use in light-weight systems. Thus, solid steel gears are used in the mechanisms of antenna satellite [1] deployment even though launching 1 lb. into space costs about $10,000.

To solve these and other problems, there is clearly a need for a qualitative improvement in the performance characteristics of general purpose machine elements. A majority of machine elements available today were developed at least 30-50 years ago and have been undergoing incremental improvements associated, mainly, with utilization of state-of-the-art materials and computational optimization techniques. In the present situation, this is not enough, and new concepts are required. This paper briefly describes one of such novel generic concepts based on the utilization of the special properties of nonlinear mechanical systems. There are numerous studies of special dynamic effects in nonlinear systems. However, much less attention has been paid to their use in design.

Besides the specific dynamic characteristics of nonlinear mechanical systems, which can be beneficial in many applications, load-carrying nonlinear elements have a built-in potential for controlled change. Accordingly, a properly designed nonlinear machine element would demonstrate the valuable property of self-adaptation. If a device incorporating a nonlinear element allows the application of internal preload, it can be used to move the "work point" along the nonlinear load-deflection characteristic. As a result, in addition to or instead of self-adaptation, the element acquires the property of adjustability in accordance with recognized design needs. These features are illustrated below.

1. Constant Natural Frequency Vibration Isolators

A "constant natural frequency" (CNF) isolator has a "hardening" load-deflection characterisitic in the vertical direction Z, such that the stiffness k_Z is proportional to the weight load W,

$$k_Z = dW/d\Delta = AW$$

where Δ is isolator deflection, A is a constant. The vertical natural frequency of the isolated object is

$$f_Z = \frac{1}{2\pi}\sqrt{\frac{k_Z g}{W}} = \frac{1}{2\pi}\sqrt{\frac{AWg}{W}} = \frac{1}{2\pi}\sqrt{Ag} = \text{const} \quad (1)$$

Such a self-adaptation to varying weight loads is very advantageous since [3], [4], [5]:

o No cumbersome calculations are required to find CG position and weight distribution between isolators. Both are needed for conventional constant stiffness (CS) isolators in order to reduce intermodel coupling.

- Nearly perfect decoupling can be achieved, leading to improvements in isolation. With CS isolators, errors in determining CG position, shifting of CG during operation, tolerances on isolator stiffness values, and a limited inventory of available isolator stiffnesses result in a strong residual coupling [3], [5].

- For a given natural frequency, CNF isolators exhibit less level change with changing load on the isolator. For a conventional CS isolator,

$$f_z = \frac{1}{2\pi}\sqrt{\frac{k_z g}{W}} \quad \text{or} \quad k_z = \frac{4\pi^2 W f_z^2}{g} .$$

For the weight change $W_2 - W_1$, the difference in the isolator deflection would be

$$\Delta_2 - \Delta_1 = \frac{W_2}{K_z} - \frac{W_1}{K_z} = \frac{g}{4\pi^2 f_z^2}\cdot\left(1 - \frac{W_1}{W_2}\right) \cdot \qquad (2)$$

For a CNF isolator,

$$\ln W = A\Delta \ \underline{\text{and}}\ \Delta_2 - \Delta_1 = \frac{g}{4\pi^2 f_z^2} \ln\frac{W_2}{W_1} . \qquad (3)$$

If $W_2 = 2.73W_1$, then for the same f_z, the value of $(\Delta_2 - \Delta_1)$ from (2) (CS isolator) would be 1.73 times larger than the value from (3) for a CNF isolator.

- Production tolerances (e.g., on rubber hardness H) lead to deviations in stiffness and natural frequency f_n from nominal values for CS isolators, as is shown in area I of Fig. 1 [6]. For CNF isolators (area II in Fig. 1) H deviations only change the limits W_{min} and W_{max} of the load range in which the isolator has CNF characteristics (shifts the range to the left for reduced and to the right for increased H). It is not importanct due to the usually large ratio $W_{max}:W_{min}$.

- A small inventory of CNF isolators is required for the installation of objects of diversified weight and design (for example, 80-90% of production machines require only one judiciously designed isolator; see [3]).

A rubber-metal CNF isolator is shown in Fig. 2a [7]. Such isolators with characteristics shown in Fig. 2b have been extensively used for the installation of many thousands of machine tools, presses, etc. Their load range W_{max}/W_{min} = 10:1 - 100:1. Smaller units have been successfully tested as car engine mounts.

The advantages of the CNF isolation concept are important also for vehicle suspensions (e.g., [8]). The most important benefits for suspension applications are uniform ride quality and smaller attitude change for front wheel drive vehicles, and for trucks where the weight on idle axles varies as 2:1 - 3:1. Specially shaped springs (such as in [9]) are used with conventional shock absorbers; thus advantages of CNF isolation are underutilized due to variations in the damping ratio with changing weight load.

Elastomeric suspension springs would alleviate this problem and, usually, would be more economical in cases where very low natural frequencies are not required, such as cargo trucks and material-handling trailers. CNF characteristics can be obtained by using elastomeric elements of special geometry (such as the spheres in [10] for truck suspensions) or of a multidisc design [11].

While spherical, toroidal, and cylindrical elastomeric elements have "natural" quasi-CNF nonlinear load-deflection characteristics in a load range $W_{max}:W_{min}$ = 6:1 - 10:1, their fatigue endurance is very much better than that of conventional elastomeric springs [10] and our tests (with B.S. Lee) have shown that their creep rate is at least 50% less than the creep rate of a rectangular elastomeric block made of the same material. The latter properties can be explained by reduced stress concentration due to the streamlined ("ideal") shapes of the elements. However, due to low horizontal stiffness, "ideal" elements can be used in suspensions only with horizontal restraint in order to enhance handling stability of the vehicles.

The multidisc suspension for an in-plant trailer shown in Fig. 3 [11] has discs with progressively increasing holes whose walls successively contact the axle surface and thus "switch out" one after another with increasing load. As a result, natural frequency is nearly constant with the load change from 2,000 N/wheel (empty) to 10,000 N/wheel (fully loaded). In addition, the damping is high and consistent, and height change from the empty to the fully loaded condition is small. The latter feature allows the assembly of trains in any combination of empty or loaded trailers, with or without suspension. The laminated design provides high transverse stiffness and good handling/ cornering characteristics. Wide implementation of such a suspension was justified by a ~20dB noise level reduction, accompanied by a 10-20 times reduction in dynamic loads between floor surface and wheel tread and a 5-10 times reduction in wheel/castor failure rate. Thus, savings from such an initially "noise abatement" measure far exceed investments for retrofitting trailers.

While a CNF isolator <u>adapts</u> itself to changing environment (weight load), the stiffness of the combination vibration isolator in Fig. 4 can be <u>adjusted</u> without disturbing the installed object. Two isolators 1 and 2, at least one of which has a nonlinear (e.g., CNF) load-deflection characteristic, can be internally preloaded with a force P_o by bolt 5. Initially isolators 1 and 2 are both subjected to load P_o, but after installation of object 4 (weight W), the total loading of the lower isolator 2 is P_o+W/2, and of the upper isolator 1 is P_o-W/2. If both 1 and 2 have identical nonlinear characteristics and $2P_o$>W, then the installation is equivalent to an isolator having stiffness $K = 2k(P_o)$, where $k(P_o)$ is the stiffness corresponding to the load P_o on the load-deflection characteristic ("working point"). Changing P_o would change K accordingly. The combination isolator thus has a <u>linear</u> (constant stiffness) load-deflection characteristic with its stiffness (i.e., natural frequency of the object) adjustable from outside. The unit can be used for specification of isolation requirements in difficult cases, for on-site optimization of isolation systems, and as a component of an active isolation system.

2. <u>"Double Nonlinearity" Effect</u>.

The combination of high damping under large vibration amplitude/resonance conditions with low damping under low amplitude/high frequency conditions is important for many isolator applications. There are many ingenious designs for active isolators or pneumatic and hydraulic isolators [6], which effectively have two sets of characteristics with "switching" between them depending on vibration amplitude or frequency.

However, the desired combination of characteristics occurs naturally in nonlinear damping devices utilizing dry (Coulomb) friction. Frictional dampers have limited use due to the instability of friction coefficients, stick-slip, etc. More reliable processes are associated with internal Coulomb friction-like interactions (wire mesh, felt, rubbers heavily filled with carbon black). These materials are characterized by the amplitude dependance of both damping and effective stiffness. Typical characteristics are shown in Fig. 5 [12]. The log decrement δ of a wire-mesh sample rapidly increases with increasing vibration amplitude a, while the dynamic stiffness coefficient $K_{dyn} = k_{dyn}/k_{st}$ decreases (although not as fast) with increasing a (here k_{st} is the static stiffness, k_{dyn} the effective stiffness in vibratory conditions). These correlations can be described by empirical formulae [12].

$$K_{dyn} = 1+A/a^{0.5}; \ \delta = Ba^{0.75} . \tag{4}$$

Felt and filled rubbers have additional mechanisms responsible for elastic and damping behavior. Accordingly, for these materials amplitude dependencies are less pronounced,

$$K_{dyn}=C + A_1/a^{0.5}, \ C>1; \ \delta=D + B_1a^{0.75} . \tag{5}$$

A, B, C, D in (5) and (6) are empirical coefficients [12].

All these materials have hardening nonlinear characteristics for static compresssion (a quasi-CNF characteristic for wire-mesh materials). However, their dynamic behavior is representative for a nonlinear mechanical system with softening characteristics. In addition, wire meshes exhibit exceptionally high damping at large amplitudes and low damping at small amplitudes (usually corresponding to high frequencies). Damping at resonances can be so high that resonance conditions can be frequently tolerated as working regimes. Thus, a passive nonlinear isolator with wire mesh combines the required large and small amplitude properties with the CNF characteristic and adapts itself to the vibration environment. However, as shown in [3], it is not suitable for isolation of precision objects from low amplitude vibrations.

3. Torsionally Flexible (TF) Couplings

The role of TF couplings in transmissions is somewhat similar to that of vibration isolators. However, TF couplings have to combine specified torsional compliance and damping with a misalignment-compensating ability and small size/low inertia. Analysis in [13] has shown that the smallest commercially available couplings are spider couplings. This makes them the most cost-effective and most versatile for design/ packaging purposes. However, their torsional and misalignment-compensating (radial) stiffness is 5-10 times higher than that of the best (but larger and more expensive) coupling designs. Due to this high torsional stiffness, influence of the coupling damping on the effective damping of the transmission system is negligible.

These shortcomings can be partly explained by the fact that the rated torque T_{cr} of the coupling has to be about 2-3 times higher than the rated torque T_{tr} of the transmission since application-related "service" factors lead to effective derating of couplings. Since a rectangular rubber block in compression can tolerate only 10-15% deformation, very hard rubber has to be used for the spider, thus resulting in very high stiffnesses in both torsional and radial directions. Due to the near-linear load-deflection characteristic of the conventional spider coupling, these high stiffnesses are the same at any loading of the transmission. Since it is well known that typical machines operate most of the time with acting torque much less than T_{tr} (lathes and milling machines use <25% of installed motor power 40-80% of their running time, and <50% for 60-90% of the time [14]), couplings utilize only a small fraction of T_{cr}. At the same time, the spider is constantly subjected itself and subjects bearings to the full amount of the misalignment-induced loads, and exhibits the high torsional stiffness associated with a high value of T_{cr}.

The nonlinear spider coupling in Fig. 6 [15] uses "ideal" shapes for the spider legs, which combine a high degree of nonlinearity with allowable compression deformations up to 40-50%. As a result, the spider can tolerate very high overload torques, equal or even exceeding those for a given size and rubber hardness (marginal torque T_{cr}) for the conventional spider. However, at the smaller torques (0.1-0.3 T_{cr}), representing actual operating torques, its torsional and radial stiffnesses would be substantially reduced. It has been confirmed experimentally with a spider coupling of L-100 Lovejoy size (T_{cr}=43.5 Nm), Fig. 7. The nonlinear spider had legs made of single rubber cylinders.

Thus, the nonlinear coupling adjusts its characteristics to the transmitted torque and, as a result, assures significant influence on the dynamic characteristics of the transmission: an increased role in the breakdown of torsional compliance; consequently, enhancement of the transmission damping; reduced loads on bearings and vibration level; a lower resonance peak for dynamic overloads due to strongly nonlinear characteristic.

4. Use of Thin-Layered Rubber-Metal Laminates

The laminates have a strong hardening nonlinearity in compression (1-2 decimal orders of magnitude change in compression modulus on displacements of 10-15 mkm [16]). This fact, together with high compressive strength (exceeding 200 MPa), and a negligible influence of compression load on shear stiffness, makes them very suitable for cushioning strong impacts [17] and for accommodation, without backlash and dead zone, of short motions while the moving bodies are transmitting large normal forces to each other. The latter feature is used in misalignment-compensating but torsionally-rigid Oldham couplings, backlash-free screw-nut connections, etc.

The most promising application of laminates is for power transmission gears. Involute profiles transmit tangential force F_t while performing a combination of rolling and sliding motion. The sliding motion (and friction force) reverses direction as the contact point crosses the pitch circle. The sliding path between the commencement of engagement and the mesh point is [18]

$$L = \frac{\pi}{4} m \cos^2 \psi \left(\frac{1}{N_1} + \frac{1}{N_2} \right), \tag{6}$$

where ψ is pressure angle; m = gear module, mm; N_1, N_2 = teeth numbers. For the reduction pair with $\psi = 20°$, $N_1 = 17$, $N_2 \geq 2N$, $L \leq 0.19m$, or about 0.1m per profile.

Conformal (Wildhaber/Novikov, W/N) gears ideally have a point contact and no geometry-related sliding. However, due to Hertzian deformations, the contact area has finite dimensions, thus causing some sliding at the periphery.

Thus, gear teeth have to endure both high bending stresses and high contact stresses in sliding contacts. This calls for expensive alloys with sophisticated treatments to provide the required internal and surface properties. Light and strong metals (Al, Ti) lack the necessary surface hardness; thus steel gears are used even for single action space applications [1]. Fiber-reinforced gear shells [19] exhibit high bending strength/ weight ratios, but poor contact properties.

Rubber-metal laminated coating on teeth profiles (Fig. 8 [8]), can accommodate payload by compression deformation and sliding-by shear deformation of the coating. The latter takes advantage of the low shear stiffness regardless of the compression (tangential) force magnitude and of low energy losses during shear deformation of some rubbers. Thus, for heavy loads, the transmission efficiency can reach 0.999-0.9999 [16]. The high nonlinearity of laminates in compression allows them to cushion dynamic loads due to pitch errors, and also to eliminate backlash by means of internal preloading of the transmission. Noise excitation due to the sliding friction force reversal at the pitch point is also eliminated, and the gear material is freed from the need to accommodate sliding and contact loads.

The surface durability of an involute gear is evaluated [21] by comparing "calculated contact stress number" S_c with allowable compression stress S_{ca},

$$S_c = C_p \sqrt{\frac{F_t}{C_v} \frac{C_s C_m C_f}{FdI}} \le S_{ca}. \qquad (7)$$

where C_v = dynamic factor, $C_v \cong 0.5$ for precision high-speed steel gears [21], $C_v \cong 1.0$ for laminate-coated gears due to cushioning effect; C_m = load distribution factor, $C_m \cong 1.5$ for high-precision steel gears [21], $C_m \cong 1.0$ for laminate-coated gears due to high local compliance and reduced effects of misalignment; I = geometry factor, ~ 15% higher for laminate-coated gears due to better teeth overlapping caused by nonlinear compliance of the coating; C_p = elastic coefficient proportional to $\sqrt{E}$, where E is the Young modulus of the surface material; for steel gears E= 2 X 10^5 MN/m^2, $C_p \cong$ 2,300, for laminate-coated gears E = 1-2 X 10^3 MN/m^2 [16], $C_p \cong 230$. All other parameters in (7) do not depend on tooth surface characteristics. Allowable contact stress S_{ca} = 600-1,300 MN/m^2 for state-of-the-art steel gears and 200-250 MN/m^2 for thin-layered rubber-metal laminates.

Accordingly, S_c for coated gears is about 16 times less than for steel gears, while S_{ca} is only 2.5-6.5 times less. Thus, the expected surface durability of the laminate-coated gears is better than of the steel gears.

For W/N gears, less shear deformation is required, and contact stresses are 2-8 times lower than in involute gears [22]. Thus, deformation conditions for the coating are significantly relaxed, resulting in even better safety margins. Coating would alleviate, due to its nonlinear compliance, a major problem of W/N gears: center distance sensitivity and intense noise generation.

The first stage of the experimental study was finished at the time of preparation of this paper; it consisted of static tests of coated involute gears. A gear pump (donated by Vickers Inc., of Troy, Michigan) was used as the test rig. Gears (N_1=N_2=10, m=6mm, c.d.=64mm, width 75mm) were reground to create allowance for coating. Two-layered laminates using 1/64 in (0.4mm) soft (H35) natural rubber layers and 0.002 in (0.051mm) tempered steel interleaves were attached to the profiles. The gears were assembled in the pump housing and subjected to oscillating torque (frequency 0.05 Hz) on an Instron testing machine.

The calculated contact and bending strengths of the tested steel gears corresponds, respectively, to F_t=14,980 N (200 N/mm) and F_t=34,960 N (466 N/mm). Loading of laminate-coated gears with F_t=45,000 N and cycling with a sinusoidal torque resulting in $(F_t)_{min}$=2,900 N and $(F_t)_{max}$=35,000 N for 100 cycles did not result in a noticeable deterioration of the coating. Further testing of involute gears and W/N gears is in progress.

5. Nonlinear Key Connection

Conventional key connections between a shaft and a sleeve (hub) can transmit large loads, but are difficult to use in many mechanisms due to their lack of adaptability. For example, increasing load intensity requires heat treatment of both connected parts, but it causes distortions in the slot shapes requiring either expensive finishing of the slots or custom fitting of keys. This, together with stress concentrations at the sharp corners in the conventional key slots, essentially precludes the use of hardened shafts and hubs. Another example is assembly of metal and ceramic parts (e.g., a metal shaft and ceramic bushing in turbines [22]), where keys are not used due to a large difference in thermal expansion of the materials creating dangerous clearances in the connection at high working temperatures, and also due to the extreme sensitivity of ceramics to stress concentration.

These problems can be helped by using a nonlinear (configuration-sensitive) key proposed in [23]. It consists of a helical spring inserted in an appropriately-shaped key slot, Fig. 9. The initial (free) diameter of the spring is larger than the inscribed diameter of the key slot opening; thus the spring has to be preloaded (pretwisted) before insertion. Before and during insertion the spring is flexible, and can accommodate a not very straight slot (e.g., distorted after heat treatment) and does not require fitting. However, after insertion, it becomes rigidized due to the cementing effect of friction forces.

Pretwisting enhances initial pressure and friction forces between the spring and slot surfaces thus reducing the danger of a buckling-like failure. Spring radius reduction caused by pretwist is associated with tensile stresses in outer fibers and compressive in inner fibers, thus increasing load-carrying capacity of the spring key; key stiffness (and the torsional stiffness of the connection) increases because of this initial stress pattern. Another effect is an increase in transverse bulging Δ_t of the pretwisted spring in relation to its deformation Δ_c in the direction of compressive payload. Due to the cementing effect of friction forces and the usually small helix angle of the spring, the latter can be modelled as a stack of rings. For a free ring, compressed by radial forces, $d=\Delta_v/\Delta_h \cong 1.08$ [24]. In a preloaded ring d is reduced and the relative importance of transverse deformation increases. As a result, a four-point contact between the key and slot surfaces can be realized instead of a two-point contact, with the corresponding sharp increase in load-carrying capacity. The degree of preload can be adjusted to control the stiffness of the connection. Thus, the spring key connection is characterized by the adaptability and adjustability typical of nonlinear mechanical systems.

An optimal design of such a connection involves a gothic arc shape of the slot surfaces to reduce contact stresses. Under an increasing transmitted torque, the wire coils would progressively envelop the gothic arc surfaces thus using a hardening nonlinear torque-twist characteristic of the connection, Fig. 10.

Conclusions

The examples described are intended to substantiate a potential for the use of nonlinear components by, first of all, utilizing their unique adaptability/adjustability properties. It is shown how the utilization of nonlinear characteristics depends both on the specifics of the nonlinear components and on the required characteristics of the machine elements.

Acknowledgement

This work is supported by NSF Grant MEA 83-08751.

REFERENCES

1. Prince, M., "Large Spacecraft Active Deployment Hinges," Proceed. of the 24th AIAA/ASME/ASCE/AMS Structures, Structural Dynamics, and Materials Conference, paper 83-0909-CP.
2. Metwalli, S.M., "Optimum Nonlinear Suspension Systems," ASME Paper 85-DET-85.
3. Rivin, E.I., "Principles and Criteria of Vibration Isolation of Production Machinery," ASME J of Mechanical Design, 1979, vol. 101, pp. 682-692.
4. Rivin, E.I., "Vibration Isolation of Production Machinery - Basic Considerations," Sound and Vibration, 1978, No. 11, pp. 14-20.
5. Rivin, E.I., "Vibration Isolation of Production Machinery - Vibration Sensitive Machines," Sound and Vibration, 1979, No. 8, pp. 18-23.
6. Rivin, E.I., "Passive Engine Mounts - Some Directions for Further Development," SAE Paper 850481.
7. U.S. Patents 3,442,475; 3,460,786.
8. Hortrich, H., "Rear Suspension Design with Front Wheel Drive Vehicles," SAE Paper 810421.
9. Vaillant, C., Ferlicca, R., "A New Design and Manufacturing Process for Suspension Coil Springs," SAE Paper 850060.
10. Schmitt, R.V., Kerr, M.L., "A New Elastomeric Suspension Spring," SAE Paper 710058.
11. Huang, B., Rivin, E.I., "Noise Abatement of In-Plant Trailers," SAE Paper 800494, U.S. Patent 4,188,048.
12. Rivin, E.I., "Energy Dissipation and Dynamic Stiffness of Elasto-Damping Materials," in "Energy Dissipation During Vibration of Elastic Systems," "Naukova Dumka", Kiev, 1968, pp. 383-389. (in Russian)
13. Rivin, E.I., "Design and Application Criteria for Connecting Couplings," ASME J of Mechanisms, Transmission, and Automation in Design, 1986, Vol. 108, pp. 96-105.
14. Components and Mechanisms of Machine Tools, ed. by D.N. Reshetov, "Mashinostroenie", Moscow, Vol. 2, 1972 (in Russian).
15. U.S. Patent 4,557,703.
16. Rivin, E.I., "Properties and Applications of Ultra-Thin-Layered Rubber-Metal Laminates," Tribology International, No. 2, 1983, pp.17-25.
17. Huang, B. Rivin, E.I., "Reduction of Impact Noise in Mechanical Presses", Proceed. of Noise-Con 83, MIT, 1983, pp. 425-432.
18. Tuplin, W.A., "Gear Load Capacity," Pitman & Sons Ltd., London, 1961.
19. Ikegami, K. Takada, M., "Strength of Fiber Reinforced Plastic Gears of Monocoque Structures," ASME Paper 84-DET-72,
20. U.S. Patent 4,184,380.
21. Gear Handbook, McGraw-Hill, N.Y., 1962, Ch. 13.
22. Brooks, A., Bellin, A.I., "The Application of Ceramics to Gas Turbines," The Leading Edge (GE Magazine), Summer 1981, pp. 21-30.
23. U.S. Patent 4,358,215.
24. Blake, A. (Editor), Handbook of Mechanics, Materials, and Structures, John Wiley, N.Y., 1985.

ПРИМЕНЕНИЕ НЕЛИНЕЙНЫХ МЕХАНИЧЕСКИХ СИСТЕМ ДЛЯ ДЕТАЛЕЙ МАШИН С УЛУЧШЕННЫМИ ХАРАКТЕРИСТИКАМИ

Описаны принципы конструирования и характеристики нескольких типов деталей машин с высокими техническими характеристиками, основанные на использовании нелинейных механических систем, которые адаптируются к окружающим условиям, а также могут регулироваться посредством изменения внутренних нагрузок (предварительного натяга). Примеры включают равночастотные виброизоляторы и подвески, виброизоляторы с переменной жесткостью, использование систем с двойной нелинейностью (жесткую при статическом нагружении и мягкую в динамических условиях), муфты использующие "идеальные" резиновые элементы, зубчатые колеса, в которых скольжение в зацеплении компенсируется сдвиговой деформацией в резиновом покрытии зубьев, шпоночное соединение, использующее элементы, поведение которых зависит от конфигурации. Приведены некоторые результаты испытаний и внедрения.

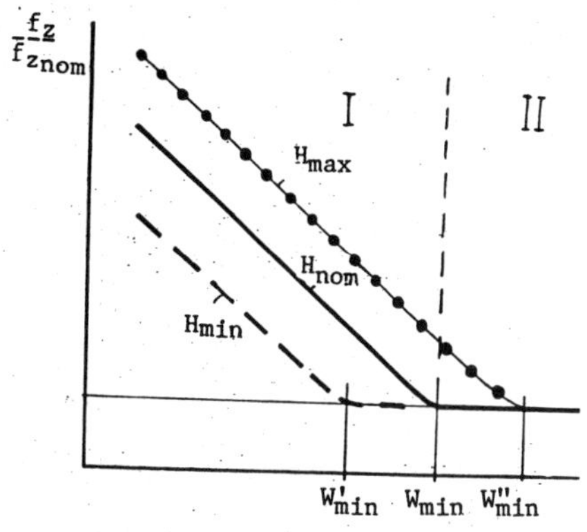

Fig. 1. Variation in rubber durometer H vs. frequency characteristics of CS (I) and CNF (II) isolator; f_{no}, H_o - nominal natural frequency, durometer; H_{max}, H_{min} - high and low value of rubber durometer.

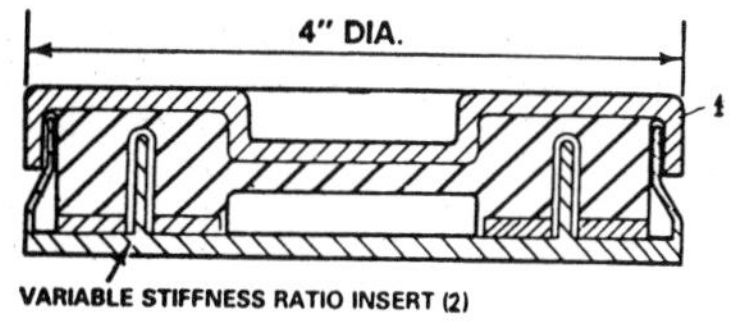

a

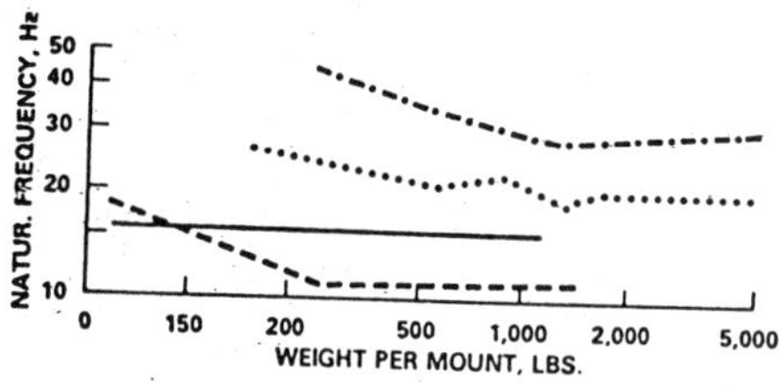

b

Fig. 2. Rubber-metal CNF isolator (a) and its frequency plots for various models (b).

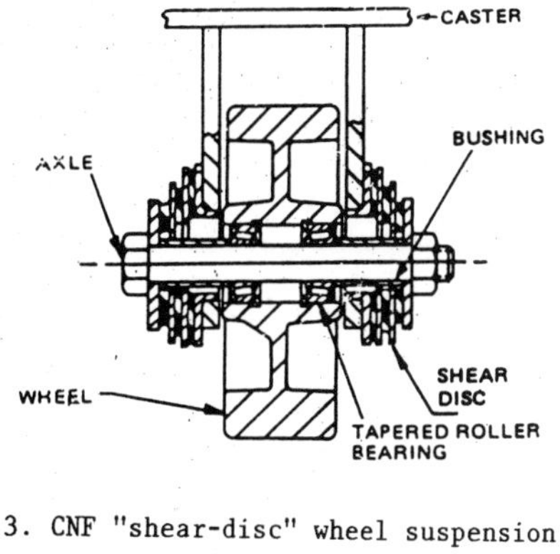

Fig. 3. CNF "shear-disc" wheel suspension.

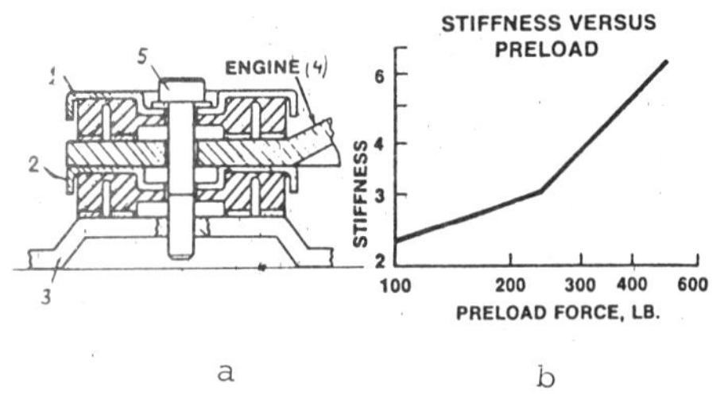

Fig. 4. Variable stiffness isolator (a) and its typical preload-stiffness plot (b).

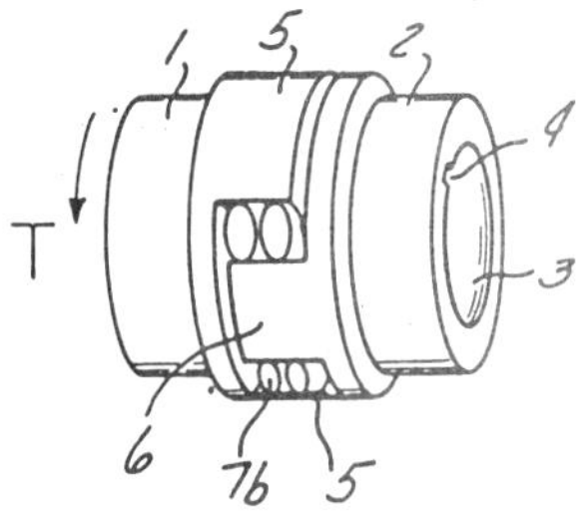

Fig. 6. Nonlinear spider coupling.

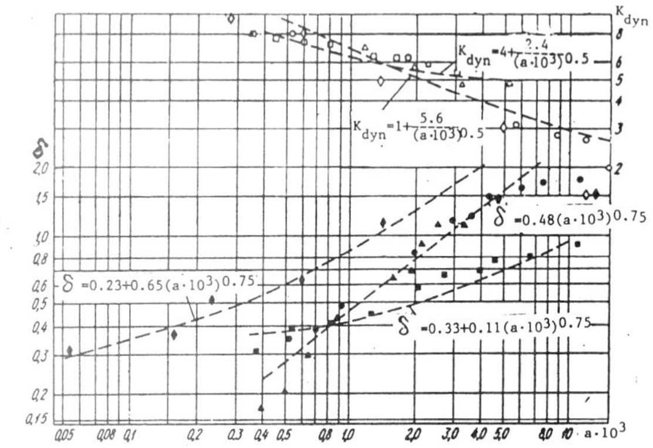

Fig. 5. Log decrement δ and dynamic stiffness coefficient K_{dyn} vs. relative amplitude a (vibration amplitude/element thickness); o, △ - wire mesh under low, high static load; ▽ - thick - fiber felt; clear symbols - K_{dyn}, black symbols - δ.

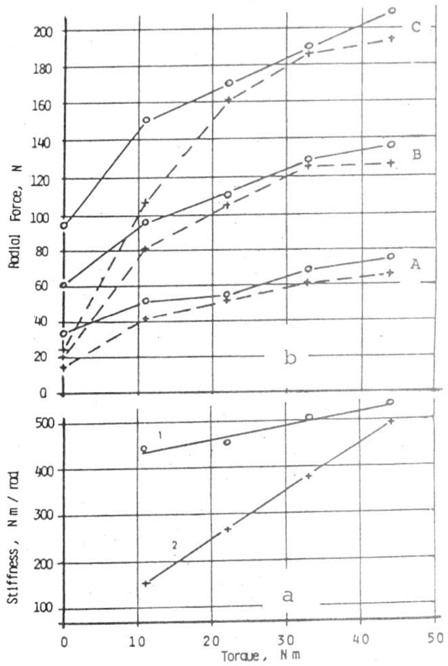

Fig. 7. Torsional (a) and radial (b) stiffness of conventional (o) and nonlinear (+) spider coupling; radial misalignment: A - 0.25 mm; B - 0.5 mm; C - 0.75 mm.

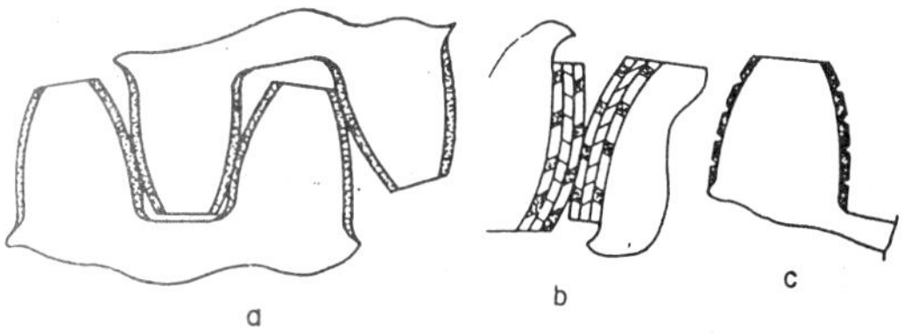

Fig. 8. Gears with elastomeric coating.

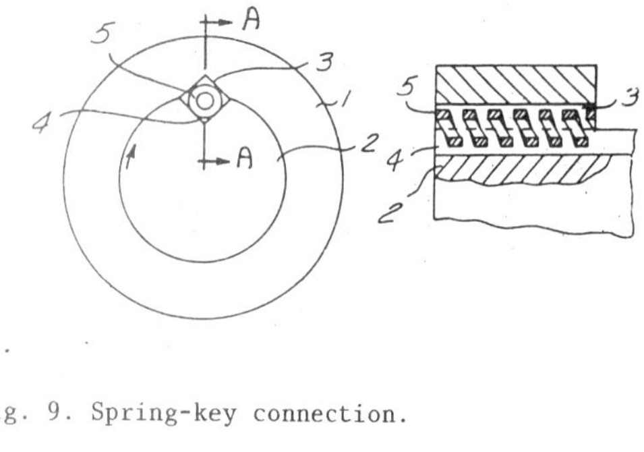

Fig. 9. Spring-key connection.

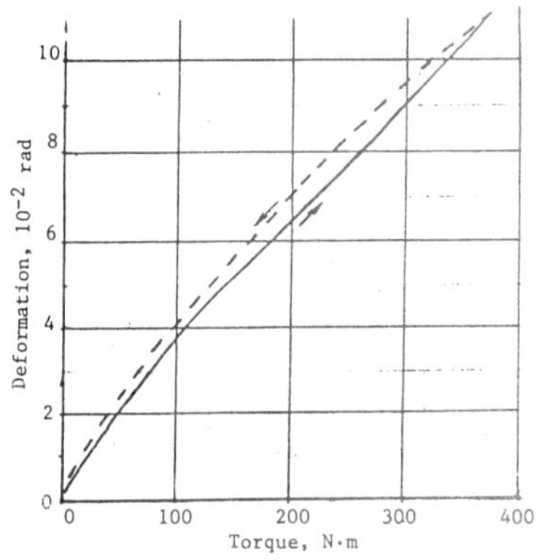

Fig.10. Load-deflection characteristic of spring-key connection (3 springs L=38mm, D=5mm, d=0.25mm, connection diameter 15mm)

Analysis and Synthesis of a Traction Drive Torque Converter

E. I. ODELL AND L. S. CAULEY

Department of Mechanical Engineering, University of South Alabama, Mobile, Alabama 36688, USA

Abstract. A mechanical torque converter that operates on the principle of gyroscopic motion of a sphere was analyzed and design methods developed. Concept analysis of the torque converter had been completed previously. The torque converter was analyzed to establish relevant criteria and design constraints. Using the established criteria and constraints a method of design was developed and applied to the torque converter.

Analysis of the traction drive contact area between the sphere and drive rings was studied and the influence of the various parameters, such as contact shape, contact stress, spin, etc., was developed. Criteria and constraints were established for each parameter. An evaluation method was developed to obtain operating characteristics for the contact area. Using results from the contact evaluations and overall converter operating limits, performance values are obtained for a specific configuration. The operational power values are charted as a function of input and output speeds.

Keywords. Torque control; variable speed gear; self-adjusting systems; gyroscopes; traction drive; torque converter; power transmission.

INTRODUCTION

A mechanical torque converter, based on traction drive of a hardened steel sphere, operates on the principle of gyroscopic torque to produce the output torque, Fig. 1. There are two inputs, the traction drive input and the rotating frame input, and one output. The concept analysis of the torque converter has been completed (Odell and Cauley, 1986), which resulted in the following input and output torque relations. The traction drive input torque, T_1, is

$$T_1 = I(A)(B)^2(\omega_0 - \omega_2)\,\omega_2 \qquad (1)$$

where I is the mass moment of inertia of the sphere, A is equal to $(N_1 N_4)/(N_2 N_3)$ which is the gear ratio parameter, and B is equal to R/r, ω_0 is output speed, and ω_2 is rotating frame speed.

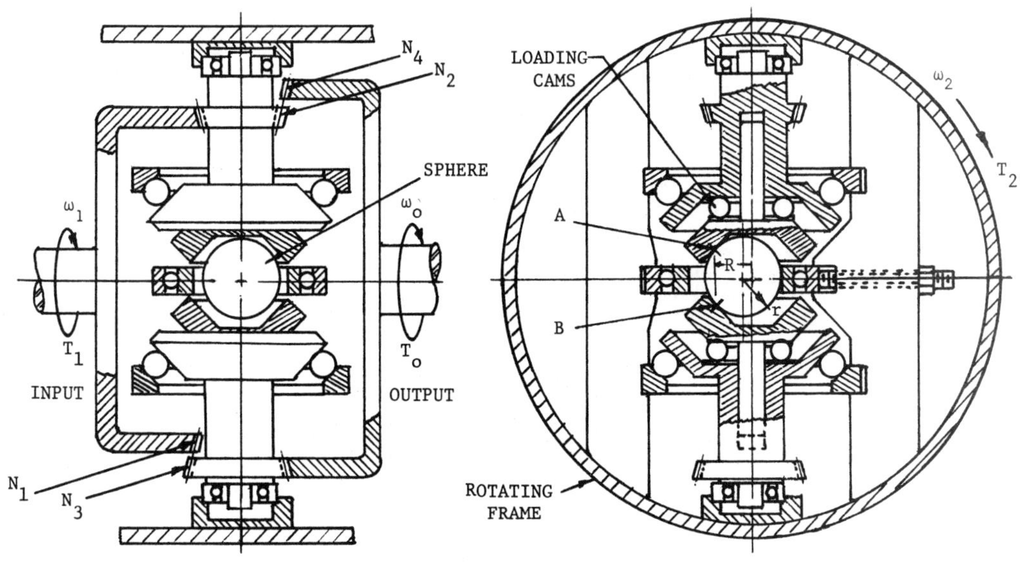

Fig. 1. Torque Converter Configuration Analyzed. Traction points: input-A, output-B. (Support bearings for input/output shafts and rotating frame not shown.)

The output torque, T_0, is

$$T_0=I(A)(B)^2(\omega_2 - \omega_1)\ \omega_2 \quad (2)$$

where ω_1 is the input speed. The frame torque, T_2 is

$$T_2=I(A)(B)^2(\omega_1 - \omega_0)\ \omega_2. \quad (3)$$

The torque converter has some unique characteristics. The input torque is independent of input speed and likewise the output torque is independent of output speed. The speeds and torques can be negative as well as positive. The maximum torques are functions of the maximum traction force at the traction points of the sphere and driving/driven surfaces. Based upon the maximum torques, a maximum/minimum speed envelope is obtained, (Odell and Cauley, 1986).

The next step in the research and development of the torque converter is the analysis and synthesis required to obtain a design method. The concept of operation has been proven and relationships obtained. A design methodology is needed in order to design the torque converter for a given application.

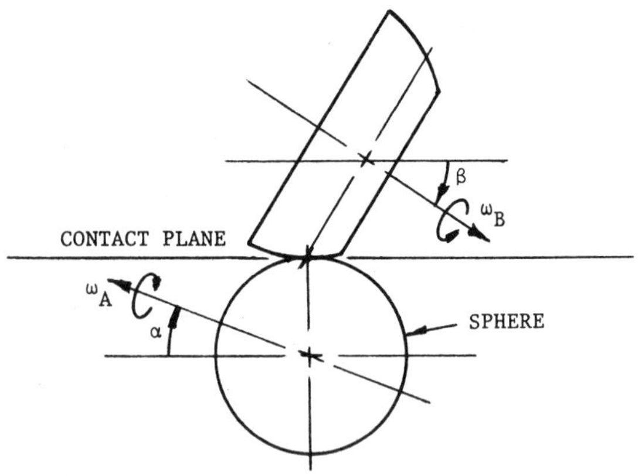

Fig. 2. Traction Contact Geometry.

ANALYSIS

The basic torque converter was reviewed to determine the critical operational and design areas. In the review, items such as frame rotation and gear speeds were checked but the critical item was the traction contact area. Structural analysis of the frame and other components was not considered.

The traction contact area between the sphere and either input or output surface was analyzed for contact stress, spin, power capacity, and fatigue life. A general contact arrangement is shown in Fig. 2. The torque converter arrangement dictates that the angle β for the input/output surfaces is fixed. The axis of rotation of the sphere shifts such that angle α varies as the input/output speeds change.

The normal force at the contact point, required for power transmission, produces an elliptical contact area. The maximum contact stress is calculated using Hertzian theory where the stress is a function of the radii of curvature and material properties of the contacting bodies. Details of this calculation can be found in the work of Rohn, Loewenthal, and Coy (1980). Typically contact stress for traction drives is in the range of 0.5 to 2.0 GPa.

The traction drive contact area was analyzed using the methods of Tevaarwerk (1979) and Loewenthal (1984). These methods consider effects of slip, spin, traction force, torque normal to the contact, and power losses, where slip is defined as the ratio of the difference in velocities of the contacting bodies to the driving body velocity, spin is the geometrical mismatch of the velocities in the contact area, angular spin velocity, ω_S, is, refer to Fig. 2,

$$\omega_S = \omega_A \sin\alpha + \omega_B \sin\beta, \quad (4)$$

and traction force is due to the shearing of the traction fluid resulting in a tangential force in the direction of motion. The required traction force to rotate the sphere was determined from the geometrical configuration of the torque converter and Eq. (1) and Eq. (2). The above method allowed for the determination of the required slip to produce the needed traction force. Power loss in the contact area due to slip and spin was calculated. The results of the above method were compared to results obtained by the method of summation of elemental values as used by Krauss (1984).

Fatigue life was evaluated using the work of Rohn, Loewenthal, and Coy (1980), which is based on 90 percent survival life. Using this method the life of each surface is calculated and then combined using the standard reciprocal formula. Since the rotational axis of the sphere is not fixed, this is a conservative fatigue life evaluation.

DESIGN PARAMETERS

Initially the design parameters considered were gear pitch line velocity, frame rotational speed, contact stress, fatigue life, traction contact surface velocity, traction force, slip, heat loss, film thickness, and wear. The use of Santotrac 50 as the lubricant was assumed. Limiting values were assigned to each parameter and the critical parameters were determined as the design study progressed. Several parameters proved not to be critical for the designs considered.

Gear pitch line velocity was limited to a maximum of 30 m/s. For the initial designs, this parameter severely limited the operational capacity and designs were modified to reduce its effect. Operational range of frame rotational speed was below the limiting value.

The remaining parameters relate to the contact area. Due to the geometry of the torque converter, the contact area was conformal which produced low contact stress. The contact stress did not exceed 0.8 GPa. The low contact stress resulted in long fatigue life and in all designs exceeds 100,000 hours.

Power transmission in the contact area is related to surface velocity, traction force, and slip. Surface velocity was limited to the range of 3 to 75 m/s. Maximum traction force is a function of the traction coefficient which varies with contact stress and surface velocity (Krauss, 1984; Heilich and Shube, 1983). The traction force was limited to 80 percent of the maximum available traction force. Slip in the contact area was limited to a 3 percent maximum.

Several empirical relationships developed by Krauss (1984) were used to evaluate contact performance. Heat loss in the contact area was limited to a maximum of 40 watts per square mm. Minimum lubricant film thickness was 0.0003 mm. Wear factor is an empirical factor relating heat loss and lubricant flow rate in the contact area. Krauss (1984) recommends a maximum value of 5.

SYNTHESIS

The torque converter was analyzed using the above analysis. The initial design configurations, not shown, had limited power capacity and operating speed range. Gear pitch line velocities were high and contact surface velocities were low. Modifications were made and studied to obtain increased performance. Other configurations developed as the design process continued.

The relationship between input and output contact surface performance and the input and output gearing was studied using computer graphics, Fig. 3. A constant input speed results in a vertical line and the output speed is the curved line. As the output speed increases the axis of the sphere, α , Fig. 2, rotates and the operating position on the graph, Fig. 3, shifts. (The angle α for the input contact increases in the negative direction and the angle decreases for the output contact). Related positions of input and output contacts have the same letter, for example input position A relates to output position A. The maximum power that can be transmitted is the smaller value of input or output capacity for the given speeds.

Computer simulation was used to study the operational characteristics of the complete torque converter. Each design configuration was simulated using the same constraint limits. Contact performance studies were used to define the speed ranges to be considered. These studies resulted in the speed-up gearing and configuration shown in Fig. 1. Other design parameters that evolved were the conformal contact geometry with relatively low contact stress.

Designs both with and without the cam loading were studied. Cam loading did increase the speed range at the slower speeds.

Evaluation of the various designs was accomplished via the use of plots similar to Fig. 4. With all design parameters, such as contact geometry, contact stress, gear ratios, etc. fixed, the power that can be transmitted is a function of the two input speeds and the output speed. It may be that for any particular combination of speeds, that one or more of the design constraints are violated and the unit should not be operated at those speeds. Figure 4 shows the allowable operating region with several power values indicated. The design parameters used to generate these results are given in the appendix. Figure 4 is for one rotating frame speed only. Increasing the frame speed results in decreased speed ranges but increased power capacity. Changing any of the design parameters influences the shape and power values of Fig. 4. Space limitation prohibit the presentation of a complete set of performance graphs.

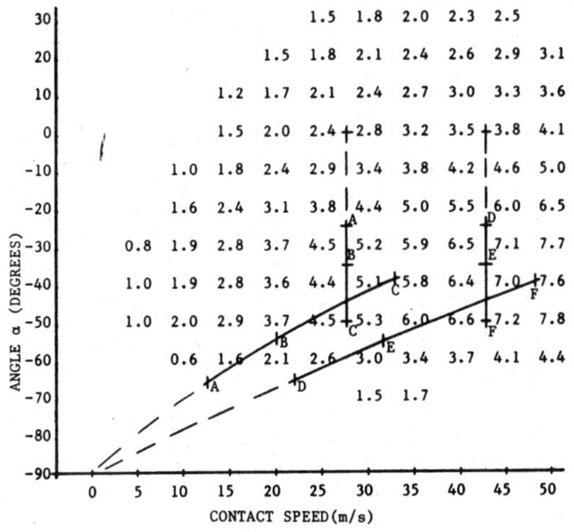

Fig. 3. Traction Contact Study Chart. Power values shown in kW.

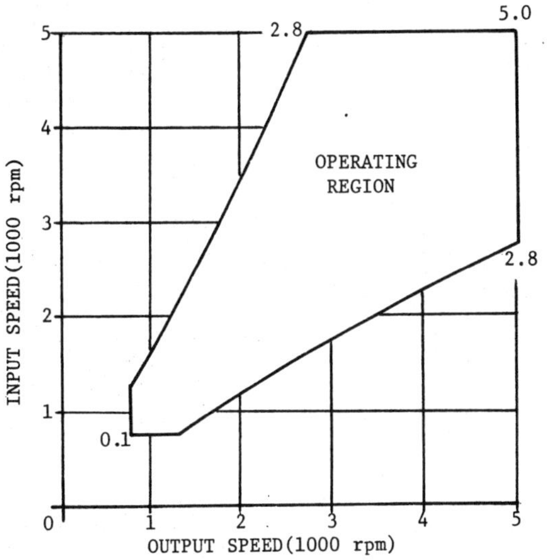

Fig. 4. Torque Converter Performance. Output power (kW) shown adjacent to operating region. Rotating frame speed is 144 rpm.

Contact performance can be evaluated for power transmission over a range of output speeds for a given input speed. The effects of contact geometry and loading are developed via the study of the allowable speed range and power transmission level. Designs that allow for high input speeds have a larger operating speed range. If input speeds are low, contact geometry and loading must be adjusted to accommodate a greater operating range. Expansion of operating range reduces the power capacity. Contact performance chart studies are valuable with regard to improving designs of the torque converter.

CONCLUSIONS

A design method has been developed for a new type of torque converter. It was based on sound engineering practice using known fundamentals. Results show that a significant amount of power can be transmitted over a usable speed range. A torque converter using a 50 mm sphere can transmit 5 kW with output speed ratios as high as 3:1. Greater power can be transmitted but the operating range is reduced.

The basic design method can be applied to any traction drive configuration. Application requires evaluation of the traction contacts and their interaction with the physical geometry of the drive configuration. If one of the surfaces in the traction contact is not spherical, then non-symmetrical curvatures must be incorporated into the design method.

The analysis and synthesis presented is one step in the overall development process of the traction drive torque converter. It is needed for designing a test model which is the next step in the development.

REFERENCES

Heilich, F. W., and E. E. Shube (1983). Traction Drives: Selection and Application. Marcel Dekker, New York.

Kraus, C. E. (1984). Bearings and Rolling Traction Analysis and Design. Excelermatic, Austin, Texas.

Loewenthal, S. H. (1984). Spin Analysis of Concentrated Traction Contacts. NASA Technical Memorandum 83713, Lewis Research Center, Cleveland, Ohio.

Loewenthal, S. H., and D. A. Rohn (1984). Elastic Model of the Traction Behavior of Two Traction Lubricants. ASLE Transactions, 27, 129-137.

Odell, E. I., and L. S. Cauley (1986). A Traction Drive Torque Converter. ASME, Paper 86-DET-82.

Rohn, D. A., S. H. Loewenthal, and J. J. Coy (1980). Simplified Fatigue Life Analysis for Traction Drive Contacts. ASME, Paper 80-C2/DET-40.

Tevaarwerk, J. L. (1979). Traction Drive Performance Prediction for the Johnson and Tevaarwerk Traction Model. NASA Technical Paper 1530, Lewis Research Center, Cleveland, Ohio.

APPENDIX

Design Parameters for Sample

Sphere diameter	50 mm
Traction ring radii of curvature	-55 mm
Traction ring rolling radius(R)	19.4 mm
Gear ratios	1:4
Traction point preload	450 N
Loading cam slope	1/70
Contact stress(max.)	0.75 GPa

Design and Analysis of Magnetic Bearings

D. K. ANAND*, J. A. KIRK*, G. E. RODRIQUEZ** AND P. A. STUDER

**Mechanical Engineering Department, University of Maryland, College Park, MD 29742, USA*
***NASA/GSFC, Greenbelt, MD 20771, USA*

The current research utilizes the approach of radially supporting a single rotor using permanent and electromagnetic in a parallel magnetic path. Static stabilization in all three translational coordinates is achieved via electromagnets that are driven by an error signal generated by radial position sensors. These sensors are located orthogonal to each other in order to obtain decoupling between motion in the x and y directions. The motor/generator is based on brushless DC-permanent magnet/ironless technology using electronic commutation. The system has been tested at over 9500 rpm with satisfactory results. The successful fabrication and testing of this has provided the necessary confidence to proceed to a larger size. More importantly it has indicated the necessity of controlling additional degrees of freedom in order to obtain stable dynamic suspension at higher rpm.

A matrix of key technologies versus various prototype designs is presented to show the current status and emphasize future directions.

Keywords: Energy storage, magnetic suspension, composite flywheel, brushless d.c. motors

INTRODUCTION

Magnetic bearings have many advantages over conventional ball bearings and therefore present unique opportunities in rotating machinery. Magnetic suspension is particularly attractive in conjuction with the use of high-strength composites for the flywheel design (Anand, Kirk and Frommer, 1985; Evans and Kirk, 1985; Kirk, Studer and Evans, 1976; Kirk and Huntington, 1977; Kirk, 1977; Kirk and Studer, 1977; Kirk, Anand, Evans and Rodriguez, 1984; Kirk, Anand and Khan, 1985). Composites allow the attainment of very high speeds and consequently high energy to weight ratios. The weight penalty due to magnetic suspension is small since the flywheel uses a relatively light flux return ring, which is a small fraction (typically 0.1 to 0.01) of the flywheel weight.

A magnetically supported bearing could theoretically have a reliable lifetime on the order of 20 years. This extraordinary lifespan is attributed to the total elimination of bearing friction. The lifetime, in fact, should be governed only by the life of the control and motor electronics.

Magnetic bearings have been considered for a wide variety of applications. Recently, magnetic bearings have been considered for applications in supporting space telescopes, vibration damping, and machine tools (Anand, Kirk and Anjanappa, 1986; Robinson, 1984; Studer, 1978).

The purpose of this work is to simulate the operation of a magnetic bearing, and apply the results of the simulation to the design and construction of a 300 WH (108 kJ) spokeless flywheel energy storage system.

BEARING CONFIGURATION

An active pancake magnetic bearing successfully developed at The University of Maryland is shown in Figs. 1 and 2. "Pancake" refers to the sandwiching of permanent magnets (PM) between ferromagnetic plates. The flux distribution from the permanent magnets support the bulk of the rotor weight. Four electromagnetic coils are located near the permanent magnets to control the rotor about its unstable equilibrium point, i.e., the point at which the air gap is constant around the stator. When the rotor displaces radially, the motion is sensed by a position transducer at the periphery of the rotor. The control system responds by sending a control current through the coils which results in an additional corrective flux distribution. This flux adds to the permanent magnet flux on the large gap side and subtracts from the permanent magnet flux on the small gap side. The net result is a corrective force which moves the rotor back to the center (nominal) position. An identical radial control system exists for the orthogonal direction. The control in the axial direction is passive.

DESIGN OF A 300 WATT-HOUR MAGNETIC BEARING

Design Considerations

The requirements for the 300 WH (108 kJ) energy storage system are given in Table 1 along with the simulation results. Geometrically, an ideal 300 WH (108 kJ) system should be as compact as possible, thus the proposed system is a stack arrangement of two magnetic bearings similar to the prototype shown in Fig. 3. Due to the 18 lb (80 N) rotor, each bearing must carry 18 lbs. (80 N) for a factor of safety of two. The center section, between the two magnetic bearings, contains a high efficiency ironless armature, electronically commutated motor/generator. The motor/generator system is described in Anand, Kirk and Anjanappa (1985), Kirk, Studer and Evans (1976), Kirk and Studer (1977), and is not discussed further in this paper.

It is assumed that, in general, the stack arrangement can be designed as two separate bearings. This arrangement allows for the control of rocking motion by independently controlling the two radial control systems. In order to have the magnetic bearing stack operate as required it is necessary to include back up (touchdown) bearings in the mechanical package.

The approach to designing the 300 WH (108 kJ) system is to compute magnetic stiffness based upon permeance modeling. The detailed modeling of the permeance are presented in Bangham (1985) and Vieira (1985). The most important characteristic is the total axial force carrying ability. Both the maximum axial force, and the curve shape (axial force vs. displacement) are important parameters. To prevent excessive sag, it is desireable to have a large slope over lower displacements. Also, a large peak axial force prevents the loss of suspension under external and axial force overloads.

The overall axial performance can be increased by altering a variety of parameters. The parameters which primarily affect the axial force in order of importance are permanent magnet size, rotor and stator diameter, and pole face size. Increasing magnet size is the easiest method to increase the axial force. However, magnet size is restricted by saturation effects. Saturation can be tolerated to some extent, so long as it does not occur within the range of radial motion; i.e., the range of operation. The range of operation is not equivalent to the actual nominal air gap since the actual air gap is constrained by touchdown bearings to prevent saturation and retain alignment. The operating range must be sufficiently large to allow the control system to correct for any external disturbances.

The total axial force is also altered by increasing the rotor and stator diameters. These dimensional changes create a larger pole face area, and thus facilitate the use of larger magnets without causing saturation to occur prematurely.

Increasing the pole face size similarly increases the bearing's ability to stall the occurrance of saturation. However, attention should be paid to the role of pole face area. As area is increased the flux density in the air gap will decrease. Since the flux term is squared in the force equation, the overall force will decrease if the pole face area is increased without correspondingly increasing the magnet size.

The beneficial effects of a large radial stiffness is seen by perturbing the flywheel position and introducing ΔB from the control coil. The correction force current about the nominal position is,

$$F = K[(B+\Delta B)^2 - (B-\Delta B)^2] \qquad K = \text{constant}$$

Simplifying,

$$F = 4K(\Delta B)B$$

This shows that a large radial flux (B) is beneficial for providing a large correcting force. This is an important and somewhat unexpected point. A large radial flux indicates the presence of a large destabilizing force. However, the nature of the control system is such that the destabilizing force is harnessed beneficially. The above procedure is also applicable when the rotor is not in the centered position. In this case the flux density will not be equivalent on both sides of the rotor, and the resulting expression is more complex.

Simulation Results

The program MAGBER was run over a wide range of parameters of interest. The input values were obtained either by iteration, experience or considerations discussed in the previous section. In this section the final results of the design are presented and discussed.

Figures 4-6 are the result of a typical design run to determine the geometry of the 300 WH (108 kJ) design. Figure 4 lists a summary of the results and the parameters of the designs attempted. Figure 5 shows axial stiffness, the most critical parameter for the design. It is so critical that all of the other parameters are of secondary importance. It was observed in earlier simulation runs that a 3 inch (76 mm) diameter system was of insufficient size to support an 18 lb (80 N) wheel with a reasonable factor of safety. A 4 inch (101 mm) stator is the smallest stator which can satisfactorily support the desired weight. From Fig. 6 the maximum axial force is ~ 30 lbs (~133 N) and the axial stiffness over small displacements is 1667 lbs/in (292 N/mm). Due to symmetry the axial force vs. displacement for each of the four separate quadrants with no radial eccentricity is identical and each quadrant supports one-quarter of the total rotor weight. Figure 6 shows the axial stiffness under a radial eccentricity of 0.003" (0.076 mm). Quadrant 3 will not support as much weight as quadrant 1. However, the increased weight carrying ability of quadrant 1 tends to cancel the effect of the decreased weight carrying ability of quadrant 3, and the effects are not seen in the resultant plot. Table 1 gives the recommended dimensions for the 300 WH (108 kJ) system based on many iterations similar to those performed above.

Control System

The control system for stabilizing the magnetic bearing is shown in Fig. 7. Basically, an eddy current transducer provides the radial position of the flywheel. The transducer signal produces an error signal that drives a current thru four electromagnets. The magnetic flux of the electromagnets adds on one side and subtracts the permanent magnetic flux on the other side, producing a corrective differential force. The control system is a nonlinear bang-bang system and is discussed in detail in Bangham (1985). The governing equations of the control system were analyzed using a program called CONTRL. This was written specifically for this application to study response and stability. Details are given in Bangham (1985). The system shown in Fig. 3 was fabricated, tested and used to stabilize the magnetic bearing.

Mechanical Design

Based upon the MAGBER analysis and control system calculations, an envelope of allowable displacement for the magnetic bearing is defined. Typically, with the flywheel in the centered position, the control system is capable of maintaining rotor stability for up to ±0.008 in (±0.203 mm) of rotor displacement. The back up bearing configuration shown in Fig. 3 is normally not in contact with the rotor except when starting the system from rest. If the flywheel movement exceeds ±0.008 inches (±0.203 mm), then the back up bearings come into operation to limit flywheel excursions until the magnetic bearing re-establishes suspension of the flywheel. Conceptually, the back up bearings can be thought of as a pair of thrust bearings with axial and radial clearances designed to allow for the normal excursions of the flywheel [±0.008 inches (0.203 mm)]. The stationary portion of the bearing set is composed of two high precision ball bearings with fittings attached to their outer rings. These fittings are designed to mate with rings attached to the flywheel. The axial separation of the ball bearing sets, as well as the axial and radial clearances between stationary and rotating members, define the range of motion through which the flywheel will move before mechanical contact occurs.

Geometric considerations require that the touchdown bearing sets should be separated by as large an axial spacing as possible. This locates the bearings adjacent and between the magnetic bearings. In addition, the rotating and non-rotating portions of the flywheel should contact at as large a diameter as possible. By locating the back up bearing sets adjacent to the magnetic bearings, making the contact diameter 3.005 inches (76.269 mm), and the axial and radial clearance gaps equal to 0.006 inches (0.152 mm), the radial motion of the flywheel at the suspension rings is 0.008 inches (0.203 mm). To avoid status of all key technologies is shown in Fig. 8.

The back up mechanical package shown in Fig. 3 has been fabricated without the motor/generator and is currently undergoing testing.

CONCLUSIONS

Results from the system parametric design, and the control system analysis, show that a 300 WH (108 kJ) energy storage flywheel can be built by incorporating characteristics of a previously built magnetic bearing into a stack arrangement. The study also indicates that a method of reducing permanent magnet unbalance must be incorporated; the effects of nonlinearities, particularly those leading to limit cycles, need further evaluation; capacitive, saturation, and hysteresis effects of the iron should be included in the model but avoided in the operating range of the bearing; and finally back up bearings are required to maintain flywheel excursions in the linear operating range of the magnetic bearing control system.

ACKNOWLEDGEMENT

The work reported on in this paper has been sponsored by the National Aeronautics and Space Administration, Goddard Space Flight Center, under grant NAG5-396.

REFERENCES

Anand, D.K., Kirk, J.A. and Anjanappa, M., "Magnetic Bearing Spindles for Enhancing Tool Path Accuracy", Advanced Manufacturing Processes, Vol. 1, No. 2, April 1986.

Anand, D.K., Kirk, J.A. and Frommer, D.A., "Design Considerations for a Magnetically Suspended Flywheel System", Proceedings of the 20th Intersociety Energy Conversion Engineering Conference, August 18-23, 1985, Miami Beach, Florida, pp. 2.449-2.453.

Bangham, M.L., "Simulation and Design of a Flywheel Magnetic Bearing", M.S. Thesis, University of Maryland, College Park, 1985.

Evans, H.E., and Kirk, J.A., "Inertial Energy Storage Magnetically Levitated Ring-Rotor", Proceedings of the 20th Intersociety Energy Conversion Engineering Conference, August 18-23, 1985, Miami Beach, Florida.

Kirk, J.A., "Flywheel Energy Storage-I: Basic Concepts", Int. J. Mech. Sci., Vol. 19, 1977, pp. 223-231.

Kirk, J.A., Anand, D.K., Evans, H.E., and Rodriguez, E.G., "Magnetically Suspended Flywheel System Study", Presented at the Integrated Flywheel Tech. 1984 Workshop, NASA Marshal Space Flight Center, Huntsville, Alabama, Feb. 7-9, 1984.

Kirk, J.A., Anand, D.K. and Khan, A.A., "Rotor Stresses in a Magnetically Suspended Flywheel System", Proceedings of the 20th Intersociety Energy Conversion Engineering Conference, August 18-23, 1985, Miami Beachg, Florida, pp. 2.454-2.462.

Kirk, J.A., and Huntington, R.A., "Energy Storage-An Interference Assembled Multiring Superflywheel", 12th Intersociety Energy Conv. Engin. Conf., Aug. 28th-Sept. 5, 1977, pp.517-524.

Kirk, J.A., and Studer, P.A., "Flywheel Energy Storage-II: Magnetically Suspended Superflywheel", Int. J. Mech. Sci., Vol. 19, 1977, pp. 233-245.

Kirk, J.A., Studer, Philip A, and Evans, Harold E., "Mechanical Capacitor", NASA TN D-8185, 1976.

Robinson, A.A., "Magnetic Bearings-The Ultimate Means of Support for Moving Parts in Space", Spacecraft Tech. Dept., ESA Tech. Directorate, ESTEC, Noordwijk, Netherlands.

Studer, Philip A., "Magnetic Bearings for Instruments in the Space Environment", NASA TM-78048,1978.

Vieira, Rogerio de Azeucdo, "Analysis of a Magnetic Bearing with Two Degrees of Freedom", Masters Thesis, University of Maryland, College Park, 1985.

Zusammenfassung: In dem hier beschriebenen Forschungs Projekt wird ein radial gelagerter Rotor mit Permanentmagnet und Electromagneten parallel unterstützt. Satische Stabilitaet wird durch Electromagneten, welche durch radiale Positionssensoren kontrolliert wreden, erreicht. Der Motor/Generator basiert auf dem bürstenlosen Gleichstrom-Permanentmagnet-Prinzip. Das System wurde bei über 9500 Umdrehungen pro Minute zufriedenstellend getested. Eine Zusammenstellung der wichtigsten Technologien mit verschiedenen Prototypenausfürungen ist dargestellt um den Stand der Entwicklung zu dokumentieren and zukünftige Enwicklungen aufzuzeigen.

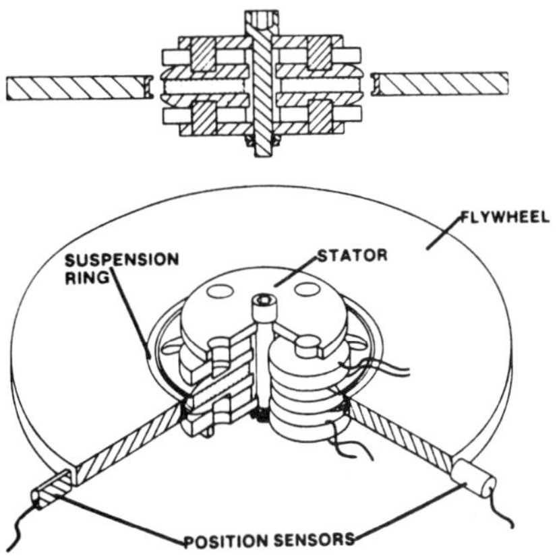

Figure 1. Pancake Magnetic Bearing

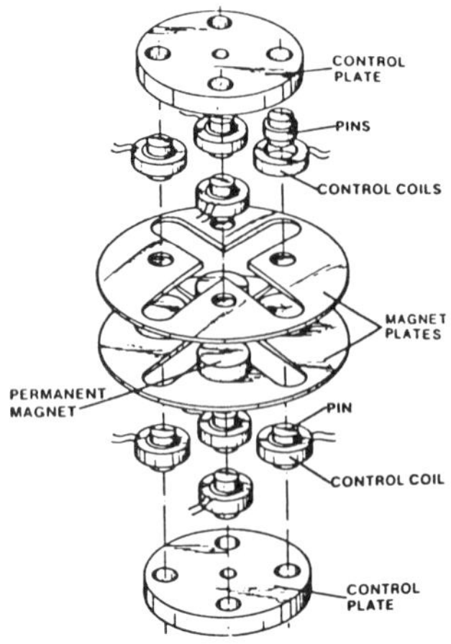

Fig. 2. Exploded View of Stator for Pancake Bearing

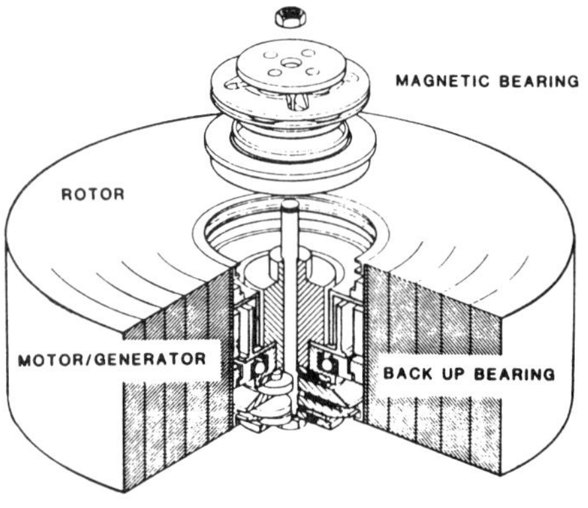

Fig. 3. Back up Bearing Arrangement for Stator-Stack

Table 1. Requirements for 300 WHR (108 kJ) Energy Storage System

ITEM	REQUIREMENT	SOURCE
ENERGY REQUIREMENT	300 Watt-Hour STORAGE CAPABILITY	DESIGN PARAMETER
TOTAL POWER CONSUMPTION	60 Watts	DESIGN PARAMETER
TOTAL ENERGY CONSUMPTION	30 Watt-Hour	DESIGN PARAMETER
ROTOR WEIGHT	18 lbs (80 1 N)	FLYWHEEL DESIGN REQUIREMENT
STATOR SIZE RANGE	3in TO 5in ID (76 1mm TO 126 9mm)	FLYWHEEL DIMENSION REQUIREMENT
STABILITY	STABLE SYSTEM UNDER 2-g RADIAL LOAD	DESIGN CRITERION
STATOR DIAMETER	4 in (101 5 mm)	SIMULATION RESULT
NOMINAL GAP	0 035 in (0 888mm)	SIMULATION RESULT
MAGNET AREA	1 45 in^2 (934 06mm^2)	SIMULATION RESULT
MAGNET LENGTH	0 3 in (7 6 mm)	SIMULATION RESULT
POLE FACE THICKNESS	0 065 in (1 650mm)	SIMULATION RESULT

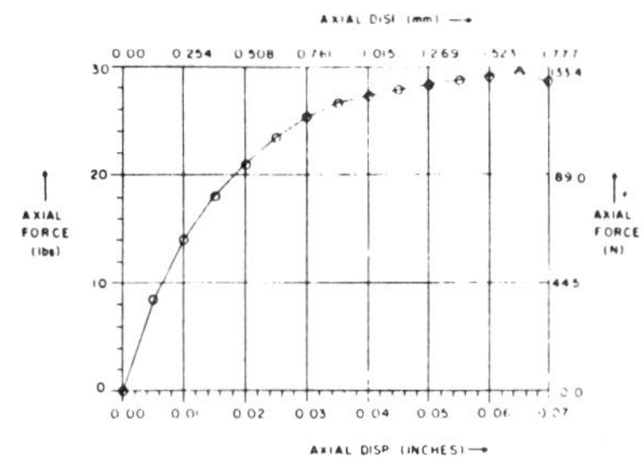

Fig. 5. Resultant Axial Stiffness/Radial Bias 0

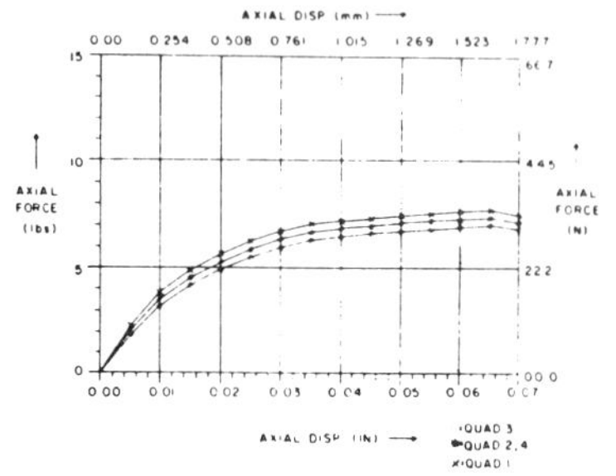

Fig. 6. Quadrant Axial Stiffness/ Radial Bias 0.003 in

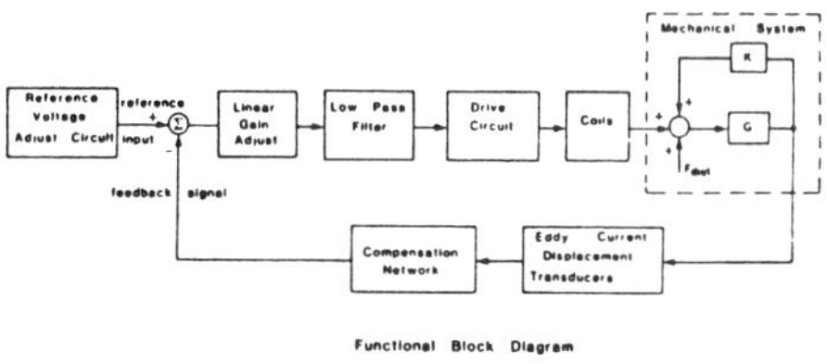

Fig. 7. Functional Block Diagram for Laboratory Prototype

	TOUCHDOWN BEARING*	MAGNETIC SUSPENSION	MOTOR/ GENERATOR	CONTROL SYSTEM*	COMPOSITE FLYWHEEL	DYNAMIC BALANCING	VAC. CHAMBER TESTING
3 inch DESIGN	C	C	C	S	D	A	T
4 inch DESIGN	C	C	D	R	D	A	-
STACK DESIGN (3 inch)	T	T	D	U	D	A	T
PROTOTYPE (4 inch deliverable)	Implementation under way		D	R	D	-	-

LEGEND

* = Major Thrust Area U = Unstable
C = Complete D = Design Complete
T = Undergoing Testing R = Technology Being Developed
S = Stable A = Analytical Evaluation Underway

Fig. 8. Status of All Key Technologies

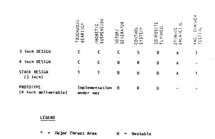

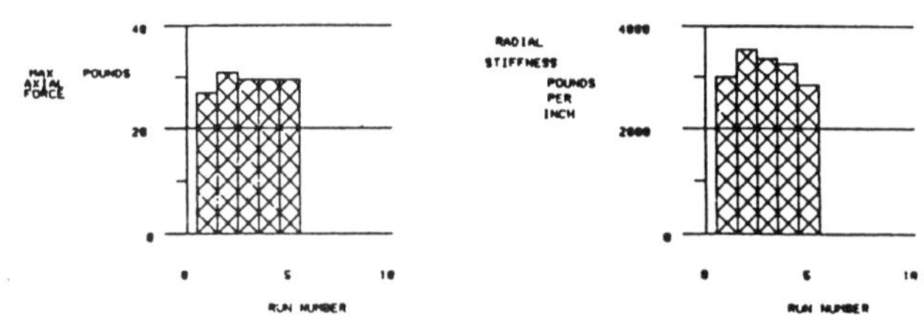

RUN	AM	DEL	KRAVE	FAMAX@H	RADS/AXF
1	1 3500	0 0000	3001 0	26 706	112 383
2	1 5000	0 0080	3532 0	30 824	114 575
3	1 4500	0 0100	3352 0	29 418	113 955
4	1 4500	0 100	3263 0	29 418	110 564
5	1 4500	0 0090	2840 0	29 454	96 436

FOR ALL RUNS LM=0 3000
T=0 0650
RI=2000
PLEAK=1 050
BSAT=1 500
H=0 0850
GZERO=0 03500

Fig. 4. Summary of Design Parameters

A Finite Element Formulation for the Hygrothermal Response of Mechanisms Fabricated with Polymeric Composite Laminates

C. K. SUNG*, B. S. THOMPSON AND M. V. GANDHI

Machinery Elastodynamics Laboratory, Mechanical Engineering Department, Michigan State University, East Lansing, MI 48824, USA
**now Associate Professor, Department of Power Engineering, National Tsing Hua University, Hsinchu, Taiwan*

ABSTRACT

The next generation of machine systems will need to incorporate high strength, high stiffness, lightweight members in order to achieve superior performance characteristics. One of the candidate materials from which these members will be fabricated is the polymeric composite laminates which possess the significant advantages of permitting their macromechanical structure to be tailored to suite each application, and in addition, they offer high strength, high stiffness, low mass members to be fabricated. The constitutive behavior of some of these advanced materials is, however, dependent upon the ambient environmental conditions and hence the impact of ambient temperature and moisture must be assessed carefully if the full potential of these materials are to be realized in practice. The subject of this paper is the development of a methodology for predicting the hygrothermoelastodynamic response of machine systems fabricated in these advanced materials. This methodology is based on a variational theorem which provides the basis for a finite element formulation. An illustrative example serves to demonstrate some preliminary work on predicting the response of a four bar linkage with a polymeric composite rocker link which is simultaneously subjected to both mechanical and hygrothermal loadings.

INTRODUCTION

The current generation of machine systems are often fabricated with articulating members possessing substantial weight and high stiffness. Such characteristics significantly constrain the performance of these systems, and this is especially true in the area of robotics where, for example, typical payload-weight/arm-weight ratios are only about 1:20 [1]. If superior performance characteristics are to be achieved, which may be measured by higher operating speeds, reduced energy consumption or a more rapid response time for example, then lightweight, high stiffness, high strength members must be designed, engineered and fabricated.

Members with these desirable properties can be developed by fabricating these machine elements with modern polymeric composite materials. Advanced composite materials offer significantly superior characteristics to the commercial metals for these machinery applications [2,3], because of their high strength and stiffness to weight ratios, superior damping properties, and furthermore, the materials may be tailored to suit each individual application. However, if the potential of these modern materials is to be realized in practice, then their sensitivity to changing environmental conditions must be assessed carefully.

In engineering practice, a robot or a piece of machinery will rarely operate under static environmental conditions such as constant temperature and constant ambient relative humidity. These environmental conditions, however, can influence the response of machine systems fabricated with polymeric composite materials because the constitutive behavior of these advanced materials is often heat and moisture dependent. Figure 1 presents a schematic of the inter-relationship between the relevant phenomena and the resulting impact on the performance of a machine system fabricated in polymeric composite materials. Polymeric composite materials absorb moisture. Thus as the ambient relative humidity increases the material absorbs more moisture which increases the mass of the structural members with an associated increase in the inertial loading and the dynamic stresses. In addition, as the humidity increases the stiffness of the structural member decreases. The increased dynamic stresses can create a more permeable medium [8,9] resulting in a further reduction in stiffness, a further increase in mass, and the subsequent diffusion of even more moisture. This coupled situation is exacerbated by the dependence of stiffness, strength and diffusivity upon the ambient temperature.

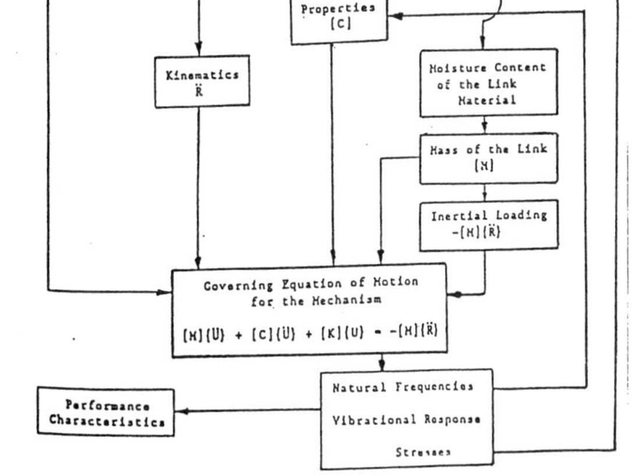

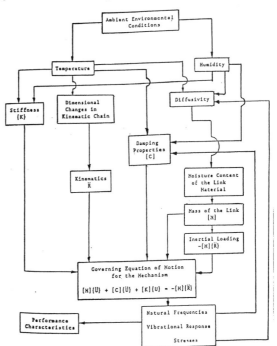

Fig. 1 Impact of Temperature/Humidity Conditions on the Dynamic Response of Mechanisms, Fabricated with Polymeric Composite Materials

If a viable methodology is to be established for the design of high-performance machine systems fabricated with polymeric composite materials, then a suite of comprehensive investigations on the hygrothermoelastodynamic response of these systems must be undertaken first. This paper represents a first step towards achieving that goal. Herein, a finite element formulation is presented for analyzing machine systems which are simultaneously subjected to both hygrothermal loading and also mechanical loading. An illustrative example is presented in which the response of a linkage mechanism with one link fabricated in a polymeric composite laminate is simulated.

FINITE ELEMENT FORMULATION

A variational theorem was presented in references [4] and [5], to provide the basis for undertaking hygrothermoelastodynamic analyses of machine systems by considering the motion of a chain of continua subjected to both mechanical and hygrothermal loadings. The first variation, yields, as stationary conditions the governing field equations and also the relevant boundary conditions for this class of problem.

The objective, herein, is to develop a "displacement" finite element formulation for a single one-dimensional finite element with two exterior nodes, each having five nodal degrees of freedom. Two nodal variables W and ϕ describe the flexural displacement and slope, respectively, which are assumed to be governed by the Bernoulli-Euler hypothesis, U describes the longitudinal displacement, T denotes the temperature field and M is the moisture concentration. Thus, the nodal "displacement" vector for the element may be defined by

$$\lfloor U \rfloor^T = \lfloor U_1, W_1, \phi_1, T_1, M_1, U_2, W_2, \phi_2, T_2, M_2 \rfloor \quad (1)$$

The general deformation displacement is represented by

$$u(x,t) = \lfloor N^s \rfloor \lfloor U \rfloor, \quad (2)$$

the temperature field is represented by

$$\theta(x,t) = \lfloor N^h \rfloor \lfloor T \rfloor, \quad (3)$$

and the moisture concentration is represented by

$$m(x,t) = \lfloor N^m \rfloor \lfloor M \rfloor \quad (4)$$

where $\lfloor N^s \rfloor$, $\lfloor N^h \rfloor$ and $\lfloor N^m \rfloor$ are the associated shape functions describing the spatial distribution of the variables.

The form of the finite element equations based on the first variation are written

$$\delta J = 0$$

$$= \int_{t_1}^{t_2} \Bigg[\int_V \lfloor \delta\gamma_{xx} \rfloor^T \Big[\lfloor \sigma_{xx} \rfloor - [C][B^s][U] + [\beta][N^h][T] + [\alpha][N^m][M] - 1/(\tilde{\rho}M_o)[D]^{-1}[G]\Omega\partial[q^{(M)}]/\partial x \Big] dV$$

$$+ \int_V [\delta s]^T \Big[[N^h][T] - \theta_o/c\,([s] - [\beta][B^s][U] + \mu[N^m][M]) \Big] dV$$

$$+ \int_V [\delta\pi]^T \Big[[N^m][M] - M_o/b([\pi] - [\alpha][B^s][U] + \mu[N^h][T]) \Big] dV$$

$$+ [\delta U]^T \Big[[M^{ss}][\ddot{U}] + [C^{ss}]\{\dot{U}\} + [K^{ss}][U] + [K^{sh}][T] + [K^{sm}][M] + [M_R^s][\dot{P}_R] - \int_V [N^s]^T[X]dV - \int_{S\sigma} [N^s]^T([\bar{g}] - [g])dS$$

$$+ \int_V 1/(\tilde{\rho}M_o)[D]^{-1}[G][B^s]^T\Omega\partial[q^{(M)}]/\partial x dV \Big]$$

$$+ [\delta T]^T \Big[[C^{hs}][\dot{U}] + [C^{hh}][\dot{T}] + [C^{hm}][\dot{M}] + [K^{hh}][T] + \int_{S_H} [N^h]^T([\bar{Q}^{(H)}] - [Q^{(M)}])dS_H \Big] \quad (5)$$

$$+ [\delta M]^T \Big[[C^{ms}][\dot{U}] + [C^{mh}][\dot{T}] + [C^{mm}][\dot{M}] + [K^{mm}][M] + \int_{S_D} [N^m]^T([\bar{Q}^{(M)}] - [Q^{(M)}])dS_D \Big]$$

$$+ \int_V [\delta\sigma]^T \Big[[\gamma] - [B^s]\,[U] \Big] dV$$

$$+ \int_V ([\delta q^{(H)}]^T[K]^{-1}/\tilde{\rho}\theta_o) \Big[[q^{(H)}] + [K][B^h][T] \Big] dV$$

$$+ \int_V ([\delta q^{(M)}]^T[D]^{-1}/\tilde{\rho}M_o) \Big[[q^{(M)}] + [D][B^m][M] + [G][B^s][\gamma] \Big] dV$$

$$- \int_V \rho[\delta P]^T \Big[[P] - [N_R][P_R] - [N][\dot{U}] \Big] dV$$

$$+ \int_{S_u} [\delta g]^T[N^s]([U] - [\bar{U}])\, dS_u$$

$$+ \int_{S_\theta} ([\delta q^{(H)}]^T/\tilde{\rho}\theta_o)[N^h]([T] - [\bar{T}])dS_\theta$$

$$+ \int_{S_M} ([\delta q^{(M)}]^T/\tilde{\rho}M_o)[N^m]([M] - [\bar{M}])dS_M \Bigg] dt$$

where,

$$[C^{ms}] = \int_V [N^m]^T M_o[\alpha][B^s]\, dV$$

$$[C^{mh}] = -\int_V [N^m]^T 2\mu M_o[N^h]\, dV$$

$$[C^{mm}] = \int_V [N^m]^T b[N^m]\, dV$$

$$[K^{mm}] = \int_V [B^m]^T\Omega[D][B^m]\, dV$$

$$[K^{ms}] = \int_V [B^m]^T\Omega[G][B^s]\, dV$$

$$[\bar{Q}^{(M)}] = -[D][B^m][M]$$

$$[C^{hs}] = \int_V [N^h]^T\theta_o[\beta][B^s]\, dV$$

$$[C^{hh}] = \int_V [N^h]^T c[N^h]\, dV$$

$$[C^{hm}] = -\int_V [N^h]^T 2\mu\theta_o[N^m]\, dV$$

$$[K^{hh}] = \int_V [B^h]^T[K][B^h]\, dV \quad (6)$$

$$[\bar{Q}^{(H)}] = -[K][B^h][T]$$

$$[M_R^s] = \int_V [N^s]^T \rho [N_R] \, dV$$

$$[M^{ss}] = \int_V [N^s]^T \rho [N^s] \, dV$$

$$[K^{ss}] = \int_V [B^s]^T [C][B^s] \, dV$$

$$[K^{sh}] = -\int_V [B^s]^T [\beta][N^h] \, dV$$

$$[K^{sm}] = -\int_V [B^s]^T [\alpha][N^m] \, dV$$

where [C] (N/m^2) is the elastic modulus matrix; [β] ($N\text{-}m/m^{3\,\circ}K$) and [α] ($N\text{-}m/m^3$) are thermal-expansion and hygroscopic-expansion coefficient matrices; [N_R] contain the shape functions approximating the rigid-body kinematic fields while [P_R] is the column vector containing the nodal rigid-body kinematic degrees of freedom. The nodal absolute velocity components are defined by [P]; the surface tractions, heat and mass fluxes are defined by [g], [$Q^{(H)}$] and [$Q^{(M)}$], respectively; [B] is the spatial derivative of shape function [N]. Space limitations have prevented a comprehensive nomenclature to be documented here. The interested reader is referred to references [4] and [5] for further details on the approach.

This completes the development of a variational formulation for undertaking hygrothermoelastodynamic analyses of flexible machine systems, and the subsequent section presents an example to illustrate the application of this formulation.

ILLUSTRATIVE EXAMPLE

The above finite element formulation was employed to investigate the hygrothermoelastodynamic response of a flexible four bar linkage operating under two radically different thermodynamic reference states. The linkage comprised a rigid crank, which was assumed to operate at a constant speed of 296 rpm in both reference states, a rigid coupler link, and a flexible rocker link which was assumed to be fabricated in a Hercules AS/3501-6 graphite epoxy laminate with a $[0/90]_s$ ply configuration.

The first reference state was characterized by an ambient temperature of 22°C and 0.5% relative humidity. Under these conditions the elastic modulus of the material was assumed to be 82.1654 x 10^3 N/mm^2 and the density 1.6 gm/cm^3. The second reference state was characterized by an ambient temperature of 80°C and a relative humidity of 90%. These extreme conditions may be encountered in many industrial environments, such as for example, where materials are being hot worked. Under these conditions of elevated temperature and high relative humidity, the elastic modulus was assumed to be adversely degraded to 78.471 x 10^3 N/mm^2 [6] and the density of the material was assumed to have increased to 1.63 gm/cm^3 [7]. The coefficients of thermal expansion of a unidirectional laminate of this material, were taken to be β_x = -0.3 (μm/m)/K° and $\beta_y = \beta_z$ = 28.1 (μm/m)/°K where the ox axis is the longitudinal fiber direction [7]. The associated swelling coefficients were assumed to be α_x = 0 and $\alpha_y = \alpha_z$ = 0.44 m/m. The coefficients of thermal expansion for the $[0/90]_s$ layup were taken to be

$$\begin{Bmatrix} \beta_1 \\ \beta_2 \end{Bmatrix}_{0^\circ} = \begin{Bmatrix} -0.3 \\ 28.1 \end{Bmatrix} \mu m/m/^\circ K$$

$$\begin{Bmatrix} \beta_1 \\ \beta_2 \end{Bmatrix}_{90^\circ} = \begin{Bmatrix} 28.1 \\ -0.3 \end{Bmatrix} \mu m/m/^\circ K \qquad (7)$$

where subscripts 1 and 2 designate the off-axis orientation. The effective coefficient of thermal expansion β_{eff}, and the effective coefficient of moisture expansion (swelling) α_{eff}, for this cross-ply laminate were assumed to be 27.8 (μm/m)/°K and 0.22 (m/m), respectively, by using equation (22) in reference [8].

The length of the crank, coupler, rocker and ground link was 63.5 mm, 308.0 mm, 310 mm and 413.0 mm respectively. The rocker link had cross-sectional dimensions of 25.4 mm perpendicular to the plane of the mechanism and 2.34 mm in the plane of the mechanism. The hygrothermal response of the flexible rocker link operating at the two reference states were simulated using a time-step of 0.0005631 seconds to numerically solve the governing finite element equations of motion. The results, which are presented in Figure 2, clearly demonstrate the large deflections associated with the system operating at the higher temperature and higher relative humidity.

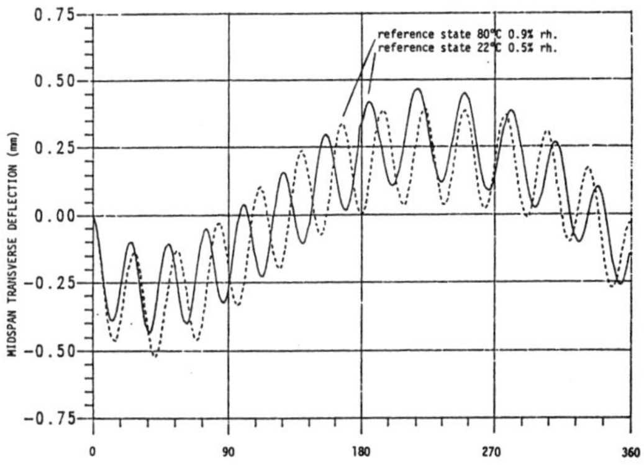

Fig. 2 Rocker Midspan Transverse Deflections in the Two Reference States

CONCLUSION

A finite element formulation has been presented for the hygrothermoelastodynamic analysis of machine systems fabricated in composite materials based on a mixed variational theorem. An illustrative example is employed to undertake a preliminary analysis of a four bar linkage simultaneously subjected to both mechanical and hygrothermal loadings in two radically different reference states.

ACKNOWLEDGEMENTS

The authors wish to acknowledge, with thanks, financial support for this work from both the National Science Foundation, under grant No. MSM-8514087, and also the Composite Materials and Structures Center at Michigan State University.

REFERENCES

1. Sadamato, K., "R&D On Intelligent Robots Being Spurted," The Japan Robot News, Vol. 1, No. 4, pp. 1-6.
2. Thompson, B.S. and Sung, C.K., "A Variational Formulation for the Dynamic Viscoelastic Finite Element Analysis of Robotic Manipulators Constructed from Composite Materials," ASME Journal of Mechanisms, Transmissions and Automation in Design, Vol. 106, No. 2, 1984, pp. 183-190.
3. Sung, C.K., Thompson, B.S., Crowley, P., and Cuccio, J., "An Experimental Study to Demonstrate the Superior Response Characteristics of Mechanisms Constructed with Composite Laminates," Mechanism and Machine Theory, Vol. 21, No. 2, 1986, pp. 103-119.
4. Sung, C.K., "A Theoretical and Experimental Investigation of the Dynamic Response of Flexible Mechanism Systems Fabricated from Fibrous Composite Materials," Doctoral Dissertation, Michigan State University, March 1986.
5. Sung, C.K. and Thompson, B.S., "A Variational Principle for the Hygrothermoelastodynamic Analysis of Mechanism Systems," ASME Paper No. 86-DET-15, ASME Journal of Mechanisms, Transmissions, and Automation in Design, (in press).
6. Springer, G.S., Environmental Effects of Composite Materials, Vol. I and II, Technomic Publishing Company, Inc., 1984.
7. Tsai, S.W. and Hahn, H.T., Introduction to Composite Materials, Technomic Publishing Co., Inc., 1980.
8. Craft, W.J. and Christensen, R.M., "Coefficient of Thermal Expansion for Composites with Randomly Oriented Fibers," Journal of Composite Materials, Vol. 15, January 1981, p. 2-20.

The Law of Braking Torque Variation Making Possible the Foil Uncoiling from a Roll of Variable Characteristic to Take Place at a Constant Speed

A. VEG, M. LUKIC AND T. L. PANTELIC

Faculty of Mechanical Engineering, University of Belgrade, Beograd, Yugoslavia

Abstract. In industrial practice, the treatment of a plastic floil takes place continually at a constant speed. The foil is usually uncoiled from a roll of variable characteristics. In order to ensure the passage of the foil at a constant speed, a braking torque is applied, in accordance with an appropriately defined function. The results obtained in this paper are used in machine development design at the Machine Mechanics Institute at the Faculty of Mechanical Engineering, University of Belgrade.

PREFACE

Modern processes in plastic industry treat the plastic foil, which moves with constant speed. To hold the process continually, it is necessary to move the foil steadily. Importance of that demand is caused by strictly defined periodical operations. Any delay in foil feeding interrupts the process.

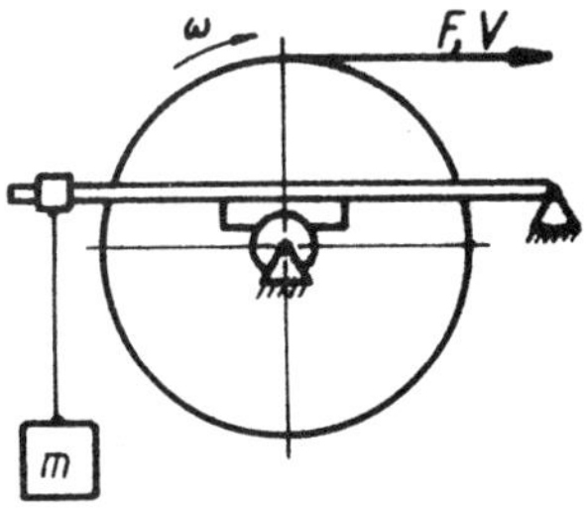

Fig. 1

The foil flow is constant, and if assumed that there is no deformation of foil (tension, wrinklage), it is imposed that the allowed speed variability is within narrow tolerances. The width of tolerated speed area is dictated by treating technology.

The roll of plastic foil unwinding,changes, it's geometric and dynamic characteristics. For that, it is necessary to compense that unsteadiness in such way, so that the resultant speed consecvently becomes constant.

There is set of various ways to solve that task. Most of them are based on backfeed controled devices. On the contrary to those devices, there is a simple construction for that purpose (Fig.1). For uncoiling speed control it is used a mechanism producing the constant braking torque. To provide the constant speed of foil throughout the process, it is necessary to variate the pulling force. This mechanism is very efficient, when applied on rolls of thin plastic foil, which unwindes slowly. When the speed uncoiling is increased, or the thickness is greater, the grade of dynamic characteristics (mass, moment of inertia, radius) changing is significant. To satisfy the demand of constant speed of uncoiling, time dependent braking torque is applied. The main idea of this paper is to define the function of of braking torque in time domain.

MECHANICAL MODEL

Model is composed of the following subassemblies:

- mechanism for braking torque application
- roll of foil
- system spring-damper-small roll
- pulling force

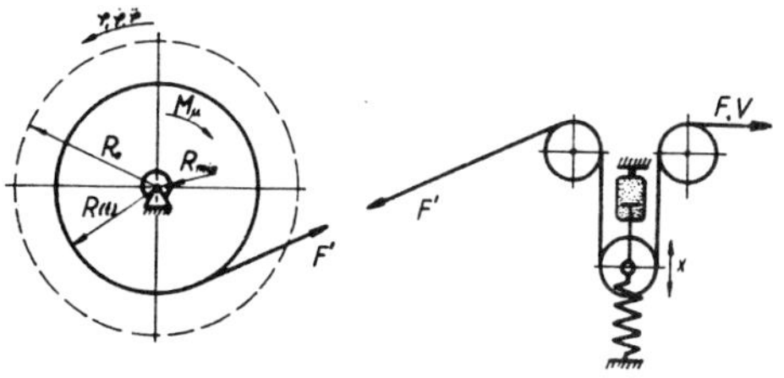

Fig. 2

Braking torque

Function defining braking torque M_μ is obtained as solution of diferential equations, depending on input quantities and initial conditions. Shape of the braking torque function determines the expected solution. That function can be applied by the mechanism based on one of the following physical phenomena:

- dry friction
- viscous friction
- electro-magnetic field.

Braking torque control depends on chosen mechanism and it's parameters. General form of braking torque function is $M_{\mu} = K \cdot P(t)$, where the K is constant defining proportional relation and P(t) is fluent variable. Due to constructiv restrains the choosing of convenient mechanism is eased. Important notice is that chosen mechanism must be able to satisfy changings of braking torque.

Roll of foil

Foil that uncoils in production process is formed as roll. Unwinding starts at maximal radius and through time decreases to minimal one, almost of neglectable cuantity. The radius of roll is time dependent. Roll mass and moment of inertia are functions of radius that changes. Following list expresses the explanation and relations of all fluent variables and constants. Those that are time dependent are marked on the list, and through text their time dependency will be omitted, for ease of writing and printing:

z - foil thickness
R_0 - max. radius
R(t) - variable radius
$\varphi(t)$ - angle coord.
$\dot{\varphi}(t)$, $\ddot{\varphi}(t)$ - angular speed and acceleration
J(t) - polar moment of inertia
F'(t) - tension force in foil
m_0 - max. mass
m(t) - variable mass

Relations: $R = R_0 - (z/2\pi)\cdot\varphi$

$$J = (m/2R_0^2)\cdot\left[R - (z/2)\right]^4$$

$$m = (m_0/R_0^2)\cdot\left[R - (z/2)\right]^2$$

System spring-damper-small roll

The subassembly spring-damper-small roll is implanted to absorb the instantenious unsteadiness that may occur, The ideal operation of all subassemblies excludes the need for such compensing system. This paper deals the motions of that subassembly, but using real initial conditions fluence of that system is annuled, especially for assumption of ideal acting of braking torque. Parameters and relations for small roll are as follows:

D - diameter
J_R - moment of inertia
β - coefficient of damping
$\psi(t)$ - angle coord.
$\dot{\psi}(t)$, $\ddot{\psi}(t)$ - angular speed and acceleration
m_R - mass
x - displacement
$\dot{x}(t)$, $\ddot{x}(t)$ - speed and acceleration
c - spring stiffness

Relations: $\dot{\varphi}R/a = V/(D - a)$

$$\dot{\psi} = \dot{\varphi}R/a = (R\dot{\varphi} + V)/D$$

$$\ddot{\psi} = (\ddot{\varphi}R + \dot{\varphi}\dot{R})/D$$

Pulling force

Uncoiling foil is transformed to new products, in treating process. Pulling force is defined in accordance to following phases in production. For the further analysis, pulling force and foil speed are considered as constant: F = conts; V = conts.

MATHEMATICAL MODEL

(1) $-d/dt(J\dot{\varphi}) = M_{\mu} - F'R$	torque equilibrium equat.
(2) $\dot{\varphi}R + \dot{x} = V$	kinematic equation
(3) $F' + F - xc - \dot{x}\beta - \ddot{x}m_R = 0$	force equilibrium-small roll
(4) $(F' - F)D/2 + d/dt(J_R\dot{\psi}) = 0$	torque equilibrium-small roll

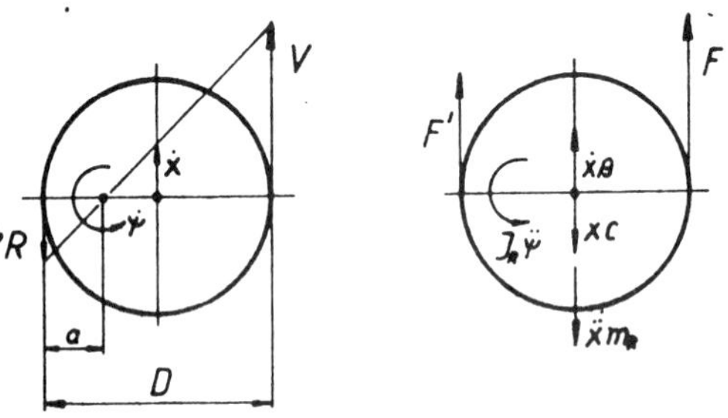

Fig. 3

(2) $x = Vt + A\varphi^2/2 - R_0\varphi + 2F/c$; $A = z/(2\pi)$, $B = 2J_R/D^2$

(3) $F' = F - (\beta\dot{\varphi} + m_R\ddot{\varphi})(R_0 - A\varphi) + \beta V + m_R A\dot{\varphi}^2 + c(Vt + A\varphi^2/2 - R_0\varphi)$

(4) $(m_R - B)(A\varphi - R_0)\ddot{\varphi} + (m_R - B)A\dot{\varphi}^2 + (A\varphi - R_0)\beta\dot{\varphi} +$

$$+\beta V + c(Vt + A\varphi^2/2 - R_0\varphi) = 0$$

replace $y = A\varphi - R_0 \Rightarrow \varphi = (y + R_0)/A$, $\dot{\varphi} = \dot{y}/A$

(4.1) $(m_R - B)(yy')'/A + \beta yy'/A + c\left[Vt + (y^2 - R_0^2)/(2A)\right] +$

$$+\beta V = 0$$

replace $w = (y^2 - R_0^2)/(2A)$, $w' = yy'/A$

(4.2) $(m_R - B)w'' + \beta w' + cw = -cVt - \beta V$ - particular linear differential equation of the second order

(4.2.1) $w_p = -Vt$ - particular solution

(4.2.2) $(m_R-B)r^2+\beta r+c=0$ - characteristic equat.

(4.2.3) $r_1=(-\beta+\sqrt{\beta^2-4c(m_R-B)})/(2m_R-2B)$

$r_2=(-\beta-\sqrt{\beta^2-4c(m_R-B)})/(2m_R-2B)$ roots

General solution for:

- real distinct roots

$$\varphi=\left[R_0-\sqrt{R_0^2-2AVt+2AC_1e^{r_1t}+2AC_2e^{r_2t}}\right]/A$$

- conjugated complex roots

$$\varphi=\left[R_0-\sqrt{R_0^2-2AVt+2Ae^{-\beta t/(2m_R-2B)}(C_1\cos Lt+C_2\sin Lt)}\right]/A$$

$$L=\sqrt{4c(m_R-B)-\beta^2}/(2m_R-2B)$$

- real repeated roots

$$\varphi=\left[R_0-\sqrt{R_0^2-2AVt+2A(C_1+C_2t)e^{-\beta t/(2m_R-2B)}}\right]/A$$

Complete solution for $\varphi(0)=0, \dot{\varphi}(0)=V/R_0$:

(5) $\varphi=(R_0-\sqrt{R_0^2-2AVt})/A$

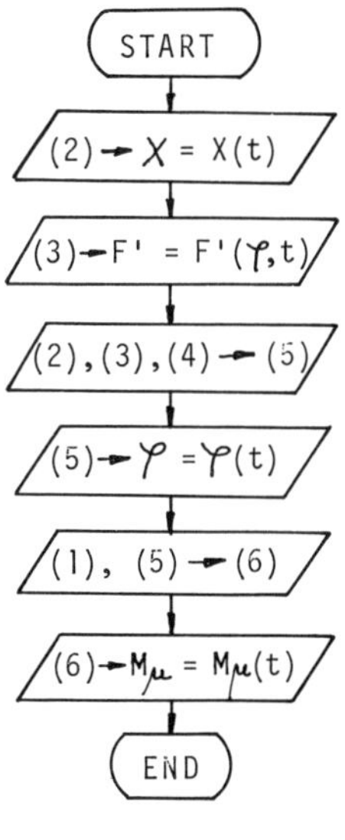

Fig. 4 - Flow of solving

General solution, which yields from characteristic equation, can be of various form, depending on mutual relation between parameters of system spring-damper-roll. For all these solutions there are expressed forms of angular coordinate changing. Initial conditions $\varphi(0)=0$, $\dot{\varphi}(0)=V/R_0$ impose the transformation of all mentioned general solutions to unique form, expressed as complete solution. That's the physical sense of mechanical model. Analysing general solutions, one by one, in the longer portion of time, regardless the initial conditions, they would convergate to the same form of complete solution (explanation follows):

- both roots r_1, r_2 are negative
- for $|\beta|>\sqrt{|\beta^2-4c(m_R-B)|}$
- under square root of solution there is e^{qt}, where $q<0$

Initial conditions, that are chosen, are of the type that will enable continual, smooth uncoiling from the very begining, causing no motion of small roll in the whole time domain. That means that the uncoiling, stable continues, and braking torque strictly follows the changing of geometric and dynamic characteristics of basic roll. Some other initial conditiones will cause oscilation of small roll which will calm down in longer or shorter period of time, depending on damping grade.

Importance of complete solution impose, to use it for the final form of braking torque.

(6) $M_\mu(t)=5Am_0V^2(R_0^2-2AVt)^{3/2}/(2R_0^4)+(\sqrt{R_0^2-2AVt})\cdot F$

EXAMPLE

R_0	= o.4 m	$z = 1o^{-4}$ m
R_{min}	= o.o2 m	V = 1 m/s
m_0	= 5oo kg	F = 2oo N
D	= o.o6 m	$c = 1o^4$ N/m
m_R	= 7 kg	$A = 16 \times 1o^{-6}$ m
J_R	$= 5.25 \times 1o^{-3}$ kgm^2	B = 3 kg

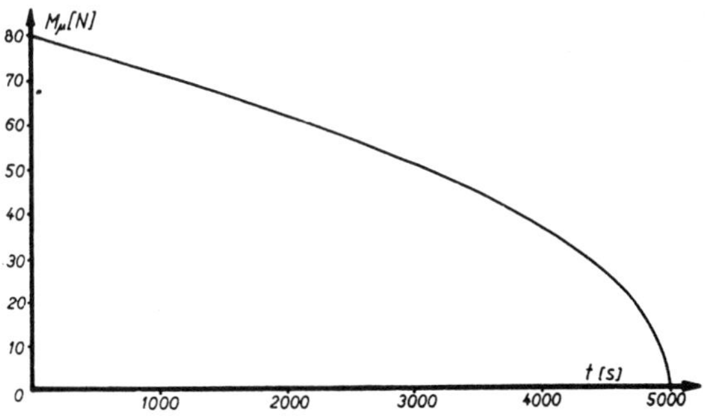

Fig. 5

SUMMARY

This paper is the result of widely spread activity on programme of developing family of machines in plastic industry. One portion of that programme delt the problems of uncoiling foil. Analysing existing solutions, and developing own one, we fould many problems and some of them are described inhere.

This was one of the first steps. Permanent engagement in this field, we hope, will bring significant progress encorporated in new devices and machines.

ZUSAMMENFASSUNG

Die Urkunde ist die Leistung einer breiten Tätigkeit auf dem Entwicklungsprogramm der Maschinen in der Plastikindustrie. Ein Teil des Programms bezieht sich auf die Folien.

Analysierend schon bekannte Losungen, bei der Entwicklung unseren eigenen, haben wir viele Problemme entdeckt. Einige von den Problemmen sind hier beschieldert.

Das war ein von ersten Schritten. Ständige Arbeit auf dem Feld, hoffen wir, bringt einen bedeutsamen Fortschritt in Verbindung der neuen Apparaten und Maschinen.

Eine Neue Realisation des Systems für Numerische Steuerung

T. PETROVIĆ*, Ž. ŽIVKOVIĆ* AND D. RANDJELOVIĆ**

*Department of Mechanical Engineering, University of Niš, Niš, Yugoslavia
**Research – Development Institute, Ei, Niš, Yugoslavia

Zusammenfassung. Aus den drei grundsätzlich möglichen Arten der Bewegungssteuerung eines Antriebssystems - Einwirkung auf Antriebselement, Übertragungselement oder Wirkelement - wird bei der vorgestellten Lösung die Beeinflussung des übertragenden Mechanismus gewählt. Das Hauptziel dabei ist, die unumgänglichen Beschleunigungseffekte möglichst nahe am Wirkelement auftreten zu lassen, um die Trägheitskräfte so gering wie möglich zu halten. Erreicht wird das durch ein Differentialgetriebe mit impulssteuerbaren und alternierend wirksamen An-bzv.Abtriebsgliedern unter Beibehaltung der kontinuierlichen Hauptantriebsbewegung. Die erreichbaren Parameter hinsichtlich Belastbarkeit und Positinierfehler sind sehr gut.

Keywords. Servomechanisms, Microprocessors, Discrete systems, Direct digital control, Function approximation, Position control, Optimisation.

EINLEITUNG

Die moderne Handhabungstechnik verlangt vielfach Linearität der Bewegungsform eines Manipulators oder seiner peripheren Einrichtungen, z.B. beim Einlegen, Palettieren oder Fügen. Jede maschinengebundene Translationsbewegung ist - im Gegensatz zu Rotationsbewegungen - notwendigerweise wegbegrenzt, muss also bei wiederholten Bewegungsvorgängen umgesteuert werden können. Hinzu kommt die Forderung nach automatisch steurbarer Positionierung mit möglichst geringem Positionierfehler.

Für den Antrieb von Positioniereinrichtungen werden überwiegend Elektromotoren eingesetzt, die eine Rotationsbewegung erzeugen. Im Falle einer zu realisierenden linearen Positionierung ist die Umwandlung von Rotation in Translation erforderlich. Viele Arbeiten sind darauf gerichtet, für diese Aufgabe neue Lösungen zu finden oder vorhandene Lösungen zu optimieren. In dem vorliegenden Beitrag wird über die optimale Steuerung eines impulsgesteuerten Schrauben-Umlaufräder-Mechanismus berichtet, der in der Maschinenbaufakultät der Universität Nis von Petrović (1981) entwickelt vorden ist und eine einsinnige Drehbewegung in eine wechselsinnige geradlinige Bewegung umwandelt. Die verbesserte Steuerung ist aus der Zusammenarbeit mit der Technischen Hochschule Ilmenau, Sektion Gerätetechnik und dem Institut Ei Niš hervorgegangen.

AUFBAU DES POSITIONIERSYSTEMS

Von den prinzipiell möglichen Bewegungswandlern wird das Schraubengetriebe aus Gründen der linearen Übertragungsfunktion, der relativ einfachen Herstellbarkeit mit hoher Genauigkeit und Spielarmut weitaus am häufigsten angewendet. Noch wenig befriedigend ist jedoch generell das Problem der Bewegungssteuerung bei Schraubengetrieben gelöst, da im allgemeinen die Steuergrössen auf die antreibende und masseintensive Spindel wirken (Bild 1a und 1b) und demzufolge keine günstigen Bedingungen für hohe Positioniergenauigkeiten und gute Reaktionsfähigkeiten gegeben sind. Im folgenden wird eine Lösung vorgestellt, die derartige Schwierigkeiten reduziert, insbesondere die Umsteuerung der Antriebsbewegung bei Richtungsänderung der Abtriebsbewegung vermeidet und die Realisierung aller charakteristischen Bewegungsabläufe gestattet (Petrović, Živković, Bögelsack, 1983).

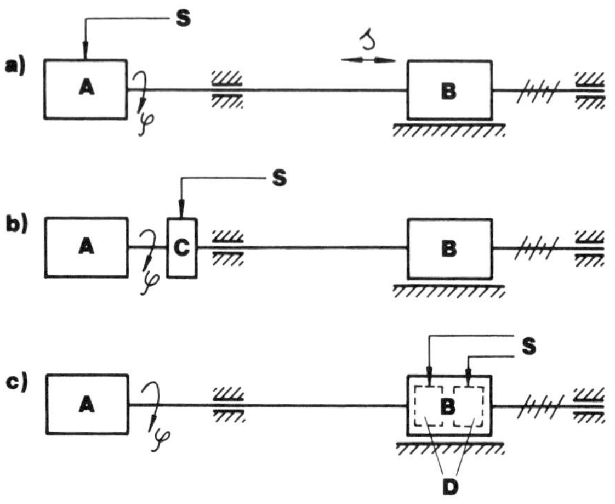

Bild 1. Arten der Bewegungssteuerung

Aus den drei grundsätzlich möglichen Arten der Bewegungssteuerung eines Antriebssystems - Einwirkung auf Antriebselement (A), Übertragungselement (C) oder Wirkelement (B) - wird bei der vorgestellten Lösung die Beeinflussung des übertragenden Mechanismus (D) gewählt.

Die Grundeinheit des Positioniersystems, das je nach Anzahl der Achsen ein-, zwie- oder dreidimensionale Positionierung zulässt, ist im Bild 2 dargestellt: Der Motor M treibt die Spindel Sp ständig und einsinnig mit der Winkelkoordinate φ an. Das Messystem MS misst die erreichte Position des Abtriebselementes mit dem darin enthaltenen Getriebe G und erzeugt das Signal

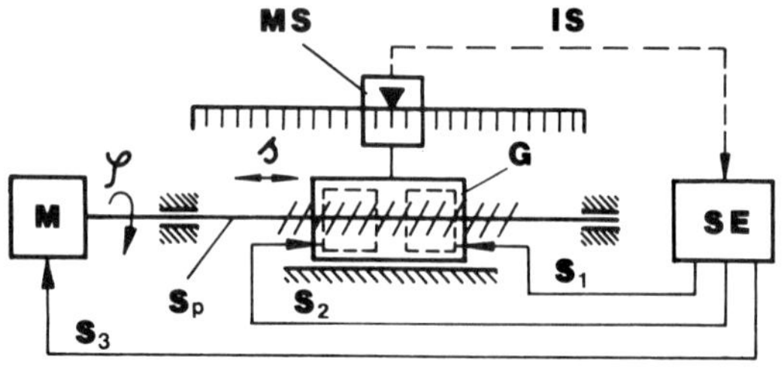

Bild 2. Grundaufbau

IS. Es wird durch die Steuereinheit SE in drei Steuersignale umgesetzt: S_1 und S_2 steuern das Abtriebselement ("Spindelmutter"), S_3 beeinflusst die Winkelgeschwindigkeit des Motors. (Bei erforderlicher konstanter Antriebswinkelgeschwindigkeit entfällt S_3). Zwischen der Antriebsbewegung und der Abtriebsbewegung besteht kein kinematisch zwangläufiger Zusammenhang, der auf das Abtriebsglied bezogene lineare inkrementale Geber ermöglicht die Kompensation aller etwaigen Fehler in den kinematischen Ketten.

STEUERUNG

Die Steuerung des Positioniersystems umfasst die Regelung der Antriebswinkelgeschwindigkeit und die Festlegung von Bewegungsrichtung bzw. Stillstand des Abtriebselementes. Üblicherweise wird in ähnlichen Positioniersystemen ausschliesslich das Antriebselement beeinflusst, z.B. wird das Antriebselement gestoppt, um den Stillstand des Abtriebselementes zu erreichen. Demgegenüber sind bei der vorgeschlagenen Lösung die Forderungen an das Antriebsregime nicht sehr streng. Es genügt eine stufenweise Steuerung im Anfahr - und Bremsbetrieb.

Mit den Steuersignalen S_1 und S_2 werden im Getriebe G jeweils Kopplungen aktiviert, die eine Bewegung des Abtriebselementes in die eine oder die andere Richtung gestattet.

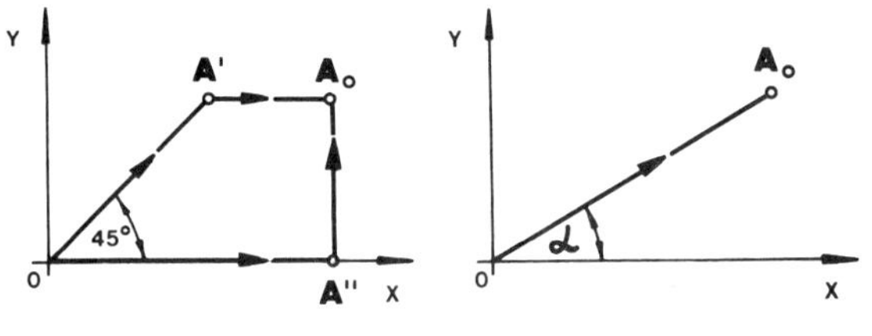

Bild 3. Trajektoriensteuerung

Bei flächenhafter Positionierung bestimmt der Quotient der Winkelgeschwindigkeiten für die x- und die y-Achse die Form der Trajektorie (Bild 3). Damit ist die Möglichkeit einer zeit- und energieoptimalen Steuerung gegeben. Beispielswieise ist das Erreichen der Position A_o im Bild 3a zeitgünstiger auf dem Weg 0-A'-A_o als auf dem Weg 0-A''-A_o.

Beim Führungssystem ist $0\text{-}A_o$ (Bild 3b) ein Segment der interpolierten Trajektorie.

An der Maschinenbau-Fakultät der Universität Niš wurden mehrere Modelle dieser Mechanismen realisiert, sowohl für die Zwecke der Leistungsübertragung bei grösser axialer Belastung als auch zur Bewegungswandlung bei Positionieraufgaben (Bild 4).

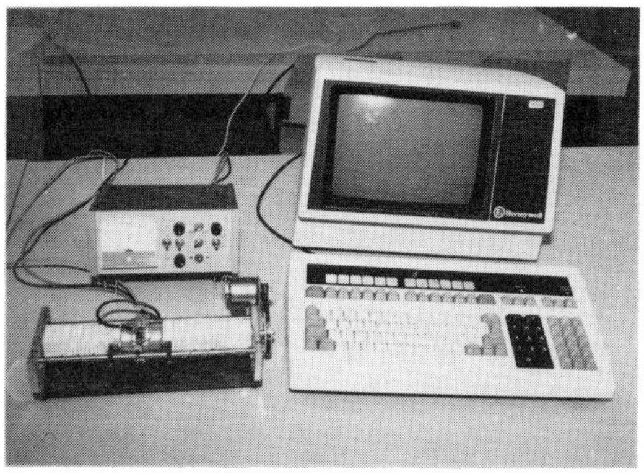

Bild 4. Versuchsaufbau zur Positionierung

Die Realisierung der Steuerung erfolgte auf der Basis des Mikrorechners COSMAC CRS II. Den zugehörigen Steueralgorithmus zeigt das Bild 5.

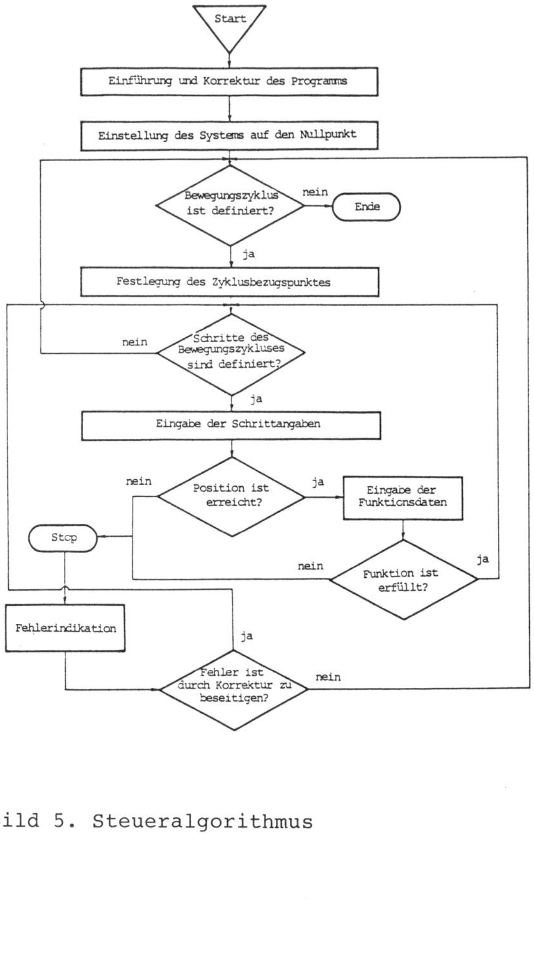

Bild 5. Steueralgorithmus

Man erkennt die drei Hauptteile: Eingabe des Programms von Hand oder externem Speicher, Festlegung des Arbeitszyklus, Kommunikation zwischen Bediener und System. Die Software besteht aus Befehlen auf der Ebene des Interpreters (Eingabe und Korrektur der Koordinaten, Eingabe des Programms usw.), so dass keine besonderen Kenntnisse in der Programmierung erforderlich sind.

LITERATUR

Bögelsack, G. (1968). Getriebe und gesteuerte Antriebe im Gerätebau, Feingerätetechnik, Heft 3, S.79-101.

Petrović, T. (1981). Zavojni mehanizam za transformaciju jednosmernog kružnog u dvosmerno pravolinijsko kretanje sa impulsnim upravljanjem, Doktorska disertacija, Mašinski fakultet Niš.

Petrović, T., Živković, Ž. und Bögelsack, G. (1983). Impulsgesteuerte Positioniereinrichtung mit kombiniertem Schrauben-Umlaufräder-Getriebe, 28. Intern. Wiss. Koll., Ilmenau, S. 165-168.

THE NEW NUMERICAL CONTROL SYSTEM REALIZATION

Summary: This paper deals with the system for numerical control based on combined screw-planetary gear mechanism, researched by Mechanical Faculty of Niš. In the paper, the system characteristics and applications are described. This system has better characteristics in comparision to the existing system under conditions that lower moving rates and larger forces at working part of the system was required. The system control by help of microcomputer is described too.

Cams

The Influence of Dimensional Accuracy of the Function Contact Face of a Cam on the Follower Motion in Cam Mechanisms

P. ŠIDLOF, Z. KOLOC, AND M. VÁCLAVÍK

Mathematical and Physical Department, ELITEX KVÚ, Liberec, ČSSR

Abstract. This paper deals with problems concerning the influence of general manufacturing imperfection of the cam contact face on the dynamic behaviour of the follower motion in cam mechanisms. On the base of experiments with the contact faces of various cams milled or grinded on CNC machine there is derived the relation approximating the power spectral density of the deviation of the acceleration of the follower center. The experiment is described and measured results are interpreted.

Keywords. Measurement of dynamic properties; cam mechanism; spectral analysis; machining, CNC-machinery.

Even the process of manufacture cams on up-to-date NC-machinery with continuous control still produces certain accuracy errors on the surface of the resulting product. Thus cross-slots with a depth of approximately 0,01 mm at the start or end step of machining are considered as typical local errors. This deviation is caused by short-time stoppage of the rotating cutting tool at a defined points and by the flexibility involved in the system workpiece-tool-machine.

Similar, but low-order errors occures in points of feed ratio changes. The wavelength value of these errors depend on the applied graduation of the angular coordinate as an independent variable, on the radius vector of the tool axis and on the tool radius. Such manufacturing errors are of general character causing surface waviness. Hence another type of general error is subject to the chip formation during milling resulting in a certain roughness of the contact face. The polar velocity of the contact point between the follower roller and the rotating cam related to the cam will be $v = 2\pi f_1 r_c$ where f_1 is the cam motion frequency and r_c is the average radius of the contact face curvature. In this situation the errors of the cam contact face produce further errors $x(t)$ of the roller center against its theoretical position or errors of acceleration $\ddot{x}(t)$ either.

Thus $x(t)$ and its derivatives are random stationary functions of time where the average value to be zero. Errors transferred from the cam surface onto the roller center within the frequency bandwidth $f \in \langle f_a, f_b \rangle$ have variance

$$\sigma_x^2(f_a, f_b) = \int_{f_a}^{f_b} G_{xx}(f)\, df \; , \qquad (1)$$

or

$$\sigma_{\ddot{x}}^2(f_a, f_b) = \int_{f_a}^{f_b} G_{\ddot{x}\ddot{x}}(f)\, df \, . \qquad (2)$$

respectively.

Using the equations (1) or (2) respectively according to (Bendat 1980) to be introduced the magnitudes of power spectral density $G_{xx}(f)$ related with the displacement error $x(t)$ of the roller center or $G_{\ddot{x}\ddot{x}}(f)$ of the corresponding acceleration $\ddot{x}(t)$ respectively. Considering the given conditions of measurement this paper deals predominantly with the magnitude $G_{\ddot{x}\ddot{x}}(f)$ related with $G_{xx}(f)$ by

$$G_{\ddot{x}\ddot{x}}(f) = (2\pi f)^4 G_{xx}(f). \qquad (3)$$

The error of acceleration $\ddot{z}$ of the working member of the mechanism corresponds to the power spectral density $G_{\ddot{z}\ddot{z}}(f)$. For the linear dynamic system is

$$G_{\ddot{z}\ddot{z}}(f) = H_{xz}^2(f)\, G_{\ddot{x}\ddot{x}}(f). \qquad (4)$$

The increase of frequency response $H_{xz}(f)$ is the absolute magnitude of the input/output ratio of the Fourrier transforms written as

$$H_{xz}(f) = \left| \frac{Z(i2\pi f)}{X(i2\pi f)} \right| . \qquad (5)$$

It is possible to record the function course of $H_{xz}(f)$ for realy mechanisms by measuring. The amplification of frequency response in mechanical systems can be determined mathematically by solving the appropriate dynamic model.

Further attention will be given now only to the power spectral density $G_{\ddot{x}\ddot{x}}(f)$ of the acceleration error of the roller center of the cam mechanism participating on $G_{\ddot{z}\ddot{z}}(f)$ as we defined in Eq.(4). The formula defining the power spectral density $G_{\ddot{x}\ddot{x}}(f)$ in radial cam mechanisms with roller followers presented in reference (Gatzen 1976) is

$$G_{\ddot{x}\ddot{x}}(f) = \frac{1}{2}(2\pi)^4 (1{,}11\, R_a)^2 f^3, \quad f \leq f_m ,$$
$$G_{\ddot{x}\ddot{x}}(f) = \frac{f_m}{f}\, G_{\ddot{x}\ddot{x}}(f_m), \quad f > f_m , \qquad (6)$$
$$f_m = \frac{f_1}{2} \sqrt{\frac{2 r_c (r_c + r_R)}{\pi r_R R_a}} ,$$

where f_1 is cam motion frequency $f_1 = \frac{v}{2\pi r_c}$, r_c is cam contact face medium radius, r_R is roller radius and R_a is cam contact face roughness (mean arithmetic deviation). Thus the expression $G_{\ddot{x}\ddot{x}}(f)$ shows in a comprehensive style the relation between this magnitude and the dispersive variance $(1{,}11\, R_a)^2$ of the cam surface, but it does not include the respective dispersive variance of the follower surface. Another insufficiency of Eq.(6) is the fact, that the dispersive variance of acceleration $\ddot{x}$, namely

$$\sigma_{\ddot{x}}^2 (f_1, \infty) = \int_{f_1}^{\infty} G_{\ddot{x}\ddot{x}}(f)\, df \qquad (7)$$

is divergent under the influence of the second portion of the formula for $G_{\ddot{x}\ddot{x}}(f)$ what does not correspond to reality in fact.

A number of experimental investigations were realized in order to acquire good knowledge about the influence of machining conditions when producing the cam contact face and the following consequence in displacement and acceleration of the roller center. Cutting of the contact face of a radial disk cam Dia.196,5 mm on a CNC-Machine, type CF-3 made by SIG is done under following conditions: workpiece was clamped with an eccentricity of 64,9 mm, processing data step were 0,5° of angular displacement round this axis. For approximation of the theoretical profile, given by segments of the spirals of Archimedes, linear interpolation was used. The tools were of 25 mm Dia and the feed was 80 mm per minute. Milling was made at the spindle speed of 700 rpm using four-teeth milling cutter with a helix angle of 65°. Grinding was made at the spindle speed of 35 000 rpm with planeting speed of 600 rpm and the planet eccentricity of less then 1 mm.For purposes of measurements the milled and grinded disk cams were mounted centric on a common shaft and fixed by means of a taper into the spindle of medium lathe. The spindle is mounted in high quality sliding bearings lubricated by a pressure oil and driven by coupled separate claw clutch by means of a flat belt. The swing-arm is made of aluminium alloy and is adapted for the use of three different rollers Dia.22, 35 and 55 mm and of a width of 12 mm. The rollers are mounted in sliding bearing Dia.10 mm. Acting force of 50 N applied to the cam by the roller is induced by means of an elastic rubber member. The influence of the acting force on the error amplitude was found important only in case of low-order forces. The lever motion was measured by means of an accelerometer, the axis of which being on the normal line of the cam at the point of contact with the follower roller. The mounting resonance frequency of the accelerometer type 224 C of ENDEVCO is 32 kHz. The accelerometer is connected to the charge amplifier KISTLER, type 5001. The cam rotation was measured by means of a special equipment for digital registration of the angular velocity; as a pickup was used an incremental rotary encoder ZEISS IGR 2000, attached to the shaft by means of a metal bellows coupling. The progression of time intervals between the pulses of the incremental encoder were recorded as measuring output. Their reciprocal values are proportional to the average angular velocity of each interval. These digital quantities were recorded on one track of an instrumentation tape recorder SE 7000 A. Voltage proportional to the acceleration $\ddot{y}$ was recorded on another track with a limit frequency 10 kHz($\pm$0,5dB) given by the chosen tape speed. During the evaluation of $\ddot{y}(t)$ the record of the angular cam velocity $\omega_1(t)$ enabled both the determination of the average values $\bar{\omega}_1 = 2\pi f_1$, and starting of the digitalization from a given angular displacement of the cam as well. Digitalization was made by using the Hewlett-Packard system voltmeter 3437 A with an eightfold deceleration of tape. The resulting digitalization frequency $f_{dig} = 40$ kHz. Each of the realizations included N = 2 048 samples. One of these realizations examining a milled disk cam with a follower roller Dia. 35 mm at f_1 = 3,16 Hz (190 rpm) is shown in Fig.1 (σ_C = 0,45 um, σ_R = 0,2 um). Before using the discrete Fourier transforms (FFT) the edge parts of the realizations of a length of 205 samples were multiplicated by the cosine function to restrain the influence of the total length of the analysed interval. Power spectral density of the measured acceleration error $\ddot{y}$ according to (Bendat 1980) will give

$$G_{\ddot{y}\ddot{y}}(f_k) = \frac{2}{N \cdot f_{dig}} \{ [\sum_{n=1}^{N} \ddot{y}_n \cos 2\pi \frac{kn}{N}]^2 + [\sum_{n=1}^{N} \ddot{y}_n \sin 2\pi \frac{kn}{N}]^2 \}. \qquad (8)$$

The resulting function $G_{\ddot{y}\ddot{y}}(f_k)$ was computed as an arithmetic mean of three realizations under constant physical experimental conditions. To eliminate the influence of frequency related effect at the contact point (x) between cam and roller and the measuring point (y) the function $H_{xy}(f)$(Fig.2) of the increased acceleration was measured by means of programmable vibration equipment LING so as to allow the computation of the power spectral density of the acceleration x of the roller center:

$$G_{\ddot{x}\ddot{x}}(f_k) = \frac{G_{\ddot{y}\ddot{y}}(f_k)}{H^2_{xy}(f_k)} \qquad (9)$$

Fig.3 illustrates the function $\xi_1(\nu)$ related to the realization conditions as shown in Fig.1:

$$\xi_1(\nu_k) = \frac{f_1 G_{\ddot{x}\ddot{x}}(f_k)}{\omega_1^4(\sigma_C^2 + \sigma_R^2)}, \quad \nu_k = \frac{f_k}{f_1} \qquad (10)$$

The expressions σ_C^2 and σ_R^2 respectively in Eq.(10) define the dispersion variance of both the cam and roller face measured by means of roughness integrator. Fig.3 shows also the shape of $\xi_1(\nu_k)$ which satisfies the expression (6) giving quantities of higher order. This might be caused by roller skipping due to low acting force during the experimental test.

From the analysis of the experimental measurements the approximation expression of the power spectral density of the roller center acceleration follows in this form:

$$G_{\ddot{x}\ddot{x}}(f) = (2\pi)^4(\sigma_C^2+\sigma_R^2)\, A(\nu)\, \frac{f_1 f^2}{\exp[(\frac{f}{f_1})^2 B(p,\nu)]},$$

$$p = \frac{r_C r_R}{10^4 (r_C + r_R)\sqrt{\sigma_C^2 + \sigma_R^2}}, \qquad (11)$$

$$A(\nu) = A_0(1 + a_1\nu + a_2\nu^2),$$

$$B(p,\nu) = B_0[1 + \beta_{01}p + \beta_{02}p^2 + (1 + \beta_{11}p + \beta_{12}p^2)b_1\nu + (1 + \beta_{21}p + \beta_{22}p^2)b_2\nu^2].$$

As the first approximation the following constant values were given:

$A_0 = 11$ $\quad a_1 = 0$ $\quad a_2 = 0{,}25\,[sm^{-1}]$
$B_0 = 6.10^{-7}$ $\quad b_1 = 0$ $\quad b_2 = 0{,}044\,[sm^{-1}]$
$\beta_{11} = \beta_{01} = 0$ $\quad \beta_{02} = 0{,}16$ $\quad \beta_{12} = \beta_{21} = 0$ $\quad \beta_{22} = 1$

This power spectral density of acceleration satisfies the general shape error of the cam contact face except those frequencies related to the steps of input coordinates for NC machining of the contact face shape. Hence these frequencies according to Fig.3 being now

$$f \in \langle 430 f_1, 500 f_1 \rangle.$$

CONCLUSION

On the base of experimental tests with cam contact faces milled or grinded on a CNC-machine, there has been developed a relation approximating the power spectral density of the acceleration error of the cam mechanism follower roller center. Further there has been stated that kinematic exciting of the follower is conspicuously influenced by manufacturing coordinates sampling.

REFERENCES

Gatzen, H. H. (1976). Bestimmung der Nutz- und Störbeschleunigungen bei Kurvengetrieben-ein Beitrag zur dynamischen Auslegung von Kurvengetrieben.Von der Fakultät für Maschinenwesen der Rheinisch-Westfalischen Technischen Hochschule Aachen zur Erlangung des akademischen Grades eines Doktor-Ingenieurs genehmigte Dissertation.

Bendat, J. S. and Piersol, A. G. (1980). Engineering Applications of Correlation and Spectral Analysis. John Wiley and Sons, Ney York, Chichester, Brisbane, Toronto, Singapore

ZUSAMMENFASSUNG

EINFLUSS DER MASSGENAUIGKEIT DER AKTIVEN FLÄCHE EINER KURVENSCHEIBE AUF BEWEGUNG DES ARBEITSGLIEDES EINES MECHANISMUS

Der vorliegende Beitrag befasst sich mit dem Einfluss der allgemeinen Fertigungs-Ungenauigkeit der Kurvenscheibe auf das dynamische Verhalten des Abtriebsgliedes eines Kurvengetriebes. Experimentell erfasste Daten von an CNC-Maschinen gefrästen oder geschliffenen Kurvenscheiben wurden untersucht und aufgrund einer Analyse die Beziehung zur Approximierung der spektralen Leistungsdichte der Beschleunigungsabweichung an der Mittelpunktsbahn des Abtriebsgliedes eines Kurvengetriebes entwickelt. Die Durchführung des Experiments wird beschrieben und die erfassten Messergebnisse erörtert.

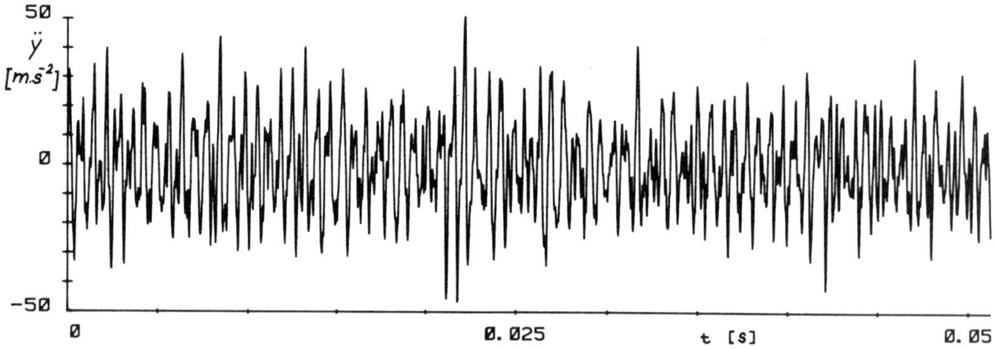

Fig.1 Acceleration of the cam mechanism follower

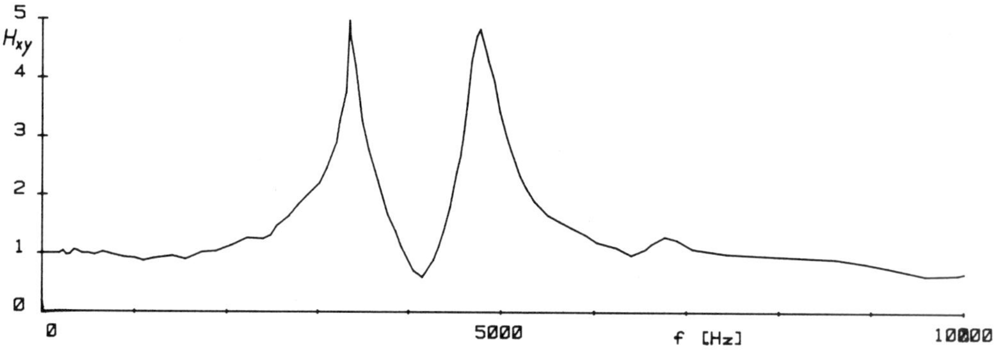

Fig.2 The frequency response between contact point of roller and accelerometer

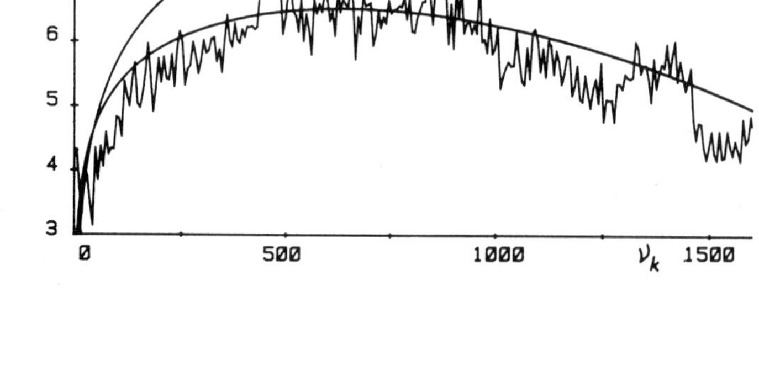

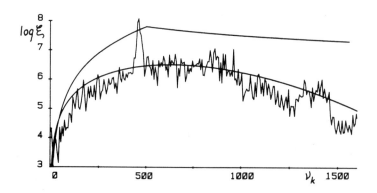

Fig.3 Non-dimensional power spectral density of acceleration of cam mechanism roller center

Optimum Design of Rigid Cam Mechanisms for Minimum Power Loss

S. A. EL-SHAKERY

Department of Production Engineering and Design, Faculty of Engineering and Technology, Menoufia University, Shibin El-Kom, Egypt

ABSTRACT : Designing the charactristics of rigid cam mechanisms for minimum power loss is formulated . The objective function which is the minimum power loss, and the constraints are formulated to be functions of cam-size, retainer spring stiffness and follower bearing dimensions . The Kuhn-Tucker optimization technique is presented . The results of such design show that the total power loss increases as either the follower offset or retainer spring stiffness increases . Moreover , this power loss decreases as the cam base circle radius increases . Furthermore, it found that this power loss is slight sensitive to variation in the dimensions of the follower bearing guid .

Keywords : Vehicles; nonlinear equations ; optimization ; power ; cam mechanisms .

INTRODUCTION

The cam-follower mechanisms are found in the vehicles, textile and printing machines, automatic machine tools etc. Therefore, several studies on design of such mechanism have been carried out . Recent study (Ghosh and Yadav, 1983) has been done to determine the minimum work loss in such mechanism, in addition, Freudenstein, Mayourian and Maki (1983) identified the nature of the functional dependence of the energy loss on such system parameters and established a minimum energy loss limit in such flixible mechanism . Moreover, El-Shakery (1985) presented the energy loss formula in such system as a function of cam contour, cam-size and flexibility of the follower train . Also El-Shakery (1985) found out the formulations to design such mechanism for high efficiency considering the flexibility of the follower train and the elastohydrodynamic phenomenon at cam-follower interface . However, the analysis of such system with Coulomb friction should prove useful in the development of a comparative measure of either the energy or the power loss in the system . Thus in this work, such analysis for frictional resistance and sliding velocity is adopted and the influence of mechanism parameters on the minimum power loss is carried out taking into consideration that the mechanism members are rigid .

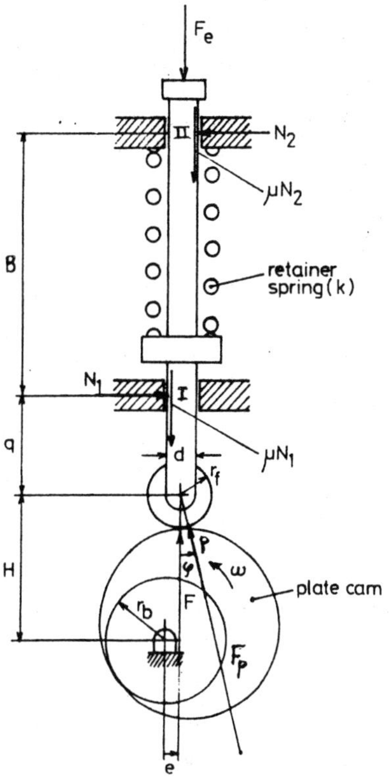

FIG.1 PLATE CAM WITH RECIPROCATING ROLLER-FOLLOWER SYSTEM

FORMULATION

The Objective Function

The rigid cam-roller-translating-follower mechanism which is shown in Fig. 1 , is taken as example . The power loss in such system is excessive at two situations, first at cam follower interface and second at guide of follower train and retainer spring compared to other situations .

Hence, the total power loss P_ℓ is given by

$$P_\ell = P_i + P_g \qquad (1)$$

Where P_i = power loss at cam-follower interface .

P_g = power loss at follower train and retainer spring guides .

P_i and P_g are estimated by

$$P_i = u_i F_p V_s \quad (2)$$

and $$P_g = u. V_c (N_1 + N_2) \quad (3)$$

Where u_i = coefficient of sliding - rolling friction at cam follower interface . This has been studies by El-Shakery (1985) using Truckman's equation and found to be approximately 0.001 for heavy loaded cam systems

F_p = normal contact force at cam follower interface .

V_s = sliding velocity .

u = coefficient of friction at follower bearing guides .

N_1 and N_2 = reaction forces at follower bearing guide .

V_c = follower velocity due to cam contour type .

These are developed in Appendix
Then, the objective function (total power loss) is computed by

$$P_e = u_i F_p V_s + | u (N_1 + N_2) V_c | \quad (4)$$

The Constraints :

The limit design equations of such system has been carried out by El-Shakery (1985). These equations are as:

1- To construct the system for safe of operation

$$r_b > e \quad \text{and} \quad r_b \geqslant d_s$$

2- To eliminate the follower jam and to minimize the thrust forces (N_1 , N_2)

$$\phi = \tan^{-1} \left(\frac{V_r - e}{H}\right) \leqslant 30^\circ \quad (\phi = \text{pressure angle}) .$$

3- To avoid contact loss at cam follower interface .

$F \geqslant 0$ (F: transmitted force from cam to the follower depends upon intertia, spring and external forces) .

4- To eliminate the undercutting of cam profile ,

$P_c > 0$ (P_c: cam curvature radius) .

5- To minimize or eliminate the wear at cam-follower interface

$\sigma^* \leqslant \sigma_d$ (σ : maximum contact stress)
(σ_d : allwable design stress)

Let, the design variables

$$X = \begin{pmatrix} x_1 \\ x_2 \\ x_3 \end{pmatrix} = \begin{pmatrix} r_b \\ e \\ k \end{pmatrix}$$

Hence, the constraints are, respectively, of the following forms ;

$$\left.\begin{aligned} g_1(x) &= x_1 - x_2 \geqslant 0 \\ g_2(x) &= 0.57\ H\ (x) - (V_r - x_2) \geqslant 0 \\ g_3(x) &= F\ (x_3) \geqslant 0 \\ g_4(x) &= P_c\ (x_1 , x_2) \geqslant 0 \\ g_5(x) &= \sigma_d - \sigma^*(X) \geqslant 0 \\ g_6(x) &= X \geqslant 0 \end{aligned}\right\} \quad (5)$$

The primary optimization problem can be stated as :

Find X which minimize P_e eqn.(4), and satisfy $g_m(x)$ eqns.(5) where $m = 1, 2, \ldots\ldots, 6$.

This problem can be solved by some optimization techniques . Here, the Kuhn - Tucker conditions are applied as

$$\left.\begin{aligned} \nabla P_e(x) - \sum_{m=1}^{6} u_m \nabla g_m(x) &= 0 \\ g_m(x) &\geqslant 0 \\ u_m \cdot g_m(x) = 0 , \quad u_m &\geqslant 0 \end{aligned}\right\} \quad (6)$$

Where u_m = slackness variables ,

$u_m\ g_m(x) = 0$: complementary slackness conditions .

IMPLEMENTATION

The following intial values are taken $r_b(=x_1) = 25$ mm , $e\ (=x_2) = 0.0$ mm , $k\ (=x_3) = 35.5\ K_p/cm$, $F_e = 1\ K_p$ $F_w = 1\ K_p$, $S_i = 8.5$ mm, $u = 0.15$, $u_i = 0.001$, $N = 200$ rpm , $\sigma_d = 32000\ K_p/cm^2$, $r_f = 20$ mm , $B_r = B_t = \pi$, cam lift $L = 1$ cm . The cam profile is cycloid .

It should be known the maximum value of any function has be determined by use of Regula-Falsi position method .

RESULTS AND DISCUSIONS

Figure 2 shows the relation between the power losses (P_i, P_g and P_e) and cam position θ and it illustrates the position of maximum absolute values of the power losses . From Figs. 3-5 , the following can be observed :

1- As either relainer spring stiffness K or follower offset e increases, the maximum power losses P_i^* , P_g^* and P_e^* increase ,

2- As cam base circle radius increases , the P_i^* increases and poth P_g^* and P_e^* decrease and

3- The power losses is sensitive to

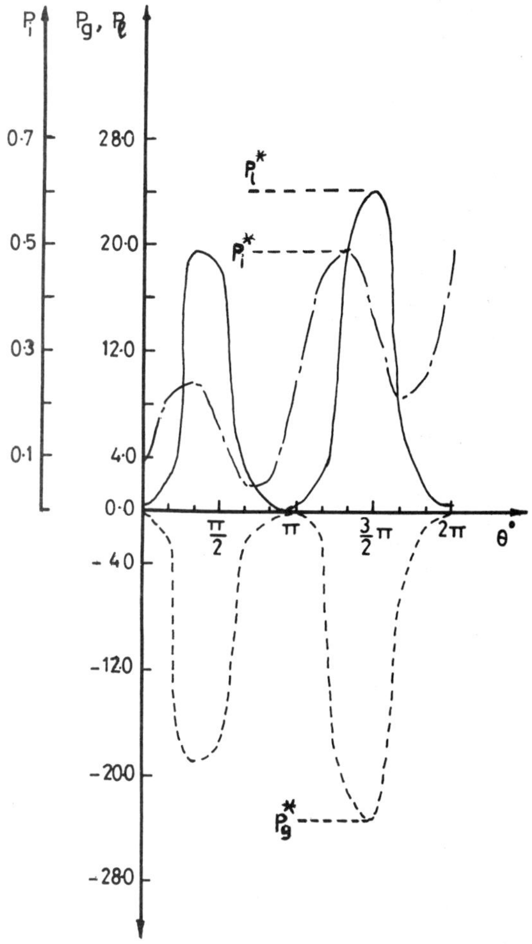

Fig. 2

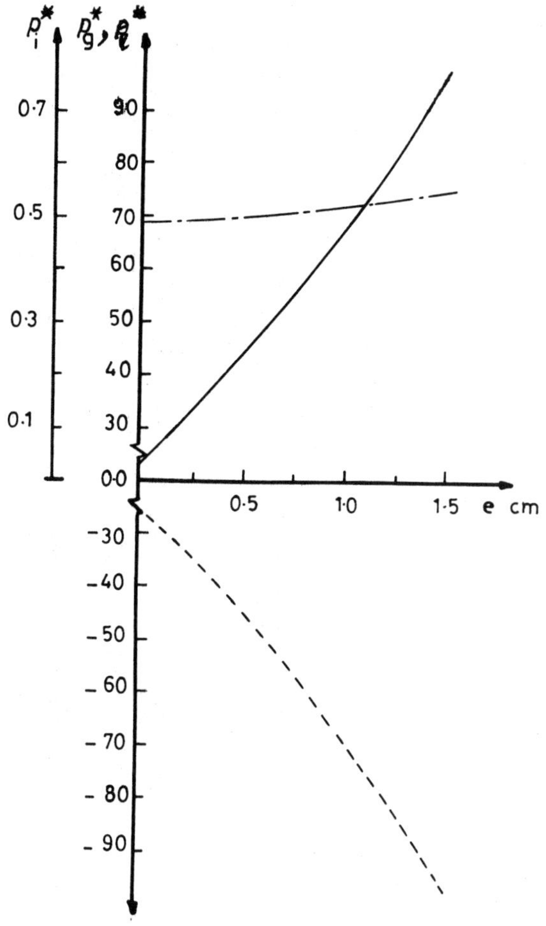

Fig. 4

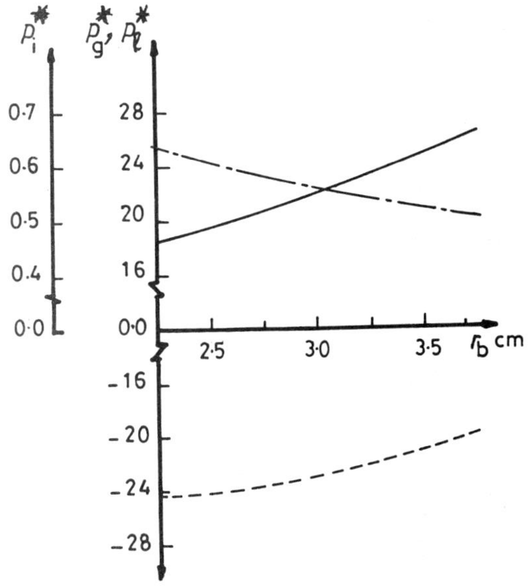

Fig. 3

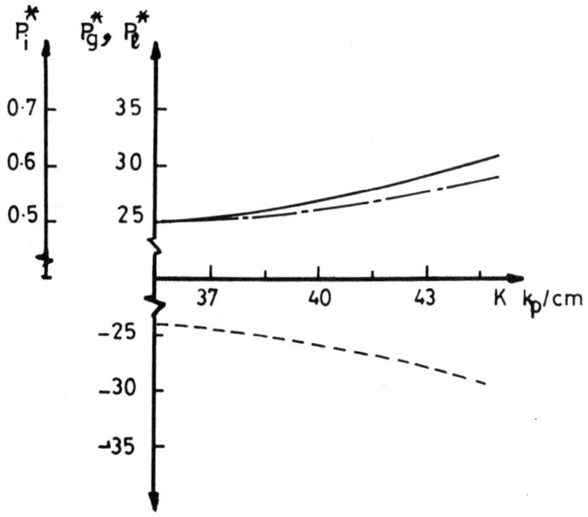

Fig. 5

In these figures P^* : maximum values

—— P_l
— · — P_i } in Kp Cm/s
- - - P_g

variation either in cam-size (r_b and e) or in retainer spring stiffness K .

In addition, the results show that the power losses is slight sensitive to the variation in the follower bearing dimensions (d and B) .

CONCLUSIONS

The power loss in cam-roller-translating-follower system is formulated and discussed . The following recommendations can be drawn :

1) To minimize the power losses in such system, the follower offset and retainer spring stiffness should be decreased and the cam base circle radius should be increased as much as possible , and

2) No significant effect of the follower guide dimensions on the power losses is seen .

REFERENCES

Ghosh, A. and Yadav, R.P. (1983). Synthesis of Cam-Follower Systems with Rolling Contact . Mechanisms And Machine Theory, 18, 49-56 .

Freudenstein, F. ; Mayourian, M. and Maki, E.R. (1983). Energy Efficient Cam - Follower Systems . ASME, Journal of Mechanisms, Transmissions and Automation in Design, 105, 681-685 .

El-Shakery, S.A. (1985). Postoptimal Cam Mechanism Design For Minimum Energy Loss part 1 . The Fourth IFTOMM Int. Symp. on Linkages, Bocharest, Romania , III-2,369-376 .

El-Shakery, S.A. (1985). Postoptimal Plate-Cam-Roller-Follower Design For High Efficiency Part 1 . 3rd Cairo Univ. Conf. on Mechanical Design and Production, Cairo, EGYPT, 2,179-186.

APPENDIX

For equations (2-4), the sliding velocity V_s , the contact force F_p and reaction forces on follower bearing guides N_1 and N_2 are given by

$$V_s = V_{tc} - V_{tf} = V_t\,(1-C) ,$$
$$V_t = V_p \cos(\beta_p + \varphi) , \quad V_p = \rho_p \cdot \omega ,$$
$$1-C = \epsilon_s : \text{sliding percentage} ,$$
$$\varphi : \text{pressure angle} = \tan^{-1}\left(\frac{V_r - e}{H}\right) ,$$
$$H = \sqrt{(r_v + r_f)^2 - e^2} + S_c$$
$$F_p = D_3/D , \quad N_1 = D_1/D , \quad N_2 = D_2/D$$

Where the Determinants D, D_1 , D_2 and D_3 are as :

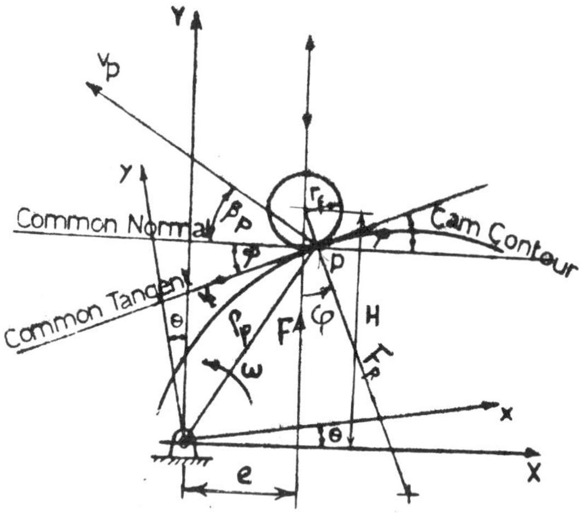

$$D = \begin{vmatrix} a_1 & a_2 & a_3 \\ b_1 & b_2 & b_3 \\ c_1 & c_2 & c_3 \end{vmatrix} , \quad D_1 = \begin{vmatrix} d_1 & a_2 & a_3 \\ d_2 & b_2 & b_3 \\ d_3 & c_2 & c_3 \end{vmatrix}$$

$$D_2 = \begin{vmatrix} a_1 & d_1 & a_3 \\ b_1 & d_2 & b_3 \\ c_1 & d_3 & c_3 \end{vmatrix} , \quad D_3 = \begin{vmatrix} a_1 & a_2 & d_1 \\ b_1 & b_2 & d_2 \\ c_1 & c_2 & d_3 \end{vmatrix}$$

Where, $a_1 = a_2 = u$, $a_3 = u_i \sin\varphi - \cos\varphi$

$b_1 = C_2 = u\,d/2$, $b_2 = B - u\,d/2$,

$b_3 = -(r_f u_i + qX)$, $C_1 = -(u\,d/2 + B)$

$C_3 = r_f u_i + (q+B)X$, $d_1 = -F$, $d_2 = d_3 = 0$

$X = u_i \cos\varphi + \sin\varphi$.

These are carried out using the equilibrium equations of the system ($\Sigma F_y = \Sigma M_I = \Sigma M_{II} = 0$, Fig. 1) .

OPTIMUM DESSEIN DE CAME MECANISME RIGIDE, POUR DE PERTE DE LA PUISSANCE MINIMA

RESUME : Le dessein de parameters de came regide mecanismes pour minimiser la perte de puissance est formule . Les fonctions objectives, de la puissance et des contraintes, sont formules en fonction de taille de came, de rigidite de retenir ressort et les dimension de desciple coussinet . "Le Kuhn-Tucker technique d'optization est presente" . Les resultats de opttmization de ce probleme mountre que la perte de puissance totale . Crios lorsque le quntile de devation de came suivi crios ou le rigidite de retenir ressort crois . De plus, cette perte de puissance crois lorsque le rayon de circule de base de came crois et le quntite de devation de came suivi decrois .

The History of Cams and Cam Mechanisms

J. MÜLLER

Lehrstuhl Getriebetechnik, Sektion Landtechnik, Wilhelm-Pieck-Universität Rostock, DDR

Abstracts. In addition to classic application in machine building, vehicle building and instrument manufacture at present there are applications in robot technology in the field of manufacturing of micro-electronic components. Technological applications of cam links are traced to the early palaeolithicum. Based on these beginnings the contribution deals with a summary of the development of Cams considering the social aspects.

Keywords. Machinery; mechanisms, cam mechanisms; history

KURVENGETRIEBLICHE SPEZIFIKA

Getriebe mit mindestens einem Kurvenglied, das mit einem benachbarten Getriebeglied durch ein Kurvengelenk verbunden ist, heißen Kurvengetriebe; wobei unter Kurvengelenk eine gelenkkige Verbindung zu verstehen ist, die infolge der geometrischen (Form) Gestalt ihrer Kontaktflächen eine allgemeine räumliche - im speziellen Fall ebene - Relativbewegung der miteinander gepaarten Glieder erlaubt.

TECHNISCH-HISTORISCHER ASPEKT

Geht man von der Definition des Begriffs "Getriebe" (Mechanismus) aus, so läßt sich gegenwärtig noch nicht gesichert die Aussage treffen, welche mechanische Einrichtung - wenn von den bei Lebewesen vorhandenen abgesehen werden soll - als erste von Menschenhand erschaffene oder verwendete anzusprechen ist; wohl läßt sich feststellen, daß der Faustkeil (Fig.1.) ein "Kurvenglied" im strengen Sinne seiner Definition darstellt. Und wenn ihn unsere Vorfahren im Altpaläolithikum zum Spalten oder Bearbeiten beispielsweise von hölzernen oder steinernen Gegenständen verwendeten, kann bereits zu diesem Zeitpunkt von der Existenz eines "Kurvengelenkes" gesprochen werden; eine Bearbeitungsmethode, die sich bis heute im Verwenden des stählernen Meißels oder Beiles fortsetzt.

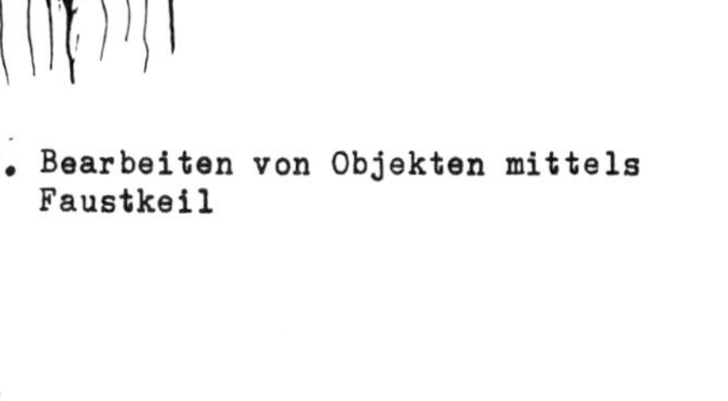

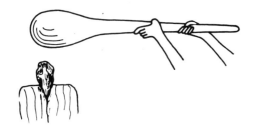

Fig.1. Bearbeiten von Objekten mittels Faustkeil

In der Periode des Mesolithikums (8000 bis 3000 v.u.Z.) begann der Übergang zum Ackerbau und damit die Notwendigkeit des Einsatzes von Bodenbearbeitungsgeräten; der in diesem Zusammenhang zu nennende hölzerne Hakenpflug "Baum des Ackerbauern" dokumentiert eine weitere prähistorische Anwendung (Fig.2.)

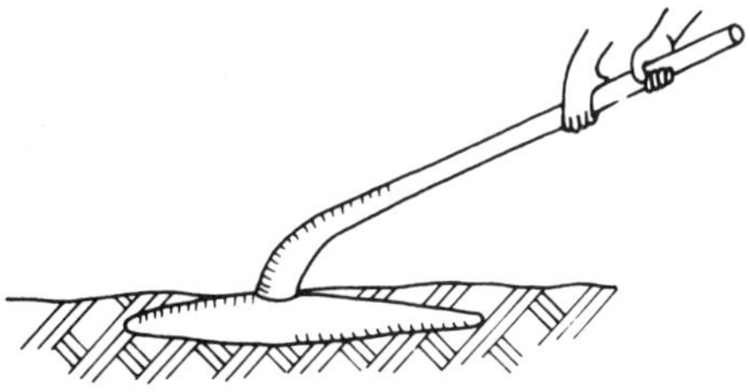

Fig.2. Hakenpflug - Baum des Ackerbauern

Die gewaltigen Bauwerke, die in den Ländern des Orients im 3. bis 2.Jahrtausend v.u.Z. entstanden, rufen noch heute unsere unumschränkte Bewunderung hervor, insbesondere wegen der dabei zu vollbringenden Leistungen beim Transport der massiven Steinblöcke bis zu 15 t, wie sie beispielsweise beim Bau der Cheopspyramide Verwendung fanden. Als technische Hilfsmittel hierbei dienten Schlitten - auf horizontalen oder geneigten Ebenen bewegt - vielfach zur Verringerung der Reibung mit unterlegten Rollen, die dafür bestimmte Personen nach dem Weiterrücken der Last aufhoben und dann erneut unter das Vorderteil des Schlittens schoben; weiterhin verdienen Winden und Flaschenzüge Erwähnung. Im 5.Jh.v.u.Z. war dann bereits der Transport auf Rädern weit verbreitet, und es fällt leicht, in diesen Hilfsmitteln den Einsatz von Kurvengelenken zu erkennen.

Griechen und Römer verstanden Windkraft zwar für die Fortbewegung ihrer Schiffe nutzbar zu machen, jedoch ist nicht bekannt geworden, ob sie auch drehende Windmühlen als Antriebseinrichtung benutzt haben. Nur das Windrad (Fig.3.) des HERON VON ALEXANDRIA (um 70 v.u.Z.) ist als einzige Einrichtung des klassischen Altertums bekannt, mit deren Hilfe Windkraft - vorschlagsweise für die Betätigung einer Orgel - nutzbar werden sollte; obwohl diese Einrichtung

Technikhistoriker als eine Art Spielzeug einschätzen, findet daran der Prototyp eines Kurvengetriebes Anwendung.

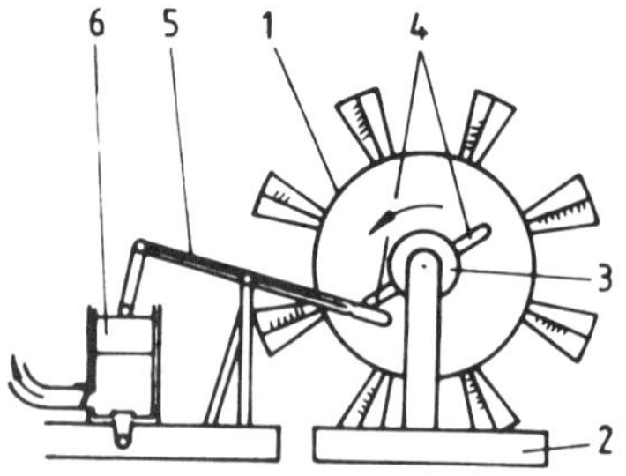

Fig. 3. HERONs Windmacher

Das Windrad (1) ist auf eigenem Fundament (2) drehbar gelagert; auf der Welle des Windrades sind ein oder mehrere Scheiben (3) parallel nebeneinander angeordnet, die radiale Stangen (4) tragen. Beim Drehen des Windrades legen sich diese Stangen gegen eine Platte am Ende des Abtriebsgliedes (5), drücken dieses nach unten und heben dadurch den Kolben (6) an, der schließlich nach Beendigung des Kontaktes zwischen radialer Stange und Platte durch seine Eigenmasse die Luft in die Orgelpfeifen drückt.

Die Notwendigkeit, zu wirtschaftlichen Zwecken größere Bodenabschnitte zu vermessen und Kanäle sowie Staudämme in den Ländern mit künstlicher Bewässerung zu errichten, führte zum Entwickeln nivellierender Instrumente; die am Nivellierinstrument des HERON VON ALEXANDRIA (1.Jh. v.u.Z.) eingesetzten, aller Wahrscheinlichkeit metallenen Schneckengetriebe, die vom Oberbegriff her wegen ihres Kurvengelenkes als kurvengetriebliche Bauform anzusprechen sind, können als deren frühzeitige Anwendung in der Feingerätetechnik angesehen werden. Nicht unerwähnt bleiben darf schließlich in diesem Zusammenhang die mit einem Schraubengetriebe - das ebenfalls ein Kurvengelenk laut Definition aufweist - ausgerüstete und für die Steigerung der Produktivität zur Weinherstellung entwickelte Schraubenpresse mit hölzerner Spindel.

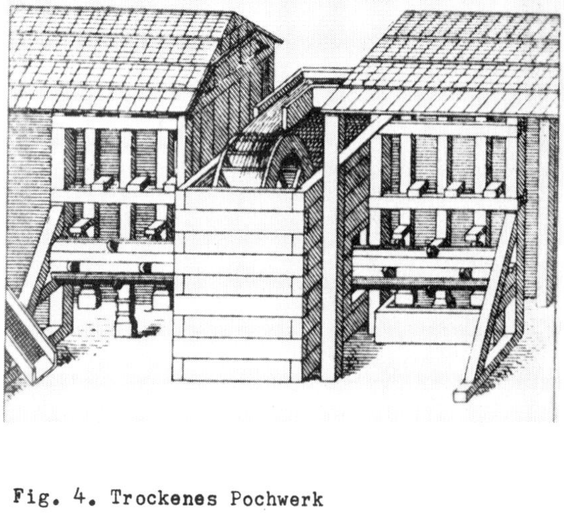

Fig. 4. Trockenes Pochwerk

Mit dem HERONschen Windmacher war schließlich das Kurvengetriebe (Nockengetriebe) erfunden worden, mit dessen Hilfe es nunmehr gelang, Wasser- oder Windkraft für die Zwecke der Verarbeitung oder Bearbeitung nutzbar zu machen, wobei erstgenannte schließlich den Vorzug erhielt. Die Tragweite dieser, in die Zeit der Sklavengesellschaft einzuordnenden Erfindung wurde aber erst später erkannt und dann im grösseren Umfang genutzt, als die gesellschaftliche Entwicklung im Feudalismus eine Steigerung der Produktivkräfte förderte - in Westeuropa sogar erst im 6.bis 1o. Jh. Als Beispiel sei der Einsatz dieses von HERON entwickelte Nockengetriebe in sogenannten Pochwerken am Ende des 15. Jahrhunderts angeführt (Fig. 4.), in denen das geförderte Erz vor dem Röst- und Schmelzprozeß zum Zwecke seiner besseren Aufbereitung zerstampft wurde. Der Antrieb erfolgt über ein Wasserrad und die hiervon angetriebene Welle trug Zapfen (Nocken), die ihren zugeordneten Stempel oder Ramme nach oben bewegte und dieser dann beim Herunterfallen durch seine Eigenmasse das Erz zerkleinerte.

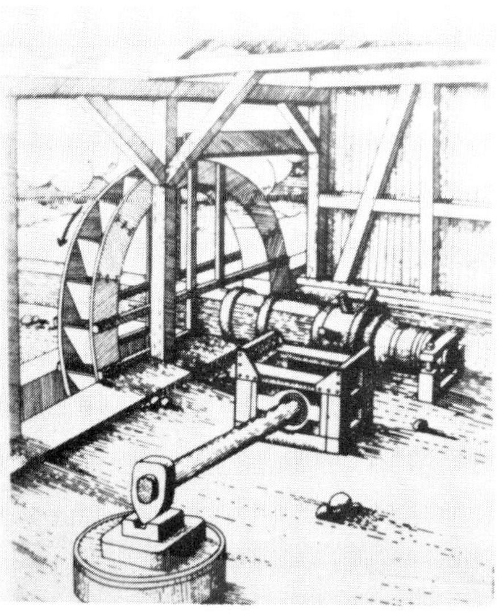

Fig.5. Schwanz- und Stielhammer des Zisterzienserklosters Oliva

Auch in anderen Produktionsbereichen der Feudalordnung fand diese Bauform eines Nockengetriebes Anwendung, wie beispielsweise im Rahmen der Papierherstellung, in der Drahtproduktion oder zum Schmieden. Ortsnamen wie Schmidmühlen bei Amberg in der Oberpfalz, der für den Anfang des 11.Jh.u.Z. (etwa 1010 bis 1028) als gesichert gilt, bieten den Indiz hierfür. Fig. 5 zeigt die Darstellung eines angetriebenen Schwanz- und Stielhammers aus dem Zisterzienserkloster Oliva bei Danzig aus dem 16.Jh.; die durch ein Wasserrad angetriebene Nockenwelle mit ihren schlagleistenartig ausgebildeten, stabilen Nocken ist unverkennbar.

Das Streben nach höherer Produktivität bedingte ein Steigern der Arbeitsgeschwindigkeiten; Konstruktionen wie nach Fig.4 und 5 erwiesen sich hierfür wenig geeignet, und man entdeckte die Möglichkeit, durch Formgebung des Nockens die kinematischen Verhältnisse beeinflussen zu können, wie Nocken mit bogenförmiger Kontur oder nach SCHMERBER mit spiralartig gekrümmter Kurvenflanke (sogenannte Daumen) dokumentieren. Weitere Anwendungen forderten nicht nur ein einfaches Bewegen des Eingriffsgliedes in seine Endlage, sondern das Realisieren einer bestimm-

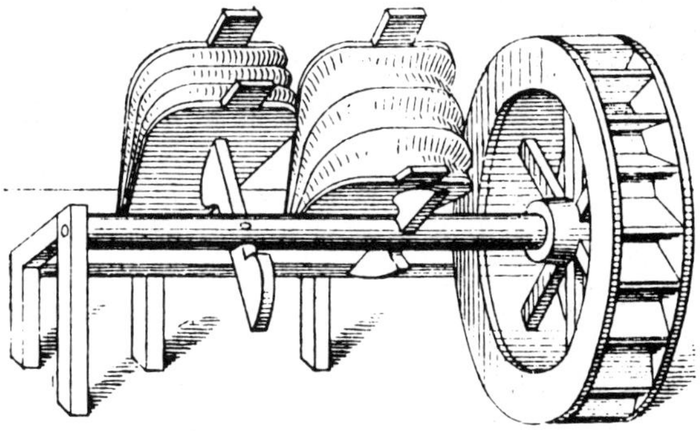

Fig. 6. Doppelgebläse

ten, vom Antriebswinkel abhängigen Funktion der Abtriebsbewegung. Beispielsweise finden sich bei LEONARDO DA VINCI Zeichnungen (Fig.7, rechts) für einen als Stackfreed bezeichneten und 1520 in Süddeutschland für Taschenuhren verwandten Mechanismus (Fig.7, links), dessen Aufgabe darin bestand, mit steigendem Drehwinkel dem Nocken eine seine Drehbewegung verlangsamende Kraft entgegenzusetzen.

In anderen Anwendungsfällen galt es, eine bestimmte Geschwindigkeit für Hin- und Rückbewegung zu verwirklichen, wie beispielsweise für das Betätigen der Kolbenstange von Gebläsen, (siehe auch Fig. 10.)

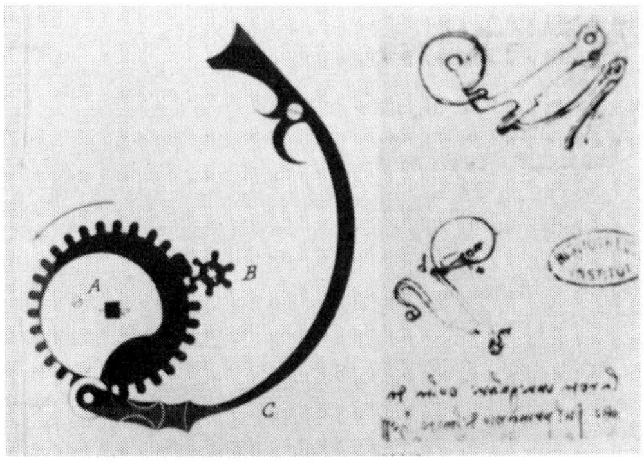

Fig. 7. Stackfreed nach LEONARDO DA VINCI

Die Erfindung des Kurvengetriebes mit umlaufender Kurvenscheibe liefert die Voraussetzung für seine breite Anwendungsvielfalt, und quasi sprunghaft erschlossen sich ihm neue Anwendungsgebiete. Der Zwang zur Mechanisierung oder Teilmechanisierung immer weiterer, bisher handwerklich durchgeführter Arbeitsvorgänge führte folgerichtig auch zum Entwickeln verschiedener räumlicher Kurvenmechanismen (Fig. 8. und 9.), so daß bereits im 16.Jh. eine Vielzahl kurvengetrieblicher Bauformen - zumindest im konzeptionellen Vorschlag - bekannt war, auf die nunmehr die Erfinder automatischer Maschinen - wie APPLEY, MERGENTHALER, JACQUARD, NARTOW u.a. - erfolgreich zurückgreifen konnten und die auch gegenwärtig als grundlegende Bauformen selbst in modernsten Technologien - wie Roboter- und Manipulatorentechnik, Fertigungseinrichtungen für elektronische und mikroelektronische Bauelemente - Anwendung finden.

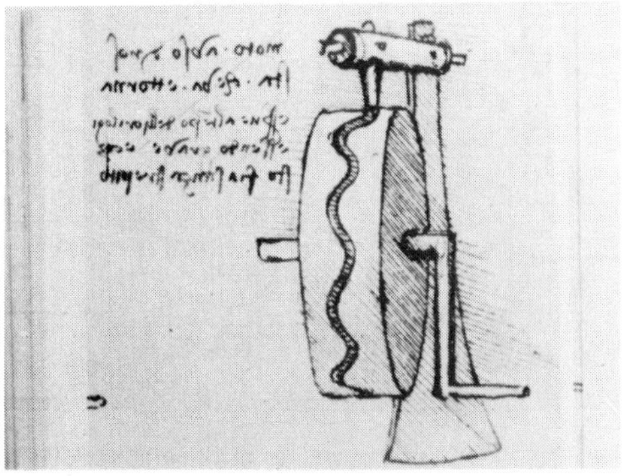

Fig. 8. Flügelregulator nach LEONARDO DA VINCI

ENTWICKLUNG WISSENSCHAFTLICHER METHODEN UND IHRE NUTZUNG FÜR KONSTRUKTION UND FERTIGUNG

Die ersten erfinderischen Lösungen für kurvengetriebliche Baugruppen dienten zum Erleichtern einfacher Lebensumstände und zum rationellen Bewältigen monotoner Arbeit, wobei kurvengetriebliche Baugruppen in Kombination mit anderen Getriebearten sogenannte "elementare Maschinen" bildeten. Bis zum Ende des Mittelalters nahmen sich die wissenschaftlichen Ansätze zur Lösung getrieblicher Aufgaben episodenhaft aus, während die Renaissance den Gedanken wachrief, menschliches Wissen zur Naturbeherrschung nutzbar zu machen, und somit den Anstoß gab zur Herausbildung naturwissenschaftlicher Methoden.

Die primären Untersuchungen über die Zusammenhänge der mechanischen Umformung von Bewegungen und Kräften an Keil und Schraube in der Antike fanden schließlich nach über anderthalb Jahrtausenden ihre Fortsetzung in den Arbeiten von LEUPOLD, beispielsweise in seinem "Theatrum machinarum"; wobei in diesen Werken Ansätze in der Entwicklung von Methoden zum Synthetisieren, also Bestimmen der Abmessungen der Kurvenscheibe für vorgelegte Bedingungen nicht übersehen werden können; auch bereits früher, bei LEONARDO, waren im Zusammenhang mit dem Entwurf des Spiralnockens erste Bemühungen zum theoretischen Entwurf des Kurvengetriebes unverkennbar. So berichtete LEUPOLD über den Entwurf verschiedener kurvengetrieblicher Lösungen, mit deren Hilfe ein Verringern der unerwünschten Ungleichförmigkeit der Drehbewegung erreicht werden soll, wie sie sich bei Antrieb durch Wind- oder Wasserkraft als unvermeidbar ergab.

Neben Methoden zum Synthetisieren von Kurvenscheiben für Kurvengetriebe bemühte sich J.N.P. HACHETTE (1769-1834) um Systematisierung von Mechanismen einschließlich kurvengetrieblicher Lösungen im Rahmen seiner Übersicht über Maschinenelemente und ihre aus dem damaligen Bedarf im Maschinenbau entspringenden Anwendungen. J.V. PONCELET (1788-1867) widmete sich gleicher Fragestellung; auf seine methodischen Angaben zur Konstruktion von Kurvengetrieben mit umlaufender Kurvenscheibe, dessen Abtriebsglied sich mit konstanter Geschwindigkeit bewegt, konnte bereits hingewiesen werden. Besondere Erwähnung verdient, daß PONCELET als erster auf das Beachten des jetzt als "Pressungswinkel" bezeichneten Winkels hinwies (Fig. 1o.) und Fragen nach geeigneten Abmessungen unter Berücksichtigen wirkender Kräfte behandelte.

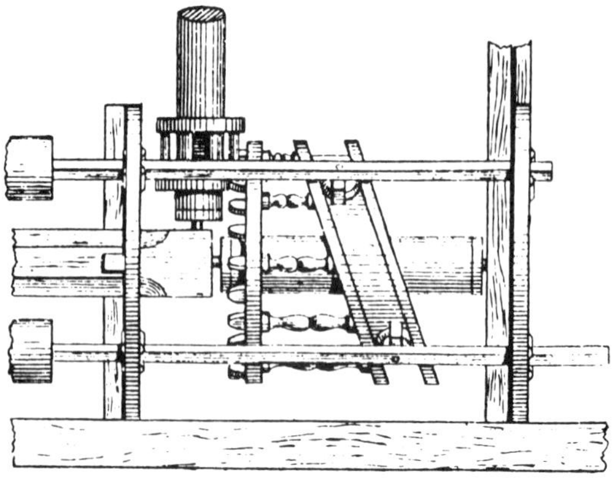

Fig. 9. Taumelscheibengetriebe

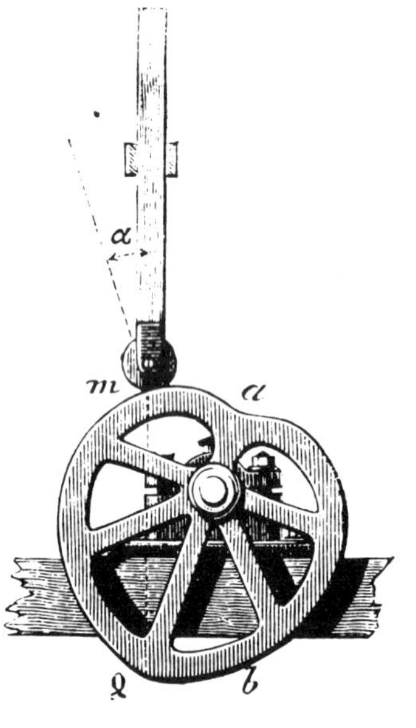

Fig. 1o. Kurvengetriebe mit umlaufender Kurvenscheibe

Ohne an dieser Stelle auf die weitere, bis in die jüngste Neuzeit fortsetzende spontane und zu leistungsfähigen CAD/CAM-Lösungen führende Entwicklung von Konstruktions- und Berechnungsmethoden für Kurvengetriebe und ihre Urheber einzugehen, darf aus Anlaß dieses historischen Rückblicks eine Bemerkung LEONARDO DA VINCI angeführt werden, die auch aus gegenwärtiger Sicht nichts an Aktualität eingebüßt hat: "Wer die Praxis ohne die Theorie liebt, ist wie ein Seemann, der auf ein Schiff steigt und nie weiß, wohin er gerät".

ZUSAMMENFASSUNG

An Hand belegter Beispiele konnten wesentliche Etappen der über viele Tausende von Jahren sich erstreckende Entwicklungsgeschichte des "Kurvengetriebes" aufgezeigt werden. Die Kürze des Beitrages gestattet nur ein skizzenhaftes Andeuten.

Die dargelegten entwicklungsgeschichtlichen Zusammenhänge beziehen sich vor allem auf Verhältnisse in Europa und angrenzende Gebiete. Weitergehende Informationen über bisher unbekannte Entwicklungen und kurvengetriebliche Lösungen in anderen Kontinenten werden vom Autor gern und dankbar entgegen genommen.

REFERENCES

Autorenkollektiv. (1976). Getriebetechnik - Kurvengetriebe (Ed: J.Volmer).VEB Verlag Technik, Berlin.

Autorenkollektiv. (1981). Allgemeine Geschichte der Technik von den Anfängen bis 187o. VEB Fachbuchverlag, Leipzig.

Fischer,H. (1914). Beiträge zur Geschichte der Werkzeugmaschine, Schmiedemaschine. BGT 6, S. 1 - 34.

Landels, J.G. (198o). Die Technik in der antiken Welt. Verlag C.H. Beck, München.

Leupold, J. (1724). Theatrum machinarum generale. Leipzig.

Mauersberger,K. (1983). Studienmaterial zur Vorlesung Geschichte der Technikwissenschaften (Teil 1). Dresdner Beiträge zur Geschichte der Technikwissenschaften, Heft 7, TU Dresden.

Nowak,H., G.Boden, J.Müller.(1984). Die Aktualität der Kurvengetriebe in der auf Mikroelektronik orientierten Technik. IFTOMM-SYMPOSIUM KURVENGETRIEBE, Karl-Marx-Stadt (DDR).

Poncelet, J.V. (1826). Cours de mechanique appliquée aux machines. Paris.

Schefold,M. (1929). Das mittelalterliche Hausbuch als Dokument für Geschichte der Technik. BGT, VDI-Verlag, Berlin.

Troitzsch,U., W.Weber. (1982). Die Technik von Anfängen bis zur Gegenwart. Verlag Westermann, Braunschweig.

Reti,L. (1974). Leonardo - Künstler, Forscher, Magier. Frankfurt/Main.

Herrn Dr.-Ing. K.Mauersberger, TU Dresden, sei für wertvolle Hinweise aus technikwissenschaftlicher Sicht herzlich gedankt.

Optimization of Cam Systems

A. MAGGIORE AND U. MENEGHETTI

Dipartimento di Ingegneria delle costruzioni meccaniche, Università degli Studi, v. Risorgimento 2, 40137 Bologna, Italia

Abstract. Geometrical parameters of cam mechanisms are usually assumed to have assigned values so they are not involved in the optimization processes. When designing automatic machines, however, some of these values may be free choices, almost within limits, so they are put in the Objective Function to be minimized. Improvements in regard of acceleration, transmitted force, hertzian pressure, and jerk, may be more consistent than that achievable by the optimum selection of the cam curve without changing the total stroke and the cam angle by introducing "extra-strokes", as defined in this paper. Particular conditions can also be introduced in the Objective Function, e.g. on velocity. Numerical examples and results obtained by this way are reported.

Keywords. Optimization of cams; objective function; extra-stroke.

INTRODUCTION

Optimization of cam systems is a leading question. Many designers have experienced how dramatically the behaviour of a mechanism can be improved due for example to a better choice of the cam curve. In a bad designed mechanism notable improvement can easily be achieved by choosing correctly its parameters but it is a very arduous task to improve a mechanism which is already well conceived. In this case, problems can not be solved only on the basis of the knowledge, experience and personal capacity of the designer. The assistence of an optimization approach is often needed and always useful if a well designed mechanism must be even improved.

Optimization can be considered from different points of view. Cam curve selection for reduced acceleration or jerk is brought forward in many textbooks (Rothbart, 1956; Neklutin, 1969; Chen, 1982; Magnani and Ruggieri, 1986). Minimum cam size has been determined by Fenton (1966). Interesting works have been done by Kwakernaak and Smit (1968), Hain (1971), Berzak (1980); quite a lot of these works are concerned with optimizing dynamical behaviour and reducing residual vibrations. The last tasks can be achieved also by suitable values of the physical parameters of the mechanism (Chew, Freudenstein and Longman, 1983).

Optimization is usually performed assuming a suitable objective function (OF) which is to be maximized or, more frequently, minimized, subject to a set of constraint equations. Geometrical parameters of the cam mechanism are usually assumed to have assigned values so they are not involved in the optimization process. When designing cams for automatic machines, however, some of these values may be free choices, so it seems to be correct to put them in the OF too. Before doing this we will spend some words to remember in which sense the total stroke and corresponding cam rotation can be sometimes considered "free choices".

BACKGROUND

We shall call "phase angle" the rotation performed by the cam during an event (or "phase") of follower's motion. The most interesting event of follower's motion is usually its rise; the related phase angle is the angle the cam rotates while the follower rises from its lowest position to the highest one.

In automatic machines, phase angles of various cams are often related each other. Consider for example Fig. 1.a. The object handled by the machine must be shifted from I to II by A, then from II to III by B and finally from III to IV by C. Obviously, B can start to displace the object only after this has reached position II. In effect, however, B can start from its dwell position before this point, because there is a gap Gb between the initial position of B and position II. What actually must happen is that B has not to travel longer than Gb before the object has reached position II. In the following of this paper we call "extra-stroke" (e.-s.) a piece of the total stroke like Gb; it may be an "initial" (like Gb) or a "final" extra-stroke, according to its position in follower's rise.

Some questions arise related to the possibility to choose an e.-s. First of all, the phase angle found in the tentative "triangular" curve of motion (Fig. 1.b) can be enlarged. The amount of the enlargement depends not only on the value of the e.-s. but also on the cam curve. The best choice of this curve is therefore connected to the value of the e.-s., the maximum permissible phase angle, and possibly other conditions. The amount of the e.-s. can be restricted by room limitations or it can be substantially unlimited.

The opportunity offered by e.-s. and enlarged cam rise angle are well known; as a rule, they are used by designers on the basis of their own insight. Present paper offers a contribution to a more systematic approach to this problem.

OBJECTIVE FUNCTION

The main point in the optimization of a cam mechanism is to build up a suitable OF. The finding of a new or particular optimization algorithm is not among our tasks, so a common library routine has been used for minimizing the OF.

A lot of suitable optimization criteria can be found in the literature. In the present paper a quite general OF was constructed by adding maximum values of acceleration, transmitted force and hertzian pressure; jerk was also taken into account. Other criteria can be easily considered without affecting the essence of the method.

The main feature of the OF is that the cam curve and the e.-s. must be considered as variables for the optimization, so a general cam curve must be used. The curve choosen is the "generalized trapezoidal" one and will be described in the next section; other curves can be used on the condition that their shape can be modified for optimization purposes.

Basic parameters of the motion must be introduced together with the extreme (minimum and maximum permissible) values of the variables, i. e. the extra-strokes and the single angles of the generalized trapezoidal curve. For each value of the e.-s. considered by the optimization algorithm the corresponding value of the phase angle is calculated by an appropriate routine; more detailed treatement of this point can be found in one paper by the second author (Meneghetti, 1983).

With completely defined values of motion parameters, OF can be calculated. The values of the variables are suitably changed by the optimization routine until optimum is reached.

The OF actually used in the subsequent examples has the form:

$$OF = W_A\ A_{MAX} + W_F\ F_{MAX} + W_P\ P_{MAX} + W_J\ J_{MAX} \qquad (1)$$

W_A, W_F, W_P, W_J, are the weights respectively of acceleration, transmitted force, hertzian pressure, and jerk; typical values are 0 or 1. A_{MAX}, F_{MAX}, P_{MAX}, J_{MAX}, are maximum values of acceleration, transmitted force, hertzian pressure, and jerk, reduced in values so that they are of the same order.

Acceleration and jerk are related only to the cam curve. Transmitted force and hertzian pressure depend also on the mechanism. Significant values of mechanism's parameters used in all the examples are reported in Table 1. Transmitted force is considered as due only to the accelerated mass (inertia force) as often occurs in automatic machines.

GENERALIZED TRAPEZOIDAL CURVE

As previously said the optimization routine must be able to change the cam curve. There are a lot of curves suitable for this purpose. The generalized trapezoidal curve was chosen because some very popular curves can be derived from this one.

Generalized trapezoidal curve was proposed by C. Cavagna and P. L. Magnani (1981). It consists of seven pieces corresponding to seven parts (in general different from each other) of the cam angle rotation (Fig. 2). The first, third, fifth

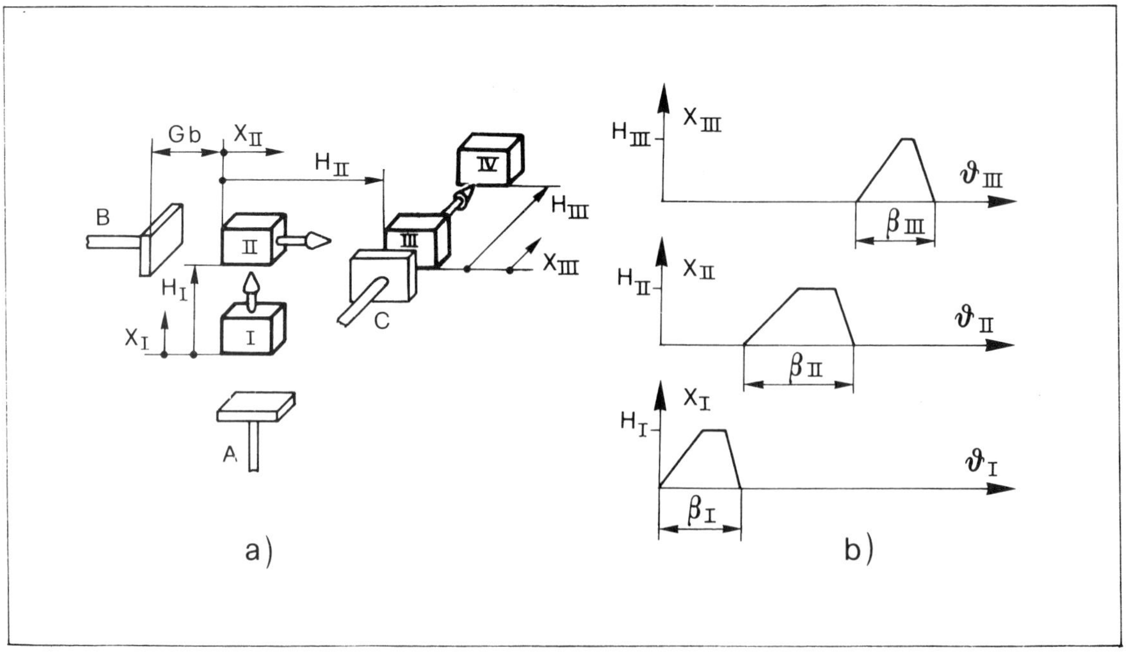

Fig. 1. a) Handling of an object from position I to IV. H_I, H_{II}, H_{III} are useful strokes respectively of A, B, and C.
b) Simplified "triangular" displacement programs of A, B, C.

and seventh piece of acceleration diagram are respectively first, second, third and last quarter of a sinusoidal curve (each part can be of different stretch). Second and sixth piece are of constant acceleration. In the fourth piece the acceleration is zero.

Many well known curves can be considered as particular cases of the generalized trapezoidal curve. For example, if we put $\beta_1 = \beta_3 = \beta_4 = \beta_5 = \beta_7 = 0$, the parabolic (constant acceleration) curve results. Putting $\beta_1 = \beta_3 = \beta_5 = \beta_7 = \beta/4$ and $\beta_2 = \beta_4 = \beta_6 = 0$, we obtain the cycloidal curve. The modified trapezoidal curve is obtained by putting $\beta_1 = \beta_3 = \beta_5 = \beta_7 = \beta/8$, $\beta_2 = \beta_6 = \beta/4$

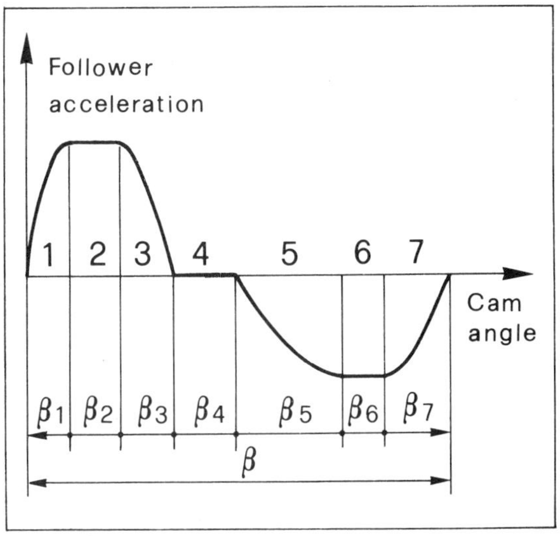

Fig. 2. Generalized trapezoidal acceleration program.

and $\beta_4 = 0$; and so on.

For simplicity purposes a less general form of the curve was used in the examples. The angle β_4 was always put zero and the part of negative acceleration was put similar (but not necessarily equal) to the positive one (that is, it was assumed $\beta_5/\beta_1 = \beta_6/\beta_2 = \beta_7/\beta_3$).

PARTICULAR CONDITIONS

Sometimes, particular conditions must be fulfilled on displacement, velocity, acceleration, and so on.

For example, with reference to Fig. 1 the velocity of B when touching the object (after displacement Gb) can be requested not to exceed a certain value in order to avoid damaging the object.

In such cases the OF becomes some more complicated. Optimization however can be performed essentially in the same way as previously described. Examples are reported in the following section.

EXAMPLES

Examples were performed in order to evaluate the method and to prove the importance of e.-s. and other conditions on the optimum cam curve.

General data of the mechanism are reported in Table 1. Useful stroke and useful angle are to be intended the values without considering e.-s. The cam is of radial type.

Twelve examples were performed; related data are reported in Table 2. In examples 1 to 4 no e.-s. has been introduced, so they are taken as reference examples. Cases 5 to 8 deal with e.-s. ≤ 1 mm without further conditions. In examples 9 to 12 a condition is introduced on velocity, that is the velocity v of the follower when touching the object (see Fig. 1 a) must be not higher than 0.5 m/s. It may be worth to note that if there is no condition on velocity or on maximum value of β, as in examples 5 to 8, the maximum permissible e.-s. always results to be the optimizing value.

Results are reported in Table 3.

In examples 1, 5, and 9, the objective has been simply to minimize the maximum acceleration. Parabolic curve results to be the best, with just a little deformation in case 9.

Examples 2, 6, and 10 deal with the minimization of the transmitted force. Optimized curves display high acceleration in initial and final part and low acceleration (i.e., low inertia forces) in the mean portion (where velocity and pressure angle are high).

In examples 3, 7, and 11 the target has been to minimize the jerk. Fairly interesting, the cycloidal curve results the best one.

Global optimization has been searched for in cases 4, 8, and 12. The best curve is quite similar to the usual modified trapezoidal one.

TABLE 1 Data of Cam Mechanism

Useful cam angle	100	deg
Useful stroke	100	mm
Base radius	200	mm
Cam angular velocity	600	rpm

TABLE 2 Data for the Examples

N.	e.-s. (mm)	W	W	W	W	v (m/s)
1	0.00	1	0	0	0	=
2	0.00	0	1	0	0	=
3	0.00	0	0	0	1	=
4	0.00	1	1	1	1	=
5	≤1.00	1	0	0	0	=
6	≤1.00	0	1	0	0	=
7	≤1.00	0	0	0	1	=
8	≤1.00	1	1	1	1	=
9	≤1.00	1	0	0	0	0.5
10	≤1.00	0	1	0	0	0.5
11	≤1.00	0	0	0	1	0.5
12	≤1.00	1	1	1	1	0.5

TABLE 3 Results

N.	β(deg)	$β_1/β$	$β_2/β$	$β_3/β$	e.-s. (mm)
1	100.0	0.50	0.00	0.00	=
2	100.0	0.00	0.35	0.15	=
3	100.0	0.25	0.00	0.25	=
4	100.0	0.18	0.14	0.18	=
5	107.5	0.50	0.00	0.00	1.00
6	107.4	0.00	0.35	0.15	1.00
7	113.0	0.25	0.00	0.25	1.00
8	112.3	0.18	0.14	0.18	1.00
9	103.8	0.00	0.48	0.02	0.27
10	103.6	0.00	0.39	0.11	0.25
11	109.6	0.25	0.00	0.25	0.45
12	108.7	0.18	0.15	0.17	0.41

In conclusion, each target seems to have its own optimizing curve, irrespective of the e.-s.; if no conditions are imposed, optimum is always found in correspondence of the maximum permissible value of e.-s.; a limited value of the e.-s. must be selected in presence of constraints, for example on follower's velocity in certain positions.

CONCLUSION

When designing a cam mechanism for an automatic machine, extra-stroke can sometimes be used. Maximum permissible extra-stroke is often the most suitable one for the optimum-performance mechanism. Limiting conditions may derive from maximum permissible cam angle or from other conditions, for example follower's velocity at the beginning or ending of the useful stroke. A suitable Objective Function can be very helpful for selecting the optimum value of the extra-stroke and the optimum shape of the cam curve. It must be emphasized that the improvement achievable by utilizing extra-stroke can be very high. For example, in case n. 4 (no extra-stroke) the value of the Objective Function is 232 whereas in case n. 12 (extra-stroke limited to 0.46 mm do to imposed maximum velocity at the beginning of the useful stroke) is 198, and in case n. 8 (extra-stroke: 1.00 mm) is 187; maximum acceleration is respectively 700, 594, and 562 m/s^2.

REFERENCES

Berzak, N. (1980). Optimization of cam follower systems with kinematic and dynamic constraints. ASME Paper 80-DET-11.

Cavagna, C., and P.L. Magnani (1981). Una generalizzazione per il progetto delle leggi di moto delle camme. Progettare, n. 12, 55-61.

Chen, F.Y. (1982). Mechanics and Design of Cam Mechanisms. Pergamon Press, New York.

Chew, M., F. Freudenstein, and R.W. Longman (1983). Application of optimal control theory to the synthesis of high-speed cam-follower systems. J. of Mech., Transm., and Aut. in Design, 105, 576-591.

Fenton, R.G. (1966). Determining minimum cam size. Machine Design, Jan. 20, 155-158.

Hain, K. (1971). Optimization of cam mechanism to give good transmissibility, maximum output angle of swing and minimal acceleration. J. of Mech., 6, 419-434.

Kwakernaak, H., and Smit, J. (1968). Minimum vibration cam profile. J. Mech. Eng. Sc., 10, 219-227.

Magnani, P.L. and G. Ruggieri (1986). Meccanismi per macchine automatiche. Utet, Torino.

Meneghetti, U. (1983). La rifasatura delle camme nelle macchine automatiche. Il Progettista Industriale, n. 6, 40-48.

Neklutin, C.N. (1969). Mechanisms and Cams for Automatic Machines. Elsevier, New York.

Rothbart, H.A. (1956). Cams. Design, Dynamics and Accuracy. John Wiley & Sons, New York.

Résumé. Les machines automatiques exigent des cames de plus en plus véloces, ainsi il est très important de concevoir celles-ci dans la façon la meilleure possible. Il est bien connu qu'un bon choix de la lois du mouvement est fondamental au but d'obténir un mécanisme efficient. Si les performances exigées sont très hautes, il est souvent nécessaire d'optimiser la lois ou même le mécanisme entier à l'aide d'un algorithme d'optimisation, ayant choisi une convénable Function-Objectif. Dans cet article on propose d'optimiser le mécanisme à came considérant comme des variables la forme de la lois de mouvement et aussi les extra-courses, c'est à dire les portions de mouvement que l'on peut introduire avant et après la course utile (voir Fig. 1). La lois utilisée est la trapézoïdale généralisée (voir Fig. 2). Ce sont souvent nécessaires des conditions spéciales, par exemple à l'égard de la vitesse du membre conduit. Les résultats remportés à titre d'exemple montrent l'importance des améliorations que l'on peut obtenir par la méthode proposée.

A New Algorithm for Automatic Assembling of Combined Cam Curves

YU-HONG LI*, WEI-HAO QI** AND WEI-QING CAO**

*Beijing Printing Machinery Research Institute, A2 Taiping Street, Xuanwu District, Beijing PRC

**Department of Mechanical Engineering, Shaanxi Mechanical Engineering Institute, Xian, PRC

Abstract. In order to simplify the assembling of combined curves of the follower motion of cam mechanisms, a model for an "Automatic Assembling Program" for computers is proposed. It is based on the "area-moment" conception and on a set of predesignated "basic acceleration curve segments" which will be selected for every part of the considered interval by the users. The computer can assemble them automatically and calculate all motion curves accurately.

Keywords. Cams; Cam curves; Transfer functions; Mechanical variables control; computer aided design.

INTRODUCTION

The combined curves of cam follower motion obtained by assembling simple function segments play an important role in cam design because of the simplicity in mathematics and the versatility in applications. However, the traditional method of assembling is to find out a number of constants in the equations of acceleration, velocity and displacement according to the boundary conditions and the internal continuity requirements. It involves a lot of manual deductions. Furthermore, for various combinations of different function segments, the deductions will be different. It is difficult to program for all combinations in a single procedure.

Kloomok and Muffley (1955) proposed a method called the "building block approach" for combined motion design, which simplified the manual deduction rather than eliminating it entirely. Neklutin (1969) offered a "universal method of cam calculation" which made use of sine and constant acceleration curves to be combined. This method later led to a so-called "Universal Curve" model where the interval is divided into seven parts: the constant acceleration is used in the 2nd and 6th parts, the zero acceleration in the 4th part, and a quarter period of sine wave in the remaining parts as transient curves; the lengths of all parts are determined by the users. We (1984) succeeded in a FORTRAN program for the model "Universal Curve" using the "area-moment" relationship, which is only 60 lines long. However, the function of the transient parts is only one type (a quarter of sine wave), therefore, its "universality" is limited.

In this paper we propose a method in which several function segments can be chosen for the transient parts of the "Universal Curve" above. Hence, its "universality" will be largely extended and the program will be able to assemble different segments automatically.

THE "AREA-MOMENT" RELATIONSHIP AND BASIC EQUATIONS

Assuming the acceleration curve A(t) is in the interval (t_o, t_n), the velocity at the end t_n is known as

$$V(t_n) = V(t_o) + \int_{t_o}^{t_n} A(t)dt \qquad (1)$$

and the displacement at t_n will be

$$S(t_n) = S(t_o) + (t_n - t_o)V(t_o) + \int_{t_o}^{t_n}\left(\int_{t_o}^{t} A(t)dt\right)dt \qquad (2)$$

The double integral above has been proved mathematically as

$$\int_{t_o}^{t_n}\left(\int_{t_o}^{t} A(t)dt\right)dt = \int_{t_o}^{t_n} A(t)(t_n - t)dt \qquad (3)$$

We can explain it as "a sum of the area elements A(t)dt under the acceleration curve A(t) multiplied by distances (t_n-t) from t to the end t_n". From this point of view, we may call it "the area-moment of curve A(t) about t_n", and denote it by D

$$D = \int_{t_o}^{t_n} A(t)(t_n - t)dt \qquad (4)$$

Consider that the interval (t_o,t_n) is divided into n parts by t_i (i=1,2,...,n-1), and the acceleration curve segments $A_i(t)$ are defined in subinterval (t_{i-1},t_i), then

$$\begin{aligned} D &= \sum_{i=1}^{n} \int_{t_{i-1}}^{t_i} A_i(t)(t_n - t)dt \\ &= \sum_{i=1}^{n} \int_{t_{i-1}}^{t_i} A_i(t)[(t_n - t_i) + (t_i - t)]dt \qquad (5) \\ &= \sum_{i=1}^{n} \left[(t_n - t_i)\int_{t_{i-1}}^{t_i} A_i(t)dt + \int_{t_{i-1}}^{t_i} A_i(t)(t_i - t)dt\right] \end{aligned}$$

Similar to D, we may define D_i "the area-moment of curve $A_i(t)$ about the end of ith part"

$$D_i = \int_{t_{i-1}}^{t_i} A_i(t)(t_i - t)dt \qquad (6)$$

And define F_i "the area under the curve $A_i(t)$" (Fig. 1)

$$F_i = \int_{t_{i-1}}^{t_i} A_i(t)\,dt \tag{7}$$

And call the ratio $\frac{D_i}{F_i}$ "the distance from the centroid of the ith area to t_i" denoted by t_{ci}

$$t_{ci} = \frac{D_i}{F_i} = \frac{\int_{t_{i-1}}^{t_i} A_i(t)(t_i - t)\,dt}{\int_{t_{i-1}}^{t_i} A_i(t)\,dt} \tag{8}$$

Then formula (5) becomes

$$\begin{aligned} D &= \sum_{i=1}^{n} \left[(t_n - t_i)F_i + D_i \right] \\ &= \sum_{i=1}^{n} \left[(t_n - t_i)F_i + t_{ci}F_i \right] \\ &= \sum_{i=1}^{n} \left[F_i (t_n - t_i + t_{ci}) \right] \end{aligned} \tag{9}$$

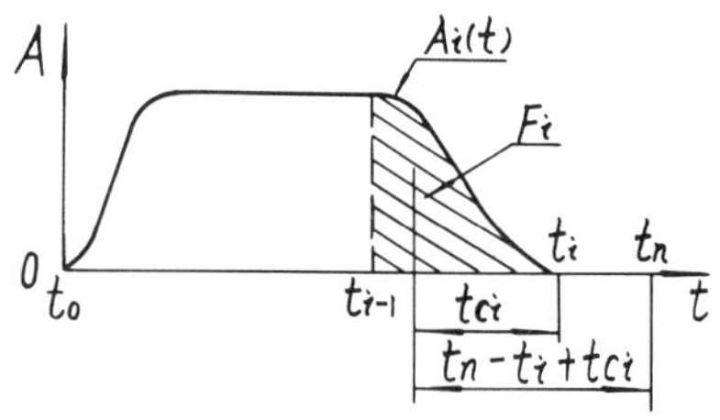

Fig.1 The "area" under $A_i(t)$ and its "centroid".

By substituting (7) and (9) into (1) and (2), a pair of "basic equations" is obtained as

$$\begin{cases} V(t_n) = V(t_0) + \sum_{i=1}^{n} F_i \\ S(t_n) = S(t_0) + (t_n - t_0)V(t_0) + \sum_{i=1}^{n} \left[F_i (t_n - t_i + t_{ci}) \right] \end{cases} \tag{10}$$

BASIC ACCELERATION CURVE SEGMENTS

Seven "basic acceleration curve segments", named by type number "j" (j=1,2,...,7), are given in Fig. 2, which will be selected by users for every part of the combined curve. For each of the upper six (j=1,2, ...,6), we note that the "area" F_j and the position of the centroid T_{cj} are propotional to the length of the segment Th and the maximum acceleration Am respectively

$$F_j = f_{vj} A_m T_h \tag{11}$$

$$T_{cj} = f_{sj} T_h \tag{12}$$

where f_{vj} is named "the factor of area", and f_{sj} "the factor of centroid".

To meet the jerk conditions J_0 and J_n at boundaries, we have developed the 7th segment, the area of which is severed into two parts, one related to the maximum acceleration Am

$$F_{7A} = \frac{1}{2} A_m T_h \tag{13}$$

and the other to the boundary jerk J_0

$$F_{7J} = \frac{1}{12} J_0 T_h^2 \tag{14}$$

Type j	Acc. Curves	Acc. Equations	Areas F_j	Centriod Position T_{cj}
1	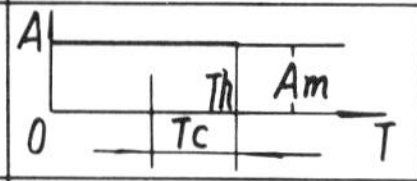	$A(T)=A_m$	$A_m T_h$ $f_v=1$	$\frac{1}{2}T_h$ $f_s=0.5$
2	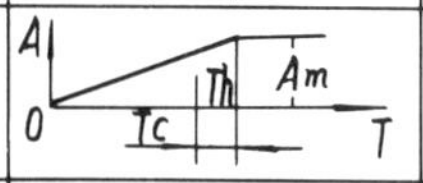	$A(T)=\frac{A_m}{T_h}T$	$\frac{1}{2}A_m T_h$ $f_v=0.5$	$\frac{1}{3}T_h$ $f_s=0.3333$
3	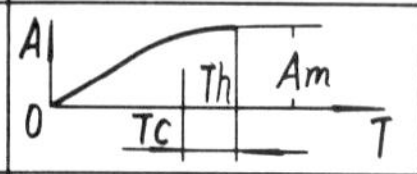	$A(T)=\frac{A_m}{T_h}(2T-\frac{T^2}{T_h})$	$\frac{2}{3}A_m T_h$ $f_v=0.6667$	$\frac{3}{8}T_h$ $f_s=0.375$
4	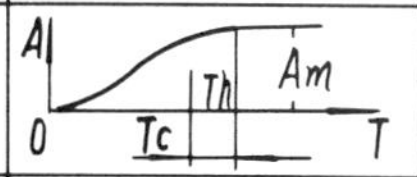	$A(T)=\frac{A_m}{T_h^2}(3T^2-\frac{2T^3}{T_h})$	$\frac{1}{2}A_m T_h$ $f_v=0.5$	$\frac{3}{10}T_h$ $f_s=0.3$
5	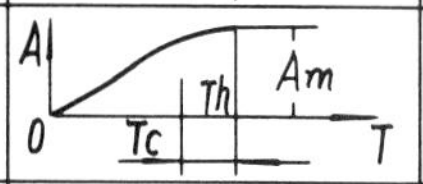	$A(T)=A_m \sin\frac{\pi T}{2T_h}$	$\frac{2}{\pi}A_m T_h$ $f_v=0.63662$	$(1-\frac{2}{\pi})T_h$ $f_s=0.36338$
6	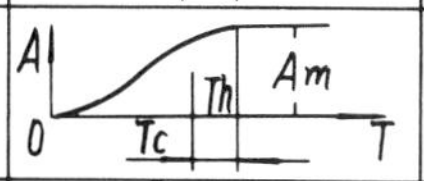	$A(T)=\frac{A_m}{2}(1-\cos\frac{\pi T}{T_h})$	$\frac{1}{2}A_m T_h$ $f_v=0.5$	$(\frac{1}{2}-\frac{2}{\pi^2})T_h$ $f_s=0.29736$
7	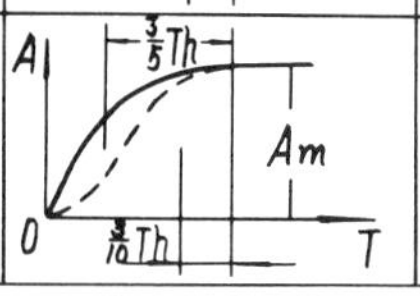	$A(T)=J_0 T + \frac{T^2}{T_h^2}(3A_m-2J_0T_h) - \frac{T^3}{T_h^3}(2A_m-J_0T_h)$	$\frac{1}{2}A_m T_h + \frac{1}{12}J_0 T_h^2$ $f_{vA}=0.5$ $f_{vJ}=0.08333$	$f_{sA}=0.3$ $f_{sJ}=0.6$

Fig. 2. Basic acceleration curve segments

Thus we can store the factors f_{vj} and f_{sj} in a computer, and use them when assembling without taking the equations into account.

Note that the segments in Fig. 2 are "monotonically increasing" functions of time T. In actual designing, we often need "monotonically decreasing" segments (in absolute value). However, we do not want another set of equations stored. To cope with this, we use a symmetry transformation—a reflection in A-axis followed by a translation in the direction of T-axis, i.e. to replace variable T in the equations of the "basic acceleration curve segments" by (Th −T). See Fig.3. Then the distance from the centroid to the end of the segment becomes $(1 - f_{sj})$ Th.

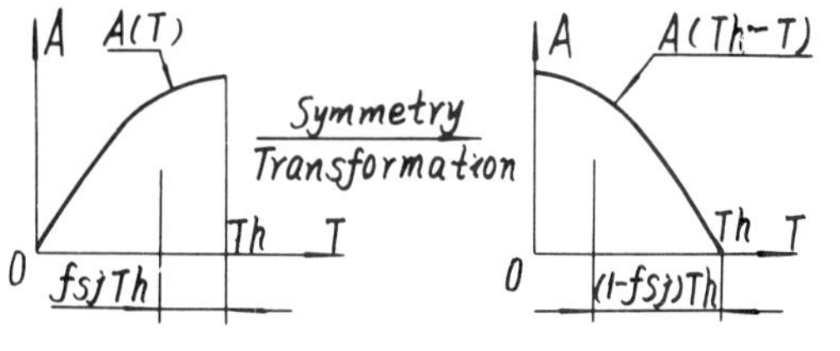

Fig. 3. Symmetry transformation of the "basic acceleration curve segments"

THE MATH MODEL FOR THE "AUTOMATIC ASSEMBLING PROGRAM"

In light of most needs of engineering, we propose a mathematical model for the assembling program of computers, which contains seven parts in the design interval (shown in Fig. 4). The constant acceleration segments (j=1) are used in the 2nd, 4th and 6th parts. The type number j and the length Th_i for each of the other parts as well as the acceleration magnitude G_4 in the 4th part will be assigned by the user. Two constant acceleration segments (j=1) are added to the first and the last parts with their magnitudes G_1 and G_7 equal to required boundary accelerations A_0 and A_n respectively. For the continuity of acceleration, the following must be satisfied:

$$\begin{aligned} Am_1 &= Aa - G_1 \;; \quad Am_2 = Aa \\ Am_3 &= Aa - G_4 \;; \quad Am_4 = G_4 \\ Am_5 &= Ab - G_4 \;; \quad Am_6 = Ab \\ Am_7 &= Ab - G_7 \end{aligned} \tag{15}$$

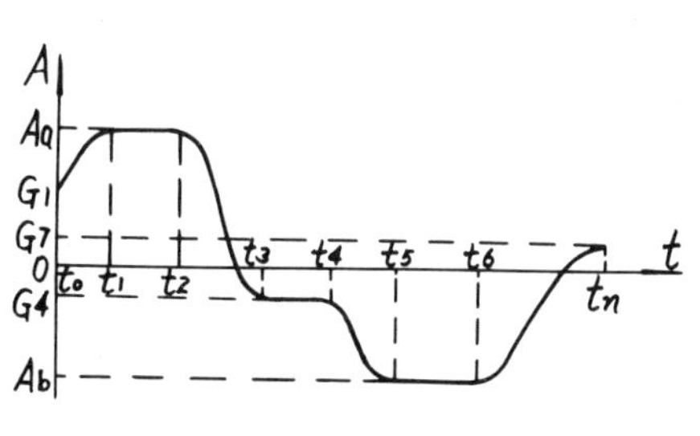

Fig.4. The math model for "automatic assembling program".

By substituting (11), (12) and (15) into (10), we finally obtain a pair of simultaneous equations.

$$\begin{cases} C_1 Aa + C_2 Ab = B_1 \\ C_3 Aa + C_4 Ab = B_2 \end{cases} \tag{16}$$

where $C_1 = \sum_{i=1}^{3} f_{vi} Th_i$ (17)

$$C_2 = \sum_{i=5}^{7} f_{vi} Th_i \tag{18}$$

$$C_3 = \sum_{i=1}^{3} \left(f_{vi} Th_i (t_n - t_i + f_{si} Th_i) \right) \tag{19}$$

$$C_4 = \sum_{i=5}^{7} \left(f_{vi} Th_i (t_n - t_i + f_{si} Th_i) \right) \tag{20}$$

$$B_1 = V_n - V_0 - G_4 Th_4 - \sum_{i=1,3,5,7} \left[(1 - f_{vi}) G_i Th_i \right] \tag{21}$$

$$\begin{aligned} B_2 = {} & S_n - S_0 - V_0 (t_n - t_0) - G_4 Th_4 \left(\frac{Th_4}{2} + t_n - t_4 \right) \\ & - \sum_{i=1,3,5,7} G_i Th_i \left[\frac{Th_i}{2} + t_n - t_i - f_{vi} (f_{si} Th_i + t_n - t_i) \right] \end{aligned} \tag{22}$$

$$G_3 = G_4 = G_5$$

In simultaneous equations (16) all C_i (i=1,2,3,4) and B_i (i=1,2) can be calculated from given data, so that the magnitude of the whole acceleration curve will be known by solving the value of Aa and Ab. At the same time, all boundary conditions will be satisfied accurately.

While designing the cam profile, the value of velocity and displacement at any time will be calculated by using corresponding equations of the "basic acceleration curve segments" through adequate coordinate transformations. All curves will be absolutely continuous because of the characteristics of integration.

For high speed cam systems, it is also important that the jerks at boundaries be matched. Now we can use the 7th type of the segments (j=7) in the first part and the last part of the interval. Then the two right-side terms of (16) become

$$B_1 = V_n - V_0 - G_4 Th_4 - \sum_{i=1,3,5,7} (1 - f_{vi}) G_i Th_i - \frac{1}{12} (J_0 Th_1^2 - J_n Th_7^2) \tag{23}$$

$$\begin{aligned} B_2 = {} & S_n - S_0 - V_0 (t_n - t_0) - G_4 Th_4 \left(\frac{Th_4}{2} + t_n - t_4 \right) \\ & - \sum_{i=1,3,5,7} G_i Th_i \left[\frac{Th_i}{2} + t_n - t_i - f_{vi} (f_{si} Th_i + t_n - t_i) \right] \\ & - \frac{J_0 Th_1^2}{12} \left(\frac{3}{5} Th_1 + t_n - t_1 \right) + \frac{1}{30} J_n Th_7^3 \end{aligned} \tag{24}$$

Notice that formulas (23) and (24) will be the same as (21) and (22) when J_0 and J_n are assigned to be zero. So that we can make a common program for the two cases above.

TWO EXAMPLES OF USAGE

Example 1. To calculate a trapezoid acceleration curve which is modified in the first and the 7th parts for continuities of jerk at boundaries.

(1) Input the boundary conditions:

Time	$t_0=0$	$t_n=1$
Displacement	$S_0=0$	$S_n=1$
Velocity	$v_0=0$	$v_n=0$
Acceleration	$A_0=0$	$A_n=0$
Jerk	$J_0=0$	$J_n=0$

(2) Input all lengths of the seven parts: Th_i= 1/8, ¼, 1/8, 0, 1/8, ¼, 1/8

(3) Input the type numbers of segment selected for them: j = 4, 1, -3, 1, 3, 1, -4

The negative sign of the type numbers means "symmetry transformation". Finally input G4=0. Immediately the computer tells:

Aa=5.27473; Ab=-5.27473; v_{max}=2.08791

And the acceleration curve is output as Fig. 5.

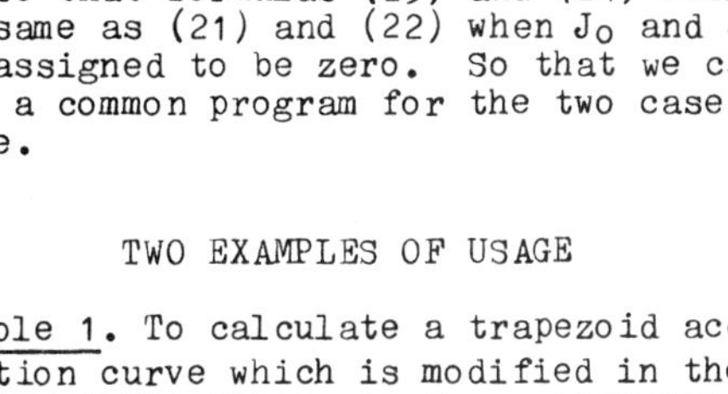

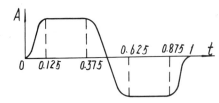

Fig. 5. Acc. curves for Example 1.

Example 2. Design the follower motion of the paper-passing mechanism of an offset machine, which needs the following boundary conditions of motion for its working (or forward) period:

Time	$t_0=0$	$t_n=1.1802$
Displacement	$S_0=4.5496$	$S_n=5.4645$
Velocity	$V_0=-0.1692$	$v_n=1.08869$
Acceleration	$A_0=0.0141$	$A_n=-0.158$
Jerk	$J_0=0.16$	$J_n=8.0$

A shape of the acceleration curve has been given roughly in Fig. 6. Now, find a combined curve which is similar to it.

According to Fig. 6, each length of the 7 parts should be:

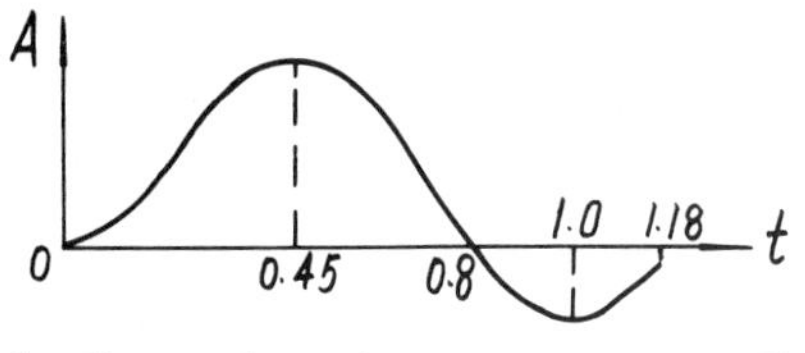

Fig.6. The original acc. curve in Ex. 2

Th_i = 0.45, 0, 0.35, 0, 0.2, 0, 0.1802
and the type numbers can be
j = 7, 1, -5, 1, 5, 1, -7.
After imputting all numbers, the computer output all of the motion curves shown in Fig.7.

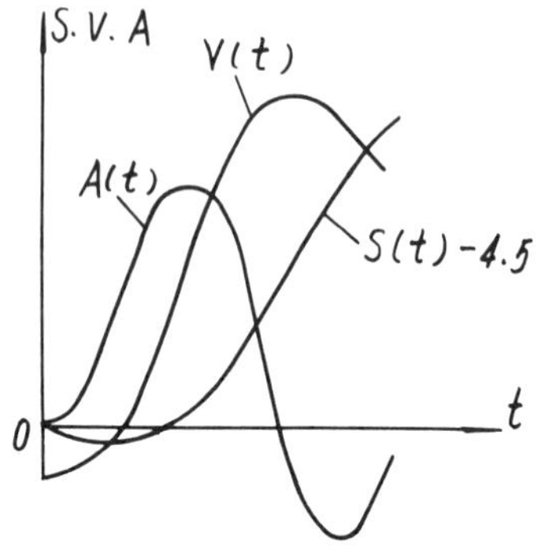

Fig. 7. Motion curves output by computer

CONCLUSION

In the mathematical model of the "Automatic Assembling Program", the "area factors" f_{vj} and "centroid factors" f_{sj} of the seven predesignated "basic acceleration curve segments" as well as their equations are stored. The design interval is divided into seven parts. For every part, the users need only to determine the length and to select a "basic acc. curve segment".

The program will use a pair of simultaneous equations based on the "area-moment" relationship to assemble them. Then calculate all curves of motion when needed.

The adventages of the algorithm are:
1. The number of unknowns involved is greatly reduced--we have only Aa and Ab.
2. Different conbinations of various curve segments are assembled by the same procedure.
3. Like playing tangram, users may simply apply the "basic segments" to form various acceleration specifications.
4. All boundary conditions (encluding jerks) will be satisfied accurately with assured continuity of all curves.
5. A short program can be used for most basic cam curves, as well as to create particular cam curves for certain purposes.

Four cams have been designed for printing machines using the program which have been proved to be satisfactory for functional requirements.

REFERENCES

Chen, F.Y. (1982). Mechanics and Design of Cam Mechanisms. Pergamon Press, New York. Chap. 7, pp.102-124.
Kloomok, M., and R. V. Muffley (1955). Plate cam design--with emphasis on dynamic effects. Prod. Eng., Feb., 156-162.
Li, Yuhong and Qi Weihao (1984). A new algorithm for "universal curve" of cam motion and its program. Machine Design (published by The Northwestern Polytechnic Institute, Xian, China), No.4,1-7.
Neklutin, C.N.(1969). Mechanisms and Cams for Automatic Machines. American Elsevier Pub. Company Inc., New York. Chap. 3, pp.91-144.
Qi, Weihao and Li Yuhong (1985). The design of cam curve from complex boundary conditions. Chinese J. Mechanical Engineering. 21, 60-65.

Eine neue Rechnungsmethode zur programmgesteuerten Kombinierung der Kurvensegmente zu einer Exzenterkurve

Dieser Artikel stellt eine neue Rechnungsmethode zur programmgesteuerten Kombinierung der Kurvensegmente zu einer Exzenterkurve vor.

------- 7 vorkonstruierte und gespeicherte "Basisbeschleunigungskurvensegmente" und deren "Flaechenfaktor" und "centroid-Faktor" ausnehmeh.
------- 2 Formelgleichungen ausnehmen, aus das gemeinsame Maximum Aa der ersten 3 Kurvensegmente und das gemeinsame Maximum Ab der zweiten 3 Kurvensegmente ausgerechnet werden koennen.
------- mit dem entsprechenden Wechsel der Koordinaten und mit den Formelgleichungen der bestimmten Kurvensegmente die Geschwindigkeiten und Bewegungen in beliebiger Zeit ausrechnen.

Mit der vorher vorgestellten Methode ist das universale Kombinietrungsprogramm in Computer ausgearbeitet, deren Vorteile sind:
1. Die Zahl der unbekannten Zahlen ist am minimum verringert worden. Nur Aa und Ab bleiben noch.
2. Die verschiedenen Kombinierungen der Kurvensegmente koennen nur mit einem gleichen Programm erledigt werden.
3. Der Anwender kann mit diesen 7 "Basisbeschleunigungs-kurvensegmenten" unterschiedliche Beschleunigungskurven formulieren.
4. Das Programm kann alle Grenzbedingen exakt erfuellen und die Kontinuierlichkeit aller Kurven gewaehrleisten.
5. Dieses einfache Programm kann in die Rechnungen der grossteiligen Standardexzenterkurven eingesetzt werden und kann nach individuellen Wuenschen spezielle Exzenterkurven ausarbeiten.

Generating Method of Cams and its Accuracy Analysis

YANG JI-HOU* AND ZHANG XI-LIANG**

*Department of Mechanical Engineering, Dalian Institute of Light Industry, Liaoning, PRC

**Department of Mechanical Fngineering, Harbin Institute of Electrical Technology, Heilongjiang, PRC

Abstract. A new principle and method of generating disk cam profile by the swing motion of the simplest four-bar linkage are proposed. This method is applicable to the extensive rise-dwell-return-dwell type of cams and can get a fully smooth motion of cam follower at comparatively high speed. The main points of this method are enumerated, such as,dimension selection of generating mechanism, adjustable parameters on the generating mechanism to preserve dimension conditions of the required cam mechanisms, types and controling method of crank motion and compensating principles of dimension error of grinding wheel when the generating method of grinding is used.

Keywords. Mechanism; disk cam; generating method; accuracy analysis.

INTRODUCTION

1. Generating method of cam profiles was studied by Nakai(1962), and was systematically discussed by Druce(1974). The generating method proposed here isn't restricted by the ordinary form of follower motions but directly looking for the smooth motion and the simple machining way. Cams can't manifest itself good characters of motion, if the accuracy of machining way isn't guaranteed enough, even the displacement diagrams are very idral. According to the experiments(Rothbart,1956), among 0.5 mm length of cam profile, if there is radial inaccuracy of 0.05 mm, a shock happen and the acceleration rule is destroied entirely. The method here was examined at a special device, and got a comparatively good result.

THE PRINCIPLE OF GENERATING METHOD OF CAM

2. we use the swing motion of rocker of a four-bar linkage to generate cam profile as shown in Fig. 1. When the crank and rocker mechanism moves from position PQRS to PQ_1R_1S, the swing angle of rocker is ψ_Δ, which is also the angle of generating cutter or grinding wheel installed on SK. The rotation of axis O of cam blank and axis P of crank must be controled by the programme of microprocessor with two coordinates. On the two extreme positions of oscilating rod,the crank must be stationary to get the max. and min. radius of cam blank. On the rise and return stage of the follower, both axes must rotate with θ_Δ and ψ_Δ respectively. At this time two axes may rotate with constant speed ratio or with modified constant ratio.

3. Obviously the following adjustable components must be provided on this generating mechanism, such as,

(1) The dimension of crank length is adjustable, to make the swing angle ψ_Δ equal to the oscilating angle β_Δ.

(2) The length L=SK of oscilating cutter arm is adjustable, to meet the length of follower arm of different cam mechanisms.

(3) The initial angular position of cutter arm $\beta_0=\angle OSK$ is adjustable, actually the angle $\angle RSK$ of the generating mechanism is adjustable, and the feed motion of cutter must also rely on this component.

(4) The fixed length l=OS is adjustable, to suit different fundamental dimensions of the cam mechanisms.

(5) The rotating speeds of two axes O,P are adjustable. Theoretically the radius of cutting tool must be equal to the radius of the follower roller.

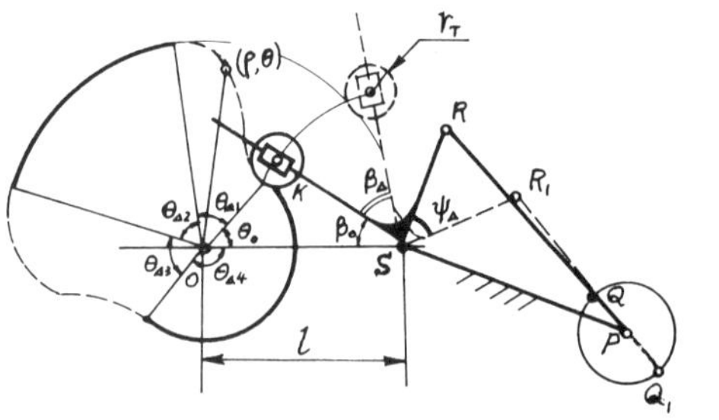

Fig. 1

THE DIMENSION SELECTION OF MECHANISM AND ITS MOTION

4. The dimensions of four-bar linkage ought to compact each other i.e. the differences between four lengths are as small as posible. The transmission angle must be in the range of $\gamma=90°\pm50°$, and the quick return angle must be less than 1° . The swing angle ψ_Δ can be changed in the range of $\psi_\Delta=0°$–$60°$, by rearangement of only crank length. After consideration and calculation according to the upper conditions, the fundamental relative lengths of links are determined as: crank a=0.38, coupler b=1.30, rocker c=0.82 and frame d=1.50, where the sum of four relative lengths is a+b+c+d=4.0. The swing angle of oscilating arm changes among 0°-60° with respect to the crank a=0-0.40.

5. In the Fig. 2, according to the formula of arm position ψ with respect to the crank angle ϕ, we knew

$$\psi=\pi-tg^{-1}\left(\frac{a\ \sin\phi}{d-a\ \cos\phi}\right)-\cos^{-1}\left(\frac{k^2+c^2-b^2}{2kc}\right) \qquad (1)$$

$$\text{but } k=\sqrt{a^2+d^2-2ad\ \cos\phi} \qquad (2)$$

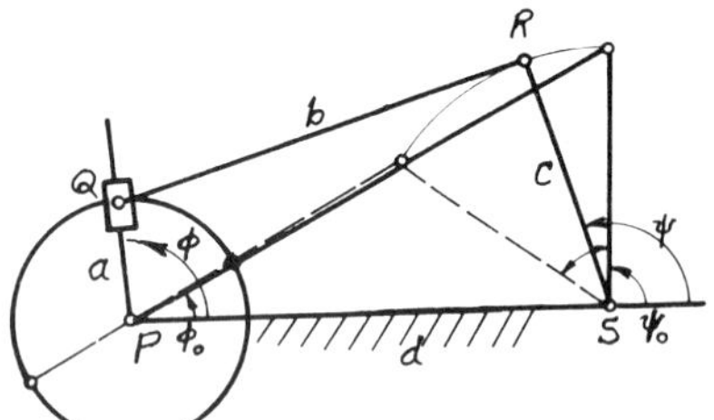

Fig. 2

When let θ represent the rotating angle of cam, and $d\theta/dt=\omega$ =const. as shown in Fig. 1, the actual angular velo. and angular acce. of the oscilating arm become to

$$\frac{d\psi}{dt}=\frac{d\psi}{d\phi}\frac{d\phi}{d\theta}\frac{d\theta}{dt}=\left(\omega\frac{d\phi}{d\theta}\right)\frac{d\psi}{d\phi} \qquad (3)$$

$$\frac{d^2\psi}{dt^2}=\left(\frac{d^2\psi}{d\phi^2}\right)\left(\frac{d\phi}{d\theta}\right)^2\left(\frac{d\theta}{dt}\right)^2+\frac{d\psi}{d\phi}\left(\frac{d^2\phi}{d\theta^2}\right)\left(\frac{d\theta}{dt}\right)^2$$
$$=\omega^2\left\{\left(\frac{d\phi}{d\theta}\right)^2\frac{d^2\psi}{d\phi^2}+\frac{d^2\phi}{d\theta^2}\frac{d\psi}{d\phi}\right\} \qquad (4)$$

By the formulas (3) we can check the base radius of cam approximately. If the speed ratio of cam axis and crank axis is const. in the rise and return stages, then $d\phi/d\theta$= const. as shown in Fig. 3(a). The formula (4) illustrate that the acceleration diagram of cam may be analog to the simple harmonic motion as shown in Fig. 4(a),a. If the speed ratio of two axes is a modifoed uniform motion as shown in Fig. 3(b),(c), the acceleration diagram of cam become to the curves b,c,d in Fig. 4(a),(b), which are near to the cycloidal motion.

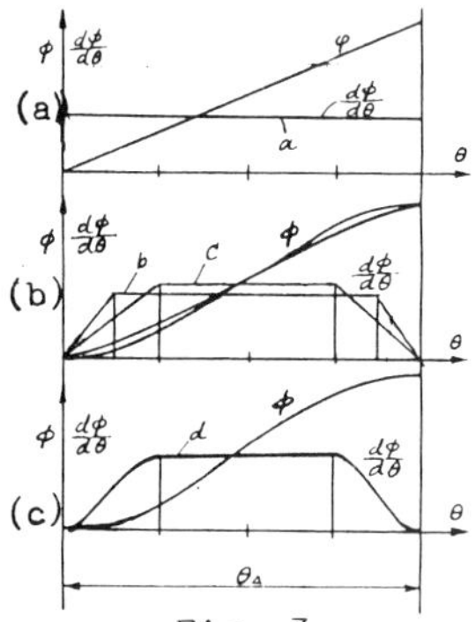

Fig. 3

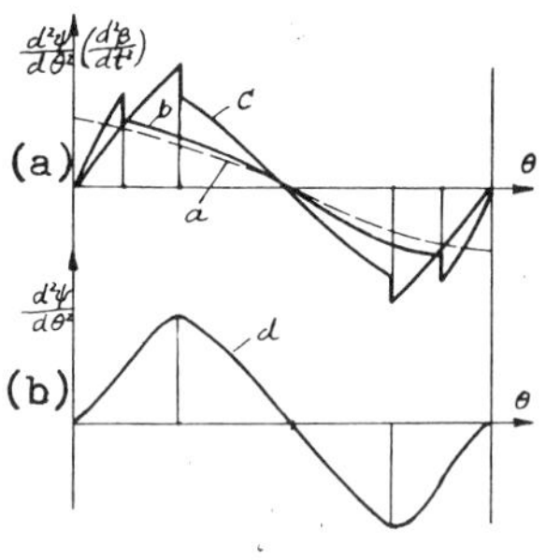

Fig. 4

THE COMPENSATING METHOD OF CUTTER ERROR

6. Theoretically the parameters of generating mechanism, such as l,L,β_0 and r_T, must equal to that of cam mechanism. Then the oscilating arm will repeat the motion of rocker and can ensure a smooth and reliable motion of the follower. But the cutter is difficult to maintain the exact dimension, for the cutter wear leads to change its own diameter, especially quicker on the grinding wheel. Therefore we ought to compensate the cutter error to get high accurate cam profiles. THis can be achieved easily by rearangement of the mechanism, and need not to use the third coordinate as ordinary control system. From Fig. 1, coordinates of cam contour (ρ,θ) are the functions of l, L, r_T, β, i.e. $\rho=\rho(r_T,l,L,\beta)$, $\theta=\theta(r_T,l,L,\beta)$, hence the error equations may be as :

$$\Delta\rho=\frac{\partial\rho}{\partial r_T}\Delta r_T+\frac{\partial\rho}{\partial l}\Delta l+\frac{\partial\rho}{\partial L}\Delta L+\frac{\partial\rho}{\partial\beta}\Delta\beta \qquad (5)$$

$$\Delta\theta=\frac{\partial\theta}{\partial r_T}\Delta r_T+\frac{\partial\theta}{\partial l}\Delta l+\frac{\partial\theta}{\partial L}\Delta L+\frac{\partial\theta}{\partial\beta}\Delta\beta \qquad (6)$$

From upper equations we can concel or compensate the error of cam contour produced by the cutter error Δr_T through adjustment of other parameters Δl, ΔL and $\Delta\beta$.

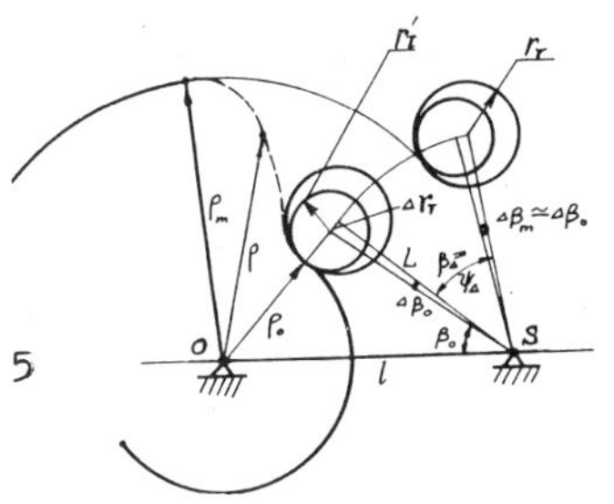

Fig. 5

7. From Fig. 5 imaginarily seeing, the better compensating parameter of the error is the initial position of oscilating arm reprisented by $\Delta\beta_o$. If the grinding wheel become smaller, at initial position of β_o, the direct and effective compensation is to diminish β_o a little,and the dimensions of l, L may be unchanged. From equ.(5), the compensating quantity of $\Delta\beta_o$ must be defined as following to get $\Delta\rho=0$.

$$\Delta\beta_o = \left[-\left(\frac{\partial\rho}{\partial r_T}\right)/\left(\frac{\partial\rho}{\partial\beta_o}\right)\right]\Delta r_T \tag{7}$$

From the results analized by Yang(1983), at the position of $\rho=\rho_o$ or ρ_m, $\beta=\beta_o$ or β_m,

$$\frac{\partial\rho}{\partial r_T}=-1,\quad \frac{\partial\rho}{\partial\beta}=\frac{Ll\ \sin\beta}{\rho+r_T} \tag{8}$$

$$\therefore\ \Delta\beta=\frac{\rho+r_T}{Ll\ \sin\beta_o}\Delta r_T \tag{9}$$

Usually when the error is compensated by equ.(9) at the base radius of a cam, the residual error of $\Delta\theta$ is small enough. By the same $\Delta\beta_o$ the error at the largest radius of a cam is also may be compensated well sometimes. If the residual error at rise position is too large due to $\Delta\beta_m \neq \Delta\beta_o$, the radius of crank may be slightly adjusted. Thus we can get exact dimensions of max. and min. radii of a cam simultaneously.

ACCURACY ANALYSIS OF CONTOUR

8. After cutter error is compensated, the residual error at the rise and return portion of cam profile are still retained. As shown in Fig. 6, when the cutter radius become smaller, $\Delta r_T = r_T' - r_T < 0$, at the position of $\beta > \beta_o$, the manufactured contour may become higher than the exact contour, even the error due to $\Delta r_T < 0$ has been compensated by $\Delta\beta_o$. Let the normal error of the contour represented by δ_K, from Fig. 6,

$AK' = r_T$, $CK' = \delta_K$, $BC = r_T'$, $AB = c\cos\alpha$

but $c = AA'$, α_o is initial pre. angle, α is pre. angle at any position, then we get

$$r_T = c\ \cos\alpha + r_T' + \delta_K \tag{10}$$

Also the following formula can be proved:

$$\cos\alpha_o = \frac{l\ \sin\beta_o}{\rho_o + r_T} - \frac{\Delta r_T}{c} \tag{11}$$

Therefore from equ.(10), we can calculate δ_K from the following formula

$$\delta_K = -\Delta r_T\left(1-\frac{\cos\alpha}{\cos\alpha_o}\right) \tag{12}$$

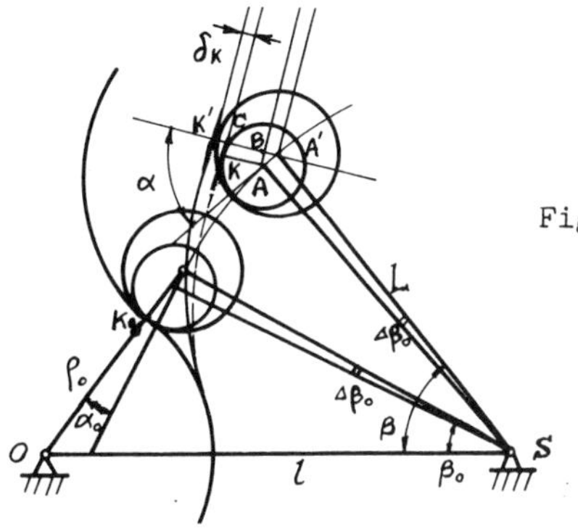

Fig. 6

Upper formula indicates that if $\Delta r_T < 0$ and $\alpha > \alpha_o$, then $\delta_K > 0$ i.e. the residual error make the contour fatter. The residual error curves of the whole contour like the form in Fig. 7, depend on the development of presure angle (Zhang,1984).

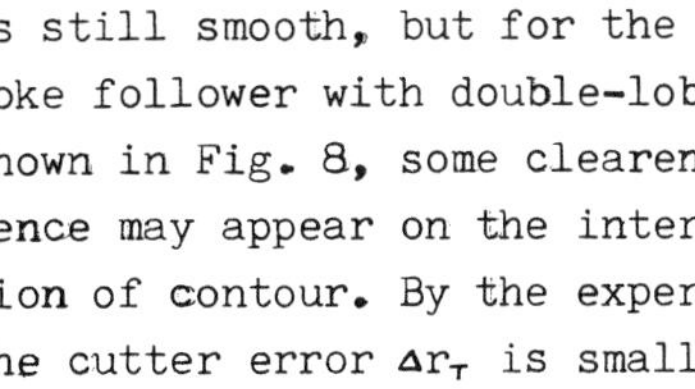

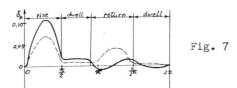

Fig. 7

9. Because the residual error changes smoothly, the motion of a force closed follower is still smooth, but for the oscilating yoke follower with double-lobed cam, as shown in Fig. 8, some clearence or interference may appear on the intermediate portion of contour. By the experience, when the cutter error Δr_T is smaller than 0.5 mm, the clearence can't exceed 0.2 mm ordinarily, and this type of cam mechanism can maintain a smooth state of working characteristics. Ordinarily allowable cutter error must be set according to the allowable clearence or interference on the cam mechanism by the theoretical calculation method, to ensure the accuracy of a cam (Yang,1984).

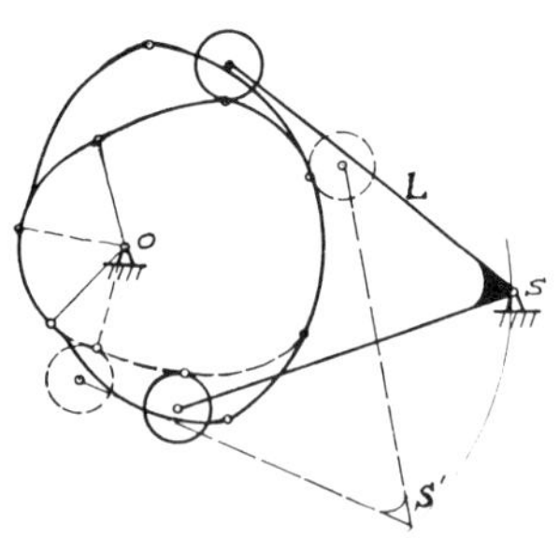

Fig. 8

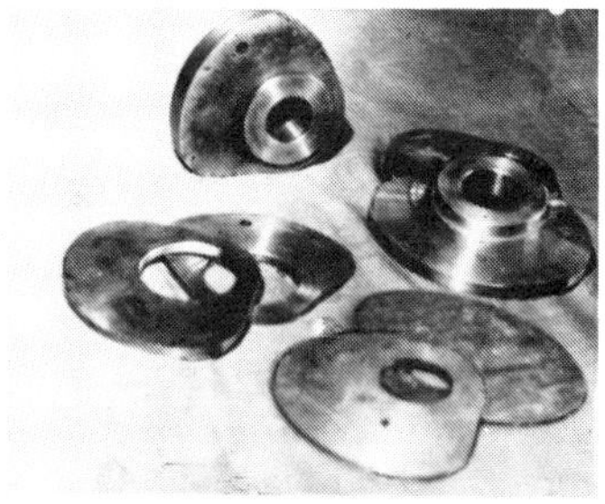
Fig. 9

Fig. 10

THE EXPERIMENTAL RESULT AND CONCLUSIONS

10. According to the former principle, a grinding device of cam generating method was made, as shown in Fig. 9. Several cams was grinded at this device as shown in Fig. 10. Four cams of them were installed at two types of automatic punching press. From the experimental data, the properties of these cams are quite better than cams made by the grinding method with template. Hence the following conclusions are obtained:

(1) The generating method of cam profile by the rocker motion of four-bar linkage can provide a continuous and smooth cam profile. It is especially adapt to the very useful rise-dwell-return type of medium and high speed cams.

(2) The generating mechanism is the simplest linkage with rotating pairs, where the clearence may be very small by using rolling bearings. Thus the cam contour become smooth enough and also the labour of template preparation can be saved.

(3) The generating mechanism likes a analog computer, which simplifies requirements to the control system of computer.

(4) For the force-closed follower, the residual error at intermediate cam profile from cutter error needn't consideration.

(5) For machining two lobed cam with yoke follower, the compensating method of cutter error must be considered to diminish the clearence or interference at the conjugated contour of rise and return stages.

REFERENCES

Druce, G. (1974). Conf. on Cam and Cam Mechanisms, London.

Nakai, Eiitsu. (1962). Machine Design, (Jap.) V6,N5.

Rothbart, H. A. (1956). Cams, John Wiley & Sons, Inc.

Yang Ji-Hou (1983). Chinese Journal of Mech. Eng., V19, N4.

Yang Ji-Hou, Zhang Xi-Liang (1984). Journal of Harbin Institute of Electrical Technology, N1.

Zhang Xi-Liang, Yang Ji-Hou (1984). Journal of Northeast Heavy Machinery Institute, N1.

МЕТОД ОБКАТКИ КУЛАЧКОВ И АНАЛИЗА ЕГО ТОЧНОСТИ

Ян Ди-Хо Жанг Ши-Лианг

Резюме. Предполагаются что новая теория и метод обкатки профилей дисковых кулачков качающимся движением простейщих шарнирных четырехзвенных механизмов. Этот метод хорошо использоватся кулачковым механизмам с качающимися толкателями, и довольное установившееся движение может быть доходиться в высокой скорости. Основные точки зтого метода представлены, т.е. длины звеньев механизма, регулированные параметры механизма обкатки, метод управления движения кривошипа четырехзвенников, точность обкатки и компенсатные методы шлифовального круга из за обкатки шлифованием.

Curvature Radius of Disk Cam Pitch Curve and Profile

SHI YONGGANG

Department of Mechanical Engineering, Zhejiang University, Hangzhou, Zhejiang, PRC

Abstract. To determine curvature radius of cam profile four new useful equations have been developed. Each equation is provided for one type of follower respectively, such as translating roller follower, translating flat-faced follower, oscillating roller follower, and oscillating flat-faced follower. In CAD of cam mechanisms by means of the equations designer can obtain the minimum of curvature radius easy. So that the curvature radius can be taken as a design condition other than as a verify condition for choosing the radius of prime circle, especially for the cam mechanisms with flat-faced follower.

Keywords. Cam; curvature; mathematical analysis; vector; differential equation; functional equation.

INTRODUCTION

The curvature radius of disk cam profile is a parameter for investigating the contact stress. For the cam mechanisms with roller follower the curvature radius of pitch curve is an important value for selecting roller radius. To avoid motion distortion of flat- faced follower, the profile must be convex at all segments. It can be distinguished by means of calculating the curvature radius. If there is any value less than zero, then ought to increase the radius of prime circle or correct the desired motion function.

Until now the calculating formula published in many text books and papers on TMM is as following

$$R=\frac{\left[\rho^2+\left(\frac{d\rho}{d\theta}\right)^2\right]^{\frac{3}{2}}}{\rho^2+2\left(\frac{d\rho}{d\theta}\right)^2-\rho\frac{d^2\rho}{d\theta^2}} \qquad (1)$$

Because the parameters ρ and θ are both transcendental functions about cam rotation angle φ, the expressions $d\rho/d\theta$ and $d^2\rho/d\theta^2$ are more complex. So that the designer could not use such equation successfully. Some authors have published that to distinguish the ratio of minimum curvature radius with prime circle by means of charts. But this method can be only used for few types of motion functions. To prevent cusp on a designed cam profile with sliding flat-faced follower, many papers show the condition of selecting prime circle radius as following

$$R_b>\left[-\left(S+\frac{d^2s}{d\varphi^2}\right)\right]_{max} \qquad (2)$$

But the equation is only adapted for the situation when the acting surface is perpendicular to the follower guide.

To deal with the problem about disk cam curvature radius perfectly, author introduces a new conception. It is that as the follower motion function and the constant parameters of cam mechanism have been specified, the pitch curve and the cam profile as well as their curvature radii are all determined simultaneously. Therefore the cam curvature radius must be a function about the specified motion function of follower and the mechanism constants. The designer can distinguish the curvature radius before obtaining the polar coordinates of pitch curve and cam profile. Because need not to differentiate the transcendental functions, the calculating procedure will be easier.

In order to obtain the equations determining curvature radius, in which the variables are only follower motion parameters and cam rotation angle, the polar vector analysis is used. As shown in Fig. 1 assume the location of any point on pitch curve or cam profile is indicated by polar vector $\mathbb{P}=\rho e^{i\theta}$. The model ρ and directional angle θ are both functions about cam rotation angle φ. The differentiation of the vector $\mathbb{P}$ with respect to φ is as following

$$\mathbb{P}'=\frac{d\rho}{d\varphi}e^{i\theta}+\rho\frac{d\theta}{d\varphi}e^{i(\theta+\pi/2)}=\tau e^{i\xi} \qquad (3)$$

where τ and ξ are the model and direction angle of vector $\mathbb{P}'$. The second differentiation is as following

$$\mathbb{P}''=\frac{d\tau}{d\varphi}e^{i\xi}+\tau\frac{d\xi}{d\varphi}e^{i(\xi+\frac{\pi}{2})} \qquad (4)$$

Therefore the curvature radius R at any consideration point can be determined by means of formula

$$R=\frac{|\mathbb{P}'|^3}{|\mathbb{P}'\times\mathbb{P}''|}=\frac{\tau}{d\xi/d\varphi} \qquad (5)$$

Equation 5 can be used to indicate bending direction of the curves too. If the value of R is positive the curve will be convex at the consideration point, and if it is negative then will be concave. The reason is that if a curve is convex at any point, the curvature center and the polar origin, i. e. the cam rotation center, must be at same side divided by tangential line drawn through the consideration point. And if the curve is concave then the curvature center and the polar origin will be at opposite sides. From Eq. 4 the normal vector of $\mathbb{P}''$, i. e. $\tau \frac{d\xi}{d\varphi} e^{i(\xi+\pi/2)}$, will locate the position of curvature center. If the model $\tau \frac{d\xi}{d\varphi}$ is positive, the direction angle of normal vector is of $(\xi+\pi/2)$, therefore the curvature center and the polar origin are at same side. And if the model is negative, the actual directional angle of normal vector is of $(\xi-\pi/2)$, then the curvature center and polar origin are at each side of trangential line. Furthermore, the sign of $\tau d\xi/d\varphi$ is identical with the sign of $\tau/(d\xi/d\varphi)$, i.e. the curvature radius R.

CURVATURE RADIUS OF DISK CAM WITH TRANSLATING ROLLER FOLLOWER

Disk cam mechanism with translating roller follower can be classified as four cases according to offset direction and contact portion of cam profile with roller as shown in Fig. 2. The cases a and b show external cams, cases c and d show internal cams. To obtain an equation which is suitable for the four cases introduces two sign coefficients η and λ, η is used to distinguish the offset direction, and λ the situation of contact. The polar vector $\mathbb{P}_B$ locating point B on pitch curve is

$$\mathbb{P}_B = \eta E e^{i(\varphi_b+\varphi)} + (S_b+S) e^{i(\varphi_b+\varphi-\frac{\pi}{2})} \tag{6}$$

where E is the offset amount, S_b and φ_b are both mechanism constants

$$S_b = \sqrt{R_b^2 + E^2}$$

$$\varphi_b = arc\,tg\,(S_b/(\eta E)),\quad 0 \leqslant \varphi_b \leqslant \pi$$

R_b is radius of prime circle, S is displacement of follower from origin position to consideration position. The first and second differentiations of vector $\mathbb{P}_B$ with respect to angle φ are as following

$$\mathbb{P}_B' = (\frac{dS}{d\varphi} - \eta E) e^{i(\varphi_b+\varphi-\frac{\pi}{2})} + (S_b+S) e^{i(\varphi_b+\varphi)} \tag{7}$$

$$\mathbb{P}_B'' = (2\frac{dS}{d\varphi} - \eta E) e^{i(\varphi_b+\varphi)} + (S_b+S-\frac{d^2S}{d\varphi^2}) e^{i(\varphi_b+\varphi+\frac{\pi}{2})} \tag{8}$$

Substituting into Eq. 5 obtain the equation of curvature radius of pitch curve

$$R_B = \frac{[(\frac{dS}{d\varphi}-\eta E)^2 + (S_b+S)^2]^{\frac{3}{2}}}{(\frac{dS}{d\varphi}-\eta E)(2\frac{dS}{d\varphi}-\eta E)+(S_b+S)(S_b+S-\frac{d^2S}{d\varphi^2})} \tag{9}$$

where $dS/d\varphi$ and $d^2S/d\varphi^2$ are reduced velocity and reduced acceleration of follower respectively. If a result of R_B is less than zero, the pitch curve is concave at the consideration point. The curvature radius of cam profile at corresponding point K, i.e. R_K, is as following

$$R_K = R_B - \lambda R_r \tag{10}$$

where R_r is radius of follower roller. If a follower is knife-edged then $R_r = 0$.

CURVATURE RADIUS OF DISK CAM PROFILE WITH TRANSLATING FLAT-FACED FOLLOWER

The cam mechanism with translating flat-faced follower is shown in Fig. 3. Where γ is the oblique angle of acting surface. The polar vector $\mathbb{P}_K$ locating the point K on cam profile is

$$\mathbb{P}_K = (R_b + S\cdot\sin\gamma) e^{i\varphi} + \frac{dS}{d\varphi}\sin\gamma\, e^{i(\varphi+\frac{\pi}{2})} \tag{11}$$

The differentiations are as following

$$\mathbb{P}_K' = (R_b + S\cdot\sin\gamma + \frac{d^2S}{d\varphi^2}\sin\gamma) e^{i(\varphi+\frac{\pi}{2})} \tag{12}$$

$$\mathbb{P}_K'' = (\frac{dS}{d\varphi} + \frac{d^3S}{d\varphi^3})\sin\gamma\, e^{i(\varphi+\frac{\pi}{2})} + [R_b + (S+\frac{d^2S}{d\varphi^2})\sin\gamma] e^{i(\varphi+\pi)} \tag{13}$$

Substituting into Eq. 5 then obtain

$$R_K = R_b + (S + \frac{d^2S}{d\varphi^2})\sin\gamma \tag{14}$$

To prevent cusp and undercutting the value of R_K must be greater than zero.

CURVATURE RADIUS OF DISK CAM WITH OSCILLATING ROLLER FOLLOWER

The four cases of such type of cam mechanism are shown in Fig. 4. The sign coefficient ς is given to distinguish the relative position of cam rotating center respect to the follower pivot. The polar vector $\mathbb{P}_B$ locating point B on the pitch curve is as following

$$\mathbb{P}_B = l_{OA} e^{i(\varphi-\varsigma\varphi_b)} + l_{AB} e^{i[\varphi-\varsigma(\varphi_b+\psi_b+\psi+\pi)]} \tag{15}$$

where l_{OA} is the distance between cam rotating center and follower pivot, l_{AB} is the follower length, φ_b and ψ_b are constant angles of which the values are

$$\varphi_b = arc\cos\frac{l_{OA}^2 + R_b^2 - l_{AB}^2}{2 l_{OA}\cdot R_b},\quad 0 \leqslant \varphi_b \leqslant \pi$$

$$\psi_b = arc\cos\frac{l_{OA}^2 + l_{AB}^2 - R_b^2}{2 l_{OA}\cdot l_{AB}},\quad 0 \leqslant \psi_b \leqslant \pi$$

And ψ is the angular displacement of follower from origin position. It is determined by means of desired motion function. The differentiations of vector $\mathbb{P}_B$ with respect to angle φ are as following

$$\mathbb{P}_B' = l_{OA} e^{i(\varphi-\varsigma\varphi_b+\frac{\pi}{2})} + l_{AB}(1-\varsigma\frac{d\psi}{d\varphi}) e^{i[\varphi-\varsigma(\varphi_b+\psi_b+\psi+\pi)+\frac{\pi}{2}]}$$

$$\mathbb{P}_B''=l_{OA}e^{i(\varphi-\varsigma\varphi_b+\pi)}+l_{AB}(1-\varsigma\frac{d\psi}{d\varphi})^2e^{i[\varphi-\varsigma(\varphi_b+\psi_b+\psi)]}$$
$$-\varsigma\frac{d^2\psi}{d\varphi^2}\cdot l_{AB}e^{i[\varphi-\varsigma(\varphi_b+\psi_b+\psi+\pi)+\frac{\pi}{2}]}$$

Assume

$$f=1-\varsigma\frac{d\psi}{d\varphi}\ ,\qquad \beta=\psi_b+\psi$$

Substituting $\mathbb{P}_B'$ and $\mathbb{P}_B''$ into Eq. 5 obtain the equation of curvature radius of pitch curve

$$R_B=\frac{l_{OA}^2+f^2l_{AB}^2-2fl_{OA}l_{AB}\cos\beta}{l_{OA}^2+f^3l_{AB}^2-l_{OA}l_{AB}[f(1+f)\cos\beta+\frac{d^2\psi}{d\varphi^2}\sin\beta]} \qquad (16)$$

where $d\psi/d\varphi$, $d^2\psi/d\varphi^2$ are reduced angular velocity and reduced angular acceleration of pivoted follower respectively.

The curvature radius of cam profile can be found from Eq. 10, and the sign coefficient λ must be selected according to the consideration case as shown in Fig. 4.

CURVATURE RADIUS OF DISK CAM PROFILE WITH PIVOTED FLAT-FACED FOLLOWER

The four cases of such type of cam mechanism are shown in Fig. 5. The sign coefficient η is used to indicate offset direction of acting surface from follower pivot. The polar vector $\mathbb{P}_K$ locating point K on profile is as following

$$\mathbb{P}_K=[l_{OA}\sin(\psi_b+\psi)+\eta E]e^{i(\varphi-\varsigma\psi)}$$
$$+l_{OP}\cos(\psi_b+\psi)e^{i(\varphi-\varsigma\psi+\frac{\pi}{2})}$$

where ψ_b is mechanism constant

$$\psi_b=\arcsin\frac{R_b-\eta E}{l_{OA}}\ ,\qquad -\frac{\pi}{2}\leq\psi_b\leq\frac{\pi}{2}$$

And ψ is angle displacement of follower from its origin position. Because the point P is the instantaneous center of velocity of cam and follower, therefore

$$l_{OP}=l_{OA}\frac{d\psi}{d\varphi}/(1-\varsigma\frac{d\psi}{d\varphi})$$

Assume

$$f=1-\varsigma\frac{d\psi}{d\varphi}\ ,\qquad \beta=\psi_b+\psi$$

The polar vector $\mathbb{P}_K$ becomes

$$\mathbb{P}_K=(l_{OA}\sin\beta+\eta E)e^{i(\varphi-\varsigma\psi)}$$
$$+(l_{OA}\frac{d\psi}{d\varphi}\cos\beta/f)e^{i(\varphi-\varsigma\psi+\frac{\pi}{2})} \qquad (17)$$

The first differentiation of vector $\mathbb{P}_K$ with respect to cam rotating angle φ is as following

$$\mathbb{P}_K'=[\eta Ef+l_{OA}(\frac{f-\varsigma\frac{d\psi}{d\varphi}}{f}\sin\beta+\frac{d^2\psi}{d\varphi^2}\cos\beta)]e^{i(\varphi-\varsigma\psi+\frac{\pi}{2})}$$

Assume the model of vector $\mathbb{P}_K'$ is equal to τ , then $\mathbb{P}_K'$ and its differentiation are

$$\mathbb{P}_K'=\tau e^{i(\varphi-\varsigma\psi+\frac{\pi}{2})}$$

$$\mathbb{P}_K''=\tau'e^{i(\varphi-\varsigma\psi+\frac{\pi}{2})}+f\tau e^{i(\varphi-\varsigma\psi+\pi)}$$

Substituting into Eq. 5 obtain the equation of curvature radius of cam profile

$$R_K=l_{OA}(\frac{f-\varsigma\frac{d\psi}{d\varphi}}{f^2}\sin\beta+\frac{\frac{d^2\psi}{d\varphi^2}}{f^3}\cos\beta)+\eta E \qquad (18)$$

To avoid cusp and undercutting the value of R_K must be greater than zero at any point of cam profile.

CONCLUSION

1. To obtain curvature radius of pitch curve and cam profile can use the Eq. 9, Eq. 10, Eq. 14, Eq. 16, Eq. 18 respectively according to the type of cam mechanism.

2. To avoid cusp and undercutting the value computed from Eq. 14 or Eq. 18 must be greater than zero.

3. To avoid undercutting the sign of R_K determined by Eq. 10 must be equal to the sign of R_B for the cam mechanism with roller follower.

REFERENCES

Mabie, H.H., and Ocvirk, F.W. (1978). Mechanisms and Dynamics of Machinery. Third Edition.

Kloomok, M., and Muffley, R.V. (Sep. 1955). Plate Cam Design-Radius of Curvature. Prod. Egn.

Rothbart, H.A. (1956). Cams.

Paul, B. (1979). Kinematics and Dynamics of Planar Machinery.

Shigley, and Uicker, (1980). Theory of Machines and Mechanism.

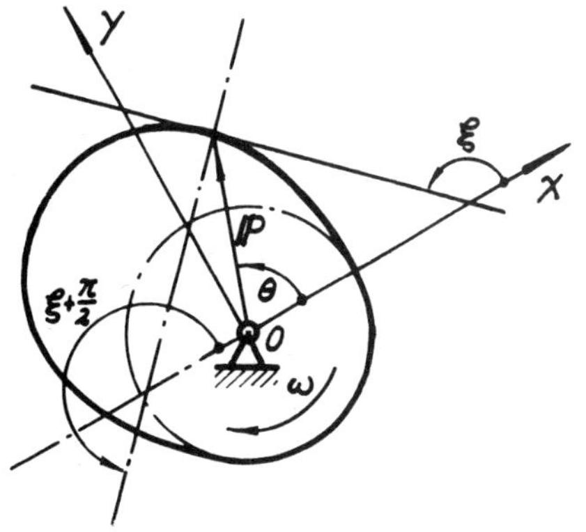

Fig. 1. Polar vector of disk cam

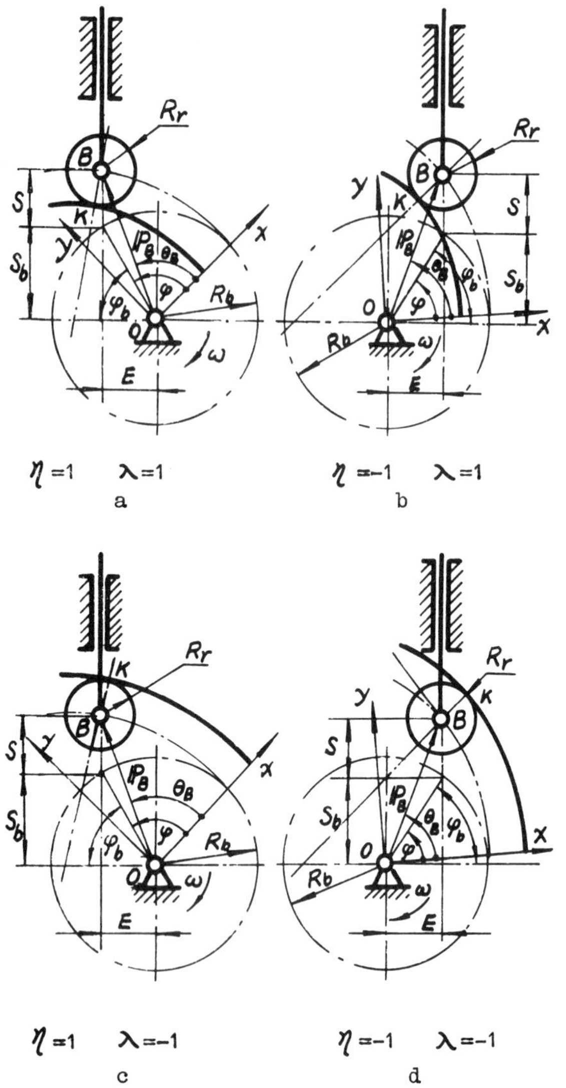

Fig. 2. Four cases of disk cam mechanism with translating roller follower

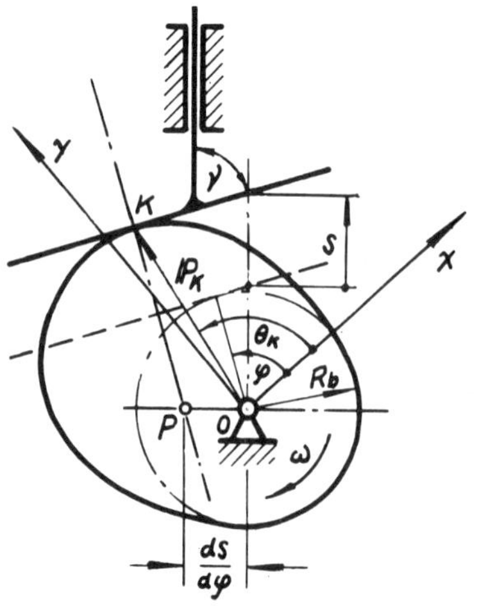

Fig. 3. Disk cam mechanism with translating flat-faced follower

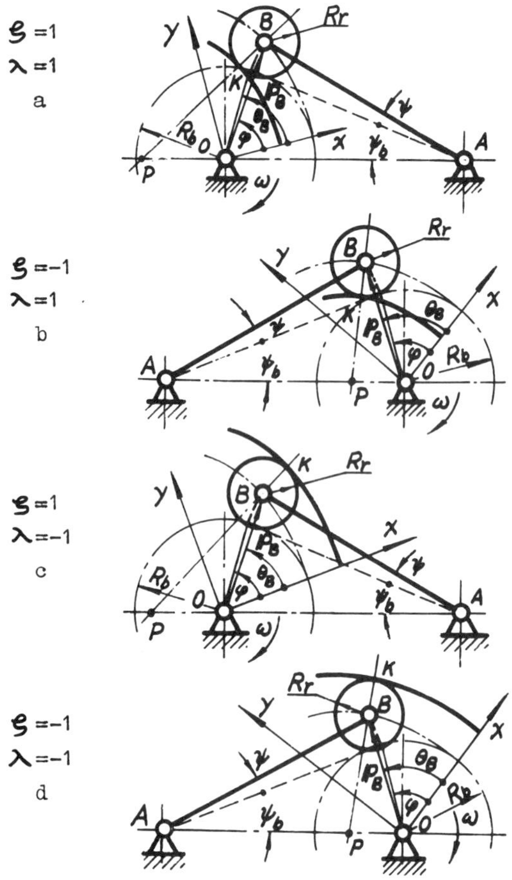

Fig. 4. Four cases of disk cam mechanism with oscillating roller follower

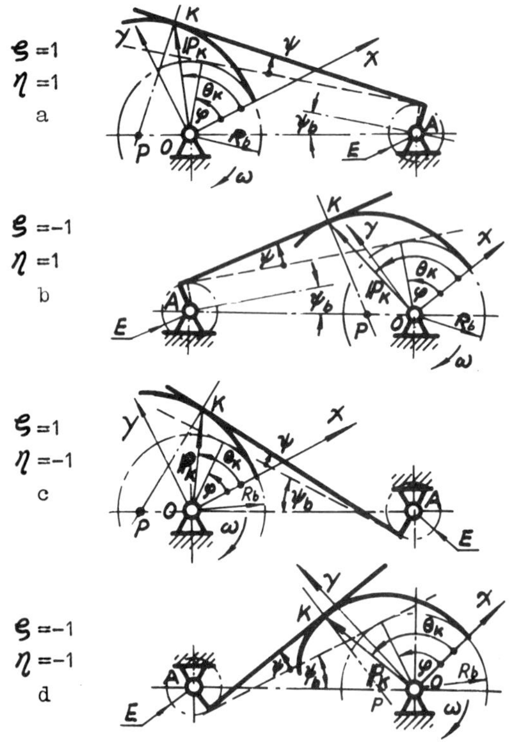

Fig. 5. Four cases of disk cam mechanism with oscillating flat-faced follower

Approximation of a Cam Profile

J. ODERFELD AND A. POGORZELSKI

Warsaw University of Technology, Warsaw 00-665, Nowowiejska 24, Poland

Abstract. Let x denote the angle of rotation of a cam, and y - a rectilinear lift of a follower. The standard method of designing cams for internal combustion engines starts with the assumption of a discrete sequence of second differences $\{y''(x_k)\}$ which is closely related to the accelerations. A suitable transformation of this sequence can be used for machining the profile of the cam. The first draft of the $\{y''(x_k)\}$ sequence is usually assumed in a graphic form, which reflects the general idea of the designer but does not meet many requirements essential to the performance of the mechanism and the engine. The paper concentrates on approximation of this sequence by a continuous spline function of third degree s"(x) with n+3 free constants. For this purpose an objective function is built in the form of a weighted sum of two components. The first one is a measure of distance between the y" and s" functions. The second is a measure of smoothness of the s" function. Minimizing the objective function in presence of various constrains yields optimal values of the constants and defines the best s" function. Then a corresponding optimal sequence of lift is being found. The paper closes with an outline of an iterative process of designing a cam profile. It is illustrated by an example.

Keywords. Cam profile; approximation; spline function; goodness of fit; smoothness of function; Lagrange´s multipliers; computer optimization.

GENERAL REMARKS ON DESIGNING CAMS

Figure 1 is a sketch of a typical timing system of an internal combustion engine. The driving link of this mechanism is a rotating cam, and its driven link is a follower which is coupled directly or indirectly with the valve. Symbols used in the sketch denote: x - angle of rotation of the cam, and y - lift of the follower. Let x=a correspond to the beginning of the follower´s motion, and x=b to its end.

A standard method of designing cams usually starts with the assumption of geometric accelerations, i.e. of a discrete sequence $\{y''(x_k)\}$, where $x_k \in \langle a,b \rangle$, k=0,1,2,...,N, and $y''(x_k)$ is the finite second difference of lifts for angle of rotation $x=x_k$. Let us remind that there exists a relation of proportionality between the terms $y''(x_k)$ and the acceleration of the follower. Hence by summing twice these terms one could get a new sequence $\{y(x_k)\}$, which would be related by affinity to the course of the lift of the follower, and which could be used for machining the cam´s profile. Various variants of this production process are well known and will not be dealt with in this paper. We intend to concentrate on generation of the sequence $\{y''(x_k)\}$ and on its modifications.

Clearly, the first draft of this sequence corresponds to the general idea of the designer as to the course of acceleration of the follower. However, this draft must be improved, until a satisfactory profile of the cam is reached. This is due to the fact that the profile has to satisfy various conditions related to the dynamics of the mechanism and to the performance of the engine.

APPROXIMATION AND SMOOTHING

We begin by substituting (Oderfeld, 1985) the original sequence $\{y''(x_k)\}$ by an approximating sequence $\{s''(x_k)\}$ which meets the following basic requirements.

Type of approximation. The approximating function s"(x) is a spline function (Fortuna, 1982) with a vector $\bar{c}$ of n+3 free constants. This function as well as its first two derivatives are continuous.

Goodness of fit. Since it can be taken for granted that the first draft has been proposed by an experienced designer it is reasonable to keep small the weighted variance

$$\sigma^2 = \frac{1}{\sum_{k=0}^{N} WL(x_k)} \sum_{k=0}^{N} WL(x_k)\cdot[s''(x_k)-y''(x_k)]^2 \qquad (1)$$

where the coefficients $WL(x_k)$ denote local weights. If all local weights are equal then formula (1) reduces simply to the variance

$$\sigma^2 = \frac{1}{N} \sum_{k=0}^{N} [s''(x_k)-y''(x_k)]^2 \qquad (2)$$

Smoothing. Rapid changes of consecutive terms $y''(x_k)$ may induce vibrations. For that reason it is advisable to keep small the expression

$$RO = \int_a^b [s^{IV}(x)]^2 dx \qquad (3)$$

This expression is used in mathematics as a measure of smoothness of the s"(x) function. It assumes large values if the function s"(x) presents a wavy outline; it is equal to zero if s"(x) is a linear function.

Constraints. All or at least some of the following constraints are desirable:

$$\left.\begin{aligned} & s'''(a) = s'''(b) = 0 \\ & s''(a) = s''(b) = 0 \\ & s'(a) = 0\ , \quad \int_a^b s''(x)dx = 0 \\ & s(a) = 0\ , \quad \int_a^b \int_a^y s''(t)dt\ dy = 0 \end{aligned}\right\} \qquad (4)$$

$$\max_{x\in\langle a,b\rangle} s(x) = H \qquad (5)$$

Constraints (4) mean that at the beginning and at the end of motion the values of jerk, acceleration, velocity and lift should be equal to zero. Constraint (5) means that the prescribed full lift H of the follower must be attained.

Criterion of optimality. Find the vector $\bar{c}$ satisfying (Laurent, 1972) the equation

$$F(\bar{c}) = N\cdot\sigma^2 + WG\cdot RO = \min! \qquad (6)$$

subjected to constraints (4) and (5). Here the coefficient WG is a global weight which takes into account importance of both: goodness of fit and smoothness.

Our solution of the optimization problem defined by eq.(6) began with a mathematical analysis of a general case, and was based on Lagrange´s method of multipliers. After some rather time-consuming preparations we did succeed in reducing the problem to a set of linear equations. It is worth mentioning that the form of these equations does not depend on the input data, which influence only some coefficients of the linear equations. It is to be noted that it was sufficient to make this analysis only once. To solve explicitly the linear equations a computer program has been written, which for any set of input data first computes the coefficients of linear equations then solves them finding components of the optimum vector $\bar{c}$. Thus the optimal approximating function s"(x) is fully defined. All the succeeding transformations and applications of this function are confided to the computer. Among them let us quote: values of $s^{IV}(x)$, s"'(x), s (x), s(x) for any desired set of values of the independent variable x, and moreover a few indices characterizing the quality of approximation e.g. σ^2 and RO.

Iterations. Roughly speaking the process of designing the cam´s profile is a series of iterations like the one just described. In each iteration the input data are kept constant and the computer optimization deals only with goodness of fit and smoothness (cfr. eq.6). The intervention of the designer is of the type "man-in the-loop". If he is satisfied with the output data he stops the process; if not - he starts the next iteration.

Two questions remain to be answered. The first one is: how to evaluate the iteration just finished? Here are a few eaxmples of traits that might be worth examining: peaks of acceleration, values of acceleration at nevralgic places; number of modes of lift; slopes of lift at start and end of motion, e.t.c.

The second question deals with modifications to be introduced between two successive iterations. Let us remind that there are three groups of input data. To the first group belong the values a, b, H, p; they are fixed beforehand by the designer, and are not open to discussion. The second group is composed of parameters n, WG, WL; they may be changed at will between two consecutive iterations. As a general rule it can be said that increasing n and/or decreasing WG results in improving goodness of fit at the cost of smoothness and vice versa. Local weights WL may be used to lay stress upon some subset of x_k values. The third and last group of input data consists of the original sequence $\{y''(x_k)\}$ which should be replaced by another one only if this seems to be absolutely necessary.

AN EXAMPLE

The input accelerations mentioned in the preceding chapter can be given in an explicit form of an equally spaced sequence or implicitly by means of a graph. The second case is much more convenient for the designer and for that reason will be treated here in some detail. Before presenting our example let us introduce suitable notations and relations.

TABLE 1 Notations

Symbol	Definition	Dimensions
$\tilde{y}''$	graphic input acceleration	arbitrary unit
$\tilde{s}''$	approximated graphic acceleration	per (p deg.)^2
$\tilde{s}$	approximated graphic lift	arbitrary unit
s"	approximated geometric acceleration (second differences)	millimetre per (p deg.)^2
s	true approximated lift	millimetre
p	width of interval	deg
ω	velocity of rotation	r.p.m
A	approximated kinematic acceleration	m·s^{-2}

Let the required full lift be H millimetres, and the maximum graphic approximated lift be max $\tilde{s}$ in arbitrary units. Then the following identites hold:

$$s'' = (H/\max \tilde{s})\tilde{s}'' \qquad (7)$$

$$s = (H/\max \tilde{s})\tilde{s} \qquad (8)$$

$$A = 10^{-3}(6\omega/p)^2 s'' \qquad (9)$$

Now we are ready to present the example. Let the broken line y" in Fig. 2 show the course of graphic input acceleration for angles of rotation $x \in \langle 0,75°\rangle$; the course for $x \in (75°,150°\rangle$ is a mirror-image. The units on the vertical axis are arbitrary. But nevertheless the y" line contains some useful information. In the first phase of motion the acceleration increases rapidly, and this gradient favours rapid starting of the mechanism important for both the inlet and outlet valves. The further ascent of acceleration from 5° to 25° is less energic in view of avoiding excessive inertia forces. Negative accelerations from 25° to 65° are dimensioned so as to keep the overall area under the curve equal to zero. As it is known, this is a necessary condition for zero velocities both at the start and at the end of the motion. For $x \in \langle 65°,75°\rangle$ i.e. for the middle part of motion the acceleration is equal to zero which should favour damping of vibrations.

On the other hand our $\tilde{y}''$ diagram displays sudden changes of acceleration which may generate shocks and vibrations. This refers to strong discontinuities c-d and e-f, and to weak discontinuities at points a, b, c, d, e, f. These defects can be easily remedied by applying our approximation scheme as shown in Fig. 2 by the $\tilde{s}''$ curve which was produced by the computer program with following parameters: n=10, WG=5, all WL=1, p=1°. Overall appearance of the $\tilde{s}''$ curve is smooth but some new defects appear. Perhaps the most serious one is the fact that for $x \in \langle 65°, 75° \rangle$ the acceleration rises considerably from -3 to 1.3 arbitrary units. A minor but characteristic demerit are small waves of negative accelerations between 35° and 60°. As we did find no modification of program parameters could remedy these drawbacks.

In this situation we decided to modify, by hand, the graphic input acceleration as shown by curve $\tilde{y}''$ in Fig. 3. Roughly speaking the modified $\tilde{y}''$ curve is free from discontinuities and is tangent to the x-axis both at x=0° and x=75°. Since we were anxious to maintain (after approximation) this smooth merging with the x-axis at x=75° we decided to introduce local weights WL as shown in Table 2.

TABLE 2 Local Weights

Angle x (deg)	75	74 76	73 77	72 78	71 79	70 80	69 81	68 82	67 83	66 84	Otherwise
WL	41	37	33	29	25	21	17	13	9	5	1

Note the very high values of WL in the nearest vicinity of x=75°.

After a few experiments with various iterations we selected the following values of parameters: n=20, WG=100 what led to a rather satisfactory approximated graphic acceleration - $\tilde{s}''$ shown in Fig. 3. Then the computer calculated approximated graphic lifts $\tilde{s}$, their maximum value max $\tilde{s}$, and applying formulas (7), (8) and (9) produced:

- a list of true approximated lifts s (with p=1° and H=8 mm); for graph presentation see Fig. 4;
- a list of approximated kinematic accelerations A (with ω=2000 r.p.m.); for graphic presentation confront the right hand vertical scale in Fig. 3.

A full documentation concerning this iteration should be carefully examined by the designer before a final approval. As an example of claims that could be raised in our case note that as seen from Fig. 4, lift of 0.1 mm is reached only after 8° of rotation. If the designer were anxious to secure a more energic opening of the timing system, one could try a next iteration with additional local weights applied to angles $x \in \langle 0°, 5° \rangle$. This would probably improve the start of motion.

We do not want to enter into a further discussion of this and similar details. Our aim was merely to show that the proposed method seems to be a flexible and useful tool in the hands of the designer of cam mechanisms.

CONCLUSION

A new computer-aided method of designing cam profile for internal combustion engines is presented. It starts with the assumption, as a first draft, of an empirical diagram of geometrical accelerations of the follower. Next we modify this draft so as to improve its dynamic quality, however without deviating much from the first draft. This plan is realized in a series of iterations. In each of them the input diagram is being approximated by means of a spline function with free constrants to be found by solution of an optimization problem.

REFERENCES

Fortuna, Z., B. Macukow and J. Wąsowski (1982). Numerical Methods. WNT, Warszawa (in Polish).

Laurent, P.J. (1972). Approximation et optimisation. Hermann, Paris (in French).

Oderfeld, J. and A. Pogorzelski (1985). Approximation of cam profile: Goodness of fit and smoothness. In Proceedings of Int. Conf. on Computer Aided Design of Machines. May 7-10,1985, Zakopane, Poland, pp.84-89 (in Polish).

Zusammenfassung

APPROXIMIEREN EINES NOCKENPROFILS

Eine typische Ausführung der Steuerung der Verbrennungsmotoren besteht aus einem Umlaufnocken (Drehwinkel - x) und einem Stössel (Hub - y). Eine weitverbreitete Methode zum Entwerfen des Nockenprofils nimmt als Ausgangspunkt den Verlauf der Beschleunigung und genauer gesagt sein Analogon in Form einer Beschleunigungsfolge der zweiten Differenzen des Hubs $\{y''(x_k)\}$.

Werden die Glieder dieser Folge entsprechend behandelt, unter Berücksichtigung der Randwerte, so erhält man die Hubfolge $\{y(x_k)\}$, die dann zur Herstellung des Nockenprofils angewendet werden kann.

Üblicherweise formuliert der Konstrukteur seinen ersten Vorschlag der Differenzfolge $\{y''(x_k)\}$ in Form einer Skizze, die zwar der Grundidee des Konstrukteurs über den Beschleunigungsverlauf entspricht, aber nicht alle Einzelheiten berücksichtigt. Das Referat behandelt vor allem Bildung der entgültigen Beschleunigungsfolge. Zuerst approximiert man die Differenzfolge $\{y''(x_k)\}$ mit einer Spline-Funktion s"(x) mit freien n+3 Konstanten. Um diesen Konstanten optimale Zahlenwerte zu geben, bildet man eine Zielfunktion in Form einer linearen Kombination mit einem globalen Gewicht WG, von zwei Summanden. Der erste davon ist das Mass der Abweichung der Funktion s"(x) von der originalen Differenzfolge. Der zweite Summand ist das Mass der Glätte der Funktion s"(x).

Nur wird das Minimum der Zielfunktion gesucht. Es sei hier bemerkt, dass dabei acht Nebenbedingungen berücksichtigt werden müssen. Das Optimierungsprinzip ist ziemlich kompliziert, aber nach einer sorgfältigen Vorbereitung konnte seine Realisierung einem Digitalrechner anvertraut werden. Nachdem die optimale Approximationsfunktion s"(x) gefunden worden ist, folgt die Druckausgabe verschiedener Listen und Diagramme, die u.a. die Beschleunigung, die Geschwindigkeit und den Hub betreffen. Es muss aber bemerkt werden, dass zu den Eingabedaten einige Parameter gehören (u.a. n und WG), welche durch den Konstrukteur frei gewählt werden können. Aus diesem Grunde wird das Verfahren stufenweise geführt. Zuerst gibt man den Parametern probeweise Zahlwerte, die der bisherigen Erfahrung entsprechen oder wenigstens scheinen, vernünftig zu sein. Dann liefert der Digitalrechner eine ausführliche Dokumentation, die dem Konstrukteur Hinweise zur eventuellen Änderung der Parameter gibt und - wenn dies notwendig ist - auch der originalfolge $\{y''(x_k)\}$.

Am Ende des Referats findet der Laser ein Beispiel des Verfahrens.

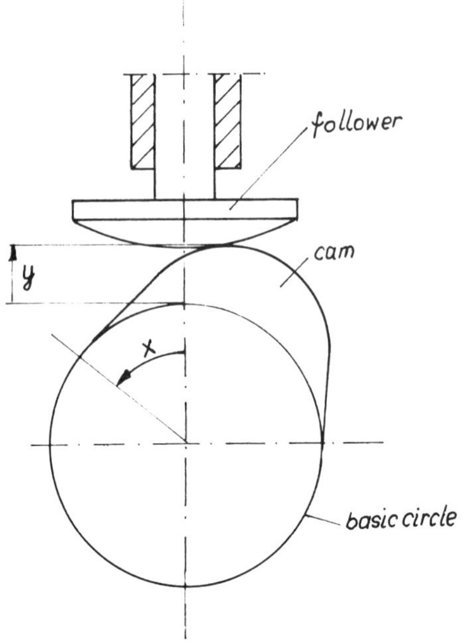

Fig. 1. General outline of a cam mechanism

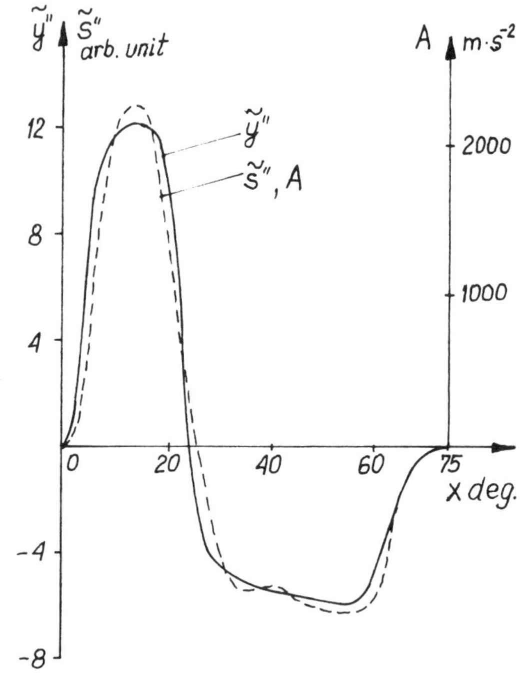

Fig. 3. Final iteration (n=20, WG=100, WL- Table 2)

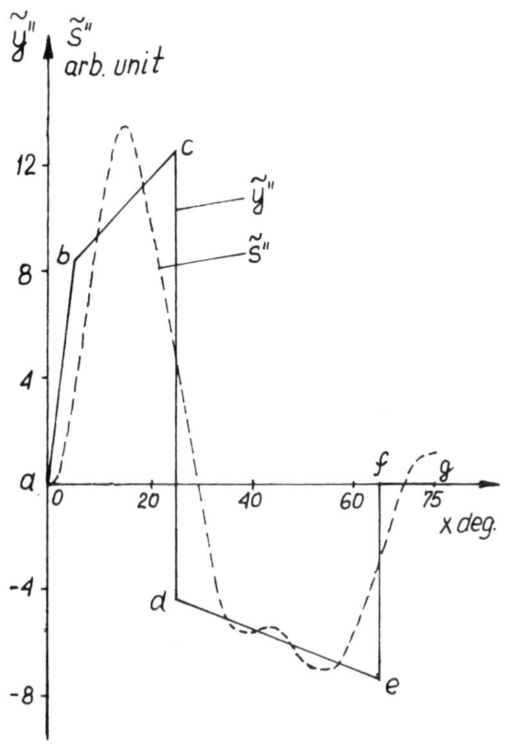

Fig. 2. First iteration (n=10, WG=5, WL=1)

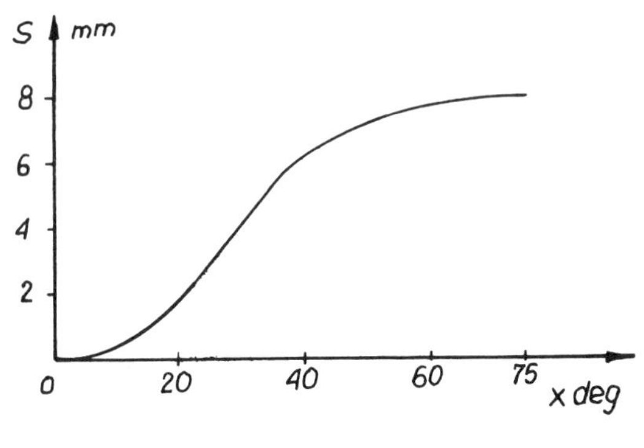

Fig. 4. Final course of lift

Mecanism for the Alternative Opening-closing of Safety Platform Doors in Mine Sinking Shafts

V. ZAMFIR, C. BOLOG AND I. ANDRAS

Department of Electromechanical Engineering Mining Institute, Petroşani, Romania

closing of safety platform doors in sinking shafts ; it is a mechanism of novel conception, applied to mine shaft sinking installations. The opening and closing of doors is made by the kibble sled which contains, in the guides plane, two double symmetrical cams whose profile is established in accordance with the desired motion laws of the doors while opening - closing using the non-dimensional coefficient method.

Keywords: synthesis of cam mechanism; non-dimensional coefficient method.

INTRODUCTION

The safety platform door mechanism in sinking shafts is shown schematically (for a single door) in fig.1.It consists of the door (1),the symmetrical crank lever (2) the elbowed articulated arms (3) and the spring (5).

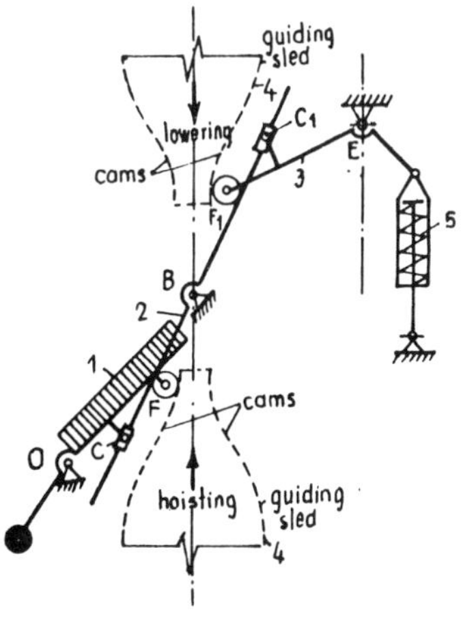

Fig.1.The safety platform door mechanism. The opening of the doors is achieved by the guiding sled (4) through the force in the winding rope, at hoisting and its deadweight at lowering. The mechanism is statically balanced and symmetrical in relation to the vertical plane of the joint B of the crank lever (2) and OC is equal and parallel to EC, OF is equal and parallel to EF_1, CB = BC_1 (these are necessary so that the mechanism should have the same kinematic parameters both at the up and down wind of the sled (4)). The door closing is achieved through the spiral spring (5).

It is desirable that the opening and closing of doors should be done smoothly without shocks or with small shocks.
In the following pages we shall give a general synthesis algorithm of the cams profile on the guiding sled (4), for meeting these demands using the non-dimensional space, velocity and acceleration coefficients, assimilating the sled (4) to a translation cam, the door (1) at hoisting and the elbowed articulated arm (3) at lowering to a rotation oscillating cam follower.

2.DEFINING THE NON-DIMENSIONAL COEFFICIENTS

We write down with s and ψ , the displacement of the cam, and the follower, respectively, at a certain moment linked by the relation $\psi = \psi(s)$.
The maximum displacements and their limits are considered to be known :

- for the cam : $s_{max} = s_f - s_o$;
- for the follower $\psi_{max} = \psi_f - \psi_o$;

in which by s_f and s_o, ψ_f and ψ_o have been written the values of the spaces for the cam, the follower respectively at the final, respectively the initial moment.
The laws of the follower motion have the following general form :

$$\psi = \psi_{(s)} ; \dot{\psi} = \frac{d\psi}{ds} ; \ddot{\psi} = \frac{d^2\psi}{ds^2} \qquad (1)$$

TABLE 1.

INTERVAL	δ_{max} ; ξ_{max}	ς	δ	ξ
I $0 \div m$	$\delta_{max} = \xi_{max}(0.25 + 0.5n)$	$\xi_{max} \frac{k^3}{6m}$	$\xi_{max} \frac{k^2}{2m}$	$\xi_{max} \frac{k}{m}$
II $m \div m+n$	$\xi_{max} = \frac{3}{A}$	$\xi_{max} \left[\frac{m^2}{6} + \frac{k(k-m)}{2}\right]$	$\xi_{max} \left(k - \frac{m}{2}\right)$	ξ_{max}
III $m+n \div 0,5$	$A = 0,5 - 0,5m + n - 2mn - n^2$	$\frac{\xi_{max}}{6} \left[3nk + 1,5(k-m-n) + \frac{(0,5-k)^2}{(0,5-m-n)}\right]$	$\frac{\xi_{max}}{2} \left\| 0,5 + n - \frac{(0,5-k)^2}{0,5-m-n} \right\|$	$\xi_{max} \frac{0,5-k}{0,5-m-n}$

The ratio is introduced :

$$K = \frac{s}{s_{max}} \tag{2}$$

which is termed cam displacement non-dimensional coefficient with the limts $0 \le k \le 1$.

The laws of the follower motion (1), taking into account (2) can be written under the form :

$$\psi = \varsigma \psi_{max} ; \dot{\psi} = \delta \frac{\psi_{max}}{t_1} ; \ddot{\psi} = \xi \frac{\psi_{max}}{t_1^2} \tag{3}$$

in which ς, δ and ξ are the non-dimensional displacement, velocity and acceleration coefficients related among themselves by the relations :

$$\delta = \frac{d\varsigma}{dk} ; \xi = \frac{d\delta}{dk} \tag{4}$$

and t_1, is the duration of the follower maximum displacement.

The non-dimensional coefficients ς, δ and ξ can be calculated for any follower motion law depending on the non-dimensional parameter of cam K displacement :

$$\varsigma = \varsigma(k) ; \delta = \delta(k) ; \xi = \xi(k) \tag{5}$$

For motion laws with two sectors symmetrical tahograms, the non-dimensional coefficients ς, δ and ξ can be calculated only for the acceleration sector $0 \le K \le 0.5$, since from the symmetry condition for the deceleration sector $0.5 \le K \le 1.0$, they are calculated from the following relations :

$$\varsigma_a(k) = 1 - \varsigma(1-k) ; \delta_a(k) = \delta(1-k) ;$$

$$\xi_a(k) = - \xi(1-k) \tag{6}$$

The coefficients ς, δ and ξ should usually be known only for the lifting interval. They can be calculated for the lowering interval with the relations :

$$\varsigma^*_{(k)} = \varsigma(1-k) ; \delta^*_{(k)} = - \delta(1-k) ;$$

$$\xi^*_{(k)} = \xi(1-k) \tag{7}$$

The trapezoid-shaped acceleration diagram traced with the non-dimensional coefficients is shown in figure 2.

In table 1 are given the relations that express the follower motion law for the trapezoid-like variation diagram expressed by the non-dimensional space, velocity and acceleration coefficients.

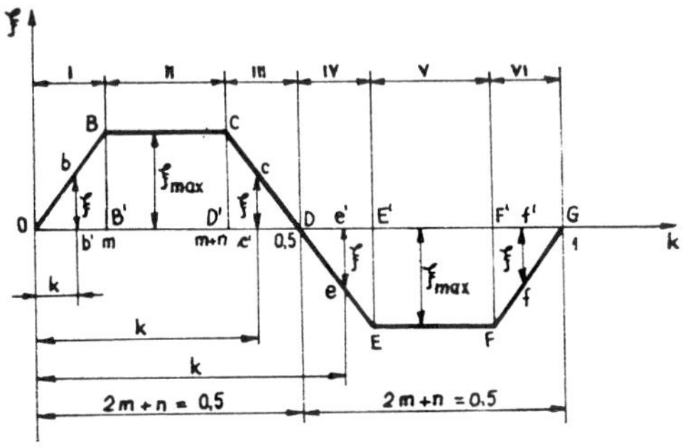

Fig.2.The trapezoid-shaped acceleration diagram

DETERMINING THE LENGTH OF THE ACTIVE

In fig.3 is shown the schematic diagram of the mechanism with translation cam and rotation oscillating cam follower. Choosing the kinematic length of the follower (element 2), equal to the unit FO = 1, the cam profiled part, L_0 will be shown in the drawing, through the re-

TABLE 2.

Nr.	THE LAWS OF THE FOLLOWER (ACCELERATION)	INTERVAL	δ_{max}; ξ_{max}	φ	δ	ξ
1.	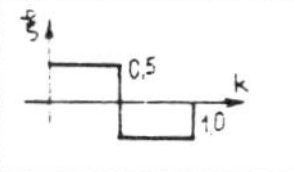	$0 \div 0{,}5$	$\delta_{max} = 2$ $\xi_{max} = 2$	$2k^2$	$4k$	4
2.	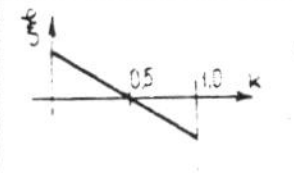	$0 \div 0{,}5$	$\delta_{max} = 1{,}5$ $\xi_{max} = 6$	$3k^2 - 2k^3$	$6k - 6k^2$	$6 - 12k$
3.	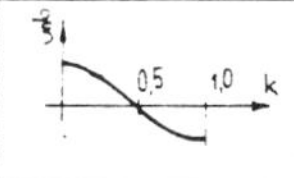	$0 \div 0{,}5$	$\delta_{max} = \frac{\pi}{2}$ $\xi_{max} = \frac{\pi^2}{2}$	$\frac{1}{2}(1 - \cos \pi k)$	$\frac{\pi}{2} \sin \pi k$	$\frac{\pi^2}{2} \cos \pi k$
4.	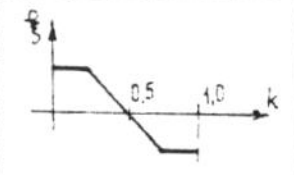	$0 \div n$ $n \div 0{,}5$	$\delta_{max} = \frac{0{,}75 + 1{,}5n}{0{,}5 + n - n^2}$ $\xi_{max} = \frac{3}{0{,}5 + n - n^2}$	$\xi_{max} \cdot \frac{k^2}{2}$ $\xi_{max}\left[-\frac{1+2n+4n^2}{24} + \frac{0{,}5+n}{2}k + \frac{(0{,}5-k)^3}{6(0{,}5-n)}\right]$	$\xi_{max} k$ $\xi_{max}\left[\frac{0{,}5+n}{2} - \frac{(0{,}5-k)^2}{2(0{,}5-n)}\right]$	ξ_{max} $\xi_{max} \frac{0{,}5-k}{0{,}5-n}$
5.	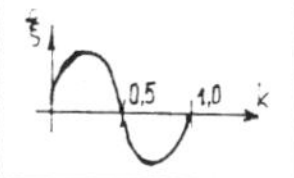	$0 \div 0{,}5$	$\delta_{max} = 2$ $\xi_{max} = 2\pi$	$k - \frac{1}{2\pi} \sin 2\pi k$	$1 - \cos 2\pi k$	$2 \sin 2\pi k$
6.	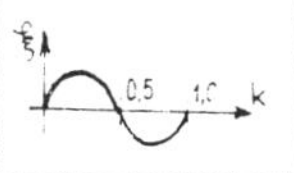	$0 \div 0{,}5$	$\delta_{max} = 2$ $\xi_{max} = 6$	$8(k^3 - k^4)$	$24k^2 - 32k^3$	$48k - 96k^2$

lative length $L_o^{(1)}$, calculated with the relation :

$$L_o^{(1)} = \frac{L_o}{FO} \qquad (8)$$

We note down with $\bar{V}_{F_2}$ the velocity of point F on the follower situated perpendicularly to FO, with $\bar{V}_{F_1}$, the velocity of the point F on the cam parallel to the translation direction (the guides axis(and with $\bar{V}_{F_2F_1}$ the velocity relative to point F on the follower relative to point F on the cam, situated on the tangent direction in F to the cam profile. These velocities meet the following vectorial equation :

$$\bar{V}_{F_2} = \bar{V}_{F_1} + \bar{V}_{F_2F_1} \qquad (9)$$

The follower position is shown by the angle,

$$\beta = \psi_o + \psi \qquad (1o)$$

in which ψ_o is the initial position angle.

The pressure angle is written down with θ and is formed between the normal nn to the cam profile and the tangent to the follower trajectory in the point F (the velocity direction $\bar{V}_{F_2}$).

Transposing graphically the relation (9), in fig.3 b and applying the sine theorem, we can write :

$$\frac{\bar{V}_{F_2}}{\bar{V}_{F_1}} = \frac{\cos(\theta + \beta)}{\cos \theta} \qquad (11)$$

If in relation (11) we replace $\bar{V}_{F_1}$ by $L_o^{(1)}/t_1$ and $\bar{V}_{F_2}$ by $\beta = \frac{d\psi}{dt}$ we obtain the following relation of the pressure angle :

$$tg\theta = \pm\, ctg\,\beta \mp \frac{\dot{\beta}\, t_1}{L_o^{(1)} \sin \beta} \qquad (12)$$

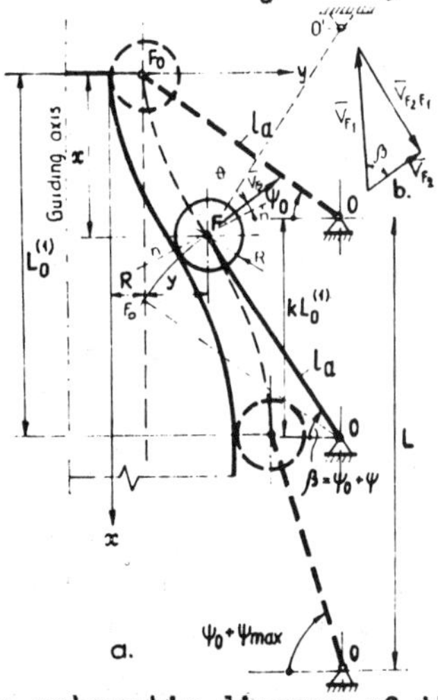

Fig.3.The schematic diagram of the mechanism

In relation (12) as well as in those that follow the inferior sign refers to the follower positioning, marked in fig.3 a by a dashed line O'F.

Annuling the derivative in relation to time in the right-hand side of relation (12) the conditions are found under

which the pressure angle Θ becomes maximum Θ_{max}.
Eliminating from equations (12) and (13) the ration $\frac{L_o^{(1)}}{\Psi_1}$, replacing Θ with a maximum quantity and taking into account (3) after transformations, is obtained :

$$tg(\Psi_o+\xi\Psi_{max}) = \mp tg(\Theta_{max}+\eta) \quad (14)$$

where,

$$tg\,\eta = \frac{\phi_1}{\delta\Psi_{max}} \quad (15)$$

in which by ϕ_1 we mean the ratio ξ/δ.
From the relation (12), in which Θ was replaced with Θ_{max}, positive or negative in order to comprise both situations of follower positioning, the relative length of the profiled part of the cam $L_o^{(1)}$ is obtained, by observing the pressure angle $\Theta = \Theta_{max}$:

$$L_o^{(1)} = \frac{\delta\Psi_1\cos\Theta_{max}}{\cos(\Psi_o+\xi\Psi_{max}\pm\Theta_{max})} \quad (16)$$

In relation (16) the sign of Θ_{max} is chosen so that the denominator should be positive.

4. DETERMINING THE CAM PROFILE

With the notation in fig.4 the coordinates of a current point of the cam profile are easily deduced :

$$x = kL + FO(\sin\Psi_o - \sin\beta)$$
$$y = FO(\cos\Psi_o - \cos\beta) \quad (17)$$

in which by kL has been noted down the follower displacement in the hypothesis of the motion reverse and by L the absolute length of the cam profile.
We notice that for k=1, $x=x_{max}$;

$$x_{max} = L + FO[\sin\Psi_o - \sin(\Psi_o+\Psi_{max})] \quad (18)$$

Therefore, for the follower to rotate with the angle Ψ_{max}, the cam has to cover a space x_{max} differing from the langth of its active part L with the value. $x_{max} - L = FO\,(\sin\Psi_o - \sin(\Psi_o + \Psi_{max}))$.
Thus, the relations 17, become :

$$x = kL + FO\{k[\sin(\Psi_o+\Psi_{max}) - \sin\Psi_o] + \sin\Psi_o - \sin\beta\}$$
$$y = FO(\cos\Psi_o - \cos\beta) \quad (19)$$

From the first relation (19) it can be noticed that not all the terms are influenced by the non-dimensional coefficient of cam K displacement and hence, at equal increases of the above, correspond unequal increases of the abscissa x.

In practice, it is much more convenient to know the ordinates of the points of the cam theoretical profile for the equidistant abscissa points. With this end in view the length of the cam active profile in divided into n intervals Δx. Replacing in the first relation (19), $L = FO \cdot L_o$ and $\beta = \Psi_o + \xi\Psi_{max}$, after ordering, is obtained :

$$K[L_o + \sin(\Psi_o+\Psi_{max}) - \sin\Psi_o] - \sin(\Psi_o+\xi\Psi_{max}) = \frac{i\Delta x}{FO} - \sin\Psi_o \; ; \; i=1,2,\ldots,n \quad (2o)$$

From relations (2o) the coefficient K can be determined corresponding to each interval that will be used in the relations (19) for determining the co-ordinates of the cam profile. The maximum ordinate is found with the relation :

$$y_{max} = FO(\cos\Psi_o - \cos(\Psi_o + \Psi_{max})) \quad (21)$$

5. EXEMPLES AND CONCLUSIONS

The clue of the above-mentioned problems is solving system (2o). With this end in view a general programme for the computer has been worked out, in FORTRAN IV language by means of which the translation cam profile can be synthesized for any motion law required to rotation oscillating follower.
The programme worked out refers to an acceleration tachogram up to 6 intervals fig.2. Nevertheless, it can be easily extended to larger numbers of intervals.
The above methodology has been applied for several motion laws, table 2, while the synthesis results have been transposed graphically by computer directly, being then used in designing.

REFERENCES

Beletki, V., Raschiot mechanismov maskin avtomatov pischevîh proizvod. stv, Viskia Skcola, Kiev, 1974.

Zamfir, V., Un nou procedeu pentru deschiderea închiderea uşilor podurilor de siguranţă din puţurile de mină în săpare, Lucrările ştiinţifice ale Institutului de Mine Petroşani Vol. XVI, fascicula 1, 1984, pp.32-35.

Patent NO.85959/1983, R.S.R.

SOMMAIRE

On présente la synthèse des cames des traineaux de guidage de la benne utilisées à l'ouverture-fermeture automatique des portes des platformes de sûreté situées dans les puits de mine en fonçage.

A Compromise Between NC Flexibility and Cam Drive High Driving Force Capacity

J. AGULLÓ AND J. VIVANCOS

Departament d'Enginyeria Mecànica, Universitat Politècnica de Catalunya, Av. Diagonal, 647, 08028 Barcelona, Spain

Abstract. To obtain periodical displacements, flexibility of conventional NC using ball-bearing screw drives makes NC attractive as a substitute of cam drives when a great number of cams is needed. However in many applications cam capacity to produce intense driving forces is difficult to beat by means of screw drives. As maximum acceleration of motor and driven mass occur simultaneously in screw drives, the motor inertia severely reduces the maximum driving torque available at high acceleration. This is true even if low inertia servomotors and optimum transmission ratio are used. NC of the rotation of a cam allows to obtain a wide variety of displacement functions while requiring a smaller driving torque than optimum screw drives. If there are single top and bottom levels for all maxima and minima, a slider-crank mechanism with numerically controlled crank rotation also allows the obtaining of a wide variety of displacements and requires lower driving torques than optimum screw drives. This approach is attractive because of its mechanical simplicity. Monotonous cam or crank rotation gives a reasonable flexibility to obtain displacement functions while keeping the number and level of maxima and minima. If rotation is allowed to reverse, new maxima and minima at intermediate levels may appear, which pushes further the potential of both approaches but at the expense of greater driving torques.

Keywords. Numerical control; machine tools; computer control; function generation; screw drives; cam drives; slider-crank mechanism.

INTRODUCTION

Cam drives are widely used in special machine tools to obtain periodic displacements as a function of piece rotation. Relationship between both movements is kinematically guaranteed and consequently it is guaranteed by constraint forces, which can easily attain high values.

When a great number of cams is needed in order to obtain a variety of displacements, their use may be costly and cumbersome. To reduce the actual number of cams it is usual practice to obtain, by means of a single cam, a family of displacements through geometrical modifications in the linkages between the cam and the element to be moved. However this approach usually lacks the flexibility required to accurately obtain displacement modifications other than a straightforward amplification or reduction by a multiplying constant.

Flexibility to obtain a wide variety of curves makes NC an attractive approach, but the straightforward introduction of conventional machine tool NC -based on screw drives- to substitute cams may prove to be unsuccessful in many applications because cam capacity to produce intense driving forces is difficult to beat by means of numerically controlled screw-drives. As maximum acceleration of motor and drived mass occur simultaneously in screw drives, the motor inertia severely reduces the maximum driving force available at high acceleration. This is true even if low inertia servomotors and optimum transmission ratio are used.

NC of the rotation of a given cam may prove to be more successful, as substantial displacement departures may require moderate motor angular acceleration while cam kinematics helps to obtain high driving forces from moderate driving torques.

If there is a single level for all maxima and a single level for all minima, the use of a slider-crank mechanism with numerically controlled crank rotation is attractive due to its inherent mechanical simplicity.

For both approaches, the kinematical equation of control and the required driving torque are obtained, and their performance is compared to that of numerically controlled screw drives by means of an illustrative example.

OPTIMUM TRANSMISSION RATIO IN A SCREW DRIVE

Lets consider the screw drive shown in Fig. 1. If the screw inertia -as well as the inertia of the gear attached to it- and frictions other than those that can be included in F_R can be neglected, the equation of motion leads to the required driving torque,

$$\Gamma = \{\ddot{y}[m + I(\frac{2\pi}{kh})^2] + F_R\}\frac{kh}{2\pi} \quad , \qquad (1)$$

where I is the moment of inertia of the motor -with the gear attached to it-, k is the transmission ratio of the gear reduction unit and h is the lead of the screw.

In order to determine the optimum transmission ratio, $\lambda \equiv kh/2\pi$, that makes minimum the driving torque, Eq. (1) must be derived and equated to zero,

$$\frac{\partial \Gamma}{\partial \lambda} = \ddot{y}(m - I\lambda^{-2}) + F_R = 0 \quad , \qquad (2)$$

from which,

$$\lambda_{op} = [\ddot{y}I/(\ddot{y}m + F_R)]^{1/2} \quad . \qquad (3)$$

Substitution of λ_{op} in Eq. (1) leads to

$$\Gamma_{op} = m\ddot{y}\ 2\left[\frac{I}{m}\ (1 + \frac{F_R}{m\ddot{y}})\right]^{1/2} \quad . \qquad (4)$$

If F_R can be neglected, Eqs. (3-4) reduce to

$$\lambda_{op} = (I/m)^{1/2} \quad , \qquad (5)$$

$$\Gamma_{op} = m\ddot{y}\ 2(I/m)^{1/2} \quad . \qquad (6)$$

For a given lead of the screw h, the optimum ratio of the gear reduction unit, which is given by

$$k_{op} = (2\pi/h)(I/m)^{1/2} \quad , \qquad (7)$$

may be > 1 if I/m is big enough and, in such a case, the gear unit is a multiplier instead of a reduction unit. The limit $I/m \to 0$ leads to the rather unrealistic case $\Gamma_{op} \to 0$, $k_{op} \to 0$.

N.C. OF A CAM DRIVE

Lets consider a cam drive that produces the displacement

$$y(\psi) = f(\psi) \quad ; \quad f(\psi + 2\pi) = f(\psi) \quad , \qquad (8)$$

where ψ is the cam angle of rotation, and lets call $y_o(\theta)$ the displacement obtained when ψ is equal to the piece angle of rotation θ,

$$y_o(\theta) = f(\theta) \quad . \qquad (9)$$

If cam rotation is numerically controlled, ψ can be made different from θ by a periodic function $g(\theta)$, of period equal to 2π, which will be called angular correction,

$$\psi = \theta + g(\theta) \quad ; \quad g(\theta + 2\pi) = g(\theta) \quad . \qquad (10)$$

In such a case, the displacement $y_g(\theta)$ obtained is

$$y_g(\theta) = f[\theta + g(\theta)] \quad . \qquad (11)$$

By means of this angular correction, points of curve y_o are moved sidewards, as it can be seen in Fig. 2 , and maxima and minima of y keep their values provided that $dg/d\theta > -1$. The path between consecutive extreme points, as path AB, is changed into another path between consecutive extreme points A'B'. If $dg/d\theta < -1$, cam rotation angle $\psi(\theta)$ is no longer monotonous increasing and extreme values at intermediate levels may appear.

For a given function $y_g(\theta)$ with the same number of extreme points as $y_o(\theta)$, angular correction $g(\theta)$ can be found by means of the inverse function

$$g(\theta) = f^{-1}\left[y_g(\theta)\right] - \theta \quad , \qquad (12)$$

for each pair of corresponding paths between consecutive extreme points, as it is illustrated in Fig. 2 . Displacements $y_g(\theta)$ can be multiplied by a constant by means of elementary kinematical resources as usual.

Displacement $y_g(\theta)$ defined by Eq. (11) leads to velocity $\dot{y}$ and acceleration $\ddot{y}$,

$$\dot{y} = \dot{\theta}(1 + g')\ f'(\theta + g) \quad , \qquad (13)$$

$$\ddot{y} = \ddot{\theta}(1 + g')\ f'(\theta + g) + \dot{\theta}^2\left[(1+g')^2 f''(\theta+g) + g''f'(\theta+g)\right] \quad , \qquad (14)$$

where ' denotes derivation relative to θ. Piece angular acceleration $\ddot{\theta}$ can usually be neglected.

DRIVING TORQUE REQUIRED BY A NUMERICALLY CONTROLLED CAM DRIVE

If it is assumed that $\ddot{\theta} = 0$ and that friction forces, other than those that can be included in F_R, as well as inertia of auxiliary elements between the cam and the moved mass m, Fig. 3, are negligible, then the equation of motion leads to the required driving torque

$$\Gamma = \dot{\theta}^2 m\left[g''(\frac{I}{m} + f'^2(\theta+g)) + (1+g')^2 f'(\theta+g) f''(\theta+g)\right] + F_R f'(\theta+g) \quad , \qquad (15)$$

where I is the moment of inertia of motor and cam -reduced to cam axis-.

N.C. OF A SLIDER-CRANK MECHANISM

In a slider-crank mechanism, if the crank to connecting-rod ratio is small enough, the displacement of the slider from its bottom death center (B.D.C.) is given by

$$y = s\left[1 - \cos(\psi)\right] \quad , \qquad (16)$$

where s is the crank length and ψ is the angle of rotation from the B.D.C.

When curve $y_g(\theta)$ to be obtained by means of NC of the crank rotation has all its maxima at the same level and all its minima equal to zero, it can be obtained by means of a monotonous increasing rotation $\psi(\theta)$. In this case, if $y_g(\theta)$ has n maxima in one period, the crank will have to rotate n revolutions per each turn of the piece. Lets call $y_o(\theta)$ the displacement obtained for $\psi = n\theta$,

$$y_o(\theta) = s(1 - \cos n\theta) \quad . \qquad (17)$$

If angle $\psi = n\theta$ is modified by means of an angular correction $g(\theta)$ -periodical with period 2π- added to θ,

$$\psi = n(\theta + g) \quad ; \quad g(\theta + 2\pi) = g(\theta) \quad , \qquad (18)$$

then the displacement $y_g(\theta)$ obtained is

$$y_g = s\left[1 - \cos n(\theta + g)\right] \quad . \qquad (19)$$

As in the case of numerically controlled cam drives, if $dg/d\theta > -1$, curves $y_g(\theta)$ are obtained from $y_o(\theta)$ by means of a point to point sidewards displacement. If $dg/d\theta < -1$, additional maxima and minima may appear because the crank reverses its rotation.

For a given $y_g(\theta)$, angular correction $g(\theta)$ to obtain it can be found in a similar way as that used in the case of cam drives.

Displacement $y_g(\theta)$ defined by Eq. (19) leads to

$$\dot{y}(\theta) = sn\dot{\theta}(1 + g')\sin n(\theta + g) \quad , \qquad (20)$$

$$\ddot{y}(\theta) = sn\ddot{\theta}(1 + g')\sin n(\theta + g) + sn\dot{\theta}^2\left[g''\sin n(\theta+g) + \frac{n}{2}(1+g')^2 \sin 2n(\theta+g)\right] . \qquad (21)$$

Piece angular acceleration $\ddot{\theta}$ can usually be neglected.

DRIVING TORQUE REQUIRED BY A NUMERICALLY CONTROLLED SLIDER-CRANK MECHANISM

Under assumptions equivalent to those considered in the case of cam drives, the equation of motion of the slider-crank drive, Fig. (4), leads to the required driving torque

$$\Gamma = \dot{\theta}^2 ms^2 n\left[g''\left(\frac{I}{ms^2} + \sin^2 n(\theta+g)\right) + (1+g')^2 \sin 2n(\theta+g)\right] + F_R s \sin n(\theta+g) \qquad (22)$$

ILLUSTRATIVE EXAMPLE OF A NUMERICALLY CONTROLLED CAM DRIVE

Lets consider a cam drive with the displacement function

$$y(\psi) = s(2 - \cos\psi - \cos 2\psi) \qquad (23)$$

Curve $y_o(\theta)$ obtained for $\psi = \theta$ is shown in Fig. 5a, and Fig. 5b shows a set of curves $y_g(\theta)$ obtained by means of the angular correction

$$g(\theta) = \varepsilon \sin\theta \quad , \qquad (24)$$

with ε values: -57.3°, -40°, -20°, 0°, 20°, 40°, 57.3°. Values $\varepsilon = \pm 57.3°$ are the threshold beyond which $\psi(\theta)$ is no more a monotonous increasing function. As it can be seen in Fig. 5b, curves $y_g(\theta)$ vary over a wide range.

The required dimensionless torque $\Gamma^* \equiv \Gamma(\theta)/ms^2\dot{\theta}^2$, for the same values of ε as in Fig. 5b and for $F_R = 0$, is shown in Fig. 6a for $I/ms^2 = 5$, and in Fig. 6b for $I/ms^2 = 200$. It can be seen that most of the torque is required by the rotational inertia as I/ms^2 increases.

The dimensionless driving torque required if the cam is substituted by a screw drive with optimum transmission ratio is shown in Figs. 7a and 7b for the same values of ε and I/ms^2 as in Fig. 6. As it can be seen, maximum values of Γ^* are far greater than those required by the cam drive.

For comparison purposes, maximum values of Γ^* are shown in Fig. 8 as a function of I/ms^2 for both approaches and for ε values of Fig. 5b. As it can be seen, maximum torque required by the numerically controlled cam drive is much smaller than that required by a numerically controlled screw drive over the wide range $0 \leq I/ms^2 \leq 200$ considered.

To correlate the dimensionless torque Γ^* with actual torques, it is worth noting that, for a given duty cycle, the maximum torque of low inertia servomotors is roughly proportional to their moment of inertia. As a reference, for intermitent cycle S3(25%) 10 min with temperature increase of 80°, it can be taken

$$\Gamma_{max}[Nm] = 16 \cdot 10^3 \, I_{motor}[Kg\ m^2] \quad , \qquad (25)$$

from which

$$\Gamma^*_{max} = \frac{\Gamma_{max}}{ms^2\dot{\theta}^2} = \frac{16 \cdot 10^3}{4\pi^2} \frac{I_{motor}}{ms^2} \frac{1}{f^2} \quad , \qquad (26)$$

where f is the rotation speed in cycles per second. So maximum dimensionless torque attainable is proportional to I_{motor}/ms^2. In Fig. 9, Γ^*_{max} is shown as a function of I/ms^2 for f = 5, 10, 15 and 20 Hz. It is assumed that $I = I_{motor}$, i.e. that cam or crank inertia can be neglected.

ILLUSTRATIVE EXAMPLE OF A NUMERICALLY CONTROLLED SLIDER-CRANK MECHANISM

To illustrate the case of a numerically controlled slider-crank mechanism, the displacement curve

$$y_o(\theta) = s(1 - \cos 2\theta) \quad , \qquad (27)$$

has been considered. In this case, n = 2. It is the same curve considered in the previous example but with the first harmonic removed -it is worth realizing that, if s≪piece mean radius, this removal is equivalent to a displacement of piece profile-.

Curve $y_o(\theta)$ is shown in Fig. 10a while Fig. 10b shows the set of curves $y_g(\theta)$ obtained by means of the angular correction $g(\theta)$ defined by Eq. (24) with ε values: -57.3°, -40°, -20°, 0°, 20°, 40°, 57.3°. As in the previous example, for $-57.3° < \varepsilon < 57.3°$ crank rotation $\psi(\theta)$ is monotonous increasing. It can be seen that, in this case, curves $y_g(\theta)$ also vary over a wide range.

The required dimensionless torque Γ^* for the same values of ε as in Fig. 10b is shown in Fig. 11a for $I/ms^2 = 5$ and in Fig. 11b for $I/ms^2 = 200$. As in the previous example, most of the torque is required by the rotational inertia for great values of I/ms^2, but now this torque tends to be twice greater than that required by a cam drive because crank angular displacement is twice the cam angular displacement.

The dimensionless driving torque required if screw drive with optimum transmission ratio is used instead of the slider-crank mechanism is shown in Figs. 12a and 12b for the same values of ε and I/ms^2 as in Fig. 11. It can be seen that maximum values of Γ^* are greater than those required by the slider-crank mechanism.

Maximum values of Γ^* for both approaches are compiled in Fig. 13 as a function of I/ms^2, in the range $0 \leq I/ms^2 \leq 200$, and for ε values of Fig. 10b. Over the full range of the considered values of I/ms^2 and ε, maximum driving torques required by the slider-crank mechanism are substantially smaller than those required by a screw drive with optimum transmission ratio. Maximum values of Γ^* available from low inertia servomotors are illustrated in Fig. 9.

CONCLUSIONS

To obtain periodic motions, the use of cam drives can be boosted by the NC of cam rotation because it allows the obtaining of a wide variety of displacement functions by means of a single cam, while requiring a lower driving torque than conventional numerically controlled screw drives with optimum transmission ratio. On top of this, to obtain small and fast periodic motions, cam drives are better suited than screw drives from mechanical design point of view.

If there is a single top level position for all maxima and a single bottom level position for all minima, the use of slider-crank mechanism with numerically controlled crank rotation is attractive because of its inherent mechanical simplicity and toughness. It also allows the obtaining of a wide variety of displacement functions while requiring a smaller torque than that required in optimum screw drives.

In both cases, monotonous cam or crank rotation gives a reasonable flexibility to obtain displacement functions while keeping the number and level of maxima and minima. If cam or crank rotation is allowed to reverse, new maxima and minima at intermediate levels may appear, which pushes further the potential of both approaches but at the expense of greater driving torques.

UN COMPROMIS ENTRE LA FLEXIBILITE DU C.N. ET LA CAPACITE DE PRODUIRE DES FORCES ELEVEES DES CAMES

Résumé. Pour l'obtention de déplacements périodiques, la flexibilité du C.N. conventionnel, utilisant des vis à billes, le rend attirant comme remplacement des cames lorsque un grand nombre de ces dernières devient nécessaire. Cependant, dans beaucoup d'applications, la capacité des cames pour produire des forces d'impulsion intenses est difficile à surpasser par des dispositifs à vis.

Dans ce cas, du fait que l'accélération maximale du moteur et celle de la masse qui subit l'impulsion se présentent de manière simultanée, l'inertie du moteur réduit sévèrement la force maximale d'impulsion pouvant être obtenue lorsque l'accélération est importante. Cela se présente même dans le cas de servomoteurs de basse inertie et un rapport de transmission optimal.

Le C.N. de la rotation d'une came déterminée permet d'obtenir une grande variété de fonctions de déplacement, en même temps que la cinématique de la came permet de produire des forces d'impulsion élevées à partir d'un couple inférieur à celui qui serait nécessaire pour un dispositif à vis.

Si tous les maxima, ainsi que tous les minima du mouvement périodique recherché ont la même valeur, l'application du mécanisme bielle-manivelle avec la rotation de la manivelle controlée numériquement présente un attrai du fait de sa simplicité et de sa robustesse mécaniques, et parce que sa cinématique permet aussi de produire des forces d'impulsion élevées à partir de couples inférieurs aux nécessaires dans les dispositifs à vis.

Dans les deux cas, si la rotation de la came ou de la manivelle est croissante monotone, le nombre de maxima et de minima se tient, de même que leur niveau. Si le sens de la rotation change dans une période, de nouveaux maxima et minima peuvent apparaître, avec des niveaux intermédiaires, mais aux prix d'utiliser des couples plus élevés.

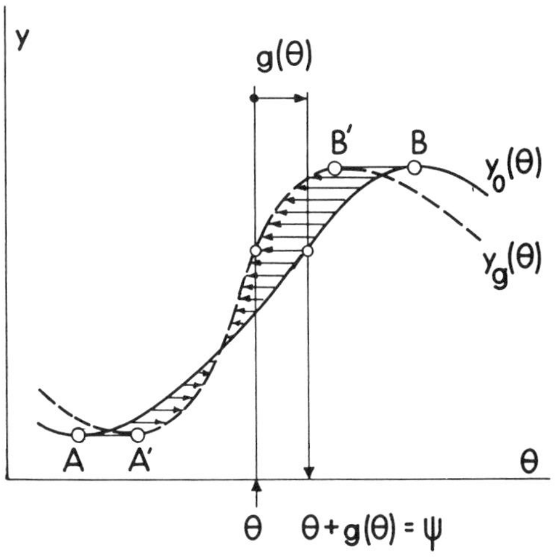

Fig. 2 Displacement modification by means of angular correction y(θ).

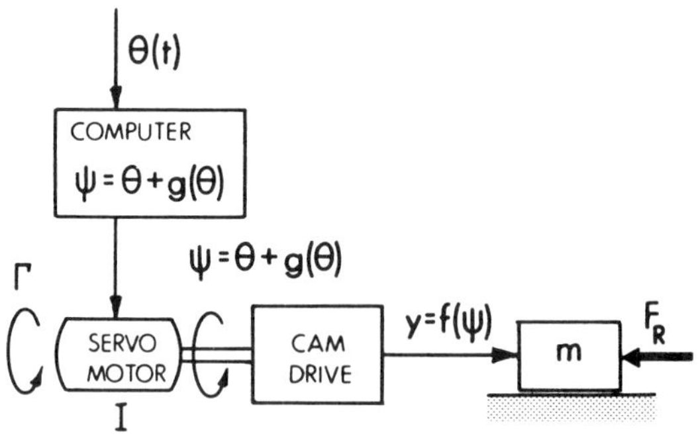

Fig. 3 NC of a cam drive.

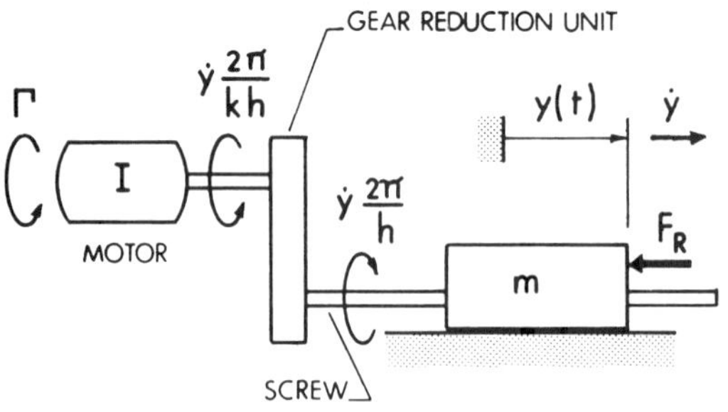

Fig. 1 Screw drive.

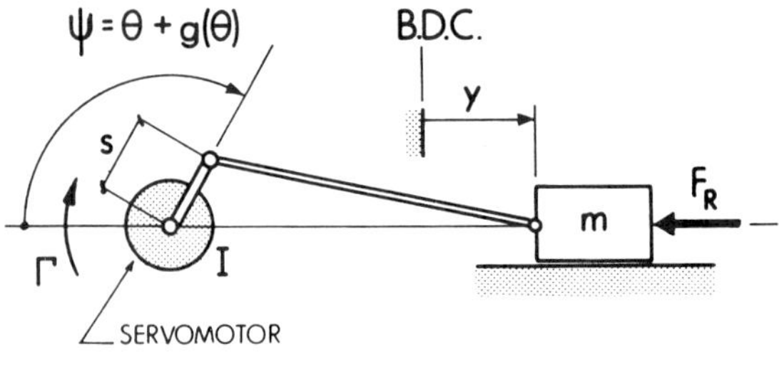

Fig. 4 NC of a slider-crank mechanism.

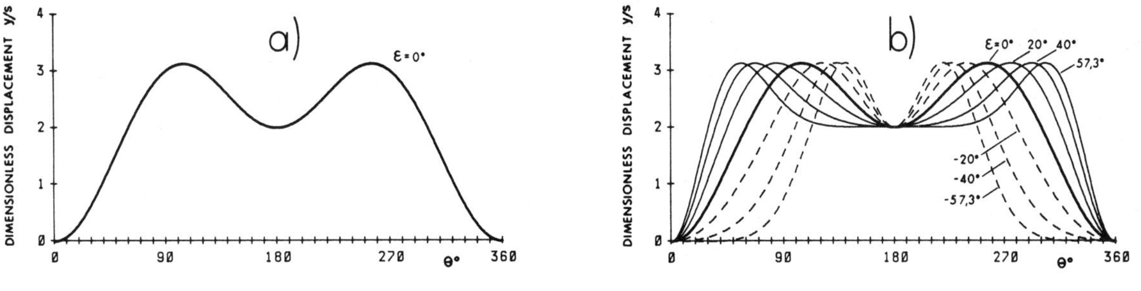

Fig. 5 Cam drive displacement functions. a) $y_o(\theta)/s$. b) $y_g(\theta)/s$.

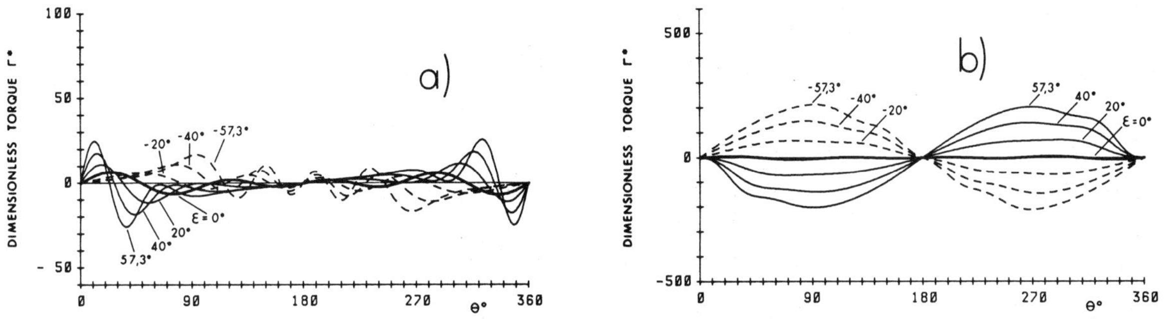

Fig. 6 Γ^* required by NC cam drive to obtain $y_g(\theta)$ of Fig. 5b for I/ms^2= : a) 5 . b) 200.

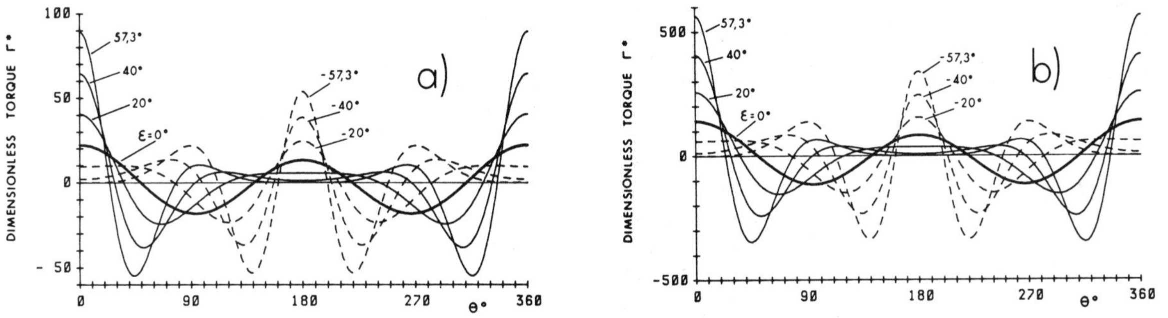

Fig.7 Γ^* required by an optimum screw drive to obtain $y_g(\theta)$ of Fig. 5b for I/ms^2= : a) 5 . b) 200.

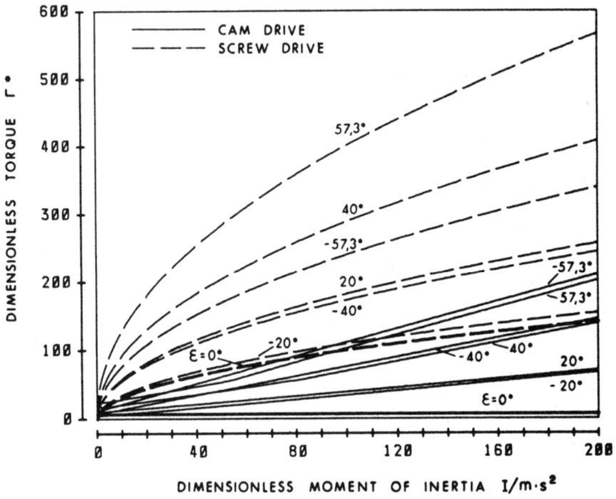

Fig. 8 Γ^*_{max} required by a NC cam drive and an optimum screw drive to obtain y_g of Fig. 5b.

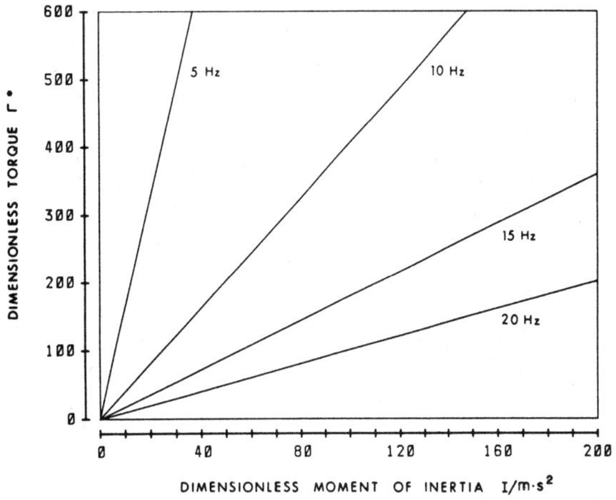

Fig. 9 Orientative Γ^*_{max} available from low inertia servomotors for a certain duty cycle.

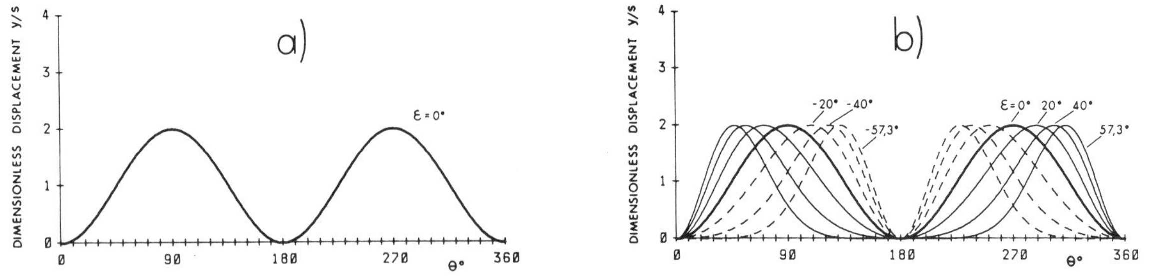

Fig. 10 Slider-crank mechanism desplacement functions. a) $y_o(\theta)/s$. b) $y_g(\theta)/s$.

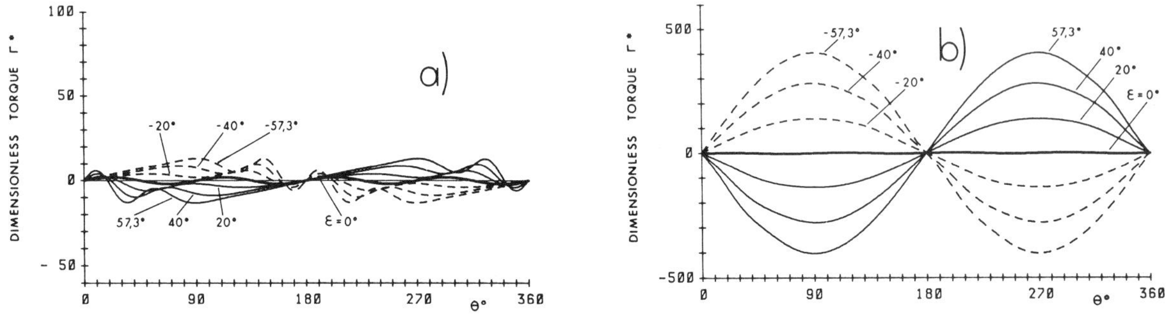

Fig. 11 Γ^* required by a NC slider crank mechanism to obtain y_g of Fig. 10b for I/ms^2 = : a) 5 . b) 200.

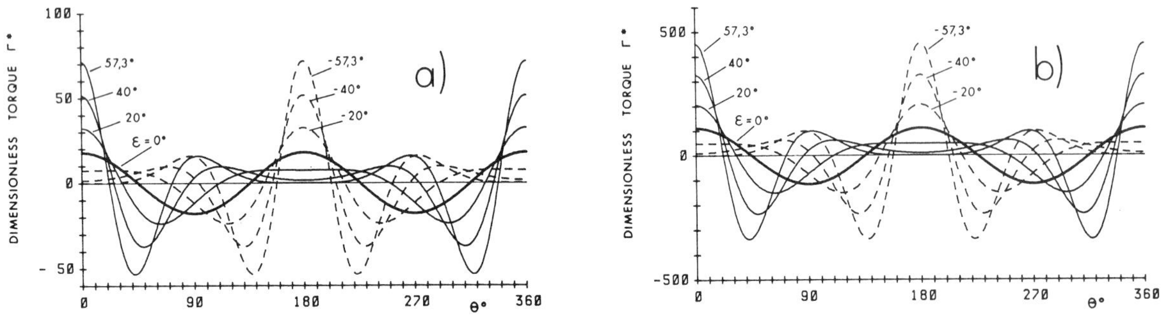

Fig. 12 Γ^* required by an optimum screw drive to obtain y_g of Fig. 10b for I/ms^2 = : a) 5 . b) 200.

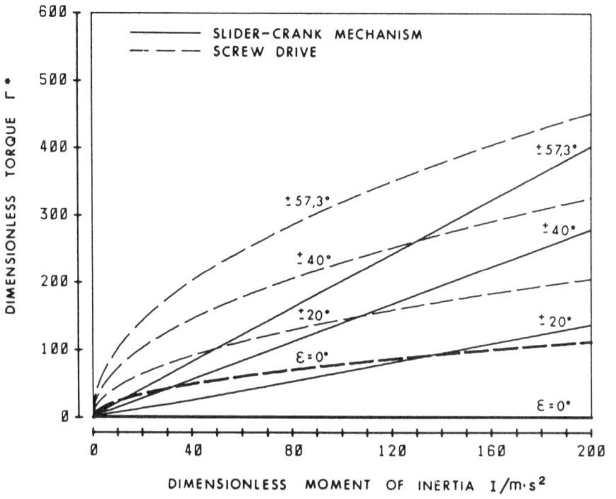

Fig. 13 Γ^*_{max} required by a NC slider-crank mechanism and an optimum screw drive to obtain y_g of Fig. 10b.

Analysis of Picking Mechanism on Shuttle Looms

M. ÇALIŞKAN*, S. T. TÜMER** AND E. F. SEVER*

*Mechanical Engineering Department, Middle East Technical University, Ankara, Turkey
**Mechanical Engineering Department, King Saud University, Riyadh, Saudi Arabia

Abstract. In weaving on shuttle looms the performance of picking largely depends on the dynamic behavior of the mechanism that projects the shuttle from one end of the loom to the other. In this study the picking mechanism of an underpick shuttle loom is modeled to study its dynamic behavior and to investigate the effects of various interactions mainly due to the characteristics of the components. The modeling is limited to the projection of the shuttle while the interactions from braking of the shuttle, the motion of the sley, and the damping exhibited by the drive are not considered. It is also assumed that the cam follower always stays in contact with the cam during picking. References are made to the experimental results by using the numerical data obtained from measurements taken on a particular loom. It is demonstrated that models of higher degree of complexity are more suitable for the analysis of the picking mechanism.

Keywords. Dynamic response; modelling; models; system analysis; textile industry.

INTRODUCTION

Conventional shuttle looms still keep their share in world cloth production as shuttleless weaving systems find more extensive use all over the world. The transition from conventional systems into shuttleless ones is particularly important for developing countries whose loom production is limited to conventional looms. Research needs to be conducted on the textile manufacturer's part to well understand the dynamics of shuttle looms before attempting to switch to shuttleless systems as well as to develop the existing technology on conventional systems.

Almost all the important setbacks of shuttle looms are related to their weft insertion system. Past analytical approaches (Vincent and Howell, 1939; Catlow, 1951a, 1951b; Hart, Patel, and Bailey 1976) which employed single degree of freedom models have proved insufficient to predict the dynamic behavior of the shuttle projection system accurately. The most detailed experimental study undertaken by Lord and Mohamed (1975) has demonstrated that the dynamic interactions experienced in the system are complicated enough not to be expressed by simple, single degree of freedom system models. On the other hand, general approaches (Chen 1975; Chen 1982; Koster 1974) have been developed to study the dynamic behavior of cam-follower systems similar to the one contained in the shuttle projection system.

In this study such general treatment techniques are applied to the weft insertion system of shuttle looms and analytical system models of higher degree of complexity have been developed to explain the experimental findings to a certain extent. Two multi-degree of freedom models along with the single degree of freedom model have been proposed to predict the dynamic behavior of the system. All three models have been tried with the parameter values obtained from Draper-X2 shuttle loom, and analytical results from these models have been compared to the experimental findings (Lord and Mohamed, 1975) qualitatively. The comparison has given an indication on the required degree of complexity of system models to predict the dynamic behavior accurately. The difference in the shuttle motion during its projection on both sides of the loom has also been investigated through these models. Furthermore, the effects of linear and cycloidal cam profiles on the motion have been determined.

SHUTTLE PROJECTION SYSTEM

A typical shuttle projection system commonly known as the picking mechanism is illustrated in Fig. 1. The bottom shaft is driven by an electric motor through a pair of gears with a reduction ratio of 1 to 2. The pick cam and its follower, the pick ball transmit the rotational motion to the shuttle as translation through a chain of components consisting of the picking shaft, the lug strap, and the picking stick in succession. The projection operation termed as picking is repeated on both sides of the loom alternatively, after the shuttle is stopped at the end of its flight following the picking operation. The basic function of the picking mechanism is to accelerate the shuttle to a certain speed within a fixed distance and a specified duration of time. The acceleration time history during picking is required to be as smooth and uniform as possible since the force acting on the shuttle is proportional to its acceleration. Typical values of maximum shuttle speed, acceleration distance, picking time interval, and maximum acceleration are around 20 m/s, 0.2 m, 20 ms, and 1000 m/s^2 respectively. In the design stage the picking mechanism is aimed to accelerate the shuttle to its maximum speed determined by the width of the woven cloth in shortest possible time and distance with the minimum attainable acceleration.

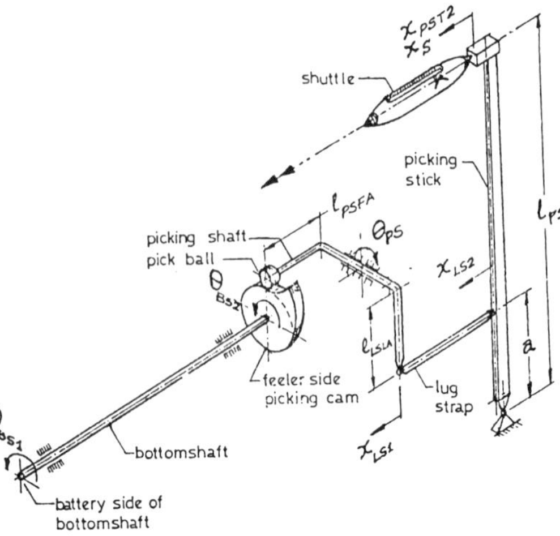

Fig. 1. Typical Shuttle Picking Mechanism

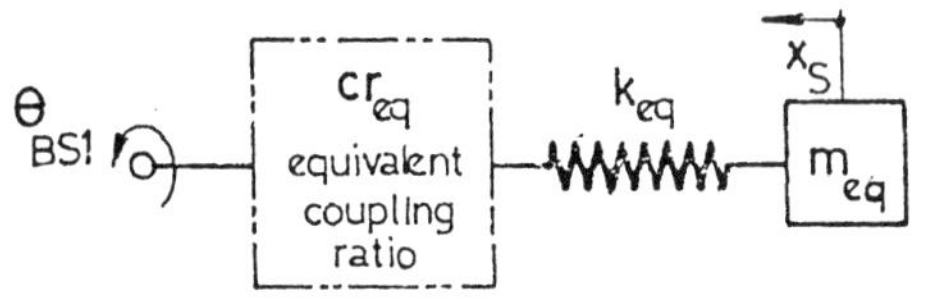

Fig. 2. Single Degree of Freedom Model

The picking mechanism exploits the elasticity of the components in the chain while accelerating the shuttle. The picking stick is the most prominent member as far as the elastic energy storage and release characteristics are concerned. If the same mechanism with the same geometry had been constructed from totally rigid components the maximum speed reached by the shuttle would have been the half of the speed given by the elastic mechanism for the same picking time interval and distance. The timing of the elastic energy release mainly from the picking stick is particularly important in accelerating the shuttle. System models which consider such effects are necessary to explain the behavior of the picking mechanism.

The inertia of the sley mechanism and the damping characteristics of the electric motor also affect the acceleration of the shuttle along with the components of the picking mechanism. The sley mechanism basically made up of a four-bar linkage causes variations in motor speed through its effective inertia as a function of position producing adverse effects on the shuttle projection.

MODELS FOR PICKING MECHANISM

In this study the existing single degree of freedom model is complemented with two four degree of freedom models given as follows:

MODEL 1 : Single degree of freedom model,
MODEL 2U: Uncoupled, four degree of freedom model,
MODEL 2C: Coupled, four degree of freedom model.

The effects of the sley mechanism and the drive motor have been neglected in the analysis.

Model 1

In the development of the single degree of freedom model, the bottom shaft is assumed to be rotated at a constant speed from the drive or "battery" side. This can be expressed as

$$\frac{d\theta_{BS1}}{dt} = \omega = \text{Constant} \tag{1}$$

which implies a very stiff drive. The model is illustrated schematically in Fig. 2. The equivalent coupling ratio is defined as the coeffient of relation between the input rotation θ_{BS1} and the output translation i.e. the shuttle motion, x_S when all the dynamical effects i.e. inertial and elastic are not considered. For a linear system model, this coefficient has to be constant. This is only possible if a linear relation exists between the motions of the cam and its follower. A linear picking cam profile has been assumed for the sake of linearity in the analysis to compare responses of different models. The equivalent spring stiffness k_{eq} is determined by equating the elastic potential energies of the equivalent system and the real system after the input θ_{BS1} is fixed ($\theta_{BS1} = 0$) and a unit displacement is given to the shuttle. The equivalent mass m_{eq} is found by equating the kinetic energies of the equivalent and the real systems for a unit shuttle speed when both systems are assumed to contain rigid components. The equation of motion for this model is given as

$$m_{eq}\ddot{x}_s + k_{eq}\,x_s = k_{eq}\,cr_{eq}\,\theta_{BS1} \tag{2}$$

Model 2C

A four degree of freedom model using θ_{BS1}, θ_{BS2}, x_{LS2}, and x_s as generalized coordinates has been developed. In this approach the kinetic energy terms due to elastic deflection of components are included at the formulation stage. This has resulted in a set of four coupled, second order ordinary differential equations. Assignment of nonlinear cam profiles is made possible although it would yield nonlinear equations of motion. In general

$$|M|\{\ddot{x}\} + |K|\{x\} = \{F\} \tag{3}$$

where $|M|$ is the mass matrix, $|K|$ is the stiffness matrix, $\{x\}$ is the vector of generalized coordinates, and $\{F\}$ is the forcing vector. The coupled equations of motion result in nondiagonal mass matrix which presents relative difficulties in the solution process. In the development procedure the bottom shaft is taken as a torsional member; the cam-follower pair as a nonlinear transducer; the pickball as a lumped mass; the picking stick as a cantilever beam; the lug strap as an axial member, and the shuttle as a lumped mass. It has also been assumed that the follower i.e. the pick ball does not leave the cam surface during picking operation.

Model 2U

Model 2U is a simplified form of Model 2C. It is obtained after contributions of the elastic displacements and velocities into the total kinetic energy have been removed. This simplification produces a set of uncoupled second order differential equations resulting in a diagonal mass matrix which is easier to handle in the solution process.

COMPARISON OF MODELS

All three models have been applied to the picking mechanism of Draper-X2 shuttle loom. A linear cam relation has been assumed to compare the models. The generalized coordinate θ_{BS1} has been specified for Model 2C and Model 2U as given in Eq.(1), thus, lowering the degree of freedom by one. The time histories obtained from three models for displacement, velocity and acceleration of the shuttle are presented in Figs. 3, 4, and 5. The time histories for a rigid picking mechanism are also illustrated in these figures. The shuttle leaves the picker when the maximum speed has been reached. The shuttle is assumed to be moving at constant speed afterwards in the analysis. The motion of the picker after the separation takes place is shown in broken lines. The time corresponding to the point of separation is the picking time whereas the distance taken by the shuttle up to that point is the picking distance.

All three models are found to yield, in general, close values for picking time and picking distance although they are different from each other. However, the 18% difference between the picking times obtained from Model 1 and Model 2C could be quite important as far as the loom timing is concerned. On the other hand, the single degree of freedom model is on the safe side when jamming of the shuttle is considered. Model 1 and Model 2U yield the same maximum shuttle speed whereas Model 2C exhibits about 7% higher speed.

The most striking difference is observed in the acceleration time histories given in Fig. 5. Maximum acceleration levels obtained from Model 1 and Model 2C differ by about 80%. The acceleration time history of Model 2C displays three distinct peaks giving a good indication on the unfavorable shock and noise conditions observed on the loom during picking operation.

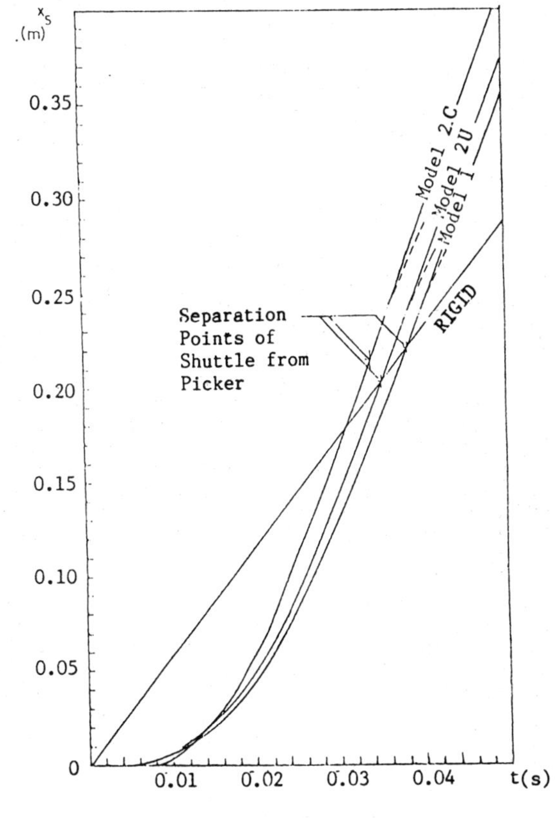

Fig. 3. Time History of Shuttle Displacement for Linear Cam

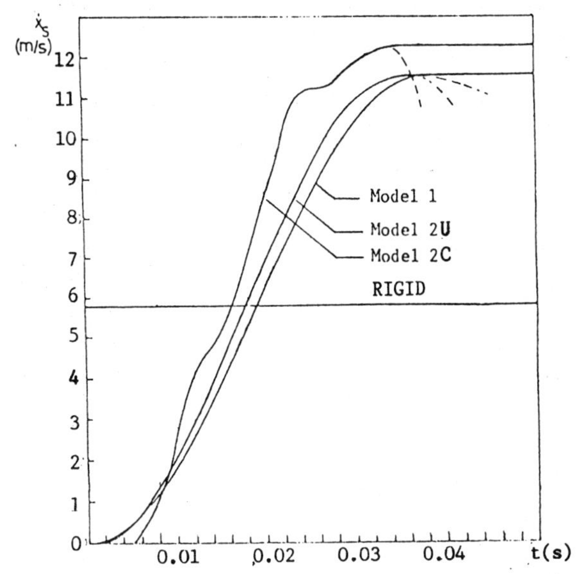

Fig. 4. Time History of Shuttle Velocity for Linear Cam

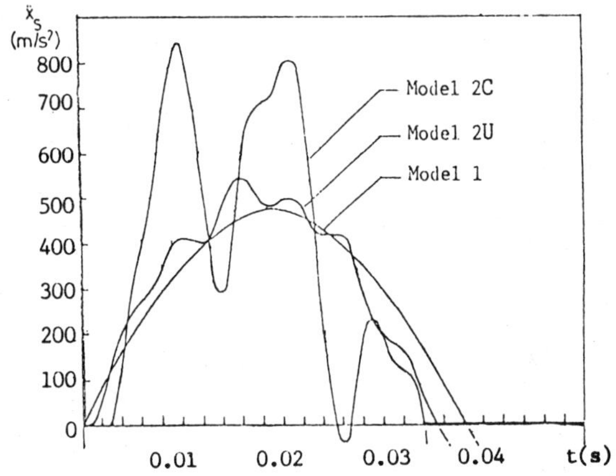

Fig. 5. Time History of Shuttle Acceleration for Linear Cam

The need for employment of multi degree of freedom models to represent the dynamic behavior of the picking mechanism becomes evident since it is possible to include several modes of vibration as opposed to less modes for lower order models. Figure 6 illustrates the experimental acceleration time history obtained on a Picanol President loom by Lord and Mohamed (1975) for a linear cam relation and constant crank speed. The experimental time history is labeled by $\ddot{x}$ whereas $a^2(s-x)$ is used to represent the analytical result from the single degree of freedom model. This figure also demonstrates the insufficiency of the single degree of freedom models to predict the dynamic behavior. A qualitative comparison of the experimental time history in Fig. 6 with the acceleration time history of Model 2C given in Fig. 5 justifies the need for multi degree of freedom system models.

The comparison of acceleration time histories for Model 2C and Model 2U given in Fig. 5 indicates the importance of the kinetic energy contribution due to elastic deformations in the system. This result does not comply with the general statement by Koster (1974) about the dismissal of such

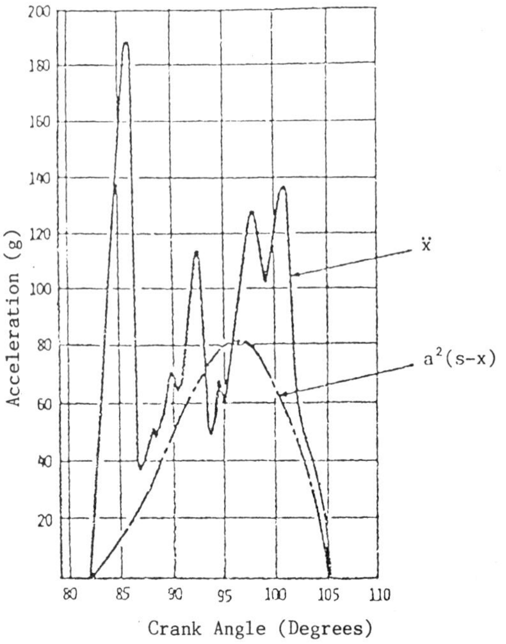

Fig. 6. Acceleration Time History by Lord and Mohamed (1975)

contributions in cam-follower systems. However, this conclusion has been reached by considering typical parameters of cam-follower systems frequently encountered in practice. The picking mechanism, on the other hand, is a special mechanism where the elasticity of components is functionally exploited, and elastic deformations in some components are almost in the same orders of magnitude with rigid body displacements. Therefore, the dynamic coupling property of equations of motion for this system appears particularly important so that it cannot be neglected in the analysis although it introduces additional difficulties in solution process.

The effect of the torsional stiffness of the bottom shaft on shuttle acceleration can also be investigated by the use of Model 2C (Sever 1985). The analysis has verified the experimental findings by Lord and Mohamed (1975) on why different cam profiles for each side of the loom must be chosen to obtain the same shuttle motion. Besides, it has been demonstrated that the cycloidal cam profile would produce higher and smoother shuttle acceleration response as opposed to linear profile (Sever 1985).

CONCLUSION

This study demonstrates the insufficiency of the popular single degree of freedom models for the picking mechanism of shuttle looms to predict the system response. The analysis for a specific loom and linear cam relation justifies the need for multi degree of freedom models where higher modes of vibration can be included. It is also determined that kinetic energy contributions due to elastic displacements have to be considered in the formulation of equations of motion.

The shuttle acceleration response obtained from coupled, multi degree of freedom model is found to correlate well to the previous experimental findings. The need for two different cam profiles for both sides of the loom is verified analytically. It is shown that it is possible to investigate the effects of cam profiles, elasticity of system components and geometric parameters on the system response. Model 2C can also be extended to include the sley mechanism and the electric motor (Sever 1985).

REFERENCES

Vincent, J.J., and W.T. Howell (1939). The mathematical theory of shuttle projection. J. Text. Inst., 30, T103.

Catlow, M.G. (1951). The force-time relations during shuttle projection, Part I: Equations of movement. J. Text. Inst., 42, T413-T442.

Catlow, M.G. (1951). The force-time relations during shuttle projection, Part II: Comparison of theoretical and experimental results. J. Text. Inst., 42, T443.

Hart, F.D., B.M. Patel, and J.R. Bailey (1976). Mechanical separation phenomena in picking mechanisms of fly-shuttle looms. ASME J. Eng. for Ind., 98, 835-839.

Lord, P.R., and M.H. Mohamed (1975). An analysis of the picking mechanism of a textile loom. ASME J. Eng. for Ind., 97, 385-390.

Chen, F.Y. (1975). A survey of the state of the art of cam system dynamics. Mech. and Mach. Theory, 12, 201-224.

Chen, F.Y. (1982). Mechanics and Design of Cam Mechanisms. Pergamon Press, N.Y.

Koster, M.P. (1974). Vibrations of Cam Mechanisms McMillan Press, London.

Sever, F.E. (1985). Modelling and Analysis of Picking Mechanism on Shuttle Looms. MSc Thesis, Middle East Technical University, Ankara.

ANALYSE EINES FADENSCHUSS-GETRIEBES FÜR DIE WEBEREI-TECHNIK

Zusammenfassung. Der Erfolg des Fadenschießens beim weben wird hauptsächlich durch das dynamische Verhalten des Getriebes, das den Schützen von einem Ende der Webmaschine bis zum anderen Ende beschleunigt, bestimmt. In dieser Arbeit wird ein derartiges Getriebe hinsichtlich seines dynamischen Verhaltens betrachted und die gegenseitigen Auswirkungen, verursacht durch das Verhalten verschiedener Teile, untersucht. Die Betrachtung bezieht sich lediglich auf die Beschleunigung des Schützens. Auswirkungen des Bremsens des Schützens, der Bewegung des Ladens, sowie der Dämpfung verursacht durch den Antrieb bleiben somit unberücksichtigt. Desweiteren wird angenommen, daß wührend des Fadenschusses, der Kontakt im kurvengetriebe gewährleistet bleibt. Vergleiche mit experimentellen Ergebnissen werden erstellt. Es wird gezeight, daß Modelle mit höherem Grad an Komplexität sich für die Analyse des Fadenschuss-Getriebes besser eignen.

Schlüsselwörter: Dynamisches Verhalten; Modellierung; Modelle; Systemanalyse; Textilindustrie.

Optimal Configurations for Parallel Shaft Indexing Mechanisms

J. REES JONES AND K. S. TSANG

Department of Mechanical, Marine and Production Engineering, Liverpool Polytechnic, Byrom Street, Liverpool L3 3AF, UK

Abstract. The power efficiency and the life of indexing boxes is determined primarily by the dimensions of the configuration and the size of the roller followers. The sensitivity of efficiency and life to configuration parameter changes is explored bringing into consideration factors such as pressure angle, sub-surface stress and fatigue, rolling velocity.

Keywords. Optimal configurations; cam profile; fatigue life; sub-surface shear stress; shakedown limit.

NOMENCLATURE

a	=	the depth of maximum principal shear stress
A	=	the centre distance between cam and turret axes
B	=	the follower length (pitch circle radius)
C	=	basic dynamic load rating
F(t)	=	the roller load as a function of time
I	=	moment of inertia of the turret
$\tilde{n}$	=	average roller speed (rpm)
n(t)	=	the roller speed as a function of time
T	=	period of time
α	=	cam rotation, zero at start of turret motion
β	=	follower rotation ($\beta=\beta_i(\alpha)+\beta_o$)
$\beta_i(\alpha)$	=	cam motion law
β_o	=	initial value of β
β'	=	$d\beta/d\alpha$
β''	=	$d\beta^2/d^2\alpha$
μ	=	pressure angle

INTRODUCTION

Indexing refers to the intermittent motion of a machine element, conveyor or table in which the dwell between periods of movement (indexing period) enables an operation to be carried out or a drive to be disengaged. Whilst such motions can be implemented in an infinite variety of ways, cams predominate in the high speed end of indexing requirements.

In the last decade the incidence of commerically produced and marketed cam indexing boxes has increased significantly, They provide a convenient and ready made component to incorporate in automated systems, a suitable cam indexing box can usually be obtained to meet the required dwell to index ratio using the suppliers technical performance literature if the input drive transmission and the output load conditions are known.

The progression to higher speeds in machines results in higher dynamic loading and consequential shorter life of indexing boxes. The question arises whether an improvement in life can be obtained within the space of an existing box enclosure by changing only the cam, roller and turret parameters in a way which will not interfere with the box dimensions. This would facilitate higher speeds without the need to go for larger boxes that might otherwise be indicated by the supplier's technical literature and avoids the question of whether a larger box would fit into the machine space available. Large boxes may not without radical changes to the layout of a current machine design.

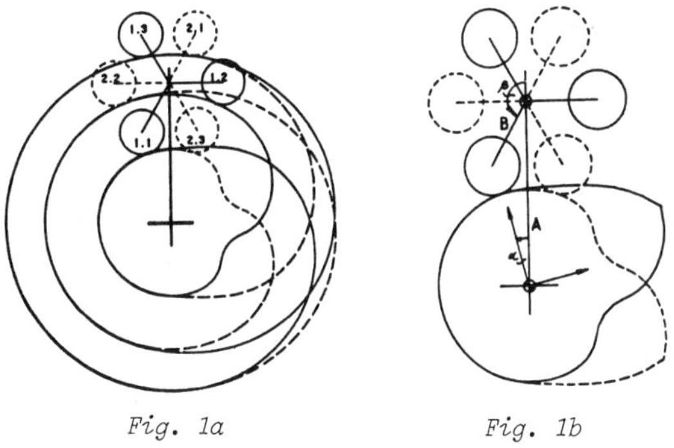

Fig. 1a *Fig. 1b*

Fig. 1. Parallel shaft indexer cam profiles, having 3 stops, 6 rollers.

The layout of cam and roller turret follower is shown in Fig.1. The configurational variables are :

i) pitch circle radius B of roller axes in the turret,
ii) roller diameters.

The effects on each of these due to variations in the pitch circle radius and the roller diameter is the subject of this paper.

The centre distance between input and output shafts is considered constant along with width of both cam and roller.
The criteria are :

i) pressure angle,
ii) roller life,
iii) fatigue endurance of cam face.

GEOMETRY

All the criteria are dependant on the geometry of the mechanism configuration and the cam profile or motion law. Commonly used motion laws in indexing applications [1] are :

i) modified trapezoidal,
ii) modified sine, modified sine constant 50/75.

The law used determines the dynamic performance of the indexer in any given machine system. The drive and driven line compliance and backlash, mass or moment of inertia and speed are machine factors associated in the determination of the dynamic performance. The commonly used laws are sufficiently distinctive to produce near optimum conditions in any given set of machine factors.

This investigation is restricted to a consideration of indexing based on the modified sine law. This was chosen on the basis that it provide good performance where the machine factors are not known and has lower pressure angles associated with it that lend to higher efficiency.

THE PROFILES

The cam profiles are calculated using the method of [2]. These are shown in Fig.1a. The resulting curves are shown to the right of Fig.1b. They are the result of the interfering envelope of the rollers in successive indexes.

The cam rotates in the anti-clockwise direction from the point shown and the indexing turret moves in the clockwise direction. Indexing of the turret is such that the roller 1.2 succeeds the roller 1.1 in the drawn position and so on.

PRESSURE ANGLE

The efficiency of force transmission through a cam and follower is determined by the geometry in particular the pressure angle. This angle is measured between the contact normal (line of action of the force) and the direction in which the roller centre moves or tends to move. Resolving this force into two components, parallel and perpendicular to the centre line of the follower gives the transmitted force that does useful work on the output and the other that is involved as a constraint force taken up by the follower pivot bearing and increasing the frictional resistance at that joint. Since the latter amounts to lost work then it accounts a significant part of efficiency loss.

The pressure angle is evaluated [3] through the relationship of Eq. (1).

$$\tan \mu = \frac{\cos \beta + B(1-\beta')/A}{\cos \beta} \qquad \ldots(1)$$

Between two and three rollers will be in nominal contact depending upon the part of the cycle. However, only one or two of these will be active in transmitting force and the actual ones involved will be determined by the direction of the load torque on the output turret. This torque may be opposite to the direction of rotation or in the same direction, as it may well be when in the overun condition associated with an inertia dominated load which is being decelerated. Where two rollers share the load the one with the least pressure angle is assumed, in the present context, to carry the entire load.

A graph of pressure angle is shown in Fig.2. In this the active pressure angle is the positive one. As long as the centre distances are fixed then only the follower length (or turret pitch circle radius) will have any effect the pressure angle. These are also represented in Fig.2.

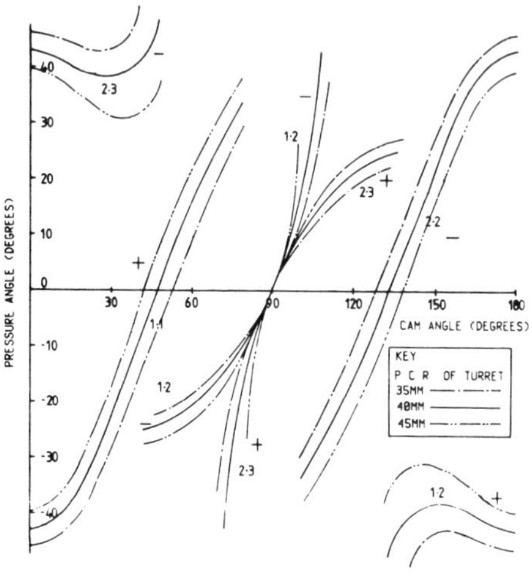

Fig. 2. Graph of pressure angle versus cam angle, with effect of changing turret pitch circle radius.

CONTACT FORCE

In this high speed consideration the inertia effects will predominate and force is calculated using Eq. (2), given in [3] as

$$\text{contact force } F = \frac{I\,\beta''}{B \cos \mu} \qquad \ldots(2)$$

To allow for the vibration effects in high speed operations, a so called 'torsion factor Ct' is introduced. It is defined in [1] as a factor that scales up the nominal contact force. The value Ct =1.4 is an average value for a dynamically well behaved system that is used in the analysis presented in this paper.

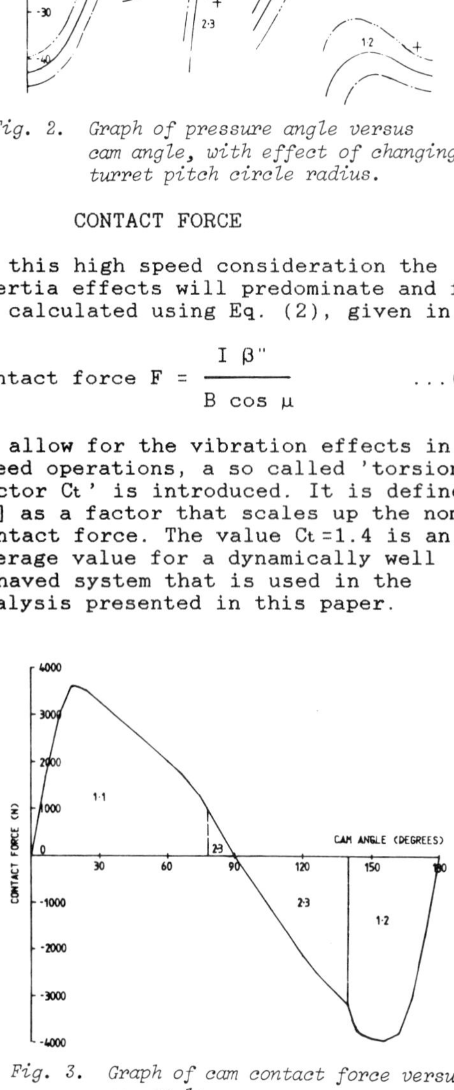

Fig. 3. Graph of cam contact force versus cam angle.

Fig.3. shows a typical graph for contact force evaluated in the above way. Only the terms in the denominator are affected by the pitch circle radius B. Although for all practical variations cos μ changes an insignificant amount. This implies that the force is inversely proportional to the pitch circle radius B.

RADIUS OF CURVATURE

The radius of curvature σ is another property that comes into the calculation of life. It is calculated [4] through the expression of Eq. (3).

$$\sigma = \frac{[A^2+B^2(1+\beta')^2+2AB(1+\beta')\cos\beta]^{3/2}}{A^2+B^2(1+\beta')^3+AB[(2+\beta')(1+\beta')\cos\beta+\beta''\sin\beta]} \qquad ...(3)$$

FATIGUE LIFE

The life of a cam box may be limited by wear or catastrophic damage through misuse or excessive preload. The wear may take place in the form of attrition of the surface or through pitting fatigue associated with rolling contact.

Failure of either the roller follower or the cam surface may occur and the prediction of such is different because of the nature of the information available.

(i) Roller Life
Advantage is taken of manufacture's data and rules which, although semi-empirical, are supported by the evidence of massed data and life histories from many sources.

This data involves a 'basic load rating' which, in dynamic conditions, is that steady load at which there is a 10% probability of failure in one million revolutions of the roller.

The particular form of the failure is not indicated by manufacturers. External contact conditions with the cam surfaces are not brought into the problem so that contact stress is not considered as an issue except between the needle or bearing rollers and the race tracks.

Roller makers assume a definite life for their rollers. This can be obtained from the dynamic loading by the semi-empirical equation for the rated life in hours L_{10h}, of a needle roller bearing given as Eq. (4) and Eq. (5).

$$L_{10h} = \frac{16666}{\tilde{n}}\left[\frac{C}{P}\right]^{10/3} \qquad ...(4)$$

$$\text{and } P = \sqrt[10/3]{\frac{\int_0^T n(t).F^{10/3}(t)\,dt}{\int_0^T n(t)\,dt}} \qquad ...(5)$$

The basic dynamic load rating C is fixed for every roller's size and these are listed in the maker's catalogue. The average roller speed $\tilde{n}$ and the equivalent roller load P are evaluated for every cam operating speed.

(ii) Surface Fatigue
Here, the evidence of rolling contact fatigue tests have been drawn upon from several sources published and unpublished. Notable amongst these are references [5] and [6]. These generally conclude that steels tend to exhibit full endurance or indefinite life if the contact stress is below a critical value referred to as the 'endurance limit'. The absence of failure up to 4 x 10^7 cycles of stress is considered to have full endurance.

The intercept of the graph of stress against cycles to failure within the 4 x 10^7 cycles ordinate established the 'endurance limit'.

The evidence of controlled tests show a good degree of correlation between the hardness of the material and the 'endurance limit'. In the theories, as they prevail the 'endurance limit' should reach the so called 'shakedown limit' that is a Hertz stress (direct) of

$$fz_{max} = 4k \qquad ...(6)$$

where $k = \dfrac{\text{Vickers Pyramid Hardness number}}{6}$

Tests show that for through hardened materials with good surface finish an 'endurance limit' of 3k is obtainable.

For surface hardened cams, which now have become most common, a modification is required. Here the hardness of the material at a depth 1.5 times the depth a of maximum sub-surface shear stress is found [7]. The 'endurance limit' is then taken as to the classical 'shakedown limit'

$$fz_{max}\Big|_{a=1.5} = 4k \qquad ...(7)$$

COMPUTER PROGRAM

A computer program has been written for parallel indexing cams in general. It is capable of analysing mechanisms that generate one or two indexes per revolution and several variations on the numbers of stops per revolution of the output. It enables the determination of

i) the effective arc of contact for each roller,
ii) the radii of curvature and pressure angles,
iii) the contact loads for different speeds and loading conditions,
iv) the direct contact stress at the surface,
v) the depth of maximum shear,
vi) the life of the roller,
vii) the endurance of the cam surface.
and viii) the graphic representation of the cam and the variables.

Inspection enables the assessment of the suitability of the profile and the effective engagement arcs are calculated. The same program is extended to include the evaluation of contact forces, roller velocities and associated roller life.

ANALYSIS

A standard parallel shaft indexing mechanism (A) selected from an indexer manufacture's catalogue was used as the basis in the analysis, these system parameters were used in the computer program. The results obtained are discussed as follows.

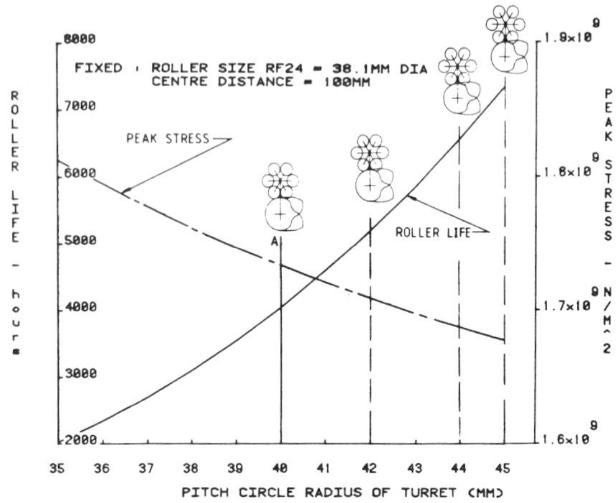

Fig. 4. Graph of roller life, peak stress versus turret pitch circle radius.

(i) Changing the Pitch Circle Radius of the Turret

The increase in pitch circle radius is beneficial for both roller life and peak stress, since the roller life goes up with increases in pitch circle radius whilst the peak stress goes down. For example, The roller life increases by 50% if the pitch circle radius is increased by 3mm and the peak stress decreases by 2.1%, see Fig.4.

The peak pressure angle is improved by approximately 1 or 2 degrees if the pitch radius is increased by 3mm. This benefits the efficiency of force transmission between the cam and the roller. The cam profile varies slightly to accommodate increases in turret pitch circle radius, see Fig.4. No detrimental effects are seen within the range considered.

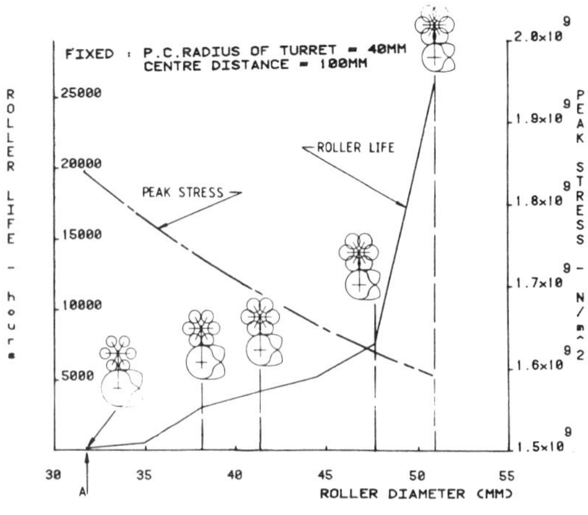

Fig. 5. Graph of roller life, peak stress versus roller diameter.

(ii) Changing the Roller Size

The roller life is calculated using a maker's semi-empirical equation. For larger roller sizes the basic dynamic loading rate C is higher, and for almost the same loading condition its equivalent bearing load P does not change very much. Also for the same input speed the roller velocity will be lower. The result of this, calculated with the above expression indicates a higher roller life Fig.5.

CONCLUSION

The analysis was based upon an arbitrary cam speed and inertia load and concentrates on the effects of configurational changes.

The geometrical configuration affects the life of rollers. The life is particularly sensitve to the roller diameter and the conclusion is that the roller should be as large in diameter as possible, subject to the constraint of rollers interfering with each other and the effective contact for three rollers is reduced to impracticable low limits.

The pitch circle radius has less influence on the roller life; nevertheless the rule is that greater is the pitch circle radius, the longer will be the roller life.

REFERENCES

1. Rees Jones, J. Reeve J.E. "Dynamic response of cam curves based on sinusoidal segments",Cams and Cam Mechanisms,MEP 1978
2. Rees Jones, J. "Mechanisms, Cam Cutting Co-ordinates",Engineering,Mar 1978 Vol 218
3. Rees Jones, J. "Mechanisms, Pressure Angles and Forces in Cams", Engineering, Jul 1978 Vol 218
4. Rees Jones, J. "Mechanisms, Cam Curvature and Interference", Engineering, May 1978 Vol 218
5. 'Fatigue in Rolling Contact' - Symposium, IMechE 1963
6. Young I.T. "A Wider Scope for Nitrided Gears", ASME May 1981
7. ESDU data item 78035 contact phenomena 1. contact stress

AUSZUG

Die Leistung und Lebensdauer der schrittegetriebe wird vorwiegend durch die Dimensionen der Anlage und die Größe der rollennachfolger bestimmt. Bei der Untersuchung der Empfindlichkeit der Leistung und Lebenslauf gegenüber Änderungen in Anlagegrößen werden solche Faktoren berücksichtigt wie Druckwinkel, unterflächliche Spannungen und Zermürbeerscheinungen, rollende Geschwindigkeit.

Effect of Initial Surface Roughness on Dynamic Performance of Plate Cams

R. L. NORTON, D. LEVASSEUR, ANN PETTIT AND SARA DILLICH

Worcester Polytechnic Institute, Worcester, MA, USA

Abstract. Previous studies (1) (2) have shown a statistically significant relationship between the method of manufacture of plate cams and the dynamic quality of the cam follower behavior. A Cam Dynamic Test Fixture (CDTF) has been designed and built to allow operation of cams made by various techniques, under controlled conditions of speed and load. Dynamic measurements have been made of follower acceleration and follower force, in both tangential and radial directions. Roller slip has also been accurately measured. Before running the test cams, the surface finish was measured with a Hommel T-20S surface roughness tester. The surface parameters were again sampled after intervals of run-in under measured conditions of force, speed, and acceleration. The data for acceleration and force were transformed to the frequency domain. All experiments were randomized and data subjected to an Analysis of Variance, to look for correlations between RMS power spectrum of acceleration, and surface finish parameters of the subject cam. Test cams for the study were made by Continuous Numerical Control manufacturing techniques using $1/4^\circ$, $1/2^\circ$ and 1° digitizing increments. Results show significant differences in the dynamic behavior of milled versus ground cams, but no differences due to digitizing increment. An alternate heat treatment method, (Ion Nitriding) shows a significant improvement in dynamic behavior over conventional through-hardening of milled cams.

Keywords. Acceleration measurement; data acquisition; fourier transforms; machining; vibration measurement; cam design, cam manufacturing; cam dynamics.

INTRODUCTION

Despite careful selection of a cam's mathematical profile contour, many variables in the manufacturing process can have significant effect on the quality of the cam's dynamic performance. In the USA, Continuous Numerical Control (CNC) manufacturing methods have largely supplanted the older techniques of analog duplication of master cams on tracer machines, for the production of small batch, non-automotive cams. The mathematical cam profile must be sampled at some increment, usually $1/4^\circ$, $1/2^\circ$ or 1°, to create the digital data for the CNC tape. The actual cam contour will then be linearly interpolated between the sampled data points by the CNC machine. (Some machines also allow circle arc interpolation.) The actual cam profile function which results may differ from the theoretical due to vagaries in the manufacturing process.

Many cams in production machinery are used "as milled," with no grinding, post-heat treat. This is often done for purely economic reasons as grinding nearly doubles the cost of a typical cam. Quite good milled finishes are now attainable. Through-hardening of high carbon steel introduces some heat distortion, which can only be eliminated by grinding. A low temperature surface hardening technique such as Ion Nitriding has the potential for reducing the thermal distortion in cams which are to be left as milled.

This experimental study investigated the effect of variation in the digitizing increment, the surface finishing method (milling vs. grinding), and the heat treatment method (through-hardening vs. Ion Nitriding). Seven pairs of cam disks were manufactured in combinations of the above factors, by a major commercial cam manufacturer. Since the surface characteristics of the cam were of principal interest, a "no-rise" cam or circular disk of 7 inch diameter was chosen in order to eliminate any dynamic effects due to cam profile design. A later paper will report on the testing of double dwell cams of various designs made with the same techniques. The test cam disk is shown in Fig. 1. All but the Ion Nitrided samples were made from AISI A-10 tool steel. The latter were made of AISI 4140 steel.

TEST EQUIPMENT

A Cam Dynamic Test Fixture (CDTF) was designed and built to serve as a dynamically quiet test bed on which to run these cams. It consists of an extremely stiff camshaft coupled to a large flywheel which is belt driven from a vibration isolated, DC, speed controlled motor (Fig. 2). Plain bearings are used throughout, including within the 2" diameter, chrome plated roller follower. The cam follower is carried on a 10" radius aluminum arm, which is instrumented to measure: radial and tangential follower acceleration, radial and tangential follower force, camshaft angular velocity and position, roller follower angular velocity, and oil bath temperature.

The test cams are clamped to a tapered and ground spindle with no keyway. The follower arm is loaded by a helical spring to a nominal force of 170 pounds. The cam is driven at 3 Hertz, while data is taken from the transducers, digitized, transformed to the frequency domain and stored. Both before and after running the cams on the CDTF, their surface profiles were measured at a particular location with a Hommel T20-S Surface Roughness Tester. This measures and records 8000 points over a 2.5 mm. length of the cam surface. Fixturing allowed the same precise location to be measured both before and after running. These surface data were digitized and transferred to a host computer for storage, analysis and plotting.

The Hommel T-20S calculates (19) surface roughness, waviness and profile parameters according to ISO and DIN standards.

TEST PROTOCOL

The fourteen test cam disks (seven pair), were made by the methods shown in Table 1. The order of testing, both for surface measurement and dynamic measurement was randomized. A "surface experiment" consisted of two replications of the surface profile of the same location on each cam (28 profiles). The cam was physically removed from the custom holding fixture (Fig. 3) between the two replications. The degree to which these two profile replications are identical is a measure of the accuracy of the fixturing and of the Hommel test apparatus. This procedure was designed to measure any differences in surface profile which might occur due to run-in wear of the cams during the dynamic testing. A "dynamic experiment" consisted of one replication of each of the cams in random order (14 tests) run on the CDTF for approximately (5) minutes each, while the acceleration (x,y) and force (x,y) measurements on the follower were taken, transformed to the frequency domain, and stored in the host computer. Follower slip and oil temperature were also monitored. The investigation consisted of (16) surface experiments and (15) dynamic experiments, which were alternated in that order.

The raw surface profile data were plotted, as were the power spectra of the acceleration and force data. Sample plots are shown in Figures 4-10. Nineteen surface parameters were calculated for each surface experiment, and the spectral power and RMS average values were calculated for the force and acceleration data from the dynamic experiments. The calculated data from all experiments were then subjected to a two-way analysis of variance, with cam type and experiment as the factors. A Newman-Keuls test was used to separate significant variables.

RESULTS

Figure 4 shows the first replication of the surface profile of cam J6L2RTG1 and the second replication of the same cam. The cam was removed and replaced on the fixture between these two measurements, but not run on the CDTF. The close duplication of these two measurements shows that the Hommel T20S and the custom fixturing are accurate enough to allow good replication. Figure 5 shows the two suface replications of the same cam after (15) dynamic experiments on the CDTF. No significant wear is evident after this short run-in period of about 75 minutes. The analysis of variance of these data also showed that there was no sigificant difference in the measured surface parameters during the (15) experiments. The factor of cam type, however, was significant at $P < .00005$. Table 2 shows the ranking of cam type for the surface parameters of average roughness (Ra) and waviness (Wt). The separation class column indicates differences which are statistically significant at $p = .01$. Cam types with the same separation class letter are not significantly different from one another. It is interesting that among the ground cams the 1° digitized samples have significantly lower roughness and waviness values than the $1/2^o$ and $1/4^o$ samples. The nitrided cams group with the milled cams as expected. Figures 6 and 7 show example profiles for milled and nitrided cam surfaces.

Figures 8, 9 and 10 show typical power spectra for ground, milled and nitrided cams respectively. The RMS average values of the spectra for (14) cams per experiment and (15) experiments were analyzed in a two-way Anova. The factor of "experiment" was not significant indicating no appreciable change in the dynamic behavior of the came over the first 75 minutes of run-in. However, the factor of "cam type" was significant at $P < .000005$. Table 3 shows the ranking and separation of the cam types tested in this series. Note that the ground cams, as a group, run quieter than the milled cams which is consistent with the surface measurement parameters in Table 2. However, the nitrided cams ran significantly quieter than the milled cams despite the similarity of their surface finishes as shown in Table 2. The digitizing increment shows no significant effect on the high frequency dynamic performance of the cams.

CONCLUSIONS

There appears to be no dynamic advantage to using a digitizing increment smaller than about 1 degree for this size cam with linear interpolated CNC. Some economic advantage may be realized by using the larger angular increments in terms of cutting speed. Grinding of the cam surface does provide a significant improvement over milling. The fact that the Ion Nitrided cams show significantly quieter dynamic behavior than the milled cams despite their identical machining and the insignificant differences in their surface roughness parameters, suggests that the lesser heat treat distortion of this lower temperature hardening process may be of benefit. Note that these data are preliminary in nature due to the fact that, at this time, only a subset (14) of a projected set of (82) cams of various designs and manufacturing methods have been tested.

ACKNOWLEDGMENTS

The authors wish to express their appreciation to the National Science Foundation, USA, for support under Grant #MSM8512913. Equipment and support was also supplied by AMP, Incorporated, Harrisburg, PA, Camco Inc., Chicago, Il, Valmet, Inc. New Britain, CT, and Dytran Inc., Canoga Park, California.

REFERENCES

1) Norton, R.L., (1985) Effect of Cam Manufacturing Methods on Dynamic Performance of Eccentric Cams, an Experimental Study Part 1. Proceedings of the 9th Applied Mechanics Conference Oklahoma State University.

2) Norton, R.L., (1985) Effect of Cam Manufacturing Methods on Dynamic Performance of Double Dwell Cams, an Experimental Study, Part II. Proceedings of the 9th Applied Mechanics Conference, Oklahoma State University

TABLE 1 Test Cam Definitions

No. Cams	Digitizing Increment	Finish	Machine	Heat Treatment	Cam Code
2	$1/4^o$	Ground	R-Theta	Through Rc57	J6L4RTGx
2	$1/2^o$	Ground	R-Theta	Through Rc57	J6L2RTGx
2	1°	Ground	R-Theta	Through Rc57	J6L1RTGx
2	1/4o	Milled	X-Y	Through Rc57	J6L4XYMx
2	$1/2^o$	Milled	X-Y	Through Rc57	J6L2XYMx
2	1°	Milled	X-Y	Through Rc57	J6L1XYMx
2	$1/4^o$	Milled	R-Theta	Ion Nitrided Rc55	J6L4RTNx

Sommaire

De précédentes études (1) (2) ont montré une relation statistique significative entre la méthode de fabrication de cames plates (a deux dimensions) et le comportement dynamique de la conduite du palpeur de came. Un appareil de test dynamique de came (CDTF) a été concu et fabriqué afin de permettre l'opération de cames produites par différentes techniques dans des conditions controlés de vitesse et de charge. Des mésures dynamiques de l'accélération et de la force du palpeur dans les directions tangentielles et radiales ont été prises. Le glissement du rouleau a également été mésuré avec précision. Avant de tester les cames, le finissage de la surface était évalué avec l'appareil de mésure de rugosité de surface, HOMMEL T-20S. Les caractéristiques de la surface étaient mesurées par la suite, après des periodes d'essais dans des conditions determinées de force, de vitesse et d'accélération. Les données d'accélération et de force ont été transformées dans le domaine des fréquences. Tous les essais ont été rendus aléatiores et toutes les données ont été soumis à une analyse de la variance pour mieux chercher des corrélations entre le spectre de puissance efficace de l'accélération et les paramètres de finissage de la surface de la came en question. Pour cette étude, les cames ont été fabriquées par des techniques de controle numérique continu utilisant des incréments d'échantillonnage de $1/4^{o}$, $1/2^{o}$, et 1°. Les résultats montrent des différences de comportment dynamique significatives entre les cames fraisées et les cames affûtées, mais aucune différence due aux incréments d'échantillonnage. Une méthode alternative de traitement thermique (Nitruration Ionique), comparée aux méthodes conventionelles de durcissement profond de cames fraissées, montre une amélioration significative dans le comportment dynamique.

Fig. 1. Circular Cam Test Disk.

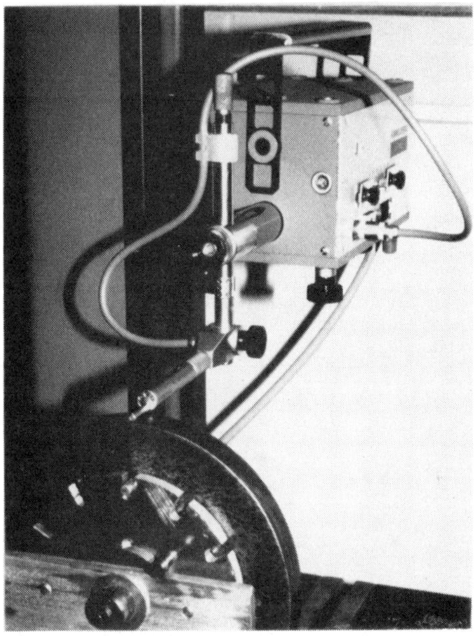

Fig. 3. Cam Holding Fixture and Hommel Surface Apparatus.

Fig. 2. Cam Dynamic Test Fixture (CDTF).

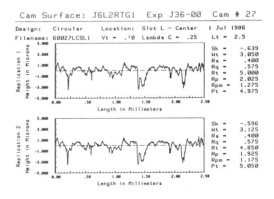

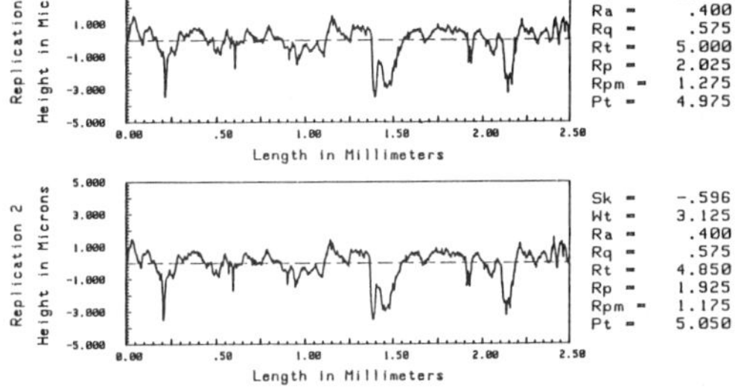

Fig. 4. Surface Profile of Cam L2-RT-G1.

TABLE 2 Surface Roughness and Waviness of (14) Test Cams

Cam Type	Average Roughness RA (microns)	Separation Class	Waviness Wt (microns)	Separation Class
1° Ground	0.189	a	1.537	a
1/4° Ground	0.347	b	3.750	b
1/2° Ground	0.380	b	3.468	b
1/2° Milled	0.700	c	5.610	c
1/4° Milled	0.705	c	5.785	c
1° Milled	0.801	d	5.810	c
1/4° Nitrided	0.803	d	5.288	c

TABLE 3 Mean Dynamic Follower RMS Acceleration (x,y) over 30 to 770 Hertz

Cam Type	Mean RMS X accel(g's)	Separation Class	Mean RMS y accel(g's)	Separation Class
1/2° Ground	0.002	a	0.012	a
1/4° Ground	0.004	a	0.016	ab
1° Ground	0.004	a	0.015	ab
1/4° Nitrided	0.008	b	0.022	b
1/2° Milled	0.017	c	0.075	c
1° Milled	0.017	c	0.071	c
1/4° Milled	0.018	c	0.076	c

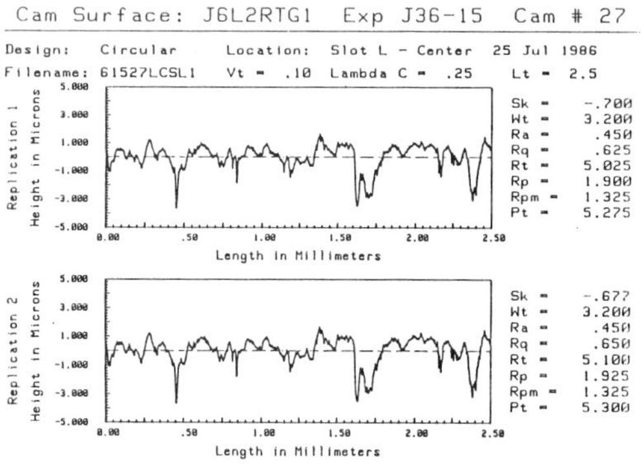

Fig. 5. Surface Profile of Cam L2-RT-G1.

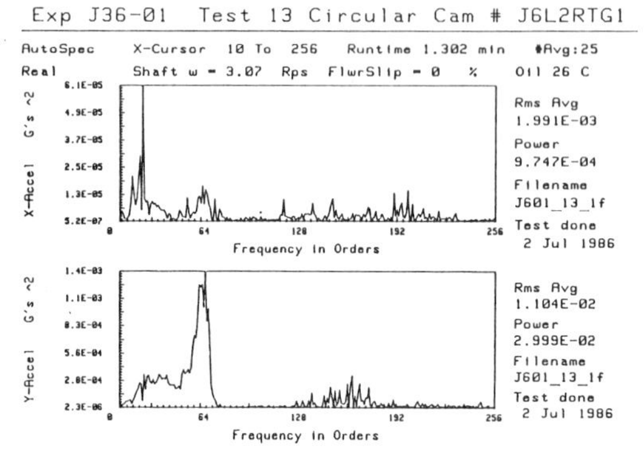

Fig. 8. Acceleration Spectrum of Cam L2-RT-G1.

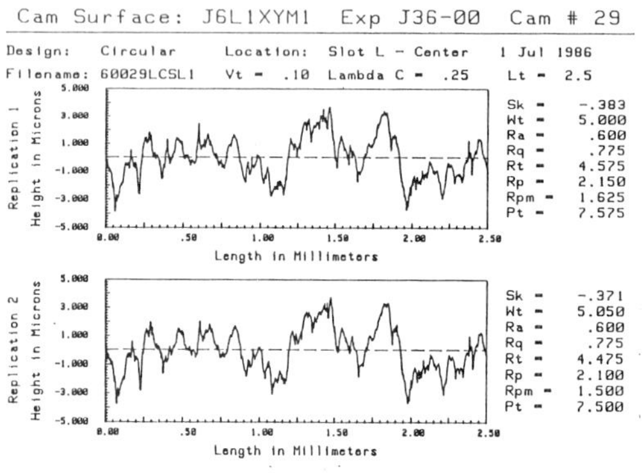

Fig. 6. Surface Profile of Cam L1-RT-M1.

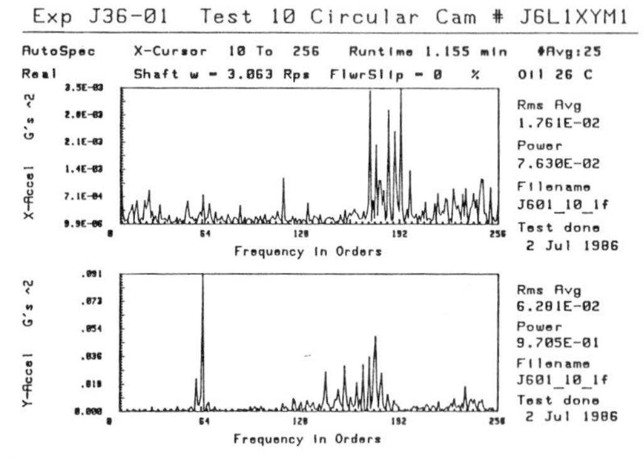

Fig. 9. Acceleration Spectrum of Cam L1-RT-M1.

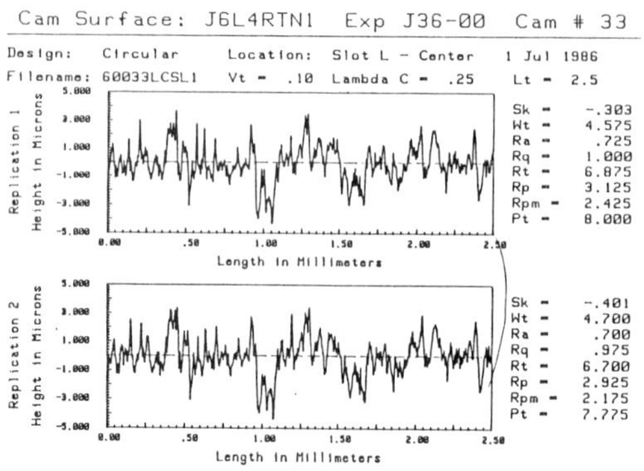

Fig. 7. Surface Profile of Cam L4-RT-N1.

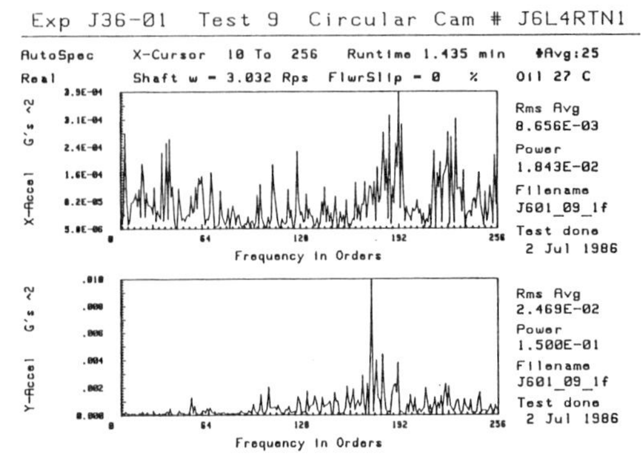

Fig. 10. Acceleration Spectrum of Cam L4-RT-N1.

Analytical Bases for Improvement of Cam Mechanisms Synthesis

JU. V. VOROBYOV* AND V. A. DUBROVSKY**

*The Institute of Chemical Engineering, Leningradskaya Street, I, Tambov, 392620, USSR

**Institute for the Study of Machines USSR Academy of Sciences, Griboedov Street, 4, Moscow 101000, USSR

Аннотация. Разработаны аналитический аппарат синтеза и критерии долговечности кулачковых механизмов, позволяющие существенно расширить область поиска оптимального решения. Аппарат синтеза содержит новые более совершенные способы образования сопряженных профилей кулачка и толкателя, позволяющие считать синтез обобщенным; предлагаемые критерии раскрывают широкие возможности учета процессов взаимодействия поверхностей контакта, связанных с их разрушением и приспособляемостью. Обобщенный синтез объединяет математические выражения, связывающие кривизны центроид и сопряженных профилей с параметрами главного критерия, в качестве которого принят критерий долговечности кулачковой пары. При разработке методов обобщенного синтеза и критериев долговечности кулачковых механизмов использованы теория множеств, методы кинематической геометрии, контактная задача теории упругости, метод создания расчетных моделей на основе триботехнических инвариантов. Рассмотрена возможность определения оптимального закона движения толкателя, исходя непосредственно из заданной совокупности критериев. Практическое значение полученных результатов заключается в увеличении ресурса, повышении качественных показателей и расширении применения кулачковых механизмов.

АНАЛИТИЧЕСКИЙ АППАРАТ ОБОБЩЕННОГО СИНТЕЗА КУЛАЧКОВЫХ МЕХАНИЗМОВ

На современном этапе развития методов синтеза кулачковых механизмов совершенствование этих методов возможно в трех направлениях: развитие аналитического аппарата синтеза, позволяющего получать оптимальные сопряженные профили; определение закона движения толкателя, исходя непосредственно из заданной совокупности критериев; дальнейшее уточнение и разработка критериев синтеза, имеющих физический смысл. Первое направление преимущественно относится к кулачковым механизмам, имеющим монолитный, то есть выполненный как одно звено, толкатель. Второе и третье - распространяются на все типы кулачковых механизмов. В число критериев, имеющих физический смысл, прежде всего входят критерии контактной долговечности кулачковой пары, позволяющие оценивать износ, усталостное выкрашивание, заедание рабочих поверхностей, а также критерии, отражающие динамические явления, среди которых особо важное место занимает динамика контакта кулачка и толкателя.

Таким образом, выявляется тенденция к усложнению методов синтеза кулачковых механизмов, связанному с необходимостью учета большого числа факторов, и к использованию критериев, имеющих конкретное физическое содержание. Разрешимость задачи синтеза в этом случае требует новых аналитических оснований, в том числе аналитических условий связи между параметрами синтеза и параметрами критериев. Отсюда возникает необходимость наиболее общей постановки задачи синтеза кулачковых механизмов и разработки математического аппарата, которые позволили бы получать все возможные решения относительно сопряженных профилей кулачка и толкателя в пределах принятых ограничений и осуществлять выбор нужного решения в соответствии с принятыми критериями. Синтез, удовлетворяющий этим требованиям будет обобщенным.

Теория обобщенного синтеза кулачковых механизмов

Требование множественности решений в пределах принятых ограничений может быть наиболее полно удовлетворено в случае, если профиль одного из звеньев представить таким общим уравнением кривой, из которого можно получить любую непрерывную кривую. Это уравнение можно интерпретировать как уравнение произвольной кривой. Записанное в полярной форме, оно имеет вид

$$r = r_0 \exp\left(\int_{\theta_0}^{\theta} \operatorname{ctg} \gamma(\theta) d\theta\right). \qquad (I)$$

Как следует из Фиг. I, положение звена 2 фиксируется радиусом-вектором r_0 и углом α_0, заключенным между r_0 и межосевой линией O_1O_2.

Функция $\gamma(\theta)$ должна обеспечивать выполнение принятых критериев. Таким образом, дальнейшее решение связано с определением функции $\gamma(\theta)$. Эту функцию можно получить из зависимостей $\gamma(PK,\beta)$ и $\theta(PK,\beta)$, где PK и β являются координатами контактной точки в полярной форме. Исходя из схемы, представленной на Фиг. I, находятся выражения $\gamma(PK,\beta)$ и $\theta(PK,\beta)$. После этого задача локально сводится к выбору параметров PK и β, удовлетворяющих принятым критериям. Решение её в общем виде возможно посредством использования области определения функции $\gamma(\theta)$ в виде неравенства $max\,\gamma(\theta) \geq \gamma(\theta) \geq min\,\gamma(\theta)$ с последующим построением искомой функции $\gamma(\theta)$. Решение целесообразно осуществлять в два этапа. Первый этап состоит в поиске области определения функции $\gamma(\theta)$, удовлетворяющей главному критерию. Второй - в сужении этой области по каждому из остальных критериев и выборе в полученной области искомой функции $\gamma(\theta)$.

В качестве главного критерия принят критерий долговечности. Этот критерий зависит от силового и кинематического взаимодействия контактирующих поверхностей, которое, в свою очередь, определяется локальными свойствами кривых, образующих сопряженные профили. Отсюда критерий долговечности имеет вид

$$Kr \Rightarrow J_0(P_n, \rho_1, \rho_2\, v_{ck}, \{a|P_a\}) \leq k_0,$$

где P_n -нормальное усилие; ρ_1 и ρ_2-радиусы кривизны сопряженных профилей; v_{ck} -скорость скольжения поверхностей; $\{a|P_a\}$ - совокупность параметров, оказывающих влияние на значения критерия. В свою очередь $P_n = P_n(\Sigma P_c, \beta)$, $v_{ck}(PK, \varphi', \psi')$, где ΣP_c -суммарная сила, оказывающая сопротивление движению толкателя, приведенная к точке контакта; φ' и ψ' -производные угловых перемещений кулачка и толкателя. Тогда имеем

$$Kr \Rightarrow J_1(PK, \beta, \rho_1, \rho_2, \{b|P_b\}, \qquad (2)$$

при этом $\{a|P_a\} \subset \{b|P_b$.
Переменные β, ρ_1 и ρ_2 определяют локальные свойства кривых, образующих сопряженные профили. Управляя этими параметрами и величиной отрезка PK можно удовлетворить любому выражению (2), заданному в виде равенства или неравенства.

Необходимый переход от выражения (2) к зависимостям $\gamma(PK,\beta)$ и $\theta(PK,\beta)$ должен осуществляться на основании решения системы уравнений, позволяющей исключить ρ_1 и ρ_2. Следовательно эта система должна содержать уравнение, устанавливающее взаимосвязь координат точки с радиусами кривизны сопряженных профилей, то есть связывать кривизны центроид с кривизнами сопряженных профилей. Подобная взаимосвязь указанных параметров имеет место в уравнении Эйлера - Савари, однако привести это уравнение к нужному виду довольно сложно. Поэтому использован новый подход, позволивший получить дифференциальное уравнение второго порядка относительно первой передаточной функции, которое зависит только от $PK, \beta, \rho_1, \rho_2$.

В итоге для механизма с качающимся толкателем имеем систему из двух уравнений и одного неравенства

$$Kr = J_1(PK, \beta, \rho_1, \rho_2, \{b|P_b\}) \leq k_0,$$

$$\frac{d^2\psi}{d\varphi^2} - \frac{d\psi}{d\varphi}\left(1-\frac{d\psi}{d\varphi}\right)tg\beta + \left(1+\frac{d\psi}{d\varphi}\right)\left(1-\frac{d\psi}{d\varphi}\right)^2 \times \frac{\rho_1\rho_2 - (PK^2 - PK(\rho_1-\rho_2))}{O_1O_2(\rho_1+\rho_2)\sin\beta} = 0,$$

$$\rho_1 = 1/|r''(\ell)|, \quad \ell = \ell(PK,\beta). \qquad (3)$$

Третье уравнение системы (3) служит для вычисления радиуса кривизны профиля ведущего звена. При численном выражении координат профиля радиус кривизны в точке контакта K_i может быть найден по координатам (y_{K_i}, x_{K_i}), $(y_{K_{i-1}}, x_{K_{i-1}})$ и положению нормали в точке K_i, заданному углом β_i. Второе уравнение системы (3) содержит указанную взаимосвязь параметров $PK, \beta, \rho_1, \rho_2$. Для поступательно движущегося толкателя аналогичное уравнение имеет вид

$$\frac{d^2S}{d\varphi^2} + \frac{dS}{d\varphi}\, ctg\beta - \frac{\rho_1\rho_2 - PK(\rho_1-\rho_2) - PK^2}{(\rho_1+\rho_2)\sin\beta} = 0.$$

Таким образом главный критерий выразится зависимостью

$$Kr \Rightarrow J(PK, \beta, \{b|P_b\}) \leq k_0,$$

на основе которой ищется область определения функции $\gamma(\theta)$. Решение заключается в выполнении операций над множествами:

1) $\left\{P|H\left(x_{P_i} = \frac{O_1O_2\frac{d\psi}{d\varphi}}{1+\frac{d\psi}{d\varphi}}\right)\right\}$, $P_i \in P$,

$$\sup H(x_P) = x_P|\left(\frac{d\psi}{d\varphi}\right)_{max}, \quad \inf H(x_P) = x_P|\left(\frac{d\psi}{d\varphi}\right)_{min}.$$

2) $M_{P_i} = \cup m_{ij}$, $m_{ij} = \{(x,y)>0 | H(x - y\,tg\beta_i - x_{P_i} = 0)\}$,

$$\sup H(\beta)_j \Rightarrow \beta_i < \pi, \quad \inf H(\beta_j) \Rightarrow \beta_j > 0.$$

3) $\mathcal{K}_{m_{ij}} = \{K_{ij} | H(K_{ij} \in m_{ij} \wedge K_{ij} | (Kr \leq k_0))\}$,

$$\sup \mathcal{K}_{m_{ij}} < \frac{O_1O_2}{1+\frac{d\psi}{d\varphi}}, \quad \inf \mathcal{K}_{m_{ij}} > 0.$$

В итоге имеем $\mathcal{K}_{P_i} = \cup \mathcal{K}_{m_{ij}}$ и точечное множество $\mathcal{K} = \cup \mathcal{K}_{P_i}$, удовлетворяющее условию $Kr \leq k_0$. Для каждого подмножества $\mathcal{K}_{P_i}$ находим значения $max\,\gamma(PK,\beta)$, $min\,\gamma(PK,\beta)$ и соответствующие им значения $\theta(PK,\beta)$.

При выборе функции $\gamma(\theta)$ следует учитывать дополнительные критерии. Эти критерии могут носить геометрический, кинематический, физический и прочий характер. Критерии, носящие геометрический характер, позволяют учесть конструктивные особенности механизма, например, условие выпуклости профиля толкателя $\gamma''(\theta) \geq 0$, а остальные критерии - обеспечить требуемые качественные показатели.

Расчет координат профиля кулачка при обобщенном синтезе

Профиль кулачка может быть найден по методу огибающей семейства профилей толкателя или методу профильных нормалей. Принципиальное решение задачи с уравнением произвольной кривой вида (I) заключается в следующем. Имеем две прямоугольные системы координат: подвижную $\Sigma = \{0, x, y\}$ и неподвижную $\Sigma_1 = \{O_1, x_1, y_1\}$ (ввиду простоты построений схему не приводим). Пусть точка O_1

является центром поворота кулачка и точка O_2, лежащая на оси, центр поворота толкателя. В начальный момент одноименные оси параллельны и одинаково направлены. Пусть точка O имеет в неподвижной системе координаты x_0, y_0. Обозначим текущий угол поворота кулачка через φ. Профиль кулачка является огибающей семейства профилей толкателя. Проверку существования огибающей при машинном способе вычисления профиля кулачка по методу обращенного движения удобно выполнять по условию: если $\varphi_i \neq \varphi_{i+1}$, то $(x_i y_i) \neq (x_{i+1} y_{i+1})$. Приняв во внимание (I), систему уравнений, выражающих профиль кулачка, получим в следующем виде

$$\sqrt{x^2+y^2} - r_0\, e^{\int_{\theta_0}^{\theta} ctg\gamma(\theta)d\theta} = 0,$$

$$x\frac{\partial x}{\partial \varphi} + y\frac{\partial y}{\partial \varphi} - \left(x\frac{\partial y}{\partial \varphi} - y\frac{\partial x}{\partial \varphi}\right)ctg\gamma(\theta) = 0. \qquad (4)$$

Значения x и y, входящие в (4), выражаются через x_1 и y_1. Второе уравнение системы (4) сводится к уравнению второго порядка относительно x_1 и y_1. Записанное по степеням y_1, оно имеет вид

$$a(x_1)y_1^2 + b(x_1)y_1 + c(x_1) = 0. \qquad (5)$$

По координатам $x_1(\varphi), y_1(\varphi)$ строится профиль кулачка и находятся параметры критерия.

Возможные упрощения решения задачи обобщенного синтеза

В некоторых случаях достаточным для практического использования может быть метод, позволяющий управлять параметрами синтеза в процессе самого синтеза. Теоретические основы такого метода следующие. Метод предполагает наращивание искомой кривой по кускам, принадлежащим известным кривым. Выбирая участки с нужными локальными свойствами и стыкуя их получаем кусочно-аналитическую функцию. Условием, которое при этом должно выполняться, является отсутствие разрыва второй производной в местах наращивания кривой. Таким образом, предлагаемый метод синтеза основан на использовании области определения функции $\gamma(\theta)$, удовлетворяющей требованию отсутствия разрыва второй производной. Для нахождения этой области используем Фиг. 2. Пусть кривая I наращивается кривой II в точке K, образуя часть профиля одного из звеньев. Потребуем, чтобы центры кривизны стыкуемых в точке K кривых совпадали и располагались в точке A, лежащей на нормали в точке K кривой I. Отрезок AK будет равен радиусу кривизны ρ этой кривой. Предполагаем функцию $\gamma(\theta)$ кусочно-аналитической в её области определения. В соответствии с уравнением (I) искомая область определения функции $\gamma(\theta)$ будет выражена множеством положений полюса кривой (I). Если полюс кривой, заданной уравнением (I), располагать на окружности, построенной на радиусе кривизны в точке стыковки кривых как на диаметре, то $\gamma(\theta)=const$, а уравнение (I) приводится к виду $r=r_0\exp(\theta\cdot ctg\gamma)$ и является уравнением логарифмической спирали (Фиг. 2, кривая III). Отсюда следует, что оперируя уравнением логарифмической спирали и перемещая её центр по окружности с диаметром ρ, можно наиболее просто наращивать кривую, у которой кривизна меняется монотонно. В частном случае, располагая центр логарифмической спирали на нормали в начальной точке касания, получаем класс окружностей различного радиуса.

Определение закона движения ведомого звена на основе принятых критериев

Представленный аналитический аппарат синтеза позволяет осуществить выбор закона движения толкателя, исходя непосредственно из совокупности принятых критериев. Решение иллюстрирует Фиг. 3. Обозначения: R -минимальный радиус-вектор кулачка; K_0-точка касания сопряженных профилей в начальном положении толкателя; β_0 -угол между линией межосевого расстояния и нормалью в точке K_0. Пусть профиль ведомого звена задан аналитически одним из способов. Зафиксируем на профиле какую-либо точку K и найдем условия, при которых она будет контактной. Для этого запишем уравнение нормали и найдем точку пересечения O нормали, проведенной в точке K, с прямой O_1K_0 и координаты этой точки. Определяем величину отрезка OK_0. Находим угол ξ между прямой O_1O и нормалью в точке K. Задаём произвольное положение полюса зацепления P на межосевой линии. Поворотом отрезка O_2P в направлении противоположном движению толкателя сносим точку P на нормаль, проведенную в точке K. Получим точку P_ξ. Определяем координаты (x,y) этой точки. Угловое перемещение ψ толкателя, при котором точка K профиля этого звена будет контактной, находим из равенства

$$\psi = arctg(|y|/(O_1O_2-|x|)).$$

Координаты точки K в том положении толкателя, при котором точка K будет контактной (в неподвижной системе координат) обозначим $x_{K\psi}, y_{K\psi}$.

Для определения угла φ поворота кулачка следует выразить координаты точки K в обращенном движении кулачка относительно толкателя (в подвижной системе координат) $x_{K\varphi}, y_{K\varphi}$.

Координаты $x_{K\psi}, y_{K\psi}$ совпадают с координатами x, y в (4), а $x_{K\varphi}, y_{K\varphi}$ с x_1, y_1 в (5). Решив уравнение (5), получим зависимость $y_1 = y_1(x_1, \varphi)$, из которой находим φ. Имея локальное значение $\psi(\varphi)$, значение $d\psi/d\varphi = O_1P/PO_2$ и координаты контактной точки $x_{K\varphi}, y_{K\varphi}$, вычисляем значения принятых критериев. Варьируя положением полюса P, добиваемся для каждой фиксированной точки профиля ведомого звена выполнения критериев. По локальным значениям $\psi_i(\varphi_i)$ восстанавливаем закон движения $\psi=\psi(\varphi)$, должным образом удовлетворяющий принятым критериям.

КРИТЕРИИ ДОЛГОВЕЧНОСТИ КУЛАЧКОВОЙ ПАРЫ

Если задана долговечность кулачковой пары, то главный критерий синтеза должен быть представлен уравнением износа. Средняя интенсивность механического износа J_M за один цикл нагружения определяется как отношение линейного износа к пути трения скольжения за время одного контакта. Так как J_M величина безразмерная, то для её выражения были использованы триботехнические инвариантные комплексы (Дроздов, 1974). В итоге уравнение интенсивности изнашивания имеет вид

$$J_M = K\left(\frac{P_{t\,max}}{HB}\right)^m \left(\frac{R_a}{h}\right)^n.$$

Обозначим первый комплекс через $\Phi_н$ и второй через $\Phi_{см}$. Комплекс $\Phi_н$ учитывает влияние на износ наибольших контактных касательных напряжений $P_{t\,max}$ и твердости материала HB. Величина контактных касательных напряжений, действующих вдоль линии контакта, получена на основе решения контактной задачи (Воробьёв, 1984). В результате имеем

$$\Phi_н \Rightarrow \frac{P_{n\,max}}{HB}\left(\frac{f^3}{2f - P_t/P_n}\right)^{1/2},$$

где $P_{n\,max}$ наибольшее контактное нормальное напряжение; f -коэффициент трения; P_t и P_n соответственно касательная и нормальная силы, действующие в контакте.

Комплекс $\Phi_{см}$ характеризует состояние смазочного слоя

$$\Phi_{см} \Rightarrow \frac{(Ra_1^2 + Ra_2^2)^{1/2}}{h_{гр} + 0{,}63\rho_n\left(\frac{\mu v_к}{P_{пн}}\right)^{0{,}7}\left(\frac{P_{пн}\beta}{\rho_n}\right)^{0{,}6}\left(\frac{\lambda}{\alpha\mu v_{ск}^2}\right)^{0{,}325} Pe_{1,2}},$$

где $P_{пн}$ -погонная нагрузка; μ -динамический коэффициент вязкости при температуре контакта; β -пьезокоэффициент смазки; λ -коэффициент теплопроводности смазки; α -коэффициент, характеризующий изменение вязкости смазки при прохождении через контакт; $Pe_{1,2}$ -среднее число Пекле в контакте; Ra -среднее арифметическое отклонение микронеровностей; $v_к = v_1 + v_2$ -суммарная скорость качения; $v_{ск}$ -скорость скольжения поверхностей; -приведенный радиус кривизны; = 0,000I мм.

Критерий заедания выражается неравенством

$$Kr \Rightarrow \Phi_{см} < 1.$$

При выполнении этого неравенства заедание отсутствует.

ЛИТЕРАТУРА

Дроздов Ю.Н. Обобщенные характеристики в анализе трения и смазки тяжелонагруженных тел. - Машиноведение, 1974, №6, с. 70-74.

Воробьёв Ю.В. Аналитические основания для оценки долговечности рабочих поверхностей при качении с проскальзыванием. - Машиноведение, 1984, №4, с. 68-76.

SUMMARY

The development of synthesis methods of cam mechanisms is connected with the necessity of taking into account a large number of factors and with using criteria which have a concrete physical content. In this connection the synthesis and criteria of contact durability of cam mechanisms are suggested. By constructive solving they allow to raise the load capacity and durability of these mechanisms by meeting demands of minimal dimensions and decreasing metal and energy capacity. In contrast to these existing methods of cam mechanisms synthesis which suppose that working profile of follower is given, suggested methods do not contain a priori information about mating profile geometry. Such synthesis may be called generalized. Its realization is based on operating local elements of mating profiles in accordance with adopted criteria. The dependences obtained which connect local elements of profiles with transfer functions are aimed at this. For cases when the motion law is not given the method or restablishment of motion law is suggested in which ambiguity of connection of contact point with transfer function is used. It follows that such a position of gear pole may be found locally and this position best satisfies given criteria. Wear and seizing criteria given as tribotechnical invariant complexes are used as the main criterion for cam mechanisms synthesis. As a result two interconnected problems are solved and used practically - the problems of generalized synthesis and securing given contact durability of cam mechanisms.

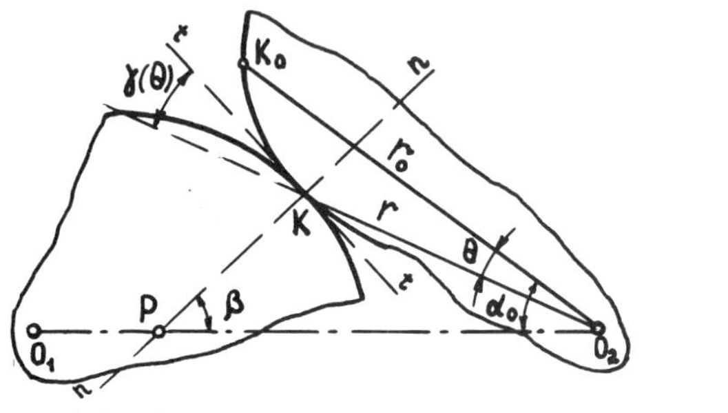

Фиг. I. Схема, характеризующая параметры произвольной кривой.

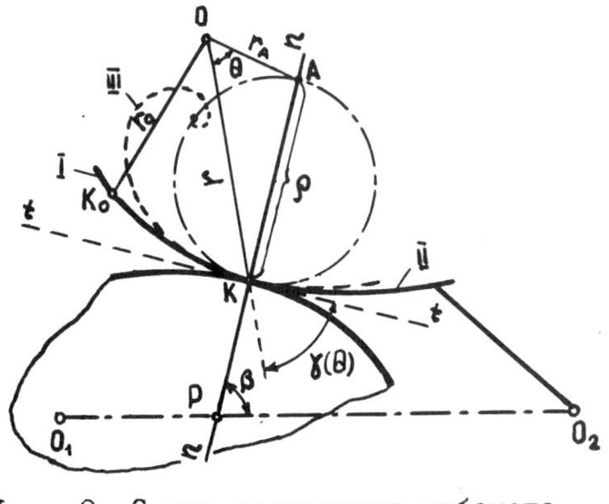

Фиг. 2. Схема построения рабочего профиля толкателя посредством его наращивания.

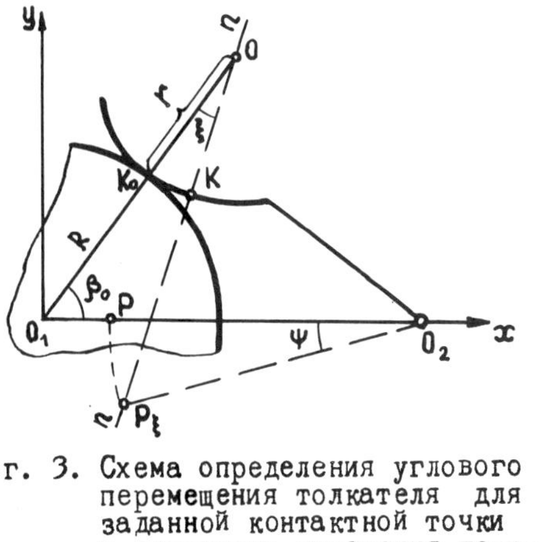

Фиг. 3. Схема определения углового перемещения толкателя для заданной контактной точки и заданного положения полюса зацепления.

Teaching Methods

Wandel in der Ausbildung zum Maschinenbauingenieur durch den zunehmenden Einsatz von CAD/CAM-Systemen

J. KNOOP

Institut für Konstructions- und Fertigungstechnik, Universität der Bundeswehr Hamburg, 2000 Hamburg 70, Postfach 70 08 22, DDR

Abstract. Die zunehmende Einbeziehung der elektronischen Datenverarbeitung sowie der CAD/CAM-Technologie hat die Anforderungen an die Mitarbeiter in Konstruktion und Fertigung erheblich verändert. Dies erfordert entsprechende Anstrengungen hinsichtlich der Ausbildung der Maschinenbauingenieure in der Zukunft. Im Aufsatz wird beschrieben, wie die klassische Ausbildung im Studium des Maschinenbaus verändert werden muß, um den neuen Gegebenheiten Rechnung zu tragen. Dabei werden insbesondere Beispiele aus der Getriebetechnik herangezogen, um die Auswirkungen der neuen Technologie in einem klassischen Fach des Maschinenbaus zu zeigen.

Keywords. CAD/CAM-Ausbildung, Maschinenbau-Ausbildung, Kinematik, Koppelgetriebe, Bewegungssimulation.

EINLEITUNG

Der Arbeitsplatz des Mitarbeiters in der Konstruktions- und Entwicklungsabteilung eines Maschinenbauunternehmens ist in den vergangenen Jahren durch die zunehmende Einbeziehung von EDV-Anlagen sowie die CAD/CAM-Technologie einem starken Wandel unterworfen. Damit haben sich auch die Anforderungen an die dort Tätigen erheblich verändert. Dies setzt hinsichtlich der Aus- und Weiterbildung der Mitarbeiter und der zukünftigen Konstrukteure erhebliche Anstrengungen der Bildungsinstitutionen im Schul- und Hochschulbereich sowie auch in der betrieblichen Weiterbildung voraus.

Bei steigendem Bedarf an Arbeitskräften mit qualifizierter Ausbildung im allgemeinen Maschinenbau und gleichzeitiger vertiefter EDV-Kenntnis zeigt sich, daß der Arbeitsmarkt diese Nachfrage in keiner Weise befriedigen kann. So sind viele Unternehmen dazu gezwungen, geeignet erscheinende Mitarbeiter mit der Rechneranwendung durch Schulungsmaßnahmen innerhalb des Unternehmens oder über Dienstleistungsunternehmen vertraut zu machen. Auch die Zahl der Absolventen von Fach-, Fachhoch- und Hochschulen mit entsprechender Ausbildung ist zu gering. Dies ist nicht weiter verwunderlich, fehlen doch in den meisten Studiengängen und Wahlfächerkatalogen EDV-spezifische Fächer wie z.B. die graphische Datenverarbeitung, strukturiertes Programmieren, Softwareengineering, CAD usw. völlig.

Im folgenden soll nun anhand der bisherigen Ausbildung des Maschinenbauingenieurs aufgezeigt werden, in welche Richtung - aus der Sicht des Konstruktions- und Fertigungsingenieurs - dieses Studium weiter entwickelt werden muß.

DER STUDIENGANG MASCHINENBAU

Es soll an dieser Stelle nicht der Streit vertieft werden, ob die Informatiker mehr Maschinenbau oder die Maschinenbauer mehr Informatik lernen müssen, um den Ingenieur in Zukunft optimal und nachfragegerecht auszubilden. Der Maschinenbauingenieur wird auch weiterhin die Schwerpunkte auf die sogenannten Kernfächer des klassischen Maschinenbaus setzen müssen, doch sollte es ebenso unbestritten sein, daß innerhalb dieser Fächer das Werkzeug EDV einen immer größeren Stellenwert einnimmt. Das bedeutet, daß EDV-nahe Fächer in den Fächerkanon des Studiums des allgemeinen Maschinenbaus mit aufgenommen werden müssen.

Denkt man speziell im Sinne des neuen Schlagwortes CIM (Computer Integrated Manufacturing) weiter voraus, so werden Qualifikationsprobleme durch den notwendigerweise steigenden Einsatz der EDV in allen Abteilungen eines Unternehmens schnell zunehmen.

In Tafel 1 sind die wesentlichen Disziplinen des Grundstudiums im Maschinenbau aufgeführt, wobei gleichzeitig der Einfluß der CAE-Anwendungen auf diese Fächer dargestellt wird. Insbesondere die Technische Mechanik und die Maschinenelemente sind davon betroffen, mittelbar allerdings auch einige andere Fächer wie die Mathematik, Thermodynamik usw. Es stellt sich also die Frage, wie die Lehrinhalte der betroffenen Gebiete an die neuen, durch rechnergestützte Methoden geänderten Anforderungen angepaßt werden können.

TAFEL 1 Einfluß der CAE-Anwendungen auf die Fächer des Grundstudiums (Maschinenbau)

Fach \ Rechneranwendung	CAD	CAM	CAT	FEM
Mathematik	◩			◩
Mechanik und Festigkeitslehre				■
Maschinenzeichnen und Darstellende Geometrie	◩			
Maschinenelemente	■	◩		◩
Thermodynamik				◩
Werkstofftechnik				
Chemie				
Physik				
Grundlagen der Elektrotechnik				
Grundlagen der Fertigungstechnik		◩	◩	

■ Anwendung unbedingt erforderlich ◩ Anwendung wünschenswert

Während im Fach Maschinenzeichnen und Darstellende Goemetrie die wichtigsten Grundlagen der Geometrie sowie die Kenntnisse im technischen Zeichnen zu vermitteln sind, was nicht notwendigerweise die Verwendung und den Einsatz einer rechnergestützten Zeichenanlage voraussetzt, sind die Auswirkungen auf die Konstruktionsübungen im Fach Maschinenelemente wesentlich tiefgreifender. So stellt sich die Frage, ob die Durchführung derartiger Übungen am Zeichenbrett nach wie vor noch sinnvoll ist und ob mit Hilfe der CAD-Anlage hier wesentliche Fortschritte erzielt werden können. Da sich die ganze Vorgehensweise beim Konstruieren am CAD-System, insbesondere bei Verwendung eines 3D-Volumenmodels, wesentlich ändert, ist dies eine ganz elementare Frage.

Dabei stellen sich erhebliche Probleme organisatorischer und infrastruktureller Natur, da der Student zum infragekommenden Zeitpunkt während seines Studiums in der Regel über keinerlei EDV-Ausbildung und Erfahrung verfügt und diese Übungen von allen Studenten eines Jahrgangs durchgeführt werden müssen, was eine leistungsfähige CAD-Anlage mit einer großen Zahl von Arbeitsplätzen voraussetzt. Um mit einem CAD-System zu arbeiten, ist zusätzlich eine relativ aufwendige Schulung von Nöten, deren Integration in die Vorlesung/Übung Maschinenelemente, praktisch unmöglich ist. Es stellt sich daher die Frage, wie man eine solche Ausbildung noch in das Grundstudium integrieren kann.

Die Technische Mechanik und insbesondere die Festigkeitslehre sind im letzten Jahrzehnt durch rechnergestützte Verfahren im besonderen Maße beeinflußt worden, wobei die Berechnungsmethode der Finiten Elemente (FEM) einen sehr großen Einfluß auf dieses Fachgebiet erhalten hat. Insbesondere die systematischen Tätigkeiten werden bei der FEM heute fast ausschließlich am Rechner erledigt (Seidel [1]). Für den Ingenieur sollte die Betonung mehr auf den kreativen Tätigkeiten liegen, wie z.B. der realistischen Auswahl eines Modells und der richtigen Umsetzung der Rechenergebnisse in die Konstruktion. Das bedeutet für die Lehrinhalte, daß Lösungsverfahren wie z.B. Seileckverfahren, Cremonaplan, Berechnung von Flächen- und Massenmomenten, Berechnung der elastischen Linie usw. nicht unbedingt im Vordergrund stehen müssen, daß aber die kreativen Fähigkeiten, wie das Erfassen einer Problemstellung, der Aufbau eines realistischen Modells, das Abschätzen und die richtige Umsetzung von Ergebnissen mehr geschult werden müssen.

Schon diese kurze und unvollständige Darstellung des EDV-Einflusses auf das Grundstudium läßt erkennen, daß eine grundlegende EDV-Ausbildung zum besseren Verständnis von FEM- und CAD-Methoden bereits im Grundstudium unerläßlich ist. Die Realisierung solcher Ideen stößt in der Praxis auf große Schwierigkeiten, da dies ja die Verdrängung anderer Studieninhalte voraussetzt und deshalb nur über eine länger andauernde Phase möglich erscheint.

Es muß durch eine erste EDV-Ausbildung zu Beginn des Studiums sichergestellt werden, daß der Student bereits im Grundstudium in die Lage versetzt wird, kleinere Programme selbst zu schreiben und vorgegebene Programmsysteme anzuwenden.

Leichter erscheint eine Erneuerung des Studienplans im Hauptstudium, ergeben sich doch hier durch Wahl der Vertiefungsrichtungen und durch die Wahlfächer mehr Möglichkeiten, neue Gebiete zu integrieren.

Tafel 2 zeigt einen Vorschlag von Funk und Knoop [2] für die Vertiefungsrichtung Konstruktions- und Produktionstechnik. Zu den Pflichtfächern sollte in jedem Fall (wenn dies nicht schon im Grundstudium erfolgt ist), das Fach "Grundlagen der Elektronischen Datenverarbeitung" gehören. Dieses Fach muß sowohl auf die EDV-Hardware als auch auf die Programmiersprachen, Betriebssysteme, Datenbanksysteme, graphische Datenverarbeitung usw. eingehen.

TAFEL 2 Hauptstudiengang Allgemeiner Maschinenbau

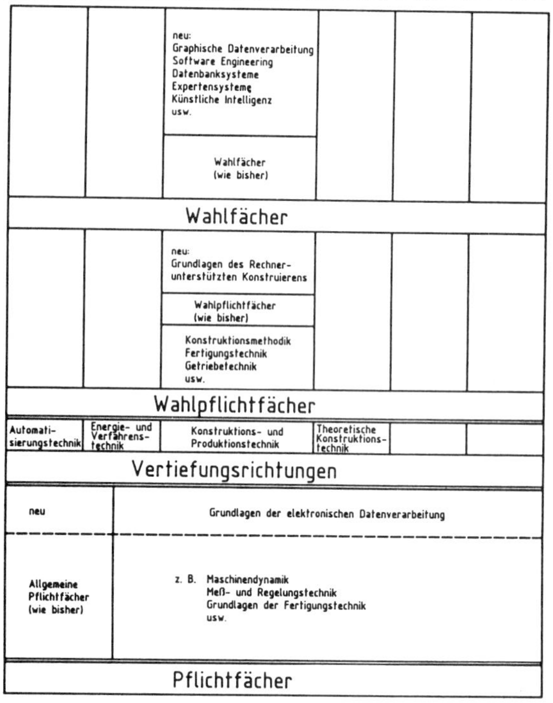
neu:
Graphische Datenverarbeitung
Software Engineering
Datenbanksysteme
Expertensysteme
Künstliche Intelligenz
usw.
Wahlfächer
(wie bisher)
Wahlfächer
neu:
Grundlagen des Rechnerunterstützten Konstruierens
Wahlpflichtfächer
(wie bisher)
Konstruktionsmethodik
Fertigungstechnik
Getriebetechnik
usw.
Wahlpflichtfächer
Automatisierungstechnik
Energie- und Verfahrenstechnik
Konstruktions- und Produktionstechnik
Theoretische Konstruktionstechnik
Vertiefungsrichtungen
neu
Grundlagen der elektronischen Datenverarbeitung
Allgemeine Pflichtfächer (wie bisher)
z. B. Maschinendynamik
Meß- und Regelungstechnik
Grundlagen der Fertigungstechnik
usw.
Pflichtfächer

Die Vertiefungsrichtungen in den allgemeinen Studienplänen unterscheiden sich naturgemäß in Abhängigkeit von gewissen Schwerpunkten, die für die spezielle Hochschule bestimmend sind, doch kann man sagen, daß die Konstruktions- und Produktionstechnik, die Automatisierungstechnik, die Energie- und Verfahrenstechnik, die Werkstofftechnik und die theoretische Konstruktionslehre fast immer angeboten werden. Auch die Strömungstechnik und die Fahrzeugtechnik wären hier noch zu nennen. Grundsätzliche Bedeutung haben in der Konstruktions- und Produktionstechnik die Fächer "Konstruktionsmethodik" und "CAD/CAM-Systeme". Diese Fächer sollten in Zukunft für diese Vertiefungsrichtung als Pflichtfächer angesehen werden.

CAD/CAM-AUSBILDUNG

Wegen der elementaren Bedeutung für die Vertiefungsrichtung Konstruktions- und Fertigungstechnik soll an dieser Stelle kurz auf den Aufbau und Inhalt einer Vorlesung "CAD/CAM-System" eingegangen werden.

- CAE-Anwendungen: Dies sind die klassischen Anwendungen der EDV im Konstruktions- und Fertigungsprozeß wie z.B. die Berechnungsverfahren, Grundlagen der FEM-Rechnung, rechnergestützte Optimierungsverfahren usw.

- CAD-Hardwarekonfigurationen: Rechnerarten, Systemkonfigurationen für CAD-Einsatz, graphische Ein- und Ausgabegeräte.

- CAD-Software: Konzepte von CAD-Systemen (2D-, 3D-Modelle, Variantenprogrammierung usw.).

- Aufbau der CAD-Systeme: Rechnerinterne Darstellung (RID), Eingabesprachen, mathematische Grundlagen der Berechnung der Geometriedaten, numerische Verfahren der Kurven- und Flächendarstellung, Hidden-Line-Algorithmen, Grundlagen der Datenbanksysteme, Schnittstellenproblematik.

- CAM (Computer Aided Manufacturing): Arbeitsplanerstellung, NC-Programmierung, Qualitätssicherung und CAT (Computer Aided Testing).

- Perspektiven: CIM (Computer Integrated Manufacturing), neue Ein- und Ausgabemedien, Standardisierung.

Vertieft wird die Vorlesung durch entsprechende Übungen, die den Studierenden mit der Arbeitstechnik mit Digitalisiertablett, Lichtstift, Menütablett usw. vertraut machen und vor allen Dingen die veränderte Vorgehensweise beim Arbeiten mit CAD-Systemen vermitteln sollen. Ergänzt werden kann eine solche Vorlesung "CAD/CAM-Systeme" durch zusätzliche Wahlfachangebote mit Themenstellungen wie z.B.:

- graphische Datenverarbeitung,
- Software-Engineering,
- strukturiertes Programmieren,
- künstliche Intelligenz,
- Expertensysteme,
- Datenbanksysteme usw.

Dabei werden sich sicherlich mit der Automatisierungstechnik Überschneidungen ergeben, die zu bereinigen bzw. aufeinander abzustimmen sind.

AUSWIRKUNGEN AUF ANDERE FÄCHER

Abschließend ein Beispiel dafür, wie durch den Einsatz der CAD-Systeme altbekannte Vorgehensweisen und Berechnungsverfahren in klassischen Fächern des Maschinenbaus beeinflußt werden können und dadurch eventuell auch wieder in den Vordergrund treten. Hier ist als besonderes Beispiel die Getriebetechnik herangezogen worden (Tolle [3]).

In Bild 1 ist ein Viergelenkgetriebe dargestellt, für dessen Koppelpunkte A,B,C und D die Geschwindigkeiten und Beschleunigungen zu ermitteln sind.

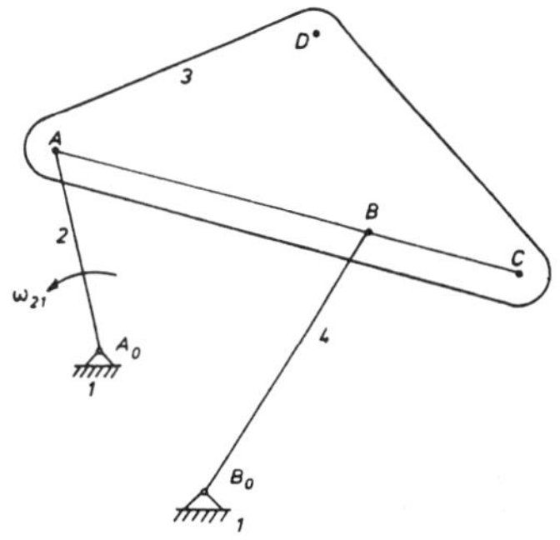

Bild 1. Viergelenkgetriebe

Bisher wurden hierfür vor allem zwei Methoden angewendet, nämlich das analytische Verfahren mit Hilfe der Vektorgleichungen und ein graphisches Verfahren mit Hilfe des Geschwindigkeitsplans. Dabei erhielt man bei der letzteren Methode nur Näherungswerte, die von der Zeichengenauigkeit abhingen. Führt man hingegen eine derartige Konstruktion mit Hilfe der CAD-Anlage durch (Bild 2), so ist für jede mögliche Getriebestellung über den am CAD-System erstellten Geschwindigkeitsplan schnell eine Lösung möglich, die der analytischen äquivalent ist, denn der Rechner vollzieht die graphische Methode durch Parallelen- und Lotbildung exakt nach.

Eine weitere wichtige Anwendung des Werkzeugs "CAD" stellt die einfach durchzuführende Bewegungssimulation eines Koppelgetriebes dar (Bild 3).

Am Beispiel einer Geradführung nach Hain [4] erkennt man zunächst die Prinzipkonstruktion und in Bild 4 ein durchkonstruiertes Hebelgetriebe, mit dem man in der Lage ist, z. B. eine Kollisonsanalyse durchzuführen.

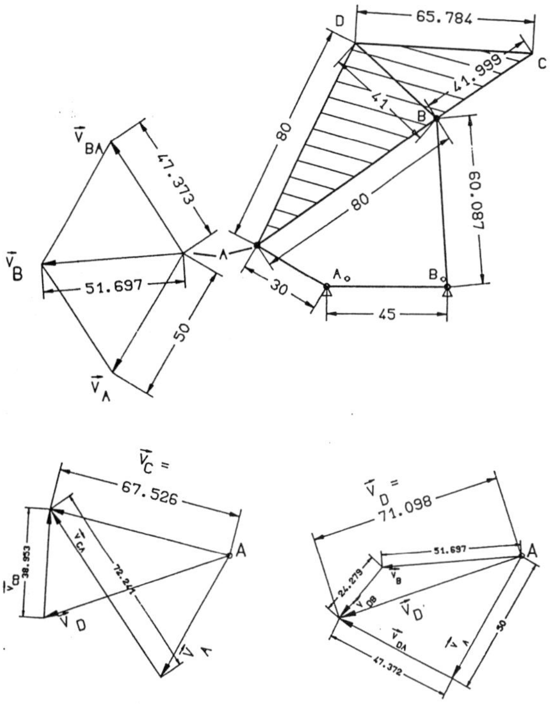

Bild 2. Aufbau des Geschwindigkeitsplanes mit Hilfe eines CAD-Systems

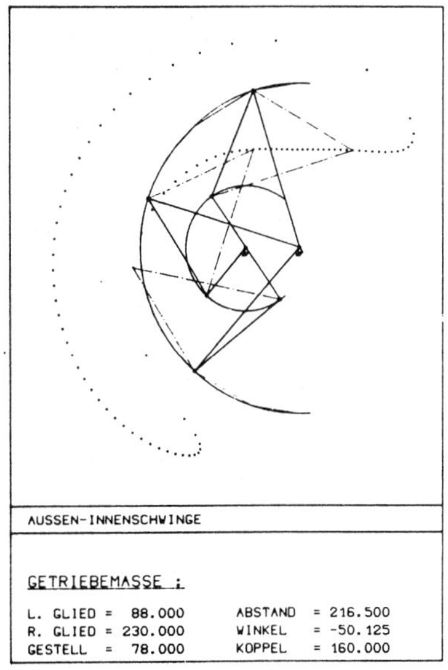

Bild 3. Viergelenkgetriebe als Geradführung

Der Fadengeber eines Fadengebergetriebes einer Nähmaschine (Bild 5) zeigt eine weitere Anwendung (nach Dittrich u. Braune [5]). Die Darstellung der Geschwindigkeitsvektoren ist jederzeit möglich (Bild 6), wie auch die hier nicht gezeigten Diagrammdarstellungen der Geschwindigkeit und Beschleunigung von beliebigen Koppelpunkten.

Die Programme zur Berechnung und Darstellung dieser Bilder sind in das CAD-System integriert und erlauben eine lückenlose interaktive Arbeitsweise am System bei der Konstruktion derartiger Getriebe.

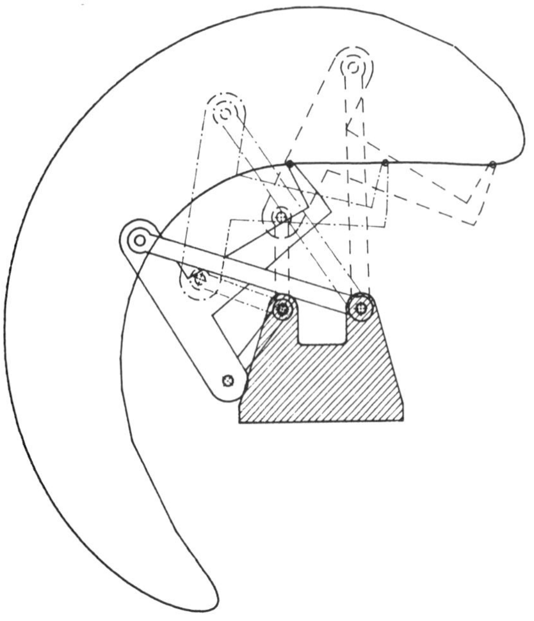

Bild 4. Viergelenkgetriebe als Geradführung [4]

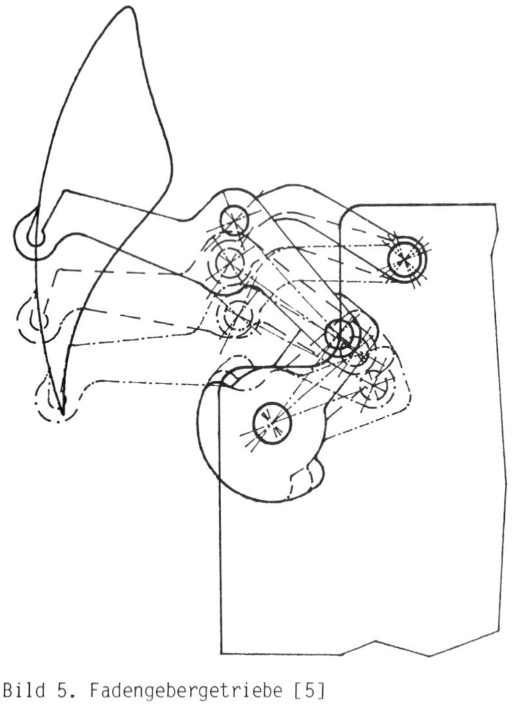

Bild 5. Fadengebergetriebe [5]

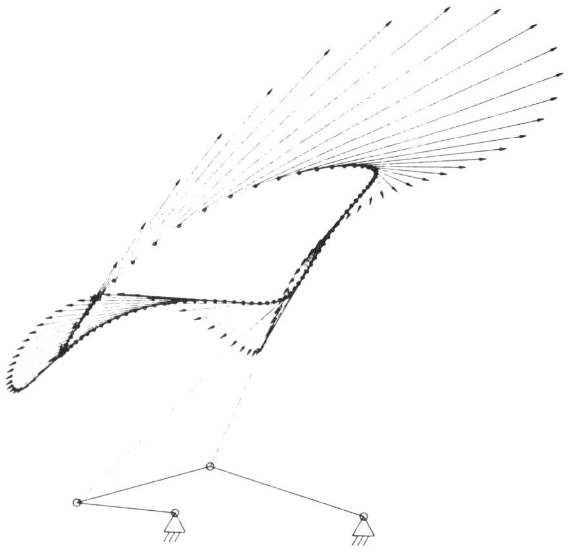

Bild 6. Geschwindigkeitshodographen

ZUSAMMENFASSUNG

Die ständig wachsende Bedeutung der EDV-Anwendung in fast allen Fächern des Maschinenbaus macht ein grundsätzliches Überdenken der Studienpläne und der Lehrinhalte dieser Fächer erforderlich. Insbesondere die unter dem Begriff CAE zusammengefaßten Rechernanwendungen in Entwicklung und Konstruktion ersetzen das seit Ende des letzten Jahrhunderts gängige Werkzeug "Zeichenbrett, Bleistift, Radiergummi, Zeichenpapier und Rechenschieber" durch ein völlig anders geartetes und zu betreibendes Instrumentarium. Das Ergebnis der Konstruktionsarbeit, die technische Zeichnung, wird dabei durch einen Datensatz ersetzt, der, einmal erstellt und in entsprechender Form abgespeichert, als Grundlage für die gesamte rechnergestützte Herstellung des Produktes dient.

Die Auswirkungen dieser in den Unternehmen zunehmend eingesetzten Technologien sind für die zum Teil völlig anderen Anforderungen an den Ingenieur verantwortlich und setzen eine vermehrte und im Studienplan frühzeitiger angesiedelte Ausbildung in EDV-spezifischen Fächern voraus.

Erwähnenswert sind die Möglichkeiten der Anwendung von CAD-Systemen in den sogenannten "klassischen Fächern" des Maschinenbaus. Hier können vor allem graphische Methoden, die bisher aufgrund von Zeichnungsungenauigkeiten als nicht exakt genug galten, neue Bedeutung erhalten. Auch das Erstellen von Modellen kann durch Simulation am Rechner ersetzt werden.

SCHRIFTTUM

Seidel, K. (1985). Maschienbaustudium - Konstruktionspraxis heute: noch kompatibel? Proceedings of ICED 85, Hamburg, Volume 2, 1027-1036

Funk, W. u. J. Knoop. (1986). Der Einfluß der CAD/CAM-Technologie auf die Hochschulausbildung im allgemeinen Maschinenbau. CAE-Journal, 5/86.

Tolle, P. (1985). Pocket Computers, CAD-Systems and Large Scale Data Processors as Design Aids for Variable-Velocity Mechanisms. Proceedings of CAD/CAM, Robotics and Automation International Conference, University of Arizona, Tucson, 629-633.

Hain, K. (1973). Getriebebeispiel-Atlas. VDI-Verlag, Düsseldorf.

Dittrich, G. u. R. Braune. (1978). Getriebetechnik in Beispielen. Oldenbourg-Verlag, München

SUMMARY

The increasing application of electronic data processing machines and CAD/CAM-technology changes requirements in qualification of mechanical engineers and designers. Therefore it is necessary to reform "classic" education to more computer influenced fields in the study of mechanical engineering. This paper shows some examples typical for the study at technical universities in Germany. Other examples in the field of kinematics demonstrate the modified procedures in mechanism analysis by application of CAD/CAM-systems.

Microcomputer Aided Teaching (μCAT) in the Field of TMM

K. KĘDZIOR AND A. OLĘDZKI

Warsaw University of Technology, Institute for Aircraft Engineering and Applied Mechanics, Nowowiejska 22/24, 00-665 Warsaw, Poland

Abstract. The recent development of unexpensive microcomputers has created out new possibilities for teachers and students. Teachers can use them as a very efficient tool for the illustration how to solve various difficult problems. They can be used in classrooms (a set of graphic monitors is necessary) to make clear the complex problems which are difficult to explain in a traditional way on blackboard. Special microcomputer programs may be used for tutoring students and even for grading them. Students can solve their home-work problems more efficiently and thoroughly while using microcomputer programs included into text-books and/or written by themselves. The paper presents an outlook of experience gained by the authors during their four years recent application of microcomputers for teaching both TMM and relative subjects.

Keywords. Teaching; education; computer-aided instruction; computer applications.

INTRODUCTION

It is obvious that the CAD and/or CAM methods and also the use of computers in theoretical and experimental investigations are of great importance in the education of contemporary student - an engineer of the future.

The last development and widespreading of cheap microcomputers make it possible to use them in teaching already in early terms. Some elements of CAD may be easily included into courses of such subjects as Machine Design and/or TMM, and the method of computer simulation may be applied in teaching Systems Dynamics and Automatic Control of Machines. Microcomputers are useful in laboratories of many subjects so as to teach students computer aided measurements and data processing.

The Institute for Aircraft Engineering and Applied Mechanics of the Warsaw University of Technology has been using microcomputers systematically for more than four years to aid teaching subjects which are not from the field of Computer Science. The paper present some examples of using microcomputers in TMM and AC. One can find here as well some conclusions coming from the hitherto made experiments.

COMPUTER AIDED LECTURING

The use of microcomputers in classes has many advantages. Due to the use of a computer monitor instead of blackboard and chalk one can faster, more precisely and fully illustrate a traditional lecture. A good example of such a use is the program "Basic Networks" made for teaching Automatic Control (authors: D.Ferdyn, K.Kędzior). The program works in a conversative way and lets the lecturer to demonstrate - in any moment of the lecture - the static and time responces of control systmes to typical test signals, as well as the frequency responses (on the complex plane or logarithmic diagrams) of one of the ten basic netweorks of automatic control for any set of parameters. The number of easy-to-show diagrams is bigger than the one usually given in the text books. Each diagram can be printed out and put into a student´s note-book. (An example of this print-out is given in Fig. 1).

The use of another program similar to the one presented above called "Four-bar linkage" illustrates classes in the subject TMM. It is one of the 25 microcomputer programs prepared by A.Olędzki (1985--1986) for TMM. In a traditional lecture in order to show the variety of possible curves, a heavy model operated by a teachers assistant was necessary. Such a model with 11 exemplary coupler curves is shown in Fig. 2. It has, which is worth mentioning, unchangeable parameters. This limit does not refer to the model written in the short "Four--bar linkage" program, which makes it much easier for lecturer to choose the proper examples (Fig. 3).

In several cases the accurate drawing by hand on a blackboard is impossible. A good example of that is given in Fig. 4, where complicated profiles obtained as envelopes of cycloidal curves were drawn by a Spectrum microcomputer using 4 lines of BASIC program only.

The other example (Fig. 5) concerns a cam profile with a flat follower. Student and/or designer can check here quickly if the given profile is a proper one (without undercutting).

Special advantages of a microcomputer aided lecture cannot be seen unless this is the only method to explain a certain problem completely. A simple example of such a use is the program "Fourier" (author: K.Nazarczuk) which visualizes the harmonics of the chosen functions of time, and also presents the process of summing up the basic harmonics (Fig.6).

A more complex example is the presentation during the lecture of the computer simulation of a physical system. CSSP program is used for this purpose (author: I.Siwicki). Thanks to the use of England´s integration procedure the program, written in PASCAL, can be used even on such a cheap microcomputer as ZX Spectrum 48k, to the simulation of any system which description does not exceed 20 ordinary differential equations. Due to properties of the used procedure the integration is stable in spite of different kinds of discontinuities or nonlinearity. It eliminates the common up to now limits which were met while discussing nonlinear systems in teaching the advanced TMM. It also makes the choice of examples more free. The program CSSP has its extension for a IBM PC. This version enables simulation experiments of almost any physical system.

MICROCOMPUTER AIDED STUDENT´S HOME WORK

The syllabus of many subjects from the group of TMM comprise some controlled students homeworks. Such works done in a traditional way will take numerous hours each. This is a real burden for a student, so overload by contemporary teaching programs. Thus the use of microcomputer shortens the time and also enriches the range and the content of the works.

In the academic year 1984/85 there was made the first use of the microcomputers ZX Spectrum 48k to this purpose. The problem was programmed as follows:
- the level of the student´s knowledge was tested first (the computer asked questions choosing them at random out of the prepared set of problems);
- in the case when the student´s knowledge was not satisfactory, the computer was giving the advice "please read the text-book again and contact me next time";
- in the case of the satisfactory responses, the student was following the questions connected with his particular problem;
- at the end of the session the computer was grading the student, relising thus the teacher of the traditional time consuming task.

Five microcomputers were prepared for this task. It took 2 weeks for 120 students to fulfil the job. The time devoted by the student to do the work was reduced from 15 hours in the past (average value) to 2 hours of his home work and 0.5 hour of the work with the microcomputer. The students were distrustful first while doing their home wirk in such a way. Yet, their later estimation was positive (some of them become even enthusiatic!).

The example presented above is not the last one to show all the possibilities of the computer aided student´s own work. Students can also apply programs used during the classes to make all the kinds of repetitions easier, to make print-outs, diagrams etc.

One can see how easily we can make tutoring programs of the future and to make the life of teachers and students much easier, as it is nowadays.

MICROCOMPUTER AIDED LABORATORY WORKS

Several students in the case of a typical laboratory, gathered in several groups do up to five one-hour exercises in one term. Each exercise usually has the following stages:
- an introductory test (checking the basic knowledge of the student);
- setting up the measuring equipment;
- measurements and registration of the obtained results;
- working out the results and making the final report;
- the final test (necessary for a grading of a student).

Each of those stages (except for the second one) may be computerized, but it seems most advisable to use the microcomputer for data acquisition and processing (the third and the fourth stage). In the most student´s practical exercises the number of the collected data is not so big, thus even microcomputers with a limited memory (48-64k) can be used. As it is easy to process the data and to present the results in the graphical form, it makes it possible to extend the range of the exercises and to pay more attention to the analysis of various aspects of the tested phenomena. We should mention here that the preparation of the software for even simple exercise is labour consimimg task (programs are most often in the computer language) and it cannot be given as the task for the students doing the exercise. Those students who make use of the ready to use programs must understand the way of data processing and the aim of the used procedures. Thus the instruction to the exercise must contain the information about that in details.

An example illustrating this problem is the software package used in the laboratory of Aerodynamics of our Institute (author: J.Wojciechowski) in the laboratory work devoted to the turbulence of the boundary layer. In Fig. 7 a measuring set is shown. The microcomputer works out the obtained results of the measurements and prints diagrams of power spectral concentration and the function of the autocorrelation of the measured signal. Getting such processed data would be impossible without the computer which not only makes it easier but also possible to perform such an interesting and instructive exercise.

We should mention here that up to now most Polish technical universities use main frame computers to teach Computer Science and to do research. We do not have enough of them, so their applicability for student experimental task - is rather questionable. In such a situation the use of microcomputers is practically the only solution.

OTHER APPLICATIONS OF MICROCOMPUTERS IN TEACHING

Due to the insufficient number of main frame computers and minicomputers in Poland, also microcomputers are more and more often used here by students to do their course works and diplomas. Our experiences show that microcomputers are quite a sufficient tool for these purposes and due to their rich graphical software they are even more comfortable in their use than bigger digital computers. For diploma works, however, the 16-bit processors and RAM above 128k is necessary, thus - the class of IBM PC and/or its clones is a proper choice.

The use of microcomputers in all the administrative works (registration of students, grading them, making lists of students, etc) helps conduct the teaching process. Since the microcomputer is not generally a device for many users, therefore it is suitable for acquisition of those data which should be under protection.

PARTICULARS ABOUT THE EQUIPMENT

Basing on our experiences we conclude that the microcomputer equipment used in teaching should:
- be reliable, since a break-down during a lecture is sometimes a loss of time that cannot be made up for;
- resistant to power supply interferences (caused by electric motors, voltage drop, power failure and others). Some of the contemporary microcomputers does not meet these requirements which causes lots of troubles;
- mechanically resistant (especially the keyboard); the "personal computer" applied in teaching stops being personal as it is used many hours a day by many people with different levels of technical and computer knowledge, so this problem should be considered seriously.

Out of the used microcomputers two proved to be particularly useful for our purposes:
- Sinclair ZX Spectrum (Plus) 48k. It has many drawbacks (its keyboard is not typical, it does

not have any floppy discs). Yet, its graphical possibilities make it mostly useful in mass teaching (lectures, classes, laboratories). Besides, it is also cheap.

- IBM PC (and other compatible ones) used in higher forms of teaching, diplomas including. It is also useful for research purposes.

FINAL CONCLUSIONS

Microcomputers make teaching more efficient by increasing the speed of the information transfer from the teacher to the student. Moreover, computer aided teaching is more attractive for students who, in this way, get used to applying the computer technique.

Many countries have already notized all these advantages. Some colleges and universities in the USA make it obligatory for students to have their own microcomputers. It is worth to emphasize that authorities of many countries pay a great attention to the use of microcomputers in teaching, which was expressed in the Soviet-American declaration published after the 1986 Geneva summit.

Microcomputer aided teaching makes it necessary for academic teachers to acquire appropriate skill in using the new equipment. For the elder members of faculty it becomes, however, a barrier difficult to overcome and therefore their attitude towards attempts of introducting microcomputer aided teaching is unfavourable. This makes a serious problem and we should realise it also.

REFERENCES

Olędzki, A. (1985). Computer aided teaching (CAT) in TMM. The Fourth Int. Symp. IFToMM SYROM´85, Bucarest, Vol. III-2, pp.315-322.

Olędzki, A. (1986). Fundamentals of Theory of Machines and Mechanisms. WNT, Warsaw (in Polish, in print).

ОБУЧЕНИЕ С ПОМОЩЬЮ МИКРОКОМПЬЮТЕРОВ (μCAT) В ОБЛАСТИ ТММ

В последние годы наблюдается все более широкое употребление недорогих микрокомпьютеров, что открывает новые возможности перед учителями и учащимися. Учителя могут применять микрокомпьютеры как чрезвычайно эффективное пособие для обучения решению различных трудных задач. Эта техника употребима в аудиториях (что требует комплекта графических дисплеев) для пояснения сложных проблем, которые трудно поддаются объяснению традиционным методом, с помощью мела и доски. Специализованные микрокомпьютеры можно применить для непосредственного обучения учащихся и для проверки их знаний, а даже для оценки. Учащиеся могут решать домашние задачи более эффективно с помощью микрокомпьютера, пользуюсь программами приведенными в учебниках или составленными самостоятельно.
В докладе авторы хотят поделиться опытом накопленным в течение четырехлетнего применения микрокомпьютеров для обучения ТММ и смежных проблем.
Накопленный опыт был употреблен для составления недавно опубликованного учебника по ТММ. В учебник включены программы иллюстрирующие большинство представленных проблем. К книге прилагается кассета с 25 программами и инструкция.

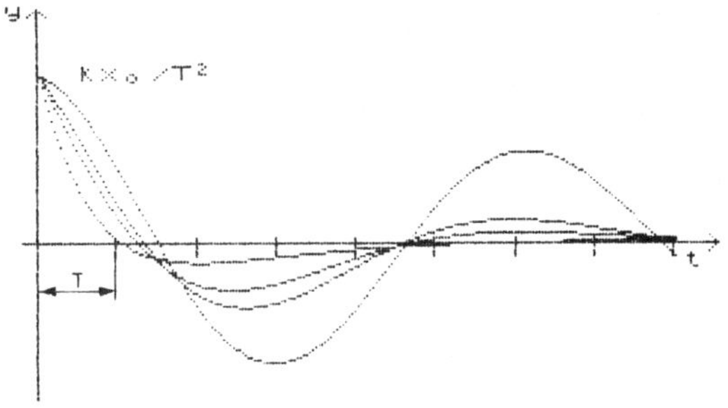

Fig. 1. Example of microcomputer print-out of the step response of the system with a transfer function $G(s)=ks^2/(T^2s^2+2\gamma Ts+1)$ for $\gamma = 0.1;\ 0.35;\ 0.7;\ 1.5$.

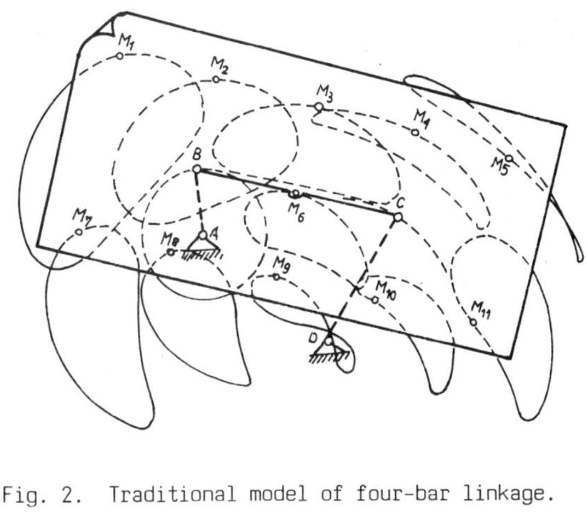

Fig. 2. Traditional model of four-bar linkage.

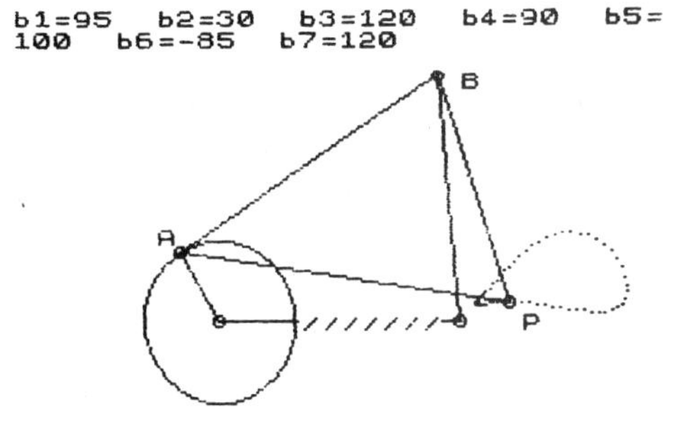

Fig. 3. Coupler curve print-out

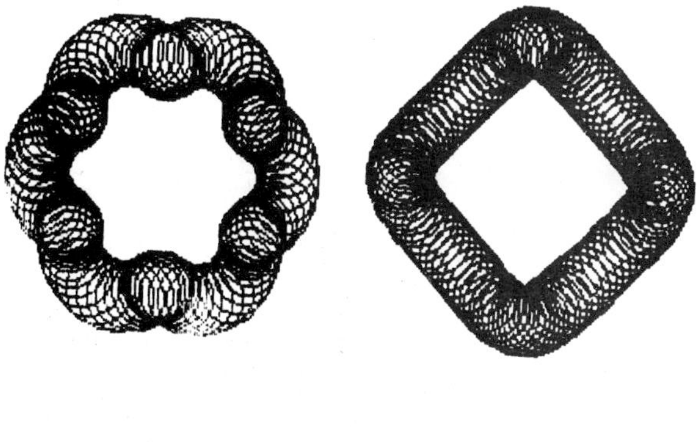

Fig. 4. Cycloidal curves print-out

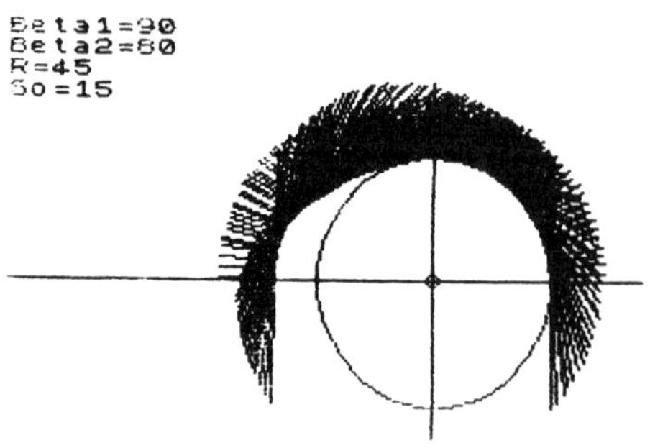

Fig. 5. Cam profile with a flat follower.

Fig. 6. Visualization of harmonics - the sum of 23 components.

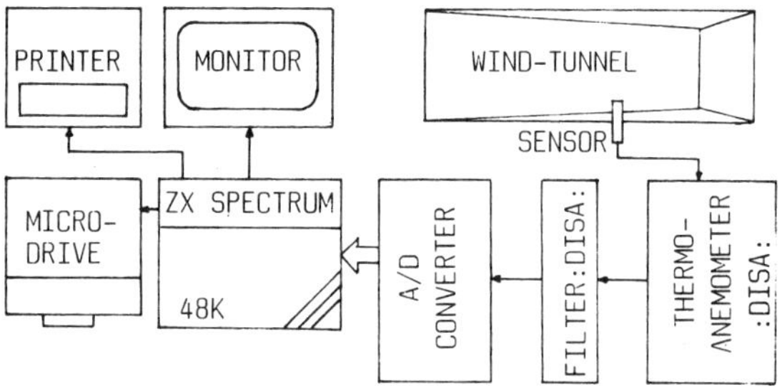

Fig. 7. Experimental set-up.

Coordonnées Spécifiques Concernant la Modernisation de la Discipline de la Théorie des Mécanismes (Analyse cinématique)

A. ORĂNESCU

Université de Galatz

Par suite d'un mandat reçu de la part de la Commission Romaine IFTOMM (International Federation for Theory of Machines and Mechanisms) de Cercle territorial de Galatz a initie l'action de moderniser la discipline de Mécanismes, à partir de 1976, avec les objectifs fondamentaux suivants:

a) Remplacer des méthodes géométriques par des méthodes numériques de calcul, en employant le calcul automatique. On a encore tenu compte de la nécessité d'employer la machine à calculer, sans programme, même pour les mécanismes d'une certaine complexité;

b) Séparer judicieusement les domaines où l'on applique les méthodes à caractère de grande généralité (pour les structures complexes) des méthodes spécifiques aux structures usuelles techniques. Pour l'élaboration de ces méthodes on a tâché d'éviter de exagérations mathématique dont les jeunes enseignants débutants subissent souvent l'influence, "le conservatorisme" qui caractérise encore les adeptes des méthodes graphiques, aussique la tendance de "généraliser à tout prix" par laquelle certaines écoles appliquent exclusivement des méthodes générales, universellement valables, mais qui, pour les structures simples, usuelles, deviennent ridiculement compliquées lors de l'application;

c) Accentuer l'importance de l'étude des mécanismes tridimensionnaux.

ANALYSE DE LA CONFIGURATION ET DE LA CINÉMATIQUE DES MÉCANISMES À ARTICULATIONS, RÉSOLUE PAR LA MÉTHODE DES BARRES

Pour faciliter la compréhension des considérations qui suivent, nous considérons des le début le mécanisme plan RRRR, comme dans la fig.1.

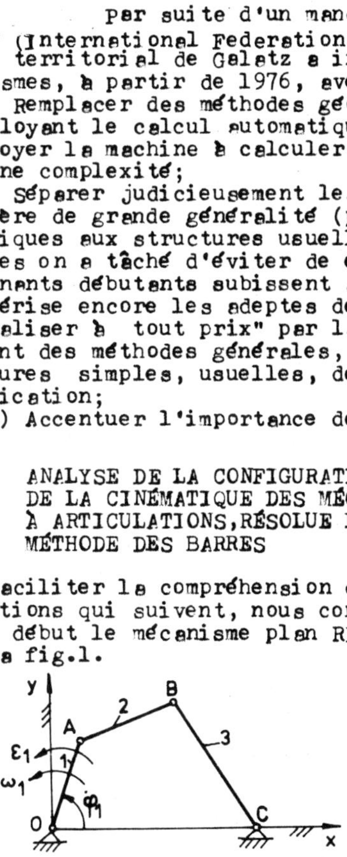

Fig.1. Mecanisme plan RRRR

Liste des paramètres d'entrée: O; C; $\ell_1 = \ell_{OA}$; $\ell_2 = \ell_{AB}$; $\ell_3 = \ell_{CB}$; φ_1, ω_1, ε_1.

Détermination des configurations:

$$\left\| \begin{matrix} x_A \\ y_A \end{matrix} \right\| = \left\| \begin{matrix} \ell_1 \cos \varphi_1 \\ \ell_1 \sin \varphi_1 \end{matrix} \right\|; \quad \begin{matrix} (x_B - x_A)^2 + (y_B - y_A)^2 - \ell_2^2 = 0 \\ (x_B - x_C)^2 + (y_B - y_C)^2 - \ell_{CB}^2 = 0 \end{matrix}$$

ayant les inconnues x_A, y_A, x_B, y_B.

Pour linéariser les équations, par une méthode de gradient ou l'algorithme de Newton, il est bien de partir des solutions approximatives, déterminées graphiquement.

Détermination des vitesses et des accélerations:

$$\|v_A\| = \omega_1 \Delta OA; \quad \|v_B\| = \|v_A\| + \omega_2 \Delta AB = \omega_3 \Delta CB \qquad (1)$$

$$\|a_A\| = -\omega_1^2 \|OA\| + \varepsilon_1 \Delta OA; \quad \|AB\| = \|a_A\| - \omega_2^2 \|AB\| + \varepsilon_2 \Delta AB = -\omega_3^2 \|CB\| + \varepsilon_3 \Delta CB \qquad (2)$$

ayant les inconnaes $\|v_{A,B}\|$, ω_2, ω_3, $\|a_{A,B}\|$, ε_2, ε_3 (où $\|v_A\|$ est la matrice colonne qui représente $\vec{v}_A$, et $\omega_2 \Delta OA$, le déterminent, $\vec{\omega}_2 \times \overrightarrow{OA}$.

Il est bin de remarquer que, le mécanisme étant de la deuxième classe - dans la classification Assur - tous les systèmes d'équations sont découplables en systèmes de deux équations à deux inconnues.

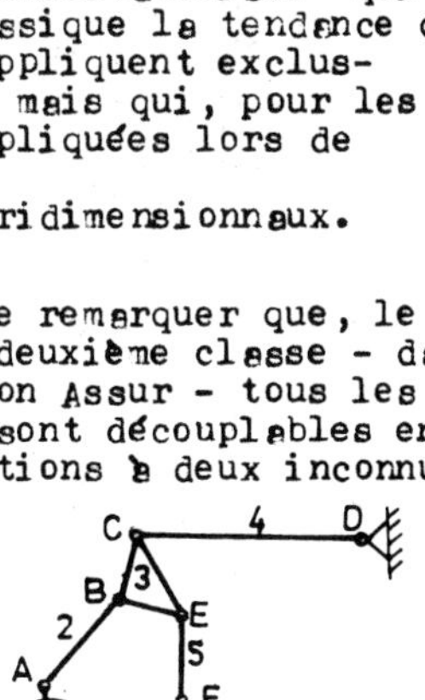

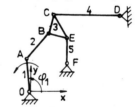

Fig. 2. Mécanisme de III-ème classe.

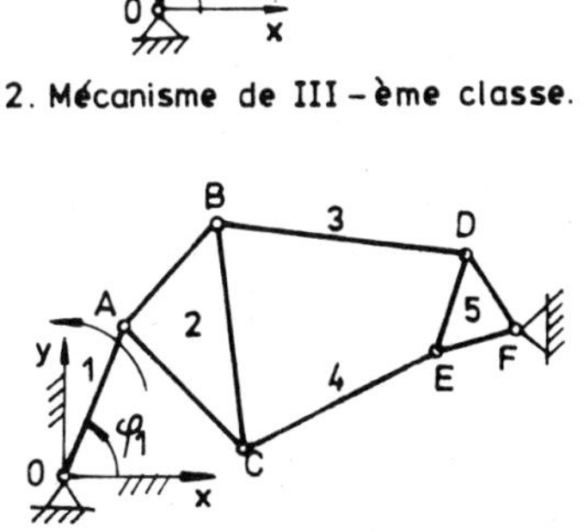

Fig.3. Mécanisme de IV-ème classe.

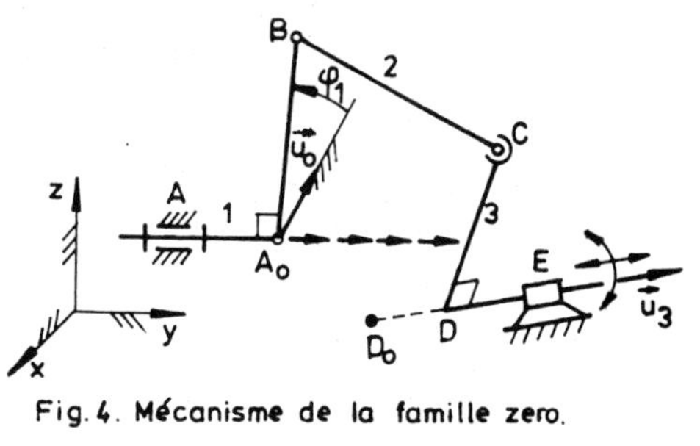

Fig.4. Mécanisme de la famille zero.

Solution de l'exemple de la figure 2 (mécanismes de III-ème classe). Après avoir déterminé les coordonnées du point A, les six coordonnées des points B, C et E se déterminent en écrivant les six équations de distance (AB), (BC), (CD), (CE), (BE) et (FE). (Les solutions graphiques approximatives peuvent être trouvées par des méthodes specifiques).
Pour le champ de vitesses on écrit les équations suivantes (type Euler):

$$\|v_B\| = \|v_A\| + \omega_2 \Delta AB = \|v_C\| + \omega_3 \Delta CB = \|v_E\| + \omega_3 \Delta EB,$$

$$\|v_B\| = \omega_1 \Delta OA + \omega_2 \Delta AB = \omega_4 \Delta DC + \omega_3 \Delta CB =$$
$$= \omega_5 \Delta FE + \omega_3 \Delta EB \qquad (3)$$

c'st - à- dire par un système de 4 équation scalaires ayant les inconnues $\omega_2, \omega_3, \omega_4, \omega_5$; après quoi il suit la détermination de $\|v_B\|$ par un remplacement simple. La détermination des accélérations suit le même schéma de calcul.
On doit remarquer que le mécanisme peut être transformè, de manière conventionnelle, dans un mécanisme de II-ème classe, si l'on change l'élément "conducteur", en admetant comme paramètre conventionnel d'entrée pour les vitesses $\omega_4^x = 1$. Le mécanisme est maintenant de deuxième classe et il peut être résolu comme dans le cas précédent. Ultérieusement les vitesses linéraires et angulaires se multiplient par le facteur de correction $k = \omega_1/\omega_1^x$
Par exemple $\|v_B\| = \|v_B^x\| \frac{\omega_1}{\omega_1^x}$; $\omega_2 = \omega_2^x \frac{\omega_1}{\omega_1^x}$ (4)

Pour les accélérations on considère le champ de vitesses réel et l'on admet, conventionnellement pour la manivelle 4, $\varepsilon_4^x = 0$ en résolvant un champ d'accélérations fictives. On obtient les accélérations réelles des expressions:

$$\|a_p\| = \|a_P^x\| + \|v_p\| \frac{\varepsilon_1 - \varepsilon_1^x}{\omega_1}; \quad \varepsilon_i = \varepsilon_i^x + \omega_i \frac{\varepsilon_1 - \varepsilon_1^x}{\omega_1} \qquad (5)$$

où, tout comme pour (4), le point P et élément i, peuvent être quelconques dans la figure 2.
On doit encore remarquer que cette méthode peut être appliquée dans de nombreux cas pratiques et qu'elle peut être utilisée, par superposition des mouvements, pour des mécanismes à degré des mobilité supérieur à l'unité.
Solution de l'exemple de la figure 3. (mécanisme de IV-ème classe).
Les configurations sont solutionnées en écrivant 8 équations de distance dont on obtient les coordonnées des points B, C, D et E. Les vitesses et les accélérations sont obtenues en écrivant des équations:

$$\|v_B\| = \|v_A\| + \omega_2 \Delta AB = \omega_5 \Delta FD + \omega_3 \Delta DB$$

$$\|v_C\| = \|v_B\| + \omega_2 \Delta BC = \|v_A\| + \omega_2 \Delta AC =$$
$$= \omega_5 \Delta FE + \omega_4 \Delta EC \qquad (6)$$

le principe de la solution correspond aussi aux accélérations-même si l'on admet de manière conventionnelle que l'élément 5 est l'élément conducteur, la classe du mécanisme se réduit à la III-ème et l'on peut appliquer les considerations précédents.

La solution du mécanisme de la fig. 4 (mécanisme tridimensionnel de la famille zéro, ayant la structure RRSC).
La liste des paramètres d'entrée est la suivante: A; A_0; D_0 (un point fixe quelconque, sur l'axe de la couple cylindrique du point E), E (le centre géométrique de la couple cylindrique du point E, ou d'un autre point considéré comme repaire sur l'axe de la couple) $l_{A_0B}, l_{BC}, l_{DC}, \vec{u}_0$ (marquant une position d'origine de la manivelle $\vec{A_0B}$), $\varphi_1, \omega_1, \varepsilon_1$.
Pour les configurations, on écrit successivement:

$$\|u_1\| = \|AA_0\|/l_{AA_0}; \quad \|u_3\| = \|D_0E\|/l_{D_0E};$$
$$(x_B - x_{A_0})^2 + (y_B - y_{A_0})^2 + (z_B - z_{A_0})^2 - l_{A_0B}^2 = 0$$
$$u_{1x}(x_B - x_{A_0}) + u_{1y}(y_B - y_{A_0}) + u_{1z}(z_B - z_{A_0}) = 0$$
$$u_{0x}(x_B - x_{A_0}) + u_{0y}(y_B - y_{A_0}) + u_{0z}(z_B - z_{A_0}) =$$
$$= l_{AB} \cos\varphi_1 \qquad (7)$$

système qui détermine les coordonnées du point B. Les 3 relations (7) peuvent être écrites d'une manière plus succinte:

$$(A_0B)^2 - l_{A_0B}^2 = 0; \quad \vec{u}_1 \cdot \vec{A_0B} = 0;$$
$$\vec{u}_0 \cdot \vec{A_0B} = l_{A_0B} \cos\varphi_1 \qquad (7')$$

Les coordonnées du point D peuvent être écrites seulement en fonction de la distance variable $d = (D_0D)$, sous la forme:

$$\left\| \begin{matrix} x_D \\ y_D \\ z_D \end{matrix} \right\| = \left\| \begin{matrix} x_{D_0} \\ y_{D_0} \\ z_{D_0} \end{matrix} \right\| + d \left\| \begin{matrix} u_{3x} \\ u_{3y} \\ u_{3z} \end{matrix} \right\| \qquad (8)$$

Il s'en suit que les points C et D incluent 4 inconnues et qu'éles se déterminent du système:

$$(BC)^2 - l_{BC}^2 = 0; \quad (\vec{u}_1, \vec{A_0B}, \vec{BC}) = 0;$$
$$\vec{u}_3 \cdot \vec{DC} = 0; \quad (DC)^2 - l_{DC}^2 = 0 \qquad (9)$$

(Le produit mixte exprimé par la deuxième relation de (9) a à la base l'observation ques les vecteurs $\vec{u}_1$, $\vec{A_0B}$ et $\vec{BC}$ sont coplanaires).
Pour les vitesses, on calcule d'abord $\|v_B\| = \omega_1 \vec{u}_1 \times \vec{A_0B}$ (10) ensuite la vitesse du point C du systeme:

$$\vec{v}_C = \vec{v}_B + (\vec{\omega}_1 + \vec{\omega}_{21}) \times \vec{BC} = \vec{v}_D + \vec{\omega}_3 \times \vec{DC} \qquad (11)$$

(Pour que les choses soient plus claires la relation (11) a été exprimée de manière vectorielle, où l'on a souligné d'une ligne chaque grandeur comme des trois grandeurs qui caractérisent le vecteur respectiv).
Dans la relation (11) le vecteur $\vec{\omega}_2 = \vec{\omega}_1 + \vec{\omega}_{21}$ a été écrit en tenant compte que $\vec{\omega}_{21} =$

$$= \omega_{21} \frac{\vec{u}_1 \times \vec{A_0B}}{l_{A_0B}} = \omega_{21} \vec{u}_{21} \qquad (12)$$

où $\vec{u}_{21}$ est situé sur l'axe de l'articulation du point B; $\vec{v}_D = v_D \vec{u}_3$; et $\vec{\omega}_3 = \omega_3 \vec{u}_3$.
Pour déterminer les accélérations on doit observer que la dérivée de l'expression (12) s'écrit successivement:

$$\vec{\varepsilon}_{21} = \frac{d\vec{\omega}_{21}}{dt} = \frac{\partial \vec{\omega}_{21}}{\partial t} + \vec{\omega}_1 \times \vec{\omega}_{21} \qquad (13)$$

ainsi que $\vec{\varepsilon}_2 = \vec{\varepsilon}_1 + \frac{\partial \omega_{21}}{\partial t} \vec{u}_{21} + \vec{\omega}_1 \times \vec{\omega}_{21}$
Il en résulte successivement

$$\|a_B\| = -\omega_1^2 \|A_0B\| + \varepsilon_1 \Delta u_1, A_0B$$
$$\vec{a}_C = \vec{a}_B + \vec{\omega}_2 \times (\vec{\omega}_2 \times \vec{BC}) + \vec{\varepsilon}_2 \times \vec{BC} =$$
$$= a_D \vec{u}_3 - \omega_3^2 \vec{DC} + \varepsilon_3 \vec{u}_3 \times \vec{DC} \qquad (14)$$

Les deux derniers membres de (14) ont pour inconnues les scalaires $\frac{\partial \omega_{21}}{\partial t}$, a_D, ε_3.

ANALYSE DE LA CONFIGURATION ET DE LA CINÉMATIQUE DES MÉCANISMES À COULISSE, RÉSOLUE PAR LA MÉTHODE DES BARRES

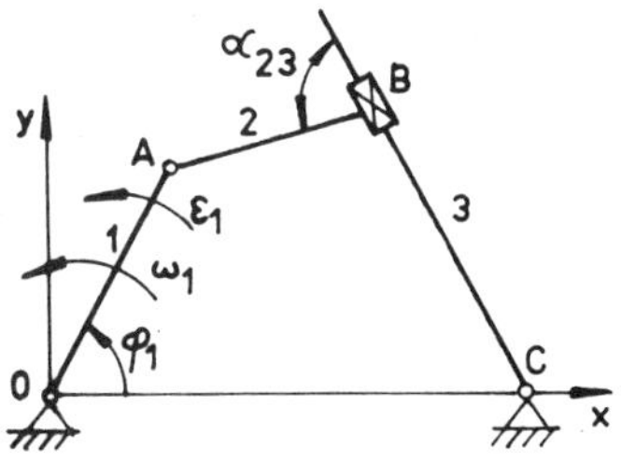

Fig.5. Mécanisme à coulisse de II - ème classe.

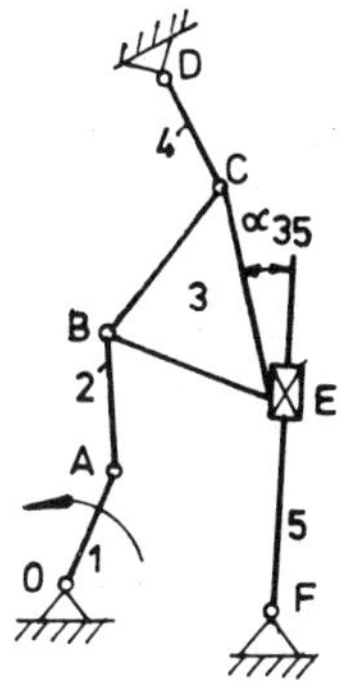

Fig. 6. Mécanisme à coulisse de III - ème classe

Solution du mécanisme de la figure 5. Pour la solution de la configuration on observe que cette fois dans la liste des paramètres d'entrée on doit inclure aussi l'angle α_{23}, de manière qu'après la détermination des coordonnées du point A pour le point B tout comme pour la longueur inconnue L_{CB}, on puisse écrire:

$$(AB)^2 - \ell^2_{AB} = 0; \quad (BC)^2 - L^2_{CB} = 0;$$
$$\vec{BA} \cdot \vec{CB} = \ell_{AB} L_{CB} \cos \alpha_{23} \qquad (15)$$

Ici encore les grandeurs approximatives se déterminent graphiquement (par des méthodes classiques).

Pour la détermination des vitesses et des accélérations on considère le point A, solidarisé sur les éléments 2 et 3 et ayant comme transporteur l'élément 3.

$$\vec{v}_A = \vec{\omega}_1 \times \vec{OA} = \vec{\omega}_3 \times \vec{CA} + v_{AA_3} \frac{\vec{CB}}{L_{CB}} \qquad (16)$$

qui a pour inconnues ω_3, v_{AA3}, puis $\vec{v}_A$ et s'écrit symboliquement:

$$\|v_A\| = \omega_1 \Delta OA = \omega_3 \Delta_{CA} + v_{AA_3} \frac{\|CB\|}{L_{CB}} \qquad (16')$$

après quoi il résulte $\vec{v}_B$ de $\|v_B\| =$

$$= \|v_A\| + \omega_2 \Delta AB \qquad (17)$$

avec la remarque $\omega_2 = \omega_3$.

Les accélérations, après le calcul de $\vec{a}_A = -\omega_1^2 \vec{OA} + \vec{\varepsilon}_1 \times \vec{OA}$, résultent de: $\vec{a}_A =$

$$= -\omega_3^2 \vec{CA} + \vec{\varepsilon}_3 \times \vec{CA} + a_{AA_3} \frac{\vec{CB}}{L_{CB}} + \vec{a}_c \qquad (17')$$

où la composante Coriolis qui dépend seulement du champ de vitesses, est obtenue de: $\vec{a}_A^{\,c} = 2\vec{\omega}_3 \times \vec{v}_{AA_3}$.

La relation (17) s'écrit symboliquement:

$$\|a_A\| = -\omega_1^2 \|OA\| + \varepsilon_1 \Delta OA = -\omega_1^2 \|CA\| + \varepsilon_3 \Delta CA +$$
$$+ a_{AA_3} \frac{\|CB\|}{L_{CB}} + 2\omega_3 v_{AA_3} \frac{\Delta CB}{L_{CB}}$$

avec les inconnues ε_3 et a_{AA3}. L'accélération $\vec{a}_B$ se détermine par une équation Euler, en observant l'égalité $\varepsilon_2 = \varepsilon_3$.

Pour la solution de la configuration du mécanisme de la figure 6, après avoir résolu le point A, les 6 coordonnées afferentes aux points B,C,E, tout comme l'inconnue L_{FE}, résultent des 6 equations de distance auxquelles on ajoute $\vec{FE} \cdot \vec{EC} = L_{FE} \cdot \ell_{EC} \cos \alpha_{35}$

Les vitesses et les accélérations se déterminent par des relations Euler entre les points B-A; B-C; B-E; auxquelles on ajoute, pour les vitesses, les relations

$$\|v_E\| = \omega_5 \Delta FE + v_{EE5} \frac{\|FE\|}{L_{FE}}, \quad \omega_5 = \omega_3 \qquad (18)$$

et pour les accélérations

$$\|a_E\| = -\omega_5^2 \|FE\| + \varepsilon_5 \Delta FE + a_{EE5} \frac{\|FE\|}{L_{FE}} + 2\omega_5 v_{EE5} \frac{\Delta FE}{L_{FE}} \qquad (19)$$

Il est, bien sûr, à noter que l'on peut appliquer dans ce cas aussi, la méthode du chagement conventionnel de l'élément conducteur, comme on a déjà montré dans la figure 2.

Pour les structures spatiaux la méthode de composition du mouvement relatif avec le mouvement de transport reste, aussi, sans modification.

ANALYSE DES CONFIGURATIONS ET DE LA CINÉMATIQUE DES MÉCANISMES, PAR LA MÉTHODE DES CONTOURS INDÉPENDANTS.

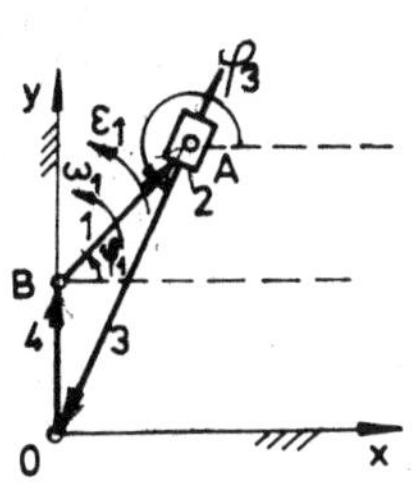

Fig.7. Mécanisme sheping.

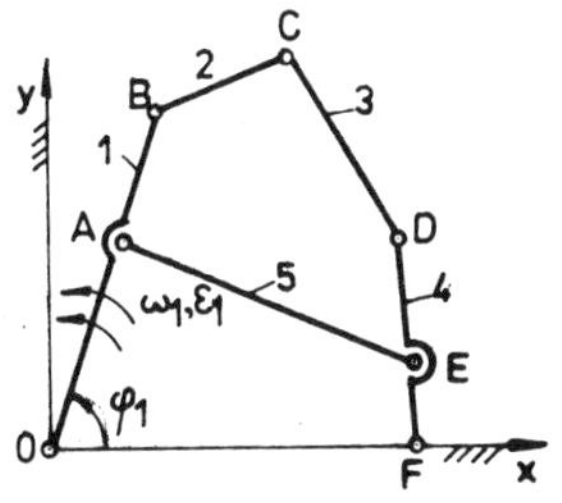

Fig.8. Mécanisme a deux contours

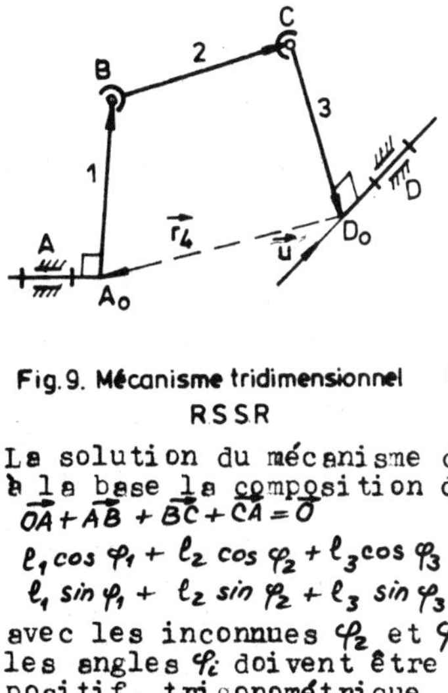

Fig.9. Mécanisme tridimensionnel RSSR

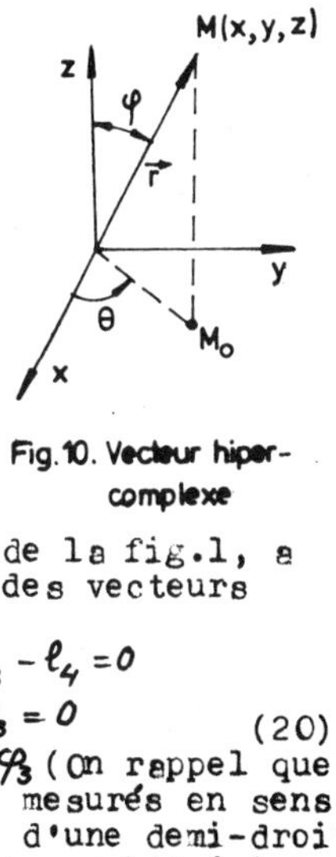

Fig.10. Vecteur hipercomplexe

La solution du mécanisme de la fig.1, a à la base la composition des vecteurs $\vec{OA} + \vec{AB} + \vec{BC} + \vec{CA} = \vec{0}$

$$\ell_1 \cos \varphi_1 + \ell_2 \cos \varphi_2 + \ell_3 \cos \varphi_3 - \ell_4 = 0$$
$$\ell_1 \sin \varphi_1 + \ell_2 \sin \varphi_2 + \ell_3 \sin \varphi_3 = 0 \qquad (20)$$

avec les inconnues φ_2 et φ_3 (On rappel que les angles φ_i doivent être mesurés en sens positif, trigonométrique, d'une demi-droite parallèle à $\vec{OX}$ qui passe par l'origine du vecteur en question).

Par des dérivations successives, par rapport au temps, on obtient:

$$\omega_1 \ell_1 \sin \varphi_1 + \omega_2 \ell_2 \sin \varphi_2 + \ldots = 0; \quad \omega_1 \ell_1 \cos \varphi_1 + \ldots = 0 \qquad (21)$$

$$\varepsilon_1 \ell_1 \sin \varphi_1 + \omega_1^2 \ell_1 \cos \varphi_1 + \ldots = 0; \quad \varepsilon_1 \ell_1 \cos \varphi_1 - \omega_1^2 \ell_1 \sin \varphi_1 + \ldots = 0 \qquad (22)$$

Pour le mécanisme de la figure 7, on écrit successivement:

$$\vec{\ell}_1 + \vec{L}_3 + \vec{\ell}_4 = \vec{0}; \quad \ell_1 \cos \varphi_1 + L_3 \cos \varphi_3 = 0; \quad \ell_1 \sin \varphi_1 + L_3 \sin \varphi_3 +$$
$$+ \ell_4 = 0; \quad -\ell_1 \omega_1 \sin \varphi_1 + v_{AA_3} \cos \varphi_3 - L_3 \omega_3 \sin \varphi_3 = 0; \quad \ell_1 \omega_1 \cos \varphi_1 +$$
$$+ v_{AA_3} \sin \varphi_3 + L_3 \omega_3 \cos \varphi_3 = 0; \qquad (23)$$
$$-\ell_1 \varepsilon_1 \sin \varphi_1 - \ell_1 \omega_1^2 \cos \varphi_1 + a_{AA_3} \cos \varphi_3 - 2\omega_3 v_{AA_3} \sin \varphi_3 + \ldots = 0$$
$$\ell_1 \varepsilon_1 \cos \varphi_1 - \ell_1 \omega_1^2 \sin \varphi_1 + a_{AA_3} \sin \varphi_3 + 2\omega_3 v_{AA_3} \cos \varphi_3 + \ldots = 0$$

Pour le mécanisme de la figure 8, on doit considérer deux contours indépendants, conformément à la relation $N = c - n = 7 - 5 = 2$ Ceux-ci pourrait être, par exemple O,F,E, A,O et A,E,D,C,B,A.

Pour le mécanisme tridimensionnel RSSR,de la figure 9, si l'on exprime le vecteur quelconque $\vec{r}$ sous la forme $\vec{r} = x + iy + jz$ où i et j sont des opérateurs complexes, on obtient de la figure 10:

$$x = r \sin\varphi \cos\theta;\ y = r \sin\varphi \sin\theta;\ z = r \cos\varphi \qquad (24)$$

Les trois projections du contour A_0 B C B_0 s'écrivent:

$$[r_1 \sin\varphi_1 \cos\theta_1] + r_2 \sin\varphi_2 \cos\theta_2 + r_3 \sin\varphi_3 \cos\theta_3 + [r_4 \sin\varphi_4 \cos\varphi_4] = 0;\ [r_1 \sin\varphi_1 \sin\theta_1] + r_2 \sin\varphi_2 \sin\theta_2 + r_3 \sin\varphi_3 \sin\theta_3 + [r_4 \sin\varphi_4 \sin\theta_4] = 0;\ [r_1 \cos\varphi_1] + r_2 \cos\varphi_2 + r_3 \cos\varphi_3 + [r_4 \cos\varphi_4] = 0 \qquad (25)$$

auxquelles on ajoute la relation de perpendicularitéu $\vec{u} \cdot \vec{r}_3 = 0$ (Fig.9):

$$u_x r_3 \sin\varphi_3 \cos\varphi_3 + u_y r_3 \sin\varphi_3 \sin\theta_3 + u_z r_3 \cos\varphi_3 = 0 \qquad (26)$$

car l'angle entre les deux vecteur peut s'exprimer ordinairement:

$$\cos\alpha = (x_1 x_2 + y_1 y_2 + z_1 z_2)/r_1 r_2 \qquad (27)$$

Dans les relations (25) on a introduit les parenthèses [ ] pour les grandeurs connues comme grandeurs d'entrée ou constantes. Si l'on prend en considération l' **axe** oy conformément à la direction du vecteur $\vec{u}$ et si l'on introduit les notations suivantes pour les termes connus:

$$r_1 \sin\varphi_1 \cos\theta_1 + r_4 \sin\varphi_4 \cos\theta_4 = a;\ r_1 \sin\varphi_1 \sin\theta_1 + r_4 \sin\varphi_4 \sin\theta_4 = b;\ r_1 \cos\varphi_1 + r_4 \cos\varphi_4 = c$$

la relation (26) donne $\theta_3 = 0$ et (25) devient:

$$a + r_2 \sin\varphi_2 \cos\theta_2 + r_3 \sin\varphi_3 = 0;\ b + r_2 \sin\varphi_2 \sin\theta_2 = 0;\ c + r_2 \cos\varphi_2 + r_3 \cos\varphi_3 = 0 \qquad (27)$$

Apres l'élimination de l'angle φ_3, par élévation au carré et remplacement, il résulte des développements simples:

$$z = ctg\,\theta_2 = \frac{ap^2 \pm c\sqrt{(a^2+c^2)(r^2 - b^2) - p^4}}{b(a^2+c^2)} \qquad (28)$$

où l'on a noté: $$p = \sqrt{\frac{a^2 + c^2 + r_2^2 - r_3^2 - b^2}{2}} \qquad (29)$$

De la relation intermédiaire

$$\sin\varphi_2 = -\frac{b}{r_2 \sin\theta_2} \qquad (30)$$

on peut maintenant calculer φ_2 et φ_3.

A partir des résultats précédents on peut déduire les implications cinématiques,mais on doit souligner, que cette méthode est surtout efficiente pour les configurations (optimisations en synthèse, etc).

Pour conclure, ci-dessus nous avons montré les principaux problèmes afférents à la configuration et à la cinématique des mécanismes par des méthodes spécifiques, comme nous avons déjà précisé. Nous n'avons pas poursuivi d'une manière systématique les applications de ces procedés.

BIBLIOGRAPHIE

Orănescu,A.,Crudu,I. (1977). Metode numerice în analiza configurației cinematicei și cinetostaticei mecanismelor. Univ.Galați.

Orănescu,A.,Bocioacă,R.,Bumbaru,S.(1979) Mecanisme,capitole de curs retrusturate. Univ.Galați.

Manolescu,N.I., Kovacs,F.,Orănescu,A. (1972). Teoria mecanismelor și a mașinilor, Ed.Did.și Ped.Buc.

SUMMARY

In this paper we suggest some numerical teaching methods of Mechanisms, trying to avoid, as much as possible,the abuse so frequently made such as: the application of the general methods by all means, as well as the use of the automatic calculation even regarding the very simple structures.

We have tried to establish some criteria in pointing out the opportunity of using the various teaching methods.

In this paper the determination of the mechanisms place was taken into consideration, as well as the kinematic field, by using the bar method regarding the jointed mechanisms and the spindled ones, both for plan and tridimensional structures.

In the end the independent contours method was considered for structures of the same types.

Creating a Mechanical Engineering Program in a Developing Nation

R. B. GAITHER

Department of Mechanical Engineering, University of Florida, Gainesville, Florida 32611, USA

Abstract. The philosophical commitments required and the challenges that have to be met in creating a new mechanical engineering program in mechanical engineering in a developing nation are discussed. Vital aspects of the program including assessments of needs, faculty selection and facilities are identified for a particular new university whose objectives required that its mechanical engineering graduates be able to compete with graduates from already established and reputable universities.

Keywords. Education, mechanical engineering, curriculum design

PHILISOPHICAL COMMITMENT

The leadership of virtually all nations whether already in advanced phases of social and technological development or only now emerging into the early phases are committed to the support of educational systems that permit their populations to both benefit from and participate in technological growth. There are many reasons for this. Among them are two that are quite pragmatic, namely: (1) to provide some assurance that they will survive in a world that today appears to be guided by a corollary of the Darwinian philosophy of survival by the fittest, and (2) to satisfy the basic human need to be creative and be active participants in a world that is not only changing rapidly but has become heavily dependent upon technology.

Attesting to the realities of this committment in those nations which are already technologically advanced is the presence of impressive public institutions of higher learning where scientific and engineering studies hold unparalled positions of visability and importance. Although many if not most of these countries often seem to be concentrating major resources on protection and defense, a close examination of national expenditures on a per capita basis reveals that education rivals and usually surpasses all other categories. In the budget of the federal government of the United States for example, direct support of education appears as a very small piece of the pie especially if one fails to note that significant portions of the budgets for other categories such as agriculture, military, and welfare contain grants and research funds to be used in the nation's universities and colleges. When these amounts are coupled with the budget outlays for education from the 50 states that comprise the United States of America and the monies that individuals and families spend on education, a final figure is arrived at that is quite staggering.

In developing nations where significant portions of the leadership have received their education in foreign lands, the opportunities for individual citizens wishing to take advantage of a large financial support structure for education are few. Moreover, the driving motives on at least a national or collective scale are different from those within technologically advanced nations. They are neither so sharply focused on maintaining technological leadership nor so deeply committed to new discovery and innovation. They are trying to "catch up". This fact, coupled with local needs and perceptions makes the planning, construction and early operation of new engineering colleges in the developing nations quite unique.

THE CHALLENGES

Today, we are witnessing an emergence and early growth of new universities and colleges that are dedicated to science and engineering among all of the developing nations of the world. Some are new directions or ventures for established universities that possess impressive histories of excellence in history, religion, philosophy and the arts. But, most are totally new. Thus, the challenges to those involved with the early development of science and engineering programs in developing countries are quite different than those confronting developers of science and engineering programs where such resources as qualified faculty, equipment, maintenance facilities and an administration that is prepared to deal with the high cost of technological education are locally accessable. The challenges, more often than not, begin at very basic levels. What follows is a description of a single case of a new university which did not exist fifteen years ago. The description is focused on a department of mechanical engineering which graduated its first class last year.

ASSESSMENT OF NEEDS

The first challenge that had to be addressed was what are the real needs of the people who were expected to participate in and receive benefits from a program in mechanical engineering within a country which had never produced locally educated graduates before. The decision to have a new university, to commit significant funds to its construction and operation and to have a high quality program in mechanical and other basic engineering fields had already been made.

For the very good reason that it is today recognized as the international language of technology, English was selected to be the language of instruction. This carried a subtle but real implication that the program was intended to produce graduates who would neither be limited in their professional work to strictly local employment nor be handicapped in any future efforts to enter upon graduate studies in all but a few of the world's engineering graduate schools. This decision also had an immediate effect upon the curriculum. Although virtually all of the students to be admitted to the new engineering programs had considerable exposure to English in high school, the ability of these students to read textbooks, understand lectures and prepare oral and written reports in English was found to be severely limited. Thus, heavy doses of English had to be inserted early in the curriculum and supplemented with summers of concentrated study and practice in English.

FACULTY SELECTION

The decision to have a program of high quality imposed a constraint on both the curriculum and the selection of faculty. For reasons that reflected a desire that the engineering graduate from this new school be able to compete with graduates of other eminent schools, the curriculum was designed to be a modern one focused upon creative design in an up-to-date technological world. The path of first creating a school of technology with a curriculum focused upon operating existing machine systems was not to be followed. This clearly required faculty who themselves were modern engineers possessing advanced degrees that demonstrated their familiarity with the borders of technological knowledge and who could teach young engineers how to deal with advanced concepts and be prepared to be productive in tomorrow's as well as today's world. The upper administration of the university had a formitable task in recruiting such a faculty. That they were eminently successful in doing so is attested to by the fact that several of the members of the first graduating class are currently enrolled in graduate study at engineering schools whose names are known to us all.

FACILITIES

Although this particular new university encountered a goodly number of growing pains and its share of the same "people problems" that all schools seem to suffer with on a periodic basis, it was not handicapped with a scarcity of funds for comfortable buildings, equipment and facilities. The entire physical plant was designed and constructed by a well known engineering and architectual firm under the direction of an architect that not only enjoys a measure of world fame but possesses a strong interest in technological education

The curriculum, selection of laboratory equipment and structure of the departments were developed initially by an advisory committee who stipulated from the beginning that they would expect major contributions to come from the new faculty. With but few exceptions (i.e. the dean who had already established himself as an authority in engineering education), the faculty were young but promising Ph.D.'s from reputable universities who shared the vision that this new university could function well and would meet the expectations of the country's leaders. The tasks that they undertook were both difficult and time consuming. To obtain the confidence of the local population in general, and the industrial leadership in particular, each of them became involved in extra curricular consulting, giving short courses on timely subjects and becoming familiar with all aspects of the nations technology. The faculty of the mechanical engineering department went so far as to insure that all of the senior level design projects addressed local needs. One was even successful in assisting his students in producing a device that could be designed and constructed with materials obtained in the local area.

DESIGN FOCUS OF THE CURRICULUM

Although incorporating the latest concepts in each subject area, the overall curriculum in mechanical engineering at this new university was constructed along traditional lines wherein emphasis was placed on the learning of fundamentals in a logical order. Temptations to bifurcate the curriculum, establish options or experiment with novel educational techniques were suppressed very early. Not only would the pursuit of such non traditional approaches to education have the potential of isolating the mechanical engineering program but it would certainly divert time and attention from efforts to achieve measurable high standards of quality.

Today, as in the beginning, the curriculum has a highly visible design focus. Fundamental courses in statics, dynamics, materials and thermodynamics are presented in a sequence such that they provide a core of knowledge upon which design courses in both the thermal and machine system stems of the curriculum are based. In the senior year of the five year program, a majority of each student's time is concentrated upon completing a design project which not only requires the integration of knowledge gained in earlier courses but calls for producing a real solution to a real problem.

CONCLUSION

This venture proved to be a truly successful one where the initial goals were achieved and the promises for the future are at this moment very good. The features of the planning for this that stand out today as having been vital are that cooperation was continually sought, that the faculty not only accepted the challenge of striving for excellence but improved upon advice on how to do so. Today, this university, in general and its mechanical engineering department in particular, are in a healthy condition. Several of the first mechanical engineering graduates together with their advisors presented papers at international conferences on design last year.

Synthesis of Mechanisms by the Function Approximation Method in the Course of the Theory of Machines and Mechanisms

S. I. PANTELEEV

Moscow Aviation Institute, Volokolamskoe Shosse, 4, Moscow 125871, USSR

КРАТКОЕ СОДЕРЖАНИЕ В докладе сообщается о разделах курса теории механизмов и машин в высших технических учебных заведениях и содержании одного из основных его разделов - синтеза механизмов. Более подробно говорится о синтезе механизмов с низшими парами по методу, основанному на теории приближения функций, созданной основоположником русской школы теории механизмов академиком П.Л.Чебышевым.

I Теория механизмов и машин является одной из основных учебных дисциплин общеинженерной подготовки специалистов широкого профиля в высших технических учебных заведениях. Она является связующим звеном между циклом общенаучных дисциплин и циклом специальных дисциплин, где изучают машины и механизмы отраслевого назначения.

В теории механизмов и машин изучаются общие методы исследования и проектирования механизмов, необходимых для создания современных машин, приборов и автоматических устройств вне зависимости от их конкретного целевого назначения. Эта дисциплина дает знания о строении механизмов, о кинематических и динамических характеристиках механизмов и управляемых кинематических цепей, о методах нахождения параметров механизмов по заданным кинематическим и динамическим свойствам, методах виброзащиты машины и человека, знания об управлении движением механизмов и машин.

2 Современная программа по теории механизмов и машин для инженерно-технических специальностей высших учебных заведений включает в себя следующие разделы:

I. Введение, где излагается: Содержание дисциплины "Теория механизмов и машин" и ее значение для инженерного образования. Связь теории механизмов и машин с другими областями знаний. История становления науки о механизмах и машинах.

2. Структура механизмов. Изучаемые темы: Основные понятия теории механизмов и машин. Основные виды механизмов. Структурный анализ и синтез механизмов.

3. Анализ механизмов. Изучаемые темы: Кинематический анализ механизмов. Трение в механизмах. Силовой анализ механизмов. Динамический анализ механизмов. Колебания в механизмах и машинах. Уравновешивание механизмов. Виброзащита.

4. Синтез механизмов. Изучаемые темы: Общие методы синтеза механизмов. Синтез механизмов с низшими парами. Синтез зубчатых механизмов. Синтез кулачковых механизмов. Синтез гидравлических механизмов.

5. Основы теории управления машин-автоматов и промышленных роботов. Изучаемые темы: Основные виды систем управления машин-автоматов и промышленных роботов. Синтез логических (релейных) систем управления промышленных роботов.

Основанием курса теории механизмов и машин являются фундаментальные положения математики и механики, которые развиваются и дополняются применительно к задачам дисциплины. Однако, в курсе решаются и такие задачи, которые не рассматриваются в изучаемой ранее дисциплине теоретической механики. И это, в первую очередь, все задачи кинематического и динамического синтеза.

3 При решении задач анализ и синтеза механизмов используется абстрактная модель механизма - его кинематическая схема. Если кинематическая схема определена правильно, то она обладает наиболее общими свойствами, относящимися к механизму данного конкретного вида, которые не зависят от конструктивного оформления его деталей и области его применения.

Синтез механизмов - один из основных разделов курса теории механизмов и машин. Главной задачей синтеза механизмов является проектирование кинематической схемы механизма - определение кинематических размеров его звеньев, которые удовлетворяют заданным основным и дополнительным условиям синтеза. От правильного выбора схемы механизма и значений его параметров зависит не только качество выполняемого процесса, но и все его прочностные и энергетические характеристики. Последующие расчеты на прочность механизмов и их деталей, установление конструктивных форм и выбор материалов уже не могут дать должного эффекта, если параметры схемы механизма не являются оптимальными.

4 Общие методы синтеза механизмов, излагаемые в курсе теории механизмов и машин, можно разделить на две группы: синтез механизмов методами оптимизации и синтез механизмов по методу приближения функций.

Оптимизационные методы дают необходимое решение задач синтеза путем многократного количественного повторения решения задач кинематического анализа механизма при различных числовых значениях его параметров. При значительном числе параметров синтеза решение задач вследствие большого числа различных вариантов механизма стало возможным только с появлением ЭВМ. Методами оптимизации в синтезе механизмов с применением ЭВМ можно получить решение любой задачи синтеза, но при этом большое значение имеет качественная оценка ожидаемых результатов решения. Оптимизационные методы дают только количественное решение задачи. Они не дают возможности анализировать влияние отдельных параметров синтеза на качество решения.

Такую возможность представляют методы синтеза механизмов, в основе которых лежит теория наилучшего приближения функций, созданная русским математиком и механиком, академиком П.Л.Чебышевым (1821-1894) /1, 2, 3/, получившая свое дальнейшее развитие в трудах ученых советской школы теории механизмов и машин /4, 5/.

5 П.Л.Чебышев преподавал в Петербургском университете математические дисциплины и прикладную механику. При изучении различных механизмов для воспроизведения прямолинейного движения без использования ползунов и направляющих он отметил, что механизм Уатта, известный под названием параллелограмма Уатта, который применялся в современных Чебышеву машинах, не дает достаточной точности прямолинейного движения. Им был предложен метод определения параметров механизма, с большой точностью воспроизводящего прямолинейное движение. Опубликованная им работа "Теория механизмов, известных под именем параллелограммов" /1/ явилась началом создания общего метода наилучшего приближения функции /2, 3/. Основанный на этом методе синтез механизмов стали называть синтезом механизмов по Чебышеву.

В русской технической литературе до Чебышева вообще не было работ, посвященных теории механизмов, и, конечно, работ, посвященных синтезу механизмов. Чебышеву принадлежат 15 работ, в которых рассматриваются труднейшие задачи синтеза механизмов. Для решения этих задач он использовал разработанную им теорию функций, наименее уклоняющихся от нуля.

Кроме синтеза приближенно направляющих механизмов, Чебышев решал задачи синтеза шарнирных механизмов с остановами, с двумя качаниями коромысла за один оборот кривошипа, с различными законами движения ведомых звеньев. Многие решенные им задачи синтеза механизмов воплощены в созданных им моделях механизмов. Чебышев на 13 лет раньше Грюблера дал формулу для определения степени подвижности плоского механизма. Поэтому с полным основанием П.Л.Чебышев является основоположником русской школы теории механизмов.

6 Теория приближения функций Чебышева основана на сопоставлении двух функций: одна из них характеризует свойства конкретного синтезируемого механизма, другая характеризует требуемые свойства этого механизма, соответствующие оптимальному выполнению его назначения.

Заданная функция, которую необходимо воспроизвести, приближенно заменяется функцией мало от нее отличающейся. П.Л. Чебышевым дано математическое решение задачи о нахождении условий, при которых эти две функции оказываются близкими, т.е. решение задачи приближения функций.

По методу приближения функций (приближенного синтеза) искомые значения выходных параметров синтеза механизма находятся из системы уравнений, составленных на основании условий минимума максимального отклонения приближающей функции от заданной, а не путем поиска, как это делается по методам оптимизации.

Приближенный синтез механизмов по П.Л.Чебышеву можно осуществлять в три этапа.

Первый этап - выбор основного условия синтеза и дополнительных ограничений.

Второй этап - упрощение аналитического выражения основного условия синтеза в виде отклонения от заданной функции. Для упрощения аналитического выражения отклонения от заданной функции можно использовать способ "взвешенной разности", впервые примененный П.Л.Чебышевым.

Третий этап - вычисление искомых параметров из условия минимума отклонения от заданной функции.

При вычислении искомых параметров механизма применяются, например, такие способы приближения функций, как интерполирование, квадратичное приближение, наилучшее приближение функций по П.Л.Чебышеву.

В упоминаемом выше механизме Уатта приближение прямой линии на отрезке длинною в 100 мм осуществлялось с точностью до 1,2 мм. В механизме, полученном Чебышевым на основании разработанной им теории наилучшего приближения функций отклонения от прямой не превышало 0,1 мм. История создания теории наилучших приближений функций П.Л.Чебышевым является примером неразрывной взаимосвязи фундаментальных и прикладных исследований.

7 Программа дисциплины "Теория механизмов и машин для инженерно-технических специальностей высших учебных заведений предусматривает изучение следующих вопросов, названных выше конкретных тем по синтезу механизмов:

Тема 1. Общие методы синтеза механизмов.

Этапы синтеза механизмов. Входные и выходные параметры синтеза. Основные и дополнительные условия синтеза. Целевые функции. Ограничения. Методы оптимизации в синтезе механизмов с применением ЭВМ. Случайный поиск. Направленный поиск. Штрафные функции. Локальный и глобальный минимумы. Комбинированный поиск. Постановка задачи приближенного синтеза механизмов по Чебышеву. Интерполирование. Квадратическое приближение функций. Наилучшее приближение функций.

Тема 2. Синтез механизмов с низшими парами.

Постановка задачи синтеза передаточного шарнирного четырехзвенника. Вычисление трех, четырех и пяти параметров синтеза. Синтез пространственного четырехзвенника. Синтез шарнирного четырехзвенника по положениям звеньев. Синтез плоских и пространственных механизмов по коэффициенту изменения средней скорости коромысла. Точные направляющие механизмы. Методы синтеза приближенных направляющих механизмов. Механизмы Чебышева. Теорема Робертса. Шарнирные механизмы с выстоем. Мальтийские механизмы. Синтез манипуляционных механизмов.

Тема 3. Синтез зубчатых механизмов.

Основная теорема зацепления. Образование сопряженных поверхностей по Оливье. Цилиндрическая зубчатая передача. Эвольвентное зацепление. Основные размеры зубьев. Кинематика изготовления сопряженных поверхностей зубьев цилиндрических эвольвентных зубчатых колес. Геометрический расчет зубчатой передачи при заданных смещениях. Построение картины зацепления. Проверка дополнительных условий при синтезе эвольвентного зацепления. Особенности внутреннего зацепления. Подрезание зубьев. Блокирующий контур. Косозубые колеса. Цилиндрическая передача Новикова. Эвольвентная коническая передача. Начальные поверхности. Последовательность синтеза зацеплений, получаемых при нарезании зубьев по методу обкатки. Виды гиперболических передач. Червячная передача. Выбор схемы планетарной передачи. Выбор числа сателлитов из условий соседства и равных углов между сателлитами. Выбор чисел зубьев в планетарных передачах. Синтез бесступенчатых передач с замкнутым дифференциалом. Планетарные коробки передачи. Дифференциальные механизмы манипуляторов и промышленных роботов.

Тема 4. Синтез кулачковых механизмов.

Виды кулачковых механизмов. Эквивалентные (заменяющие) механизмы. Выбор допускаемого угла давления. Определение основных размеров из условия ограничения угла давления. Определение основных размеров из условия выпуклости кулачка. Выбор закона движения ведомого звена с учетом его упругости. Определение профиля кулачка по заданному закону движения ведомого звена. Выбор радиуса качения ролика. Условие качения ролика. Выбор замыкающей пружины.

Тема 5. Синтез гидравлических механизмов.

Типовая схема объемного гидропривода. Уравнение движения гидравлического механизма. Определение геометрических параметров тормозного устройства (регулируемого дросселя) из условия воспроизведения заданного закона торможения.

8 Изложение этих вопросов различных тем синтеза механизмов, в том числе и синтеза по методу приближения функций представлено в изданных в Советском Союзе учебниках и учебных пособиях /6, 7/.

Метод приближения функций по Чебышеву применяется также и в курсовом проектировании по теории механизмов и машин /8/.

Метод приближения функций П.Л.Чебышева, предназначенный для решения задач теории механизмов и машин, сейчас применяется во многих областях техники, применяется в физических и даже биологических исследованиях, где требуется сопоставление функций.

CONCLUSION

Sections of the program have been treated containing the course of the theory of machines and mechanisms delivered at higher technical establishments, as well as the subject-matter of every section and the program of the subject - theory of machines and mechanisms - for engineering specialities followed one of the basic sections of the given course, and namely, the synthesis of mechanisms. The application of P.L.Chebyshev function approximation method have been also considered for the synthesis of mechanisms with lowest pairs.

ЛИТЕРАТУРА

I. Чебышев П.Л. Теория механизмов, известных под именем параллелограммов. Соч., т.I, с.I09-I43, СПб, I899.
2. Чебышев П.Л. Вопросы о наименьших величинах, связанные с приближенным представлением функций. Соч., т.I, с.27I-378.
3. Чебышев П.Л. О функциях, наименее уклоняющихся от нуля. Соч., т.I, с.I87-2I5, Приложение к XXII тому Записок Академии Наук, № I, I873.
4. Левитский Н.И. Синтез механизмов по Чебышеву, Москва-Ленинград, Изд-во Академии Наук СССР, I946.
5. Левитский Н.И. Проектирование плоских механизмов с низшими парами. Москва-Ленинград, Изд-во Академии Наук СССР, I950.
6. Левитский Н.И. Теория механизмов и машин. Москва. Изд-во "Наука", I979.
7. Левитская О.Н., Левитский Н.И. Курс теории механизмов и машин. 2-е издание. Москва. Изд-во "Высшая школа", I985.
8. Левитская О.Н. Синтез гидравлического механизма в курсовом проекте по теории механизмов и машин. Сборник научно-методических статей по теории механизмов и машин. Выпуск 6. Изд-во "Высшая школа", I978.

Rotor Dynamics

Residual Vibration Amplitude on Influence Coefficient Balancing Procedure

D. L. ZORATTO* AND A. G. CASTRO**

*Department Mechanical Engineering, COPPE/Universidade Federal do Rio de Janeiro, Rio de Janeiro, Brazil
**Universidad Industrial de Santander, Colombia

Abstract. The residual vibration amplitude after balancing a rotor by the influence-coefficient procedure depends on the combinations of errors of amplitude and phase shift measurement at the balancing planes. This paper deals with a simulation procedure to determine the pattern distribution that should be expected for the residual vibration amplitudes, as function of the standard deviation of measurement errors. Taking one set of measurement as an "exact" one, we generate a number of new sets of fictitious measurement, by adding some random variables ε_1 and ε_2 representing the error on amplitude and phase shift respectivelly. Each new set is treated by the influence-coefficient procedure determining at each location of the rotor a balancing weight ω_i. The vectorial difference between this weight and the one calculated for the reference set, determines, by applying the influence-coefficient procedure in reverse way, a residual vibration amplitude that should result from one particular combination of measurement errors. Some parametric curves representing the medium value(and also the medium value plus one standard deviation of the pattern distribution) as function of ε_s are presented.

Keywords. Balancing; flexible rotors; influence coefficients; residual vibration.

INTRODUCTION

The influence of measurement errors on the residual vibration amplitude after balancing a flexible rotor is of great interest to the practicioners devoted to this subject.

As regard to the influence-coefficient method,even if it is known the standard deviation of amplitude and phase shifts measurement it is not possible to determine prior to balance the actual value of the residual amplitude vibration. The aim in this case is to predict the pattern distribution of residuals. This was done by, Iwatsubo (1976) by using a theoretical statistical treatment. Alternatively we present here a simulation method for determining this pattern by treating statistically a number of artificial sets obtained from one experimental set taked as reference.

THE INFLUENCE-COEFFICIENT PROCEDURE

The well-known influence-coefficient procedure for balancing flexible rotors (Tessarzik J. M., R. H. Badgley and W. J. Anderson, 1972) is based on a set of amplitude and phase shift measurement: at a number of measurement planes in original configuration and in configurations that are modified by aggregating some test masses in a number of balancing planes (usually the same as the measurements planes).

The unbalancing masses are determined, by the equation:

$$\{U\} = |\alpha|^{-1} \{Y\} \tag{1}$$

where {Y} is the original vibration amplitudes vector, {U} is the unbalancing mass vector and $|\alpha|$ is the influence-coefficient matrix, determined by calculating:

$$\alpha_{ij} = \frac{Y_{i(1)} - Y_{i(0)}}{T_i} \tag{2}$$

where $Y_{i(0)}$ is the original vibration measured at plane i and $Y_{i(1)}$ is vibration measured after the aggregation of a test mass (that produces a moment T) at the same plane i.

We can see by the expression (1) that the error in calculating the unbalancing masses depends essencially of the errors of measurements of amplitude and phase shifts of vibrations, but depends also from the propagation of these errors in calculating the α_s by expression (2).

Taking expression (1) in reverse way, we can see that the residual unbalance ΔU, resulted after the aggregation of the calculated balancing masses will produce residual amplitudes that depends from the same errors mentioned above. Even if we know the standard deviation of the measurement of amplitude and phase shifts the α_s and also the residual vibrations depend on the combinations of errors in the several measurements that have to be done.

RESIDUAL AMPLITUDE VIBRATION

We suppose to known the matrix α_* of the influence coefficients that represents "exactly" the rotor. This matrix could be obtained by applying the influence-coefficient procedure to a set of measurements amplitude and phase shift without errors. In this case it is possible to determine the "exact" balancing weights $-U_*$ by the expression:

$$\{U_*\} = |\alpha|_*^{-1} \{Y_*\} \tag{3}$$

where Y_* are the "exact" values of the amplitude and phase shifts at each location.

The matrix $|\alpha|_*$ and the vector $\{U_*\}$ are taked as references in comparing with the ones containing errors. Amy other set of measurement with errors would lead to

$$\{U\}' = |\alpha|^{-1}.\{Y\}' \quad (4)$$

where $U_i' = U_{*i} \pm \Delta U_i$ and $|\alpha|'$ is the influence coefficient matrix obtained from the values Y_i' with errors.

If we apply the influence coefficient procedure in reverse way to the vector $\{U_* - U'\} = \{\Delta U\}$ we obtain the residual vibration amplitude vector $\{\Delta Y\}$ for this particular set of measurements which is:

$$\{\Delta Y\} = |\alpha|_* . \{U_* - U'\} \quad (5)$$

By developing the first equation of (5) we have:

$$\Delta Y_1 = (\alpha_{11*} \pm \Delta U_1) + (\alpha_{12*} \pm \Delta U_2) + (\alpha_{13*} \pm \Delta U_3) + + (\alpha_{1n*} \pm U_n) \quad (6)$$

that shows the dependency of ΔY_1 from the ΔU_S.

GENERATING NEW SETS OF MEASUREMENTS BY SIMULATION

The idea of simulating artifically a set of fictitious measurement, taking one experimental set of measurement as reference is here developed. The standard deviation of amplitude and phase shifts measurement possible to obtain by the equipment, is supposed to be known. The actual value of the error for each measurement would follow in principle a normal curve distribution. The combination of these errors in one set, would give, by the reasoning presented above, a particular value of residual vibration amplitude and we suppose that, if we have a number of different sets, they will give us a kind of pattern distribution for the residual vibration amplitude. This pattern distribution,of course, will more or less represent the balancing system, depending of the number of sets at disposal. For this purpose, to have many experimental sets would be troublesome and the "articial" simulation of new sets could satisfy this requirement if we take some cares.

Recognizing the statiscal caracteristic of an actual measurement, the generation of an "artificial" measurement is done by associating a random number (between 0 and 1) to the area under the normal curve that represents the probability of occurrence of an specific deviation from the medium value of a particular measurement. Actually, for one measurement we have to associate two random numbers, one for amplitude and one for phase shift.

SIMULATION RESULTS

The simulation was made taking as reference one experimental set performed by Castro (1986) in a flexible rotor with four measurement and balancing planes i.e. twenty measurements of amplitude and phase shifts. In practice the resulting residual amplitude vibration for this set when the balancing masses are applied is 20% of the original vibration amplitude.

Of course, since we use as reference, one set that contains some errors the resulting pattern distribution for the residual vibration amplitude, is higher than should be.For this reason. The results here presented by the figure 1 to 4 have to be interpreted as a tendency of the residual vibration amplitude, pattern distribution as function of the deviation standard.

The medium value of the residual vibration amplitude (in %) is showed by Figure 1 as function only of the standard deviation of the amplitude measurement (in %).

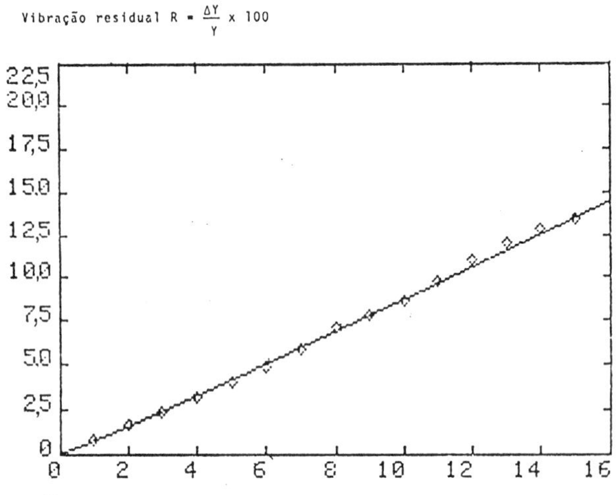

Fig. 1 - Medium value of RVA as function of amplitude errors

We can see that in the range simulated this medium value is quite linear with the standard deviation.

The medium value of RVA as function of the standard deviation (in degrees) of the fase shift error only, is presented in Figure 2. We can see that the balancing procedure is quite more sensible to phase shift errors than for amplitude errors, what is in agreement with the theory.

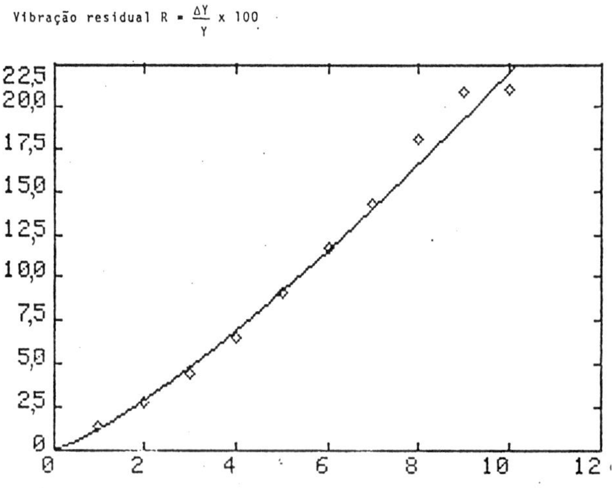

Fig. 2 - Medium value of RVA as function of phase shift errors

The medium value of RVA as function of combinations of standard deviations in amplitude and phase shift is presented in Figure 3.

The Figure 4 presents the curves of medium value of RVA (μ_{RVA}) and $\mu_{RVA} + \sigma_{RVA}$ for 4 degrees phase shift standard deviation which permits to see the strong increase of residual vibration amplitude with increasingly errors.

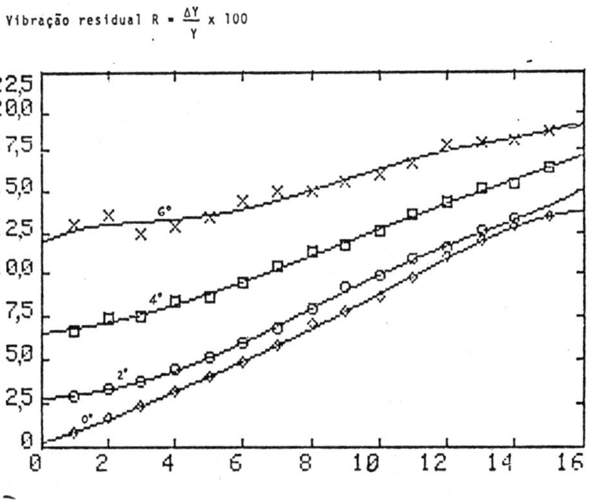

Fig. 3 - Medium value of RVA as function of amplitude an phase shifts errors

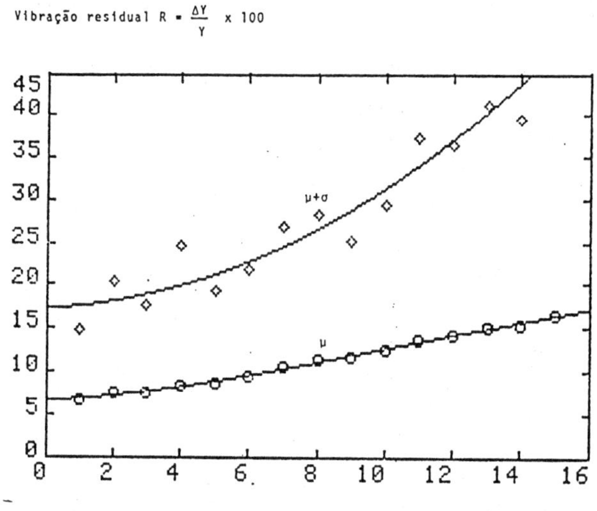

Fig. 4 - Medium value of RVA and $\mu_{RVA} + \sigma_{RVA}$ for 4 degrees phase shift error

CONCLUSIONS

Considering the resulting pattern distribution of the residual vibration amplitude as a normal curve we can estimate that in 68% of the cases the residual vibration amplitude after balancing would be less than the value predicted by curves $\mu + \sigma$ as in Figure 4.

Since the results was obtained, taking as reference a matrix $|\alpha|$ that contains errors, we should expect that the curves here presented, are higher than the real ones.

The method shows a potential application to evaluate the exactitude of a set of measurement prior to agregate the calculated balancing masses. By our experience a bad set taked as reference in applying the method described, leads to higher curves than the ones showed, and because of this it can be let a side prior to agregate the balancing masses.

REFERENCES

Iwatsubo, T. (1976). Error analysis of vibration of rotor/bearing system. *Inst. of Mech. Eng. London.*

Tessarzik, J. M., R. H. Badgley and W. J. Anderson (1973). Flexible rotor balancing by exact point -speed influence-coefficient method. *Journal of Engineering for Industry (ASME)*, Feb., 148-157.

Castro, A. G. (1986). Balanceamento de Rotores Flexíveis pelo Método dos Coeficientes de Influência. *M.Sc. Thesis, Federal University of Rio de Janeiro (COPPE-UFRJ).*

Resumen. Vibración residual en el método de balanceo por coeficients de influencia. La vibración residual despues del balanceo de un rotor flexible por el metodo de coeficients de influencia es funcion de la combinación de los errores de medición de la amplitud e ângulo de fase en los planos de balanceo. Este trabajo se relaciona con el desenvolvimiento de um procedimiento de simulación para determinar la forma de distribuición estadistica que se debe esperar para las amplitudes de vibracion residuales en funcion de de los errores de medición. Partiendo de un conjunto de mediciones consideradas sin errores, se construye um grau número de conjuntos artificiales de mediciones por asociación de variables randomicas ε_1 y ε_2 que representan los errores en amplitud y ângulo de fase respectivamente. Cada nuevo conjunto es tratado por el método de los coeficientes de influencia determinândose en cada plano uma masa de balanceo ω_i. La diferencia vectorial entre esta masa y aquella calculada para el conjunto de referencia determina por la aplicación del procedimiento de los coeficientes de influencia en sentido inverso, uma vibración residual que resultaria de uma particular combinación de errores. Las curvas presentadas muestram la variación de la media de vibracion residual (y tambien la media mas una deviación padron) en funcion de los errores de medición.

A Contribution to Analysis of Continuous Models of Drive Systems

A. MIODUCHOWSKI AND M. G. FAULKNER

Department of Mechanical Engineering, University of Alberta, Edmonton, Alberta, Canada T6G 2G8

INTRODUCTION

In the present paper, a method is proposed for calculation of velocities, strains and displacements in an arbitrary cross-section of shafts of a multi mass driven system with a torsional absorber. We assume that the model representing the drive system consists of shafts, segments of which have different but constant diameters and rigid disks connected with them. These rigid disks have constant mass moments of inertia with respect to axis of rotation and are loaded by external moments which are in general arbitrary. The torsional absorber is attached to the free end of the system and consists of an auxiliary mass elastically connected to the main system.

It should be noted that the analysis of torsional deformations in multi mass drive systems based on continuous models as proposed in this paper was first used in the case of a single rotor and shaft [1] and a two mass drive system [2]. The technique was further developed to analyze a multi mass drive system with uniform shafts [3] and later the case of stepped shafts was discussed as well [4].

In order to describe the dynamical behavior of the continuous model of a drive system the theory of propagation of one dimensional torsional elastic waves is employed together with appropriate initial and boundary conditions. Internal and external damping of the system is described by an equivalent damping. This approach leads to a system of linear ordinary retarded type differential equations with constant coefficients. An effective and stable numerical procedure is proposed for the solution of this system of differential equations and some numerical results are presented in graphical form.

THREE MASS DRIVE SYSTEM

Consider a system consisting of 3 rigid bodies and two shafts. The rigid bodies have mass moments of inertia J_i, $i = 1,2,3$ and the disk denoted by number 2 is acted upon by the external moment M_2. The shafts are deformable only in torsional manner and the centroidal axes of all elements coincide with the main axis of the drive system, Fig. 1. The ith shaft is characterized by the length ℓ, density ρ, shear modulus G and polar moment of inertia J_{0i}, $i = 1,2$.

Torsional dynamic absorbers are usually, but not necessarily, connected to the free end of the drive system. Fundamentally a torsional absorber assembly consists of an auxiliary mass elastically connected to the main system, Fig 1 (two configurations are shown by the dash line). The moment of inertia of the auxiliary mass is denoted by J_4 and the polar moment of inertia of the shaft by J_{03}.

For the continuous model of the system discussed here the investigation of angular displacements $\theta_i(x,t)$, velocities $\partial\theta_i(x,t)/\partial t$ and strains $\partial\theta_i(x,t)/\partial x$ of the shafts is reduced to the solution of 3 wave equations

$$\frac{\partial^2\theta_i(x,t)}{\partial t^2} - c^2 \frac{\partial^2\theta_i(x,t)}{\partial x^2} = 0, \quad i = 1,2,3 \qquad (1)$$

where $c^2 = G/\rho$, with the following boundary conditions:

$$- J_1 \frac{\partial^2\theta_1(x,t)}{\partial t^2} + GJ_{01} \frac{\partial\theta_1(x,t)}{\partial x} - D_1 \frac{\partial\theta_1(x,t)}{\partial t} = 0, \quad x = 0$$

$$M_2 - J_2 \frac{\partial^2\theta_2(x,t)}{\partial t^2} + GJ_{02} \frac{\partial\theta_2(x,t)}{\partial x} - GJ_{01} \frac{\partial\theta_1(x,t)}{\partial x} - D_2 \frac{\partial\theta_2(x,t)}{\partial t} = 0, \quad x = \ell$$

$$- J_3 \frac{\partial^2\theta_3(x,t)}{\partial t^2} + GJ_{03} \frac{\partial\theta_3(x,t)}{\partial x} - GJ_{02} \frac{\partial\theta_2(x,t)}{\partial x} - D_3 \frac{\partial\theta_2(x,t)}{\partial t} = 0, \quad x = 2\ell$$

$$- J_4 \frac{\partial^2\theta_3(x,t)}{\partial t^2} - GJ_{03} \frac{\partial\theta_3(x,t)}{\partial x} - D_4 \frac{\partial\theta_3(x,t)}{\partial t} = 0, \quad x = 2.5\ell$$

$$\theta_1(x,t) = \theta_2(x,t) \quad x = \ell \qquad (2)$$

$$\theta_2(x,t) = \theta_3(x,t) \quad x = 2\ell$$

and initial conditions

$$\theta_i(x,t) = \frac{\partial\theta_i(x,t)}{\partial t} = 0, \quad i = 1,2,3 \qquad (3)$$

The damping moments which are assumed to be applied to the rigid bodies are assumed in the form

$$M_i = - D_i \frac{\partial\theta_i(x,t)}{\partial t}$$

where D_i is a coefficient of equivalent viscous damping. These damping moments take into account internal as well as external damping. Additionally it was assumed that both shafts are of equal length e and the length of the damper shaft is $0.5e$. It should be stressed however, that the analysis as outlined above is easily applicable to other lengths of shafts and, or different geometry and location of the torsional damper, e.g. damper located inside the rotor, dash-dot line in Fig. 1.

Upon the introduction of non-dimensional quantities:

$$\bar{x} = \frac{x}{e}, \quad \tau = \frac{ct}{e}, \quad \bar{\theta}_i = \frac{\theta_i}{\theta_0}, \quad \bar{M}_2 = M_2 \frac{e^2}{J_1 h_0 c^2} \tag{4}$$

$$\bar{D}_i = \frac{D_i e}{J_1 c}, \quad K_i = \frac{J_{01} \rho e}{J_i}, \quad A_i = \frac{J_1}{J_i}$$

the relations (1), (2) and (3) become

$$\frac{\partial^2 \theta_i}{\partial t^2} - \frac{\partial^2 \theta_i}{\partial x^2} = 0, \quad i = 1,2,3 \tag{5}$$

$$-\frac{\partial^2 \theta_1}{\partial t^2} + K_1 \frac{\partial \theta_1}{\partial x} - D_1 \frac{\partial \theta_1}{\partial t} = 0, \quad x = 0$$

$$A_2 M_2 - \frac{\partial^2 \theta_2}{\partial t^2} + K_2 \frac{\partial \theta_2}{\partial x} - \frac{A_2}{A_1} K_1 \frac{\partial \theta_1}{\partial x}$$

$$- A_2 D_2 \frac{\partial \theta_2}{\partial t} = 0, \quad x = 1$$

$$-\frac{\partial^2 \theta_3}{\partial t^2} + K_3 \frac{\partial \theta_3}{\partial x} - \frac{A_3}{A_2} K_2 \frac{\partial \theta_2}{\partial x}$$

$$-A_3 D_3 \frac{\partial \theta_3}{\partial t} = 0, \quad x = 2$$

$$-\frac{\partial^2 \theta_3}{\partial t^2} - \frac{A_4}{A_3} K_3 \frac{\partial \theta_3}{\partial x} - A_4 D_4 \frac{\partial \theta_3}{\partial t} = 0$$

$$x = 2.5 \tag{6}$$

$$\theta_1 = \theta_2, \quad x = 1$$

$$\theta_2 = \theta_3, \quad x = 2$$

and

$$\theta_i = \frac{\partial \theta_i}{\partial t} = 0, \quad \text{for } t = 0, \quad i = 1,2,3 \tag{7}$$

where for convenience all $\theta_i(x,t)$ functions are written as θ_i and all bars are omitted.

The solutions of equations (5) are sought in the nondimensional form:

$$\theta_i = f_i[(t-t_{0i}) - (x-x_{0i})] + g_i[(t-t_{0i}) + (x-x_{oi})]$$

where the function f_i represents a wave propagating to the right and g_i represents a wave propagating to the left side of the i-th shaft as a result of application of the external moment M_2. It is assumed that the functions f_i and g_i are continuous and for negative arguments equal zero. By taking into account that the first perturbation occurs at the instant $t_{0i} = 0$ one gets:

$$\theta_1 = f_1(t-x) + g_1(t+x), \quad x_{01} = 0$$

$$\theta_2 = f_2(t-x+1) + g_2(t+x-1), \quad x_{02} = 1 \tag{8}$$

$$\theta_3 = f_3(t-x-2) + g_3(t+x-2), \quad x_{03} = 2$$

By substituting (8) into the boundary conditions (6) a system of equations is obtained for functions f_i and g_i. One easily observes that there occur simple relationship between arguments of the functions appearing in the same equation. Upon denoting the largest argument in each equation by the variable ζ the arguments of the remaining functions are then shifted by one or two. This procedure leads to the following final system of six linear, ordinary differential equations of the first and second order for the unknown functions $f_i(\zeta)$ and $g_i(f)$, $i = 1,2,3$:

$$g_1(\zeta) = f'_2(\zeta-1) + g'_2(\zeta-1) - f'_1(\zeta-2)$$

$$g'_2(\zeta) = f'_3(\zeta-1) + g'_3(f-1) - f'_2(\zeta-2)$$

$$g''_3(\zeta) + r_4 g'_3(\zeta) = - f''_3(\zeta-1) + s_4 f'_3(\zeta-1)$$

$$f''_1(\zeta) + r_1 f'_1(\zeta) = - g''_1(\zeta) + s_1 g'_1(\zeta)$$

$$f''_2(\zeta) + r_2 f'_2(\zeta) = A_2 M_2 - g''_2(\zeta) + s_2 g'_2(\zeta)$$

$$+ t_2 f'_1(\zeta-1)$$

$$f''_3(\zeta) + r_3 f'_3(\zeta) = - g''_3(\zeta)$$

$$+ s_3 g'_3(\zeta) + t_3 f'_2(\zeta-1) \tag{9}$$

where:

$$r_1 = K_1 + D_1 \qquad r_3 = K_3 + A_3 K_2/A_2 + A_3 D_3$$

$$r_2 = K_2 + A_2 K_1/A_1 + A_2 D_2 \qquad r_4 = A_4 K_3/A_3 + A_4 D_4$$

$$s_1 = K_1 - D_1 \qquad t_2 = 2A_2 k_1/A_1$$

$$s_2 = K_2 - A_2 K_1/A_1 - A_2 D_2 \qquad t_3 = 2A_3 K_2/A_2$$

$$s_3 = K_3 - A_2 K_2/A_2 - A_3 D_3$$

$$s_4 = A_4 K_3/A_3 - A_4 D_4$$

The system of equations (9) should be solved in the given sequence in the successive intervals of the argument ζ. Because functions f_i and g_i equal zero for negative arguments, hence when solving equations (9) in this sequence the right hand sides of these equations are always known if moment $M_2(\zeta)$ is a known function of time.

MULTI MASS DRIVE SYSTEM

Consider a drive system consisting of (N) rigid bodies and (N-1) shafts, Fig. 2. The rigid bodies have mass moments of inertia J_i, $i = 1,2,\ldots,N$ and are acted upon by external moments M_i. The shafts are deformable only in a torsional manner and centroidal axes of all elements coincide with the main axis of the drive system. The ith shaft, $i = 1,2,\ldots,N-1$ is characterized by the length e_i, density ρ, shear modulus G and polar moment of inertia $J_{o,i}$.

For the continuous model of the drive system discussed here the investigation of angular displacements, velocities and strains of the shafts is reduced to the solution of (N-1) wave equations:

$$\frac{\partial^2 \theta_i(x,t)}{\partial t^2} - c^2 \frac{\partial^2 \theta_i(x,t)}{\partial x^2} = 0, \quad i = 1,2,\ldots N-1 \tag{10}$$

where $c^2 = G/\rho$, with the following boundary conditions:

$$M_1(t) - J_1 \frac{\partial^2\theta_1(x,t)}{\partial t^2} + GJ_{0,1}\frac{\partial\theta_1(x,t)}{\partial x} - D_1\frac{\partial\theta_1(x,t)}{\partial t} = 0 , \quad \text{for } x = 0$$

$$M_i(t) - J_i \frac{\partial^2\theta_i(x,t)}{\partial t^2} + GJ_{0,i}\frac{\partial\theta_i(x,t)}{\partial x} - GJ_{0,i-1}\frac{\partial\theta_{i-1}(x,t)}{\partial x} - D_i\frac{\partial\theta_i(x,t)}{\partial t} = 0 ,$$

$$\text{for } x = \sum_{j=1}^{i-1} e_j , \quad i = 2,3\ldots N-1 \tag{11}$$

$$M_N(t) - J_N \frac{\partial^2\theta_{N-1}(x,t)}{\partial t^2} - GJ_{0,N-1}\frac{\partial\theta_{N-1}(x,t)}{\partial x} - D_N\frac{\partial\theta_{N-1}(x,t)}{\partial t} = 0$$

$$\text{for } x = \sum_{j=1}^{N-1} e_j$$

$$\theta_{i-1}(x,t) = \theta_i(x,t), \text{ for } x = \sum_{j=1}^{i-1} e_j ,$$

$$i = 2,3\ldots N-1$$

and initial conditions:

$$\theta_i(x,t) = \frac{\partial\theta_i(x,t)}{\partial t} = 0, \text{ for } t = 0,$$

$$i = 1,2,\ldots N-1$$

Upon the introduction of non-dimensional quantities (4), the solutions of eqs. (10) are sought in the form

$$\theta_i(x,t) = f_i(t - x + \sum_{j=1}^{i-1} e_j) + g_i(t + x - \sum_{j=1}^{i-1} e_j \tag{12}$$

$$i = 1,2,\ldots,N-1$$

In the formula (12) it is taken already into account that the first perturbation in the i-th shaft occurs in the cross-section $x = e_1 + \ldots + e_{i-1}$ at the instant $t = 0$. Furthermore it is assumed that the functions f_i and g_i are continuous and for negative argument equal zero. Following the same procedure as before we get the final system of eqns. for the unknown functions f_i and g_i:

$$g_i(\zeta) = f_{i+1}(\zeta - e_i) + g_{i+1}(\zeta - e_i) - f_i(\zeta - 2e_i),$$

$$i = 1,2,\ldots N-2$$

$$g''_{N-1}(\zeta) + r_N g'_{N-1}(\zeta) = A_N M_N(\zeta - e_{N-1}) - f''_{N-1}(\zeta - 2e_{N-1}) + s_N f'_{N-1}(\zeta - 2e_{N-1})$$

$$f''_1(\zeta) + r_1 f'_1(\zeta) = M_1(\zeta) - g''_1(\zeta) + s_1 g'1(\zeta)$$

$$f''_i(\zeta) + r_i f'_i(\zeta) = A_i M_i(\zeta) - g''_i(\zeta) + s_i g'_i(\zeta) + t_i f'_{i-1}(\zeta - e_{i-1}) \tag{13}$$

$$i = 2,3,\ldots N-1$$

where:

$$r_1 = K_1 + D_1 , \quad s_1 = K_1 - D_1$$

$$r_i = K_i + \frac{A_i}{A_{i-1}} K_{i-1} + A_i D_i ,$$

$$s_i = K_i - \frac{A_i}{A_{i-1}} K_{i-1} - A_i D_i$$

$$t_i = 2A_i K_{i-1}/A_{i-1} \text{ for } i = 2,3,\ldots N-1$$

$$r_N = A_N(D_N + K_{N-1}/A_{N-1}) ,$$

$$s_N = A_N(K_N - 1/A_{N-1} - D_N)$$

Consider now a multi mass drive system and a torsional dynamic absorber which is attached to it. It is usually, but not necessarily, connected to the free end of the drive system. Fundamentally a torsional absorber assembly consists of an auxiliary mass elastically connected to the main system, Fig. 2 (the absorber is shown by the dash line). Consequently the absorber system can now be regarded as a special case of the mass (N+1) connected to the main system by means of the elastic shaft (N). Hence it is easy to see that the dynamic theory of a multi mass drive system as presented above applies directly to the case of the multi mass drive system with the torsional absorber along with all the assumptions and results. In particular, one can write down the final system of governing equations for this case by simply replacing (N) with (N+1) in eqns. (13), namely:

$$g_i(\zeta) = f_{i+1}(\zeta - e_i) + g_{i+1}(\zeta - e_i) - f_i(\zeta - 2e_i),$$

$$i = 1,2,\ldots N-1$$

$$g''_N(\zeta) + r_{N+1} g'_N(\zeta) = - f''_N(\zeta - 2e_N) + s_{N+1} f'_N(\zeta - 2e_N)$$

$$f''_1(\zeta) + r_1 f'_1(\zeta) = M_1(\zeta) - g''_1(\zeta) + s_1 g'_1(\zeta) \tag{14}$$

$$f''_i(\zeta) + r_i f'_i(\zeta) = A_i M_i(\zeta) - g''_i(\zeta) + s_i g'_i(\zeta) + t_i f'_{i-1}(\zeta - e_{i-1})$$

$$i = 2,3,\ldots N$$

where constants r_i, s_i and t_i are similar to those given by (13), with (N) replaced by (N+1).

It should be stressed that the same numerical procedure applies whether the drive system has a torsional absorber attached to it or not. The system of eqns. (14), and (13), should be solved in the given sequence in the successive intervals of argument ζ where lengths of those intervals depend upon the physical dimensions of shafts. Because functions f_i and g_i equal zero for negative arguments, hence when solving eqns. (14), or (13), in the given sequence the right-hand sides of these eqns. are always known as the moments $M_i(\zeta)$, (usually an oscillating type) are known functions of time.

NUMERICAL PROCEDURE AND EXAMPLE

One notes that all second order differential equations in the set (14) have the form

$$f''(\zeta) + rf'(\zeta) = - g''(\zeta - e) + sg'(\zeta - e) + h(\zeta - k)$$

where $g'(\zeta - e)$ and $h(\zeta - k)$ are given functions, and whose solution for $\xi \geq \xi o$ is as follows

$$f'(\xi) = e^{-r(\xi-\xi_o)} \left[\int_{\xi_o}^{\xi} [-\,g''(\xi-e) + sg'(\xi-e) + h(\xi-k)]\, e^{r(x-\xi_o)} dx + f'(\xi_o) \right]$$

after integrating by parts this finally gives

$$f'(\xi) = f'(\xi_o)e^{-r(\xi-\xi_o)} + g'(\xi_o-e)e^{-r(\xi-\xi_o)} - g'(x-e) + e^{-r(\xi-\xi_o)}$$

$$[(r+s) \int_{\xi_o}^{\xi} g'(x-e)e^{r(x-\xi_o)} dx + \int_{\xi_o}^{\xi} h(x-k)e^{r(x-\xi_o)} dx]$$

Since both integrals in equation (15) can be easily evaluated because $g'(x-e)$ and $h'(x-k)$ are known functions, one can now apply this equation to transform all second order differential equations in (14) into equations suitable for numerical integration.

As an example of the multi mass drive system a two mass drive system with non dimensional lengths of shafts equal to one is considered and the results are compared with the situation when a torsional absorber is attached to the free end of the drive system.

In Fig. 3 angular displacements $\theta(x,t)$ are shown for the cross-section x = 0 of the two mass drive system with the parameters: $\kappa = 0.1$, $D_1 = D_2 = 1.0$, $A_2 = 1.0$. The system is acted upon by the external moment $M_1(t) = 7 * 10^{-5} \sin \omega t$ only, where $\omega = 25/6 * \pi * 10^{-3}$ (n = 750 rpm), $\omega = \pi/1.8 * 10^{-2}$ (n = 1000 rpm), $\omega = 2\pi/1.8 * 10^{-2}$ (n = 2000 rpm) and $\omega = 4\pi/1.8 * 10^{-2}$ (n = 4000 rpm). The torsional absorber (dash line) is characterized by $A_3 = 10/3$ and $\kappa_2 = 0.0177$ [5]. It can be noticed that for these two cases the maximum difference in displacement occurs for $\omega = 25/6 * \pi * 10^{-3}$, and this difference decreases with the increase of angular velocity.

REFERENCES

Mioduchowski, A. and Nadolski, W. (1983). Torsional Deformations of a Shaft with Rotor Having Time-Dependent Moment of Inertia. Meccanica, No. 18, pp. 182-184.

Mioduchowski, A., Pierlorz, A. and Nadolski, W. (1983). Torsional Deformations of a Two Mass Drive System. Proceedings of the 6th World Congress TMM, pp. 543-545.

Nadolski, W. Pielorz, A. and Mioduchowski, A. (1984). Dynamic Investigation of Multi Mass Drive System by Means of Torsional Waves. ZAMM, Vol. 64, No. 9, pp. 427-431.

Nadolski, W., Pielorz, A. and Mioduchowski, A. (1985). Multi Mass Drive Systems With Stepped Shafts. Meccanica, No. 20, pp. 164-170.

Ker Wilson, W. (1968). Practical Solutions of Torsional Vibration Problems. Vol. IV, Devices for Controlling Vibration. 3rd ed., Chapman and Hall, London.

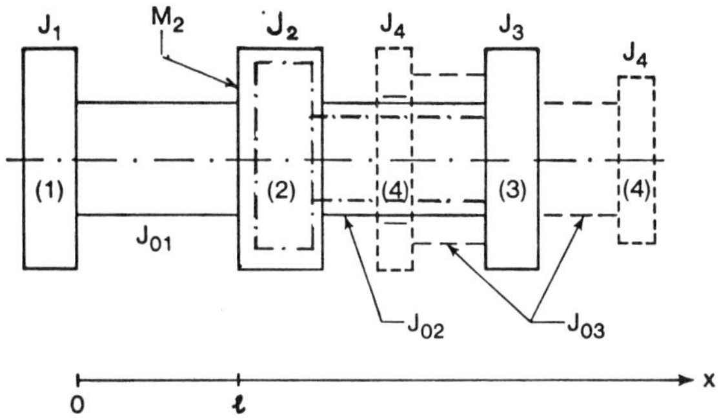

Fig. 1 Three Mass Drive System

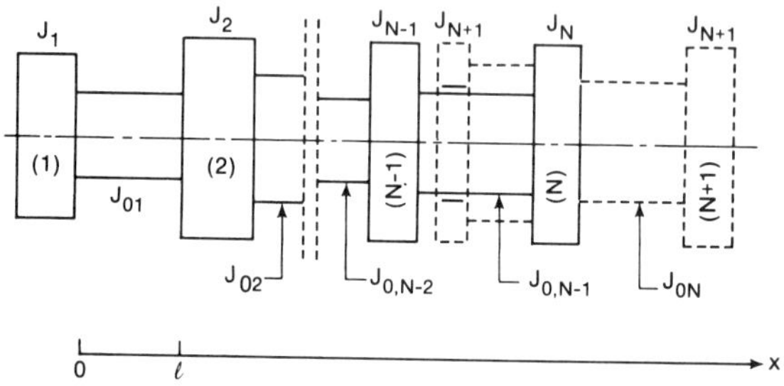

Fig. 2 Multi Mass Drive

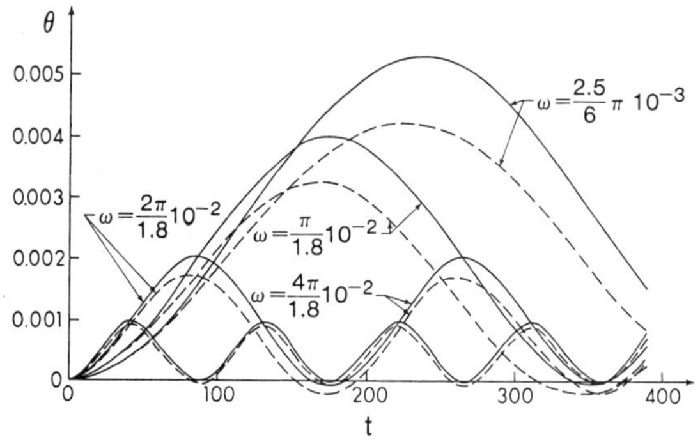

Fig. 3 Angular displacements $\theta(0,t)$ of the two mass drive system without and with an absorber (dash line)

Übersicht: In diser Arbeit wird das kontinuierliche Modell eines Mehrmassen - Antriebssystems betrachtet. Die Methode der dynamischen Analyse beruht auf der Theorie der eindemensionalen Ausbreitung von elastischen Torsionswellen. Es wird ein System dikutiert, das aus N Spindeln und N+1 Starrkörpern besteht. Numerische Ergebnisse werden für ein System von 2 Spindeln und 3 Starrkörpern in grafischer Form dargestellt.

On the Energy Dissipation of a Single Rotor

F. P. J. RIMROTT AND A. FIGUEROA

Mechanical Engineering, University of Toronto, Toronto, Ontario, Canada M5S 1A4

Abstract. When a single solid rotor rotates in a fashion where angular momentum vector and angular velocity vector are not collinear, there is inevitably energy dissipation due to hysteresis. This circumstance makes the attitude of a torque-free rotor unstable, and causes drifting until collinearity is achieved. Or, if an external energy source is available, then its torque must be of just such magnitude and direction, that the work done is equal to the hysteresis losses, in order to check the drifting. If the rotor is constrained such that permanent noncollinearity is impressed on the system, there is consequently a continuous supply of external power needed to replenish the energy dissipated through hysteresis.

In the present paper, the energy dissipation function is established, and shown to be proportional to an innate dissipativity of the rotor system and to the rotor's attitude, i.e. the angle between angular momentum vector the rotor's symmetry axis.

Subsequently, it is shown what deformations of the rotor are associated with energy dissipation. The deformations are found to be a function of the innate dissipativity and the attitude angle only, thus making the deformed rotor configuration constant, provided a suitable floating coordinate system is chosen to describe the deformations. As a consequence of the constant configuration, the concept of rigidity, so essential in rotor dynamics, need not be relinquished.

Keywords. Gyroscopes; space vehicles; stability.

NOMENCLATURE

A, B, C	Principal inertia moments of gyro in undeformed state, Ws^3
C	Centre of mass
$Cuvz$	Floating coordinate system along principal axes of body in undeformed state
$Cxyz$	Coordinate system along principal axes of body in undeformed state
D	Dissipation energy, Ws
$\mathbf{H}$	Angular momentum, Ws^2
$[I]$	Inertia tensor matrix, Ws^3
I_{uz} etc.	Inertia products of gyro in deformed state in principal coordinates of the undeformed state, Ws^3
$\mathbf{M}$	Torque, Ws
U	Work, Ws
$\mathbf{e}_u$ etc.	Standard basis vectors
$\boldsymbol{\Omega}$	Angular velocity of $Cuvz$ coordinate system, rad/s
β	Dissipation coefficient, Ws^6
ν	Nutation (cone) angle ("attitude"), rad
$\dot{\nu}$	Attitude drift, rad/s
$\dot{\sigma}$	Spin, rad/s
$\boldsymbol{\omega}$	Angular velocity of $Cxyz$ coordinate system, rad/s

THE COLLINEARITY THEOREMS

Energy dissipation due to hysteresis is the rule for any solid rotor when subjected to load alternation. Based on this fact, three collinearity theorems have been stated by Rimrott (1985), of which the first refers to the practically important case of the attitude drift of a torquefree gyro, while the second and third are of a more general nature and include the first as a special case. Summerized the collinearity theorems say the following:

There is no energy dissipation, when angular velocity and angular momentum are collinear.

When angular velocity and angular momentum are not collinear, there will be internal energy dissipation (1)
(a) with the energy drawn from the body's kinetic energy in case of a torquefree gyro,
(b) with the energy drawn from the work done by the external torque in case of a driftfree gyro, and
(c) with the energy drawn from a combination of the above in other cases.

The term 'angular velocity' used in the theorems stated above refers to the angular velocity of the principal coordinate system fixed to the body in its *undeformed* state and originating at its mass centre.

NEAR-RIGID SOLID

Rimrott and Yu (1987) have shown, that a near-rigid solid whose inertia tensor has the form

$$\begin{bmatrix} A & 0 & 0 \\ 0 & A & 0 \\ 0 & 0 & C \end{bmatrix} \qquad (2)$$

in the undeformed state, assumes the form

$$[I] = \begin{bmatrix} A & 0 & I_{uz} \\ 0 & A & I_{vz} \\ I_{uz} & I_{vz} & C \end{bmatrix} \qquad (3)$$

in the deformed state, when one and the same floating $Cuvz$ coordinate system, associated with the principal axes in the undeformed state is used. The inertia products are, of course, much smaller in magnitude than the inertia moments.

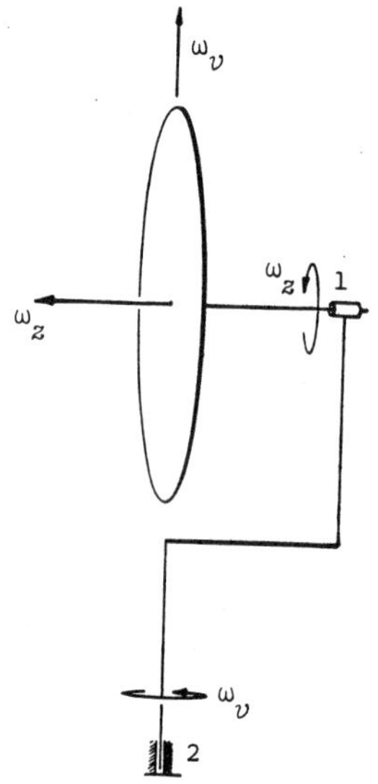

Fig. 1. Disk gyro in Undeformed State

INTERNAL ENERGY DISSIPATION AND ASSOCIATED DEFORMATION

In a viscously damped oscillator, the rate $\dot{d}$ of energy dissipation follows a relationship $\dot{d} = c_1 \omega^2 X^2$, where X is the vibration amplitude, and ω the frequency. The 'frequency' with which a gyro body oscillates is the spin $\dot{\sigma}$, the vibration 'amplitude' is proportional to the alternating accelerations of the gyro body which contain terms with coefficients $\omega_v^2 \omega_z^2$. Thus for a whole gyro body

$$\dot{D} = \beta \omega_v^2 \omega_z^2 \dot{\sigma}^2 \qquad (4)$$

On the other hand, there is a deformation, whose elastic part is proportional to the Coriolis accelerations $\omega_v \omega_z$, and whose hysteretic part is proportinal to the Coriolis accelerations and the spin $\dot{\sigma}$ (Zhang and Ling, 1985). When expressed by an inertia product,

$$I_{uz} = -\beta \omega_v \omega_z \dot{\sigma} \qquad (5)$$

A CONSTRAINED ROTOR

In Figure 1 a flexible rotor, constrained to move at a constant ω_v and a constant ω_z, is shown.

There will be elastic deformation as shown in Figure 2, leading to an I_{vz}. The effects of the presence of I_{vz} can be shown to be insignificant (Rimrott and Kessaris, 1987) and we shall therefore disregard the elastic deformations by setting $I_{vz} = 0$.

More significant are the deformations due to the hysteresis of the material, which lead to an I_{uz}. Since the spin of the gyro in the present example is $\dot{\sigma} = \omega_z$, we obtain, from equation (5)

$$I_{uz} = -\beta \omega_v \omega_z^2 \qquad (6)$$

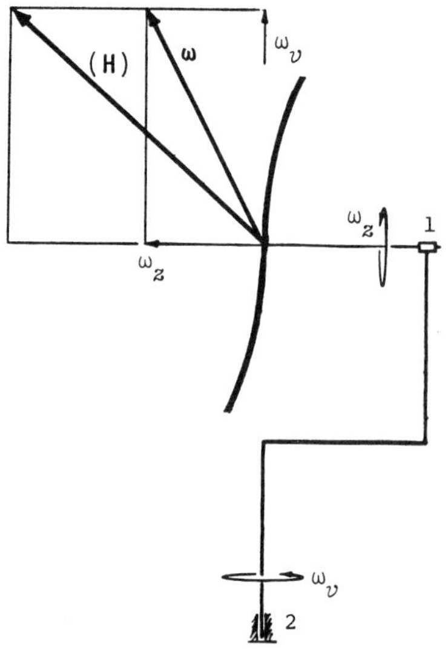

Fig. 2. Deformed State, with I_{vz}

Note that I_{uz} = constant, since ω_v = constant and ω_z = constant.

With hysteresis present, and elasticity disregarded,

$$[I] = \begin{bmatrix} A & 0 & I_{uz} \\ 0 & A & 0 \\ I_{uz} & 0 & C \end{bmatrix} \qquad (7)$$

$$\boldsymbol{\omega} = [\mathbf{e}_u \quad \mathbf{e}_v \quad \mathbf{e}_z] \begin{bmatrix} 0 \\ \omega_v \\ \omega_z \end{bmatrix} \qquad (8)$$

$$\boldsymbol{\Omega} = [\mathbf{e}_u \quad \mathbf{e}_v \quad \mathbf{e}_z] \begin{bmatrix} 0 \\ \omega_v \\ 0 \end{bmatrix} \tag{9}$$

$$\mathbf{H} = \{\mathbf{e}\}^T[I]\{\omega\} \tag{10}$$

$$\mathbf{H} = [\mathbf{e}_u \quad \mathbf{e}_v \quad \mathbf{e}_z] \begin{bmatrix} I_{uz}\omega_z \\ A\omega_v \\ C\omega_z \end{bmatrix} \tag{11}$$

$$T = \frac{1}{2}\mathbf{H}\cdot\boldsymbol{\omega} = \frac{1}{2}(A\omega_v^2 + C\omega_z^2) \tag{12}$$

The torque that is applied to the gyro is

$$\mathbf{M} = \dot{\mathbf{H}} = \overset{0}{\mathbf{H}} + \boldsymbol{\Omega}\times\mathbf{H} \tag{13}$$

$$\mathbf{M} = [\mathbf{e}_u \quad \mathbf{e}_v \quad \mathbf{e}_z] \begin{bmatrix} C\omega_v\omega_z \\ 0 \\ -I_{uz}\omega_v\omega_z \end{bmatrix} \tag{14}$$

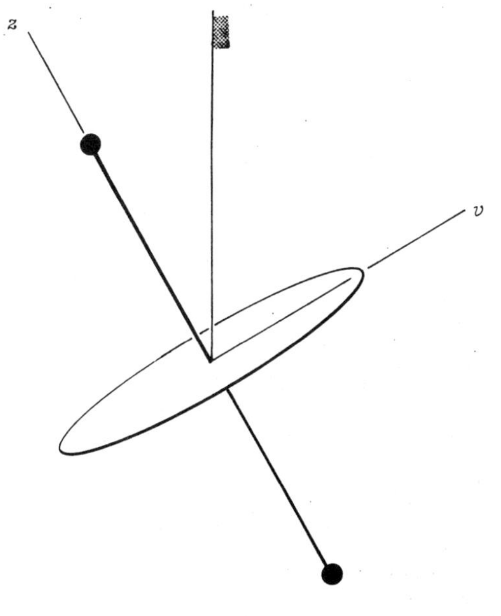

Fig. 3. Torquefree Gyro in Undeformed State

The power that must be supplied to the gyro (via the electric motors 1 and 2) is

$$\dot{U} = \mathbf{M}\cdot\boldsymbol{\omega} = -I_{uz}\omega_v\omega_z^2 \tag{15}$$

Invoking equation (5) we find that

$$\dot{U} = \beta\omega_v^2\omega_z^4 \tag{16}$$

which equals the dissipation rate $\dot{D}$ of equation (4), since $\dot{\sigma} = \omega_z$ in the present application.

From equation (6), the inertia product is seen to be constant. As a consequence the equations derived in the dynamics of rigid bodies apply, in spite of the fact that the gyro body is deformed, leading to the conclusion that a term such as 'dynamics of bodies of constant configuration' might be more approprite than 'dynamics of rigid bodies'.

A TORQUEFREE ROTOR

In a torquefree rotor, such as the one shown in Figure 3, there will again be elastic deformations leading to an I_{vz} (Figure 4) which can be shown to have only insignificant effects on the motion. Hysteresis, on the other hand, represented by an I_{uz}, has a significant effect, in that it causes the attitude ν to drift.

In the present case the spin is

$$\dot{\sigma} = \frac{A - C}{A}\omega_z \tag{17}$$

thus, from equation (5),

$$I_{uz} = -\beta\frac{A - C}{A}\omega_v\omega_z^2 \tag{18}$$

Since

$$\omega_v = \frac{H}{A}\sin\nu \tag{19a}$$

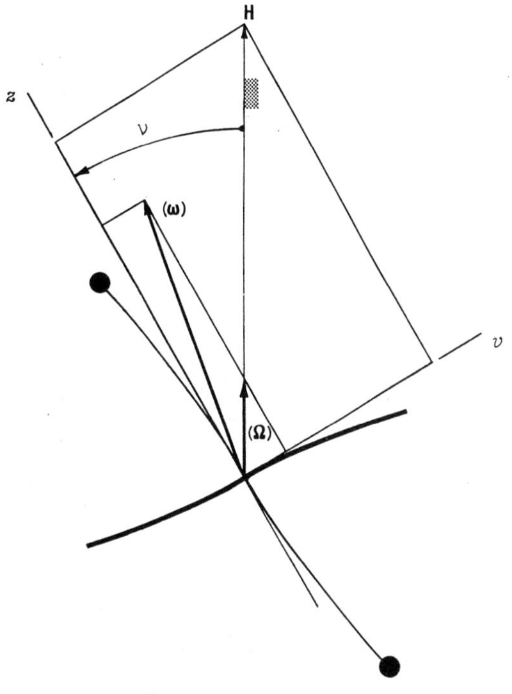

Fig. 4. Deformed State, with I_{vz}

and

$$\omega_z = \frac{H}{C}\cos\nu \tag{19b}$$

one obtains

$$I_{uz} = -\beta\frac{(A - C)H^3}{2A^2C^2}\cos\nu\sin 2\nu \tag{20}$$

and an indication how the deformation I_{uz} associated with hysteresis is dependent upon the attitude angle ν.

With hysteresis present, and elasticity disregarded,

$$[I] = \begin{bmatrix} A & 0 & I_{uz} \\ 0 & A & 0 \\ I_{uz} & 0 & C \end{bmatrix} \tag{21}$$

$$\boldsymbol{\omega} = [\mathbf{e}_u \quad \mathbf{e}_v \quad \mathbf{e}_z] \begin{bmatrix} \dot{\nu} \\ \omega_v \\ \omega_z \end{bmatrix} \tag{22}$$

$$\boldsymbol{\Omega} = [\mathbf{e}_u \quad \mathbf{e}_v \quad \mathbf{e}_z] \begin{bmatrix} \dot{\nu} \\ \frac{H}{A}\sin\nu \\ \frac{H}{A}\cos\nu \end{bmatrix} \tag{23}$$

$$\mathbf{H} = [\mathbf{e}_u \quad \mathbf{e}_v \quad \mathbf{e}_z] \begin{bmatrix} 0 \\ H\sin\nu \\ H\cos\nu \end{bmatrix} = \text{constant} \tag{24}$$

with $H\sin\nu = A\omega_v$ (25)

and $H\cos\nu = C\omega_z + I_{uz}\dot{\nu} \simeq C\omega_z$ (26)

since

$$H_u = A\dot{\nu} + I_{uz}\omega_z = 0 \tag{27}$$

the drifting becomes

$$\dot{\nu} = -\frac{I_{uz}H}{AC}\cos\nu \tag{28}$$

Or, with the help of equation (20)

$$\dot{\nu} = \beta\frac{(A - C)H^4}{2A^3C^3}\cos^2\nu\sin 2\nu \tag{29}$$

i.e. the attitude is drifting; towards $\nu = 90°$ if $A > C$, and towards $0°$ if $A < C$.

There is no torque applied, thus $\mathbf{M} = \dot{\mathbf{H}} = \mathbf{0}$, leading with the help of equation (13) to

$$\dot{\omega}_v = H\dot{\nu}\cos\nu/A \tag{30a}$$

$$\dot{\omega}_z = -H\dot{\nu}\sin\nu/C \tag{30b}$$

There is also no power supplied to the gyro. The energy D dissipated is fed directly by a loss of kinetic energy T.

With

$$T = \frac{1}{2}\mathbf{H}\cdot\boldsymbol{\omega} = \frac{H^2}{2}\left(\frac{\sin^2\nu}{A} + \frac{\cos^2\nu}{A}\right) \tag{31}$$

the rate of kinetic energy change is

$$\dot{T} = -\frac{(A - C)H^2\sin 2\nu}{2AC}\dot{\nu} \tag{32}$$

Since there is a one-to-one relationship

$$\dot{T} = -\dot{D} \tag{33}$$

we find, involving equation (29),

$$\dot{D} = \beta\frac{(A - C)^2H^2}{4A^4C^4}\cos^2\nu\sin^2 2\nu \tag{34}$$

which can be reduced to

$$\dot{D} = \beta\omega_v^2\omega_z^2\dot{\sigma}^2 \tag{35}$$

In the present example, I_{uz} is strictly speaking, not constant with time. Its change, however, is so small that it can be considered a constant without loss of accuracy.

CONCLUSION

The effect of elastic material hysteresis on rotor behaviour in case of noncollinearity of angular momentum and angular velocity is demonstrated to lead to an inertia product of constant or almost constant magnitude which allows the treatment of the problem by the methods of dynamics of bodies of constant configuration.

REFERENCES

Rimrott, F.P.J. (1985). Dissipative Rigids in Gyrodynamics. ZAMM, 65, 7, 287-300

Rimrott, F.P.J., Yongxi Yu (1987). The Near-Rigid Solid in Gyrodynamics. In G. Faulkner (Ed.), Proceedings CANCAM 87, Edmonton

Rimrott, F.P.J., Anastasios Kessaris (1987). Elasticity and Hysteresis Effects on the Behaviour of Axisymmetric Solid Gyros. In G. Faulkner (Ed.), Proceedings CANCAM 87, Edmonton

Zhang, W., F.H. Ling (1986). Dynamic Stability of the Rotating Shaft made of Boltzmann Visco-elastic Solid. ASME Journal of Applied Mechanics, 53, 424-429

ZUSAMMENFASSUNG

Wenn Drallvektor und Winkelgeschwindigkeitsvektor eines Rotors nicht kollinear sind, dann findet unweigerlich eine Dissipation von mechanischer Energie innerhalb des Rotorsystems statt. Als Folge dieser Energiedissipation driften drehmomentfreie Rotoren, bei gleichbleibendendem Drall und bei sinkender kinetischer Energie, bis in eine Winkellage, bei der Kollinearität und ein Minimum an kinetischer Energie erreicht sind. Nur wenn ein genau abgestimmtes äußeres Drehmoment vorhanden ist, welches dem Rotorsystem laufend die verlorengehende Energie ersetzt, kann dem Driften Einhalt geboten werden.
Wenn ein Rotorsystem durch andauernde Nichtkollinearität gekennzeichnet ist, dann läßt sich automatisch sagen, daß dem System dauernd mechanische Energie zugeführt werden muß.

An Hand einer an der Nabe fest eingebauten rotierenden Scheibe wird gezeigt, welche Gestaltsänderungen auftreten und welchen Einfluß diese auf das Rotorverhalten haben. Es stellt sich heraus, daß die Gestaltsänderungen konstant sind, wenn sie in einem geeigneten Zwischenkoordinatensystem ausgedrückt werden, d. h. der Rotor ist zwar verformt, kann aber weiterhin als Körper konstanter Gestalt angesehen werden, ein Umstand von grundlegender Wichtigkeit in der Dynamik. Nicht ganz so günstig ist die Lage beim zweiten Beispiel, bei dem an Hand eines drehmomentfreien Kreisels gezeigt wird, daß die Gestaltsänderung nur annähernd konstant ist. Jedoch geht das Driften typischerweise so langsam vor sich, daß auch hier mit einer konstanten Gestaltsänderung gerechnet werden kann, ohne zu große Fehler zu verursachen.

Application D'une Methode Pseudo-Modale a la Determination du Comportement Dynamique des Monorotors et des Multirotors Coaxiaux

P. BERTHIER, G. FERRARIS, J. DER HAGOPIAN AND M. LALANNE

Laboratoire de Mécanique des Structures, U.A. C. N. R. S. 862, Institute National des Sciences Appliquées, 20, avenue Albert Einstein, 69621 Villeurbanne Cédex, France

Résumé : Le calcul des vitesses critiques et de la réponse au balourd de monorotors ou multirotors coaxiaux est effectué à l'aide de deux méthodes, l'une directe, l'autre pseudo-modale en utilisant une base réduite réelle. Les temps de calcul et la précision des deux méthodes sont comparés à l'aide de deux exemples industriels, une turbine à gaz et un moteur d'avion moderne. La méthode pseudo-modale permet avec une excellente précision de réduire largement les temps de calcul.

Mots clés : Rotor, dynamique, méthodes numériques.

Keywords : Rotor, dynamic, numerical methods.

INTRODUCTION

Le calcul du comportement dynamique des machines tournantes s'effectue fréquemment à partir d'une modélisation par éléments finis, (Childs, 1976 ; Glasgow, 1980 ; Li, 1981, Nelson, 1976, 1981 ; Rouch, 1980 ; Zorzi ; Berthier, 1983). Les équations du mouvement comprennent souvent plusieurs centaines de degrés de liberté et les matrices du système ne sont pas toutes symétriques par suite de l'effet de Coriolis et des caractéristiques de raideur et d'amortissement des paliers et roulements utilisés. Les vitesses critiques sont déduites d'un calcul de valeurs propres en général complexes et la réponse au balourd est déduite d'un balayage en vitesse de rotation.

Deux méthodes sont comparées ici, une directe et une pseudo-modale :

- la méthode directe utilise un processus itératif pour déterminer les premières valeurs propres et vecteurs propres complexes du système et la résolution linéaire du système complet pour calculer la réponse au balourd.
- la méthode pseudo-modale nécessite un changement de base réduit réel, premiers modes du système rendu symétrique et non amorti, afin d'obtenir un système différentiel réduit. Le système est alors utilisé pour calculer directement les valeurs propres complexes et la réponse au balourd.

Les temps de calcul et les précisions sont comparés sur deux exemples industriels : une turbine à gaz et un moteur d'avion qui est un multirotor coaxial.

MISE EN EQUATIONS

Monorotor

La mise en équations par éléments finis d'un monorotor a déjà été présentée en détail par Berthier, Ferraris et Lalanne, 1983. Il est rappelé que les disques sont rigides, que les arbres sont non amortis et schématisés par des éléments de poutres avec effets secondaires de cisaillement et d'inertie de rotation, que les paliers symétriques ou dissymétriques sont caractérisés par des termes constants. Dans ces conditions, les équations ont la forme générale :

$$M\,\ddot{\delta} + C(\Omega)\,\dot{\delta} + K\delta = F(\Omega,t) \qquad (1)$$

Avec :

- δ , vecteur des N coordonnées généralisées
- M , matrice masse symétrique
- C , matrice comportant une partie antisymétrique due à l'effet de Coriolis et une partie non symétrique correspondant à l'amortissement visqueux des paliers.
- K , matrice raideur dont la dissymétrie est due à la raideur des paliers.
- F , vecteur des forces généralisées représentant ici l'effet de balourd.
- Ω , vitesse de rotation.
- t , temps.

Multirotor coaxial

La mise en équations des multirotors coaxiaux a été présentée par Berthier, Ferraris et Lalanne, 1986. Chaque rotor est caractérisé par des éléments finis du même type que ceux d'un monorotor, mais possède une vitesse propre de rotation. Les différents rotors sont liés par des paliers inter-arbres et les vitesses propres de rotation ne sont pas indépendantes. Par souci de simplification, les équations sont présentées ici en supposant deux rotors et un seul élément de liaison. Les équations de chaque rotor en utilisant les notations précédentes sont les suivantes :

Rotor 1 (vitesse angulaire Ω_1) :

$$M_1\,\ddot{\delta}_1 + C_1(\Omega_1)\dot{\delta}_1 + K_1\delta_1 = F_1(\Omega_1,t) - R_1 \qquad (2)$$

Rotor 2 (vitesse angulaire Ω_2) :

$$M_2\,\ddot{\delta}_2 + C_2(\Omega_2)\dot{\delta}_2 + K_2\delta_2 = F_2(\Omega_2,t) - R_2 \qquad (3)$$

avec R_1, R_2 les actions des rotors 1 et 2 sur le palier inter-arbre.

L'équilibre de l'élément de liaison s'écrit en négligeant sa masse :

$$R_1 + R_2 = 0 \qquad (4)$$

avec R_1 et R_2 dépendant des déplacements et vitesses des noeuds de chaque rotor en regard du palier.

En remplaçant R_1 et R_2 fonctions des déplacements

et vitesses dans (2) et (3) et en utilisant (4), il vient :

$$M\delta^{\bullet\bullet} + C(\Omega_1,\Omega_2)\delta^{\bullet} + K\delta = F(\Omega_1,\Omega_2,t) \quad (5)$$

qui est une équation différentielle formellement identique à (1). Les rotors sont de plus couplés aérodynamiquement et il faut associer à (5) la loi reliant les vitesses de rotation du type :

$$\Omega_1 = a\,\Omega_2 + b \quad (6)$$

où a et b sont des constantes.

RESOLUTION DES EQUATIONS

La recherche des vitesses critiques passe par celle des valeurs propres et vecteurs propres complexes du système sans second membre de forme générale :

$$M\delta^{\bullet\bullet} + C\delta^{\bullet} + K\delta = F \quad (7)$$

La réponse au balourd est la solution en régime permanent de (7) avec second membre.

Méthode directe

La méthode bi-itérative utilisée est celle de Borri et Mantegazza, 1977. Elle conduit à la recherche des valeurs propres complexes d'un système réduit calculées par la méthode du Q.R. présentée par Smith, 1976. Les valeurs propres sont de la forme a + jω. Le terme ω caractérise la pulsation et a est représentatif de l'amortissement. Par analogie avec un système à un degré de liberté :

$$a + j\omega = \alpha\omega_o + j\omega_o\sqrt{1 - \alpha^2} \quad (8)$$

où ω_o est la pulsation du système non amorti et α le facteur d'amortissement visqueux. Les intersections des courbes, évolution des pulsations propres ω en fonction de la vitesse de rotation Ω avec la droite $\omega = \Omega$, ou $\omega = \Omega_1$ et $\omega = \Omega_2$ dans le cas des multirotors, permettent de déterminer les vitesses de rotation critiques.
Dans le cas du calcul de la réponse au balourd, le second membre étant harmonique, les déplacements nodaux δ sont recherchés sous forme harmonique pour chaque valeur de la vitesse de rotation Ω et obtenus par identification des termes en sinus et cosinus dans les deux membres de (7). Cette méthode a été préférée à une méthode modale utilisant la base complexe calculée à l'occasion de la recherche des valeurs propres car elle est plus rapide et exacte.

Méthode pseudo-modale

Dans un premier temps, on recherche une base modale réelle réduite d'un système voisin de (7). Les effets de Coriolis, l'amortissement des paliers sont négligés et la matrice raideur est arbitrairement symétrisée. Le changement de base :

$$\delta = \phi\, p \quad (9)$$

est déduit de la solution de :

$$M\delta^{\bullet\bullet} + K^{*}\delta = 0 \quad (10)$$

où K^* est symétrique.
Les valeurs propres et vecteurs propres de (10) sont réels et donc facilement calculables par la méthode itérative présentée par Bathe et Wilson, 1973. La base modale du système (10) est prise ensuite comme base pseudo-modale du système (7). Ceci conduit après prémultiplication de (7) par la transposée de ϕ au système réduit :

$$\phi^t M\phi\, p^{\bullet\bullet} + \phi^t C\phi\, p^{\bullet} + \phi^t K\phi\, p = \phi^t F \quad (11)$$

soit :

$$m\, p^{\bullet\bullet} + c\, p^{\bullet} + k\, p = f \quad (12)$$

avec m, matrice symétrique
c,k, matrices dissymétriques
f, vecteurs des forces pseudo-modales.
Les solutions de (12) sans second membre sont cherchées sous la forme :

$$p = P\, e^{rt} \quad (13)$$

soit :

$$(r^2 m + rc + k)\, P = 0 \quad (14)$$

ou bien :

$$\begin{vmatrix} 0 & I \\ -k^{-1}m & -k^{-1}c \end{vmatrix} \begin{vmatrix} rP \\ P \end{vmatrix} = \frac{1}{r} \begin{vmatrix} rP \\ P \end{vmatrix} \quad (15)$$

avec I matrice unité et -1 symbole d'inversion.
Les valeurs propres de (15) sont alors obtenues par la méthode du Q.R. Lors du calcul de la réponse au balourd, les variables pseudo-modales p de (12) sont cherchées sous forme harmonique pour chaque valeur de la vitesse de rotation et l'on revient aux variables réelles par (9).

APPLICATIONS

Turbine à gaz

Le rotor et sa modélisation par éléments finis est présenté sur la Fig.1. Les paliers sont différents et dissymétriques. Le diagramme de Campbell et la réponse au balourd ont déjà été présentés par Berthier, 1983.

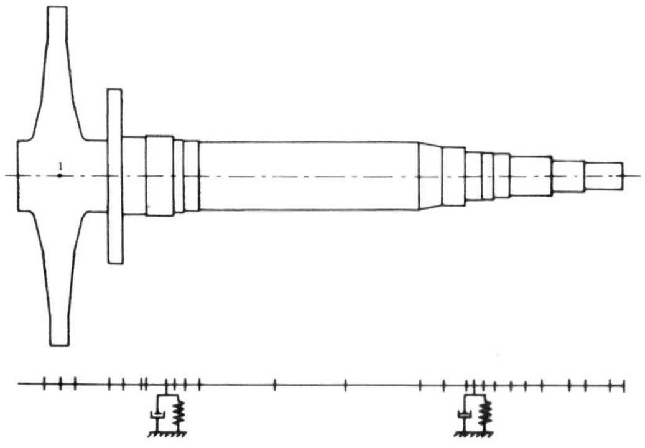

Fig.1. Turbine à gaz

Les calculs par la méthode directe sont effectués avec 10 + 4 vecteurs propres et la précision au cours des itérations est testée sur les dix premiers. Les calculs par la méthode pseudo-modale sont effectués avec 2 + 4, 4 + 4, 6 + 4, 8 + 4, 10 + 4 vecteurs avec la précision testée sur les 2, 4, 6, 8, 10 premiers vecteurs.
Les vitesses critiques sont présentées dans le tableau 1, Les fréquences propres $N = \omega/2\pi$ et les facteurs d'amortissements α à 30.000 tr/mn sont présentés dans le tableau 2. Les temps de calculs sont indiqués en secondes et les signes + ou - indiquent le sens de precession directe ou inverse. Les fréquences obtenues par la méthode directe sont exprimées en Hertz et les vitesses critiques en tours/minute. Les valeurs obtenues par la méthode pseudo-modale sont exprimées en valeur relative par rapport à celles de la méthode directe.
La méthode pseudo-modale conduit à des résultats identiques à ceux de la méthode directe en un

temps de calcul beaucoup moins important puisque le rapport des temps de calcul est approximativement de vingt à un.
Les réponses au balourd sont présentées dans le tableau (3). Seules sont comparées les amplitudes maximum à la résonance et les vitesses de rotation correspondantes. Les temps de calculs non indiqués sur le tableau sont là encore favorables à la méthode pseudo-modale dans un rapport de 2 à 1 car le temps de calcul de la base pseudo-modale est de l'ordre de celui de cinquante pas de calcul de réponse au balourd.

	Méthode	Méthode pseudo-modale				
	directe	2+4	4+4	6+4	8+4	10+4
N_1	10600.	.005	.0	.0	.0	.0
N_2	16980.	.019	.002	.0	.0	.0
N_3	28357.		.005	.0	.0	.0
N_4	30212.		-.001	-.001	-.002	-.002
				.001	.0	.0

Tableau 1 : Turbine à gaz - Vitesses critiques

	Méthode directe		Méthode pseudo-modale									
			2+4		4+4		6+4		8+4		10+4	
Temps	73.3		1.1		1.6		2.2		2.8		3.5	
	N	α	N	α	N	α	N	α	N	α	N	α
N_1^-	116.61	.0011	.030	.454	.003	-.182	.0	.0	.0	.0	.0	.0
N_1^+	310.27	.0080	.032	.075	.004	.087	.0	.012	.0	.0	.0	.0
N_2^-	470.22	.0080	.017	.250	.006	.150	.0	.0	.0	.0	.0	.0
N_3^-	503.77	.0173	.027	.289	-.002	.029	-.002	.023	-.002	.006	-.002	.006
N_2^+	595.73	.0118	.012		.006	-.085	.0	.017	.0	.017	.0	.017
N_3^+	726.18	.0172		-.398	.015	.099	.003	-.023	-.005	-.017	.002	-.017
N_4^-	872.74	.0188		-.372	.0	.005	.0	.0	-.001	.0	.001	.0
N_4^+	1261.34	.0150			.030	-.573	.006	-.080	.0	-.027	.0	-.027
N_5^-	1415.13	.0168					.007	.030	.003	.048	.003	.042
N_5^+	1662.75	.0163					.010	-.050	.006	-.049	.005	-.025

Tableau 2 : Turbine à gaz - Fréquences propres et facteurs d'amortissement à 30.000 tr/mn.

	Méthode directe		Méthode pseudo-modale									
			2+4		4+4		6+4		8+4		10+4	
	Ω	Max.	Ω	Max.	Ω	Max.	Ω	Max.	Ω	Max.	Ω	Max.
N_1	10585.	.147E-3	.005	-.075	.0	.068	.0	.007	.0	.0	.0	.0
N_2	16963.	.590E-3	.019	-.017	.002	-.075	.0	-.019	.0	-.003	.0	.0
N_3	28360.	.540E-4	.016	-.213	.006	-.095	.001	.018	.001	.007	.0	.010
N_4	30296.	.294E-4	.028	.088	.001	-.017	.001	-0.20	.0	-.007	.0	-.007
N_5	36350.	.133E-4	.021	.060	.010	.015	.003	-.001	.001	.0	.001	.0

Tableau 3 : Turbine à gaz - Réponse au balourd.

Moteur d'avion

Le moteur d'avion civil, Fig.2, est représenté schématiquement sur la Fig.3. On y distingue :
- un rotor haute pression (HP),
- un rotor basse pression (BP),
- une enveloppe fixe, stator (ST),
- un roulement inter-arbre (RI),
- des paliers ou éléments de liaison (EL).

Les ensembles disques-aubes sont pris en compte dans le calcul, mais ne sont pas portés sur le schéma par souci de clarté.
La modélisation de ce système est complexe. Certains éléments du moteur sont représentés par des poutres et des disques rigides ; d'autres éléments à géométrie compliquée sont pris en

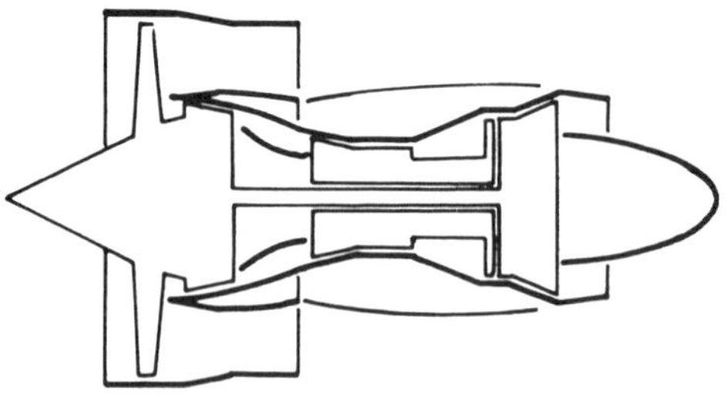

Fig.2. Moteur d'avion

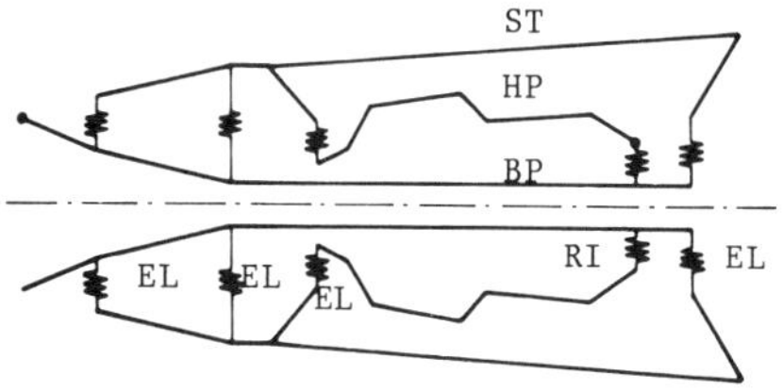

Fig.3. Moteur d'avion. Schématisation

	Méthode directe	Méthode pseudo-modale 26+4	30+4	32+10
Temps	485	21	28	30
N_1^+				
N_2^+	1154.26	.0001	.0001	.0
N_3^+	2506.98	.0004	.0004	.0001
N_4^+	3456.79	-.0034	.0001	.0
N_5^+	4254.84	.0006	.0006	.0003
N_6^+	5707.17	.0019	.0019	.0008
N_{10}^+	2058.42	.0047	.0030	.0006
N_{11}^+	3256.65	.0021	.0020	.0009
N_{12}^+	3315.82	.1138	.1127	.0091

Tableau 4 : Moteur - Vitesses critiques.

compte par l'introduction de termes de raideur et de masse résultant de la condensation d'un modèle élément fini.
Les résultats relatifs à ce moteur ont déjà été présentés en détail par Berthier, 1986. Les calculs par la méthode directe sont effectués avec 32 +10 vecteurs propres et les calculs par la méthode pseudo-modale avec 26 + 4, 30 + 4, 32 + 10 vecteurs propres. Les vitesses critiques du moteur sont présentées dans le tableau 4. Les fréquences propres, lorsque la vitesse de rotation du rotor B.P. est de 5.000 tr/mn sont présentées dans le tableau 5. Les unités employées et la forme des résultats sont les mêmes que pour la turbine à gaz. A précision égale, la méthode pseudo-modale améliore les temps de calcul dans un rapport de quinze à un.

	Méthode directe	Méthode pseudo-modale 26+4	30+4	32+10
Temps	485	21	28	30
N_1^-	3.5676	.0001	.0001	.0
N_1^+	3.8945	.0001	.0001	.0
N_2^-	16.497	.0001	.0001	.0
N_2^+	19.870	.0001	.0001	.0
N_3^-	29.118	.0002	.0001	.0
N_3^+	42.026	.0007	.0007	.0002
N_4^-	40.504	.0012	.0012	.0004
N_4^+	60.685	.0001	.0001	.0
N_5^-	50.631	.0016	.0013	.0006
N_5^+	71.108	.0006	.0006	.0003
N_6^-	74.540	.0002	.0001	.0
N_6^+	94.485	.0017	.0016	.0007
N_7^-	81.019	.0009	.0007	.0002
N_7^+	108.17	.0006	.0004	.0
N_8^-	103.33	.0016	.0015	.0005
N_8^+	140.05	.0011	.0006	.0001
N_9^-	121.66	.0004	.0003	.0
N_9^+	179.38	.0014	.0006	.0
N_{10}^-	159.60	.0286	.0284	.0049
N_{10}^+	206.57	.0040	.0023	.0003
N_{11}^-	175.36	.0238	.0240	.0031
N_{12}^-	213.09	.0	.0	.0
N_{11}^+	214.28	.0031	.0027	.0004
N_{12}^+	220.28	.0365	.0360	.0034

Tableau 5 : Moteur - Fréquences propres à 5.000 tr/mn.

La réponse au balourd est effectuée sans amortissement. La réponse du moteur à un balourd porté par le rotor BP est tracée sur la Fig.4. Le tracé ne laisse pas apparaître de différence entre les deux méthodes car seuls les premiers modes du système interviennent dans la réponse. La réponse à un balourd porté par le rotor HP est tracée sur la Fig.5. Les courbes sont superposées si le nombre de vecteurs choisis dans la méthode pseudo-modale est (32 + 10). Elles ne le sont plus lorsque le nombre de vecteurs de la réponse pseudo-modale est plus faible (32 + 4) car la précision sur la vitesse critique N_{12}^+ n'est pas suffisante comme l'indique le tableau 4. Les temps de calcul sont là encore favorables à la méthode pseudo-modale.

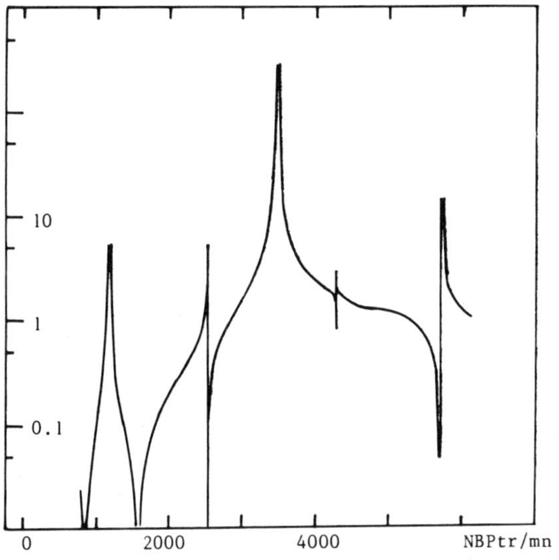

Fig.4. Moteur d'avion. Réponse au balourd BP

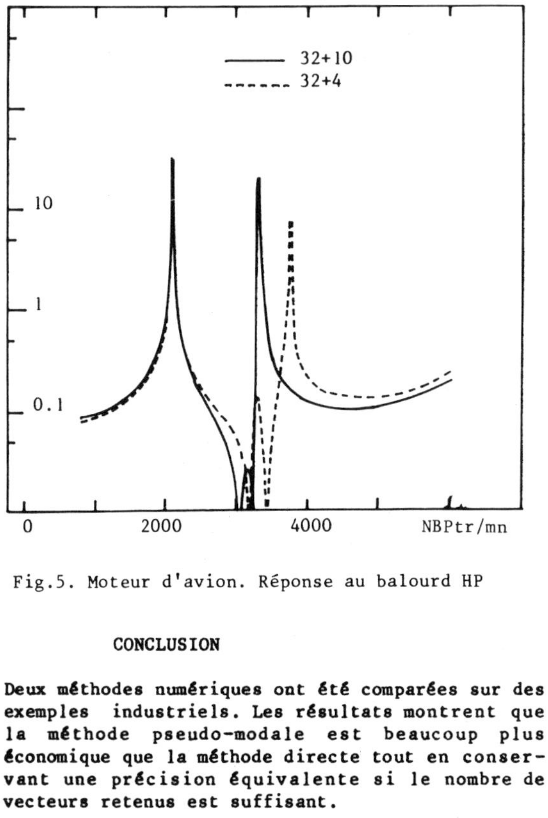

Fig.5. Moteur d'avion. Réponse au balourd HP

CONCLUSION

Deux méthodes numériques ont été comparées sur des exemples industriels. Les résultats montrent que la méthode pseudo-modale est beaucoup plus économique que la méthode directe tout en conservant une précision équivalente si le nombre de vecteurs retenus est suffisant.

REMERCIEMENTS

Ces recherches ont bénéficié de l'aide de la D.R.E.T., de la S.N.E.C.M.A. et de la Société CREUSOT-LOIRE (FRAMATOME). Une partie des moyens de calcul utilisés ont été attribués par le Conseil Scientifique du Centre de Calcul Vectoriel pour la Recherche.

REFERENCES

Bathe, K.J. and E.L. Wilson (1973). Solution methods for eigenvalue problems in structural mechanics. Int. J. Numer. Methods Eng., vol.6, pp.213-226.

Berthier, P., G. Ferraris, and M. Lalanne (1983). Prediction of critical speeds, unbalance and nonsynchronous forced response of rotors. The Shock and Vibration Bulletin. Bulletin 53, Part.4, pp.103-111.

Berthier, P., G. Ferraris, et M. Lalanne (1986). Prédiction du comportement dynamique des moteurs d'avion : vitesses critiques - effets de balourds. Journal de Mécanique Théorique et Appliquée, vol.5, n°4, pp.1-14.

Borri, M. and P. Mantegazza (1977). Efficient solution of quadratic eigen problems arising in dynamic analysis of structures. Comp. Meth. Appl. Mech. Eng., vol.12, pp.19-31.

Childs, D.W. (1976). A modal transient simulation model for flexible axisymetric rotor. J. Eng. Ind., vol.98, serie B, n°1, pp.312-319.

Glasgow, D.A. and H.D. Nelson (1980). Stability analysis of rotor bearing systems using component mode analysis. J. Mech. Des., vol.102, n°2, pp.352-359.

Li, D.F. and E.J. Gunther (1981). Component mode synthesis of large rotor systems. A.S.M.E. Paper 81 GT 147.

Nelson, H.D. and J.M. Mc. Vaugh (1976). The dynamic of rotor bearing systems using finite elements. J. Eng. Ind., vol.98, serie B, n°2, pp.593-600.

Nelson, H.D. and W.L. Meacham (1981). Transient analysis of rotor bearing systems using component mode analysis. A.S.M.E. Paper 81 GT 110.

Rouch, K.E. and J.S. Rao (1980). Dynamic reduction in rotor dynamics by the finite element method. J. Mech. Des., vol.102, n°2, pp.360-368.

Smith, B.T. (1976). Matrix eigensystem routines. Lectures notes in Computer Science, vol.6, Goos and Hartmanis, Springer Verlag.

Zorzi, E.S. and H.D. Nelson (1980). The dynamic of rotor bearing system with axial torque using finite element approach". J. Mech. Des., vol.102, n°1, pp.158-161.

SUMMARY

At first the equations, based on a finite element modelisation, of the dynamic behavior of a monorotor and of a coaxial multirotor are presented. The equations are solved using a direct and a pseudo-modal method.

- The direct method is based on an iterative procedure which gives the lowest complex eigenvalues, eigenvectors and thus the critical speeds and the response to unbalance.
- The pseudo-modal method uses a transformation from physical coordinates to pseudo-modal coordinates by using the first lowest modes of an undamped non rotating symmetric system of equations coming from the initial equations. The new pseudo-modal system is solved as above.

Computer time and accuracy of results have been compared for two industrial examples : a gaz turbine and a jet engine. The results show the pseudo-modal method is much more economic than the direct method and gives, when properly used, results with the same accuracy.

Modelling and Active Vibration Control of Flexible Rotors

H. ULBRICH AND S. FÜRST

Institute of Mechanics, Technical University Munich, Arcisstraße 21, D-8000 München 2, FRG

Abstract. The appearance of number of resonances and even instabilities in turbomachines under working conditions is a well known fact. Optimizing the dynamical behavior by variations of system parameters in a certain range (passive parameter optimization) often doesn't yield sufficient results. In most cases further improvement is only possible by an active control of rotor vibrations. Limited space for the actuators, however, is one of the main problems in implementing a control system in practice. For this reason it is sometimes better to use actuators acting via bearing housings (i.e. ball-journal bearings) and thereby indirectly on the rotor in opposition to magnetic bearings, which act directly on the rotor. Describing the rotor as a "hybrid multibody system" the consideration of all mass- and gyroscopic effects is easily achieved. Based on different optimization strategies a number of controllers have been designed and tested by computer simulations. Some theoretical results are corroborated by experiments. Results demonstrate impressively the improvement of the system's dynamical behavior compared to systems without active vibration control.

Keywords. Rotorsystems; dynamics; modeling; hybrid multi-body system; controllability/observability; optimal control; electromagnetic actuators; state feedback; output feedback control.

INTRODUCTION

Improvement of the dynamic behavior of rotors which are supported by conventional bearings by use of controlled forces has been dealt with in numerous works for instance Schweitzer, 1974; Ulbrich and Kleemann, 1983; Gondhalekar and others, 1979; Burrows and Sahinkaya, 1983; used control forces acting directly on the rotor without contact whilst in Nonami (1985) as well as in this work they act upon the bearing housings. This latter method yields the advantage that the actuators do not have to be installed at sometimes hardly accessible places but, instead, design changes may be made at the bearing locations.Good results can only be expected if one can guarantee sufficient observability and controllability of the natural vibrations that are to be influenced. In order to check on these system properties and to design an effective control adequate modeling is necessary.

The description of the system as a "hybrid multi-body system" is, in current standards, presumably the most effective way of modeling, especially under consideration of control aspects. The method implies the subdivision of the whole system into rigid and flexible parts. These substructures are to be coupled by special elements characterized by adequate force functions. Description of the system under this modeling concept enables all mass and gyroscopic effects (even those of the shaft) to be considered in a very simple way. Also, a modular structure of the system equations without loss of physical transparency is achieved. Even the consideration of control aspects (e.g. low system order or simple adjustment) in an adequate way is made possible.

In the following this method is therefore taken as the basis for discussion of the usefulness of a number of different control concepts for rotor systems. The equations of motion for a concrete example are given. The investigation of the natural behavior yields hints on the choice of the system order. After checking observability and controllability several control concepts are designed in order to decrease resonance and to remove occuring instabilities.

MODELING AND MATHEMATICAL DESCRIPTION OF ROTOR SYSTEMS

The derivation of the equations of motion for large systems (consisting of rigid and elastic elements) can be achieved by direct evaluation of d'Alemberts principle (in the form of Lagrange). For the rotor systems under consideration the result may be written as:

$$\delta \underline{y}^T \sum_{i=1}^{k} \int_{m_i} \{ (\frac{\partial \underline{v}}{\partial \dot{\underline{y}}})^T [dm \cdot \underline{a} - d\underline{f}^e] + (\frac{\partial \underline{\omega}}{\partial \dot{\underline{y}}})^T [d\underline{\bar{I}}\,\dot{\underline{\omega}} + \underline{\tilde{\omega}}\, d\underline{\bar{I}}\,\underline{\omega} - d\underline{l}\ ] + (\frac{\partial \pi}{\partial \underline{y}})^T \}_i = 0 , \tag{1}$$

with k_1 elastic beam-like bodies and k_2 rigid bodies ($k=k_1=k_2$), where

$\underline{y}$:	vector of generalized coordinates
$\underline{a}$:	acceleration of center of mass m
$\underline{\omega}$	angular velocity
$\underline{I}$:	tensor of moments of inertia
$\underline{J}_T = \frac{\partial \underline{v}}{\partial \dot{\underline{y}}}$	jacobian matrix of translation
$\underline{J}_R = \frac{\partial \underline{\omega}}{\partial \dot{\underline{y}}}$	jacobian matrix of rotation
π	potential function of elastic beam
$\underline{f}^e$:	non-potential forces acting on a beamelement or body
$\underline{l}^e$:	non-potential torques acting on a beamelement or body

A description of the system in such a way implies the subdivision of the generalized minimal coordinates as "hybrid coordinates" into two types: "only" time dependent degrees of freedom (e.g. motions of the rigid bearing units) and degrees of freedom which represent elastic deformations (e.g. bending and torsional vibrations of the elastic shafts). The functions of the latter type, taken to describe the motion of a shaft element are separated using Ritz' method:

$$u(z,t)=\underline{u}^T(z)\underline{c}_u(t),\ \underline{u},\ \underline{c}_u \in \underline{R}^{f_u} \quad (2)$$

with f_u being the number of considered admissible shape-functions per continuous coordinate (analogue for v(z,t)). Here it is useful to choose eigenfunctions of the non-rotating rotor as admissible shape-functions, Bremer (1983), thereby allowing for low system order even at high rotation frequencies The determination of these functions is carried out in a separate calculation, e.g. by use of cubic spline functions, hermite polynomials, by exact methods within the current model (e.g. transfer functions), or experimentally. The eigen behavior in connection with the expected excitation frequencies reveals the commendable system order (number of elastic degrees of freedom). In the following only small deviations from a given reference system are considered. Therefore linearization is feasable with terms caused by so-called cinematic coupling being neglected as small of second order. Now the equations of motion for the separate moduls (substructures) can be given in the usual form of mechanical systems:

$$\begin{bmatrix} \underline{M}_{xx} & \underline{O} \\ \underline{O} & \underline{M}_{yy} \end{bmatrix}_i \ddot{\underline{y}}_i + \begin{bmatrix} \underline{D}_{xx} & \underline{G}_{xy} \\ -\underline{G}_{yx} & \underline{D}_{yy} \end{bmatrix}_i \dot{\underline{y}}_i + \begin{bmatrix} \underline{K}_{xx} & \underline{O} \\ \underline{O} & \underline{K}_{yy} \end{bmatrix}_i \underline{y}_i = \underline{h}_i . \quad (3)$$

The coordinates that are necessary for the description of the separate moduls are comprised in the vector $\underline{y}_i$. The force vector $\underline{h}_i$ containes the forces acting on the i-th substructure. They are not only owing to outside the system but can also be caused by coupling effects between the subsystems. In Table 1 the appearing submatrices in Eq. (3) for an elastic subsystem are given with respect to the inertial system I. Systematically the

submatrices in Eq. (3) for an elastic subsystem are given with respect to the inertial system I. Systematically the substructures are combined to synthesize the overall system. The jacobian matrices $\underline{J}_T$ and $\underline{J}_R$ appearing in Eq.(1) serve as relating elements. They may be interpreted as transformation of all force and moment relations into the common basis of the chosen minimal coordinates. Changes in modeling certain substructures can be performed without having to re-work the whole system. The set of equations of the overall system, as well, exhibits the structure of ordinary mechanical system equations

$$\underline{M}\ddot{\underline{y}} + (\underline{D}+\underline{G})\dot{\underline{y}} + (\underline{K}+\underline{N})\underline{y} = \underline{h} \quad (4)$$

with the configuration vector of the whole system

$$\underline{y} = [\underline{y}_1^T, \underline{y}_2^T, \ldots, \underline{y}_K^T]^T;\ \underline{y} \in R^f. \quad (5)$$

Introducing the state space vector

$$\underline{x}(t) = [\underline{y}^T(t),\ \dot{\underline{y}}^T(t)]^T;\ \underline{x} \in R^n \quad (6)$$

Eq.(4) becomes

$$\dot{\underline{x}}(t)=\underline{A}\underline{x}(t)+\underline{B}\underline{u}(t);\quad (\underline{A} \in R^{n,n}, \underline{B} \in R^{n,r}), \quad (7)$$

with system order n=2f, f being the number of degrees of freedom, r the number of actuators and $\underline{u}(t)$ being the control vector. A comprehensive description of the kinematic and kinetic relations may be found in Ulbrich (1986).

CONTROLLERS FOR ROTOR SYSTEMS

The design of linear and nonlinear controllers both analogue and digital has been treated in great detail in literature. There are numerous methods and concepts of design and realisation of controllers. The choice of a concept is mainly determined by the aim and the physical realities of the plant (i.e. the open loop). In the following a purely time invariant and linear system

$$\dot{\underline{x}}(t) = \underline{A}\ (\Omega)\underline{x}(t) + \underline{B}\underline{u}(t),$$
$$\underline{y}(t) = \underline{C}\ \underline{x}(t);\ \underline{C} \in R^{m,n} \quad (8)$$

is assumed where

$\underline{A}$ = system matrix, $\underline{B}$ = control matrix (depending on the actuator locations and on the kind of actuators), $\underline{C}$ = measurement matrix (related to the measurement locations and devices), m = number of measurements (measurement coordinates).

Based on the structure of the control law two concepts may be distinguished:

- state feedback

$$\underline{u}(t)= \underline{K}_S\ \underline{x}(t);\ \underline{K}_S \in R^{r,n}, \quad (9)$$

- output feedback

$$\underline{u}(t)= \underline{K}_O\ \underline{y}_M(t);\ \underline{K}_O \in R^{r,m}. \quad (10)$$

This means that the control vector $\underline{u}(t)$ is a linear function of either the system state vector $\underline{x}(t)$ or of the system output, the measurement vector $\underline{y}_M(t)$, respectively.

State feedback control

In designing a state feedback control the following methods are mostly used:

a) optimisation according to the quadratic integral criterion

$$J = \int_0^\infty (\underline{x}^T \underline{Q}\, \underline{x} + \underline{u}^T \underline{R} \underline{u})dt \rightarrow \min. \qquad (11)$$

This approach reduces the design problem to the task of solving the algebraic Riccati equation. The weak point in this way of procedure is the adaption of the weighting matrices $\underline{Q}$ and $\underline{R}$ for the specific problem.

b) choice of the eigenvalues λ_i of the closed control loop (pole assignement)

$$[\lambda_i \underline{E} - (\underline{A}-\underline{BK})]\underline{q}_i = \underline{O}. \qquad (12)$$

In contrary to the previous method the choice of suitable eigenvalues (poles) is the problem of this approach. By the placement of eigenvalues a specific system behavior is enforced so that in case of poorly adapted values extremely high control forces may appear; in dealing with rotor systems disadvantageous influence on gyroscopic effects may even deteriorate the system dynamics. Hints for the choice of suitable poles are given in the work Ackermann (1977).

c) modal state control

is a method that allows for a determined shift of one or more eigenvalues. This can be usefull especially for rotor systems which, for whatsoever reasons, are run in the near of resonances. With the control concept given in the following, however, one can always shift only as many poles as actuators are available, Korn and Wilfert (1982). In the control design stage the system represented by Eq. (8) is transformed into

$$\dot{\underline{\bar{x}}}(t) = \underline{\Lambda}\, \underline{\bar{x}}(t) + T^{-1}\underline{B}\underline{u}(t) = [\underline{\Lambda} - \underline{K}_M]\underline{\bar{x}}. \qquad (13)$$

The matrix $\underline{\Lambda}$ is real with a diagonal block structure. The gain matrix $\underline{K}_s$ in (9) results from

$$\underline{K}_s = \underline{B}^+ \underline{T}\, \underline{K}_M\, \underline{T}^{-1} \qquad (14)$$

with $\underline{B}^+ = (\underline{B}^T\underline{B})^{-1}\underline{B}^T$ being the pseudo inverse of the column regular control matrix $\underline{B}$ from Eq. (8).

d) combined state feedback control

This kind of control is useful in cases where the application of a controller designed according to a) does not yield sufficient damping of some natural vibrations. Here the additional application of a modal feedback control according to c) can improve the system behavior significantly. The additional effort in realisation for this case is negligible since a summation of gain matrices only effects the gain coefficients whereas the structure of the control remains the same.

Output feedback control

As the determination of all system state coordinates is often very difficult most controllers are based on output feedback only. According to the reduction of measurement expense the possible influence on system behavior is restricted. However, the obtainable results are still sufficient for most applications. Significant improvement by means of state feedback often requires enormous expense which in most cases is not tenable under economic aspects.

The design of an output feedback control always implies a parameter optimisation, i.e. a optimal tuning of the coefficients of the feedback matrix in Eq. (10). Designing an output feedback using the quadratic quality criterion the quality funtional can be acquired by solving the Ljapunov matrix equation. Utilizing all symmetry characteristics reduces the number of variable parameters significantly thereby decreasing the calculation expence, as well. It is important to note that in contrary to the Riccati controller different initial conditions $\underline{x}(t_O)=\underline{x}_O$ will result in different optimal matrices $\underline{K}_O$. The initial conditions may be set explicitly in such a way as to enforce consideration of some critical natural mode shapes (e.g. by equating $\underline{x}_O$ to a specific eigenvector characterizing a natural mode shape).

Under certain circumstances (i.e. complete observability and complete controllability) the output feedback control allows for determined shift of specific eigenvalues, as well. Considering the modal description of the system as in Eq. (13) the feedback matrix $\underline{K}_O$ in Eq. (11) becomes

$$\underline{K}_O = \underline{B}^+ \underline{T}\, \underline{K}_M \underline{T}^{-1} \underline{C}^+ \qquad (15)$$

with $\underline{C}^+ = \underline{C}^T(\underline{C}\underline{C}^T)^{-1}$ being the pseudo inverse of the row regular measurement matrix $\underline{C}$ in Eq. (8). In general the matrix $\underline{C}$ is not quadratic so that the calculation in Eq. (15) cannot be inverted; this means that all other poles are shifted, as well. This requires a check on stability and system behavior after the shift of the poles (Korn and Wilfert, 1982).

Example of application

In the previous chapters the theoretical fundamentals for the description of hybrid rotor systems and for the design of suitable controllers have been supplied. Using the example of a horizontally mounted rotor their utility is now tested both by computer simulation and experimentally. Figure 1 displays a photograph of the experimental setup. The mechanical model can be seen in Fig. 2. It essentially consists of an elastic rotor mounted in two bearings each of them represented by a discrete mass. The coupling to the shaft is modelled by the stiffness matrices $(\underline{C}_L)_i$ and the damping matrices $(\underline{D}_L)_i$. The stiffness matrices $(\underline{C}_a)_i$ represent the outer bearing support. In the existing experimental setup there are two possibilities of applying determined forces to the rotor, namely contactless forces by a magnetic bearing or indirectly using electro-magnetic actuators acting upon the roller-bearing housings. In this latter case the magnetic bearing is not used as a control element but simulates additional distortion such as non-conservative forces. Using the vector of generalized coordinates as in Eq. (5)

$$\underline{y}(t)=[\underline{c}_u^T(t),\underline{c}_v^T(t),x_1,y_2,x_2,y_2]^T \quad \underline{y}\in R^f, \qquad (16)$$

the equations of motion for the overall system consisting of three substructures (i.e. the elastic rotor, bearing 1 and bearing 2) can be specified systematically. The dimension of the fx1 vector $\underline{y}$ equals the number of degrees of freedom of the system with vectors $c_u(t)$ and $c_v(t)$ of dimension f_u and the rigid body coordinates x_1, y_1 for bearing 1 and x_2, y_2 for bearing 2. The contribution of the elastic rotor (subsystem 1) to the system matrix can be composed from the submatrices given in Table 1. The coupling of the bearings to the shaft is carried out using the jacobian matrices of translation (no rotation present). They can be written as

$$(\underline{J}_T)_2 = \begin{bmatrix} \underline{0}_1 & \begin{array}{cc} 1 & 0 \\ 0 & 1 \\ 0 & 0 \end{array} & \underline{0}_3 \end{bmatrix} \qquad (17)$$

for bearing 1 (subsystem 2) and

$$(\underline{J}_T)_3 = \begin{bmatrix} \underline{0}_1 & \underline{0}_2 & \begin{array}{cc} 1 & 0 \\ 0 & 1 \\ 0 & 0 \end{array} \end{bmatrix} \qquad (18)$$

for bearing 2 (subsystem 3). For example, the discrete bearing forces acting upon the rotor from bearing 1 at location z_1 may be stated as

$$\underline{f}_{11}(z_1,t)=(\underline{C}_L)_1[(\underline{J}_T)_2-(\underline{J}_T(z_1))_1]\underline{y}(t)+$$
$$+ (\underline{D}_L)_1[(\underline{J}_T)_2-(\underline{J}_T(z_1))_1]\dot{\underline{y}}(t). \qquad (19)$$

In order to derive the system equations explicitly, according to the separation method introduced in Eq.(2) adequate admissible shape-functions (Bremer, 1983) for the elastic coordinates have to be chosen.

Controllability and observability

To a high degree controllability and observability are determined by natural mode shapes (eigenvectors) of the system. Using magnetic bearings - direct control action - all natural mode shapes of the rotor can be influenced sufficiently depending on the choice of the bearing locations (Ulbrich, 1986). On the other hand natural mode shapes and therefore the location of nodes as well depend on the frequency of rotation. This may even cause the loss of controllability of specific natural modes at certain frequencies, (Ulbrich and Kleemann, 1983).

If the system is mounted conventionally, i.e. in roller-journal bearings, and the vibration behavior has to be improved by controlled forces, the use of indirect action of control forces is an alternative to magnetic bearings. The effectiveness of this method is, to a great extent, characterized by the coupling of the bearing housings to the rotor shaft. For the second synchronous natural mode in Fig. 3 a measure for the controllability is shown as a function of the bearing mass m_L and the bearing-shaft coupling stiffness c_k respectively. In Figures 3 it can be seen that controllability decreases with reduced c_k since with softer coupling the effectiveness of the actuator forces declines. The behavior with respect to the outer bearing support is vice versa, i.e. sufficient mobility has to be supplied. For the example under consideration good controllability of the important natural mode shapes in the current frequency range is guarranteed if the conditions

$$\frac{c_k}{c_w} > 1 \quad \text{and} \quad \frac{c_a}{c_w} < 1 \qquad (20)$$

hold (where c_w is a measure for the overall stiffness of the rotor structure). The influence of the bearing mass on the controllability is small within the considered frequency range. Similar statements are achieved for the observability if actuator location and measurement location are identical.

RESULTS

For the rotor system shown in Fig. 1 a number of different controllers according to chapter 3 have been designed. The numerical investigations proved that the control goals could easily be achieved when using a system state feedback optimized according to Eq.(11) and/or combined with modal control (Ulbrich, 1986). The determination of the complete system state is, in most cases, too costly especially when "elastic degrees of freedom" are considered. Therefore in the current example only different output feedback controllers were tested. The actuator forces acting upon the bearing at location (S) (Fig. 1) are controlled by m_M measurement vectors $\bar{y}_i$ (indicates measurement location) which are composed of displacement and velocity signals in ^Ix- and ^Iy-direction. The forces are determined by

$$\begin{bmatrix} F_x \\ F_y \end{bmatrix} = \sum_{i=1}^{m_M} \begin{bmatrix} k_{xx} & k_{xy} & d_{xx} & d_{xy} \\ k_{yx} & k_{yy} & d_{yx} & d_{yy} \end{bmatrix} \bar{y}_i . \qquad (21)$$

The control design at first is performed by means of the root-locus analysis. The result can then be optimized using a quadratic quality criterion. Figure 4 shows the final results achieved for three different cases in form of frequency response functions versus rotor frequency. The design of the different controllers is based on the consideration of the first bending natural mode of the elastic rotor (natural frequency at approximately 20 Hz). Only measurements of displacement (x_1, y_1) at location (M1) (coinciding with the actuator location S) for the feedback were used to achieve the results displayed in Fig. 5. A significant reduction of resonance peaks can be observed. Addition of another displacement signal (taken at measurement location (M2)) with positive sign (positive feedback) leads to a still sharper decrease of resonance amplitudes Consideration of the velocities measured at (M2) in the feedback loop does not cause any additional improvement.

CONCLUSION

Improvement of the dynamics of rotor systems often fails because of the necessary expense. The present paper aimed for a systematic computer-adapted description of rotor systems with elastic substructures and for a determined change in vibration behavior by means of controlled forces acting via the bearing housings. The results show that by development of adequate actuators in connection with suitable control concepts facilities can be supplied to improve the vibration behavior of both existing rotors and systems in the design stage.

REFERENCES

Ackermann, J. (1977). Entwurf durch Polvorgabe. Regelungstechnik, Heft 6, S. 173-179 und S. 209-215.

Bremer, H. (1983). Kinetik starr-elastischer Mehrkörpersysteme. Fortschr.-Ber. VDI-Z., Reihe 11, Nr. 53.

Burrows, C.R.; Sahinkaya, M.N. (1983). Vibration Control of Multi-Mode Rotor-Bearing Systems. Proc. Royal Society of London A386, pp. 77-93.

Gondhalekar, V.M.; Nikolajsen, J.; Jayawant, B.V. (1979). Electromagnetic Control of Flexible Transmission Shaft Vibration. Proc. IEE, Vol. 126, No. 10, pp. 1008-1010.

Korn, U.; Wilfert, H. (1982). Mehrgrößenregelungen - Moderne Entwurfsprinzipien im Zeit- und Frequenzbereich. VEB-Verlag Technik, Berlin.

Nonami, K. (1985). Vibration Control of Rotor Shaft Systems by Active Control Bearings. ASME-Paper 85-DET-126.

Schweitzer, G. (1974). Stabilization of Self-exited Rotor Vibration by an Active Damper. Dynamics of Rotors-Symp. Lyngby/Denmark, pp. 472-493.

Ulbrich, H.; Kleemann, U. (1983). Untersuchung von Instabilitäten eines aktiv beeinflußten Rotors mit elastischer Struktur. ZAMM, Bd. 63, S. 120-122.

Ulbrich, H. (1986). Dynamik und Regelung von Rotorsystemen. Habilitationsschrift, Technische Universität München.

Table 1. Submatrices of elastic substructures with respect to the inertial system I.

submatrix	influence
${}^{I}(\underline{M}_{xx})_i = (\int_0^L \rho[A(z)\underline{u}(z)\underline{u}^T(z)+I_x(z)\underline{u}'(z)\underline{u}'^T(z)]dz)_i$	inertia
${}^{I}(\underline{D}_{xx})_i = (d_i\int_0^L EI_x(z)\underline{u}''(z)\underline{u}''^T(z)dz)_i$	material damping
${}^{I}(\underline{G}_{xy})_i = (2\Omega\int_0^L \rho I_x(z)\underline{u}'(z)\underline{u}'^T(z)dz)_i$	gyroscopic effects
${}^{I}(\underline{K}_{xx})_i = (\int_0^L EI_x(z)\underline{u}''(z)\underline{u}''^T(z)dz)_i$	elastic forces

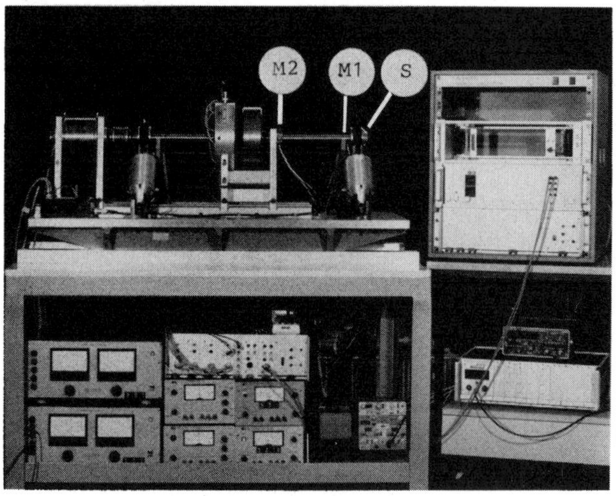

Fig. 1. Experimental set-up.

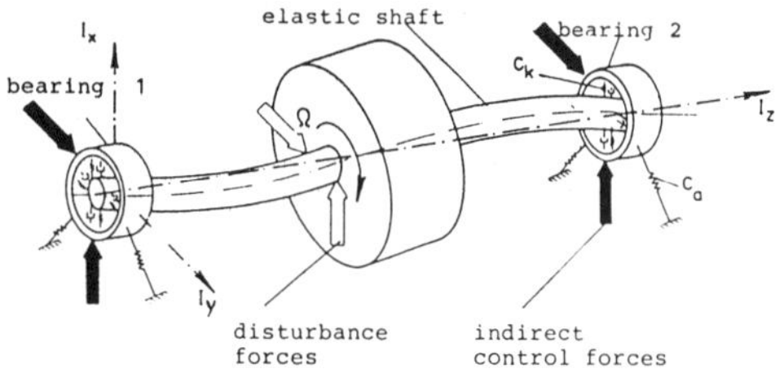

Fig. 2. Mechanical model of rotor system.

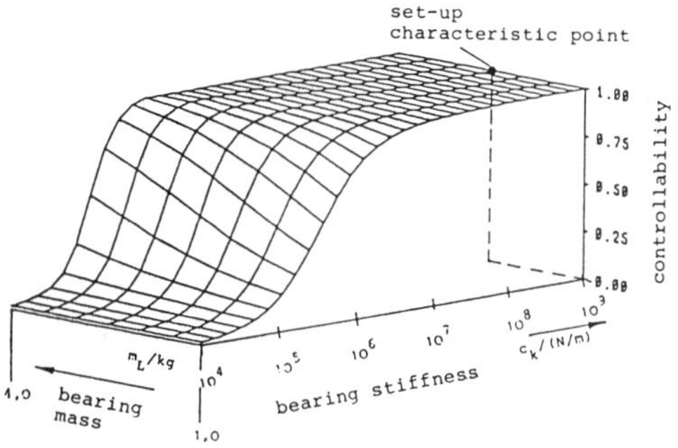

Fig. 3. Controllability for the second natural mode.

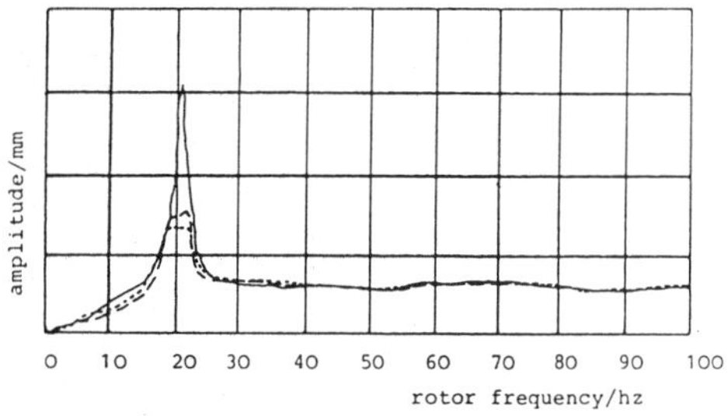

Fig. 4. Measured response function.

—— a) velocity feedback only, 1 measurement location.
– – – b) displacement and velocity feedback, 1 measurement location.
----- c) displacement and velocity feedback, 2 measurement locations.

An Investigation of Unstable Vibrations of a Geared Torsional System

B. KISHOR

Institute of Technology, Banaras Hindu University, Varanasi-221005, India

Abstract. Mathematical model of a typical two stage reduction gear train employed in turbomachines is given in terms of coupled nonlinear differential equations with variable coefficients. Parametric resonances and other nonlinear responses are studied. A condition of zero width of parametric resonance zone is obtained which can be used in design. Cause of occurrence of peculiar vibrations as reported in a real turbomachine system is also explained.

Keywords. Locked train type reduction gear; gear error; variable mesh stiffness; jump phenomenon; zone of synchronization; parametric resonance.

INTRODUCTION

Geared-torsional rotor systems employed in turbomachines are prone to vibration instabilities. Moreover, current trends in the design of aircraft engines, space vehicles and other power transmission machines have introduced new risks of unrecognised vibration instabilities. In the past investigators have attempted to solve some of these problems notable amongst whom are Tondl(1965); Kishor(1979) Kubo & Kiyono(1980); Mitchell & David (1985)and Kumar, Sankar & Osman(1985). Yamada & Mitsui(1979) reported a vibration phenomenon in a marine steam turbine which was different from oil whip, oil whirl etc. Moreover, there was a sudden rise of the amplitude of vibrations and the vibrations were accompanied by abnormal sound. Present paper is devoted to the study of this phenomenon through a mathematical model.

MATHEMATICAL MODEL

A locked train type reduction gear and turbine is schematically shown in Fig.1. Gear errors due to runout which may be the result of manufacture, installation or operation are incorporated in the model. Torsional deformations of meshing gears are given by the following

$$\theta_i = r_{ij}\,\theta_j + r_{ij}\,b_j \sin\theta_j \tag{1}$$

where $r_{ij} = R_j/R_i$; $b_j = a_j/R_j$; R_j , R_i are the radii of base cylinders.

Gear teeth are considered to have equivalent variable mesh stiffness represented by K_{e1} and K_{e2} at the pitch points along the tangent lines of pitch circles. The K_{ei} (i=1,2), following Tordion & Gauvin (1977), are

$$K_{ei} = K_m + K_{vi} \tag{2a}$$

$$K_{vi} = a_o/2 + \sum_{m=1}^{\infty}(a_m \sin mvt + b_m \cos mvt) \tag{2b}$$

where $a_o = 4K_a(2-c_i)$; $i = 1,2$;

$a_m = (2K_a/m\pi)\,(\cos 2m\pi\,(c_i - 1) - 1)$

$b_m = (2K_a/m\pi)\,\sin\,2m\pi\,(1 - c_i)$;

K_a = amplitude of variable stiffness;

c_i = contact ratio and K_m = amplitude of constant mesh stiffness.

Motion of the wheel of the first stage is considered to have significant effects on that of the pinion of the second stage only through torsional vibrations. Both the mass and mass moment of inertia of the wheel are much larger than those of the pinion, it is , therefore, assumeed to neglect the transverse motions of the wheel.

Bearings are modelled as mass-spring-dashpot systems. Deflections u_x and u_y are taken to be small with respect to the equilibrium position of the journal. Oil film reaction forces F_i(i= x,y) are

$$F_i = K_{ij}\,u_j + C_{ij}\,\dot{u}_j \tag{3}$$

where a repeated subscript denotes sum.

Equations of motion of the system are

$$M\ddot{u}_x + F_x = 0 \tag{4}$$

$$M\ddot{u}_y + I_1/R_1\ddot{\theta}_1 + A_{11}/R_1\theta_1 - A_{12}/R_1\,\theta_1^3 - K_{t1}/R_1\,\theta_2 + C_{t1}(\dot{\theta}_1 - \dot{\theta}_2) + F_y = 0 \tag{5}$$

$$I_2\ddot{\theta}_2 + B_{11}\theta_2 - K_{t1}\theta_1 - B_{12}\theta_2^3 + C_{t1}(\dot{\theta}_2 - \dot{\theta}_1) \quad (6)$$

where $A_{11} = K_{t1} + (2R_1 + a_1)^2 K_{e1}$

$A_{12} = (4/3R_1 + 2/3\ a_1)\ a_1 K_{e1}$

$B_{11} = K_{t1} + K_{t2} r_{23}^2 (1 + b_2)^2 + (2R_2 + a_2)^2 K_{e2}$

$B_{12} = 2/3 K_{t2} r_{23}^2 b_2 (1 + b_2) + (4/3R_2 + 2/3a_2) a_2 K_{e2}$

and I_1, I_2 = mass moments of inertia of gears 1 and 2; K_{t1}, K_{t2} = torsional shaft stiffnesses; C_{t1}, C_{t2} = torsional damping; M = mass of the pinion. $(\dot{})$ over a letter represents differentiation w.r.t.time.

METHOD OF SOLUTION

Exact solution of equations(1) to(6) is difficult to obtain but important informations can be had even if particular cases are solved. Of particular interest is the case when vibrations are mainly torsional and the effects of the bearings are neglected. Approximate expression for meshing stiffness is obtained from equation(2b) by retaining first two terms only in the expansion. Incorporating these simplifications in equations(1) to (6) and upon simplification, the following set of equations is obtained:

$$I_1\ddot{\theta}_1 + K_{t1}(\theta_1 - \theta_2) + C_{t1}(\dot{\theta}_1 - \dot{\theta}_2) + (2R_1 + a_1)^2 K'(1 - b\cos vt)\ \theta_1 - A'_{12}\theta_1^3 = 0 \quad (7)$$

$$I_2\ddot{\theta}_2 + K_{t1}(\theta_2 - \theta_1) + C_{t1}(\dot{\theta}_2 - \dot{\theta}_1) - B'_{12}\theta_2^3 + (2R_2 + a_2)^2 K'(1 - b\cos vt)\ \theta_2 = 0 \quad (8)$$

where $b = 1/(1 + 2c - 4\pi)$; $K' = 2K_a b/\pi$;

$A'_{12} = (4/3\ R_1 + 2/3\ a_1) a_1 K_m$;

$B'_{12} = (4/3\ R_2 + 2/3\ a_2)\ a_2 K_m$

Equations(7) and(8) are susceptible to computer simulation but approximate analytical solutions are obtained instead.

Modified asymptotic method, suited to the present case, has been used to arrive at the solution. According to the essence of the method it is possible to construct an equivalent equation, to a first order of approximation, for the present system(Minorsky, 1962). Following the procedure, an approximate solution of equations (7) and (8) is sought in the form given below:

$$\theta_i = \phi_i\ A\cos(v/2\ t + n);\ (i=1,2) \quad (9)$$

where ϕ_i (i=1,2) are fundamental mode functions of the corresponding linear case, A and n are slowly varying functions of time.

Using (9) in equations (7) and (8), following the procedure of the asymptotic method and upon simplification, equivalent equation of the system so obtained is given below :

$$\ddot{z} + w^2(1 - h\cos vt)\ z + 2c\,\dot{z} - ez^3 = 0 \quad (10)$$

where $z = A\cos(v/2\ t + n)$; $2c = C_1/m_1$;

$w^2 = (K_1 + K_2)/m_1$; $h = bK_2/(K_1 + K_2)$;

$e = E_1/m_1$; $C_1 = C_{t1}(\phi_1 - \phi_2)^2$;

$E_1 = A'_{12}\ \phi_1^4 + B'_{12}\ \phi_2^4$

and $m_1 = I_1\phi_1^2 + I_2\phi_2^2$; $K_1 = K'_{b1}(\phi_1 - \phi_2)^2$;

$K_2 = (2R_1 + a_1)^2 K'\phi_1^2 + (2R_2 + a_2)^2 K'\phi_2^2$

Asymptotic method further requires that A and n of equation (9) must satisfy, to a first order approximation, the following set of equations :

$$\frac{dA}{dt} = -\ cA - \frac{Ahw^2}{2v}\sin 2n \quad (11)$$

$$\frac{dn}{dt} = w - v/2 - \frac{3eA^2}{4v} - \frac{hw^2}{2v}\cos 2n \quad (12)$$

Stationary values of the amplitude and the phase of the vibrations is obtained by equating the right hand sides of equations (11) and (12) to zero. Eliminating the phase n , following relationship between the amplitude (A) and the modulation frequency (v) is obtained :

$$A^2 = (4/3e)\left[(v/2)^2 - w^2 \mp D/2\right] \quad (13)$$

$$\text{where}\quad D = (\ h^2w^4 - 4v^2c^2)^{1/2} \quad (14)$$

With the help of equation (13) a resonance curve is plotted which is shown in Fig. 2. The branch AB is unstable and CB is stable. The phenomenon exhibits the usual hysteresis character. For determining the limits of the zone of synchronisation, A is equated to zero and then the following is obtained :

$$w^2 - D/2 < (v/2)^2 < w^2 + D/2 \quad (15)$$

Hence, the width of the synchronization zone (D) is given by equation (14).

Experimental results of Yamada & Mitsui (1979) are also plotted in Fig.2. The jump (AD) fits into the predicted curve. It is interesting to note that the presence of damping reduces the interval AC (Fig.2) inside which the parametric resonance arises. The minimum modulation percentage which is necessary for parametric resonance, when damping is present, is given by the inequality $h < 4\ c/w$.

CONCLUSIONS

The cause of loss of rotor motion stability and the rise of self-excited vibrations is one of the most important aspects of investigation in the field of rotor dynamics. Yamada and Mitsui (1979) observed certain unusual vibration phenomena accompanied by abnormal sound. The present investigation stems from the idea that the gear trains used in high power machines (e.g. turbomachines) should affect the onset of unstable vibrations. Equations of motion of one such gear train system are obtained which are coupled non-linear differential equations with variable coefficients. Complete solution of such a system of equations remains to be obtained mathematically. An approximate solution of the system is obtained using an asymptotic method of non-linear vibration analysis. Analytical model as well as solutions show the presence of parametric resonances as well as softening spring response. It is interesting to note that the experimental findings of Yamada & Mitsui (1979) and the predicted response are shown in Fig. 2. The experimental plot of the response shows a softening spring type response. The jump phenomenon observed during experimentation should cause a sudden loss of contact. Subsequently this will lead to hammering action of one tooth on to the other. This hammering action is the probable cause of generation of abnormal sound.

It was reported by Benton & Seireg (1978) that the operation of a gear system in unstable speed region leads to excessive dynamic loads and may eventually lead to complete destruction of teeth. Present investigation provides a scale for minimum modulation percentage necessary for parametric resonance (one of the principal causes of rotor instability).

Equation(14) predicts that the width of synchronization zone(D) becomes zero for $h = 4c/w$. This implies zero width of parametric resonance zone. Hence, damping when introduced in geared systems will reduce the risks of rotor instability arising out of the variable mesh stiffness of the gear teeth.

REFERENCES

Benton,M. and A.Seireg(1978). Factors influencing instability and resonances in geared systems. Am.Soc. Mech.Engrs. J. Mech.Design,103, 372

Kishor,B.(1979). Effect of gear errors on nonlinear vibrations of a gear train system. Proc. Fifth IFToMM Congress, 1122-1125

Kubo,A. and S.Kiyono(1980). Vibrational excitation of cylindrical involute gears due to tooth form errors. Bull. Japan Soc. Mech.Engrs., 23, 1536-1543

Kumar,A.S., T.S.Sankar and M.O.M.Osman (1985). On dynamic tooth load and stability of a spur gear system using the state space approach. Trans. Am. Soc. Mech.Engrs. J.Mech.Transmissions and Automation in Design, 107,54-60

Mitchell,L.D. and J.W.David(1985). Proposed solution methodology for the dynamically coupled nonlinear geared rotor mechanics equations. Trans. Am. Soc. Mech.Engrs. J.Vib. Acoustics, Stress and Reliability in Design, 107, 112-116

Minorsky, N.(1962) . Nonlinear Oscillations. D.Van Nostrand Co.Inc., Newyork

Tondl,A.(1965). Some problems of rotor dynamics. Chapman & Hall, London

Tordion, G.V. and R.Gauvin(1977). Dynamic stability of a two stage gear train under the influence of variable meshing stiffness. Trans. Am. Soc. Mech.Engrs. J.Engg. Ind., 785-791

Yamada, T. and J.Mitsui(1979). A study on the unstable vibration phenomena of a reduction gear system, including the lightly loaded journal bearings, for a marine steam turbine. Bull. Japan Soc. Mech.Engrs. , 22, 98-106

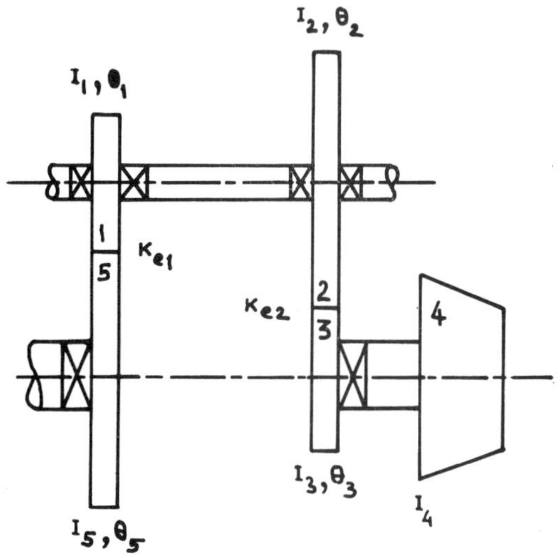

FIG.1: LOCKED TYPE GEAR TRAIN

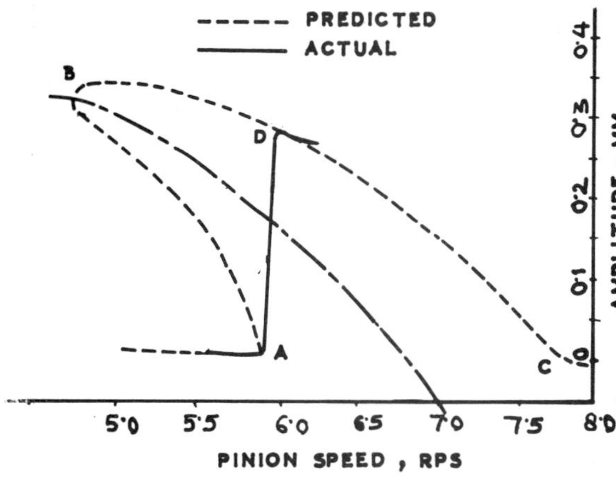

FIG.2: VIBRATION PHENOMENON

The Effect of Material Damping on the Unbalanced Response and Stability of a Flexible Rotor Supported in Journal Bearings

C. RAJALINGHAM AND B. S. PRABHU

Department of Applied Mechanics, Indian Institute of Technology, Madras, India

Abstract. Taking into account the stiffness asymmetry associated with the material damping, its influence on the unbalanced response and stability of a flexible rotor supported in journal bearings is investigated. The critical speeds of the system are taken as those speeds at which the energy dissipation in the bearing is a maximum. This investigation shows that in the case of excessive material damping the stiffening effect predominates over the damping effect. Thus an optimum material damping improves the dynamic performance of the system.

Keywords. Material damping, Kelvin solid, Stiffness asymmetry, Journal Bearing, Critical Speed, Threshold speed, Dissipation coefficient.

INTRODUCTION

The material damping is capable of creating unstable rotor vibrations at speeds above the first critical speed. Kimball (1925) investigated the influence of the material damping on the vibrations of rotors supported in rigid bearings. Based on experimental observations, Tondl (1924) reported the fact that the material damping has an unfavourable influence on the transitional vibrations of rotors when passing quickly through the critical speed and when rapidly reducing the speed after the beginning of selfexcited vibrations.

Neglecting stiffness asymmetry, Rao et al (1981) incorporated the material damping as an additional damping for the investigation of the unbalanced response of a flexible rotor supported in hydrodynamic bearings. Since the stiffness asymmetry is the characteristic aspect of the material damping, it is more appropriate to include this aspect also in the investigations on the stability and the unbalanced response of such flexible rotor-bearing system. The present investigation deals with the influence of the material damping on the critical and threshold speeds of a rotor-bearing system.

Rajalingham et al (1986) introduced the energy dissipation in the bearing per revolution as a measure to assess the severity of rotor vibration and the critical speeds based on such criteria can be measured directly from deceleration-speed characteristic during the coast down phenomena. The above criteria is used in this investigation also for the definition of the critical speeds.

NOMENCLATURE

$a_{jx},\ldots$	defined in equations (7-10)
$\tilde{a}_{jx},\ldots$	$\tilde{a}_{jx} = a_{jx} - i\, b_{jx},\ldots$
b	bearing width
$b_{jx},\ldots$	defined in equations (7-10)
c	radial clearance
c_s	shaft damping coefficient
$\bar{c}_{xx},\ldots$	dimensionless damping coefficients
[C]	damping coefficient matrix
$[\bar{C}]$	$[\bar{C}] = (2c\omega/M_s g([C]$
C_D	dissipation coefficient
d	journal diameter
g	acceleration due to gravity
i	$i = \sqrt{(-1)}$
J_1	$J_1 = \bar{k}_{xx} + \bar{k}_{yy}$
J_2	$J_2 = \bar{c}_{xx} + \bar{c}_{yy}$
J_3	$J_3 = \bar{k}_{xx}\bar{k}_{yy} - \bar{k}_{xy}\bar{k}_{yx}$
J_4	$J_4 = \bar{c}_{xx}\bar{c}_{yy} - \bar{c}_{xy}\bar{c}_{yx}$
J_5	$J_5 = \bar{k}_{xx}\bar{c}_{yy} + \bar{k}_{yy}\bar{c}_{xx} - \bar{k}_{xy}\bar{c}_{yx} - \bar{k}_{yx}\bar{c}_{xy}$
k_s	shaft stiffness
$\bar{k}_{xx},\ldots$	dimensionless stiffness coefficient
$\tilde{k}_{xx},\ldots$	$\tilde{k}_{xx} = \bar{k}_{xx} + i\,\bar{c}_{xx},\ldots$
[K]	stiffness coefficient matrix
$[\bar{K}]$	$[\bar{K}] = (2c/M_s g)[K]$
$[\tilde{K}]$	$[\tilde{K}] = [\bar{K}] + i\,[\bar{C}]$

M_s	mass of rotor
So_o	$So_o = (M_s g\psi^2/2bd\eta\omega_o)$
t	time
(x_j, y_j)	co-ordinates of journal center
(x_r, y_r)	co-ordinates of rotor center
(x_s, y_s)	$x_s = x_r - x_j,\ y_s = y_r - y_j$
$(\bar{x}_j, \bar{y}_j)$	$\bar{x}_j = x_j/c,\ \bar{y}_j = y_j/c$
$(\bar{x}_s, \bar{y}_s)$	$\bar{x}_s = (x_s\ \delta_s \cos^2\phi)/c,$ $\bar{y}_s = (y_s - \delta_s \sin\phi\cos\phi)/c$
β	$\beta = c_s/(2.\sqrt{(k_s M_s)})$
δ_o	unbalance eccentricity
$\bar{\delta}_o$	$\bar{\delta}_o = \delta_o/c$
δ_s	$\delta_s = M_s g/k_s$
ϵ	eccentricity ratio
η	viscosity
ω	journal speed
$\bar{\omega}$	$\bar{\omega} = \omega/\omega_o$
ω_o	$\omega_o = \sqrt{(g/c)}$
ω_s	$\omega_s = \sqrt{(k_s/M_s)}$
$\bar{\omega}_s$	$\bar{\omega}_s = \omega_s/\omega_o$
ω_b	threshold speed
ω_c	critical speed
ω_{bo}	ω_b when $\beta \to 0$
$\omega_{b\infty}$	ω_b when $\beta \to \infty$
ϕ	$\phi = \tan^{-1}(c_s\omega/k_s)$
ψ	clearance ratio, $\psi = 2c/d$
$'$	denotes derivative with respect to ωt

THEORY

Governing Equations

Assuming the material of the shaft carrying a central rotor and supported in identical journal bearings to be a Kelvin solid, the equations governing the motion of the system can be written in dimensionless form as,

$$\bar{\omega}^2 \bar{x}_s'' + 2\beta\bar{\omega}\bar{\omega}_s \bar{x}_s' + \bar{\omega}_s^2 \bar{x}_s + 2\beta\bar{\omega}\,\bar{\omega}_s \bar{y}_s = -\bar{\omega}^2 \bar{x}_j + \bar{\omega}^2 \bar{\delta}_o \cos\omega t \quad (1)$$

$$\bar{\omega}^2 \bar{y}_s'' + 2\beta\bar{\omega}\,\bar{\omega}_s \bar{y}_s' + \bar{\omega}_s^2 \bar{y}_s - 2\beta\bar{\omega}\,\bar{\omega}_s \bar{x}_s = -\bar{\omega}^2 \bar{y}_j + \bar{\omega}^2 \bar{\delta}_o \sin\omega t \quad (2)$$

$$2\beta\bar{\omega}\,\bar{\omega}_s \bar{x}_s' + \bar{\omega}_s^2 \bar{x}_s + 2\beta\bar{\omega}\,\bar{\omega}_s \bar{y}_s = \bar{k}_{xx}\bar{x}_j + \bar{k}_{xy}\bar{y}_j + \bar{c}_{xx}\bar{x}_j' + \bar{c}_{xy}\bar{y}_j' \quad (3)$$

$$2\beta\bar{\omega}\,\bar{\omega}_s \bar{y}_s' + \bar{\omega}_s^2 \bar{y}_s - 2\beta\bar{\omega}\,\bar{\omega}_s \bar{x}_s = \bar{k}_{yx}\bar{x}_j + \bar{k}_{yy}\bar{y}_j + \bar{c}_{yx}\bar{x}_j' + \bar{c}_{yy}\bar{y}_j' \quad (4)$$

Stability

The threshold speed and the frequency of whirling at the threshold can be evaluated numerically from the characteristic equation using the Rouths criteria. The expressions for the threshold speeds for the cases $\beta \to 0$ and $\beta \to \infty$ can readily be obtained as,

$$(\omega_{bo}/\omega_o) = [J_2^2 J_4 J_5/\{(J_2 + J_5/\bar{\omega}_s^2)(J_2^2 J_3 - J_1 J_2 J_5 + J_5^2)\}]^{1/2} \quad (5)$$

$$(\omega_{b\infty}/\omega_o) = [J_2 J_4 J_5/(J_2^2 J_3 - J_1 J_2 J_5 + J_5^2)]^{1/2} \quad (6)$$

Since J_1, J_2, J_3, J_4 and J_5 are positive quantities, equations (5,6) have meaning only when $(J_2^2 J_3 - J_1 J_2 J_5 + J_5^2) > 0$. At sufficiently high values of ϵ, $(J_2^2 J_3 - J_1 J_2 J_5 + J_5^2) \leq 0$ and consequently the system remains stable at that value of ϵ for all values of $\bar{\omega}$. Equations (5,6) reveal $\omega_{b\infty} > \omega_{bo}$, in addition to the fact that as $\bar{\omega}_s \to \infty$, $\omega_{b\infty} \to \omega_{bo}$. Thus excessive material damping can improve the stability of a flexible rotor-bearing system.

Unbalanced Response

Assuming the system to be stable, its synchronous unbalanced response could be expressed as,

$$\bar{x}_j = \bar{\delta}_o(a_{jx}\cos\omega t + b_{jx}\sin\omega t) \quad (7)$$

$$\bar{y}_j = \bar{\delta}_o(a_{jy}\cos\omega t + b_{jy}\sin\omega t) \quad (8)$$

$$\bar{x}_s = \bar{\delta}_o(a_{sx}\cos\omega t + b_{sx}\sin\omega t) \quad (9)$$

$$\bar{y}_s = \bar{\delta}_o(a_{sy}\cos\omega t + b_{sy}\sin\omega t) \quad (10)$$

Substituting equations (7-10) in equations (1-4) and simplifying gives

$$(-\bar{\omega}^2 + 2i\beta\bar{\omega}\,\bar{\omega}_s + \bar{\omega}_s^2)\tilde{a}_{sx} + (2\beta\bar{\omega}\,\bar{\omega}_s)\tilde{a}_{sy} = \bar{\omega}^2(\tilde{a}_{jx} + 1) \quad (11)$$

$$(-2\beta\bar{\omega}\,\bar{\omega}_s)\tilde{a}_{sx} + (-\bar{\omega}^2 + 2i\beta\bar{\omega}\,\bar{\omega}_s + \bar{\omega}_s^2)\tilde{a}_{sy} = \bar{\omega}^2(\tilde{a}_{jy} - i) \quad (12)$$

$$(2i\beta\bar{\omega}\,\bar{\omega}_s + \bar{\omega}_s^2)\tilde{a}_{sx} + (2\beta\bar{\omega}\,\bar{\omega}_s)\tilde{a}_{sy} = (\tilde{k}_{xx}\tilde{a}_{jx} + \tilde{k}_{xy}\tilde{a}_{jy}) \quad (13)$$

$$(-2\beta\bar{\omega}\,\bar{\omega}_s)\tilde{a}_{sx} + (2i\beta\bar{\omega}\,\bar{\omega}_s + \bar{\omega}_s^2)\tilde{a}_{sy}$$

$$= (\tilde{k}_{yx}\tilde{a}_{jx} + \tilde{k}_{yy}\tilde{a}_{jy}) \qquad (14)$$

Eliminating $\tilde{a}_{sx}, \tilde{a}_{sy}$ from equations (13,14) using equations (11,12) and solving for $\tilde{a}_{jx}, \tilde{a}_{jy}$ yields,

$$\tilde{a}_{jx} = \frac{1}{\tilde{D}}\left[\left\{\frac{1}{\bar{\omega}^2} - \frac{1}{\bar{\omega}_s^2} + \frac{4i\beta}{\bar{\omega}\,\bar{\omega}_s}\right\}\{i\tilde{k}_{xy} + \tilde{k}_{yy}\} - \left\{1 + \frac{4i\beta\bar{\omega}}{\bar{\omega}_s}\right\}\right] \qquad (15)$$

$$\tilde{a}_{jy} = \frac{1}{\tilde{D}}\left[\left\{\frac{1}{\bar{\omega}^2} - \frac{1}{\bar{\omega}_s^2} + \frac{4i\beta}{\bar{\omega}\,\bar{\omega}_s}\right\}\{-i\tilde{k}_{xx} - \tilde{k}_{yx}\} + i\left\{1 + \frac{4i\beta\bar{\omega}}{\bar{\omega}_s}\right\}\right] \qquad (16)$$

where

$$\tilde{D} = \left[1 + \frac{4i\beta\bar{\omega}}{\bar{\omega}_s} + \frac{2\beta\bar{\omega}}{\bar{\omega}_s^3}\right][\tilde{k}_{xy} - \tilde{k}_{yx}] - \left[\left(\frac{1}{\bar{\omega}^2} - \frac{1}{\bar{\omega}_s^2} + \frac{4i\beta}{\bar{\omega}\,\bar{\omega}_s}\right) - \frac{2i\beta\bar{\omega}}{\bar{\omega}_s^3}\right][\tilde{k}_{xx} + \tilde{k}_{yy}]$$

$$+ \left[\left(\frac{1}{\bar{\omega}^2} - \frac{1}{\bar{\omega}_s^2} + \frac{4i\beta}{\bar{\omega}\,\bar{\omega}_s}\right)\left(\frac{1}{\bar{\omega}^2} - \frac{1}{\bar{\omega}_s^2}\right)\right][\tilde{k}_{xx}\tilde{k}_{yy} - \tilde{k}_{xy}\tilde{k}_{yx}] \qquad (17)$$

Dissipation Coefficient

The energy dissipation in the bearings due to the unbalanced response of the journal center per revolution of the rotor can be expressed as (Rajalingham, **et al**, 1986)

$$E = \pi . C_D . M_s g c . \bar{\delta}_o^2 \qquad (18)$$

where

$$C_D = [(a_{jx}^2 + b_{jx}^2)\bar{c}_{xx} + (a_{jx}a_{jy} + b_{jx}b_{jy})(\bar{c}_{xy} + \bar{c}_{yx}) + (a_{jy}^2 + b_{jy}^2)\bar{c}_{yy} - (a_{jx}b_{jy} - b_{jx}a_{jy})(\bar{k}_{xy} - \bar{k}_{yx})] \qquad (19)$$

The dissipation coefficient, C_D is a measure of the severity of rotor vibration and hence for this investigation the critical speeds are defined as those speeds at which C_D is a maximum.

COMPUTATION

Computations were carried out in double precesion for rotors supported in full cylindrical journal bearings of $b/d=0.75$ at $\epsilon = 0.10, 0.15, \ldots 0.90$ for $\bar{\omega}_s = 0.25, 0.50, 0.75, 1.0, 2.0$ and $\beta = 0.025, 0.050, 0.075, 0.10, 0.25, 0.50, 0.75, 1.0, 2.0$.

For the selected values of ϵ, $\bar{\omega}_s$ and β, (ω/ω_s) was varied from 0. by steps of 1. and the interval in which ω_b/ω_s lies was noted. This interval was then scanned using steps of 0.1 and the smaller interval containing ω_b/ω_s was noted. This procedure was repeated using successive steps of 0.01, 0.001, 0.0001 and 0.00001 and the value of ω_b/ω_s was noted.

For the selected values of ϵ, $\bar{\omega}_s$ and β, ω/ω_s was increased from 0. to 1.5 by a step of 0.001 and the corresponding ω/ω_s for which C_D takes its maximum values were noted.

RESULTS AND DISCUSSION

Figure 1 shows that the threshold speed first decreases and then increases as the material damping increases. Thus

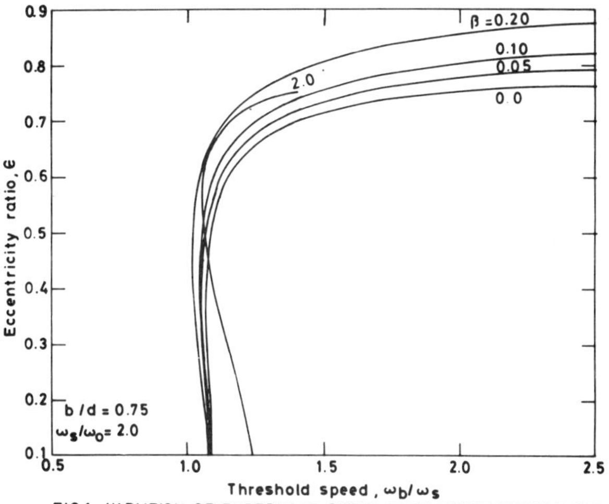

FIG.1. VARIATION OF THRESHOLD SPEED WITH ECCENTRICITY RATIO.

for excessive material damping the stiffness asymmetry associated with the material damping has predominant influence on the stability of flexible rotors supported in journal bearings. The improvement of the stability due to excessive material damping suggests a stiffening effect of the stiffness anisotropy.

Figure 2 shows that the major critical speed decreases as the material damping increases. However, the minor critical speed always increases with the material damping and for the case of excessive material damping the minor critical speed becomes larger than the corresponding major critical speed. Figure 3 shows the peak values of the dissipation coefficients at these critical speeds.

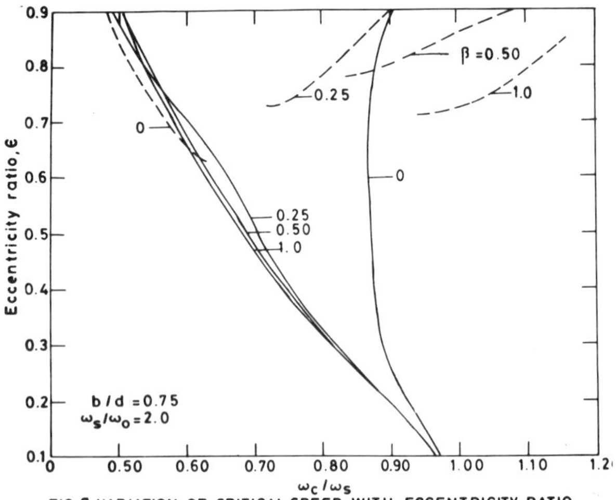

FIG. 2. VARIATION OF CRITICAL SPEED WITH ECCENTRICITY RATIO.

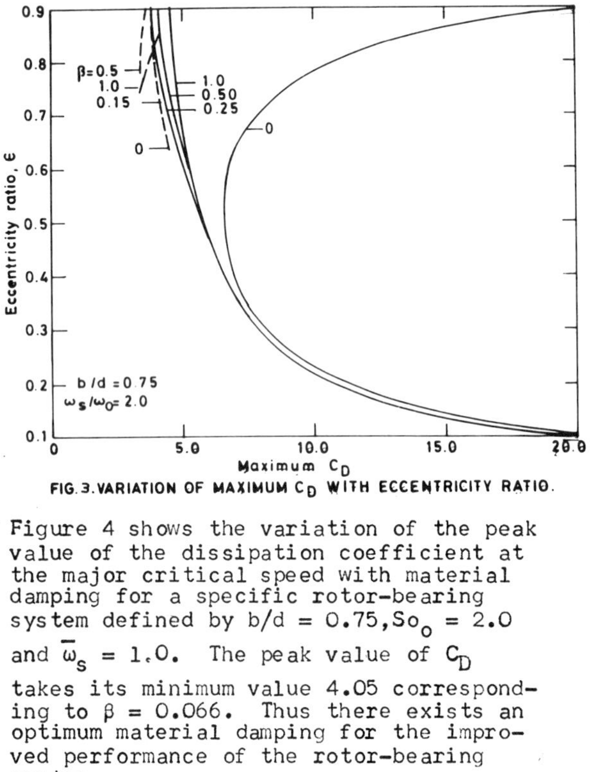

FIG. 3. VARIATION OF MAXIMUM C_D WITH ECCENTRICITY RATIO.

Figure 4 shows the variation of the peak value of the dissipation coefficient at the major critical speed with material damping for a specific rotor-bearing system defined by b/d = 0.75, So_o = 2.0 and $\bar{\omega}_s$ = 1.0. The peak value of C_D takes its minimum value 4.05 corresponding to β = 0.066. Thus there exists an optimum material damping for the improved performance of the rotor-bearing system.

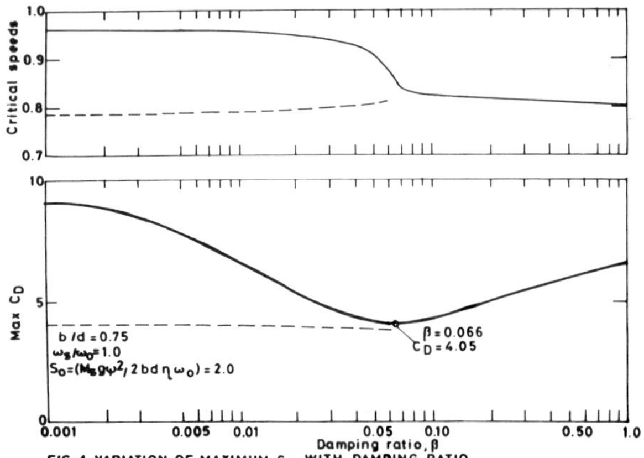

FIG. 4. VARIATION OF MAXIMUM C_D WITH DAMPING RATIO.

CONCLUSIONS

1. Low material damping destabilises the system whereas excessive material damping enhances its stability.

2. The severity of rotor vibration at the major critical speed can be minimised by suitably optimising the material damping.

REFERENCES

Kimball, A.L. (1925). Internal Friction as a Cause of Shaft Whirling. Phil. Mag., 49, 724-727.

Tondl, A. (1924). Some Problems in Rotor Dynamics, Chapman and Hall, London, p.69.

Rao, J.S., Bhat, R.B. and Sankar, T.S. (1981). Effect of Damping on Synchronous Whirl of a Rotor in Hydrodynamic Bearings. Trans. CSME, 6, 155-161.

Rajalingham, C., Ganesan, N. and Prabhu, B.S. (1986). The Dissipation Coefficient and Its Application to Flexible Rotor-Bearing System, Wear, 107, 343-354.

Zusammenfassung. Die Rotorunwucht erzwungenen Schwingungen und Stabilitatsgrenze des elastischen Rotoren in Gleitlagern mit Berucksichtigung des Innerer Damfung wurde untersucht. Die Biege Kritische Drehzahlen eines System sind die Drehzahlen wenn die Energie verlost dissipation in Gleitlagern ist ein a maximale. Die Untersuchungen zeigten, das mit grosse Innerer Damfing die Federung effeckte ist grosser als der Damfung. Das heisst, es gibt eine optimale Innerer Damfung fur die System dynamische eigenschaften.

Influence of Modal Variation on Flexible Rotor Balancing

J. S. BRAR* AND K. N. GUPTA**

*Department of Mechanical Engineering, Delhi College of Engineering, Delhi-110016, India
**ITMMEC, Indian Institute of Technology, New Delhi-110016, India

Abstract. Modal balancing of flexible rotors requires mode shapes, which can be influenced by the flexibility of supports. In this paper modal variation is affected by changing the pedestal stiffness. Besides, the effect on balancing is also studied by making use of mode shapes corresponding to rigid supports instead of actual mode shapes. Balancing is performed for first three modes only by using N-plane modal balancing method and the effect of modal variation on the residual unbalance response of the rotor is studied. Two classes of rotors are considered: rotors having flexibility compared to that of bearing (a) comparable and (b) sufficiently large. It is found that consideration of actual mode shapes and provision of flexible supports in the installation result in better balancing of the rotor.

Keywords. Balancing; flexible rotor.

INTRODUCTION

Many rotors operate above the critical speeds, and hence call for flexible rotor balancing for their smooth operation. The effectiveness of balancing is judged by the residual unbalance response at critical speeds. Support stiffness play a role in influencing this effectiveness. They also affect the mode shapes. This aspect is being studied in this paper by varying the support stiffness in comparison to rotor stiffness.

Gladwell and Bishop (1959) had shown how the modal theory of balancing could be applied to a rotor on flexible bearings at its two ends but the bearings were assumed as massless springs. Bishop and Parkinson (1972) discussed the effect of support flexibility in reference to use of balancing machines for flexible rotor balancing. The N-plane balancing method makes use of a linear combination of normal modes of the rotor in approximating its deflection caused by unbalance. This way of approximation considerably simplifies the balancing procedure. However, it introduces the dependence of the rotor balance on its normal modes which are being influenced by boundary conditions including support stiffness.

ROTOR DETAILS

The rotor under investigation is shown in Fig. 1. It consists of five discs. The rotor is supported at its ends (station 2 and 10) on two identical bearings which are housed in flexible pedestals. The support stiffness can be varied by varying pedestal stiffness.

Two sets of disc masses have been considered. In oneset (Rotor A) each disc weights 91.0 Kg. In the second set (Rotor B) the extreme discs weigh 4.0 Kg. each while three intermediate discs are of weight 8.0 Kg. each. In the first set, the diameter of the shaft between the bearings is 10.0 cm with a span of 1.7 m and in the second set, it is 1.8 cm with a span of 70.0 cm.

Rotors A and B belong respectively to the classes (a) rotor with flexibility comparable to that of bearing, and (b) rotor with flexibility sufficiently higher than that of bearing.

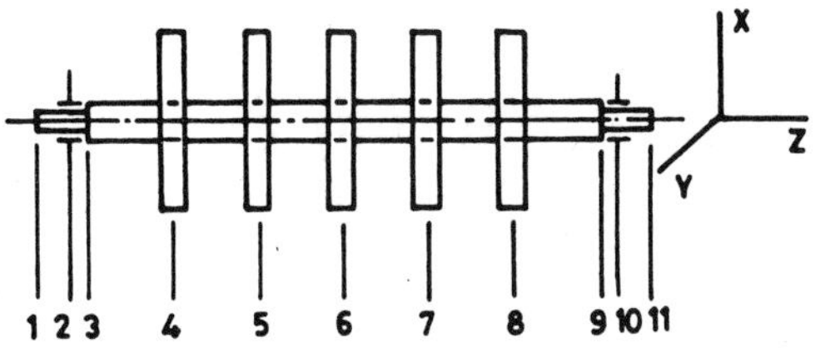

Fig. 1 Rotor and stations

BALANCING STUDIES

Balancing Procedure

N-plane balancing method (Bishop, 1959,

1972; Kellenberger, 1972) has been used for balancing the rotor. Practical rotors generally have two to three critical speeds within the range of their operation. So the rotors A and B are balanced for their first three modes. An arbitrary unbalance has been assumed as given in Table 1. For balancing three modes, three balance planes are selected preferably at antinodes, i.e. at stations 4,6 and 8 as shown in Fig. 1. The equations determining the balancing masses can be expressed as follows:

$$\sum_{j=1}^{3} \phi_n(z_j) U_j = -\sum_{i=1}^{M} \phi_n(z_i) u_i \qquad (1)$$

$$n = 1,2 \text{ and } 3$$

where u_i refers to assumed unbalance in i_{th} plane, z_i is the coordinate of i_{th} station, U_j and ϕ_n refer respectively to balancing mass at j_{th} plane and n_{th} modal vector.

A computer programme has been developed to calculate balancing masses at the required balance planes in x and y directions. It first calculates the eigen values and eigen vectors (modal vectors), which are substituted in eq.(1) for evaluating balancing masses U_j. These masses are incorporated in the system and the programme determines maximum displacement response due to residual unbalance at three critical speeds. Transfer matrix approach has been used for finding eigen values, eigen vectors and unbalance.

Modal Variation

The support characteristics affect eigen values and eigen vectors which are required in modal balancing method. Stiffness and damping characteristics of the journal bearing (Rao, 1983, Swamy, 1975) depend on Sommerfeld number or eccentricity ratio of the bearings, which further depends on bearing parameters and viscosity of lubricant. For introducing modal variation, the pedestal stiffness is varied in relation to the rotor stiffness. To account for this variation, a factor called stiffness ratio 'C' is introduced which is defined as follows:

$$C = \frac{K_p}{f_1^2 \; M} \qquad (2)$$

where K_p is the pedestal stiffness in N/m, f_1 is frequency of first mode with rigid support in rad/sec and M is rotor mass at the bearing support in Kg.

For modal variation C is being varied from 0.1 to 10.0. The variation of eccentricity ratio from 0.4 to 0.6 is considered for evaluating the bearing characteristics. These values can be obtained simply on variation of clearance ratio from 0.001 to 0.0015.

For further studies on influence of modal variation on rotor balance two cases are considered. In the first case, normal modes are determined by assuming the rotor to be supported on rigid supports and they are applied to the actual rotor for determining the balancing masses and subsequently the residual unbalance response after incorporating them. In the second case, normal modes are determined corresponding to the actual situation and are used to carry out the same exercise of determining balancing masses and residual unbalance response. Figs 2 and 3 display first three modes of the rotor corresponding to actual support and rigid support conditions respectively.

TABLE 1 Rotor Unbalance

Station No.	Unbalance gm cm			
	Rotor A		Rotor B	
	X-plane	Y-plane	X-plane	Y-plane
4	100.0	200.0	10.0	20.0
5	100.0	100.0	10.0	10.0
6	-100.0	100.0	-10.0	10.0
7	100.0	-100.0	10.0	-10.0
8	-100.0	-100.0	-10.0	-10.0

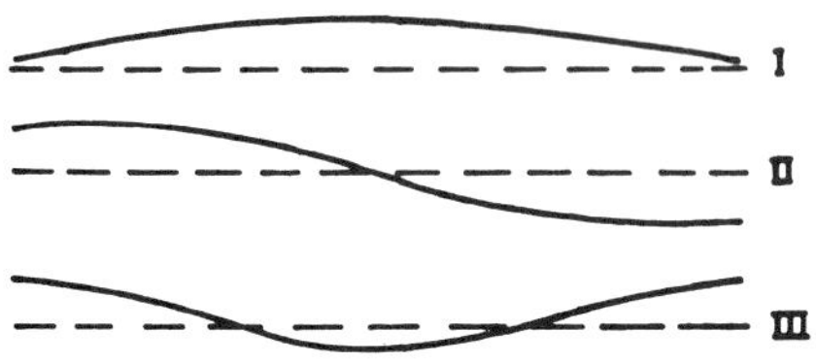

Fig. 2 Mode shapes of rotor with flexible support.

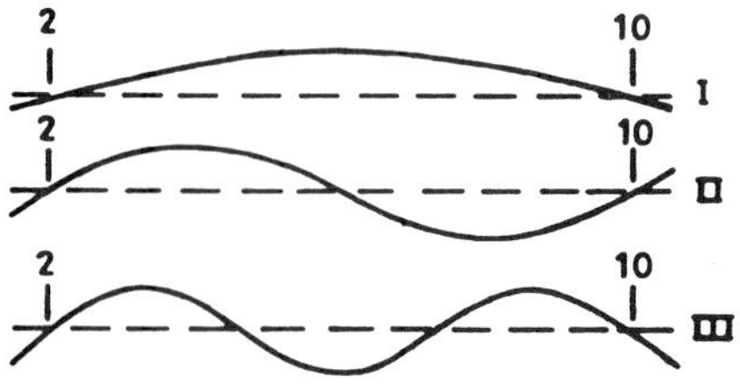

Fig. 3 Mode shapes of rotor with rigid support (Stations 2 and 10 are bearing supports).

RESULTS AND DISCUSSION

The residual unbalance response at first three critical speeds as obtained for rotors A and B after incorporating balancing masses is plotted against stiffness ratio in Figs. 4 and 5. The curves with dotted lines correspond to rigid support conditions, while those with firm lines refer to actual support conditions. It can be noted that there is a wide gap in the values of unbalance response for these two sets of curves, but the gap reduces with the increase in stiffness ratio value. This is obvious, because with increase in C value the actual support conditions tend towards rigid support conditions. Further, it is marked that the gap at C=10 is less for rotor B than for rotor A. The reason is that in case of rotor B the journal bearing stiffness is nearly ten times larger than that of rotor, but in case of rotor A both are nearly of the same order. So in former case, the actual support conditions are closer to rigid support conditions. It is also noticed that the gap is becoming less and less as one moves to responses at higher critical speeds. In case of rotor A, the response curves corresponding to actual support conditions become somewhat flat for value of C greater than 1 but this trend is not observed in case of rotor B. The bearing flexibility which is higher in the former case becomes dominating for C greater than 1, and thus it influences the response curve.

Another important feature, which is noted in the set of response curves corresponding to actual support conditions is that the response is low for smaller values of C, thus indicating that flexible supports provide better balancing of the rotor.

CONCLUSIONS

Following conclusions can be drawn on the basis of these studies:

(a) Mode shapes corresponding to actual support conditions should be taken into account for performing modal balancing exercise on a flexible rotor. This provides better balancing. However, when supports are sufficiently stiff (C>5) compared to the rotor, mode shapes corresponding to rigid supports may be used without much loss in the balancing.

(b) Flexible supports provide better balancing of the rotor.

REFERENCES

Bishop, R.E.D. and Parkinson, A.G. (1972). On the use of balancing machines for flexible rotors. Engineering for Industry J., 561 pp.

Gladwell, G.M.L. and Bishop, R.E.D. (1959). The Vibration of Rotating shafts supported in flexible bearings. Mechanical Engineering Science J., Vol 3, 195 pp.

Kellenberger, W. (1972). Should a flexible rotor be balanced in N or (N+2) planes? Engineering for Industry J., 548 pp.

Rao, J.S. (1983). Rotor Dynamics. Wiley Eastern Ltd. (India), First edition.

Swamy, S.T.N., Prabhu, B.S. and Rao, B.V.A. (1975) Stiffness and damping characteristics of finite width journal bearing with a non-Newtonian film and their application to instability prediction. Wear, 32, 379 pp.

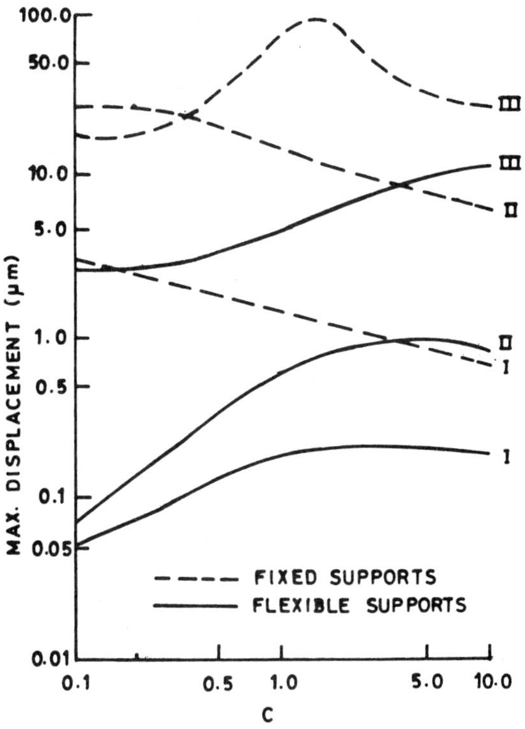

Fig. 4 Unbalance response for rotor A

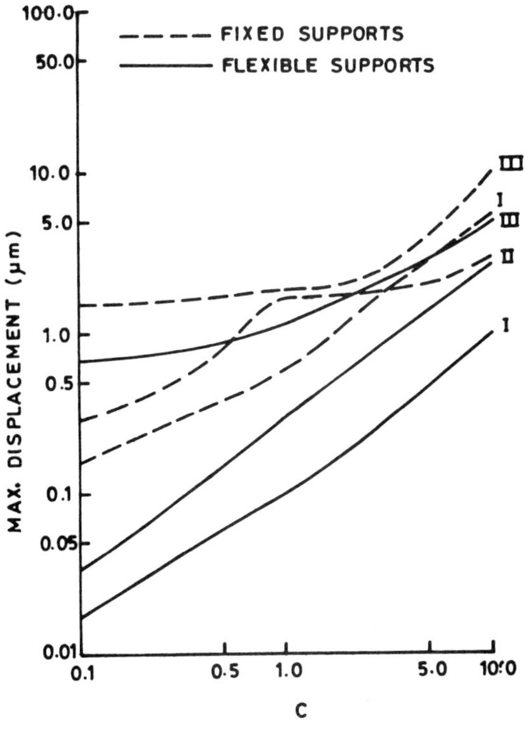

Fig. 5 Unbalance response for rotor B.

Zusammenfassung. In der vorliegenden Arbeit wird die Methode der Eigenfunktionen zum Rotorauswuchten verwendet, um den Einfluss der Schwingungsformänderung über die Rotorauswucht zu untersuchen. Schwingungsformen werden durch Änderung der Lagerbocksteifigkeit verändert. Es ist auch untersucht, wie das verhalten wegen gebliebener Unwucht durch Einführung der Eigenfunktionen entsprechend der starren Lagerungen statt der aktuellen Schwingungsformen beeinflusst wird. Zwei Arten von Rotoren sind betrachtet. In einer ist die Rotornachgiebigkeit im Verhältnis zu der Gleitlagersfederung vergleichbar, während in der Zweiten sie wesentlich gross ist. Man hat beschlossen, dass die Betrachtung der aktuellen Schwingungsformen und das Versorgen der nachgiebigen Lagerbocken ergeben ein besseres Rotorauswuchten.

On the Vibrations of Rotor Due to Parametric Excitation and Internal Friction

KAZUTOYO KONO*, AKIHIKO HOSOI**, ERIYA NAKAUCHI** AND SHINICHI NAGAYAMA**

*Department of Mechanical Engineering, Naruto University of Education, Naruto, Tokushima, Japan
**Faculty of Education, Tokushima University, Tokushima, Japan

Abstract. In the case of a rotating shaft system supported by ball bearing (LM stroke ST), gas bearing, or magnetic bearing, etc., the longitudinal oscillation of a shaft with free ends occurs. Equations of motion are then introduced. Three spring constants of this shaft system change simultaneously as the function of time t, respectively. And the parametric excitation occurs in the shaft system. The width of the unstable region at major critical angular velocity becomes wider as the amplitude of longitudinal displacement and the frequency of parametric oscillation increase, respectively. On the other hand the self-induced vibration due to internal friction of a rotating shaft system supported by the same ball bearing can be expected. And the conclusions derived are the same as those in case of parametric excitation.

Keywords. Vibration; oscillation; parametric excitation; internal friction; self-induced vibration; spring constants of shaft.

INTRODUCTION

The lateral vibrations of the system carrying a rotor have been usually treated as a vibratory system of four degrees-of-freedom system. If a force P and a couple M_t applied to the rotor on a circular shaft result in a deflection r and an inclination angle θ of the rotor, the following linear relations $P = \alpha r + \gamma\theta$, $M_t = \gamma r + \delta\theta$ hold, where α, γ, and δ are all spring constants of the shaft. Now, in the case of a rotating shaft system supported by ball bearing (LM stroke ST), gas bearing, or magnetic bearing, etc., the longitudinal oscillation of a shaft with free ends occurs. So we introduce the equations of motion of rotating shaft system having lateral and longitudinal vibration at the same time. The three spring constants of this shaft system α, γ, and δ change simultaneously as the function of time t, respectively. As the result of the aforesaid phenomenon the parametric excitation occurs in the shaft system. This excitation has great influence upon the natural frequencies $p_i (i = 1\text{--}4)$ versus rotating speed of shaft ω diagram. The p_2 curve which introduces the major critical angular velocity ω_c, where the relation $p_2 = \omega$ holds, becomes the lobed curve p_2 which is more distinguished than p_1, p_3, and p_4 ones. And as the result of this lobed curve p_2, the major critical angular velocity ω_c has the unstable region $(\omega_{c1}, \omega_{c2})$. Generally the width $(\omega_{c1}, \omega_{c2})$ of major critical angular velocity becomes wider as the amplitude of longitudinal displacement ε and the frequency of parametric oscillation Ω increase, respectively. So we have to decrease the value of ε and/or Ω, respectively.

Next, the self-induced vibration due to internal friction of the rotating shaft system supported by ball bearing (LM stroke ST) can be expected, and hence we examined the theoretical and experimental treatment of this vibration. Conclusions obtained are the same as those in case of parametric excitation.

EQUATIONS OF MOTION AND FREQUENCY EQUATION DUE TO PARAMETRIC EXCITATION

On the vibrations of a shaft carrying a rotor the most general equations of motion are given as follows:

$$\omega = \dot{\phi} + \dot{\psi} = \text{constant}$$

$$M\ddot{x} + c_1\dot{x} + \alpha x + \gamma\theta_x - \Delta\alpha\{x\cos2(\omega t + \zeta) + y\sin2(\omega t + \zeta)\} - \Delta\gamma\{\theta_x\cos2(\omega t + \zeta) + \theta_y\sin2(\omega t + \zeta)\} = 0,$$

$$M\ddot{y} + c_1\dot{y} + \alpha y + \gamma\theta_y - \Delta\alpha\{x\sin2(\omega t + \zeta) - y\cos2(\omega t + \zeta)\} - \Delta\gamma\{\theta_x\sin2(\omega t + \zeta) - \theta_y\cos2(\omega t + \zeta)\} = 0,$$

$$I\ddot{\theta}_x + I_p\omega\dot{\theta}_y + c_2\dot{\theta}_x + \gamma x + \delta\theta_x - \Delta I d/dt(\dot{\theta}_x\cos2\omega t + \dot{\theta}_y\sin2\omega t) - \Delta\gamma\{x\cos2(\omega t + \zeta) + y\sin2(\omega t + \zeta)\} - \Delta\delta\{\theta_x\cos2(\omega t + \zeta) + \theta_y\sin2(\omega t + \zeta)\} = 0,$$

$$I\ddot{\theta}_y - I_p\omega\dot{\theta}_x + c_2\dot{\theta}_y + \gamma y + \delta\theta_y - \Delta I d/dt(\dot{\theta}_x\sin2\omega t - \dot{\theta}_y\cos2\omega t) - \Delta\gamma\{x\sin2(\omega t + \zeta) - y\cos2(\omega t + \zeta)\} - \Delta\delta\{\theta_x\sin2(\omega t + \zeta) - \theta_y\cos2(\omega t + \zeta)\} = 0 \qquad \text{-----------(1)}$$

where M is the mass of the rotor; I_p is the polar moment of inertia; $I = (I_1 + I_2)/2$ is the mean value of the diametral moments of inertia of the rotor; $\Delta I = (I_1 - I_2)/2$ the asymmetry of the rotor; $\Delta\alpha$, $\Delta\gamma$, and $\Delta\delta$ the inequalities in stiffness; α, γ, and δ the mean values of the spring constants of the shaft; x and y the deflection of the rotor; $\theta_x = \theta\cos\phi$ and $\theta_y = \theta\sin\phi$ the components of the inclination angle; $\omega = \dot{\phi} + \dot{\psi}$ the angular velocity of the shaft; c_1 and c_2 the coefficients of damping; t time. Eulerian angles θ, ϕ, ψ and orientation ζ are shown in Fig.1.

In this paper, we treat the simple case, i.e., a vibratory system consisting of a light elastic shaft with circular shaft and a symmetrical rotor is treated, in which the deflections x, y and the inclination angles θ_x, θ_y of the rotor couple each other through gyroscopic terms.

Putting $\Delta I = 0$, $\Delta\alpha = \Delta\gamma = \Delta\delta = 0$, $c_1 = c_2 = 0$, and for convenience sake, the following dimensionless quantities are introduced:

$$i_p = I_p/I,\quad x' = x(M/I)^{1/2},\quad y' = y(M/I)^{1/2},$$
$$t' = t(\alpha/M)^{1/2},\quad \omega' = \omega(M/\alpha)^{1/2}, \tag{2}$$
$$\gamma' = (\gamma/\alpha)(M/I)^{1/2},\quad \delta' = (\delta/\alpha)(M/I)$$

Substituting Eq.(2) into Eq.(1) and omitting primes on the dimensionless quantities the frequency equation is derived as follows;

$$f(p) = (1 - p^2)(\delta + i_p\omega p - p^2) - \gamma^2$$
$$= (p - P_1)(p - p_2)(p - p_3)(p - p_4) = 0 \qquad \text{-----------(3)}$$

where p_1, p_2, p_3, and p_4 are the natural frequencies of the rotating shaft system and relation $p_1 > 1 > p_2 > 0 > p_3 > -1 > p_4$ holds. Natural frequencies p_i of the system versus rotating angular velocity ω of the shaft diagram ($p - \omega$ diagram) is shown in Fig. 2. Putting $p = \omega$ in Eq.(3), we obtain the value of major critical angular velocity ω_c.

UNSTABLE REGION NEAR MAJOR CRITICAL ANGULAR VELOCITY

When the shaft is supported by ball bearing (MS stroke ST), etc., the shaft fluctuates periodically to the longitudinal direction as the results of whirling. It is assumed here that the longitudinal sliding length Δs fluctuates sinusoidally with frequency Ω, and is given by

$$\Delta s = \nu\varepsilon \cos \nu\Omega t \tag{4}$$

in which ε is a small quantity and ν is a certain constant, and Ω is a frequency of longitudinal oscillation of the shaft.

The spring constants $\alpha(t)$, $\gamma(t)$, and $\delta(t)$ in the case of free-free supported system are derived from beam theory, respectively, neglecting any higher power than 3rd order of small value Δs.

$$\alpha(t) = 3\ell EI_0\{\ell^2 - 3ab - 6(a - b)\varepsilon\nu\cos\nu\Omega t + 12\varepsilon^2\nu^2\cos^2\nu\Omega t\}/\{a^3b^3 + 6a^2b^2(a - b)\cdot \varepsilon\nu\cos\nu\Omega t + 12ab(b^2 - 3ab + a^2)\varepsilon^2\nu^2\cos^2\nu\Omega t\}$$

$$\gamma(t) = 3\ell EI_0(a - b - 4\varepsilon\nu\cos\nu\Omega t)/\{a^2b^2 + 4ab(a - b)\varepsilon\nu\cos\nu\Omega t + 4(b^2 - 4ab + a^2)\varepsilon^2\nu^2\cos^2\nu\Omega t\}$$

$$\delta(t) = 3\ell EI_0/\{ab + 2(a - b)\varepsilon\nu\cos\nu\Omega t - 4\varepsilon^2\nu^2\cdot \cos^2\nu\Omega t\} \qquad \text{-----------(5)}$$

where E is Young's modulus of elasticity, I_0 is moment of inertia of the area of cross section of shaft. A horizontal shaft with a diameter of d, a length of ℓ is supported by ball bearings (LM stroke ST) with a bore of $\phi 10$ at its right and left shaft ends, and carries a symmetrical rotor at the position where the rotor is at a and b ($a < b$) distances from the right and left shaft ends, respectively.

By substituting Eq.(5) into Eq.(2) and moreover Eq.(2) into Eq.(1), the frequency equation Eq.(3) is derived easily. And the natural frequencies p_i versus rotating speed of shaft ω diagrams obtained by means of a digital computer are shown in Fig.3 (a), (b), (c). The first intersecting point of p_2 curve (lobed curve) and $p = \omega$ straight line gives the point ω_{c1}, and the last one gives ω_{c2}, respectively.

For obtaining the exact values of the width of the unstable region $\Lambda = \omega_{c2} - \omega_{c1}$, and the boundaries ω_{c1} and ω_{c2} for the unstable vibrations near ω_c, it is necessary to calculate the root of $p = \omega$ in Eq.(3) by means of a digital computer. The major critical angular velocity ω_c, however, is provided easily by Eq.(6).

$$\omega_c = \left[\left[(i_p - 1) - \delta + \left[\{(i_p - 1) - \delta\}^2 + 4(i_p - 1)(\delta - \gamma^2)\right]^{1/2}\right]/2(i_p - 1)\right]^{1/2} \qquad \text{-----------(6)}$$

In Fig.3, the width Λ is affected by the value of ε and/or Ω. Computer-aided $p_2 - \omega$ diagrams in which the upper and lower envelopes of lobed curve p_2 are denoted respectively by dotted lines, are shown in Fig.4(a), (b), (c). The larger the value of ε and/or Ω, the wider the width Λ as shown in Fig.5, i.e., the non-dotted area shown in Fig.5 indicates the unstable region. The width Λ becomes narrower as the value of ε decreases.

With the aim of observing the unstable vibrations appearing in the neighborhood of ω_c, the experimental apparatus shown in Fig.6 was used, and a horizontal circular shaft supported with ball bearing (LM stroke ST) at the both shaft ends connected to universal joint (SP-10-00A) at the both shaft ends, severally, and a rotor with I_p larger than I, were adopted. The horizontal shaft with a diameter of $\phi 10.2$ and a length of 654.0 mm carries a symmetrical rotor at the position where the rotor is at $a = 174$ and $b = 480$ mm distances from the right and left shaft ends, respectively. Dimensions of the rotor are as follows:

$$W = 7.71 \text{ kgf},\quad I_p = 2.4585 \text{ kgf cm s}^2,$$
$$I = 1.2294 \text{ kgf cm s}^2 \tag{7}$$

In this paper, the experimental apparatus has only one major critical angular velocity ω_c at $\omega = p_2$ as shown in Eq.(6) because $I_p > I$.

Response curves and the unstable regions near the major critical speed ω_c are shown in Fig.7. Amplitudes of steady vibration with frequency ω are indicated by symbols ○, □, △, for $\Omega = 31$ rad/s, Δs = 0.5, 0.3, o.1 cm, respectively, which were obtained by experiments. The actual unstable regions come slightly lower than the analytical values which are denoted respectively by the vertical one-dotted chain lines in Fig.7.

EQUATIONS OF MOTION DUE TO INTERNAL FRICTION

The self-induced vibration due to internal friction i.e., hysteresis whirl of the rotating shaft system supported by ball bearing (LM stroke ST) is treated as the following equations, simultaneously. Equations of motion are given as follows:

$$M\ddot{x} + c_1\dot{x} + \alpha x + \gamma\theta_x = (F_{isx} + F_{irx}) = \xi_x F_{ix}$$
$$M\ddot{y} + c_1\dot{y} + \alpha y + \gamma\theta_y = (F_{isy} + F_{iry}) = \xi_y F_{iy}$$
$$I\ddot{\theta}_x + I_p\omega\dot{\theta}_y + c_2\dot{\theta}_x + \gamma x + \delta\theta_x = (F_{is\theta x} + F_{ir\theta x}) = \xi_{\theta x}F_{i\theta x} \tag{8}$$
$$I\ddot{\theta}_y - I_p\omega\dot{\theta}_x + c_2\dot{\theta}_y + \gamma y + \delta\theta_y = (F_{is\theta y} + F_{ir\theta y}) = \xi_{\theta y}F_{i\theta y}$$

The longitudinal and the lateral deflection with internal friction or hysteresis are considered. Right hand sides in Eq.(8) $\xi_x F_{ix}$, $\xi_y F_{iy}$, $\xi_{\theta x}F_{i\theta x}$, $\xi_{\theta y}F_{i\theta y}$ indicate a whirling force in x, y, θ_x, θ_y directions, severally. And ξ_x, ξ_y, $\xi_{\theta x}$, $\xi_{\theta y}$ are the coefficients of whirling force. The hysteresis loops for various deflections of rotor due to thrust and radial forces are geometrically similar. In this case the area of hysteresis is proportional to the square of the deflection as shown in Fig. 8 (a), (b).

Putting the right hand sides in Eq.(8) $\xi_x F_{ix} = -\xi_1 y$, $\xi_y F_{iy} = \xi_1 x$, $\xi_{\theta x} F_{i\theta x} = -\xi_2 \theta_y$, $\xi_{\theta y} F_{i\theta y} = \xi_2 \theta_x$, and the following dimensionless quantities Eq.(9), and Eq.(2) are introduced.

$$p' = p(M/\alpha)^{1/2},\quad c_1' = c_1/(M\alpha)^{1/2},\quad c_2' = c_2(M/\alpha)^{1/2}/I,$$
$$\xi_1' = \xi_1/\alpha,\quad \xi_2' = \xi_2/(I\alpha/M) \qquad \text{--------(9)}$$

Inserting Eq.(9), (2) into Eq.(8) and omitting primes on the dimensionless quantities, and considering a rotating rectangular coordinate system o-x'y'z which rotates about the z-axis with an angular velocity of p and assuming that the x'-axis coincides with the x-axis at t = 0, we have the following relationships:

$$\begin{matrix} x \\ y \end{matrix} = x' \begin{matrix} \cos \\ \sin \end{matrix}\, pt \begin{matrix} - \\ + \end{matrix}\, y' \begin{matrix} \sin \\ \cos \end{matrix}\, pt,$$
$$\begin{matrix} \theta_x \\ \theta_y \end{matrix} = \theta'_x \begin{matrix} \cos \\ \sin \end{matrix}\, pt \begin{matrix} - \\ + \end{matrix}\, \theta'_y \begin{matrix} \sin \\ \cos \end{matrix}\, pt \qquad (10)$$

Substituting Eq.(10) into Eq.(8), and assuming solutions of equation to be denoted as follows:

$$\begin{matrix} x' \\ y' \end{matrix} = \begin{matrix} A \\ B \end{matrix} e^{st},\quad \begin{matrix} \theta_x' \\ \theta_y' \end{matrix} = \begin{matrix} C \\ D \end{matrix} e^{st} \qquad (11)$$

Inserting Eq.(11) into equations of motion, we have the following characteristic equation:

$$s^8 + K_7 s^7 + K_6 s^6 + K_5 s^5 + K_4 s^4 + K_3 s^3 + K_2 s^2 + K_1 s + K_0 = 0 \qquad \text{------------(12)}$$

where
$$K_7 = I' + E',$$
$$K_6 = J' + E'I' + F',$$
$$K_5 = K' + E'J' + F'I' + G',$$
$$K_4 = M' + E'K' + F'J' + G'I' + H',$$
$$K_3 = E'M' + F'K' + G'J' + H'I',$$
$$K_2 = F'M' + G'K' + H'J',$$
$$K_1 = G'M' + H'K',$$
$$K_0 = H'M'$$

in which
$$E' = 2c_1,\quad F' = 2p^2 + c_1^2 + 2,\quad G' = 2c_1 p^2 + 2c_1 - 4p\xi_1,\quad H' = (1 - P^2)^2 + (c_1 p - \xi_1)^2;\quad I' = 2c_2,$$
$$J' = 2(\delta + i_p \omega p - p^2 - \gamma^2) + c_2^2 + (2p - i_p\omega)^2;$$
$$K' = 2(\delta + i_p\omega p - p^2 - \gamma^2)c_2 + 2(2p - i_p\omega)(c_2 p - \xi_2),\quad M' = (\delta + i_p\omega p - p^2 - \gamma^2)^2 + (c_2 p - \xi_2)^2$$

The unstable region due to internal friction in which vibrations mount up exponentially is derived by solving Eq.(12). Namely, on the region where Eq.(12) has roots of complex with a positive real part the rotating system becomes unstable.

CONCLUSION

The width of the unstable region denoted by $(\omega_{c1}, \omega_{c2})$ where ω_{c2}, ω_{c1} are the upper and lower boundaries of the unstable region in the neighborhood of major critical angular velocity ω_c increases with the magnitude of longitudinal displacement ε and/or the frequency of parametric oscillation Ω.

The self-induced vibration due to internal friction of rotating shaft system with ball bearing (LM stroke ST) can be expected. Conclusions derived are the same as those in case of parametric excitation.

The experimental results obtained by the simulated apparatus are in perfect agreement with the analytical ones.

REFERENCES

Ota, H., and K. Kono (1971a). Unstable Vibrations Induced by Rotationally Unsymmetric Inertia and Stiffness Properties. Bulletin of JSME, 14, 29-38.

Yamamoto, T., H. Ota, and K. Kono (1968). On the Unstable Vibrations of a Shaft With Unsymmetrical Stiffness Carrying an Unsymmetrical Rotor. Trans. ASME, Ser.E, J. Appl. Mech., 35, 313-321.

Yamamoto, T., and K. Kono (1970). On Vibrations of a Rotor with Variable Rotating Speed. Bulletin of JSME, 13, 757-765.

Yamamoto, T., and K. Kono (1971b). Forced Vibrations of a Rotor with Rotating Inequality. Bulletin of JSME, 14, 1059-1068.

Yamamoto, T., H. Ota, and K. Kono (1972). On Vibrations of a Rotor with Rotating Inequality and with Variable Rotating Speed. Memoirs of the Faculty of Engineering, Nagoya Univ., 24, 1-80.

Über Rotorschwingungen infolge parametrischer Anregung und innerer Reibung

Die lateralen Schwingungen eines einen Rotor tragenden Systems werden gewöhnlich als ein Schwingungssystem von vier Freiheitsgraden beschrieben. Wenn eine Kraft P und ein Kräftepaar M_t auf einen auf einem Kreisschaft arbeitenden Rotor einwirken, so ergeben sich in Bezug auf die Beugung r und den Neigungswinkel θ die folgenden linearen Beziehungen: $P = \alpha r + \gamma\theta$, $M_t = \gamma r + \delta\theta$. Dabei sind α, γ und δ Federkonstanten des Schaftes. Wenn nun ein rotierendes Drehschaft-System auf einem Kugellager (LM stroke ST), einem magnetischen Lager o. ä. aufliegt, entsteht an einem Schaft mit freien Enden eine longitudinale Oszillation. Daher benötigen wir Bewegungsgleichungen für ein Drehschaft-System mit gleichzeitig lateralen und longitudinalen Schwingungen. Die drei Federkonstanten α, γ, und δ dieses Schaft-Systems verändern sich jeweils gleichzeitig als Funktion der Zeit t. Als Folge ergibt sich in dem Schaft-System eine parametrische Anregung. Diese hat auf die Eigenfrequenzen p_i (i = 1 —4), die gegen die Rotationsgeschwindigkeit ω des Schaftes gerichtet sind, großen Einfluß. Im Diagramm erscheint die Kurve p_2, welche die Hauptgrenzwinkelgeschwindigkeit ω_c darstellt (hier ist $p_2 = \omega$), viel stärker ausgebuchtet als kurven p_1, p_3 und p_4. Infolge dieser Ausbuchtungen ergibt sich für die Hauptgrenzwinkelgeschwindigkeit ω_c der labile Bereich $(\omega_{c1}, \omega_{c2})$. Die Bandbreite $(\omega_{c1}, \omega_{c2})$ der Hauptgrenzwinkelgeschwindigkeit ω_c nimmt generell zu mit einer Vergrößerung der Amplitude der longitudinalen Verschiebung ε beziehungsweise der Frequenz der parametrischen Schwingung Ω. Daher ist jeweils der Wert von ε und/oder Ω möglichst zu verringern.

Da infolge der inneren Reibung eines auf einem Kugellager (LM stroke ST) aufsitzenden Drehschaft-Systems selbstverursachte Schwingungen zu erwarten sind, haben wir diese theoretisch und experimentell untersucht. Dabei ergaben sich die gleichen Schlußfolgerungen wie im Fall der parametrischen Anregung. Die in Simulationsversuchen erzielten experimentellen Resultate stimmen mit den analytischen gut überein.

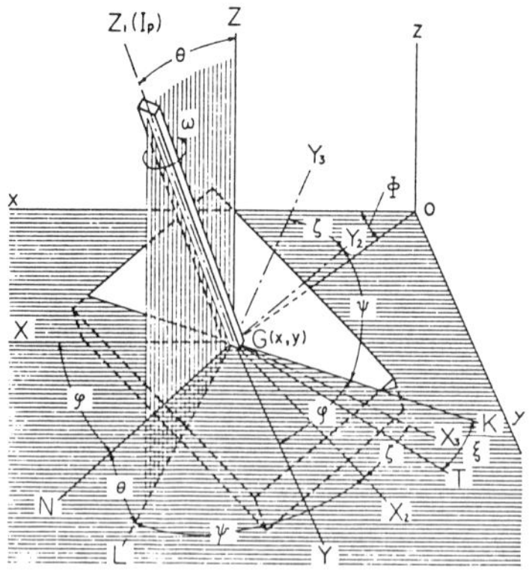

Fig.1 Rotating shaft system carrying a rotor

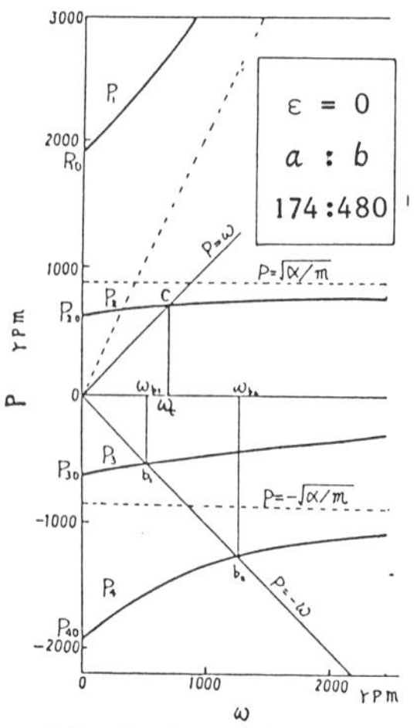

Fig.2 P-ω diagram

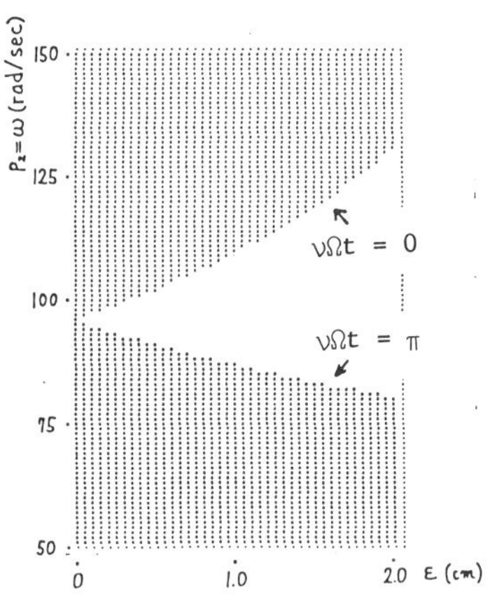

Fig.5 Unstable region

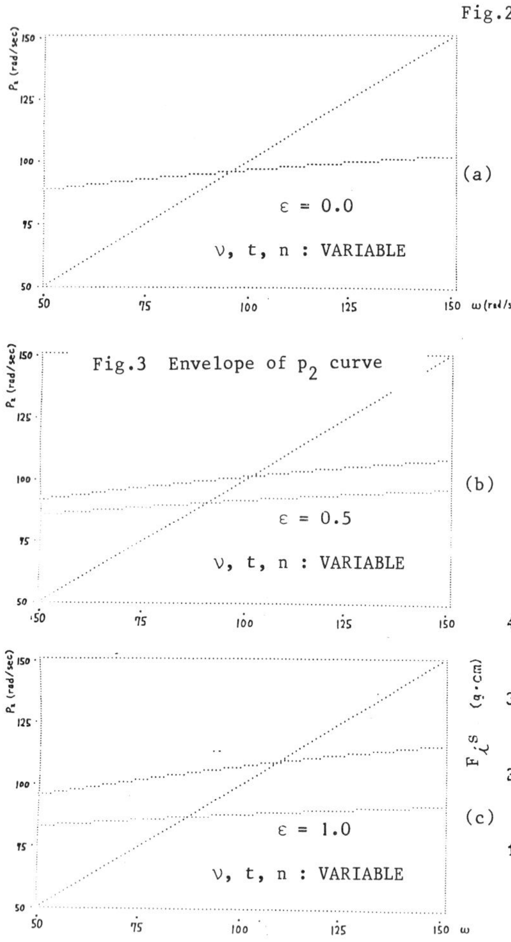

Fig.3 Envelope of p_2 curve

1 rotor
2 strain gauge
3 universal joint
4 guard ring
5 slip-ring
6 motor
7 displacement pickup
8 magnetic exciter
9 dynamic strain meter
10 recorder
11 analyser

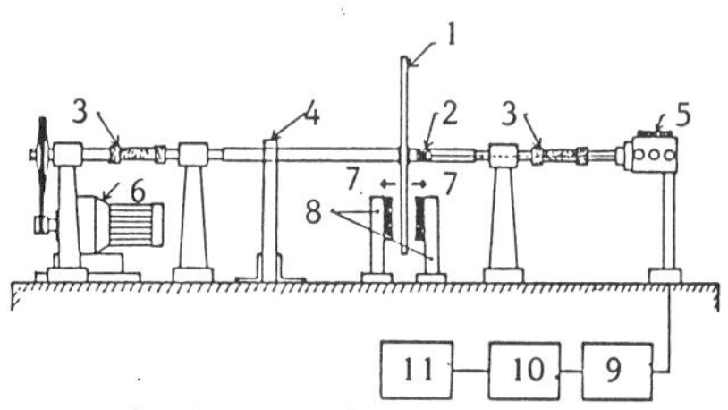
Fig.6 Experimental apparatus

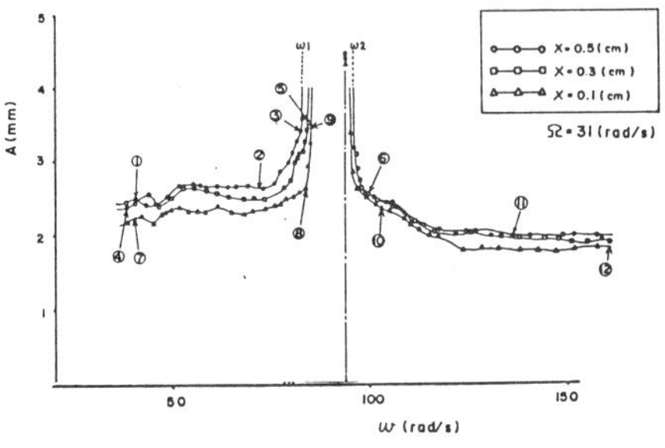

Fig.7 Response curves (ω_c)

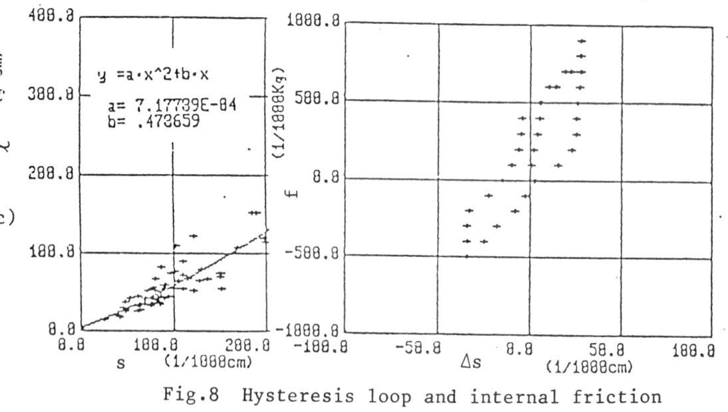

Fig.8 Hysteresis loop and internal friction

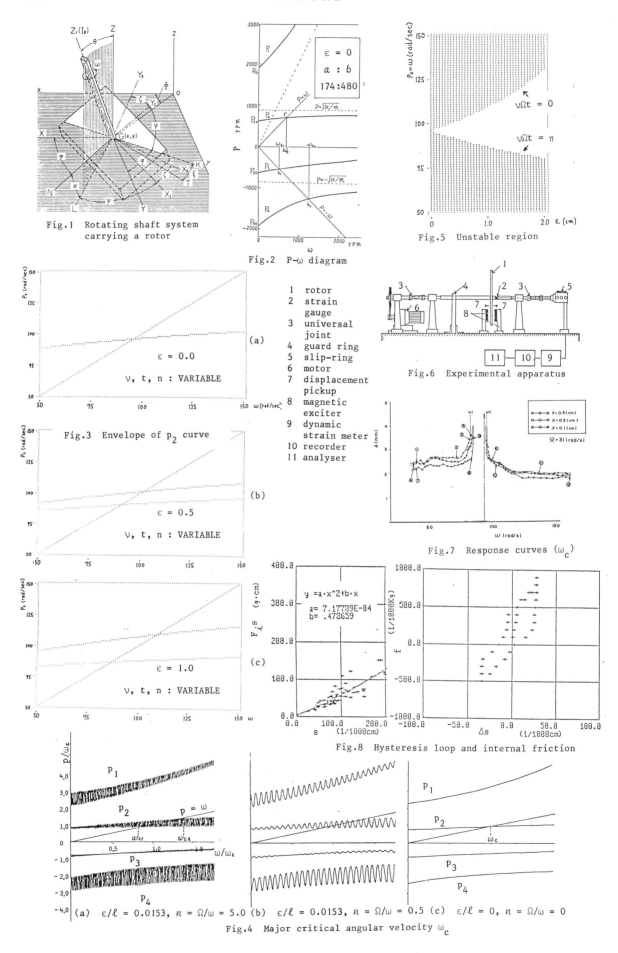

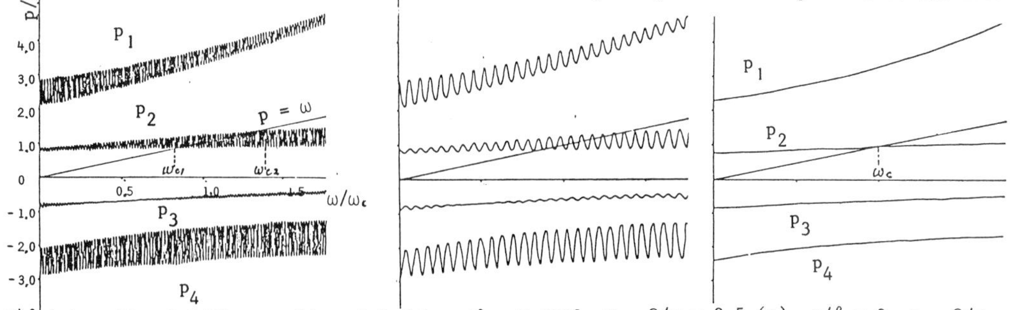

(a) $\varepsilon/\ell = 0.0153$, $n = \Omega/\omega = 5.0$ (b) $\varepsilon/\ell = 0.0153$, $n = \Omega/\omega = 0.5$ (c) $\varepsilon/\ell = 0$, $n = \Omega/\omega = 0$

Fig.4 Major critical angular velocity ω_c

Axial Vibration of Metal Flexible Couplings

S. YANABE*, Y. ONDA* AND T. OKAZAKI**

*Department of Mechanical Engineering, Technological University of Nagaoka, Niigata, Japan 940-21
**Department of Mechanical Engineering, Kyusyu Engineering University, Fukuoka, Japan 804

Abstract. A metal flexible coupling, which is composed of two packs of leaf springs and a spacer, is widely used in connecting rotating shafts of machines. Concerning this coupling, however, an axial vibration is often observed because of the low axial stiffness of the leaf springs. In this paper, the effects of three different misalignment (axial, parallel, and angular) conditions as well as the number and the thickness of the leaf spring on the coupling axial vibration are experimentally investigated. The results show that the axial misalignment strengthens the axial stiffness and the parallel or the angular misalignment strongly increases the resonance amplitude. Further a formula for evaluating the axial stiffness of the leaf springs is theoretically derived, and the predicted results are in good agreement with the experimental ones.

Keywords. Vibration of rotating body, Axial vibration, Coupling, Leaf spring, Misalignment, Axial stiffness, Resonance amplitude, Damping ratio

INTRODUCTION

A metal flexible coupling is widely used in connecting rotating shafts of machines, since it has a simple structure, needs no lubrication, accommodates large misalignments of shaftings and so on. A coupling of this type, however, tends to generate an axial vibration because a coupling spacer lies between two packs of leaf springs with low axial stiffness. In this paper, an experimental apparatus is constructed to investigate characteristics of the coupling axial vibration under various operating conditions. Effects of the thickness and the number of leaf spring together with three different misalignments (axial, parallel, and angular) are experimentally investigated. Further, a formula for evaluating an axial stiffness of the leaf springs is theoretically derived.

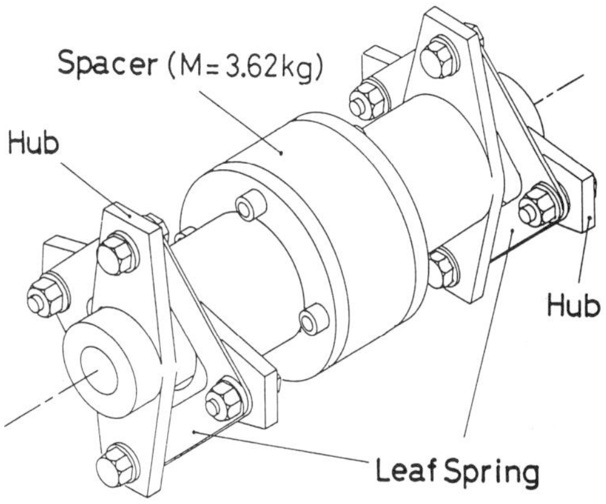

Fig.1 Metal flexible coupling

EXPERIMENTAL APPARATUS AND METHOD

A perspective view of the tested coupling is shown in Fig.1. A mass of a spacer is M=3.62kg. A leaf spring has a figure of a hollow square with 90 mm in side and 20 mm in width. The thickness of the leaf spring is 0.4, 0.5, 0.65, and 0.8 mm. An outline of experimental apparatus is shown in Fig.2. A speed-variable motor drives a driven shaft through the tested coupling. The driven shaft is supported on radial ball bearings and its axial movement is tightly restrained by thrust bearings and thrust collars. Figure 3 shows the three different misalignment conditions treated here, and the corresponding leaf spring deformations are shown in Fig.4. The assembled apparatus is operated in the speed range of 300-2700 rpm and the axial vibration of the coupling spacer is measured by an eddy current type gap sensor A(see Fig.2). Preparatory tests show that amplitudes of both axial and bending vibrations of the driven shaft were small enough in all over the tested speed range and that torsional vibrations do not affect the coupling axial vibration.

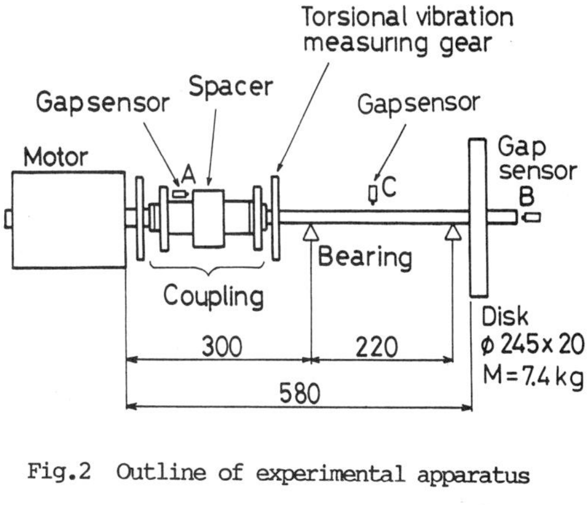

Fig.2 Outline of experimental apparatus

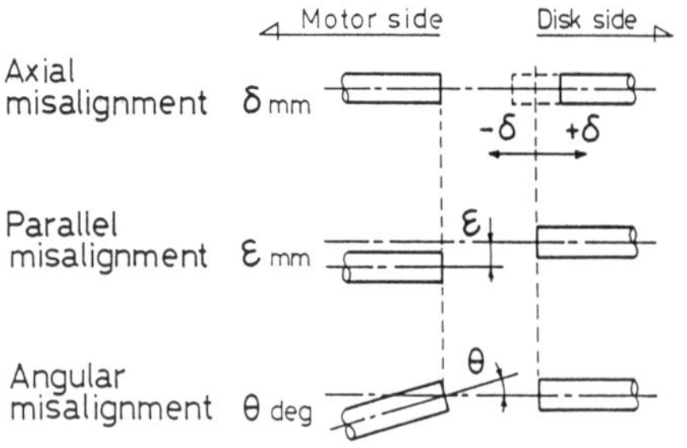

Fig.3 Three different misalignments

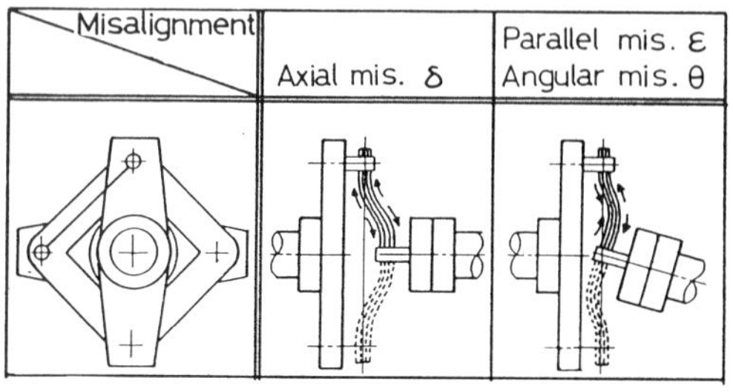

Fig.4 Deformations of leaf springs

EXPERIMENTAL RESULTS

Axial Vibration under Good Alignment Conditions

Effects of the number of the leaf spring n. Axial vibration responses obtained by changing the number of the leaf spring with 0.65mm in thickness are shown in Fig.5. Damping ratios obtained by free vibration tests are written in the same figure. The vibration wave forms show that the main component of the axial vibration is synchronous with the shaft rotation but there exist higher harmonic resonances. Figure 5 shows that both the natural frequency f_n and the damping ratio ζ become remarkably larger and the resonance amplitude A becomes remarkably smaller, with increase in n.

Effects of the thickness of the leaf spring t. Figure 6 shows the variation of the axial vibration response due to t. The solid lines in Fig.6 denote the response curves for n=1 and the dashed lines for n=3. Figure 6 shows that the natural frequency f_n becomes higher with increase in t but the damping ratio ζ and the resonance amplitude A are hardly affected by the thickness.

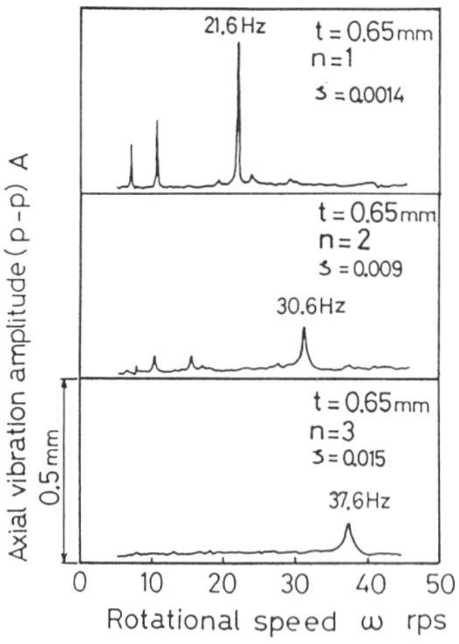

Fig.5 Variation of axial vibration response due to the number of the leaf spring n

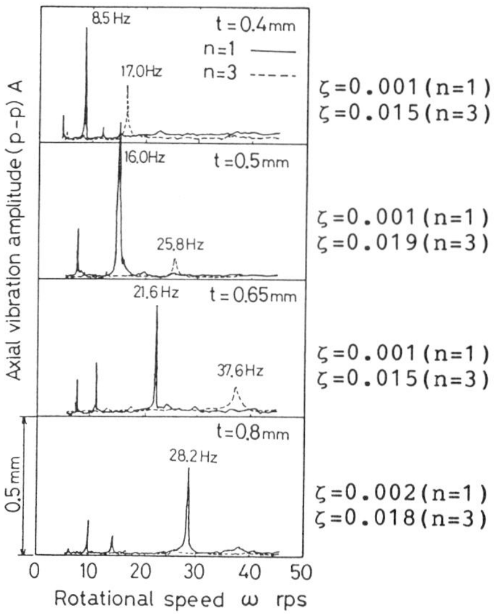

Fig.6 Variation of axial vibration response due to the thickness of the leaf spring t

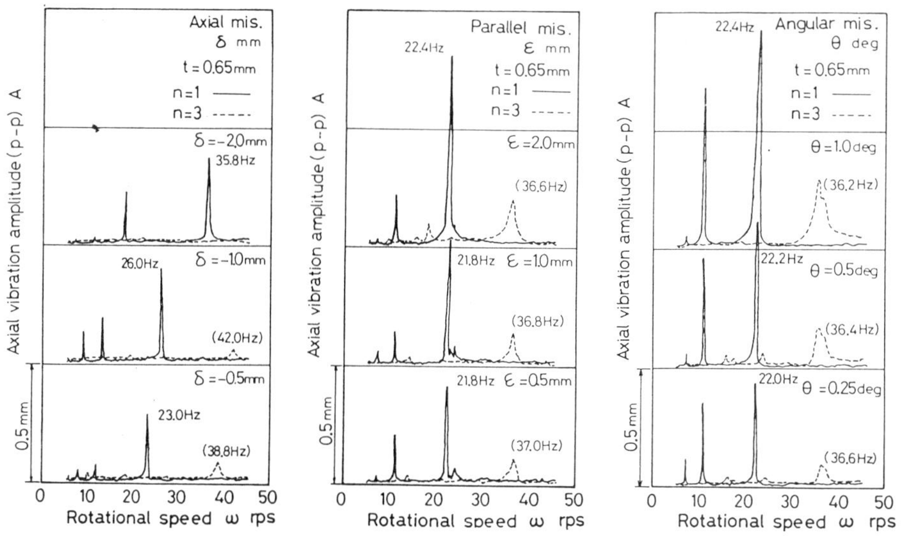

(a) Axial misalignment (b) Parallel misalignment (c) Angular misalignment

Fig.7 Variations of axial vibration response due to three different misalignments

Axial Vibration under Misalignment Conditions

Figures 7(a)-(c) show the axial vibration responses obtained by changing the levels of the three different misalignments δ , ε and θ . The distinction between the solid and the dashed lines in Fig.7 is the same as in Fig.6. Rearranging the results shown in Fig.7, the variations of the natural frequency f_n, the damping ratio ζ and the resonance amplitude A are plotted against each misalignment level in Figs.8(a)-(c). Symbols ○ and □ denote the cases for n=1 and n=3 respectively. From Figs.7 and 8, the followings are found.

(1) The f_n becomes remarkably higher with increase in δ but it is not hardly affected by ε and θ .
(2) The A becomes strongly larger with increase in ε and θ, but it is not so much affected by δ.
(3) These tendencies are always observed irrespective of the number of the leaf n.
(4) For n=1, the ζ hardly changes irrespective of the type and the level of misalignment. For n=3, ζ becomes large as ε and θ increase or δ decreases.

Each misalignment has different effects on the axial vibration and these are comming from the difference of the leaf spring deformation. It will be discussed later.

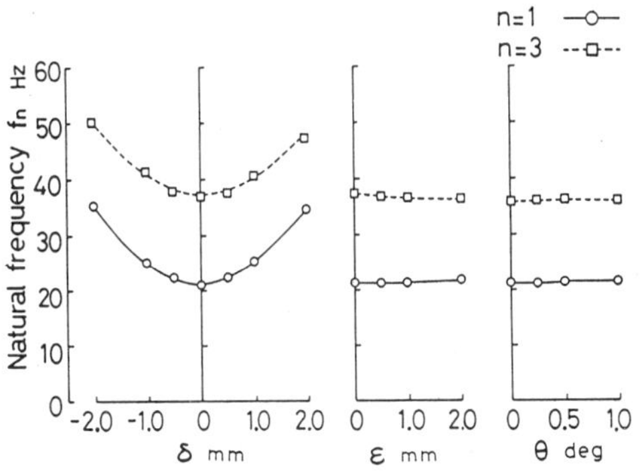

(a) Variation of natural frequency

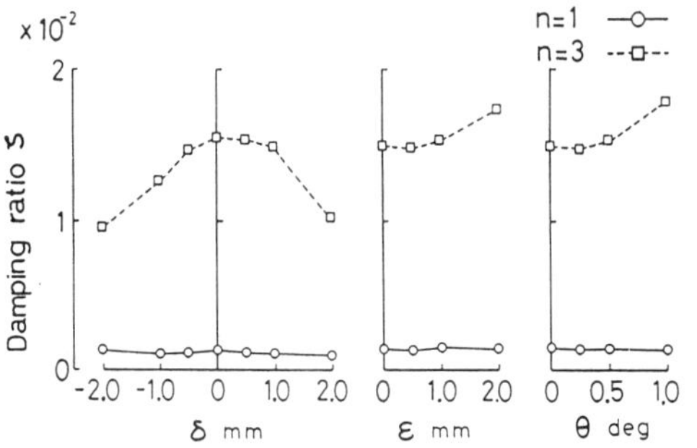

(b) Variation of damping ratio

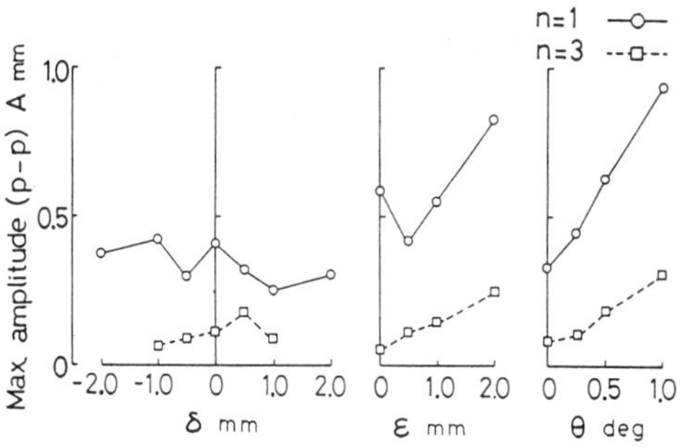

(c) Variation of resonance amplitude

Fig.8 Effects of misalignments on the axial vibration

THEORETICAL ANALYSIS OF AXIAL STIFFNESS OF THE COUPLING

Under Good Alignment Conditions

Figure 9 shows a deformation of the leaf springs due to the axial displacement of the coupling spacer x_o, which is produced by an axial force P_o. The axial stiffness of the leaf springs k_b can be expressed by the equation

$$k_b=\partial P_0/\partial x_0=\partial^2 U_b/\partial x_0{}^2 \quad (1)$$

Where U_b is an increment of a strain energy of the leaf springs due to bending deformation. A deflection curve of a leaf can be written as

$$y=x_0(3s^3-2s^2), \quad s=x/\ell \ ,0\leq s\leq 1 \quad (2)$$

Then the U_b can be calculated as follows.

$$U_b=4n\int_0^\ell (M^2/2EI)dx=24EInx_0{}^2/\ell^3 \quad (3)$$

Substituting Eq.(3) into Eq.(1), we obtain

$$k_b=48EIn/\ell^3 \quad (4)$$

In the following calculation, ℓ is replaced by ℓ_b (= ℓ -d, d=15 mm:diameter of a washer which fixes the leaf springs). Equation (4) can be also derived by another method.

Under Misalignment Conditions

The experimental results show that only the axial misalignment δ reinforces the axial stiffness of the leaf springs. This is mainly because δ produces a tensile deformation together with the bending one in the leaf spring and an increment of the strain energy due to this tensile deformation reinforces the axial stiffness. It should be noted here that all members of the leaf springs are not subjected to the same tensile deformation because of position errors of bolt holes on each leaf spring. We assume only one leaf is subjected to the deformation.

Based on the above considerations, the increment of the axial stiffness k_t due to axial misalignment $\delta(=2x_o)$ can be derived as follows. The increment of the strain energy U_t due to the tensile deformation can be calculated as

$$U_t=4\int_0^\ell \{AE(\Delta\ell/\ell)^2\}/2dx$$
$$=2AE\int_0^\ell \{(dy/dx)^2/2\ell\}^2dx \quad (5)$$

where y is expressed by Eq(2). The k_t can be obtained as

$$k_t=\partial^2 U_t/\partial x_0{}^2=216AEx_0{}^2/25\ell^3 \quad (6)$$

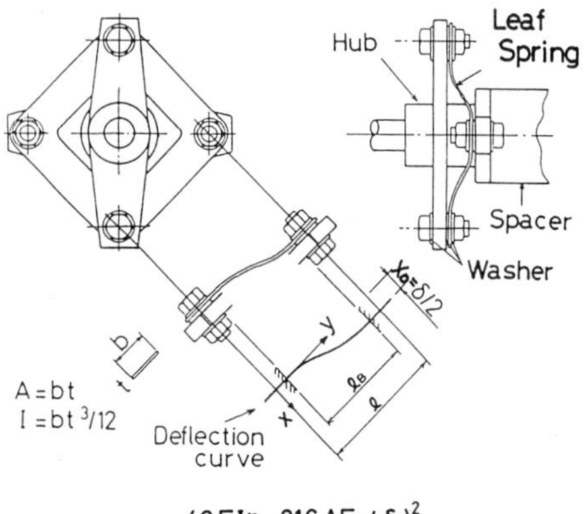

$$K=\frac{48EIn}{\ell_B^3}+\frac{216AE}{25\ \ell^3}\left(\frac{\delta}{2}\right)^2$$

Fig.9 Deformation of leaf spring due to axial displacement of spacer

The axial stiffness of the leaf springs taking account of the effects of the axial misalignment can be written as

$$K=k_b+k_t=48EIn/\ell_b^{\,3}+54AE\delta^2/25\ell^3 \qquad (7)$$

DISCUSSIONS

Natural Frequency of Coupling Axial Vibration

The natural frequency of the coupling axial vibration f_n ($=\sqrt{K/M}/2\pi$) are calculated by using the axial stiffness K in Eq.(7) and the results are compared with the experimental ones in Figs.10 (a) and (b). Both results have a good correspondence and the maximum difference is 17%.

Damping Ratio of Coupling Axial Vibration

A damping force is comming from the friction forces acting between the leaf springs. When n=1, the damping ratio hardly changes irrespective of the type and the magnitude of the misalignment. When the number of the leaf spring increases, the contact area between the leaf springs becomes large and then the damping ratio remarkably increases. As the parallel ε or the angular θ misalignment increases, the contact forces becomes larger. While, the axial misalignment δ decreases the damping ratio because both the contact forces and the contact areas between the leaf springs seem to decrease due to the axial misalignment.

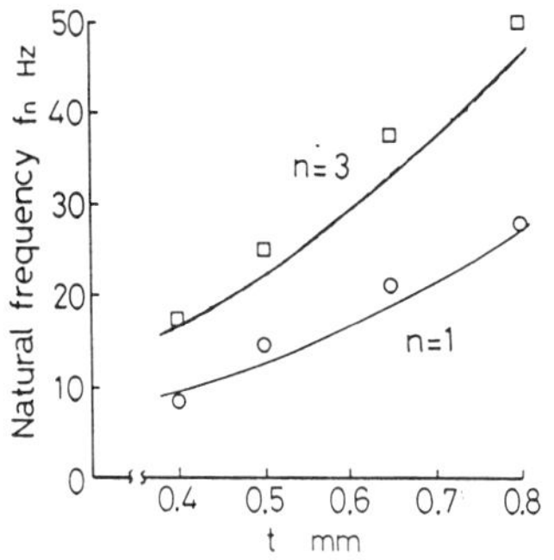

(a) Under good alignment conditions

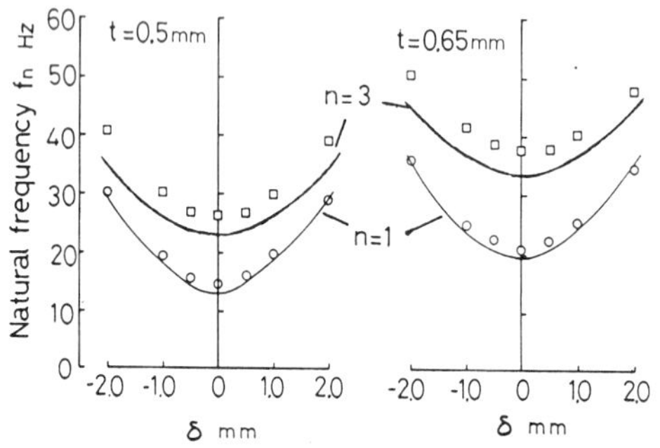

(b) Under axial misalignment conditions

Fig.10 Comparison of natural frequencies obtained by experiments and calculations

Amplitude at Resonance and Excitation Force of Coupling Axial Vibration

For good alignment conditions, the resonance amplitude seems to be almost inversely proportional to the damping ratio. When ε or θ increases, the resonance amplitude strongly gets larger, though the damping ratio leaves constant or slightly increases. This means an increase of an excitation force. The deformation of the leaf springs due to ε or θ periodically changes with shaft rotation and it excites the axial vibration of the coupling spacer. While, the deformation due to δ does not influence on the excitation force.

CONCLUSIONS

Conclusions are summarized as follows.

(1) The increase in the number of the leaf spring reinforces both the axial stiffness and the damping ratio. The increase in the thickness of the leaf spring reinforces the axial stiffness but hardly affects the damping ratio.

(2) The axial misalignment increases the axial stiffness of the leaf springs because of large tensile stress produced in leaf spring, but hardly influences the resonance amplitude.

(3) The parallel or the angular misalignment produces the deformation in leaf springs, which varies alternately with the shaft rotation. So the average stiffness of the leave springs does not change but the axial excitation force becomes very large. This means that the natural frequency does not change but the resonance amplitude becomes very large.

(4) The formula, which predicts the axial stiffness of the coupling taking account of the effects of the axial misalignment, is theoretically derived. The natural frequencies calculated by using the formula have a good agreement with the experimental results.

REFERENCES

(1) Landon,F.K. and Counter,L.F., Proc. 5th Turbomach. Symp., (1976), 125-131.
(2) Mancuso,J.R., ASME Paper, 77-DET-131(1977), 10.

Zusammenfassung. Eine metallische bewegliche Kupplung, die aus zwei Päcke Blattfedern und einem Distänzstuck besteht wird, ist um drehenden Wellen zu verbinden weit benuzt. Betreffs dieser Kupplung, aber, axiale Schwingungen wegen der wenigen Axialsteifheit oft beobachtet werden. In dieser Abhandlung, die Einflüsse der verschiedenen Misaufstellungs bedingungen (axiale, parallele, winklige) soeben wie so der Nummer und Dicke der Blattfedern versuchsweise untersucht werden. Die Ergebnise zeigen daβ die axiale Misaufstellung die Steifheit stürkt und daβ die parallele order winklige Misaufstellung die Resonanzamplitude stäklich gröβt. Weiter ein Formular um die Axialsteifheit der Blattfeders abzuschätzen theoretisch hergeleitet wird, und die Vorhersage beim Formular mit experimentallen Ergebnisse gut übereinstimmt.

Modal Balancing of a Flexible Rotor Using Response Gradient Method

C. W. LEE, S. W. HONG, N. Y. KIM AND Y. D. KIM

Department of Mechanical Engineering, Korea Advanced Institute of Science and Technology, Seoul, Korea

Abstract. Conventional flexible balancing methods necessitate trial runs in the vicinity of the critical speeds to be balanced, endangering the rotor systems due to excessive vibrations. In the present paper, a new balancing method for flexible rotor bearing systems, not requiring trial runs near at critical speeds, is deveolped by determining the influence coefficients based on the derivatives of the unbalance response away from the critical speeds. Theoretical and experimental results prove that the suggested method permits safe operation during balancing without much loss of the quality in balancing.

Keywords. Flexible rotor; Modal balancing; Influence coefficient; Derivative of unbalance response.

INTRODUCTION

It is well known that vibrations of high speed flexible rotor bearing systems such as compressors and turbine generators can be effectively suppressed by the flexible balancing which eliminates the modal unbalances [Parkinson,1980; Saito, 1985]. Most of the conventional methods inevitably necessitate trial runs just at or close to the critical speeds to be balanced, so as to effectively extract the modal unbalance components from the response signals generally associated with many modes. The trial operations near at the critical speeds, however, are likely to cause excessive vibrations.

In this paper, a new balancing method for flexible rotor bearing systems, not requiring trial runs near at critical speeds, is developed by determining the influence coefficients based on the derivatives of the unbalance responses away from the critical speeds. This method is derived theoretically, and its effectiveness is evaluated by experiments.

ANALYSIS

Equation of Motion

The discretized equation of motion of a typical rotor bearing system is of the form [Kim,1986]

$$\underline{M}\,\ddot{\tilde{q}} + \underline{C}\,\dot{\tilde{q}} + \underline{K}\,\tilde{q} = \tilde{f}(t) \tag{1}$$

where $\underline{M}$ is the mass matrix, $\underline{C}$ is the matrix incorporating the damping and gyroscopic effects, and $\underline{K}$ is the stiffness matrix. $\underline{C}$ and $\underline{K}$ are nonsymmetric and indefinite, while $\underline{M}$ is symmetric and positive definite. The matrices are of the order NxN, N being the dimension of the displacement and force vectors, $\tilde{q}$ and $\tilde{f}$.

Equation (1) may be rewritten as, in state space form,

$$\underline{M}^*\dot{\tilde{x}} + \underline{K}^*\tilde{x} = \tilde{F}(t) \tag{2}$$

where
$$\underline{M}^* = \begin{bmatrix} \underline{0} & \underline{M} \\ \underline{M} & \underline{C} \end{bmatrix} \qquad \underline{K}^* = \begin{bmatrix} -\underline{M} & \underline{0} \\ \underline{0} & \underline{K} \end{bmatrix}$$

$$\tilde{x} = \begin{Bmatrix} \dot{\tilde{q}} \\ \tilde{q} \end{Bmatrix} \qquad \tilde{F} = \begin{Bmatrix} \tilde{0} \\ \tilde{f} \end{Bmatrix}$$

Unbalance Response by Modal Analysis

The eigenvalue problem and the adjoint problem associated with Eq.(2) yield the 2N eigenvalues, λ_i, and the corresponding 2N right and left eigenvectors represented by

$$\tilde{u}_i = \begin{Bmatrix} \lambda_i \tilde{y}_i \\ \tilde{y}_i \end{Bmatrix} \qquad \tilde{v}_i = \begin{Bmatrix} \lambda_i \tilde{z}_i \\ \tilde{z}_i \end{Bmatrix} \qquad i=1,2,\ldots,2N \tag{3}$$

where $\tilde{y}_i$ and $\tilde{z}_i$ are the displacement right and left eigenvectors, respectively, corresponding to λ_i. Using the biorthogonality conditions

$$\underline{V}^T\underline{M}^*\underline{U} = \underline{I}$$
$$\underline{V}^T\underline{K}^*\underline{U} = -\underline{\Lambda} \tag{4}$$

Eq.(2) becomes

$$\dot{\tilde{r}} - \underline{\Lambda}\,\tilde{r} = \underline{V}^T\tilde{F} \tag{5}$$

where
$$\underline{U} = [\tilde{u}_1, \tilde{u}_2, \ldots, \tilde{u}_{2N}]$$
$$\underline{V} = [\tilde{v}_1, \tilde{v}_2, \ldots, \tilde{v}_{2N}]$$
$$\underline{\Lambda} = \mathrm{Diag}(\lambda_1, \lambda_2, \ldots, \lambda_{2N})$$

and $\underset{\sim}{r}$ is the generalized modal coordinate vector.

In the presence of unbalance, the force vector takes the form

$$\underset{\sim}{F}(t) = \begin{Bmatrix} \underset{\sim}{0} \\ \underset{\sim}{W} \end{Bmatrix} \Omega^2 e^{j\Omega t} \tag{6}$$

where $\underset{\sim}{W}$ denotes the system unbalance vector, and Ω the rotational speed. Assuming the steady state solution of the form

$$r_i = r_{oi} e^{j\Omega t} \tag{7}$$

we obtain, by substituting Eqs.(6) and (7) into Eq.(5),

$$r_{oi} = \frac{\underset{\sim}{z}_i^T \underset{\sim}{W}}{j\Omega - \lambda_i} \Omega^2 \tag{8}$$

Thus, the unbalance response vector q of the system becomes

$$q_s = \sum_{i=1}^{2N} \frac{\underset{\sim}{z}_i^T \underset{\sim}{W}}{j\Omega - \lambda_i} y_{is} \Omega^2, s=1,2,\ldots,N \tag{9}$$

where $\underset{\sim}{q} = [q_1, q_2, \ldots, q_N]^T$

$\underset{\sim}{y}_i = [y_{i1}, y_{i2}, \ldots, y_{iN}]^T$

New Balancing Method

It will prove convenient to rewrite Eq.(9) as

$$b_s = q_s/\Omega^2 = \sum_{i=1}^{2N} \frac{\underset{\sim}{z}_i^T \underset{\sim}{W}}{j\Omega - \lambda_i} y_{is}, s=1,2,\ldots,N \tag{10}$$

Here, b_s implies the system response to a spin speed independent constant force. Differentiation of Eq.(10) with respect to Ω up to order n yields

$$D^n b_s = j^n n! \sum_{i=1}^{2N} R_i^{n+1} a_{is} \tag{11}$$

where ! denotes factorial and

$$R = \frac{1}{j\Omega - \lambda_i}, \quad a_{is} = \underset{\sim}{z}_i^T \underset{\sim}{W} y_{is}$$

$$D^n = \frac{d^n}{d\Omega^n}$$

When the system is run near at the r-th critical speed, where

$$|R_r|^{k+1} >> |R_i|^{k+1} \quad ; i = 1,\ldots,2N, \ i \neq r \tag{12}$$

one can find n, for every small $\varepsilon > 0$, such that

$$\left| \frac{D^m b_{rs}}{j^m m!} - R_r^{m+1} a_{rs} \right| < \varepsilon \quad ; \text{for all } m > n \tag{13}$$

Thus, Eq.(11) can be approximated as

$$D^n b_{rs} = j^n n! R_r^{n+1} a_{rs} \tag{14}$$

Equation (14) implies that any desired, dominant, modal unbalance component can be effectively extracted from the response signals, which are generally associated with many modes, by means of differentiation of b with respect to Ω.

In order to identify experimentally the modal unbalances without a priori knowledge on the system, two test runs are required; with the original unbalance and then with the known trial unbalance attached to the system. When a trial unbalance w_k is positioned at the location k of the system and the system is run in the vicinity of the r-th critical speed, the trial unbalance vector $\underset{\sim}{W}^k$ may be expressed by

$$\underset{\sim}{W}^k = w_k \underset{\sim}{G}^k \tag{15}$$

where $\underset{\sim}{G}^k$ is the force vector due to a unit unbalance attached at the location k of the system. The resulting response becomes

$$b_{rs}^k = \sum_{i=1}^{2N} R_i \underset{\sim}{z}_i^T (\underset{\sim}{W} + \underset{\sim}{W}^k) y_{is} \tag{16}$$

Differentiating Eq.(16) with respect to Ω up to order n yields

$$D^n b_{rs}^k = j^n n! R_r^{n+1} (a_{rs} + a_{rs}^k), \quad s=1,2,\ldots,N \tag{17}$$

where $a_{rs}^k = \underset{\sim}{z}_r^T \underset{\sim}{W} y_{rs}$

From Eqs.(14) and (17), one obtains the influence coefficients defined as

$$\alpha_{rs}^k = \frac{D^n b_{rs}^k - D^n b_{rs}}{w_k} \tag{18}$$

where r,s,k denote the mode considered, the location of response measurement, and the trial unbalance (or correction) plane, respectively. The correction unbalance, w_k^*, to be attached at the location k, can be determined from

$$w_k^* = \frac{D^n b_{rs}}{D^n b_{rs} - D^n b_{rs}^k} w_k \tag{19}$$

Equation (19) implies that modal balancing can be conducted effectively with a sufficiently large n. Small n may often be sufficient especially when the system critical speeds are well separated and the system is run at the speed as close to the r-th critical speed as possible.

EXPERIMENTS AND DISCUSSIONS

Experiments are conducted through the first critical speed with the rotor bearing system in the laboratory.

Test Rig

Figure 1 shows the general assembly of the test rig. The test shaft with six

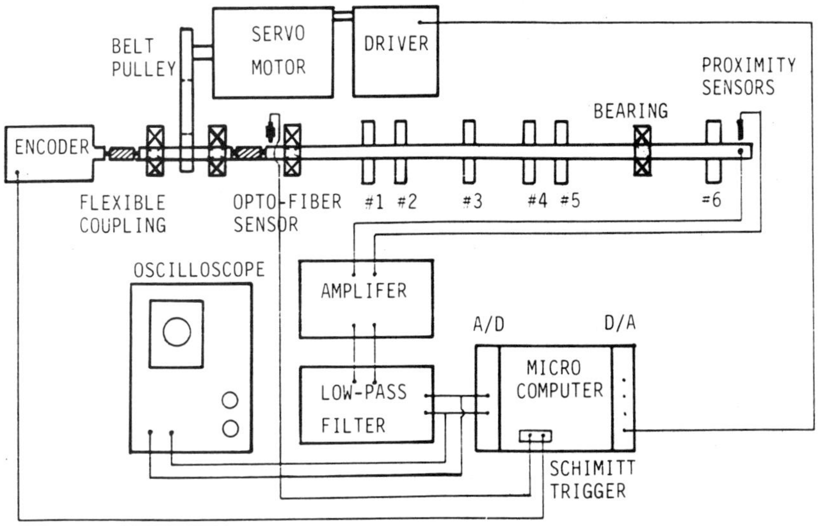

Fig. 1 Experimental set-up

disks is connected to the intermidiate shaft through a flexible coupling. The test shaft supported by two identical ball bearings is driven by a computer controlled D.C. servo motor through a flat belt and pulley system.

Two proximity sensors measure the horizontal and vertical shaft vibrations near one end of shaft as shown in Fig.1. The measured signals are sampled and analysed by a microcomputer which calculates the major and minor radii and the inclination angle of the whirl. An optofiber sensor located at the other end of the shaft generates one pulse per revolution, which is used to monitor the rotational speed and to provide the reference angular position of the shaft, initiating the A/D sampling every revolution. The sampling interval is controlled by an encoder.

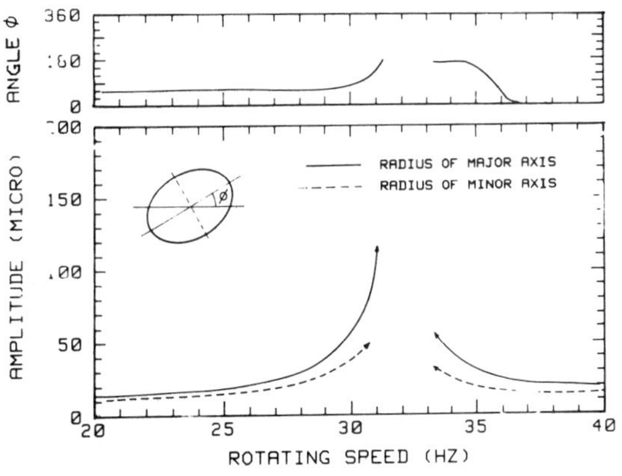

Fig. 2 Unbalance response before balancing

Experiments and Results

The proposed balancing method is conducted with the system having the first critical speed of about 32 Hz. The typical unbalance response before balancing is presented in Fig.2. The unbalance response data with and without a trial unbalance in the range of 24 to 28 Hz are taken for least square polynomial curve fitting, and the influence coefficient at 27.5 Hz is obtained by taking the first order derivatives of the polynomials of order 5 fitted to the data. The orders of the polynomials and the derivatives are strictly limited by noises contaminated during data acquisition or processing. In the present experiment, the second or higher order derivatives do not yield better results than the first, and the polynomial of order 5 is found to be optimal within the considered speed range.

Experiments are conducted under various unbalance conditions. The results of the proposed balancing at 27.5 Hz, are shown in Fig.3 when an unbalance of 30 g-cm is attached to disk 4. Similar results are also obtained for other cases.

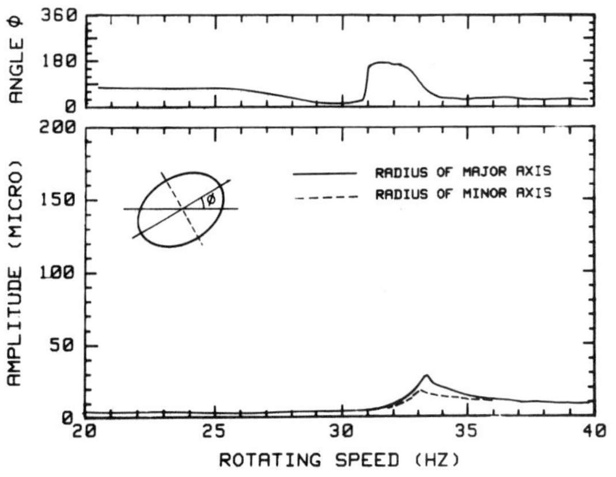

Fig. 3 Unbalance response after balancing

CONCLUSION

A new balancing method for flexible rotor bearing systems, not requiring trial runs at the critical speeds to be balanced, is developed by determining the influence coefficients based on the derivatives of the unbalance responses below the critical speeds. This method is then successfully applied to the balancing of a flexible rotor in the laboratory. The proposed method permits safe operation during balancing without much loss of the quality in balancing.

REFERENCES

Parkinson, A. G., Darlow, M. S., and Smalley, A. J. (1980), A theoretical introduction to the development of a unified approach to flexible rotor balancing. Journal of Sound and Vibration, 48(4), 489-506.

Saito, S., Azuma, T. (1985), Balancing of flexible rotors by the complex modal methods. Trans. of the ASME, Journal of Vibration, Acoustics, Stress, and Reliability in Design, 105, 94-100.

Kim, Y. D., Lee, C. W. (1986), Finite element analysis of rotor bearing systems using modal transform. to appear in Journal of Sound and Vibration, 111(2).

SUMMARY

Conventional balancing methods necessitate trial runs near at the critical speeds to be balanced, so as to effectively extract the required modal unbalance component from the response signals, which are generally associated with many modes. The trial runs, however, often cause excessive vibration, possibly damaging the systems. In this paper, a new balancing method is suggested which does not require trial runs near at the critical speeds. The method suggests a new way of determining the influence coefficients based on the derivatives of the unbalance responses. It is shown theoretically that any dominant modal unbalance component can be effectively extracted from the response signals by using the derivatives of the curve fitted unbalance responses. Actual experiments are also performed to verify the feasibility of the method. The experimental results prove that the proposed method permits safe operation during balancing without much loss of the balancing quality.

Dynamic Analysis and Experiments of Rotor – Bearing – Foundation System for Large Steam Turbine – Generator Set

GONG HANSHENG, CHANG HANYING, HE YI, HUANG WENZEN*
AND ZHENG ZHAOCHANG**

Shanghai Power Plant Equipment Research Institute,
***Tsinghua University, Beijing, PRC*

Abstract. On large steam turbine-generator sets, the effect of dynamic characteristics of foundation on the static and dynamic behaviors of rotor-bearing-foundation system can't be ignored. In this paper, the component mode systhesis method is used in the dynamic analysis of the system, which is divided into two substructures, the rotor and the foundation, the oil film bearing is treated as a connector. As a numerical example, an undamped critical speed of the 300MW turbo-generotor are calculated and compared with the experiments.

KEYWORDS Rotor dynamics; vibration analysis; component mode synthesis; rotating machinery;integral foundation;frame foundation.

NOMENCLATURE

C	damping matrix
Cij	elements of bearing damping matrix (i=y,z j=y,z)
F	bearing force vector
f	external force vector
K	stiffness matrix
Kij	elements of bearing stiffness matrix (i=y,z j=y,z)
M	mass matrix
Q	general variable vector
q	unknown vetor
x y z	cartesian coordinate system
X	displacement vector
s	eigenvalue

INTRODUCTION

For the dynamic analysis of many rotating machines, it is necessary to take into account the influence of dynamic behaviors of the foundation. In this case the method of component synthesis is more available and powerful. The total system is divided into two substructures, the internal damping and gyroscopic effects are neglected. The concrete foundation is commonly regarded as having proportional material damping. Thus the component modes of both substructures can be considered as real modes. The characteristics of fluid film bearing are directly assembled into the global stiffness and damping matrix of the complete system. They are, therefore, matrixes of non-symmetric and are dependent on the shaft speed. It is known, that the global modes of the complete system are in general complex. It leads to a solution of a complex eigenvalue for the calculation of the damped critical speeds. On the other hand, the undamped critical speed and their corresponding real modes of complete system can also be found, when the damping of the complete system is negligible, and the symmetry global system stiffness matrix is formed.

* Postgraduate student for doctor degree

EQUATIONS OF MOTION IN PHYSICAL COORDINATES

The rotor is subdivided into 3-d beam elements by the finite element method, and the foundation is also discreted in same manner. Let X(i) denotes the displacements or physical coordinates of the ith substructure (i=1,2) then, the eq. of motion for ith substructure can be expressed in the form

$$M^{(i)}\ddot{X}^{(i)} + C^{(i)}\dot{X}^{(i)} + K^{(i)}X^{(i)} = f^{(i)} + F^{(i)} \qquad (1)$$

$$X^{(i)} = \begin{Bmatrix} X_I \\ X_B \end{Bmatrix}^{(i)} \qquad (2)$$

Where I denotes the internal
B denotes the interface
and leads to the corresponding partition of matrix.

$$\begin{bmatrix} M_{II} & M_{IB} \\ M_{BI} & M_{BB} \end{bmatrix}^{(i)} \begin{Bmatrix} \ddot{X}_I \\ \ddot{X}_B \end{Bmatrix}^{(i)} + \begin{bmatrix} C_{II} & C_{IB} \\ C_{BI} & C_{BB} \end{bmatrix}^{(i)} \begin{Bmatrix} \dot{X}_I \\ \dot{X}_B \end{Bmatrix}^{(i)} + \begin{bmatrix} K_{II} & K_{IB} \\ K_{BI} & K_{BB} \end{bmatrix}^{(i)} \begin{Bmatrix} X_I \\ X_B \end{Bmatrix}^{(i)}$$

$$= \begin{Bmatrix} f_I \\ f_B \end{Bmatrix}^{(i)} + \begin{Bmatrix} 0 \\ F_B \end{Bmatrix}^{(i)} \qquad (3)$$

F represents the force of the 2nd substructure to the 1st substrucure through the bearings, hence $F_B^{(1)} = -F_B^{(2)}$ (4)

It is well known that the bearing force is characterized by 8 dynamic coefficients and are dependent on the rotating speed W. It is obviously expressed by

$$\begin{Bmatrix} Fy \\ Fz \end{Bmatrix} = -\begin{bmatrix} Kyy & Kyz \\ Kzy & Kzz \end{bmatrix} \begin{Bmatrix} Y^{(1)} - Y^{(2)} \\ Z^{(1)} - Z^{(2)} \end{Bmatrix} - \begin{bmatrix} Cyy & Czz \\ Czy & Czz \end{bmatrix} \begin{Bmatrix} Y^{(1)} - Y^{(2)} \\ Z^{(1)} - Z^{(2)} \end{Bmatrix} \qquad (5)$$

Where, in general, kyz = kzy, cyz = czy
It is assumed that the complete system has l bearings, then the eq (4) can be extended to the general form

$$F_B^{(1)} = -K_b\,(X_B^{(1)} - X_B^{(2)}) - C_b\,(\dot{X}_B^{(1)} - \dot{X}_B^{(2)}) \tag{6}$$

Where $K_b = \text{diag}\left(\begin{bmatrix} K_{YY} & K_{YZ} \\ K_{ZY} & K_{ZZ} \end{bmatrix}_i\right)$ (7)

$$C_b = \text{diag}\left(\begin{bmatrix} C_{YY} & C_{YZ} \\ C_{ZY} & C_{ZZ} \end{bmatrix}_i\right) \tag{8}$$

$$F_B^{(1)} = [(fy, fz)_1 \ldots\ldots (fy, fz)_L]^T \tag{9}$$

Generally $K_b \neq K_b^T$ $\quad C_b \neq C_b^T$

Since $X_B^{(1)} - X_B^{(2)} = [I : -I]\begin{Bmatrix} X_B^{(1)} \\ X_B^{(2)} \end{Bmatrix}$ (10)

Where I is the identity matrix of order (21*21).

$$F_B^{(1)} = -[K_b : -K_b]\begin{Bmatrix} X_B^{(1)} \\ X_B^{(2)} \end{Bmatrix} - [C_b : -C_b]\begin{Bmatrix} X_B^{(1)} \\ X_B^{(2)} \end{Bmatrix} \tag{11}$$

EQUATIONS OF MOTION IN MODAL COORDINATES

Now, the coordinate transformation of equation (2) takes the form

$$\begin{Bmatrix} X_I \\ X_B \end{Bmatrix} = [a]\begin{Bmatrix} q_I \\ q_B \end{Bmatrix} \tag{12}$$

$$a = \begin{bmatrix} T_I^{(i)} & T_B^{(i)} \\ 0 & I \end{bmatrix}$$

T is the eigenvectors of the conservative eigenproblem of the ith substructure, ie.

$$(-[\,{}^\frown W^2_\smile\,]^{(i)} M_{II}^{(i)} + K_{II}^{(i)})\, T^{(i)} = 0 \tag{13}$$

Within the interval (0, $W_{max}^{(1)}$) $W_{max}^{(1)}$ is choised as 2 times of operating speed. T is the solution of the static problem, ie.

$$T_B^{(i)} = -(K_{II}^{(i)})^{-1} K_{IB}^{(i)} \tag{14}$$

inserting eq.(12) into (3). We get the ith substructure equation of motion in model coordinates

$$\begin{bmatrix} I & M_{12} \\ M_{21} & M_{22} \end{bmatrix}^{(i)} \begin{Bmatrix} \ddot{q}_I \\ \ddot{X}_B \end{Bmatrix}^{(i)} + \begin{bmatrix} C_{11} & C_{12} \\ C_{21} & C_{22} \end{bmatrix}^{(i)} \begin{Bmatrix} \dot{q}_I \\ \dot{X}_B \end{Bmatrix}^{(i)} + \begin{bmatrix} W^2 & 0 \\ 0 & K_{22} \end{bmatrix}^{(i)} \begin{Bmatrix} q_I \\ X_B \end{Bmatrix}^{(i)}$$

$$= \begin{Bmatrix} T_I^T f_I \\ T_B^T f_I + f_B \end{Bmatrix}^{(i)} + \begin{Bmatrix} 0 \\ F_B \end{Bmatrix}^{(i)} \tag{15}$$

EQUATIONS OF MOTION OF THE COMPLETE SYSTEM

The eq of motion of both substructures can be assembled as:

$$M\ddot{Q} + C\dot{Q} + KQ = F \tag{16}$$

$$Q = \{\, q_I^{(1)}\ q_I^{(2)}\ X_B^{(1)}\ X_B^{(2)}\,\}^T \tag{17}$$

$$M = \begin{bmatrix} I & 0 & M_{12}^{(1)} & 0 \\ 0 & I & 0 & M_{12}^{(2)} \\ M_{21}^{(1)} & 0 & M_{22} & 0 \\ 0 & M_{22} & 0 & M_{22}^{(2)} \end{bmatrix} \tag{18}$$

$$C = \begin{bmatrix} C_{11}^{(1)} & 0 & C_{12}^{(1)} & 0 \\ 0 & C_{11}^{(2)} & 0 & C_{12}^{(2)} \\ C_{21}^{(1)} & 0 & C_b + C_{22}^{(1)} & -C_b \\ 0 & C_{21} & -C_b & C_b + C_{22}^{(2)} \end{bmatrix} \tag{19}$$

$$K = \begin{bmatrix} W^2 & & & \\ & W^2 & & \\ & & K_b + K_{22}^{(1)} & -K_b \\ & & -K_b & K_b + K_{22}^{(2)} \end{bmatrix} \tag{20}$$

$$F = \begin{Bmatrix} T_I^{(1)} f_I^{(1)} \\ \vdots \\ \vdots \end{Bmatrix} \tag{21}$$

SOLUTION OF THE EQUATIONS OF MOTION

In order to analysis the dynamic behavior of the complete system, the homogeneous equation of (16) must be considered. So we have

$$M\ddot{Q} + C\dot{Q} + KQ = 0 \tag{22}$$

Equation (22) can be converted, using the substitutions $q_1 = Qe^{st}$, $q_2 = sq_1$, to the linear generalized form of eigenproblem

$$Aq = sBq \tag{23}$$

$$A = \begin{bmatrix} M & 0 \\ 0 & -K \end{bmatrix} \quad B = \begin{bmatrix} 0 & M \\ M & C \end{bmatrix} \quad q = \begin{Bmatrix} q_1 \\ q_2 \end{Bmatrix}$$

If M,C and K are all symmetric, the supermetrices of (23) are symmetric. But this is not always the case because of the effects of the oil bearings and the flow induced aerodynamic cross forces in the sealings. In this situation there is the possibility that complex eigenvalues and by-orthogonal complex eigenvectors occur in conjugat pairs. Eigensolution procedures for this kind of problem are more complicated. Equation (23) can be solved in two ways. By premultiplying B^{-1} (if B is nonsingular) to either side of (23), we get the standard form of eigenproblem

$$\tilde{A}q = sq\ ,\ \tilde{A} = B^{-1}A \tag{24}$$

which can be solved by well known QR method. If B is singular or quasi-singular, QR method is no longer available. In this case, an alternative method called QZ method proposed by Moler and Stewart, which need not perform any matrix invertion, is more powerful for generalized form of eigenproblem (23).

Another advantage of the QZ method is that, with certain improvement, all the eigenvector and eigenrow as well as eigenvalues can be obtained simultaneously, so it can be used to analysis dynamic sensitivity of the system if needed. The QZ algorithm proceed in four stages. In the first, which is a generalization of Household reduction of a single matrix to Hessenberg form, A (in equation (23)) is reduced to upper Hessenberg form and at the same time B is reduced to upper triangular form. the second step, which is a generalization of the Francis implicit double shift QR algorithm, A is reduced to quasi-triangular form while the triagular form of B is mantained. In the third stage all the eigenvalues are extracted. In the last stage eigenvectors are obtained from the quasi-triangular matrices and then transformed back into the original coordinate system.

Through back substitution using (12), (17), all the complex mode shaps of the system in phisical coordinate can be obtained from the eigensolutions of equation (23). Damped critical speeds can also be calculated by iterations using complex eigenvalues.

EXAMPLE

The calculated results of the undamped critical speeds for a 300 MW turbine-generator set are shown in TABLE 1. The structure of the rotor and the foundation are shown in Fig. 1 and Fig. 2. Fig. 3-8 present some of the mode shaps. The influence of the foundation has been considered in the R.B.F.system (rotor-bearing-foundation system), while it has not involved in the R.B.system (rotor-bearing system).

TABLE 1 Results of Calculation and Experiment

	R.-B. System	R.-B.-F. System	Measured Value
Critical Speed(1st)	838 r.p.m.	840 - 910 r.p.m.	915 r.p.m.
Critical Speed(2nd)	2513 r.p.m.	2011-2434 r.p.m.	2410 r.p.m.
Critical Speed(3rd)	2678 r.p.m.	2069-2754 r.p.m.	
Critical Speed(4th)	3200 r.p.m.	3167-3563 r.p.m.	

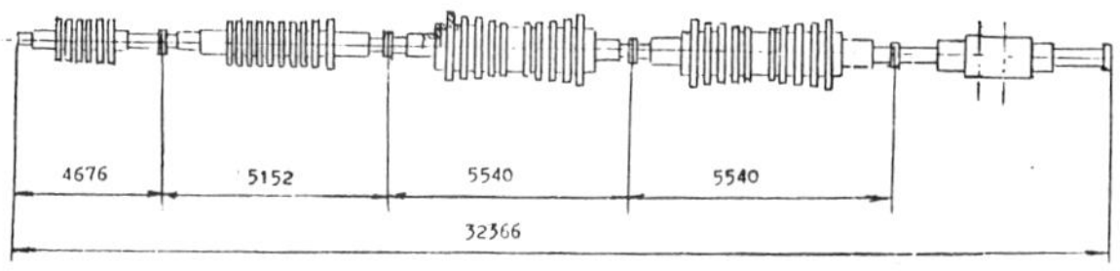

Fig. 1 Structure of the 300 MW turbine-generator rotor

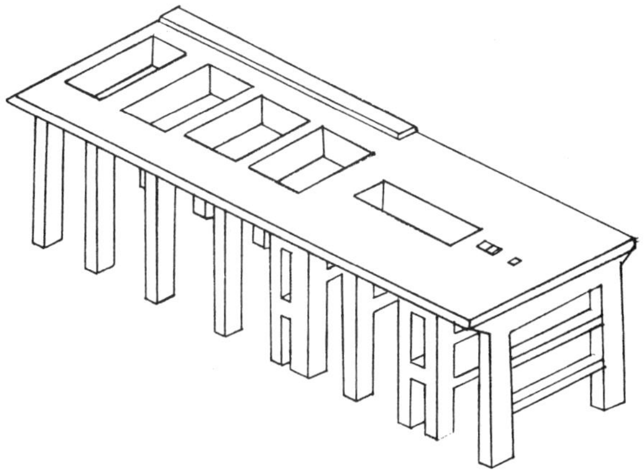

Fig. 2 The structure of the foundation

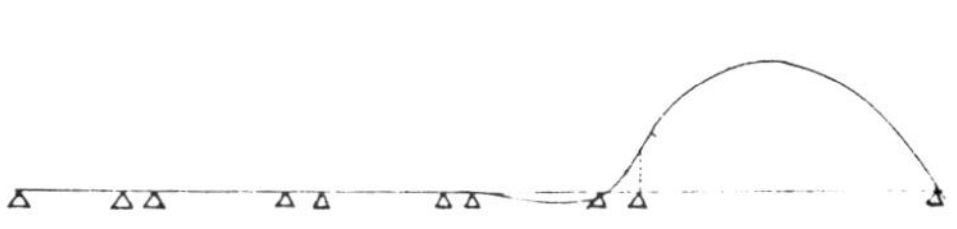

Fig. 3 The mode shape(1st) of the rotor-bearing system (838 r.p.m)

Fig. 4 The mode shape(1st) of the rotor-bearing-foundation system (840 r.p.m.)

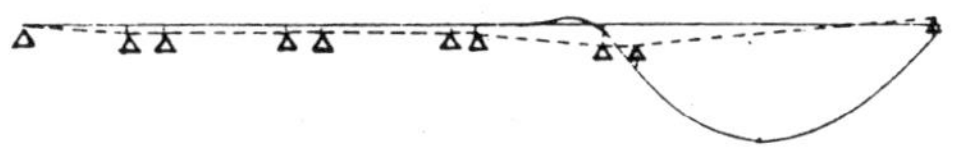

Fig. 5 The mode shape(1st) of the rotor-bearing-foundation system (910 r.p.m)

Fig. 6 The mode shape(2nd) of the rotor-bearing system (2513 r.p.m.)

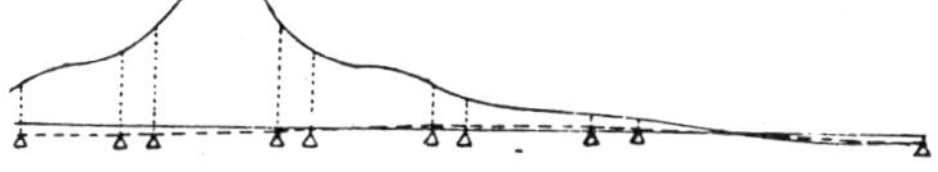

Fig. 7 The mode shape(2nd) of the rotor-bearing-foundation system (2011 r.p.m.)

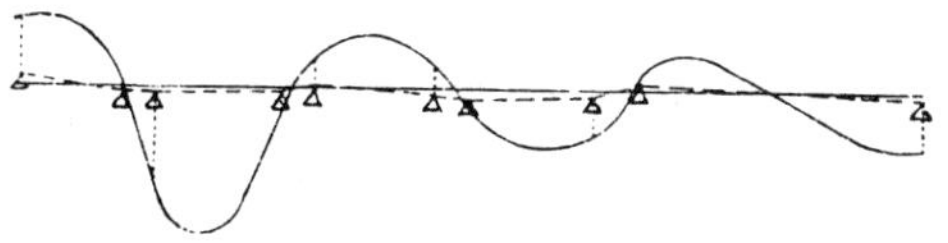

Fig. 8 The mode shape (2nd) of the rotor-bearing-foundation system (2434 r.p.m.)

EXPERIMENT

In ordor to compare the influence of the flexibility of foundation on the critical speed of the rotor-bearing system, the experiments were carried out on two testing rigs, one with an integral foundation and the other with a frame one. The results in Fig. 9 and Fig. 10 show that there is a very narrow critical speed range for the rotor on the integral foundation, while the range is much wider for the one with the more flexible frame foundation. In addition to rig testing, an on-situ experiment has also been performed on a 300 MW turbine-generator set, the results are given in TABLE 1.

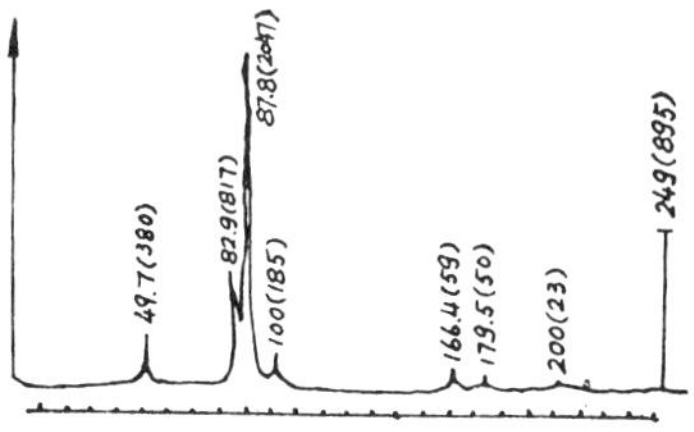

Fig. 9 The measured value of critical speed on the integral foundation

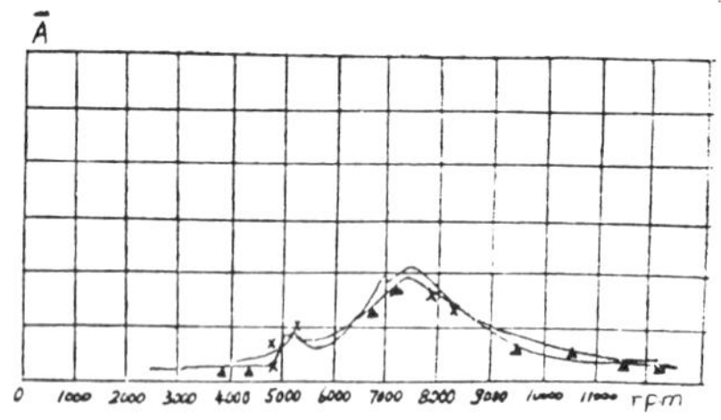

Fig. 10 The measured value of critical speed on the flexible frame foundation

CONCLUSIONS

1. The influence of the foundation of a large turbine-generator set on the dynamic behaviors of the rotor-bearing system can not be ignored.
2. The critical speed range is wider for the rotor on flexible frame foundation than the one on rigid integral foundation.
3. The calculation results presents a good agreement with the experiment.

REFERENCES

Jaker,M. (1980). Vibration analysis of large rotor-bearing-foundation systems using a model condensation for the redution of unknowns. I Mech E Conference Publication c280/80.

Kuang, J.H., and Tsuei, Y.G., (1985). A more general method of substruture mode synthesis for dynamic analysis. AIAA. J., 23, 618-623.

Gong, H.S., and Ding, K.Y., (1985). Mode synthesis method for rotor-bearing-foundation systems. In chinese.

Gong, H.S., and Zheng, Z.C., (1985). Calculation and experiment on the model of the large turbine-generator-foundation system. In chinese.

A. Pons., Study of some effects of hydrodynamic bearings--non-linear behaviour on rotating machine operation.

Zheng Zhaochang, (1985) et all. Gyroscopic mode systhesis in the dynamic analysis of a multi-shaft rotor-bearing system. ASME Paper 85-lgt-73

Lund, J.W., (1974) Model response of a flexible rotor in fluid bearing. ASME Trans. Journal of Eng. for Industry, vol 96, No 2, May 1974

The Optimal Balancing of Flexible Rotors

A. J. URBANEK* AND E. WITTBRODT**

*Department of Mechanical Engineering, Technical University of Koszalin, Koszalin, Poland

**Machine Design Department, Technical University of Gdańsk, Gdańsk, Poland

Abstract. The modal optimal balancing method of flexible rotors is discussed. The following conditions are taken into account in the method:

- effectivity /only two startings are needed, so the balancing time is shorter,and costs - smaller/,
- elimination of prospective reasons of turbo-machine failures /backlashes between rotor and stator are included into analysis/,
- technical usefulness of the method /a portable microcomputer set for rotor balancing in its own bearings of a turbo-machine should be introduced/.

The computer method for calculation of the optimal balancing parameters of correction is described. The rotor system is discretized by the rigid finite element method, and the modal unbalances are identified using experimental-analytical techniques. The optimal balancing isconnected with a minimization of the bearings reaction forces when lateral deflections of flexible rotors are limited.

Keywords. Computer - aided design; Computer applications; Discrete systems; Optimisation; Vibration control.

INTRODUCTION

The main aim of flexible rotors balancing, in according to known methods, is liquidation /or reduction to the necessary minimum/ of the bearings reaction forces. The flexible rotor balanced in a such way can deflect more than backlashes between rotor and stator are. It is known from the flexible rotors balancing and exploitation practice, that it is the main reason of damages. It can often to cause: mechanical damages of rotors, increase of temperatures dued by friction, and also unbalances dued by thermo-deformation of rotors. From this point of view it is necessary to take into calculation the limit of lateral deflection in some cross-sections of rotors. There is also very important the balancing time and costs. It is possible to make the time shorter and costs smaller by :

- reduction of necessary startings of rotors,
- reduction of number of measurements,
- providing of correction realization in the design proccess of turbo-machines,
- increasing of measurement possibilities.

Taking into account described above conditions, the autors propose the new flexible rotors balancing methods. The method has advantages, and does not have disavantages of the known methods. A schema of the proposed modal flexible rotors balancing methodis shown in fig. 1.

There are introduced the three stages:

i/ to discretize a flexible rotor system by the rigid finite element method,

ii/ to identify modal unbalances of a flexible rotor,

iii/ to calculate the optimal correction parameters.

A DISCRETE MODEL OF A FLEXIBLE ROTOR SYSTEM

A discrete model is introduced for numerical calculation of natural frequencies and modes of a flexible rotor. The discrete

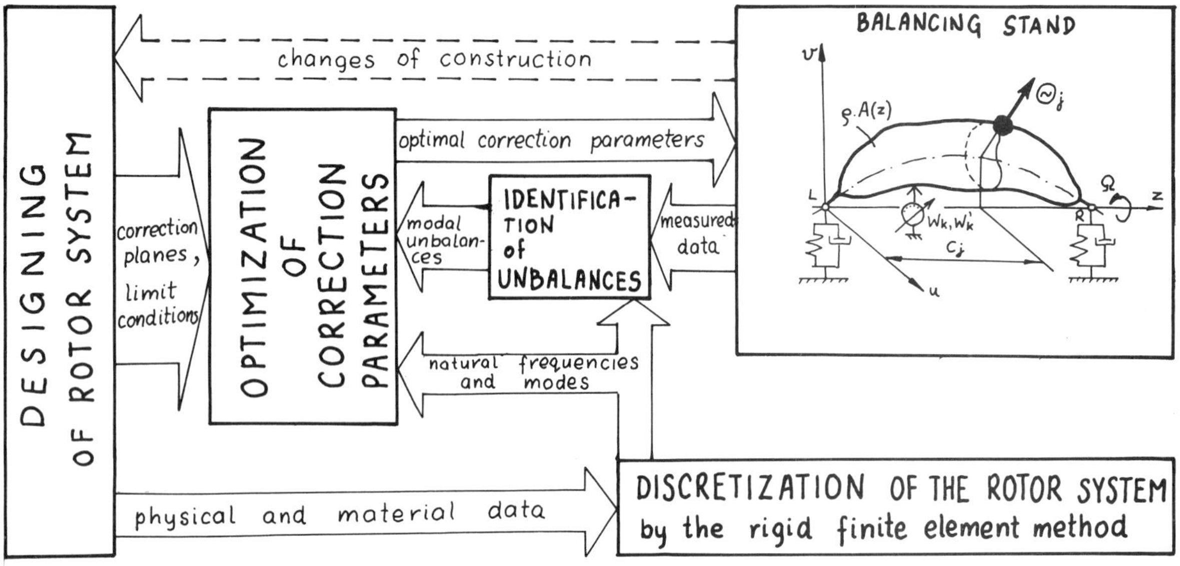

Fig. 1. A schema of the modal flexible rotors balancing method.

model contains /Kruszewski,1975/: the rigid finite elements /RFE/, and the spring-damper elements /SDE/. An example of the discrete model is shown on fig.2. Parameters of the discrete model should be calculated as it is described in detail in /Kruszewski,1975/.

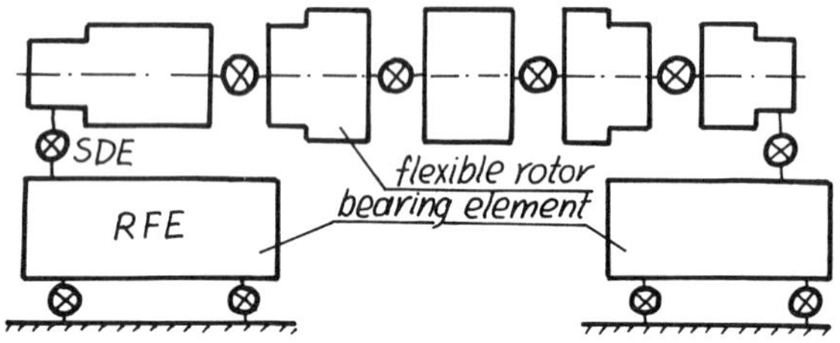

Fig. 2. The discrete model.

MODAL UNBALANCES

A method of identification of modal unbalances bases on the influences coefficients theory, and on the measured modal deflections of a flexible rotor/Parkinson,1980/. The calculated natural frequencies and modes of the discrete model are also used. It is necessary to add an identify mass. The cross-section vibrations of the rotor are measured twice; with, and without the identify mass. A rotational speed should be approximately equale the critical rotational speed no. k of the rotor. The measured deflection vectors are shown on fig. 3.

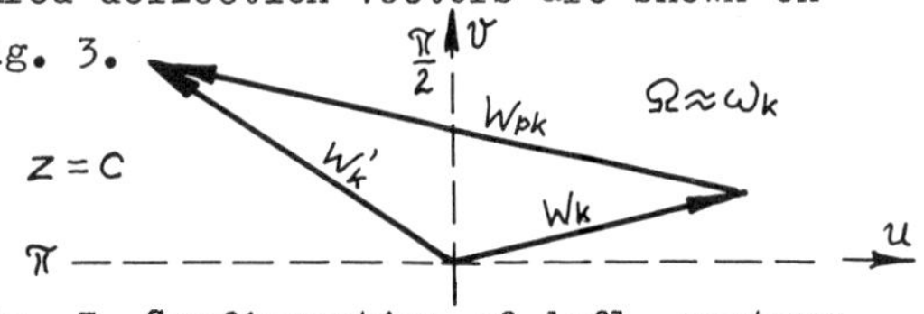

Fig. 3. Configuration of defl. vectors.

The k-mode unbalance vector is calculated from the equation

$$D_k = \frac{W_k}{W_{pk}} \cdot \frac{D_p}{Z_k} \cdot \Phi_k(c) \,, \qquad (1)$$

where:

W_k - k-mode deflection vector /complex number/,

W_{pk} - k-mode deflection vector dued by the identify mass /complex number/,

D_p - unbalance moment dued by the identify mass /complex number/,

C - coordinate of the measuring plane,

and

$$Z_k = \int_0^l \Phi_k^2(z)\,dz \qquad (2)$$

is the k-mode normalize function, and $\Phi_k(z)$ is the k-mode of vibration. The z axis is the one of the bearing supports of the rotor. A distribution of the modal unbalances can be expressed as it is written below

$$D_k(z) = D_k \cdot \Phi_k(z) \qquad (3)$$

THE COMPUTER PROGRAMME FOR OPTIMAL CORRECTION PARAMETERS CALCULATIONS

The computer programme for an optimal balancing of flexible rotors is written in BASIC, and introduced on the ZX Spectrum Plus 64 K microcomputer.The programme consists of the three subroutines:

i/ rotor parameters subroutine,

ii/ objective function subroutine,

iii/ optimization subroutine.

The rotor parameters subroutine is used for:

- input data of the discrete model,
- eigen values and modes calculation,

- numerical integration using the Gauss procedure.

The objective function subroutine is employed to generate the modyfied criterial function /the objective function/, which is adapted adequately to the following limit conditions :

- the dynamic reactions in bearings can not exceed permissible values / $|R_L| \leqslant R_{perm}$, $|R_R| \leqslant R_{perm}$ /,
- rotor deflections in the chossen planes can not exceed the backlash between the rotor and the stator / $w(z_w) \leqslant w(z_w)_{perm}$/.

The k-mode deflection in the chossen plane of the rotor, coordinate of which is z_w /fig. 4/is equale/Bishop,1972/

$$W_k(z_w)=\frac{(\xi_k \cdot \omega_k)^2 \cdot [D_k + \frac{1}{Z_k}\sum_{j=1}^{J}\Theta_j \Phi(c_j)]}{\{[\omega_k^2(1-\xi_k^2)]^2 + 4 \cdot \mu_k^2 \cdot \omega_k^3 \cdot \xi_k^2\}^{\frac{1}{2}}} \cdot \Phi_k(z_w) \quad (4)$$

where:

ω_k - k-mode frequency,

ξ_k - modal damping coefficient,

μ_k - modal mistuning coefficient.

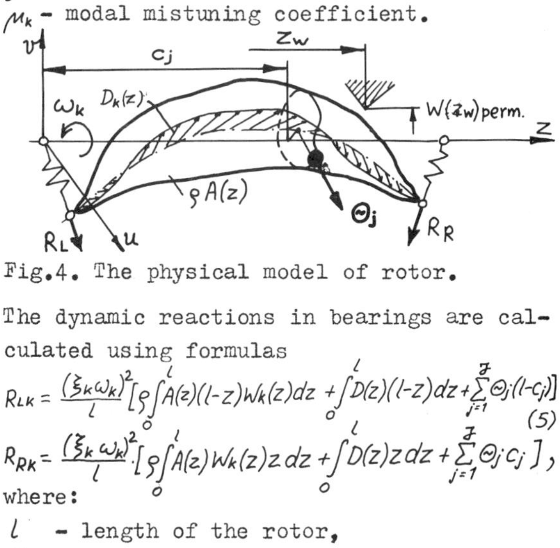

Fig.4. The physical model of rotor.

The dynamic reactions in bearings are calculated using formulas

$$R_{LK} = \frac{(\xi_k\omega_k)^2}{l}[\varrho\int_0^l A(z)(l-z)W_k(z)dz + \int_0^l D(z)(l-z)dz + \sum_{j=1}^{J}\Theta_j(l-c_j)]$$

$$R_{RK} = \frac{(\xi_k\omega_k)^2}{l} \cdot [\varrho\int_0^l A(z)W_k(z)z\,dz + \int_0^l D(z)z\,dz + \sum_{j=1}^{J}\Theta_j c_j], \quad (5)$$

where:

l - length of the rotor,

Θ_j - unbalance of the correction mass,

c_j - coordinate of the correction mass,

J - total number of the correction masses.

The objective function has the form

$$F = (R_L + R_P)^{\frac{1}{2}} + K(\Theta, \delta_0) = (R_L + R_P)^{\frac{1}{2}} \cdot \left\{1 + \frac{[W(z_w) - W(z_w)_{perm}]^2}{\delta_0}\right\}, \quad (6)$$

and it is accomodated to its minimization by the penalty function method. The penalty coefficient δ_0 is choosen experimentaly to convergence the objective function. The penalty function $K(\Theta,\delta_0)$ is a continuous one for $\Theta \in R^N$, and such that

$$K(\Theta,\delta_0) \begin{cases} = 0 & , \Theta \in \mathbf{\Theta} \\ > 0 & , \Theta \notin \mathbf{\Theta} \end{cases}$$

where $\mathbf{\Theta}$ - the set of permissible parameters.

The optimization subroutine is employed to solve an optimization problem by the Davidon - Fletcher - Powell method /Findeisen, 1980/.

The results of calculations will be presented on the Congress.

CONCLUSIONS

1. The proposed method needs only two startings, so the balancing time can be shorter, and costs - smaller.
2. A discretization of the rotor system by the rigid finite element method makes possible to take into account dynamics of all rotating machine, so the method is more general.
3. The objective function and limit conditions may be formulated in according to the type of a flexible rotor and its working conditions.
4. Using for calculation a personal computer gives possibility to construct the portable set for optimal flexible rotors balancing in their own bearings.

REFERENCES

Kruszewski, J., W. Gawroński, E. Wittbrodt, F. Najbar, and S. Grabowski /1975/. The rigid finite element method. Arkady, Warsaw. 291 pp.

Parkinson A.G., M.S. Darlow, and A.J. Smalley /1980/. A theoretical introduction to the development on unified approach to flexible rotor balancing. Journal of Sound and Vibration, 64, Vol. 4, 489 - 506 pp.

Bishop, R.E.D., and A.G. Parkinson /1972/. On the use of balancing machines for flexible rotors. Journal of Engineering for Industry., 5 , 561 - 571 pp.

Urbanek, A.J., and Z. Gosiewski /1982/. The method of identification of unbalance of a flexible rotors. Works of Institute of Machine Design., no. 5, Series B. Koszalin,148 - 158 pp.

Findeisen, W., J. Szymanowski, and A. Wierzbicki /1980/. Theory and optimal calculation methods. Państwowe Wydawnictwo Naukowe, Warszawa. 707 pp.

SUMMARY

The modal optimal balancing method flexible rotors is propossed. Advantages of the method are:

- effectivity /two startings are needed, and balancing time is shorter/,
- reduction of prospective reasons of turbo - machine failures /backlashes between rotor and stator are taken into analysis to calculate admissible deflections, and
- technical usefulness /the optimal correction planes and a way of fixing correction masses are selected at a turbo - - machine design stage/.

A rotating system is descretized by the rigid finite element method, and the natural frequencies and modes are calculated numerically. It gives possibility : to take into account dynamics of a whole rotor system, to reduce the number of measurements, to use the method for balancing rotors in their own bearings, and to make standstills of a turbo - machine shorter and costs smaller.

The modal unbalances are identified using experimental - analytical techniques. Deflections of a rotor are measured in one plane and using only one trial mass. The deflections are measured twice: without and with the trial mass, when the rotor rotates with angular speeds equale its critical speeds.

The optimal correction parameters /Θ_j, c_j / are calculated in according to obtain the minimum dynamic reaction forces in bearings, and when the lateral deflections in choosen planes are limited /fig. 5/. The deflections are limited by values of backlashes. The Davidon - Fletcher - Powell algorithm is employed to calculate the optimal parameters.

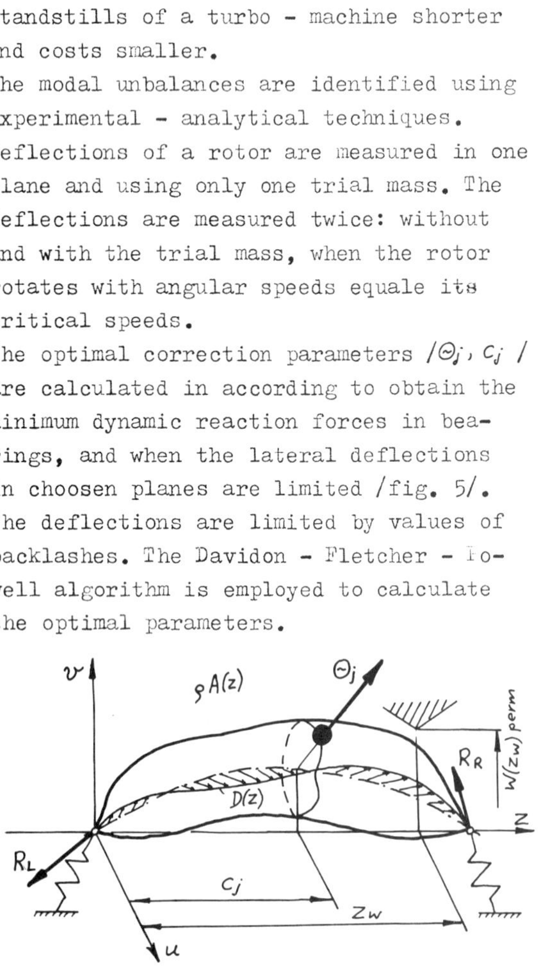

Fig. 5. The model of the rotor.
Рис. 5. Модель уравновешиваемого ротора.

The problem is solved using a simple microcomputer /ZX Spectrum Plus 64K/. It gives the possibility to construct a portable balancing set for the optimal balancing of any kind flexible rotors.

СОДЕРЖАНИЕ

В реферате предлогается модальный метод оптимального уравновешивания гибких роторов, имеющий следующие преимущества:

- эффективность /возможность снижения числа измерений, сокращения времени уравновешивания и материальных затрат/,
- снижение аварийного риска /возможность учёта конструкционных зазоров между ротором и статором/,
- техническая пригодность метода/определение оптимальных плоскостей коррекции и способа закрепления уравновешивающих грузов ещё в процессе проектирования роторной системмы/.

Благодаря компьютерному способу вычисления форм собственных колебаний и собственных частот /метод жёстких конечных элементов/, имеется возможность учёта динамических свойств целой роторной системы, снижения числа пробных пусков ротора, применения метода для уравновешивания роторов в собственных подшипниках. Всё это сокращает время уравновешивания и снижает материальные затраты.

Определение форм неуравновешенности совершается экспериментально - расчётным путём, при использовани одной измерительной плоскости и одного пробного груза. Колебания ротора измеряются два раза: без пробного груза и после его закрепления на роторе, при скоростях вращения близких критическим. Оптимальные коррекционные параметры /   / вычисляются минимизацией динамических сил в подшипниках при ограничённых упругих деформациях в некоторых поперечных сечениях ротора /рис.5/. Эти деформации связаны с конструкционными зазорами между ротором и статором. Оптимизация проводится по методу Давидона - Флетшера - Повелла. Применение миникомпьютера дает возможность создания переносной аппаратуры для уравновешивания некоторых типов гибких роторов.

Increase the Rotor Stability by Anisotropic Suspension

J.-H. WANG AND M.-T. TSAI

Department of Power Mechanical Engineering, National Tsing Hua University, Taiwan, PRC

Abstract. The instability caused by clearance excitation leads to a limit performance of turbomachines. This instability may be improved by using flexible pedestals with anisotropic stiffness. The effect of anisotropic supports is investigated by simplified model under the influences of different parameters, such as shaft stiffness, the ratio of running speed to critical speed, and pedestal stiffness. The stability behavior of two kinds of bearing and antisymmetric support condition are also studied. The dynamic characteristics of complex rotor-bearing system are investigated by finite element method. An optimum anisotropic support condition of a typical 900 MW turbogenerator system is found, and the steady state vibrations with this optimum support condition are compared with rigid support condition.

Keywords. Stability; dynamic response; turbomachine; steam whirl.

INTRODUCTION

In modern power generation system, it is a tendency to increase the ratio of power generated per unit mass of rotating elements. With higher running speed and output, turbomachines may experience instability, which leads to an output limit of power generation unit. The causes of instability normally are due to steam whirl, oil whip, and hysteretic damping for symmetric rotors. Steam whirl (also called clearance excitation) is a self-excited nonsynchronous vibration which is induced by leakage steam flow in clearance areas between stationary and moving parts (Alford, 1965; Pollman, 1978). Oil whip is also a self-excited nonsynchronous vibration which is due to characteristic of fluid-film bearing (Glienicke, 1980; Yukio Hori, 1985). The instability due to hysteretic damping is less important when it compares with oil whip and steam whirl.

A lot of studies (Bansal, 1975; Lund 1974; Rao, 1985) about stability of rotor-bearing system were concerned with oil whip. Kurohashi (1984) investigated the effect of anisotropic property of bearing on the stability of rotor subjected to clearance excitation. Gunter (1959) showed that instability caused by internal friction may be improved with anisotropic supports. Ebner (1985) used a model of symmetric Laval-rotor sypported by flexible pedestals to study the effect of anisotropic pedestal. Basically the self-excited vibration is induced by certain mechanism of energy transfer. In the case of instability caused by clearance excitation, many investigator (e.g. Ebner, 1985; Walter, 1984) believed that the running stability could be increased by anisotropic pedestal, because the energy transferred to the rotor will be decreased due to the fact that the whirl orbit becomes more elliptical. But it is necessary to know whether the results obtained by Ebner (1985) using a simplified model are still valid for complex rotor-bearing system and whether the optimum anisotropic pedestals are still suitable for steady state operation.

In this paper a simplified lumped mass-model is used to study the threshold performance of rotor-bearing system supported by flexible pedestals, and then the stability of a typical 900 MW turbogenerator supported with anisotropic pedestals is studied by finite element method to verify the validity of the results obtainted from simplified model. Finally the steady state responses of rotors supported with optimum anisotropic pedestals are investigated to compare with results from rigid pedestals.

MODEL AND EQUATION OF MOTION

Both simplified model and finite element model are used in this section to derive the equations of motion.

Simplified model

Fig.1 shows the simplified model for the purpose of investigating stability behavior of a rotor-bearing system supported by the flexible pedestals. This model includes a Laval-rotor mounted between two bearings which are supported by damped flexible pedestals represented by springs and dampers. The mass of shaft and rotor is assumed to concentrate at central points of journals and rotor individually. The inertial properties of rotor contain mass and rotary inertial while the journals are treated as point mass, each of them is connected by a massless shaft. The forces acting on the system are unbalance force, clearance exciting force, and interactive forces between bearing and journal.

The steam whirl effect is characterized by change in boundary losses at circumference of the blading resulting from rotor displacement which is caused by any perturbation. The variation of blade force at circumference causes a resultant force which acts on rotor in a direction perpendicular to the rotor displacement :

$$\begin{Bmatrix} F_x \\ F_y \end{Bmatrix} = \begin{bmatrix} 0 & -q \\ q & 0 \end{bmatrix} \begin{Bmatrix} u_d \\ v_d \end{Bmatrix} , \qquad (1)$$

where u_d and v_d are displacements of rotor in x and y directions, q is clearance excitation factor or aerodynamic cross coupling stiffness which can be evaluated for different machines theoretically and experimentally. With the increasing of output torque, the off diagonal terms of Eq.(1) increase and lead to instability.

The nonlinear characteristics of the journal bearing can be linearlized at static equilibrium position under the assumption of small vibrations. The dynamic characteristics of journal bearing are represented by eight stiffness and damping coefficients which are functions of Sommerfeld number (Glienicke, 1980). The model of bearing supported on flexible pedestal is shown in Fig.2. The forces act on journal and bearing can be expressed as :

$$\begin{Bmatrix} F_{xJ} \\ F_{yJ} \\ F_{xB} \\ F_{yB} \end{Bmatrix} = -\begin{bmatrix} C_{xx} & C_{xy} & -C_{xx} & -C_{xy} \\ C_{yx} & C_{yy} & -C_{yx} & -C_{yy} \\ -C_{xx} & -C_{xy} & C_{xx}+C_x & C_{xy} \\ -C_{yx} & -C_{yy} & C_{yx} & C_{yy}+C_y \end{bmatrix} \begin{Bmatrix} \dot{u}_J \\ \dot{v}_J \\ \dot{u}_B \\ \dot{v}_B \end{Bmatrix}$$

$$-\begin{bmatrix} K_{xx} & K_{xy} & -K_{xx} & -K_{xy} \\ K_{yx} & K_{yy} & -K_{yx} & -K_{yy} \\ -K_{xx} & -K_{xy} & K_{xx}+K_x & K_{xy} \\ -K_{yx} & -K_{yy} & K_{yx} & K_{yy}+K_y \end{bmatrix} \begin{Bmatrix} U_J \\ V_J \\ U_B \\ V_B \end{Bmatrix} . \qquad (2)$$

The off-diagonal terms in stiffness matrix play roles of destablizing forces, and lead to instability of oil whip as running speed increased. Due to the limitation of space the complete equations of motion are not derived here in detail, only the matrix form is expressed in following :

$$[M]\{\ddot{q}\} + ([C] - \Omega[G])\{\dot{q}\} + [K]\{q\} = \{F\} \qquad (3)$$

where [M][C][G][K] are mass, damping, gyroscpic, and stiffness matrices of order 12x12, as shown in appendix.

Finite Element Model

The simplified model can't exactly describe the behaviors of a complex rotor-bearing system like turbogenerator system in power plant. For complex system the dynamic behavior normally can be studied only by approximate methods, for example by finite element method (Nelson, 1976). Figure 3 is a typical turbogenerator system which will be analyzed by finite element method. The finite element model used here follows Rayleigh beam theory and includes the effects of flexibility of pedestal and clearance exciting force. The complex rotor-bearing system consists of rigid disks, rotors and bearings, the corresponding equations of motion are

1. rigid disk

$$([M_T^d]+[M_R^d])\{\ddot{q}^d\} - \Omega[G^d]\{\dot{q}^d\}+[K^d]\{q^d\}=\{F^d\} \qquad (4)$$

2. rotor element

$$([M_T^e]+[M_R^e])\{\ddot{q}^e\} - \Omega[G^e]\{\dot{q}^e\}+[K^e]\{q^e\}=\{F^e\} \qquad (5)$$

3. bearing

$$[M_T^b]\ \{\ddot{q}^b\}+[C^b](\{\dot{q}^b\}-\{\dot{q}^e\}) + [K^b](\{q^b\} - \{q^e\}) = \{F^b\}. \qquad (6)$$

The assembled system equation of motion is

$$[M]\{\ddot{q}\}+([C]-\Omega[G])\{\dot{q}\}+[K]\{q\}=\{F\} \qquad (7)$$

which has same matrix form as Eq.(3) except that the order of matrices is NxN, where N=4(n+1)+2b with n=number of rotor elements, b=number of bearings.

To get the eigenvalues of Eqs.(3) and (7), the Eqs. are tranformed to generlized eigenvalue problem :

$$\lambda[M^*]\{\Phi\}+[K^*]\{\Phi\}=\{0\}, \qquad (8)$$

where

$$[M^*]=\begin{bmatrix} [0] & [M] \\ [M] & [C]-\Omega[G] \end{bmatrix} \quad [K^*]=\begin{bmatrix} -[M] & [0] \\ [0] & [K] \end{bmatrix}.$$

λ is eigenvalue, $\{\Phi\}$ is eigenvector of state varible.
Premultiply Eq.(8) by $[K^*]^{-1}$, then one gets a standard eigenvalue problem :

$$[D]\{\Phi\}=\frac{1}{\lambda}\ \{\Phi\} \qquad (9)$$

where dynamic matrix

$$[D]=\begin{bmatrix} [0] & [I] \\ -[K]^{-1}[M] & -[K]^{-1}([C]-\Omega[G]) \end{bmatrix}.$$

Because of nonsymmetry of [K], [C] and [G], eigenvalues of Eq.(8) are N sets of complex conjugate :

$$\lambda_i=\sigma_i \pm j\omega_i \qquad i = 1,2,\cdots;N \qquad (10)$$

It is well known that the system becomes unstable if there is any positive σ. The instability threshod is defined as a state which has at least one of the eigenvalues with zero real part while the others with negative real part.

The steady state response due to mass unbalance can be written as particular solutions of Eqs.(3) and (7). The unbalance force can be written as :

$$\{F\} = \{F_c\}\cos\Omega t + \{F_s\}\sin\Omega t. \qquad (11)$$

Then the steady state response has the same form as Eq.(11) :

$$\{q\} = \{q_c\}\cos\Omega t + \{q_s\}\sin\Omega t. \qquad (12)$$

One can then solve Eq.(3)(7) by Gauss elimination:

$$\begin{Bmatrix} \{X_c\} \\ \{X_s\} \end{Bmatrix} = \begin{bmatrix} [K]-\Omega^2[M] & \Omega([C]-\Omega[G]) \\ -\Omega([C]-\Omega[G]) & [K]-\Omega^2[K] \end{bmatrix}^{-1} \begin{Bmatrix} \{F_c\} \\ \{F_s\} \end{Bmatrix}. \qquad (13)$$

RESULTS OF STABILITY ANALYSIS

Results of Simplified Model

The simplified model of Fig.1 is usedto investigate the stability by consideration of many system parameters variation, especially the variation of anisotropic pedestal. All the dimensionless parameters are listed in Table 1. The maximum clearance excitation, S_{max}, which the rotor can sustain is investigated under the influences of anisotropic stiffness, speed ratio, and different bearing types. Some of typical results are shown in Figs.4 to 7. The bearings used here are five-pad tilting-pad bearing and two-lobed bearing with the ratio of width to diameter B/D=0.5, and Sommerfeld number S_0=0.1. The external damping is completely neglected, so U=0.0, T=1.0. Another assumption is symmetric rotor with $\alpha_a=\alpha_b=0.5$, $\mu^L=\mu^R=\mu$. Supercript L and R indicate left and right.

Figure 4 shows that the threshold performance is improved by anisotropic pedestal stiffness with five-pad tilting-pad bearings. The support condition is symmetric, i.e. $Q^L=Q^R=Q=5.0$ and $P^L=P^R=P$. The system with stiff shaft (μ=0.6) can sustain higher clearance excitation in rigid support region (P>1.0) owing to the fact that there is smaller amplitude and energy of vibration. The threshold performance decreases with increasing speed ratio as a result of descreasing of bearing damping in high speed. An optimum support condition is found in a region about anisotropic pedestal stiffness P=0.05 for all case in study. This interesting result from simplified model indicates that the instability due to clearance excitation of large turbogenerator perhaps can also be improved by anisotropic pedestal. This question will be discussed later. The existence of an optimum support condition can be explained as following. The vibration of rotor can be considered as a superposition of the vibration of bearing and the vibration of rotor relative to the bearing. For more rigid condition (P=100-1.0), the vibrations of the bearings are small and have little influence on vibration of rotor. With decreasing of horizontal stiffness, the vibrations of bearings grow, especially in the soft direction, and make the rotor whirl orbits more elliptical. Due to the elliptical orbit of rotor, the work done by clearance exciting force is decreased, so that the threshold performance is improved. But if the stiffness in horizontal direction decreases further, the system becomes too soft and less stable.

Figure 5 shows the same stability map with two-lobed bearing. In this case, the threshold performance decreases as the increasing of anisotropic degree of pedestal stiffness for stiff shaft (μ=0.6). The reason is due to the fact that the vibration of rotor relative to bearing is small, and an increase of vibration of bearing leads to increasing of vibration of rotor. With low speed ratio (W=0.8), the two-lobed bearing has better threshold performance than five-pad tilting-pad bearing at least in the area of rigid support condition. But the five-pad tilting-pad bearing which is free from instability of oil whip has better stability characteristics with high speed ratio (W=1.6). This is an important result that one should first know the main source of instability (bearing force or clearance exciting force), then one can correctly select the best bearing type.

Figure 6 shows the case of more rigid vertical stiffness of pedestal (i.e. $Q=Q^L=Q^R=50.0$), a little improvement of stability can be observed. The optimum support condition shifts to P=0.005. The same trend exists for two-lobed bearing. Figure 7 shows an antisymmetric support condition in which the horizontal stiffness of left pedestal is equal to the vertical stiffness of right pedestal, and vi'ce ver'sa ($P=P^L=P^{R'}$, $Q=Q^L=Q^{R'}$). In more rigid support region (P=100-1.0), the difference between antisymmetric and symmetric support is small. But with increasing of anisotropic degree of pedestal stiffness, the threshold performance decreases monotonously. The same situation is happened to the case of two-lobed bearing.

Results of Finite Element Model

The results of simplified model show that usually there is an optimum anisotropy of pedestal stiffness, but it is still necessary to see whether such characteristic also exists for large turbogenerator system. A typical 900 MW turbine-generator system, as shown in Fig.3, is used here to check this question by finite element method. The system with length 45 m and weight 485 tons is mounted on eight bearings with flexible pedestals. The clearance exciting force is assumed here to act only in high pressure rotor. Fig.8 shows a stability map with same pedestal stiffness $K_y=9.95\times10^9$ kg/m for all bearings. The rotor system mounted on two-lobed bearing behaves like a system with stiff shaft in simplified model, the maximum instability threshold happens in the case of rigid support. If the same system is mounted on five-pad tilting-pad bearing, there is a maximum instability threshold with the anisotropic stiffness ratio P=0.1. From this result one point should be emphasized here, namely, the maximum clearance excitation S_{max} of rotor system with rigid supported two-lobed bearing is anyway larger than that of five-pad bearing. It is surprising to find that the five-pad tilting-pad bearing is free from oil whip, but this type bearing is not so good to sustain clearance exciting force as usually thought.

EFFECT OF ANISOTROPIC PEDESTAL STIFFNESS

The result in Fig.8 shows that there is an optimum point in p≈0.1 for five-pad tilting-pad bearing. We are interested to know how the steady state vibration of rotor system supported with optimum pedestal is.

The complex rotor system mounted on five-pad tilting-pad bearing with two support conditions ($K_y=9.95\times10^{10}$kg/m, P=1.0 and $K_y=9.95\times10^9$kg/m, P= 0.1) are compared here. The unbalance mass found in generator, low pressure turbine, and high pressure turbine is assumed to 1.5, 1.0, 0.5 kg-m respectively. Figure 9 shows the steady state response at the first, the sixth critical speed and rated speed with rigid support condition (K_y=9.95 $\times10^{10}$ kg/m, P=1.0), and Fig.10 shows the same response of anisotropic support condition (K_y=9.95x 10^9 kg/m, P=0.1). The vibration of rotor with anisotropic support condition, which yields the greatest stability, is larger than the vibration with rigid support.

CONCLUSION

The instability due to clearance excitation leads to a limit performance of turbogenerators. This instability may be improved by using flexible pedestals with anisotropic stiffness. In present work, the effect of anisotropic stiffness on the stability due to clearance excitation is investigated by simplified model and finite element model. The steady state vibrations with optimum support condition are also compared with those with rigid support condition. Some conclusions can be emphasize here :

1. The stability of system usually can be improved by anisotropic pedestal stiffness for less stiff shaft.
2. The rotor system mounted on two-lobed bearing is more stable than the system mounted on five-pad tilting-pad bearing in lower operation speed range, and less stable in higher speed range.
3. It is well known that the five-pad tilting-pad bearing is free from oil whip, but when the pedestal is rigid as today usually used, the five-pad tilting-pad bearing is not so good as two-lobed bearing to sustain clearance exciting force.
4. Although the instability of rotor system due to clearance exciting force may be improved by using anisotropic pedestal, the steady state vibrations due to mass unbalance may become worse.
5. So far the results show that the optimum range of anisotropy of pedestal stiffness, P, is

strongly dependent on the parameters of rotor system, there is no simple rule to find this optimum range. For some systems, there may be even no optimum range.

REFERENCE

Alford, J.S. (1965). Protecting turbomachinery from self-excited rotor whirl. ASME Journal of Engineering for Power, 333-344.

Bansal, P.N., R.G. Kirk (1975). Stability and damped critical speeds of rotor-bearing system. ASME Journal of Engineering for Industry, 1325-1332.

Ebner, F.L. (1985). Stability behavior of clearance-excited turborotors with external anosotropy. ASME Paper No. 85-DET-149, ASME Design Engineering Division Conference, Cincinnati, Ohio, Spe. 10-13.

Glienicke, J., D.C. Han and M. Leonhard (1980) Practical determination and use of bearing dynamic coefficients. Tribology International, 297-309.

Gunter, E.J., P.R. Trumpler (1959). The influence of internal friction on the stability of high speed rotors with anisotropic supports. ASME Journal of Engineering for Industry, 1105-1113.

Kurohashi, M. and co-workers (1984). Stability Analysis of rotor-bearing system subjected to cross coupling force. Paper no. C 266/84 Third International Conference on Vibrations in Rotating Machinery 11-13 Stp. 1984 University of York.

Lund, J.W. (1974). Stability and critical speeds of a flexible rotor in fluid-film bearing. ASME Journal of Engineering for Industry, 509-517.

Nelson, H.D. and J.M. MacVaugh, (1976). The dynamics of rotor-bearing systems using finite elements. ASME Journal of Engineering for Industry, 98, 593-600.

Pollman, E., H. Schwerdtfeger and H. Termuehlen (1978). Flow excited vibrations in high pressure turbines (steam whirl). ASME Journal of Engineering for Power, 100, 219-228.

Roa, J.S. (1985). Instability of rotors mounted in fluid film bearings with a negative cross-coupled stiffness coefficient. Mechanism and Machine Theory, 20, 181-187.

Walter Tranpel (1982). Thermische Turbomaschinen, Springer-Verlag, Berlin.

Yukio Hori (1959). A theory of oil whip. ASME Journal of Applied Mechanics, 189-198.

APPENDIX

The constituent of Eq. (3) are:

displacement vector

$$\{q\}^T=[u_d \;\; u_B^L \;\; u_B^R \;\; u_J^L \;\; u_J^R \;\; \beta l \;\; v_d \;\; v_B^L \;\; v_B^R \;\; v_J^L \;\; v_J^R \;\; dl],$$

mass matrix

$$[M]=\mathrm{diag}[m_d \;\; m_B^L \;\; m_B^R \;\; m_J^L \;\; m_J^R \;\; \frac{I_T}{l^2} \;\; m_d \;\; m_B^L \;\; m_B^R \;\; m_J^L \;\; m_J^R \;\; \frac{I_T}{l^2}],$$

damping and gyrosopic matrix

$$[C]-\Omega[G]=\begin{bmatrix}
0 & 0 & 0 & 0 & 0 & 0 & 0 & 0 & 0 & 0 & 0 & 0 \\
0 & C_x^L+C_{xx}^L & 0 & -C_{xx}^L & 0 & 0 & 0 & C_{xy}^L & 0 & -C_{xy}^L & 0 & 0 \\
0 & 0 & C_x^R+C_{xx}^R & 0 & -C_{xx}^R & 0 & 0 & 0 & C_{xy}^R & 0 & -C_{xy}^R & 0 \\
0 & -C_{xx}^L & 0 & C_{xx}^L & 0 & 0 & 0 & -C_{xy}^L & 0 & C_{xy}^L & 0 & 0 \\
0 & 0 & -C_{xx}^R & 0 & C_{xx}^R & 0 & 0 & 0 & -C_{xy}^R & 0 & C_{xy}^R & 0 \\
0 & 0 & 0 & 0 & 0 & 0 & 0 & 0 & 0 & 0 & 0 & \Omega\frac{I_p}{l^2} \\
0 & 0 & 0 & 0 & 0 & 0 & 0 & 0 & 0 & 0 & 0 & 0 \\
0 & C_{yx}^L & 0 & -C_{yx}^L & 0 & 0 & 0 & C_y^L+C_{yy}^L & 0 & -C_{yy}^L & 0 & 0 \\
0 & 0 & C_{yx}^R & 0 & -C_{yx}^R & 0 & 0 & 0 & C_y^R+C_{yy}^R & 0 & -C_{yy}^R & 0 \\
0 & -C_{yx}^L & 0 & C_{yx}^L & 0 & 0 & 0 & -C_{yy}^L & 0 & C_{yy}^L & 0 & 0 \\
0 & 0 & -C_{yx}^R & 0 & C_{yx}^R & 0 & 0 & 0 & -C_{yy}^R & 0 & C_{yy}^R & 0 \\
0 & 0 & 0 & 0 & 0 & -\Omega\frac{I_p}{l^2} & 0 & 0 & 0 & 0 & 0 & 0
\end{bmatrix}$$

stiffness mamtrix

$$
[K]=\begin{pmatrix}
3EIA & 0 & 0 & -\frac{3EI}{a^3} & -\frac{3EI}{b^3} & -3EIB & q & 0 & 0 & 0 & 0 & 0 \\
0 & K_x^L+K_{xx}^L & 0 & -K_{xx}^L & 0 & 0 & 0 & K_{xy}^L & 0 & -K_{xy}^L & 0 & 0 \\
0 & 0 & K_x^R+K_{xx}^R & 0 & -K_{xx}^R & 0 & 0 & 0 & K_{xy}^R & 0 & -K_{xy}^R & 0 \\
-\frac{3EI}{a^3} & -K_{xx}^L & 0 & K_{xx}^L+\frac{3EI}{a^3} & 0 & \frac{3EI}{a^2 l} & 0 & -K_{xy}^L & 0 & K_{xy}^L & 0 & 0 \\
-\frac{3EI}{b^3} & 0 & -K_{xx}^R & 0 & K_{xx}^R+\frac{3EI}{b^3} & -\frac{3EI}{b^2 l} & 0 & 0 & -K_{xy}^R & 0 & K_{xy}^R & 0 \\
-3EIB & 0 & 0 & \frac{3EI}{a^2 l} & -\frac{3EI}{b^2 l} & \frac{3EI}{abl} & 0 & 0 & 0 & 0 & 0 & 0 \\
-q & 0 & 0 & 0 & 0 & 0 & 3EIA & 0 & 0 & -\frac{3EI}{a^3} & -\frac{3EI}{b^3} & 3EIB \\
0 & K_{yx}^L & 0 & -K_{yx}^L & 0 & 0 & 0 & K_y^L+K_{yy}^L & 0 & -K_{yy}^L & 0 & 0 \\
0 & 0 & K_{yx}^R & 0 & -K_{yx}^R & 0 & 0 & 0 & K_{yy}^R+K_y^R & 0 & -K_{yy}^R & 0 \\
0 & -K_{yx}^L & 0 & K_{yx}^L & 0 & 0 & -\frac{3EI}{a^3} & -K_{yy}^L & 0 & K_{yy}^L+\frac{3EI}{a^3} & 0 & -\frac{3EI}{a^2 l} \\
0 & 0 & -K_{yx}^R & 0 & K_{yx}^R & 0 & -\frac{3EI}{b^3} & 0 & -K_{yy}^R & 0 & K_{yy}^R+\frac{3EI}{b^3} & \frac{3EI}{b^2 l} \\
0 & 0 & 0 & 0 & 0 & 0 & 3EIB & 0 & 0 & -\frac{3EI}{a^2 l} & \frac{3EI}{b^2 l} & \frac{3EI}{abl}
\end{pmatrix}
$$

$$A=\left(\frac{1}{a^3}+\frac{1}{b^3}\right) \quad B=\left(\frac{1}{a^2 l}-\frac{1}{b^2 l}\right) \quad K_d=3EI\left(\frac{1}{a^3}+\frac{1}{b^3}\right)$$

force vector

$$\{F\}^T=m_d e\,\Omega^2 \lfloor \cos(\Omega t+\varphi)\ 0\ 0\ 0\ 0\ 0\ \sin(\Omega t+\varphi)\ 0\ 0\ 0\ 0\ 0 \rfloor$$

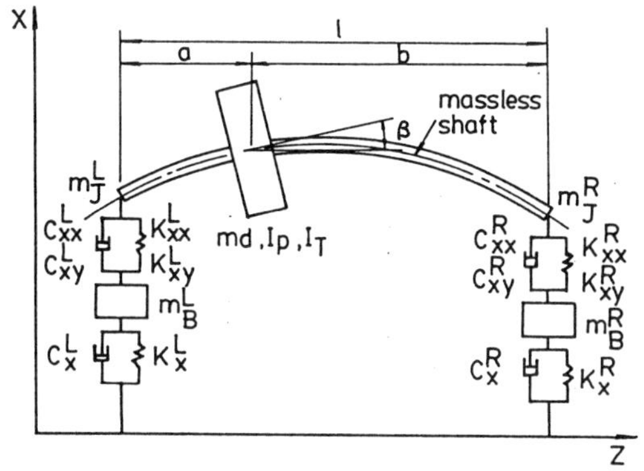

Fig. 1 Rotor model. Only projection in x-z plane is shown.

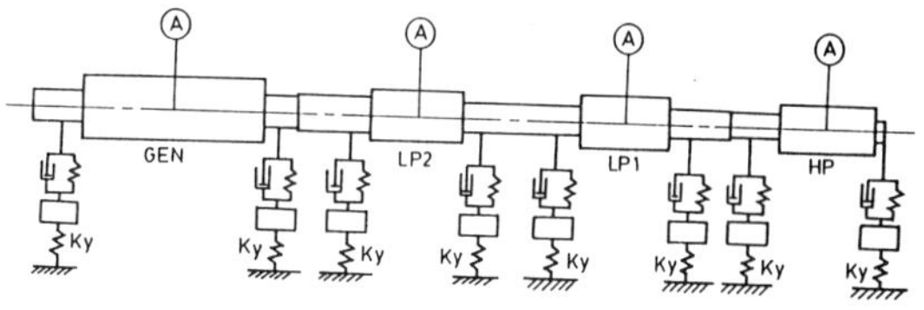

Fig. 3 A typical large turbogenerator system. The positions labeled A represent positions of unbalance.

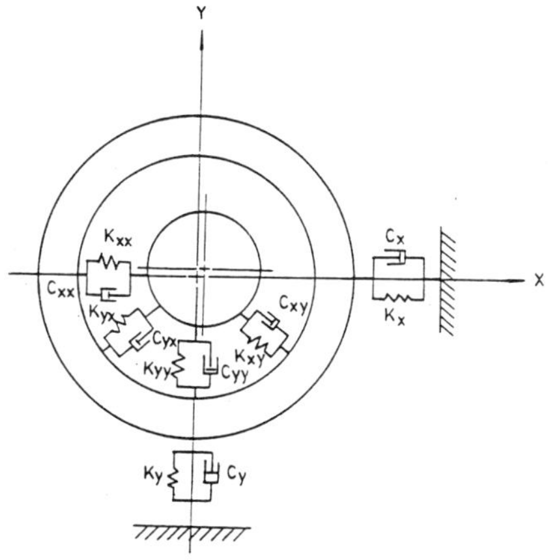

Fig. 2 Model of bearing supported on a flexible pedestal.

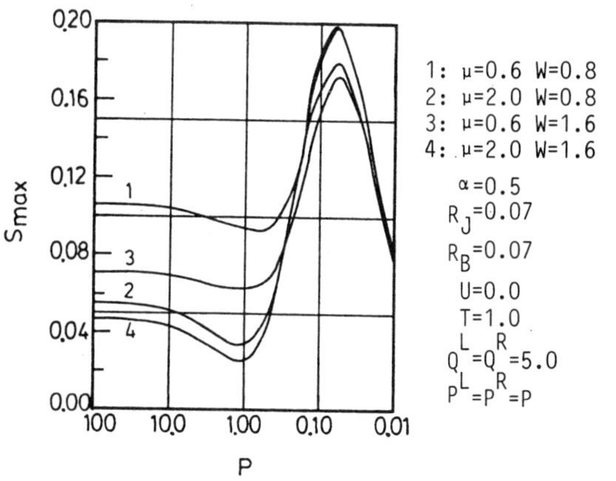

Fig. 4 Threshold performance as a function of anisotropic degree. (Five-pad tilting-pad bearing B/D=0.5;So=0.1)

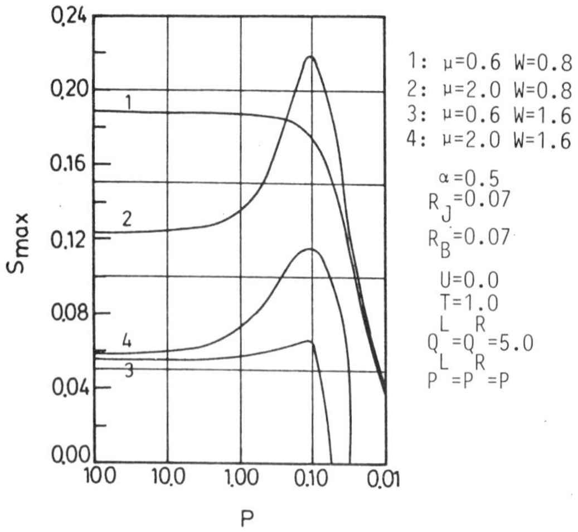

Fig. 5 Threshold performance as a function of anisotropic degree. (Two-lobed bearing B/D=0.5;So=0.1)

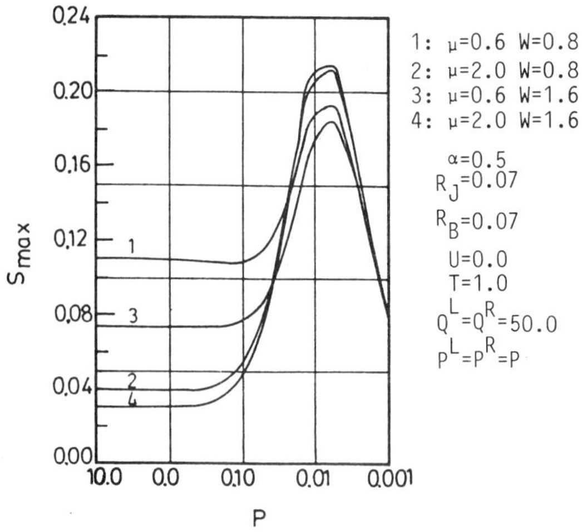

Fig. 6 Threshold performance as a function of anisotropic degree. (Five-pad tilting-pad bearing B/D=0.5; So=0.1)

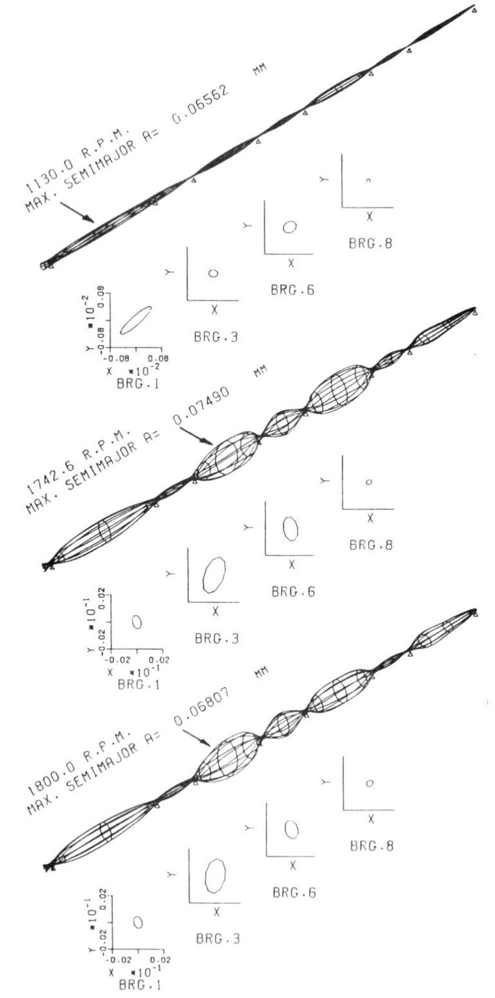

Fig. 9 The steady state response due to mass unbalance with rigid pedestal. (K_y=9.95×10^{10}Kg/m;P=1.0)

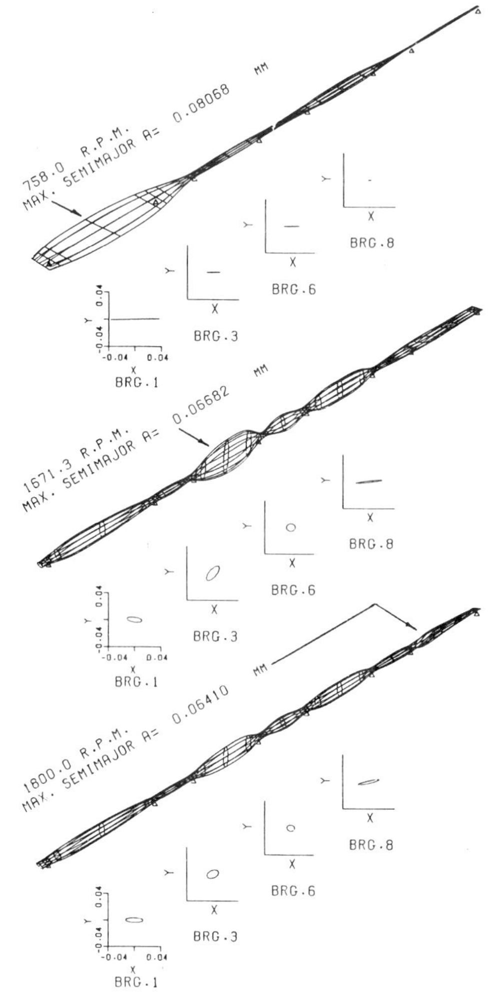

Fig. 10 The steady response due to wass unbalance with flexible pedestal. (K_y=9.95×10^{9}Kg/m; P=0.1)

TABLE 1 Dimensionless Parameters

dimensionless parameter	symbol	defnition
relative shaft stiffness	μ	$g/\omega_k^2\ \Delta R_{min}$
mass ratio bearing/rotor	R_B	m_B/m_d
mass ratio journal/rotor	R_J	m_J/m_d
dimensionless moment of inertial	R_a	$I_T/m_d\ l^2$
dimensionless polor moment of inertial	R_p	$I_p/m_d\ l^2$
length ratio	α_a	a/l
length ratio	α_b	b/l
ratio of running to crtical speed	W	Ω/ω_k
stiffness ratio pedestal/shaft	Q	K_y/K_d
stiffness ratio pedestal/shaft	Q'	K_x/K_d
anisotropic degree of pedestal stiffness	P	K_x/K_y
anisotropic degree of pedestal stiffness	P'	K_y/K_x
damping ratio pedestal/bearing	U	C_y/C_{yy}
anisotropic degree of pedestal damping	T	C_x/C_y
leakage excitation factor	S	q/K_d

$\omega_k = \sqrt{K_d/m_d}$; K_d:shaft stiffness

ΔR_{min}:smallest radial bering clearance

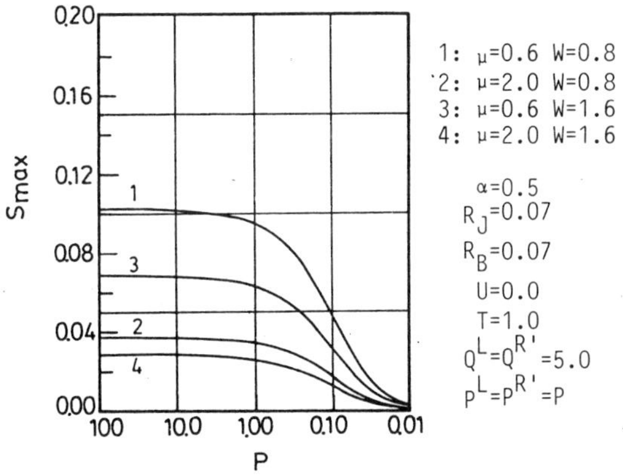

Fig. 7 Threshold performance as a function of anisotropic degree with antisymmetric support. (Five-pad tilting-pad bearing B/D=0.5; So=0.1)

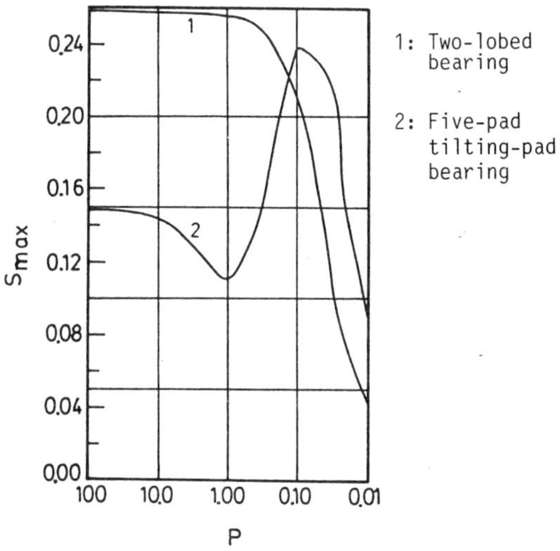

Fig. 8 Threshold performance as a function of anisotropic degree. (K_y=9.95×10⁹Kg/m; $P=K_x/K_y$.)

Abstrakte Eine Ursache der Instabilität in Turbomaschinen ist die Strömungsvorgänge in den Labyrinthdichtungen in Beschaufelungsteil der Rotoren und in den Wellendichtungen. Daher gibt is eine Grenzleistung einer Turbomaschine. Die Stabilität kann durch die Einsetzung eines anisotropen Lagergehäuses verbessert werden. In dieser Arbeit werden die Einflüsse der Systemparameter eines Lavalläufers auf die Stabilität ermittelt. Um die Gültigkeit der Ergebnisse aus dem Lavalläufer zu überprüfen, werden die Einflüsse der Anistropie auf das Stabilitätsverhalten eines typischen 900 MW Turbogenerator untersucht, Außerdem werden die stationäre unwuchterzwungenen Schwingungen des anisotrop gelagerten Turbogenerators gerechnet.

Geometrical Non-Linearities in the Vibrations of Circular Plate-Shaped Rotors

G. GOGU* AND M. G. MUNTEANU**

*Department of Machine Elements and Mecanisms, University of Braşov, Romania

**Department of Mechanics and Strength of Materials, University of Braşov, Romania

Summary. The calculation of natural frequencies of circular plate-shaped rotors is presented taking into consideration the geometrical non-linearities due to membrane stresses. These stresses are produced by the forces of inertia, by thermal gradients by different axial or radial loadings e.a. The equation of motion is solved by the finite element method. Specialized semi-analytical finite element, conceived by the authors, are used in the calculation of the eigenvalues. The calculation method proposed in the paper is applicable both to the rotors having the form of disks, with constant or variable thickness, and to the rotors in the form of curved plates. The method allows to take into consideration the masses distributed on the external boundary of the rotor.

Keywords. Dynamic response; eigenvalues, eigenmodes, finite element method; semi-analytical finite elements.

INTRODUCTION

The circular plate-shaped rotors are frequently used in practice. Rotors in the form of disks with constant or variable thickness are used in direct-current electric machines and in gas turbines. Rotors in the form of curved plates are used in centrifugal pumps, turbines, centrifugal filters or separators.

The high rotational speeds produce centrifugal forces and membrane stresses in rotors with considerable values. Besides these stresses, during running, stresses due to radial and/or axial pressures and to thermal gradients, actting on radius and/or on thickness, can appear in the body of rotors

The state of stress in the circular plate-shaped rotors introduces geometrical non-linearities in their dynamic behavior (Zienckiewicz, 1975). The state of stress can produce elastic stability loss by bifurcation (buckling) or by reaching the critical speeds. Thus, to point out the effect of the state of stresses on the dynamic behavior of circular plate-shaped rotors second order calculations must be performed. To this end the paper advances the use of specialized semi-analytical finite elements.

The transverse vibrations of circular plate-shaped rotors are discussed by using the method of calculation proposed by the autors of this paper. The eigenvalues and eigenfrequencies for eigenmodes with nodal diameters and nodal circles are calculated.

The calculations performed allow the determination of the critical speeds and the critical state of stress at which bifurcation buckling occurs.

The paper presents the effects of membrane stresses produced by the forces of inertia, during running, by the thermal gradients due to the nonuniform distribution of temperature on the radius and by hooping, in the case of hooped rotors. The effect of these stresses are illustrated by diagrams plotted on the basis of numerical results obtained by the authors.

EQUATION OF MOTION IN THE CASE OF CIRCULAR PLATE-SHAPED ROTORS

By means of the finite element method the partial differential equations, which describes the dynamic behavior of circular plate - shaped rotors, is replaced by a system of linear equations:

$$\left([K] + [K_G] \right) \{u\} + [M]\{\ddot{u}\} = \{P\} , \quad (1)$$

which includes the boundary conditions. The following notations were used: $[K]$ is the elastic rigidity matrix; $[K_G]$ - the geometric rigidity matrix, which depends on the membrane stresses; $[M]$ - the mass matrix; $\{u\}$ and $\{\ddot{u}\}$ - the vector of nodal displacement and the vector of nodal accelerations, respectively; $\{P\}$ - the vector of nodal load.

For an element the rigidity matrices and mass matrix, respectively, have the expressions:

$$[k] = \int_{Ae} [B]^T [D] [B] \, dA, \quad (2)$$

$$[k_G] = h \int_{Ae} [G]^T [\sigma_o][G] \, dA, \quad (3)$$

$$[m] = \rho h \int_{Ae} [N]^T [N] \, dA, \quad (4)$$

where: A_e is the surface of the finite element; $[D]$ - the Hooke's law matrix; $[\sigma_o]$ - the vector of stresses in the median plane of the rotor; $[B]$ and $[G]$ -the matrices relating the curvature and the torssion vector $\{\varkappa\}$ and the slope vector $\{w'\}$, respectively, to the nodal displacemente vector of the element $\{w_e\}$; $[N]$ - the matrix of the interpolation function relating the displacement of any-point to the nodal displacements of the element.

The matrix calculated using Eg.(4) represents the consistent mass matrix. In the case of rotors with slashes on their external boundary the concentrated masses matrix is also performed.

Based upon Eq (1), two eigenvalue problems can be defined:

$$([K] + [K_G] - \omega^2 [M]) \{u\} = 0, \qquad (5)$$

$$([K] + \lambda [K_G]) \{u\} = 0, \qquad (6)$$

where λ is a proportionality factor of the membrane stress field. Equation (5) enables the eigenfrequencies ω to be determined, taking into account the presence of the different fields of membrane stresses. By Eq.(6) we get the value of the critical membrane stress, at which bifurcation buckling occurs, as well as the coresponding eigenmode.

SPECIALIZED SEMI-ANALYTICAL FINITE ELEMENT FOR PLANAR ROTORS

To solve the problem of eigenvalues (5)in the case of planar rotors having constant or variable thickness it is sufficient to study only a circular sector inserted between two consecutive nodal diameters (Fig.1). The nodal diameters being skew - symmetry axes, the coresponding sides are considered articulated. We also consider that on the circumference of a circle of radius r, the displacement w has a sinusoidal variation. Thus, a unidimensional finite element was developed (Gogu and Munteanu, 1985). Each element has two nodes. Each node introduces two degrees of freedom, the displacement w along the z axes and the rotation in the zr plane, considered positive in the sense of rotation of the r axes upon the z axes using the shortest way. The finite element interpolation function is a polynomial

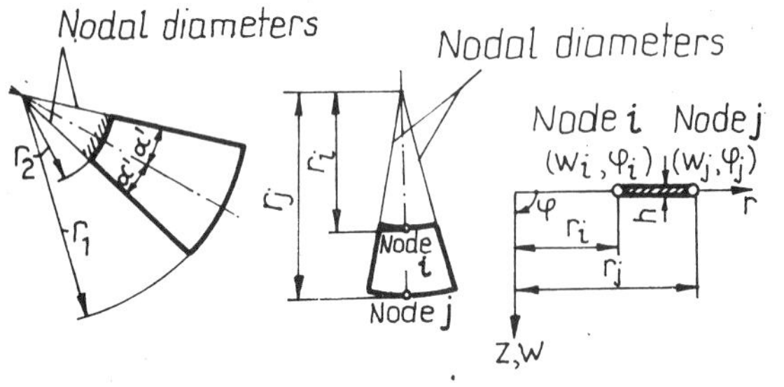

Fig.1. Specialized semi-analytical finite element for planar rotors.

function of the form:

$$w_{qe}(r,\theta)=(c_{1q}+c_{2q}r+c_{3q}r^2+c_{4q}r^3)\cos q\theta, (7)$$

where $\theta \in [-\pi/2q, \pi/2q]$, q being the number of nodal diameters. The constants $c_{kq}(k=1,2,3,4)$ are determined from the condition that, in nodes Eq.(7) leads to the nodal displacements.

SPECIALIZED SEMI-ANALYTICAL FINITE ELEMENT FOR CURVED ROTORS

To solve the problem of eigenvalues (5) in the case of curved rotors, a specialized semi-analytical finite element was developed having two nodes and four unknown displacements per node. Each node introduces the displacements: $\bar{u}$, $\bar{v}$, $\bar{w}$ and $\partial\bar{w}/\partial s$. The nodal displacements $\bar{u}$ and $\bar{w}$ are related to a local reference point attached to the finite element (Fig.2) and $\bar{v}$ is to circumferential displacement.

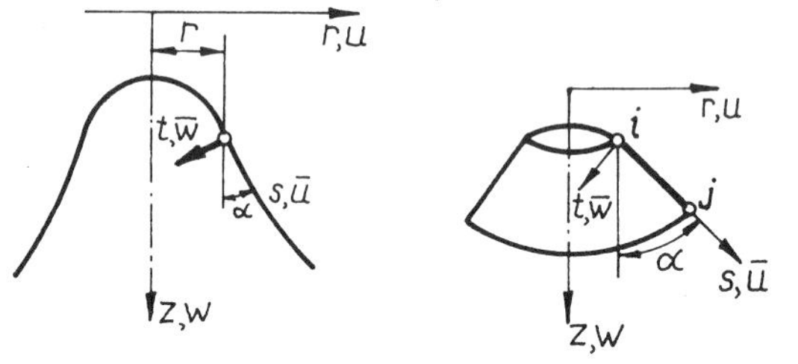

Fig.2. Specialized semi-analytical finite element for curved rotors.

In this case the finite element interpolation function is a polynomial function of the form:

$$\begin{aligned} \bar{u}_{qe} &= (c'_{1q}+c'_{2q}s)\cos q\theta , \\ \bar{v}_{qe} &= (c'_{3q}+c'_{4q}s)\sin q\theta , \\ \bar{w}_{qe} &= (c'_{5q}+c'_{6q}s+c'_{7q}s^2+c'_{8q}s^3)\cos q\theta , \end{aligned} \qquad (8)$$

where $\theta \in [-\pi/2q, \pi/2q]$. The constants $c'_{kq}(k=1,2,\dots,8)$ are determined according to the nodal displacements, as for any type of finite element.

INFLUENCE OF MEMBRANE STRESSES ON THE EIGENFREQUENCIES OF CIRCULAR PLATE-SHAPED ROTORS

The problem of eigenvalues (5) may be written in the form:

$$([K] + \lambda[K_G] - \omega^2[M]) \{u\} = 0 \qquad (9)$$

The solution of Eq.(9) may be determined by starting form the solutions of the following two eigenvalue problems:

$$([K] - \omega^2 [M]) \{u\} = 0 , \qquad (1o)$$

$$([K] + \Psi[K_G]) \{u\} = 0. \qquad (11)$$

In the case of rotors with an axial-symmetrical membrane stresses field there exist an approximate relation between the eigenvalues of equations (9),(1o) and (11) having the form (Munteanu, Gogu and Radu, 1987):

$$\Omega_q = \omega_q \sqrt{1 - \frac{\lambda}{\Psi_q}} , \qquad (12)$$

where Ω_q, ω_q and Ψ_q represent the eigenvalues of equations (9),(1o) and (11), for the eigenmodes with q nodal diameters and $l = 0$ nodal circles, modes of practical interest for planar rotors especially. For these eigenmodes the lowest natural frequencies are obtained.

The subprograms for the calculation of eigenvalues Ω_q, ω_q and Ψ_q is based on an iterative method that permits the simultaneous determination of several eigenvalues and eigenvectors (Corr and Jennings, 1976). The method is rapidly convergent, 5 to lo iteration cycles being in general, sufficient to obtain satisfactory accuracy.

Based on relation (12), the influence of various fields of membrane stress on the natural frequencies can be determined without solving problem (9) several times for different values of parameter λ. The computer time is reduced considerably due to the fact that the eigenvalues are not calculated for each new solution of Eq. (5).

The eigenvalues Ψ_q represent the critical values of rotor bifurcation buckling. From Eq.(12) we can observe that if the factor λ increases the natural frequencies Ω_q decrease. If $\lambda = \Psi_q$, then the natural frequencies become zero. If the eigenvalues Ψ_q are positive, then the natural frequencies decrease in the presence of membrane stresses. If the eigenvalues Ψ_q are negative, then the natural frequencies increase.

For the same boundary conditions, dimensions and membrane stress field some of eigenvalues Ψ_q increase and other ones decrease. Thus, for example, in the case of planar rotors, considered clamped at the internal boundary of radius r_2, the membrane stresses produced during rotation lead to higher natural frequences for all the eigenmodes. Stresses produced by the thermal gradients on radius corresponding to a positive difference of temperature between the external and internal boundary of rotor ($\Delta t = t_1 - t_2 > 0$) lead to higher natural frequencies for the eigenmodes with q = 0 and q = 1 nodal diameter and to lower natural frequencies for the eigenmodes with $q \geqslant 2$ nodal diameters, respectively. The membrane stresses produced during hooping, in the case of hooped rotors, lead to lower natural frequencies for the eigenmodes with q = 0 and q = 1 nodal diameter and to higher natural frequencies for the eigenmodes with $q \geqslant 2$ nodal diameters, respectively.

The eigenmodes with the lowest positive values of Ψ_q are called critical eigenmodes. The critical eigenmodes depend on the boundary conditions and on the membrane stress field, being independent on the rotor dimensions. For example, in the case of planar rotors considered clamped at the internal boundary of radius r_2, the critical eigenmodes depend only on the ratio $\varrho_s = r_2/r_1$, r_1 being the external radius of rotor. Thus, the critical eigenmodes are characterized by l =0 nodal circles and q = 2 nodal diameters for $\varrho_s \leqslant 0{,}26$, q = 3 nodal diameters for $0{,}26 < \varrho_s \leqslant 0{,}46$, q= 4 nodal diameters for $0{,}46 < \varrho_s \leqslant 0{,}58$, e.a.

Therefore, for the planar rotors, the critical eigenmodes are those with l =0 nodal circles and $q \geqslant 2$ nodal diameters.

For the aforementioned eigenmodes the bifurcation buckling occurs and the minimum critical speeds are reached.

The membrane stresses produced by the thermal gradients on radius ($\Delta t > o$), leading to decreasing of natural frequencies for the eigenmodes with $q \geqslant 2$ nodal diameters, have nagative effect on the elastic stability of plate-shaped rotors.

The membrane stresses produced in the hooped rotors during hooping leading to increasing of natural frequencies for the eigenmodes with $q \geqslant 2$ nodal diameters have a positive effect on the elastic stability of plate-shaped rotors. This accounts for the use of hooped rotors particularly in the case of high thermal gradients on radius.

A problem analogous with the one presented above is that of determining the critical stresses in the presence of several fields of membrane stresses, when the critical values are known for each field in turn, i.e. that of solving the problem:

$$([K] + \sum_i \lambda_i [K_{Gi}])\{w\} = 0 , \qquad (13)$$

when the solutions of the below problems are known:

$$([K] + \lambda_i [K_{Gi}])\{w\} = 0. \qquad (14)$$

By following the same calculation method it is found that the stability loss occurs when the condition

$$\sum_i \frac{\lambda_i}{\Psi_{qi}} = 1 \qquad (15)$$

is satisfied. Thus the relation (12) may be generalized

$$\Omega_q = \omega_q \sqrt{1 - \sum_i \frac{\lambda_i}{\Psi_{qi}}} . \qquad (16)$$

Relations (12), (15) and (16) agree with the theoretical results obtained by the finite element method applied with problems (5) and (9), as well as with the experimental results, as presented in Figs.3 and 4. The rotor presented in Figs. 3 and 4 has the following features: external radius r_1= 185 mm, clamping radius r_2= 62.5 mm, hooping radius r_3=14o mm, thickness h=2 mm and hooping radial pressure p_r= 5,5 MPa.

On the basis of frequency-rotational speed diagrams the critical speeds may be determined taking into consideration the influence of the different membrane stress fields. It is considered that the critical speed is reached if the natural frequency of running rotor becomes equal with the product q.n therefore when the backward wave natural frequency (f_q^b) becames zero (Tobias and Arnold, 1957), as illustrated in Fig.4, n being the rotor speed.

CONCLUSIONS

The study of plate-shaped rotor natural frequencies must be realized taking into consideration the membrane stress fields. These stresses are produced especially by the centrifugal forces and, in some cases,

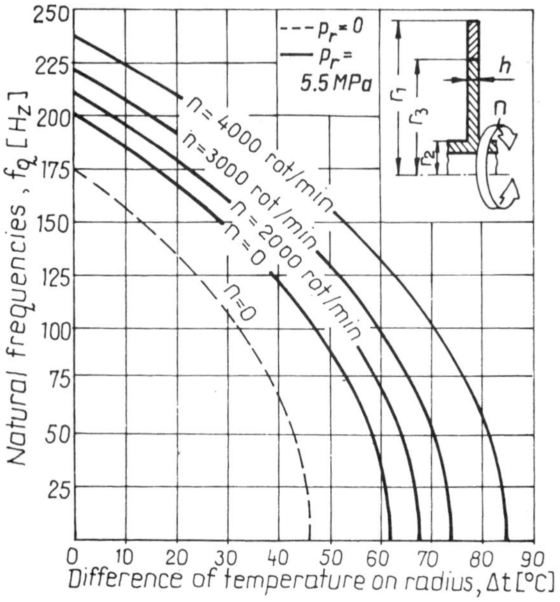

Fig.3. Influence of rotational speed, difference of temperature on radius (t 0) and hooping on rotor natural frequencies.

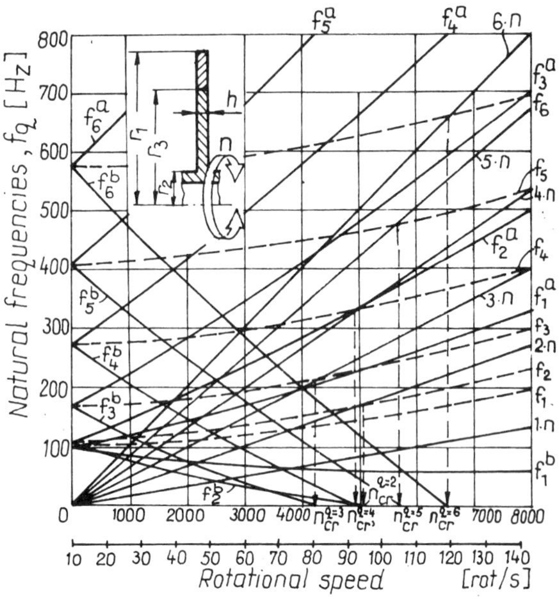

Fig.4. Frquency-rotational speed diagram for a hooped rotor with difference of temperature on radius(Δt=2o°C).

by the thermal gradients or by the hooping pressure.

In order to calculate the natural frequencies and, implicitly, the rotor critical speeds taking into consideration the membrane stress fields any general program, based on the finite element method, e.g. NONSAP (Bathe, Wilson, Iding, 1974) may be used. However this envolves long computer time. Under such conditions the specialized programs prove more favorable. Such a program was conceived by the authors of this paper on the basis of the semi-analytical finite elements. The teoretical results presented in the paper are in good agreement with the experimental data.

REFERENCES

Bathe, K-J., Wilson, E., and Iding, R.H. (1974). NONSAP: A Structural Analysis Program for Static and Dynamic Response of Nonlinear Systems. University of California, Berkeley.

Corr, R.B. and Jennings, A. (1976). Simultaneous iteration algorithm for symmetric eigenvalue problems, Int.Jour. Numerical Meth.in Eng., p.647-663.

Gogu, G. and Munteanu, M.G. (1985). Specialized semianalytical finite element use to study the vibration of the circular plates. IV-th IFToMM. Int.Symp., Bucharest, Romania, Vol.III-1, p.193-2oo.

Munteanu, M.G., Gogu, G. and Radu, G. (1987). On displacements and eigenfrequencies of thin circular plates in the presence of membrane stresses. St. Resear.Appl. Mechanics (under point).

Tobias, S.A. and Arnold, R.N. (1957). The influence of dynamical imperfection on the vibration of rotating disks. Proc.Inst.Mech.Eng., p.669 - 69o.

Zienkiewicz., O.C. (1975). The Finite Element Method in Engineering Science. Mc. Graw-Hill, London.

NON-LINÉARITÉS GÉOMÉTRIQUES DANS LES VIBRATIONS DE ROTORS EN FORME DE PLAQUES CIRCULAIRES

- Résumé -

Les rotors en forme de plaques circulaires sont fréquemment utilisés dans la construction des machines électriques à courant continu, des turbines, des pompes, des centrifugeuses etc. Dans cet ouvrage on présente le calcul des fréquences propres aux rotors en forme de plaques circulaires avec la prise en considération des non-linéarités géométriques dues aux tensions de leur corps. Ces tensions sont produites par les forces centrifuges, par les gradients thermiques, par les diverses sollicitation axiales ou radiales. L'équation du mouvement est résolue par la méthode des éléments finis. Pour le calcul des valeurs propres sont utilisés les éléments finis semianalytique spécialisés, conçus par les auteurs. La méthode de calcul proposée dans l'ouvrage est applicable aux rotors en forme de disques d'une épaisseur constante ou variable ainsi qu'aux rotors en forme de plaques courbes. La méthode permet également la prise en considération des masses distribuées sur le contur extérieur du rotor.

Analysis and Evaluation of Fatigue in Turbogenerators Shafts Due to Different Perturbations of The Electric Network

L. RUBIO, J. JOSÉ AND M. G. JULIO

E.T.S.I. Industriales de la Universidad Politécnica de Madrid, Departamento de Ingeniería Mecánica y Fabricación, c/ José Gutierrez Abascal 2, 28006 Madrid, Spain

INTRODUCCION.

In the present article they are considered mechanical effects which are produced in turbogenerators as a consequence of disturbances of the Electric Network. That mechanical effects will be characterized by mechanical fatigue that can produce in the turbogenerator shafts. This study is inserted in a wider work, whom final objective is the - design of a Torsional Stress Analyzer for turbogenerators of large power stations.

For knowing the fatigue caused by some particular Network disturbance, it is first necessary to assess the originated transient electromagnetic torque, and then to calculate transient torsional torques caused in each part of shaft as a consequence of that electromagnetic torque: known the torsional torques it is now possible to deduce fatigue caused by these torques and to obtain the shaft lifetime loss. In this article it is considered a transient electromagnetic torque measurement method,substantially different that habitually employed; it is - about dip voltage method, that avoids use of special measurement devices for transient currents,and it enables calculation of electromagnetic torque - only with the mediums in power stations and a few - more additional devices. In this article they are followed all necessary steps to reach the fatigue, beginning from the dipo tensión method. We will use notations fust like Reference /i/.

1. Transient Electromagnetic Torque: Source and - Measurement.

We can suppose the turbonerator on a power - station, that is, supplying power to the Electric Network. If suddenly, it is originated a disturbance in the Network, e.g.: shortcircuit in a line, - all points of this Network will more or less appreciate that disturbance and this will too affect to the generator; in the generator will be produce a variation on the working conditions, which will -- give as result a transient electromagnetic torque.

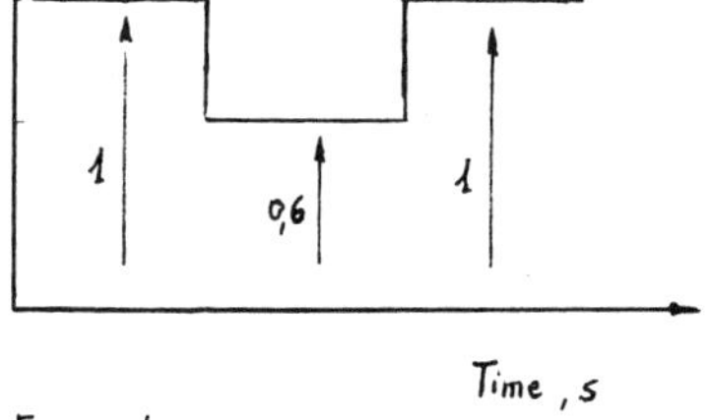

Figure 1

It is possible to obtain the expression of - transient electromagnetic torque caused after an electric disturbance, only knowing a few parameters of generator and with the measure of the duration and the deep of reduction voltage, produced during electric fault, but without knowing transient currrent; the reduction of voltage produced after - - fault will be named by dip voltage. In figure 1 it can be seen a modellized voltage dip, with a 40% - deep and a duration of 100 ms. Although in this -- paper we will not prove the deduction of electromagnetic couple formules as function of dip voltage - value (that can be seen on reference /1/, it is - convenient to point out that it results on 4 expressions of the electromagnetic couple, depending on type As a illustrative example, the expression of the electromagnetic couple for the case of a monophasic voltage dip is:

$$T_{em} = Te_{al} + Te_{u-i} + Te_{u-r}$$

2. Mechanical Response of a Turbogenerator.

In the mechanical system of the turbonegerator electromagnetic torque is a resistant couple in -- opposite to the motor couples developped by different parts of the turbine: in normal conditions -- electromagnetic couple balances the motor couple. After an alteration on resistant couple, it is an alteration on electromagnetic couple when it has - been produced a dip voltage in synchronous generator, the mechanical system reponds and, as consequence of this, torsional torque are developped -- which tend to compensate the original alteration of electromagnetic torque.

In this section we study mechanical response produced in the turbogenerator group as consequence of a voltage dip in the Electric Network; for this, it is modellized the turbogenerator rotor and on - this they are applied all acting torques, obtaining in this manner the torsional couples as reply.

2.1. Mechanical Model

From mechanical point of view, the rotor of a turbogenerator is a complex system; this has a - - variable section shaft, where are coupled or jointed the different masses, because of this the mathematical model which represents mechanical system is difficult for determining. A several authors have concluded on your investigations that it is possible to simplify mechanical model responding to electromagnetic couple originated by electrical disturbances so that a lumped mass model is enough for - study of the shaft /2/.

On the other hand, number of masses in the model is a parameter that will cause repercussión on the precission of results. At this respect,IEEE -

Working Group propose taking multimass model with an inertia by rotor to get a fiable results; in - this models, each main element (generator, high - pression stage turbine...) is considered as a - - rigid body with determinated mass which is linked to the rest of elements by means of elastic shafts without mass.

In the present work, we'll take six masses - that is a proposed number for fatigue studies. -- Nevertheless, generalization of the result to a .. number of masses have not difficulty.

2.2. Mechanical Damping

There is a factor to be considered, not mentioned until now: it is the damping factor; it can be considered two types of mechanical damping: -- from internal or external origin.

Internal Damping (called usually Di) is originated by mechanical hysteresis which contains material of the turbogenerator shaft: it can be including coupled parts.

Internal Damping (called usually Di) is originated by mechanical hysteresis, which contains material of the turbogenerator shaft; it can be including coupled parts.

In spite of being completely definated, it is very difficult for large turbogenerators to obtain real values of two quoted damping factors. In the event of taking into account the mechanical - -- damping and not being availaible, a real values - are possible to make use of the recommended values of I.E.E.E. for its Second Benchmark Model /3/. - Two parameters, Di and De, have a very little influence in the peak values of the torsor torques resulting of a Electric Network disturbance, - -- whenever the subsynchronous Resonance would not - Present, that occurs in all Spanish Electric - - Network. However, to assess the fatigue beared by material, it is important to keep in mind the - - damping of torsional vibrations /4/.

2.3. Calculation of Torsional Torques

Once the model and its parameters are set up, it is necessary for calculating torsional torques as response to a disturbance to obtain torsional angles of mechanical model and then, to deduce - that torques. We have done the calculation by -- planning a differential equations system which -- shows consideration of all actuating mechanical - phenomenes. Bearing in mind that it is a six - -- masses model, there are six equations with six -- unknowns angles. For instance, equations corresponding to the part of the generator is:

$$2.H_g.D^2.\theta_g = K_{BPB,g}.(\theta_{BPB} - \theta_g) - K_{g,e}(\theta_g - \theta_e) -$$

$$- dg.D.\theta_g - Tel$$

To solve the system of six equations it have been employed a method of numerical calculus, by use of computing techniques; concretely, we have been done use of CSMP (Continuous System Model - Program), which gives excellent results for this type of mechanical couple originated by a triphasic short-circuit on the part included between the two low pression stages of a turbine. A wider information about application of CSMP to solve this type of problems may be obtained in the Reference /5/.

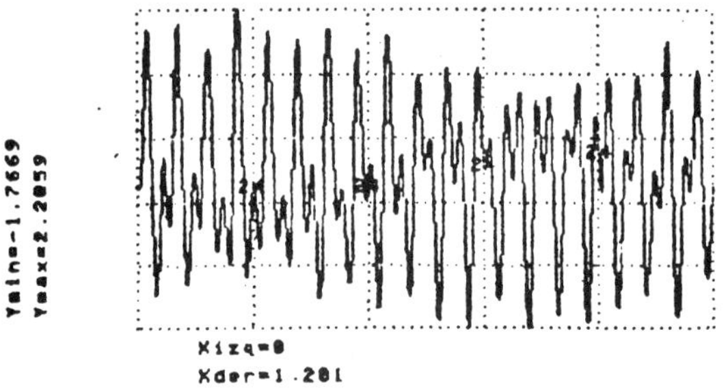

3. Fatigue and Loss of useful life in the turbogenerator shaft.

IEEE Subsynchronous Resonance Working Group defines the fatigue as the process of progressive localized, permanent and structural change - -- occuring in a material sujected to conditions -- which produce fluctuating stresses and strains at some point or points and which may culminate in - cracks or complete fracture after a sufficient -- number of fluctuations. It is important to note - that fatigue is a cumulative process and it is not physically observable until it is formed a crack in the material.

(On the turbogenerator shaft in this case).

In our case and with torsor torques known, - mechanical stresses which support each part of the shaft is calculated, and from there it is already possible to get the fatigue produced. The methods to calculate this fatigue are well knowns; in reference /6/ you can find some of these methods.

4. Conclusion

Bearing in mind that fatigue failures are -- most of produced in the shaft of large turbogenerators and, therefore, the more importants when - studying these shafts, it is comprehensive the -- importance for the supervision of functional state of the shafts. With proposed method, that is, dip voltage method, it is possible to know for - each electric disturbance produced in Electric - Network the transient electromanetic torque caused by use of a ver simple measurement equipment: - - later, it is possible to evaluate mechanical fatigue caused. It is important to note that Mechanical model has been chosen as a lumped model which is - giving an excellent results. Resolution of Mechanical Equations is done by using a Computer Simulation Program, called CSMP. The simplicity of dip voltage method suggest the design of a equipment - control for accumulated fatigue in the shaft of - turbogenerators, whereupon it is nor necessary the use of special measurement equipments. With this fatigue control equipment (called also Torsional - Stress Analyzer), it is possible the use of devices and systems existents as normal equipment in the large units of electric power generation. That Torsional Stress Analyzer would have two important characteristics: simplicity and economy of means.

5. References.

/1/ Lasso, J.J.
"Influencia de los fenomenos transitorios en los sistemas eléctricos sobre los esfuerzos mecánicos en los ejes de los grupos turbina-generador" - - E.T.S.I.I. , Noviembre 1985.

/2/ Ramey, D.G; Kung, G.C.
"Important parameters in considering transient torques on turbine-generators shaft systems" IEEE Trans. on Power Apparatus and Systems. Vol. PAS-99 No.1, 1980.

/3/ IEEE Subsynchronous Resonance Working -- Group:
"Second Benchmark Model for computer simulation of subsynchronous resonance" IEEE Trans. on Power - - Apparatus and Systems. vol. PAS-104 No.5, 1985.

/4/ Goosens, J.F.;Calvaer, A.j.; Soenen, L.J. "Full-scale short-circuit and other test on the dynamic torsional response of Rodenhuize Nr 4-300 MW-3000 RPM Turbogenerator" IEEE Trans. on Power Apparatus and Systems. Vol. PAS-100 No.9 1981.

/5/ López Martinez, J.M.:
"Estudio del comportamiento vibracional de árboles de Turbogenerador a los, esfuerzos torsionales por defectos en la Red Eléctrica" Proyecto fin de carrera, E.T.S.I.I. Madrid 1985.

/6/ Shigley, J.E.
"Diseño en Ingeniería Mecánica" Ed. Mc.Graw-Hill, 1978.

/7/ Vioque Vioque,J.
" Procedimiento sistemático para análisis por - - ordenador de vibraciones en máquinas" Proyecto fin de carrera E.T.S.I.I. Madrid 1985

Résumé. On a analysé le comportement mécanique des générateurs de l'electricité à consequence des troubles électriques dans le résau. Ces effects mécaniques rendt fatigue mécanique dans l'axe du générateur.

Keyword: fatigue in turbogenerators: perturbation of the electric network.

Transient and Steady-State Torsional Response of Internal Combustion Engine Crankshaft with Viscous Damping and Periodic Excitation

V. KARADAĞ

Mechanical Engineering Faculty, Technical University of Istanbul, Istanbul, Turkey

Abstract. In the present investigation, the resonant and nonresonant steady-state and transient response of the crankshaft of the internal combustion engines have been reexamined by the digital computing procedures. The periodic forces in accordance with the firing order of the engine, several external equivalent viscous dampings of bearings and pistons, and the hysteretic damping of crankshaft have been considered in the analysis. The complex eigenvalue and modal analysis procedures and average acceleration method have been incorporated in the analysis. It has been shown that, the transient vibrational stresses acting on the crankshaft due to reciprocating masses and gas forces during, and through resonance and steady state conditions, can reach to the allowable limits, and can lead to the ultimate fatigue failure of the crankshaft. The contribution of the transient response to the failure is, also, experimentally verified. Comparisons, have been, also, made between several digital computing procedures of transient and steady state vibration analysis.

Keywords. Dynamic response; internal combustion engines; damping; digital computer applications; stress control; vibration control.

INTRODUCTION

Recently, the vibrations of rotating machinery trains, gear train systems, and crankshafts have been reexamined by digital computing procedures. The literature dealing with the subject is, usually, based on the hard won experiences (Harris and Crede, 1976; Nestorides, 1958; Wilson, 1958, 1963,1965). In the recent publications; Doughty (1985), has developed the complete steady-state torsional response of free-free machine trains to periodic excitation. Kashiwagi (1985), has defined the engine damping magnifications on shafting both with and without an end-damper. The complex eigenvalue problem related to the subject is, also, a subject of investigation as in (Doughty and Vafaee, 1985; Songyuan, 1985; Zorzi and Lee, 1985). However, the transient vibrations of the crankshafts of multi-cylinder internal combustion engines have not been analysed in the recent literature. The works of (Evans and others, 1985; Horvath, 1963; Hizume, 1976; Pollard, 1967,1972) related to the subject on the rotating shaft systems are worth to mention. An introductive work is by Paul (1979), in which the transient performance of one cylinder engine with the resisting torque of the velocity-squared type has been analysed on a one degree of freedom model.

In the present work, first, a multi degree of freedom torsional system model of a multi-cylinder inline engine has been constructed, employing equivalent viscous dampings of bearing and pistons, and hysteretic damping of the crankshaft. This model is similar to the model used frequently in the literature (Doughty and Vafaee, 1985; Harris and Crede, 1976; Nestorides, 1958; Wilson, 1958, 1963,1965). The equivalent viscous damping constants related to the bearings of the practical examples have been experimentally determined. The equivalent viscous damping constant of the hysteretic damping of crankshaft, has been determined by using the natural damping of the crankshaft material. Then, the transient response of the crankshafts of several internal combustion engines have been analysed in addition to the common type of torsional analysis of steady-state response evaluation for average speed of operation and the associated periodic excitation torques. Several comparisons, have been made between the modal analysis procedures and the average acceleration method including the digital computing cpu times. The importance of the subject have been experimentally verified on a special test engine.

THE EQUATION OF MOTION AND THE SOLUTION METHODS

First, the lumped mass model of the crankshaft of the in-line, four-stroke cycle, single-acting, reciprocating engine with the station-to-station and station-to-the ground viscous damping has been constructed (Fig. 1). This type of modeling is the most common application and adequate for most practical purposes. This modeling is, also, a suitable and efficient method to investigate the transient response by digital computers. For the present purposes, only systems reducible to equivalent single shaft systems and a special system with a cogged belt drive at the flywheel of the test engine have been considered. Then, the general excitation forces in accordance with the firing order of engine is incorporated in the analysis. The variation of the mass polar moment of inertia of the reciprocating engine mechanism is, separately included in the analysis, since, the problem is nonlinear in character and computing is time consuming. It is, also, assumed that operating conditions are identical in all cylinders and there is no coupling between the axial, bending and torsional modes. The equations of the motion of the system have been obtained by making use of the damped torsional system model,

$$|M|\{\ddot{\psi}\} + |C|\{\dot{\psi}\} + |K|\{\psi\} = \{M(t)\} \qquad (1)$$

$|M|$, $|C|$, and $|K|$ are the mass, damping, and

stiffness matrices, respectively. This equations can be reduced to a set of 2n simultaneous first-order equations as,

$$\begin{vmatrix} 0 & M \\ M & C \end{vmatrix} \{\dot{y}\} + \begin{vmatrix} -M & 0 \\ 0 & K \end{vmatrix} \{y\} = \begin{Bmatrix} 0 \\ \{M(t)\} \end{Bmatrix} \qquad (2)$$

This complex eigenvalue problem can be solved by either the double QR method, transfer matrix methods and the Faddeev-Leverrier iteration method. Double QR and Faddeev-Leverrier method has been used in the analysis. Then, the following uncoupled equations can be easily obtained,

$$a_{ii}\dot{q}_i + b_{ii}q_i = m_{ii}(t) \quad (i=1,2n) \qquad (3)$$

These uncoupled equations of the initial value problem are solved applying the fourth order Runge-Kutta integration procedure.

A transient torsional analysis program based on the average acceleration technique of time step numerical integration, which is a second order implicit method, is, also, used in the analysis. The equations of motion at time t_{i+1} of the system, and

$$\{\psi_{i+1}\} = \{\psi_i\} + h\{\dot{\psi}_i\} + h^2/4\{\ddot{\psi}_i\} + h^2/4\{\ddot{\psi}_{i+1}\} \qquad (4)$$

$$\{\dot{\psi}_{i+1}\} = \{\dot{\psi}_i\} + h/2\{\ddot{\psi}_i\} + h/2\{\ddot{\psi}_{i+1}\} \qquad (5)$$

can be used to solve $\{\ddot{\psi}_{i+1}\}$ as,

$$\{\ddot{\psi}_{i+1}\} = |D|^{-1}\{\{M(t_{i+1})\} - |C|\{\{\dot{\psi}_i\} + \frac{h}{2}\{\ddot{\psi}_i\}\} - |K|.\{\{\psi_i\} + h\{\dot{\psi}_i\} + \frac{h^2}{4}\{\ddot{\psi}_i\}\}\} \qquad (6)$$

where,

$$|D| = |M| + \frac{h^2}{4}|K| + \frac{h}{2}|C| \qquad (7)$$

The numerical integration may be performed with the average acceleration method at each time interval by using the equations (4) to (6). But, the resonant frequencies should be, separately, determined for the critical rotational speeds of the crankshaft.

THE NUMERICAL RESULTS AND DISCUSSIONS

Several examples have been used to investigate the subject. The first example is a 16 cylinder Diesel engine-generator system. The eigenvalues calculated by the double QR method which is used in the present work in the modal analysis application, are compared with the values obtained by the transfer matrix method of (Doughty and Vafaee, 1985), in Table 1, (Fig. 2). The calculated eigenvalues and eigenvectors are the same for this case.

The second example is a branched torsional system of two cylinder engine-generator-blower with a free-free damper (Doughty, 1985). The average acceleration method have been used in the analysis. Engine speed is 1800 rpm. The typical displacements of transient response and steady-state response have been shown in Table 2, for the non-resonant case. The maximum value of the ratio of maximax response to the steady-state response amplitude is 1.39 for step torque excitation and 1.83 for coexistant step and full-cycle sinusoidal torque excitation at the free-free damped mass of the system. These values are about the same with the values, generally, obtained in one degree of freedom undamped systems with the same type excitation of (Harrison and Crede, 1976), and four degrees of freedom system of (Pollard, 1972).

TABLE 1 Eigenvalues of 16 Cylinder Diesel Engine-Generator System

M	undamped	Eigenvalues Doughty and Vafaee(damped)	present (damped)
0	0.0+j0.0	0.0 +j0.0	0.0 +j0.0
1	0.0+j67.424	-1.637+j67.404	-1.637+j67.404
2	0.0+j376.685	-0.332+j376.687	-0.332+j376.687
3	0.0+j743.288	-8.009+j743.208	-8.009+j743.208

TABLE 2 Two Cylinder Engine-Generator-Blower System

Average acceleration method is used in the present case with 2.5° integration steps, with zero initial conditions. ($x10^{-3}$ rd.).

	The sole effect of step torque				Coexistant sinusoidal and step torques		
	present				present		
i	2 cyc.	3 cyc.	steady	Doughty	2 cyc.	3 cyc.	Doughty
1	0.0	0.0	0.0	0.0	-0.776	-0.730	-0.723
2	-0.662	-0.587	-0.624	-0.624	-1.906	-0.674	-0.623
3	-1.308	-1.375	-1.356	-1.355	-2.703	-1.346	-1.357
4	-2.466	-4.817	-3.929	-3.929	-3.803	-4.835	-3.929
5	-2.486	-4.820	-3.929	-3.929	-3.821	-4.841	-3.929
6	-0.654	-0.687	-0.678	-0.678	-1.352	-0.673	-0.678
7	-1.333	-2.374	-2.243	-2.272	-2.022	-2.375	-2.270
8	-2.034	-4.058	-3.859	-3.867	-2.685	-4.048	-3.863
I	-	-	-	-	-0.196	-0.216	-0.215

TABLE 3 6 Cylinder Engine-Generator System

One-node torsional vibrational stresses imposed on shafting by gas forces with correction for inertia. Third order stresses at resonance cases. Zero initial conditions; in two cycles.

		Nonresonant cases		Resonance cases			
		r.p.m.		r.p.m.			
maximax response	$\frac{N}{cm^2}$	300	350	210	280	840	840 steady initial cond.
transient	$\tau_{tr_{max}}$	1269	794	1826	1478	2636	9495
steady-st.	$\tau_{tr_{max}}$	731	472	1778	681	7897	2465
3 cycles		-	-	-	-	10313	-
Wilson	$\tau_{st_{max}}$	683	487	2137	3515	9997	9997
Lewis	$\tau_{st_{max}}$	-	-	1305	3030	22108	22108
Kashiwagi	$\tau_{st_{max}}$	-	-	-	-	9366	9366

The third example, is six-cylinder, in-line, four stroke cycle, single-acting, oil engine-generator set of (Wilson, 1963), (Fig. 4). The maximax response and steady-state response one-node stresses imposed on shafting by torsional vibration at resonant cases have been shown in Table 3. The firing order is 1-3-5-6-4-2. The maximum value of the maximax response stresses, act during first 120° crankshaft rotation. The maximum value of the ratio of the maximax response stresses to the steady-state response stresses is 2.17 with zero initial conditions of the steady state response. The ratio of the torque due to the gas forces acting on a piston, to the torque due to the inertia forces of reciprocating masses, is 2.02 at the rated speed of 300 rpm for this example Initial conditions used in determining the transient response may, also, change the obtained stresses. The transient and residual response stresses in this case explain the widely divergent conclusions when estimating such stresses, (Table 3).

The fourth example is a FIAT 124, 4 cylinder automobile engine (Fig. 5). In this case, the mass polar moments of inertia are determined experimentally from the engine crankshaft. The equivalent viscous damping coefficients of the several bearings and pistons are, also, determined experimentally (30°C conditions). The equivalent viscous damping for the hysteretic damping of the crankshaft has been calculated by using the natural damping of crankshaft material which is, also, determined experimentally (Steidel, 1979). This damping is 1% of critical damping, where the typical industry values are assumed, ranging from 1 to 4% (Evans and others, 1985). The engine damping determined by the Carter's empirical formula in (Wilson, 1963), is 0.15 Nms/rd. The external damping determined by Holzer formula is 0.05, and by Shannon formula (Kozesnik, 1962), is 0.04 Nms/rd, for this example. Experimentally determined equivalent viscous damping of the crankshaft bearing is 0.22 Nms/rd (30°C, per cylinder), for this case. The values of the harmonic components of the tangential effort due to gas forces of a four stroke cycle, spark ignition oil-engine are taken from (Nestorides, 1958; Wilson, 1963), with the correction for inertia. The torsional vibration stresses of the crankshaft obtained by various methods have been shown in Table 4. The testing values obtained by Faddeev-Leverrier method are the same with the values obtained in (Tse, Morse and Hinkle, 1978), for a three degrees of freedom system. The differences between the steady state response stresses are mainly due to the different damping values. An increase have been, generally, observed when the stresses have been calculated by using the steady state response as initial conditions, instead of zero initial conditions (Table 5), (Fig. 6). The effect of the

TABLE 4 F 124 Engine Crankshaft Stresses.
1-3-4-2 firing order. Engine speed 4000 rpm, second order gas forces with inertia correction, with zero initial conditions. The maximum shear stresses after two rotations of the crankshaft.

Between	N/cm^2	Modal Analysis Faddeev 10° step size	Modal Analysis d.QR 10° step size	Modal Analysis av.ac. 2.5° step size	lewis	wilson (equivalent energy)	wilson (practical formula)	kashiwagi
Fly. 4thP.	τ_{tr}	366	289	332	--	--	--	--
	τ_{st}	211	256	271	2693	10178	4060	126
4thP. 3rdP.	τ_{tr}	265	241	262	--	--	--	--
	τ_{st}	161	201	201	--	--	3836	119

variation of the mass polar moment of inertia of the reciprocating parts of the reciprocating engine mechanism on the transient response and steady state response, has been shown in Table 5, (apparent damping effect). The same effect, has been, also, shown in Table 6, including a rectangular step pulse of 5 times the torque acting on

TABLE 5 F 124 Engine Crankshaft Stresses. (continued from the table 4). The $steady_2$ initial conditions are the conditions of the steady state response of 1-4-2-3 firing order for the disturbed case; var. int., is variation of inertia.

Between N/cm^2	Faddeev 10° step size zero int. cond.	Faddeev 10° step size steady state cond.	D.QR 10° step size steady state cond.	Average acceleration 2.5° step size zero int. cond.	Average acceleration 2.5° step size $steady_2$ state cond.	Average acceleration 2.5° step size $steady_2$ state cond. var.int.
Fly. τ_{tr}	366	452	377	437	532	468
4thP. τ_{tr}	211	289	271	271	287	287
4thP. τ_{tr}	265	351	321	351	365	563
3rdP. τ_{st}	161	211	201	211	211	211
2ndP. τ_{tr}	50	121	121	171	244	199
1stP. τ_{st}	90	60	70	70	70	70

TABLE 6 F 124 Engine with a rectangular pulse acting on the crankshaft
The magnitude of the pulse=-98.984 Nm during first 10° of crankshaft rotation, =0 otherwise. Rotation speed 4000 rpm, average acceleration method with 2.5° step sizes, in two full crankshaft rotation. Initial conditions of $steady_2$ of Table 5. var. inrt.=variation of inertia.

Between N/cm^2	no pulse	pulse on the flywheel no var. inertia	pulse on the flywheel with var. inertia	pulse on 4th P. no var. inrt.	pulse on the 4th piston steady state initial cond. no var. inrt.	pulse on the 4th piston steady state initial cond. with var. inrt.
Fly. τ_{tr}	532	547	460	686	1068	1098
4thP. τ_{st}	287	287	271	287	316	318
4thP. τ_{tr}	365	361	374	724	799	807
3rdP. τ_{st}	211	201	201	231	241	245
2ndP. τ_{tr}	244	292	216	153	420	423
1stP. τ_{st}	70	70	70	80	91	72

crankshaft due to gas forces with inertial correction, during first 10° rotation of the crankshaft. First and third order one-node stresses have been shown in Table 7. The effect of the firing order has been, also, shown in Table 7. The coefficient of fluctuation of speed due to torsional vibration of crankshaft during transient and steady state response is δ_{tr}=0.01 and δ_{st}=0.0069 when the variation of polar moment of inertia is included and δ_{tr}=0.0099, δ_{st}=0.0068 otherwise, at the flywheel. The same coefficient is δ_{tr}=0.012, δ_{st}=0.0051 and δ_{tr}=0.011, δ_{st}=0.0050, respectively, on the other end of the crankshaft. This coefficient during steady state is the same with the values of (Nestorides, 1958; Wilson, 1963).

TABLE 7 F 124 Engine Crankshaft Stresses
Other cases, continued from Table 4. Average acceleration method.

$\frac{N}{cm^2}$ Between	4000 rpm first order gas+inrt. forces steady state initial cond.	4000 rpm third order gas+inrt. forces zero initial cond.	1-4-2-3 firing order only inertia forces acting zero initial values r p m 3151	4000	5252
Fly. τ_{tr}	262	44	137	160	46
4^{th}P. τ_{st}	6	5	125	103	38
4^{th}P. τ_{tr}	293	85	136	331	431
3^{rd}P. τ_{st}	170	63	134	259	426

TABLE 8 Digital Computation Time Consumed By Modal Analysis Methods and Average Acceleration Method
IBM 370 CPU time. The displacements and velocities obtained in two full rotation of crankshaft.

Dimensions	Modal analysis Faddeev 10° step size	Modal analysis d.QR 10° step size	Average acceleration 2.5^o step size	Average acceleration 1.25^o step size	Average acceleration with var. inrt. 2.5^o step size
5 x 5	32	30	10	15	42
7 x 7	809	168	16	26	-

The stresses acting on the crankshaft are not changed considerably, when a rectangular step torque is applied on the flywheel (Table 6). But, the stresses produced by the same rectangular pulse applied on the other end of crankshaft are, considerably, increased (Table 6). The effect of the variation of polar moment of inertia for this shock excitation is very small in normal conditions.

Digital computation time consumed by the modal analysis methods and average acceleration method have been shown in Table 8. The average acceleration method consumes less digital computing time with the increasing degrees of freedom of the system and is, also, effective in all cases.

The fifth example is a special test engine-generator set which has been shown in Fig. 1. All equivalent viscous damping coefficients in this case, have been determined experimentally (30^oC conditions). In this case, there is a cogged special belt drive between the flywheel and generator. This cogged belt drive, change the torsional vibration characteristics of the crankshaft, considerably. First, the effect of the belt drive has been neglected for comparison reasons, then, included in the analysis (Table 9). In this case; the stresses produced by the reciprocating masses are in steadily increasing character with the zero

TABLE 9 4 Cylinder Engine-Generator Set
Rotation speed 2000 rpm, second order inertia forces acting on each pistons. Average acceleration method, 2.5^o step size.

$\frac{N}{cm^2}$ Between	7 x 7 no belt drive case: present (zero init. cond.)	lewis	wilson (equivalent energy)	wilson (practical formula)	kashiwagi	8x8 with drive belt case: zero init. cond.	steady init. cond.
Fly. τ_{tr}	-	-	-	-	-	0	85
4^{th}P. τ_{st}	8843	24224	12270	12100	4117	57	8
3^{rd}P. τ_{tr}	-	-	-	-	-	0	174
2^{nd}P. τ_{st}	-	-	-	-	-	154	104
2^{nd}P. τ_{tr}	-	-	-	-	-	0	218
1^{st}P. τ_{st}	3389	-	5078	5008	1704	107	143

initial conditions, as is the case in some of the other examples. But, these stresses decrease considerably with the initial conditions of the steady state response. The transient response of the system, when the system is excited during the steady-state operating conditions, can effect the resulting residual stresses produced during steady state operating conditions (coincident dual modes). In this test case, the maximum transient stresses calculated on the crankshaft are produced between the first and second piston, and the maximum stresses calculated during steady-state operating conditions, are produced between the third and fourth pistons (160 N/cm^2). The fatigue failure of the crankshaft has occurred on the crankshaft between the first and second pistons, in a relatively short duration of the resonant working conditions. Therefore, the contribution of the transient response to the failure is unquestionable, and is experimentally verified (Table 9).

CONCLUSIONS

It has been shown that, the transient torsional vibration stresses produced by the gas forces and inertia forces acting on each of the cylinders of reciprocating engines, can reach to the allowable limits, and can lead to the ultimate fatigue failure of the crankshafts, in general. The contribution of the transient response to the failure is, also, experimentally verified. It has been shown that the steady-state response residual stresses acting on the crankshaft during operating conditions can be increased by the transient response of the crankshaft, when the crankshaft is excited during the steady-state operating conditions. This phenomenon, also, explains one of the reason of the widely divergent conclusions when estimating such stresses acting on the crankshafts.

REFERENCES

Doughty, S., and Vafaee, G. (1985). Transfer matrix eigensolutions for damped torsional system. J. Vibration, Acoustics, Stress, and Reliability in Desing, Vol. 107, pp. 128-132.

Doughty, S. (1985). Steady-state torsional response with viscous damping. J. Vibration, Acoustics, Stress, and Reliability in Design. Vol. 107, pp. 123-127.

Evans, B.F., Smalley, A.J., Simmons, H.R., and P.E. (1985). Startup of synchronous motor drive trains: The application of transient torsional analysis to cumulative fatigue assesment. ASME paper no. 85-DET-122, ASME Design Engineering Conference, Cincinnati, Ohio.

Harris, C.M., and Crede, E. (1976). Shock and Vibration Handbook. Sec. ed., McGraw-Hill Book Com.

Horvath, A. (1963). Fluctuating torsional vibrations when running up a turbo-compressor asynchronously by means of salient-pole synchronous motor. The Brown Bovery Review. Vol. 50, No. 6/7, June/July, pp. 417-429.

Hizume, A. (1976). Transient torsional vibration of steam turbine and generator shaft due to high speed reclosing of electric power lines. J. Engineering for Industry. Trans. ASME, Vol. 98, pp. 968-979.

Kashiwagi, A. (1985). Evaluation method of internal damping in torsional vibration of engine shafting. Bulletin of JSME. Vol. 28, No. 244, pp. 2386-2393.

Kozesnik, J. (1962). Dynamics of Machines. Sntl-Publishers of Technical Literature, Prague, E.P. Noordhoff Ltd. Groningen-The Netherhands.

Nestorides, E.J. (1958). A Handbook on Torsional Vibration. B.I.C.E.R.A., University Press, Cambridge.

Pollard, E.I. (1967). Torsional response of systems. J. Engineering for Power. Trans. ASME, Vol. 89, No. 3, pp. 316-324.

Pollard, E.I. (1972). Transient torsional vibration due to suddenly applied torque. J. Engineering for Industry. Trans. ASME, Vol. 94, pp. 595-602.

Paul, B. (1979). Kinematics and Dynamics of Planar Machinery. Prentice-hall, Inc., Englewood Cliffs, New Jersey, USA.

Songyuan, Lu (1985). A method for calculating damped critical speed and stability of rotor-bearing systems. ASME paper No. 85-DET-116, ASME Design Engineering conference. Cincinnati, Ohio.

Steidel, R.F. (1979). An Introduction to Mechanical Vibrations. Sec. ed., John Wiley and Sons.

Tse, F.S., Morse, I.E., and Hinkle, R.T. (1978). Mechanical Vibrations Theory and Applications. Sec. ed., Allyn and Bacon Inc..

Wilson, W.K. (1958,1963,1965). Practical Solution of Torsional Vibration Problems. Vol. 1, Vol. 2, Vol. 3, third ed., Chapman Hall Ltd. London.

Zorzi, E.S., and Lee, C.C. (1985). An efficient approach for calculating the damped natural frequencies of rotating machinery. ASME paper No. 85-DET-118, ASME Design Engineering Conference. Cincinnati, Ohio.

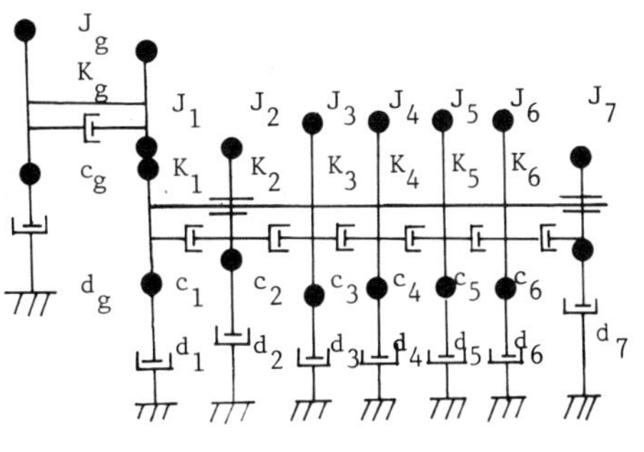

Fig. 1. 4 cylinder engine-generator set.

J_g=0.02734	kgm^2	c_g=0.00353	Nms/rd
J_1=0.2468	"	c_1=0.1499	"
J_2=2.04 10^{-5}	"	c_2=0.00247	"
J_3=5.72 10^{-4}	"	c_3=0.00213	"
J_4=J_5=5.05 10^{-4}	"	c_4=c_3	"
J_6=5.84 10^{-4}	"	c_5=0.00229	"
J_7=2.04 10^{-5}	"	c_6=0.00309	"
K_1=22777.1	Nm/rd	d_1=0.0	"
K_2=4091.0	"	d_2=d_7=0.0085	"
K_3=K_4=K_5=2252.6	"	d_3=0.0165	"
K_6=4084.1	"	d_4=0.0168	"
K_g=0.024	"	d_5=0.0167	"
		d_6=0.0157	"
		d_g=0.05	"

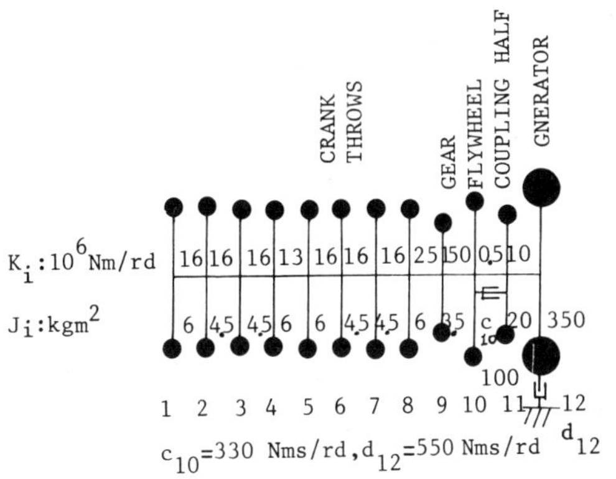

Fig. 2. 16 cylinder Diesel engine-generator system.

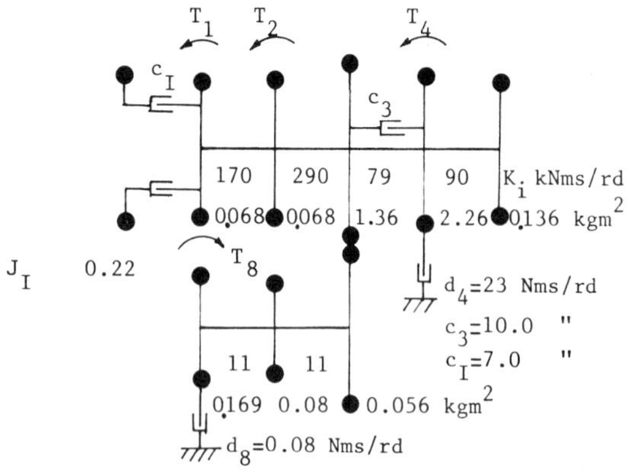

Fig. 3. Two cylinder engine-generator-blower system.

$T_1=106.06+123\cos(188.5t)+235\ \sin(188.5t)$

$T_2=106.06-123\cos(188.5t)-235\ \sin(188.5t)$

$T_4=-160.0$, $T_8=-10.0$, $T_3=T_5=T_6=T_7=T_I=0$,

$n_{1_{cr}}=1664$ rpm, $n_{2_{cr}}=2794$ rpm

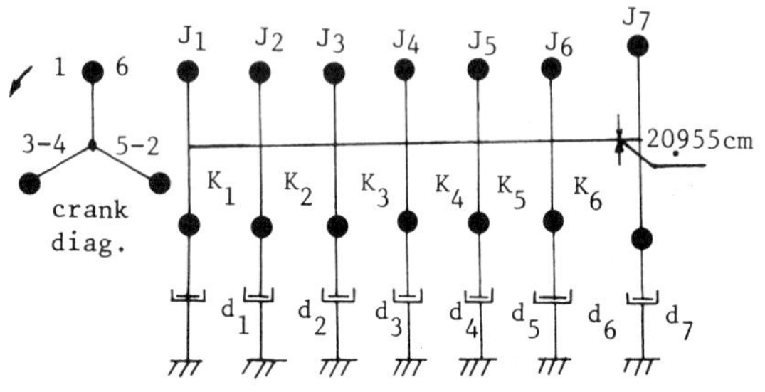

Fig. 4. 6 cylinder Diesel engine-generator system.

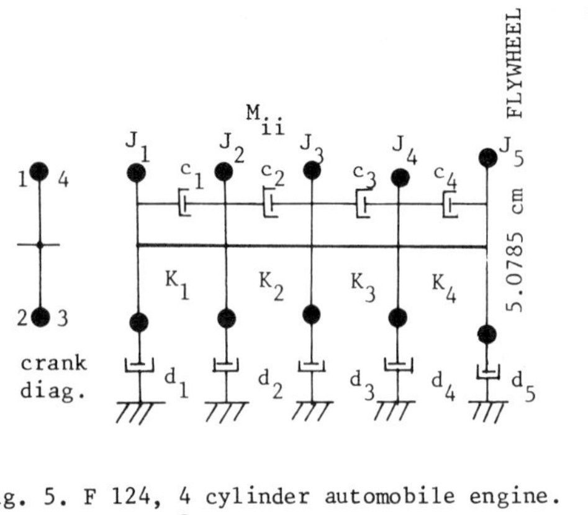

Fig. 5. F 124, 4 cylinder automobile engine.

$J_1=0.000393$ kgm^2, $J_2=J_3=J_4=J_1$, $J_5=0.0709$ kgm^2,

$K_1=K_2=K_3=258208$ Nm/rd, $K_4=387799$ Nm/rd,

$c_1=c_2=c_3=0.06367$ Nms/rd, $c_4=0.33163$ Nms/rd,

$d_1=0.97207$ Nms/rd, $d_2=d_3=1.08237$ Nms/rd,

$d_4=1.02722$ Nms/rd, $d_5=0.16546$ Nms/rd

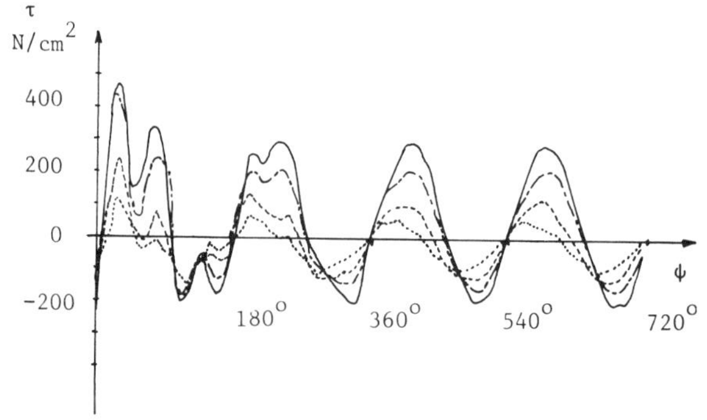

Fig. 6. F 124, 4 cylinder engine crankshaft stresses. Second order gas forces with inertia correction, rotation speed is 4000 rpm.
———: Flywheel-4th piston,
– – – –: 4th piston-3rd piston,
-------: 3rd piston-second piston,
.......: second piston-first piston.

ZUSAMMENFASSUNG

In dieser Arbeit wird die stationaere und transiente Antwort der Kurbelwelle von Verbrennungskraftmaschinen im Resonanz-und Nichtresonanzfall numerisch untersucht. In der Analysis werden die von der Zündfolge abhängigen periodischen Zwangskraefte, verschiedene äquivalente viskose Dämpfungen von Lagern und Kolben, sowie die hystheretische Krankwellendämpfung mitberücksichtigt. Neber der Modalanalyse mit komplexen Eigenwerten wird das Verfahren der mittleren Beschleunigung angewandt.

Es wird gezeigt, dass die transienten Spannungen, die infolge der hin-und hergehenden Massen und der Gaskräfte im Resonanzfall und im stationaeren Betrieb erhebliche Werte annehmen und durch Werkstoffermüdung zu Krankwellenbrüchen führen können. Der von der transienten Antwort herrührende Beitrag zum Versagen wird auch experimentel nachgewiesen. Daneben werden verschiedene numerische Rechenverfahren für transiente und stationaere Schwingungen miteinander verglichen.

Spectral Features of the Response of a Rigid Rotor Mounted on Discontinuously Nonlinear Supports

R. D. NEILSON AND A. D. S. BARR

Department of Engineering, University of Aberdeen[1], Aberdeen, Scotland, UK

Abstract. The spectral features of the response of a rigid rotor mounted in discontinuously nonlinear supports are discussed. The mathematical model of such a system and the numerical procedures developed to permit its response to be obtained are presented. The resulting response is found, for some speed ranges to be aperiodic and to contain sidebands in its spectrum. These sidebands are found to diverge continuously with increasing speed resulting in a fan-like structure of frequencies if viewed as a waterfall plot. An experimental rig which corresponds physically to the model is described and comparison made between experimental and theoretical results. Good agreement is found.

INTRODUCTION

In recent years there has been a tendency to increase the power and efficiency of rotating machinery. This is particularly true of gas turbine aero-engines. Alongside this desire for increased output, the smooth running of such machinery is often also of paramount importance both for mechanical reliability and consumer satisfaction and confidence. Consequently rotordynamic effects, which in some cases may be nonlinear, become increasingly important in the design and operation of such machinery.

Apart from known nonlinearities such as occur, for instance, in squeeze film bearings, other effects such as clearances in the bearings or rotor/stator rubs may also influence the vibration response. In the past some aspects of these phenomena have been considered [1, 2, 3, 4, 5, 6]. In most cases the presence of subharmonic motion has been predicted. The presence of aperiodic nonlinear motion in a Jeffcott shaft model with a bearing clearance and aperiodic whirl of the stator in a rotor/stator system have been shown in [2] and [6] respectively. These analyses do not however, explain certain characteristics occasionally seen in the vibration response of aeroengines undergoing engine roughness testing prior to acceptance. A response from such a test is shown in Fig. 1. The diagram is a plot of frequency content, with magnitude depicted on a grey scale basis, against % of the maximum speed of the higher speed rotor. It shows spectral sidebands fanning out from the first engine order of the high speed rotor at about 85% of maximum speed.

The present work attempts to explain the source of these unusual spectral characteristics by the use of a relatively simple rigid rotor model with bearing clearance effects.

ROTOR MODEL

The model consists of an axisymmetric rigid rotor whose motion is described by the coordinate system of Fig. 2 and which is supported at both ends in nonlinear elastic supports with radial force/deflection characteristics of the form depicted in Fig. 3. This stiffness characteristic results in restoring force vectors of the form

$$\bar{F}_A = kr_A(-\bar{e}_r) \qquad \text{if } r_A < g_A \quad (1)$$

$$\bar{F}_A = [kr_A+K_A(r_A-g_A)](-\bar{e}_r) \qquad \text{if } r_A > g_A \quad (2)$$

at end A with similar terms at end B, where $\bar{e}_r$ is a unit vector in the radial direction.

By taking components of these forces and applying a free body analysis and angular momentum theory, the equations of motion for small amlitude (linear) motion, ie where $r_A < g_A$ and $r_B < g_B$ may be shown to be

$$M\ddot{X} + C_X\dot{X} + 2kX = m_R\Omega^2 r\cos\Omega t \quad (3)$$

$$M\ddot{Y} + C_Y\dot{Y} + 2kY = m_R\Omega^2 r\sin\Omega t \quad (4)$$

$$A\ddot{\alpha} + C_\alpha\dot{\alpha} + C\Omega\dot{\beta} + \frac{kl^2}{2}\alpha = -m_T\Omega^2 r\ell_T\sin(\Omega t+\phi) \quad (5)$$

$$A\ddot{\beta} + C_\beta\dot{\beta} - C\Omega\dot{\alpha} + \frac{kl^2}{2}\beta = m_T\Omega^2 r\ell_T\cos(\Omega t+\phi) \quad (6)$$

where M, A and C are the mass and two principal moments of inertia of the rotor and m_R and m_T refer to equivalent imbalance masses for cylindrical and conical motion respectively. Other systems of equations govern the motion if $r_A > g_A$ and/or $r_B > g_B$. For the sake of brevity only the case for $r_A > g_A$ when $K_B = 0$ (i.e. where there is no discontinuity at end B) will be considered. The equations of motion governing this system are

$$M\ddot{X} + C_X\dot{X} + (2k+K_A)X + \tfrac{1}{2}K_A\ell\beta - K_A g_A\left\{\frac{(X+\beta\frac{\ell}{2})}{[(X+\beta\frac{\ell}{2})^2 + (Y-\alpha\frac{\ell}{2})^2]^{\frac{1}{2}}}\right\} = m_R\Omega^2 r\cos\Omega t \quad (7)$$

$$M\ddot{Y} + C_Y\dot{Y} + (2k+K_A)Y - \tfrac{1}{2}K_A\ell\alpha - K_A g_A\left\{\frac{(Y-\alpha\frac{\ell}{2})}{[(X+\beta\frac{\ell}{2})^2 + (Y-\alpha\frac{\ell}{2})^2]^{\frac{1}{2}}}\right\} = m_R\Omega^2 r\sin\Omega t \quad (8)$$

[1] This work was undertaken at the Department of Mechanical Engineering, Dundee University, Scotland under M.O.D. grant No. 2205/021/NGTE.

$$A\ddot{\alpha} + C_\alpha\dot{\alpha} + C\Omega\dot{\beta} + \frac{\ell^2}{4}(2k+K_A)\alpha - \frac{1}{2}K_A\ell Y + \frac{1}{2}K_A\ell g_A\left\{\frac{(Y-\alpha\frac{\ell}{2})}{[(X+\beta\frac{\ell}{2})^2 + (Y-\alpha\frac{\ell}{2})^2]^{\frac{1}{2}}}\right\} = -m_T\Omega^2 r\ell_T\sin(\Omega t+\phi) \quad (9)$$

$$A\ddot{\beta} + C_\beta\dot{\beta} - C\Omega\dot{\alpha} + \frac{\ell^2}{4}(2k+K_A)\beta + \frac{1}{2}K_A\ell X - \frac{1}{2}K_A\ell g_A\left\{\frac{(X+\beta\frac{\ell}{2})}{[(X+\beta\frac{\ell}{2})^2 + (Y-\alpha\frac{\ell}{2})^2]^{\frac{1}{2}}}\right\} = m_T\Omega^2 r\ell_T\cos(\Omega t+\phi) \quad (10)$$

The two systems of equations, 3 to 6 and 7 to 10 are valid only for motion in the ranges given. Since, in general, the motion of the system may be unsteady, i.e. the motion of the end A of the rotor may oscillate accross the boundary $r_A = g_A$ in a complicated way, its solution in time must proceed in a piece-wise manner with the equations of motion changing as necessary.

NUMERICAL INTEGRATION TECHNIQUES

Due to the discontinuous nature of the equations of motion of the system (Eqs. 3 to 10), solution by analytical or approximate methods is either unwieldy or impractical and recourse must be made to a numerical means of solution. This, however necessitates a piece-wise solution of the equations. Consequently the transitions between segments of the response calculated using the linear model and those computed using the equations of motion appropriate when in contact with the snubber spring(s) must be accurately located in time. This has previously been shown necessary to prevent collapse of the order of the solution (see [7]).

The methods used for this piece-wise solution have been developed from those presented in [7] and are based on the classical 4th. order Runge Kutta algorithm. An additional computation is also included to give an estimate of the solution at the mid-point of the time step as well as at the end point. This approximation is of the form

$$y_{n+\frac{1}{2}} = y_n + \frac{h}{4}(k_1+k_2) \quad (11)$$

Although of lower order than the main integration (error of order $O(h^3)$ compared to $O(h^5)$) this approximation permits the use of a quadratic interpolation algorithm for locating points of discontinuity within the time step.

The detection and interpolation for points of discontinuity was accomplished using discontinuity functions. These are functions calculated from the response of the system having a zero at the point of discontinuity. The detection procedures check during each time step if the values of the discontinuity functions at the beginning and end of the step straddle a zero or indicate the presence of a local extremum close to the zero. In most cases a quadratic interpolation scheme was used to locate the discontinuity, but this was backed up by a linear interpolation scheme and a bisection routine in the case of failure. Bisection was also used exclusively for finding zeros in the case of a local extremum of the discontinuity functions.

Subsequent to the computation of the response, a coordinate transformation was applied to give the motion in the x and y directions at the two bearing supports from the modal coordinates of Eqs. 3 to 10.

SPECTRAL ANALYSIS

The response of the system was subjected to standard analysis techniques to obtain its spectral content. In order to reduce the problem of leakage the Hanning window was applied to all time records prior to performing the F.F.T.. The spectra themselves were calculated on the basis of the true amplitude of the components, rather than the more commonly used r.m.s. value or the Power Spectral Density. This has the advantage that the response amplitudes may be compared directly with such physical quantities as the radial clearance. The computed spectra were subsequently plotted in the form of waterfall plots.

RESPONSE OF THE MODEL

A range of numerical tests were undertaken on the model using different levels of imbalance and different discontinuous stiffnesses. It was found that three types of response could be elicited from the system, linear response, nonlinear synchronous response and aperiodic response. It is this last type of response which is of most interest to the present work.

Figure 4 shows a typical time history, spectrum and orbit for the model driven at 1800rpm with $K_A = 3.46$MN/m and $g_A = 0.216$mm. As can be seen the motion is aperiodic and the orbit does not repeat itself within the time plotted, but instead seems gradually to rotate in space. The effect of this on the spectrum is the presence of sidebands, which are not spaced at a subharmonic of the shaft speed. Figure 5 shows a waterfall plot for the same discontinuous stiffness but including the responses in the speed range 1000 to 3000rpm (16.7 to 50.0Hz.). This plot shows clearly the sidebands in the speed range 1600 to 2200rpm (26.7 to 36.7Hz.). The sidebands diverge from the shaft speed with increasing shaft speed and show a form somewhat similar to that depicted in Fig. 1. It should be noted that none of the sideband frequencies corresponds to either of the natural frequencies of the linear model which are 20.3Hz. (cylindrical mode) and 28.1Hz. (conical mode).

Further investigation suggests that although the response at the bearing supports is aperiodic in the x and y directions, it may be periodic in the radial direction.

DESCRIPTION OF THE EXPERIMENTAL RIG

To permit comparison with the theoretical results, a mechanical rig was designed and built. The rig (Figs. 6 and 7) consisted basically of an elastically supported rigid rotor, driven by a 1.5kW. variable speed motor with a thyristor control system along with various instrumentation systems. The rotor assembly itself comprised a 90mm diameter mild steel rotor running in two single row roller bearings (Ransome, Hoffman and Marples XLLRJ3"). The elastic support for this was provided by two sets of flexural elements (rods). These were attached at one end to the housings in which the bearing outer races were mounted and at the other to large support blocks. These blocks were in turn bolted to a cast iron bed plate. The stiffness of the rods was chosen to provide the two natural frequencies of the rig (cylindrical mode and conical mode) well within the operating range of the drive system (300-3000rpm).

Holes were drilled and tapped at either end of the rotor inboard of the bearings to allow imbalance weights to be added. To ensure that the level of unbalance being applied was accurately known, the rotor was dynamically balanced prior to incorporation in the rig.

The discontinuous stiffness was effected by placing an elastically supported snubber ring around the bearing housing. A radial gap between the two parts permitted some motion of the rotor assembly prior to contacting the ring. The stiffness of the snubber ring support assembly was substantially greater than that of the bearing housing supports, thus giving a hardening discontinuous stiffness characteristic. The elastic support for the snubber ring was provided by rods in a manner similar to that used for the bearing housing supports. Snubber ring assemblies could be fitted to either or both bearing housings and two stiffnesses of rods and two rings with different radial clearances were manufactured to allow a variety of discontinuity conditions to be modelled. It was assumed that the natural frequencies of the snubber assemblies would be sufficiently far removed from the excitation of the rotor system to permit them to be modelled by massless springs.

The response of the rig was measured at four points simultaneously by two pairs of non-contacting eddy current probes located at the bearing housings. A pair of l.e.d. opto-switches and a disk with holes drilled in it were used to monitor the speed of the rotor and the position of the unbalance mass. Due to the large quantity of data which could be generated from a single run of the rig, viz 4 vibration responses, a speed signal and a phase marker, it was found to be more convenient to record the data and process it subsequently than to use "on-line" monitoring. A Racal Store 7DS recorder was used for this purpose. The recorded vibration response data were analysed subsequently for spectral content using a Hewlett Packard 5423A Structural Dynamics Analyser controlled by a Hewlett-Packard 9826A micro-computer and were plotted in the form of waterfall plots.

RESPONSE OF THE RIG

The rig was run with the same discontinuous stiffness characteristics and similar imbalance levels to the model. An example of the resulting responses is shown in Fig. 8. This compares very favourably with the predicted spectral response characteristics of Fig. 5 with sidebands being present over a similar range and with comparable spacing and amplitude. The earlier drop out of the nonlinear synchronous resonance is the only major discrepancy and is probably due to the effect of the inertia of the snubber ring which was neglected in the theoretical model. In all other aspects the responses are the same and this tends to validate the phenomenon as being genuine.

CONCLUSIONS

Good agreement is found between computation from the mathematical model of the rotor system and the responses obtained from the experimental rig. Both model and rig simulate the type of vibration response, depicted in Fig. 1, which occasionally occurs in full scale aeroengine roughness testing. The results of the tests indicate that in rotor systems having symmetric motion dependent stiffness discontinuities, aperiodic responses characterised by sidebands may occur in certain speed ranges. These sidebands may give a divergent or fan like structure of frequencies if plotted in the form of a waterfall plot and this might be used as a diagnostic test for clearance effects of this type. There is some evidence that for certain parameters and over a narrow speed range continuous bifurcation of the fan structure may occur for increasing speed as a transition into chaotic motion of the system.

REFERENCES

1. Ehrich, F. F. Subharmonic vibration of rotors in bearing clearance. A.S.M.E. Paper 66-MD-1, presented at the Design Engineering Conference and Show, 9-12 May, 1966, pp. 1-4.
2. Ehrich, F. F. and O'Connor, J. J. Stator whirl with rotors in bearing clearance. *Trans. A.S.M.E., J. Eng. Ind. 89* (1967), 381-389.
3. Bently, D. E. Forced subrotative speed dynamic action of rotating machinery. A.S.M.E. Paper 74-PET-16, presented at the Petroleum Mechanical Engineering Conference, Dallas, Texas, 15-18 Sept., 1974, pp. 1-8.
4. Childs, D. W. Fractional frequency rotor motion due to nonsymmetric clearance effects. *Trans. A.S.M.E., J. Eng. Power 104* (1982), 533-541.
5. Muszynska, A. Partial lateral rotor to stator rubs. Proceedings of the I. Mech. E. 3rd. Intl. Conf. on Vibrations in Rotating Machinery, Univ. of York, 11-13 Sept., 1984, pp. 327-335.
6. Day, W. B. Nonlinear rotordynamics analysis. *N.A.S.A. Report CR 171425* (1985), .
7. Borthwick, W. K. D. The numerical solution of discontinuous structural systems. Proceedings of the Second Int. Conf. on Recent Advances in Structural Dynamics, University of Southhampton, 9-13 April, 1984, pp. 307-316.

Резюме

Статья описывает результаты научного исследования в котором жесткий вал, поддерживаемый с обеих концов эластически опертыми роликовыми подшипниками, управляется внешним силовым источником и вибрирует под влиянием насаженного дебаланса. Эластичные несущие опоры имеют высоко нелинейную характеристику. Они состоят из линейной жесткости до сдвига равному радиальному зазору, при котором вторичная жесткость входит в действие. Эта вторичная жесткость намного больше первой на регулируемый фактор около десяти. Система симулирует условия при которых подшипник в системе вала пересекает свой зазор и соприкасается с поддерживающей конструкцией.

Установлены формулы движения системы, допуская что пружины опор подшипника не имеют массы. Формулы не линейные когда зазор пересечен и имеют четыре степени свободы. Включена гироскопическая пара между угловыми видами. Линейные нормальные формы не вращательного движения с маленькой амплитудой использованы как координаты.

Временная характеристика системы вычислена посредством численного интегрирования включая алгоритм, который, для сохранения порядка процесса, повторяется, чтобы точно найдти по времени точки соприкосновения со вторичной пружиной. Интегрированная временная характеристика подвергается анализу Фурье для того чтобы установить величину присуствующих частотных компонентов. В некоторых областях скорости вала спектр проявляет многократную структуру "вентиляторного" типа, которая, по-видимому, является типичной для этого вида системы.

Также представлены экспериментальные результаты, полученные от описанного устройства. Результаты, полученные от устройства, находятся в хорошем согласии с теоретическим прогнозом. Работа имеет отношение к контролированию шероховатости при испытаниях газовых турбин в натуральную величину, при которых иногда наблюдаются подобные спектральные характеристики.

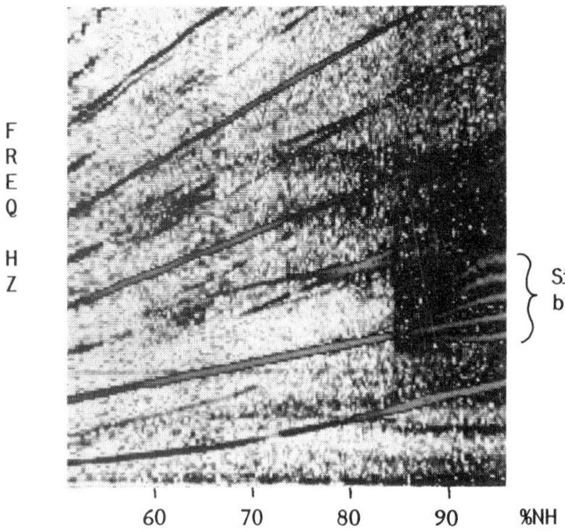

Fig. 1 Aeroengine vibration response.

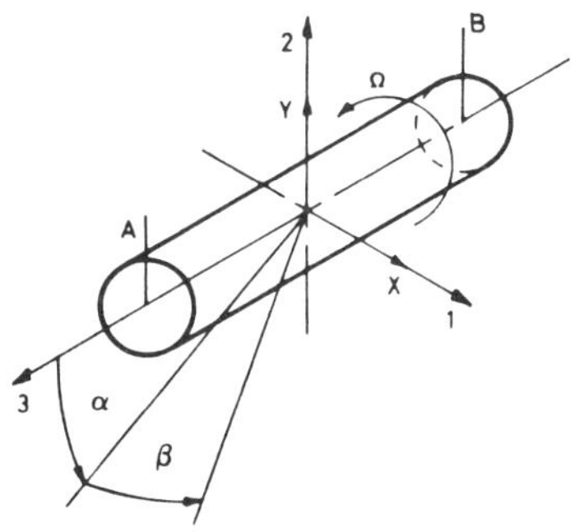

Fig. 2 Coordinate system of the model.

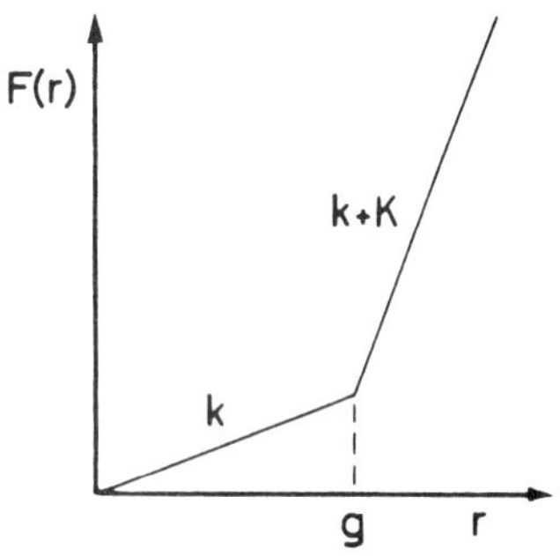

Fig. 3 Force/deflection characteristic.

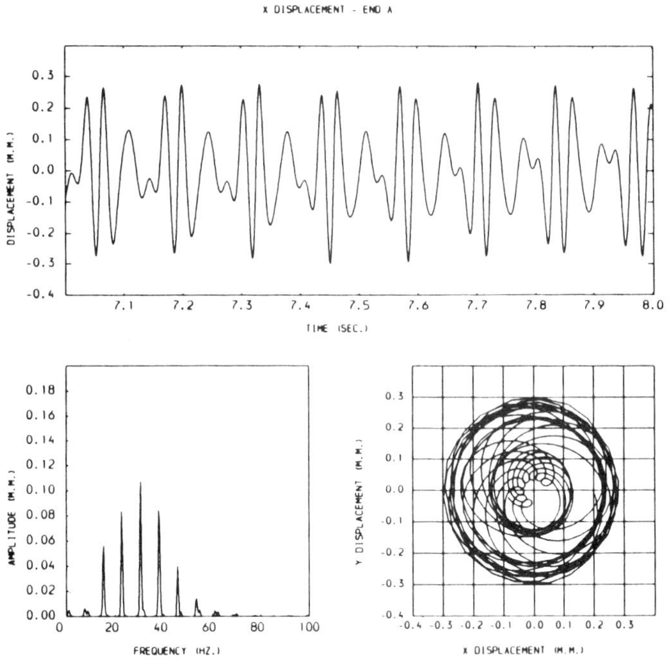

Fig. 4 Time history, orbit and spectrum obtained from the model.

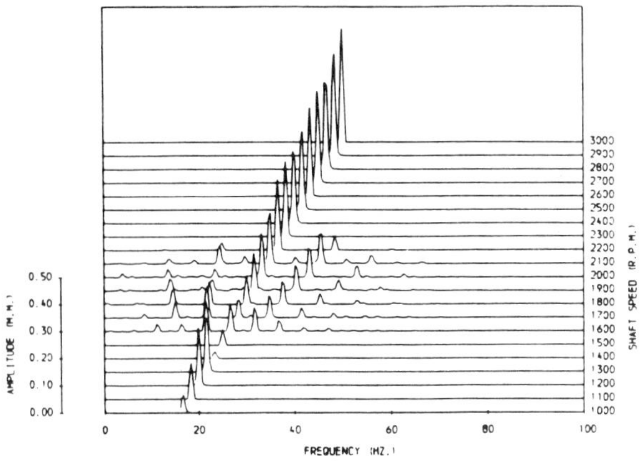

Fig. 5 Spectral response of the model.

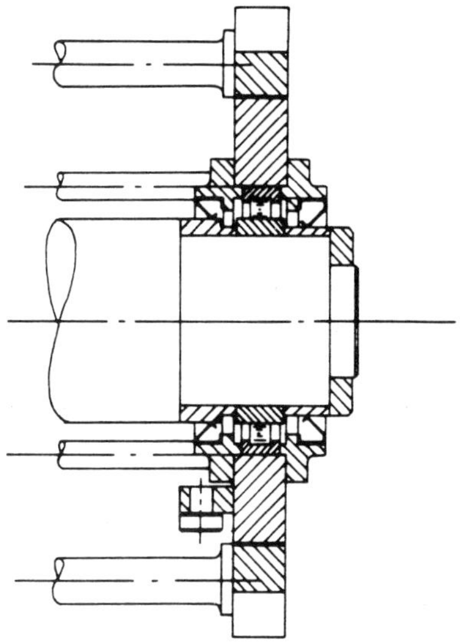

Fig. 6 Arrangement of the bearing assembly.

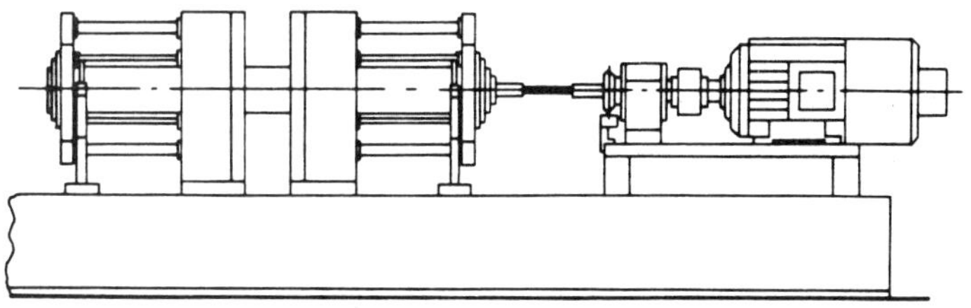

Fig. 7 General arrangement of the rig.

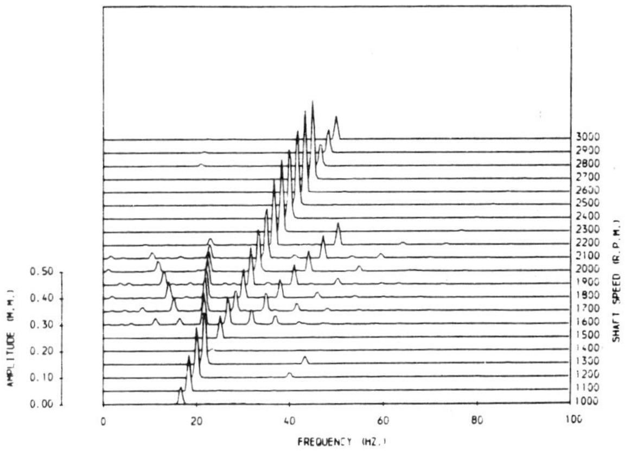

Fig. 8 Spectral response of the rig.

Resolution of a Paradox Concerning the Instability of Unsymmetric Rotors

S. H. CRANDALL

Massachusetts Institute of Technology, Cambridge, MA 02139, USA

Abstract. It has long been known that a rotor with unsymmetric stiffness and inertia properties can be unstable for steady rotation at rate Ω over certain speed ranges. Unsymmetric stiffness was analyzed by Stodola (1924) and unsymmetric inertia was discussed by Smith (1933). Combined stiffness and inertia asymmetries were studied by Crandall and Brosens (1961). A convenient measure of the degree of asymmetry in the cases of pure elastic asymmetry or pure inertia asymmetry is the divergence between the natural frequencies ω_1 and ω_2 of the nonrotating rotor ($\Omega = 0$) associated with the principal diametral axes of the elastic or inertia asymmetry. For either pure elastic or pure inertia asymmetry the greater the divergence between ω_1 and ω_2 the stronger the instability where the instability strength is measured by either the extent of the unstable speed range $\Omega_2 - \Omega_1$ or by the rate of exponential growth at the midpoint of the unstable range. In the case of combined elastic and inertia asymmetries there is an additional parameter ϕ which is the angle between the principal axis of maximum inertia and the principal axis of maximum stiffness. As ϕ increases from 0 to $\pi/2$ the divergence between ω_1 and ω_2 increases, but according to Crandall and Brosens (1961) the strength of the instability decreases. This anomalous result has been independently verified, analytically and experimentally by Yamamoto, Ota and Kono (1968). The present paper supplies an elementary physical explanation of the paradox.

Keywords. Rotor dynamics; dynamic stability; gyroscopic coupling; asymmetric inertia; asymmetric stiffness.

INTRODUCTION

A transparent physical analysis is developed in stages. First elastic asymmetry in a simple planar model is considered and the relationships between the natural frequencies ω_i of the nonrotating system and the location and strength of the unstable rotation speed range are developed (Crandall, 1983). Then using an analogy between the centrifugal and inertia forces acting on a particle in a rotating plane and the inertia torques acting on a rotating disk the corresponding relationships between nonrotating natural frequencies and the instability speed range are developed for a rotor with unequal diametral moments of inertia. It is observed that elastic asymmetry and inertial asymmetry have nearly parallel effects except for an important inversion in the unequal nonrotating natural frequencies induced by the introduction of the two types of asymmetry. Finally when elastic and inertial asymmetries are combined in the same rotor the reason why the spread between the nonrotating natural frequencies $\omega_2 - \omega_1$ varies in the opposite sense to the extent of the unstable speed range $\Omega_2 - \Omega_1$ as the orientation between principal elastic axes and principal inertial axes is changed is clarified.

PLANAR MODEL WITH ELASTIC ASYMMETRY

The fundamental whirling mode of the classical Föppl-Jeffcott model may be described by the planar model shown in Fig. 1. The ideal bearings at the ends of the shaft are imagined to be moved into the same central plane and the central disk is modeled by a mass particle supported by elastic springs representing the bending compliance of the shaft. When there is elastic asymmetry the principal stiffnesses k_1 and k_2 are unequal. Introducing ϵ_s as a dimensionless measure of the stiffness inequality ($0 \leq \epsilon_s < 1$) we write

$$k_1 = k(1 + \epsilon_s)$$
$$k_2 = k(1 - \epsilon_s) \tag{1}$$

where $k = m\omega_0^2$ is the average stiffness. When there is no rotation ($\Omega = 0$) the system has two orthogonal modes of vibration along the principal elastic axes with natural frequencies ω_i given by

$$\omega_1^2 = \omega_0^2(1 - \epsilon_s) \quad \text{and} \quad \omega_2^2 = \omega_0^2(1 + \epsilon_s) \tag{2}$$

When rotating steadily the system has somewhat complicated natural whirling motions (Crandall, 1962) but there are two simple synchronous whirls. When $\Omega = \omega_1$ the mass particle is in neutral equilibrium under the action of centrifugal force and the restoring spring force, anywhere along the 2-axis. When rotating with $\Omega = \omega_2$ steady synchronous whirl is possible with the mass displaced anywhere along the 1-axis.

To examine the behavior of the system for a rotation rate Ω between ω_1 and ω_2 consider the case $\Omega = \omega_0$. Fig. 2 shows the forces acting on the mass under the assumption that a synchronous whirl with fixed angle ψ is possible. Note that for $\Omega = \omega_1$ synchronous whirl is possible with $\psi = 0$ and that for $\Omega = \omega_2$ synchronous whirl is possible with $\psi = \pi/2$. For Ω between these speeds it is natural to expect a ψ-value between the corresponding limits. When $\Omega = \omega_0$ the restoring force kr due to the average stiffness balances the centrifugal force $mr\Omega^2$ for any runout magnitude r. The force contributions from the stiffness inequality in (1) have the resultant $\epsilon_s kr$ acting perpendicular to the runout. The effect of a tangential force whose magnitude is proportional to the runout acting on a whirling particle which is otherwise in equilibrium under the influence of a central linear restoring force and centrifugal force is to drive the orbit into a logarithmic spiral (Crandall, 1983). The runout increases (decreases) exponentially with time when the tangential force is in the (opposes the) direction of the rotation. It will shortly be shown that for small ϵ_s the angle ψ in Fig. 2 when $\Omega = \omega_0$ is nearly $\pi/4$ and the exponential growth rate is $\epsilon_s\omega_0/2$ correct to second order in ϵ_s. The energy input to the orbit from the tangential force is obtained directly from the external agent which maintains the rotation rate Ω.

The force analysis in Fig. 2 can be repeated in the other quadrants of the rotating coordinate system. In the first and third quadrants the unbalanced tangential force due to the stiffness inequality drives the whirl while in the second and fourth quadrants the unbalanced tangential force opposes the whirl. The general solution for the orbit of m when $\omega_1 < \Omega < \omega_2$ for arbitrary initial conditions is a linear combination of an outgoing spiral and an ingoing spiral. When Ω is outside this range orbits with fixed ψ are no longer possible. The natural whirls are orbits with primary frequencies close to ω_0 as seen by a nonrotating observer. Since the rotation is now at a rate Ω which differs considerably from ω_0 the orbit moves more or less regularly around through the four quadrants of the rotating frame. It is alternately urged forward and then backward by the resultant of the asymmetric spring forces but with no net input of energy in the steady state natural whirls. The system of Fig. 1 thus has stable free whirling motions for all rotation rates Ω except those in the range $\omega_1 < \Omega < \omega_2$. The extent of the unstable range is $\omega_2 - \omega_1 \approx \epsilon_s\omega_0$. The exponential growth rate of the unstable whirl varies smoothly from zero at the end points $\Omega = \omega_1$ and $\Omega = \omega_2$ to a maximum of approximately $\epsilon_s\omega_0/2$ close to the speed $\Omega = \omega_0$. A final comment on the system of Fig. 1 is that the mode with *lower* nonrotating natural frequency ω_1 is along the axis of the *softer* spring and that the neutral equilibrium whirl at the *lower* speed end of the unstable range has $\psi = 0$ so that the runout is along the same *softer* spring axis.

PARTICLE DYNAMICS IN A ROTATING PLANE

To establish a basis for an analogy with the nutational motion of a rotating disk we next consider the equations of motion for a mass particle in a rotating plane. The equations may also be used to provide rigorous confirmation of the results outlined in the preceding section. In Fig. 3 the mass particle m has coordinates ξ, η in a coordinate system which rotates at the rate Ω in the inertial x, y plane. When a force with components f_ξ, f_η acts on the particle the dynamic equations are

$$\frac{1}{m}\begin{Bmatrix} f_\xi \\ f_\eta \end{Bmatrix} = \begin{Bmatrix} \ddot{\xi} \\ \ddot{\eta} \end{Bmatrix} + 2\Omega\begin{bmatrix} 0 & -1 \\ 1 & 0 \end{bmatrix}\begin{Bmatrix} \dot{\xi} \\ \dot{\eta} \end{Bmatrix} - \Omega^2\begin{Bmatrix} \xi \\ \eta \end{Bmatrix} \tag{3}$$

where the terms on the right are the relative acceleration, the Coriolis acceleration, and the centripetal acceleration. In the case where the forces f_ξ, f_η are the restoring forces of the springs in Fig. 1 with stiffnesses given by (1) the equations may be put in the form

$$\begin{Bmatrix} \ddot{\xi} \\ \ddot{\eta} \end{Bmatrix} + 2\Omega\begin{bmatrix} 0 & -1 \\ 1 & 0 \end{bmatrix}\begin{Bmatrix} \dot{\xi} \\ \dot{\eta} \end{Bmatrix} = (\Omega^2 - \omega_0^2)\begin{Bmatrix} \xi \\ \eta \end{Bmatrix} - \epsilon_s\omega_0^2\begin{bmatrix} 1 & 0 \\ 0 & -1 \end{bmatrix}\begin{Bmatrix} \xi \\ \eta \end{Bmatrix} \tag{4}$$

The complete solution of this system is a superposition of four modes of the form

$$\begin{Bmatrix} \xi \\ \eta \end{Bmatrix} = \begin{Bmatrix} a \\ b \end{Bmatrix} e^{st} \tag{5}$$

where the eigenvalues for s are the roots of

$$s^2 = -(\omega_0^2 + \Omega^2) \pm \omega_0(4\Omega^2 + \epsilon^2\omega_0^2)^{1/2} \tag{6}$$

and the corresponding amplitude ratios are

$$\frac{b}{a} = \frac{s^2 - \Omega^2 + \omega_0^2(1+\epsilon_s)}{2\Omega s} \tag{7}$$

For example, when $\Omega = \omega_0$, and ϵ_s is small, the roots for s are close to $\pm 2i\omega_0$ (with b/a close to $\pm i$) and $\pm\epsilon_s\omega_0/2$ [with b/a close to $\pm(1 + \epsilon_s/4)$]. The former pair describe the backward whirl which appears to have a frequency of $2\omega_0$ when viewed from the coordinate frame rotating forward at rate ω_0. The latter pair describe the exponential growth (decay) of runout with fixed angle ψ in the unstable speed range.

An interesting aproximate procedure for obtaining the latter results can be based on the observation that when ϵ_s is small the motion in the unstable speed range will be very nearly a synchronous whirl with slow movement with respect to the rotating frame. For such slow movements it is appropriate to neglect the relative accelerations on the left of (4) in comparison with the Coriolis acceleration. When this is done the approximate solution, useful in the speed range $\omega_1 < \Omega < \omega_2$, is a superposition of two modes of the form (5) where the eigenvalues for s are now

$$s = \pm\frac{\left[(\Omega^2 - \omega_1^2)(\omega_2^2 - \Omega^2)\right]^{1/2}}{2\Omega} \tag{8}$$

and the corresponding amplitude ratios are

$$\frac{b}{a} = \pm\left(\frac{\omega_2^2 - \Omega^2}{\Omega^2 - \omega_1^2}\right)^{1/2} \tag{9}$$

When $\Omega = \omega_0$ which to first order in ϵ_s is the central speed of the unstable speed range $\omega_1 < \Omega < \omega_2$, the growth rates (8) reduce exactly to $s = \pm\epsilon_s\omega_0/2$ which to first order in ϵ_s are the maximum and minimum growth rates and the amplitude ratios (9) become $b/a = \pm 1$ which implies that $\psi = \pm\pi/4$ in Fig. 2.

SMALL NUTATIONS OF A ROTATING DISK

We consider next a rigid disk performing small nutations while rotating about its fixed center of mass at rate ω. A possible example is the second mode of the Föppl-Jeffcoat model shown in Fig. 4. Let I_1, I_2, and I_3 be the principal moments of inertia with the 3-axis along the axis of rotation and 1- and 2-axes along perpendicular diameters. As a dimensionless measure of the inertial asymmetry we introduce the parameter ϵ_i $(0 \le \epsilon_i < 1)$

$$\begin{aligned} I_1 &= I(1+\epsilon_i) \\ I_2 &= I(1-\epsilon_i) \end{aligned} \tag{10}$$

where I is the average diametral moment of inertia. The magnitude of the axial moment of inertia I_3 can approach the value $2I$ for a very thin disk or it can be very much smaller than I if the rotor shape is a long spindle of small diameter. Introducing the axial inertia parameter $A\,(0 < A < 1)$ we write

$$I_3 = A(I_1 + I_2) = 2AI \tag{11}$$

If the torque acting on the rotor has components τ_1 and τ_2 along the principal diametral axes and an axial rotation rate Ω is maintained, Eulers angular momentum equations (see, for example, Crandall and colleagues, 1968) are

$$\begin{aligned} \tau_1 &= I_1\dot{\omega}_1 - (I_2 - I_3)\omega_2\Omega \\ \tau_2 &= I_2\dot{\omega}_2 + (I_1 - I_3)\omega_1\Omega \end{aligned} \tag{12}$$

where ω_1 and ω_2 are the components of the absolute angular velocity of the rotor along the diametral axes. For small nutation (tipping of the diametral plane) the orientation of the diametral plane can be described by small tip angles θ_1 and θ_2 about the principal diameters as shown in Fig. 5. Because these diameters rotate with the rotor at rate Ω there are two components to the total time derivatives $\dot{\theta}_1$ and $\dot{\theta}_2$. One component is due to the change in orientation of the tipped plane and the other is due to the fact that the rotation Ω transports the diameters to new positions in the already tipped plane. The first component is due to the angular velocity of the diametral plane, the second is due to convection. The total time derivatives of the tip angles can thus be written from inspection of Fig. 5 as

$$\begin{aligned} \dot{\theta}_1 &= \omega_1 + \Omega\theta_2 \\ \dot{\theta}_2 &= \omega_2 - \Omega\theta_1 \end{aligned} \tag{13}$$

from which we obtain the absolute angular velocity components and their derivatives as functions of the tip angles and their derivatives

$$\begin{aligned} \omega_1 &= \dot{\theta}_1 - \Omega\theta_2 \qquad & \dot{\omega}_1 &= \ddot{\theta}_1 - \Omega\dot{\theta}_2 \\ \omega_2 &= \dot{\theta}_2 + \Omega\theta_1 \qquad & \dot{\omega}_2 &= \ddot{\theta}_2 + \Omega\dot{\theta}_1 \end{aligned} \tag{14}$$

When these are substituted in Euler's equations (12) along with the notations of (10) and (11) the result can be written in the form

$$\frac{1}{I}\begin{Bmatrix} \tau_1 \\ \tau_2 \end{Bmatrix} = \begin{bmatrix} 1+\epsilon_i & 0 \\ 0 & 1-\epsilon_i \end{bmatrix}\begin{Bmatrix} \ddot{\theta}_1 \\ \ddot{\theta}_2 \end{Bmatrix} + 2(1-A)\Omega\begin{bmatrix} 0 & -1 \\ 1 & 0 \end{bmatrix}\begin{Bmatrix} \dot{\theta}_1 \\ \dot{\theta}_2 \end{Bmatrix} - \Omega^2\begin{bmatrix} 1-2A-\epsilon_i & 0 \\ 0 & 1-2A+\epsilon_i \end{bmatrix}\begin{Bmatrix} \theta_1 \\ \theta_2 \end{Bmatrix} \tag{15}$$

It may be noted that there is a rough similarity between these nutation equations and the dynamic equations for a particle in a rotating plane (3). The analogy becomes perfect in the limiting case when $\epsilon_i \to 0$ and $A \to 0$.

Equation (15) can be applied to several classical stability problems for a rotating rigid body. For example, when $\tau_1 = \tau_2 = 0$ the rotor is torque free and standard stability analysis of (15) shows that steady rotation at rate Ω about the axis with moment of inertia I_3 is stable except when I_3 is the intermediate moment of inertia; i.e., when $I_2 < I_3 < I_1$ $(1-\epsilon_i < 2A < 1+\epsilon_i)$. In the case where the torques in (15) are supplied by axially symmetric torsional springs so that $\tau_1 = -k\theta_1$ and $\tau_2 = -k\theta_2$, standard stability analysis (Crandall and Brosens, 1961) shows that steady rotation about the axis with moment of inertia I_3 is stable for all speeds $0 < \Omega < \infty$ if I_3 is the largest moment of inertia $(2A > 1+\epsilon_i)$. If I_3 is the intermediate moment of inertia then steady rotation about the I_3 axis is stable up to a critical speed Ω_1 and then unstable for all speeds Ω greater than Ω_1. Finally if I_3 is the smallest moment of inertia $(2A < 1-\epsilon_i)$ steady rotation about the I_3 axis is stable up to a first critical speed Ω_1, unstable for a range of speeds extending from Ω_1 up to a second critical speed Ω_2, and then stable for all higher speeds. Since only in the last case is the stability behavior for inertia asymmetry parallel to that for stiffness asymmetry we shall limit the further discussion of inertia asymmetry to the case where I_3 is the smallest moment of inertia $(2A < 1-\epsilon_i)$.

STABILITY OF ROTOR WITH ASYMMETRIC INERTIA

We now consider the preceding case in greater detail. The system has the general appearance of Fig. 4. The straightening tendency of the bent shaft is modelled by an axially symmetric spring with torsional spring constant $k = I\omega_0^2$. Inserting $\tau_i = -k\theta_i$ in (15) we rewrite it in a form parallel to (4)

$$\begin{bmatrix} 1+\epsilon_i & 0 \\ 0 & 1-\epsilon_i \end{bmatrix} \begin{Bmatrix} \ddot{\theta}_1 \\ \ddot{\theta}_2 \end{Bmatrix} + 2(1-A)\Omega \begin{bmatrix} 0 & -1 \\ 1 & 0 \end{bmatrix} \begin{Bmatrix} \dot{\theta}_1 \\ \dot{\theta}_2 \end{Bmatrix} = \left[\Omega^2(1-2A) - \omega_0^2\right] \begin{Bmatrix} \theta_1 \\ \theta_2 \end{Bmatrix} - \epsilon_i \Omega^2 \begin{bmatrix} 1 & 0 \\ 0 & -1 \end{bmatrix} \begin{Bmatrix} \theta_1 \\ \theta_2 \end{Bmatrix} \tag{16}$$

Note that the inertia inequality term on the right of (16) plays a role similar to the stiffness inequality term on the right of (4).

When the system is not rotating ($\Omega = 0$) the natural frequencies of free vibration are given, directly from (16), by

$$\omega_1^2 = \frac{\omega_0^2}{1+\epsilon_i} \quad \text{and} \quad \omega_2^2 = \frac{\omega_0^2}{1-\epsilon_i} \tag{17}$$

associated with θ_1-vibration and θ_2-vibration respectively. Note that in contrast with the asymmetric stiffness case, the *smaller* natural frequency ω_1 is associated with the axis of *greater* moment of inertia and the *larger* natural frequency ω_2 is associated with the axis of *lesser* moment of inertia.

At the ends of the unstable speed range, $\Omega_1 < \Omega < \Omega_2$, there are neutrally stable synchronous whirls with constant θ_1 or θ_2. With the left hand side of (16) set equal to zero we find the critical speeds Ω_1 and Ω_2 given by

$$\Omega_1^2 = \frac{\omega_0^2}{1-2A+\epsilon_i} \quad \text{and} \quad \Omega_2^2 = \frac{\omega_0^2}{1-2A-\epsilon_i} \tag{18}$$

associated with constant tip angles θ_2 and θ_1 respectively. Note that for the limiting case $A \to 0$ (a needle-like rotor) the critical speeds (18) are the same as the nonrotating natural frequencies (17) as was the case for stiffness asymmetry but for rotors with $A > 0$ the critical speeds (18) are higher and farther apart than the nonrotating natural frequencies (17). For small inertia inequality ϵ_i both the separation $\omega_2 - \omega_1$ between the nonrotating natural frequencies and the extent $\Omega_2 - \Omega_1$ of the unstable speed range grow in direct proportion to ϵ_i.

The growth rate of the unstable solution may be obtained from an exact solution of (16). A useful approximate estimate can be obtained by assuming that small ϵ_i implies small growth rate s which in turn implies that the accelerations $\ddot{\theta}_i$ will be of higher order than the velocities $\dot{\theta}_i$. If the acceleration terms are neglected in (16) it has solutions of the form (5) with growth rate

$$s = \pm \frac{\left\{\left[\Omega^2(1-2A+\epsilon_i) - \omega_0^2\right]\left[\omega_0^2 - \Omega^2(1-2A-\epsilon_i)\right]\right\}^{1/2}}{2\Omega(1-A)} \tag{19}$$

and corresponding amplitude ratios

$$\frac{b}{a} = \pm \left[\frac{\omega_0^2 - \Omega^2(1-2A-\epsilon_i)}{\Omega^2(1-2A+\epsilon_i) - \omega_0^2}\right]^{1/2} \tag{20}$$

These results are correct to first order in the inertia inequality ϵ_i for Ω in the unstable speed range $\Omega_1 < \Omega < \Omega_2$. They are somewhat more complicated than the corresponding results, (8) and (9), for stiffness inequality because of the gyroscopic coupling controlled by the axial inertia parameter A. In the limiting case $A \to 0$ the similarity between (19), (20) and (8), (9) is much stronger. The maximum and minimum growth rates (19) are very nearly

$$s = \pm \frac{\epsilon_i \Omega_0}{2(1-A)} \tag{21}$$

which are the values given by (19) when the speed Ω has the value Ω_0 given by

$$\Omega_0^2 = \frac{\omega_0^2}{1-2A} \tag{22}$$

To first order in ϵ_i the speed Ω_0 is the central frequency of the unstable speed range $\Omega_1 < \Omega < \Omega_2$. At speed Ω_0 the amplitude ratios (20) reduce to $b/a = \pm 1$. These results for inertia inequality are parallel to those obtained for stiffness inequality.

The parallelism could be further emphasized by constructing a diagram similar to Fig. 2 for the inertia inequality case. In place of the forces which act to produce relative velocity and acceleration with respect to the rotating frame we would show the torques which produce relative angular velocity and angular acceleration. In place of the average spring force kr would be the elastic torque $k\theta$ where $\theta^2 = \theta_1^2 + \theta_2^2$. In place of the centrifugal force $mr\Omega^2$ would be the average inertia torque $(I - I_3)\Omega^2\theta$ and in place of the tangential force $\epsilon_s kr$ due to the stiffness inequality there would be a tangential torque $\epsilon_i I\Omega^2\theta$ due to the inertial inequality. The basis for this analogy lies in the similarity of the right hand sides of (4) and (16). The two diagrams behave similarly in both cases as the speed Ω advances through the unstable speed range. At the onset of instability $\psi = 0$. There is neutral equilibrium with the mass displaced along the softer elastic axis in the stiffness inequality case and with the rotor tipped about the diametral axis with smallest moment of inertia in the inertia inequality case. In the former case the lower critical speed Ω_1 is given by $\Omega_1^2 = k_2/m = k(1-\epsilon_s)/m$ while in the latter case it is given by $\Omega_1^2 = k/(I_1 - I_3) = k/I(1-2A+\epsilon_i)$. As the speed increases the angle ψ increases until the situation of maximum growth rate is achieved with $\psi \approx \pi/4$ as pictured in Fig. 2. With further increase in speed the angle ψ continues to increase until at the end of the unstable range $\psi = \pi/2$ where there is neutral equilibrium with the mass displaced along the stiffer elastic axis in the stiffness inequality case and with the rotor tipped about the diametral axis with largest moment of inertia in the inertia inequality case. In the former case the higher critical speed Ω_2 is given by $\Omega_2^2 = k_1/m = k(1+\epsilon_s)m$ while in the latter case it is given by $\Omega_2^2 = k/(I_2 - I_3) = k/I(1-2A-\epsilon_i)$. A rotor with inertia inequality thus behaves in a parallel fashion to a rotor with stiffness equality as the two rotors pass through their respective unstable speed ranges. The major difference between the two cases is due to the inversion of the nonrotating natural modes with lower and higher frequencies. In the case of stiffness inequality the axis which has the runout at the onset of instability is also the axis of vibration for the nonrotating mode with *lower* natural frequency while in the case of inertia inequality the axis about which the rotor tips at the onset of instability is the axis of vibration for the nonrotating mode with *higher* natural frequency. This inversion is responsible for the paradoxical behavior when stiffness inequality and inertia inequality are combined in the same rotor. The underlying reason for the inversion is that in the case of stiffness inequality it is the *same* inequality term on the right hand side of (4) which governs both the divergence in the nonrotating natural frequencies and the extent of the unstable speed range while in the case of inertia inequality these two results depend on *different* inequality terms. In (16) the divergence in the nonrotating natural frequencies is governed by the inertia inequality in the first term on the left while the extent of the unstable speed range is governed by the last term on the right. The nonrotating natural frequency ω_2 about the 2-axis depends on I_2 through $\omega_2^2 = k/I_2 = k/I(1-\epsilon_i)$ while the critical speed Ω_1 which involves tip about the 2-axis depends on I_1 through $\Omega_1^2 = k/(I_1 - I_3) = k/I(1-2A+\epsilon_i)$.

ROTOR WITH ASYMMETRIC STIFFNESS AND ASYMMETRIC INERTIA

Consider the system of Fig. 4 where now both the rigid disk and the elastic shaft are asymmetric. Let the orientation of the elastic principal axes a, b with respect to the inertial principal axes 1,2 be as indicated in Fig. 6 where ϕ is the angle between the axis with diametrical moment of inertia $I_1 = I(1+\epsilon_i)$ and the axis with torsional stiffness $k_a = k(1+\epsilon_s)$. The angle between the axis with diametral moment of inertia $I_2 = I(1-\epsilon_i)$ and the axis with torsional stiffness $k_b = k(1-\epsilon_s)$ is also ϕ. Consider the case where $\phi = 0$. In this case the nonrotating natural frequencies for tipping about the 1 and 2 axes are given by

$$\omega_1^2 = \frac{k(1+\epsilon_s)}{I(1+\epsilon_i)} \quad \text{and} \quad \omega_2^2 = \frac{k(1-\epsilon_s)}{I(1-\epsilon_i)} \tag{23}$$

and the critical speeds for neutral equilibrium with tip angles about the 1 and 2 axes are given by

$$\Omega_2^2 = \frac{k(1+\epsilon_s)}{I(1-2A-\epsilon_i)} \quad \text{and} \quad \Omega_1^2 = \frac{k(1-\epsilon_s)}{I(1-2A+\epsilon_i)} \tag{24}$$

respectively. Note that for a given inertia inequality ϵ_i, introducing a stiffness inequality $\epsilon_s > 0$ tends to *increase* the extent of the unstable speed range between Ω_1 and Ω_2 and *decrease* the divergence between ω_2 and ω_1. In this configuration the two inequalities *reinforce* each other in the unstable speed range, increasing the growth rate of the unstable solution and widening the unstable speed range. However they *oppose* one another in their effects on the nonrotating natural frequencies (23). Conversely when $\phi = \pi/2$ in Fig. 6 the nonrotating natural frequencies for tipping about the 1 and 2 axes are given by

$$\omega_1^2 = \frac{k(1-\epsilon_s)}{I(1+\epsilon_i)} \quad \text{and} \quad \omega_2^2 = \frac{k(1+\epsilon_s)}{I(1-\epsilon_i)} \tag{24}$$

while the critical speeds for neutral equilibrium with tip angles about the 1 and 2 axes are given by

$$\Omega_1^2 = \frac{k(1-\epsilon_s)}{I(1-2A-\epsilon_i)} \quad \text{and} \quad \Omega_2^2 = \frac{k(1+\epsilon_s)}{I(1-2A+\epsilon_i)} \tag{25}$$

In this configuration the two inequalities *oppose* one another in the unstable speed range, decreasing the growth rate of the unstable solution and narrowing the unstable speed range while at the same time *reinforcing* the tendency of the nonrotating natural frequencies to separate.

For intermediate angles ϕ the stiffness matrix in the 1,2 frame is obtained by the tensor transformation

$$\begin{bmatrix} \cos\phi & -\sin\phi \\ \sin\phi & \cos\phi \end{bmatrix} \begin{bmatrix} k(1+\epsilon_s) & 0 \\ 0 & k(1-\epsilon_s) \end{bmatrix} \begin{bmatrix} \cos\phi & \sin\phi \\ -\sin\phi & \cos\phi \end{bmatrix} = k\begin{bmatrix} 1 & 0 \\ 0 & 1 \end{bmatrix} + \epsilon_s k \begin{bmatrix} \cos 2\phi & \sin 2\phi \\ \sin 2\phi & -\cos 2\phi \end{bmatrix} \tag{26}$$

When the torques obtained from this stiffness matrix are inserted in (15) the resulting equations can be put in the form of (16) with identical terms except for the final inequality term which now becomes

$$-\begin{bmatrix} \epsilon_s\omega_0^2\cos 2\phi + \epsilon_i\Omega^2 & \epsilon_s\omega_0^2\sin 2\phi \\ \epsilon_s\omega_0^2\sin 2\phi & -\epsilon_s\omega_0^2\cos 2\phi - \epsilon_i\Omega^2 \end{bmatrix} \begin{Bmatrix} \theta_1 \\ \theta_2 \end{Bmatrix} \tag{27}$$

Note that (27) reduces to the final term in (16) when $\epsilon_s = 0$ and to the final term in (4) when $\epsilon_i = 0$ and $\phi = 0$. The equations of motion for small nutations of a rotating asymmetric disk supported by an asymmetric elastic shaft as indicated in Fig. 4 are thus obtained by substituting (27) for the final term in (16). Here we use these equations to obtain graphs of the unstable speed range $\Omega_1 < \Omega < \Omega_2$ and the nonrotating natural frequencies ω_1 and ω_2 as functions of the orientation angle ϕ. Fig. 7 shows the results for the limiting case $A \to 0$ when $\epsilon_s = \epsilon_i = 0.1$. In this case with no gyroscopic coupling and equal inequalities the opposing effects of the inequalities with respect to the extent of the unstable range $\Omega_2 - \Omega_1$ and the divergence between the nonrotating natural frequencies ω_1 and ω_2 is seen in its purest form. At $\phi = 0$ we have $\omega_1 = \omega_2$ but the extent of the unstable speed range $\Omega_2 - \Omega_1$ is maximum, while at $\phi = \pi/2$ the instability has disappeared but the divergence between ω_1 and ω_2 is maximum. A more realistic case with $A = 1/4$ and $\epsilon_s = 0.1$, $\epsilon_i = 0.05$ is shown in Fig. 8. Here the effect of the axial inertia is to raise the speeds of the unstable speed range and to make the maximum extent of the unstable speed range larger than the maximum divergence between nonrotating natural frequencies. The general tendency of $\Omega_2 - \Omega_1$ to decrease when $\omega_2 - \omega_1$ increases (and vice-versa) still remains. These diagrams suggest that an asymmetric rotor with a finite range of unstable speeds could be made stable at all speeds by deliberately introducing additional asymmetry at the proper orientation.

REFERENCES

Crandall, S.H. and Brosens, P.J. (1961). On the stability of rotation of a rotor with rotationally unsymmetric inertia and stiffness properties, *J. Appl. Mech.* **28**, 567–570.

Crandall, S.H. (1962). Rotating and reciprocating machines. In W. Flugge (Ed.), *Handbook of Engineering Mechanics*, McGraw-Hill Book Co., New York. Chap. 58.

Crandall, S.H., Karnopp, D.C., Kurtz, Jr., E.F., and Pridmore-Brown, D.C. (1968). *Dynamics of Mechanical and Electromechanical Systems*, McGraw-Hill Book Co., New York. pp. 226–228.

Crandall, S.H. (1983). The physical nature of rotor instability mechanisms. In M.L. Adams (Ed.), *Rotor Dynamical Instability*, ASME Publication AMD, Vol. 55, pp. 1–18.

Smith, D.M. (1933). The motion of a rotor carried by a flexible shaft in flexible bearings. *Proc. Roy. Soc. London*, **A142**, 92–118.

Stodola, A. (1924). *Dampf-und Gasturbinen*, Springer Verlag.

Yamamoto, T., Ota, H., and Kono, K. (1968). On the unstable vibrations of a shaft with unsymmetrical stiffness carrying an unsymmetrical rotor. *J. Appl. Mech.* **35**, 313–321.

FIGURES

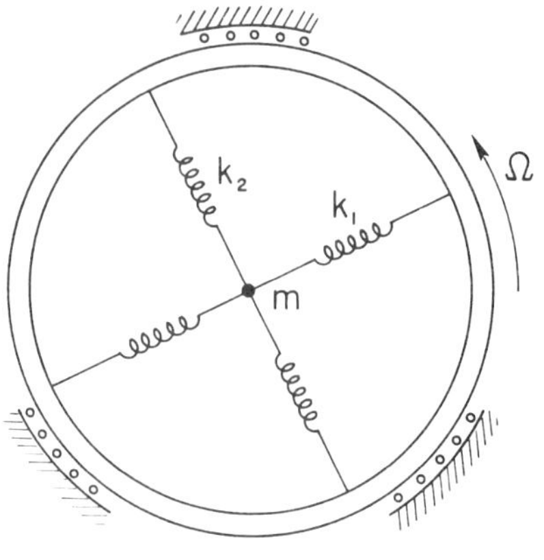

Fig. 1 Planar model of fundamental mode of Föppl-Jeffcott rotor with asymmetric stiffness.

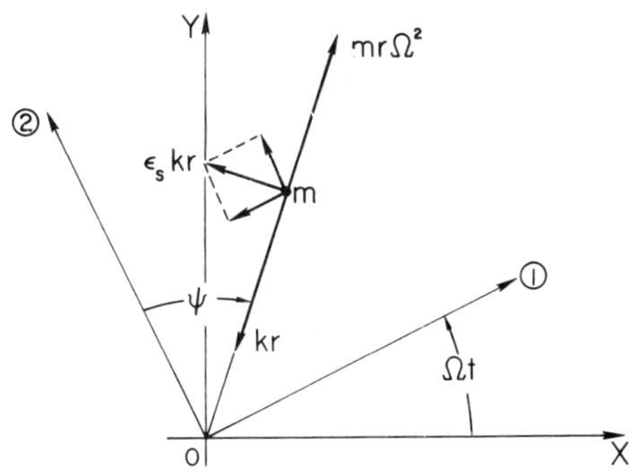

Fig. 2 Forces acting on mass whirling with fixed angle ψ and slowly varying runout r.

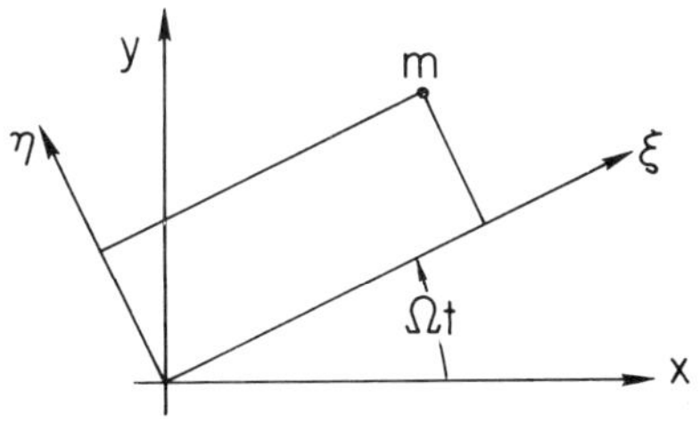

Fig. 3 Mass particle has displacements ξ and η with respect to rotating coordinate system.

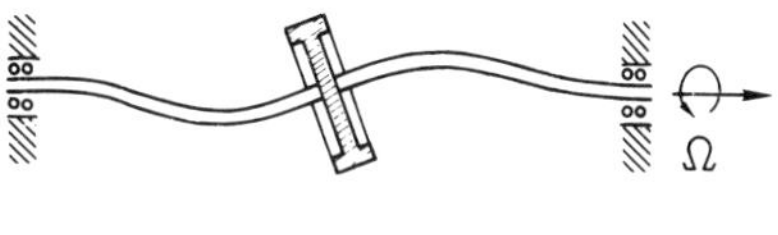

Fig. 4 Second mode of Föppl-Jeffcott rotor in which center of disk remains fixed while diametral plane of disk tips.

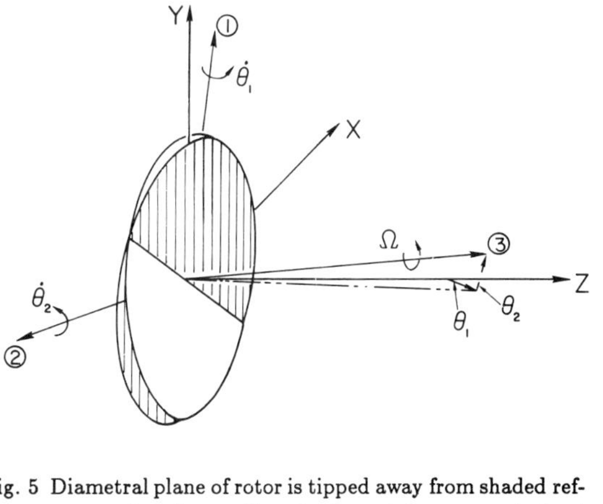

Fig. 5 Diametral plane of rotor is tipped away from shaded reference plane by small angles θ_1 and θ_2 measured about principal inertia axes fixed in rotor which rotates at rate Ω.

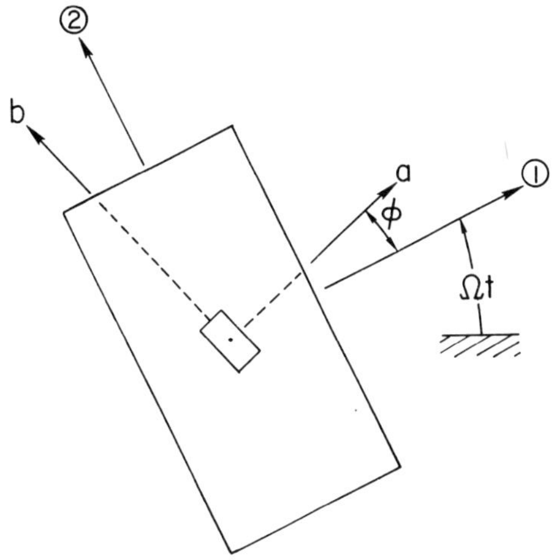

Fig. 6 Asymmetric disk with principal diametral inertia axes 1,2 mounted on asymmetric shaft with principal bending axes a, b.

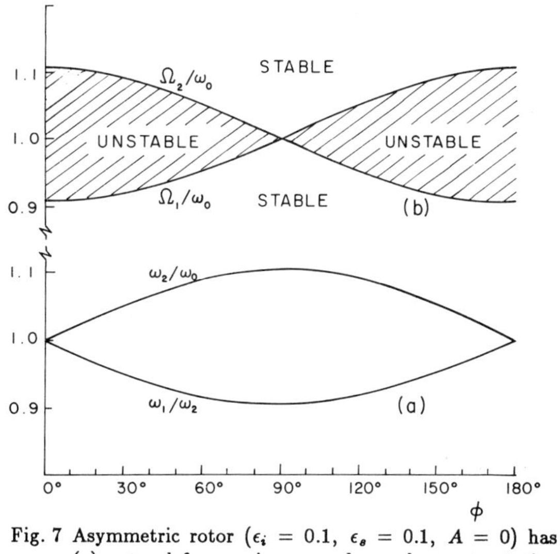

Fig. 7 Asymmetric rotor ($\epsilon_i = 0.1$, $\epsilon_s = 0.1$, $A = 0$) has (a) natural frequencies ω_1 and ω_2 when not rotating ($\Omega = 0$) and has (b) stability limits Ω_1 and Ω_2 when rotating.

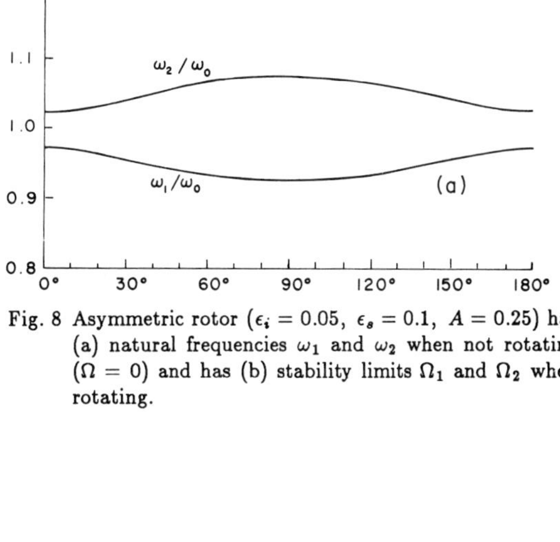

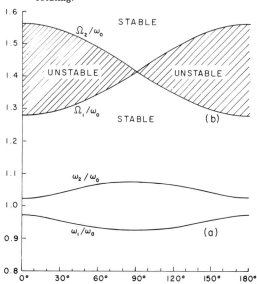

Fig. 8 Asymmetric rotor ($\epsilon_i = 0.05$, $\epsilon_s = 0.1$, $A = 0.25$) has (a) natural frequencies ω_1 and ω_2 when not rotating ($\Omega = 0$) and has (b) stability limits Ω_1 and Ω_2 when rotating.

SOMMAIRE

Une Résolution du Paradoxe de l'instabilité des Rotors asymetriques

On sait despuis longtemps qu'un rotor avec des propriétés de raideur et d'inertie asymetriques peut être instable pour une rotation constante à un taux de vitesse angulaire Ω pour une gamme de vitesses particuliere. Un rotor avec la raideur asymetrique a été analysé par Stolola (1924) et l'inertie asymetrique a été discutée par Smith (1933). La combinaison de la raideur et l'inertie asymetriques a été étudiée par Crandall et Brosens (1961). Une mesure convenable du degré d'asymetrie dans la cas de la raideur asymetrique ou dans les cas de l'inertie asymetrique est la divergence entre les frequences propres ω_1 et ω_2 du rotor sans rotation ($\Omega = 0$) associées avec les axes principaux de la raideur ou de l'inertie. Dans chaque cas separé la plus grande est la divergence entre ω_1 et ω_2 la plus forte est l'instabilité, où la force de l'instabilité est mesurée par l'éntendue de la gamme de vitesses instables $\Omega_2 - \Omega_1$ ou par le degré de l'accroîssement exponentiel au centre de la gamme de vitesses instables. Dans le cas avec la combinaison de la raideur et l'inertie asymetriques, il y a un paramètre supplementaire ϕ qui est l'angle entre l'axe principal de l'inertie et l'axe principal de la raideur. Quand ϕ accroît de 0 à $\pi/2$ la divergence entre ω_1 et ω_2 s'accroît mais, conformément à Crandall and Brosens (1961), la force de l'instabilité décroît. Ce résultat anomal a été indépendement verifié, analytiquement et par experience (Yamamoto, Ota and Kono, 1968). Ce rapport présente une expliquation elémentaire du paradoxe.

Dynamics of Unbalanced Rotor in the Interaction with the Other Physical Fields

E. R. RACHIMOV, A. B. KIDERBEK-ULI, A. RACHMATULLAEV AND
T. Sh. IMANKULOV

Kazakh State University, Alma-Ata, USSR

Аннотация. В работе рассматривается движение роторной системы, имеющей в первом случае полость, частично заполняемую несколькими несмешивающимися жидкостями, во втором случае - с учетом подвижности станины. Задача решается с применением метода теории пограничного слоя, асимптотического метода Боголюбова-Митропольского.

Ключевые слова. Ротор, вязкая жидкость, устойчивость, станина, критическая скорость.

КОЛЕБАНИЯ РОТОРНОЙ СИСТЕМЫ, ЧАСТИЧНО ЗАПОЛНЕННОЙ МНОГОСЛОЙНОЙ ЖИДКОСТЬЮ

Рассмотрим гибкий ротор, вращающийся с постоянной угловой скоростью, расположенный симметрично относительно упругих опор. Цилиндрическая полость ротора частично заполняется n несжимаемыми несмешивающимися жидкостями. Уравнения движения рассматриваемой системы имеют вид

$$\left.\begin{aligned} m\ddot{x}+(n_e+n_i)\dot{x}+\Omega_0 n_i y+c_x x &= me\Omega_0^2\cos\Omega_0 t+Rh\int_0^{2\pi}\sigma_{r/r=q}\cos(\Omega_0 t+\varphi)d\varphi, \\ m\ddot{y}+(n_e+n_i)\dot{y}-\Omega_0 n_i x+c_y y &= me\Omega_0^2\sin\Omega_0 t+Rh\int_0^{2\pi}\sigma_{r/r=q}\sin(\Omega_0 t+\varphi)d\varphi, \end{aligned}\right\} \quad (1)$$

$$\left.\begin{aligned} &\frac{\partial U_i}{\partial t}+U_i\frac{\partial U_i}{\partial r}+\frac{V_i}{r}\frac{\partial U_i}{\partial \varphi}-\frac{V_i}{r}-2\Omega_0 V_i-\frac{\nu_i}{r}\frac{\partial f_i}{\partial \varphi}=-\frac{1}{\rho_i}\frac{\partial P_i}{\partial r}-\ddot{x}\cos(\Omega_0 t+\varphi)-\ddot{y}\sin(\Omega_0 t+\varphi), \\ &\frac{\partial V_i}{\partial t}+U_i\frac{\partial V_i}{\partial r}+\frac{V_i}{r}\frac{\partial V_i}{\partial r}+\frac{U_i V_i}{r}+2\Omega_0 U_i+\nu_i\frac{\partial f_i}{\partial r}=-\frac{1}{\rho_i r}\frac{\partial P_i}{\partial \varphi}+\ddot{x}\sin(\Omega_0 t+\varphi)-\ddot{y}\cos(\Omega_0 t+\varphi), \end{aligned}\right\} \quad (2)$$

$$\frac{\partial (rU_i)}{\partial r}+\frac{\partial V_i}{\partial \varphi}=0, \quad (3)$$

где $f=\frac{1}{r}\frac{\partial U_i}{\partial \varphi}-\frac{\partial V_i}{\partial r}-\frac{V_i}{r}$,

с граничными условиями:

на стенке ротора при $r=R$

$$\left.\begin{aligned} U_{n/r=R}&=0, \\ V_{n/r=R}&=0; \end{aligned}\right\} \quad (4)$$

на границе раздела жидкостей при $r=r_i$

$$\left.\begin{aligned} U_i&=U_{i+1}, \\ V_i&=V_{i+1}; \end{aligned}\right\} \quad (5)$$

условие равенства касательных и нормальных напряжений

$$\nu_i\rho_i\left(\frac{1}{r}\frac{\partial U_i}{\partial \varphi}+\frac{\partial V_i}{\partial r}-\frac{V_i}{r}\right)=\nu_{i+1}\rho_{i+1}\left(\frac{1}{r}\frac{\partial U_{i+1}}{\partial \varphi}+\frac{\partial V_{i+1}}{\partial r}-\frac{V_{i+1}}{r}\right), \quad (6)$$

$$\left[P_i-\frac{1}{2}\rho_i\Omega_0^2(r^2-r_i^2)+2\rho_i\nu_i\frac{\partial U_i}{\partial r}\right]_{/r=r_i+\xi_i(\varphi,t)}= \left[P_{i+1}-\frac{1}{2}\rho_{i+1}\Omega_0^2(r^2-r_{i+1}^2)+2\rho_{i+1}\nu_{i+1}\frac{\partial U_{i+1}}{\partial r}\right]_{/r=r_i+\xi_{i+1}(\varphi,t)}, \quad (7)$$

$$\partial\xi_i(\varphi,t)/\partial t=U_i=U_{i+1} \text{ при } r=r_i , \quad (8)$$

на свободной поверхности жидкости

$$\nu_1\rho_1\left[\frac{1}{r}\frac{\partial U_1}{\partial r}+\frac{\partial V_1}{\partial r}-\frac{V_1}{r}\right]_{/r=r_0}=0, \quad (9)$$

$$\left[P_1-\frac{1}{2}\rho_1(r^2-r_0^2)+2\nu_1\rho_1\frac{\partial U_1}{\partial r}\right]_{/r=r_0+\xi(\varphi,t)}=0, \quad (10)$$

$$\frac{\partial\xi(\varphi,t)}{\partial t}=U_1 \text{ при } r=r_0 , \quad (11)$$

где $r_0\le r\le R$, $i=\overline{1,n-1}$

Решение уравнений (2) с граничными условиями (4)-(II) начнем методом малых параметров, с помощью которого получим уравнения нулевого, первого и т.д. порядков. Решением уравнения нулевого порядка является

$$u_{i0}=0,\ v_{i0}=0,\ P_{i0}=0 \qquad (12)$$

учитывая которое уравнения движения жидкостей представляются в виде

$$\left.\begin{aligned}&\frac{\partial u_i}{\partial t}-2\Omega_0 v_i-\frac{\nu_i}{r}\frac{\partial f_i}{\partial \varphi}=-\frac{1}{\rho_i}\frac{\partial P_i}{\partial r}-\\&\quad-\ddot{x}\cos(\Omega_0 t+\varphi)-\ddot{y}\sin(\Omega_0 t+\varphi),\\&\frac{\partial v_i}{\partial t}+2\Omega_0 u_i+\nu_i\frac{\partial f_i}{\partial \varphi}=-\frac{1}{\rho_i r}\frac{\partial P_i}{\partial \varphi}+\\&\quad+\ddot{x}\sin(\Omega_0 t+\varphi)-\ddot{y}\cos(\Omega_0 t+\varphi).\end{aligned}\right\} \qquad (13)$$

Уравнения (13),(3) с граничными условиями (4)-(11) решаются вблизи твердой стенки $r=q_n-\delta_{2n}^c\varepsilon_n$, $\delta_{2n}^c=(q_n-r)/\varepsilon_n$, границы раздела жидкостей $r=q_i+\varepsilon_{i+1}\delta_j^c$, $r=q_i-\varepsilon_i\delta_\kappa^c$ и свободной поверхности жидкости $r=1+\delta_1\varepsilon_1$, $\delta_1^c=(r-1)/\varepsilon_1$, где δ_i^c - толщина пограничных слоёв $(i=\overline{1,2n}),(j=\overline{1,2n-1};\ \kappa=\overline{1,2n-2})$. Вводя потенциалы Ламба можно составляющие скоростей частиц жидкостей представить в виде

$$\left.\begin{aligned}u_i&=\frac{\partial \Phi_i}{\partial r}+\frac{1}{r}\frac{\partial \Psi_i}{\partial \varphi},\\v_i&=\frac{1}{r}\frac{\partial \Phi_i}{\partial \varphi}-\frac{\partial \Psi_i}{\partial r}.\end{aligned}\right\} \qquad (14)$$

Тогда уравнения (13) и (3) принимают соответственно вид

$$\partial\Psi_i/\partial t-\varepsilon_i^2\Delta\Psi_i=0, \qquad (15)$$

$$\Delta\Phi_i=0. \qquad (16)$$

В случае вынужденного движения функции Φ_i и Ψ_i можно представить как

$$\left.\begin{aligned}\Phi_i&=R_i(z)\exp(i(\sigma t-\varphi)),\\\Psi_i&=R_i^*(z)\exp(i(\sigma t-\varphi)).\end{aligned}\right\} \qquad (17)$$

и решая уравнения движения жидкостей вне и внутри пограничных слоёв имеем $3n$ уравнений, откуда определяются составляющие скоростей частиц жидкостей

$$u_i=f_i(C_\kappa^\ell,\delta_i^c,q_i),$$

$$v_i=f_i(C_\kappa^\ell,\delta_i^c,q_i),$$

и $4n$ уравнений, получаемых из граничных условий (4)-(11), для определения постоянных интегрирования. Из второго уравнения системы (13) определяется давление и сила реакции жидкости, и, далее - характеристическое уравнение системы для определения зон неустойчивости. Представим решение для двухслойной жидкости, частично заполняющей полость ротора, совершающего прецессионное движение. Уравнение движения ротора в комплексной форме записывается в виде

$$m\ddot{z}+n\dot{z}+cz=F, \qquad (18)$$

где F есть сила реакции жидкости на стенки полости ротора

$$F=Rh\int_0^{2\pi}\sigma_{r/r=q}\exp(i(\Omega_0 t+\varphi)), \qquad (a)$$

σ есть нормальное напряжение силы жидкости на стенке полости ротора

$$\sigma_{r/r=q}=P-2\nu_2\rho_2\frac{\partial u_2}{\partial r}\Big/_{r=q}. \qquad (б)$$

Уравнения движения жидкостей имеют вид

$$\left.\begin{aligned}&\frac{\partial u_i}{\partial t}-2\Omega_0 v_i-\frac{\nu_i}{r}\frac{\partial f_i}{\partial \varphi}=-\frac{1}{\rho_i}\frac{\partial P_i}{\partial r}+\omega^2 e\exp i(\sigma t-\varphi),\\&\frac{\partial v_i}{\partial t}+2\Omega_0 u_i-\nu_i\frac{\partial f_i}{\partial r}=-\frac{1}{\rho_i r}\frac{\partial P_i}{\partial \varphi}-\omega^2 e\exp(i(\sigma t-\varphi)),\end{aligned}\right\} \qquad (19)$$

решая которые вне и внутри пограничных слоёв с учетом (14)-(17) будем иметь

$$\left.\begin{aligned}&\Phi_1=\left(C_1^{(1)}r+\frac{C_2^{(1)}}{r}\right)\exp(i(\sigma t-\varphi)),\\&\Phi_2=\left(C_1^{(1)}r+\frac{C_2^{(2)}}{r}\right)\exp(i(\sigma t-\varphi)),\\&\Psi_1(\delta_1,\varepsilon_1,\varphi)=C_4^{(1)}\Big[1-\frac{\delta_1\varepsilon_1}{2}+\frac{3\varepsilon_1^2}{8}\Big(\delta_1^2-\\&-\frac{\delta_1}{\sqrt{\lambda}}\Big)\Big]\exp(-\sqrt{\lambda}\,\delta_1)\exp i(\sigma t-\varphi),\\&\Psi_1(\delta_2,\varepsilon_1,\varphi)=C_3^{(1)}\Big[1+\frac{\delta_2\varepsilon_1}{2q_1}+\frac{3\varepsilon_1^2}{8q_1^2}\Big(\delta_2^2-\\&-\frac{\delta_2}{\sqrt{\lambda}}\Big)\Big]\exp(-\sqrt{\lambda}\,\delta_2)\exp i(\sigma t-\varphi),\\&\Psi_2(\delta_4,\varepsilon_2,\varphi)=C_3^{(2)}\Big[1+\frac{\delta_4\varepsilon_2}{2q}+\frac{3\varepsilon_2^2}{8q^2}\Big(\delta_4^2-\\&-\frac{\delta_4}{\sqrt{\lambda}}\Big)\Big]\exp(-\sqrt{\lambda}\,\delta_4)\exp i(\sigma t-\varphi),\\&\Psi_2(\delta_2,\varepsilon_2,\varphi)=C_4^{(2)}\Big[1-\frac{\delta_3\varepsilon_2}{2q_1}+\frac{3\varepsilon_2^2}{8q_1^2}\Big(\delta_3^2-\\&-\frac{\delta_3}{\sqrt{\lambda}}\Big)\Big]\exp(-\sqrt{\lambda}\,\delta_3)\exp i(\sigma t-\varphi).\end{aligned}\right\} \qquad (20)$$

С учётом формул (20) из (14) определяются составляющие скоростей частиц жидкостей, давление и σ_r из (б)

$$\sigma_{r/r=q}=\Big[-\frac{2C}{q}\Big(\lambda+\frac{\sqrt{\lambda}}{q}\varepsilon_2+\frac{3}{4q^2}\varepsilon_2^2\Big)+$$

$$+\frac{9\rho_2}{q^6\lambda}\varepsilon_2^5\Big]C_2^{(2)}\exp(i(\sigma t-\varphi)),$$

и выражение для силы реакции жидкости из (а), подставив которое в уравнение движения ротора (18) получим характеристическое уравнение системы. Его решение представим в виде

$$\omega=\omega_0+\varepsilon_1\omega_1+\varepsilon_2\omega_2+\dots$$

с помощью которого получаются уравнения

нулевого, первого, второго и т.д. порядков. Уравнением нулевого порядка является полином 6-ой степени. Уравнения для ω_1 и ω_2 являются линейными, их коэффициенты зависят от ω_0 и параметров системы.

ДИНАМИКА СИСТЕМЫ "РОТОР-ФУНДАМЕНТ" С ПЕРЕМЕННЫМИ КОЭФФИЦИЕНТАМИ

Уравнения движения гибкого ротора с переменной массой, несимметрично расположенного относительно упругих опор, совместно с двигателем установленного на изолированной станине в общем виде имеют вид:

$$\left.\begin{array}{l}
1.\ m(\tau)\ddot{x}+p_1 x-q_1\alpha-p_1 x_\kappa-\sigma_1\theta-h_1 p_1 \Phi=\mu\left[-n_{11}(\dot{x}-\dot{x}_\kappa)-n_{12}(\dot{\alpha}-\dot{\theta})-n_{11}L_p\dot{\theta}+\right.\\
\left.\quad +m(\tau)e\dot{\varphi}^2\cos\varphi+h_1 n_{11}\dot{\Phi}\right],\\
2.\ m(\tau)\ddot{y}+p_2 y-q_2\beta-p_2 y_\kappa-\sigma_2\psi=\mu\left[-n_{11}(\dot{y}-\dot{y}_\kappa)-n_{12}(\dot{\beta}-\dot{\psi})-n_{11}L_p\dot{\psi}+m(\tau)e\dot{\varphi}^2\cos\varphi,\right.\\
3.\ J(\tau)\ddot{\alpha}+J_0(\tau)\dot{\varphi}\dot{\beta}-\tau_1 x+s_1\alpha-\gamma_1 x_\kappa-\delta_1\theta=\mu\left[-n_{12}(\dot{x}-\dot{x}_\kappa)-n_{12}(\dot{\alpha}-\dot{\theta})-n_{12}L_p\dot{\theta}(J_0(\tau)-\right.\\
\left.\quad -J(\tau))\delta\dot{\varphi}^2\sin(\varphi-\varkappa)\right],\\
4.\ J(\tau)\ddot{\beta}+J_0(\tau)\dot{\varphi}\dot{\alpha}-\tau_2 y+s_2\beta-\gamma_2 y_\kappa-\delta_2\psi=\mu\left[-n_{12}(\dot{y}-\dot{y}_\kappa)-n_{22}(\dot{\beta}-\dot{\psi})-n_{12}L_p\dot{\psi}-(J_0(\tau)-\right.\\
\left.\quad -J(\tau))\delta\dot{\varphi}^2\cos(\varphi-\varkappa)\right],\\
5.\ M\ddot{x}_\kappa+p_x x_\kappa+n_1 x+n_2 x_\kappa+k_1\alpha+k_2\theta+H_1\Phi=\mu\left[-n_{11}(\dot{x}_\kappa-\dot{x})-n_n\dot{x}_\kappa-n_{12}(\dot{\theta}-\dot{\alpha})+n_0\theta-H_2\dot{\Phi}\right],\\
6.\ M\ddot{y}_\kappa+p_y+\bar{n}_1 y+n_2 y_\kappa+k_1\beta+k_2\psi+p_y L_3\Phi=\mu\left[-n_{11}(\dot{y}_\kappa-\dot{y})-n_n y_\kappa-n_{12}(\dot{\psi}-\dot{\beta})-n_0\psi\right],\\
7.\ J_1\ddot{\theta}+s_x\theta+n_3 x+n_4 x_\kappa+k_3\alpha+k_4\theta=\mu\left[-(k_0+k_6)\dot{\theta}-n_5(\dot{x}_\kappa-\dot{x})-n_n L\dot{x}_\kappa-k_5(\dot{\theta}-\dot{\alpha})\right],\\
8.\ J_2\ddot{\psi}+s_y\psi+\bar{n}_3 y+\bar{n}_4 y_\kappa+k_3\beta+k_4\psi=\mu\left[-(k_0+k_6)\dot{\psi}-n_5(\dot{y}_\kappa-\dot{y})-n_n L_\kappa\dot{y}_\kappa-k_5(\dot{\psi}-\dot{\beta})\right],\\
9.\ J_3\ddot{\Phi}+s_\varphi\Phi+H_3\Phi+H_6\dot{x}_\kappa-H_5 x-H_7\theta=\mu\left[-k_\varphi\dot{\Phi}-H_4\Phi-H_8\dot{x}_\kappa-H_9\dot{x}+M_c(\dot{\varphi}-\dot{\Phi})\right],\\
10.\ J_0(\tau)\dfrac{d\omega}{dt}=M_д(\varphi)-M_c(\dot{\varphi}-\dot{\Phi})-m(\tau)e(\ddot{y}\cos\varphi-\ddot{x}\sin\varphi)-J_0(\tau)\left[\ddot{\alpha}\beta+\dot{\alpha}\dot{\beta}-\right.\\
\quad -2\delta\dot{\beta}\dot{\varphi}\cos(\varphi-\varkappa)+\delta\dot{\varphi}^2\beta\sin(\varphi-\varkappa)+J(\tau)\delta\left[\ddot{\alpha}\cos(\varphi-\varkappa)+\ddot{\beta}\sin(\varphi-\varkappa)\right]-J_0(\tau)\delta\ddot{\alpha}\cos(\varphi-\varkappa),
\end{array}\right\}\quad (1)$$

где μ – малый параметр; $H_1=ph_1-p_x h_2$, $H_2=h_1 n_{11}-h_2 n_n$, $H_3=h_1^2 p_1+p_x h_2^2+p_4 L_3^2-p_\kappa h_2$, $H_4=h_1^2 n_{11}+h_2^2 n_n$, $H_5=h_1 p_1$, $H_7=-h_1 p_1$, $H_6=-p_x h_2+h_1 p_1$, $H_8=h_2 n_n+h_1 n_{11}$, $H_9=h_1 n_{11}$.

Так как обобщённая математическая модель рассматриваемой системы является системой нелинейных дифференциальных уравнений с переменными коэффициентами, то здесь удобнее всего использовать асимптотический одночастотный метод Боголюбова-Митропольского, позволяющий исследовать как переходные, так и стационарные режимы работы систем с широким применением ЭВМ. Решение первых девяти уравнений системы (I), описывающих колебательное движение данной системы, соответствующее одной из собственных частот λ_κ будем иметь в виде

$$q_j^{(\kappa)}=a_\kappa C_{1\kappa}e^{i(\lambda_\kappa t+\psi_\kappa)}+a_\kappa C_{j\kappa}e^{-i(\lambda_\kappa t+\psi_\kappa)}, \qquad (2)$$

где q_j – обобщённые координаты $x, y, \alpha, \beta, x_\kappa, y_\kappa, \theta, \psi, \varphi$, причём $s\lambda_1\neq r\lambda_\kappa$; s, r – взаимно простые числа; a_κ и ψ_κ – амплитуда и фаза вынужденных колебаний; $C_{j\kappa}(\tau), C^*_{j\kappa}(\tau)$ собственные функции. Введя обозначения $C_{1\kappa}(\tau)=1+i$, $C^*_{1\kappa}(\tau)=1-i$, $C_{2e,\kappa}/C_{1\kappa}=$ $=C_{2e,\kappa}/C^*_{1\kappa}=i\hat{C}_{2e,\kappa}$, $C_{2e+1,\kappa}/C_{1\kappa}=C^*_{2e+1,\kappa}/C^*_{1\kappa}=\hat{C}_{2e+1,\kappa}$

получим систему независимых алгебраических уравнений относительно $\hat{C}_{j\kappa}$ $(j=2,9)$. Тогда решая характеристическое уравнение при заданной угловой скорости можно построить график $\lambda_\kappa=\lambda_\kappa(\omega)$. Заметим, что собственная частота системы будет зависеть от угловой скорости и переменности параметров системы. Принимая $\omega=\lambda_\kappa$ и решая характеристическое уравнение можно найти критические скорости системы $\lambda=\lambda_{\kappa p}$. В зависимости от характеристик опор, вала и других параметров в системе возможны несколько критических скоростей $\lambda_1, \lambda_2, \lambda_3, \lambda_4$. Предполагая, что в системе отсутствует внутренний резонанс, в работе рассматриваются поочередные переходы через возможные критические зоны λ_κ. Амплитуда a_κ и фаза ψ_κ нестационарных колебаний определяются из уравнений:

$$\left.\begin{array}{l}
da_\kappa/dt=-D_\kappa(\tau,\omega)a_\kappa-E_{1\kappa}\cos(\psi_\kappa-\\
\quad -\pi/4)+E_{2\kappa}\sin(\psi_\kappa+\varkappa-\pi/4),\\
d\psi_\kappa/dt=\lambda_\kappa(\tau)-\omega(\tau)+E_{1\kappa}\sin(\psi_\kappa-\\
\quad -\pi/4)/a_\kappa+E_{2\kappa}/a_\kappa\cos(\psi_\kappa+\varkappa-\pi/4),
\end{array}\right\}\quad (3)$$

где

$$E_{1\kappa}(\tau,\omega)=\frac{1/\sqrt{2}\,m(\tau)e\omega^2(\tau)[1-\hat{C}_{2\kappa}(\tau)]}{m_1(\tau)[\lambda_\kappa+\omega(\tau)]+m_2(\tau)},$$

$$E_{2\kappa}(\tau,\omega)=\frac{1/\sqrt{2}\,[J_0(\tau)-J(\tau)]\delta\omega^2(\tau)[C_{3\kappa}(\tau)-C_{4\kappa}(\tau)]}{m_1(\tau)[\lambda_\kappa+\omega(\tau)]+m_2(\tau)},$$

$$\mathcal{D}_\kappa(\tau,\omega)=\frac{m_3(\tau)+\frac{d}{dt}(m_1\lambda_\kappa)+2\lambda_\kappa N(\tau)}{2m_1(\tau)\lambda_\kappa+m_2(\tau)}.$$

В последнем уравнении системы (I), представим ω как суперпозицию плавно меняющегося члена Ω_κ и суммы малых вибрационных членов. Тогда ввиду малости последних, угловую скорость ω примем в первом приближении равной Ω_κ, а затем определим её из десятого уравнения (I). Так как $a_\kappa, \psi_\kappa, \hat{C}_{\ell,\kappa}$ являются медленно меняющимися параметрами, то считая их для одного периода постоянными, находим $\dot{x}$, $\dot{y}, \dot{\alpha}, \dot{\beta}, \ddot{x}, \ddot{y}, \ddot{\alpha}, \ddot{\beta}$ и подставляя их в правую часть десятого уравнения системы (I) и усредняя его по φ за время равное одному периоду, получим уравнение относительно Ω_κ:

$$(\tau)d\Omega_\kappa/dt=M(\Omega_\kappa)-M_c(\Omega_\kappa-\dot{\varphi})-$$
$$-L(\tau,a_\kappa,\psi_\kappa,\Omega_\kappa). \qquad (4)$$

Для исследования особенностей нестационарных колебаний обобщённой динамической модели "ротор-фундамент" в области одной из критических скоростей необходимо уравнения первого приближения (3) и (4) проинтегрировать с применением ЭВМ и построить амплитудно-частотную характеристику нестационарных колебаний, изменяя обобщённые координаты по времени и другие параметры системы. Для стационарных режимов должно быть

$$\frac{da_\kappa}{dt}=0,\ \frac{d\psi_\kappa}{dt}=0,\ \frac{d\Omega_\kappa}{dt}=0\ . \qquad (5)$$

Соотношение (5) можно рассматривать как условия существования стационарных режимов движения. Для всех расчетных схем уравнения первого приближения определяются уравнениями (3) и (4), но необходимо для каждого конкретного случая определить коэффициенты уравнения движения, вычислять собственные частоты, собственные функции. Отметим, что при исследовании частных задач нет необходимости каждый раз решать уравнения движения этих систем тем или иным математическим методом, а достаточно вносить соответствующие изменения в выражения для коэффициентов системы. Это позволяет при исследовании динамики роторных систем легко переходить от одной расчетной модели к другой, провести вычислительный эксперимент и получить сравнительные качественные характеристики рассматриваемых систем с применением ЭВМ.

ЛИТЕРАТУРА

Гробов, В.А.(I96I). Асимптотические методы расчёта изгибных колебаний валов турбомашин. Изд-во АН СССР, I66 с.

Дерендяев, Н.В., Сандалов, В.М.(I982). Об устойчивости вращения ротора, заполненного слоисто-неоднородной вязкой несжимаемой жидкостью. Тезисы докл. II Всесоюзного съезда по ТММ, Одесса, с.I40.

Митропольский, Ю.А.(I964). Проблемы асимптотической теории нестационарных колебаний. Изд-во "Наука", 437 с.

SUMMARY

The work is devoted to the investigation of rotor systems. For the case, when rotor has a cylindrical cavity, partially filled with a few types of unmixed fluid the equalization of movement of the system "rotor-fluid" was composed and it was offered the methodics of solution with the application of method of small parameters, the theory of boundary layer, giving an opportunity the characteristic equation for the determination of the zones of instability under the variation of parameters. In the case, when the rotor with a modified mass in time is settled on the isolated foundation, the movement of the system "rotor-foundation" in common case is described with the help of ten differential equatation with variable factors. This task is solved with the application of Bogolyubov-Mitropolsky method, wihich allows to investigate stationary and instationary regimes of the system movement.

The Vibrations of the Rotor for the Case of Winding up of the Band with Constant Velocity

L. CVETICANIN AND M. ZLOKOLICA

Faculty of Technical Sciences, 21000 Novi Sad, V. Vlahovica 3, Yugoslavia

Abstract. In the paper the vibrations of the rotor on which the band is winding up with constant velocity are analysed. The rotor is assumed as a shaft-disc system. The shaft is supported in two bearings. The mass of the shaft can be neglected in comparation to the mass of the disc. The elastic force in the shaft is non-linear. The disc is settled in the middle of the shaft. The mass of the disc is varying during the time. It is caused by winding up of the band. The mass center position is also varying. Thus, the mathematical model of the rotor is a system of non-homogenius nonlinear differential equations with varyable parameters. This system can be solved by the use of the method of multiple scales. The numerically got results are shown and discussed.

Keywords. Rotor; difference equations; nonlinear equations; mathematical analysis; vibrations.

INTRODUCTION

In the paper (Cveticanin, 1984) the rotor on which the band is winding up with constant velocity is analysed. But, in the paper the varying of the position of the mass center to the rotating center is neglected. So, it is supposed that the mass center and the center of rotating are the same points. It is not the case for the real rotors.

In this paper the vibrations for this kind of rotors are analysed, but taking into account of that phenomena.

The rotor is analysed as a shaft-disc system. The shaft is supported in two bearings. The mass of the shaft can be neglected. The elastic force in the shaft is non-linear. The disc is settled in the middle of the shaft. The varying of the mass of the disc is,

$$m = m_o + \rho L v h \qquad (1)$$

the varying of the radius is

$$R = (R_o{}^2 + vth/\pi)^{1/2} \qquad (2)$$

and the varying of the position of the mass center is

$$r = (2R^2 hL\rho/m_o)\sin\theta/2 \qquad (3)$$

where R_o is the radius of the empty disc, m_o is the mass of the empty disc, v is the velocity of winding up, ρ is the density of the band, L is the width of the band, h is the thickness of the band and

$$\theta = 2\pi R/h \qquad (4)$$

the angle of winding up of the rotor. The model of the rotor is shown in Fig. 1.

The vibrations are analysed in the plane which is normal to the axle of the rotor.

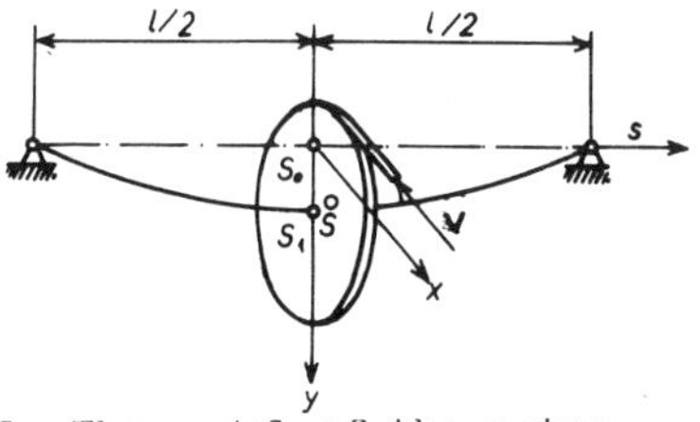

Fig. 1. The model of the rotor

MATHEMATICAL MODEL OF THE ROTOR

Using the Impulse method the differential equation of the moving of the rotor is

$$m(d^2\vec{r}_{s1}/dt^2 + d\vec{\Omega}/dt \times \vec{r} + 2\vec{\Omega} \times d\vec{r}/dt + \vec{\Omega} \times (\vec{\Omega} \times \vec{r}) + d^2\vec{r}/dt^2) = \vec{F} + \vec{\Phi} \qquad (5)$$

where is $d^2\vec{r}_{s1}/dt^2$ the absolute acceleration of S_1, $\vec{\Omega} \times (\vec{\Omega} \times \vec{r})$ is the normal component of the acceleration, $2\vec{\Omega} \times d\vec{r}/dt$ is the Coriolis acceleration, $d\vec{\Omega}/dt \times \vec{r}$ is the tangential acceleration, $d^2\vec{r}/dt^2$ is the relative acceleration, $\vec{F}$ is the active force and $\vec{\Phi}$ the reactive force.

In the Fig. 2. the positions of mass center S, the position of rotating center S_1 are plotted.

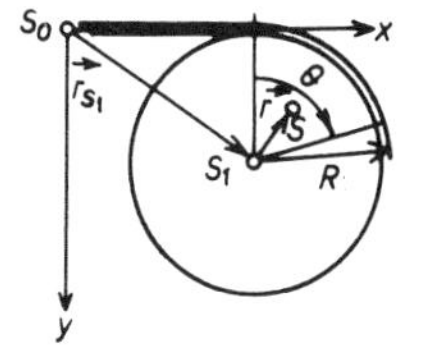

Fig. 2. The position vectors of S and S_1

The reactive force is

$$\vec{\Phi} = -(dm/dt\ dR/dt)\vec{j} \qquad (6)$$

where $\vec{j}$ is the unit vector in y axle direction.

Substituting the term (6) into eq. (5) and after projecting the equations on the axes of the coordinate system xS_oy, it is

$$m(d^2x/dt^2-\Omega^2 r\sin\theta/2+d\Omega/dt\ r\cos\theta/2+ d^2r/dt^2\sin\theta/2+2\Omega dr/dt\cos\theta/2)=X$$
$$m(d^2y/dt^2+\Omega^2 r\cos\theta/2+d\Omega/dt\, r\sin\theta/2- d^2r/dt^2\cos\theta/2+2\Omega dr/dt\ \sin\theta/2)=Y- dm/dt\ dR/dt \quad (7)$$

where x and y are the coordinates of position vector of mass center, X and Y are the projections of the force.

For complex deflection

$$z=x+iy \quad (8)$$

and complex force

$$Z=X+iY \quad (9)$$

where $i=\sqrt{-1}$ is the imaginary unit the system of differential equations (7) is

$$m\ d^2z/dt^2=Z-i\ dm/dt\ dR/dt-9mR^2\varrho Lh\Omega^2 e^{i\theta}/(4m_o)+mR^2\varrho Lh\Omega^2/(4m_o)+imR^2\varrho Lh(d\Omega/dt)/m_o (3e^{i\theta}/4-1/2)+imh^2L\varrho v\Omega(3e^{i\theta}-1)/(\pi m_o) \quad (10)$$

The elastic force in the shaft is after Bolotin (1964)

$$Z=-b_1z+b_2z\,|z|^2 \quad (11)$$

where b_1 is the linear coefficient of elasticity and b_2 is the nonlinear coefficient of elasticity.

To simplify the analysis, the dimensionless coefficients are introduced

$$T=\omega_o t;\ \omega_o^2=b_1/m_o;\ Z=z/L;\ \mu=\varrho Lvh/(m_o\omega_o);\ p=h/(R_o\pi);\ \tau=\mu T;\ A=m_o/(2\varrho\pi L^2R_o);\ \alpha=\varrho_o/\varrho;\ k=9v/(4\omega_o L);\ b=b_2L/(\varrho vh\omega_o)$$

where ϱ_o is the density of the disc. The dimensionless differential equation is

$$d^2Z/dT^2+\omega^2(\tau)Z=-\mu bZ|Z|^2\omega^2(\tau)+\mu k/9-\mu ke^{i\theta}-\mu^2A\omega^2(\tau)/(1+\alpha\tau)^{1/2}i\omega^2(\tau)+\mu^2 2A\pi i(21/4e^{i\theta}-3/2)/(1+\alpha\tau)^{1/2} \quad (12)$$

where:

$$\omega^2(\tau)=1/(1+\tau) \quad (13)$$

$$\theta(\tau)=2(1+\alpha\tau)^{1/2}/p \quad (14)$$

PARAMETER ANALYSIS

Using the parameter analysis it can be concluded that μ is a small parameter and τ is a slow varying time. The nonlinearity is small, too. After neglecting the terms with μ^2 as the small values of the second order it is

$$d^2Z/dT^2+\omega^2(\tau)Z=\mu bZ|Z|^2\omega^2(\tau)+\mu k/9-\mu ke^{i\theta} \quad (15)$$

In eq. (15) it is possible to neglect the term $(\mu k/9)$ as a small value, because it is smaller than other terms ten of more times. Now, it is

$$d^2Z/dT^2+\omega^2(\tau)Z=\mu bZ|Z|^2\omega^2(\tau)-\mu ke^{i\theta} \quad (16)$$

For the case when the mass center and the rotating center have the same positions the eq. (16) becomes

$$d^2Z/dT^2+\omega^2(\tau)Z=bZ|Z|^2\omega^2(\tau) \quad (17)$$

This eq. (17) is already analysed in the paper of Cveticanin (1984).

In this paper the resonance case for the eq.(17) will be analysed.

THE ANALYTICAL SOLVING METHOD

The eq. (17) is a differential equation with small non-linearity and with slow varying parameters and because of that for solving the asymptotic method of Bogolubov-Mitropolski (1958) can be applied.

The solution can be supposed in the form

$$Z=ae^{i(\theta+\vartheta)} \quad (18)$$

where

$$da/dT=\mu A_1(\tau,a,\vartheta) \quad (19)$$

$$d\vartheta/dT=\omega(\tau)-\nu(\tau)+\mu B_1(\tau,a,\vartheta) \quad (20)$$

$$\nu(\tau)=\mu\alpha/p(1+\alpha\tau) \quad (21)$$

The coefficients A_1 and B_1 will be denoted. After substituting of the eqs. (18-20) into eq.(17) it is

$$2\mu A_1 i\omega(\tau)+ai\mu(d\omega/d\tau)-2a\mu\omega(\tau)B_1+\mu\partial A_1/\partial\vartheta(\omega-\nu)+\mu ai(\omega-\nu)\partial B_1/\partial\vartheta=\mu ba^3\omega^2(\tau)-\mu ke^{-i\vartheta} \quad (22)$$

Separating the real and imaginary parts it is

$$\partial A_1/\partial\vartheta(\omega-\nu)-2a\omega B_1=ba^3\omega^2-k\cos\vartheta$$
$$a(\omega-\nu)\partial B_1/\partial\vartheta+2\omega B_1=-a\partial\omega/\partial\tau+k\sin\vartheta \quad (23)$$

The solutions of the eq. (23) are

$$A_1=-a/(2\omega)\ (d\omega/d\tau)+k\sin\vartheta/(\omega+\nu) \quad (24)$$

$$B_1=-ba^2\omega/2+k\cos\vartheta/a(\omega+\nu) \quad (25)$$

After substituting (24) and (25) into (19) and (20) it is

$$da/dT=4k^2p\mu_1\sin\vartheta/9\alpha(\omega+\nu)+kp\mu_1\omega^2a/(9\alpha)$$
$$a(d\vartheta/dT)=a(\omega-\nu)+4k^2p\mu_1\cos\vartheta/9\alpha(\omega+\nu)-2\mu a^3\omega b^*/9 \quad (26)$$

where

$p=h/(R_o\pi)$ - the geometric parameter of the band of the band

$\alpha=\varrho_o/\varrho$ - the parameter of density

$k=9^ov/(4\omega_o L)$ - parameter of velocity

$\mu_1=L/R_o$ - the geometric parameter of the rotor

$b^*=b_2/(4\omega_o^2\varrho_oR_o\pi)$ - the parameter of non-linearity

THE NUMERICAL SOLVING METHOD

The system of differential equations (26) cannot be solved analytically, but numerically only. The results are plotted.

In Fig. 3. the amplitede as a function of time for some values of parameter μ_1 is plotted.

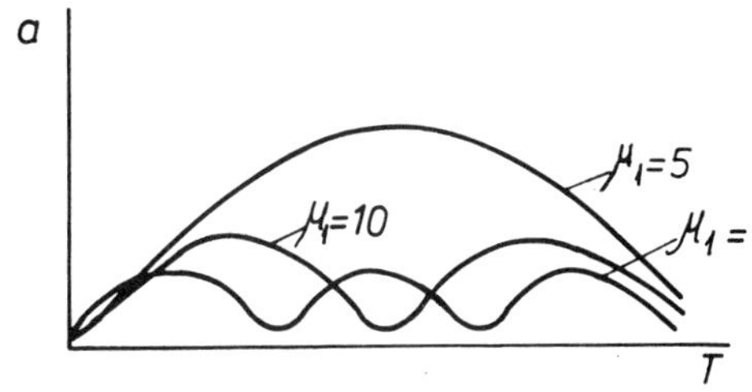

Fig.3. Amplitude-time curves for some values of μ_1

In Fig. 4. the amaplitude as a function of time for some values of parameter p is plotted.

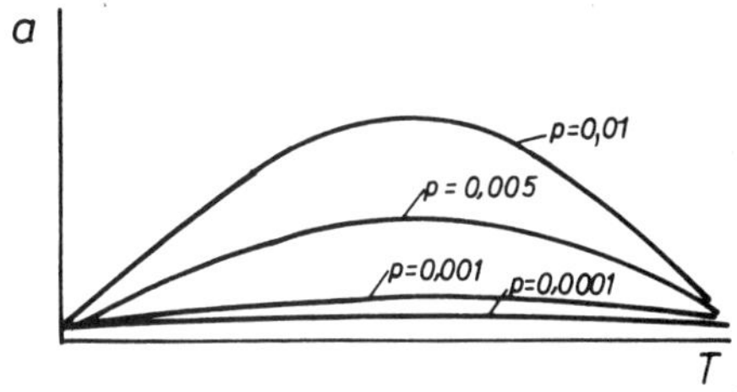

Fig. 4. Amplitude-time curves for some values of p

The amplitude as the function of time for some parameters of b^* is plotted in Fig. 5.

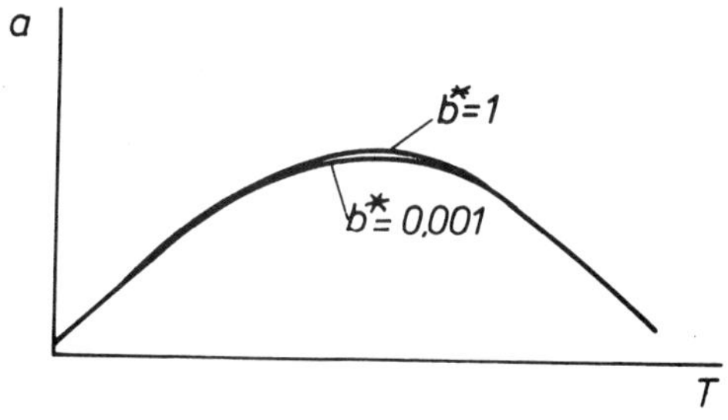

Fig. 5. Amplitude-time curves for some values of b^*

In Fig. 6. the amplitude as a function of time for some values of parameter α is plotted.

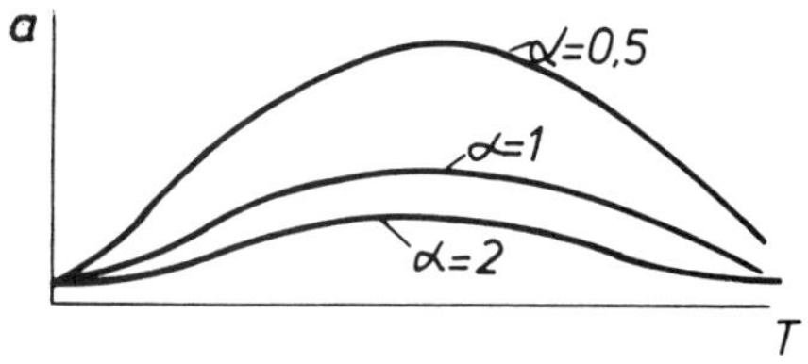

Fig. 6. Amplitude-time curves for some values of α

In Fig. 7. the amplitude as a function of time for some values of parameter k is plotted.

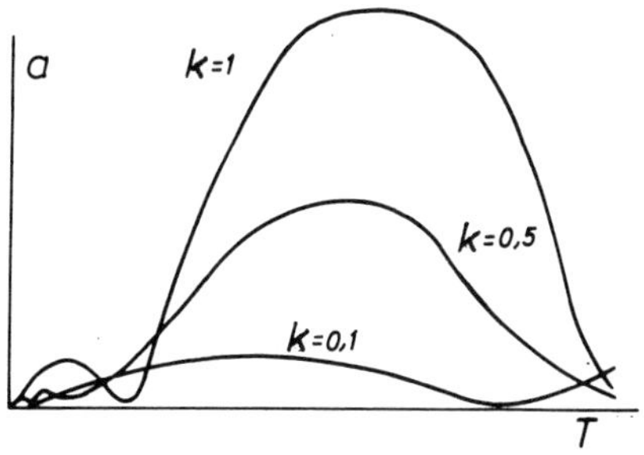

Fig. 7. Amplitude-time curves for some values of k

CONCLUSION

The amplitude of vibrations of the rotor is a function of the geometric parameter of the rotor. The width of the disc smaller in comparation to the radius of the empty disc, the amplitude of vibration is lareger. So, varying the dimensions of the disc of the rotor it is possible to influence the amplitude of vibrations.

The vibrations of the rotor are the functions of the characteristics of the band which is winding up. The thinner the band, the vibrations will be smaller. For the band with larger density of material, the amplitude of vibrations is larger.

For this kind of rotor with such parameters the influence of the nonlinearity is neglectible.It means, that it is enough to analyse the rotor as a linear model.

The velocity of winding up of the band has a large importane for the productivity of these machines. The greater the velocity the higher the productivity. But, the velocity cannot be increased very much, because the larger the velocity, the amplitudes of vibrations are larger. So, the optimal value of the velocity of the band must be denoted for which the vibrations are under a limited value, but the productivity is enough high.

REFERENCES

Bogolubov, N.N., and Y.A. Mitropolski(1958). Asimptoticheskii metodi v teorii nelinejnih kolebanij. Fizmatgiz,Moscow.
Bolotin, V.V. (1964). Dynamic stability of elastic systems. Holden-Day, San Francisco.
Cveticanin, L. (1984). Vibrations of a textile machine rotor. J. of Sound and Vibration, 97, 181-187.

DIE ROTORSCHWINGUNGEN BEIM WICKELN DES BANDES MIT DER KONSTANTEN GESCHWINDIGKEIT

L. Cveticanin und M. Zlokolica

Kurzfassung. In dieser Abhandlung wurden Rotorschwingungen analysiert auf dem ein Band mit der konstanten Geschwindigkeit gewickelt wird. Rotor besteht aus Welle-Disk-System. Die Welle ist in zwei Lagern gelagert. Das Wellengewicht ist gering in Verhältnis auf dem Diskgewicht. Das Diskgewicht verändert sich zeitlich infolge der Wicklung des dünnen Band. Die Lage des Rotorsschwerpunkts im Bezug auf dem Drehenszentrum ändert sich auch. Das mathematische Modell des Rotors ist also ein System der unhomogenen unlinearen Differentialgleichungen mit wechselnden Koefizienten. Zur Lösung verwendet man Method der mehrskalaren Zerlegung, das für diesen Fall geeigent ist. Es wurden die Schwingungen für den Resonanzfall analysiert. Die analytisch bekomenen Lösungen wurden mit den numerischen Ergebnissen vergleichen.

Bio-mechanics

Spatial Six Degree of Freedom Motion Measurement with Application to Human Motion Analysis

T. J. CHALKO* AND C. E. WILLIAMS**

*Department of Mechanical and Industrial Engineering, University of Melbourne, Parkville, Vic 3052, Australia
**11 St. Leonards Road, Devenport, Auckland 9, New Zealand.

ABSTRACT . Measurement of instantaneous relative spatial position and motion of two rigid bodies was obtained by means of the 6 degree of freedom spatial mechanism developed by authors. Angular displacements of all joints of the mechanism were measured by a set of precise potentiometers and digitized as a function of time. Conversion to linear cartesian and angular coordinates involved a computer algorithm based on the direct kinematics approach. Calibration was carried out for the assembled mechanism ,in a given workspace. Joint angles were calculated for a number of defined positions of the mechanism using the inverse kinematics computer algorithm. Calculated angles were then compared with the corresponding sensor readings and the characteristic of each joint sensor was determined individually by the least square fit method. Differentiation of the obtained position trajectories was performed to find linear and angular velocities as well as accelerations. Digital filtering was applied to eliminate the noise caused by the data acquisition system as well as by the digital differentiation procedure. The device has been applied to measure the relative position of the violin bow with respect to the violin itself. This enabled bowing mechanics and bowing skills analysis. An example of data obtained is presented.

KEYWORDS.Mechanism kinematics,motion measurement,spatial mechanism, human motion,violin bowing

INTRODUCTION

Three-dimensional human motion in cartesian coordinates has been successfully measured using accelerometry, using the Doppler effect with an ultrasonic transducer (Nadler and Goldman 1958), and using ultrasonic pulsing (Fleischer and Lange 1983). Other researchers have measured three degrees of freedom of the motor cycle rider body motion using potentiometers as the joints in a simple mechanism (Weir , Zellner and Teper 1978 , Prem 1983).

A task analysis of violin bowing entails a description of instrument handling characteristics and delineation of the patterns of appropriate kinematic and kinetic bowing variables. In the study described so far in the literature the analysis of bowing has been two-dimensional, using mainly cinematography (with limited accuracy). Literature study also shows that researchers consider at least 5 out of 6 bow coordinates relative to the violin to be significant in the change of direction of bowing. Therefore the aim of the project was established to measure simultaneously all six spatial coordinates of the bow , enabling their analysis.

The application of an open kinematic chain of mobility 6 (6 degrees of freedom) was found the most effective way to measure all six coordinates of the relative motion of two bodies in space (i.e. three cartesian coordinates of linear displacement and three angular coordinates).

Applied to measure the violin bow motion , the device was installed between the violin and the end of the bow and was found to provide a precise and continuous measurement of the position and motion of the bow relative to the violin , without significantly affecting its function and with minimal interference to the player.

MEASURING DEVICE

A sketch of the device , its attachment to the violin and the system of coordinates used are shown in Fig.1.

The mechanism contains six revolute joints (constructed from potentiometers), connecting the links .

The basic criterion for the device design for the particular application discussed here was to achieve minimal disturbance to the player during bowing. This was achieved primarily by constructing the mechanism links from lightweight materials (hollow aluminium tubes 9.5 mm outer diameter with 1 mm walls) and selection of optimum lengths of the links, giving minimal motion of the centre of gravity of the mechanism (i.e. minimizing the work done to overcome gravity forces) in the range of bowing space.

The weight of the device was found by violinists to incur minimal restriction to their normal bowing action. However, an obvious improvement to the mechanism would be to further decrease the device weight.

Helipot 10-turn 5k Ω potentiometers were chosen to function as and measure angular positions of the mechanism joints. The mechanical resistance of the potentiometers was not found to be significant during bowing. The potentiometer spindles were fitted to the tubular links by means of additional bushes. Wiring was attached along the linkage and lead away so it did not obstruct the player's motion.

The electrical diagram of the measuring device is presented in Fig.2. All connections were shielded in order to reduce the electrical noise.

DATA ACQUISITION AND TRANSFORMATION

The diagram of the data acquisition procedure is shown if Fig.3. During the experiment , the transducers signals (voltages) were digitized as a function of time and stored on the disc. Following the collection of data, a kinematic analysis of the bow motion was performed by means of

conversion of the potentiometer signals into 6 spatial coordinates. Rotation of each potentiometer determined the angles between adjoining members of the mechanism. Calculation of these joint angles from the potentiometer voltages enabled the position of the bow to be ascertained, by transformation of coordinates.
Initially, voltages were registered (V_1 -V_6 - potentiometer voltages, and V_7 - sound level signal). Then, the angular rotations of potentiometers (i.e. joint angles $\phi_1 - \phi_6$) were calculated from the voltage output according to the following formula:

$$\phi_i = \frac{V_i - V_{0i}}{AS_i} \tag{1}$$

where :
i = 1...6 is the potentiometer number
V_i is the voltage from the i-th potentiometer,
V_{0i} is the reference or zero angle voltage for the i-th potentiometer
AS_i is the angular scale of the i-th potentiometer.

V_{0i} and AS_i are the constants determined individually for each potentiometer in the calibration procedure. It was found that linear representation of the potentiometer of the form (1)(or V_i = ϕ_i AS_i +V_{0i}) gives sufficient accuracy.
The potentiometer angles computed from the above formula were then used to calculate the relative position of the two ends of the mechanism. The algorithm for this operation is presented in Fig.4. The coordinate systems used are shown in Fig.1a-d.
Position of a rigid body is determined if the cartesian coordinates of three points on it are known. Hence, coordinates of three points need to be transformed from ObXbYbZb (system of coordinates attached to one end of the mechanism - the bow) into OXYZ (attached to the other end of the mechanism - the violin) to enable calculation of angular orientation of the bow relative to the violin . In the particular application reported in this paper the following angles were used to define the bow angular orientation with respect to the violin (see Fig.1b 1c and 1d) :
α : the angle between the bow projection on the YZ plane and the Y axis,
β : the angle between the bow projection on the XY plane and the Y axis,
γ : the angle of bow twist - between the bow symmetry plane and the X axis.

The computer algorithm in FORTRAN, showing a single point coordinates transformation from ObXbYbZb into OXYZ , given sines and cosines of the joint angles is shown in Fig.5.
The complete transformation of a set of 6 joint angles (ϕ_1 ... ϕ_6) into six spatial coordinates (X,Y,Z, α, β, γ) takes about 40 ms when performed by a PDP 11/23+ computer (1 MHz clock) under RT-11 operating system.

CALIBRATION PROCEDURE

In order to calibrate and test the accuracy of the device, a calibration rig was constructed, consisting of a straight metal bar to which the device was clamped (Fig.7). Measurements of a number of configurations of the mechanism on the calibration rig were carried out to calibrate and test accuracy of the device and also check the performance of the transformation software. The three cartesian and three angular coordinates of the bow (X,Y,Z, α, β, γ) were measured in relation to the origin in different positions along the calibration bar within the normal bowing trajectory limits. A Keithley 172 autoranging digital multimeter was used to record the potentiometer voltages with an accuracy of .1 mV. Measurement of the spatial coordinates was carried out with an accuracy better than 1 mm (cartesian coordinates) and 1 degree (angular coordinates).
A programme was devised, using inverse kinematics, to calculate the theoretical value of the angular rotation of each potentiometer (joint angles ϕ_1 .. ϕ_6) given the spatial coordinates measured using the calibration rig. Then the calculated joint angles for each position were plotted against the corresponding potentiometer voltages. The relationship between the joint angles and the voltages of each potentiometer was found to be fairly linear, although different for each potentiometer, and best straight lines were fitted using the least square fit method. The average deviation of the potentiometer signals about the fitted lines shown in Table 1 represent the average error of the device in the considered range of the potentiometers.
All deviation are given in Volts i.e equivalent voltage errors for corresponding potentiometers and compared with data acquisition accuracy.
The example calibration plots are shown in Fig.7. The angular scale (AS_i) and the zero angle voltage (V_{0i}) of each potentiometer were calculated being the slope and intercept of the fitted i-th straight line . Finally, the transformation software was tested for single positions by inserting voltages (V_1 - V_6) ,measured on the rig , and comparison of the computed spatial coordinates with the coordinates measured directly on the rig. The calibration and testing procedure is shown in Fig.8.

The repeatability of the device measurements was tested for single positions on the calibration rig. The mechanism was moved away from and then returned to the selected positions and the differences in potentiometer voltage readings were recorded. The standard deviations from the mean values of repeated trials are listed in Table 1.
The accuracy and repatability of the device was tested also on the device (with both ends clamped on the calibration rig) being flicked manually. The potentiometer voltages were then recorded after a steady state was achieved. This test was repeated , and the standard deviations from the average are also listed in Table 1.
The signal-to-noise ratio in the data acquisition system was optimized by careful shielding, especially at the A-D input in the computer, and by boosting the potentiometer supply voltage to 40 volts from a very stable and 'noise-free' (i.e. having minimal AC electrical noise components in DC output) power supply (HP 6116A).
The accuracy of the A-D converter was tested with a calibration power supply (HP 6116A) and showed an average error of .1% (1 mV error with 1 V input). Consequently, the accuracy of the data acquisition system was within the resolution of the transducer system.
The resultant accuracy of spatial coordinates measurement using the device was found better than 2 mm for the cartesian coordinates and better than 1 degree for the angular coordinates. The main source of errors were mechanical and electrical inaccuracies of the potentiometers (backlash and local nonlinearities for small rotations). This accuracy was obtained for static position measurement and motion of low frequency (0-4 Hz) which is below the 1-st natural frequency of the mechanism clamped into the calibration rig.

APPLICATION TO VIOLIN BOWING MEASUREMENT

Example results of the violin bow motion measurement using the above described device are

presented in Fig.9. Six coordinates of the bow with respect to the violin as well as the recorded sound level generated during the test are shown as functions of time.
Differentiation of the bow coordinates gives corresponding (cartesian and angular) components of the bow velocity and acceleration.
The digital filter based on the fast Fourier transformation technique (FFT) was used to filter the raw data as well as the velocity and acceleration plots. The filter was set to be low pass with the cut off frequency 10 Hz to eliminate the high frequency noise caused by the data acquisition system as well as by the digital differentiation procedure. Filtering technique was applied to the complete bowing cycles (periods) since the Fourier transformation requires periodic input function.

CONCLUSIONS

Described device presents a unique solution for the measurement of motion in six degrees of freedom. In the particular application presented , the resulting data provided a vast amount of precise and useful information about the motion of the bow as manipulated by skilled and less skilled violinists.
The device may have many applications, being easily adaptable and attachable to a wide range of equipment and instruments. A large variety of motion patterns (within a sphere of about one metre) may be measured with an accuracy of 1-2 mm, and recorded either directly by a computer or using a data recorder.
Analysis of data provides opportunity for skill analysis, modelling, monitoring and assessment of many skills. The mechanism provides data that is readily accessible for analysis, and that is a continuous measurement of the motion under consideration.
The main limitation of the device, apart from its limited resolution , is the restriction regarding the frequency of motion being measured (below 4 Hz). This may be improved by increasing the natural frequencies of the device i.e. decreasing its mass (weight) and increasing its stiffness.
The accuracy and resolution of measurements may be improved by the use of more precise angular resolvers in the joints.
Further developments may include also the real time data processing so that the required spatial coordinates could be observed and recorded during the experiment.
The relationship between the performer or operator and the measuring tool and analyst is an important feature in the analysis of human skills. An advantage of filming procedures is that data can be collected with no physical encumbrance to the performer, and even without his knowledge. However the limitations of filming (e.g. circumstantial data acquisition and processing, limited accuracy in measuring three-dimensional motion , difficulty to extend measurements to more than 3 coordinates) perhaps override this advantage.
There are certain disadvantages in the measurement of a human motion using a mechanical measuring device. One is restriction (although a minimal one) to the operator's 'normal' movement. The analyst needs to ascertain whether the restrictions enable reliable measurements to be made of the skill parameters in question. In this study the criterion for skilled performance was the sound produced. Since the violinists who took part in the experiment were able to produce results that were consistent with those of their 'normal' bowing, then the means of producing these results were also considered to be consistent with 'normal' bowing.

REFERENCES

[1] Nadler, G., and Goldman, J., 'The UNOPAR', The Journal of Industrial Engineering, 9, 1, 1958, pp. 58-65.
[2] Fleischer, A.G., and Lange, W.,'Analysis of hand movements during the performance of positioning tasks', Ergonomics, 26, 6, 1983, 555-564.
[3] Weir, D.H., Zellner, J.W., and Teper, G.L., 'Motorcycle handling', Systems Technology, Inc., Report TR-1086-1, vol. II, 1978.
[4] Prem, H., Motorcycle Rider Skills Assessment, Ph.D. thesis, University of Melbourne, 1983.
[5] Schmidt, R.A., Zelaznik, H., Hawkins, B., Frank, J.S., and Quinn, J.T. Jr., 'Motor-output variability: a theory for the accuracy of rapid motor acts', Psychological Review, 36, 5, 1979, pp. 415-452.

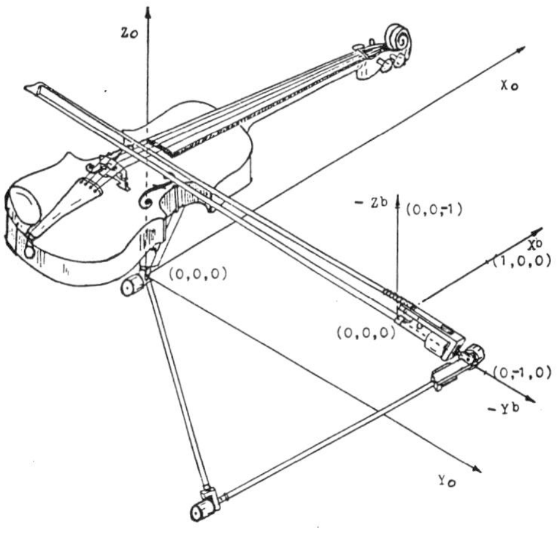

Fig.1a Six degrees of freedom motion measuring device attached to the violin and the bow.

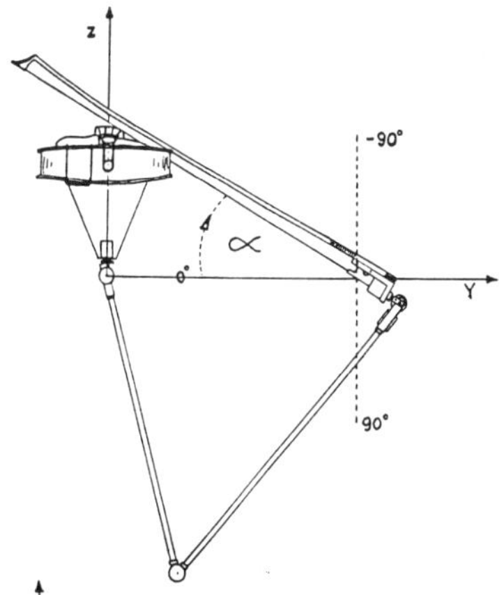

Fig.1a System of coordinates (α)

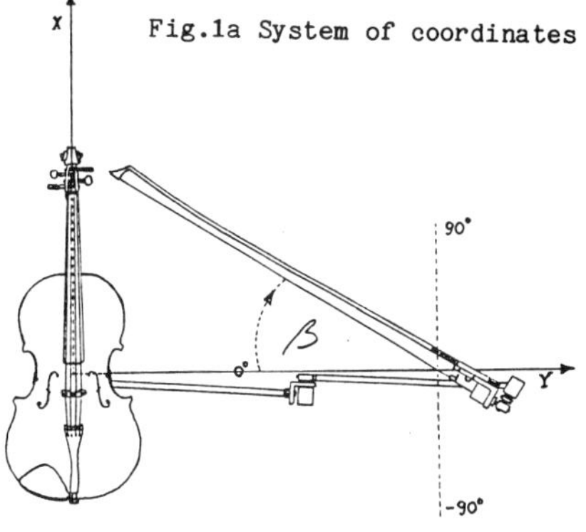

Fig.1b System of coordinates (β)

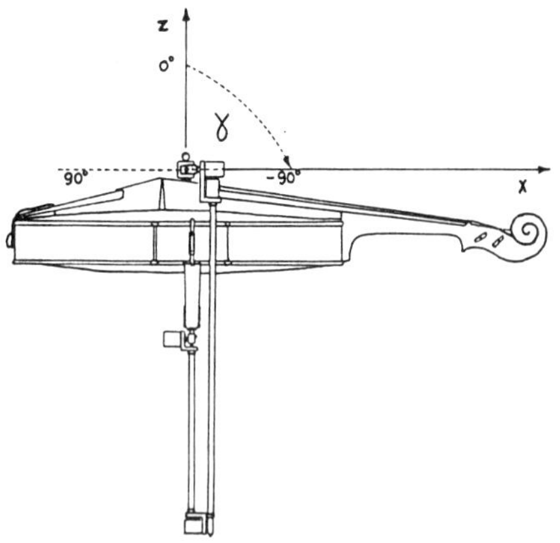

Fig.1c System of coordinates (γ)

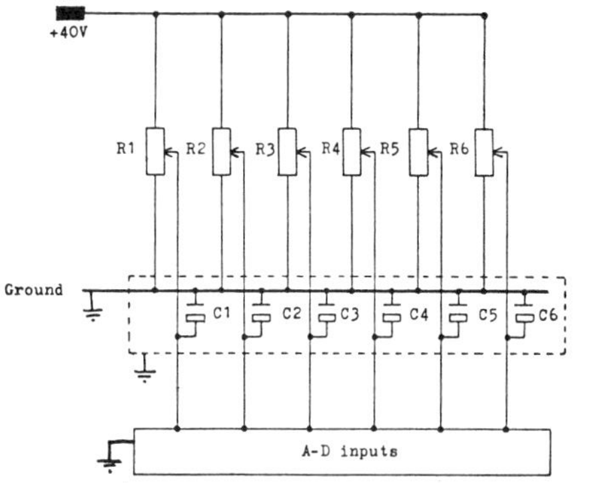

Fig.2 Electrical diagram of the device

```
--------------------------------------------------
| Input 6 potentiometer voltages V_i              |
|       1  2  3  4  5  6                          |
--------------------------------------------------
        |  |  |  |  |  |
---------------------------------   -------------------------
| Calculate 6 joint angles      |-| input calibration data |
|       1  2  3  4  5  6        | |     AS_i  , V_0i       |
---------------------------------   -------------------------
        |  |  |  |  |  |
-------------------------------------   ---------------------------
| TRANS3 subroutine                 | | |                         |
|                                   |--| Input constants (dimen-  |
| Transformation of coordinates of the | | sions of the mechanism,etc) |
| point (0,0,0) in (ObXbYbZb) into OXYZ |  ---------------------------
| coordinate system (X0,Y0,Z0) (TRANS) |
|                                   |
| Transformation of coordinates of the |
| point (C,1,0) in (ObXbYbZb) into OXYZ |
| coordinate system (X1,Y1,Z1) (TRANS) |
|                                   |
| Transformation of coordinates of the |
| point (0,0,1) in (ObXbYbZb) into OXYZ |
| coordinate system (X2,Y2,Z2) (TRANS) |
|                                   |
| Calculate angles  α, β, γ         |
|          α =-atan (Z1-Z)/(Y1-Y)   |
|          β = atan (X1-X)/(Y1-Y)   |
|          γ = atan (Z2-Z)/(X2-X)   |
-------------------------------------
        |  |  |  |  |  |
-------------------------------------
| RETURN TO CALLING PROGRAMME       |
|    X,Y,Z, α, β, γ                 |
| relative cartesian and angular    |
|    coordinates                    |
-------------------------------------
```

Fig.4 Flow diagram of coordinate transformation programme

```
--------------------------------------------
| VIOLIN BOWING VARIABLES                  |
| (bow motion and sound)                   |
| 1  2  3  4  5  6  Sound                  |
--------------------------------------------
  |  |  |  |  |  |   |
--------------------------------------------
|           TRANSDUCERS                    |
| ( 6 potentiometers and microphone)       |
|  V1 V2 V3 V4 V5 V6   V7                  |
--------------------------------------------
  |  |  |  |  |  |   |
--------------------------------------------
|      SIGNAL CONDITIONERS                 |
| (filtering capacitors, AC/DC             |
|     converter for sound level)           |
--------------------------------------------
  |  |  |  |  |  |   |
--------------------------------------------
| A/D 8-CHANNEL CONVERTER (12 BIT)         |
| (100 Hz sampling frequency)              |
--------------------------------------------
  |  |  |  |  |  |   |
--------------------------------------------
|      COMPUTER STORAGE                    |
|        (Hard disk)                       |
| V1 ..V7  - transducer voltages           |
--------------------------------------------
```

Fig.3 Data acquisition process

```
C-----------------------------------------------------------------
      SUBROUTINE TRANS(S,C,XB,YB,ZB,X,Y,Z)
      DIMENSION S(6),C(6)
      COMMON A1,A2,A3,A4,A5
C-----------------------------------------------------------------
C     S(6) ,C(6) - MATRICES OF SINS AND COS'S OF JOINT ROTATION
C                  ANGLES  F1.....F6
C     A1 ...A6   - DIMENSIONS OF LINKS
C-----------------------------------------------------------------
C-----ROTATION OF THE SYSTEM OF COORDINATES ABOUT CURRENT Y AXIS
      CALL ROT(S(6),C(6),YB,ZB,XB,Y1,Z1,X1)
      Y1=Y1+A1
      X1=X1+A5
C-----ROTATION OF THE SYSTEM OF COORDINATES ABOUT CURRENT X AXIS
      CALL ROT(S(5),C(5),X1,Y1,Z1,X2,Y2,Z2)
      Y2=Y2+A4
C-----ROTATION OF THE SYSTEM OF COORDINATES ABOUT CURRENT Y AXIS
      CALL ROT(S(4),C(4),Y2,Z2,X2,Y3,Z3,X3)
      X3=X3-A3
C-----ROTATION OF THE SYSTEM OF COORDINATES ABOUT CURRENT X AXIS
      CALL ROT(S(3),C(3),X3,Y3,Z3,X4,Y4,Z4)
      Y4=Y4+A2
C-----ROTATION OF THE SYSTEM OF COORDINATES ABOUT CURRENT X AXIS
      CALL ROT(S(2),C(2),X4,Y4,Z4,X5,Y5,Z5)
C-----ROTATION OF THE SYSTEM OF COORDINATES ABOUT CURRENT Z AXIS
      CALL ROT(S(1),C(1),Z5,X5,Y5,Z,X,Y)
      RETURN
      END
C-----------------------------------------------------------------
      SUBROUTINE ROT(S,C,X0,Y0,Z0,X,Y,Z)
C-----------------------------------------------------------------
C     ROTATION OF THE SYSTEM OF COORDINATES ABOUT AXIS
C     CORRESPONDING TO THE FIRST COORDINATE GIVEN
C     S,C - SIN AND COS OF THE ANGLE OF ROTATION
C-----------------------------------------------------------------
      Y=Y0*C+Z0*S
      Z=-Y0*S+Z0*C
      X=X0
      RETURN
      END
```

Fig.5 FORTRAN program to perform transformation of coordinates of a single point from ObXbYbZb into OXYZ

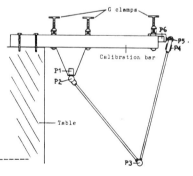

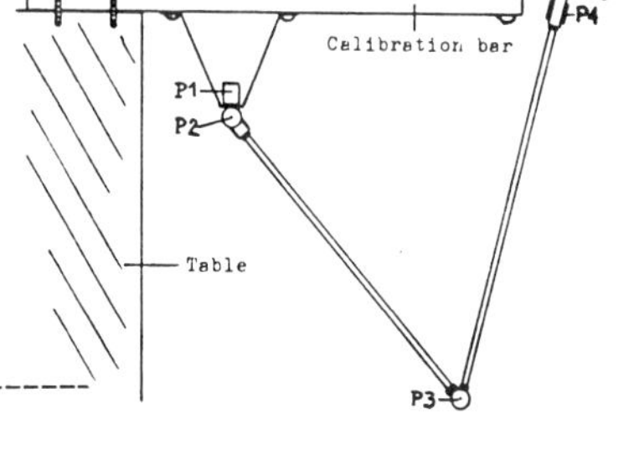

Fig.6 Calibration rig

pot. no.	CONSTANTS AS [V/rad]	V_0 [V]	ERRORS - STANDARD DEVIATIONS [V] straight line fit	position repeat.	flick test	A/D error per V input
1	-.471127	1.22702	.005	.002	.002	.001
2	-.575039	1.48378	.002	.002	.003	.001
3	.597536	1.74281	.002	.002	.001	.001
4	.530554	.60344	.005	.002	.002	.001
5	.598873	1.46045	.003	.006	.0001	.001
6	-.644545	.95625	.022	.003	.002	.001

TABLE 1. Potentiometers constants and errors - standard deviations (Volts)

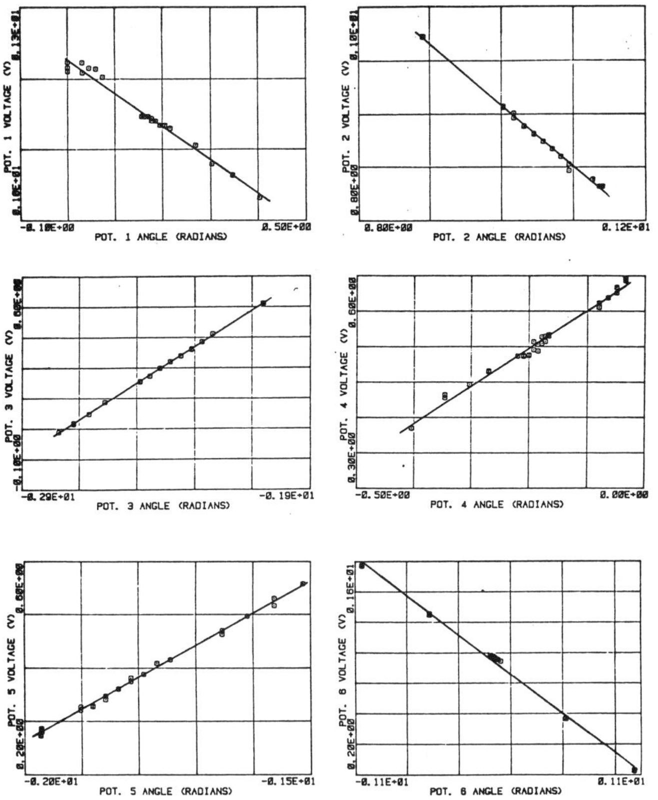

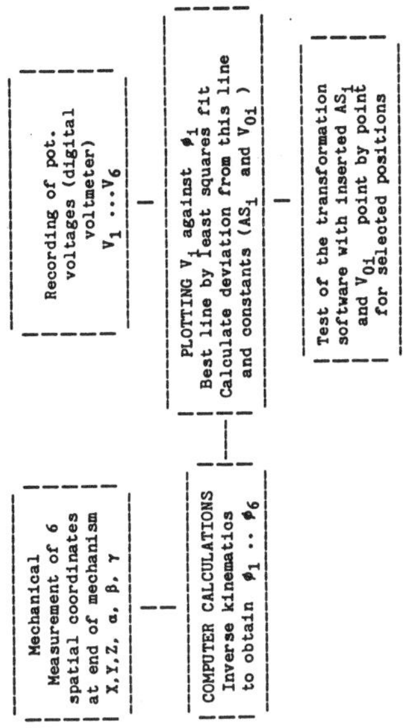

Fig.8 Flow diagram of calibration and testing procedures

Fig.7 Relationship between joint angles and potentiometer voltages for six potentiometers. Shown are fitted lines.

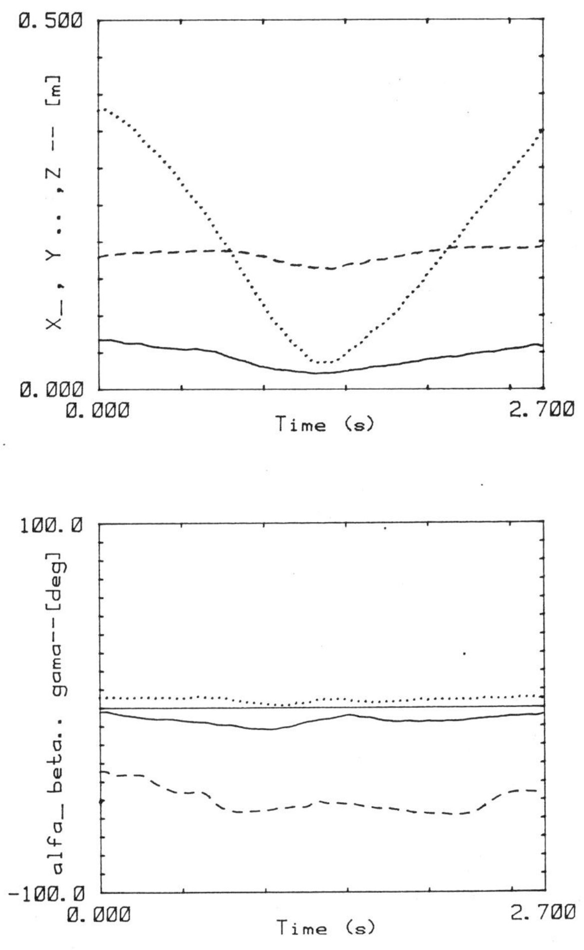

Fig.9 Example results of bow motion measurement

ПРОСТРАННОЕ ИЗМЕРЕНИЕ ДВИЖЕНИЯ О ШЕСТИ КООРДИНАТАХ С ПРИМЕНЕНИЕМ ДЛЯ АНАЛИЗА ЧЕЛОВЕЧЕСКОГО ДВИЖЕНИЯ

РЕЗЮМЕ.Измерение мгновенного взаимно-относительного пространственного положения и движения двух негибких тел создаётся при помощи пространственного механизма о 6 степенях свободы совершенным авторами. Угловые перемещения всех шарниров механизма измерялись при помощи аккуратных потенциометров и в цифровой форме вводились в компьютер.

Трансформирование этих измерений в линейные картезианские и угловые координаты создавалось при помощи вычислительного алгоритма и компьютера.

Калибрование проводилось для смонтированного комплектного механизма в данной области рабочего пространства.Угловые перемещения шарниров были вычислены для нескольких данных положений механизма при помощи обратного компьютерного алгоритма. Эти теоретические углы сравнивались с соответствующими показаниями потенциометров и характеристики каждого шарнирного потенциометра определялись индивидуально, методом малейшего квадратного отклонения. Дифференциация полученных трайекторий совершалась для получения линейных и угловых скоростей и ускорений.Цифровое фильтрование применялось к результатам для избавления от шума в электрической системе, а также отклонений по поводу цифровой дифференциации.

Механизм применялся для измерения взаимно-относительного положения и движения скрипичного смычка и самой скрипки во время игры. Это создало возможность анализа механики человеческого движения , а также анализа и сравнения техники скрипачей.

Примерные результаты измерений представляются.

On the Design of a Mechanical Arm for a Wheel-chair

A. HEMAMI

Department of Mechanical Engineering, Concordia University, Montreal, Canada, H3G 1M8

Abstract. For a mechanical arm to be fixed to a wheel-chair, or a similar device used by the handicapped, decision on the number, type (revolute or prismatic) of joints and the order they are arranged is the first question to be answered. Many other problems of theoretical or practical nature, such as the mechanical design, joint actuation, how to command and control the arm motion, development of the necessary hardware and software is to be resolved before such a device can be successfully employed.

Such an arm must be capable of fulfilling the needs of a user as much as possible. In this sense, it must have a better manoeuverability than the industrial robot manipulators; this calls for additional degrees of freedom which make an arm redundant and consequently more complicated for motion control.

On the other hand, for the class of application this arm will be used for, the limitations on the size of various parts and the computer size and power impose certain constraints that cannot be avoided. The paper presents the results on structure, its kinematic solutions and proposals to deal with redundancy.

Keywords. Robots; Manipulation; Optimization; Kinematic Equations; Wheel-chair.

INTRODUCTION

This work is part of the study towards design of a mechanical arm that may be added to a mobile robot or be fixed to a wheel-chair or a table to be used by the handicapped. More emphasis has been given for the latter application, though at this stage only part of the problems involved in such a design are considered. For instance, the way a user is going to command the arm for the motion he/she desires is a secondary problem which is not studied at this stage.

For most of the manufacturing jobs that are accomplished by industrial robots, 4, 5 and 6 degrees of freedom suffice. Six degrees of freedom are enough in order that any desired position and orientation be achieved by the end effector within the working envelope of an arm. Any extra degree of freedom in this sense will be redundant. On the other hand, a human arm is highly redundant (Whitney 1986); this allows for better flexibility in operation, for instance in the presence of obstacles. However, if the hand is excluded, the arm and wrist only may be regarded as having seven degrees of freedom (Hemami, 1986), that is, with one redundant degree of freedom, although not always all the seven degrees of freedom are used (Hemami, 1986).

Furthermore, in doing a task a human may bend or turn his body to the side, that is, additional degrees of freedom are brought into effect. This is less likely to be the case of a wheel-chair user. But, for a mechanical arm that has to be moved around, simplicity and compactness of a controller is of primary importance. In other words, anything that adds to the complexity must be avoided, as much as possible.

STRUCTURE

By structure, we mean the number of links and the way they are jointed together. A first candidate may be thought of as an elbow type industrial robot (such as PUMA arm, but without the offsets) with six degrees of freedom, as shown in Fig. 1. This structure seems to be reasonably good in reaching the points around the user, in particular on the right hand side. Also, it can be conveniently set in a "park" position by rotating the first joint 180° away from the position in which it is shown, when not in use. However, from the stability point of view and the fact that the forearm can move only in the vertical plane passing through the upper arm (and this may not be so convenient for performing some tasks), the design illustrated in Fig. 2 is preferrable.

This is because the points in front of the user may be reached more easily without the upper arm obstructing the user's field of view. The difference between the two structures is in joint 2 (in additio to a 90° rotation of the first joint axis, being horizontal instead of vertical). In the latter, the axis of rotation θ_2 is along the link length whereas in the former it is perpendicular to it. This arrangement better resembles a human arm. In both of the above arms (and in what follows) a spherical wrist is used. A spherical wrist has three rotary joints all of which axes concide in the wrist point (point Q in the figures).

In order to increase the flexibility of the arm, in particular in reaching the points that are outside the reach of the arm in Fig. 2, a prismatic joint may be added to the forearm. This will enhance the ability of the arm in tasks such as picking up something from the floor or off the shelf. The arm-wrist combination in this case has seven degrees of freedom, as illustrated in Fig. 3.

KINEMATICS REVIEW

In programming and manipulation of industrial robots, the position and orientation of an end effector attached to the robot wrist is known with respect to a fixed reference coordinate system which is usually at the base of the arm. This is shown in the form of a 4 x 4 matrix T which results from the multiplication of the homogeneous transformation matrices A_i (i = 1, 2, ..., number of degrees of freedom) (see for example Paul 1981, Lee 1982, Wolovich 1987) which define the relationship between successive reference frames attached to each joint according to (Denavit and Hartenberg 1955). Thus if there are n degrees of freedom

$$T = A_1 A_2 \ldots A_n = \left[\begin{array}{c|c} R & \bar{p} \\ \hline 000 & 1 \end{array}\right] = \begin{bmatrix} n_x & s_x & a_x & p_x \\ n_y & s_y & a_y & p_y \\ n_z & s_z & a_z & p_z \\ 0 & 0 & 0 & 1 \end{bmatrix} \qquad (1)$$

where the 3 x 3 matric R denotes the orientation of the effector and p implies the coordinates of the end effector position. The three orthogonal vectors $\bar{a}$, $\bar{s}$ and $\bar{n}$ are the approach, slide and normal vectors, respectively, as shown in Fig. 4-a. Movement of the arm will be manipulated, therefore, by finding the inverse kinematic solution of the joint variables for various configurations of the arm.

For applications such as prosthetic arm, and the arm under study, it is more desirable (Whitney 1969) to express the required motion with respect to the hand (the end effector in an industrial robot) coordinate system rather than the base coordinate system. In this sense six independent motions, three translations along and three rotations about the x, y and z axes attached to the hand, namely reach, sweep, lift, turn, tilt and twist, as illustrated in Fig. 4-b, may be defined.

In order to execute any one or combinations of these motions, either the velocities or the position relations may be used as follows. For the former resolved motion rate control is suggested (Whitney 1972); if the components of a vector $\bar{q}$ represent the 3 translations and 3 rotations, respectively, as functions of $\bar{\theta}$ the n joint variables in the form of

$$\bar{q} = f(\bar{\theta}) \qquad (2)$$

then, the corresponding linear and angular velocities may be defined as

$$\dot{\bar{q}} = \begin{bmatrix} \bar{V} \\ \bar{\Omega} \end{bmatrix} = J(\bar{\theta})\,\dot{\bar{\theta}} \qquad (3)$$

where $\bar{V}$ represents the vector of velocities for lift, sweep and reach, and $\bar{\Omega}$ represents the angular velocities for turn, tilt and twist, and J is a 6 x n Jacobian matrix whose elements reflect the rate of change of $\bar{V}$ and $\bar{\Omega}$ with respect to $\bar{\theta}$, that is,

$$\left[J(\bar{\theta})\right]_{ij} = \frac{\partial f_i}{\partial \theta_j} \qquad (4)$$

For a nonredundant arm there are only 6 joints, and J is square; knowing $\bar{V}$ and $\bar{\Omega}$ then allows to find the rate of change of joint variables $\dot{\bar{\theta}}$ from equation (3) for nondegerate cases where J is not singular. This solution is symbolically denoted by

$$\dot{\bar{\theta}} = \left[J(\bar{\theta})\right]^{-1} \begin{bmatrix} \bar{V} \\ \bar{\Omega} \end{bmatrix} \qquad (5)$$

though the inverse of Jacobian matrix by itself is not necessary (Hollerbach and Sahar 1983). Equation (5) indicates that the knowledge of the joint angles is essential for the calculation of their rate of change at any momentary configuration of an arm. In other words, it has to be continuously updated.

On the other hand, bearing in mind that in grasping a stationary object, a person stops his/her hand motion when the required position/orientation is reached, the execution of the six motions in Fig. 4-b may be achieved in the following manner.

Supposing that the current position/orientation of hand is known in the form of the T-matrix in equation (1), and it is required that the hand is moved along its x-axis x_H (lift). If x_H is increased successively by a small value Δ_x in a number of steps (until a stop command is issued) then for each intermediate step the new position of the hand with respect to its former position coordinate system is defined by

$$\Delta T = \begin{bmatrix} & & & \Delta_x \\ & I & & 0 \\ & & & 0 \\ 0 & 0 & 0 & 1 \end{bmatrix} \qquad (6)$$

and with respect to the base coordinate system is defined by

$$T_0 = T \cdot \Delta T = \left[\begin{array}{c|c} R & R\begin{bmatrix} \Delta_x \\ 0 \\ 0 \end{bmatrix} + \bar{p} \\ 0\;0\;0 & 1 \end{array}\right] \qquad (7)$$

in light of equation (1). Equation (7) implies that the orientation R of the hand has not changed (as expected) but the value of its position vector has changed by

$$R \begin{bmatrix} \Delta_x \\ 0 \\ 0 \end{bmatrix} = \begin{bmatrix} n_x \Delta_x \\ n_y \Delta_x \\ n_z \Delta_x \end{bmatrix} = \bar{n}\,\Delta_x \qquad (8)$$

The corresponding joint variables may now be calculated from inverse kinematic solutions. The calculations have to be repeated as many times as necessary or until any of the joint angles has reached its limit. Similarly, movement of the hand in its y_H and z_H directions or a combination of lift, sweep and reach can be performed in the same manner.

A similar approach is used for turn, tilt and twist. In this sense, if $\Delta\theta_x$, $\Delta\theta_y$ and $\Delta\theta_z$ are the small rotations about x_H, y_H and z_H, then

$$\Delta T = \begin{bmatrix} 1 & 0 & 0 & 0 \\ 0 & \text{Cos}\,\Delta\theta_x & -\text{Sin}\,\Delta\theta_x & 0 \\ 0 & \text{Sin}\,\Delta\theta_x & \text{Cos}\,\Delta\theta_x & 0 \\ 0 & 0 & 0 & 1 \end{bmatrix} \quad \text{for turn} \qquad (9\text{-a})$$

$$\Delta T = \begin{bmatrix} \text{Cos}\,\Delta\theta_y & 0 & \text{Sin}\,\Delta\theta_y & 0 \\ 0 & 1 & 0 & 0 \\ -\text{Sin}\,\Delta\theta_y & 0 & \text{Cos}\,\Delta\theta_y & 0 \\ 0 & 0 & 0 & 1 \end{bmatrix} \quad \text{for tilt} \qquad (9\text{-b})$$

and

$$\Delta T = \begin{bmatrix} \text{Cos}\,\Delta\theta_z & -\text{Sin}\,\Delta\theta_z & 0 & 0 \\ \text{Sin}\,\Delta\theta_z & \text{Cos}\,\Delta\theta_z & 0 & 0 \\ 0 & 0 & 1 & 0 \\ 0 & 0 & 0 & 1 \end{bmatrix} \quad \text{for twist} \qquad (9\text{-c})$$

Equations (9) imply that the position of the hand does not change with respect to the coordinate system whereas a new orientation is obtained in each step as far as the rotating motion is continued. The step size Δ_x, $\Delta\theta_x$ and so on must be experimentally determined such that the hand motion is smooth enough while we have a real time control.

Computation of a manipulator's Jacobian matrix is time consuming (Lenarčič 1984, Orin and Schrader 1984). Computation and updating the T-matrix depends on the step size for Δ_x, etc. We have not yet compared the computation time and their relation for the proposed arm.

DEALING WITH REDUNDANCY

In an arm with r redundant joints, neither the Jacobian matrix is square nor the inverse kinematic equations can be solved without either assigning arbitrary values to r number of joints or bringing into effect some conditions that result in r additional equations. These conditions, contain the criterion to optimize a desired performance index, or the existance of some restrictions due to, say, obstacles that must be avoided. Stanišiś and Pennock have used an extra degree of freedom to overcome the degeneracy; but this cannot be always the case.

There exist two problems involved that must be resolved. The first one is to find an appropriate performance index for optimization, or the way the restrictive conditions, in case they exist, be embedded into the kinematic equations. The second one concerns the fact that the solution may not be determined in real time with a small microcomputer that is going to be used for this purpose.

Two alternative ways to deal with redundancy in our particular problem having only one redundant degree of freedom have been thought of as:

1- Use one of the joints - here the prismatic joint in particular - as an auxiliary joint that remain fixed until one of the other joints has reached its lower or upper limit.

2- By a series of lock/unlock switching, motion of arm is broken into smaller segments for each of which one of the joints is locked and the rest form a nonredundant arm.

In both of the above proposals it will become necessary to know the kinematic solutions for different classes of arms that result from eliminating one of the joints, or if the resolved motion rate approach is employed knowledge of the Jacobian matrix for those reduced structure arms becomes essential. If for simplicity the three last joints associated with the wrist are excluded from taking part in the substitution process in "1" or lock/unlock process in "2", there are classes of arms with four different structure to be dealt with.

In number 2, the order in which the lock/unlock process should switch between the participating joints may be determined on the basis of inducing certain characteristics (trajectory, planning or optimization for instance) in the motion. However, this can be done off-line, whence the results are transferred into the computer memory as a look-up table, etc.

For all cases discussed above, further work is necessary in order to find the most appropriate technique. For this reason, it was decided at this time to use only a nonredundant arm with six degrees of freedom including a prismatic joint, as shown in Fig. 5. This is similar to the arm in Fig. 3 when joint 2 is fixed at 90°degrees. However, a passive joint 2 may be added to rotate the elbow and forearm by 90°degrees, for parking purposes.

KINEMATIC SOLUTIONS

In this section, inverse kinematic solutions are found for the arm in Fig. 5. The arm itself is shown in Fig. 6. The coordinate frame attached to the first joint is $x_0y_0z_0$ and the rotation θ_1 takes place about z_0-axis which is the same as x_b-axis of the body frame $x_by_bz_b$ indicated in Fig. 5. The geometric parameters associated with the joints using the Denavit-Hartengerg (1955) representation is shown in Table 1. The joint transformation matrices are shown in Table 2, where S_i and C_i $(i=1,\dots,6)$ stand for Sin θ_i and Cos θ_i, respectively.

Table 1 - Geometric Parameters

Joint number	1	2	3	4	5	6
a_i (length)	ℓ	0	0	0	0	0
d_i (offset)	0	0	variable	0	0	e
α_i (twist)	90°	90°	0	90°	90°	0

Table 2 - Transformation Matrices

$$A_1 = \begin{bmatrix} C_1 & 0 & S_1 & \ell C_1 \\ S_1 & 0 & -C_1 & \ell S_1 \\ 0 & 1 & 0 & 0 \\ 0 & 0 & 0 & 1 \end{bmatrix} \qquad A_2 = \begin{bmatrix} C_2 & 0 & S_2 & 0 \\ S_2 & 0 & -C_2 & 0 \\ 0 & 1 & 0 & 0 \\ 0 & 0 & 0 & 1 \end{bmatrix}$$

$$A_3 = \begin{bmatrix} 1 & 0 & 0 & 0 \\ 0 & 1 & 0 & 0 \\ 0 & 0 & 1 & d \\ 0 & 0 & 0 & 1 \end{bmatrix} \qquad A_4 = \begin{bmatrix} C_4 & 0 & S_4 & 0 \\ S_4 & 0 & -C_4 & 0 \\ 0 & 1 & 0 & 0 \\ 0 & 0 & 0 & 1 \end{bmatrix}$$

$$A_5 = \begin{bmatrix} C_5 & 0 & S_5 & 0 \\ S_5 & 0 & -C_5 & 0 \\ 0 & 1 & 0 & 0 \\ 0 & 0 & 0 & 1 \end{bmatrix} \qquad A_6 = \begin{bmatrix} C_6 & -S_6 & 0 & 0 \\ S_6 & C_6 & 0 & 0 \\ 0 & 0 & 1 & e \\ 0 & 0 & 0 & 1 \end{bmatrix}$$

In tables 1 and 2, ℓ is the length of the upper arm, d is the length of the forearm (variable), and e is the effective length of the hand, from wrist point Q to finger tip point (The tool point) P. In a spherical wrist (Featherston 1983, Paul and Zhang 1986) when the position/orientation of hand is known in the form of the T-matrix in equation (1) with respect to the arm base coordinate system $x_0y_0z_0$, or any other coordinate frame like $x_by_bz_b$ in Fig. 5, then the wrist point position is readily found from

$$\bar{q} = \bar{p} - \bar{a}\, e \tag{10}$$

where $\bar{a}$ is the approach vector. The x, y and z components of point Q are known now and will be denoted by Q_x, Q_y and Q_z, respectively.

Figure 7-a and 7-b illustrate two views of the arm in Fig. 6 up to the wrist point when looked at along the z_0-axis (in the negative direction) and normal to the plane of the upper and forearms, respectively.

Figure 7-a depicts that

$$\theta_1 = \tan^{-1} \frac{Q_y}{Q_x} \tag{11}$$

and it is clear from Fig. 7-b that

$$\theta_2 = \tan^{-1} \frac{Q_z}{\sqrt{Q_x^2 + Q_y^2} - \ell} + \frac{\pi}{2} \tag{12}$$

Also, in the right angle triangle $Q Q_M E$

$$d^2 = Q_z^2 + (\sqrt{Q_x^2 + Q_y^2} - \ell)^2 \tag{13}$$

SOMMAIRE

Dans la conception et la fabrication d'un bras mécanique destiné à être fixé sur une chaise roulante ou sur une table, les deux problèmes principaux consistent tout d'abord à choisir une structure appropriée - numéros de série des joints et leur disposition - et ensuite à définir un mode de fonctionnement. Puis, parmi les problèmes de manipulation devant être résolus à un stade ultérieur, il y a, au niveau de la conception mécanique, le système de commande et d'asservissement du bras, l'élaboration de logiciels, etc. Certains aspects tant théoriques qu'expérimentaux de cette conception exigeront beaucoup de recherche.

Les deux problèmes fondamentaux, celui de la structure ainsi que la manipulation de ce bras, sont examinés dans cet exposé. Nous avons opté pour la fabrication d'un bras offrant sept degrés de liberté et comprenant un joint prismatique afin de pouvoir atteindre, à l'aide de l'avant-bras, un point relativement éloigné. Cependant, la complexité de la cinématique du bras étant relative au nombre de joints et également en raison des limites qu'imposent la puissance et les dimensions d'un ordinateur pour la supervision et l'asservissement de ce bras, seuls six degrés de liberté peuvent être incorporés. Des solutions cinématiques sont appliquées pour ce modèle réduit, au cas où sa manipulation devrait dépendre du contrôle plutôt que de la vitesse de déplacement. La décision finale reposera néanmoins sur les résultats des recherches en cours dans ce domaine.

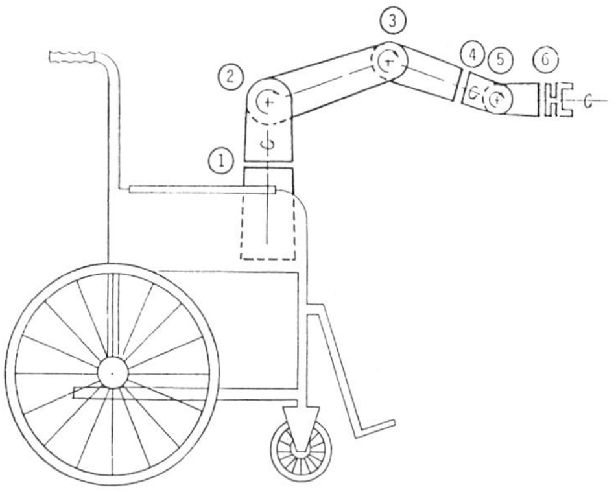

Fig. 1. PUMA Type Arm

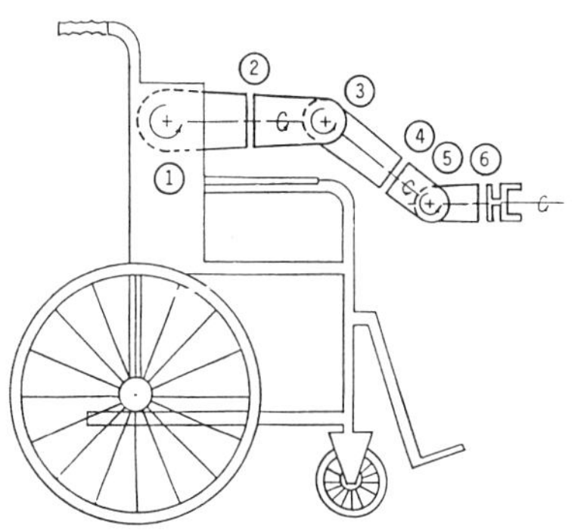

Fig. 2. A Mechanical Arm that Better Resembles a Human Arm

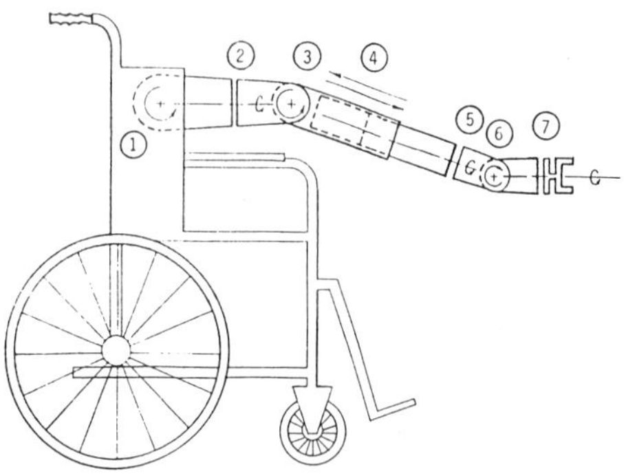

Fig. 3. Arm in Figure 2 Plus a Prismatic Joint

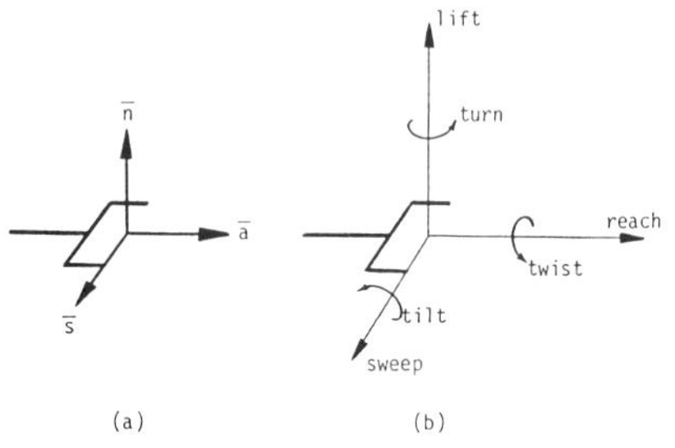

Fig. 4. Linear and Angular Motions in Hand Coordinates

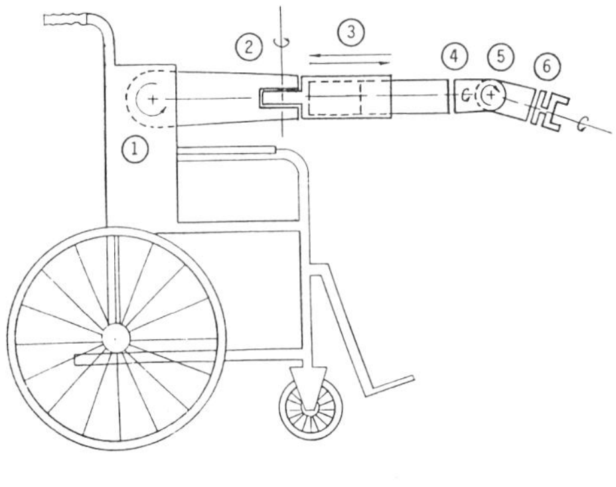

Fig. 5. A 6-Degree of Freedom Arm with a Prismatic Joint

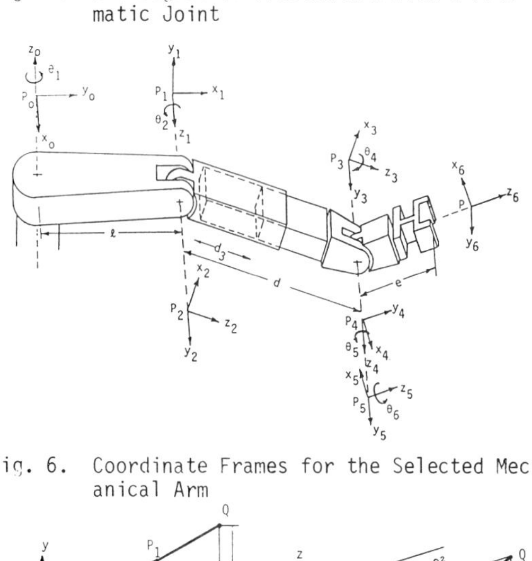

Fig. 6. Coordinate Frames for the Selected Mechanical Arm

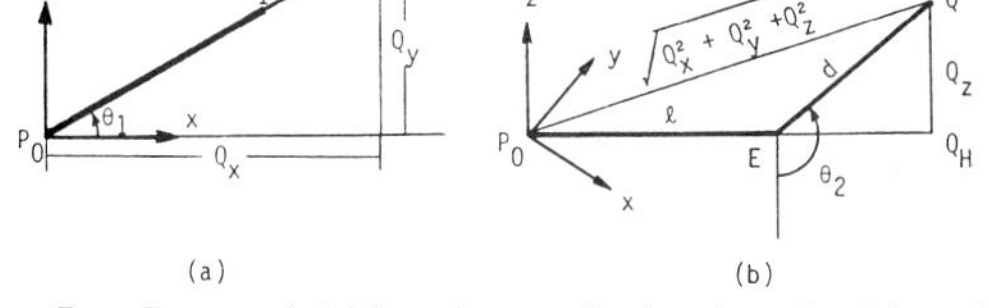

Fig. 7. Top and Side Views of the Arm in Fig. 6

which leads to

$$d = \sqrt{Q_x^2 + Q_y^2 + Q_z^2 + \ell^2 - 2\ell\sqrt{Q_x^2 + Q_y^2}} \qquad (14)$$

When the joint variables θ_1, θ_2 and d are found, the rest of joint variables corresponding to the wrist can be determined as follows (Hollerback and Sahar 1983, Hemami 1985):

from equation (1), the rotation part of the T-matrix, that is the 3 x 3 matrix R, is obtained from

$$R = R_1 R_2 R_3 R_4 R_5 R_6 \qquad (15)$$

where R_i (i = 1,...,6) are the rotation parts of transformation associated with the joints. In light of equation (15), and of the knowledge of θ_1, θ_2 and d, we have

$$R_4 R_5 R_6 = (R_1 R_2 R_3)^{-1} R \qquad (16)$$

If the right-hand side of equation (16) is denoted by W, then

$$R_4 R_5 R_6 = W = \begin{bmatrix} w_{11} & w_{12} & w_{13} \\ w_{21} & w_{22} & w_{23} \\ w_{31} & w_{32} & w_{33} \end{bmatrix} \qquad (17)$$

and it follows from Table 2 and equation (17) that

$$R_5 R_6 = \begin{bmatrix} C_5 C_6 & -C_5 S_6 & S_5 \\ S_5 S_6 & -S_5 S_6 & -C_5 \\ S_6 & C_6 & 0 \end{bmatrix} = R_4^{-1} W$$

$$= \begin{bmatrix} w_{11}C_4 + w_{21}S_4 & w_{12}C_4 + w_{22}S_4 & w_{13}C_4 + w_{23}S_4 \\ w_{31} & w_{32} & w_{33} \\ w_{11}S_4 - w_{21}C_4 & w_{12}S_4 - w_{22}C_4 & w_{13}S_4 - w_{23}C_4 \end{bmatrix} \qquad (18)$$

except for a degenerate case for which in equation (17)

$$\begin{cases} w_{33} = -1 \\ w_{13} = w_{23} = w_{31} = w_{32} = 0 \end{cases} \qquad (19)$$

and corresponds to $\theta_5 = 0$, in general comparison of equations (18) and (19) leads to the following results

$$\theta_4 = \tan^{-1} \frac{w_{23}}{w_{13}} \qquad (20)$$

$$\theta_5 = \tan^{-1} \frac{\sqrt{w_{13}^2 + w_{23}^2}}{-w_{33}} \qquad (21)$$

$$\theta_6 = \tan^{-1} \frac{w_{11} \sin\theta_4 - w_{21} \cos\theta_4}{w_{12} \sin\theta_4 - w_{22} \cos\theta_4} \qquad (22)$$

In equation (16) it is not necessary to invert the product matrix $R_1 R_2 R_3$, since the inverse of the resulting matrix will be equal to its transpose because of orthogonality properties. The inverse kinematic solutions discussed here will be necessary if the manipulation of arm is going to be on the displacement bases, as discussed in the preceding section. If an extra joint (joint 2 in Fig. 3) is added, this will only provide part of the solution for the kinematic equations.

CONCLUSIONS

The primary problems concerning the design and development of a mechanical arm that may be used on a wheel-chair or be fixed to a table, are first to select an appropriate structure in the sense of the type, number and arrangement of joints, and second to determine the way it can be manipulated. The mechanical design; how the arm is commanded and controlled, software development, etc. are among other problems that have to be resolved in later stages. Some considerable research work is required to be carried out for various matters that are involved in such a design, both theoretical and experimental.

In this paper, the two fundamental problems of the structure and the manipulation are discussed. A seven degree of freedom arm, including a prismatic joint to enhance the capability of reaching some relatively further points, by stretching the forearm, has been decided on for this purpose. However, because the redundancy in the number of joints increases the complexity of the kinematics of an arm, and in view of the limitations associated with the power and size of a computer to supervise and control the manipulation of such arm, at this stage only six degrees of freedom are considered. The kinematic solutions for this reduced model are obtained in case the manipulation will be based on displacement control and not the velocity. Nevertheless, the final decision depends on the results of further research that is currently in progress on the subject.

REFERENCES

Denavit, J. and R.S. Hartenberg (1955). A kinematic notation for lower-pair mechanisms based on matrices, J. Applied Mechanics, 22, 215-221.
Featherstone, R. (1983). Position and velocity transformations between robot end-effector coordinates and joint angles, Int. J. Robotics Research, 2, No. 2, 35-45.
Hemami, A. (1985). Control and programming of a two-arm robot, Proc. Robots 9 Conf., Detroit, 16-38/16-58.
Hemami, A. (1986). On a human-arm-like mechanical manipulator, Robotica, to appear.
Hollerbach, J.M. and G. Sahar (1983). Wrist partitioned, inverse kinematic acceleration and manipulator dynamics, Int. J. Robotics Research, 2, No. 4, 65-76.
Lee, C.S.G. (1982). Robot arms kinematics, dynamics and control, (IEEE) Computer, December, 62-80.
Lenarčič, J. (1984). A new method for calculating the Jacobian for a robot manipulator, Robotica, 1, 205-209.
Orin, D.E. and W.W. Schrader (1984). Efficient computation of the Jacobian for robot manipulators, Int. J. Robotics Research, 3, No. 4, 66-75.
Paul, R.P. (1981). Robot manipulators: mathematics, programming and control, MIT Press, Cambridge.
Paul, R.P. and H. Zhang (1986). Computationally efficient kinematics for manipulators with spherical wrists based on the homogeneous transformation representation, Int. J. Robotics Research, 5, No. 2, 32-44.
Stanišić, M.M. and G.R. Pennock (1985). A non-degenerate kinematic solution of a seven-jointed robot manipulator, Int. J. Robotics Research, 4, No. 2, 10-20.
Whitney, D.E. (1969). Resolved motion rate control of manipulators and human prostheses, IEEE Trans. Man-Machine Systems, MMS-10, No. 2, 47-53.
Whitney, D.E. (1972). The mathematics of Coordinated control of prosthetic arms and manipulators (ASME) J. Dynamic Systems, Measurement and Control, 94, No. 4, 303-309.
Wolovich, W.A. (1987). Robotics: basic analysis and design, HRW, N.Y.

Innovative Pantograph to make a Robot Skate

A. V. S. BHASKAR AND T. NAGARAJAN

Precision Engineering and Instrumentation Laboratory, Department of Mechanical Engineering, Indian Institute of Technology Madras, Madras 600 036, India

Abstract. A robot with mobility on legs constitutes the central idea of this paper. The robot drives itself using 4 skates. Each skate is supporting the body through a double pantograph. What follows presents the need for the innovative pantograph model, how the mobility is achieved using it, and the identification of the optimum point of contact and lift-off of the skate to achieve best kick. Acceleration plots of the skates have been used in the study. The peak velocity region in the stroke of its legs has been proposed for optimum landing of the skates.

Key words. Mobile robots; legged locomotion; obstacle clearance; gait; pantograph.

INTRODUCTION

Legged robots have proved to be very successful on uneven terrain (Raibert, 1982). They are able to sense and avoid obstacles. They can maintain their stability even if one or few of their leg-tips are not in plane with the others. They can also climb steps(Fischetti, 1985). The only marked disadvantage is that they are slow. On the contrary, the wheeled robots are considerably faster but cannot crossover obstacles or climb steps. Clearly robots which could have speed, capacity to climb steps and above all, "Walking over" obstacles while moving at atleast the speed of a wheeled robot, would be a rich blend of the two orthodox designs. It opens up a new avenue in robot locomotion- Robots that have legs in conjunction with wheels at their end.

GENERAL STRUCTURE OF THE ROBOT

The robot, named Grass Hopper, has 4 "bays" in its main body (Fig. 1) each of which houses a "double pantograph" mechanism (Fig. 2). An ankle, powered by an electric motor, is coupled to the bottom of each leg. The ankle carries a skate below it. This roller skate can be indexed through 360 degrees using the motor, similar to twisting an ankle in a human foot.

Legged locomotion is a science, which, try however it can to resist the temptation to copy human orthopaedics, kinematics, cybernatics and bionics, is carried away by the simplicity in movement and maneuverability. The present paper is an example of such a case, wherein the human motions involved in the use of skates while roller skating, is applied to the locomotion of the robot. One slight change is that it is extended to 4 legs.

The Muscle Behind The Robot

At the upper end of the pantograph mechanism are two pneumatic rodless cylinders, one placed horizontally and the other vertically (Fig. 5). The basic rule of a pantograph applies here also, i.e., keeping one of the three points fixed, if the second is moved, the third follows the path faithfully, but for the variation in the velocity (all the three points are always collinear). Likewise, here the

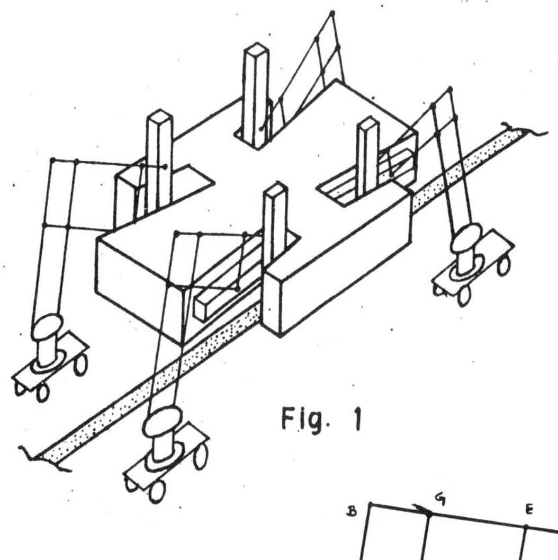

Fig. 1

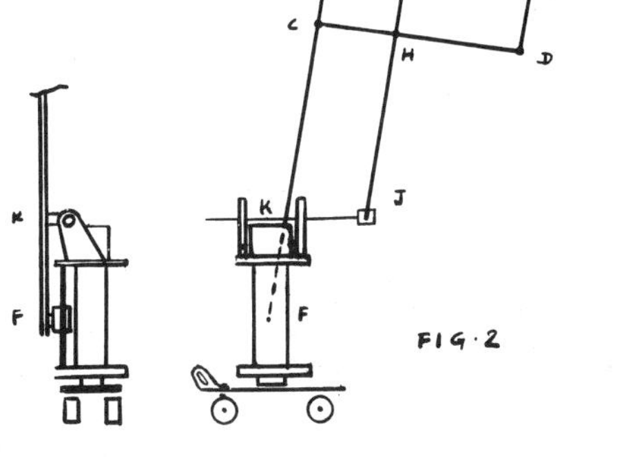

FIG·2

vertical cylinder joint is arrested and on moving the piston of the horizontal rodless cylinder, the skate moves perfectly parallel to the ground. And when the vertical cylinder alone is actuated, the skate purely travels vertical. It is left to the user to judiciously combine the two motions to obtain a particular foot motion (Fig. 5) which may be described as 'Horizontal backward kick; kick and slow lift; reaching highest point; drawing-in and slow lowering; final touch down' corresponding to the fully drawn-in condition of the horizontal cylinder.

Each of the bays are at an angle 20 degrees to the main body axis. The pantograph kicks in a plane of the same angle as the bay. If the skates are oriented so as to be parallel to the the direction of the kick, the wheels will rotate freely without any friction; the leg and the skate alone slide backward on the ground, the body stays as it is. The body is unable to move forward because the foot does not give any support against which it can kick-off to overcome its inertia.

The same mode of operation is applied to the legs of the robot also. The combined use of wheels and legs is an important breakthrough aimed at exploiting the advantage of wheels fully by alienating the drive mechanism (pneumatic) from them. The wheels are as such free to rotate in any direction.

Order of Foot Placements

According to Todd (1985) a gait is a repetitive pattern of foot placements (Fig. 4). It is usual to assume that each leg is sufficiently specified as a two state device, the states being on the ground and off it. The legs on the ground are supporting and propelling the mobile, and those in the air are being retracted. The concept of gait assumes a regular progression forwards or backwards. Clearly there are times when an animal stands still or shuffles in an irregular way and the idea of gait has no utility at these times. Any of the 4-legged animals have a 1-2-4-3 gait

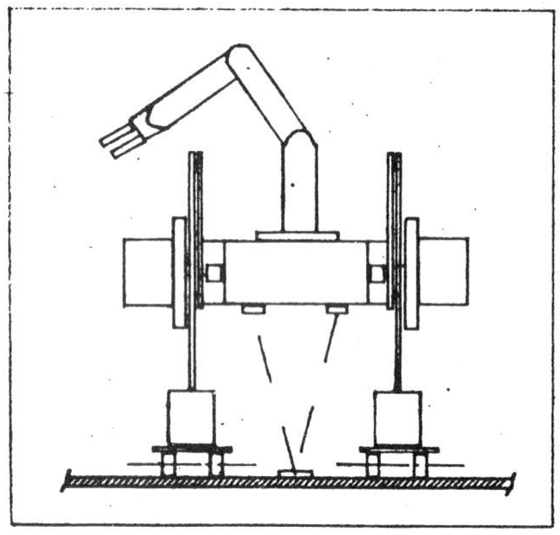

Fig. 3

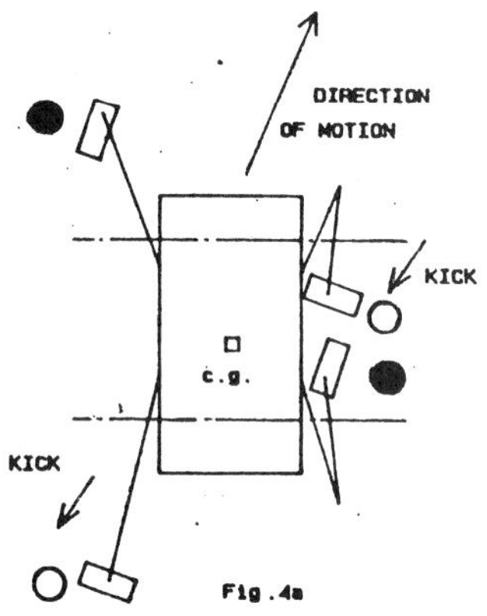

Fig.4a

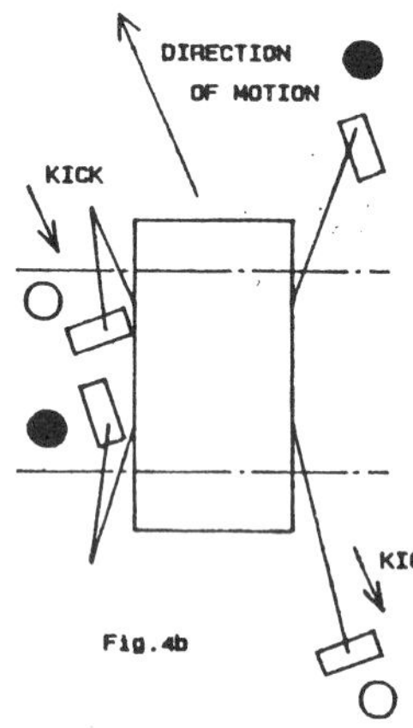

Fig.4b

To solve this problem one can do away with the skates and its wheels and use the tip of the pantograph to rest on the ground. But this reduces its speed and capacity to turn.

The Skater's Secret

A human being also faces the same problem described above,while skating. This is very obvious with a beginner. An experienced skater, to start with, keeps one skate and the corresponding foot (say the left foot) in the direction of motion and kicks-off with the other. The second foot, (the one which is kicking, i.e., the right foot) is twisted at the ankle and the length of the foot is at right angles to the first. The wheels of the right skate do not rotate and provide for the necessary support. At the end of the kick the right leg is lifted off the ground, the foot twisted back to normal (parallel to the direction of motion) and placed in front. The left leg now kicks as the other did previously.

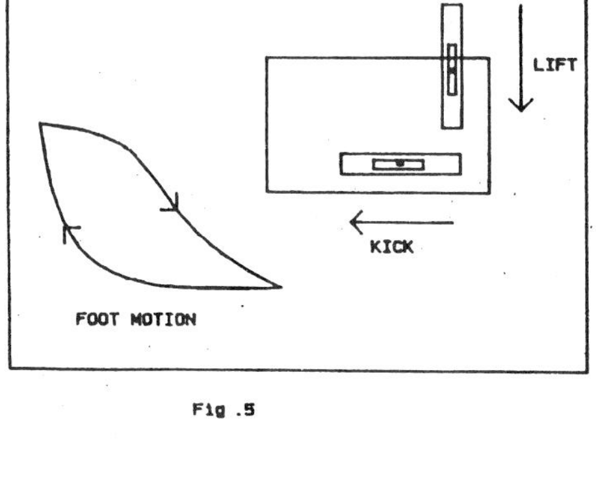

Fig.5

(legs numbered anticlockwise in plan view starting from the right rear leg). This gait assures that at any time only one leg is in the air and the rest form a triangle within which falls the center of gravity.

The double pantograph in conjunction with the four wheels on each leg gives a very good stability. Because of this the minimum number of legs to be on the ground can be reduced to two. As a result its gait is (1,3)-(2,4).

Two diagonally opposite legs when placed on the ground, form a rectangle with the extreme four wheels as corners. The center of gravity falls within this area. Moreover the longitudinal axis of the rectangle, if by careful arrangement of the legs passes through the C.G., maximum stability is

being in a horizontal plane is to be satisfied through out the kick of the leg. It is permitted to go out of the horizontal plane only while doing other motions such as lift and lowering.

In order to correct the plane of the skate continuously, the second pantograph AEGHDJ is superimposed on the first pantograph. Both have the same ratios. A and D are correspondingly the vertical and horizontal cylinder connecting points. Both J and F move on a horizontal line. With respect to world coordinates, J and F are moving (Translating) along X, with only the distance between then along X varying. Hence vertical drop between them is always constant. If that is equal to KF then the skate is horizontal and will continue to be so through out. Only KJ varies and as that

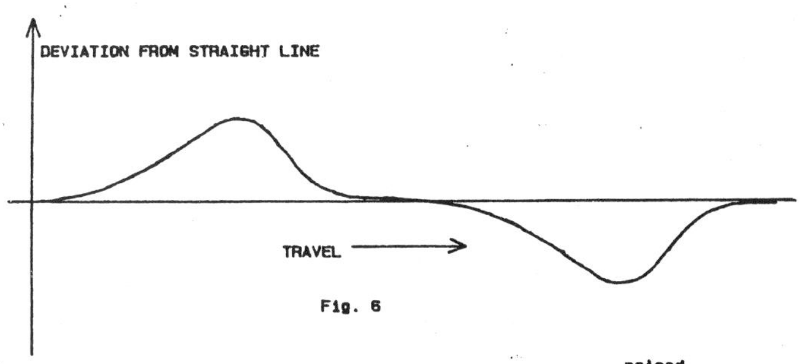

Fig. 6

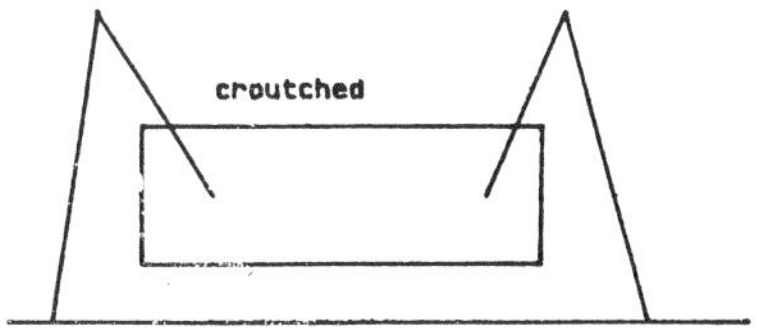

Fig. 7a

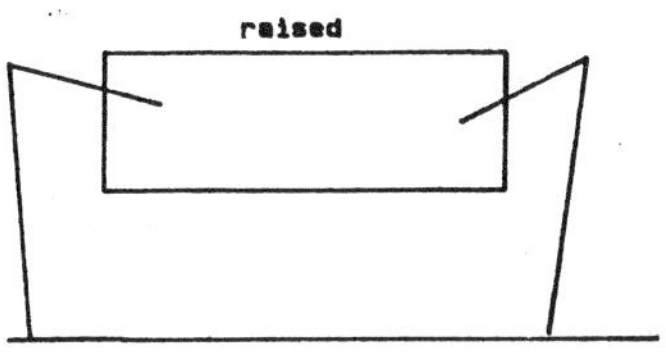

Fig. 7b

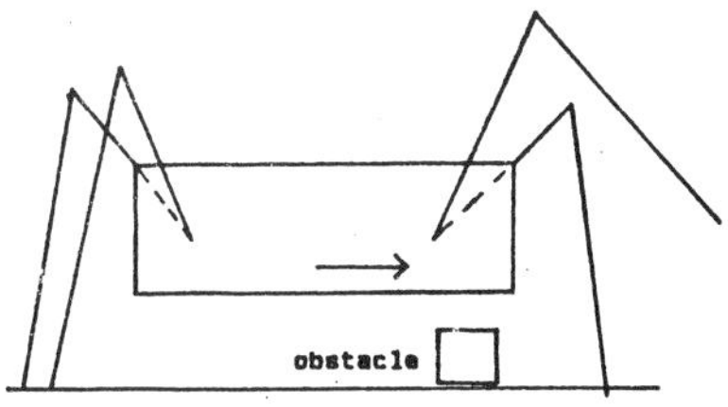

Fig. 8a

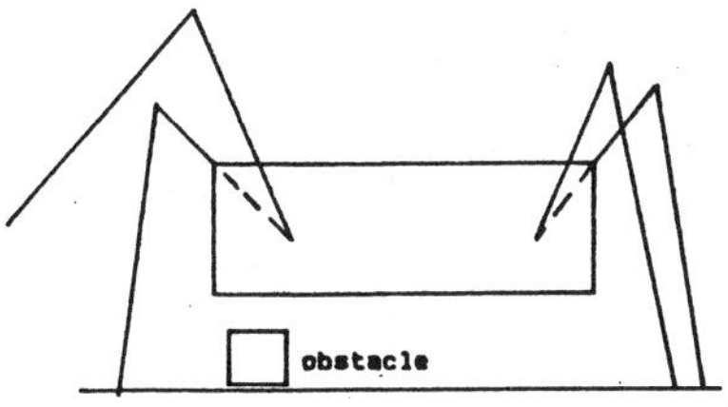

Fig. 8b

achieved. And as the legs once on the ground, do not move relative to the body, this position is unchanged till the time comes and the other two diagonal legs have to be placed on the ground and the first pair will have to be used for kicking. It can also switch over to the 1-2-4-3 animal gait if need be.

The Double Pantograph

A single pantograph A B C D E F (Fig.2) would normally be required if the robot were of the pure legged type. But now the requirement of using skates necessiates modification of the same. The 4 wheels on each skate are to be parallel to the ground for all of them to touch, in order to avoid unnecessary stress developed otherwise. Moreover this condition of all 4 wheel axes

variation is being accomodated by sliding of KJ in J, KF is vertical.

OPTIMUM LEG LANDING POINT

An acceleration pickup clamped on to the skate gave the actual acceleration plots of the leg (Fig.9). The sharp spikes are the acceleration peaks of the horizontal forward and reverse leg motions. This figure represents entire cycle of foot motion.

The pickup was placed with its sensitivity plane horizontal and the leg reciprocation was recorded with the vertical cylinder remaining idle (Fig. 10 a). The first two peaks are correspondingly the forward acceleration (during kick) and the reverse

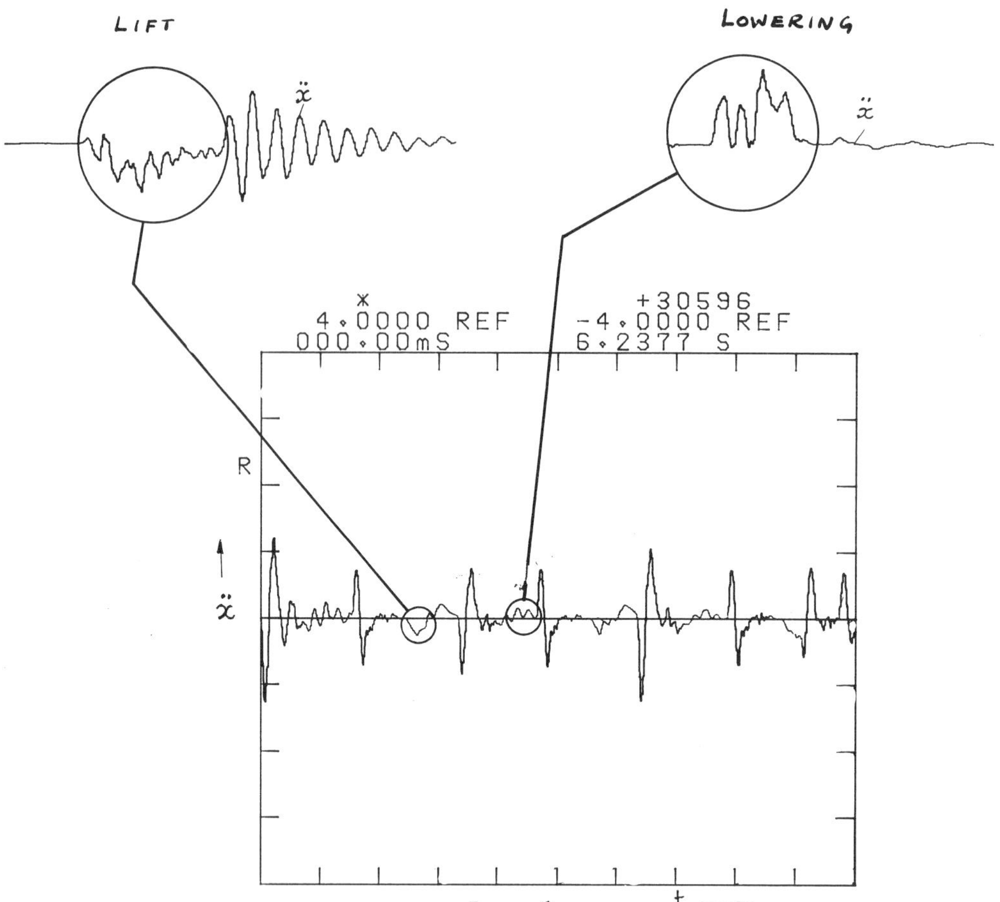

FIG. 9

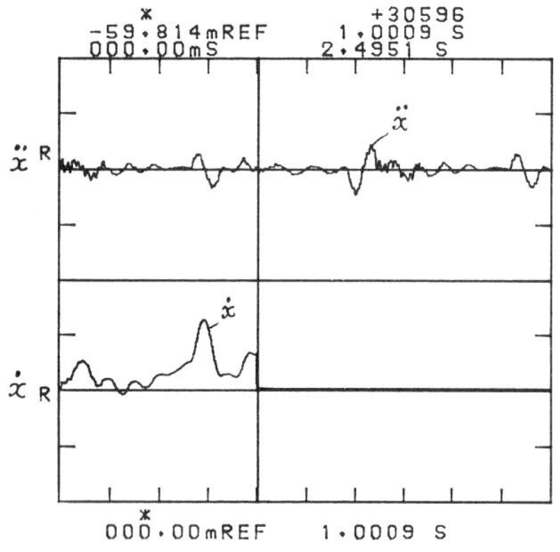

FIG. 10 A

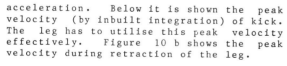

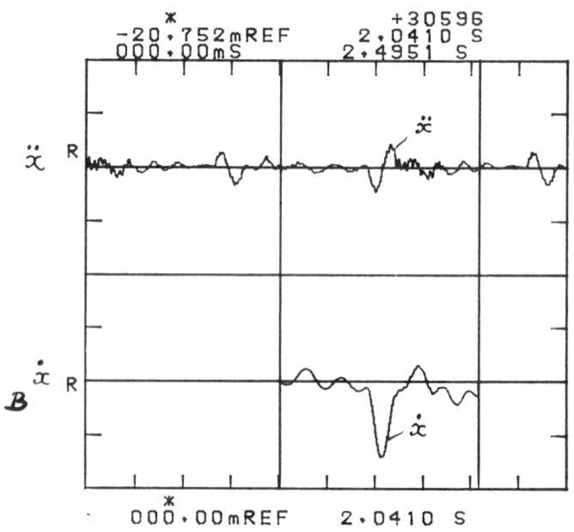

FIG. 10 B

acceleration. Below it is shown the peak velocity (by inbuilt integration) of kick. The leg has to utilise this peak velocity effectively. Figure 10 b shows the peak velocity during retraction of the leg.

Conclusion

For effective utilization of the kick, the leg has to touch down at a point before the peak velocity region. This point of contact is 15% of the stroke from the fully retracted position in the case of this robot 'Grass hopper'. This is unique in the sense that it combines both wheeled and legged versions effectively.

REFERENCES

Bhaskar, A.V.S. and Nagarajan, T. (1986). A mobile 4 legged robot for unmanned material handling in an industrial environment. AIMTDR Conference, IIT Delhi.

Fischetti, Mark. A. (1985). Robots do the dirty work. IEEE Spectrum, April, p.65-72.

Raibert, Marc.H and Sutherland, Ivan. E. (1982). Machines that walk. Scientific American, p.32-41.

Todd, T.J. (1985). Walking machines - an introduction to legged robots. Kogen Page.

A Study of Dynamics in Initiation and Stopping of Human Gait

T. YAMASHITA*, T, TANIGUCHI* AND R. KATOH**

*Kyushu Institute of Technology, Tobata, Kitakyushu 804, Japan
**Toa University, Shimonoseki 751, Japan

Abstract. The transient characteristics of the human gait, observed in the initiation and stopping, were analyzed. In the experiment, two force plates which detected only the vertical force component were used. A simple model consisting of a rigid body and two massless legs was used in our model analysis. Both the supporting force and the point of application behave differently compared to the steady walking. Our model analysis clearly showed a mechanism of these specific characteristics from the view point of the mechanics. The following results were obtained. Many of time parameters had correlation with the cadence. The comparisons of simulated results with experimental results showed that the simulation method was appropriate.

Keywords. Gait initiation; Gait stopping; Gait experiment; Simulation; Human walking; Modeling; Regression analysis.

INTRODUCTION

Many researchers have studied human level walking by various methods which can be classified into the experiment and the model analysis (Cunningham, 1958; Murray, 1967; Frank, 1970). Tsuchiya and others (1971) have shown experimentally that some unique phenomena are observed in the characteristic of both reaction force and its point of application in gait initiation process. These characteristics are often referred to "inverse" because the motion takes place in the direction that is reverse to a direction which is anticipated from the steady state. The mechanisms of the inverse phenomenon, or how the inverse characteristic relates the body motion in transient process, is not studied deeply. The model simulation method is useful for the analysis of the phenomena. Although Cook and Cozzens (1976) have experimentally analyzed the characteristics in gait initiation process, there are few studies about the gait stopping process.

This paper analyzes the transient characteristics of the human gait, observed in the initiation and stopping by an experimental method and a modeling method. An experimental method and a modeling method for both the initiation and the stopping process will be presented. In the experiment, force plates method was adopted to detect the force acting between floor and leg. As for the human body motion, most important characteristic is the motion of point of application acting the legs. We measured only the vertical force component so that the motions of the point of application can be calculated from the measured force data. The results of both the experiment and model simulation are given and discussed.

EXPERIMENT

Two force plates which detected only the vertical force component were used to measure separately the force acting to each leg. They were placed in parallel together with some auxiliary platforms to provide an experimental walkway. Parameter values characterizing the process, that is, the positions of the point of application, supporting periods of each leg, and step lengths and widths were also measured.

These data were collected manually and through the dynamic strain meter, A/D converter, and the scanner and were processed by a desktop computer (YHP 9845B), which was also used in the model simulation. Our subjects were ordered to walk at various speeds, from a very slow walking to high speed, which were indicated by a metronome for every test. Ten cadences ranged from 48 to 152 (steps/min) were adopted to direct walking speed in both gait initiation and stopping.

Figure 1 shows a time history of reaction force and a trajectory of point of application on horizontal plane as examples of experimental results. The unique characteristics in transient state of walking clearly appears in the figure, that is, both the supporting force and the point of application behave differently compared to the steady walking. These characteristics are often referred to "inverse" as mentioned above. Amount of inverse motion of reaction force and of point of application are shown respectively F_p, F_m, X_1 and Y_1 in Fig. 1.

Parameters mentioned above were also analyzed by the regression method to find correlations between these parameters and walking speeds.

MODEL SIMULATION

A simple model consisting of a rigid body and two massless legs, which have given satisfactory results for the steady walking (Yamashita and Taniguchi, 1979), was used in our model analysis. A mathematical model was derived in a suitable form for the analysis of the transient process. The equations are derived provided that the body moves in a three dimensional space and the leg length is constant. The following differential equations result from linealization (Yamashita and Taniguchi, 1986a,1986b).

$$m \ddot{X}_E = W (X_E - X_F)/(1 + h_z) \quad (1)$$

$$m \ddot{Y}_E = W (Y_E - Y_F)/(1 + h_z) \quad (2)$$

where X,Y position in the earth-fixed rectangular

coordinate system,
W body weight (= mg),
h_z vertical distance between hip joint and center of gravity,
g gravitation constant,
l length of the leg,
m mass of the body,
subscript
E center of gravity of the body mass,
F position of the point of application of resultant force.

In the simulation the following conditions were assumed: a steady walking is achieved after the contact of second-swinging leg in gait initiation; the stopping process requires two steps following the steady state; the differential equations were solved to find the amounts of the inverse motion in two directions. Parameter values obtained in our experiment were utilized in the simulation in order to characterize a real human walking.

RESULTS

Table 1 shows some of the result of the regression analysis, where T_i means supporting time in i-th phase (i=1,2...,6). It became clear that the time parameters and some movements of the point of application, including the inverse motion X_1, are related to the cadence. On the other hand, the inverse amount Y_1 is related to the inverse amount of reaction force F_p (Yamashita and Taniguchi, 1986a).

The characteristic of inverse motion of both the reaction force and its point of application were also measured in gait stopping process. Tendency of the result of regression analysis are also similar to that of initiation (Yamashita and Taniguchi, 1986c).

Our model analysis clearly showed a mechanism of these specific characteristics from the view point of the mechanics. Also, this model was useful to explain the difference between the motion of the center of mass of the body and that of the point of application in transient phase of walking because the equation included these two as variables. The nondimensional time $\tau = t\sqrt{g/(l+h_z)}$ was introduced and lengths were normalized by the leg length. So expressions of time and lengths are nondimensional in following figures.

Both the initiation and the stopping process were successfully simulated for a wide range of walking speed. Figure 2 compares the amount of inverse motion with simulated and measured, and shows that the simulated values exist within the limit of measured data. Figures 3 and 4 show the time history of the center of gravity and point of application in gait initiation and gait stopping, respectively. Although the point of application moves initially inverse direction, the center of mass moves in the reasonable direction (Fig.3): the inverse motion of the point of application in the lateral direction makes the body move toward the leg that supports the body first; the inverse motion of the point of application in the sagittal direction makes the body accelerate in the walking direction.

In gait stopping process, although the point of application moves reverse direction to the walking direction in last double supporting phase, the center of mass stops the motion by moving forward constantly in the sagittal direction (Fig. 4).

Figure 5 shows a time history of sagittal reaction force simulated from initial stage to the steady state. The sagittal forces in initial stage is always positive, which means that the force makes the body accelerate to forward. It's quite reasonable to initiate the gait. These positive forces are related to the inverse motion of point of application.

DISCUSSION

Our model consisting of one body and two massless legs (just as birds — flamingo or crane), is very simple as compared with the form of human being. People who want to manufacture and control a biped walking robot, will insist that it is too simple, because
(1) mechanical model such as ours can't be manufactured in practice, and
(2) this model guarantees that the swing leg can be controlled without any restrictions and this is not realistic.

On the other hand, results simulated by the model coincide with the experimental results, as shown in Figs. 2 and 3, also in the paper already reported (Yamashita and Taniguchi, 1979). This seems to be due to following both causes :
(i) The control strategy to maintain the body at the vertical position is adequate,
(ii) our model can provide characteristics mathematically or functionally equivalent to the human walking, though the model seems to be never similar to the structure of man.

The fact (i) is often used strategy on controlling the biped walking machine. (ii) is significant for the robot designers. This suggests that human leg has the form satisfying such condition, which has been given by Katoh and Mori (1984), the motion of the swinging leg has little influence on the motion of the supporting leg. This can contribute to the simplification of control strategy of human biped locomotion or biped walking machine.

CONCLUSION

The transient characteristics of the human gait, observed in the initiation and stopping, were analyzed. The experimental method and the modeling method for both the initiation and the stopping process are presented.

The following results were obtained.
(1) Many of time parameters have correlation with the cadence.
(2) The inverse characteristics have important relation to both the human gait initiation and stopping: the inverse change of the reaction force makes the body move toward the leg that supported the body first; that of the point of application in the sagittal direction makes the body accelerate in the walking direction.
(3) Inverse motion of the point of application of resultant force is related to that of reaction force.
(4) Both the initiation and the stopping process are successfully simulated for a wide range of walking speed.
(5) So, the simulation method is appropriate, and this model is useful to explain the difference between the motion of the center of mass of the body and that of the point of application in transient phase of walking.

These results suggest that the assumed control strategy, that is, to keep the body upright is right and proper to simulate the human walking characteristics.

REFERENCES

Cook, T. and B. Cozzens (1976). The initiation of gait. In R. M. Herman and others (Ed.), Neural Control of Locomotion, Prenum Press. pp. 65-75.

Cunningham D. M. (1958). Components of floor reaction during walking, Univ. of Calif., Berkley.

Frank, A. A. (1970). An approach to the dynamic analysis and synthesis of biped locomotion mechanics, Med. and Biol. Engg., 8, 465-476.

Katoh, R. and M. Mori (1984). Control method of biped locomotion giving asymptotic stability of trajectory, Automatica, 20, 405-414.

Murray, M. P. (1967). Gait as a total pattern of movement. Am. J. Phys. Med., 46, 290-333.

Tsuchiya, K. and others (1971). Gait analysis by foot print. In SOBIM Japan (Ed.), Proceedings of The Second Domestic Symposium on Biomechanisms. pp.147-162.

Yamashita, T. and T. Taniguchi (1979). Simulation of human walking characteristics by simple model. Proceedings of the Fifth World Congress on Theory of Machines and Mechanisms. 851-854.

Yamashita, T. and T. Taniguchi (1986a). Study of human gait initiation —— Experiment and model analysis. Trans. of the Society of Instrument and Control Engineering, 22, 176-202.

Yamashita, T. and T. Taniguchi (1986b). Study of human gait initiation —— Regression analysis of experimental data and simulation. Trans. of the Society of Instrument and Control Engineering, 22, 303-309.

Yamashita, T. and T. Taniguchi (1986c). Study of human gait stopping. Trans. of the Society of Instrument and Control Engineering, 22, 411-416.

ZUSAMMENFASSUNG

Die übergangsverhalten Charactereigenschaften des menschlichen Gehens, beim Gangbeginn und im Ganghalt, wurde untersucht. Im Experiment wurden zwei biomechanische Meßplattformen angewandt, um die ledigliche Vertikalkomponente der Kräfte zu messen. Ein einfaches Modell, das aus ein festener Körper und zwei masslosen Beine besteht, wurde in unsere Modellanalyse. Die beide Bodenkräfte und Angriffspunkte verschieden verhalten unterschiedlich im Vergleich mit den Eigenschaften des beständigen Gehens. Die vorschlägte Modellanalyse zeitgte deutlich ein Mechanismus der bestimmten Charactereigenschaft hinsichtlich mit der Mechanik. Die nachfolgende Ergebnisse wurde durch erhaltet. Die viele Zeitparameter haben die Beziehung mit dem Gleichschritt. Durch im Vergleich mit den simulieren Ergebnissen und experimentallen Ergebnissen die Methode ist gültig geworden.

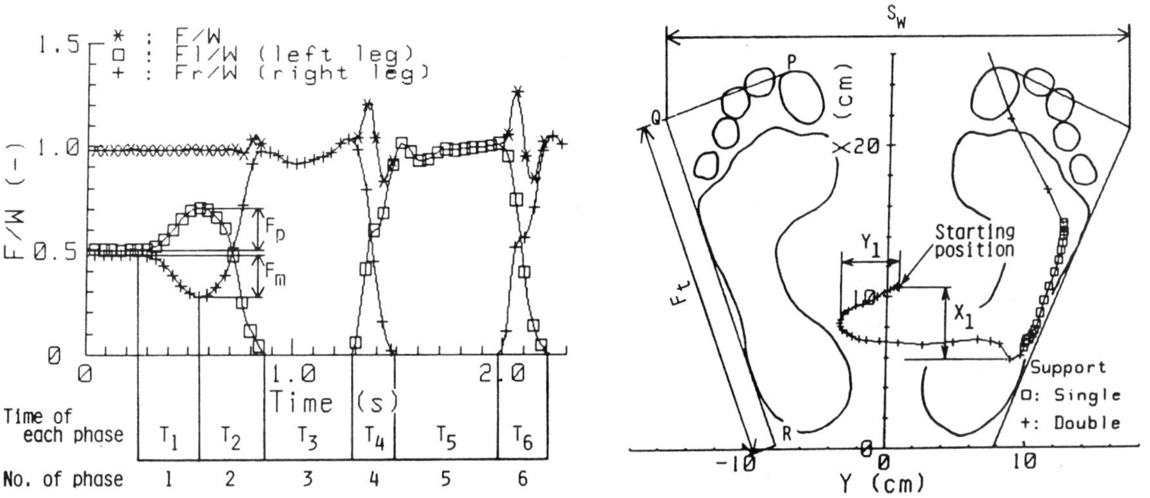

(1) Time history of the vertical reaction force.

(2) Trajectory of the point of application on horizontal plane.

Fig. 1. Examples of measured data (cadence: 80 steps/min).

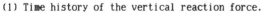

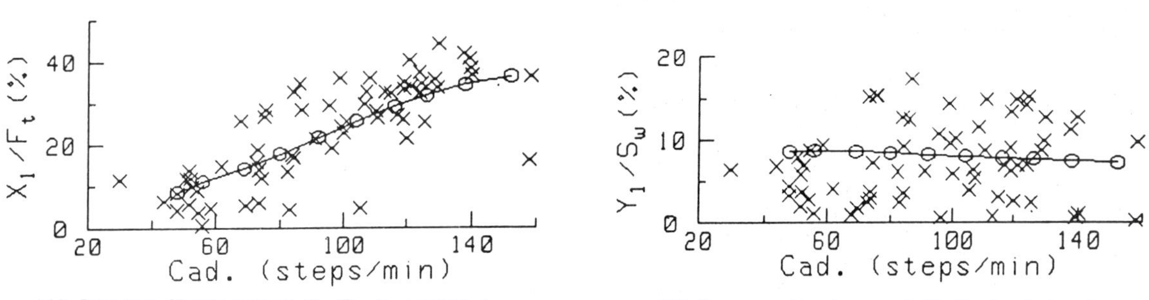

(1) Inverse displacement X_1/F_t in sagittal.

(2) Inverse displacement Y_1/S_w in lateral.

Fig. 2. Comparison of simulated results with experimental results (X:experimental data).

Table 1 Results of regression analysis of parameters. X_L and Y_L indicate the moving amount in steady state

Parameter	Regression Coefficients A_0	Regression Coefficients A_1	Correlation Coefficient r^2	Type
Cad. vs X_1	−1.2799	0.0710	0.6091	1st
Cad. vs X_L	36.9658	0.2021	0.5085	1st
Cad. vs Y_1	−1.7069	0.9540	0.0409	Log
Cad. vs Y_L	7.9030	−0.0004	0.0099	Exp
Cad. vs T_1	0.2648	−0.0015	0.0058	Exp
Cad. vs T_2	0.6889	−0.0094	0.3062	Exp
Cad. vs T_3	0.7813	−0.0069	0.8066	Exp
Cad. vs T_4	0.5581	−0.0111	0.7712	Exp
Cad. vs T_5	1.1087	−0.0096	0.8893	Exp
Cad. vs T_6	1.5827	−0.2999	0.6926	Log
Cad. vs F_p	−0.1415	0.0560	0.0855	Log
Cad. vs F_m	−0.1818	0.0666	0.1015	Log
Cad. vs F_p+F_m	−0.0344	0.1226	0.0940	Log
X_1 vs F_p	0.0344	0.0137	0.2901	1 st
Y_1 vs F_p	0.0151	0.0353	0.7962	1 st

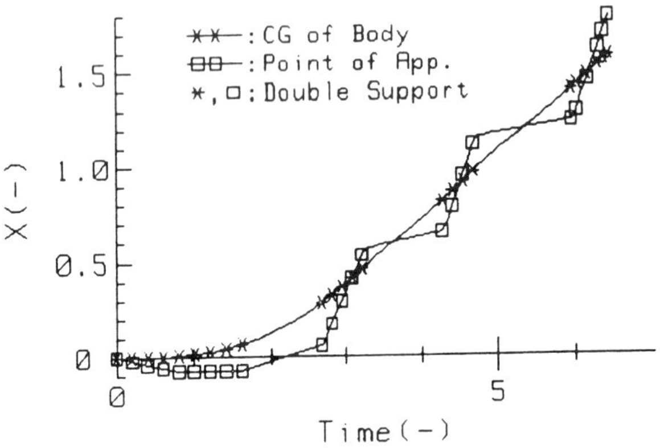

(1) Time history of sagittal motion.

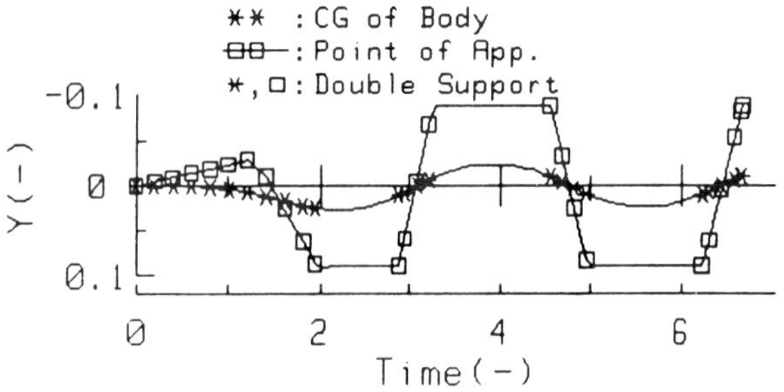

(2) Time history of lateral motion.

Fig. 3. Examples of simulated kinematics in gait initiation (cadence: 104 steps/min).

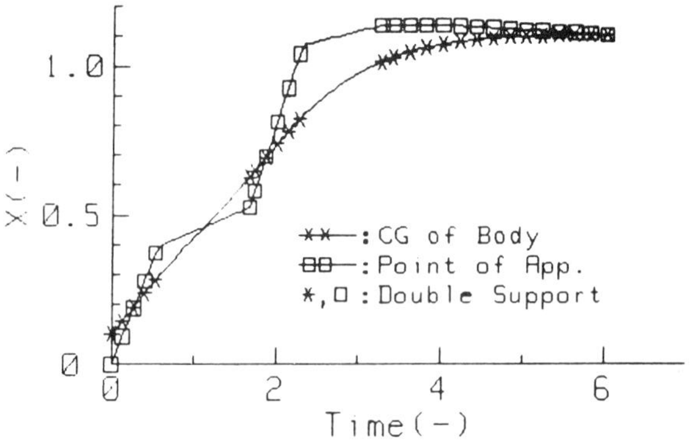

(1) Time history of sagittal motion.

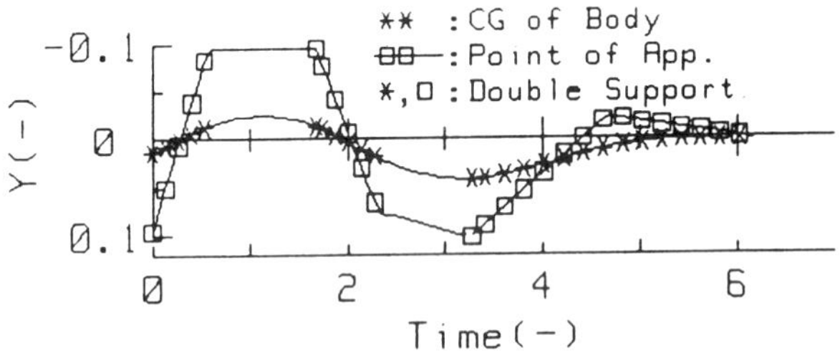

(2) Time history of lateral motion.

Fig. 4. Examples of kinematics in gait stopping (cadence: 104 steps/min).

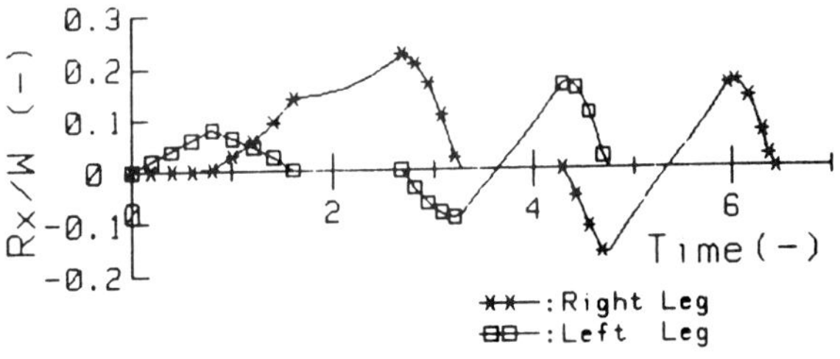

Fig. 5. An example of the time history of sagittal reaction force simulated.

The Development of a Microcomputer-controlled Electrical Prosthesis with six Degrees of Freedom

TAKANORI HIGASHIHARA*, YUKIO SAITO* AND HIROSHI ITOH**

*Tokyo Denki University, Faculty of Science and Engineering, Hatoyama-cho, Hiki-gun, Saitama, Japan
**Tokyo Denki University, First Faculty of Engineering, Nisiki-cho, Chiyoda-ku, Tokyo, Japan

Abstract. Although we have been developed a total-arm electrical artificial limb with 11 degrees of freedom under the control of a microcomputer, but not yet put into practical use. The principal reason is that the conventional shoulder mechanism is not rigid enough because the combinations of gear mechanism caused considerable moment at the joint. In addition, the axes of two rotation degrees in a shoulder does not intersect. It gives an unnatural shoulder motion. This study consists of using a space linkage shoulder mechanism while aiming at an artificial limb with simplified shoulder having six degrees of freedom as designed by the CAD system. The hand part of this artificial limb can be readily changed to either a cosmetic hand type or hook type according to the purpose. The control equipment of the artificial limb has been made compactly using microcomputer and new ICs. Thereby we achived an intelligent electrical artificial limb for a person with both arms amputated. Moreover the use of microcomputer allows a coordinated motor control and thus the artificial limb may be operated under the simple command with a chin switch.

Keywords. Artificial limbs; prosthetic; computer control; computer-aided design; microprocessors; biomedical

INTRODUCTION

The total-arm electrical artificial limb with 11 degrees of freedom under the control of microcomputer has been developed, but not yet put into practical use. It is attributed to following factors;
1) The conventional shoulder mechanism uses a gear train which includes independent rotation around X and Y axes or X and Z axes during 2 degrees of freedom operation in the shoulder. These axes are located in an offset position, resulting in unnatural shoulder motion.
2) The gear mechanism causes considerable moment at the joint and thus is not rigid enough.
3) Due to reasons described, the appearance and shape of actual human arm is not achieved.

In a view of above problem factors, this study is intended to develop the shoulder artificial limb with 6 degrees of freedom with the space linkage mechanism in the shoulder. This artificial limb is for a person with both arms amputated and has 6 degrees of freedom including double space linkage mechanism for shoulder, external and internal rotation of upper arm, extension and flexion of elbow, inner and outer rotation of forearm, and opening and closing of hand.
The appearance and shape of the artificial limb has been CAD designed on the basis of data obtained from the measurment of the arm of the healthy person with a three-dimensional coordinate measuring machine. The hand part is readily changeable to either a cosmetic hand type and hook type according to the kind of operation required.
The control system is designed with a microcomputer (Z-80) for the coordinated control of 6 degrees of freedom. The another arm (i.e. dummy arm) includes a microcomputer block, motor driver block, and battery. Operation is made basically with a touch switch operated by a chin, and classified into a coordinated operation mode for specified motion and a manual mode for independent operation of each motor. In addition, a voice controller, joystick, moving according to the motion of shoulder, etc. may be used as an input device.

MECHANISM OF ARTIFICIAL LIMB

Fig.1 (a) and (b) is a skeleton diagram indicating the mechanism of a total-arm electrical artificial limb with 11 degrees of freedom.
For the design of a new total-arm artificial limb with 6 degrees of freedom, there is no need to change the conventional arm mechanism for flexion and extension of elbow, and inner and outer rotation of forearm. The new mechanism is applied for the shoulder and hand. 2 degrees of freedom of shoulder of Bs and Au type total-arm artificial limbs have axes offset and a gear mechanism (low rigidity). Accordingly, as shown in Fig. 1 (c), a universal joint is newly used in the shoulder, and the double linkage mechanism incorporated contributes to increase rigidity. The hand part, it has a structure which is ready to change to a cosmetic hand type or hook type according to the task required. This hand part can thus offer a grip force appropriate to the work.

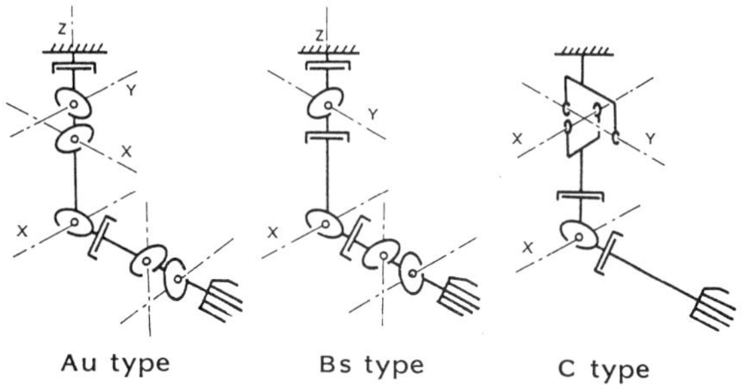

Fig. 1. Skeleton diagram of a shoulder mechanism

SPACE LINKAGE SHOULDER MECHANISM

The space linkage shoulder mechanism is a three-point supporting joint mechanism consists of columns supporting the load on the front end. The shoulder and upper arm are connected with a universal joint to obtain two degrees of freedom of front and side upward motions. The upper arm part has 2 motors (each for the front and side upward motion). Each motor is the gear reduced by a harmonic drive and drives the ball screw. Front and side upward motion links are mounted at the shoulder so that they are orthogonal for each nut part. Driving the screw part causes the nut to move, and makes the arm to swing upward via a linkage mechanism. As shown in Fig. 2, an origin is set at a shoulder projection(acromion), and the operation of upper arm with the length of a radius 'R' causes the elbow to move on a spherical surface.

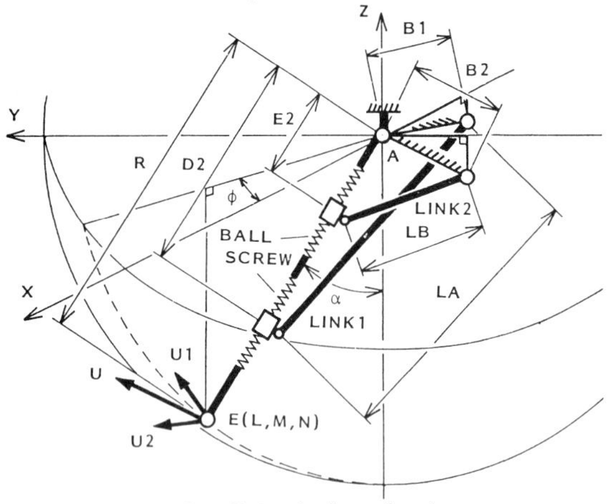

Fig. 2. Principal mechanism

The coordinate system of the elbow is

$$\begin{pmatrix} L \\ M \\ N \end{pmatrix} = \begin{pmatrix} |R \cdot SIN\ \alpha| \cdot COS\ \phi \\ |R \cdot SIN\ \alpha| \cdot SIN\ \phi \\ -R \cdot COS\ \alpha \end{pmatrix} \quad (1)$$

where, 'α' is an upward motion angle, 'φ' an angle of rotation around Z axis, and 'R' a length of upper arm.

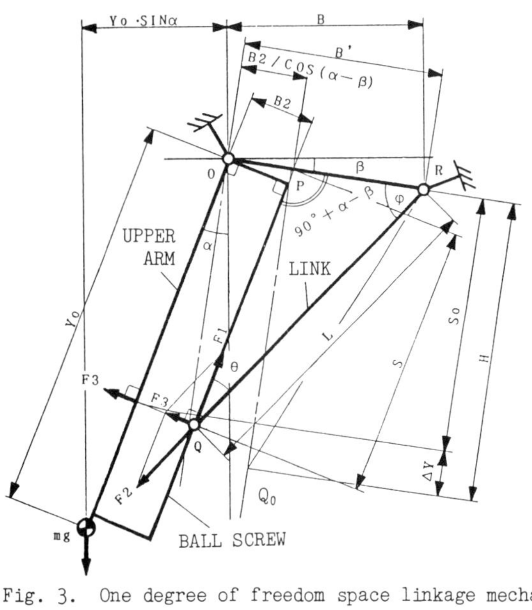

Fig. 3. One degree of freedom space linkage mechanism

Fig. 3 shows a conceptual view of one degree of freedom space linkage mechanism with the arm stretched. The travel ΔY of the nut when the arm is swung upward for an angle of 'α' is, $\Delta Y = H - S_0$

$$= H - \frac{L \cdot COS(\beta - \theta + \alpha)}{COS(\alpha - \beta)} - B2 \cdot TAN(\alpha - \beta) \quad (2)$$

where,

H; Length from point O to the first nut position(Q_0)
L; Length of the connecting rod
B2; Dislocation from the point O to the ball screw
α; Angle of upward motion
β; Connecting rod end angle θ; ∠PQR

For example, ΔY is 39 mm when the arm is swung upward for 90°, when the arm is swung side upward for 60° is 35 mm.
The force F1 to pull up the nut is;

$$F1 = \frac{mg \cdot Y_0 \cdot SIN\ \alpha \cdot COS(\alpha - \beta)}{L \cdot SIN\ \psi - B2 \cdot SIN(\alpha - \beta)} \cdot \frac{1}{TAN\ \theta} \quad (3)$$

where,

Y_0; Length from point O to the center of gravity of arm mg; Weight ψ; ∠PRQ

The moment at the elbow is calculated as follows.

$$U1 = D2/R \cdot Mt \cdot TAN(COS^{-1}((LA^2 + D2^2 - B1^2)/(2 \cdot LA \cdot D2)))$$
$$U2 = E2/R \cdot Mt \cdot TAN(COS^{-1}((LB^2 + E2^2 - B2^2)/(2 \cdot LB \cdot E2))) \quad (4)$$

where,

U1; Movement of the shoulder flexion
U2; Movement of the shoulder abduction
D2; Length from the original point A to the nut of shoulder flexion
E2; Length from the original point A to the nut of shoulder abduction
B1,B2; Distance of each heel size
LA,LB; Length of each connection rod
Mt; Transfer torque by own actuator
R; Length of upper arm

Fig. 4 (a) shows a resultant of moment obtained from the formula (4), with a front upward angle plotted on the axis of an abscissa and the swing angle taken as parameter. As is evident from the figure, in all swing angles as parameters, the maximum force of action is obtained at around the front upward angle of 40°. Location for the maximum force is set at this position because this is a center of an operation range in a daily life with artificial limb.
Since the space linkage mechanism supports the arm with two links, the load is distributed and the arm structure withstands larger external force at the front end of arm.

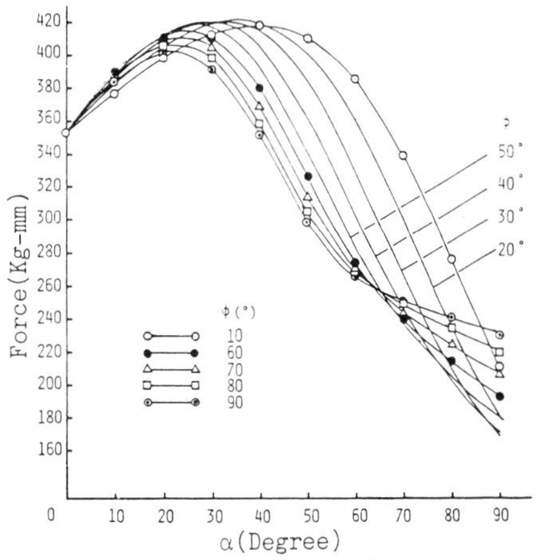

(a) Resultant of force for the upward angle (Parameter; The swing angle)

Fig. 4. Resultant force of shoulder

SPECIFICATIONS OF THE SHOULDER ARTIFICIAL LIMB WITH SIX DEGREES OR FREEDOM

Fig. 5 shows an overall view of the shoulder artificial limb with 6 degrees of freedom includes front and side upward motion of the shoulder, external and internal rotation of upper arm, flexion and extension of elbow, inner and outer rotation of forearm, and opening and closing of hand. The total length of arm is 586 mm and the weight 2.3Kg. The artificial limb is made mainly of aluminum alloys and additionally composite material (carbon fiber and duralumin) for reduction of weight.
The motor used is a rare earth coreless DC motor solid with a reduction gear. Five motors (other than that for fingers) are directly coupled to a photoelectric encoder. The shoulder movable range is divied into the front upward (front 90°- rear 30°) and side upward (outward 60°- inward 0°) because daily operation with artificial limb includes followings;
(1) Eating (2) Shifting things on the table
(3) Turning the page (4) Writing characters
(5) Picking up the receiver
Accordingly, the shoulder need not have a wide movable range.
The hand part is of a quick-change type enabling selection of a cosmetic hand type for decorative use and a hook type for functional use according to the purpose. The fingers of cosmetic hand type fit well to a shape to be gripped because of elastic characteristic of the MP joint.

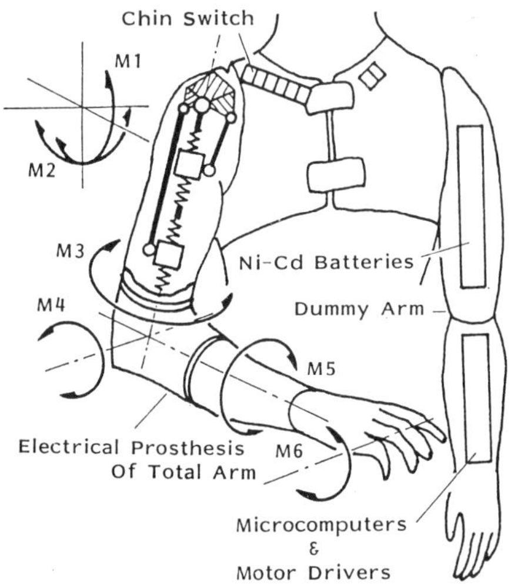

Fig. 5. Overall view of the artificial limb

APPEARANCE DESIGN WITH CAD

The appearance and shape of artificial limb has been designed on the basis of measurement of the arm of healthy person with the three-dimensional coordinate measuring machine and modified to fit to the shape of actual arm.
Designing with CAD offers following advantages;
1) Arm appearance fits for the shape of artificial arm is obtained on the basis of measurement data of healthy person
2) Simulation of arm motion is possible with a measured arm shape because the center of rotation of each joint of artificial limb is known
3) It is possible to determine the motor position properly because any section of the arm is obtained.
Fig. 6 shows the result of data processing after measurement of the shoulder and arm of healthy person.

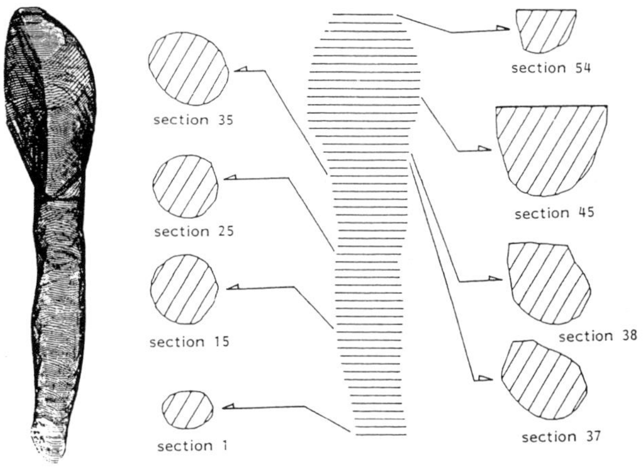

Fig. 6. Measured shape of the arm of healthy person

CONTROL SYSTEM

The shoulder artificial limb has 5 degrees of freedom excluding the hand part and it is difficult for the person with both arms amputated to undertake coordinated operation even though he utilizes his remaining physical function at its highest. It is therefore necessary to control 5 motors by a microcomputer program. The use of microcomputer also facilitates calculation such as coordinate transformation and speed change, safety alarm, and connection with various input/output devices. Use of the motor as a software servo with a microcomputer helps reducing the size of the control equipment.
Fig. 7 shows a system diagram of the controller for the artificial limb. The microcomputer, motor driver, and battery are built into the left (dummy) arm.
The person with both arms amputated gives command with a touch switch using remained function of the shoulder, joystick, tongue switch, chin switch, voice, and/or EMG. However, the touch switch and EMG can extract a limited number of independent signals while the tongue switch, when used, makes eating difficult. On the other hand, the present art of voice recognition device is unreliable as an operation command. The artificial limb in this study uses a chin switch. This switch consists of 7 key switches (mode selection and switches from #1 to #6) on a socket.
The control section consists of CPU (CMOS Z-80), PROM 8Kbyte, RAM 8Kbyte, and I/O (8255-2) for command input and output to the motor driver. The motor driver section consists of 2 set of circuit same as the control section. One CPU performs software-based positioning control of 3 motors. The positioning control is made possible by the signal from the encoder directly coupled to the motor. Namely, this driver section can control 6 motors simultaneously. The motor driver is a full-bridge driver for changing the motor running direction and can control 4 modes of forward, reverse, stop, and brake.

COORDINATED CONTROL

There must be two modes for operation of this shoulder artificial limb: a manual mode to drive motors independently and a coordinated motion mode under distribution of DDA (digital differential analyzer) pulse. MICOM1 is a microcomputer of control section and MICOM2 and 3 are microcomputers of motor driver section. To drive the motor, MICOM1 of the main commands the motor running direction pass a motor control bus line to MICOM2 and 3 of the subsystem which in turn gives an accept signal to the driver to drive the motor. While the

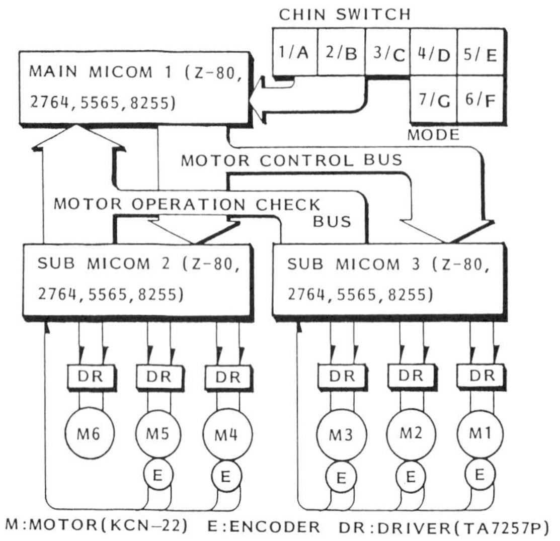

Fig. 7. System diagram of the controller

motor is running, MICOM2 and 3 keep the motor operation check bus in busy state to reject any subsequent motor running direction command from MICOM1 till all motors assigned are driven by one pulse. MICOM1 waits till the motor operation check bus becomes ready and, upon confirmation of the ready state, outputs the motor running direction to MICOM2 and 3. In the manual mode, pressing the switch corresponding to the motor after setting of the command causes motion of any arbitrary joint. The coordinated operation mode allows the fixed sequence operation (travel to any arbitrary grid point, water drinking) through combination of switches after setting of the command. In this mode, travel to a target position is mode through pulse distribution according to the DDA calculation method. Intersections are assumed as grid points in an interval of 10 cm in X, Y, and Z directions to enable the artificial limb to move in an arbitrary direction to any point in the space. When the artificial limb is to move to any arbitrary grid point, the number of motor pulses equivalent to 5 degrees of freedom is stored beforehand as a table on ROM. During practical operation, the number of pulses of each motor to the grid point are accessed from ROM and distributed and output to each motor. The fixed sequence operation performs a series of consecutive operations by reading sequentially data of grid points. The command program as described above is shown in Table 1. To ensure safety of the artificial limb, each joint is equipped with limit switch. In terms of software, development is being made on a motor output check program which checks if the motor is running in a command direction and a safety range check program which checks if the artificial limb is being operated within a movable range of each joint.

TABLE 1 Command program

COMMAND 1ST	COMMAND 2ND		ACTION OF PROGRAM
A			RETURN TO ORIGINAL POSITION
B	A		TEMPORARY STOP, RESUME
B	B	0-25	THE FIXED SEQUENCE OPERATION
C	0-6		INDEPENDENT ACTION OF EACH MOTOR
D	GRID POINT		TRAVEL TO ANY GRID POINT
E	0-3		SPEED CONVERSION

CONCLUSIONS

The total-arm electrical artificial limb with 6 degrees of freedom incorporated with a space linkage mechanism in the shoulder and the design with CAD system ensures function and visual satisfaction. This new artificial limb offers following features.

(1) Space linkage shoulder mechanism ensures natural shape and motion of the shoulder while offering highly rigid structure.

(2) Appearance design of the shoulder artificial limb with CAD achives a shape close to the arm of healthy person.

(3) Hand part can be changed to a cosmetic hand type or hook type enabling kind of a hand appropriate to the work.

(4) The control equipment of artificial limb is made compact by a recent progress of LSI, thereby it contributed to the realization of intelligent motor-driven artificial limb for the person with both arms amputated.

(5) Microcomputer incorporated enables coordinated control of motors. Simple operation command with a chin switch is enough to move the artificial limb.

This study has been supported by The Ministry of Education and Tokyo Denki University which have made possible the laboratory work in the paper.

REFERENCES

H.FUNAKUBO,et.al (1980). Total-Arm Prosthesis Driven by 12 Micro-Motor, Pocketable Microcomputer and Voice and Look-Sight Microcommanding System. I.C.R.E.,pp. 39-42

Y.SAITO, T.HIGASHIHARA (1985). A mechanism of electrical powered shoulder arm with a double linkage. RESNA 8th ANNUAL CONFERENCE, pp. 76-77

RÉSUMÉ

La communication présente la conception du tout bras prothèse electrique nouveau utilisant l'espace linkage mécanisme. Nous avons mesuré le bras au moyen d'un appareil de mesure tridimensionnel et transformé à la forme de prothèse. Nous avons mese un appareil a microordinateur compact utilisant un interrupteur de la mâchoire comme le contrôle.

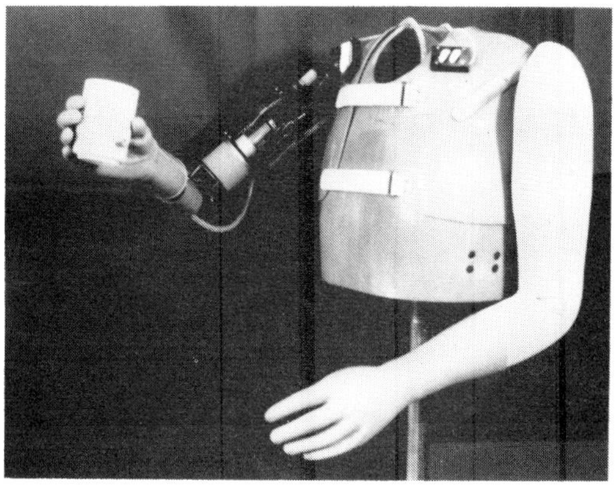

Fig. 8. The total-arm electrical artificial limb with 6 degrees of freedom

Three Dimensional Coordinate Measuring Machine and Data Processing Method for a Functional Cosmetic Arm

YUKIO SAITO*, MASANOBU SUGIYAMA* AND TORU OHSHIMA**

*Department of Industrial Machinery Engineering, Faculty of Science and Technology, Tokyo Denki University, Saitama, Japan
**Department of Medical Engineering, Tokyo Metropolitan Prosthetic and Orthopaedic Research Institute, Tokyo, Japan

ABSTRACT

The object of this study is described hereunder.
We have already developed a new three-dimensional coordinate measurement system for the human body and have actually performed the measurement of natural hands. The value obtained from the measurement is expressed as a three-dimensional coordinate measuring machine in terms of point to point positioning.
So that, there are many technical difficulties to detect edge of device, smooth curved shape, inner diameter, slender steep slope and soft shape.
This paper also discussed the fundamental methods of these data processing we followed to process the observed data.

Keywords, Computer-aided design; Artificial limb; Computer graphics; Machining; Medical systems; Numerical,control;Biomedical.

INTRODUCTION

The measurement of different parts of human body presupposes the existence of hardware as well as software of general applicability. We have already developed a new three dimensional coordinate measurement system of the surface of human arm and foot.

In this study reported hereunder, we have developed a method of automatic measurement and data processing for any desired profile on human body and industrial devices. The value. obtained from the measurement is expressed as a three dimensional coordinate measuring machine in terms of point to point positioning. So that, generaly, we have many technical difficulties as follows;

(a) Edge detection of machine parts
(b) Detection of smooth curved shape
(c) Detection of inner diameter as deep hole
(d) Detection of slender steep slope
(e) Detection of soft shape as a raw egg

The above detections are the fundation of automatic measurement by a three dimensional coordinate measuring machine and difficult detective items.

Accordingly, we have developed a three dimensional measuring machine for the detection of soft living body shape, a data processing methods to the manufacturing of a functional cosmetic hand.

ALGORITHM OF A LIVING BODY MEASUREMENT

The algorithm of measuring living body is structured by ;

(a) Automatic measurement programming of human hand and foot
(b) Programming of data processing for the smooth curve and soft shape profile
(c) Programming for machining on 3-D coordinate measuring machine as shown in Figure 1.

And this study includes a display program to carry out a dialog style extensively.
We made a module formation so that one program performs only one processing and then, simplified the formation of new program and the operation such as debug, addition and amendment etc.
These modules are composed of data processing and common area which definite a data format of input/output among modules.

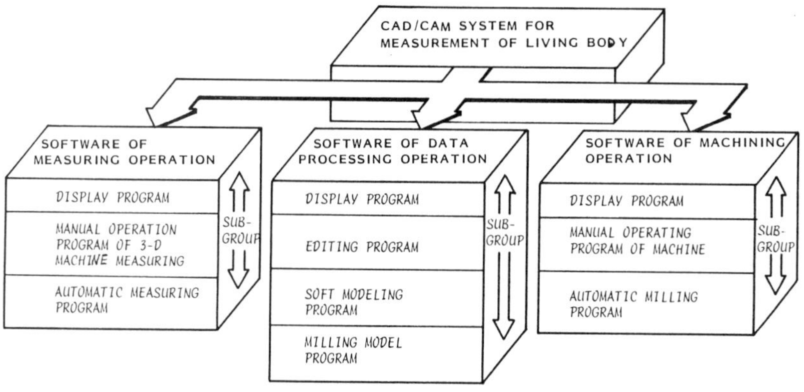

Figure 1 CAD/CAM Programming for the living body shape

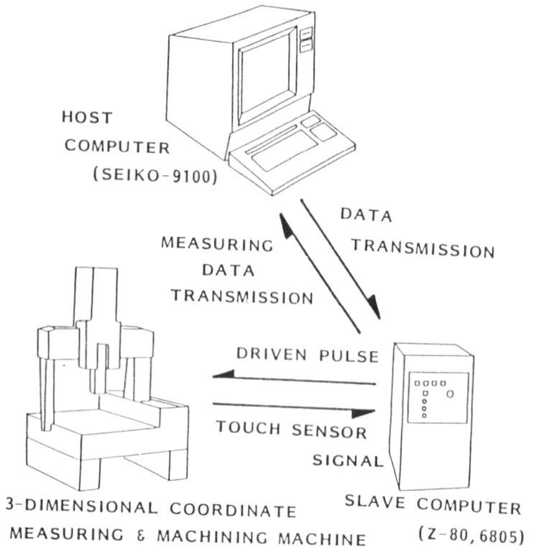

Figure 2 Measuring and machining system

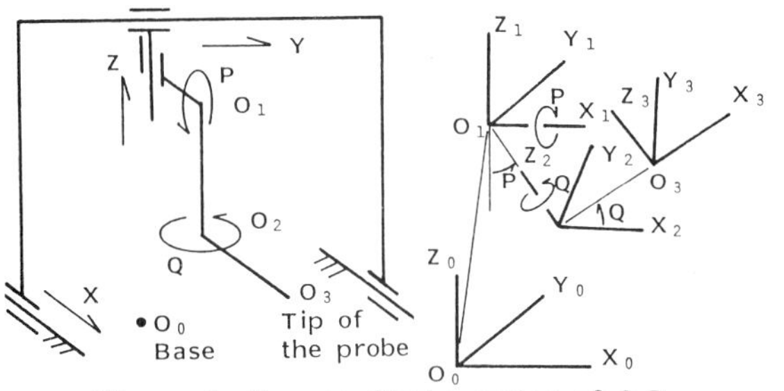

Figure 3 The coordinate system of 3-D coordinate measuring machine

HARDWARE STRUCTURE AND COORDINATE SYSTEM

Figure 2 shows a measuring and machining system. The host computer we used for a data processing is a personal microcomputer system (SEIKO-9100). The total system includes a Z-80 CPU as a motor driven motor control, and 6805 CPU is a numerical control system for processing of sensor signals and allowing pulse distribution control of the 3 degrees of freedom of 3-D measuring machine.

For the measurement of foot, we have used a two-axes rotary mechanism for the movement along a z-axis.
Namely, the coordinate system is composed of five axes, and is based on an orthogonal coordinate system with three degrees of freedom. The value obtained from the measurement is expressed as displacements of 3-D coordinate measuring machine, in terms of position.

$A_s(X_s, Y_s, Z_s)$ of z-axis and angles of rotation P and Q are shown in Figure 3.
And then, we can obtain a position $A_h(X_h, Y_h, Z_h)$ of the tip of probe.
If we express the conversion matrix as R_P(P-rotation), R_Q(Q-rotation), T_r(distance between P and Q axes), and T_s(probe length) then we have

$$A_h = R_P \cdot R_Q \cdot T_r \cdot T_s \cdot A_s \quad \cdots\cdots (1)$$

also the corresponding inversion is given by

$$A_s = R_P^{-1} R_Q^{-1} T_r^{-1} T_s^{-1} A_h^{-1} \cdots\cdots (2)$$

BASIC SHAPE MODEL FOR AUTOMATIC MEASUREMENT

A sensor for measuring the living body is a capacitance type touch sensor (length 50mm, diameter 0.5mm, sensitivity 10PF) which sends an ON/OFF signal to the host computer.

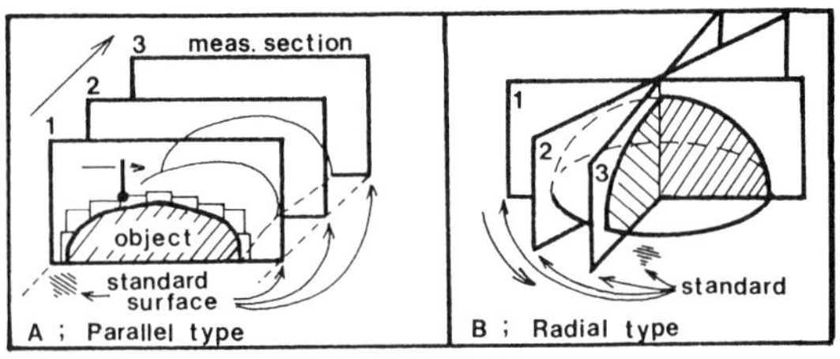

Figure 4 Basic path of probe

Figure 4 shows a basic path of the probe which has cross section of parallel type and radial type.
And then, Figure 5 shows basic structures of measuring cross section. The measurement of almost object profile becomes possible by the combination of basic structures.

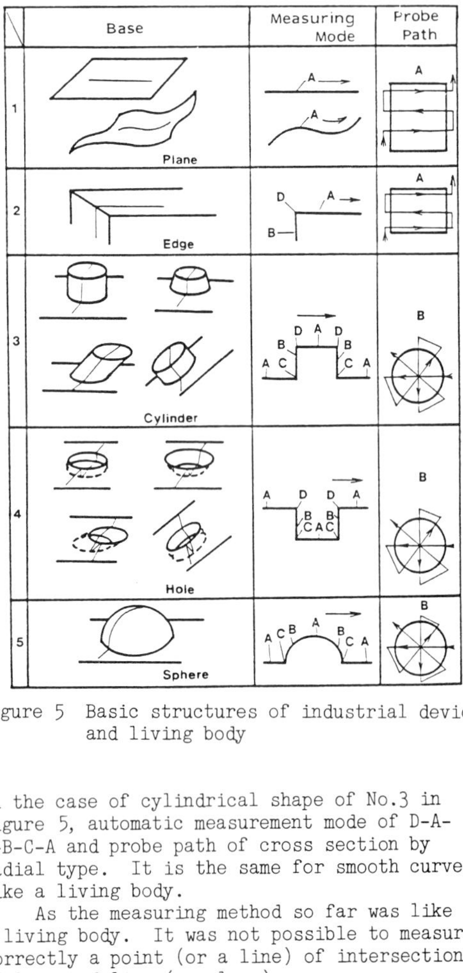

Figure 5 Basic structures of industrial device and living body

In the case of cylindrical shape of No.3 in Figure 5, automatic measurement mode of D-A-D-B-C-A and probe path of cross section by radial type. It is the same for smooth curves like a living body.

As the measuring method so far was like a living body. It was not possible to measure correctly a point (or a line) of intersection of line and line (or plane).
When the probe moves from the measurement point P_{n-1} to P_n in Figure 6, the condition deciding next point P_f is decided by the direction of measuring unit vector A, B, C, D.
When the actual movement amounts for measuring unit vector are $P_q(A)$, $P_r(B)$, $P_s(C)$, and $P_t(D)$ and measuring point P_f exists on A-curve or B-curve,

$$P_{n+1} \neq P_f$$

The probe path to obtain a next measuring point P_f is able to classify into seven mode by the direction of measuring unit vectors, a relation between measured point P_n and former measured point P_{n-1}.
The conditions of classification are:

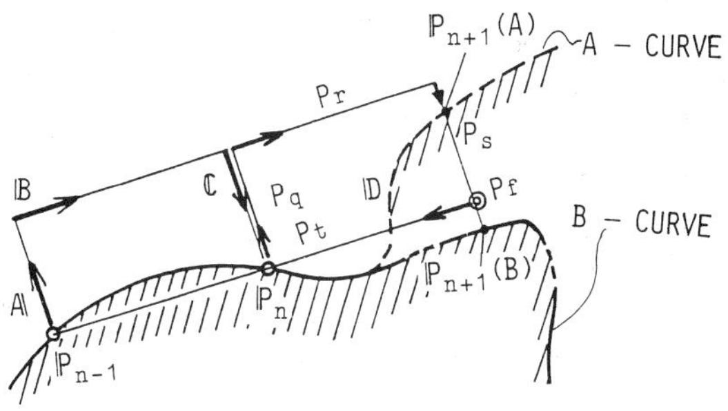

Figure 6 Measuring vector

1) The direction cosines of a cross section
2) $P_{n+1} \neq P_f$
3) $P_n - P_{n-1} > \sqrt{Pq'^2 + Pr'^2}$ (= a minimum setting pitch)
4) Measuring unit vector when P_n is obtained

Where, Pq and Pr are the moved distance Pq'(A), Pr'(B), Ps'(C), and Pt'(D) which removed from point P_n.

Figure 7 shows an example of edge detection, (a) (b) show the movement of probe which moves at regular intervals on smooth surface shape ,and (c) shows this result of edge detection and automatic pitch transformation.

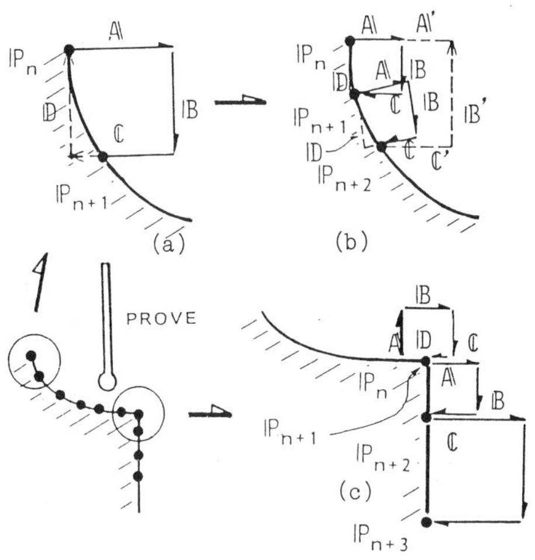

Figure 7 An example of edge detection

DATA PROCESSING AFTER THE MEASUREMENT ON LIVING BODY

Figure 8 shows a stae of measuring hand indicated on CRT display of personal computer. A subject extends his arm on a reclining chair, with elbows and forearm in a relaxed pose. These process of mesurement and data processing are shown in Figure 9.

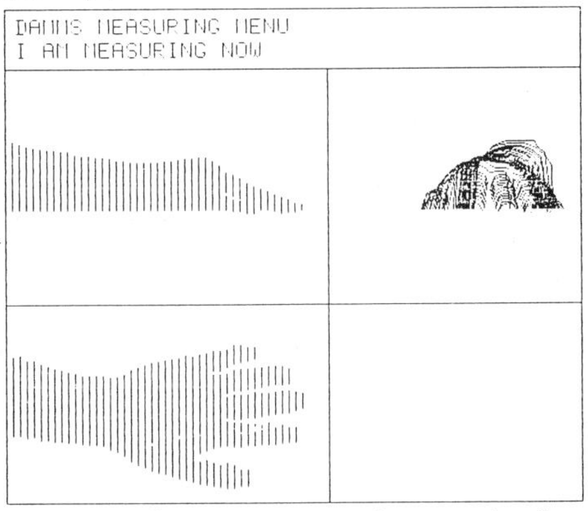

*GPR Figure 8 CRT display of measuring hand

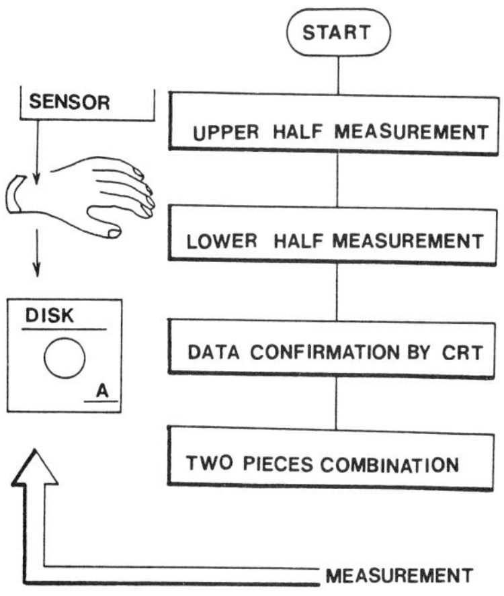

Figure 9 Measuring process of hand

The data processing referred to over here consists principally of:

(1) Joining together the different parts into which the living body was divided and measured.
(2) Correcting the measured data for part of the profile.
(3) Smooth interpolation of measured data.

For the hand, it was possible to join together the different parts with the back as the reference, expanding or contracting the palm as necessary.
For the foot, however, all the parts measured had a direct relation on the shape of shoes or that of prosthetic devices.
So that, the simple method of data processing for the hand could not be applied on foot directly. The foot is measured after sectioning it into four areas given by the three-dimensional coordinates, and connected together to generate the total profile of the complete foot.

SMOOTH INTERPOLATION FOR LIVING BODIES

We have used B-spline and ellipse interpolation in order to obtain a smooth curve between sections. It is very difficult to interpolate a complicated object like a hand contains five fingers,but it is possible with the use of a new algorithm developed in this study utilizing B-spline interpolation.
An ellipse interpolation is used to get a suitable roundness in the finger tips. This method of interpolation is simple and suitable for the hand and other living bodies.

MEASUREMENT AND PROCESSING RESULTS

As shown in Figure 10, we automatically measured about 800 points in each section interval (3-8mm) at a speed of 0.8s/point.
Figure 11 shows a plastic doll measured 4197 points in 2mm pitch.

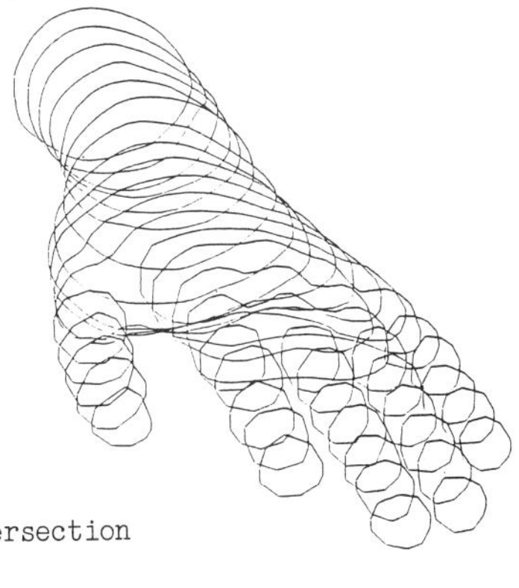

Figure 10
Measured intersection

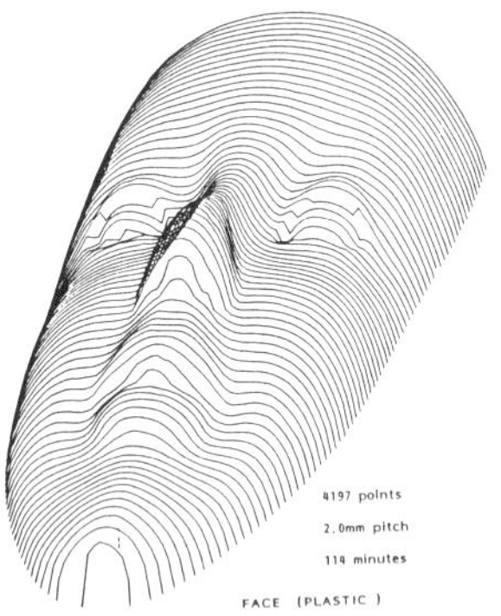

Figure 11 Measured smooth shape model

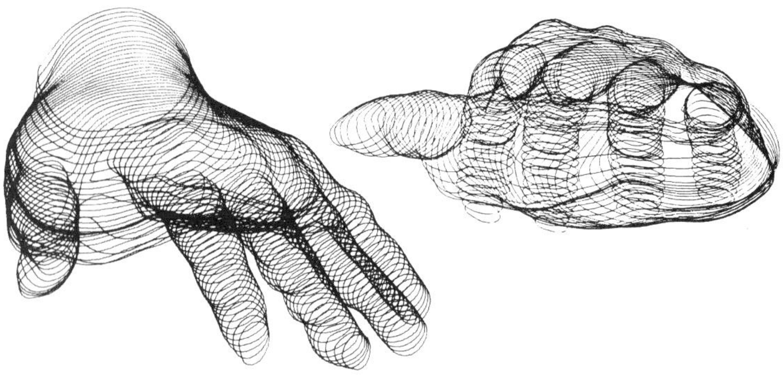

Figure 12 Smooth interpolation & motion analysis

By the movement analysis of natural hand, it is cleared that the rate of angle change of three joints is not constant.
When we used a experimental result practically, the connection of MP-PIP and MP-DIP joints of extension and flexion can be obtain next equation.

Extension movement;

$$\theta_2 = 4.091\theta_1 - 0.106\theta_1^2 + 0.0009\theta_1^3$$
$$\theta_3 = 3.317\theta_1 - 0.049\theta_1^2 + 0.0001\theta_1^3 \qquad \cdots\cdots (3)$$

Flexion movement;

$$\theta_2 = 1.468\theta_1 - 0.017\theta_1^2 + 0.0002\theta_1^3$$
$$\theta_3 = 1.344\theta_1 - 0.009\theta_1^2 + 0.0001\theta_1^3 \qquad \cdots\cdots (4)$$

Figure 12 shows the result of a simulation introduced by equation (3) and (4),
Figure 13 shows a mother model of a glove manufactured by this system.
The mother model of the cosmetic glove is divided into an upper and lower side to mill a cube wax. The respective milling data are transported from the host to the slave computer. So that, a cutter program of mother model is written in BASIC language and can be checked a tool correction and interference.
Figure 14 shows the functional cosmetic arm manufactured by this system. This prosthesis has the movements of extension and flexion for each finger. This is because of the reaction of a positioning spring plate used in the joint part of inner mechanism.

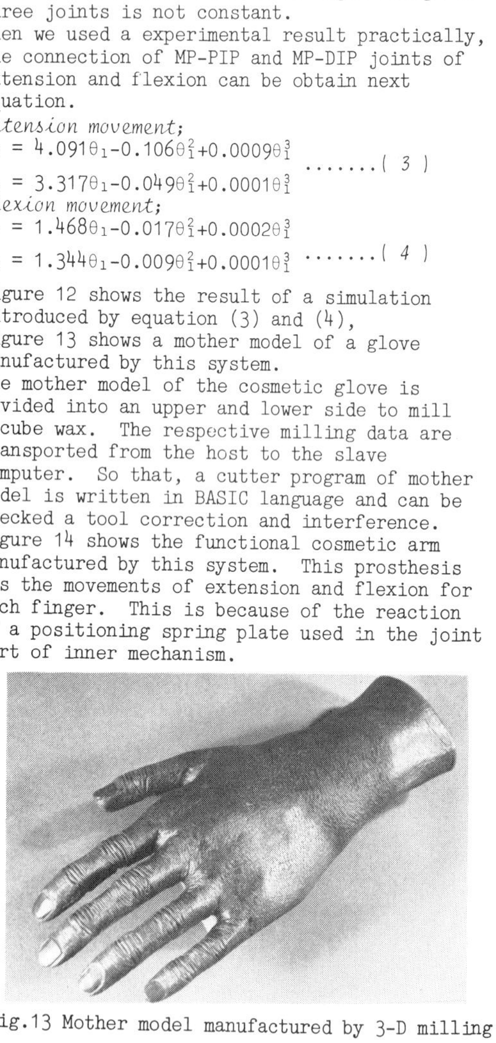

Fig.13 Mother model manufactured by 3-D milling

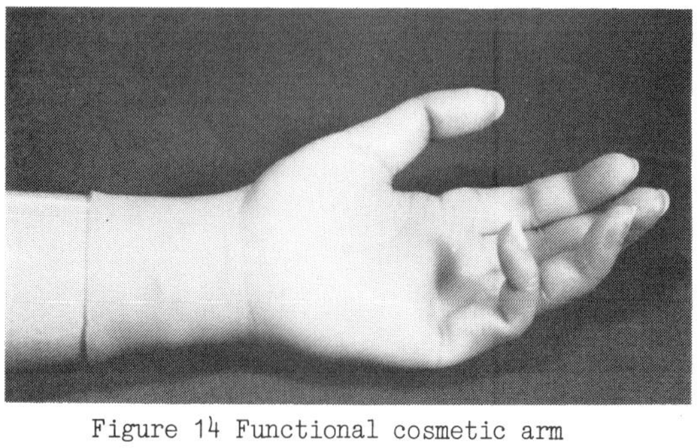

Figure 14 Functional cosmetic arm

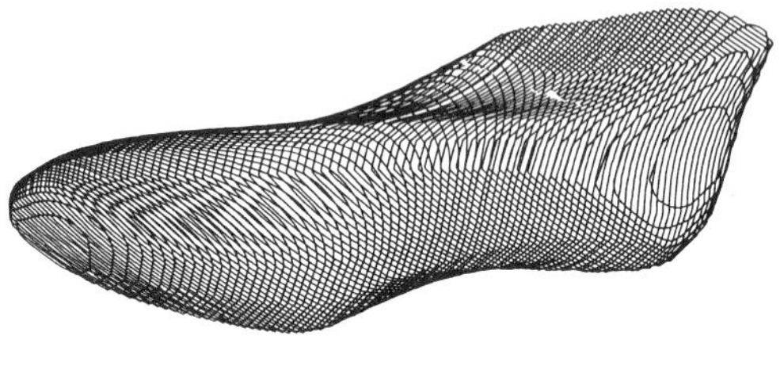

Figure 15 Measured last model and its smooth interpolation

CONCLUSION

We derived a soft model with the primary purpose of abstracting the characteristic features of profiles and also a conversion of the functionally measured data. As a result, a functional cosmetic glove fitted for the handicapped person is manufactured by this system.
Accordingly we developed a three dimensional coordinate measuring machine with five degrees of freedom and the software necessary for the measurement of hand and foot profile.
And then, it was possible to apply the same method on the measurement of industrial devices and soft material object as a raw egg.
A complicated and soft object like a foot can be measured automatically, exactly and fully.
These systems are quite useful in shortning the design time, because designs are fine-tuned numerically, they can be optimized without constructing test model.

REFERENCES

Ohshima,T and Saito,Y (1984).Studies on a Three-Dimensional Automatic Measuring and Milling system. Seimitu Kikai., 50, 830-835.
Saito, Y and Ohshima,T (1984). A New Tree-Dimensional Measuring System for the Living Body. 2nd. I.C.R.E., 151-152

RÉSUMÉ

La présente étude est décrit ci-après.
L'équipment tridimensionnel été développé dans le cadre de cette étude permet de mesurer la main handicapés physiques, de gérer les données measurées et de façonner un modèle de gant facilement et automatiquement.
Donc, il y a beaucoup de difficultés d'ordre technique pour détecter le coin, la courbe douce, le diamètre interne, le pente rapide et la forme molle.
Nous avons fait face à les problèmes.

Elasto-Dynamics Synthesis of Six-legged Walking Mechanism

SUN HAN-XU* AND GAN DONG-YING**

*Beijing Institute of Aeronautics and Astronautics, Beijing, PRC
**Changchun Institute of Optics and Fine Mechanics, Academia Sinica, PRC

INTRODUCTION

The real motion of mechanism may be divided into two parts: one is the motion of rigid body and the other is the vibration of mechanism related to the static position of the rigid body. The former may be achieved by kinematics analysis and the latter by bivration analysis. In the process of its walking, the walking mechanism is "frozen" at instantaneous position in motion and assumed as a "transient construction", then, it may be analysed with analytical construction methods. In this way, the dynamic properties of the walking mechanism can be analysed druing walking. We analyze mechanism for its synthetic uses. The dynamic properties of a robot are determined by its construction parameters and electric parameters of control circuits. The construction parameters of a robot are its construction sizes. This paper has introduced the optimizing method that makes the robot greater in stiffness and lighter in weight, which the authors have used in the design.

THE ESTABLISHMENT OF ELASTO-DYNAMIC MODEL OF SIX-LEGGED WALKING MECHANISM

This walking mechanism has six legs, each leg having three active articulations. The linking form of this walking mechanism and related coordinates of each leg are shown in Fig. 1.

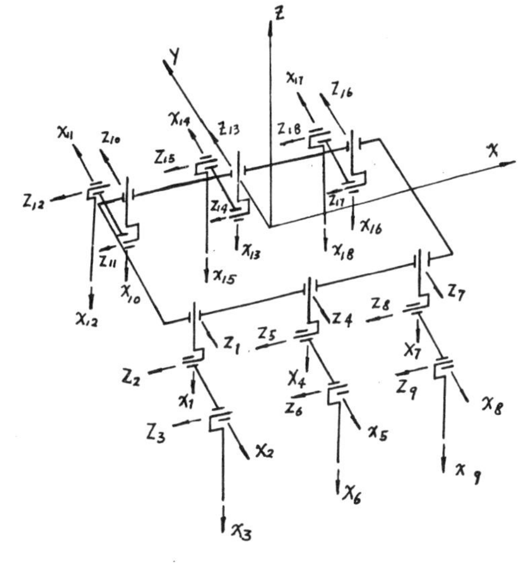

Fig. 1. The linking form of the walking mechanism.

Kinematic Expressions of Each Bar

According to different uses, various demands have been asked on the walking mechanism. The frame of the walking mechanism must have large stiffness because it must take the load of various devices. Such as manipulators, camera. So that, we analyse the frame as a rigid body and the other eighteen bars as elastic bodies. Each bar is considered as a bar-unit, so we have eighteen bar-units. The n-th bar's condition is represented in coordinates of self-natural system as Fig. 2.

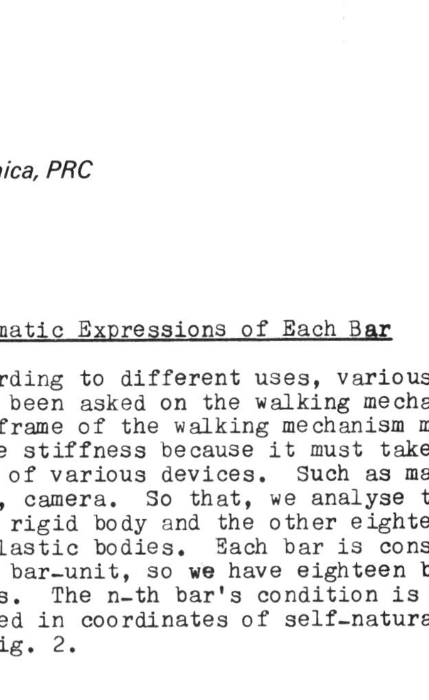

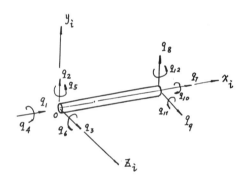

Fig. 2. The n-th bar's condition is represented in coordinates of self-natural system.

Each end of the bar has three displacements and three rotation coordinates, so that, each bar can be expressed with twelve coordinates. There are twelve external forces corresponding to the twelve coordinates.

$$\{q^n\} = (q_1^n, q_2^n, \ldots, q_{12}^n)^T$$

$$\{p^n\} = (p_1^n, p_2^n, \ldots, p_{12}^n)^T$$

In which, $\{q^n\}$ is generalized coordinates of the n-th bar. $\{p^n\}$ is generalized force of the n-th bar. Let $[m]_{12x12}$ is generalized mass matrix of bar unit and $[k]_{12x12}$ is generalized stiffness matrix of bar unit. Matrix $[m]$ and $[k]$ are traditional matrixs. They are from reference (6).

In the case that damping is omitted, the motion equation of the n-th bar has been given in coordinates of self natural system as:

$$[m_n]\{\ddot{q}^n\} + [k_n]\{q^n\} = \{p^n\} \qquad (1)$$

Transition of Coordinates

In the above, each motion equation has been derived from the coordinates of unitself system, but the relative position between units is described in the same coordinates of the general system. In order to link the six legs, the coordinates of each leg are transformed into the coordinates of the frame system. The zero position of relative rotating angles between bars is defined in Fig. 3.

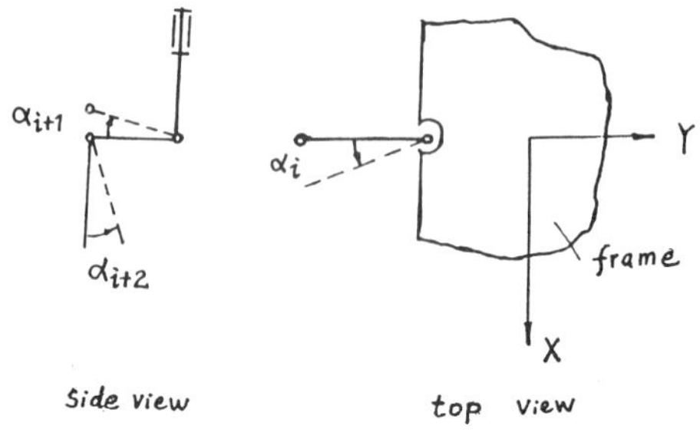

Fig. 3. The zero position of relative rotating angles between bars.

Let $\{\bar{q}^i\}$ (i = 1, 2,...,18) is the twelve-dimension' vector of the i-th bar in the coordinates in frame's system. The transformation matrix from coordinates of each bar's system to the frame's is gotten. In following C expresses cosine and S expresses sine .

$$[t_i] = \begin{bmatrix} 0 & 0 & -1 \\ C\theta_i & -S\theta_i & 0 \\ -S\theta_i & -C\theta_i & 0 \end{bmatrix}$$

$$[t_{i+1}] = \begin{bmatrix} -C\theta_{i+1}\cdot S\theta_i & C\theta_{i+1}\cdot C\theta_i & S\theta_{i+1} \\ S\theta_{i+1}\cdot S\theta_i & S\theta_{i+1}\cdot C\theta_i & S\theta_{i+1} \\ -C\theta_i & S\theta_i & 0 \end{bmatrix}$$

$$[t_{i+2}] = \begin{bmatrix} -S\theta_{i+2}C\theta_{i+1}S\theta_i-C\theta_{i+2}S\theta_{i+1}S\theta_i & -S\theta_{i+2}C\theta_{i+1}C\theta_i-C\theta_{i+2}S\theta_{i+1}C\theta_i & S\theta_{i+2}S\theta_{i+1}-C\theta_{i+2}C\theta_{i+1} \\ -C\theta_{i+2}C\theta_{i+1}S\theta_i+S\theta_{i+2}S\theta_{i+1}S\theta_i & -C\theta_{i+2}C\theta_{i+1}C\theta_i+S\theta_{i+2}S\theta_{i+1}S\theta_i & C\theta_{i+2}S\theta_{i+1}+S\theta_{i+2}C\theta_{i+1} \\ -C\theta_i & S\theta_i & 0 \end{bmatrix}$$

$$[t_j] = \begin{bmatrix} 0 & 0 & -1 \\ -C\theta_j & S\theta_j & 0 \\ S\theta_j & C\theta_j & 0 \end{bmatrix}$$

$$[t_{j+1}] = \begin{bmatrix} C\theta_{j+1}S\theta_j & C\theta_{j+1}C\theta_j & -S\theta_{j+1} \\ -S\theta_{j+1}S\theta_j & S\theta_{j+1}C\theta_j & C\theta_{j+1} \\ -C\theta_j & S\theta_j & 0 \end{bmatrix}$$

$$t_{j+1} = \begin{bmatrix} S\theta_{j+2}C\theta_{j+1}S\theta_j-C\theta_{j+2}S\theta_{j+1}S\theta_j & S\theta_{j+2}C\theta_{j+1}C\theta_j+C\theta_{j+2}S\theta_{j+1}C\theta_j & -S\theta_{j+2}S\theta_{j+1}+C\theta_{j+2}C\theta_{j+1} \\ -C\theta_{j+2}C\theta_{j+1}S\theta_j-S\theta_{j+2}S\theta_{j+1}S\theta_j & -C\theta_{j+2}C\theta_{j+1}C\theta_j+S\theta_{j+2}S\theta_{j+1}C\theta_j & C\theta_{j+2}S\theta_{j+1}+S\theta_{j+2}C\theta_{j+1} \\ -C\theta_j & S\theta_j & 0 \end{bmatrix}$$

(i=1,4,7); (j=10,13,16)

$$[T_n]_{12x12} = \begin{bmatrix} [t_n]_{3x3} & & & \\ & [t_n]_{3x3} & & \\ & & [t_n]_{3x3} & \\ & & & [t_n]_{3x3} \end{bmatrix}$$

$$\{q^n\} = [T_n]\{\bar{q}^n\} \quad (n=1,2,\ldots,18) \qquad (2)$$

Substituting (2) into (1) and multipling (1) with $[T_n]^T$ to the left of equation (1), we get equation (3).

$$[T_n]^T[m_n][T_n]\{\ddot{\bar{q}}^n\}+[T_n]^T[k_n][T_n]\{\bar{q}^n\} = [T_n]^T\{p^n\} \qquad (3)$$

Let $[\bar{m}_n] = [T_n]^T[m_n][T_n]$

$[\bar{k}_n] = [T_n]^T[k_n][T_n]$

$\{\bar{p}_n\} = [T_n]^T\{p^n\}$

So that, Eq. (3) is represented as:

$$[\bar{m}_n]\{\ddot{\bar{q}}^n\} + [\bar{k}_n]\{\bar{q}^n\} = \{\bar{p}^n\} \qquad (4)$$

Eq. (3) is leg's motion equation in the coordinates of frame's system.

Linking

The linking between bars in a leg and the linking of lets with the frame are considered as rigid linking. Displacement between two bars at the linking point must be equal to each other. The sum of generalized forces that act on six-legs is equal to external force that acts on the frame. Pile the motion equations of three bars in the i-th leg, we get:

$$\begin{bmatrix} \bar{m}_n^i & & \\ & \bar{m}_{n+1}^i & \\ & & \bar{m}_{n+2}^i \end{bmatrix} \begin{Bmatrix} \ddot{\bar{q}}^n \\ \ddot{\bar{q}}^{n+1} \\ \ddot{\bar{q}}^{n+2} \end{Bmatrix} + \begin{bmatrix} \bar{k}_n^i & & \\ & \bar{k}_{n+1}^i & \\ & & \bar{k}_{n+2}^i \end{bmatrix} \begin{Bmatrix} \ddot{\bar{q}}^n \\ \ddot{\bar{q}}^{n+1} \\ \ddot{\bar{q}}^{n+2} \end{Bmatrix} = \begin{Bmatrix} \bar{p}^n \\ \bar{p}^{n+1} \\ \bar{p}^{n+2} \end{Bmatrix} \qquad (5)$$

Let $\{Q_i\}_{24x1} = (\bar{q}_1^n, q_2^{-n}, \ldots, \bar{q}_{12}^n, \bar{q}_7^{n+1}, \ldots,$

$$\bar{q}_{12}^{n+1},\ \bar{q}_{7}^{n+2}, \ldots, \bar{q}_{12}^{n+2})^T$$

The linking matrix:

$$[D]=\begin{bmatrix} [I]_{6x6} & & & \\ & [I]_{6x6} & & \\ & [I]_{6x6} & & \\ & & [I]_{6x6} & \\ & & [I]_{6x6} & \\ & & & [I]_{6x6} \end{bmatrix}_{36x24}$$

$$\begin{Bmatrix} \bar{q}^{n} \\ \bar{q}^{n+1} \\ \bar{q}^{n+2} \end{Bmatrix}_{36x1} = [D]_{36x24}\{Q_i\}_{24x1} \qquad (6)$$

Substituting Eq. (6) into Eq. (5) and multipling Eq. (5) with matrix $[D]^T$ to the left of Eq. (5), we obtain Eq. (7).

$$[M_i]_{24x24}\{\ddot{Q}_i\}+[K_i]_{24x24}\{Q_i\}=\{P_i\} \qquad (7)$$

In which:

$$[K_i]_{24x24}=[D]^T\begin{bmatrix} [K_n^i] & & \\ & [K_{n+1}^i] & \\ & & [K_{n+2}^i] \end{bmatrix}[D]$$

$$[M_i]_{24x24}=[D]^T\begin{bmatrix} [\bar{m}_n^i] & & \\ & [\bar{m}_{n+1}^i] & \\ & & [\bar{m}_{n+2}^i] \end{bmatrix}[D]$$

$$\{P_i\}=[D]^T\begin{Bmatrix} \{\bar{p}^n\} \\ \{\bar{p}^{n+1}\} \\ \{\bar{p}^{n+2}\} \end{Bmatrix} \qquad (i=1,\ldots,6)$$

Because the frame is considered as a rigid body the displacements of linking points between six lets and the frame are equal to the displacements of the rigid frame.

Let $\bar{Q}_i=(\bar{q}_7^i,\ldots,\bar{q}_{12}^i)$; $\bar{\bar{Q}}_i=(\bar{q}_1^i,\ldots,\bar{q}_{12}^i)$

and $\{\bar{Q}\}_{114x1}=(\bar{\bar{Q}}_1, \bar{Q}_2, \ldots, \bar{Q}_{18})^T$

The complete linking matrix is given as:

$$[S]=\begin{bmatrix} [I]_{6x6} & & & & & & \\ & [I]_{18x18} & & & & & \\ [I]_{6x6} & & & & & & \\ & & [I]_{18x18} & & & & \\ [I]_{6x6} & & & [I]_{18x18} & & & \\ [I]_{6x6} & & & & & & \\ & & & & [I]_{18x18} & & \\ [I]_{6x6} & & & & & & \\ & & & & & [I]_{18x18} & \\ [I]_{6x6} & & & & & & \\ & & & & & & [I]_{18x18} \end{bmatrix}_{144x114}$$

So that:

$$\begin{Bmatrix} Q_1 \\ Q_2 \\ \vdots \\ Q_{18} \end{Bmatrix} = [S]\{\bar{Q}\}_{114x1} \qquad (8)$$

Pile motion equations of six legs, we get

$$\begin{bmatrix} [M_1] & & & & & \\ & [M_2] & & & & \\ & & [M_3] & & & \\ & & & [M_4] & & \\ & & & & [M_5] & \\ & & & & & [M_6] \end{bmatrix}\begin{Bmatrix} \{\ddot{Q}_1\} \\ \{\ddot{Q}_2\} \\ \{\ddot{Q}_3\} \\ \{\ddot{Q}_4\} \\ \{\ddot{Q}_5\} \\ \{\ddot{Q}_6\} \end{Bmatrix} +$$

$$\begin{bmatrix} [K_1] & & & & & \\ & [K_2] & & & & \\ & & [K_3] & & & \\ & & & [K_4] & & \\ & & & & [K_5] & \\ & & & & & [K_6] \end{bmatrix}\begin{Bmatrix} \{Q_1\} \\ \{Q_2\} \\ \{Q_3\} \\ \{Q_4\} \\ \{Q_5\} \\ \{Q_6\} \end{Bmatrix} = \begin{Bmatrix} \{P_1\} \\ \{P_2\} \\ \{P_3\} \\ \{P_4\} \\ \{P_5\} \\ \{P_6\} \end{Bmatrix} \qquad (9)$$

Substituting Eq. (8) into Eq. (9) and multipling Eq. (9) with $[S]^T$ to the left of Eq. (9), we get:

$$[\bar{M}]\{\ddot{\bar{Q}}\} + [\bar{K}]\{\bar{Q}\} = \{\bar{P}\} \qquad (10)$$

In which: (M, K as $M_{114x114}$, $K_{114x114}$; Q, P as Q_{114x1}, P_{114x1})

$$[\bar{M}]=[S]^T\begin{bmatrix} [M_1] & & & & & \\ & [M_2] & & & & \\ & & [M_3] & & & \\ & & & [M_4] & & \\ & & & & [M_5] & \\ & & & & & [M_6] \end{bmatrix}[S]$$

$$[\bar{K}]=[S]^T\begin{bmatrix} [K_1] & & & & & \\ & [K_2] & & & & \\ & & [K_3] & & & \\ & & & [K_4] & & \\ & & & & [K_5] & \\ & & & & & [K_6] \end{bmatrix}[S]$$

$$[\bar{P}]=[S]^T\begin{Bmatrix} \{P_1\} \\ \{P_2\} \\ \{P_3\} \\ \{P_4\} \\ \{P_5\} \\ \{P_6\} \end{Bmatrix}$$

Eq. (10) is the motion equation of the six-legged walking mechanism in the coordinates of the frame system.

ELASTO-DYNAMICS SYNTHESIS

The dynamic properties of robot are determined by its construction. In order to design a robot which has the smallest motion deviation and moving activity, the optimal size of robot's construction is solved with optimization method. After determing the form of robot construction according to walking demands, the cross-section magnitudes of bars are variables in the process of optimizing.

The First Optimizing Goal

A robot is made of many bars and the frame. There are relative motions between bars, and between the frame and the six legs. Because the width and length of the frame and the lengths of bars are determined by the authors in kinematics synthesis, they are not variables in the process of dynamic optimizing, only the crosssection magnitudes of bars are variables. It is difficult for us to explain quanttatively how much contribution each bar has to the system's stiffness, but natural frequency, especially, the lowest order natural frequency of system reflects system's stiffness. The bigger the lowest order frequency of system, the greater the system's stiffness. The first optimizing goal is to let the walking vehicle have maximum basic frequency. In section II, we get the complete motion equation in the coordinates of the frame's system, and transform it into the coordinates of system fixed with the earch. Let angles between frame coordinates axes and fixed coordinates axes are θ_{xx}, θ_{xy}, θ_{xz}, θ_{yx}, θ_{yy}, θ_{yz}, θ_{zx}, θ_{zy}, θ_{zz}. We get transformation matrix as:

$$[b]_{3x3} = \begin{bmatrix} C\theta_{x\bar{x}} & C\theta_{x\bar{y}} & C\theta_{x\bar{z}} \\ C\theta_{y\bar{x}} & C\theta_{y\bar{y}} & C\theta_{y\bar{z}} \\ C\theta_{z\bar{x}} & C\theta_{z\bar{y}} & C\theta_{z\bar{z}} \end{bmatrix}; \quad [B] = \begin{bmatrix} [b] & & & \\ & [b] & & \\ & & \ddots & \\ & & & [b] \end{bmatrix} 38$$

In which, $S\theta$ express Sineθ ; $C\theta$ express cosineθ . The relation between the generalized coordinates $\{\tilde{Q}\}_{114x1}$ of the system fixed with the earth and the generalized coordinates $\{\bar{Q}\}_{114x1}$ of the system fixed with the frame is expressed as:

$$\{\bar{Q}\}_{114x1} = [B]_{114x114}\{\tilde{Q}\}_{114x1} \quad (11)$$

Substituting Eq. (11) into Eq. (10) and multipling Eq. (10) with $[B]^T$ to the left of Eq. (8), we get Eq. (12).

$$[\tilde{M}]_{114x114}\{\ddot{\tilde{Q}}\}_{114x1}+[\tilde{K}]_{114x114}\{\tilde{Q}\}_{114x1} = \{\tilde{P}\}_{114x1} \quad (12)$$

in which: $[\tilde{M}] = [B]^T[\bar{M}][B]$

$[\tilde{K}]=[B]^T[\bar{K}][B]$ $\{P\}=[B]^T\{P\}$

After got characteristic value of matrix $[K]^T\cdot[M]$,λ,we get the natural frequency of the system ω :

$$\omega = \sqrt{1/\lambda} \quad (13)$$

The bigger value ω , the smaller value $\sqrt{\lambda}$. For the first optimizing gool function, we write

$$F1\ (x) = \sqrt{\lambda_{max}} \quad (14)$$

In which, the variable x is n-dimensional column vector of crossection size of bars.

The Second Optimizing Goal

We not only demand the robot has bigger stiffness, but also moves actively and saves power. In order to make the robot act dexterously, the second optimizing goal is to make the walking vehicle lightest in weight. In order to make the walking vehicle steady in motion and symmetric in construction, leg I, III, IV, VI are designed with the same construction, and leg II, V are designed with the other construction. As an example, we choose circle-crosssection bar so that the weight of leg I can be written as:

$$W1= \pi\,\xi((R_1^2-r_1^2)\cdot L_1+(R_2^2-l_2^2)\cdot L_2+(R_3^2-r_3^2)\cdot L_3); \quad (15)$$

The weight of leg II can be written as:

$$W2=\pi\,\xi((R_4^2-r_4^2)L_4+(R_5^2-r_5^2)L_5+(R_6^2-r_6^2)L_6) \quad (16)$$

In which: ξ is the density of material; R_i is external radius of the i-th bar's crosssection; r_i is inner radius of the i-th bar's crosssection; L_i is length of the i-th bar. The weight of the frame is determined by the working demands, which is a constant. The second optimizing goal is written as:

$$F2(X) = C + 4W1 + 2W2 \quad (17)$$

The overall goal function written as:

$$F(X) = D1\cdot F1(X) + D2\cdot F2(X) \quad (18)$$

In which, D1,D2 are the "power" of these division goal function. In process of optimizing goal function must satisfy a series of restrictions. These restrictions include:

$$r_i \geqslant 0 \quad (i=1,\ldots,6) \quad (19)$$

These restrictions prevent the inner radius of bar from negative value.

$$R_i - r_2 > 0 \quad (i=1,\ldots,6) \quad (20)$$

These restrictions prevent area of crosssection from negative value. In order to prevent area of crosssection from zero and prevent bars from distress, we especially replenish the following restriction:$\omega_{min} > 9$. In order to prevent the walking robot from collapse, we use mono-direction transformation mechanism in each articulator.

RESULTS

By calculation, we adopt D1=0.6 and D2=0.4. The length of bars are calculated with kinematic optimizing by authors as:

L_1=10 cm, L_2=15 cm, L_3=23 cm, L_4=10 cm, L_5=31 cm, L_6=30 cm. In order to use the same crosssection bars material we let $R_1=R_2=R_3=R_4=R_5=R_6$ and $r_1=r_2=r_3=r_4=r_5=r_6$. we select the initial numeral values of variables as:

R_i=19 mm; r_2=9 mm;(i=1,...,6)

Using changeable derivation multiplan body method, we calculated and got values of variables for design as shown in the following:

R_i=25.3 mm; r_i=5.3 mm;(i=1,...,6)

CONCLUSION

1) In order to make the walking robot get bigger stiffness and move actively, the best way, we consider is to increase the elasticity of its construction and use the method presented by the authors.
2) We can depose the restriction about $R_1=R_2=R_3=R_4=R_5=R_6$ and $r_1=r_2=r_3=r_4=r_5=r_6$.
3) The method in this paper is a method of computer aid design.

REFERENCES

1. Zhang Ce: "An Introduction to Kineto-Elasto-dynamics" design of mechinery. 1985. Vol.1.
2. M. Vukobrativic: Dynamics of Manipulation Robot Theory and Application. 1982.
3. Jiang Ping: "Elasto-Dynamic analysis and control of robot with mult-articulations —six-legged walking vehicle", defend paper for mastor degree, 1985.
4. Charles M, Close and and Dean K. Frederick: Modeling and Analysis of Dynamic Systems.
5. Fourth Edition: Vibration Problems in Engineering.
6. Yang Di,Tang Heng Ling and Liao Bo Yu: "Dynamics of Machine Tool". 1983.

A Mathematical Model of a Flexible Manipulator of the Elephant's Trunk Type

G. MALCZYK* AND A. MORECKI**

*CBKO Pruszkow, ul. Staszica 1, Informatics Division
**ITLiMS PW Warszawa, al. Niepodleglosci 222, Team for Robotics and Biomechanics, Technical University of Warsaw

Summary. The paper describes the physical model of an elephant's trunk as well as the mathematical model of its kinematics in static conditions. The model is based upon anatomical and physiological studies of the trunk.
Motion capabilities of the trunk and its muscular structure, forming the power transmission system, has been taken into account.

INTRODUCTION

The technical biomechanics workgroup at ITLiMS PW is occupied with elastic manipulators and robots. This paper describes the kinematics of the elephant's trunk under static conditions. The created model may be used as an aid in the construction of manipulator's arm. The mathematical description is given for the physical model introduced in [6] and [7]. The creation of the physical model has been preceded by extensive studies of anatomy and physiology of an elephant's trunk [1,2,6,7] . The aim of the paper is to describe a method of construction of an elephant's - trunk - type manipulator. The literature refering to the subject of snake-like or spine-like constructions [3,4,8,9,10,11] and elastic structures [5] has been investigated.

THE DESCRIPTION OF ANATOMICAL AND PHYSIOLOGICAL PROPERTIES OF THE ELEPHANT'S TRUNK

The description of anatomy will be limited to the motion system (muscular system) and the description of physiology will take into account only the motion ability of the elephant's trunk. The muscular system can be subdivided into two subsystems, namely: the internal subsystem consisting of inner muscles, fat and connective tissue and the external subsystem composed of external muscular layer. The trunk does not contain a skeleton, so the structure of movement is continous. The internal subsystem is functioning as a "skeleton" and the external subsystem works as a drive mechanism. The internal subsystem functions as a cylindrial core. It is surounded by a circular muscular layer of variable thickness. The internal subsystem becomes narrower along the trunk and the thickness of the muscular layer increases. The trunk , on its circumference hasthree callosities, located: one in front and the other two at both sides of the rear part of the trunk. Along the trunk the callosities become stronger. The structure of the trunk's cross-section at different hights does not change. The fibers in the external muscular layer are placed along the trunk and the fibers of the muscles in the internal subsystem are located radially. In the lower end of the trunk the fibres of the rear callosities, located in the external muscular layer, are placed askew downwords towards the back of the trunk. The cross-section of the trunk is shown in figure1, and its model is pictured in figure 2.

Certain coordinate systems and nomenclature have been introduced to simplify the physiological studies. The position of the trunk when it is pointing straight down is considered as its normal position (fig.3).The figure shows the axis and planes of reference. The trunk may be subdivided into three parts:the base, the stem and the tip.
The ability of trunk's movement is immense. Nevertheless some schematocs of the trunk's motion can be introduced. The base of the trunk can move only in sagittal plane. Rotary motion or movement from side not take place or is negligible. The stem move in sagittal plane as well as in frontal plane. Here rotary motion is also not possible. The tip of the trunk can move in all three planes. The motion in saggital plane is called the forward and backward movement. The motion in frontal plane is called left and right bending. The motion in transverse planeis called turning movement.This nomen-

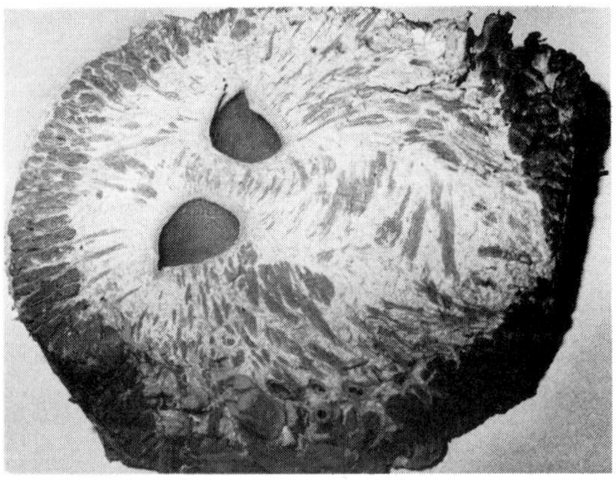

Fig. 1 Cross-section of elephant trunk at a level of the base

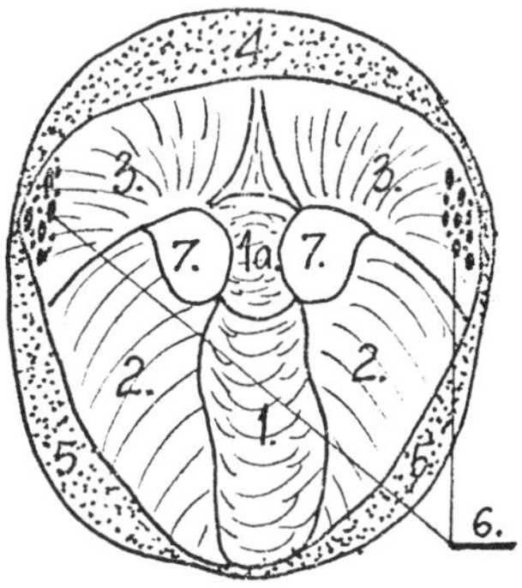

Fig.2 Model of the cross-section 1-5 conventional regions 6-nerves, 7-snout tubes

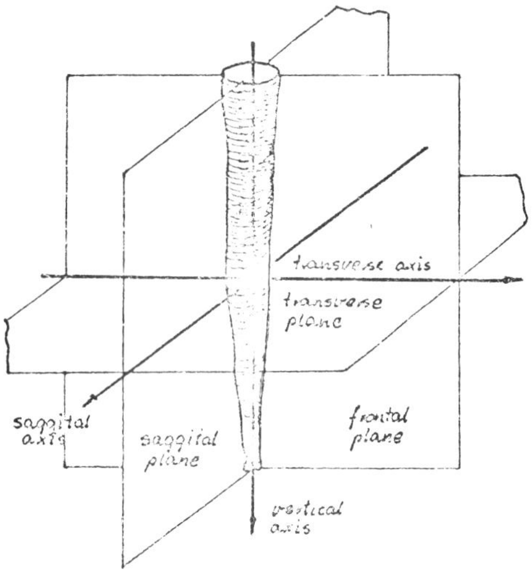

Fig.3 Normal position of elephant trunk and reference coordinates

clature pertains to the normal position of the trunk . The possibilities of the trunk's motion are shown in fig.4.

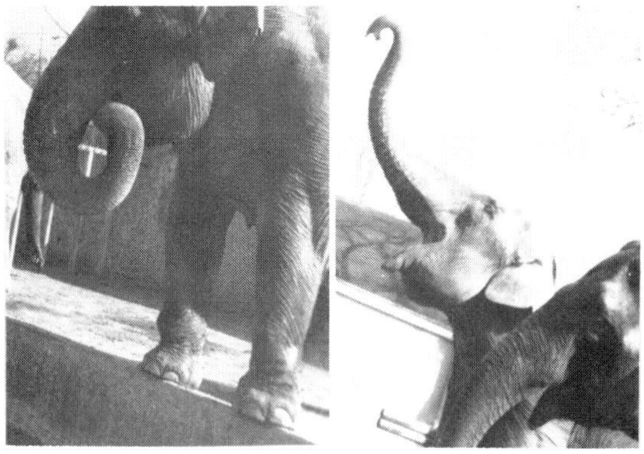

Fig.4 Some examples of the elephant trunk movements

The relation between muscular structure and the nervous system has been described in [6].

PHYSICAL MODEL OF THE ELEPHANT'S TRUNK CONSIDERED AS AN ELASTIC STRUCTURE

The model is based upon anatomical and physiological studies of the trunk. Motion capabilities of the trunk and its muscular structure, forming the power transmission system, has been taken into

Fig.4 Continued

account [6,7]. The model (Fig.5) consists of concentrated masses connected with weightless links. Each link is considered to have different rigidity in two directions and torsional rigidity about the vertical axis. The connections between the masses and the links are rigid. The model is divided into three parts corresponding to the three parts of the elephant's trunk. The rigidity of each links has been selected so as to represent precisely possibilities of the trunk's motion In the base part of the trunk the rigidity in the direction of the transverse axis and the torsional rigidity are much larger than the rigidity in the direction of saggital axis. Torsional rigidity is much geater than the other two rigidities of the stem. All three rigidities of the tip are of the same order . The rigid connections between mases and links assure a continous structure of motion. The model is equiped with a power transmission system of muscular type (with unilateral action). The power transsmission consists of 43 actuators. The two chains located at both sides of the rear contain 21 actuators each. The last actuator is located in front, along the full lenght of the trunk. These locations correspond to the positions of three callosities in the external muscular layer. In the tip of the trunk the actuators are positioned askew to the vertical axis. This anables the tip to perform rotary motion . The structure with such power transmission can execute 39 independent movements.

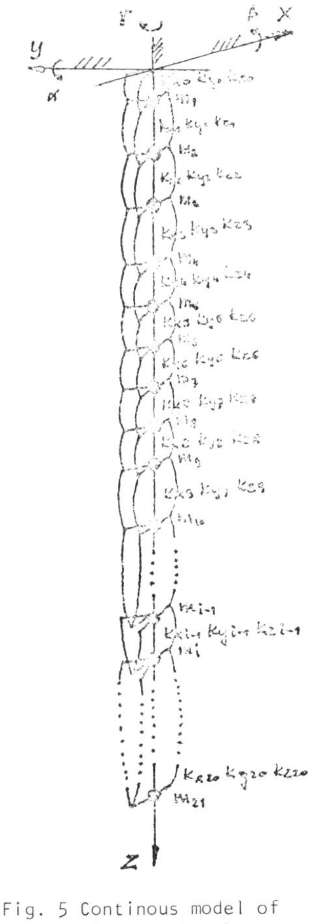

Fig. 5 Continous model of elephant trunk movements

A MATHEMATICAL DESCRIPTION OF THE KINEMATICS OF ELEPHANT'S TRUNK

This description pertains to the elastic structure modelling the elephant's trunk, not to the trunk

itself.
The following assumptions have been made: the segments, corresponding to the elements of the structure, are described by a vector function

$$r_i = r_i(l_j) = x_i(l_j), y_i(l_j), z_i(l_j) \;;$$

each of its components is a function of class C^3 for their argument in the range $l_j(0,l_o)$; in transition points between segment i and i+1 the derivatives r_i', r_i'', r_i''' are bilaterally equal: each segment has a constant lenght.
Several coordinate frames have been introduced to simplify the discription (fig.6). The inertial coo-

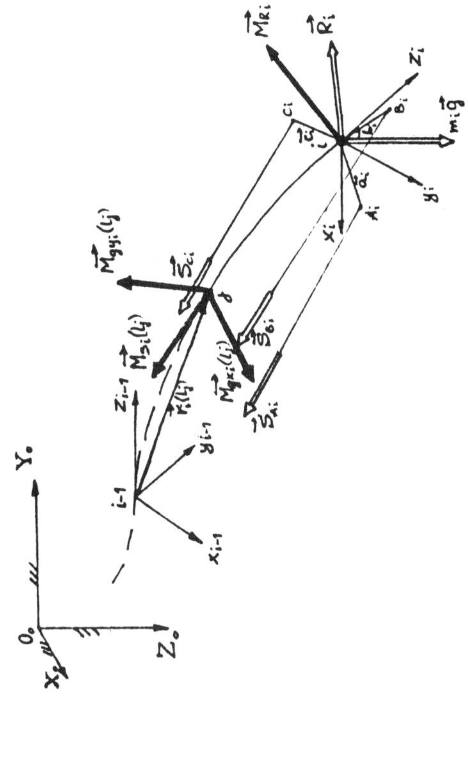

Fig.6 Statical model of the basic element

rdinate frame $X_oY_oZ_oO_o$, having its origin in O_o, is located so that: axis Z_o is colinear with the axis of the model in its normal position, axis X_o is defined by a vector with its origin in O and its end at the point of attachment of the front actuator to the base A_o, axis Y_o is constructed in such a way, that vectors X_o, Y_o, Z_o will form a dextrorotatory orthogonal coordinate frame. Moreover 21 movable coordinate frames have been fixed to the elements of the model. The origin of each coordinate frame coincides with the origin of each element 1,2,3,...,21. The axis of each frame are defined by Frenet's trihedron located at the origin in such a way that: normal versor is colinear with axis x_i, binormal versor is colinear with axis y_i and tangential versor is colinear with axis z_i.
The axis z_o and Z_o of frames $x_oy_oz_o$ and $X_oY_oZ_o$ are colinear and the angle of rotation about Z_o between x_o and X_o or y_o and Y_o depends on the operation of the actuators.
Knowing the configurations of all elements in local coordinate frames the configuration of the model can be found. Because all the elements are similar, we will describe only one of them in a local coordinate frame. The 3rd Sarett-Frenet's equation describes the curvature of the element in a local frame:

$$\frac{d\vec{n}_i(l_j)}{dl_j} = \tau_i(l_j)\cdot \vec{b}_i(l_j) - |\varkappa_i(l_j)|\cdot \vec{t}_i(l_j), \qquad (1)$$

where:
t_i, n_i, b_i -are the vectors: tangential, normal and binormal to the curvature of the element at the point delimited by the length l_j.
τ_i -is the relative torsion at the point delimited by the length l_j.
$\varkappa_i$ -is the relative curvature of the element at the point delimited by the length l_j.
l_j -is the independent variable - the length of the element measured along the curvature, $l_j \in (0, l_{oi})$
l_{oi} -is the total length of the element.

Now, $r_i(l_j)$ can be defined as a vector function, where $r_i(l_j)$ is the vector joining the origin of the local coordinate frame and the consecutive points on the curve (fig.6).

$$\vec{t}_i(l_j) = \frac{dr_i(l_j)}{dl_j} \qquad (2)$$

$$\vec{n}_i(l_j) = \frac{d^2 r_i(l_j)}{dl_j^2} \left| \frac{dl_j^2}{d^2\vec{r}_i(l_j)} \right| \qquad (3)$$

$$\vec{b}_i(l_j) = \vec{t}_i(l_j) \times \vec{n}_i(l_j) \qquad (4)$$

where x denotes the vector cross product.
The curvature and torsian can be derived from (1):

$$\tau_i(l_j) = \frac{(\vec{M}_{si}(l_j))\,dl_j}{G_i \cdot J_{oi}} \qquad (5)$$

$$|\varkappa_i(l_j)| = \frac{|\vec{M}_{gi}(l_j)|}{E_i J_i} \qquad (6)$$

where:
J_{oi} -axial moment of inertia of the cross section of the elephant's trunk in segment i.
J_i -moment of inertia of the cross-section in the direction of axis x_i or y_i (Axial symetry of the cross-section is assumed $J_{xi}=J_{yi}=J_i$. Moreover J_i is considered constant along the length of the element).
G_i -modulus of rigidity (coefficient of transverse elasticity).
E_i -longitudinal modulus of elasticity (Young's modulus)
(E_i and G_i are assumed constant along the length of the trunk).
$\vec{M}_{si}(l_j)$ -torque moment acting on an element at the point delimited by the length l_j. The value of the function $m_{si}(l_j)$ describes the distribution of torque moment along the elements of the model. The direction and sense of the vector are determined by the tangential versor at the point l_j.
$\vec{M}_{gi}(l_j)$ - bending moment in two planes acting on an element at the point delimited by the length l_j. The value of the function $M_{gi}(l_j)$ determines the distribution of bending moment in space along the model's element. The direction and sense of the vector are determined by the sum of normal and binormal vectors.

Substituting (2,3,4,5,6) into (1) we obtain vector differential equation describing the vector function $r_i(l_j)$

$$\frac{d}{dl_j}\frac{\vec{r}_i''(l_j)}{|r_i''(l_j)|}=\frac{\vec{M}_{si}(l_j)dl_j}{G_iJ_{oi}}\vec{r}'(l_j)\times$$
$$\times\frac{\vec{r}''(l_j)}{\vec{r}''(l_j)}-\frac{\vec{M}_{gi}(l_j)}{E_iJ_i}\vec{r}'(l_j) \qquad (7)$$

Now, the torque and bending moment will be determined from simple relations supplied by statics (fig.6). The resultant moment about point j can be denoted as:

$$\vec{M}_{wi}=\vec{S}_{A1}(\vec{a}+\vec{ij})+\vec{S}_{B1}\times(\vec{b}+\vec{ij})+\vec{S}_{C1}\times(\vec{c}+\vec{ij})+$$
$$+\vec{M}_{Ri}+(m_i\vec{g}+\vec{R}_i)\times\vec{ij} \qquad (8)$$

where: $\vec{S}_A,\vec{S}_B,\vec{S}_C$ -are the muscular forces in the front and both actuators located at the rear.

a,b,c -the distance of the points of application of forces S from the point i+1.

m_i -the mass of element i.

$\vec{R}_i$ - the force of reaction at the point i+1 caused by the elements with a number greater than i.

$\vec{M}_{Ri}$ -the moment of reaction at the point i+1 caused by the elements with a number greater than i+1.

ij -vector $\vec{ij}=\vec{r}_i(l_j)-\vec{r}_i(l_j)$

The resultant moment M_{wi}, expressed in a mobile coordinate frame x_i,y_i,z_i, can be decomposed into components expressed in coordinate frame fixed to the Frenet's trihedron constructed at point "j".

$$M_{wi}=\left[M_{gx_i},\ M_{gy_i},\ M_{gs_i}\right] \qquad (9)$$

where: M_{gx_i},M_{gy_i} - are the bending moments in two planes at the point j of element i

M_{s_i} - torque moment at the point j of element i.

$$M_{g_i}=M_{gx_i}+M_{gy_i} \qquad (10)$$

Using equation (7), the following problem can be solved: in the configuration of consecutive elements of the model is given as

$$\vec{r}_i(l_j)=\left[x_i(l_j),y_i(l_j),z_i(l_j)\right];$$
$$i=0,\ldots,20; \qquad (11)$$
$$j=0,1,\ldots,n;$$

where: i - the number of elements of the model

j - the number of points at which the element's configuration has been determined

in discrete form, the forces S (which have to be exerted by the actuators, so that this configuration can be obtained) can be determined.
The inverse problem can not be solved using this discription, because the boundary conditions are not defined.

CONCLUSIONS

A mathematical description of the elephant's trunk, considered as an elastic structure, has been given. It can be seen, from the analitical description, that the inverse problem can not be solved, if it is started by the above method, because the boundary conditions are not defined. As a continuation of of this work, the verification of the assumptions and the m thod of description will be performed. The following conclusions can be derived from the work we have been doing. The continous mass must be introduced to the mathematical description of the model's kinematics and statics. Moreover global coordinate frames must be employed to determine the boundary conditions. After the verification the model will be simulated on a computer.

ACKNOWLEDGEMENT

This project is supported by the Polish Academy of Sciences (Central Programme for Fundamental Research No. 7.1).

REFERENCES

Boas J.E.V., S.Pauli: The elephant head, part 2. Published at the ccst of the Calsberg-fund., Copenhagen 1925.

Чеботарев И.Т.: Взаимоотношение венозных образований и мышц в хоботе слона. Труды Московской Ветеринарной Академии. Том 27. Москва, 1963.

Hirose S., Ikuna K., Umetani Y.: A new design method of servo-actuators based on the shape memory effect. Tokyo Institute of Techn., Tokyo 1984.

Hirose S., Umetani T.: An active cord mechanism with oblique sivilvel joints and its control. Proc. of the 4th Ro.man.sy-81, Zaborow, Poland 1981, pp. 327-340.

Hemani A., Studies on a high Weight and Flexible Robot Manipulator. "Robotics" No 1. Novoth-Holand 1985.

Malczyk G.: Model i propozycja rozwiazania manipulatora typu traba slonia. Praca magisterska, Warszawa, 1984.

Malczyk G., Morecki A.: Ealstyczny manipulator typu traba slonia. Praca Naukowa Instytutu Cybernetyki Technicznej Politechniki Wroclawskiej, I Krajowa Konferencja Robotyki, Tom 2, Wroclaw 1985.

Prospekt - Spine - Robotics, Sweden 1983.

Prospekt - Spine News, Sweden 1983.

Prospect - Spine News, Sweden 1984.

Roth B., Rastegar J., Scheinman V.: On design of computer controlled manipulators. Proc. 1st CISM-IFToMM "Ro.man.sy" Udine, Italy Sept. 5-8, 1973, Springer Verlag 1974.

Sokolowska-Pietuchowa J.: Anatomia czlowieka, PZWL , W-wa, 1983.

Краткое содержание

В работе описана физическая модель хобота слона и на этой основе предложена математическая модель для статических условий. Проведенные соотвествующие анатомические и физиологические исследования хобота слона, которые выявили разные функциональные и рабочие способности этого интересного "манипулятора". На базе этого анализа предложены две модла, а именно дискретная и силошная.
Для описания класических свойств структуры хобота использовано третие уровнение Соррета-Френета.
Работа продолжается в двух напровленях, а именно машинной симульяции и построения действительного манипульятора.

Some Problems of the Design and Control of a Quadruped Walking Machine MK-4

K. JAWOREK, A. MORECKI, W. POGORZELSKI AND T. ZIELINSKA

Institute of Aircraft Engineering and Applied Mechanics, Warsaw University of Technology, Warsaw, Poland

Abstract. This paper describes the design of mechanical and control systems of a machine, which be able to perform different locomotion activities, such as walking or running in diverse terrain. A mathematical model describing such walking machines from the point of view of their gait, is proposed. The model includes the mechanical part of the system - executing the movements, as well as the control part of the system, which is used for planning the movement. Some results obtained from computer simulation are presented.

Keywords. Control system analysis; modelling; robots; walking machines.

MECHANICAL STRUCTURE AND CONTROL SYSTEM OF THE MACHINE MK-4

The first investigations performed by our team in the field of quadruped locomotion started in 1983 (Jaworek, Pogorzelski, 1983). The main goal of this project was to design a machine, which will be able to perform different locomotion activity which proper velocity and external loading even in a diverse terrain.

It was assumed that the legs of this machine will imitate the limbs of such mammals as rabits, dogs or horses - that is the fore legs are different from the hind legs. At present the machine can only move in the plain OXZ (Fig.1). Each leg has two degrees (Fig.2 and Fig.3) of freedom and is operated by bilateral hydraulic actuators. In the next version one degree of freedom more will be added to each leg. The actuators are equiped with limit microswitches of the on-off type. The structure of each leg ensures a one-to-one relationship between the displacement of hydraulic actuators and the position of the leg's end. The dimensions of the platform and the legs were assumed. The machine with such structure was named MK-4. The simplest kind of movement gait of the machine can be divided into phases with three and four legs supporing the platform. Consecutive phases, during which three legs are supporting the platform, can be distinguished by the lifted leg: left fore leg, right hind-, right fore- and hind left leg. The movement is quasi-static , that is the projection of the center of gravity of the machine must be inside the polygon - vertices of which are located in the points of machine support.

In the beginning simple system with open loop control was elaborated (Jaworek, Pogorzelski, 1984). Next the motion on slanting terrain was discussed (Jaworek, Pogorzelski, Zielinska , 1985). Inertialess model of the machine as well as ideal drive system were assumed. Implementation of microprocessor system made possible the open loop control (Jaworek, Pogorzelski, and Zielinska 1984,1985). In the further investigations the hierarchical structure of on adaptive control system will be assumed. Digital control (partly concurrent-Fig.4), based on the concept of indirect control, which application of extermal regulation on the executive level, was anticipated. The executive level was ealborated in details. In Fig.5 the main part of the solution is presented, namely the digital control system. P.Bieniaszewski has tested this system by computer simulation. The functioning of the executive level for one leg was tested on an example of control of the two-link open kinematic chain with hydraulic actuators. An ideal model of the drives was assumed for the simulation process. A special algorithm for finding on extremum of a function with two arguments was elaborated and tested. The visualisation of the chains motion confirms possibility indirect control of the mechanism with several degrees of freedom using extremal regulation and indirect control principle. Fig. 6 presents the simulation of the motion caused by the executive level.

MATHEMATICAL MODEL OF THE MK-4 WALKING MACHINE

Considering the walking machine as a system, which has to realise the process of motion, its mathematical model was elaborated. The model was subdivided into two parts: the mechanical part and the control part.

The motion of the machine is possible, when:
- the motion of each limb is possible, that is
a) the legs' design does not limit the motion,
b) the legs do not colidate,
- the motion's stability conditions are met, that is the machine does not lean or turn over.

We introduce the concept of discrete real state of the walking machine - $\mathbf{p}^j$ - at time t_j

$$\mathbf{p}^j = \mathbf{p}(t_j) = (p_1^j, p_2^j, p_3^j, p_4^j) \qquad (1)$$

The state p^j is defined by the coordinates of the leg's end. The values of the leg's coordinates are limited by the leg's design. Each state of the machine the position of the legs' ends can be attained by moving the hydraulic actuators fixed to the legs. Each leg is propelled by two actuators. The positions of the actuators are related to the output $\mathbf{u}^j = \mathbf{u}(t_j)$ of the control system.

$$\mathbf{u}^j \in \mathbf{U}, \quad \text{and} \qquad (2)$$

$$\mathbf{U} = U_{11} \times U_{21} \times U_{12} \times U_{22} \times U_{13} \times U_{23} \times U_{14} \times U_{24},$$

where U_{ki} ($k = 1,2$; $i = 1,2,3,4$) is the set of values attained by the control systems k-th actuator connected to the i-th leg. The discrete real state of the walking machine is a function of the output.

$$\mathbf{p}^j = f(\mathbf{u}^j) \qquad (3)$$

The control part of the system computes the values of the state variables. This values are the values of the discrete theoretical states $\mathbf{p}^{*j}$ ($\mathbf{p}^{*j} \in \mathbf{P}^*$). The value of each output $\mathbf{u}^j$ is the function of the discrete theoretical state.

$$\mathbf{u}^j = G(\mathbf{p}^{*j}) \qquad (4)$$

The values from the set $\mathbf{P}^{*j}$ of the state $\mathbf{p}^{*j}$ must be such that the motion will be possible.
Let:

- $\chi^* = (p^{*0}, p^{*1}, \dots, p^{*N})$, $N = 0,1,2,\dots$
  - any sequence of theoretical states of the machine,
- $S^* = S^*(t_j)$ - the factors effecting the stability of motion,
- $H(\chi^*, S^*)$ - the function describing the conditions of stable motion.

A subset of theoretical states, for which the stability of motion is assured, can be denoted by D_s^*. Now:

$$H(\chi^*, S^*) \in D_s^*, \text{ for } D_s^* \subset P^*, D_s^* \neq 0 \tag{5}$$

When S^* = const., the functioning of the control part can be modeled as:

$$p^{*j+1} = F(p^{*j}, j) \tag{6}$$

WALKING MACHINE'S DESCRIPTION

Interconnecting the description of the control and mechanical parts leads to a walking machine's model.

The relationship (4) is realized as a function:

$$u^{*j} = G'(p^{*j}), \tag{7}$$

$$u^j = G''(u^{*j}), \tag{8}$$

where u^{*j} is the theoretical - computed in the control part - displacement of the actuators. This displacement is computed for the instant t_j. u^j is real displacement at time t_j.

The relationship (7) describes the functioning of the control part and the relation (8) describes the cooperation of the control and mechanical part. The model of the walking machine is shown in Fig. 7.

COMPUTER SIMULATION OF THE MACHINE'S MOTION

The deduced mathematical model was the basis for synthesis of the control system. The functioning of the system was tested by computer simulation. It was assumed that the control system has to permit the motion of the machine in the terrain with slopes of variable angles. The simulation was performed on RIAD-35 (IBM 370 compatible) and SM-4 (PDP-11 compatible) computers. The obtained results were satisfactory. The simulation system was implemented in FORTRAN IV. The results of the simulation were obtained in numeric form, that is the numerical values of the leg end's coordinates in discrete instants of time and the angle between the platform and the plane of the ground were displayed on the screen. The motion of the simplified image of the machine could also be displayed on the computer screen.

CONCLUSION

At present a hardware model of the hind leg with tree degrees of freedom is under construction. This model requires some modifications of the optimal control algorithm. The execute level must be equipment with learning ability to improve its functioning. In the other words future developments will supply the quadruped with a certain level of machine intelligence, which will allow it to accept external stimulus such as operators instructions the speed and the direction of motion signals carrying information about the position of the platform relative to the ground or the state of limit switches in the actuators.
This system will evolve into a closed loop multilevel loop. The improvement of the indirect control system, which is relying on the assigment of a goal by external controller, will enable the control of the legs under real conditions, in which there exist: friction, clearences and variable resistance of the ground. The implementation of such a system needs a specialized computer. This computer must have the ability to perform concurrent computations and must be equipment with a language for communication with the operator.

ACKNOWLEDGEMENTS

This project is supported by the Polish Academy of Sciences (Central Programme for Fundamental Research No. 7.1). This work is conduced by the Robotics and Biomechanics Group at the Institute of Aircraft Engineering and Applied Mechanics. Its memebers are: Prof.A.Morecki, Ph.D.(head of the group), K.Jaworek, Ph.D.,W.Pogorzelski, Ph.D., Dr. habil., T.Zielinska, M.Sc.. The indirect control was simulated on a computer by P.Bieniaszewski. It was the topic of his M.Sc.theses directed by W.Pogorzelski.

REFERENCES

Bieniaszewski, P.(1986). Extremal control of the walking machine MK-4.(In Polish). M.Sc. theses, Warszawa.

Jaworek,K., W. Pogorzelski(1983). On multilevel control of four legged walking machine. (In Polish). 6-th National Conference on Biocybernetics and Biomedical Engineering. Warszawa. pp. 331-332.

Jaworek,K., Pogorzelski W. (1984). Gait of the four legged walking machine and its open loop control. (In Polish). 10-th Polish Conference on Theory of Machines and Mechanisms. Warszawa pp. 67-75.

Jaworek,K., W. Pogorzelski and T. Zielinska(1985). Motion over sloping terrain of the four legged walking machine MK-4. (In Polish). 7-th Conference on Biocybernetics and Biomedical Engineering. Gdansk. pp. 364-366.

Jaworek,K., W. Pogorzelski (1985). Target oriented hierarchic locomotion robot LR-4. Proceedings of the 3-rd International Conference: Control Problems of Industrial Robots. Varna.

Jaworek,K. (1985). Modern algorithms and microprocessor control systems of the anthropomorphic robots and walking machines. (In Polish). Proceedings of the 6-th Conference on: Advances in Construction of Precise Electronic and Mechanical Equipment MIKRONIKA '85.Warszawa.

Jaworek,K., W. Pogorzelski(In printing). Robotics and related fields of konwledge.(In Polish). Polish Academy of Sciences, Warszawa.

Waldron K.J., A.Perry, V.V.Vohnout and R.B.McGhee (1984). Configuration design of the Adaptive Suspension Vehicle. The Inter. Journal of Robotics Research, 3, No. 2, 37-47.

Zielinska,T. (1985). Mathematical model of the walking machine. (In Polish). 1-st National Conference on Robotics.Wroclaw, pp. 109-112.

Zielinska,T. (1986). Modelling of the gait of the four legged machine. (In Polish). Ph.D. theses, Warszawa.

Некоторые проблемы проектирования и управления четырехножного шагающего аппарата МК-4.

Резюме

Работы выполняемые коллективом в области четырехножной локомоции начались предложением структуры и размеров шагающего аппарата - машины МК - 4. Для цели управления шагающим аппаратом придумано специальный метод управления , так называемый метод посредственного управления. Следуя из этого проведено разработку нижнего уровня - уровня исполнительного - для многоуровневой системы управления аппаратом. Используя микропроцессор Z- 80 А проведено симуляцию исполнительного уровня,в том числе действия цифрового экстремального регулятора.

В теоретическом плане вступительно разработано математическую модель, которая представляет походку аппарата. Походка возможна тогда, когда исподнены определенные геометрические и кинематические

условия.
В описании выделено часть которая планирует движения /управляющая часть / и часть которая исполняет движение /механическая часть/.

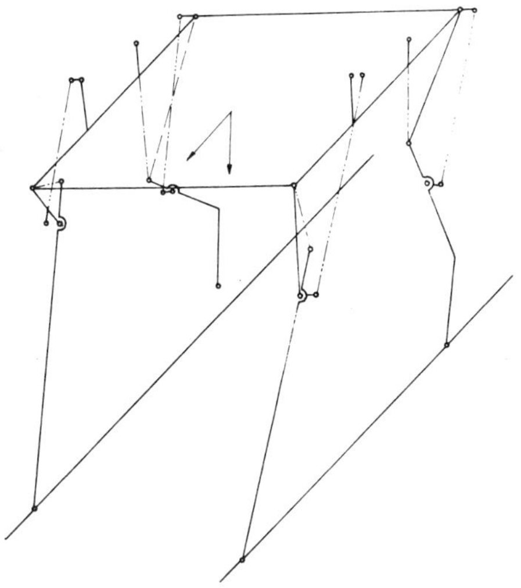

Fig. 1. Model of the walking machine MK- 4.

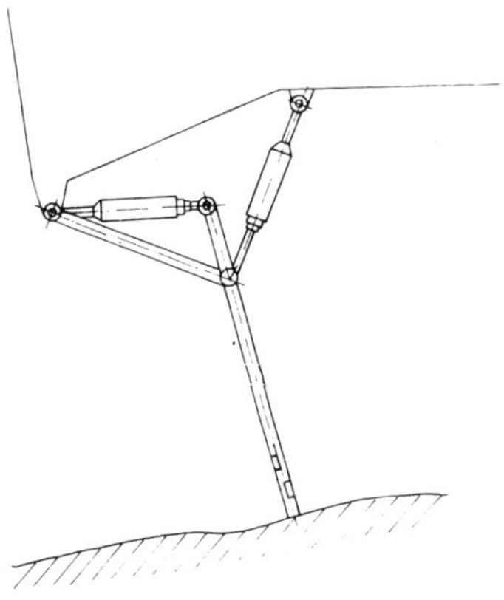

Fig. 2. Construction of the foreleg .

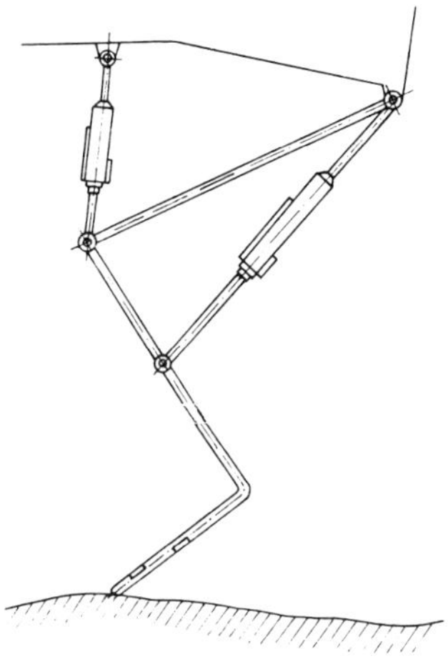

Fig. 3. Construction of the hindleg .

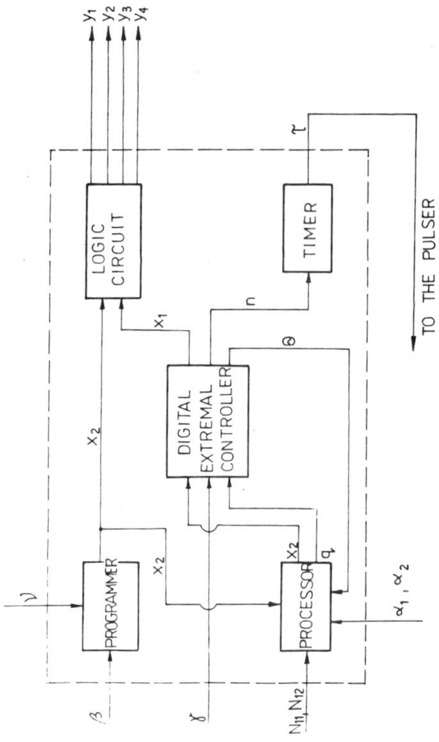

Fig. 4. Structure of the control system.

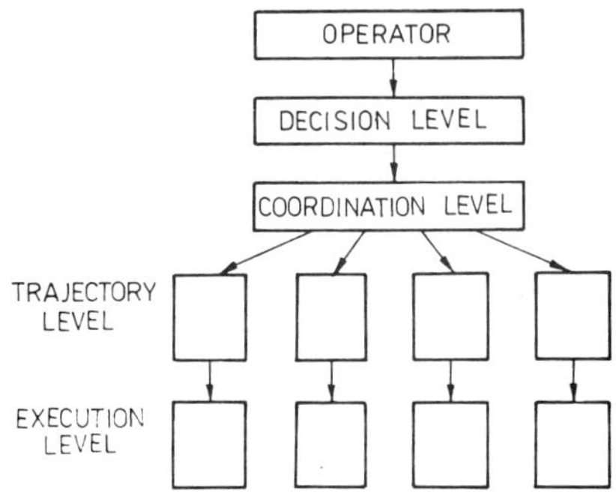

Fig. 5. The main part of the digital control system.

a.)

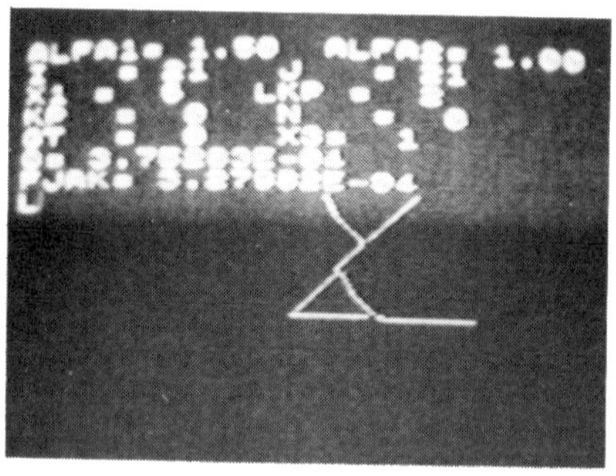

b.)

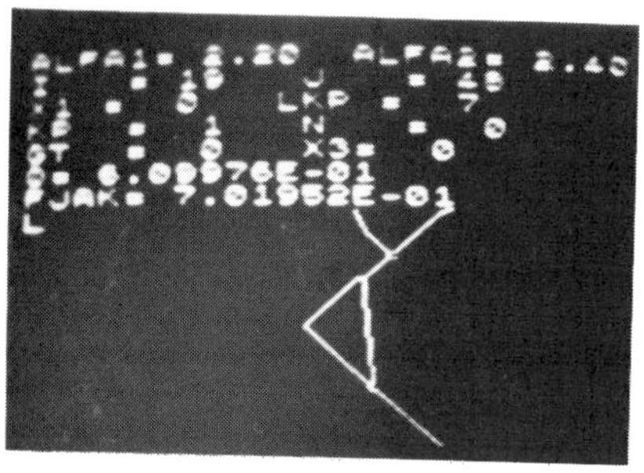

Fig. 6. Computer simulation of two-link model of the walking machine's leg displayed in two position. ALFA 1, ALFA 2- the variable values of the initial and final position.

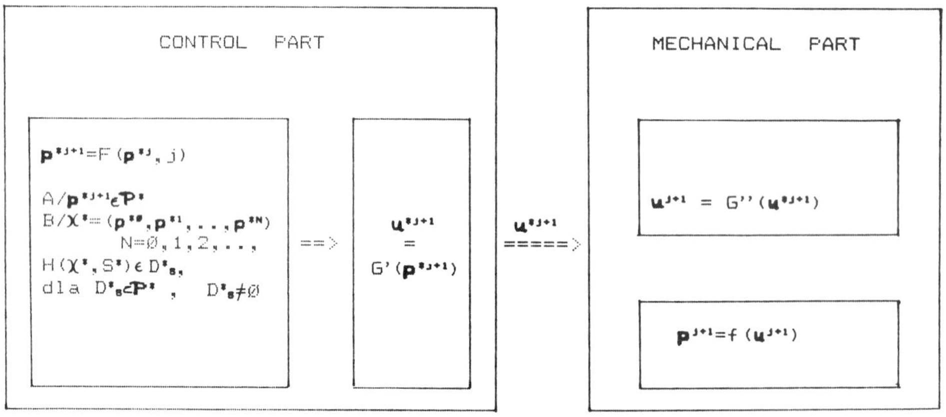

Fig. 7. Mathematical model of the walking machine.

On the Mechanical Properties of Lumbar Spine

M. DIETRICH AND P. KUROWSKI

Warsaw University of Technology, Warsaw, Poland

Abstract. In an etiology of lumbar spine diseases, which produce very popular low-back pain a mechanical factors are more or less important. How great is influence of mechanical factors it can be investigated primary using modelling methods. The theoretical multioptimum model which allows to investigate forces acting in the system of lumbar spine was proposed. The influence of intra-abdominal pressure as well as vertebrae and muscles sizes upon the forces were discussed. Distributions of stresses in vertebrae due to the forces were experimentally investigated using three-dimensional photoelasticity (frozen stress) method. The difference between spongy and cortical bone were considered in two dimensional FEM model of sagittal plane of two vertebrae bodies and intervertebral disc.

Keywords. Mechanics of spine; spondylolysis; osteophyte; model analysis; FEM; photoelasticity.

INTRODUCTION

Considerable part of the human population suffers from so called low-back pain. The pain is generated by one of diseases of lumbar part of spine, like spondylolysis, spondylolysthesis, slip disc, osteophyte formation and others. Spondylolysis means fractures in vertebrae arches which sometimes leads to forward slip of vertebral body that is called spondylolysthesis. Slip disc means such deformation of intervertebral disc (displacement of nucleus pulposus, fracture of annulus fibrosus) which produces pressure on nervous system. Osteophyte formation that is a kind of bone remodelling process which changes the shape of vertebra what in consequence can limit motion of the spine and produces pressure on nervous system.

All mentioned diseases are related to mechanical condition of the spine, first of all to loadings of spine elements. In a medical literature one can find a lot of publications (number of them is so big that it is not possible to indicate all of them in references) discussed the nature of mentioned diseases. At present there are no opinions neglecting the mechanical factors in etiology of these diseases. There are discussions however, how great is the influence of mechanical factors in comparison with biological and chemical ones. For instance the authors opinion about the importance of mechanical factors in the etiology of spondylolysis is given in |1|. The opinion about osteophyte formation is given in |2|.

It can be presume that spondylolytic cracs have fatige nature. High number of repetitions of external loading (for instance during intensive exercises in some kinds of sport) produces fracture in the place in vertebra of high magnitude of stress.

The osteophyte formation is one of typical aging-related bone remodelling processes. Growtgh of osteophytes is likely to be induced by resultant changes in stresses in lumbar vertebrae that occur in the course of aging. But the first to show symptoms of aging are usually intervertebral discs.

This short introduction shows how important is knowledge of mechanical conditions of lumbar spine for expanation of the nature of the disease mentioned. A lot of papers in biomechanical literature concern this problem, but in our opinion there is still lack of enough reliable information about the mechanical properties of lumbar spine. Some of the results connected with the authors investigations will be described in this paper. Due to difficult and limited access to structure of a spine for in vivo investigations the authors employed modelling methods only.

THE SIMPLE MODEL OF LUMBAR SPINE

The lumbar spine is operated by highly complicated system of muscles and these are under permenent control of the nervous system. The number of muscles far exceeds the number of degrees of freedom of this biomechanism so the nervous system can control muscles in the way which allows to perform any given task in the most convenient way. A study on the mechanics of the lumbar spine leads to conclusion that this process can be considered as a multioptimum problem. Except of conditions resulting from the maximum values of muscle forces, the nervous system can tend to minimize the total muscle effort or minimize an intervertebral reaction or to perform the task in the shortest time ect.

As low-back pain concerns practically all groups of the human population, conclusions on the influence of mechanical factors to the disease should be drawn from studies of a typical spine subjected to typical loads. In our opinion a typical case is the spine of a healthy subject performing a common motion activity that is known to cause high loads in the lower back. Forward and backward bending was chosen as an example of such motion activity. The mechanical model of the lumbo-sacral spine corresponding to this movement has been proposed |1,3|. It consists of seven rigid bodies representing the upper trunk, five lumbar vertebrae and pelvis, all connected by hinges simulating intervertebral discs (Fig. 1). Muscles action is introduced to the model as forces acting along straight lines between points where muscles are connected with the skeleton. Two big groups of muscles are considered in the model: erector spinae

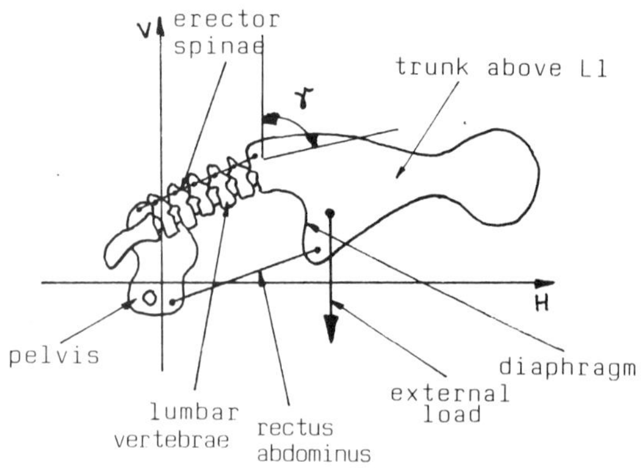

Fig. 1. The mechanical model of human lumbar spine.

divided into six segments and muscles of the frontal wall of abdomen.

The corresponding mathematical model consists of 18 equations of equilibrium and the set of inequality constrains: muscle forces, normal components of intervertebral reactions and abdominal pressure should be positive. The equations of equilibrium and inequality conditions must be satisfied by 20 unknown values of forces: 7 muscle forces, 6 normal components and 6 tangential components of intervertebral reactions and the force of abdominal pressure.

This problem has no unique solution until we define the criterion of muscles control. In other words, we should recognize how in nature does the nervous system operate on subconscious level. The selection of criteria govering the muscles action is perhaps the most difficult part of the study. To enable the reasonable selection we confined the analysis of the model to simple motion activity; statical bending the trunk forward and backward in the sagittal plane. Three different objective functions were considered: minimization of the scalar sum of all muscle forces 1 or of the scalar sum of all intervertebral reactions 2 and mini-max criterion requiring that absolute maximum value of the shear force 3 (tangential component of intervertebral reaction) in any of intervertebral joints is minimum. It is obvious that many other reasonable objective functions can be created. The chosen three are comparatively simple and can represent demands of different kinds. The first formula represents demand of the minimum muscle effort or minimum energy losses. The second one represents condition of minimum averaging loading of the spine. The third one corresponds with the prevention of intervertebral discs from high shear stresses |4|.

This results for forward bending and upright position show that the optimum solutions corresponding to all the functions are identical. The optimum solution takes place for the value of abdominal pressure supporting the upper part of the trunk with possibly big force however not producing contradictory action of abdomen muscles.

The analysis of backward bending shows three different optimum results slightly differing in values. The example result of the calculation is given in Fig. 2.

All above results prove that there are no essential differences in the work of lumbar spine according to criteria described by the three different objective functions. The results show that this simple mathematical model is not very sensive to the reasonable change of objective function.

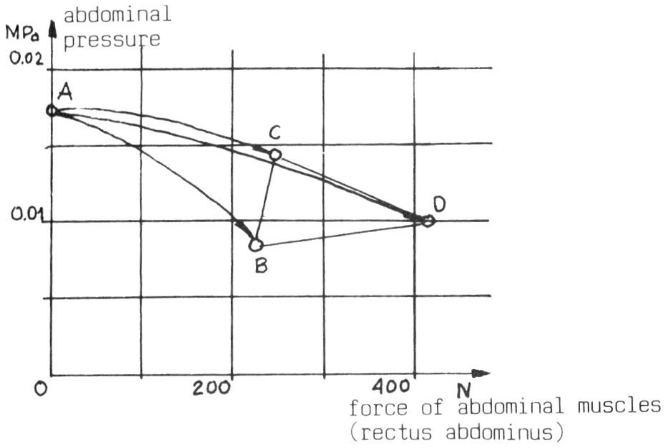

Fig. 2. The optimum solutions for three different objective functions:
A: all functions, $0 \leq \gamma \leq 80°$
B: function 1
C: function 2 } $\gamma = -20°$
D: function 3

INFLUENCE OF SOME PROPERTIES OF THE LUMBAR SPINE TO THE LOADS OF ELEMENTS OF THE SPINE

A lot of parameters of the spine system strongly influences the loading of the elements of the system. Basing on the results of our investigations some of them can be discussed: vertebra size, dimension of cross section of erector spine (or wider of muscles on the back side of spine) and magnitude of intra-abdominal pressure.

Analysis of the mechanical model shows that a small difference in the size of vertebrae results in a difference of the magnitude and direction of the muscle forces, and it can qualitatively change the load on the vertebrae. As an example Fig. 3 shows the change in direction of muscle force acting on the spinal process of vertebra L5. This change results from the change in dimensions of the os sacrum or more precisely of vertebra S1. The magnitude of F_1 is approximatively twice as much as magnitude F_2.

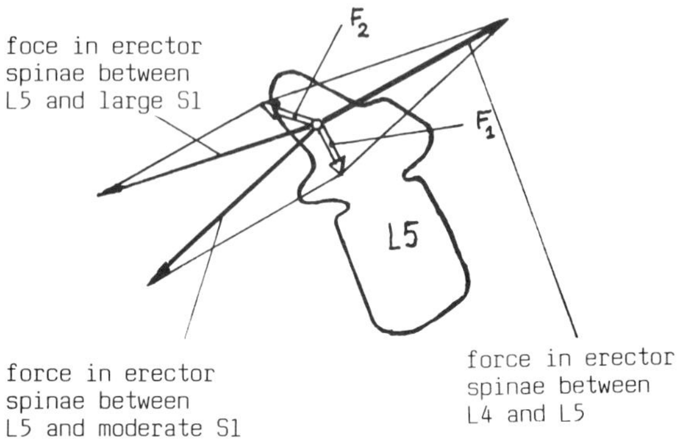

Fig. 3. The change of the resultant force exerted on the vertebral arch due to the change of size of vertebra S1.

The difference in size of cross section of erector spinae also results in strong difference of loading of the system. The place and area of the insertion of erector spinae to the spinous process of vertebrae depends on the magnitude of cross section of erector spinae (Fig. 4). For better developed musculature when the muscles, including erector spinae are greater, the point of insertion is more shifted in back side then for less developed muscu-

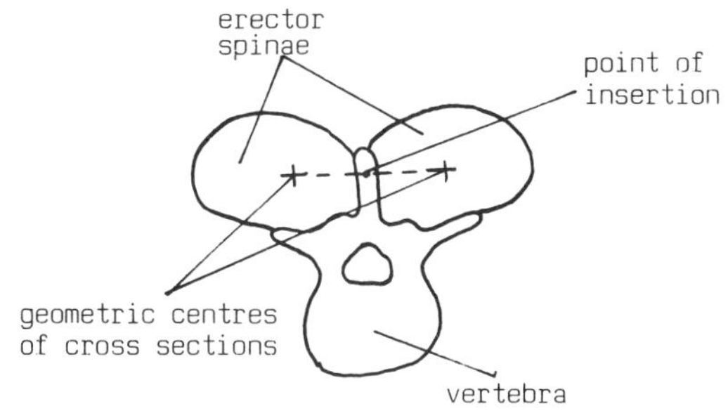

Fig. 4. The location of point of insertion of erector spinae depending on the position of geometric centres of its cross sections.

lature. As the distance between the point of insertion and central point of nucleus pulposus is small in comparison with another dimensions of spine system, even small difference in this distance produces great differences of loadings acting in the spine system.

The investigations confirmed a high relieving effect of intra-abdominal pressure upon the loadings in the lumbar spine. The exemplary results are shown in Fig. 5 and 6.

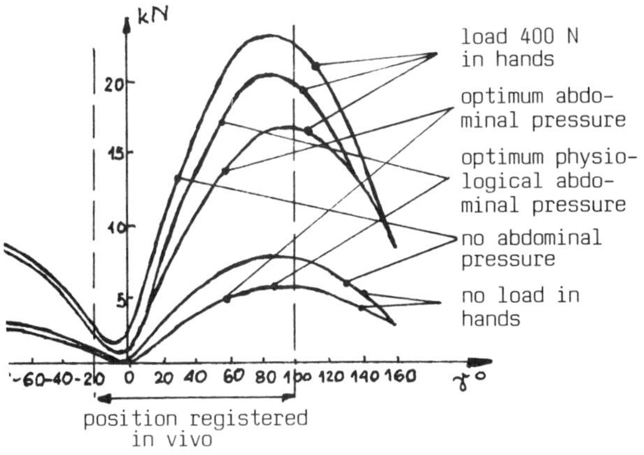

Fig. 5. Values of objective function 1 (sum of forces in muscles)

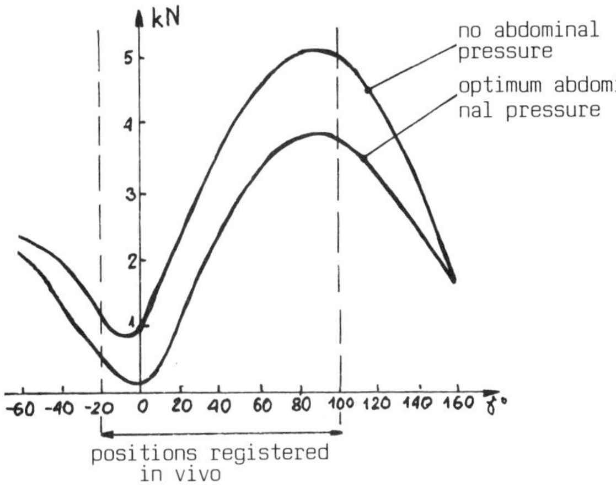

Fig. 6. Force acting between vertebrae body L5 and S1 when load 400 N is kept in hands

In the case of optimum pressure, but not exceeding physiological value (in the investigations the physiological value for normal man was assumed to be equal to 0.024 MPa) the force acting between vertebrae are about 20% lower then in the case of no pressure. For the man in very good condition the physiological pressure can be higher, for instance equal to 0.03 MPa |4|. In such case the relieving effect will be ever greater. This relieving effect can be decisive for a safety of spine. As it is seen in the Fig. 6 the force acting between L5 and S1 reaches 5 kN.

As the compressive strength of lumbar vertebrae is of the order 6500 N, one can see that lifting a load of 400 N in the position of deep forward bending may be dangerous for the spine, and the relieving effect of intra-abdominal pressure may be important in preventing damage to the vertebral calumn.

From the results obtained, qualitatively described here, it is evident that within natural variation of geometry of spine as well as of development of the muscle system, there are spines specially prone to failure due to external loading and ones comparatively resisten to loadings.

EXPERIMENTAL MODEL FOR STRESS ANALYSIS IN LUMBAR VERTEBRAE

To get an information about stress distribution in the vertebrae an experimental model of lower part of lumbar spine was proposed. It consists of vertebrae L4, L5 and S1 connected by intervertabral discs and joints. The models of wertebrae were made from epoxy resin; discs and articular cartilage were made from silikon rubber. Investigations of such model can give only rough estimation of stress distribution, first of all because of difference between unisotropic structure of bone and isotropic material of model. The geometry of the model vertebrae was identical to the geometry of the autopsy specimen vertebrae. The model was described in |1|, where analog criteria were discussed.

The loads, found with the help of the mathematical model and multiplied by suitable factors to obtain the best optical results, were applied to the photoelastic models in a special loading frame. When loaded in this frame, the models were stress-frozen using standard procedure for stress-freezing. The stress-frozen models were tested in two ways: some were cut into thin slices (Fig. 7) to find stress concentrations, others into small cubes to find effective stresses according to the Huber-Van Mises criterion as described in |5|. Cubes were taken from vertebra body, pedicles and parts interarticularis. The application of the Huber-Van Mises criterion to the material of vertebra is justified by the elastoplastic properties of cortical bone. The cortical bone is the main component of parts of vertebra where high stress can be expected so each of the cracks in these places is connected with the damage of cortical bone.

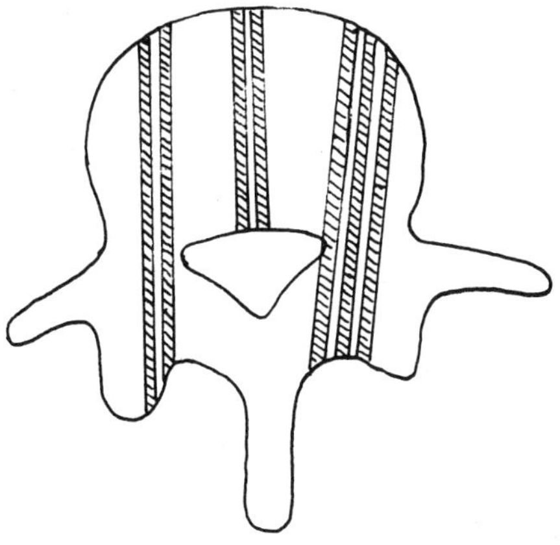

Fig. 7. Locations of investigated slices.

Maximum values (in MPa) of the stresses in different parts of the model of vertebra L5 are shown beneth; the loading of the vertebra corresponds to the situation when external load 400 N is kept in hands.

TABLE 1 Maximum effective stress in vertebra

	body	pedicle	part interarticularis
Upright posture	1.3	0.6	1.3
Deep forward bending	6.5	4.5	5.7
Backward bending	1.7	1.1	1.3

The results obtained show that high effective stresses are located in vertebrae body and in parts interarticularis. In the last place the spondylolytic fractures are clinically observed. The thin slices show the greatest concentration of the stresses in parts interarticularis. Therefore in this place the maximum value of local stress can be expected and in this place development of fatige crack can be expected.

Damage tests of models of vertebrae were also made. The results show different modes of fractures in arches, in parts interarticularis in particular, due to small differences of the loadings and geometry of vertebrae tested.

FINITE ELEMENT MODELS

As the experimental investigations are expensive and time consuming, for more wide investigations we used theoretical modelling of vertebrae and Finite Element Method for stress calculations.

Most simple FEM model is a plane one. A plane strain model of lumbar motion segment (Fig. 8) was developed for investigations the effective stress distribution in sagittal plane of vertebrae body |2|.

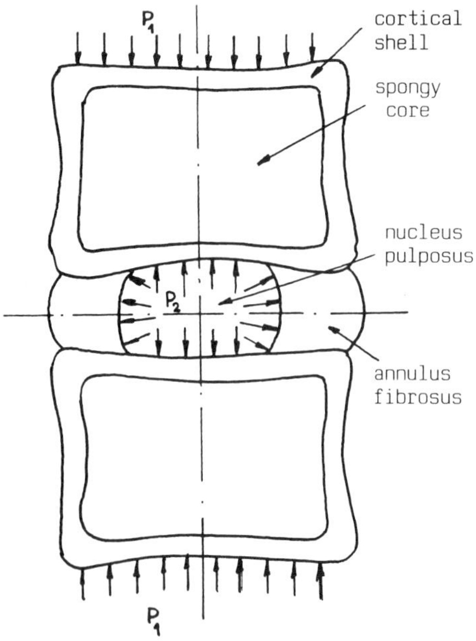

Fig. 8. The plain model representing two vertebrae bodies and intervertebral disc.

In model both spongy and cortical bone were modelled as well as an intervertabral disc being in different conditions. The last one was simulated by using different relationship between average pressure on the related end-plate p_1 and interdiscal pressure p_2. For the case of healthy disc high effective stresses were found on the inner side of vertebral body shell in the central part of the end-plate. For the degenerated disc high stresses were found in inner part of the vertebral body wall and in vertebral body rims. The locations well correspond with clinically observed modes of failure. The model allowed also study of osteophytes growth, the important kind of bone remodelling process.

To simulate the formation of osteophytes a shape optimization of the model was done. The results of the process, in the case of degenerated discs showed remodelation of the rims shape in the way corresponding with clinically onserved one. There were no remodelation in the case of the healthy disc. The results show strong relationship between disc degeneration and osteophyte formation in the lumbar spine.

The three dimensional model of vertebrae consisting of 116 elements has been also proposed. Elements having 13, 15 and 20 nodes have been used. Investigations of this model are now in progress.

REFERENCES

1. Dietrich, M., P. Kurowski (1985). The importance of mechanical factors in the etiology of spondylolysis. Spine, 10, No.6.
2. Kurowski, P., S. Tsutsumi, A. Kubo (1986). Computer simulation of the osteophyte growth in the lumbar spine. European Soc. of Biomech. Fifth Meeting, West Berlin.
3. Dietrich, M., P. Kurowski (1983). The model of the human lumbar spine. Proc. of the Sixth Congress on Theory of Machines and Mechanisms, New Delhi.
4. Eie, N., P. Wehn (1962). Measurements of the intra-abdominal pressure in relation to weight bearing in the lumbo sacral spine. J.Oslo City Hosp.
5. Dally, J.W., W.F. Riley (1965). Experimental Stress Analysis, McGraw Hill, New York.

О МЕХАНИЧЕСКИХ СВОЙСТВАХ ПОЯСНИГНОГО ПОЗВОНОЧНИКА

Большая часть человечества страдает болью кресца, вызванной разными болезнями пояснигной части позвоночника. В зарождении и развитии этих болезней более или менее существенную роль играют механические факторы. Чтобы выяснить значение этих факторов исследуется механическое состояние (расположение, нагрузки, деформации, напряжения) позвоночника. Так как общие исследования in vivo практически невозможны, наиболее доступными являются методы моделирования.
В работе описана простая физическая и математическая модель пояснигного позвочника, на основании которой были определены силы, действующие в системе позвоночника - силы развиваемые мышцами, межпозвоночные силы а также воздействие давления брюшной полости. Для отображения способа управления мышцами с помощью нервной системы, была применена концепция полиоптимизации. На основе результатов исследований этой модели можно было сделать выводы, касающиеся влияния на величину сил, действующих в системе пояснигного позвоночника таких величин как величина позвонков, развитие мускулатуры и величина давления брюшной полости.
Распределение напряжений в позвонках исследовалось экспериментальным путем с помощью трехмерного эластооптического метода (замораживание напряжений). Были определены места концентрации напряжений, в которых можно ожидать повреждения позвонков. Эти места соответствуют клинически наблюдаемым местам трескания позвоночных дуг.
Предлагается также двухмерная дискретная теоретическая модель, моднлирующая разные свойства коркового вещества и губчатой кости. Исследование этой модели, проведенные с помощью метода законченного элемента, позволили определить влияние состояния межпозвоночного диска на распределение напряжений в позвоночном столбе и на процесс приспособления костей позвонка (образование и увеличение отростков на позвоночном столбу).

The Design of Prosthesis for Vertebral Body Replacement of the Spine due to Metastatic Malignant Tumors

E. ALICI* AND Ö. Z. ALKU**

*Faculty of Medicine, Edge University, Izmir, Turkey
**Department of Civil Engineering, Dokuz Eylül University Izmir, Turkey

Abstract. Patients suffering from metastatic malignant tumors of the spine generally have a short life, suffer severe pain and their daily activities are certainly limited. The lesion region of the spine must be firmly fixed for the restoration of spinal stability, as well as anterior decompression on the spinal cord.

In previous work bone cement and bone grafts have been used for fixation purposes and therefore freedom of movement in some regions of the spine is limited. Also if bone grafting is used, union takes a long time. If bone cement is used for fixation, the region will be subjected to severe stress which would loosen the cement from the bone. All these undesired results have prompted the necessity of designing a prosthesis to fulfill the required functions of vertebral and intervertabral discs.

In this study, it is intended to design prosthesis with shapes and dimensions as identical to those of real vertebrae as possible. The objective of the design is also to develop prosthesis that will perform the normal vertebral functions of the human body.

Keywords. Prosthesis; vertebrae; uniaxial compression; load-carrying capacity; freedom of movement; rotation.

INTRODUCTION

The limited life spans of patients suffering from metastatic malignant tumors of the spine necessitate orthopedic surgical measures besides chemotherapy and radiotherapy. It is important in this case to recover the lost functions of the vertebrae so that the patient retrieves his normal daily activities. The spinal instability in these patients have been successfully restored by bone grafts (Fielding at al., 1979). However, this method not only limits the freedom of movement to a great extent but also reguires a long time until the union is completed. Thus, the patient has to suffer these problems of the external fixation. Therefore, specialists are urged to search for new methods to decrease the severe pain and limited daily activities of patients during their already shortened life as a consequence of malignant tumors. This new method must perform two functions, first to decrease pain and provide freedom of movement, and second to restore the usual functions of the spine. These two objectives can be realized by biomechanics and by the cooperation of the medical doctor and the engineer.

DIMENSIONS OF THE VERTEBRAE, FREEDOM OF MOVEMENT AND THE DESIGN OF PROSTHESIS

Prosthesis must have dimensions identical to those of real vertebrae for proper fixation. For this purpose, dimensions of 147 vertebrae have been measured with a precision of 0.1 mm using seven fresh spines. In this way, the average dimensions of dorsal, cervical and lumber vertebrae are determined (Alıcı, 1982).

In literature, the height of the intervertebral disc is given as 3 mm in the cervical, 5 mm in the dorsal and 9 mm in the lumber regions (Kapanjı,1974). The intervertebral disc permits forward, backward and sideward movements about the vertical axis with angles of $6^{o}-10^{o}$ in the cervical and lumber regions and of $3^{o}-6^{o}$ in the dorsal region (Kapanjı, 1974; Yumasev and Furman, 1976). In this case, the disc serves the same purpose as a hinge by enabling the vertebrae move backward, forward and sideward within certain limits.

The prosthesis, designed to be identical to real vertebrae discs, consists of two parts. The two heads, which are used to fix the prosthesis to the healthy vertebrae, are made of cobalt, chromium and molybdenum. Between the two heads is a body made of polyethylene with high molecular density (Fig.1). The prosthesis head has a spherical head, cylindrical neck, flat plate, and a conical stem, whereas the body consists of two grooves at both ends where the spherical heads rest.

The angle that the prosthesis head above the body makes with the vertical axis is greater than the rotation of a normal intervertebral disc. This is to prevent the neck and the plate from touching the body and thereby to eliminate the forces that may lead to the loosening of the stem in the vertebra. Consequently, the length of the neck is so selected that the plate makes $5^{o}-10^{o}$ movements with respect to regions on the body. The diameters and the locations of the spherical heads and the grooves are determined according to those of the nucleus populus in the intervertebral disc (Fig.2). These characteristics may be observed in Fig.1 for the dorsal region. The diameters of heads determined for the cervical, dorsal and lumber regions are given in Table 2 in the next section.

SIZING OF THE PROSTHESIS

Table 1 gives the physical characteristics of the prosthesis materials, the cobalt-chromium-molybdenum mixture, high molecular density polyethylene and the bone cement used to fix the stem to the vertebra.

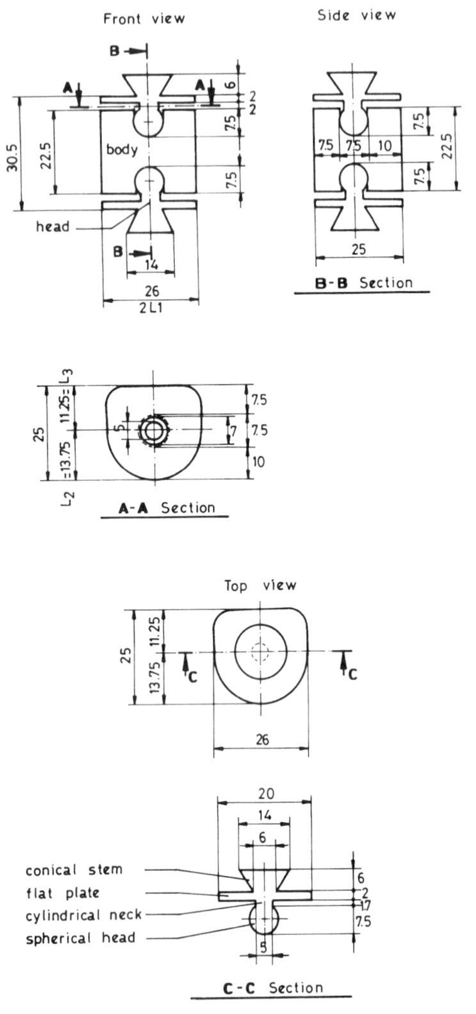

Fig. 1. Model of the dorsal region prothesis.

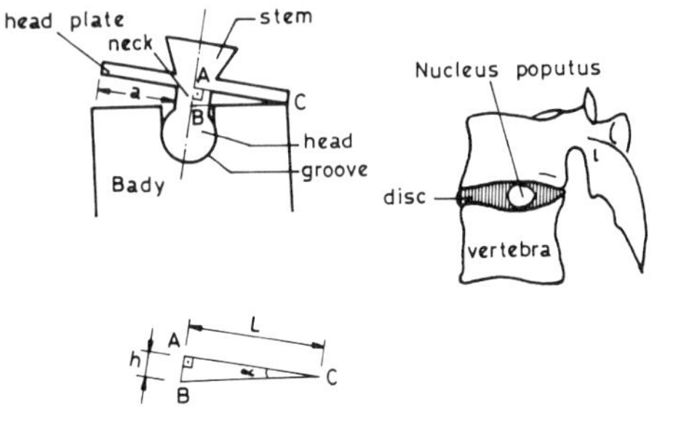

Fig. 2. Rotation of the prosthesis head above the body, with characteristics of the head and the groove similar to those of the nucleus populus.

Table 1 Physical properties of materials used in prosthesis design

Physical property	Material		
	High mol. density poly.	Bone cement	Co-Cr-Mo mixture
Specific weight (gp/cm^3)	0.94		8.3
Modulus of elasticity (kp/cm^2)	4000-12700	21000	5154300
Elongation rate (%)	450		2
Tensile strength (kp/cm^2)		700	7800
Compressive strength (kp/cm^2)	220	1050	6000

The mixture is 62% cobalt, 27% chromium, 5% molybdenum, 3% manganese, 1% silicon, 1% carbon, and 1% iron. The elasticity modulus of polyethylene with high molecular weight varies within a wide range. Therefore, an attempt is made to determine the modulus through experiments. Laboratory tests thus performed revealed the value of the elasticity modulus to be in the order of 5800 kg/cm^2. This figure then is used throughout the study. Biomechanical analyses have shown that, in the spine, effects of bending (tensile stresses) are received by tissues around the vertebrae and those of compression by the disc and the vertebrae.

Figure 3 shows the prosthesis proposed by Saegesser and Roark. The uniaxial compression force is determined by:

$$P = \left(\frac{\sigma_{max}}{0.9774}\right)^3 \left(\frac{1}{E1} + \frac{1}{E2}\right)^2 / \left(\frac{D2 - D1}{D1 . D2}\right)^2 \qquad (1)$$

which is developed form the general Hertz equation for applications of prosthesis as described in this study (Saegesser at al, 1975).

In the above, σ max denotes the yield strength of high molecular weight polyethylene under compression; E1 and E2, the modulus of elasticity of the Co-Cr-Mo. mixture (head) and of polyethylene (grooves) respectively; D1 and D2, diameters of the spherical head and the groove; D2-D1, freedom of movement and P, the uniaxial load that the prosthesis can carry within elastic limits. When these values are known Hertz formulation is then used to determine the theoretical compression forces to which the prosthesis will be subjected. Table 2 shows the calculation of P values for E1=5154300 kp/cm^2, E2=5800 kp/cm^2 and σ_{max} = 220 kp/cm.

Table 2 Loads that vertebral prosthesis can carry

Location of vertebra	D1	D2	D2-D1	P
	cm	cm	cm	kp
5th cervical	0.45	0.46	0.01	145.60
7th dorsal	0.75	0.77	0.02	283.30
3rd lumber	1.20	1.23	0.03	822.80

The calculated forces are much greater than those that the vertebrae are subject to during normal daily activities. The discs in the middle cervical, upper dorsal and middle lumber regions of an 80 kp person in standing position receives forces in the order of 3 kp, 14 kp and 47 kp, respectively (Leonardi, 1966). The surface of the bead plate must conform to the shape and the top and bottom surface dimensions of the body. Once the location and the diameter of the groove is determined, the lengths of the right and left sides (L1), the front (L2) and back (L3) of the plate may be easily specified. The only thing to do then is to determine the thickness of the plate, the height and diameter of the neck. For different regions, the height h of the neck may be calculated using Fig. 2 and simple trigonometric relations of a straight triangle while providing freedom of movement (rotation) in the order of 5^0 - 10^0 about the vertical axis. The h values can then be used to determine the angular freedom of movement (rotation) for forward, backward and sideward movements (Table 3).

Table 3 Characteristics of the head and the plate

Location of vertebrae		5th cervical	7th dorsal	3rd lumber
L1 (mm)		6.75	13.00	20.00
L2 (mm)		6.75	13.75	18.00
L3 (mm)		6.75	11.25	12.00
h (mm)		1.20	2.00	2.00
	To the sides	10.08	8.75	5.71
	Forward	10.08	8.28	6.34
	Backward	10.08	10.08	9.46

The forces that the prosthesis will theoretically receive are previously given in Table 1. The diameter of the circular neck can be determined by increasing these forces 10% for safety purposes. Using the general formula for compression, the diameter d is found by

$$\frac{\pi d^2}{4} = \frac{1.10P}{\sigma_{max}} \qquad (2)$$

for different locations of the vertebrae and the results are given in Table 4.

Table 4 Characteristics of the neck

Location of vertebrae	1.10P	Required cross-section of the neck	Diameter	Proposed diameter
	kp	mm^2	mm	mm
5th cervical	160	2.67	0.92	2.50
7th dorsal	310	5.17	2.56	5.00
3rd lumber	900	15.00	4.37	8.00

The vertebrae which are fixed to the head plate by bone cement distribute the stresses they receive on to the head plates. The critical section due to this distributed load is the side of the prosthesis stem. The head plate with respect to the side of the stem may be approximated as a cantilever beam. This approach is fairly realistic and simple as a computational method (Fig. 3).

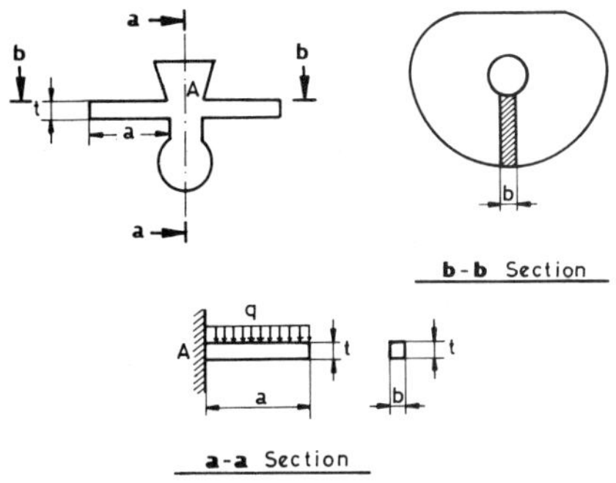

Fig. 3. Simple computational model for determination of plate thickness, where the head plate is approximated as a cantilever beam.

The bending moment Ma occurs at the very end of the prosthesis stem:

$$Ma = \frac{q.a^2}{2} \qquad (3)$$

where q is the distributed load conveyed from the vertebrae to the prosthesis plate with unit width. The maximum tensile stress that occurs at section a-a is

$$\sigma = \frac{Ma}{W} \qquad (4)$$

W is the section modulus of the plate with width b and thickness t:

$$W = \frac{bt^2}{6} \qquad (5)$$

The thickness of the plate can then be determined using (3), (4) and (5):

$$t = \sqrt{\frac{6\ Ma}{b\sigma}} \quad \text{or} \quad t = a\sqrt{\frac{3q}{b\sigma}} \qquad (6)$$

Computation of the thickness of the head plate is given in Table 5 for the 5th cervical, 7th dorsal and 3rd lumber regions. In this computation, the loads determined in Table 2 are increased 10%, and the assumption of a cantilever beam as a computational method is used to find the required thickness of plate. Finally, proposed thickness of plate is given for each region, according to the required thickness and geometric considerations.

Table 5 Computation of thickness of the head plate

Location of vertebrae		5th cervical	7th dorsal	3rd lumber
1.10 P	kp	160	310	900
Area of plate	mm^2	300	540	960
Distributed Load	kp/mm	0.5333	0.5741	0.9375
Length of Cantilever Beam	mm	8.75	10.00	14.00
Moment of Cantilever Beam	kp.mm	20.415	28.705	91.875
Required thickness of plate	mm	1.25	1.49	2.66
Proposed thickness of plate	mm	2.00	2.00	3.00

Dimensions of prosthesis in all three regions are given in the above tables; only the dorsal prosthesis is presented in Fig. 1.

CONCLUSIONS

1. Authentic dimensions and shapes of vertebrae are determined to develop a prosthesis model that fits the properties of the vertebrae.
2. The prosthesis designed can perform the functions of the disc and the vertebrae. The spherical head and the groove act as a hinge, thereby eliminating the forces that may loosen up the prosthesis from the vertebrae where it is fixed. The energy absorption and load-dampening characteristics of intervertebral discs are also found in high molecular weight polyethylene.
3. Theoretical loads that the prosthesis may be subject to are much smaller than those they receive in normal daily activities.

REFERENCES

İnan, M. (1973). Cisimlerin mukavemeti (Strength of materials). Ofset Matbaacılık, 135-140.

Leonardi, A. (1966). Biomecanica del disco intervertebrale Congresso della societa Italiana di ortopedia e traumatologia. Catania, 22-25 Octobre, 117-119.

Fielding, H.F., R.N. Pyle, V.G. Fietti, (1979). Anterior cervical vertebral body resection and bone-grafting for benign and malignant tumors. J.Bone Joint Surg., 61-A, 251-253.

Kapanji, I.A.,(1974). Fisiologia articolore tronco e rachide, edizione Italiana a Cura del Prof Gui ?., Soc. Editrice Demi-Roma., Vol. III, 43.

Saegesser, M.,C. Burri, C.H. Herforth, M.Jöger, (1979). Michael ungethum tecnologische und biomechanische aspecte der hüft und klineallo-arthroplastik. Aktuelle probleme in chirurgie und orthopadie. 9, 27-29.

Yumashev, G.S.,M.E. Furman, (1976). Osteochondrosis of the spine. Mir Publishers, Moskow. 26-30.

Roark, R.J., (1954). Formulas for stress and Strain. Mc. Graw-Hill, pp. 287.

Alıcı, E. (1982). Omurganın biomekanik özellikleri ve omurların kötü huylu urlarında uygulanabilecek omur protezleri (Biomechanical properties of vertebrae and prosthesis as a treatment method in case of malignant tumor metastases). Thesis for Associate professor ship, Ege University, 16-39.

Panjabi, M. (1977). Experimental determination of spinal motion segment behavior. Orthop. Clin. North Am., 8, 169.

Charnley, J. (1965). A biomechanical analysis of the use of cement to anchor the femoral head prosthesis. J. Bone Joint Surg., 47-B, 354.

Vertebral Prosthesis and their Biomechanical Properties

E. ALICI* AND Ö. Z. ALKU**

*Faculty of Medicine, Ege University, Izmir, Turkey
**Department of Civil Engineering, Dokuz Eylül University Ismir, Turkey

Abstract. Vertebral prosthesis are considered as a treatment method for medullary pressure and instabilities which occur at malignant tumor metastases. The prosthesis are designed to replace vertebra body where the malignant tumors are usually situated. The developed samples of prosthesis have been fixed to the vertebrae and the mechanical behavior of the system has been studied using 70 fresh vertebrae. Similar tests are applied to samples without prosthesis for purposes of comparison. The part of the study presented herein is totally experimental and all results given refer to actual tests.

Keywords. Prosthesis; vertebrae; compression; load-deformation; load-carrying capacity; energy absorbtion capacity.

INTRODUCTION

The prosthesis, which are designed to conform to the function of the vertebrae and for which the dimensions together with the load carrying capacity are theoretically determined, need to be investigated under versatile conditions and effects. This experimental study is carried out on vertebrae with and without fixed prothesis heads. The series of experiments are so designed that eventually comparisons between the two cases can be made.

EXPERIMENTAL STUDY AND RESULTS

All experiments are carried out by an Instron 1114 testing machine with a compression rate of 0.2 cm/min and recording rate of 2 cm/min.

Cervical, dorsal and lumber prosthesis, composed of two heads of a cobalt-chromium-molybdenum mixture and a polyethylene body with a high molecular weight, have been seven times compression tested in the elastic region. Load-displacement curves are thus obtained for cases of loading and unloading (Fig. 1). The average energy absorption rates of the cervical, dorsal and lumber prosthesis have been found to be in the order of 12.4 kp.cm, 56.8 kp.cm, and 137.7 kp.cm, respectively.

In order to determine the reduction in the load carrying capacity of the spine due to the placement of the prosthesis heads to the vertebrae, 308 vertebrae obtained from 14 fresh spines have been uniaxially compression tested under two cases of with and without fixed prosthesis heads. A group of 154 vertebrae obtained by segmenting 7 spines have had polyester plates fixed at the top and bottom surfaces. The other group of 154 vertebrae have had flat polyester plates fixed at one surface and metallic prosthesis heads fixed with bone cement on the other surface. The load-displacement curves obtained from two experimental series are then compared. It is found that the load carrying capacity of the vertebrae without prosthesis heads increases gradually from the cervical region downwards, and at the lower dorsal vertebra a sudden increase occurs, finally reaching a maximum value at the fourth lumber vertebra. The results of the experiments show that the vertebrae with the fixed prosthesis heads lead to a lower load carrying capacity in comparison to vertebra without the prosthesis heads under uniaxial compression (Fig. 2). Although this capacity in the case of fixed prosthesis heads is observed to be 9.33 % lower than it is in the case without the prosthesis heads, it seems adequate for the daily functions of the human body in terms of forces acting on the intervertebral discs (Leonardi, 1966; Nachemson and Morris, 1964).

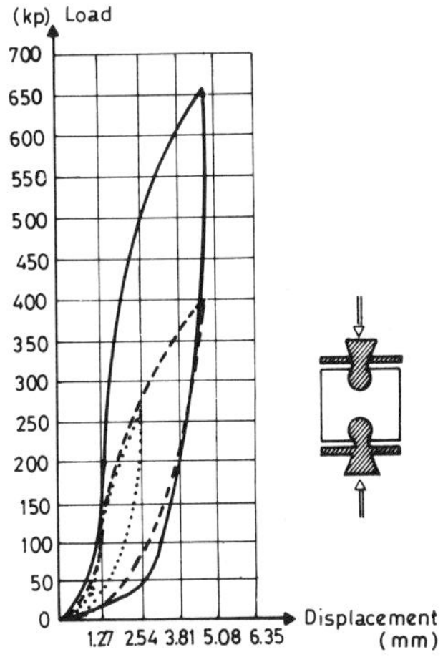

Fig. 1. Load-displacement and recovery curves for prosthesis under compression, with full curves standing for lumber, dashed curves for dorsal and dotted curves for cervical prosthesis.

Taking into consideration that the system may be affected differently by the replacement of prosthesis to different regions of the spine, a group

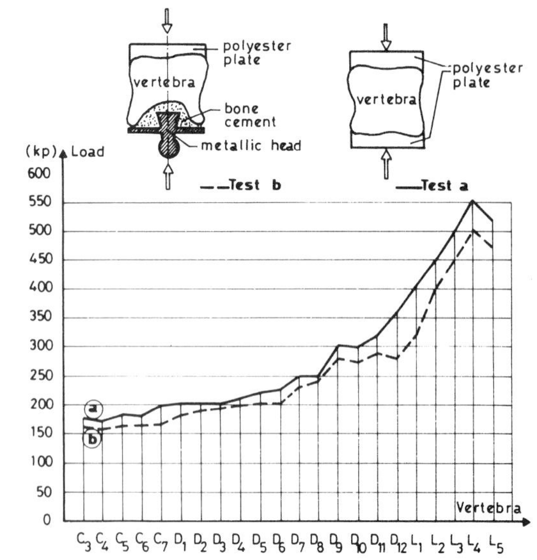

Fig. 2. Compressive strengths, in the elastic region of vertebrae with and without fixed prosthesis heads.

of 56 fresh spines divided into cervical, dorsal and lumber regions and 168 regional samples with and without prosthesis have been subjected to uniaxial compression and bending test during which the system has been bent forwards, backwards and to the sides up to an angle of 10^{0} off the vertical axis. In the cervical region the 5th, in the dorsal region the 7th, and in the lumber region the 3rd, vertebrae have been replaced by the prosthesis.

The tests are carried out in the elastic region, using samples with and without prosthesis heads. As a result, the load carrying capacities of the samples under compression are obtained through load-displacement curves, together with values of elasticity modulus and of absorbed energy during loading and unloading (Fig. 3, a, b, c).

The results of the test using regional samples have shown that the samples without prosthesis have 9.38 % more load carrying capacity than the ones with prosthesis. It is noticed that samples with and without the prosthesis exhibit greater bearing capacity in axial compression than in the combined bending at all three regions. As to the bearing capacity in combined bending, they show greater capacity in bending about y-y axis than they do about x-x axis. The same characteristics result respectively in the energy absorption capacity of the system.

The average energy absorption values are 64.19 kp. cm and 52.35 kp.cm for the samples without and with the prosthesis respectively. The energy absorption of the samples without prosthesis is 18.44 % higher than the value of the ones with prosthesis. The modulus of elasticity has been found indifferent for the samples with and without prosthesis.

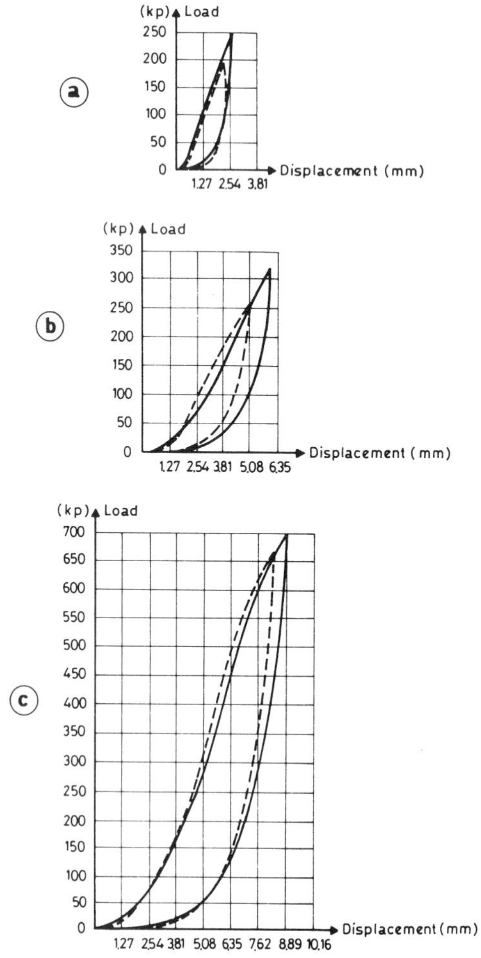

Fig. 3. Load-displacement curves for uniaxially compression tested samples with (dashed curves) and without (full curves) prosthesis fixed in the: ⓐ_cervical region ⓑ_dorsal region ⓒ_lumber region

CONCLUSION

Tests carried out on samples with and without prosthesis heads have shown that the former lead to losses in the order of 9 % in terms of their load-carrying capacities and of 18 % in terms of their energy absorption capacities when compared to the latter. However, it is also observed that the prosthesis heads can safely meet a great portion of forces acting on the vertebrae.

REFERENCES

İnan, M. (1973). Cisimlerin mukavemeti (Strength of materials). Ofset Matbaacılık, 72-83.

Leonardi, A. (1966). Biomecanica del disco intervertebrale Congresso della societa Italiana di ortopedia e traumatologia. Catania, 22-25 Octobre, 117-119.

Nachemson, A. (1960). Lumber intradiscal pressure. Acta orthop., Scand. Suppl. 43.

Nachemson, A., J.M. Morris (1964). İnvivo measurements of intradiscal pressure. J. Bone Joint Surg., 46, 1077.

Alıcı, E. (1982). Omurganın biomekanik özellikleri ve omurların kötü huylu urlarında uygulanabilecek omur protezleri (Biomechanical properties of vertebrae and prosthesis as a treatment method in case of malignant tumor metastases). Thesis for associate professorship, Ege University, 59-72.

Alku, Ö.Z. (1977). Çok katlı çerçevelerin elastik stabilitesinin açı metodu ile incelenmesi (Analysis of elastic stability in multi-story frames by the slope-deflection method). M.Sc. Thesis, Ege University, 1-8.

A Drink Transport Mechanism for Disabled People

C. W. STAMMERS* AND A CLARKE**

*School of Mechanical Engineering, University of Bath, Bath, UK
**Royal National Hospital for Rheumatic Diseases, Bath, UK

Abstract. A six bar chain mechanism of the Watt I type has been built to aid handicapped persons with drinking of hot fluids. The drink transporter is driven via a crank-rocker chain.

The analysis of pin forces and motor torque demand is given, along with that of cup velocity and acceleration.

Operational factors such as transporter range, cup speed, and cup refill are discussed.

Keywords. Medical, disabled, drinking, chain mechanism.

NOMENCLATURE

a	length of AD
a_o	length of AB_o
b	length of AB
b_o	length of B_oC_o
c	length of BC
c_o	length of C_oD_o
d	length of DC
d_o	length of AD_o
e	length of BE
f	length of DF
F_1,F_2	pin forces at C
F_{10}, F_{20}	pin forces at D_o
G_1,G_2	pin forces at B
G_{10},F_{20}	pin forces at C_o
H_1,H_2	pin forces at A
J_1,J_2	pin forces at G
K_1,K_2	pin forces at F
ℓ_{13}	length of BE
ℓ_{25}	length of CF
ℓ_{36}	length of EG
ℓ_{37}	length of EH
$P_1.P_2$	pin forces at D
v	velocity
W	weight of cup
W_b	weight of link AB
W_e	weight of link BE
W_f	weight of link DF
W_{36}	weight of link EG
W_{37}	weight of link EH
W_{56}	weight of link FG
Z_1,Z_2	pin forces at E
β	angle between BE and EH
ε	angle of main chain to horizontal
ε_o	angle of input chain to horizontal
η	coupler angle, main chain
η_o	coupler angle, input chain
μ	transmission angle, main chain
μ_o	transmission angle, input chain
θ	crank angle, main chain
θ_o	crank angle, input chain
ρ	angle DFG
τ	angle FGE

INTRODUCTION

A survey (Harris, Cox and Smith, 1971) estimated that in the UK there were some 157,000 'very severely' disabled and 357,000 'severely' handicapped persons. For many of these, those suffering from arthritis, multiple sclerosis, motor neurone disease, stroke, Parkinson's disease or cerebral palsy for instance, the handicap involves difficulty with feeding and drinking.

In a survey (Clarke and colleagues, 1985) performed on a population of 45 such patients, the most important requirement so far as mechanical devices were concerned was one which would provide a hot drink while the person was alone.

There has been considerable interest in robotic devices in a feeding role: Paeslack and Roesler (1977), Mason and Peizer (1978), Schneider and Seamone (1980), Kingma and colleagues (1981) but the costs are considerable, although Semple (1981) and Davies and Semple (1981) have built a microprocessor controlled arm for around \$600.

In collaboration with Bath Institute of Medical Engineering and the Royal National Hospital for Rheumatic Diseases in Bath, a low cost (below \$100) mechanical device has been developed. Plane or spatial n link mechanisms offer a relatively inexpensive solution. Such devices cannot be steered, but then not all disabled persons are able, or wish to, steer a robotic arm.

THE DRINK TRANSPORTER

A photograph of the transporter developed is shown in Fig 1.

The programme began in the form of a final year undergraduate project (Cleland and Lisak, 1985). A four bar plane chain was studied initially as this can exhibit 'pick and place' coupler paths. However, a more compact solution was found to be a six bar chain of the Watt I type used in a crossed configuration (Fig 2a). Member FG is doubled and passes each side of BE. Aluminium bar, 20 mm by 6 mm, was used.

The six bar chain cannot rotate through 360°; a crank rocker input chain was employed (Fig 2b). This not only removes the necessity for micro-switches but also reduces the torque demanded by the load (see below). Microswitches have in fact been subsequently fitted, but this was to provide semi-automatic operation if desired.

Cup speed has to be low (less than 100 mm/s) (a) for psychological reasons and (b) to make the 'stop' decision a fairly casual one.

A 1.6 w 6v dc motor is employed turning at 6000 rev/min and geared down by 2560:1 (a standard accessory).

A coffee maker is used, previously loaded by a helper. When coffee is first required, the coffee maker is turned on via a 40 mm x 20 mm button (which can be operated by hand, elbow, head or suck/blow). The coffee is directed into the cup held in the jaws of the transporter alongside. (Fig 1).

The delivery of coffee takes only a few seconds. Lift of the cup is via a large button which is biased off. Cup motion is halted whenever pressure is removed from the button. A microswitch stops the cup at maximum extension if a stop has not been previously demanded.

Extension of up to 424 mm is achieved. The horizontal and vertical range can be adjusted by tilting of the mechanism. In the configuration employed, maximum horizontal reach was 372 mm and vertical 203 mm.

The desired range was obtained by patient consultation.

On touching a 'home' button the cup returns to the start position without further demand, stop being achieved by a second microswitch.

Factors of particular interest are the torque required by the load (typical 5N, maximum 10N) and also the cup velocity and acceleration.

ANALYSIS

Because motor speed is low (about 0.25 rad s^{-1}), the mass of the links modest (approx 0.1 kg) and accelerations below 0.05 ms^{-2}, inertia terms can be neglected by comparison with the static forces.

The geometry of the chain (Fig 2) was determined by the constraint method.

The pin forces (Fig 3) are obtained from link moment and force equilibrium. EG is length $\ell 36$, EH length $\ell 37$.

$$(J_1 \sin\zeta - J_2 \cos\zeta)\ell 36 - (W + 0.5W_{37})\ell 37 \cos(\varepsilon - \beta + \eta) = 0 \quad (1)$$

$$Z_1 \cos\beta - Z_2 \sin\beta + J_1 \cos\zeta + J_2 \sin\zeta + (W + W_{37})\sin(\beta - \eta - \varepsilon) = 0 \quad (2)$$

$$Z_1 \sin\beta + Z_2 \cos\beta + J_1 \sin\zeta - J_2 \cos\zeta - (W + W_{37})\cos(\varepsilon - \beta + \eta) = 0 \quad (3)$$

$$K_1 \sin\rho + K_2 \cos\rho + 0.5W_{56}\cos(\rho + \phi + \varepsilon) = 0 \quad (4)$$

$$J_1 = K_1 \cos\rho - K_2 \sin\rho + W_{56}\sin(\rho + \phi + \varepsilon) \quad (5)$$

$$J_2 = K_1 \sin\rho + K_2 \cos\rho - W_{56}\cos(\rho + \phi + \varepsilon) \quad (6)$$

$$F_2 d = K_2 f + 0.5 f W_f \cos(\phi + \varepsilon) \quad (7)$$

$$cG_1 \sin(\theta - \eta) - cG_2 \cos(\theta - \eta) + (e - c)Z_2 - W_e(c - 0.5e)\cos(\eta - \varepsilon) = 0 \quad (8)$$

$$G_1 \cos(\theta - \eta) + G_2 \sin(\theta - \eta) - W_e \sin(\eta + \rho) + F_1 \cos\mu + F_2 \sin\mu - Z_1 = 0 \quad (9)$$

$$G_1 \sin(\theta - \eta) - G_2 \cos(\theta - \eta) - W_e \cos(\eta + \varepsilon) + F_1 \sin\mu - F_2 \cos\mu - Z_2 = 0 \quad (10)$$

$$d_o F_{10} \sin\mu_o + 0.5W_b b\cos(\theta + \varepsilon) = G_2 b$$

The weight of the small links B_oC_o, C_oD_o is neglectd

$$F_{20} = 0$$

$$G_{10}\sin(\theta_o - \eta_o) = G_{20}\cos(\theta_o - \eta_o) \quad (11)$$

$$F_{10} = G_{10}\cos(\theta_o - \eta_o) + G_{20}\sin(\theta_o - \eta_o) \quad (12)$$

$$T = b_o G_{20} \quad (13)$$

$$F_{10} d_o \sin\mu + 0.5W_b b\cos(\theta + \varepsilon) = bG_2 \quad (14)$$

giving for demanded motor torque

$$T = [bG_2 - 0.5W_b b\cos(\theta + \varepsilon)]b_o \sin(\theta - \varepsilon)/d_o \sin\mu_o \quad (15)$$

Solution of equations (1) to (10) allows the 10 pin forces $J_1, J_2, Z_1, Z_2, K_1, K_2, F_1, F_2, G_1$ and G_2 to be obtained for a given cup load W and link weight.

Motor torque T can then be obtained from equation (15). The factor bo $\sin(\theta_o - \eta_o)/d_o \sin\mu_o$ in the latter shows that when the input coupler is roughly in line with the input crank, as happens at the two extremes of cup travel, demanded torque is much lower than would occur for drive at the main chain crank.

Velocity and Acceleration

As with the resolution of chain geometry, successive pairs of links are studied. The analysis for the case in which the end of one link is fixed is given by Molian (1982).

Two links, length d and c, joined at (X_2, Y_2) have ends at (X_o, Y_o) and (X_1, Y_1) respectively.

$$(X_2 - X_1)^2 + (Y_2 - Y_1)^2 = c^2$$

$$(Y_2 - X_o)^2 + (Y_2 - Y_o)^2 = d^2$$

Differentiating

$$\begin{bmatrix} (X_2 - X_1) & (Y_2 - Y_1) \\ (X_2 - X_o) & (Y_2 - Y_o) \end{bmatrix} \begin{bmatrix} \dot{X}_2 \\ \dot{Y}_2 \end{bmatrix} = \begin{bmatrix} (X_2 - X_1)\dot{X}_1 + (Y_2 - Y_1)\dot{Y}_1 \\ (X_2 - X_o)\dot{X}_o + (Y_2 - Y_o)\dot{Y}_o \end{bmatrix}$$

or $A\mathbf{v}_2 = \mathbf{b}$

Differentiating again $A\dot{\mathbf{v}}_2 = \dot{\mathbf{b}} - \dot{\mathbf{A}}\mathbf{v}_2$

giving the acceleration at point 2.

Cup velocity and acceleration are obtained by successive applications of the dyad analysis

RESULTS

Link dimensions (see Fig 2) are

AB_o = 145 mm;	B_oC_o = 53 mm;	C_oC_o = 114 mm;
AD_o = 110 mm;	AB = 305 mm;	BC = 60 mm;
CD = 290 mm;	AD = 55 mm;	BE = 285 mm;
CF = 50 mm;	EG = 90 mm;	FG = 290 mm;
EH = 365 mm.		

Motor (geared) output speed is 0.25 $rads^{-1}$.

Cup velocity is presented in Fig 4 as a function of the motor angle permitted by the microswitches. (Pick up of the cup occurs at 50° and maximum extension at 230°). A roughly sinusidal velocity profile is evident. Velocity falls as the cup approaches the mouth; the decision to initiate a manual stop (if desired) can be taken in a leisurely manner. As peak total velocity is only 63 mms^{-1} (at mid extension) the mechanism does not appear threatening to the user.

The corresponding cup accelerations were calculated. Horizontal acceleration is of particular interest because of the possibility of spillage. However, levels are so low that there is no problem with this. With peak acceleration around 30 mms^{-2} or 0.003 g, dynamic effects are not significant.

The motor torque required for a 10 N load (max load considered) is compared in Fig 5 with that for no cup (self loading only). Self loading constitutes about 20% of the peak torque demand. Peak power demand is 0.35 watt which is largely due to raising the 10 N load at peak vertical velocity of 29 mms^{-1}.

The torque demand (Fig 6) is reduced at the ends of the range since the angle between the input crank and coupler is close to 180° at the beginning of lift and around zero at the end of lift. If the drive were applied at the main chain crank considerably greater torque could be needed.

DEVELOPMENT

When a mechanism is developed, as here, by analysis rather than synthesis, it is likely that some improvement can be made in the parameters. Factors to consider are:

(a) horizontal range
(b) vertical range
(c) cup velocity profile
(d) peak motor torque demand

Increased horizontal range means that the device can be further removed from the user, allowing more space for books and personal effects. Vertical range can be increased, at the expense of horizontal by tilting the transporter, although this increases the motor torque demand.

Although the coffee maker can make up to 10 cups, in the original model only one cup of coffee could be used since there was no way of repeatedly filling the cup from the coffee maker jug.

A jug was constructed having a hole covered by a disc. As the transporter arm returns to the 'home' position a rack on the arm engages the disc, rotating it and allowing coffee to enter the cup by gravity feed. When the desired quantity of coffee has been drawn, the arm is moved forward, closing off the jug aperture.

A new 'home' position is now required forward of the cup refill position to allow the cup to be returned without refill being initiated.

Having overcome the basic engineering and design problems it is now proposed to mount studies involving several disabled people. Initially these will be in the laboratory to allow final design changes to be made and to check safety. This will be followed by field studies, using disabled people in their own homes, to assess reliability and practicality for day to day use.

REFERENCES

Clarke, A K, Clay, T, Hillman, M, and Orpwood, R, (1985). The extension of environmental control by the use of robotics. Autumn Conference, Tissue Viability Society, Bath, England.

Cleland, A C, and Lisak, Z, (1985). A drink robot for the handicapped. School of Engineering Report 738, University of Bath.

Davies, B L, and Semple E C, (1981). A simple low cost microprocessor controlled medical manipulator. Conference on External Control of Human Extremities. Dubrovnic. pp 55-64.

Harris, A I, Cos, E, and Smith, R W, (1971). Handicapped and Impaired in Great Britain, HMSO, London.

Kingma, Y J, Chambers, M M, and Durdle, N, Robotic Feeding Device for Quadraplegics (1983) 16th Annual Hawaii International Conference on System Science. pp 495-499.

Mason, C P, and Peizer, E, (1978). Medical Manipulator for Quadraplegics. Proceedings of the International Conference on Telemanipulators for the Physically Handicapped. Roquencourt, France. pp 309-317.

Molian, S, (1982) The constraint method of kinematic analysis. In S Molian, Mechanism Design. Cambridge University Press, Cambridge, England. pp 108-109.

Paeslack V, and Roesler, H, Design and control of a manipulator for tetrapelgics. (1977). Mechanism and Machine Theory, 12, 413-423.

Schneider, W, and Seamone, W, (1980). A microprocessor controlled robotic arm allows self feeding for a quadraplegic IEEE New York Workshop on the Application of Personal Computing to Aid the Handicapped. pp 31-35.

Semple E C, (1981). A robot arm and workable system for the assistance of the physically handicapped' Engineering in Medicine, 10, 143-147.

Stepourjine, R, (1979). A hydraulic modular robot to help seriously motor handicapped people. Proceedings Fifth World Congress on Theory of Machines and Mechanisms Montreal. pp 1527-1531.

RESUME

Suivant un sondage auprès de personnes handicapées, un mécanisme permettant à une telle personne de se servir d'une boisson chaude a été conçu.

Une cafétière automatique dispense du café dans une tasse retenue par un mécanisme à six barres entraîné par un moteur électrique à courant continu et un engrenage. Le coût total étant de l'ordre de $100.

Les vitesses, accélérations ainsi que les efforts et le couple moteur sont étudiés.

Le perfectionnement d'un tel système qui permettra a la personne handicapée de se servir successivement de tasses de café est présenté.

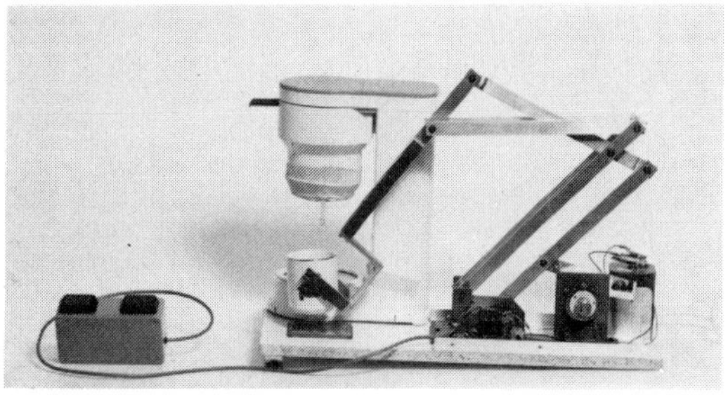

FIG 1 : DRINK TRANSPORTER

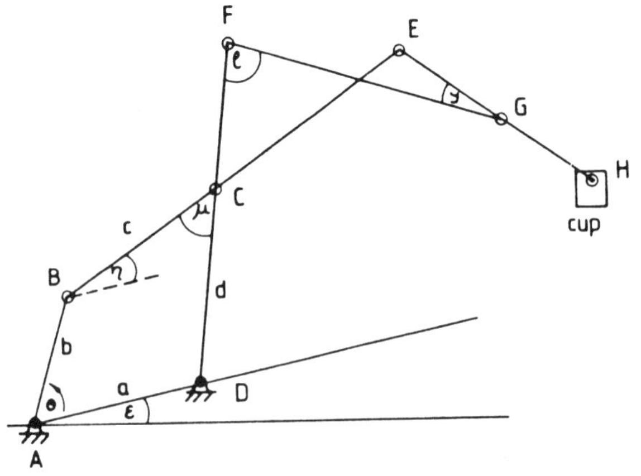

Fig. 2a : Watt I six bar chain

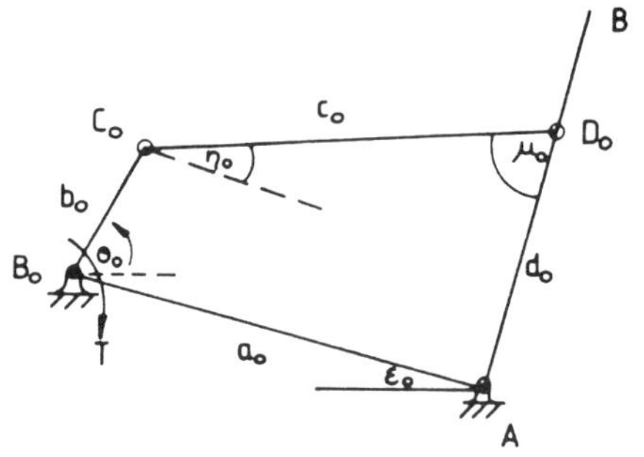

Fig. 2b Input chain

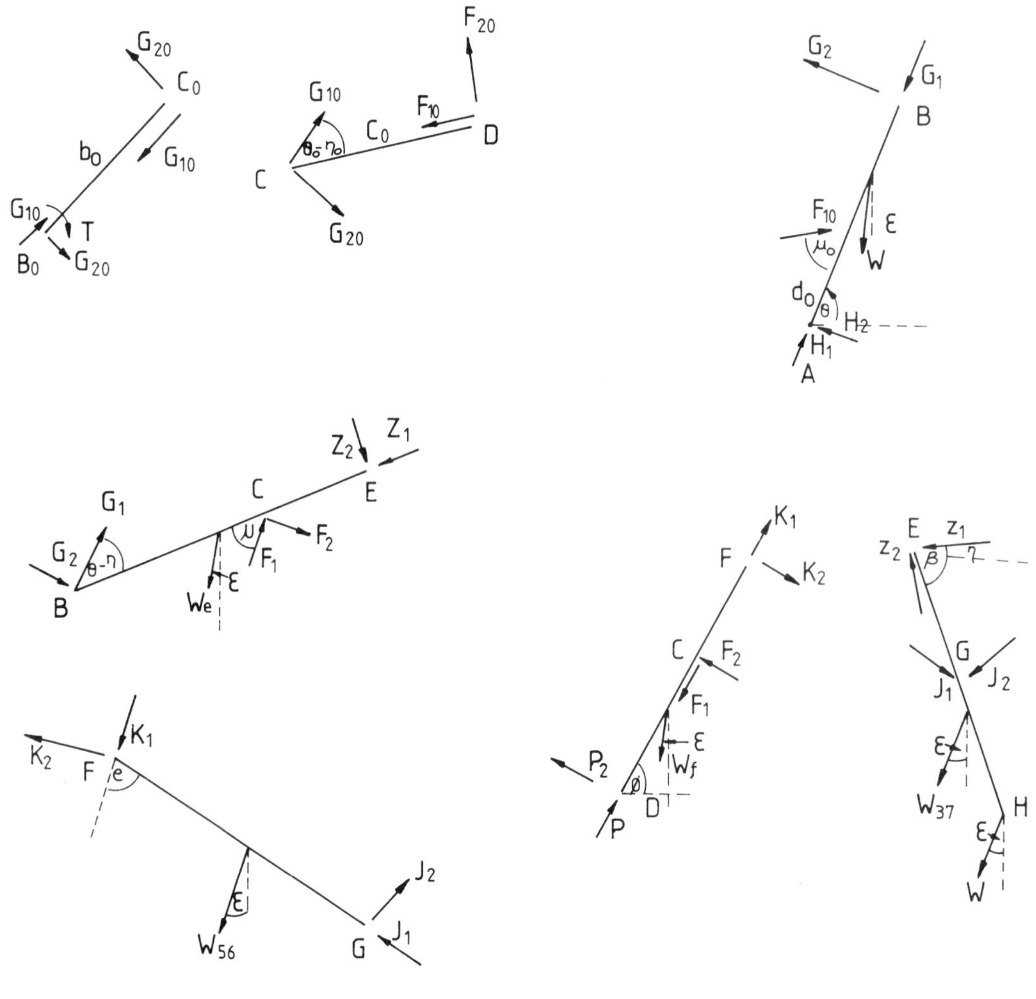

Fig. 3 : Pin forces

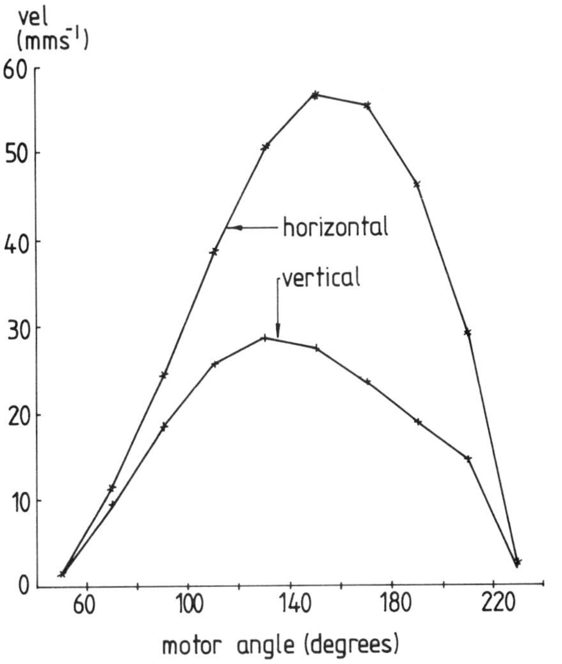

Fig. 4: Cup velocity

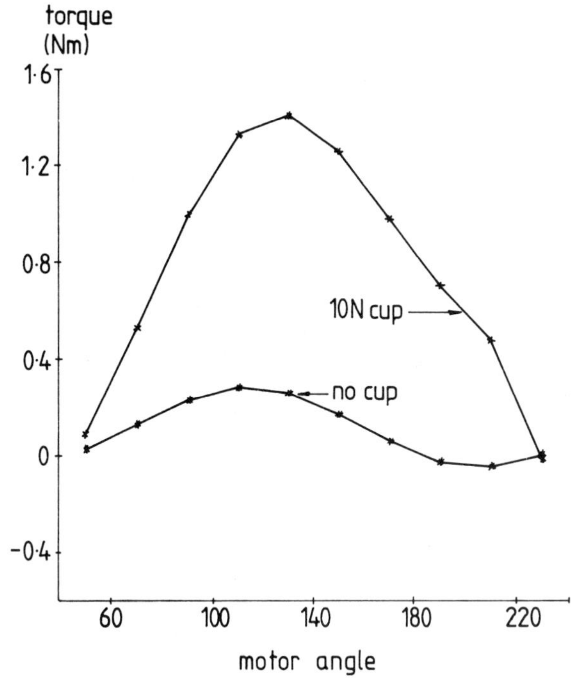

Fig. 5: Motor torque demand

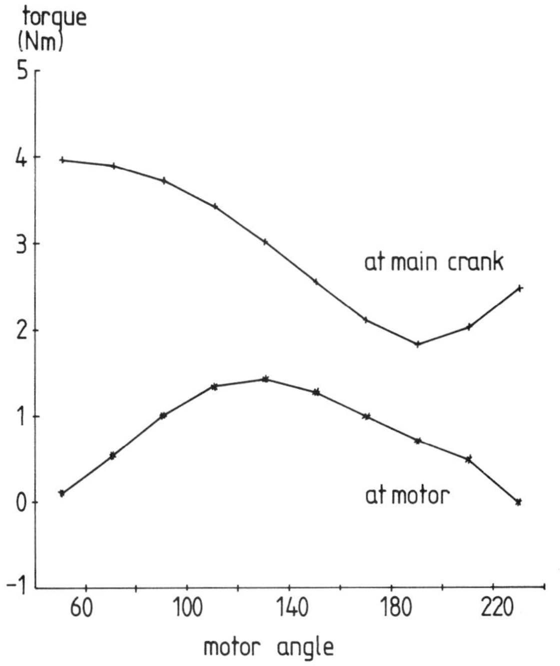

Fig. 6: Torque demand, 10 N cup

Arthropods: 350 Million Years of Successful Walking Machine Design

E. F. FICHTER*, B. L. FICHTER** AND S. L. ALBRIGHT***

*Department of Industrial and General Engineering, Oregon State University, Corvallis, OR 97331, USA
**Department of Entomology, Oregon State University, Corvallis, OR 97331, USA
***Department of Mechanical Engineering, Oregon State University, Corvallis, OR 97331, USA

Abstract. Sophisticated legged vehicles can be of advantage for such tasks as climbing steep inclines, manouvering around obstacles and traversing narrow beams by holding on with grippers. Leg designs capable of handling these challenges would permit use of legged vehicles in construction and maintenance of lattice work structures as those proposed for space stations and those already used for underwater structures. Arthropods provide excellent examples of more complex leg structures. For 350 million years these animals have been evolving efficient and effective means for dealing with a great variety of habitats including rough surfaces, cliff faces, web structures and crevices. Study of the mechanics of these different solutions and the fitness of each solution for the habitat of the arthropod should advance our design concepts for legged vehicles.

Keywords. Arthropod; legged vehicle; manipulation; robot; walking machine.

INTRODUCTION

Legged vehicles or walking machines have many potential applications. Most walking machines currently being used for development of control systems, navigation systems, and leg mechanisms are designed for walking on relatively smooth and nearly horizontal surfaces. However, there are many other situations in which we would like to use walking machines. Legged vehicles would be much better than wheeled vehicles for manouvering around or climbing over obstacles. From observing animals we know that legs are very good for climbing steep inclines which may also be strewn with obstacles. There have been a few experimental efforts with machines which climb walls and with machines which move between walls which are very close together. There are many potential uses for such machines in hazardous environments resulting from chemical processing, nuclear energy applications, and underground mining. Legged vehicles have been suggested for use in construction and maintenance of space stations. A similar environment in some respects is building and maintaining underwater structures. These environments are similar because the effect of gravity is minimized so that legs must be able to hold on to the substrate in addition to providing support. Machines used for building high-rise open-work structures on the earth would also need to hold on to the substrate on which they are working.

Different work environments will demand different performance from the legs of these vehicles. To work in a gravity field, legs must resist the gravity loads on the vehicle. Legs must be flexible enough to move around or over obstacles. When working in confined spaces, legs must be flexible enough to manouver without colliding with walls. The feet must be able to hold on to the substrate when the machine is required to climb steep or vertical surfaces or traverse ceilings. Feet which hold on are also necessary when the machine is to operate where gravity has little effect.

In addition to these requirements which change with different applications, there is a universal requirement: the kinematic constraints imposed by the mechanical system consisting of body and legs must be incorporated in the controller design. If feet slide, energy is wasted and both substrate and feet may be damaged. Similarly, if legs cannot exert forces and moments necessary to continuously support the body, the system collapses. The solutions to both of these control problems must originate in the kinematics of the system.

CURRENT ENGINEERING RESEARCH

A long history of literature on walking machines is reviewed by Raibert (1986). He remarks that only about 50% of the earth's land surface is accessible to vehicles with wheels or tracks; walking machines could make much of the remaining 50% accessible. The same versatility which makes walking machines attractive for manouvering in inaccessible parts of earth's surface also gives them advantages in hazardous environments (Ichikawa, Qzaki, and Sadakane, 1983).

Most of the present work on walking machines is based on studies of animal gait (McGhee, 1968). One result has been to treat leg motion of walking vehicles as a finite state machine. This approach is adequate only for very simple environments.

Most legged machines now under development have one leg (Raibert, 1986), two legs (Hemami, and Chen, 1984; Miura, and Shimoyama, 1984), four legs (Hirose, 1984; Raibert, 1986), or six legs (Klein, Olson, and Pugh, 1983; Orin, Tsai, and Cheng, 1985; Ozguner, and Tsai, 1985; Petternella, and Salinari, 1974; Raibert, and Sutherland, 1983; Song, Waldron, and Kinzel, 1985). Ichikawa, Qzaki, and Sadakane (1983), however, describes a five legged machine. These are only a few examples of papers on walking machines.

One area of primary concern in current walking machine research is balance. The walking machine must maintain a balance of both static and dynamic forces if it is not to fall over. Of major concern in determining gait for machines with more than two legs is maintaining static balance (Peternella, and Salinari, 1974). There has been little research beyond this except for Raibert's work on dynamic balance of hopping machines (Raibert, and Sutherland, 1983; Raibert, 1986). There has also been a large amount of work on dynamic balance in biped machines (Hemami, and Chen, 1984; Miura, and Shimoyama, 1984).

The general area which has received most attention is control, particularly control which results in efficient movement. Orin, Tsai, and Cheng (1985) examined dynamic control and Song, Waldron, and Kinzel (1985) reported on leg geometry. Hirose (1984) tried to minimize energy consumption by keeping the body at a fixed height.

Another aspect of control is reduction of computation time for the walking algorithm. Raibert, and Sutherland (1983) discussed hardwiring leg movements into control circuitry. In this solution the control computer only has to provide necessary parameters to the circuitry for each step. This approach resembles part of the arthropod solution to this same problem.

Control of body attitude and load distribution on the feet is discussed by Klein, Olson, and Pugh (1983). The legs can adjust to correct for roll and pitch of the body and to equalize the load born by each leg.

An aspect of control that borders on navigation is how and where the feet are placed. Hirose (1984) discussed use of tactile sensors to locate appropriate fool holds. This method has the disadvantage that the sensor must feel for the next foot hold. Another approach is to use a vision system to locate the next foot hold (Ozguner, and Tsai, 1985).

Most engineering research on current, first generation walking machines is limited to balance control, sensing, and navigation. This is quite appropriate for walking machines which travel over relatively simple substrates. As substrates become more complex, more sophisticated legged vehicles will be required. More will need to be known about mechanics of leg movement, how this movement interacts with movement of the body, and how environment influences leg design.

Legs on current machines are fairly simple; most have only three degrees-of-freedom. They use a point contact foot or its equivalent, e.g. a foot connected to the leg with a ball joint which is not actively controlled. More actively controlled degrees-of-freedom will be required for more versatile leg movement.

While in current foot designs some foot slippage is tolerated, in legged vehicles which will move on steep or lattice work substrates, foot slippage is unacceptable. A better understanding of the kinematics of the system will make it possible to avoid foot slippage and also avoid singular configurations of the mechanical system. Foot slippage can be avoided by considering constraints that multiple closed loops of the mechanism place on motion and by quantifying forces that the foot-substrate interface can sustain without slipping. The same kinematic knowledge of the system can be used to avoid configurations in which forces exerted by actuators do not constrain the motion of the system; these are the mechanically singular configurations.

Novel and effective solutions to these problems may come from analyzing existing models of legged vehicles. Millions of different legged vehicles are immediately available in the form of arthropods.

CURRENT BIOLOGICAL RESEARCH

Arthropods are magnificient models for investigating walking machine design. Insects, spiders, crabs and lobsters are typical arthropods: they travel on articulated, multi-segmented legs with an external skeleton permitting muscle attachments at nearly any point to optimize efficiency or speed. Their habitats include smooth and very rough surfaces; up-right, vertical and up-side-down surfaces; unyielding rock and supple web surfaces and vast open spaces as well as narrow, twisted crevices. Each species has survived by minimizing locomotory energy demands for routine necessities such as grazing for food while retaining the quick burst of speed many use for evading predators.

A strong argument has been presented that the hexapodous state is of such advantage that it has been independently evolved in five major groups of arthropods (Manton, 1972, 1979). Certainly hexapodous running is used by arthropods as diverse as some prawns and crabs, certain spiders and most insects. Each of these classes of hexapods uses their three pairs of legs in a distinctive manner with different leg actions, gaits, fields of movement and mechanisms connecting the body and leg.

Virtually all biologists working with any aspect of leg motion (see Tables 1 and 2 for a partial listing of work done) have noted differences in action among the matched pairs of legs on a single arthropod as well as contrasts between different arthropods but have had insufficient mathematical tools to analyse or explain these differences.

They have recognized that arthropods walk not only without foot slippage (Manton, 1967) but also without loss of claw grip (Manton, 1958a). Rotation of the entire leg is also used in different ways. Centipedes use leg rotation to strengthen the leg and allow more rapid leg motion while spiders use leg rotation to lengthen the stride (Manton, 1958a).

Because of their small size and the rigorous competition involved in survival, arthropods have had to become efficient machines. One trade-off in joint development involves using a more restrictive skeletal framework such as the hinge joint instead of the three degree-of-freedom ball-and-socket joint. The hinge does not allow the freedom of movement available with the ball-and-socket but requires much less control musculature. In an animal of very restricted size, less control musculature means more muscle mass to power locomotion (Manton, 1958b). One centipede species which has a ball and socket joint at each leg base has 34 muscles per leg; millipedes employ all hinge joints in their legs and move using 2 muscles per leg (Manton, 1973).

Arthropods are responsive to forces which scaling laws may allow us to ignore in walking machines. Nearly all arthropods walk with their body slung between leg arches. This lowering of the center of gravity gives stability against forces such as the wind, a factor probably not as critical to a larger animal or machine (Manton, 1951). However, stability required by the arthropod may also be desirable for the machine; for example, a low center of gravity may be advantageous for work on steep terrains.

Nearly all exploration of walking in arthropods has centered on gaits or on muscle and neural control (Tables 1 and 2 mention many arthropod species studied to date but omit most multiple papers on the same animal). The gait and forces acting on the stick insect, the arthropod most frequently used for such studies, are quite well understood. A diversity of arthropods has proved amenable to investigations on walking; these works will become increasingly pertinent once leg motion is understood.

One final note is that we expect to need to move beyond arthropod solutions to mobility problems for a walking machine design. Over the millions of years of evolution, species will be optimized to their habitat within their own genetic constraints. For example, if 6 legs are really the best for a spider in a given situation, this option is not a possibility because spider genetics dictate an animal with 8 legs. Similarly, if any change in design is not continuously advantageous, the competition involved in evolution would be expected to eliminate that solution. A novel and highly advantageous option which could only be achieved through this disadvantageous intermediate step would thus never occur. We expect to grasp the patterns in leg mobility of arthropods and then move beyond these solutions to others pertinent to walking machines.

REFERENCES

Barnes, W.J.P. (1975). Leg co-ordination during walking in the crab, *Uca pugnax*. J. comp. Physiol., 96:237-256.

Bowdan, E. (1978a). Walking and rowing in the water strider, *Gerris remigis*. I. A cinematographic analysis of walking. J. comp. Physiol. 123: 43-49.

Bowdan, E. (1978b). Walking and rowing in the water strider, *Gerris remigis*. II. Muscle activity associated with slow and rapid mesothoracic leg movement. J. comp. Physiol. 123:51-57.

Bowerman, R.F. (1975). The control of walking in the scorpion. I. Leg movements during normal walking. J. comp. Physiol. 100:183-196.

Burns, M.D. (1973). The control of walking in Orthoptera. I. Leg movements in normal walking. J. exp. Biol. 58:45-58.

Burrows, M., and G. Hoyle (1973). The mechanism of rapid running the the ghost crab, *Ocypode ceratophthalma*. J. exp. Biol. 58:327-349.

Chasserat, C., and F. Clarac (1983). Quantitative analysis of walking in a decapod crustacean, the rock lobster *Jasus lalandii*. J. exp. Biol. 107:219-243.

Cruse, H. (1976a). The function of the legs in the free walking stick insect, *Carausius morosus*. J. comp. Physiol. 112:235-262.

Cruse, H. (1976b). The control of body position in the stick insect (*Carausius morosus*), when walking over uneven surfaces. Biol. Cybern. 24:25-33.

Cruse, H., F. Clarac, and C. Chasserat (1983). The control of walking movements in the leg of the rock lobster. Biol. Cybern. 47:87-94.

Cruse, H., and U. Muller (1986). Two coupling mechanisms which determine the coordination of ipsilateral legs in the walking crayfish. J. exp. Biol. 121:349-369.

Delcomyn, F. (1971). The locomotion of the cockroach *Periplaneta americana*. J. exp. Biol. 54:443-452.

Franklin, R.F. (1985). The locomotion of hexapods on rough ground. In M. Gewecke and G. Wendler (Eds.) Insect locomotion. Verlag Paul Parey, Berlin. pp.69-78.

Graham, D. (1972). A behavioural analysis of the temporal organisation of walking movements in the 1st instar and adult stick insect (*Carausius morosus*). J. comp. Physiol. 81:23-52.

Graham, D. (1977). Unusual step patterns in the free walking grasshopper *Neoconocephalus robustus*. II. A critical test of the leg interactions underlying different models of hexapod co-ordination. J. exp. Biol. 73:159-172.

Graham, D. (1978). Unusual step patterns in the free walking grasshopper *Neoconocephalus robustus*. I. General features of the step patterns. J. exp. Biol. 73:147-157.

Gray, P.T.A., and P.J. Mill (1983). The mechanics of the predatory strike of the praying mantid *Heirodula membranacea*. J. exp. Biol. 107:245-275.

Hemami, H., and B.R. Chen (1984). Stability analysis and input design of a two-link planar biped. Int. J. Robotics Res. 3(2):93-100.

Herreid, C.F.II, and R.J. Full (1986). Locomotion of hermit crabs (*Coenobita compressus*) on beach and treadmill. J. exp. Biol. 120:283-296.

Hirose, S. (1984). A study of design and control of a quadraped walking vehicle. Int. J. Robotics Res. 3(2):113-133.

Hughes, G.M. (1952) The co-ordination of insect movements. I. The walking movements of insects. J. exp. Biol. 29:267-285.

Ichikawa, Y., N. Qzaki, and K. Sadakane (1983). A hybrid locomotion vehicle for nuclear power plants. IEEE Trans. Syst. Man Cybern. SMC-13(6):1089-1093.

Klein, C.A., K.W. Olson, and D.R. Pugh (1983). Use of force and attitude sensors for locomotion of a legged vehicle over irregular terrain. Int. J. Robotics Res. 2(2):3-17.

Macmillan, D.L. (1975). A physiological analysis of walking in the American lobster (*Homarus americanus*). Phil. Trans. Roy. Soc. Lond. B 270:1-59.

Manton, S.M. (1951). The evolution of arthropodan locomotory mechanisms. Part 2. General introduction to the locomotory mechanisms of the Arthropoda. J. Linn. Soc. (Zool.) 42:93-117.

Manton, S.M. (1958a). Hydrostatic pressure and leg extension in arthropods, with special reference to arachnids. Ann. Mag. nat. Hist. 13(1):161-182.

Manton, S.M. (1958b). Habits of life and evolution of body design in Arthropoda. J. Linn. Soc. (Zool.) 44:58-72.

Manton, S.M. (1967). The polychaete *Spinther* and the origin of the Arthropoda. J. nat. Hist. 1:1-22.

Manton, S.M. (1972). The evolution of arthropodan locomotory mechanisms. Part 10. Locomotory habits, morphology and evolution of the hexapod classes. J. Linn. Soc. (Zool.) 51:203-400.

Manton, S.M. (1973). The evolution of arthropodan locomotory mechanisms. Part 11. Habits, morphology and evolution of the Uniramia (Onychophora, Myriapoda, Hexapoda) and comparisons with the Arachnida, together with a functional review of uniramian musculature. J. Linn. Soc. (Zool.) 53:257-375.

Manton, S.M. (1977). The Arthropoda: Habits, functional morphology, and evolution. Clarendon Press, Oxford. 527pp.

Manton, S.M. (1979). Functional morphology and the evolution of the hexapod classes. In A.P. Gupta (Ed.) Arthropod phylogeny. Van Nostrand Reinhold Co., New York. pp. 387-465.

McGhee, R.B. (1968). Some finite state aspects of legged locomotion. Math. Biosci. 2(1):67-84.

Miura, H., and I. Shimoyama (1984). Dynamic walk of a biped. Int. J. Robotics Res. 3(2):60-74.

Moffett S., and G. S. Doell (1980). Alternation of locomotor behavior in wolf spiders carrying normal and weighted egg cocoons. J. exp. Zool. 213:219-226.

Orin, D., C.K. Tsai, and F.T Cheng (1985). Dynamic computer control of a robot leg. IEEE Trans. Ind. Electron. IE-32(1):19-25.

Ozguner, F., and S.J. Tsai (1985). Design and implementation of a binocular-vision system for locating footholds of a multi-legged walking robot. IEEE Trans. Ind. Electron. IE-32(1):26-31.

Parry, D.A., and R.H.J. Brown (1959). The jumping mechanism of salticid spiders. Exp. Biol. 36:654-664.

Pearson, K.G., and R. Franklin (1984). Characteristics of leg movements and patterns of coordination in locusts walking on rough terrain. Int. J. Robots Res. 3:101-112.

Petternella, M., and S. Salinari (1974). Feasibility study on six-legged walking robots. Proc. 4th Int. Symp. Industrial Robots, Tokyo: Japan Industrial Robot Assoc. 33-42.

Pond, C.M. (1975). The role of the walking legs in aquatic and terrestrial locomotion in the crayfish *Austropotamosius pallipes* (Lereboullet). J. exp. Biol. 62:447-454.

Raibert, M.H. (1986). Legged robots. Comm. ACM 29(6):499-514.

Raibert, M.H., and I.E. Sutherland (1983). Machines that walk. Scientific American 248(2):44-53.

Sherman, E., M. Novotny, and J.M. Camhi (1977). A modified walking rhythm employed during righting behavior in the cockroach *Gromphadorhina portentosa*. J. comp. Physiol. 113:303-316.

Song, S.M., K.J. Waldron, and G.L. Kinzel (1985). Computer-aided geometric design of legs for a walking vehicle. Mech. Mach. Theory 20(6):587-596.

Takahata, M., H. Komatsu, and M. Hisada (1984). Positional orientation determined by the behavioural context in *Procambarus clarkii* Girard (Decapoda: Macrura). Behav. 88:240-265.

Wendler, G. (1966). The co-ordination of walking movements in arthropods. Symp. Soc. Exp. Biol. 20:229-249.

Wilson, D.M. (1967). Stepping patterns in Tarantula spiders. J. exp. Biol. 47:133-151.

Zanker, J.M., and T.S. Collett (1985). The optomotor system on the ground: on the absence of visual control of speed in walking ladybirds. J. comp. Physiol. 395-402.

TABLE 1 Spatial and Temporal Stepping Patterns of Arthropods

INSECTS		
stick insect	*Carausius morosus*	Graham, 1972
cockroach	*Gromphadorhina portentosa*	Sherman, Novotny and Camhi, 1977
	Periplaneta americana	Delcomyn, 1971
	Blatta orientalis	Hughes, 1952
grasshopper	*Neoconocephalus robustus*	Graham, 1977, 1978
locust	*Locusta migratoria*	Pearson, and Franklin, 1984
water strider	*Gerris remigis*	Bowdan, 1978a
beetle	*Cantharis fusca*	Wendler, 1966
	Dytiscus marginalis	Hughes, 1952
	Hydrophilus piceus	Hughes, 1952
	Carabus violaceus	Hughes, 1952
	Chrysolema orichalcea	Hughes, 1952
	Blaps mucronata	Hughes, 1952
ARACHNIDS		
spider	*Dugesiella hentzi*	Wilson, 1967
daddy-long-legs	unidentified species	Franklin, 1985
CRUSTACEANS		
lobster	*Jasus lalandii*	Chasserat, and Clarac, 1983
	Homarus americanus	Macmillan, 1975
crab	*Ocypode ceratophthalma*	Burrows, and Hoyle, 1973
	Coenobita compressus	Herreid, and Full, 1986
	Uca pugnax	Barnes, 1975
crayfish	*Austropotamobius pallipes*	Pond, 1975

TABLE 2 Control of Arthropod Leg Motions

INSECTS		
stick insect	*Carausius morosus*	Cruse, 1976a, 1976b
praying mantis	*Heirodula membranacea*	Gray, and Mill, 1983
cockroach	*Gromphadorhina portentosa*	Sherman, Novotny and Camhi, 1977
locust	*Locusta migratoria*	Pearson, and Franklin, 1984
	Schistocerca gregaria	Burns, 1973
water strider	*Gerris remigis*	Bowdan, 1978b
beetle	*Coccinella septempunctata*	Zanker, and Collett, 1985
ARACHNIDS		
scorpion	*Hadrurus arizonensis*	Bowerman, 1975
spider	*Sitticus pubescens*	Parry, and Brown, 1959
	Pardosa tristis	Moffett, and Doell, 1980
CRUSTACEANS		
lobster	*Jasus lalandii*	Cruse, Clarac and Chasserat 1983
	Homarus americanus	Macmillan, 1975
crab	*Ocypode ceratophthalma*	Burrows, and Hoyle, 1973
crayfish	*Astacus leptodactylus*	Cruse, and Muller, 1986
	Procambarus clarkii	Takahata, Komatsu and Hisada 1984

Arthropoden: 350 millionen Jahre erfolgreicher Laufmaschinenentwurf

Hochentwickelte Laufmaschinen können, für Aufgaben wie klettern an weichen Abhängen, manövrieren um Hindernisse oder überqueren schmaler Träger, von Vorteil sein. Eine Beinkonstruktion, die diese Herausforderungen bewältigen kann, würde den Gebrauch von Laufmaschinen in Konstruktion und Instandhaltung von Gittertragwerken, wie sie für Raumstationen vorgeschlagen, und für Unterwasserstrukturen bereits benutzt werden, zulassen. Diese könnten auch in Minen oder Stütztragwerken eingesetzt werden.

Derzeitige Laufmaschinen sind Testgerate der ersten Generation für den Gebrauch von sensoren und die kontrolle der Bewegung. Ihr Beinmechanismus ist unkompliziert. Wenn das Potential, dass uns Laufmaschinen bieten besser ausgenutzt werden soll, muss unsere Anstrenung in Richtung höher entwickelter Beinkonstuktionen, und Optimirung von Mobilität und Kontrolle gerichtet werden. Das vorhandensein von Modellen würde diese NeuentwichLungen vereinfachen und beschleunigen. Wir denken, dass durch die Benutzung von Arthropoden als diese Modelle viel gewonnen werden kann.

Arthropoden sind excellente Beispiele für komplexere Beinstrukturen. Seit 350 milliolnen Jahre entwickeln diese Tiere leistungsfähige und wirksame Mittel um eine grosse Vielfalt von Umweltbedingungen einschliesslich rauhen Oberfläichen, felsigen Flachen, Gewebestrukturen und Felsspalten handhaben zu können. Diene unterschiedlichen Anforderungen führten zum Gebrauch von 6 multi-segment Beinen zur Fortbewegung, und zur Ausbildung von 5 Haupt gruppen von Arthropoden. Überflüssig zu sagen, dass diese 5 Lösungen nicht identisch sind, doch die Unterschiede wurden bisher nur in Bezug auf Gangart und Nerven kontrolle untersucht. Das Studiun der Beinmechanik dieser verschiedenen Lösungen, unter Berücksichtigung der dazugehörigen Umweltbedingungen, sollte unser Entwurfskonzept für Laufmaschinen sehr verbessern.

A Comparative Evaluation of Energetic Efficiency of the Walking Propelling Agent

A. A. ANARKULOV, A. P. BESSONOV AND N. V. UMNOV

Mechanical Engineering Research Institute, USSR Academy of Sciences, Moscow, USSR

Краткое содержание. Дискретность колеи шагающего транспортного робота является предпосылкой меньших затрат энергии на перемещение по сравнению с колесным движителем. Однако с другой стороны, то обстоятельство, что каждая нога часть времени находится в фазе переноса и не участвует в поддержании веса машины, увеличивает величину вертикальной реакции на ногу по сравнению с реакцией на колесо при одном и том же числе колесных и шагающих движителей. Разработанная быстрая процедура прямого определения реакций многоногого транспортного средства позволила сравнить походки по величине затрат энергии на ходьбу по идеально пластичному грунту и выявить распределение максимальных реакций многоногих машин в зависимости от числа ног и режима ходьбы.

Keywords. Six-legged walking robot; symmetric wave gait; soil; energetic efficiency; mutually exclusive tendencies.

В литературе неоднократно указывалось, что оснащение подвижного робота шагающим движителем существенно расширяет транспортные возможности робота главным образом за счет повышения его проходимости. Причем в этом повышении основной вклад вносят адаптивные характеристики робота, улучшающие профильную проходимость, позволяющие преодолевать препятствия типа уступов, ям, щелей, лестниц и т.п. Однако для транспортного робота, работающего в условиях открытой местности, важно и то, что шагающий движитель обладает повышенной грунтовой проходимостью, т.е. должен меньше тратить энергии на единицу пути, чем традиционный колесный движитель.

Исследования, подтверждающие это положение, обычно проводились теоретически на идеализированной модели одиночного шагающего движителя [I]. При этом обычно указывалось и на то обстоятельство, что при ходьбе шагающая опора только часть цикла взаимодействует с грунтом, внедряясь в него, а другая часть переносится по воздуху, затрачивая на это существенно меньше энергии. Видимым следствием такого типа передвижения будет цепочка следов на грунте, оставшаяся после прохода шагающего робота в противовес сплошной колее, остающейся после проезда колесного или гусеничного робота. Эта дискретность колеи шагающего движителя чаще всего и приводится в качестве доказательства его энергетической эффективности.

С нашей точки зрения здесь упускается существенное обстоятельство. Дело в том, что после того, как какая-либо нога отрывается от грунта для совершения переноса, вертикальная реакция на всех остальных ногах от силы тяжести машины должна увеличиться. А поскольку глубина деформации грунта и работа, затрачиваемая на это, зависят от реакции, то следовательно должна увеличиться энергия, затрачиваемая на единицу пути. Иными словами след от каждой ноги будет глубже, чем глубина колеи колесного робота с тем же количеством опор. Налицо явление кажущегося утяжеления шагающего робота по сравнению с колесным. Покажем это простым примером. Пусть шестиногий робот с невесомыми ногами идет походкой "трешки", т.е. нога ровно полцикла находится в фазе опоры, и ровно через полцикла происходит смена одной опорной триады ног на другую. Из элементарных соотношений можно показать, что максимальная реакция в каждой опоре равна половине веса машины, при этом у средних ног реакция постоянна, у передних она увеличивается от нуля до максимальной, у задних - уменьшается от максимальной до нуля. Сумма всех максимальных реакций, т.е. тех величин, которые определяют максимальную деформацию грунта, для робота с этой походкой равна трем весам машины, и следовательно каждый след в три раза глубже колеи шестиколесного робота.

Таким образом, в оценке энергетических затрат выявляются две противоречивые тенденции, которые удобно сравнить на идеализированной модели взаимодействия машин с грунтом. Причем идеализация должна касаться не только машины, но и грунта. Модель шагающей машины традиционна - невесомые ноги, бесконечно жесткий корпус, геометрический центр которого совпадает с центром тяжести, одинаковая упругость у всех ног, достаточно медленная ходьба. Для модели же грунта нам показалось удобным использовать модель идеально-пластичного грунта. Под ним мы понимаем такой грунт, приращение деформации которого пропорционально приращению нагрузки на него, если приращение нагрузки положительно, т.е. нагрузка увеличивается. В то же время приращение деформации равно нулю, т.е. деформация не изменяется, если приращение нагрузки отрицательно (нагрузка уменьшается). В этой модели грунта величина остаточной деформации - глубина следа пропорциональна максимальной нагрузке вне зависимости от формы закона нагружения. Отсюда получается простой геометрический критерий затрат энергии на ходьбу - сумма глубины следов всех ног.

Здесь хочется обратить внимание на то, что хотя сумма глубин следов пропорциональна сумме максимальных реакций всех ног, эти последние должны находиться при непосредственном моделировании ходьбы по пластичному грунту, т.е. для этих целей нельзя использовать существующее распределение максимальных реакций, полученное при ходьбе по жесткой опорной поверхности. Причиной этого является статическая неопределимость системы корпус-ноги-грунт, приводящая к зависимости реакций в ногах от величины деформации грунта. Иными словами, если при ходьбе по жесткой поверхности опорные концы ног всегда находятся в одной плоскости, то при ходьбе по деформируемому грунту компланарность опорных концов ног не сохраняется, и следовательно меняется распределение реакций. Для многих походок величины максимальных реакций некоторых ног также меняются.

Однако прежде, чем получить такое новое распределение реакций, необходимо ответить на вопрос о единственности этого распределения. Такой вопрос возникает в связи с тем, что по-существу распределение реакций зависит от предистории ходьбы, от того какое было распределение реакций, и следовательно деформаций, на предыдущем цикле и т.д. Нет априорной уверенности, что, во-первых, при достаточно длительной ходьбе с произвольными начальными условиями по деформациям ног система выйдет на установившийся режим, и во-вторых, этот режим, если он будет, сохранится при смене начальных условий.

Ответ на эти вопросы был получен математическим моделированием. Для этого изучалось распределение реакций в ногах после перехода границы между абсолютно жестким и идеально пластичным грунтом и достаточно длительной ходьбы по последнему. При этом варьировались походки, а в пределах одной походки - расстояние до границы в момент начала движения, т.е. по-существу начальная фаза в момент начала переходного процесса ходьбы по идеально-пластичному грунту.

Результаты моделирования позволяют обоснованно утверждать, что при любой начальной фазе установившийся цикл ходьбы существует, и не зависит от нее.

Теперь процедура определения распределения максимальных реакций упрощается, так как можно исследовать переходный процесс при произвольной начальной фазе.

Нашей задачей исследования было сравнить походки по их энергетической эффективности. Для этой цели все статически устойчивые волновые симметричные походки шестиногих роботов сравнивались по вышеустановленному критерию - суммарная глубина следа всех ног. Результаты такого сравнения приведены на рис. I. На нем показано распределение глубины суммарного следа в зависимости от походок.

Нами рассматривалось движение шестиногого шагающего робота с невесомыми ногами и массой корпуса, равной 6 единицам. Деформация Δ грунта под ногой была принята пропорциональной нагрузке F на ногу. Коэффициент пропорциональности принят равным 0,25.

$$\Delta = 0.25 \cdot F.$$

На рисунке в области статической устойчивости волновых походок шестиногих роботов построены расчетные линии равной глубины суммарного следа за цикл ходьбы. На нем видны неблагоприятные области суммарной глубиной, равной 4,5 единицы - это соответствует случаю, когда каждая нога в цикле ходьбы имеет максимально возможную максимальную реакцию, равную 3 единицам (половина веса машины). При равномерном распределении веса между шестью ногами суммарная деформация грунта была бы I,5 единицы - в три раза меньше. Такая ситуация возможна только теоретически (точка 0,0) и приведена лишь для демонстрации величины возможного разброса энергетических затрат при ходьбе.

Этот рисунок в целом подтверждает энергетическую неравноценность походок и позволяет выбрать те походки, затраты при ходьбе которыми по мягкому грунту будут меньше. Это будут те походки, которые расположены в областях, в которых величина суммарного следа будет меньше. Таким образом выявился еще один критерий для сравнения походок.

ЛИТЕРАТУРА

I. Беккер М.Г., Введение в теорию систем местность-машина, -М.: Машиностроение, I973. 520 с.

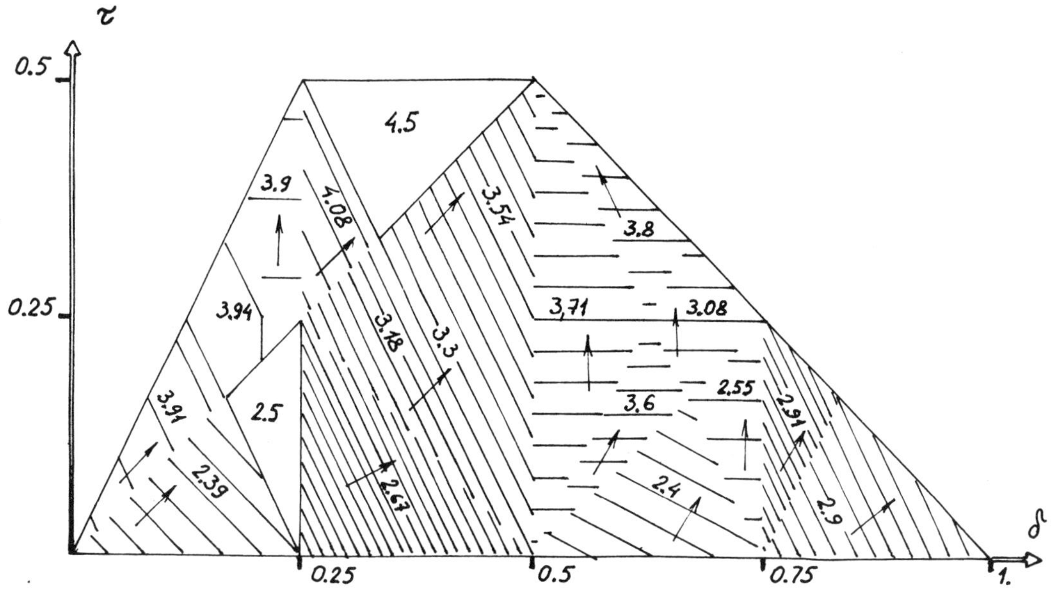

Рис. I. Распределение суммарной за цикл ходьбы глубины следа.
τ - относительное время переноса ног;
δ - относительный сдвиг в фазе работы ног одного борта;
Линии равной глубины построены через 0.04; направление возрастания глубины отмечено в каждой зоне стрелкой.

SUMMARY

When evaluating energetic efficiency of the work of a robot walking propelling agent on the soil one should take into account two mutually exclusive tendencies. On one hand, the intermittence of the walking robot's track decreases power wastes on the resistance of the soil in comparison with a wheeled robot and on the other hand, the more specific load on the supporting leg than on the wheel, increases these wastes.

Understanding that the quantity of wastes of power themselves on walking depend, first of all, on the type of the soil but not on the design of legs or the organization of movement we have put the task to show only inequal value of the gaits by this criterion.

For this purpose a numerical experiment was carried out for the walking of the six-legged walking robot with massless legs upon the perfectly flexible soil. As a result the zones with anamalously big wastes of power on walking and more favourable zones on which one should orient himself when choosing gaits for six-legged ones in the area of existence of the wave statically stable gaits were found. In particular, the well-known gait "triples" is one of the most unfavourable types of gaits for the six-legged walking robots.

History

Some Remarks on and Around Burmester's Work at the Occasion of the Centenary of His "Lehrbuch der Kinematik"

T. KOETSIER

Department of Mathematics and Computer Science, Free University Amsterdam, Postbox 7161, 1007 MC, Amsterdam, The Netherlands

Abstract. The third and last installment of *Ludwig Burmester*'s *Lehrbuch der Kinematik* appeared in 1887. The book contains the first major attempt at a synthesis of *theoretical kinematics* and *kinematics of mechanisms*. The present paper discusses the book from this point of view. The book is also considered within the wider context of the development of 19th century mechanical engineering.

Keywords. History of kinematics, history of geometry.

INTRODUCTION

The development of *kinematics of mechanisms* as an independent area of scientific investigation started in 1794 with the foundation of the École Polytechnique in Paris (Ferguson, 1962). During the first half of the 19th century also *theoretical kinematics*, which deals with the more general kinematical properties of motion, appeared as an independent area of research (Koetsier, 1986). By means of certain fundamental theorems concerning the instantaneous center of rotation, centrodes, etc. theoretical kinematics soon obtained considerable coherence as a field. Kinematics of mechanisms remained a mainly descriptive area until the work of the engineer *Franz Reuleaux* (1829-1905), author of the influential book *Theoretische Kinematik* (Reuleaux, 1875). Reuleaux' work is responsible for the way we look at mechanisms today. The essential element in Reuleaux' theory is the fact that the pedestall or analogous body in a mechanism is seen as just a *link*. Thus Reuleaux reached the more general point of view, where not the motions of the particular links with respect to the fixed link, but instead the relations between the relative motions of the links (including the fixed link) with respect to each other characterize a mechanism. By means of notions like *kinematic pair, kinematic chain, kinematic analysis and synthesis* Reuleaux turned kinematics of mechanisms into a coherent field. Before Reuleaux the application of theoretical kinematics in kinematics of mechanisms was essentially limited to the analysis of the instantaneous motion of a moving link with respect to the fixed link. The instantaneous center of rotation was determined and tangents to and radii of curvature of trajectories were determined. After Reuleaux' work the time was ripe for a much further reaching application of theoretical kinematics in kinematics of mechanisms. In this respect Ludwig Burmester's book *Lehrbuch der Kinematik, für Studirende der Maschinentechnik, Mathematik und Physik, geometrisch dargestellt, Erster Band, Die Ebene Bewegung* (Burmester, 1888) was a major leap forward[1]. Burmester's book, consisting of 941 pages, accompagnied by an atlas containing 863 figures (Burmester, 1888a) contains the first serious attempt at a synthesis of *theoretical kinematics* and *kinematics of mechanisms*. In his book Burmester gave an extensive treatment of theoretical kinematics and applied the general results thus obtained to a wealth of examples from kinematics of mechanisms. Because a complete description and evaluation of Burmester's book here is out of the question we shall restrict ourselves in this paper to two of his major contributions: his work on *synthesis of mechanisms* and his work on *velocity analysis of mechanisms*. Burmester's treatment of these subjects in his Lehrbuch der Kinematik is based upon three fundamental papers which he wrote in 1876, 1877 and 1880. These papers are a very good illustration of Burmester's work. While working on these papers Burmester must have developed the project for his Lehrbuch der Kinematik to which he dedicated the whole period between 1880 and 1887. In the present paper we shall restrict ourselves to these papers.

SYNTHESIS OF MECHANISMS

In 1872 *Louis Ernst Hans Burmester* (1840-1927) (Fig. 1) became ordinary professor of descriptive geometry in Dresden. Probably his three year younger colleague *Trajan Rittershaus*, a great admirer of Reuleaux' work, who became in 1874 professor of theoretical and applied kinematics at the same polytechnical school, turned Burmester's existing interest in theoretical kinematics towards kinematics of mechanisms. In 1876 Kirsch offered a new method to design four-bar straight-line linkages (Kirsch, 1876): i) Take a four-bar linkage and consider a large number of *discrete* positions of the coupler-plane, ii) Draw for each position the inflexion circle, iii) The point in the coupler-plane which is on or very close to as many of the inflexion circles as possible, describes approximately a straight line in the positions considered. Burmester did not appreciate the method. From his point of view it was very uncertain. Yet Kirsch' paper suggested to Burmester a question that was new: Are there points in the coupler plane of a given four-bar mechanism, whose homologous points in, say, 3,4,5,... given discrete positions are on a straight line? In (Burmester, 1876) the question was answered: In general the set of points in a moving plane that are in three given discrete positions on a straight line is a circle and there exists in general only one point that is in four given positions on a straight line. In (Burmester, 1877) Burmester turned to more general problems considering points in an arbitrary moving plane whose homologous points in 3,4,5,... given discrete positions are on

[1] The third and last installment of the book appeared in 1887. The title page of the three installments bound together gives 1888 as the year of publication. The planned second volume on spatial movement never appeared.

a circle. For four positions Burmester introduced and proved the main properties of the *circle-point curves*[2]) and the *center-curves* and he showed that there are in general four points (nowadays called Burmester-points) whose homologous points are in five positions on a circle. It is characteristic that Burmester immediately applied the theory in different ways in kinematics of mechanisms. We shall briefly consider one example. In order to show that his theory was also applicable to other than four-bar linkages Burmester applied his theory to determine length and position of link CE in Stephenson's link mechanism (Fig. 2a). D should approximately move on a straight line. Taking realistic dimensions for the other links from literature Burmester considered four positions of AB with the homologous points $D_1=D_3$, D_2 and D_4 on the straight line k, such that moreover $D_1D_4=D_1D_2$, $A_1D_1 // A_2D_2$, $A_3D_3 // A_4D_4$ (Fig. 3). Burmester's theory had yielded that if the four poles P_{13}, P_{14}, P_{23}, P_{24} form a rhomb, the center-point curve degenerates and consists of the line at infinity and the two diagonals of the rhomb (Burmester, 1877, p.331). This happens to be the case in the situation under consideration. To an arbitrary center-point E on the diagonal ℓ corresponds a circle-point C with homologous points $C_1=C_4=P_{14}$, $C_2=C_3=P_{23}$. This very special situation makes it possible to choose an arbitrary fifth position of AB with D_5 on k. Then first C_5 is determined (being the point in the fifth position corresponding to P_{14} in first and fourth position) and E is the center of the circle through C_5, P_{14} and P_{23}.

ROBERT STEPHENSON'S LINK MECHANISM

The application of Burmester's theory to Stephenson's link mechanism shows that Stephenson's linkage had become subject of theoretical interest. In 1842 Robert Stephenson and Co. brought out the *William-Howe link motion*[3]) (Fig. 2a). Valve V moves to and fro parallel to the cylinder and alternately lets steam in and off in the cylinder at both ends of the piston. The two excentrics HF and HG (angle FHG is fixed) rotate about an axis at H driven directly by the piston. The engine driver can change the position of E by means of a handle and thus put the engine in reverse or change the point of cut-off of steam. Stephenson's motion, as it is usually called, was tremendously succesfull. It was for many years in almost universal use on locomotives (Johnson, 1944, p.239).

The first to study Stephenson's linkage theoretically was *Edouard Phillips* (1821-1889). By means of the instantaneous center of rotation Y of AB with respect to the "fixed link" (Fig. 2a, Fig. 2b) Phillips analytically determined the relation between the shift of valve V and the angle of rotation of the excentrics HF and HG. Obviously Y is on the line EC. Phillips essentially reasoned as follows. Let $w_H \overset{\text{def}}{=}$ the angular velocity of the two excentrics HF and HG. Then w_Z ($\overset{\text{def}}{=}$ the angular velocity of FB about Z) $= w_H\cdot(HF/FZ)$ and w_Y ($\overset{\text{def}}{=}$ the angular velocity of BA about Y) $= w_H\cdot(HF/FZ)\cdot(ZB/YB)$. Similarly, $w_Y=w_H\cdot(HG/GU)\cdot(UA/YA)$. Equating the two expressions for w_Y yields $(HF/FZ)\cdot(ZB/YB)= =(HG/GU)\cdot(UA/YA)$ and this means that the lines BF, AG and YH are concurrent. This immediately yields a construction for Y (Phillips, 1854). Phillips' paper very well shows the approach of the "old school of kinematics" (Rittershaus, 1875, p.428). The paper also shows how the determination of instantaneous centers in linkages consisting of more than four links became a problem. Burmester would essentially solve that problem.

VELOCITY ANALYSIS OF MECHANISMS

At the request of Reuleaux the mathematician *Siegfried H. Aronhold* (1819-1884) in Berlin started teaching a course on theoretical kinematics in 1866-67 (Rittershaus, 1875, p.425). Aronhold considered chains of rigid systems $S_0,S_1,S_2,\ldots,S_n$ such that S_{i+1} moves in a prescribed way with respect to S_i. One of Aronholds first and for applications important results was the Aronhold-Kennedy three centers theorem (Aronhold, 1872, p.136).[4]) In 1875 Rittershaus applied the Aronhold-Kennedy theorem in a velocity analysis of the Quintenz-scales (Rittershaus, 1875a, p.47). The Quintenz-scales are kinematically equivalent to Stephenson's linkage and in 1877 Burmester applied the same construction to check the straight-line mechanism that we discussed above by drawing the tangents to the trajectory of D (Fig. 2) (Burmester, 1877, p.332).

Yet the full possibilities of the Aronhold-Kennedy theorem in velocity analysis first became clear when Burmester published his (Burmester, 1880)[5]). In that paper Burmester systematically investigated the general motion of n rigid systems with respect to each other in the plane. We shall restrict ourselves to two remarks on this paper. Burmester showed how in general the instantaneous centers of rotation can be determined graphically. Those very simple methods have become classical. It is remarkable that Burmester's work in this respect stimulated *Martin Grübler* (1851-1935) to start his well-known investigations concerning the number of degrees of freedom of an arbitrary mechanism (Grübler, 1883). Burmester had reasoned as follows. If we have n systems, then we have $\binom{n}{2}$ instantaneous centers. In general, for $n > 3$, if the centers of (n-2) systems with respect to the two other systems are given, all other centers can be constructed easily. Very often, however, less centers are needed. Burmester showed that for *even* n the position of a set of $\frac{3}{2}n-2$ well-chosen centers (e.g. no three on one line) determines the positions of the other centers. Grübler, confronted with this result, reasoned as follows (Grübler, 1883, p.168 and 187). If all centers can be uniquely determined then the system as a whole has one (internal) degree of freedom. Burmester's result implies that if we have a kinematic chain consisting of n links, connected by means of $j = \frac{3}{2}n-2$ turning-joints, that do not have special positions (e.g. three on one line), the chain has *one* internal degree of freedom. It may have been this particular line of thought that brought Grübler to his much more general research on degrees of freedom[6]). In his paper Burmester also showed how to determine graphically in a simple way the relative instantaneous velocities. The constructions are based upon the theorem, proved in 1876 by *Schadwill*, which says that the endpoints of the orthogonal velocities of the points of a line ℓ are on a straight line parallel to ℓ

---

[2]) In 1877 Burmester still used the name "Angriffskurve", presumably because it can be used to determine a point of application ("Angriffspunkt"). The name circle-point curve ("Kreispunktkurve") appears in (Burmester, 1888).

[3]) For the detailed history of that invention I refer to the excellent account in (Warren, 1923, pp.359-370).

[4]) The theorem was found independently by Kennedy and published in 1886 (Ferguson, 1962, p.215).

[5]) Ferguson wrote: A.B.W. Kennedy and R.H. Smit added "to Reuleaux' work the elements that would give kinematic analysis essentially its modern shape" (Ferguson, 1962, p.215). Ferguson here forgot to mention Burmester, who came up earlier with similar results and whose work was much more influential.

[6]) Grübler's formula in its simplest form $f=3n-2j-3$ should be called the Chebyshev-Grübler formula. It was known to Sylvester (1873/75, p.10) and to Chebyshev (1869, p.86). Chebyshev also gave a simple proof.

(Schadwill, 1876). Characteristically, in (Burmester, 1880) the theory is applied to a large number of mechanisms, including, of course, Stephenson's linkage.

REULEAUX' CRITICISM OF BURMESTER

In 1830 Ampère had introduced the word *kinematics* (cinématique) for the science "in which motions are considered by themselves as observed in the bodies surrounding us, and specially in those systems of apparatus which we call machines" (Reuleaux, 1875, p.12). In 1875 Reuleaux defined kinematics as "the study of those arrangements of the machine by which the mutual motions of its parts, considered as changes of position, are determined" (Ibid., p.40). In his book Burmester gave this definition: "the geometrical theory of motion and its application to machines" (Burmester, 1888, p.2). Burmester explicitly criticised Reuleaux for deviating from Ampère's intensions and excluding the geometrical theory of motion from kinematics.

In 1890 Reuleaux reacted to Burmester's definition that had in the mean time been accepted by several other authors. Reuleaux wrote, referring to Burmester's statements concerning Ampère: "They show to which completely unforseeable dangers a developing science is exposed, *because* they are *completely fairytales, phantasy from A to Z*" (Reuleaux, 1890, p.244). He wrote:"While one, in Germany, I emphasize, recently says, writes, teaches "*kinematics*" instead of "*geometrical theory of motion*", one also started spreading .... that for Ampère himself *kinematics* meant *geometrical theory of motion*. Such a far going deviation from truth in scientific affairs has, as far as I know, never occurred before, neither in Germany, nor in any other civilized country" (Reuleaux, 1890, p.247). Reuleaux was angry. He defended kinematics against aggressive conquerers who tried to annex an independent state of kinematics and turn it into a subordinate province of the empire of geometry: "Gleichzeitig gedachte die Geometrie, d.h. gedachten deren selbstgewählte eroberungsfeurige Vorkämpfer, die zahlreichen, inzwischen von und in der Kinematik aufgestellten Lehrsätze ganz einfach mit dem Rechte des Eroberers abzuspannen und wie geduldige Lämmer einzutreiben in ihren Pferch. Das kann aber nicht statthaben, wie ich bewiesen zu haben überzeugt bin" (Reuleaux, 1890, p.248). Reuleaux' emotional reaction can only be understood against the background of the development of mechanical engineering as a whole in the second half of the 19th century. In the 80s mechanical engineering methodologically had not yet become an established discipline (Braun, 1977). Reuleaux' kinematics can be considered as an attempt to found such a discipline independent of mechanics and mathematics. Reuleaux' ideas had already been attacked in the 70s and in 1890 Reuleaux was still facing considerable opposition. In the course of the 90s Reuleaux would be confronted with the so-called engineer-movement, led by *Alois Riedler*. For Riedler and his followers Reuleaux represented a far too theoretical approach to mechanical engineering. In 1895 the engineer-movement demanded and eventually obtained (although temporarily) radical changes in the curriculae of the German technological universities. Most of mathematics and Reuleaux' kinematics were removed from the program. In 1896 Riedler would succeed in throwing Reuleaux over. Reuleaux then stopped lecturing (Stäckel, 1915, p.44). In 1890 Reuleaux needed a synthesis of engineering practice and kinematics of mechanisms, instead, he got Burmester's synthesis of the geometry of motion and kinematics of mechanisms. In retrospect we can say that at the time of its appearance Burmester's book was mainly a theoretical exercise that missed the essential link with engineering practise. Only in the course of the 20th century the book would begin to prove its value in practice.

HARTMANN'S TENDER COUPLING

It is easy to find 19th century examples of mechanisms invented by engineers that stimulated the theoreticians. Stephenson's linkage is only one example. Watt's straight-line linkage is another example (Koetsier, 1983 and 1983a). It is much harder to find 19th century examples of influence in the other direction: theoretical work solving practical design problems. Yet such examples do exist. The potential value of the synthesis of kinematics of mechanisms and theoretical kinematics was clearly proved by Reuleaux' pupil *Wilhelm Hartmann* when he designed his locomotive-tender coupling (Hartmann, 1884) (Brauer, 1885) (K., 1885). In principle locomotive and tender can be connected by means of a simple ball-joint. The joint coincides with the point of intersection F of the longitudinal axes of tender and locomotives in curves. Sometimes, however, more complex solutions are required. Hartmann reasoned as follows. Instead of a ball-joint at F one could connect locomotive and tender by means of a bar PO and two gear profiles m and n touching at F (Fig 4). Of course the radii of the circles r_m and r_n should satisfy: $1/r_m + 1/r_n = 1/OF + 1/PF$. Hartmann's solution consists of the following. He replaced the two profiles by an equivalent connection. By means of the Euler-Savary formula he determined the centers of curvature O_1 and O_2 of resp. the points P_1 and P_2 (Fig. 5) and instead of the two profiles he used the bars P_1O_1 and P_2O_2 (together with PO) to connect locomotive and tender. In order to take up compressive forces during oscillatory movements he added two rollers and tooth profiles on which the rollers could roll (Fig. 5).

FINAL REMARKS

Burmester's Lehrbuch has been very influential. Much later contributions in kinematics can be considered as an extension of his work. Yet at the time of its appearance the gap between theory and engineering practice was wide. In particular as for synthesis the engineer had good reasons to be sceptical. Still in 1930 Burmester's pupil *R. Müller* wrote on Burmester's examples of straight-line mechanisms designed by means of the Burmester theory: "Everyone, who considers the figures drawn by Burmester's masterhand, must indeed become excited about the amazingly high degree of approximation [...], however, we do not know how many tedious and vain attempts the author needed to find fitting data by trying" (Müller, 1930, p.10). Kinematics could help the engineer, it could not solve all his (kinematical) problems. One only has to consider e.g. three phases in the evolution of Stephenson's link mechanism (reconstructed by Warren) to see how true this is (Fig. 6).

REFERENCES

Aronhold, S.H. (1872). Grundzüge der kinematischen Geometrie. *Verhandlungen des Vereins zur Beförderung des Gewerbfleisses in Preussen*, 29, pp. 129-155.

Braun, H-J. (1977). Methodenprobleme der Ingenieurwissenschaft, 1850 bis 1900, *Technikgeschichte*, 44, pp. 1-18.

Brauer, E.A. (1885). Review of (Hartmann, 1884), *Zeitschrift des Vereins Deutscher Ingenieure*, 29, pp. 213-216.

Burmester, L. (1876). Ueber die Geradführung durch das Kurbelgetriebe, *Der Civilingenieur*, 22, pp. 597-606 (Mit Tafel 30).

Burmester, L. (1877). Ueber die Geradführung durch das Kurbelgetriebe, *Der Civilingenieur*, 23, pp. 227-250 and pp. 319-342 (Mit Tafeln 10,11, 14,15,16)

Burmester, L. (1880). Ueber die momentane Bewegung ebener kinematischer Ketten, *Der Civilingenieur*, 26, pp. 247-286 (Mit Tafeln 14-18).

Burmester, L. (1888). *Lehrbuch der Kinematik, Band 1, Die Ebene bewegung*, Leipzig.

Burmester, L. (1888a). *Atlas zu Lehrbuch der Kinematik, Band 1, Die Ebene Bewegung*, Leipzig.

Chebyshev, P.L. (1869). Sur les parallélogrammes. *Oeuvres de P.L. Tchebychef*, Tom.II, N.Y., pp. 83-107.

Ferguson, E.S. (1962). *Kinematics of Mechanisms from the Time of Watt*, United States National Museum, Bulletin 228, Washington, D.C.

Grübler, M. (1883). Allgemeine Eigenschaften der zwangläufigen ebenen kinematischen Kette, *Der Civilingenieur*, 29, pp. 167-200 (Mit Tafel 21).

Hartmann, W. (1884). *Theorie der Locomotiv-Tender-Kupplungen*, Berlin.

Johnson, R.P. (1944). *The Steam Locomotive*, New York.

K., (1885). Ueber den Einfluss der Locomotivtender-Kupplungen auf die Betriebssicherheit von Eisenbahnen, *Organ für die Fortschritte des Eisenbahnwesens*, 22 (Neue Folge), pp. 30-31.

Kirsch, (1876). Zur Theorie der Geradführungen, *Der Civilingenieur*, 22, pp. 321-334 (Mit Tafel 17).

Koetsier, T. (1983). A Contribution to the History of Kinematics-I, *Mechanism and Machine Theory*, 18, pp. 37-42.

Koetsier, T. (1983a). A Contribution to the History of Kinematics-II, *Mechanism and Machine Theory*, 18, pp. 43-48.

Koetsier, T. (1986). From kinematically generated curves to instantaneous invariants: episodes in the history of instantaneous planar kinematics, *Mechanism and Machine Theory*, 21, to appear.

Müller, R. (1930). Ludwig Burmester, *Jahresbericht der Deutschen Mathematiker Vereinigung*, 39, pp. 1-21.

Phillips, E. (1854). Theorie der variabeln Expansion mittelst Stephenson's Coulisse, *Der Civilingenieur*, 1, pp. 164-182 (Mit Tafel 18).

Reuleaux, F. (1875). *Theoretische Kinematik: Grundzüge einer Theorie des Maschinenwesens*, Berlin (References to 1963 Dover edition in English).

Reuleaux, F. (1890). Ueber das Verhältnis zwischen Geometrie, Mechanik und Kinematik, *Zeitung des Vereins Deutscher Ingenieure*, 34, pp. 217-225, 243-248.

Rittershaus, T. (1875). Zur heutigen Schule der Kinematik, *Der Civilingenieur*, 21, pp. 425-450.

Rittershaus, T. (1875a) Zur Theorie der Quintenz-Waage, *Der Civilingenieur*, 21, p. 45-50.

Schadwill, S. (1876). Das Gliedervierseit als Grundlage der Kinematik, *Verhandlungen des Vereins zur Förderung des Gewerbfleisses in Preussen*, 55, pp. 378-447.

Stäckel, P. (1915). *Die Mathematische Ausbildung der Architekten, Chemiker und Ingenieure an den Deutschen Technischen Hochschulen*, Leipzig und Berlin.

Sylvester, J.J. (1873/75). On Recent Discoveries in Mechanical Conversion of Motion, *The Collected Mathematical Papers of J.J. Sylvester*, Vol.III, N.Y. 1973, pp. 7-25.

Warren, J.G.H. (1923). *A Century of Locomotive-Building by Robert Stephenson & Co. 1823-1923*, London (Reprinted in 1970).

Zeuner, G. (1904). *Die Schiebersteuerungen*, Leipzig (6th edition).

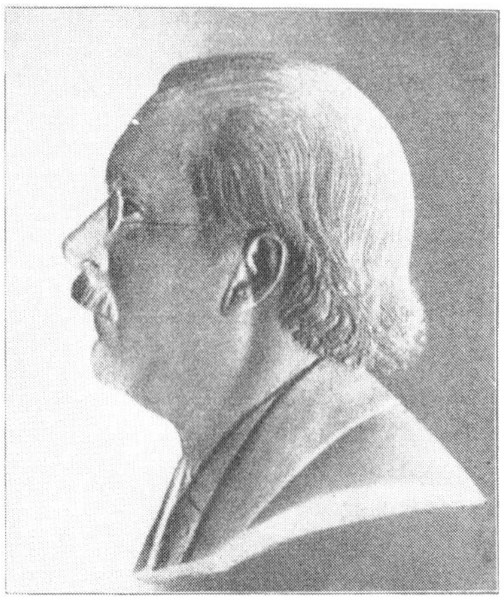

Fig. 1: Ludwig Burmester (Müller, 1930).

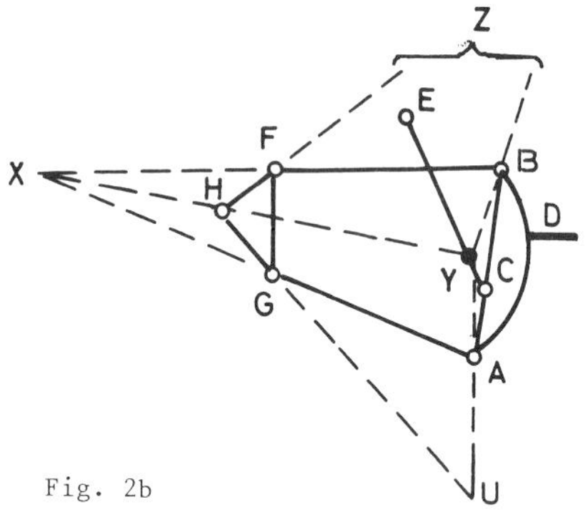

Fig. 2b

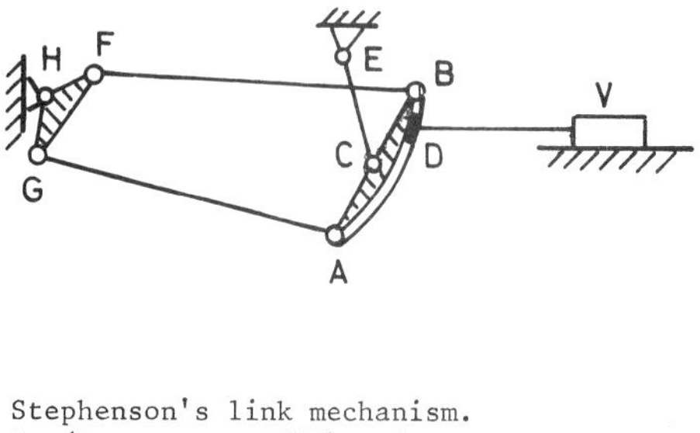

Fig. 2a: Stephenson's link mechanism.
In (Burmester, 1877) D is treated as a fixed point of link AB.

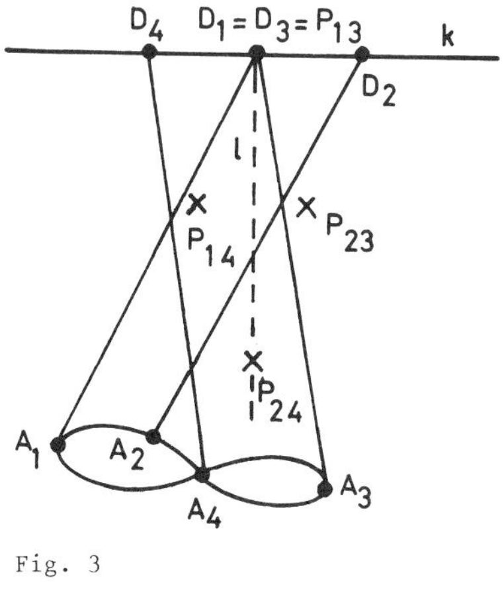

Fig. 3

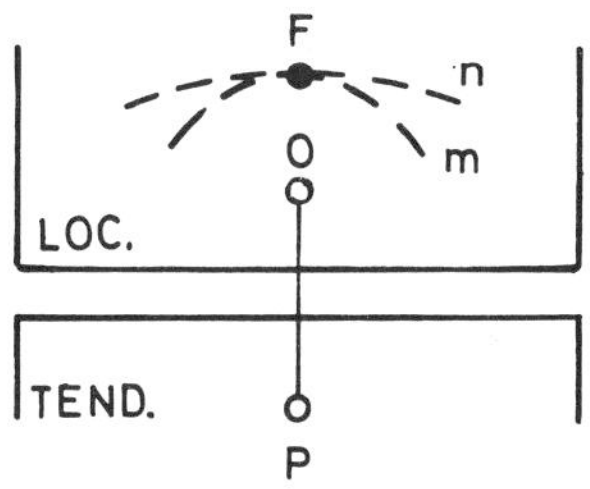

Fig. 4

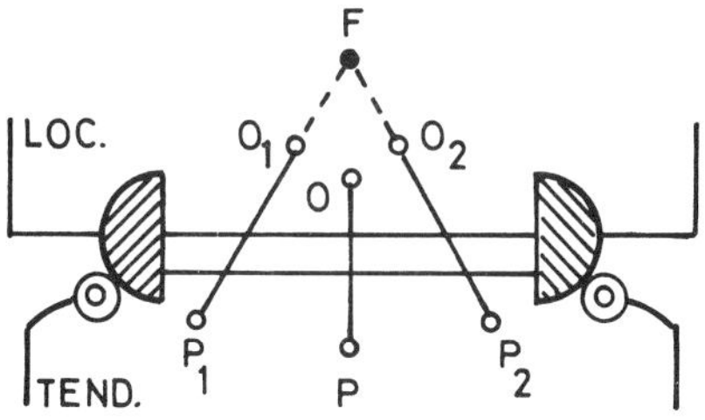

Fig. 5: Hartmann's Locomotiv-tender coupling

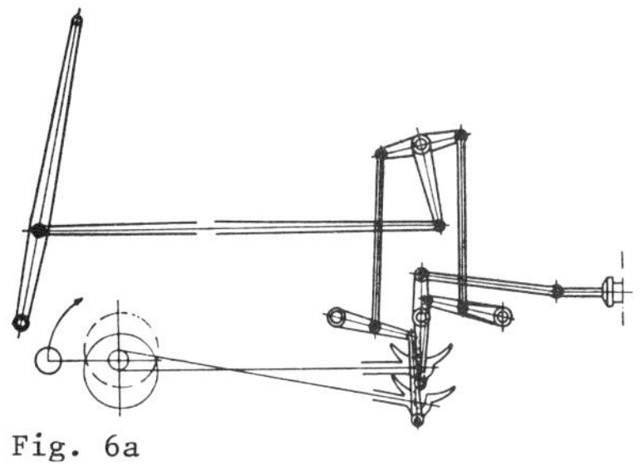

Fig. 6a

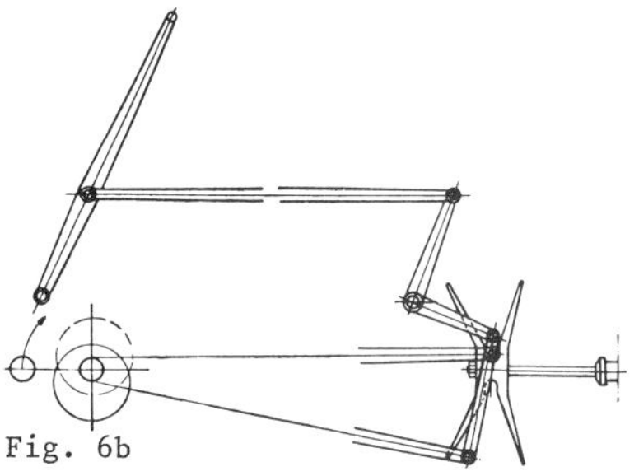

Fig. 6b

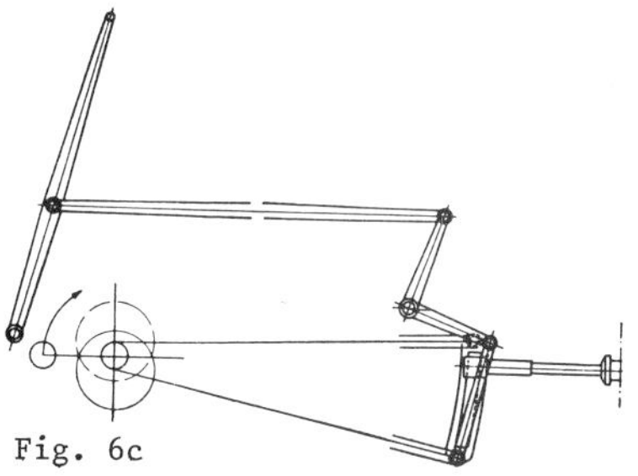

Fig. 6c

Fig. 6a: Valve-gear with forked gabs and single reversing lever, as applied by Stephenson (Warren, 1923, p.371)

Fig. 6b: Valve-gear with gabs combined on valve spindle, as applied by Stephenson, 1841 (Warren, 1923, p.371)

Fig. 6c: Valve-gear with slotted link, by Williams, and developed by Howe, 1842 (Warren, 1923, p.371)

Kurzfassung. Die dritte und letzte Lieferung *Ludwig Burmesters* Lehrbuch der Kinematik erschien in 1887. Das Buch war der erste umfassende Versuch einer Synthesis der *Theoretischen Kinematic* und der *Getriebekinematik*. Im vorliegenden Aufsatz wird das Buch in diesem Hinsicht besprochen. Ausserdem wird das Buch in grösseren Rahmen der Entwicklung der Maschinenlehre im 19ten Jahrhundert betrachtet.

History of Machines and Mechanisms Theory in Works of Soviet Scientists

M. K. USKOV AND A. A. PARKHOMENKO

Mechanical Engineering Research Institute, USSR Academy of Sciences, Moscow, USSR

Достижения советских ученых в разработке научных основ механики машин общепризнаны. Значителен их вклад и в изучение историко-научной проблематики. Именно этому вопросу посвящена настоящая статья

Научные исследования в области теории машин и механизмов получили значительное развитие в трудах русских ученых XIX столетия и в работах ученых и специалистов СССР. Их творческая деятельность стала основой формирования и развития советской школы механики машин, занимающей авторитетное место в мировой науке.

Расширение и углубление научных исследований по теории машин и механизмов, вовлечение в эту работу многих ученых и инженеров, накопление внушительного научного потенциала в этой области привели к необходимости изучения исторических путей развития науки о машинах, анализа основных направлений исследований, выявления творческого вклада ученых и специалистов.

Однако далеко не сразу необходимость исторических исследований была теоретически осознана и практически реализована. Длительное время исторический аспект в работах ученых-машиноведов либо совсем отсутствовал, либо не играл сколько-нибудь заметной роли. Ситуация изменилась в 30-40-х годах, когда историко-научная проблематика стала составной частью работ ряда видных ученых и, прежде всего, академика И.И.Артоболевского. Именно ему принадлежит заслуга в постановке многих исторических изысканий, в серьезной разработке научного наследия ученых-механиков и машиноведов. Помимо этого, он сумел "привить вкус" к истории своим ученикам и коллегам.

Изучение историко-научной проблематики было необходимо И.И.Артоболевскому в силу того, что в теории механизмов и машин утверждение и развитие новых идей не могло осуществляться без глубокого анализа классического наследия корифеев русской и мировой науки. Достаточно сказать, что идеи и методы, положенные И.И.Артоболевским в разработанную им систему классификации механизмов, возникли в результате всестороннего анализа и творческого развития ряда теоретических положений П.Л.Чебышева и Л.В.Ассура.

Уже в первых своих учебных курсах И.И.Артоболевский приводит краткие исторические очерки развития механики машин, а в изданном в 1940 г. учебнике для университетов "Теория механизмов и машин" он уделяет значительное место историческому анализу всех основных направлений этой науки /I/. Целый ряд специальных работ посвятил И.И.Артоболевский исследованию научного и технического творчества акад. П.Л.Чебышева (см., напр.: /2,3/). Особое место занимает подготовленное им совместно с Н.И.Левитским большое исследование "Механизмы П.Л.Чебышева", вошедшее в один из томов научного наследия знаменитого математика и механика /4/. И.И.Артоболевский показал П.Л.Чебышева как основоположника отечественной школы теории механизмов; он написал биографию ученого, использовав при этом многие архивные материалы, введя в научный оборот ряд новых историко-научных фактов.

Сложную историко-техническую работу выполнил И.И.Артоболевский по выявлению и подбору всех механизмов П.Л.Чебышева (а их оказалось более 40), по составлению их кинематических схем, определению размеров звеньев, анализу отдельных механизмов, составлению таблиц и графиков, построению траекторий важнейших точек и описанию всех механизмов. Надо сказать, что к анализу трудов П.Л.Чебышева И.И.Артоболевский возвращался неоднократно, в том числе в 60-70-е годы.

В работах, посвященных отечественной школе теории машин и механизмов, зародившейся в середине XIX в., И.И.Артоболевский охарактеризовал большой вклад, который был сделан в ее создание и дальнейшее развитие видными учеными России: математиком и механиком акад. П.О.Сомовым, видным специалистом по гидродинамической теории трения акад. Н.П.Петровым, известным ученым в области теории регулирования машин проф. И.А.Вышнеградским, крупнейшим ученым-механиком проф. Н.Е.Жуковским. В частности, И.И.Артоболевским была освещена малоизвестная роль Н.Е.Жуковского в развитии прикладной механики, показано, как велик был диапазон интересов ученого в этой области - от общих задач геометрии механизмов до гидродинамической теории смазанных тел и общих уравнений движения механизмов и машин /5/.

В ряде научных работ в 30-40-е, а затем в 50-60-е годы И.И.Артоболевский, отталкиваясь от исторических прецедентов, намечает актуальные задачи развития как традиционных, так и новых направлений науки о машинах. Он выступает по этим вопросам на общих собраниях Академии наук СССР, на заседаниях Отделения технических наук, на различных конференциях, публикует проблемные материалы на эти темы. Характерны в этой связи такие его работы, как "Значение проблемы теории машин и механизмов в машиностроении, ее современное состояние и очередные задачи" (Известия АН СССР, ОТН, 1940, № 1); "Теория машин и механизмов" (в кн.: Советская техника за 25 лет; 1945);"Развитие науки о машинах" (Вестник АН СССР, 1958, № 6) и др.

Одной из последних работ И.И.Артоболевского, посвященных историческим истокам теории машин, ее состоянию и перспективам, явился его доклад "Прошлое, настоящее и будущее теории машин и механизмов". С этим докладом он выступил в сентябре 1975 г. на IУ Международном конгрессе по теории машин и механизмов в Англии. Все прогнозные оценки И.И.Артоболевского неразрывно связаны с исторически закономерными направлениями развития мирового машиностроения, тенденциями научно-технического прогресса, развитием механики машин и теории управления машинами.

Вполне закономерным для историко-научных исследований И.И.Артоболевского было его обращение к творчеству крупнейшего итальянского ученого и инженера эпохи Возрождения Леонардо да Винчи. Известно, что работы Леонардо - классический пример сочетания теории с практикой. Это и привлекло Артоболевского к изучению его жизни и трудов. Он собирает многочисленные фотокопии работ Леонардо, фотоснимки спроектированных им механизмов и машин, тщательно изучает и анализирует их и на основе большой работы с этими материалами публикует несколько очерков, посвященных научному и инженерному наследию Леонардо да Винчи (см., напр. /6/).

Крупный ученый-исследователь, возглавивший большую научную школу и ставший в 60-70-х годах первым президентом Международной Федерации ИФТОММ, - академик И.И.Артоболевский всю свою жизнь интересовался историей науки и техники, считая это научное направление весьма важным как для повышения эрудиции и общей культуры ученых и инженеров, так и непосредственно для практических целей.

Значительный вклад в историю науки о машинах и механизмах внес видный советский исследователь профессор А.Н.Боголюбов. Его историко-научное творчество обширно и многообразно; оно включает в себя анализ крупных научных школ, изучение классических и современных направлений теории машин и механизмов, исследование жизни и трудов видных ученых и инженеров России и СССР, Западной Европы и США. Широкую известность приобрели такие монографии А.Н.Боголюбова, как "История механики машин" и особенно - "Теория механизмов и машин в историческом развитии ее идей" /7/.

В этих работах дан широкий и всесторонний анализ многих направлений мировой научно-технической мысли в области теории машин и механизмов. Причем, анализом охвачены люди, идеи и события, начиная с далеких античных времен и вплоть до середины XX века. Специальную монографию посвятил А.Н.Боголюбов истории становления и развития советской школы механики машин, проследив ее истоки от трудов видных русских ученых до работ советских исследователей 60-70-х годов нынешнего столетия /8/. Затруднительно перечислить всех отечественных и иностранных ученых, жизни и творчеству которых посвятил свои статьи, очерки и биографические книги проф. А.Н.Боголюбов. Одной из последних его работ является книга об акад. И.И.Артоболевском как крупнейшем деятеле советской и мировой науки /9/. В недавно изданном А.Н.Боголюбовым научно-биографическом справочнике "Математики и механики" широко отражены жизнь и деятельность многих ученых-механиков и машиноведов /10/.

Историко-научное направление в исследованиях по теории машин и механизмов получило плодотворное развитие в трудах профессоров В.В.Добровольского, Н.И.Левитского, А.П.Бессонова и других советских ученых. В частности, В.В.Добровольский не только написал ряд исторических очерков об известных ученых - П.Л.Чебышеве, Л.В.Ассуре, Н.И.Мерцалове, но уделил немало внимания историко-техническим проблемам в своих капитальных работах по общей теории механизмов /11/ и синтезу механизмов /12/. Большое число докладов, статей, аналитических обзоров по развитию методов анализа и синтеза механизмов подготовил проф. Н.И.Левитский. Ему же принадлежит ряд работ по истории русской науки о машинах. Известный ученый-машиновед проф. А.П.Бессонов активно работает в области анализа достижений советской школы механики машин, а также пропаганды мировых научных достижений в этой области, особенно в сфере деятельности Международной Федерации ИФТОММ.

Изучению развития теоретической и прикладной механики от античности до наших дней (в том числе механики машин) посвящены некоторые публикации академика А.Ю.Ишлинского и многочисленные работы историков науки профессоров А.Т.Григорьяна и И.Б.Погребысского. Немало сделано в области истории машиностроительной науки в России и СССР известным историком техники проф. А.А.Чекановым.

Нельзя не отметить такой стимулирующий фактор для исторических работ, как многолетняя деятельность всесоюзного Семинара по теории механизмов и машин, а также его многочисленных филиалов в различных городах СССР. При поддержке и содействии И.И.Артоболевского, В.В.Добровольского и других ученых на Семинаре и в его филиалах, на различных всесоюзных совещаниях и конференциях неоднократно заслушивались и обсуждались доклады по историко-научной тематике, намечались новые направления работ в этой области.

Видное место заняли историко-научные исследования в творчестве крупного ученого-машиноведа академика А.А.Благонравова, много лет возглавлявшего Институт машиноведения Академии наук СССР (ИМАШ). А.А. Благонравов справедливо считал, что внимание к истории науки не только содействует более глубокому анализу научно-технических проблем, но и является необходимым условием расширения кругозора и эрудиции научных работников. Глубокий исторический подход к исследованиям связан, по его мнению, с высокой культурой труда ученого, педагога, инженера. И не удивительно, что на протяжении многих лет научной деятельности А.А.Благонравов систематически публиковал статьи и выступал с докладами по вопросам истории науки и техники. Известен он и как редактор ряда фундаментальных историко-технических изданий, член Комитета Советского национального объединения историков естествознания и техники.

В 50-60-е годы А.А.Благонравовым были опубликованы научно-биографические статьи, посвященные крупным деятелям науки и техники - Леонардо да Винчи, академику В.С.Кулебакину, академику А.И.Бергу, профессорам М.М.Хрущову, И.В.Крагельскому и др. В этот период и позднее, в 70-е годы, А.А.Благонравов подготовил ряд исторических обзоров, раскрывающих развитие общей теории машин, теории трения, других направлений деятельности Института машиноведения. Особое место в творческом наследии А.А.Благонравова заняли работы, посвященные развитию ракетно-космической науки и техники. Он проявлял большую заинтересованность в изучении, анализе и прогнозировании ряда важных направлений космических исследований, раскрывал исторический опыт международного сотрудничества в этой области (см., напр.: /I3, I4,I5/).

При подготовке капитального научного труда об опыте изучения космического пространства А.А.Благонравов был назначен главным редактором издания. Эта книга /I6/ - плод совместных усилий большого коллектива ученых, конструкторов, специалистов - вышла в свет к первому десятилетию космической эры. В дальнейшем,когда была задумана такая же книга о втором космическом десятилетии, работа над ней начиналась под руководством А.А.Благонравова. Фундаментальная монография, обобщившая колоссальный опыт исследовательских работ в космосе за период I967-I977 годов,вышла в свет, когда Благонравова уже не стало /I7/. Издание завершилось под руководством академика С.Н.Вернова.

Активное участие принял А.А.Благонравов в работе XIII Международного конгресса по истории науки, который состоялся в Москве в августе I97I г. Он был избран в президиум конгресса, участвовал в работе секции истории машиностроения и ряда других секций, представил на конгресс доклад об основных направлениях исследований в области истории авиационной и космической техники.

Историко-научная традиция в деятельности видных ученых и руководителей Института машиноведения продолжена ныне в ряде работ академика К.В.Фролова, возглавившего с 1975 года ИМАШ. Им опубликованы статьи, доклады и очерки о видных советских ученых-машиноведах, дан ретроспективный анализ научных исследований в области теории колебаний, в развитии вибрационной техники и технологии, показано становление научной школы биомеханики систем "человек-машина". Академиком К.В. Фроловым издана (в соавторстве) монография, посвященная жизни и творческой деятельности замечательного ученого нашей эпохи - акад. А.А.Благонравова /I8/. К.В. Фролов является главным редактором международных научных сборников "Проблемы машиностроения и автоматизации", в которых публикуется немало исторических и прогнозных материалов о развитии современного машиноведения.

Определенный вклад в историю науки о машинах и механизмах внесли авторы данного доклада. Ими опубликованы в научных журналах и сборниках статьи о развитии машиноведческих исследований, о творческой деятельности видных ученых, подготовлена и издана монография "Развитие теории и практики советского машиноведения" /I9/. В книге анализируется большой исторический путь советской науки о машинах, при этом особое внимание уделено таким направлениям исследований, как теория механизмов и машин, теория трения и износа в машинах, проблемы прочности материалов и конструкций, теория точности и др.

Закончим этот доклад словами, которые произнес академик И.И.Артоболевский, открывая пленарное заседание IУ Международного конгресса ИФТОММ в 1975 г., в Англии: "Мне предстоит, - говорил он, - нелегкая задача осветить историю развития теории машин и механизмов, дать обзор ее современного состояния и, что самое трудное, сделать попытку прогноза на будущее. Вы можете спросить: какой смысл говорить о прошлом науки, когда так широк спектр будущего? В подкрепление своей постановки вопроса я сошлюсь на слова одного из наших современников, крупнейшего физика, создателя квантовой механики В.Гейзенберга: "Чтобы обозреть прогресс науки в целом, полезно сравнить современные проблемы науки с проблемами предшествующей эпохи и исследовать те специфические изменения, которые претерпевала та или иная важная проблема в течение десятилетий или даже столетий" /20, с.I28/. Этими словами И.И.Артоболевский подчеркнул значимость историко-научных исследований и еще раз обратил внимание ученых и специалистов на то важное обстоятельство, что основные направления современной науки о машинах и механизмах тесно и неразрывно связаны с тенденциями развития классического машиноведения.

ЛИТЕРАТУРА

I. Артоболевский И.И. Теория механизмов и машин. М.-Л.: ОНТИ, I940.

2. Артоболевский И.И. Роль и значение П.Л.Чебышева в истории развития теории механизмов. - Известия АН СССР. ОТН, I945, № 4-5.

3. Артоболевский И.И., Левитский Н.И. П.Л.Чебышев и русская теория механизмов. - В кн.: Труды Семинара по ТММ. Том 2, вып. 5, М.: Изд-во АН СССР, I947.

4. Артоболевский И.И., Левитский Н.И. Механизмы П.Л.Чебышева. - В кн.: Научное наследие П.Л.Чебышева. Вып. 2. Теория механизмов. М.: Изд-во АН СССР, I945.

5. Артоболевский И.И. Работы Н.Е.Жуковского по прикладной механике. - Труды по истории техники. Вып. 4. М.: Изд-во АН СССР, I954.

6. Артоболевский И.И. Леонардо да Винчи - ученый и инженер. - В кн.: Труды Семинара по ТММ. Том I3, вып. 5I. М.: Изд-во АН СССР, I953.

7. Боголюбов А.Н. Теория механизмов и машин в историческом развитии ее идей. М.: Наука, I976.

8. Боголюбов А.Н. Советская школа механики машин. М.: Наука, I975.

9. Боголюбов А.Н. Иван Иванович Артоболевский. М.: Наука, I982.

I0. Боголюбов А.Н. Математики и механики. Киев: Наукова думка, I984.

II. Добровольский В.В. Теория механизмов. М.: Машгиз, I95I; Изд. 2-е перераб. и доп. М.: Машгиз, I953.

I2. Артоболевский И.И., Добровольский В.В. Синтез механизмов. М.: Гостехиздат, I944.

I3. Благонравов А.А. Исследование верхних слоев атмосферы при помощи высотных ракет. - Вестник АН СССР, I957, № 6.

I4. Благонравов А.А., Крошкин М.Г. Геофизические исследования с помощью ракет и искусственных спутников. - Вестник АН СССР, I960, № 7.

I5. Благонравов А.А. Исследования космического пространства. - Вестник АН СССР, I970, № 9.

I6. Успехи СССР в исследовании космического пространства. (Первое космическое десятилетие. I957-I967). М.: Наука, I968.

I7. Успехи СССР в освоении космического пространства. (Второе космическое десятилетие. I967-I977). М.: Наука, I978.

I8. Фролов К.В., Пархоменко А.А., Усков М.К. Анатолий Аркадьевич Благонравов. М.: Наука, I982.

I9. Усков М.К., Пархоменко А.А. Развитие теории и практики советского машиноведения. М.: Наука, I980.

20. Академик И.И.Артоболевский. Воспоминания современников. М.: Знание, I983.

Rational Mechanics

Dynamics of a Flexible Rotor

B. P. ZHURAVLIEV

Institute of Mechanics, USSR Academy of Sciences, Moscow, USSR

Аннотация. В работе рассматривается пространственный вариант установленного ранее[1] эффекта инертности упругих волн в симметричных твердых телах. Предполагается, что упругое тело обладает сферической симметрией и под действием внешних сил вращается с произвольно меняющейся во времени и в пространстве угловой скоростью. Показано, что изменение пространственной картины упругих колебаний тела может быть описано следующим образом. Для каждой собственной формы колебаний существует такая система координат, в которой можно наблюдать стоячую волну, соответствующую данной форме. Получено уравнение Пуассона, определяющее положение этой системы координат относительно тела, показывающее, что компоненты угловой скорости упомянутой стоячей волны относительно тела в проекциях на связанные с ней оси пропорциональны компонентам угловой скорости тела относительно пространства в проекциях на оси тела. Этот результат был получен исходя из принципа Даламбера-Лагранжа, записанного для упругого твердого тела. Обобщенные координаты Лагранжа были выбраны при помощи введения в конфигурационном пространстве задачи базиса из полной ортонормированной системы собственных функций соответствующей задачи о свободных колебаниях невращающегося тела. Далее, уравнения, записанные в собственных трехмерных подпространствах преобразовывались при помощи ортогональных, зависящих от времени преобразований, к самосопряженной форме. Показано, что матрица, определяющая такое приведение всегда существует и находится единственным образом. Рассмотрен пример тонкой сферической оболочки, для которой приведены числовые данные.

Ключевые слова. Gyroscopes; Navigation; Wave Theory.

I. Явление прецессии стоячих волн в тонком упругом кольце, вращающемся с постоянной угловой скоростью ω, впервые обсуждалось Брайаном в[2]. Им было показано, что система координат, в которой может наблюдаться стоячая волна упругих колебаний, вращается с постоянной угловой скоростью Ω относительно абсолютного пространства:

$$\Omega = \frac{k^2-1}{k^2+1}\,\omega \tag{I.I}$$

где k - номер формы колебаний.

В [3] были опубликованы результаты эксперимента с тонкой полусферической оболочкой. В первоначально неподвижной оболочке возбуждалась стоячая волна упругих колебаний, соответствующая основной форме с четырьмя узлами на окружности. Затем оболочка вокруг оси симметрии поворачивалась на угол 90° и останавливалась. Было отмечено, что стоячая волна также поворачивалась, не изменяя своей формы (как твердое тело), и останавливалась. При этом угол поворота волны относительно неподвижного основания составлял 63°. В качестве теоретического объяснения наблюдаемого эффекта автор [3] ссылается на результат (I.I) Брайана. Однако, описанные два факта никакого отношения друг к другу не имеют. В теоретической модели Брайана кольцо вращается с постоянной угловой скоростью и прецессия волны (I.1) описывается в рамках спектральной теории линейных систем с постоянными коэффициентами. В[3] скорость вращения оболочки существенно переменна и наблюдаемый эффект представляет собой качественно новый факт в свойствах упругих колебаний симметричных тел.

Первое теоретическое объяснение и описание этого эксперимента дано в[1]. В частности было показано, что результат Брайана допускает широкое обобщение: формула (I.I) является точной в рамках рассматриваемой модели не только для постоянной угловой скорости ω, но и для угловой скорости, зависящей произвольным образом от времени:

$$\Omega(t)=\frac{k^2-1}{k^2+1}\,\omega(t) \qquad (I.2)$$

Если обе части этого соотношения проинтегрировать, то получается аналогичное соотношение для углов поворота тела и волны, что и дает объяснение эксперимента[3]. В[4] этот же факт был теоретически установлен для тонкой полусферической оболочки.

Если соотношение (I.2) продифференцировать, то получим, что угловое ускорение волны пропорционально угловому ускорению кольца. Момент внешних сил, ускоряющих кольцо, вызывает и ускорение волны, что и позволяет говорить об инертных свойствах волн в симметричных упругих системах. И в случае кольца и в случае оболочки эффект инертности упругих волн имеет одномерный характер: угловая скорость $\omega(t)$ есть скаляр, характеризующий вращение упругого твердого тела вокруг неподвижной в пространстве оси. Ниже рассматривается обобщение этого эффекта на произвольный пространственный случай.

2. Рассмотрим упругое сферически симметричное твердое тело со свободной границей на которое действуют массовые силы плотности f.

Главный вектор сил, действующих на тело $\int_V f\,dm$ без ограничения общности будем полагать равным нулю. Под действием главного момента $\int_V r\times f\,dm$ тело меняет свою ориентацию в пространстве (r - радиус-вектор произвольной точки тела, dm - элемент массы, V - область, занятая телом). Для описания упругих деформаций тела введем систему координат $x_1x_2x_3$ связанную с телом так, чтобы выполнялись условия:

$$\int_V x\,dm=0\,,\qquad \int_V r\times x\,dm=0 \qquad (2.I)$$

где $x=(x_1,x_2,x_3)$ - упругое смещение точки в недеформированном состоянии занимавшей положение r. Условия (2.I) характеризуют координатный трехгранник, относительно которого тело в среднем (по всем частицам) не перемещается и не поворачивается.

Ставится следующая задача: зная абсолютную угловую скорость трехгранника $x_1x_2x_3$ в проекциях на его же оси $\omega(t)$ определить как ведут себя волны упругих деформаций.

3. Запишем принцип Даламбера-Лагранжа для рассматриваемого тела:

$$\int_V [\ddot{x}+\omega\times(\omega\times(r+x))+\dot{\omega}\times(r+x)+2\omega\times\dot{x}+ \\ +\frac{1}{\rho}\nabla\Pi-f]\cdot\delta x\,dm=0 \qquad (3.I)$$

здесь ρ - плотность, зависящая лишь от $|r|$; $\nabla\Pi$ - градиент квадратичного функционала линейной теории упругости. Координаты, определяющие угловое положение тела, как целого, не варьируются, предполагается, что угловая скорость $\omega(t)$ - известная функция времени.

Для выбора обобщенных координат рассмотрим случай $\omega=0$. В[5] было показано, что спектр собственных колебаний свободного твердого тела при условиях (2.I) дискретен. Дискретность спектра означает следующее. Возрастающая последовательность частот собственных колебаний $\nu_1\leqslant\nu_2\leqslant\ldots$ неограничена, а собственные элементы $h_1(r)$, $h_2(r)$,... соответствующие этим частотам образуют ортонормированную систему функций, полную в конфигурационном пространстве задачи:

$$\int_V h_n(r)h_l(r)\,dm=\delta_n^l \qquad (3.2)$$

Это позволяет ввести независимые лагранжевы координаты, описывающие все степени свободы при деформировании тела, в общем случае $\omega(t)\not\equiv 0$ следующим образом:

$$x=\sum_{n=1}^{\infty} q_n(t)h_n(r) \qquad (3.3)$$

Задача о собственных колебаниях сферически симметричного свободного тела допускает группу $SO(3)$, поэтому спектр собственных частот вырожден и он состоит из последовательности, по крайней мере, трехкратных частот: $\nu_1 = \nu_2 = \nu_3 \leqslant \nu_4 = \nu_5 = \nu_6 \leqslant \ldots$
Конфигурационное пространство при этом представляет собой прямое произведение трехмерных собственных подпространств:

$$\{h_1, h_2, h_3\} \times \{h_4, h_5, h_6\} \times \ldots$$

Фиксируем номер m произвольного собственного подпространства и введем обозначения для соответствующих обобщенных координат: $q_{3m-2} = u$, $q_{3m-1} = v$, $q_{3m} = w$ (m =1,2,...). Подставляя (3.3), а также $\delta x = \sum_{n=1}^{\infty} \delta q_n h_n(r)$ в (3.I) и приравнивая нулю коэффициенты при независимых вариациях δq_n, получаем бесконечную систему обыкновенных дифференциальных уравнений второго порядка относительно q_n вида:

$$\ddot{u} + au + bv + cw - (\dot{\omega}, \varkappa_3)v + (\dot{\omega}, \varkappa_2)w - 2\dot{v}(\omega, \varkappa_3) + 2\dot{w}(\omega, \varkappa_2) + F_1 + L_1 = 0$$

$$\ddot{v} + bu + dv + ew + (\dot{\omega}, \varkappa_3)u - (\dot{\omega}, \varkappa_1)w - 2\dot{w}(\omega, \varkappa_1) + 2\dot{u}(\omega, \varkappa_3) + F_2 + L_2 = 0$$

$$\ddot{w} + cu + ev + fw - (\dot{\omega}, \varkappa_2)u + (\dot{\omega}, \varkappa_1)v - 2\dot{u}(\omega, \varkappa_2) + 2\dot{v}(\omega, \varkappa_1) + F_3 + L_3 = 0 \qquad (3.4)$$

в которой скалярные коэффициенты имеют вид:

$$a = \int_V (h_{3m-2}, \omega)^2 dm - \omega^2, \quad b = \int_V (h_{3m-2}, \omega)(h_{3m-1}, \omega)\, dm,$$

$$c = \int_V (h_{3m-2}, \omega)(h_{3m}, \omega)\, dm, \quad d = \int_V (h_{3m-1}, \omega)^2 dm - \omega^2,$$

$$e = \int_V (h_{3m-1}, \omega)(h_{3m}, \omega)\, dm, \quad f = \int_V (h_{3m}, \omega)^2 dm - \omega^2,$$

$$F_1 = \int_V (\nabla\Pi, h_{3m-2})\, dV, \quad F_2 = \int_V (\nabla\Pi, h_{3m-1})\, dV, \quad F_3 = \int_V (\nabla\Pi, h_{3m})\, dV$$

L_1, L_2, L_3 – представляют собой линейные функции обобщенных координат, соответствующих другим собственным подпространствам. Присутствие этих членов характеризует тот факт, что системы типа (3.4) для различных подпространств не являются независимыми друг от друга.

При получении уравнений (3.4) было предположено ради простоты, что массовые силы f ортогональны всем собственным функциям: $\int_V f \cdot h_n(r)\, dm = 0$. Это означает, что в f присутствует лишь постоянная составляющая $(\int_V r \times f\, dm \neq 0)$, обеспечивающая вращение тела со скоростью $\omega(t)$.

Векторные коэффициенты $\varkappa_1, \varkappa_2, \varkappa_3$ имеют вид:

$$\varkappa_1 = \int_V h_{3m-1} \times h_{3m}\, dm, \quad \varkappa_2 = \int_V h_{3m} \times h_{3m-2}\, dm,$$

$$\varkappa_3 = \int_V h_{3m-2} \times h_{3m-1}\, dm$$

В силу сферической симметрии выбор собственных векторов h_{3m-2}, h_{3m-1}, h_{3m} можно осуществить так, чтобы

$$\varkappa_1 = \varkappa(1,0,0), \quad \varkappa_2 = \varkappa(0,1,0), \quad \varkappa_3 = \varkappa(0,0,1)$$

где

$$\varkappa = \pm|\varkappa_1| = \pm|\varkappa_2| = \pm|\varkappa_3| = \pm\left|\int_V h_{3m-1} \times h_{3m}\, dm\right| \qquad (3.5)$$

Очевидно

$$0 \leqslant |\varkappa| \leqslant 1$$

Если ввести обозначения

$$z = \begin{pmatrix} u \\ v \\ w \end{pmatrix}, \quad \omega = \begin{pmatrix} p \\ q \\ r \end{pmatrix}, \quad L = \begin{pmatrix} L_1 \\ L_2 \\ L_3 \end{pmatrix}, \quad G = \begin{pmatrix} 0 & -r & q \\ r & 0 & -p \\ -q & p & 0 \end{pmatrix}$$

то уравнения (3.4) можно переписать в векторной форме

$$\ddot{z} + Az + \varkappa\dot{G}z + 2\varkappa G\dot{z} + L = 0 \qquad (3.6)$$

где A – симметрическая матрица позиционных сил, состоящая из коэффициентов упругих сил F_1, F_2, F_3 и коэффициентов a, b, c, d, e, f.

4. Уравнение (3.6) определяет эволюцию m-ой формы колебаний свободного твердого тела, вызванную наличием вращения. Эта эволюция определяется двумя обстоятельствами. Во-первых, сама форма колебаний непосредственно реагирует на вращение тела, что определяется наличием в уравнении (3.6) членов с G и $\dot{G}$. Во-вторых, рассматриваемая форма подвергается воздействию со стороны других форм. Сразу заметим, что это воздействие является незначительным, поскольку, к примеру, при решении уравнений (3.6) методом осреднения все члены, определяемые L, в пер-

вом приближении исчезают.

Имеет место следующий факт. Существует такая система координат $z \to y$: $z = \mathcal{M}y$ где $\mathcal{M}$ - зависящая от времени ортогональная матрица преобразования координат, в которой уравнение (3.6) при $b \equiv 0$ имеет самосопряженную форму. В этой системе координат уравнение (3.6) допускает решения типа стоячей волны. Покажем это. Подставляя $z = \mathcal{M}y$ в (3.6), найдем:

$$\ddot{y} + 2\mathcal{M}'(\dot{\mathcal{M}} + \varkappa G\mathcal{M})\dot{y} + \mathcal{M}'(\ddot{\mathcal{M}} + 2\varkappa G\dot{\mathcal{M}} + \varkappa\dot{G}\mathcal{M} + A\mathcal{M})y = 0 \qquad (4.1)$$

Потребуем

$$\dot{\mathcal{M}} = \varkappa G\mathcal{M} \qquad (4.2)$$

Получим $\ddot{\mathcal{M}} = -\varkappa\dot{G}\mathcal{M} + \varkappa^2 G^2\mathcal{M}$, и, подставляя в (4.1), найдем:

$$\ddot{y} + \mathcal{M}'(A - \varkappa^2 G^2)\mathcal{M}y = 0$$

Таким образом, если в неподвижном теле возбудить стоячую волну колебаний с каким-нибудь чистым током, и после этого привести тело во вращение с произвольной угловой скоростью, то стоячая волна будет поворачиваться относительно тела по закону (4.2). Уравнение (4.2) есть уравнение Пуассона. Сравним его с уравнением Пуассона для самого твердого тела $\dot{\mathcal{N}} = -G\mathcal{N}$, в котором ортогональная матрица $\mathcal{N}$ определяет положение твердого тела в инерциальном пространстве. Откуда и видно, что угловая скорость стоячей волны относительно тела пропорциональна угловой скорости тела относительно пространства: $\Omega_0(t) = -\varkappa\omega(t)$, или же для скорости волны относительно пространства имеем:

$$\Omega(t) = (1 - \varkappa)\omega(t) \qquad (4.3)$$

Соотношение (4.3) и представляет собой обобщение скалярного соотношения типа (1.2) на пространственный случай.

ЛИТЕРАТУРА

1. Журавлев В.Ф., Климов Д.М. Волновой твердотельный гироскоп. М.: Наука, 1985, 126 с.
2. Bryan G.H. (1890). On the beats in the vibrations of a revolving cylinder or bell.- Proc. Cambridge Philos. Soc. Nath.Phys.Sci.,vol.7, p.101-111.
3. Scott W.B. (1982). Delco makes low-cost gyro prototype.- Aviat. Week, vol. 117 , p.64-72.
4. Журавлев В.Ф., Попов А.Л. (1983). О прецессии собственной формы колебаний сферической оболочки при ее вращении. - Изв.АН СССР. МТТ, № 5, с.17-23.
5. Weil H. (1915). Das asymptotische Verteilungsgesetz der Eigenschwingungen eines beliebig gestalteten elastischen Körpers, Rend. Circolo mat. Palermo, 39,
6. Чернина В.С. (1973). Свободные колебания тонкой замкнутой сферической оболочки. Теория оболочек и пластин. М.:Наука.

SUMMARY

On étudie le comportement des ondes dans le corps élastique, ayant une symétrie sphérique, dans le cas ou ce corps tourne autour d'un pointfixe avec une vitesse angulaire arbitrairement dépendante du temps. On établi que ce comportement peut être décrit de la façon suivante. Il existe un repère qui tourne par rapport au corps avec une vitesse bien déterminée et dans lequel il existe une onde immobile, correspondante à l'une des formes propre. La configuration de cette onde ne dépend pas du temps, cette onde immobile ce comporte comme un corps rigide que glisse le long du corps considéré. La vitesse de la rotation de ce repère (ou de l'onde immobile) par rapport au corps est proportionnele à la vitesse de la rotation du corps par rapport aux étoiles.

Subject Index